T0238799

Lecture Notes in Computer Science 9135

Commenced Publication in 1973
Founding and Former Series Editors:
Gerhard Goos, Juris Hartmanis, and Jan van Leeuwen

Advanced Research in Computing and Software Science

Subline of Lecture Notes in Computer Science

More information about this series at http://www.springer.com/series/7407

Magnús M. Halldórsson · Kazuo Iwama
Naoki Kobayashi · Bettina Speckmann (Eds.)

Automata, Languages, and Programming

42nd International Colloquium, ICALP 2015
Kyoto, Japan, July 6–10, 2015
Proceedings, Part II

 Springer

Editors
Magnús M. Halldórsson
Reykjavik University
Reykjavik
Iceland

Kazuo Iwama
Kyoto University
Kyoto
Japan

Naoki Kobayashi
The University of Tokyo
Tokyo
Japan

Bettina Speckmann
Technische Universiteit Eindhoven
Eindhoven
The Netherlands

ISSN 0302-9743 ISSN 1611-3349 (electronic)
Lecture Notes in Computer Science
ISBN 978-3-662-47665-9 ISBN 978-3-662-47666-6 (eBook)
DOI 10.1007/978-3-662-47666-6

Library of Congress Control Number: 2015941869

LNCS Sublibrary: SL1 – Theoretical Computer Science and General Issues

Springer Heidelberg New York Dordrecht London

Printed on acid-free paper

Springer-Verlag GmbH Berlin Heidelberg is part of Springer Science+Business Media
(www.springer.com)

Preface

ICALP 2015, the 42nd edition of the International Colloquium on Automata, Languages and Programming, was held in Kyoto, Japan during July 6–10, 2015. ICALP is a series of annual conferences of the European Association for Theoretical Computer Science (EATCS), which first took place in 1972. This year, the ICALP program consisted of the established track A (focusing on algorithms, automata, complexity, and games) and track B (focusing on logic, semantics, and theory of programming), and of the recently introduced track C (focusing on foundations of networking).

In response to the call for papers, the Program Committee received 507 submissions, the highest ever: 327 for track A, 115 for track B, and 65 for track C. Out of these, 143 papers were selected for inclusion in the scientific program: 89 papers for Track A, 34 for Track B, and 20 for Track C. The selection was made by the Program Committees based on originality, quality, and relevance to theoretical computer science. The quality of the manuscripts was very high indeed, and many deserving papers could not be selected.

The EATCS sponsored awards for both a best paper and a best student paper for each of the three tracks, selected by the Program Committees. The best paper awards were given to the following papers:

- Track A: Aaron Bernstein and Clifford Stein. "Fully Dynamic Matching in Bipartite Graphs"
- Track B: Jarkko Kari and Michal Szabados. "An Algebraic Geometric Approach to Nivat's Conjecture"
- Track C: Yiannis Giannakopoulos and Elias Koutsoupias. "Selling Two Goods optimally"

The best student paper awards, for papers that are solely authored by students, were given to the following papers:

- Track A: Huacheng Yu. "An Improved Combinatorial Algorithm for Boolean Matrix Multiplication"
- Track A: Radu Curticapean. "Block Interpolation: A Framework for Tight Exponential-Time Counting Complexity"
- Track B: Georg Zetzsche. "An Approach to Computing Downward Closures"

Track A gave out two student paper awards this year because of the very high quality of the two winning papers.

The conference was co-located with LICS 2015, the 30th ACM/IEEE Symposium on Logic in Computer Science.

Apart from the contributed talks, ICALP 2015 included invited presentations by Ken-ichi Kawarabayashi, Valerie King, Thomas Moscibroda, Anca Muscholl, Peter O'Hearn, of which the latter two were joint with LICS. Additionally, it contained tutorial sessions by Piotr Indyk, Andrew Pitts, and Geoffrey Smith, all joint with LICS, and a masterclass on games by Ryuhei Uehara. Abstracts of their talks are included in these proceedings as well. The program of ICALP 2015 also included presentation of the EATCS Award 2015 to Christos Papadimitriou.

This volume of the proceedings contains all contributed papers presented at the conference in Track A. A companion volume contains all contributed papers presented in Track B and Track C together with the papers and abstracts of the invited speakers. The following workshops were held as satellite events of ICALP/LICS 2015:

HOPA 2015 — Workshop on the Verification of Higher-Order Programs

LCC 2015 — 16th International Workshop on Logic and Computational Complexity

NLCS 2015 — Third Workshop on Natural Language and Computer Science

LOLA 2015 — Workshop on Syntax and Semantics for Low-Level Languages

QCC 2015 — Workshop on Quantum Computational Complexity

WRAWN 2015 — 6th Workshop on Realistic Models for Algorithms in Wireless Networks

YR-ICALP 2015 — Young Researchers Forum on Automata, Languages and Programming

We wish to thank all authors who submitted extended abstracts for consideration, the Program Committees for their scholarly effort, and all referees who assisted the Program Committees in the evaluation process.

We thank the sponsors (ERATO Kawarabayashi Large Graph Project; MEXT Grant-in-Aid for Scientific Research on Innovative Areas "Exploring the Limits of Computation"; Research Institute for Mathematical Sciences, Kyoto University; and Tateisi Science and Technology Foundation) for their support.

We are also grateful to all members of the Organizing Committee and to their support staff.

Thanks to Andrei Voronkov and Shai Halevi for writing the conference management systems EasyChair andWeb-Submission-and-Review software, which were used in handling the submissions and the electronic Program Committee meeting, as well as in assisting in the assembly of the proceedings.

Last but not least, we would like to thank Luca Aceto, the president of EATCS, for his generous advice on the organization of the conference.

May 2015

Magnús M. Halldórsson
Kazuo Iwama
Naoki Kobayashi
Bettina Speckmann

Organization

Program Committee

Track A

Peyman Afshani	Aarhus University, Denmark
Hee-Kap Ahn	POSTECH, South Korea
Hans Bodlaender	Utrecht University, The Netherlands
Karl Bringmann	Max-Planck Institut für Informatik, Germany
Sergio Cabello	University of Ljubljana, Slovenia
Ken Clarkson	IBM Almaden Research Center, USA
Éric Colin de Verdière	École Normale Supérieure Paris, France
Stefan Dziembowski	University of Warsaw, Poland
David Eppstein	University of California at Irvine, USA
Dimitris Fotakis	National Technical University of Athens, Greece
Paul Goldberg	University of Oxford, UK
MohammadTaghi Hajiaghayi	University of Maryland at College Park, USA
Jesper Jansson	Kyoto University, Japan
Andrei Krokhin	Durham University, UK
Asaf Levin	Technion, Israel
Inge Li Gørtz	Technical University of Denmark, Denmark
Pinyan Lu	Microsoft Research Asia, China
Frédéric Magniez	Université Paris Diderot, France
Kazuhisa Makino	Kyoto University, Japan
Elvira Mayordomo	Universidad de Zaragoza, Spain
Ulrich Meyer	Goethe University Frankfurt am Main, Germany
Wolfgang Mulzer	Free University Berlin, Germany
Viswanath Nagarajan	University of Michigan, USA
Vicky Papadopoulou	European University Cyprus, Cyprus
Michał Pilipczuk	University of Bergen, Norway
Liam Roditty	Bar-Ilan University, Israel
Ignaz Rutter	Karlsruhe Institute of Technology, Germany
Rocco Servedio	Columbia University, USA
Jens Schmidt	TU Ilmenau, Germany
Bettina Speckmann	TU Eindhoven, The Netherlands
Csaba D. Tóth	California State University Northridge, USA
Takeaki Uno	National Institute of Informatics, Japan
Erik Jan van Leeuwen	Max-Planck Institut für Informatik, Germany
Rob van Stee	University of Leicester, UK
Ivan Visconti	University of Salerno, Italy

Track B

Andreas Abel	Chalmers and Gothenburg University, Sweden
Albert Atserias	Universitat Politècnica de Catalunya, Spain
Christel Baier	TU Dresden, Germany
Lars Birkedal	Aarhus University, Denmark
Luís Caires	Universidade Nova de Lisboa, Portugal
James Cheney	University of Edinburgh, UK
Wei Ngan Chin	National University of Singapore, Singapore
Ugo Dal Lago	University of Bologna, Italy
Thomas Ehrhard	CNRS and Université Paris Diderot, France
Zoltán Ésik	University of Szeged, Hungary
Xinyu Feng	University of Science and Technology of China, China
Wan Fokkink	VU University Amsterdam, The Netherlands
Shin-ya Katsumata	Kyoto University, Japan
Naoki Kobayashi	The University of Tokyo, Japan
Eric Koskinen	New York University, USA
Antonín Kučera	Masaryk University, Czech Republic
Orna Kupferman	Hebrew University, Israel
Annabelle McIver	Macquarie University, Australia
Dale Miller	Inria Saclay, France
Markus Müller-Olm	University of Münster, Germany
Andrzej Murawski	University of Warwick, UK
Joel Ouaknine	University of Oxford, UK
Prakash Panangaden	McGill University, Canada
Pawel Parys	University of Warsaw, Poland
Reinhard Pichler	TU Vienna, Austria
Simona Ronchi Della Rocca	University of Turin, Italy
Jeremy Siek	Indiana University, USA

Track C

Ioannis Caragiannis	University of Patras, Greece
Katarina Cechlarova	Pavol Jozef Safarik University, Slovakia
Shiri Chechik	Tel Aviv University, Israel
Yuval Emek	Technion, Israel
Sándor Fekete	TU Braunschweig, Germany
Pierre Fraigniaud	CNRS, Université Paris Diderot, France
Leszek Gąsieniec	University of Liverpool, UK
Aristides Gionis	Aalto University, Finland
Magnús M. Halldórsson	Reykjavik University, Iceland
Monika Henzinger	Universität Wien, Austria
Bhaskar Krishnamachari	University of Southern California, USA
Fabian Kuhn	University of Freiburg, Germany
Michael Mitzenmacher	Harvard University, USA
Massimo Merro	University of Verona, Italy

Gopal Pandurangan   University of Houston, USA
Pino Persiano   University of Salerno, Italy
R. Ravi   Carnegie Mellon University, USA
Ymir Vigfusson   Emory University, USA
Roger Wattenhofer   ETH Zürich, Switzerland
Masafumi Yamashita   Kyushu University, Japan

Organizing Committee

Masahito Hasegawa   Kyoto University, Japan
Atushi Igarashi   Kyoto University, Japan
Kazuo Iwama   Kyoto University, Japan
Kazuhisa Makino   Kyoto University, Japan

Financial Sponsors

ERATO Kawarabayashi Large Graph Project
MEXT Grant-in-Aid for Scientific Research on Innovative Areas: "Exploring the Limits of Computation"
Research Institute for Mathematical Sciences, Kyoto University
Tateisi Science and Technology Foundation

Additional Reviewers

Abboud, Amir
Abdulla, Parosh
Abed, Fidaa
Abraham, Ittai
Ailon, Nir
Ajwani, Deepak
Albers, Susanne
Almeida, Jorge
Alt, Helmut
Alur, Rajeev
Alvarez, Victor
Alvarez-Jarreta, Jorge
Ambainis, Andris
Aminof, Benjamin
Anagnostopoulos, Aris
Andoni, Alexandr
Angelidakis, Haris
Anshelevich, Elliot
Antoniadis, Antonios

Arai, Hiromi
Aronov, Boris
Asada, Kazuyuki
Aspnes, James
Aubert, Clément
Augustine, John
Auletta, Vincenzo
Austrin, Per
Avin, Chen
Avni, Guy
Baelde, David
Baillot, Patrick
Bansal, Nikhil
Banyassady, Bahareh
Barnat, Jiri
Barth, Stephan
Barto, Libor
Basavaraju, Manu
Bassily, Raef

Baswana, Surender
Bateni, Mohammadhossein
Batu, Tugkan
Baum, Moritz
Béal, Marie-Pierre
Beigi, Salman
Beimel, Amos
Ben-Amran, Amr
Berenbrink, Petra
Bernáth, Attila
Berthé, Valérie
Bes, Alexis
Besser, Bert
Bevern, René Van
Bi, Jingguo
Bienstock, Daniel
Bille, Philip
Bilò, Vittorio
Bizjak, Ales
Björklund, Henrik
Blais, Eric
Bläsius, Thomas
Blömer, Johannes
Bogdanov, Andrej
Bojanczyk, Mikolaj
Bollig, Benedikt
Bonfante, Guillaume
Bonnet, Edouard
Bourhis, Pierre
Bousquet, Nicolas
Boyar, Joan
Bozzelli, Laura
Bradfield, Julian
Brandes, Philipp
Brandt, Sebastian
Braverman, Vladimir
Bresolin, Davide
Brzuska, Christina
Brânzei, Simina
Bucciarelli, Antonio
Buchbinder, Niv
Buchin, Kevin
Bulatov, Andrei
Cai, Jin-Yi
Cai, Zhuohong
Canonne, Clement

Cao, Yixin
Carayol, Arnaud
Carmi, Paz
Caron, Pascal
Caskurlu, Bugra
Cassez, Franck
Castagnos, Guilhem
Castellani, Ilaria
Castelli Aleardi, Luca
Cenzer, Douglas
Chakrabarty, Deeparnab
Chalermsook, Parinya
Chan, T.-H. Hubert
Chan, Timothy M.
Chattopadhyay, Arkadev
Chekuri, Chandra
Chen, Ho-Lin
Chen, Wei
Chen, Xi
Chen, Xujin
Chitnis, Rajesh
Chlamtac, Eden
Chlebikova, Janka
Cho, Dae-Hyeong
Chonev, Ventsislav
Christodoulou, George
Cicalese, Ferdinando
Cimini, Matteo
Clairambault, Pierre
Claude, Francisco
Clemente, Lorenzo
Cleve, Richard
Cloostermans, Bouke
Cohen-Addad, Vincent
Columbus, Tobias
Cording, Patrick Hagge
Coretti, Sandro
Cormode, Graham
Cornelsen, Sabine
Cosentino, Alessandro
Coudron, Matthew
Crouch, Michael
Cygan, Marek
Czerwiński, Wojciech
Czumaj, Artur
Dachman-Soled, Dana

Dahlgaard, Søren
Dalmau, Victor
Dantchev, Stefan
Daruki, Samira
Das, Anupam
Dasler, Philip
Datta, Samir
Daum, Sebastian
Dawar, Anuj
De Bonis, Annalisa
De Caro, Angelo
De, Anindya
Dehghani, Sina
Deligkas, Argyrios
Dell, Holger
Demangeon, Romain
Demri, Stéphane
Denzumi, Shuhei
Diakonikolas, Ilias
Dibbelt, Julian
Dietzfelbinger, Martin
Dinsdale-Young, Thomas
Dinur, Itai
Disser, Yann
Dobrev, Stefan
Doerr, Carola
Döttling, Nico
Dotu, Ivan
Doty, David
Dräger, Klaus
Drucker, Andrew
Duan, Ran
Dubslaff, Clemens
Duetting, Paul
van Duijn, Ingo
Duncan, Ross
Durand, Arnaud
Durand-Lose, Jérôme
Dürr, Christoph
Dvorák, Wolfgang
Dyer, Martin
Efthymiou, Charilaos
Eirinakis, Pavlos
Elbassioni, Khaled
Elmasry, Amr
Emanuele, Viola

Emmi, Michael
Emura, Keita
Englert, Matthias
Epelman, Marina
Epstein, Leah
Ergun, Funda
Erickson, Alejandro
Esfandiari, Hossein
Fahrenberg, Uli
Farinelli, Alessandro
Faust, Sebastian
Fawzi, Omar
Fefferman, Bill
Feldman, Moran
Feldmann, Andreas Emil
Feng, Yuan
Fernique, Thomas
Ferraioli, Diodato
Fijavz, Gasper
Filinski, Andrzej
Filmus, Yuval
Filos-Ratsikas, Aris
Find, Magnus Gausdal
Firsov, Denis
Fleiner, Tamas
Foerster, Klaus-Tycho
Fomin, Fedor
Fontes, Lila
Forbes, Michael A.
Forejt, Vojtech
Formenti, Enrico
François, Nathanaël
Fränzle, Martin
Frascaria, Dario
Friedrich, Tobias
Fu, Hongfei
Fuchs, Fabian
Fuchsbauer, Georg
Fukunaga, Takuro
Fuller, Benjamin
Funk, Daryl
Fürer, Martin
Gabizon, Ariel
Gaboardi, Marco
Gacs, Peter
Gaertner, Bernd

Galanis, Andreas
Galčík, František
Ganguly, Sumit
Ganor, Anat
Ganty, Pierre
Garg, Naveen
Gaspers, Serge
Gawrychowski, Pawel
Gazda, Maciej
Gehrke, Mai
Gemsa, Andreas
Georgiadis, Loukas
Gerhold, Marcus
van Glabbeek, Rob
Göller, Stefan
Goncharov, Sergey
Göös, Mika
Gopalan, Parikshit
Gorbunov, Sergey
Gouveia, João
Grandjean, Etienne
Grandoni, Fabrizio
Green Larsen, Kasper
Grigoriev, Alexander
Grohe, Martin
Groote, Jan Friso
Grossi, Roberto
Grunert, Romain
Guessarian, Irène
Guiraud, Yves
Guo, Heng
Gupta, Anupam
Hadfield, Stuart
Hague, Matthew
Hahn, Ernst Moritz
Haitner, Iftach
Halevi, Shai
Hamann, Michael
Hampkins, Joel
Hansen, Kristoffer Arnsfelt
Har-Peled, Sariel
Harrow, Aram
Hastad, Johan
Hatano, Kohei
Haverkort, Herman
He, Meng

Heindel, Tobias
Hendriks, Dimitri
Henze, Matthias
Hermelin, Danny
Herranz, Javier
Heunen, Chris
Heydrich, Sandy
Hlineny, Petr
Hoffmann, Frank
Hoffmann, Jan
Hofheinz, Dennis
Hofman, Piotr
Holm, Jacob
Holmgren, Justin
Hong, Seok-Hee
Houle, Michael E.
Høyer, Peter
Hsu, Justin
Huang, Shenwei
Huang, Zengfeng
Huang, Zhiyi
Hwang, Yoonho
van Iersel, Leo
Im, Sungjin
Immerman, Neil
Inaba, Kazuhiro
Iovino, Vincenzo
Ishii, Toshimasa
Italiano, Giuseppe F.
Ito, Takehiro
Ivan, Szabolcs
Iwata, Yoichi
Izumi, Taisuke
Jaberi, Raed
Jaiswal, Ragesh
Jancar, Petr
Janin, David
Jansen, Bart M.P.
Jansen, Klaus
Jayram, T.S.
Jeavons, Peter
Jeffery, Stacey
Jerrum, Mark
Jeż, Łukasz
Jhanwar, Mahabir Prasad
Johnson, Matthew

Johnson, Matthew P.
Jones, Mark
Jones, Neil
Jordan, Charles
Jørgensen, Allan Grønlund
Jovanovic, Aleksandra
Jukna, Stasys
Kakimura, Naonori
Kalaitzis, Christos
Kamiyama, Naoyuki
Kanade, Varun
Kanazawa, Makoto
Kane, Daniel
Kanellopoulos, Panagiotis
Kantor, Erez
Kanté, Mamadou Moustapha
Kaplan, Haim
Karhumaki, Juhani
Kari, Jarkko
Kärkkäinen, Juha
Kashefi, Elham
Katajainen, Jyrki
Katz, Matthew
Kawachi, Akinori
Kazana, Tomasz
Kelk, Steven
Keller, Barbara
Keller, Orgad
Kenter, Sebastian
Kerenidis, Iordanis
Khan, Maleq
Khani, Reza
Khoussainov, Bakhadyr
Kida, Takuya
Kiefer, Stefan
Kijima, Shuji
Kim, Eun Jung
Kim, Heuna
Kim, Min-Gyu
Kim, Ringi
Kim, Sang-Sub
Kishida, Kohei
Kiyomi, Masashi
Klauck, Hartmut
Klavík, Pavel
Klima, Ondrej

Klin, Bartek
Knauer, Christian
Kobayashi, Yusuke
Kollias, Konstantinos
Kolmogorov, Vladimir
Komusiewicz, Christian
König, Barbara
König, Michael
Konrad, Christian
Kontogiannis, Spyros
Kopczynski, Eryk
Kopelowitz, Tsvi
Kopparty, Swastik
Korman, Matias
Kortsarz, Guy
Korula, Nitish
Kostitsyna, Irina
Kotek, Tomer
Kothari, Robin
Kovacs, Annamaria
Kozen, Dexter
Kraehmann, Daniel
Kral, Daniel
Kralovic, Rastislav
Kratsch, Dieter
Kratsch, Stefan
Krcal, Jan
Krenn, Stephan
Kretinsky, Jan
Kreutzer, Stephan
van Kreveld, Marc
Kriegel, Klaus
Krinninger, Sebastian
Krishna, Shankara Narayanan
Krishnaswamy, Ravishankar
Krizanc, Danny
Krumke, Sven
Krysta, Piotr
Kulkarni, Raghav
Kumar, Amit
Kumar, Mrinal
Künnemann, Marvin
Kuperberg, Greg
Kuroda, Satoru
Kurz, Alexander
Kyropoulou, Maria

Labourel, Arnaud
Lachish, Oded
Łącki, Jakub
Lagerqvist, Victor
Lamani, Anissa
Lammich, Peter
Lampis, Michael
Lanese, Ivan
Lange, Martin
Lasota, Sławomir
Laudahn, Moritz
Laura, Luigi
Laurent, Monique
Lauriere, Mathieu
Lavi, Ron
Lazic, Ranko
Le Gall, Francois
Le, Quang Loc
Le, Ton Chanh
Lecerf, Gregoire
Lee, James
Lee, Troy
Lengler, Johannes
Leonardos, Nikos
Leung, Hing
Levy, Paul Blain
Lewenstein, Moshe
Lewis, Andrew E.M.
Li, Guoqiang
Li, Jian
Li, Liang
Li, Yi
Li, Yuan
Li, Zhentao
Liaghat, Vahid
Lianeas, Thanasis
Liang, Hongjin
Liu, Jingcheng
Liu, Shengli
Liu, Zhengyang
Livnat, Adi
Lodi, Andrea
Löding, Christof
Loff, Bruno
Löffler, Maarten
Lohrey, Markus

Lokshtanov, Daniel
Lopez-Ortiz, Alejandro
Lovett, Shachar
Lucier, Brendan
Luxen, Dennis
Mahabadi, Sepideh
Mahmoody, Mohammad
Makarychev, Konstantin
Makarychev, Yury
Maneth, Sebastian
Manlove, David
Manokaran, Rajsekar
Manthey, Bodo
Manuel, Amaldev
Mardare, Radu
Martens, Wim
Masuzawa, Toshimitsu
Matsuda, Takahiro
Matulef, Kevin
Matuschke, Jannik
May, Alexander
Mayr, Richard
McGregor, Andrew
Megow, Nicole
Meier, Florian
Meir, Or
Mertzios, George
de Mesmay, Arnaud
Mestre, Julian
Michail, Othon
Michalewski, Henryk
Mignosi, Filippo
Mihalák, Matúš
Misra, Neeldhara
Mitsou, Valia
Mnich, Matthias
Mogelberg, Rasmus
Mohar, Bojan
Moitra, Ankur
Monemizadeh, Morteza
Montanaro, Ashley
Morihata, Akimasa
Morin, Pat
Morizumi, Hiroki
Moruz, Gabriel
Moseley, Benjamin

Mousset, Frank
Mucha, Marcin
Mueller, Tobias
Müller, David
Müller-Hannemann, Matthias
Murakami, Keisuke
Murano, Aniello
Musco, Christopher
Mustafa, Nabil
Nadathur, Gopalan
Nagano, Kiyohito
Nakazawa, Koji
Nanongkai, Danupon
Narayanan, Hariharan
Navarra, Alfredo
Navarro, Gonzalo
Nayyeri, Amir
Nederhof, Mark-Jan
Nederlof, Jesper
Newman, Alantha
Nguyen, Huy
Nguyen, Kim Thang
Nguyen, Viet Hung
Niazadeh, Rad
Nicholson, Patrick K.
Niedermann, Benjamin
Nielsen, Jesper Buus
Nielsen, Jesper Sindahl
Nies, André
Nikolov, Aleksandar
Nishimura, Harumichi
Nitaj, Abderrahmane
Nöllenburg, Martin
Nordhoff, Benedikt
Novotný, Petr
Obremski, Maciej
Ochremiak, Joanna
Oh, Eunjin
Okamoto, Yoshio
Oliveira, Igor
Onak, Krzysztof
Ordóñez Pereira, Alberto
Oren, Sigal
Orlandi, Claudio
Otachi, Yota
Ott, Sebastian

Otto, Martin
Oveis Gharan, Shayan
Ozeki, Kenta
Ozols, Maris
Padro, Carles
Pagani, Michele
Pagh, Rasmus
Paluch, Katarzyna
Panagiotou, Konstantinos
Panigrahi, Debmalya
Paolini, Luca
Parter, Merav
Pasquale, Francesco
Paul, Christophe
Pedersen, Christian Nørgaard Storm
Pelc, Andrzej
Penna, Paolo
Perdrix, Simon
Perelli, Giuseppe
Persiano, Giuseppe
Pettie, Seth
Peva, Blanchard
Philip, Geevarghese
Phillips, Jeff
Piccolo, Mauro
Pietrzak, Krzysztof
Pilaud, Vincent
Piliouras, Georgios
Pilipczuk, Marcin
Pinto, Joao Sousa
Piterman, Nir
Place, Thomas
Poelstra, Andrew
Pokutta, Sebastian
Polak, Libor
Polishchuk, Valentin
Pountourakis, Emmanouil
Prencipe, Giuseppe
Pruhs, Kirk
Prutkin, Roman
Qin, Shengchao
Quas, Anthony
Rabehaja, Tahiry
Räcke, Harald
Raghavendra, Prasad
Raghothaman, Mukund

Raman, Rajiv
Raskin, Jean-Francois
Razenshteyn, Ilya
Regev, Oded
Rehak, Vojtech
Reis, Giselle
van Renssen, André
Reshef, Yakir
Reyzin, Leonid
Reyzin, Lev
Riba, Colin
Richerby, David
Riely, James
Riveros, Cristian
Robere, Robert
Robinson, Peter
Roeloffzen, Marcel
Röglin, Heiko
Rote, Günter
Rotenberg, Eva
Roth, Aaron
Rothvoss, Thomas
de Rougemont, Michel
Rümmele, Stefan
Sabel, David
Sabok, Marcin
Sacchini, Jorge Luis
Sach, Benjamin
Saha, Ankan
Saha, Chandan
Saitoh, Toshiki
Sakavalas, Dimitris
Salvati, Sylvain
Sanchez Villaamil, Fernando
Sangnier, Arnaud
Sankowski, Piotr
Sankur, Ocan
Saptharishi, Ramprasad
Saraswat, Vijay
Satti, Srinivasa Rao
Saurabh, Saket
Sawant, Anshul
Scharf, Ludmila
Schieber, Baruch
Schlotter, Ildikó
Schneider, Stefan

Schnitger, Georg
Schoenebeck, Grant
Schrijvers, Okke
Schweitzer, Pascal
Schweller, Robert
Schwitter, Rolf
Schöpp, Ulrich
Scquizzato, Michele
Seddighin, Saeed
Segev, Danny
Seidel, Jochen
Seiferth, Paul
Sekar, Shreyas
Sen, Siddhartha
Senizergues, Geraud
Serre, Olivier
Seshadhri, C.
Seto, Kazuhisa
Seurin, Yannick
Shepherd, Bruce
Sherstov, Alexander
Shi, Yaoyun
Shinkar, Igor
Shioura, Akiyoshi
Siebertz, Sebastian
Singh, Mohit
Sitters, Rene
Sivignon, Isabelle
Skorski, Maciej
Skrzypczak, Michał
Skutella, Martin
Smith, Adam
Soares Barbosa, Rui
Sobocinski, Pawel
Solan, Eilon
Sommer, Christian
Son, Wanbin
Sorensen, Tyler
Sorge, Manuel
Sottile, Frank
Spalek, Robert
Spoerhase, Joachim
Srba, Jiri
Srivastava, Piyush
Staals, Frank
Stampoulis, Antonis

Staton, Sam
Stefankovic, Daniel
Stein, Clifford
Stein, Yannik
Stenman, Jari
Stephan, Frank
Stirling, Colin
Stokes, Klara
Stolz, David
Strasser, Ben
Streicher, Thomas
Sun, He
Sun, Xiaorui
Suomela, Jukka
Svendsen, Kasper
Sviridenko, Maxim
Swamy, Chaitanya
Takahashi, Yasuhiro
Takazawa, Kenjiro
Talebanfard, Navid
Tamaki, Suguru
Tan, Li-Yang
Tan, Tony
Tang, Bo
Tanigawa, Shin-Ichi
Tasson, Christine
Tavenas, Sébastien
Teillaud, Monique
Telelis, Orestis
Thaler, Justin
Thapper, Johan
Thomas, Rekha
Ting, Hingfung
Tiwary, Hans
Torán, Jacobo
Tov, Roei
Tovey, Craig
Treinen, Ralf
Triandopoulos, Nikos
Trung, Ta Quang
Tsukada, Takeshi
Tulsiani, Madhur
Tuosto, Emilio
Tzamos, Christos
Uchizawa, Kei
Ueno, Shuichi

Uitto, Jara
Ullman, Jon
Ullman, Jonathan
Umboh, Seeun
Unno, Hiroshi
Uno, Yushi
Uramoto, Takeo
Urrutia, Florent
Vagvolgyi, Sandor
Vahlis, Yevgeniy
Valiron, Benoît
Vanden Boom, Michael
Vdovina, Alina
Veith, David
Venkatasubramanian, Suresh
Venkitasubramaniam,
    Muthuramakrishnan
Ventre, Carmine
Vereshchagin, Nikolay
Vidick, Thomas
Vijayaraghavan, Aravindan
Vildhøj, Hjalte Wedel
Vinayagamurthy, Dhinakaran
Vishnoi, Nisheeth
Vitanyi, Paul
Vivek, Srinivas
Vondrak, Jan
Voudouris, Alexandros
Wahlström, Magnus
Walter, Tobias
Walukiewicz, Igor
Wasa, Kunihiro
Watanabe, Osamu
Wee, Hoeteck
Wegner, Franziska
Wei, Zhewei
Weichert, Volker
Weinberg, S. Matthew
Weinstein, Omri
Wenner, Alexander
Werneck, Renato
Wexler, Tom
White, Colin
Wichs, Daniel
Wiese, Andreas
Willard, Ross

Williams, Ryan
Williamson, David
Wilson, David
Wimmer, Karl
Winslow, Andrew
Woeginger, Gerhard J.
Wojtczak, Dominik
de Wolf, Ronald
Wolff, Alexander
Wong, Prudence W.H.
Woodruff, David
Wootters, Mary
Worrell, James
Wrochna, Marcin
Wu, Xiaodi
Wu, Zhilin
Xiao, Tao
Xie, Ning
Xu, Jinhui
Yamakami, Tomoyuki
Yamamoto, Masaki
Yamauchi, Yukiko
Yang, Kuan

Yaroslavtsev, Grigory
Yehudayoff, Amir
Yodpinyanee, Anak
Yogev, Eylon
Yoon, Sang-Duk
Yoshida, Yuichi
Yun, Aaram
Yuster, Raphael
Zampetakis, Emmanouil
Zanuttini, Bruno
Zemor, Gilles
Zhang, Chihao
Zhang, Jialin
Zhang, Qin
Zhang, Shengyu
Zhou, Gelin
Zhou, Yuan
Živný, Stanislav
Zois, Georgios
Zorzi, Margherita
van Zwam, Stefan
Zwick, Uri

Towards the Graph Minor Theorems
for Directed Graphs

Ken-ichi Kawarabayashi[1]([⊠]) and Stephan Kreutzer[2]

[1] National Institute of Informatics, 2-1-2 Hitotsubashi, Chiyoda-ku, Tokyo, Japan
k_keniti@nii.ac.jp
[2] Technical University Berlin, Sekr TEL 7-3, Ernst-Reuter Platz 7, 10587
Berlin, Germany
stephan.kreutzer@tu-berlin.de

Abstract. Two key results of Robertson and Seymour's graph minor theory are:

1. a structure theorem stating that all graphs excluding some fixed graph as a minor have a tree decomposition into pieces that are almost embeddable in a fixed surface.

2. the *k-disjoint paths problem* is tractable when k is a fixed constant: given a graph G and k pairs $(s_1, t_1), \ldots, (s_k, t_k)$ of vertices of G, decide whether there are k mutually vertex disjoint paths of G, the ith path linking s_i and t_i for $i = 1, \ldots, k$.

In this talk, we shall try to look at the corresponding problems for digraphs.

Concerning the first point, the grid theorem, originally proved in 1986 by Robertson and Seymour in Graph Minors V, is the basis (even for the whole graph minor project). In the mid-90s, Reed and Johnson, Robertson, Seymour and Thomas (see [13,26]), independently, conjectured an analogous theorem for directed graphs, i.e. the existence of a function $f : \mathbb{N} \to \mathbb{N}$ such that every digraph of directed treewidth at least $f(k)$ contains a directed grid of order k. In an unpublished manuscript from 2001, Johnson, Robertson, Seymour and Thomas give a proof of this conjecture for planar digraphs. But for over a decade, this was the most general case proved for the conjecture.

We are finally able to confirm the Reed, Johnson, Robertson, Seymour and Thomas conjecture in full generality. As a consequence of our results we are able to improve results in Reed et al. in 1996 [27] to disjoint cycles of length at least l. This would be the first but a significant step toward the structural goals for digraphs (hence towards the first point).

Concerning the second point, in [19] we contribute to the disjoint paths problem using the directed grid theorem. We show that the following can be done in polynomial time:

K.-I. Kawarabayashi—This work was supported by JST ERATO Kawarabayashi Large Graph Project and by Mitsubishi Foundation.

S. Kreutzer—This project has received funding from the European Research Council (ERC) under the European Unions Horizon 2020 research and innovation programme (grant agreement No 648527).

Suppose that we are given a digraph G and k terminal pairs $(s_1, t_1), (s_2, t_2), \ldots, (s_k, t_k)$, where k is a fixed constant. In polynomial time, either

- we can find k paths $P_1, \ldots, P_k$ such that P_i is from s_i to t_i for $i = 1, \ldots, k$ and every vertex in G is in at most four of the paths, or
- we can conclude that G does not contain disjoint paths $P_1, \ldots, P_k$ such that P_i is from s_i to t_i for $i = 1, \ldots, k$.

To the best of our knowledge, this is the first positive result for the general directed disjoint paths problem (and hence for the second point). Note that the directed disjoint paths problem is NP-hard even for $k = 2$. Therefore, this kind of results is the best one can hope for.

We also report some progress on the above two points.

Dynamic Graphs: Time, Space and Communication

Valerie King[(⊠)]

University of Victoria, Victoria, Canada
val@uvic.ca

Abstract. A dynamic graph is a graph which experiences a sequence of local updates, typically in the form of edge insertions and deletions. A dynamic graph algorithm for a graph property is a data structure which processes a sequence of updates while answering queries about a property. The concern of dynamic graph algorithms is primarily time, to minimize the update and query time. A graph streaming algorithm is a one-time computation of a graph property, where the input is reported one edge or one node and adjacent edges at a time. The concern of a streaming algorithm is space, to use memory nearly linear in the number of nodes and sublinear in the number of edges. In distributed computing, each node in a graph has only local information about the graph structure. Each must make a decision which may affect a global property of the graph, based on this information and messages received from its neighbors. The concern is communication, to minimize the number of bits communicated and time in terms of the number of communication rounds.

This talk will report on a convergence of these approaches, resulting in a dynamic data structure for answering connectivity queries in a graph in worst case update and query time polylogarithmic in the size of the graph and sublinear space, and the first distributed algorithm to build a spanning forest which requires substantially less communication than edges in the graph. The talk will also discuss future directions for work in this area.

Automated Synthesis of Distributed Controllers

Anca Muscholl[(✉)]

LaBRI, University of Bordeaux, Bordeaux, France
anca@labri.fr

Abstract. Synthesis is a particularly challenging problem for concurrent programs. At the same time it is a very promising approach, since concurrent programs are difficult to get right, or to analyze with traditional verification techniques. This paper gives an introduction to distributed synthesis in the setting of Mazurkiewicz traces, and its applications to decentralized runtime monitoring.

Incentive Networks

Thomas Moscibroda[✉]

Microsoft Research and Tsinghua University, Beijing, China
moscitho@microsoft.com

1 Extended Abstract

Crowdsourcing and human-computing systems that mobilize people's work have become the method of choice to quickly and efficiently solve many tasks. Commercial offerings such as Gigwalk or Amazon's Mechanical Turk allow users to recruit people to complete tasks. Crowdsourcing is commonly used to obtain large-scale user data, such as environmental data, application traces, to generate maps, or for labelling. A key challenge in successfully deploying any such system is *how to incentivize people to participate and contribute* as much as possible. In fact, this challenge is very common in systems that rely on user contributions. For instance, social forums, file-sharing services, public computing projects (e.g. SETI@Home), or collaborative reference works often suffer from the well-known bootstrapping problem. These systems can become self-sustaining once the scale of participation exceeds a certain threshold, but below this threshold, they may not by themselves provide sufficient inherent benefit for users.

In my talk, I will discuss the algorithmic foundations of two types of network-based incentive structures that can be used to encourage users to participate in and contribute to a system: *Incentive Trees* [1,4] and *Incentive Networks* [5].

Incentive Trees. Incentive Trees are tree-based mechanisms in which (i) each participant is rewarded for contributing to the system, and (ii) a participant can make referrals and thereby solicit new participants to also join the system and contribute to it. The mechanism incentivizes solicitations by making a solicitor's reward depend on the contributions (and recursively also on their further solicitations, etc) made by such solicitees. An *Incentive Tree mechanism* is an algorithm that determines how much reward each participant receives based on all the participants' contributions as well as the structure of the solicitation tree. Incentive Trees have been widely used in various domains and under different names, e.g., in referral trees, multi-level marketing schemes [2], affiliate marketing, MIT's winning strategy in the Red Balloon Challenge [6], and even in the form of the infamous illegal Pyramid Schemes.

Using an axiomatic approach, we seek to understand the possibilities and limitations of incentive trees. Our goal is to characterize what desirable properties are achievable; and to design incentive tree mechanisms that achieve a best possible set of such properties. The key challenge is to find mechanisms that simultaneously guarantee contribution and solicitation incentive, while also preventing strategic attacks, such as multi-identity attacks. As it turns out, the set of desirable properties that are mutually satisfiable is robust with regard to a wide set of modeling assumptions. For the two most basic models, I will present key impossibility results, as well as mechanisms that are "optimal" in the sense that they achieve a maximally satisfiable subset of desirable properties. Interestingly, the algorithmic structure of these optimal solutions is unusual and reveals new insights into the structure of incentive trees.

Incentive Networks. Incentive Networks are a different concept: They can be used to maximize the users' contribution when participation has already been established. Consider a basic economic incentive system, in which each participant receives a reward according to his own contribution to the system. Examples of such basic systems are endless: Jobs with hourly wages, membership savings rewards (buy 10 coffee, get one free), loyalty or airplane mileage programs, etc. In each of these systems, a user receives a reward based on his own contribution.

In this talk, I will discuss an alternative to the above basic economic system: *Incentive Networks.* In an incentive network, a participant's reward depends not only on his *own* contribution; but also in part on the contributions made by his social contacts or friends. The concept is exceedingly natural and practical: Instead of receiving pay only for your own work, in an incentive network you are rewarded for your own work plus your friend's work. So, for example, each worker is paid per hour of his own work, plus an additional amount for each hour of his friends' work. Or, in a membership rewards program, a coffee shop offers rewards to a customer whenever she consumes a cup of coffee, as well as whenever one of her designated friends does.

I will show that the key parameter effecting the efficiency of such an *Incentive Network-based reward system* depends on the participant?s *directed altruism* [3]. Directed altruism is the extent to which someone is willing to work if his work results in a payment to his friend, rather than to himself. Specifically, we characterize the condition under which an Incentive Network-based economy is more efficient than the basic "pay-for-your-contribution" economy, and we quantify the savings when using incentive networks. I will discuss the impact of the network topology and exogenous parameters on the efficiency of incentive networks. The results suggest that in many real-world practical settings, Incentive Network-based reward systems or compensation structures could be more efficient than the ubiquitous "pay-for-your-contribution" schemes.

References

1. Douceur, J., Moscibroda, T.: Lottery Trees: motivational deployment of networked systems. In: Proc. of SIGCOMM (2007)
2. Emek, Y., Karidi, R., Tennenholtz, M., Zohar, A.: Mechanisms for multi-level marketing. In: Proc. of 12th ACM Conference on Electronic Commerce (EC) (2011)
3. Leider, S., Mobius, M., Rosenblat, T., Do, Q.: Directed altruism and enforced reciprocity in social networks. The Quarterly Journal of Economics (2009)
4. Lv, Y., Moscibroda, T.: Fair and resilient incentive tree mechanisms. In: Proc. of 32nd ACM Symposium on Principles of Distributed Computing (PODC) (2013)
5. Lv, Y., Moscibroda, T.: Incentive networks. In: Proc. of 29th AAAI Conference on Artificial Intelligence (AAAI) (2014)
6. Pickard, G., Pan, W., Rahwan, I., Cebrian, M., Crane, R., Madan, A., Pentland, A.: Time Critical Social Mobilization. Science (2011)

Fast Algorithms for Structured Sparsity

Piotr Indyk[(✉)]

Massachusetts Institute of Technology, Cambridge, USA
indyk@mit.edu

Abstract. Sparse representations of signals (i.e., representations that have only few non-zero or large coefficients) have emerged as powerful tools in signal processing theory, algorithms, machine learning and other applications. However, real-world signals often exhibit rich structure beyond mere sparsity. For example, a natural image, once represented in the wavelet domain, often has the property that its large coefficients occupy a *subtree* of the wavelet hierarchy, as opposed to arbitrary positions. A general approach to capturing this type of additional structure is to model the support of the signal of interest (i.e., the set of indices of large coefficients) as belonging to a particular family of sets. Computing a sparse representation of the signal then corresponds to the problem of finding the support from the family that maximizes the sum of the squares of the selected coefficients. Such a modeling approach has proved to be beneficial in a number of applications including compression, de-noising, compressive sensing and machine learning. However, the resulting optimization problem is often computationally difficult or intractable, which is undesirable in many applications where large signals and datasets are commonplace.

In this talk, I will outline some of the past and more recent algorithms for finding structured sparse representations of signals, including piecewise constant approximations, tree-sparse approximations and graph-sparse approximations. The algorithms borrow several techniques from combinatorial optimization (e.g., dynamic programming), graph theory, and approximation algorithms. For many problems the algorithms often run in (nearly) linear time, which makes them applicable to very large datasets.

Computational Complexity of Puzzles and Games

Ryuhei Uehara[✉]

School of Information Science, Japan Advanced Institute of Science and Technology,
Asahidai 1-1, Nomi, Ishikawa, Japan
uehara@jaist.ac.jp

Abstract. A computation consists of algorithm of basic operations. When you consider an algorithm, you assume, say, the standard RAM model, that has "usual" arithmetic operations. On the other hand, when you consider an algorithm on a DNA computer, your basic operations are duplication and inversion on a string. Then you need to consider completely different algorithms, and their computational complexity also changes. That is, when we discuss computational complexity of a problem, it strongly depends on the set of basic operations you use. When you enjoy a puzzle, you have to find an algorithm by combining reasonable basic operations to its goal. (Some puzzles require to find the basic operations themselves, but we do not consider such puzzles in this talk.) From the viewpoint of theoretical computer science, puzzles give us some insight to computation and computational complexity classes in various way.

Some puzzles and games give reasonable characterizations to computational complexity classes. For example, "pebble game" is a classic model that gives some complexity classes in a natural way, and "constraint logic" is recent model that succeeds to solve a long standing open problem due to Martin Gardner that asks the computational complexity of sliding block puzzles. Such puzzles gives us "typical" and characterization and "intuitive" understanding for some computational complexity classes.

On the other hand, there are some puzzles and games that give non-trivial interesting aspects of computational complexity classes. For example, consider "14-15 puzzle" which is classic well known sliding puzzle. By parity, we can determine if one arrangement can be slid to the other in linear time. Moreover, we can always find a way for sliding between them in quadratic time. However, interestingly, finding the optimal solution is NP-complete in general. I also introduce a relatively new notion of the reconfiguration problem. This series of new problems will give some new notion of computational complexity classes.

Contents – Part II

Invited Talks

Towards the Graph Minor Theorems for Directed Graphs 3
    Ken-Ichi Kawarabayashi and Stephan Kreutzer

Automated Synthesis of Distributed Controllers 11
    Anca Muscholl

Track B: Logic, Semantics, Automata and Theory of Programming

Games for Dependent Types 31
    Samson Abramsky, Radha Jagadeesan, and Matthijs Vákár

Short Proofs of the Kneser-Lovász Coloring Principle 44
    James Aisenberg, Maria Luisa Bonet, Sam Buss, Adrian Crãciun,
    and Gabriel Istrate

Provenance Circuits for Trees and Treelike Instances 56
    Antoine Amarilli, Pierre Bourhis, and Pierre Senellart

Language Emptiness of Continuous-Time Parametric Timed Automata 69
    Nikola Beneš, Peter Bezděk, Kim G. Larsen, and Jiří Srba

Analysis of Probabilistic Systems via Generating Functions and Padé
Approximation 82
    Michele Boreale

On Reducing Linearizability to State Reachability 95
    Ahmed Bouajjani, Michael Emmi, Constantin Enea, and Jad Hamza

The Complexity of Synthesis from Probabilistic Components 108
    Krishnendu Chatterjee, Laurent Doyen, and Moshe Y. Vardi

Edit Distance for Pushdown Automata 121
    Krishnendu Chatterjee, Thomas A. Henzinger, Rasmus Ibsen-Jensen,
    and Jan Otop

Solution Sets for Equations over Free Groups Are EDT0L Languages 134
    Laura Ciobanu, Volker Diekert, and Murray Elder

Limited Set quantifiers over Countable Linear Orderings 146
    Thomas Colcombet and A.V. Sreejith

Reachability Is in DynFO 159
Samir Datta, Raghav Kulkarni, Anish Mukherjee, Thomas Schwentick,
and Thomas Zeume

Natural Homology 171
Jérémy Dubut, Éric Goubault, and Jean Goubault-Larrecq

Greatest Fixed Points of Probabilistic Min/Max Polynomial Equations,
and Reachability for Branching Markov Decision Processes 184
Kousha Etessami, Alistair Stewart, and Mihalis Yannakakis

Trading Bounds for Memory in Games with Counters................. 197
Nathanaël Fijalkow, Florian Horn, Denis Kuperberg,
and Michał Skrzypczak

Decision Problems of Tree Transducers with Origin.................. 209
Emmanuel Filiot, Sebastian Maneth, Pierre-Alain Reynier,
and Jean-Marc Talbot

Incompleteness Theorems, Large Cardinals, and Automata
over Infinite Words.. 222
Olivier Finkel

The Odds of Staying on Budget 234
Christoph Haase and Stefan Kiefer

From Sequential Specifications to Eventual Consistency 247
Radha Jagadeesan and James Riely

Fixed-Dimensional Energy Games Are in Pseudo-Polynomial Time 260
Marcin Jurdziński, Ranko Lazić, and Sylvain Schmitz

An Algebraic Geometric Approach to Nivat's Conjecture 273
Jarkko Kari and Michal Szabados

Nominal Kleene Coalgebra 286
Dexter Kozen, Konstantinos Mamouras, Daniela Petrişan,
and Alexandra Silva

On Determinisation of Good-for-Games Automata 299
Denis Kuperberg and Michał Skrzypczak

Owicki-Gries Reasoning for Weak Memory Models 311
Ori Lahav and Viktor Vafeiadis

On the Coverability Problem for Pushdown Vector Addition Systems
in One Dimension... 324
Jérôme Leroux, Grégoire Sutre, and Patrick Totzke

Compressed Tree Canonization. 337
  Markus Lohrey, Sebastian Maneth, and Fabian Peternek

Parsimonious Types and Non-uniform Computation 350
  Damiano Mazza and Kazushige Terui

Baire Category Quantifier in Monadic Second Order Logic 362
  Henryk Michalewski and Matteo Mio

Liveness of Parameterized Timed Networks. 375
  Benjamin Aminof, Sasha Rubin, Florian Zuleger, and Francesco Spegni

Symmetric Strategy Improvement. 388
  Sven Schewe, Ashutosh Trivedi, and Thomas Varghese

Effect Algebras, Presheaves, Non-locality and Contextuality 401
  Sam Staton and Sander Uijlen

On the Complexity of Intersecting Regular, Context-Free,
and Tree Languages 414
  Joseph Swernofsky and Michael Wehar

Containment of Monadic Datalog Programs via Bounded Clique-Width 427
  Mikołaj Bojańczyk, Filip Murlak, and Adam Witkowski

An Approach to Computing Downward Closures 440
  Georg Zetzsche

How Much Lookahead Is Needed to Win Infinite Games?. 452
  Felix Klein and Martin Zimmermann

Track C: Foundations of Networked Computation: Models, Algorithms and Information Management

Symmetric Graph Properties Have Independent Edges 467
  Dimitris Achlioptas and Paris Siminelakis

Polylogarithmic-Time Leader Election in Population Protocols 479
  Dan Alistarh and Rati Gelashvili

Core Size and Densification in Preferential Attachment Networks 492
  Chen Avin, Zvi Lotker, Yinon Nahum, and David Peleg

Maintaining Near-Popular Matchings 504
  *Sayan Bhattacharya, Martin Hoefer, Chien-Chung Huang,
  Telikepalli Kavitha, and Lisa Wagner*

Ultra-Fast Load Balancing on Scale-Free Networks   516
Karl Bringmann, Tobias Friedrich, Martin Hoefer, Ralf Rothenberger,
and Thomas Sauerwald

Approximate Consensus in Highly Dynamic Networks: The Role
of Averaging Algorithms   528
Bernadette Charron-Bost, Matthias Függer, and Thomas Nowak

The Range of Topological Effects on Communication   540
Arkadev Chattopadhyay and Atri Rudra

Secretary Markets with Local Information   552
Ning Chen, Martin Hoefer, Marvin Künnemann, Chengyu Lin,
and Peihan Miao

A Simple and Optimal Ancestry Labeling Scheme for Trees   564
Søren Dahlgaard, Mathias Bæk Tejs Knudsen, and Noy Rotbart

Interactive Communication with Unknown Noise Rate   575
Varsha Dani, Mahnush Movahedi, Jared Saia, and Maxwell Young

Fixed Parameter Approximations for k-Center Problems in Low Highway
Dimension Graphs   588
Andreas Emil Feldmann

A Unified Framework for Strong Price of Anarchy in Clustering Games   601
Michal Feldman and Ophir Friedler

On the Diameter of Hyperbolic Random Graphs   614
Tobias Friedrich and Anton Krohmer

Tight Bounds for Cost-Sharing in Weighted Congestion Games   626
Martin Gairing, Konstantinos Kollias, and Grammateia Kotsialou

Distributed Broadcast Revisited: Towards Universal Optimality   638
Mohsen Ghaffari

Selling Two Goods Optimally   650
Yiannis Giannakopoulos and Elias Koutsoupias

Adaptively Secure Coin-Flipping, Revisited   663
Shafi Goldwasser, Yael Tauman Kalai, and Sunoo Park

Optimal Competitiveness for the Rectilinear Steiner Arborescence
Problem   675
Erez Kantor and Shay Kutten

Normalization Phenomena in Asynchronous Networks   688
   Amin Karbasi, Johannes Lengler, and Angelika Steger

Broadcast from Minicast Secure Against General Adversaries   701
   Pavel Raykov

Author Index   713

Contents – Part I

Track A: Algorithms, Complexity and Games

Statistical Randomized Encodings: A Complexity Theoretic View 1
  Shweta Agrawal, Yuval Ishai, Dakshita Khurana,
  and Anat Paskin-Cherniavsky

Tighter Fourier Transform Lower Bounds 14
  Nir Ailon

Quantifying Competitiveness in Paging with Locality of Reference 26
  Susanne Albers and Dario Frascaria

Approximation Algorithms for Computing Maximin Share Allocations 39
  Georgios Amanatidis, Evangelos Markakis, Afshin Nikzad,
  and Amin Saberi

Envy-Free Pricing in Large Markets: Approximating Revenue
and Welfare 52
  Elliot Anshelevich, Koushik Kar, and Shreyas Sekar

Batched Point Location in SINR Diagrams via Algebraic Tools 65
  Boris Aronov and Matthew J. Katz

On the Randomized Competitive Ratio of Reordering Buffer Management
with Non-uniform Costs 78
  Noa Avigdor-Elgrabli, Sungjin Im, Benjamin Moseley, and Yuval Rabani

Serving in the Dark Should Be Done Non-uniformly 91
  Yossi Azar and Ilan Reuven Cohen

Finding the Median (Obliviously) with Bounded Space 103
  Paul Beame, Vincent Liew, and Mihai Pătraşcu

Approximation Algorithms for Min-Sum k-Clustering
and Balanced k-Median 116
  Babak Behsaz, Zachary Friggstad, Mohammad R. Salavatipour,
  and Rohit Sivakumar

Solving Linear Programming with Constraints Unknown 129
  Xiaohui Bei, Ning Chen, and Shengyu Zhang

Deterministic Randomness Extraction from Generalized and Distributed
Santha-Vazirani Sources   143
  Salman Beigi, Omid Etesami, and Amin Gohari

Limitations of Algebraic Approaches to Graph Isomorphism Testing   155
  Christoph Berkholz and Martin Grohe

Fully Dynamic Matching in Bipartite Graphs   167
  Aaron Bernstein and Cliff Stein

Feasible Interpolation for QBF Resolution Calculi   180
  Olaf Beyersdorff, Leroy Chew, Meena Mahajan, and Anil Shukla

Simultaneous Approximation of Constraint Satisfaction Problems   193
  Amey Bhangale, Swastik Kopparty, and Sushant Sachdeva

Design of Dynamic Algorithms via Primal-Dual Method   206
  Sayan Bhattacharya, Monika Henzinger, and Giuseppe F. Italiano

What Percentage of Programs Halt?   219
  Laurent Bienvenu, Damien Desfontaines, and Alexander Shen

The Parity of Set Systems Under Random Restrictions with Applications
to Exponential Time Problems   231
  Andreas Björklund, Holger Dell, and Thore Husfeldt

Spotting Trees with Few Leaves   243
  Andreas Björklund, Vikram Kamat, Łukasz Kowalik, and Meirav Zehavi

Constraint Satisfaction Problems over the Integers with Successor   256
  Manuel Bodirsky, Barnaby Martin, and Antoine Mottet

Hardness Amplification and the Approximate Degree
of Constant-Depth Circuits   268
  Mark Bun and Justin Thaler

Algorithms and Complexity for Turaev-Viro Invariants   281
  Benjamin A. Burton, Clément Maria, and Jonathan Spreer

Big Data on the Rise? – Testing Monotonicity of Distributions   294
  Clément L. Canonne

Unit Interval Editing Is Fixed-Parameter Tractable   306
  Yixin Cao

Streaming Algorithms for Submodular Function Maximization   318
  Chandra Chekuri, Shalmoli Gupta, and Kent Quanrud

Multilinear Pseudorandom Functions.   331
  Aloni Cohen and Justin Holmgren

Zero-Fixing Extractors for Sub-Logarithmic Entropy.   343
  Gil Cohen and Igor Shinkar

Interactive Proofs with Approximately Commuting Provers   355
  Matthew Coudron and Thomas Vidick

Popular Matchings with Two-Sided Preferences and One-Sided Ties   367
  Ágnes Cseh, Chien-Chung Huang, and Telikepalli Kavitha

Block Interpolation: A Framework for Tight Exponential-Time
Counting Complexity   380
  Radu Curticapean

On Convergence and Threshold Properties of Discrete Lotka-Volterra
Population Protocols   393
  Jurek Czyzowicz, Leszek Gąsieniec, Adrian Kosowski,
  Evangelos Kranakis, Paul G. Spirakis, and Przemysław Uznański

Scheduling Bidirectional Traffic on a Path.   406
  Yann Disser, Max Klimm, and Elisabeth Lübbecke

On the Problem of Approximating the Eigenvalues of Undirected Graphs
in Probabilistic Logspace.   419
  Dean Doron and Amnon Ta-Shma

On Planar Boolean CSP.   432
  Zdeněk Dvořák and Martin Kupec

On Temporal Graph Exploration.   444
  Thomas Erlebach, Michael Hoffmann, and Frank Kammer

Mind Your Coins: Fully Leakage-Resilient Signatures
with Graceful Degradation.   456
  Antonio Faonio, Jesper Buus Nielsen, and Daniele Venturi

A $(1+\varepsilon)$-Embedding of Low Highway Dimension Graphs into Bounded
Treewidth Graphs   469
  Andreas Emil Feldmann, Wai Shing Fung, Jochen Könemann,
  and Ian Post

Lower Bounds for the Graph Homomorphism Problem   481
  Fedor V. Fomin, Alexander Golovnev, Alexander S. Kulikov,
  and Ivan Mihajlin

Parameterized Single-Exponential Time Polynomial Space Algorithm
for Steiner Tree 494
  *Fedor V. Fomin, Petteri Kaski, Daniel Lokshtanov, Fahad Panolan,
  and Saket Saurabh*

Relative Discrepancy Does not Separate Information
and Communication Complexity.................................... 506
  *Lila Fontes, Rahul Jain, Iordanis Kerenidis, Sophie Laplante,
  Mathieu Laurière, and Jérémie Roland*

A Galois Connection for Valued Constraint Languages of Infinite Size..... 517
  Peter Fulla and Stanislav Živný

Approximately Counting H-Colourings Is #BIS-Hard 529
  Andreas Galanis, Leslie Ann Goldberg, and Mark Jerrum

Taylor Polynomial Estimator for Estimating Frequency Moments......... 542
  Sumit Ganguly

ETR-Completeness for Decision Versions of Multi-player (Symmetric)
Nash Equilibria.. 554
  Jugal Garg, Ruta Mehta, Vijay V. Vazirani, and Sadra Yazdanbod

Separate, Measure and Conquer: Faster Polynomial-Space Algorithms
for Max 2-CSP and Counting Dominating Sets 567
  Serge Gaspers and Gregory B. Sorkin

Submatrix Maximum Queries in Monge Matrices Are Equivalent
to Predecessor Search 580
  Paweł Gawrychowski, Shay Mozes, and Oren Weimann

Optimal Encodings for Range Top-k, Selection, and Min-Max........... 593
  Paweł Gawrychowski and Patrick K. Nicholson

2-Vertex Connectivity in Directed Graphs 605
  *Loukas Georgiadis, Giuseppe F. Italiano, Luigi Laura,
  and Nikos Parotsidis*

Ground State Connectivity of Local Hamiltonians 617
  Sevag Gharibian and Jamie Sikora

Uniform Kernelization Complexity of Hitting Forbidden Minors 629
  *Archontia C. Giannopoulou, Bart M.P. Jansen, Daniel Lokshtanov,
  and Saket Saurabh*

Counting Homomorphisms to Square-Free Graphs, Modulo 2 642
  Andreas Göbel, Leslie Ann Goldberg, and David Richerby

Approximately Counting Locally-Optimal Structures   654
  Leslie Ann Goldberg, Rob Gysel, and John Lapinskas

Proofs of Proximity for Context-Free Languages and Read-Once
Branching Programs (Extended Abstract).   666
  Oded Goldreich, Tom Gur, and Ron D. Rothblum

Fast Algorithms for Diameter-Optimally Augmenting Paths.   678
  Ulrike Große, Joachim Gudmundsson, Christian Knauer,
  Michiel Smid, and Fabian Stehn

Hollow Heaps   689
  Thomas Dueholm Hansen, Haim Kaplan, Robert E. Tarjan,
  and Uri Zwick

Linear-Time List Recovery of High-Rate Expander Codes.   701
  Brett Hemenway and Mary Wootters

Finding 2-Edge and 2-Vertex Strongly Connected Components
in Quadratic Time.   713
  Monika Henzinger, Sebastian Krinninger, and Veronika Loitzenbauer

Improved Algorithms for Decremental Single-Source Reachability
on Directed Graphs.   725
  Monika Henzinger, Sebastian Krinninger, and Danupon Nanongkai

Weighted Reordering Buffer Improved via Variants of Knapsack
Covering Inequalities   737
  Sungjin Im and Benjamin Moseley

Local Reductions   749
  Hamid Jahanjou, Eric Miles, and Emanuele Viola

Query Complexity in Expectation.   761
  Jedrzej Kaniewski, Troy Lee, and Ronald de Wolf

Near-Linear Query Complexity for Graph Inference   773
  Sampath Kannan, Claire Mathieu, and Hang Zhou

A QPTAS for the Base of the Number of Crossing-Free Structures
on a Planar Point Set.   785
  Marek Karpinski, Andrzej Lingas, and Dzmitry Sledneu

Finding a Path in Group-Labeled Graphs with Two Labels Forbidden.   797
  Yasushi Kawase, Yusuke Kobayashi, and Yutaro Yamaguchi

Lower Bounds for Sums of Powers of Low Degree Univariates   810
  Neeraj Kayal, Pascal Koiran, Timothée Pecatte, and Chandan Saha

Approximating CSPs Using LP Relaxation   822
  Subhash Khot and Rishi Saket

Comparator Circuits over Finite Bounded Posets   834
  Balagopal Komarath, Jayalal Sarma, and K.S. Sunil

Algebraic Properties of Valued Constraint Satisfaction Problem   846
  Marcin Kozik and Joanna Ochremiak

Towards Understanding the Smoothed Approximation Ratio
of the 2-Opt Heuristic   859
  Marvin Künnemann and Bodo Manthey

On the Hardest Problem Formulations for the 0/1 Lasserre Hierarchy   872
  Adam Kurpisz, Samuli Leppänen, and Monaldo Mastrolilli

Replacing Mark Bits with Randomness in Fibonacci Heaps   886
  Jerry Li and John Peebles

A PTAS for the Weighted Unit Disk Cover Problem   898
  Jian Li and Yifei Jin

Approximating the Expected Values for Combinatorial Optimization
Problems Over Stochastic Points   910
  Lingxiao Huang and Jian Li

Deterministic Truncation of Linear Matroids   922
  Daniel Lokshtanov, Pranabendu Misra, Fahad Panolan,
  and Saket Saurabh

Linear Time Parameterized Algorithms for Subset Feedback Vertex Set   935
  Daniel Lokshtanov, M.S. Ramanujan, and Saket Saurabh

An Optimal Algorithm for Minimum-Link Rectilinear Paths
in Triangulated Rectilinear Domains   947
  Joseph S.B. Mitchell, Valentin Polishchuk, Mikko Sysikaski,
  and Haitao Wang

Amplification of One-Way Information Complexity via Codes
and Noise Sensitivity   960
  Marco Molinaro, David P. Woodruff, and Grigory Yaroslavtsev

A $(2+\varepsilon)$-Approximation Algorithm for the Storage Allocation Problem   973
  Tobias Mömke and Andreas Wiese

Shortest Reconfiguration Paths in the Solution Space
of Boolean Formulas   985
  Amer E. Mouawad, Naomi Nishimura, Vinayak Pathak,
  and Venkatesh Raman

Computing the Fréchet Distance Between Polygons with Holes   997
  Amir Nayyeri and Anastasios Sidiropoulos

An Improved Private Mechanism for Small Databases   1010
  Aleksandar Nikolov

Binary Pattern Tile Set Synthesis Is NP-Hard.   1022
  *Lila Kari, Steffen Kopecki, Pierre-Étienne Meunier, Matthew J. Patitz,
  and Shinnosuke Seki*

Near-Optimal Upper Bound on Fourier Dimension of Boolean Functions
in Terms of Fourier Sparsity   1035
  Swagato Sanyal

Condensed Unpredictability   1046
  Maciej Skórski, Alexander Golovnev, and Krzysztof Pietrzak

Sherali-Adams Relaxations for Valued CSPs   1058
  Johan Thapper and Stanislav Živný

Two-Sided Online Bipartite Matching and Vertex Cover: Beating
the Greedy Algorithm   1070
  Yajun Wang and Sam Chiu-wai Wong

The Simultaneous Communication of Disjointness with Applications
to Data Streams   1082
  Omri Weinstein and David P. Woodruff

An Improved Combinatorial Algorithm for Boolean Matrix Multiplication. . . .   1094
  Huacheng Yu

Author Index   1107

Invited Talks

Towards the Graph Minor Theorems
for Directed Graphs

Ken-Ichi Kawarabayashi[1]($\boxtimes$) and Stephan Kreutzer[2]

[1] National Institute of Informatics, 2-1-2 Hitotsubashi, Chiyoda-ku, Tokyo, Japan
k_keniti@nii.ac.jp
[2] Technical University Berlin, Sekr TEL 7-3, Ernst-Reuter Platz 7, 10587 Berlin, Germany
stephan.kreutzer@tu-berlin.de

Abstract. Two key results of Robertson and Seymour's graph minor theory are:
1. a structure theorem stating that all graphs excluding some fixed graph as a minor have a tree decomposition into pieces that are almost embeddable in a fixed surface.
2. the *k-disjoint paths problem* is tractable when k is a fixed constant: given a graph G and k pairs (s_1, t_1), ..., (s_k, t_k) of vertices of G, decide whether there are k mutually vertex disjoint paths of G, the ith path linking s_i and t_i for $i = 1, \ldots, k$.
In this talk, we shall try to look at the corresponding problems for digraphs.

Concerning the first point, the grid theorem, originally proved in 1986 by Robertson and Seymour in Graph Minors V, is the basis (even for the whole graph minor project). In the mid-90s, Reed and Johnson, Robertson, Seymour and Thomas (see [13,26]), independently, conjectured an analogous theorem for directed graphs, i.e. the existence of a function $f : \mathbb{N} \to \mathbb{N}$ such that every digraph of directed treewidth at least $f(k)$ contains a directed grid of order k. In an unpublished manuscript from 2001, Johnson, Robertson, Seymour and Thomas give a proof of this conjecture for planar digraphs. But for over a decade, this was the most general case proved for the conjecture.

We are finally able to confirm the Reed, Johnson, Robertson, Seymour and Thomas conjecture in full generality. As a consequence of our results we are able to improve results in Reed et al. in 1996 [27] to disjoint cycles of length at least l. This would be the first but a significant step toward the structural goals for digraphs (hence towards the first point).

Concerning the second point, in [19] we contribute to the disjoint paths problem using the directed grid theorem. We show that the following can be done in polynomial time:

Suppose that we are given a digraph G and k terminal pairs $(s_1, t_1), (s_2, t_2), \ldots, (s_k, t_k)$, where k is a fixed constant. In polynomial time, either

– we can find k paths $P_1, \ldots, P_k$ such that P_i is from s_i to t_i for $i = 1, \ldots, k$ and every vertex in G is in at most four of the paths, or

K.-I. Kawarabayashi—This work was supported by JST ERATO Kawarabayashi Large Graph Project and by Mitsubishi Foundation.
S. Kreutzer—This project has received funding from the European Research Council (ERC) under the European Unions Horizon 2020 research and innovation programme (grant agreement No 648527).

M.M. Halldórsson et al. (Eds.): ICALP 2015, Part II, LNCS 9135, pp. 3–10, 2015.
DOI: 10.1007/978-3-662-47666-6_1

– we can conclude that G does not contain disjoint paths $P_1, \ldots, P_k$ such that P_i is from s_i to t_i for $i = 1, \ldots, k$.

To the best of our knowledge, this is the first positive result for the general directed disjoint paths problem (and hence for the second point). Note that the directed disjoint paths problem is NP-hard even for $k = 2$. Therefore, this kind of results is the best one can hope for.

We also report some progress on the above two points.

Keywords: Directed graphs · Grid minor · The directed disjoint paths problem

1 Introduction

One of the deepest and the most far-reaching theories of the recent 20 years in discrete mathematics (and theoretical computer science as well) is Graph Minor Theory developed by Robertson and Seymour in a series of over 20 papers spanning the last 20 years [28]. Their theory leads to "structural graph theory", which has proved to be a powerful tool for coping with computational intractability. It provides a host of results that can be used to design efficient (approximation or exact) algorithms for many NP-hard problems on specific classes of graphs that occurs naturally in applications.

Two key results of Robertson and Seymour's graph minor theory are:

1. a structure theorem stating that all graphs excluding some fixed graph as a minor have a tree decomposition into pieces that are almost embeddable in a fixed surface.
2. the *k-disjoint paths problem* is tractable when k is a fixed constant: given a graph G and k pairs $(s_1, t_1), \ldots, (s_k, t_k)$ of vertices of G, decide whether there are k mutually vertex disjoint paths of G, the ith path linking s_i and t_i for $i = 1, \ldots, k$.

In order to solve these two problems, of particular importance is the concept of *treewidth*, introduced by Robertson and Seymour. Treewidth has gained immense attention ever since, especially because many NP-hard problems can be handled efficiently on graphs of bounded treewidth [1]. In fact, all problems that can be defined in monadic second-order logic are solvable on graphs of bounded treewidth [4].

A keystone in the proof of the above two results (and many other theorems) is a grid theorem [29]: any graph of treewidth at least some $f(r)$ is guaranteed to have the $r \times r$ grid graph as a minor. This gird theorem played a key role in the k-disjoint paths problem [17,30]. It also played a key role for some other deep applications (e.g., [12,21,22]).

This grid theorem has also played a key role for many algorithmic applications, in particular via bidimensionality theory (e.g., [6–8]), including many approximation algorithms, PTASs, and fixed-parameter algorithms. These include feedback vertex set, vertex cover, minimum maximal matching, face cover, a series of vertex-removal parameters, dominating set, edge dominating set, R-dominating set, connected dominating set, connected edge dominating set, connected R-dominating set, and unweighted TSP tour.

The grid theorem of [29] has been extended, improved, and re-proved by Robertson, Seymour, and Thomas [31], Reed [25], Diestel, Jensen, Gorbunov, and Thomassen [10], Kawarabayashi and Kobayashi [16] and Leaf and Seymour [23]. Very recently, this has been improved to be polynomial [3]. On the other side, the best known lower bound is $\Omega(r^2 \log r)$.

A linear upper bound has been shown for planar graphs [31] and for bounded genus graphs [7]. Recently this min-max relation is also established for graphs excluding any fixed minor H: every H-minor-free graph of treewidth at least $c_H r$ has an $r \times r$ grid minor for some constant c_H [9]. The bound is now explicitly described as $|H|^{|H|}$ [16] This bound leads to many powerful algorithmic results on H-minor-free graphs [7,9] that are previously not known.

2 What about Digraphs?

The structural techniques discussed in graph minor theory all relate to undirected graphs. What about directed graphs? Given the enormous success for problems of width parameters (c.f., treewidth) defined on undirected graphs, it is quite natural to ask whether they can also be extended to analyze the structure of digraphs. In principle by ignoring the direction of edges, it is possible to apply many techniques for undirected graphs to directed graphs. However, we would have an information loss and might fail to properly distinguish between simple and hard input instances. For example, the k-disjoint paths problem for digraphs is NP-complete even when we consider the fixed value $k = 2$ (Fortune, Hopcroft and Wylie [11]), but it is polynomially solvable for all fixed k for undirected graphs [15,30]. Hence, for computational problems whose instances are directed graphs, many methods for undirected graphs may be less useful.

As a first step (but also a significant step) towards overcoming such a difficulty, Reed in 1999 and Johnson, Robertson, Seymour and Thomas [13] proposed a concept of *directed treewidth* and showed that the k-disjoint paths problem is solvable in polynomial time for any fixed k on any class of graphs of bounded directed treewidth [13]. Reed and Johnson et al. also conjectured a directed analogue of the grid theorem.

Conjecture 1 (Reed; Johnson, Robertson, Seymour, Thomas [13]). There is a function $f : \mathbb{N} \to \mathbb{N}$ such that every digraph of directed treewidth at least $f(k)$ contains a cylindrical grid of order k as a butterfly minor.

Actually, according to [13], this conjecture was formulated by Robertson, Seymour and Thomas, together with Alon and Reed at a conference in Annecy, France in 1995. Here, a *cylindrical grid* consists of k concentric directed cycles and $2k$ paths connecting the cycles in alternating directions. A *butterfly minor* of a digraph G is a digraph obtained from a subgraph of G by contracting edges which are either the only outgoing edge of their tail or the only incoming edge of their head. All details for these notations can be found in appendix.

Let us now report progress on the conjecture. In an unpublished manuscript, Johnson et al. [14] proved the conjecture for planar digraphs. In [18], this result was generalised to all classes of directed graphs excluding a fixed undirected graph as an undirected minor. For instance, this includes classes of digraphs of bounded genus. Another related result was established in [19], where a half-integral grid theorem was proved (for the definition of a "half-integral directed grid", we refer the reader to [19]).

Very recently, we finally confirm this conjecture [20]. We believe that this is a first but an important step towards a more general structure theory for directed graphs based on directed treewidth, similar to the grid theorem for undirected graphs being the basis of more general structure theorems (including the main graph minor structure theorem).

3 Algorithmic Contributions

Our main algorithmic interest is the directed k-disjoint paths problem. Recall that for undirected graphs the problem is solvable in polynomial time for any fixed number k. For directed graphs, the situation is much worse since the problem is NP-complete even for only two such pairs.

Theorem 1 (Fortune, Hopcroft, and Wyllie [11]). *The following problem is NP-complete even for $k = 2$:*

> DIRECTED DISJOINT PATHS
>   *Input:* A digraph G and terminals $s_1, t_1, s_2, t_2, \ldots, s_k, t_k$.
>   *Problem:* Find k (vertex) disjoint paths $P_1, \ldots, P_k$ such that P_i is from s_i to t_i for
> $i = 1, \ldots, k$.

Therefore much work has gone into finding polynomial time algorithms for solving this problem on restricted classes of digraphs. See e.g. [5,32] for work in this direction.

In this talk, we are not so much interested in solving disjoint paths problems on special classes of digraphs, but rather in obtaining algorithms working on all directed graphs. We therefore have to relax some of the conditions. Indeed, we allow each vertex of the graph to be contained in small number of paths linking the source/terminal pairs.

Using the directed grid minor, the following is shown in [19].

Theorem 2. *For every fixed $k \geq 1$ there is a polynomial time algorithm for deciding the following problem.*

> QUARTER- INTEGRAL DISJOINT PATHS
>   *Input:* A digraph G and terminals $s_1, t_1, s_2, t_2, \ldots, s_k, t_k$.
>   *Problem:*
>   – Find k paths $P_1, \ldots, P_k$ such that P_i is from s_i to t_i for $i = 1, \ldots, k$
>   and every vertex in G is in at most four of the paths, or
>   – conclude that G does not contain disjoint paths $P_1, \ldots, P_k$ such that
> P_i is from s_i to t_i for $i = 1, \ldots, k$.

As far as we are aware, this is the first result that establishes a positive result, and gives a polynomial time algorithm for the variant of the disjoint paths problems on the class of all digraphs. Note that this result is best possible in a sense. Indeed, Slivkins [33] proved that the directed disjoint paths problem is W[1]-hard already on acyclic digraphs and it is not hard to extend this result to the half- or quarter-integral case. Hence in terms of running time our algorithm is optimal in the sense that it cannot be improved to $\mathcal{O}(f(k)n^c)$ for any fixed constant c.

As we said, the key is to use a cylindrical grid. The following theorem tells us why a "directed" grid minor is important.

Theorem 3. *Let $s_1, \ldots, s_k, t_1, \ldots, t_k$ be (not necessarily distinct) $2k$ vertices in a digraph G. Suppose that G has a cylindrical grid W of order $8k^3$. Let $S = \{s_1, \ldots, s_k\}$ and $T = \{t_1, \ldots, t_k\}$. Suppose furthermore that*

1. *there is no separation (A_1, B_1) of order at most k such that A_1 contains S and B_1 contains all but at most k vertices Q_1 of in-degree or out-degree at least two in W, and there is no path from S to Q_1 in $G - (A_1 \cap B_1)$, and*
2. *there is no separation (A'_1, B'_1) of order at most k such that A'_1 contains T and B'_1 contains all but at most k vertices Q_2 of in-degree or out-degree at least two in W, and there is no path from Q_2 to T in $G - (A'_1 \cap B'_1)$.*

Then in polynomial time, we can find k paths $P_1, \ldots, P_k$ in G such that endpoints of P_i are s_i, t_i for $i = 1, \ldots, k$, and moreover each vertex in G is used in at most two of these paths.

Using Theorem 3, we are currently working on the following conjecture.

Conjecture 2. For a fixed constant k, there is a polynomial time algorithm for the following problem:

DIRECTED HALF-DISJOINT PATHS
    Input: A digraph G and terminals $s_1, t_1, s_2, t_2, \ldots, s_k, t_k$.
    Problem: Find k paths $P_1, \ldots, P_k$ such that P_i connects from s_i to t_i for $i = 1, \ldots, k$ and every vertex in G is in at most two of the paths.

As pointed out above, this is the best we can hope.

4 Additional Notations

An $r \times r$ *grid* is a graph which is isomorphic to the graph W_r obtained from Cartesian product of paths of length $r - 1$, with vertex set $V(W_r) = \{(i, j) \mid 1 \le i \le r, 1 \le j \le r\}$ in which two vertices (i, j) and (i', j') are adjacent if and only if $|i - i'| + |j - j'| = 1$.

The (4×5)-grid, as well as the (8×5)-wall (which can be defined in a similar way) are shown in Figure 1.

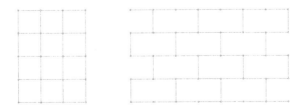

Fig. 1. The (4×5)-grid and the (8×5)-wall

A *tree decomposition* of a graph G is a pair $(T, \mathcal{W})$, where T is a tree and $\mathcal{W}$ is a family $\{W_t \mid t \in V(T)\}$ of vertex sets $W_t \subseteq V(G)$, such that the following two properties hold:

(1) $\bigcup_{t \in V(T)} W_t = V(G)$, and every edge of G has both ends in some W_t.

(2) If $t, t', t'' \in V(T)$ and t' lies on the path in T between t and t'', then $W_t \cap W_{t''} \subseteq W_{t'}$.

The *width* of a tree decomposition $(T, \mathcal{W})$ is $\max_{t \in V(T)} |W_t| - 1$. The *treewidth* of a graph G is the minimum width over all possible tree decompositions of G.

Robertson and Seymour developed the first polynomial time algorithm for constructing a tree decomposition of a graph of bounded width [30], and eventually came up with an algorithm which runs in $O(n^2)$ time, for this problem. Reed [24] developed an algorithm for the problem which runs in $O(n \log n)$ time, and then Bodlaender [2] developed a linear time algorithm.

Directed Treewidth We briefly recall the definition of directed treewidth from [13].

By an *arborescence* we mean a directed graph R such that R has a vertex r_0, called the *root* of R, with the property that for every vertex $r \in V(R)$ there is a unique directed path from r_0 to r. Thus every arborescence arises from a tree by selecting a root and directing all edges away from the root. If $r, r' \in V(R)$ we write $r' > r$ if $r' \neq r$ and there exists a directed path in R with initial vertex r and terminal vertex r'. If $e \in E(R)$ we write $r' > e$ if either $r' = r$ or $r' > r$, where r is the head of e. Let G be a digraph, and let $Z \subseteq V(G)$. We say that a set $S \subseteq (V(G) - Z)$ is Z-*normal* if there is no directed walk in $G - Z$ with the first and the last vertex in S that uses a vertex of $G - (Z \cup S)$. It follows that every Z-normal set is the union of the vertex-sets of strong components of $G - Z$. As one readily checks, a set S is Z-normal if and only if the vertex-sets of the strong components of $G - Z$ can be numbered $S_1, S_2, \ldots, S_d$ in such a way that

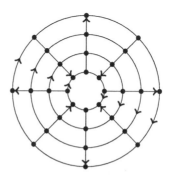

Fig. 2. Cylindrical grid G_4.

1. if $1 \leq i < j \leq d$, then no edge of G has head in S_i and tail in S_j, and
2. either $S = \emptyset$, or $S = S_i \cup S_{i+1} \cup \cdots \cup S_j$ for some integers i, j with $1 \leq i \leq j \leq d$.

Definition 1. *A directed tree-decomposition of a digraph G is a triple (R, X, W), where R is an arborescence, and $X = (X_e : e \in E(R))$ and $W = (W_r : r \in V(R))$ are sets of vertices of G that satisfy*
1. *$(W_r : r \in V(R))$ is a partition of $V(G)$ into nonempty sets, and*
2. *if $e \in E(R)$, then $\bigcup (W_r : r \in V(R), r > e)$ is X_e-normal.*
The width *of (R, X, W) is the least integer w such that for all $r \in V(R)$, $|W_r \cup \bigcup_e X_e| \leq w + 1$, where e is taken over all edges incident to r. The* directed treewidth *of G is the least integer w such that G has a directed tree-decomposition of width w.*

Sometimes, we call W_r or X_e a *bag* for $r \in V(R)$ and $e \in E(R)$. It is easy to see that the directed tree-width of a subdigraph of G is at most the tree-width of G.

References

1. Arnborg, S., Proskurowski, A.: Linear time algorithms for NP-hard problems restricted to partial k-trees. Discrete Appl. Math. **23**, 11–24 (1989)
2. Bodlaender, H.L.: A linear-time algorithm for finding tree-decomposition of small treewidth. SIAM J. Comput. **25**, 1305–1317 (1996)
3. Chekuri, C., Chuzhoy, J.: Polynomial bounds for the grid-minor theorem. In: Symp. on Theory of Computing (STOC), pp. 60–69 (2014)
4. Courcelle, B.: Graph rewriting: An algebraic and logic approach, in Handbook of Theoretical Computer Science 2, pp. 194–242. Elsevier (1990)
5. Cygan, M., Marx, D., Pilipczuk, M., Pilipczuk, M.: The planar directed k-vertex-disjoint paths problem is fixed-parameter tractable. In: 54th Annual IEEE Symposium on Foundations of Computer Science (FOCS), pp. 197–206 (2013)
6. Demaine, E.D., Fomin, F.V., Hajiaghayi, M., Thilikos, D.M.: Bidimensional parameters and local treewidth. SIAM J. Discrete Mathematics **18**, 501–511 (2004)
7. Demaine, E.D., Fomin, F.V., Hajiaghayi, M., Thilikos, D.M.: Subexponential parameterized algorithms on graphs of bounded genus and H-minor-free graphs. J. ACM **52**, 866–893 (2005)
8. Demaine, E.D., Hajiaghayi, M.: Bidimensionality: new connections between FPT algorithms and PTASs. In: Proc. 16th Annual ACM-SIAM Symposium on Discrete Algorithms (SODA), pp. 590–601 (2005)
9. Demaine, E.D., Hajiaghayi, M.: Linearity of grid minors in treewidth with applications through bidimensionality. Combinatorica **28**, 19–36 (2008)
10. Diestel, R., Gorbunov, K.Y., Jensen, T.R., Thomassen, C.: Highly connected sets and the excluded grid theorem. J. Combin. Theory Ser. B **75**, 61–73 (1999)
11. Fortune, S., Hopcroft, J.E., Wyllie, J.: The directed subgraph homeomorphism problem. Theor. Comput. Sci. **10**, 111–121 (1980)
12. Grohe, M., Kawarabayashi, K., Marx, D., Wollan, P.: Finding topological subgraphs is fixed-parameter tractable. In: The 43rd ACM Symposium on Theory of Computing (STOC 2011), pp. 479–488
13. Johnson, T., Robertson, N., Seymour, P.D., Thomas, R.: Directed tree-width. J. Comb. Theory, Ser. B **82**(1), 138–154 (2001)
14. Johnson, T., Robertson, N., Seymour, P.D., Thomas, R.: Excluding a grid minor in digraphs (2001). (unpublished manuscript)
15. Kawarabayashi, K., Wollan, P.: A shorter proof of the graph minors algorithm - the unique linkage theorem. In: Proc. 42nd ACM Symposium on Theory of Computing (STOC 2010), pp. 687–694 (2010). A full version of this paper http://research.nii.ac.jp/~k_keniti/uniquelink.pdf
16. Kawarabayashi, K., Kobayashi, Y.: Linear min-max relation between the treewidth of h-minor-free graphs and its largest grid. In: Dürr, C., Wilke, T. (eds.) STACS, volume 14 of LIPIcs, pp. 278–289. Schloss Dagstuhl - Leibniz-Zentrum fuer Informatik (2012)
17. Kawarabayashi, K., Kobayashi, Y., Reed, B.: The disjoint paths problem in quadratic time. J. Combin. Theory Ser. B **102**, 424–435 (2012)
18. Kawarabayashi, K., Kreutzer, S.: An excluded grid theorem for digraphs with forbidden minors. In: ACM/SIAM Symposium on Discrete Algorithms (SODA) (2014)
19. Kawarabayashi, K., Kobayashi, Y., Kreutzer, S.: An excluded half-integral grid theorem for digraphs and the directed disjoint paths problem. In: Proc. of the ACM Symposium on Theory of Computing (STOC), pp. 70–78 (2014)
20. Kawarabayashi, K., Kreutzer, S.: The directed excluded grid theorem. In: STOC 2015. arXiv:1411.5681 [cs.DM]

21. Kawarabayashi, K., Reed, B.: A nearly linear time algorithm for the half-integral dis-joint paths packing. In: ACM-SIAM Symposium on Discrete Algorithms (SODA 2008), pp. 446–454
22. Kleinberg, J.: Decision algorithms for unsplittable flows and the half-disjoint paths problem. In: Proc. 30th ACM Symposium on Theory of Computing (STOC), pp. 530–539 (1998)
23. Leaf, A., Seymour, P.: Treewidth and planar minors (2012)
24. Reed, B.: Finding approximate separators and computing tree width quickly. In: The 24th ACM Symposium on Theory of Computing (STOC 1992)
25. Reed, B.: Tree width and tangles: a new connectivity measure and some applications, in Surveys in Combinatorics, London Math. Soc. Lecture Note Ser. 241, pp. 87–162. Cambridge Univ. Press, Cambridge (1997)
26. Reed, B.: Introducing directed tree-width. Electronic Notes in Discrete Mathematics **3**, 222–229 (1999)
27. Reed, B.A., Robertson, N., Seymour, P.D., Thomas, R.: Packing directed circuits. Combi-natorica **16**(4), 535–554 (1996)
28. Robertson, N., Seymour, P.D.: Graph minors I - XXIII, 1982–2010. Appearing in Journal of Combinatorial Theory, Series B from 1982 till 2010
29. Robertson, N., Seymour, P.D.: Graph minors. V. Excluding a planar graph. J. Combin. Theory Ser. B **41**, 92–114 (1986)
30. Robertson, N., Seymour, P.: Graph minors XIII. The disjoint paths problem. Journal of Combinatorial Theory, Series B **63**, 65–110 (1995)
31. Robertson, N., Seymour, P.D., Thomas, R.: Quickly excluding a planar graph. J. Combin. Theory Ser. B **62**, 323–348 (1994)
32. Schrijver, A.: Finding k disjoint paths in a directed planar graph. SIAM Jornal on Computing **23**(4), 780–788 (1994)
33. Slivkins, A.: Parameterized tractability of edge-disjoint paths on directed acyclic graphs. In: Di Battista, G., Zwick, U. (eds.) ESA 2003. LNCS, vol. 2832, pp. 482–493. Springer, Heidelberg (2003)

Automated Synthesis of Distributed Controllers

Anca Muscholl[(✉)]

LaBRI, University of Bordeaux, Bordeaux, France
`anca@labri.fr`

Abstract. Synthesis is a particularly challenging problem for concurrent programs. At the same time it is a very promising approach, since concurrent programs are difficult to get right, or to analyze with traditional verification techniques. This paper gives an introduction to distributed synthesis in the setting of Mazurkiewicz traces, and its applications to decentralized runtime monitoring.

1 Context

Modern computing systems are increasingly distributed and heterogeneous. Software needs to be able to exploit these advances, providing means for applications to be more performant. Traditional concurrent programming paradigms, as in Java, are based on threads, shared-memory, and locking mechanisms that guard access to common data. More recent paradigms like the reactive programming model of Erlang [4] and Scala [34,35] replace shared memory by asynchronous message passing, where sending a message is non-blocking.

In all these concurrent frameworks, writing reliable software is a serious challenge. Programmers tend to think about code mostly in a sequential way, and it is hard to grasp all possible schedulings of events in a concurrent execution. For similar reasons, verification and analysis of concurrent programs is a difficult task. Testing, which is still the main method for error detection in software, has low coverage for concurrent programs. The reason is that bugs in such programs are difficult to reproduce: they may happen under very specific thread schedules and the likelihood of taking such corner-case schedules is very low. Automated verification, such as model-checking and other traditional exploration techniques, can handle very limited instances of concurrent programs, mostly because of the very large number of possible states and of possible interleavings of executions.

Formal analysis of programs requires as a pre-requisite a clean mathematical model for programs. Verification of sequential programs starts usually with an abstraction step – reducing the value domains of variables to finite domains, viewing conditional branching as non-determinism, etc. Another major simplification consists in disallowing recursion. This leads to a very robust computational model, namely *finite-state automata* and *regular languages*. Regular languages of words (and trees) are particularly well understood notions. The deep connections between logic and automata revealed by the foundational work of Büchi, Rabin, and others, are the main ingredients in automata-based verification.

© Springer-Verlag Berlin Heidelberg 2015
M.M. Halldórsson et al. (Eds.): ICALP 2015, Part II, LNCS 9135, pp. 11–27, 2015.
DOI: 10.1007/978-3-662-47666-6_2

In program synthesis, the task is to turn a specification into a program that is guaranteed to satisfy it. Synthesis can therefore provide solutions that are correct by construction. It is thus particularly attractive for designing concurrent programs, that are often difficult to get right or to analyze by traditional methods. In distributed synthesis, we are given in addition an architecture, and the task is to turn the specification into a distributed implementation over this architecture.

Distributed synthesis proves to be a real challenge, and there are at least two reasons for this. First, there is no canonical model for concurrent systems, simply because there are very different kinds of interactions between processes. Compare, for example, multi-threaded shared memory Java programs, to Erlang or Scala programs with asynchronous function calls. This issue is connected with another, more fundamental reason: techniques for distributed synthesis are rather rare, and decidability results are conditioned by the right match between the given concurrency model and the kind of questions that we ask.

Mazurkiewicz traces were introduced in the late seventies by A. Mazurkiewicz [31] as a simple model for concurrency inspired by Petri nets. Within this theory, Zielonka's theorem [45] is a prime example of a result on distributed synthesis.

This paper gives a brief introduction to Mazurkiewicz traces and to Zielonka's theorem, and describes how this theory can be used in the verification and the design of concurrent programs. We focus on the synthesis of concurrent programs and its application to decentralized runtime monitoring.

Monitoring is a more lightweight alternative to model-checking and synthesis. The task is to observe the execution of a program in order to detect possible violations of safety requirements. Monitoring is a prerequisite for control, because it can gather information about things that went wrong and about components that require repair actions. In programming, monitoring takes the form of assertions: an invalidation of an assertion is the first sign that something has gone wrong in the system. However, concurrent programs often require assertions concerning several components. A straightforward but impractical way to verify such an assertion at runtime is to synchronize the concerned components and to inquire about their states. A much better way to do this is to write a distributed monitor that deduces the required information by recording and exchanging suitable information using the available communication in the program. Mazurkiewicz trace theory and Zielonka's theorem can provide a general, and yet practical method for synthesizing distributed monitors.

Overview of the paper. Section 2 sets the stage by describing some classical correctness issues for concurrent programs. Section 3 introduces Mazurkiewicz traces, and Section 4 presents some applications to decentralized monitoring.

2 Distributed Models: Some Motivation

Concurrent programming models usually consist of entities, like processes or threads, that evolve in an asynchronous manner and synchronize on joint events, such as access to shared variables, or communication. We start with some illustrating examples from multi-threaded programming, and with some typical correctness properties. This will allow us to present the type of questions that we want to address.

A multi-threaded program consists of an arbitrary number of concurrently executing threads. We will assume that there is a fixed set $\mathcal{T}$ of threads. There is no global clock, so threads progress asynchronously. Threads can either perform local actions or access the global memory, consisting of shared variables from a fixed set X. Possible actions of a thread $T \in \mathcal{T}$ include reads $r(T, x, v)$ and writes $w(T, x, v)$ on a shared variable $x \in X$ (for some value v) and acquiring $acq(T, L)$, resp. releasing $rel(T, L)$ a lock L. More complex forms of access to the shared memory, such as compare-and-set (CAS), are commonly used in lock-free programming. We will not use CAS in the remaining of this section, but come back to it in Section 3.

Partial orders are a classical abstraction for reasoning about executions of multi-threaded programs. The computation on each thread is abstracted out by a set of events, and the multi-threaded execution is abstracted in form of a partial order on these events. An early example is Lamport's *happens-before* relation [27], originally described for communicating systems. This relation orders the events on each thread, and the sending of a message before its receive. In multi-threaded programs with shared memory, where locks guard the access to shared variables, the happens-before relation orders two events if they are performed by the same thread or they use the same lock.

A more formal, general definition of the happens-before relation for programs goes as follows. Let Σ be the set of actions in a program. We will assume throughout the paper that Σ is finite. Depending on the problem that we consider, we will assume that there is a binary *conflict* relation $D \subseteq \Sigma \times \Sigma$ between the actions of the program. For example, we will have $a\ D\ b$ if a and b are performed by the same thread. Starting with a linear execution $a_1 \cdots a_n \in \Sigma^*$ of a program, the *happens-before* relation is the partial order on positions defined as the reflexive-transitive closure of the relation $\{i \prec j \mid i < j \text{ and } a_i\ D\ a_j\}$. As we will see in Section 3, if the conflict relation is symmetric, this partial order is a *Mazurkiewicz trace*.

In the remaining of this section we outline two frequently considered correctness issues for concurrent programs, that will be used as examples for decentralized monitoring in Section 4.

2.1 Race Detection

Race detection is one of the widely studied problems of concurrent software. Informally, a race occurs whenever there are conflicting accesses to the same variable without proper synchronization. Detecting races is important since executions with races may yield unexpected behaviors, caused by the outcome of the computation depending on the schedule of threads.

In order to define races for multi-threaded programs with lock synchronization we need to introduce the happens-before relation for such programs. Let Σ

be the set of actions in a program, for instance:

$$\Sigma = \{w(T, x), r(T, x), acq(T, L), rel(T, L) \mid T, x, L\}.$$

Two actions from Σ are in *conflict* if

- they are performed by the same thread, or
- they acquire or release the same lock.

A *race* occurs in an execution if there are two accesses to the same shared variable such that

- they are *unordered* in the happens-before relation, and
- at least one of them is a write.

Example 1. Figure 1 illustrates a race problem due to locking that is too fine-grained. Two threads have access to a list pointed to by head. Thread 1 adds an element to the head of the list, while Thread 2 deletes the head element. The two instructions protected by the lock are ordered in the happens-before relation. However, t1.next = head and head = head.next are unordered. Since the first instruction is a read, and the second a write, this situation is a race condition.

```
type list {int data; list *next}
list *head
```

Thread 1

```
1: t1 = new(list);
2: t1.data = 42;
3: t1.next = head;
4: ack(lock)
5:    head = t1
6: rel(lock)
```

Thread 2

```
7:    t2 = head;
8:    ack(lock)
9:      head = head.next
10: rel(lock)
```

Fig. 1. A race condition in the execution $1, 2, 3, 7, 8, 9$: events 3 and 9 are unordered in the happens-before relation

2.2 Atomicity

Atomicity, or conflict serializability, is a high-level correctness notion for concurrent programs that has its origins in database transactions. A transaction consists of a block of operations, such as reads and writes to shared memory variables, that is marked as atomic. An execution of the transaction system is serial if transactions are scheduled one after the other, without interleaving them. A serial execution reflects the intuition of the programmer, about parts of the code marked as transactions as being executed atomically.

In order to define when a multi-threaded program is conflict-serializable, we need first the notion of equivalent executions. Two executions are *equivalent* if they define the same happens-before relation w.r.t. the following conflict relation: Two actions from $\Sigma = \{w(T, x), r(T, x) \mid T, x\}$ are in conflict if

- they are performed by the same thread, or
- they access to the same variable, and at least one of them is a write.

The above conflict relation has a different purpose than the one used for the race problem: here, we are interested in the values that threads compute. Two executions are considered to be equivalent if all threads end up with the same values of (local and global) variables. Since a write $w(T, x)$ potentially modifies the value of x, its order w.r.t. any other access to x should be preserved. This guarantees that the values of x are the same in two equivalent executions.

A program is called *conflict-serializable* (or *atomic*) if every execution is equivalent to a serial one. As we will explain in Section 3 this means that every execution can be reordered into an equivalent one where no transaction is interrupted.

Example 2. Figure 2 shows a simple program with two threads that is not conflict-serializable. The interleaved execution where `Thread` 2 writes after the read and before the write of `Thread` 1, is not equivalent to any serial execution.

```
Thread 1

1: atomic {
2:    read(x);
3:    write(x)
4:    }
```

```
Thread 2

5: atomic {
6:    write(x);
7:    }
```

Fig. 2. A program that is not conflict-serializable: the execution $1, 2, 5, 6, 7, 3, 4$ is not equivalent to any serial execution

3 Mazurkiewicz Traces and Zielonka's Theorem

This section introduces Mazurkiewicz traces [31], one of the simplest formalisms able to describe concurrency. We will see that traces are perfectly suited to describe dependency and the happens-before relation. The notion of conflicting actions and the happens-before relation seen in the previous section are instances of this more abstract approach.

The definition of traces starts with an alphabet of actions Σ and a *dependence relation* $D \subseteq \Sigma \times \Sigma$ on actions, that is reflexive and symmetric. The idea behind this relation is that two dependent actions are always ordered, for instance because the outcome of one action affects the other action. For example, the actions of acquiring or releasing the same lock are ordered, since a thread has to wait for a lock to be released before acquiring it.

Example 3. Coming back to the problems introduced in Sections 2.1 and 2.2, note that the conflict relations defined there are both symmetric. For example, we can define the dependence relation D over the alphabet $\Sigma = \{r(T, x), w(T, x) \mid T \in \mathcal{T}, x \in X\}$ of Section 2.2, by letting $a \, D \, b$ if a, b are in conflict.

While the dependence relation coincides with the conflict relation, the happens-before relation is the *Mazurkiewicz trace* order. A Mazurkiewicz trace is the labelled partial order $T(w) = \langle E, \preceq \rangle$ obtained from a word $w = a_1 \ldots a_n \in \Sigma^*$ in the following way:

- $E = \{e_1, \ldots, e_n\}$ is the set of *events*, in one-to-one correspondence with the positions of w, where event e_i has label a_i,
- $\preceq$ is the reflexive-transitive closure of $\{(e_i, e_j) \mid i < j, \ a_i \ D \ a_j\}$.

From a language-theoretical viewpoint, traces are almost as attractive as words, and several results from automata and logics generalize from finite and infinite words to traces, see e.g. the handbook [10]. One of the cornerstone results in Mazurkiewicz trace theory is based on an elegant notion of finite-state distributed automata, *Zielonka automata*, that we present in the remaining of the section.

Informally, a Zielonka automaton [45] is a finite-state automaton with control distributed over several *processes* that synchronize on shared actions. Synchronization is modeled through a distributed action alphabet. There is no global clock: for instance between two synchronizations, two processes can do a different number of actions. Because of this, Zielonka automata are also known as *asynchronous automata*.

A *distributed action alphabet* on a finite set $\mathbb{P}$ of processes is a pair (Σ, dom), where Σ is a finite set of *actions* and $dom : \Sigma \to (2^{\mathbb{P}} \setminus \emptyset)$ is a *domain function*. The domain $dom(b)$ of action b comprises all processes that synchronize in order to perform b. The domain function induces a natural dependence relation D over Σ by setting $a \ D \ b$ if $dom(a) \cap dom(b) \neq \emptyset$. The idea behind is that executions of two dependent actions affect at least one common process, so their order matters. By contrast, two *independent actions* a, b, i.e., where $dom(a) \cap dom(b) = \emptyset$, can be executed either as ab or as ba, the order is immaterial.

Example 4. We reconsider Example 3. The dependence relation D defined there can be realized by a distributed alphabet (Σ, dom) on the following set of processes:

$$\mathbb{P} = \mathcal{T} \cup \{\langle T, x \rangle \mid T \in \mathcal{T}, x \in X\}.$$

Informally, each thread $T \in \mathcal{T}$ represents a process; in addition, there is a process for each pair $\langle T, x \rangle$. The process $\langle T, x \rangle$ stands for the cached value of x in thread T.

The domain function defined below satisfies $a \ D \ b$ iff $dom(a) \cap dom(b) \neq \emptyset$:

$$dom(a) = \begin{cases} \{T, \langle T, x \rangle\} & \text{if } a = r(T, x) \\ \{T, \langle T', x \rangle \mid T' \in \mathcal{T}\} & \text{if } a = w(T, x). \end{cases}$$

The intuition behind $dom(a)$ is as follows. A read $r(T, x)$ depends both on the internal state of thread T and the cached value of x, and will affect the state of T. A write $w(T, x)$ depends on the internal state of thread T and will affect not only the state of T, but also the cached values of x on other threads using x, since the new value will be written into these caches.

A *Zielonka automaton* $\mathcal{A} = \langle (S_p)_{p \in \mathbb{P}}, (s_p^{init})_{p \in \mathbb{P}}, \delta \rangle$ over (Σ, dom) consists of:

- a finite set S_p of (local) states with an initial state $s_p^{init} \in S_p$, for every process $p \in \mathbb{P}$,
- a transition relation $\delta \subseteq \bigcup_{a \in \Sigma} \left(\prod_{p \in dom(a)} S_p \times \{a\} \times \prod_{p \in dom(a)} S_p \right)$.

For convenience, we abbreviate a tuple $(s_p)_{p \in P}$ of local states by s_P. An automaton is called *deterministic* if the transition relation is a partial function.

Before explaining the semantics of Zielonka automata, let us comment the idea behind the transitions and illustrate it through an example. The reader may be more familiar with synchronous products of finite automata, where a joint action means that every automaton having this action in its alphabet executes it according to its transition relation. Joint transitions in Zielonka automata follow a *rendez-vous* paradigm, meaning that processes having action b in their alphabet can exchange information via the execution of b: a transition on b depends on the states of all processes executing b. The following example illustrates this effect:

Example 5. The CAS (compare-and-swap) operation is available as atomic operation in the JAVA package java.util.concurrent.atomic, and supported by many architectures. It takes as parameters the thread identifier T, the variable name x, and two values, old and new. The effect of the instruction y = CAS(T,x,old,new) is conditional: the value of x is replaced by new if it is equal to old, otherwise it does not change. The method returns true if the value changed, and false otherwise.

A CAS instruction can be seen as a synchronization between two processes: P_T associated with the thread T, and P_x associated with the variable x. The states of P_T are valuations of the local variables of T. The states of P_x are the values x can take. An instruction of the form y = CAS(T,x,old,new) becomes a synchronization action between P_T and P_x with the two transitions of Figure 3 (represented for convenience as Petri net transitions).

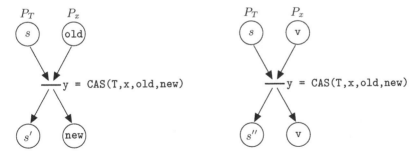

Fig. 3. CAS as transitions of a Zielonka automaton. On the left side of the figure we have the case when the value of x is old; on the right side v is different from old. Notice that in state s' the value of y is true, whereas in s'', it is false.

A Zielonka automaton can be seen as a usual finite-state automaton, whose set of states $S = \prod_{p \in \mathbb{P}} S_p$ is given by the global states, and transitions $s \xrightarrow{a} s'$ if $(s_{dom(a)}, a, s'_{dom(a)}) \in \delta$, and $s_{\mathbb{P} \setminus dom(a)} = s'_{\mathbb{P} \setminus dom(a)}$. Thus states of this automaton are tuples of states of the processes of the Zielonka automaton. As a language acceptor, a Zielonka automaton $\mathcal{A}$ accepts a *trace-closed language* $L(\mathcal{A})$, that is, a language closed under commuting adjacent independent symbols. Formally, a language L is trace-closed when $uabv \in L$ if and only if $ubav \in L$, for all $u, v \in \Sigma^*$ and all independent actions a, b.

A cornerstone result in the theory of Mazurkiewicz traces is a construction transforming a sequential automaton into an equivalent deterministic Zielonka automaton. This beautiful result is one of the rare examples of distributed synthesis with broader scope.

Theorem 1. *[45] Given a distributed alphabet (Σ, dom), and a regular trace-closed language $L \subseteq \Sigma^*$ over (Σ, dom). A deterministic Zielonka automaton $\mathcal{A}$ such that $L(\mathcal{A}) = L$ can be effectively constructed.*

The only assumption of the theorem above is that the language of the automaton is trace-closed, but this is unavoidable. Moreover, trace closure can be checked easily, e.g. on the minimal DFA of the given language.

The construction behind Theorem 1 is technically involved, but also very fascinating. The crux is to show how to put together distributed information using additional memory that is *finite*[1]. Many researchers contributed to simplify the construction and to improve its complexity, see [8, 16, 19, 33] and references therein. The most recent construction [16] produces deterministic Zielonka automata of size that is exponential in the number of processes (and polynomial in the size of a DFA for L). The exponential dependence on the number of processes is necessary, modulo a technical assumption (that is actually required for monitoring).

Theorem 2 ([16]). *There is an algorithm that takes as input a distributed alphabet (Σ, dom) over n processes and a DFA $\mathcal{A}$ accepting a trace-closed language over (Σ, dom), and computes an equivalent deterministic Zielonka automaton $\mathcal{B}$ with at most $4^{n^4} \cdot |\mathcal{A}|^{n^2}$ states per process. Moreover, the algorithm computes the transitions of $\mathcal{B}$ on-the-fly in polynomial time.*

4 Distributed Monitoring

The construction of deterministic Zielonka automata opens interesting perspectives for monitoring concurrent programs. In order to monitor a concurrent program at runtime, the monitor has to be distributed (or decentralized). This means that there is a local monitor on each thread, and these local monitors can exchange information. The exchange can be implemented by allowing local

---

[1] Vector clocks [30] are a similar notion in distributed computing, but they do not require a finite domain of values.

monitors to initiate extra communication, or, more conservatively, by using the available communication in the program in order to share monitoring-relevant information. We follow the latter setting here, since adding communication can reduce the concurrency, and it is very difficult to quantify how much performance is lost by adding communication.

Apart from detecting violations of safety properties at runtime, the information gathered by such monitors can be also used to recover from an unsafe state. Of course, this can be done only at runtime, and not offline, by inspecting sequential executions a posteriori.

Our general approach for this kind of distributed monitoring is simple: we have some trace-closed, regular property ϕ that should be satisfied by every execution of a given program or system. To detect possible violations of ϕ at runtime, we construct a monitor for ϕ and run it in parallel with the program. Consider the scenario where the program P is modeled by a Zielonka automaton $\mathcal{A}_P$. If a monitor is also a Zielonka automaton $\mathcal{A}_M$, then running the monitor M in parallel to P amounts to build the usual product automaton between $\mathcal{A}_P$ and $\mathcal{A}_M$ process-wise.

Interestingly, many properties one is interested to monitor on concurrent programs can be expressed in terms of the happens-before relation between specific events, as the following example illustrates.

Example 6. Consider the *race detection problem* from Section 2.1. A race occurs when two conflicting accesses to the same variable are unordered in the happens-before relation. Therefore, a violation of the "no-race" property is monitored by looking for two *unordered* accesses to the same variable, at least one of them being a write.

Monitoring a violation of *atomicity* (recall Section 2.2) is done by checking for every transaction on some thread T, that no action c of some thread $T' \neq T$ happened after the beginning $a = beg(T)$ of the transaction on T (cf. instruction 1 of Example 2) and before its matching end $b = end(T)$ (cf. instruction 4). In other words, the monitor looks for events c on $T' \neq T$ satisfying $a \prec c \prec b$ in the happens-before relation.

Determining the partial ordering between specific events is closely related to the kernel of all available constructions behind Zielonka's theorem. This is known as the *gossip automaton* [33], and the name reflects its rôle: it computes what a process knows about the knowledge of other processes. Using *finite-state* gossiping, processes can put together information that is distributed in the system, hence reconstruct the execution of the given DFA.

The gossip automaton is already responsible for the exponential complexity of Zielonka automata, in all available constructions. A natural question is whether the construction of the gossip automaton can be avoided, or at least simplified. Perhaps unsurprisingly, the theorem below shows that gossiping is not needed when the communication structure is acyclic.

The *communication graph* of a distributed alphabet (Σ, dom) with unary or binary action domains is the undirected graph where vertices are the processes, and edges relate processes $p \neq q$ if there is some action $a \in \Sigma$ such that $dom(a) = \{p, q\}$.

Theorem 3 ([23]). *Let (Σ, dom) be a distributed alphabet with acyclic communication graph. Every regular, trace-closed language L over Σ can be accepted by a deterministic Zielonka automaton with $O(s^2)$ states per process, where s is the size of the minimal DFA for L.*

The theorem above can be useful to monitor programs with acyclic communication if we can start from a small DFA for the trace-closed language L representing the monitoring property. However, in some cases the DFA is necessarily large because it needs to take into account many interleavings. For example, monitoring for some unordered occurrences of b and c, requires a DFA to remember *sets* of actions. In this case it is more efficient to start with a description of L by partial orders. We discuss a solution for this setting in Section 4.1 below.

We need to emphasize that using Zielonka automata for monitoring properties in practice does not depend only on the efficiency of the constructions from the above theorems. In addition to determinism, further properties are desirable when building monitoring automata. The first requirement is that a violation of the property to monitor should be detectable locally. The reason for this is that a thread that knows about the failure can start some recovery actions or inform other threads about the failure. Zielonka automata satisfying this property are called *locally rejecting* [16]. More formally, each process p has a subset of states $R_p \subseteq S_p$; an execution leads a process p into a state from R_p if and only if p knows already that the execution cannot be extended to a trace in $L(\mathcal{A})$. The second important requirement is that the monitoring automaton $\mathcal{A}_M$ should not block the monitored system. This can be achieved by asking that in every global state of $\mathcal{A}_M$ such that no process is in a rejecting state, every action is enabled. A related discussion on desirable properties of Zielonka automata and on implementating the construction of [16] is reported in [2].

4.1 Gossip in Trees

In this section we describe a setting where we can compute efficiently the happens-before relation for selected actions of a concurrent program. We will first illustrate the idea on the simple example of Section 2.2. The program there has two threads, T_1 and T_2, and one shared variable x. For clarity we add actions $beg(T_i), end(T_i)$ denoting the begin/end of the atomic section on T_i. Thus, the alphabet of actions is:

$$\Sigma = \{beg(T_1), end(T_1), w(T_1, x), r(T_1, x), beg(T_2), end(T_2), w(T_2, x)\}.$$

The dependence relation D includes all pairs of actions of the same thread, as well as the pairs $(r(T_1, x), w(T_2, x))$ and $(w(T_1, x), w(T_2, x))$. Following Example 4, the Zielonka automaton has processes $\mathbb{P} = \{T_1, T_2, \langle T_1, x \rangle, \langle T_2, x \rangle\}$ and the domains of actions are:

$$dom(beg(T_i)) = dom(end(T_i)) = \{T_i\},$$
$$dom(r(T_i, x)) = \{T_i, \langle T_i, x \rangle\}$$
$$dom(w(T_i, x)) = \{T_i, \langle T_i, x \rangle, \langle T_{3-i}, x \rangle\}$$

Note that we can represent these four processes as a (line) tree in which the domain of each action spans a connected part of the tree, see Figure 4.

Fig. 4. A tree of processes T. The dashed part is the domain of the read action, whereas the dotted parts are the domains of the two writes.

The Mazurkiewicz trace in Figure 5, represented as a partial order, shows a violation of conflict-serializability: event c at step 6 satisfies $a \prec c \prec b$, where a represents step 1, and b step 4. The happens-before relation can be computed piecewise by a Zielonka automaton, by exchanging information via the synchronization actions. Figure 6 illustrates how processes update their knowledge about the partial order. Note how the two partial orders represented by thick lines, are combined together with the action $w(T_2, x)$ of step 6, in order to produce the partial orders of processes $\langle T_1, x \rangle$, $\langle T_2, x \rangle$ and T_2 in the last column. Thus, after step 6 these processes know that action $w(T_2, x)$ happened after $beg(T_1)$. Executing then steps 3 and 4 will inform process T_1 about the violation of conflict-serializability.

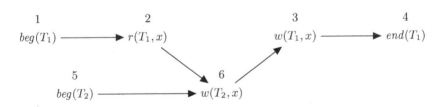

Fig. 5. Violation of conflict-serializability (partial order)

Gossiping in trees of processes works more generally as follows. We call a distributed alphabet (Σ, dom) over process set $\mathbb{P}$ *tree-like*, if there exists some tree T with $\mathbb{P}$ as set of vertices, such that the domain of each action is a *connected* part of T.

Note that the tree T is uniquely defined by action domains in the special case where the actions have at most binary domains. Otherwise, there can be several trees as above, but we will assume that the distributed alphabet comes with a suitable tree representation.

We are also given a set of monitoring actions $\Gamma \subseteq \Sigma$. The task is to compute for each process $p \in \mathbb{P}$ the *happens-before relation w.r.t.* Γ, in other words the happens-before relation between the most recent occurrences of actions from Γ that are known to process p. This information is a DAG where the set of nodes is a subset of Γ. Figure 6 below gives an example of such a computation.

Theorem 4. *Given a tree-like distributed alphabet* (Σ, dom) *with tree* $\mathcal{T}$, *and a set* $\Gamma \subseteq \Sigma$ *of actions. The happens-before relation w.r.t.* Γ *can be computed by a Zielonka automaton where every process p maintains two DAGs of size* $|\Gamma| + out(p)$, *with* $out(p)$ *the out-degree of p in* $\mathcal{T}$. *Each update of the DAGs can be done in linear time in their size.*

Theorem 4 provides a rather simple way of reconstructing the happens-before relation with *finite* additional memory, and in a distributed way. Each synchronization action b will update the DAGs maintained by the processes executing b, by combining these DAGs and selecting the most recent knowledge about actions of Γ. As an example, suppose that processes p and q are neighbors in the tree, say, q is the father of p. As long as there is no synchronization involving both p and q, process p has the most recent knowledge about occurrences of Γ-actions belonging to processes in the subtree of p. As soon as some action synchronizes p and q, process q will be able to include p's knowledge regarding these actions, in its own knowledge.

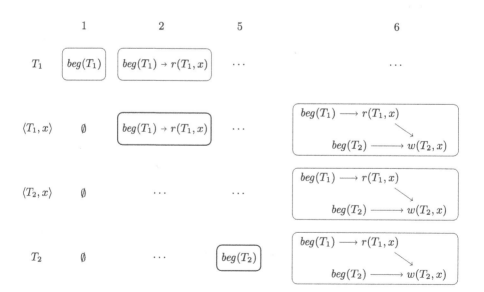

Fig. 6. Computing the partial order. Numbers in the first line stand for the program lines in Example 2. Dots mean that the information does not change.

5   Related Work

This brief overview aimed at presenting the motivation behind distributed synthesis and how Mazurkiewicz trace theory can be used for monitoring concurrent programs. To conclude we point out some further related results.

Our discussion turned around distributed synthesis in a simple case where the program evolves without external actions from an environment. Synthesis of open systems, i.e., systems with an environment, is a more complex problem. Synthesis started as a problem in logics, with Church's problem asking for an algorithm to construct devices that transform sequences of input bits into sequences of output bits, so that the interaction conforms to a given logical formula [7]. Later, Ramadge and Wonham proposed the *supervisory control* formulation [42], where a plant and a specification are given; a controller should be designed such that its product with the plant satisfies the specification, no matter what the environment does. Synthesis is a particular case of control where the plant allows for every possible behavior. Rabin's result on the decidability of monadic second-order logic over infinite trees answered Church's question for MSO specifications [41].

Synthesis without environment. The problem of distributed synthesis without environment was first raised in the context of Petri nets. The task there is to decide whether an automaton, viewed as a graph, is isomorphic to the marking graph of a net. Ehrenfeucht and Rozenberg introduced the notion of regions, that determines how to decompose a graph for obtaining a net [11].

Zielonka's algorithm has been applied to solve the synthesis problem for models that go beyond Mazurkiewicz traces. One example is synthesis of communicating automata from graphical specifications known as *message sequence charts*. Communicating automata are distributed finite-state automata communicating over point-to-point FIFO channels. As such, the model is Turing powerful. However, if the communication channels are bounded, there is a tight link between execution sequences of the communicating automaton and Mazurkiewicz traces [22]. Actually we can handle even the case where the assumption about bounded channels is relaxed by asking that they are bounded for *at least one* scheduling of message receptions [18]. Producer-consumer behaviors are captured by this relaxed requirement.

Multiply nested words with various kinds of bounds on stack access [25, 26, 40], are an attractive model for concurrent programs with recursion, because of decidability properties and expressiveness. In [5] the model is extended to nested Mazurkiewicz traces and Zielonka's construction is lifted to this setting.

For runtime monitoring, a similar approach as ours is advocated in [43], that proposes an epistemic temporal logic for describing safety properties. A distributed implementation of a monitor for such properties is obtained, based on a variant of vector clocks.

Synthesis with environment. One way to lift Church's problem to the distributed setting was proposed by Pnueli and Rosner [39]. They showed that synthesis is decidable for very restricted architectures, namely pipelines. General undecidability was already known from the framework on multi-player games [38]. Subsequently, [28] showed that pipelines are essentially the only decidable case. Some interesting extensions of Pnueli and Rosner's setting have been considered in [13,15,24].

Alternatively, a distributed formulation of Church's problem can be formulated in Ramadge and Wonham's supervisory control setting. This problem, when plants and controllers are Zielonka automata, has been considered in [14,17,29,34]. In this formulation, local controllers exchange information when synchronizing on shared actions. In this way, the arguments for undecidability based on hidden information, as in [38,39], do not apply. Decidability of distributed control for Zielonka automata is still open. It is known though that this formulation admits more decidable architectures: the control problem for local parity specifications is decidable over acyclic architectures [34], thus in cases where Pnueli and Rosner's model is undecidable.

Yet another distributed version of Ramadge and Wonham's problem is considered in [9], this time for Petri nets. The problem is to compute local controllers that guarantee basic properties like e.g. deadlock avoidance. The limitation of this approach is that the algorithm may fail to find local controllers, although they exist.

Game semantics and asynchronous games played on event structures are introduced in [32]. Such games are investigated in [21] from a game-theoretical viewpoint, showing a Borel determinacy result under some restrictions.

Verification. As we already mentioned, automated verification of concurrent systems encounters major problems due to state explosion. One particularly efficient technique able to addresses this problem is known as *partial order reduction* (POR) [20,37,44]. It consists of restricting the exploration of the state space by avoiding the execution of similar, or equivalent runs. The notion of equivalence of runs used by POR is based on Mazurkiewicz traces. The efficiency of POR methods depends of course on the precise equivalence notion between executions. Recent variants, such as dynamic POR, work without storing explored states explicitly and aim at improving the precision by computing additional information about (non)-equivalent executions [1].

There are other contexts in verification where analysis gets more efficient using equivalences based on Mazurkiewicz traces. One such setting is counterexample generation based on partial (Mazurkiewicz) traces instead of linear executions [6]. Previous work connecting concurrency issues and Mazurkiewicz trace theory concerns atomicity violations [12], non-linearizability and sequential inconsistency [3].

Acknowledgments. Very special thanks to Jérôme Leroux, Gabriele Puppis and Igor Walukiewicz for numerous comments on previous versions of this paper.

References

1. Abdulla, P., Aronis, S., Jonsson, B., Sagonas, K.: Optimal dynamic partial order reduction. In: POPL 2014, pp. 373–384. ACM (2014)
2. Akshay, S., Dinca, I., Genest, B., Stefanescu, A.: Implementing realistic asynchronous automata. In: FSTTCS 2013, LIPIcs, pp. 213–224. Schloss Dagstuhl - Leibniz-Zentrum fuer Informatik (2013)
3. Alur, R., McMillan, K., Peled, D.: Model-checking of correctness conditions for concurrent objects. In: LICS 1996, pp. 219–228. IEEE (1996)
4. Armstrong, J.: Programming Erlang: Software for a Concurrent World. Pragmatic Bookshelf (2007)
5. Bollig, B., Grindei, M.-L., Habermehl, P.: Realizability of concurrent recursive programs. In: de Alfaro, L. (ed.) FOSSACS 2009. LNCS, vol. 5504, pp. 410–424. Springer, Heidelberg (2009)
6. Černý, P., Henzinger, T.A., Radhakrishna, A., Ryzhyk, L., Tarrach, T.: Efficient synthesis for concurrency by semantics-preserving transformations. In: Sharygina, N., Veith, H. (eds.) CAV 2013. LNCS, vol. 8044, pp. 951–967. Springer, Heidelberg (2013)
7. Church, A.: Logic, arithmetics, and automata. In: Proceedings of the International Congress of Mathematicians (1962)
8. Cori, R., Métivier, Y., Zielonka, W.: Asynchronous mappings and asynchronous cellular automata. Information and Computation 106, 159–202 (1993)
9. Darondeau, P., Ricker, L.: Distributed control of discrete-event systems: A first step. Transactions on Petri Nets and Other Models of Concurrency 6, 24–45 (2012)
10. Diekert, V., Rozenberg, G. (eds.): The Book of Traces. World Scientific, Singapore (1995)
11. Ehrenfeucht, A., Rozenberg, G.: Partial (set) 2-structures: Parts i and ii. Acta Informatica 27(4), 315–368 (1989)
12. Farzan, A., Madhusudan, P.: Monitoring atomicity in concurrent programs. In: Gupta, A., Malik, S. (eds.) CAV 2008. LNCS, vol. 5123, pp. 52–65. Springer, Heidelberg (2008)
13. Finkbeiner, B., Schewe, S.: Uniform distributed synthesis. In: LICS 2005, pp. 321–330. IEEE (2005)
14. Gastin, P., Lerman, B., Zeitoun, M.: Distributed games with causal memory are decidable for series-parallel systems. In: Lodaya, K., Mahajan, M. (eds.) FSTTCS 2004. LNCS, vol. 3328, pp. 275–286. Springer, Heidelberg (2004)
15. Gastin, P., Sznajder, N.: Fair synthesis for asynchronous distributed systems. ACM Transactions on Computational Logic 14(2), 9 (2013)
16. Genest, B., Gimbert, H., Muscholl, A., Walukiewicz, I.: Optimal Zielonka-type construction of deterministic asynchronous automata. In: Abramsky, S., Gavoille, C., Kirchner, C., Meyer auf der Heide, F., Spirakis, P.G. (eds.) ICALP 2010. LNCS, vol. 6199, pp. 52–63. Springer, Heidelberg (2010)
17. Genest, B., Gimbert, H., Muscholl, A., Walukiewicz, I.: Asynchronous games over tree architectures. In: Fomin, F.V., Freivalds, R., Kwiatkowska, M., Peleg, D. (eds.) ICALP 2013, Part II. LNCS, vol. 7966, pp. 275–286. Springer, Heidelberg (2013)
18. Genest, B., Kuske, D., Muscholl, A.: A Kleene theorem and model checking algorithms for existentially bounded communicating automata. Inf. Comput. 204(6), 920–956 (2006)
19. Genest, B., Muscholl, A.: Constructing exponential-size deterministic Zielonka automata. In: Bugliesi, M., Preneel, B., Sassone, V., Wegener, I. (eds.) ICALP 2006. LNCS, vol. 4052, pp. 565–576. Springer, Heidelberg (2006)

20. Godefroid, P., Wolper, P.: Using partial orders for the efficient verification of deadlock freedom and safety properties. Formal Methods in System Design **2**(2), 149–164 (1993)
21. Gutierrez, J., Winskel, G.: On the determinacy of concurrent games on event structures with infinite winning sets. J. Comput. Syst. Sci. **80**(6), 1119–1137 (2014)
22. Henriksen, J.G., Mukund, M., Kumar, K.N., Sohoni, M., Thiagarajan, P.S.: A Theory of Regular MSC Languages. Inf. Comput. **202**(1), 1–38 (2005)
23. Krishna, S., Muscholl, A.: A quadratic construction for Zielonka automata with acyclic communication structure. Theoretical Computer Science **503**, 109–114 (2013)
24. Kupferman, O., Vardi, M.Y.: Synthesizing distributed systems. In: LICS 2001, pp. 389–398. IEEE (2001)
25. La Torre, S., Madhusudan, P., Parlato, G.: A robust class of context-sensitive languages. In: LICS 2007, pp. 161–170. IEEE (2007)
26. La Torre, S., Parlato, G.: Scope-bounded multistack pushdown systems: fixed-point, sequentialization, and tree-width. In: FSTTCS 2012, LIPIcs, pp. 173–184. Schloss Dagstuhl - Leibniz-Zentrum fuer Informatik (2012)
27. Lamport, L.: Time, clocks, and the ordering of events in a distributed system. Operating Systems **21**(7), 558–565 (1978)
28. Madhusudan, P., Thiagarajan, P.S.: Distributed controller synthesis for local specifications. In: Orejas, F., Spirakis, P.G., van Leeuwen, J. (eds.) ICALP 2001. LNCS, vol. 2076, pp. 396–407. Springer, Heidelberg (2001)
29. Madhusudan, P., Thiagarajan, P.S., Yang, S.: The MSO theory of connectedly communicating processes. In: Sarukkai, S., Sen, S. (eds.) FSTTCS 2005. LNCS, vol. 3821, pp. 201–212. Springer, Heidelberg (2005)
30. Mattern, F.: Virtual time and global states of distributed systems. In: International Workshop on Parallel and Distributed Algorithms, pp. 215–226. Elsevier (1989)
31. Mazurkiewicz, A.: Concurrent program schemes and their interpretations. DAIMI Rep. PB 78, Aarhus University, Aarhus (1977)
32. Melliès, P.-A.: Asynchronous games 2: The true concurrency of innocence. TCS **358**(2–3), 200–228 (2006)
33. Mukund, M., Sohoni, M.A.: Keeping track of the latest gossip in a distributed system. Distributed Computing **10**(3), 137–148 (1997)
34. Muscholl, A., Walukiewicz, I.: Distributed synthesis for acyclic architectures. In: FSTTCS 2014, LIPIcs, pp. 639–651. Schloss Dagstuhl - Leibniz-Zentrum fuer Informatik (2014)
35. Odersky, M., Rompf, T.: Unifying functional and object-oriented programming with Scala. Communications of the ACM **57**(4), 76–86 (2014)
36. Odersky, M., Spoon, L., Venners, B.: Programming in Scala. Artima (2010)
37. Peled, D.: All from one, one for all: on model checking using representatives. In: Courcoubetis, C. (ed.) CAV 1993. LNCS, vol. 697, pp. 409–423. Springer, Heidelberg (1993)
38. Peterson, G.L., Reif, J.H.: Multi-person alternation. In: FOCS 1979, pp. 348–363. IEEE (1979)
39. Pnueli, A., Rosner, R.: Distributed reactive systems are hard to synthesize. In: FOCS 1990. IEEE (1990)
40. Qadeer, S., Rehof, J.: Context-bounded model checking of concurrent software. In: Halbwachs, N., Zuck, L.D. (eds.) TACAS 2005. LNCS, vol. 3440, pp. 93–107. Springer, Heidelberg (2005)
41. Rabin, M.O.: Automata on Infinite Objects and Church's Problem. American Mathematical Society, Providence (1972)

42. Ramadge, P., Wonham, W.: The control of discrete event systems. Proceedings of the IEEE **77**(2), 81–98 (1989)
43. Sen, K., Vardhan, A., Agha, G., Rosu, G.: Decentralized runtime analysis of multithreaded applications. In: International Parallel and Distributed Processing Symposium (IPDPS 2006). IEEE (2006)
44. Valmari, A.: Stubborn sets for reduced state space generation. In: Rozenberg, G. (ed.) Advances in Petri Nets 1990. LNCS, vol. 483, pp. 491–515. Springer, Heidelberg (1991)
45. Zielonka, W.: Notes on finite asynchronous automata. RAIRO-Theoretical Informatics and Applications **21**, 99–135 (1987)

Track B: Logic, Semantics, Automata and Theory of Programming

Games for Dependent Types

Samson Abramsky[1], Radha Jagadeesan[2], and Matthijs Vákár[1(✉)]

[1] University of Oxford, Oxford, UK
Matthijsvakar@gmail.com
[2] DePaul University, Chicago, USA

Abstract. We present a game semantics for dependent type theory (DTT) with Π-, Σ-, intensional Id-types and finite inductive type families. The model satisfies Streicher's criteria of intensionality and refutes function extensionality. The principle of uniqueness of identity proofs is satisfied.

The model is fully and faithfully complete at the type hierarchy built without Id-types. Although definability for the hierarchy with Id-types remains to be investigated, the notions of propositional equality in syntax and semantics do coincide for (open) terms of the Id-free type hierarchy.

1 Introduction

Dependent Type theory (DTT) can be seen as the extension of the simple λ-calculus along the Curry-Howard correspondence from a proof calculus for (intuitionistic) propositional logic to one for predicate logic. It forms the basis of many proof assistants, like NuPRL, LEGO and Coq, and is increasingly being considered as a more expressive type system for programming, as implemented in e.g. ATS, Cayenne, Epigram, Agda and Idris. [1] A recent source of enthusiasm in this field is homotopy type theory (HoTT), which refers to an interpretation of DTT into abstract homotopy theory [2] or, conversely, an extension of DTT that is sufficient to reproduce significant results of homotopy theory [3]. In practice, the latter means DTT with Σ-, Π-, Id-types, a universe satisfying the **univalence axiom**, and certain higher inductive types. The univalence axiom is an extensionality principle which implies, in particular, the axiom of function extensionality [3].

Game semantics provides a unified framework for intensional, computational semantics of various type theories, ranging from pure logics [4] to programming languages [5–7] with a variety of features (e.g. non-local control [8], state [9], dynamically generated local names [10]) and evaluation strategies [11]. A game semantics for DTT has, surprisingly, so far been absent. Our hope is that such a semantics will provide an alternative analysis of the implications of the subtle shades of intensionality that arise in the analysis of DTT [12,13]. Moreover, the game semantics of DTT is based on very different, one might say orthogonal intuitions to those of the homotopical models: temporal rather than spatial, and directly reflecting the structure of computational processes. One goal, to which we hope this work will be a stepping stone, is a game semantics of HoTT doing

© Springer-Verlag Berlin Heidelberg 2015
M.M. Halldórsson et al. (Eds.): ICALP 2015, Part II, LNCS 9135, pp. 31–43, 2015.
DOI: 10.1007/978-3-662-47666-6_3

justice to both the spacial and temporal aspects of identity types. Indeed, such an investigation might even lead to a computational interpretation of the univalence axiom which has long been missing, although a significant step in this direction was recently taken by the constructive cubical sets model of HoTT [14].

Our game theoretic model of DTT is inspired in part by the domain model of DTT [15]. This model views a type family as a continuous function to a domain of domains, a witness of a Π-type $\Pi_{x:A}B$ as a continuous (set theoretic) dependent function and interprets identity types via a kind of intersection. We follow this recipe for modelling type families and identity types. We adapt the viewpoint of the game semantics of system F [7] to describe the Π-type to capture the intuitive idea that the specialisation of a term at type $\Pi_{x:A}B$ to a specific instance $B[a/x]$ is the responsibility solely of the context that provides the argument a of type A; in contrast, any valid term of $\Pi_{x:A}B$ has to operate within the constraints enforced by the context. Our definition draws its power from the fact that, in a game semantics, these constraints are enforced not only on completed computations, but also on the incomplete computations that arise when a term interacts with its context. Thus, while we follow the formal recipes of [15], the temporal character of game semantics results in strikingly different properties of the resulting model.

In the rest of this paper, we describe a game theoretic model of DTT with Σ-, Π- and Id-types, where AJM-games interpret types and history-free winning strategies interpret terms. Our model has the following key properties.

- The place of the Id-types in the intensionality spectrum compares as follows with the domain semantics and with HoTT.

	Domains	HoTT	Games
Failure of Equality Reflection	✓	✓	✓
Streicher Intensionality Criteria ($I1$) and ($I2$)	✓	✓	✓
Streicher Intensionality Criterion ($I3$)	✗	✗	✓
Failure of Function Extensionality (FunExt)	✗	✗	✓
Failure of Uniqueness of Identity Proofs (UIP)	✗	✓	✗

- We show that the model satisfies a full and faithful completeness result with respect to the terms of a version of DTT with Σ-, Π- and Id-types and finite inductive type families, at the type hierarchy built without Id. In contrast, the domain theoretic model of [15] is not (fully) complete or faithful.
- The notions of propositional equality of these (open) terms coincide in syntax and semantics.

2   A Category of Games

The idea behind game semantics is to model a computation by an alternating sequence of interactions (the play) between a program (Player) and its environment (Opponent), following some rules specified by its datatype (the game). In this translation, programs become Player strategies, while termination corresponds to a strategy being winning or beating all Opponents. The charm of this interpretation is that it not only fully captures the intensional aspects of a program but that it combines this with the structural clarity of a categorical model, thus interpolating between traditional operational and denotational semantics.

We assume the reader has some familiarity with the basics of categories of AJM-games and strategies, as described in [16], and only briefly recall the definitions. We define a category **Game** which has as objects AJM-games.

Definition 1 (Game). *A **game** A is a tuple $(M_A, \lambda_A, P_A, \approx_A, W_A)$, where*

- M_A *is a countable set of **moves**;*
- $M_A \xrightarrow{\lambda_A = \langle \lambda_A^{OP}, \lambda_A^{QA} \rangle} \{O, P\} \times \{Q, A\}$ *is a function which indicates if a move is made by **Opponent** (O) or **Player** (P) and if it is a **Question** (Q) or an **Answer** (A), for which we write $\overline{O} = P$, $\overline{P} = O$ and $M_A^O := \lambda_A^{OP^{-1}}(O)$,* $M_A^P := \lambda_A^{OP^{-1}}(P)$, $M_A^Q := \lambda_A^{QA^{-1}}(Q)$ *and* $M_A^A := \lambda_A^{QA^{-1}}(A)$;
- $P_A \subseteq M_A^{\circledast}$ *is a non-empty prefix-closed set of **plays**, where $M_A^{\circledast}$ is the set of finite sequences of uniquely occurring moves, with the properties*
  (p1) $s = at \Rightarrow a \in M_A^O$;
  (p2) $\forall_i \lambda_A^{OP}(s_{i+1}) = \overline{\lambda_A^{OP}(s_i)}$, *where we write s_i for the i-th move in s;*
  (p3) $\forall_{t \leq s} |t \upharpoonright_{M_A^A}| \leq |t \upharpoonright_{M_A^Q}|$.
  *Here, $\leq$ denotes the prefix order and $|s|$ the length of a sequence. Write $\mathsf{j}_{A,s}(m)$ for the last unanswered question preceding an answer m in a play s, which we say m answers. $\mathsf{j}_{A,s}$ will be used to enforce **stack discipline**.*
- $\approx_A$ *is an equivalence relation on P_A, satisfying*
  (e1) $s \approx_A t \Rightarrow \lambda_A^*(s) = \lambda_A^*(t)$;
  (e2) $s \approx_A t \wedge s' \leq s \wedge t' \leq t \wedge |s'| = |t'| \Rightarrow s' \approx_A t'$;
  (e3) $s \approx_A t \wedge sa \in P_A \Rightarrow \exists_b sa \approx_A tb$.
  Here, λ_A^* is the extension of λ_A to sequences.
- $W_A \subseteq P_A^{\infty}$ *is a set of **winning plays**, where P_A^{∞} is the set of infinite plays, i.e. infinite sequences of moves such that all their finite prefixes are in P_A, such that W_A is closed under $\approx_A$ in the sense that*

$$(s \in W_A \wedge t \notin W_A) \Rightarrow \exists_{s_0 \leq s, t_0 \leq t} |s_0| = |t_0| \wedge s_0 \not\approx_A t_0.$$

Our notion of morphism will be defined in terms of strategies on games.

Definition 2 (Strategy). *A **strategy on** A is a subset $\sigma \subseteq P_A^{\text{even}}$ satisfying*

(Causal Consistency): $sab \in \sigma \Rightarrow s \in \sigma$;
(Representation Independence): $s \in \sigma \wedge s \approx_A t \Rightarrow t \in \sigma$;
(Determinacy): $sab, ta'b' \in \sigma \wedge sa \approx_A ta' \Rightarrow sab \approx_A ta'b'$.

We write $\mathsf{str}(A)$ for the set of strategies on A. We sometimes identify σ with the subset of P_A that is obtained as its prefix closure. In fact, we restrict to history-free strategies, as we are modelling computation without mutable state.

Definition 3 (History-Free Strategy). *We call a strategy $\sigma \in \mathsf{str}(A)$ **history-free**, if there exists a non-empty causally consistent subset $\phi \subseteq \sigma$ (called a **history-free skeleton**) such that*

(Uniformization): $\forall_{sab\in\sigma} s \in \phi \Rightarrow \exists!_{b'} sab' \in \phi;$
(History-Freeness 1): $sab, tac \in \phi \Rightarrow b = c;$
(History-Freeness 2): $(sab, t \in \phi \ \wedge \ ta \in P_A) \Rightarrow tab \in \phi.$

Then, ϕ is induced by a partial function on moves and $\sigma = \{t \mid \exists_{s\in\phi} t \approx_A s\}$.

From now on, we assume strategies to be history-free. Winning conditions give rise to the notion of a winning strategy, the semantic equivalent of a normalising or total term. A winning strategy always has a response to any valid O-move. Furthermore, if the result of the interaction between a strategy and Opponent is an infinite play, then this is a member of the set of winning plays.

Definition 4 (Winning Strategy). *A strategy $\sigma \in \mathsf{str}(A)$ is **winning** if it satisfies*

(Finite Wins): If s is $\leq$-maximal in σ, then s is $\leq$-maximal in P_A.
(Infinite Wins): If $s_0 \leq s_1 \leq \dots$ is an infinite chain in σ, then $\bigcup_i s_i \in W_A$.

We write $\mathsf{wstr}(A)$ for the set of winning strategies on A. Next, we define some constructions on games, starting with their symmetric monoidal closed structure.

Definition 5 (Tensor Unit). *We define the game $I := (\emptyset, \emptyset, \{\epsilon\}, \{(\epsilon, \epsilon)\}, \emptyset)$.*

Definition 6 (Tensor). *For games A, B, we define*
$A \otimes B := (M_A + M_B = \Sigma_{i\in\{A,B\}} M_i, [\lambda_A, \lambda_B], P_{A\otimes B}, \approx_{A\otimes B}, W_{A\otimes B})$ *with*

- $P_{A\otimes B} = \{s \mid s \upharpoonright_A \in P_A \wedge s \upharpoonright_B \in P_B \wedge \mathsf{fst}^*(\mathsf{j}^*_{A\otimes B,s}(s \upharpoonright_{M^A_{A\otimes B}})) = \mathsf{fst}^*(s \upharpoonright_{M^A_{A\otimes B}})\};$
- $s \approx_{A\otimes B} t := s \upharpoonright_A \approx_A t \upharpoonright_A \ \wedge \ s \upharpoonright_B \approx_B t \upharpoonright_B \ \wedge \ \forall_{1\leq i\leq |s|} s_i \in M_A \Leftrightarrow t_i \in M_A;$
- $W_{A\otimes B} := \{s \in P^\infty_{A\otimes B} \mid (s \upharpoonright_A \in P^\infty_A \Rightarrow s \upharpoonright_A \in W_A) \wedge (s \upharpoonright_B \in P^\infty_B \Rightarrow s \upharpoonright_B \in W_B)\}.$

Definition 7 (Linear Implication). *For games A, B, we define*
$A \multimap B := (M_A + M_B = \Sigma_{i\in\{A,B\}} M_i, [\overline{\lambda_A}, \lambda_B], P_{A\multimap B}, \approx_{A\multimap B}, W_{A\multimap B})$ *with*

- $P_{A\multimap B} = \{s \mid s \upharpoonright_A \in P_A \wedge s \upharpoonright_B \in P_B \wedge \mathsf{fst}^*(\mathsf{j}^*_{A\multimap B,s}(s \upharpoonright_{M^A_{A\multimap B}})) = \mathsf{fst}^*(s \upharpoonright_{M^A_{A\multimap B}})\};$
- $s \approx_{A\multimap B} t := s \upharpoonright_A \approx_A t \upharpoonright_A \ \wedge \ s \upharpoonright_B \approx_B t \upharpoonright_B \ \wedge \ \forall_{1\leq i\leq |s|} s_i \in M_A \Leftrightarrow t_i \in M_A;$
- $W_{A\multimap B} := \{s \in P^\infty_{A\multimap B} \mid s \upharpoonright_A \in W_A \Rightarrow s \upharpoonright_B \in W_B\}.$

Note that the definitions of λ_- imply that in $A \otimes B$ only Opponent can switch between A and B, while in $A \multimap B$ only Player can. These definitions on objects extend to strategies, e.g. for (winning) strategies $\sigma \in \mathsf{str}(A), \tau \in \mathsf{str}(B)$, we can define a (winning) strategy $\sigma \otimes \tau = \{s \in P^{\mathsf{even}}_{A\otimes B} \mid s \upharpoonright_A \in \sigma \ \wedge \ s \upharpoonright_B \in \tau\} \in \mathsf{str}(A\otimes B)$. This gives us a model of multiplicative intuitionistic linear logic, with all structural morphisms consisting of appropriate variants of copycat strategies.

Theorem 1 (Linear Category of Games). *We define a category **Game** by*

- $\mathsf{ob}(\mathbf{Game}) := \{A \mid A \text{ is an AJM-game}\};$
- $\mathbf{Game}(A, B) := \mathsf{wstr}(A \multimap B);$
- $\mathsf{id}_A := \{s \in P_{A\multimap A} \mid s \upharpoonright_{A^{(1)}} \approx_A s \upharpoonright_{A^{(2)}}\}$, *the **copycat strategy** on A;*

– for $A \xrightarrow{\sigma} B \xrightarrow{\tau} C$, the composition (or **interaction**) $A \xrightarrow{\sigma;\tau} C$ is defined from parallel composition $\sigma \| \tau := \{s \in M^{\circledast}_{(A \multimap B) \multimap C} \mid s \restriction_{A,B} \in \sigma \land s \restriction_{B,C} \in \tau\}$ plus hiding: $\sigma ; \tau := \{s \restriction_{A,C} \mid s \in \sigma \| \tau\}$.

Then, $(\mathbf{Game}, I, \otimes, \multimap)$ is, in fact, a symmetric monoidal closed category.

To make this into a model of intuitionistic logic, a Cartesian closed category (ccc), through the (first) Girard translation, we need two more constructions on games, to interpret the additive conjunction and exponential, respectively.

Definition 8 (With). We define the game
$A \& B := (M_A + M_B, [\lambda_A, \lambda_B], P_A + P_B, \approx_A + \approx_B, W_A + W_B)$.

Definition 9 (Bang). We define $!A := (\mathbb{N} \times M_A, \lambda_A \circ \mathsf{snd}, P_{!A}, \approx_{!A}, W_{!A})$ with

– $P_{!A} = \{s \mid \forall_{i \in \mathbb{N}} s \restriction_i \in P_A \land \mathsf{fst}^*(\mathsf{j}^*_{!A,s}(s \restriction_{M^A_{!A}})) = \mathsf{fst}^*(s \restriction_{M^A_{!A}})\}$;
– $s \approx_{!A} t := \exists_{\pi \in S(\mathbb{N})} \forall_{i \in \mathbb{N}} s \restriction_i \approx_A t \restriction_{\pi(i)} \land (\pi \circ \mathsf{fst})^*(s) = \mathsf{fst}^*(t)$;
– $W_{!A} := \{s \in P^{\infty}_{!A} \mid \forall_i s \restriction_i \in P^{\infty}_A \Rightarrow s \restriction_i \in W_A\}$.

Next, we note that $!$ can be made into a co-monad by defining, for $A \xrightarrow{\sigma} B$,

$$!\sigma := \{s \in P^{\mathsf{even}}_{!A \multimap !B} \mid \exists_{\pi \in S(\mathbb{N})} \forall_{i \in \mathbb{N}} s \restriction_{(\pi(i),A),(i,B)} \in \sigma\},$$

and natural transformations

$$!A \xrightarrow{\mathsf{der}_A} A := \{s \in P^{\mathsf{even}}_{!A \multimap A} \mid \exists_{i \in \mathbb{N}} s \restriction_{!A} \restriction_i \approx_A s \restriction_A\} \quad \text{and}$$

$$!A \xrightarrow{\delta_A} !!A := \{s \in P^{\mathsf{even}}_{!A \multimap !!A} \mid \exists_{p:\mathbb{N} \times \mathbb{N} \rightarrowtail \mathbb{N}} \forall_{i,j \in \mathbb{N}} s \restriction_{!A} \restriction_{p(i,j)} \approx_A s \restriction_{!!A} \restriction_i \restriction_j\}.$$

This allows us to define the co-Kleisli category $\mathbf{Game}_!$, which has the same objects as $\mathbf{Game}$, while $\mathbf{Game}_!(A,B) := \mathbf{Game}(!A, B)$. We have a composition $(f,g) \mapsto f^{\dagger}; g$, where we write $f^{\dagger} := \delta_{\mathsf{dom}(f)}; !(f)$, for which the strategies der_A serve as identities. We can define finite products in $\mathbf{Game}_!$ by I and $\&$ and write

$$\mathsf{diag}_A := \{s \in P^{\mathsf{even}}_{!A \multimap (A \& A)} \mid \exists_{i \in \mathbb{N}} (s \restriction_{!A} \restriction_i \approx_A s \restriction_{A^{(1)}} \neq \epsilon) \lor (s \restriction_{!A} \restriction_i \approx_A s \restriction_{A^{(2)}} \neq \epsilon)\}$$

for the diagonal $!A \longrightarrow A \& A$. Moreover, we have Seely-isomophisms $!I \cong I$ and $!(A \& B) \cong !A \otimes !B$, so we obtain a linear-non-linear adjunction $\mathbf{Game} \leftrightarrows \mathbf{Game}_!$, hence a model of multiplicative exponential intuitionistic linear logic. In particular, by defining $A \Rightarrow B := !A \multimap B$, we get a ccc. We write $\mathsf{comp}_{A,B,C}$ for the internal composition $((A \Rightarrow B) \& (B \Rightarrow C)) \longrightarrow A \Rightarrow C$ in $\mathbf{Game}_!$.

Theorem 2 (Intuitionist Category of Games). $(\mathbf{Game}_!, I, \&, \Rightarrow)$ is a ccc.

Note that $W_A = \emptyset$ for the hierarchy of intuitionistic types A that are formed by operations I, $\&$ and $\Rightarrow$ from finite games, so winning strategies are the total strategies - strategies which respond to any O-move - for which infinite chattering does not occur in any interaction.

3   Dependent Games

The previous section sketched how **Game$_!$** models simple intuitionistic type theory. Next, we show how it comes equipped with a notion of dependent type. This leads to an indexed ccc **DGame$_!$** of dependent games and strategies.

We define a poset **Game$_\trianglelefteq$** of games with $A \trianglelefteq B := (M_A = M_B) \wedge (\lambda_B|_{M_A} = \lambda_A) \wedge (P_A \subseteq P_B) \wedge (s \approx_A t \;\Leftrightarrow\; s \in P_A \wedge s \approx_B t) \wedge (W_A = W_B \cap P_A^\infty)$. Given a game C, we define the cpo $\mathsf{Sub}(C)$ as the poset of its $\trianglelefteq$-subgames. We note that, for $A, B \in \mathsf{Sub}(C)$, $A \trianglelefteq B \Leftrightarrow P_A \subseteq P_B$.

For a game A, we define the set $\mathsf{ob}(\mathbf{DGame_!}(A))$ of **games with dependency on** A as the set of continuous functions $\mathsf{str}(A) \xrightarrow{B} \mathsf{Sub}(\bigcup B)$ for some other game $\bigcup B$. In practice, when defining a dependent game, we often leave $\bigcup B$ implicit as $\bigcup_{\sigma \in \mathsf{str}(A)} B(\sigma)$. Define I, $!$, $\otimes$, $\multimap$ and $\&$ pointwise on dependent games B, also performing the operation on $\bigcup B$. Writing $s \mapsto \bar{s}$ for the function $P_{!A} \longrightarrow \mathcal{P}(P_A)$ inductively defined on the empty play, Opponent moves and Player moves, respectively, as $\epsilon \mapsto \emptyset$, $s(i,a) \mapsto \bar{s}$, $s(i,a)(i,b) \mapsto \overline{s(i,a)} \cup \{t \mid \exists_{s' \in \bar{s}} t \approx_A s'ab\}$, we define the dependent function space as follows.

Definition 10 (Π-Game). *Given $B \in \mathsf{ob}(\mathbf{DGame_!}(A))$, we define a subgame $\Pi_{!A}B \;\; \trianglelefteq \;\; !A \multimap \bigcup B$ of dependent functions from A to B, by*

$$P_{\Pi_{!A}B} := \{\epsilon\} \bigcup$$

$$\{sa \mid s \in P_{\Pi_{!A}B}^{\mathsf{even}} \wedge ((\exists_{\tau \supseteq \overline{sa\restriction_{!A}}} \tau \in \mathsf{wstr}(A)) \Rightarrow \exists_{\overline{sa\restriction_{!A}} \subseteq \tau \in \mathsf{wstr}(A)} sa \restriction_{\bigcup B} \in P_{B(\tau)})\} \bigcup$$

$$\{sab \mid sa \in P_{\Pi_{!A}B}^{\mathsf{odd}} \wedge \forall_{\overline{sab\restriction_{!A}} \subseteq \tau \in \mathsf{wstr}(A)} sa \restriction_{\bigcup B} \in P_{B(\tau)} \Rightarrow sab \restriction_{\bigcup B} \in P_{B(\tau)}\}.$$

Following the mantra of game semantics for quantifiers [7], in $\Pi_{!A}B$, Opponent can choose a winning strategy τ on A while Player has to play in a way that is compatible with all choices of τ that have not yet been excluded. Similarly to the approach taken in the game semantics for polymorphism [7], we do not specify all of τ in one go, as this would violate "Scott's axiom" of continuity of computation. Instead, τ is gradually revealed, explicitly so by playing in $!A$ and implicitly by playing in B. That is, unless the play in $!A$ is such that it does not extend to define a winning strategy on A. Then, any further play in $!A \multimap \bigcup B$ is permitted. This can occur in two scenarios: either Opponent has not played along ($!\sigma$ for) a (partial) strategy σ on A, or the play in $!A$ defines a strategy on A but none of its extensions are winning. The latter, for instance, occurs when A models an uninhabited type, like the propositions 0 or $\mathsf{Id}_{\mathsf{Things}}(\mathsf{Jelly\ beans}, \mathsf{Marsupials})$. For an example, let $\mathsf{days}(n) := \{m \mid \text{there are} > m \text{ days in the year } n\}$ and define $\widetilde{\mathsf{days}}_*(\bot) = \widetilde{\emptyset}_*$, $\widetilde{\mathsf{days}}_*(n) := \widetilde{\mathsf{days}(n)}_*$ to obtain a game depending on $\widetilde{\mathbb{N}}_*$ (with $\widetilde{\mathsf{days}}(n) = \widetilde{\mathbb{N}_{<365*}}$ or $\widetilde{\mathbb{N}_{<366*}}$). Here, $\widetilde{X}_*$ signifies the game with $P_{\widetilde{X}_*} = \{\epsilon, *\} \cup \{*x \mid x \in X\}$ and $\approx_X = \mathsf{id}_X$ Then, the following are valid strategies. The fourth example is especially important, as it generalises to a (derelicted) B-copycat on $\Pi_{!A}(!B \multimap B)$ for arbitrary B, denoted $\mathbf{v}_{[A],[B]}$ in section 4. This motivates why Opponent can narrow down the fibre of B freely, while Player can only

$!\widetilde{\mathbb{N}}_*$	$\widetilde{days}_*$	$!\widetilde{\mathbb{N}}_*$	$\widetilde{days}_*$	$!\widetilde{\mathbb{N}}_*$	$\widetilde{days}_*$	$!\widetilde{\mathbb{N}}_*$	$!\widetilde{days}_*$	$\widetilde{days}_*$	
	$*$		$*$		$*$			$*$	O
364		$(i, *)$		$(i, *)$		$(i, *)$			P
		$(i, 1984)$		$(i, 1984)$		(i, m)			O
				$(i+1, *)$			m		P
				$(i+1, 1985)$					O
			365		365				P

Fig. 1. Three strategies on $\Pi_{!\widetilde{\mathbb{N}}_*}\widetilde{days}_*$ and one on $\Pi_{!\widetilde{\mathbb{N}}_*}!\widetilde{days}_* \multimap \widetilde{days}_*$. The first as all years have > 364 days, the second as 1984 was (among other things) a leap year, the third as Player can play any move in $\bigcup \widetilde{days}_* = \widetilde{\mathbb{N}}_{<366*}$ after Opponent has not played along a strategy on $\widetilde{\mathbb{N}}_*$ and the fourth as Opponent makes the move m first, after which Player can safely copy it. In the paired moves, Player chooses an (irrelevant) index i.

play without narrowing down the fibre further. To see that Player should not be able to narrow down the fibre of B, note that we do not want $f := \{\epsilon, *365\}$ to define a strategy on $\Pi_{!\widetilde{\mathbb{N}}_*}\widetilde{days}_*$, as 1983; $f = \{\epsilon, *365\} \notin \mathrm{str}(\widetilde{days}_*(1983))$.

Theorem 3. *We obtain a strict indexed ccc* $\mathbf{Game}_!^{op} \xrightarrow{(\mathbf{DGame}_!, -\{-\})} \mathbf{Cat}$ *of dependent games, if we define*

- *fibrewise hom-sets* $\mathbf{DGame}_!(A)(B,C) := \mathrm{wstr}(\Pi_{!A}(!B \multimap C))$;
- *fibrewise identities* $\mathrm{der}_B := \{s \in P_{\Pi_{!A}(!B \multimap B)} \mid \exists_i s \lceil_{!B} \lceil_i \approx_B s \lceil_B\}$;
- *if* $B \xrightarrow{\tau} C \xrightarrow{\tau'} D \in \mathbf{DGame}_!(A)$, $\tau^\dagger;_A \tau' := \mathrm{diag}_A^\dagger; \tau^\dagger \otimes \tau'; \mathrm{comp}_{\bigcup B, \bigcup C, \bigcup D}^1$;
- *given* $f \in \mathbf{Game}_!(A', A)$, *we define* $B\{f\} \in \mathrm{ob}(\mathbf{DGame}_!(A'))$ *by* $B\{f\}(\sigma) := B(!(\sigma); f)$ *and* $\bigcup B\{f\} := \bigcup B$ *and* $\tau\{f\} := f^\dagger; \tau|_{\Pi_{!A'}(!B\{f\} \multimap C\{f\})}$, *where we write* $(-)|_X$ *for the restriction of (the plays of) a strategy to* $P_X{}^2$.

Seeing that $\mathbf{Game}_!$ additionally has a terminal object I to interpret the empty context, we are well on our way to producing a model of dependent type theory [17]: we only need to interpret context extension. This takes the form of the comprehension axiom for $\mathbf{DGame}_!$, which states that for each $A \in \mathrm{ob}(\mathbf{Game}_!)$ and $B \in \mathrm{ob}(\mathbf{DGame}_!(A))$ the following functor is representable

$$x \mapsto \mathbf{DGame}_!(\mathrm{dom}(x))(I, B\{x\}) : (\mathbf{Game}_!/A)^{op} \longrightarrow \mathsf{Set}.$$

Unfortunately, this fails, as $\mathbf{Game}_!$ does not yield a sound interpretation dependent contexts. Essentially, the problem is that we do not have **additive Σ-types**, appropriate generalisations $\Sigma_A^\&B$ of $\&$ to interpret dependent context extension in $\mathbf{Game}_!$.

---

[1] To be precise, we can interpret τ and τ' as partial (history-sensitive) strategies on $!A \multimap (!\bigcup B \multimap \bigcup C)$ and $!A \multimap (!\bigcup C \multimap \bigcup D)$, respectively, and note that $\tau^\dagger;_A \tau'$ defines a winning (history-free) strategy on $\Pi_{!A}(!B \multimap D)$. Similarly for $\tau\{f\}$.

[2] In fact, we can note that, we only need to restrict O-moves to $\Pi_{!A'}(!B\{f\} \multimap C\{f\})$, in which case P-moves automatically respect the rules of the game.

Theorem 4. DGame₁ *does not satisfy the comprehension axiom.*

4   A Category with Families of Context Games

All is not lost, however. In fact, we have almost translated the structural core of the syntax of DTT into the world of games and strategies. The remaining generalisation, necessitated by the lack of additive Σ-types, is to dependent games depending on multiple (mutually dependent) games. We can produce a categorical model of DTT out of the resulting structure by applying a so-called **category of contexts (Ctxt) construction**, which is precisely how one builds a categorical model from the syntax of dependent type theory [13,18]. Alternatively, this construction can be seen as a universal way of making our indexed category satisfy the comprehension axiom, extending its base category by (inductively) adjoining strong Σ-types formally, analogous to the Fam-construction of [11] which adds formal co-products.

The problem which needs to be addressed is how to interpret dependent functions of more variables. For this purpose, we define a **context game** to be a finite list $[X_i]_{1 \leq i \leq n}$ where X_i is a game with dependency on $[X_j]_{j < i}$, i.e. a continuous function $\Sigma(\mathsf{str}(X_1), \ldots, \mathsf{str}(X_{i-1})) \xrightarrow{X_i} \mathbf{Game}_{\lhd}$, where we write $\Sigma(\ldots)$ for the usual iterated Σ-type of domains [15], i.e. the set-theoretic Σ-typed induced with the product order. For a game X_{n+1} depending on $[X_i]_{i \leq n}$, we define $\Pi_{!X_n} X_{n+1}$ depending on $[X_i]_{i \leq n-1}$ by $\Pi_{!X_n} X_{n+1}(\sigma_1, \ldots, \sigma_{n-1}) := \Pi_{!X_n(\sigma_1, \ldots, \sigma_{n-1})} X_{n+1}(\sigma_1, \ldots, \sigma_{n-1}, -)$ and $\bigcup \Pi_{!X_n} X_{n+1} := !(\bigcup X_n) \multimap \bigcup X_{n+1}$.

For illustration, define a game $\widetilde{\mathsf{RA}}_*$ depending on the context game $[\widetilde{\mathbb{N}}_*, \widetilde{\mathsf{days}}_*]$ by $\mathsf{RA}(n, m) := \{\text{Rick Astley lyrics from songs released before day } m \text{ of year } n\}$. Then, the following two strategies illustrate that a dependent function may query its arguments in unexpected order or may not query some at all.

$!\widetilde{\mathbb{N}}_*$	$!\widetilde{\mathsf{days}}_*$	$\widetilde{\mathsf{RA}}_*$		$!\widetilde{\mathbb{N}}_*$	$!\widetilde{\mathsf{days}}_*$	$\widetilde{\mathsf{RA}}_*$		
		$*$				$*$		O
$(i,*)$				$(i,*)$				P
	$(i, m > 206)$			$(i, n > 1987)$				O
$(j,*)$						Never Gonna Let You Down		P
$(j, 1987)$								O
		Never Gonna Give You Up						P

Fig. 2. Two examples of (partial) strategies on the iterated Π-game $\Pi_{!\widetilde{\mathbb{N}}_*} \Pi_{!\widetilde{\mathsf{days}}_*} \widetilde{\mathsf{RA}}_*$

We define a category $\mathsf{Ctxt}(\mathbf{DGame}_!)$ with objects context games and morphisms which are defined inductively as (dependent) lists of winning strategies on appropriate Π-games. We show that this has the structure of a category with families (CwF) [13], a canonical notion of model of DTT. This gives a more concise presentation of the resulting indexed category with comprehension, where we also add formal Σ-types in the fibres.

Definition 11 (CwF). *A CwF is a category $\mathcal{C}$ with a terminal object $\cdot$, for all objects Γ a set $\mathsf{Ty}(\Gamma)$, for all $A \in \mathsf{Ty}(\Gamma)$ a set $\mathsf{Tm}(\Gamma, A)$, for all $\Gamma' \xrightarrow{f} \Gamma$ in $\mathcal{C}$ functions $\mathsf{Ty}(\Gamma) \xrightarrow{-\{f\}} \mathsf{Ty}(\Gamma')$ and $\mathsf{Tm}(\Gamma, A) \xrightarrow{-\{f\}} \mathsf{Tm}(\Gamma', A\{f\})$, such that*

$A\{\mathrm{id}_\Gamma\} = A$	*(Ty-Id)*	$A\{g \circ f\} = A\{g\}\{f\}$	*(Ty-Comp)*
$t\{\mathrm{id}_\Gamma\} = A$	*(Tm-Id)*	$t\{g \circ f\} = t\{g\}\{f\}$	*(Tm-Comp)*,

for $A \in \mathsf{Ty}(\Gamma)$ a morphism $\Gamma.A \xrightarrow{\mathbf{p}_{\Gamma,A}} \Gamma$ of $\mathcal{C}$ and $\mathbf{v}_{\Gamma,A} \in \mathsf{Tm}(\Gamma.A, A\{\mathbf{p}_{\Gamma,A}\})$ and, finally, for all $t \in \mathsf{Tm}(\Gamma', A\{f\})$ a morphism $\Gamma' \xrightarrow{\langle f,t\rangle} \Gamma.A$ such that

$\mathbf{p}_{\Gamma,A} \circ \langle f,t\rangle = f$	*(Cons-L)*	$\mathbf{v}_{\Gamma,A}\{\langle f,t\rangle\} = t$	*(Cons-R)*
$\langle \mathbf{p}_{\Gamma,A}, \mathbf{v}_{\Gamma,A}\rangle = \mathrm{id}_{\Gamma.A}$	*(Cons-Id)*	$\langle f,t\rangle \circ g = \langle f \circ g, t\{g\}\rangle$	*(Cons-Nat)*.

Theorem 5. *We have a CwF* $(\mathsf{Ctxt}(\mathbf{DGame}_!), \mathsf{Ty}, \mathsf{Tm}, \mathbf{p}, \mathbf{v}, -.-, \langle -,-\rangle)$.

We define the required structures. All equations follow trivially from the definitions and the two lemmas stated. We define $\mathsf{Ty}([X_i]_i)$ as the set of **context games with dependency** on $[X_i]_i$: $[Y_j]_j \in \mathsf{Ty}([X_i]_i)$ iff $[X_i]_i.[Y_j]_j := [X_1, \ldots, X_n, Y_1, \ldots, Y_m]$ is a context game, while $\cdot := []$ is the terminal object.

Next, $\mathsf{mor}(\mathcal{C})$ and $-\{-\}_{\mathsf{Ty}}$ (and a special case of $-\{-\}_{\mathsf{Tm}}$) are defined through a mutual induction, where, in the last clause, we consider $\langle \sigma_1, \ldots, \sigma_n\rangle$ as a partial (history-sensitive) strategy on $!I \multimap \bigcup X_1 \& \ldots \& \bigcup X_n \cong I \multimap \bigcup X_1 \& \ldots \& \bigcup X_n$:

$$\mathsf{Ctxt}(\mathbf{DGame}_!)([X_i]_{i \leq n}, [Y_j]_{j \leq m}) := \{[f_j]_{j \leq m} \mid f_j \in \mathsf{wstr}(\Pi_{!X_1} \ldots \Pi_{!X_n} Y_j\{[f_k]_{k<j}\})\}$$

$$Y_j\{[f_k]_{k<j}\}(\sigma_1, \ldots, \sigma_n) := Y_j(f_1\{[\sigma_i]_{i \leq n}\}, \ldots, f_{j-1}\{[\sigma_i]_{i \leq n}\}), \quad \bigcup Y_j\{[f_k]_{k<j}\} = \bigcup Y_j$$

$$f_k\{[\sigma_i]_{i \leq n}\} := (\langle \sigma_1, \ldots, \sigma_n\rangle^\dagger; f_k)|_{Y_k\{[f_l]_{l<k}\}(\sigma_1, \ldots, \sigma_n)}.$$

The identities are defined as lists of derelicted copycats. Let us define a strategy $\mathsf{der}_{[X_j]_j, X_i}$ which plays the derelicted copycat on the whole image of X_i: $\mathsf{der}_{[X_j]_j, X_i} := \{s \in P_{\Pi_{!X_1} \ldots \Pi_{!X_n} X_i} \mid \exists_k s \upharpoonright_{!X_i} \upharpoonright_k \approx \bigcup X_i s \upharpoonright_{X_i}\}$. We then define $\mathsf{id}_{[X_i]_i} := [\mathsf{der}_{[X_j]_j, X_i}]_i$ and $\mathbf{p}_{[X_i]_i, [Y_j]_j} := [\mathsf{der}_{[X_i]_i.[Y_j]_j, X_k}]_k$. Let us define

$$\mathsf{Tm}([X_i]_{i \leq n}, [Y_j]_{j \leq m}) := \{[f_j]_j \mid [X_i]_i \xrightarrow{[\mathsf{der}_{[X_i]_i, X_1}, \ldots, \mathsf{der}_{[X_i]_i, X_n}, f_1, \ldots, f_m]} [X_i]_i.[Y_j]_j\}.$$

Then, we can define $\mathbf{v}_{[X_i]_i, [Y_j]_j} := [\mathsf{der}_{[X_i]_i.[Y_j]_j, Y_k}]_k$. Note that these are well-defined because of the following lemma.

Lemma 1. $\mathsf{der}_{[X_j]_j, X_i} \in \mathsf{wstr}(\Pi_{!X_1} \cdots \Pi_{!X_n} X_i\{[\mathsf{der}_{[X_j]_j, X_k}]_{k \leq i-1}\})$.

We define $\langle [f_j]_{j \leq m}, [g_k]_{k \leq l}\rangle := [f_1, \ldots, f_m, g_1, \ldots, g_l]$. We inductively define the composition of $[X_i]_{i \leq n} \xrightarrow{[f_j]_j} [Y_j]_{j \leq m} \xrightarrow{[g_k]_k} [Z_k]_k$ in $\mathsf{Ctxt}(\mathbf{DGame}_!)$ as follows

$$[f_j]_j ; [g_k]_k := [\langle f_1, \ldots, f_m\rangle^\dagger; g_k|_{\Pi_{!X_1} \cdots \Pi_{!X_n} Z_k\{[f_j]_j; [g_{k'}]_{k'<k}\}}]_k,$$

where f_j are considered as partial (history-sensitive) strategies on $!(\bigcup X_1 \& \cdots \& \bigcup X_n) \multimap \bigcup Y_j$.

Lemma 2. $[f_j]_j ; [g_k]_k$ *is a list of winning strategies if $[g_k]_k$ and $[f_j]_j$ are.*

Note that for $[X_i]_i \xrightarrow{[f_j]_j} [Y_j]_j$ and $[g_k]_k \in \mathsf{Tm}([Y_j]_j, [Z_k]_k)$,

$$[f_j]_j; \langle [\mathsf{der}_{[Y_{j'}]_{j'}, Y_j}]_j, [g_k]_k \rangle = \langle [f_j]_j, [h_k]_k \rangle,$$

for some $[h_k]_k$. We use this to define $-\{-\}_{\mathsf{Tm}}$: $[g_k]_k \{[f_j]_j\} := [h_k]_k$.

Remark 1. Note that, in $\mathsf{Ctxt}(\mathbf{DGame}_!)$, $[A, B] \cong [A\&B]$ if A and B are games (without mutual dependency) and $[] \cong [I]$.

5   Semantic Type Formers

We show that our CwF supports Σ-, Π-, and Id-types. We characterise some of the properties of the Id-types, marking their place in the intensionality spectrum.

Theorem 6. *Our CwF supports Σ- and Π-type with their β- and η-rules.*

Σ-types are just interpreted by concatenation of lists. We define a Σ-type $\Sigma_{[Y_j]_j}[Z_k]_k \in \mathsf{Ty}([X_i]_{i \leq n})$ as $[Y_j]_j.[Z_k]_k$ for $[Z_k]_{k \leq l} \in \mathsf{Ty}([X_i]_{i \leq n}.[Y_j]_{j \leq m})$.

We have already seen Π-types $\Pi_{[X_i]_{i \leq n}}[Y] := [\Pi_{!X_1} \cdots \Pi_{!X_n} Y]$ of dependent games. What remains to be defined are Π-types $\Pi_{[X_i]_i}[Y_j]_j$ of general dependent context games, which can now be reduced to the former, as we have that $\Sigma_{f:\Pi_{x:A}B}\Pi_{x:A}C[f(x)/y]$ satisfies the rules for $\Pi_{x:A}\Sigma_{y:B}C$.

Corollary 1. *Note that this means that $\mathsf{Ctxt}(\mathbf{DGame}_!)$ is in particular a ccc.*

We turn to identity types next, which are essentially defined as those of the domain semantics of DTT [15]. Interestingly, due to the more intensional nature of game semantics, they acquire a more intensional character, refuting FunExt.

Let us define $[X_i]_{i \leq n} \trianglelefteq [Y_i]_{i \leq n}$ for context games if $X_i\{[\sigma_j]_{j<i}\} \trianglelefteq Y_i\{[\sigma_j]_{j \leq i}\}$ for all $1 \leq i \leq n$ and $[\sigma_j]_{j<i} \in \Sigma(\mathsf{str}(X_1), \ldots, \mathsf{str}(X_{i-1}))$. For $[Y_j]_j \in \mathsf{Ty}([X_i]_i)$, define $\mathsf{Id}_{[Y_j]_j} \in \mathsf{Ty}([X_i]_i.[Y_j]_j.[Y_{j'}]_{j'})$ through the intersection of subgames[3]

$$\mathsf{Id}_{[Y_j]_j}([\sigma_i]_i, [\tau_j]_j, [\tau'_j]_j) := [\tau_j \cap \tau'_j]_j \trianglelefteq [Y_j]_j\{[\sigma_i]_i\}.$$

Theorem 7. *This definition satisfies the I-, E- and β-rules for Id-types.*

For Id-I, $x : A \vdash \mathsf{refl}_t : \mathsf{Id}_B(t, t)$ can be interpreted as the list of strategies $[\![t]\!]$ but at $\Pi_{[\![A]\!]}\mathsf{Id}_{[\![B]\!]}([\![t]\!], [\![t]\!]) \trianglelefteq \Pi_{[\![A]\!]}[\![B]\!]$, where we write $[\![-]\!]$ for the interpretation of DTT in our model. We can interpret Id-E such that the interpretation of its conclusion does not depend on the particular proof of identity[4].

In addition to being non-extensional (i.e. refuting the principle of equality reflection), these identity types can be said to be intensional in a positive sense.

Theorem 8. *Streicher's Criteria of Intensionality [12] are satisfied, i.e.*

---

[3] Here, we identify a strategy σ on X with the subgame $\{s \in P_X \mid \exists_{t \in \sigma} s \leq t\} \trianglelefteq X$.

[4] In fact, we need this rule in the syntax for a **faithful** interpretation in the model.

(I1) there exist ⊢ A type *such that* $x, y : A, z : \mathsf{Id}_A(x, y) \not\vdash x \equiv y : A$;

(I2) there exist ⊢ A type *and* $x : A \vdash B$ type *such that* $x, y : A, z : \mathsf{Id}_A(x, y) \not\vdash$ $B \equiv B[y/x]$ type;

(I3) for all ⊢ A type, ⊢ $p : \mathsf{Id}_A(t, s)$ *implies* ⊢ $t \equiv s : A$.

(I1) relies on the interpretation of terms carrying intensionality. For instance, we can take $[\![A]\!] = \widetilde{\mathbb{B}}_*$, where $\mathbb{B} := \{\mathsf{tt}, \mathsf{ff}\}$, and evaluate the first and second projections on $[\![x]\!] = [\![z]\!] = \bot$ and $[\![y]\!] = \mathsf{tt}$. (I2) relies on semantic types having intensional features. We can use $[\![B]\!] := (\bot, \mathsf{ff} \mapsto I, \mathsf{tt} \mapsto \widetilde{\mathbb{B}}_*)$ on the data of (I1). (I3) follows as $[p_i]_i \in \mathsf{Ctxt}(\mathbf{DGame}_!)([], \mathsf{Id}_{[X_i]_i}([f_i]_i, [g_i]_i) := [f_i \cap g_i]_i)$ implies that $p_i = f_i = g_i$ for all i, as winning strategies are maximal.

The proofs of (I1) and (I2) also work for the domain model of DTT. (I3) relies on the crucial difference between the domain and games models in their interpretation of identity types of open terms. Indeed, FunExt is seen to fail in the games model: note that for strict and non-strict constantly tt functions f and g, we have $[f] \in \mathsf{Tm}([\widetilde{\mathbb{B}}_*], \mathsf{Id}_{[\widetilde{\mathbb{B}}_*]}([f], [g]))$, while $\mathsf{Tm}([], \mathsf{Id}_{\Pi_{[\widetilde{\mathbb{B}}_*]}[\widetilde{\mathbb{B}}_*]}([f], [g])) = \emptyset$.

Theorem 9. FunExt *is refuted: for* ⊢ $f, g : \Pi_{x:A}B$, *we do not generally have* $z : \Pi_{x:A}\mathsf{Id}_B(f(x), g(x)) \vdash \mathsf{FunExt}_{f,g} : \mathsf{Id}_{\Pi_{x:A}B}(f, g)$.

On the other hand, it turns out that we have the principle of uniqueness of identity proofs UIP, by playing copycats between (the first) $[\![A]\!]$ and $\mathsf{Id}_{\mathsf{Id}_{[\![A]\!]}}$.

Theorem 10. *We have* $x, y : A, p, q : \mathsf{Id}_A(x, y) \vdash \mathsf{UIP}_A : \mathsf{Id}_{\mathsf{Id}_A(x,y)}(p, q)$.

6   Ground Types and Completeness Results

We illustrate how our model of dependent games and winning strategies satisfies a completeness result with respect to the syntax of DTT with Σ-, Π- and Id-types and finite inductive type families, at the Id-free hierarchy of types.

We describe a scheme for inductively defining finite type families. Let A be a type. Then, we specify a finite inductive definition of a type family $x : A \vdash B$ type by specifying finitely many closed terms $a_1, \ldots, a_n : A$ and distinct symbols b_{ij}, $1 \leq i \leq n$, $1 \leq j \leq m_i$. The idea is that B is a type family, such that $B[a_i/x]$ contains precisely the distinct closed terms $b_{i,1}, \ldots, b_{i,m_i}$. These type families are more limited than general inductive definitions as they are freely generated by (finitely many) **closed** terms, while one would allow open terms in the general case. This means that we precisely get the inductive type families that have finitely many non-empty fibres which are all finite types.

We interpret such a definition as specifying *I*- and *E*-rules for B:

$$\frac{}{\vdash b_{i,j} : B[a_i/x]}\, B\text{-}I_{i,j} \qquad \frac{x : A, y : B \vdash C\ \text{type}}{\vdash \mathsf{case}_B : \Pi_{x:A,y:B,z_{11}:C[a_1/x,b_{1,1}/y],\ldots,z_{nm_n}:C[a_n/x,b_{n,m_n}/y]}C}\, B\text{-}E,$$

together with the β- and η-rules, commutative conversions and a rule[5] defining a falsum eliminator from $\mathsf{Id}_B(b_{i,j}, b_{i',j'})$ for distinct constructors $b_{i,j}$, $b_{i',j'}$ of B.

---

[5] Note that this rule is derivable in presence of a universe.

Let $\mathsf{Ctxt}(\mathbf{DGame}_!)_{\mathrm{fin}\Sigma\Pi}$ be the full subcategory of $\mathsf{Ctxt}(\mathbf{DGame}_!)$ on the hierarchy generated by Σ- and Π-types and finite inductive dependent games (and substitution), as below. Then we have the following results.

Theorem 11 (Finite Inductive Dependent Games). *Finite inductive type families B over a type A, where $B(a_i)$ is generated by $\{b_{ij} \mid j\}$, have a sound interpretation in* $\mathsf{Ctxt}(\mathbf{DGame}_!)_{\mathrm{fin}\Sigma\Pi}$: $[\![B]\!] : [\![a_i]\!] \mapsto [\{\widehat{b_{ij} \mid j}\}_*]$, else $\longmapsto [\widetilde{\emptyset}_*]$.

Theorem 12 (Full and Faithful Completeness). *All morphisms in* $\mathsf{Ctxt}(\mathbf{DGame}_!)_{\mathrm{fin}\Sigma\Pi}$ *are faithfully definable in DTT with Σ-, Π- and Id-types with an identity proof irrelevant eliminator and finite inductive type families.*

By embedding our fragment of DTT in finitary PCF, where we allow larger types at the cost of non-termination, faithfulness follows from the corresponding result for PCF [6]. Definability is proved along the lines of the template of [19] and hinges on the decomposition lemma for PCF-games.

Although the completeness properties of the model at the hierarchy with Id-types remain to be studied in detail, we do have the following.

Theorem 13. *The semantic propositional equality of (open) terms of the Id-type free hierarchy does agree with that of the syntax.*

If $p \in \mathsf{Tm}([\![A]\!], \mathsf{Id}_{[\![B]\!]}([\![f]\!], [\![g]\!]))$, it follows that $[\![f]\!](a') = [\![g]\!](a')$ for all $a' \in \mathsf{Ctxt}(\mathbf{DGame}_!)_{\mathrm{fin}\Sigma\Pi}([], [\![A]\!])$. Because of theorem 12, this implies that $f(a) = g(a)$ for all $\vdash a : A$. As there are only finitely many such a, we can perform a case analysis on these to construct $x : A \vdash \text{'case'}(a, \overrightarrow{\mathsf{refl}_{f(a)}}) : \mathsf{Id}_B(f, g)$.

7   Future Work

Ultimately, the main goal is a thorough intensional, computational analysis of HoTT [3]. Obvious concrete directions for future work are the following:

- modifying the model to break UIP, perhaps through nominal games;
- examining the phenomena of function extensionality and univalence;
- study of universes and a more intensional notion of type family;
- study of (higher) inductive type families and their definability results;
- establishing completeness results for the type hierarchy with Id-types;
- constructing models of DTT with side effects;
- synthesising strategies from a dependently typed specification;
- study of a possible embedding of the model in the co-Eilenberg-Moore category **Game**$^!$, which might simplify its presentation.

Acknowledgments. Samson Abramsky was supported by the EPSRC, AFOSR and the John Templeton Foundation. Radha Jagadeesan acknowledges support from the NSF. Matthijs Vákár was supported by the EPSRC and the Clarendon Fund.

References

1. Altenkirch, T., McBride, C., McKinna, J.: Why dependent types matter. Manuscript, 235 (2005). http://www.cs.nott.ac.uk/txa/publ/ydtm.pdf
2. Awodey, S., Warren, M.A.: Homotopy theoretic models of identity types. Mathematical Proceedings of the Cambridge Philosophical Society **146**(01), 45–55 (2009)
3. HoTTbaki, U.: Homotopy Type Theory: Univalent Foundations of Mathematics. Institute for Advanced Study (2013). http://homotopytypetheory.org/book
4. Abramsky, S., Jagadeesan, R.: Games and full completeness for multiplicative linear logic. The Journal of Symbolic Logic **59**(02), 543–574 (1994)
5. Hyland, J.M.E., Ong, C.H.: On full abstraction for PCF: I, II, and III. Information and Computation **163**(2), 285–408 (2000)
6. Abramsky, S., Jagadeesan, R., Malacaria, P.: Full abstraction for PCF. Information and Computation **163**(2), 409–470 (2000)
7. Abramsky, S., Jagadeesan, R.: A game semantics for generic polymorphism. Annals of Pure and Applied Logic **133**(1), 3–37 (2005)
8. Laird, J.: Full abstraction for functional languages with control. In: Proceedings of the 12th Annual IEEE Symposium on Logic in Computer Science, LICS 1997, pp. 58–67. IEEE (1997)
9. Abramsky, S., Honda, K., McCusker, G.: A fully abstract game semantics for general references. In: Proceedings of the Thirteenth Annual IEEE Symposium on Logic in Computer Science, pp. 334–344. IEEE (1998)
10. Abramsky, S., Ghica, D.R., Murawski, A.S., Ong, C.H., Stark, I.D.: Nominal games and full abstraction for the nu-calculus. In: Proceedings of the 19th Annual IEEE Symposium on Logic in Computer Science, pp. 150–159. IEEE (2004)
11. Abramsky, S., McCusker, G.: Call-by-value games. In: Nielsen, M., Thomas, W. (eds.) Computer Science Logic. Lecture Notes in Computer Science, vol. 1414, pp. 1–17. Springer, Berlin Heidelberg (1998)
12. Streicher, T.: Investigations into intensional type theory (1993). http://www.mathematik.tu-darmstadt.de/streicher/HabilStreicher.pdf
13. Hofmann, M.: Syntax and semantics of dependent types. In: Extensional Constructs in Intensional Type Theory, pp. 13–54. Springer (1997)
14. Bezem, M., Coquand, T., Huber, S.: A model of type theory in cubical sets. In: 19th International Conference on Types for Proofs and Programs, TYPES 2013, vol. 26, pp. 107–128 (2014)
15. Palmgren, E., Stoltenberg-Hansen, V.: Domain interpretations of Martin-Löf's partial type theory. Annals of Pure and Applied Logic **48**(2), 135–196 (1990)
16. Abramsky, S., Jagadeesan, R.: Game semantics for access control. Electronic Notes in Theoretical Computer Science **249**, 135–156 (2009)
17. Vákár, M.: A categorical semantics for linear logical frameworks. In: Pitts, A. (ed.) FOSSACS 2015. LNCS, vol. 9034, pp. 102–116. Springer, Heidelberg (2015)
18. Pitts, A.M.: Categorical logic. In: Abramsky, S., Gabbay, D., Maibaum, T., (eds.) Handbook of Logic in Computer Science, vol. 5, pp. 39–128. OUP (2000)
19. Abramsky, S.: Axioms for definability and full completeness. In: Proof, Language and Interaction: Essays in Honour of Robin, pp. 55–75. MIT Press (2000)

Short Proofs of the Kneser-Lovász Coloring Principle

James Aisenberg[1], Maria Luisa Bonet[2], Sam Buss[1(✉)],
Adrian Crăciun[3], and Gabriel Istrate[3]

[1] Department of Mathematics, University of California, San Diego,
La Jolla, CA 92093-0112, USA
jaisenberg@math.ucsd.edu, sbuss@ucsd.edu
[2] Computer Science Department, Universidad Politécnica de Cataluña,
Barcelona, Spain
bonet@cs.upc.edu
[3] West University of Timişoara, and the e-Austria Research Institute,
300223 Timişoara, Romania
acraciun@ieat.ro, gabrielistrate@acm.org

Abstract. We prove that the propositional translations of the Kneser-Lovász theorem have polynomial size extended Frege proofs and quasi-polynomial size Frege proofs. We present a new counting-based combinatorial proof of the Kneser-Lovász theorem that avoids the topological arguments of prior proofs for all but finitely many cases for each k. We introduce a miniaturization of the octahedral Tucker lemma, called the *truncated Tucker lemma*: it is open whether its propositional translations have (quasi-)polynomial size Frege or extended Frege proofs.

1 Introduction

This paper discusses proofs of Lovász's theorem about the chromatic number of Kneser graphs, and the proof complexity of propositional translations of the Kneser-Lovász theorem. We give a new proof of the Kneser-Lovász theorem that uses a simple counting argument instead of the topological arguments used in prior proofs, for all but finitely many cases. Our arguments can be formalized in propositional logic to give polynomial size extended Frege proofs and quasi-polynomial size Frege proofs.

Frege systems are sound and complete proof systems for propositional logic with a finite schema of axioms and inference rules. The typical example is a "textbook style" propositional proof system using *modus ponens* as its only

J. Aisenberg—Supported in part by NSF grants DMS-1101228 and CCF-1213151.
M.L. Bonet—Supported in part by grant TIN2013-48031-C4-1.
S. Buss—Supported in part by NSF grants DMS-1101228 and CCF-1213151, and Simons Foundation award 306202.
A. Crăciun and G. Istrate—Supported in part by IDEI grant PN-II-ID-PCE-2011-3-0981 "Structure and computational difficulty in combinatorial optimization: an interdisciplinary approach".

© Springer-Verlag Berlin Heidelberg 2015
M.M. Halldórsson et al. (Eds.): ICALP 2015, Part II, LNCS 9135, pp. 44–55, 2015.
DOI: 10.1007/978-3-662-47666-6_4

rule of inference, and all Frege systems are polynomially equivalent to this system [7]. Extended Frege systems are Frege systems augmented with the extension rule, which allows variables to abbreviate complex formulas. The *size* of a Frege or extended Frege proof is measured by counting the number of symbols in the proof [7]. Frege proofs are able to reason using Boolean formulas; whereas extended Frege proofs can reason using Boolean circuits (see [9]). Boolean formulas are conjectured to require exponential size to simulate Boolean circuits; there is no known direct connection, but by analogy, it is generally conjectured that there is an exponential separation between the sizes of Frege proofs and extended Frege proofs. This is one of the important open questions in proof complexity; for more on proof complexity see e.g. [2,4,5,7,10,13].

As discussed by Bonet, Buss and Pitassi [2] and more recently by [1,6], we have hardly any examples of combinatorial tautologies, apart from consistency statements, that are conjectured to exponentially separate Frege and extended Frege proof size. These prior works discussed a number of combinatorial principles, including the pigeonhole principle and Frankl's theorem. Istrate and Crăciun [8] recently proposed the Kneser-Lovász principle as a candidate for exponentially separating Frege and extended Frege proof size. In this paper we give quasi-polynomial size Frege proofs of the propositional translations of the Kneser-Lovász theorem for all fixed k. Thus they do not provide an exponential separation of Frege and extended Frege proof size.

Our proof is also interesting because it gives a new method of proving the Kneser-Lovász theorem. Prior proofs use (at least implicitly) a topological fixed-point lemma. The most combinatorial proof is by Matoušek [12] and is inspired by the octahedral Tucker lemma; see also Ziegler [14]. Our new proofs mostly avoid topological arguments and use a counting argument instead. These counting arguments can be formalized with Frege proofs. Indeed, one of the important strengths of Frege proofs is that they can reason about integer arithmetic. These techniques originated in polynomial size Frege proofs of the pigeonhole principle [3] which used carry-save-addition representations for vector addition and multiplication in order to express and prove properties about integer operations in polynomial size. For the Kneser-Lovász theorem, the counting arguments reduce the general case to "small" instances of size $n \leq 2k^4$. For fixed k, there are only finitely many small instances, and they can be verified by exhaustive enumeration. As we shall see, this leads to polynomial size extended Frege proofs, and quasi-polynomial size Frege proofs, for the Kneser-Lovász principles.

It is surprising that the topological arguments can be largely eliminated from the proof of the Kneser-Lovász theorem. The only remaining use of topological arguments is to establish the "small instances". It would be interesting to give an additional argument that avoids having to prove the small instances separately. One possibility for this would be to adapt the proof based on the octahedral Tucker lemma to quasi-polynomial size Frege proofs. The first difficulty with this is that the octahedral Tucker lemma has exponentially large propositional translations. To circumvent this, we present a miniaturized version of the octahedral Tucker lemma called the *truncated Tucker lemma*. The truncated Tucker lemma has polynomial size propositional translations. We prove that the

Kneser-Lovász tautologies have polynomial size constant depth Frege proofs if the propositional formulas for the truncated Tucker lemma are given as additional hypotheses. However, it remains open whether these truncated Tucker lemma principles have (quasi-)polynomial size Frege or extended Frege proofs.

The (n, k)-Kneser graph is defined to be the undirected graph whose vertices are the k-subsets of $\{1, \ldots, n\}$; there is an edge between two vertices iff those vertices have empty intersection. The Kneser-Lovász theorem states that Kneser graphs have a large chromatic number:

Theorem 1 (Lovász [11]). *Let $n \geq 2k > 1$. The (n, k)-Kneser graph has no coloring with $n - 2k + 1$ colors.*

It is well-known that the (n, k)-Kneser graph has a coloring with $n - 2k + 2$ colors (see e.g. the appendix to the arXiv version of this paper), so the bound $n - 2k + 1$ is optimal. For $k = 1$, the Kneser-Lovász theorem is just the pigeonhole principle.

Istrate and Crǎciun [8] noted that, for fixed values of k, the propositional translations of the Kneser-Lovász theorem have polynomial size in n. They presented arguments that can be formalized by polynomial size Frege proofs for $k = 2$, and by polynomial size extended Frege proofs for $k = 3$. This left open the possibility that the $k = 3$ case could exponentially separate the Frege and extended Frege systems. It was also left open whether the $k > 3$ case of the Kneser-Lovász theorem gave tautologies that require exponential size extended Frege proofs. As discussed above, the present paper refutes these possibilities. Theorems 4 and 5 summarize these results.

Let $[n]$ be the set $\{1, \ldots, n\}$; members of $[n]$ are called *nodes*. We identify $\binom{n}{k}$ with the set of k-subsets of $[n]$, the *vertices* of the (n, k)-Kneser graph.

Definition 2. *An m-coloring of the (n, k)-Kneser graph is a map c from $\binom{n}{k}$ to $[m]$, such that for $S, T \in \binom{n}{k}$, if $S \cap T = \emptyset$, then $c(S) \neq c(T)$. If $\ell \in [m]$, then the* color class P_ℓ *is the set of vertices assigned the color ℓ by c.*

The formulas Kneser$_k^n$ are the natural propositional translations of the statement that there is no $(n - 2k + 1)$-coloring of the (n, k)-Kneser graph:

Definition 3. *Let $n \geq 2k > 1$, and $m = n - 2k + 1$. For $S \in \binom{n}{k}$ and $i \in [m]$, the propositional variable $p_{S,i}$ has the intended meaning that vertex S of the Kneser graph is assigned the color i. The formula* Kneser$_k^n$ *is*

$$\bigwedge_{S \in \binom{n}{k}} \bigvee_{i \in [m]} p_{S,i} \;\rightarrow\; \bigvee_{\substack{S,T \in \binom{n}{k} \\ S \cap T = \emptyset}} \bigvee_{i \in [m]} (p_{S,i} \wedge p_{T,i}).$$

Theorem 4. *For fixed parameter $k \geq 1$, the propositional translations* Kneser$_k^n$ *of the Kneser-Lovász theorem have polynomial size extended Frege proofs.*

Theorem 5. *For fixed parameter $k \geq 1$, the propositional translations* Kneser$_k^n$ *of the Kneser-Lovász theorem have quasi-polynomial size Frege proofs.*

When both k and n are allowed to vary, it is open whether the Kneser_k^n tautologies have quasi-polynomial size (extended) Frege proofs, or equivalently, have proofs with size quasi-polynomially bounded in terms of n^k.

2   Mathematical Arguments

Section 2.1 gives the new proof of the Kneser-Lovász theorem; this is later shown to be formalizable with polynomial size extended Frege proofs. Section 2.2 gives a slightly more complicated but more efficient proof, later shown to be formalizable with quasi-polynomial size Frege proofs. The next definition and lemma are crucial for Sects. 2.1 and 2.2.

Any two vertices in a color class P_ℓ have non-empty intersection. One way this can happen is for the color class to be "star-shaped":

Definition 6. *A* color class P_ℓ *is* star-shaped *if* $\bigcap P_\ell$ *is non-empty. If P_ℓ is star-shaped, then any $i \in \bigcap P_\ell$ is called a* central element *of* P_ℓ.

The next lemma bounds the size of color classes that are not star-shaped. It will be used in our proof of the Kneser-Lovász theorem to establish the existence of star-shaped color classes. The idea is that non-star-shaped color classes are too small to cover all $\binom{n}{k}$ vertices.

Lemma 7. *Let c be a coloring of $\binom{n}{k}$. If P_ℓ is not star-shaped, then*

$$|P_\ell| \;\leq\; k^2 \binom{n-2}{k-2}.$$

Proof. Suppose P_ℓ is not star-shaped. If P_ℓ is empty, the claim is trivial. So suppose $P_\ell \neq \emptyset$, and let $S_0 = \{a_1, \ldots, a_k\}$ be some element of P_ℓ. Since P_ℓ is not star-shaped, there must be sets $S_1, \ldots, S_k \in P_\ell$ with $a_i \notin S_i$ for $i = 1, \ldots, k$.

To specify an arbitrary element S of P_ℓ, we do the following. Since S and S_0 have the same color, $S \cap S_0$ is non-empty. We first specify some $a_i \in S \cap S_0$. Likewise, $S \cap S_i$ is non-empty; we second specify some $a_j \in S \cap S_i$. By construction, $a_i \neq a_j$, so S is fully specified by the k possible values for a_i, the k possible values for a_j, and the $\binom{n-2}{k-2}$ possible values for the remaining members of S. Therefore, $|P_\ell| \leq k^2 \binom{n-2}{k-2}$. □

2.1   Argument for Extended Frege Proofs

Let $k > 1$ be fixed. We prove the Kneser-Lovász theorem by induction on n. The base cases for the induction are $n = 2k, \ldots, N(k)$ where $N(k)$ is the constant depending on k specified in Lemma 8. We shall show that $N(k)$ is no greater than k^4. Since k is fixed, there are only finitely many base cases. Since the Kneser-Lovász theorem is true, these base cases can all be proved by a fixed Frege proof of finite size (depending on k). Therefore, in our proof below, we only show the induction step.

Lemma 8. *Fix $k > 1$. There is an $N(k)$ so that, for $n > N(k)$, any $(n-2k+1)$-coloring of $\binom{n}{k}$ has at least one star-shaped color class.*

Proof. Suppose that a coloring c has no star-shaped color class. Since there are $n - 2k + 1$ many color classes, Lemma 7 implies that

$$(n - 2k + 1) \cdot k^2 \binom{n-2}{k-2} \geq \binom{n}{k}. \tag{1}$$

For fixed k, the left-hand side of (1) is $\Theta(n^{k-1})$ and the right-hand side is $\Theta(n^k)$. Thus, there exists an $N(k)$ such that (1) fails for all $n > N(k)$. Hence for $n > N(k)$, there must be at least one star-shaped color class.   □

To obtain an upper bound on the value of $N(k)$, note that (1) is equivalent to

$$(n - 2k + 1)k^3(k - 1) \geq n(n - 1). \tag{2}$$

Since $2k - 1 \geq 1$, (2) implies that $(n - 1)k^4 > n(n - 1)$ and thus that $n < k^4$. Thus, (1) will be false if $n \geq k^4$; so $N(k) < k^4$.

We are now ready to give our first proof of the Kneser-Lovász theorem.

Proof (of Theorem 1, except for base cases). Fix $k > 1$. By Lemma 8, there is some $N(k)$ such that for $n > N(k)$, any $(n - 2k + 1)$-coloring c of $\binom{n}{k}$ has a star-shaped color class. As discussed above, the cases of $n \leq N(k)$ cases are handled by exhaustive search and the truth of the Kneser-Lovász theorem. For $n > N(k)$, we prove the claim by infinite descent. In other words, we show that if c is an $(n - 2k + 1)$-coloring of $\binom{n}{k}$, then there is some c' which is an $((n - 1) - 2k + 1)$-coloring of $\binom{n-1}{k}$.

By Lemma 8, the coloring c has some star-shaped color class P_ℓ with central element i. Without loss of generality, $i = n$ and $\ell = n - 2k + 1$. Let

$$c' = c \upharpoonright \left(\tbinom{n-1}{k}\right)$$

be the restriction of c to the domain $\binom{n-1}{k}$. This discards the central element n of P_ℓ, and thus all vertices with color ℓ. Therefore, c' is an $((n - 1) - 2k + 1)$-coloring of $\binom{n-1}{k}$. This completes the proof.   □

2.2   Argument for Frege Proofs

We now give a second proof of the Kneser-Lovász theorem. The proof above required $n - N(k)$ rounds of infinite descent to transform a Kneser graph on n nodes to one on $N(k)$ nodes. Our second proof replaces this with only $O(\log n)$ many rounds, and this efficiency will be key for formalizing this proof with quasi-polynomial size Frege proofs in Sect. 3.2.

We refine Lemma 8 to show that for n sufficiently large, there are many (i.e., a constant fraction) star-shaped color classes. The idea is to combine the upper bound of Lemma 7 on the size of non-star-shaped color classes with the trivial upper bound of $\binom{n-1}{k-1}$ on the size of star-shaped color classes.

Lemma 9. *Fix $k > 1$ and $0 < \beta < 1$. Then there exists an $N(k, \beta)$ such that for $n > N(k, \beta)$, if c is an $(n - 2k + 1)$-coloring of $\binom{n}{k}$, then c has at least $\frac{n}{k}\beta$ many star-shaped color classes.*

Proof. The value of $N(k, \beta)$ can be set equal to $\frac{k^3(k-\beta)}{1-\beta}$. Let $n > \frac{k^3(k-\beta)}{1-\beta}$, and suppose c is an $(n - 2k + 1)$-coloring of $\binom{n}{k}$. Let α be the number of star-shaped color classes of c. It is clear that an upper bound on the size of each star-shaped color class is $\binom{n-1}{k-1}$. There are $n - \alpha - 2k + 1$ many non-star-shaped classes, and Lemma 7 bounds their size by $k^2\binom{n-2}{k-2}$. This implies that

$$\binom{n-1}{k-1}\alpha + k^2\binom{n-2}{k-2}(n - \alpha - 2k + 1) \geq \binom{n}{k}. \tag{3}$$

Assume for a contradiction that $\alpha < \frac{n}{k}\beta$. Since $n > \frac{k^3(k-\beta)}{1-\beta}$, $0 < \beta < 1$, and $k \geq 2$, we have $n - 1 > k^3(k-1) > k^2(k-1)$. Therefore, $\binom{n-1}{k-1} > k^2\binom{n-2}{k-2}$, and if α is replaced by the larger value $\frac{n}{k}\beta$, the left hand side of (3) increases. Thus,

$$\binom{n-1}{k-1}\frac{n}{k}\beta + k^2\binom{n-2}{k-2}\left(n - \frac{n}{k}\beta - 2k + 1\right) > \binom{n}{k}.$$

Since $\binom{n-1}{k-1}\frac{n}{k} = \binom{n}{k}$ and $n - \frac{n}{k}\beta - 2k + 1 = \frac{k-\beta}{k}n - 2k + 1$,

$$k^2\binom{n-2}{k-2}\left(\frac{k-\beta}{k}n - 2k + 1\right) > (1 - \beta)\binom{n}{k}. \tag{4}$$

We have $\frac{k-\beta}{k}(n - 1) > \frac{k-\beta}{k}n - 2k + 1$. Therefore, (4) gives

$$k^3(k - 1)\frac{k-\beta}{k}(n - 1) > (1 - \beta)n(n - 1).$$

Dividing by $n - 1$ gives $k^3(k - \beta) > (1 - \beta)n$, contradicting $n > \frac{k^3(k-\beta)}{1-\beta}$. $\square$

We now give our second proof of the Kneser-Lovász theorem.

Proof (of Theorem 1, except for base cases). Fix $k > 1$. By Lemma 9 with $\beta = 1/2$, if $n > N(k, 1/2)$ and c is an $(n - 2k + 1)$-coloring of $\binom{n}{k}$, then c has at least $n/2k$ many star-shaped color classes. We prove the Kneser-Lovász theorem by induction on n. The base cases are for $2k \leq n \leq N(k, 1/2)$, and there are only finitely of these, so they can be exhaustively proven. For $n > N(k, 1/2)$, we structure the induction proof as an infinite descent. In other words, we show that if c is an $(n - 2k + 1)$-coloring of $\binom{n}{k}$, then there is some c' that is an $((n - \frac{n}{2k}) - 2k + 1)$-coloring of $\binom{n-\frac{n}{2k}}{k}$. For simplicity of notation, we assume $\frac{n}{2k}$ is an integer. If this is not the case, we really mean to round up to the nearest integer $\lceil \frac{n}{2k} \rceil$.

By permuting the color classes and the nodes, we can assume w.l.o.g. that the $\frac{n}{2k}$ color classes P_ℓ for $\ell = n - \frac{n}{2k} - 2k + 2, \ldots, n - 2k + 1$ are star-shaped, and each such P_ℓ has central element $\ell + 2k - 1$. That is, the last $\frac{n}{2k}$ many color classes are star-shaped and their central elements are the last $\frac{n}{2k}$ nodes in $[n]$. (It is possible that some star-shaped color classes share central nodes; in this case, additional nodes can be discarded so that $n/2k$ are discarded in all.)

Define c' to be the coloring of $\binom{n - n/2k}{k}$ which assigns the same colors as c. The map c' is a $(\frac{2k-1}{2k}n - 2k + 1)$-coloring of $\binom{\frac{2k-1}{2k}n}{k}$, since $n - \frac{n}{2k} = \frac{2k-1}{2k}n$. This completes the proof of the induction step. □

When formalizing the above argument with quasi-polynomial size Frege proofs, it will be important to know how many iterations of the procedure are required to reach the base cases, so let us calculate this.

After s iterations of this procedure, we have a $((\frac{2k-1}{2k})^s n - 2k + 1)$-coloring of $\binom{(\frac{2k-1}{2k})^s n}{k}$. We pick s large enough so that $(\frac{2k-1}{2k})^s n$ is less than $N(k, 1/2)$. In other words, since k is constant,

$$s = \log_{\frac{2k}{2k-1}}\left(\frac{n}{k^3(2k-1)}\right) = O(\log n)$$

will suffice, and only $O(\log n)$ many rounds of the procedure are required.

We do not know if the bound in Lemma 9 is optimal or close to optimal. An appendix in the arXiv version of this paper will discuss the best examples we know of colorings with large numbers of non-star-shaped color classes.

3   Formalization in Propositional Logic

3.1   Polynomial Size Extended Frege Proofs

We sketch the formalization of the argument in Sect. 2.1 as a polynomial size extended Frege proof, establishing Theorem 4. We concentrate on showing how to express concepts such as "star-shaped color class" with polynomial size propositional formulas. For space reasons, we omit the straightforward details of how (extended) Frege proofs can prove properties of these concepts.

Fix values for k and n with $n > N(k)$. We describe an extended Frege proof of Kneser$_k^n$. We have variables $p_{S,j}$ (recall Definition 3), collectively denoted just $\vec{p}$. The proof assumes Kneser$_k^n(\vec{p})$ is false, and proceeds by contradiction. The main step is to define new variables $\vec{p}'$ and prove that Kneser$_k^{n-1}(\vec{p}')$ fails. This will be repeated until reaching a Kneser graph over only $N(k)$ nodes.

For this, let Star(i, ℓ) be a formula that is true when $i \in [n]$ is a central element of the color class P_ℓ; namely,

$$\text{Star}(i, \ell) := \bigwedge_{S \in \binom{n}{k}, \, i \notin S} \neg p_{S,\ell}.$$

We use Star$(\ell) := \bigvee_i \text{Star}(i, \ell)$ to express that P_ℓ is star-shaped.

The extended Frege proof defines the instance of the Kneser-Lovasz principle Kneser_k^{n-1} by discarding one node and one color. The first star-shaped color class P_ℓ is discarded; accordingly, we let

$$\text{DiscardColor}(\ell) := \text{Star}(\ell) \wedge \bigwedge_{\ell' < \ell} \neg\text{Star}(\ell').$$

The node to be discarded is the least central element of the discarded P_ℓ:

$$\text{DiscardNode}(i) := \bigvee_\ell \left[\text{DiscardColor}(\ell) \wedge \text{Star}(i, \ell) \wedge \bigwedge_{i' < i} \neg\text{Star}(i', \ell)\right].$$

After discarding the node i and color class P_ℓ, the remaining nodes and colors are renumbered to the ranges $[n-1]$ and $[n-2k]$, respectively. In particular, the "new" color j (in the instance of Kneser_k^{n-1}) corresponds to the "old" color $j^{-\ell}$ (in the instance of Kneser_k^n) where

$$j^{-\ell} = \begin{cases} j & \text{if } j < \ell \\ j+1 & \text{if } j \geq \ell. \end{cases}$$

And, if $S = \{i_1, \ldots, i_k\} \in \binom{n-1}{k}$ is a "new" vertex (for the Kneser_k^{n-1} instance), then it corresponds to the "old" vertex $S^{-i} \in \binom{n}{k}$ (for the instance of Kneser_k^n), where $S^{-i} = \{i'_1, i'_2, \ldots, i'_k\}$ with

$$i'_t = \begin{cases} i_t & \text{if } i_t < i \\ i_t + 1 & \text{if } i_t \geq i. \end{cases}$$

For each $S \in \binom{n-1}{k}$ and $j \in [n-1]$, the extended Frege proof uses the extension rule to introduce a new variable $p'_{S,j}$ defined as follows

$$p'_{S,j} \equiv \bigvee_{i,\ell} \left(\text{DiscardNode}(i) \wedge \text{DiscardColor}(\ell) \wedge p_{S^{-i}, j^{-\ell}}\right).$$

As seen in the definition by extension, $p'_{S,j}$ is defined by cases, one for each possible pair i, ℓ of nodes and colors such that the node i is the least central element of the P_ℓ color class, where P_ℓ is the first star-shaped color class. The extended Frege proof then shows that $\neg\text{Kneser}_k^n(\vec{p})$ implies $\neg\text{Kneser}_k^{n-1}(\vec{p}')$, i.e., that if the variables $p_{S,j}$ define a coloring, then the variables $p'_{S,j}$ also define a coloring. For this, it is necessary to show that there is at least one star-shaped color class; this is provable with a polynomial size extended Frege proof (even a Frege proof) using the construction of Lemma 8 and the counting techniques of [3].

The extended Frege proof iterates this process of removing one node and one color until it is shown that there is a coloring of $\binom{N(k)}{k}$. This is then refuted by exhaustively considering all graphs with $\leq N(k)$ nodes.   □

3.2 Quasi-Polynomial Size Frege Proofs

This section discusses some of the details of the formalization of the argument in Sect. 2.2 as quasi-polynomial size Frege proofs, establishing Theorem 5. First we will form an extended Frege proof, then modify it to become a Frege proof. As before, the proof starts with the assumption that $\mathrm{Kneser}_k^n(\vec{p})$ is false. As we describe next, the extended Frege proof then introduces variables $\vec{p}'$ by extension so that $\mathrm{Kneser}_k^{n-n/2k}$ is false. This process will be repeated $O(\log n)$ times. The final Frege proof is obtained by unwinding the definitions by extension.

For a set X of formulas and $t > 0$, let "$|X| < t$" denote a formula that is true when the number of true formulas in X is less than t. "$|X| < t$" can be expressed by a formula of size polynomially bounded by the total size of the formulas in X, using the construction in [3]. "$|X| = t$" is defined similarly.

The formulas $\mathrm{Star}(i, \ell)$ and $\mathrm{Star}(\ell)$ are the same as in Sect. 3.1. A color ℓ is now discarded if it is among the least $n/2k$ star-shaped color classes.

$$\mathrm{DiscardColor}(\ell) \ := \ \mathrm{Star}(\ell) \wedge (|\{\mathrm{Star}(\ell') : \ \ell' \le \ell\}| \le n/2k)$$

The discarded nodes are the least central elements of the discarded color classes.

$$\mathrm{DiscardNode}(i) \ := \ \bigvee_\ell \left[\mathrm{DiscardColor}(\ell) \wedge \mathrm{Star}(i, \ell) \wedge \bigwedge_{i' < i} \neg\mathrm{Star}(i', \ell) \right].$$

The remaining, non-discarded colors and nodes are renumbered to form an instance of $\mathrm{Kneser}_k^{n-n/2k}$. For this, the formula $\mathrm{RenumNode}(i', i)$ is true when the node i' is the ith node that is not discarded; similarly $\mathrm{RenumColor}(j', j)$ is true when the color j' is the jth color that is not discarded.

$$\mathrm{RenumNode}(i', i) := (|\{\neg\mathrm{DiscardNode}(i'') : i'' < i'\}| = i - 1) \wedge \neg\mathrm{DiscardNode}(i')$$

$$\mathrm{RenumColor}(j', j) := (|\{\neg\mathrm{DiscardColor}(j'') : j'' < j'\}| = j - 1) \wedge \neg\mathrm{DiscardColor}(j')$$

For each $S = \{i_1, \ldots, i_k\} \in \binom{n-n/2k}{k}$ and $j \in [(n - n/2k) - 2k + 1]$, we define by extension

$$p'_{S,j} \equiv \bigvee_{i'_1, \ldots i'_k, j'} \left(\bigwedge_{t=1}^k (\mathrm{RenumNode}(i'_t, i_t)) \wedge \mathrm{RenumColor}(j', j) \wedge p_{\{i'_1, \ldots, i'_k\}, j'} \right).$$

The Frege proof then argues that if the variables $p_{S,j}$ define a coloring, then the variables $p'_{S,j}$ define a coloring, i.e., that $\neg\mathrm{Kneser}_k^n(\vec{p}) \rightarrow \neg\mathrm{Kneser}_k^{n-n/2k}(\vec{p}')$. The main step for this is proving there are at least $n/2k$ star-shaped color classes by formalizing the proof of Lemma 9; this can be done with polynomial size Frege proofs using the counting techniques from [3]. After that, it is straightforward to prove that, for each $S \in \binom{n-n/2k}{k}$ and $j \in [(n - n/2k) - 2k + 1]$, the variable $p'_{S,j}$ is well-defined; and that the $\vec{p}'$ collectively falsify $\mathrm{Kneser}_k^{n-n/2k}$.

This is iterated $O(\log n)$ times until fewer than $N(k, 1/2)$ nodes remain. The proof concludes with a hard-coded proof that there are no such colorings of the finitely many small Kneser graphs.

To form the quasi-polynomial size Frege proof, we unwind the definitions by extension. Each definition by extension was polynomial size; they are nested to a depth of $O(\log n)$. So the resulting Frege proof is quasi-polynomial size.   □

4   The Truncated Tucker Lemma

This section introduces the truncated Tucker lemma. The usual (octahedral) Tucker lemma implies the truncated Tucker lemma and the truncated Tucker lemma implies the Kneser-Lovász theorem. The truncated Tucker lemma is of particular interest, since its propositional translations are only polynomial size; in contrast, the propositional translations of the usual Tucker lemma are of exponential size. Additionally, there are polynomial size constant depth Frege proofs of the Kneser-Lovász tautologies from the truncated Tucker tautologies.

Our definition and proof of the truncated Tucker lemma borrows techniques and notation from Matoušek [12].

Definition 10. *Let* $n \geq 1$. *The* octahedral ball $\mathcal{B}^n$ *is:*

$$\mathcal{B}^n := \{(A, B) : A, B \subseteq [n] \text{ and } A \cap B = \emptyset\}.$$

Let $1 \leq k \leq n$. *The* truncated octahedral ball $\mathcal{B}_k^n$ *is:*

$$\mathcal{B}_k^n := \left\{(A, B) : A, B \in \binom{n}{k} \cup \{\emptyset\}, \ A \cap B = \emptyset, \ \text{and} \ (A, B) \neq (\emptyset, \emptyset)\right\}.$$

Definition 11. *Let* $n > 1$. *A mapping* $\lambda : \mathcal{B}^n \to \{1, \pm 2, \ldots, \pm n\}$ *is* antipodal *if* $\lambda(\emptyset, \emptyset) = 1$, *and for all other pairs* $(A, B) \in \mathcal{B}^n$, $\lambda(A, B) = -\lambda(B, A)$.

Let $n \geq 2k > 1$. *A mapping* $\lambda : \mathcal{B}_k^n \to \{\pm 2k, \ldots, \pm n\}$ *is* antipodal *if for all* $(A, B) \in \mathcal{B}_k^n$, $\lambda(A, B) = -\lambda(B, A)$.

Note that -1 is not in the range of λ, and $(\emptyset, \emptyset)$ is the only member of $\mathcal{B}^n$ that is mapped to 1 by λ.

For $A \subseteq [n]$, let $A_{\leq k}$ denote the least k elements of A. By convention $\emptyset_{\leq k} = \emptyset$, but otherwise the notation is used only when $|A| \geq k$.

The Tucker lemma uses the subset relation $\subseteq$ on $[n]$, but the truncated Tucker lemma uses instead a stronger partial order $\preceq$ on $\binom{n}{k}$.

Definition 12. *Let* $\preceq$ *be the partial order on sets in* $\binom{n}{k} \cup \{\emptyset\}$ *defined by* $A_1 \preceq A_2$ *iff* $(A_1 \cup A_2)_{\leq k} = A_2$.

Lemma 13. *The relation* $\preceq$ *is a partial order with* $\emptyset$ *its least element.*

Proof. It is clearly reflexive. For anti-symmetry, $A_1 \preceq A_2$ and $A_2 \preceq A_1$ imply that $A_1 = (A_1 \cup A_2)_{\leq k} = (A_2 \cup A_1)_{\leq k} = A_2$. For transitivity: Suppose $A_1 \preceq A_2$ and $A_2 \preceq A_3$. Then $(A_1 \cup A_2)_{\leq k} = A_2$ and $(A_2 \cup A_3)_{\leq k} = A_3$. This implies that

$$A_3 = (A_2 \cup A_3)_{\leq k} = ((A_1 \cup A_2)_{\leq k} \cup A_3)_{\leq k} = (A_1 \cup (A_2 \cup A_3)_{\leq k})_{\leq k} = (A_1 \cup A_3)_{\leq k}.$$

Therefore $A_1 \preceq A_3$. That $\emptyset$ is the least element is clear from the definition.   □

Definition 14. *Two pairs* (A_1, B_1) *and* (A_2, B_2) *in* $\mathcal{B}^n$ *are complementary w.r.t. an antipodal map* λ *on* $\mathcal{B}^n$ *if* $A_1 \subseteq A_2$, $B_1 \subseteq B_2$ *and* $\lambda(A_1, B_1) = -\lambda(A_2, B_2)$.

For (A_1, B_1) *and* (A_2, B_2) *in* $\mathcal{B}_k^n$, *write* $(A_1, B_1) \preceq (A_2, B_2)$ *when* $A_1 \preceq A_2$, $B_1 \preceq B_2$, *and* $A_i \cap B_j = \emptyset$ *for* $i, j \in \{1, 2\}$. *The pairs* (A_1, B_1) *and* (A_2, B_2) *are* k-complementary *w.r.t. an antipodal map* λ *on* $\mathcal{B}_k^n$ *if* $(A_1, B_1) \preceq (A_2, B_2)$ *and* $\lambda(A_1, B_1) = -\lambda(A_2, B_2)$.

Theorem 15 (Tucker lemma). *If* $\lambda : \mathcal{B}^n \rightarrow \{1, \pm 2, \dots, \pm n\}$ *is antipodal, then there are two elements in* $\mathcal{B}^n$ *that are complementary.*

Theorem 16 (Truncated Tucker). *Let* $n \geq 2k > 1$. *If* $\lambda : \mathcal{B}_k^n \rightarrow \{\pm 2k \dots, \pm n\}$ *is antipodal, then there are two elements in* $\mathcal{B}_k^n$ *that are* k-*complementary.*

For a proof of Theorem 15, see [12]. An appendix to the arXiv version of this paper proves Theorem 16 from Theorem 15.

The truncated Tucker lemma has polynomial size propositional translations. For each $(A, B) \in \mathcal{B}_k^n$, and for each $i \in \{\pm 2k, \dots, \pm n\}$, let $p_{A,B,i}$ be a propositional variable with the intended meaning that $p_{A,B,i}$ is true when $\lambda(A, B) = i$. The following formula $\text{Ant}(\vec{p})$ states that the map is total and antipodal:

$$\bigwedge_{(A,B)\in\mathcal{B}_k^n} \bigvee_{i\in\{\pm 2k,\dots,\pm n\}} (p_{A,B,i} \wedge p_{B,A,-i}).$$

The following formula $\text{Comp}(\vec{p})$ states that there exists two elements in $\mathcal{B}_k^n$ that are k-complementary:

$$\bigvee_{\substack{(A_1,B_1),(A_2,B_2)\in\mathcal{B}_k^n, \\ (A_1,B_1)\preceq(A_2,B_2) \\ i\in\{\pm 2k,\dots,\pm n\}}} (p_{A_1,B_1,i} \wedge p_{A_2,B_2,-i}).$$

The truncated Tucker tautologies are defined to be $\text{Ant}(\vec{p}) \rightarrow \text{Comp}(\vec{p})$. (We could add an additional hypothesis, that for each A, B there is at most one i such that $p_{A,B,i}$, but this is not needed for the Tucker tautologies to be valid.) There are $< n^{2k}$ members (A, B) in $\mathcal{B}_k^n$. Hence, for fixed k, there are only polynomially many variables $p_{A,B,i}$, and the truncated Tucker tautologies have size polynomially bounded by n. On the other hand, the propositional translation of the usual Tucker lemma requires an exponential number of propositional variables in n, since the cardinality of $\mathcal{B}^n$ is exponential in n.

Proof (Theorem 1 from the truncated Tucker lemma). Let $c : \binom{n}{k} \rightarrow \{2k, \dots, n\}$ be a $(n - 2k + 1)$-coloring of $\binom{n}{k}$. We show that this implies the existence of an antipodal map λ on $\mathcal{B}_k^n$ that has no k-complementary pairs. Let $\leq$ be a total order on $\binom{n}{k} \cup \{\emptyset\}$ that refines the partial order $\preceq$. Define $\lambda(A, B)$ to be $c(A)$ if $A > B$ and $-c(B)$ if $B > A$. We argue that there are no k-complementary pairs in $\mathcal{B}_k^n$ with respect to λ. Suppose there are, say (A_1, B_1) and (A_2, B_2). Since λ must assign these opposite signs, either $A_1 < B_1 \leq B_2 < A_2$ or $B_1 < A_1 \leq A_2 < B_2$. In the former case it must be that, $c(B_1) = c(A_2)$ and in the latter case that $c(A_1) = c(B_2)$. Since $B_1 \cap A_2$ and $A_1 \cap B_2$ are empty in either case we have a contradiction, since c was assumed to be a coloring. $\qquad\square$

The above proof of the Kneser-Lovász theorem from the truncated Tucker lemma can be readily translated into polynomial size constant depth Frege proofs.

Question 17. Do the propositional translations of the Truncated Tucker lemma have short (extended) Frege proofs?

References

1. Aisenberg, J., Bonet, M.L., Buss, S.R.: Quasi-polynomial size Frege proofs of Frankl's theorem on the trace of finite sets (201?) (to appear in Journal of Symbolic Logic)
2. Bonet, M.L., Buss, S.R., Pitassi, T.: Are there hard examples for Frege systems? In: Clote, P., Remmel, J. (eds.) Feasible Mathematics II, pp. 30–56. Birkhäuser, Boston (1995)
3. Buss, S.R.: Polynomial size proofs of the propositional pigeonhole principle. Journal of Symbolic Logic **52**, 916–927 (1987)
4. Buss, S.R.: Propositional proof complexity: An introduction. In: Berger, U., Schwichtenberg, H. (eds.) Computational Logic, pp. 127–178. Springer, Berlin (1999)
5. Buss, S.R.: Towards NP-P via proof complexity and proof search. Annals of Pure and Applied Logic **163**(9), 1163–1182 (2012)
6. Buss, S.R.: Quasipolynomial size proofs of the propositional pigeonhole principle (2014) (submitted for publication)
7. Cook, S.A., Reckhow, R.A.: The relative efficiency of propositional proof systems. Journal of Symbolic Logic **44**, 36–50 (1979)
8. Istrate, G., Crăciun, A.: Proof complexity and the Kneser-Lovász theorem. In: Sinz, C., Egly, U. (eds.) SAT 2014. LNCS, vol. 8561, pp. 138–153. Springer, Heidelberg (2014)
9. Jeřábek, E.: Dual weak pigeonhole principle, boolean complexity, and derandomization. Annals of Pure and Applied Logic **124**, 1–37 (2004)
10. Krajíček, J.: Bounded Arithmetic. Propositional Calculus and Complexity Theory. Cambridge University Press, Heidelberg (1995)
11. Lovász, L.: Kneser's conjecture, chromatic number, and homotopy. Journal of Combinatorial Theory, Series A **25**(3), 319–324 (1978)
12. Matoušek, J.: A combinatorial proof of Kneser's conjecture. Combinatorica **24**(1), 163–170 (2004)
13. Segerlind, N.: The complexity of propositional proofs. Bulletin of Symbolic Logic **13**(4), 417–481 (2007)
14. Ziegler, G.M.: Generalized Kneser coloring theorems with combinatorial proofs. Inventiones Mathematicae **147**(3), 671–691 (2002)

Provenance Circuits
for Trees and Treelike Instances

Antoine Amarilli[1][✉], Pierre Bourhis[2], and Pierre Senellart[1,3]

[1] Institut Mines-Télécom, Télécom ParisTech, CNRS LTCI, Paris, France
antoine.amarilli@telecom-paristech.fr
[2] CNRS CRIStAL, Université Lille 1, INRIA Lille, Lille, France
pierre.bourhis@univ-lille1.fr
[3] National University of Singapore, CNRS IPAL, Singapore, Singapore
pierre.senellart@telecom-paristech.fr

Abstract. Query evaluation in monadic second-order logic (MSO) is tractable on trees and treelike instances, even though it is hard for arbitrary instances. This tractability result has been extended to several tasks related to query evaluation, such as counting query results [2] or performing query evaluation on probabilistic trees [8]. These are two examples of the more general problem of computing augmented query output, that is referred to as *provenance*. This article presents a provenance framework for trees and treelike instances, by describing a linear-time construction of a circuit provenance representation for MSO queries. We show how this provenance can be connected to the usual definitions of semiring provenance on relational instances [17], even though we compute it in an unusual way, using tree automata; we do so via intrinsic definitions of provenance for general semirings, independent of the operational details of query evaluation. We show applications of this provenance to capture existing counting and probabilistic results on trees and treelike instances, and give novel consequences for probability evaluation.

1 Introduction

A celebrated result by Courcelle [9] has shown that evaluating a fixed monadic second-order (MSO) query on relational instances, while generally hard in the input instance, can be performed in linear time on input instances of *bounded treewidth* (or *treelike* instances), by encoding the query to an automaton on tree encodings of instances. This idea has been extended more recently to monadic Datalog [14]. In addition to query evaluation, it is also possible to *count* in linear time the number of query answers over treelike instances [2,22].

However, query evaluation and counting are special cases of the more general problem of capturing *provenance information* [7,17] of query results, which describes the link between input and output tuples. Provenance information can be expressed through various formalisms, such as *provenance semirings* [17] or Boolean formulae [26]. Besides counting, provenance can be exploited for practically important tasks such as answering queries in incomplete databases, maintaining access rights, or computing query probability [26]. To our knowledge,

M.M. Halldórsson et al. (Eds.): ICALP 2015, Part II, LNCS 9135, pp. 56–68, 2015.
DOI: 10.1007/978-3-662-47666-6_5

no previous work has looked at the general question of efficient evaluation of expressive queries on treelike instances while keeping track of provenance.

Indeed, no proper definition of provenance for queries evaluated via tree automata has been put forward. The first contribution of this work (Section 3) is thus to introduce a general notion of *provenance circuit* [11] for tree automata, which provides an efficiently computable representation of all possible results of an automaton over a tree with uncertain annotations. Of course, we are interested in the provenance of *queries* rather than automata; however, in this setting, the provenance that we compute has an intrinsic definition, so it does not depend on which automaton we use to compute the query.

We then extend these results in Section 4 to the provenance of queries on treelike relational instances. We propose again an intrinsic definition of provenance capturing the subinstances that satisfy the query. We then show that, in the same way that queries can be evaluated by compiling them to an automaton on tree encodings, we can compute a provenance circuit for the query by compiling it to an automaton, computing a tree decomposition of the instance, and performing the previous construction, in linear time overall in the input instance. Our intrinsic definition of provenance ensures the provenance only depends on the logical query, not on the choice of query plan, of automaton, or of tree decomposition.

Our next contribution in Section 5 is to extend such definitions of provenance from Boolean formulae to $\mathbb{N}[X]$, the *universal provenance semiring* [17]. This poses several challenges. First, as semirings cannot deal satisfactorily with negation [1], we must restrict to *monotone* queries, to obtain monotone provenance circuits. Second, we must keep track of the multiplicity of facts, as well as the multiplicity of matches. For this reason, we restrict to unions of conjunctive queries (UCQ) in that section, as richer languages do not directly provide notions of multiplicity for matched facts. We generalize our notion of provenance circuits for automata to instances with unknown multiplicity annotations, using arithmetic circuits. We show that, for UCQs, the standard provenance for the universal semiring [17] matches the one defined via the automaton, and that a provenance circuit for it can be computed in linear time for treelike instances.

Returning to the non-monotone Boolean provenance, we show in Section 6 how the tractability of provenance computation on treelike instances implies that of two important problems: determining the probability of a query, and counting query matches. We show that probability evaluation of fixed MSO queries is tractable on probabilistic XML models with local uncertainty, a result already known in [8], and extend it to trees with event annotations that satisfy a condition of having bounded *scopes*. We also show that MSO query evaluation is tractable on treelike block-independent-disjoint (BID) relational instances [26]. These tractability results for provenance are achieved by applying message passing [20] on our provenance circuits. Last, we show the tractability of counting query matches, using a reduction to the probabilistic setting, capturing a result of [2].

2   Preliminaries

We introduce basic notions related to trees, tree automata, and Boolean circuits.

Given a fixed *alphabet* Γ, we define a Γ-*tree* $T = (V, L, R, \lambda)$ as a set of *nodes* V, two partial mappings $L, R : V \to V$ that associate an internal node with its left and right child, and a *labeling function* $\lambda : V \to \Gamma$. Unless stated otherwise, the trees that we consider are rooted, directed, ordered, binary, and full (each node has either zero or two children). We write $n \in T$ to mean $n \in V$. We say that two trees T_1 and T_2 are *isomorphic* if there is a bijection between their node sets preserving children and labels (we simply write it $T_1 = T_2$); they have *same skeleton* if they are isomorphic except for labels.

A *bottom-up nondeterministic tree automaton* on Γ-trees, or Γ-*bNTA*, is a tuple $A = (Q, F, \iota, \delta)$ of a set Q of *states*, a subset $F \subseteq Q$ of *accepting states*, an *initial relation* $\iota : \Gamma \to 2^Q$ giving possible states for leaves from their label, and a *transition relation* $\delta : Q^2 \times \Gamma \to 2^Q$ determining possible states for internal nodes from their label and the states of their children. A *run* of A on a Γ-tree $T = (V, L, R, \lambda)$ is a function $\rho : V \to Q$ such that for each leaf n we have $\rho(n) \in \iota(\lambda(n))$, and for every internal node n we have $\rho(n) \in \delta(\rho(L(n)), \rho(R(n)), \lambda(n))$. A run is *accepting* if, for the root n_{r} of T, $\rho(n_{\mathrm{r}}) \in F$; and A *accepts* T (written $T \models A$) if there is some accepting run of A on T. Tree automata capture usual query languages on trees, such as MSO [27].

A *Boolean circuit* is a directed acyclic graph $C = (G, W, g_0, \mu)$ where G is a set of *gates*, $W \subseteq G \times G$ is a set of *wires* (edges), $g_0 \in G$ is a distinguished *output gate*, and μ associates each *gate* $g \in G$ with a *type* $\mu(g)$ that can be inp (*input gate*, with no incoming wire in W), $\neg$ (NOT-gate, with exactly one incoming wire in W), $\wedge$ (AND-gate) or $\vee$ (OR-gate). A *valuation* of the *input gates* C_{inp} of C is a function $\nu : C_{\mathrm{inp}} \to \{0, 1\}$; it defines inductively a unique *evaluation* $\nu' : C \to \{0, 1\}$ as follows: $\nu'(g)$ is $\nu(g)$ if $g \in C_{\mathrm{inp}}$ (i.e., $\mu(g) = \mathrm{inp}$); it is $\neg\nu'(g')$ if $\mu(g) = \neg$ (with $(g', g) \in W$); otherwise it is $\bigodot_{(g', g) \in W} \nu'(g')$ where $\odot$ is $\mu(g)$ (hence, $\wedge$ or $\vee$). Note that this implies that AND- and OR-gates with no inputs always evaluate to 1 and 0 respectively. We will abuse notation and use valuations and evaluations interchangeably, and we write $\nu(C)$ to mean $\nu(g_0)$. The *function* captured by C is the one that maps any valuation ν of C_{inp} to $\nu(C)$.

3   Provenance Circuits for Tree Automata

We start by studying a notion of provenance on trees, defined in an uncertain tree framework. Fixing a finite alphabet Γ throughout this section, we view a Γ-tree T as an *uncertain tree*, where each node carries an unknown Boolean annotation in $\{0, 1\}$, and consider all possible *valuations* that choose an annotation for each node of T, calling $\overline{\Gamma}$ the alphabet of annotated trees:

Definition 3.1. *We write* $\overline{\Gamma} := \Gamma \times \{0, 1\}$. *For any* Γ-*tree* $T = (V, L, R, \lambda)$ *and valuation* $\nu : V \to \{0, 1\}$, $\nu(T)$ *is the* $\overline{\Gamma}$-*tree with same skeleton where each node* n *is given the label* $(\lambda(n), \nu(n))$.

We consider automata on *annotated* trees, namely, $\overline{\Gamma}$-bNTAs, and define their *provenance* on a Γ-tree T as a Boolean function that describes which valuations of T are accepted by the automaton. Intuitively, provenance keeps track of the dependence between Boolean annotations and acceptance or rejection of the tree.

Definition 3.2. *The* provenance *of a $\overline{\Gamma}$-bNTA A on a Γ-tree $T = (V, L, R, \lambda)$ is the function* $\mathrm{Prov}(A, T)$ *mapping any valuation* $\nu : V \to \{0, 1\}$ *to 1 or 0 depending on whether* $\nu(T) \models A$.

We now define a *provenance circuit* of A on a Γ-tree T as a circuit that captures the provenance of A on T, $\mathrm{Prov}(A, T)$. Formally:

Definition 3.3. *Let A be a $\overline{\Gamma}$-bNTA and $T = (V, L, R, \lambda)$ be a Γ-tree. A* provenance circuit *of A on T is a Boolean circuit C with $C_{\mathsf{inp}} = V$ that captures the function* $\mathrm{Prov}(A, T)$.

An important result is that provenance circuits can be tractably constructed:

Proposition 3.1. *A provenance circuit of a $\overline{\Gamma}$-bNTA A on a Γ-tree T can be constructed in time* $O(|A| \cdot |T|)$.

The proof is by creating one gate in C per state of A per node of T, and writing out in C all possible transitions of A at each node n of T, depending on the input gate that indicates the annotation of n. In fact, we can show that C is treelike for fixed A; we use this in Section 6 to show the tractability of tree automaton evaluation on probabilistic XML trees from $\mathsf{PrXML}^{\mathsf{mux,ind}}$ [19].

It is not hard to see that this construction gives us a way to capture the provenance of any *query* on trees that can be expressed as an automaton, no matter the choice of automaton. A *query* q is any logical sentence on $\overline{\Gamma}$-trees which a $\overline{\Gamma}$-tree T can *satisfy* (written $T \models q$) or *violate* ($T \not\models q$). An automaton A_q tests query q if for any $\overline{\Gamma}$-tree T, we have $T \models A_q$ iff $T \models q$. We define $\mathrm{Prov}(q, T)$ for a Γ-tree T as in Definition 3.2, and run circuits for queries as in Definition 3.3. It is immediate that Proposition 3.1 implies:

Proposition 3.2. *For any fixed query q on $\overline{\Gamma}$-trees for which we can compute an automaton A_q that tests it, a provenance circuit of q on a Γ-tree T can be constructed in time* $O(|T|)$.

Note that provenance does not depend on the automaton used to test the query.

4 Provenance on Tree Encodings

We lift the previous results to the setting of *relational instances*.

A *signature* σ is a finite set of *relation names* (e.g., R) with associated *arity* $\mathrm{arity}(R) \geqslant 1$. Fixing a countable domain $\mathcal{D} = \{a_k \mid k \geqslant 0\}$, a *relational instance* I over σ (or σ-instance) is a finite set I of *ground facts* of the form $R(\mathbf{a})$ with $R \in \sigma$, where $\mathbf{a}$ is a tuple of $\mathrm{arity}(R)$ elements of $\mathcal{D}$. The *active domain* $\mathrm{dom}(I) \subseteq \mathcal{D}$ of I is the finite set of elements of $\mathcal{D}$ used in I. Two instances I

and I' are *isomorphic* if there is a bijection φ from $\mathrm{dom}(I)$ to $\mathrm{dom}(I')$ such that $\varphi(I) = I'$. We say that an instance I' is a *subinstance* of I, written $I' \subseteq I$, if it is a subset of the facts of I, which implies $\mathrm{dom}(I') \subseteq \mathrm{dom}(I)$.

A *query* q is a logical formula in (function-free) first- or second-order logic on σ, without free second-order variables; a σ-instance I can *satisfy* it ($I \models q$) or *violate* it ($I \not\models q$). For simplicity, unless stated otherwise, we restrict to *Boolean queries*, that is, queries with no free variables, that are *constant-free*. This limitation is inessential for *data complexity*, namely complexity for a fixed query: we can handle non-Boolean queries by building a provenance circuit for each possible output result (there are polynomially many), and we encode constants by extending the signature with fresh unary predicates for them.

As before, we consider unknown Boolean annotations on the facts of an instance. However, rather than annotating the facts, it is more natural to say that a fact annotated by 1 is kept, and a fact annotated by 0 is deleted. Formally, given an instance σ, a *valuation* ν is a function from the facts of I to $\{0,1\}$, and we define $\nu(I)$ as the subinstance $\{F \in I \mid \nu(F) = 1\}$ of I. We then define:

Definition 4.1. *The* provenance *of a query q on a σ-instance I is the function* $\mathrm{Prov}(q, I)$ *mapping any valuation $\nu : I \to \{0,1\}$ to 1 or 0 depending on whether $\nu(I) \models q$. A* provenance circuit *of q on I is a Boolean circuit C with $C_{\mathrm{inp}} = I$ that captures $\mathrm{Prov}(q, I)$.*

We study provenance for treelike instances (i.e., bounded-treewidth instances), encoding queries to automata on tree encodings. Let us first define this. The *treewidth* $w(I)$ of an instance I is a standard measure [23] of how close I is to a tree: the treewidth of a tree is 1, that of a cycle is 2, and that of a k-clique or k-grid is $k-1$; further, we have $w(I') \leqslant w(I)$ for any $I' \subseteq I$. It is known [9,12] that for any fixed $k \in \mathbb{N}$, there is a finite alphabet Γ_σ^k such that any σ-instance I of treewidth $\leqslant k$ can be encoded in linear time [4] to a Γ_σ^k-tree T_I, called the *tree encoding*, which can be decoded back to I up to isomorphism (i.e., up to the identity of constants). Each fact in I is encoded in a node for this fact in the tree encoding, where the node label describes the fact.

The point of tree encodings is that queries in *monadic second-order logic*, the extension of first-order logic with second-order quantification on sets, can be encoded to automata which are then evaluated on tree encodings. Formally:

Definition 4.2. *For $k \in \mathbb{N}$, we say that a Γ_σ^k-bNTA A_q^k* tests *a query q for treewidth k if, for any Γ_σ^k-tree T, we have $T \models A_q^k$ iff T decodes to an instance I such that $I \models q$.*

Theorem 4.1 [9]. *For any $k \in \mathbb{N}$, for any MSO query q, one can compute a Γ_σ^k-bNTA A_q^k that tests q for treewidth $\leqslant k$.*

Our results apply to any query language that can be rewritten to tree automata under a bound on instance treewidth. Beyond MSO, this is also the case of *guarded second-order logic* (GSO). GSO extends first-order logic with second-order quantification on arbitrary-arity relations, with a semantic restriction to *guarded tuples* (already co-occurring in some instance fact); it captures MSO (it has the same expressive power on treelike instances [16]) and many common database query languages, e.g., *frontier-guarded Datalog* [3]. We use GSO in the sequel as our choice of query language that can be rewritten to automata. Combining the result above with the results of the previous section, we claim that provenance for GSO queries on treelike instances can be tractably computed, and that the resulting provenance circuit has treewidth independent on the instance.

Theorem 4.2. *For any fixed $k \in \mathbb{N}$ and GSO query q, for any σ-instance I such that $w(I) \leqslant k$, one can construct a provenance circuit C of q on I in time $O(|I|)$. The treewidth of C only depends on k and q (not on I).*

The proof is by encoding the instance I to its tree encoding T_I in linear time, and compiling the query q to an automaton A_q that tests it, in constant time in the instance. Now, Section 3 worked with $\overline{\Gamma_\sigma^k}$-bNTAs rather than Γ_σ^k-bNTAs, but the difference is inessential: we can easily map any $\overline{\Gamma_\sigma^k}$-tree T to a Γ_σ^k-tree $\epsilon(T)$ where any node label $(\tau, 1)$ is replaced by τ, and any label $(\tau, 0)$ is replaced by a dummy label indicating the absence of a fact; and we straightforwardly translate A to a $\overline{\Gamma_\sigma^k}$-bNTA A' such that $T \models A'$ iff $\epsilon(T) \models A$ for any $\overline{\Gamma_\sigma^k}$-tree T. The key point is then that, for any valuation $\nu : T \to \{0, 1\}$, $\epsilon(\nu(T))$ is a tree encoding of $\nu(I)$ (defined in the expected way), so we conclude by applying Proposition 3.1 to A' and T. As in Section 3, our definition of provenance is intrinsic to the query and does not depend on its formulation, on the choice of tree decomposition, or on the choice of automaton to evaluate the query on tree encodings.

Note that tractability holds only in data complexity. For combined complexity, we incur the cost of compiling the query to an automaton, which is nonelementary in general [21]. However, for some restricted query classes, such as *unions of conjunctive queries* (UCQs), the compilation phase has lower cost.

5   General Semirings

In this section we connect our previous results to the existing definitions of *semiring provenance* on arbitrary relational instances [17]:

Definition 5.1. *A commutative semiring $(K, \oplus, \otimes, 0_K, 1_K)$ is a set K with binary operations $\oplus$ and $\otimes$ and distinguished elements 0_K and 1_K, such that $(K, \oplus)$ and $(K, \otimes)$ are commutative monoids with identity element 0_K and 1_K, $\otimes$ distributes over $\oplus$, and $0_K \otimes a = 0_K$ for all $a \in K$.*

Provenance for semiring K is defined on instances where each fact is annotated with an element of K. The provenance of a query on such an instance is an element of K obtained by combining fact annotations following the semantics of the query, intuitively describing how the query output depends on the annotations (see exact definitions in [17]). This general setting has many specific applications:

Example 5.1. For any variable set X, the monotone Boolean functions over X form a semiring $(\text{PosBool}[X], \vee, \wedge, 0, 1)$. On instances where each fact is annotated by its own variable in X, the $\text{PosBool}[X]$-provenance of a query q is a monotone Boolean function on X describing which subinstances satisfy q. As we will see, this is what we defined in Section 4, using circuits as compact representations.

The *natural numbers* $\mathbb{N}$ with the usual $+$ and $\times$ form a semiring. On instances where facts are annotated with an element of $\mathbb{N}$ representing a multiplicity, the provenance of a query describes its number of matches under the bag semantics.

The *tropical semiring* [11] is $(\mathbb{N} \sqcup \{\infty\}, \min, +, \infty, 0)$. Fact annotations are costs, and the tropical provenance of a query is the minimal cost of the facts required to satisfy it, with multiple uses of a fact being charged multiple times.

For any set of variables X, the *polynomial semiring* $\mathbb{N}[X]$ is the semiring of polynomials with variables in X and coefficients in $\mathbb{N}$, with the usual sum and product over polynomials, and with $0, 1 \in \mathbb{N}$.

Semiring provenance does not support negation well [1] and is therefore only defined for *monotone* queries: a query q is *monotone* if, for any instances $I \subseteq I'$, if $I \models q$ then $I' \models q$. Provenance circuits for semiring provenance are *monotone* circuits [11]: they do not feature NOT-gates. We can show that, adapting the constructions of Section 3 to work with a notion of *monotone* bNTAs, Theorem 4.2 applied to monotone queries yields a *monotone* provenance circuit:

Theorem 5.1. *For any fixed $k \in \mathbb{N}$ and monotone GSO query q, for any σ-instance I such that $w(I) \leqslant k$, one can construct in time $O(|I|)$ a monotone provenance circuit of q on I whose treewidth only depends on k and q (not on I).*

Hence, for monotone GSO queries for which [17] defines a notion of semiring provenance (e.g., those that can be encoded to *Datalog*, a recursive query language that subsumes UCQs), our provenance $\text{Prov}(q, I)$ is easily seen to match the provenance of [17], specialized to the semiring $\text{PosBool}[X]$ of monotone Boolean functions. Indeed, both provenances obey the same intrinsic definition: they are the function that maps to 1 exactly the valuations corresponding to subinstances accepted by the query. Hence, we can understand Theorem 5.1 as a tractability result for $\text{PosBool}[X]$-provenance (represented as a circuit) on treelike instances.

Of course, the definitions of [17] go beyond $\text{PosBool}[X]$ and extend to arbitrary commutative semirings. We now turn to this more general question.

$\mathbb{N}[X]$*-provenance for UCQs.* First, we note that, as shown by [17], the provenance of Datalog queries for *any* semiring K can be computed in the semiring $\mathbb{N}[X]$, on instances where each fact is annotated by its own variable in X. Indeed, the provenance can then be *specialized* to K, and the actual fact annotations in K, once known, can be used to replace the variables in the result, thanks to a *commutation with homomorphisms* property. Hence, *we restrict to $\mathbb{N}[X]$-provenance* and to instances of this form, which covers all the examples above.

Second, in our setting of treelike instances, we evaluate queries using tree automata, which are compiled from logical formulae with no prescribed execution plan. For the semiring $\mathbb{N}[X]$, this is hard to connect to the general definitions of provenance in [17], which are mainly designed for positive relational algebra operators or Datalog queries. Hence, to generalize our constructions to $\mathbb{N}[X]$-provenance, *we now restrict our query language to UCQs*, assuming without loss of generality that they contain no equality atoms, We comment at the end of this section on the difficulties arising for richer query languages.

We formally define the $\mathbb{N}[X]$-*provenance* of UCQs on relational instances by encoding them straightforwardly to Datalog and using the Datalog provenance definition of [17]. The resulting provenance can be rephrased as follows:

Definition 5.2. *The $\mathbb{N}[X]$-provenance of a UCQ $q = \bigvee_{i=1}^{n} \exists \mathbf{x}_i\, q_i(\mathbf{x}_i)$ (where q_i is a conjunction of atoms with free variables $\mathbf{x}_i$) on an instance I is defined as:*
$$\mathrm{Prov}_{\mathbb{N}[X]}(q, I) := \bigoplus_{i=1}^{n} \bigoplus_{f:\mathbf{x}_i \rightarrow \mathrm{dom}(I) \text{ such that } I \models q_i(f(\mathbf{x}_i))} \bigotimes_{A(\mathbf{x}_i) \in q_i} A(f(\mathbf{x}_i)).$$
In other words, we sum over each disjunct, and over each match of the disjunct; for each match, we take the product, over the atoms of the disjunct, of their image fact in I, identifying each fact to the one variable in X that annotates it.

We know that $\mathrm{Prov}_{\mathbb{N}[X]}(q, I)$ enjoys all the usual properties of provenance: it can be specialized to $\mathrm{PosBool}[X]$, yielding back the previous definition; it can be evaluated in the $\mathbb{N}$ semiring to count the number of matches of a query; etc.

Example 5.2. Consider the instance $I = \{F_1 := R(a,a), F_2 := R(b,c), F_3 := R(c,b)\}$ and the CQ $q : \exists xy\, R(x,y)R(y,x)$. We have $\mathrm{Prov}_{\mathbb{N}[X]}(q, I) = F_1^2 + 2F_2 F_3$ and $\mathrm{Prov}(q, I) = F_1 \vee (F_2 \wedge F_3)$. Unlike $\mathrm{PosBool}[X]$-provenance, $\mathbb{N}[X]$-provenance can describe that multiple atoms of the query map to the same fact, and that the same subinstance is obtained with two different query matches. Evaluating in the semiring $\mathbb{N}$ with facts annotated by 1, q has $1^2 + 2 \times 1 \times 1 = 3$ matches.

Provenance circuits for trees. Guided by this definition of $\mathbb{N}[X]$-provenance, we generalize the construction of Section 3 of provenance on trees to a more expressive provenance construction, before we extend it to treelike instances as in Section 4.

Instead of considering $\overline{\Gamma}$-trees, we consider $\overline{\Gamma}^p$-trees for $p \in \mathbb{N}$, whose label set is $\Gamma \times \{0, \ldots, p\}$ rather than $\Gamma \times \{0, 1\}$. Intuitively, rather than uncertainty about whether facts are present or missing, we represent uncertainty about the *number of available copies of facts*, as UCQ matches may include the same fact multiple times. We impose on $\overline{\Gamma}$ the partial order $<$ defined by $(\tau, i) < (\tau, j)$ for all $\tau \in \Gamma$ and $i < j$ in $\{0, \ldots, p\}$, and call a $\overline{\Gamma}^p$-bNTA $A = (Q, F, \iota, \delta)$ *monotone* if for every $\tau < \tau'$ in $\overline{\Gamma}^p$, we have $\iota(\tau) \subseteq \iota(\tau')$ and $\delta(q_1, q_2, \tau) \subseteq \delta(q_1, q_2, \tau')$ for every $q_1, q_2 \in Q$. We write $\mathrm{Val}^p(T)$ for the set of all p-*valuations* $\nu : V \rightarrow \{0, \ldots, p\}$ of a Γ-tree T. We write $|\mathrm{aruns}(A, T)|$ for a $\overline{\Gamma}^p$-tree T and $\overline{\Gamma}^p$-bNTA A to denote the number of accepting runs of A on T. We can now define:

Definition 5.3. *The $\mathbb{N}[X]$-provenance of a $\overline{\Gamma}^p$-bNTA A on a Γ-tree T is*
$$\mathrm{Prov}_{\mathbb{N}[X]}(A, T) := \bigoplus_{\nu \in \mathrm{Val}^p(T)} |\mathrm{aruns}(A, \nu(T))| \bigotimes_{n \in T} n^{\nu(n)}$$
where each node $n \in T$ is identified with its own variable in X. Intuitively, we

sum over all valuations ν of T to $\{0, \ldots, p\}$, and take the product of the tree nodes to the power of their valuation in ν, with the number of accepting runs of A on $\nu(T)$ as coefficient; in particular, the term for ν is 0 if A rejects $\nu(T)$.

This definition specializes in PosBool[X] to our earlier definition of Prov(A, T), but extends it with the two features of $\mathbb{N}[X]$: multiple copies of the same nodes (represented as $n^{\nu(n)}$) and multiple derivations (represented as $|\text{aruns}(A, \nu(T))|$). To construct this general provenance, we need *arithmetic circuits*:

Definition 5.4. *A K-circuit for semiring $(K, \oplus, \otimes, 0_K, 1_K)$ is a circuit with $\oplus$- and $\otimes$-gates instead of OR- and AND-gates (and no analogue of NOT-gates), whose input gates stand for elements of K. As before, the constants 0_K and 1_K can be written as $\oplus$- and $\otimes$-gates with no inputs. The element of K captured by a K-circuit is the element captured by its distinguished gate, under the recursive definition that $\oplus$- and $\otimes$-gates capture the sum and product of the elements captured by their operands, and input gates capture their own value.*

We now show an efficient construction for such provenance circuits, generalizing the monotone analogue of Proposition 3.1. The proof technique is to replace AND- and OR-gates by $\otimes$- and $\oplus$-gates, and to consider possible annotations in $\{0, \ldots, p\}$ instead of $\{0, 1\}$. The correctness is proved by induction via a general identity relating the provenance on a tree to that of its left and right subtrees.

Theorem 5.2. *For any fixed $p \in \mathbb{N}$, for a $\overline{\Gamma}^p$-bNTA A and a Γ-tree T, a $\mathbb{N}[X]$-circuit capturing $\text{Prov}_{\mathbb{N}[X]}(A, T)$ can be constructed in time $O(|A| \cdot |T|)$.*

Provenance circuit for instances. Moving back to provenance for UCQs on bounded-treewidth instances, we obtain a linear-time provenance construction:

Theorem 5.3. *For any fixed $k \in \mathbb{N}$ and UCQ q, for any σ-instance I such that $w(I) \leqslant k$, one can construct a $\mathbb{N}[X]$-circuit that captures $\text{Prov}_{\mathbb{N}[X]}(q, I)$ in time $O(|I|)$.*

The proof technique is to construct for each disjunct q' of q a $\overline{\Gamma}^p$-bNTA $A_{q'}$, where $\Gamma := \Gamma_\sigma^k$ is the alphabet for tree encodings of width k, and p is the maximum number of atoms in a disjunct of q. We want $A_{q'}$ to test q' on tree encodings over Γ, *while preserving multiplicities*: this is done by enumerating all possible self-homomorphisms of q', changing σ to make the multiplicity of atoms part of the relation name, encoding the resulting queries to automata as usual [9] and going back to the original σ. We then apply a variant of Theorem 5.2 to construct a $\mathbb{N}[X]$-circuit capturing the provenance of $A_{q'}$ on a tree encoding of I but for valuations that sum to the number of atoms of q'; this restricts to bag-subinstances corresponding exactly to matches of q'. We obtain a $\mathbb{N}[X]$-circuit that captures $\text{Prov}_{\mathbb{N}[X]}(q, I)$ by combining the circuits for each disjunct, the distinguished gate of the overall circuit being a $\oplus$-gate of that of each circuit.

Remember that an $\mathbb{N}[X]$-circuit can then be specialized to a circuit for an arbitrary semiring (in particular, if the semiring has no variable, the circuit can be used for evaluation); thus, this provides provenance for q on I for any semiring.

Going beyond UCQs. To compute $\mathbb{N}[X]$-provenance beyond UCQs (e.g., for monotone GSO queries or their intersection with Datalog), the main issue is fact multiplicity: multiple uses of facts are easy to describe for UCQs (Definition 5.2), but for more expressive languages we do not know how to define them and connect them to automata.

In fact, we can build a query P, in guarded Datalog [15], such that the smallest number of occurrences of a fact in a derivation tree for P cannot be bounded independently from the instance. We thus cannot rewrite P to a fixed finite bNTA testing multiplicities on all input instances. However, as guarded Datalog is monotone and GSO-expressible, we can compute the PosBool$[X]$-provenance of P with Theorem 4.2, hinting at a difference between PosBool$[X]$ and $\mathbb{N}[X]$-provenance computation for queries beyond UCQs.

6   Applications

In Section 5 we have shown a $\mathbb{N}[X]$-provenance circuit construction for UCQs on treelike instances. This construction can be specialized to any provenance semiring, yielding various applications: counting query results by evaluating in $\mathbb{N}$, computing the cost of a query in the tropical semiring, etc. By contrast, Section 4 presented a provenance construction for arbitrary GSO queries, but only for a Boolean representation of provenance, which does not capture multiplicities of facts or derivations. The results of both sections are thus incomparable. In this section we show applications of our constructions to two important problems: *probability evaluation*, determining the probability that a query holds on an uncertain instance, and *counting*, counting the number of answers to a given query. These results are consequences of the construction of Section 4.

Probabilistic XML. We start with the problem of probabilistic query evaluation, beginning with the setting of *trees*. We use the framework of *probabilistic XML*, denoted PrXML^{fie}, to represent probabilistic trees as trees annotated by propositional formulas over independent probabilistic events (see [19] for the formal definitions), and consider the *data complexity* of the *query evaluation* problem for a MSO query q on such trees (i.e., computing the probability that q holds).

This problem is intractable in general, which is not surprising: it is harder than determining the probability of a single propositional annotation. However, for the less expressive *local* PrXML model, PrXML$^{\text{mux,ind}}$, query evaluation has tractable data complexity [8]; this model restricts edges to be annotated by only one event literal that is only used on that edge (plus a form of mutual exclusivity).

We can use the provenance circuits of Section 4 to justify that query evaluation is tractable for PrXML$^{\text{mux,ind}}$ and capture the data complexity tractability result of [8]. We say that an algorithm runs in *ra-linear time* if it runs in linear time assuming that arithmetic operations over rational numbers take constant time and rationals are stored in constant space, and runs in polynomial time without this assumption. We can show:

Theorem 6.1 [8]. *MSO query evaluation on* PrXML$^{\text{mux,ind}}$ *has ra-linear data complexity.*

We can also show extensions of this result. For instance, on PrXML^{fie}, defining the *scope* of event e in a document D as the smallest subtree in the left-child-right-sibling encoding of D covering nodes whose parent edge mentions e, and the *scope size* of a node n as the number of events with n in their scope, we show:

Proposition 6.1. *For any fixed $k \in \mathbb{N}$, MSO query evaluation on* PrXML^{fie} *documents with scopes assumed to have size $\leqslant k$ has ra-linear data complexity.*

BID instances. We move from trees to relational instances, and show another bounded-width tractability result for *block-independent disjoint* (BID) instances (see [26]). We define the *treewidth* of a BID instance as that of its underlying relational instance, and claim the following (remember that query evaluation on a probabilistic instance means determining the probability that the query holds):

Theorem 6.2. *For any fixed $k \in \mathbb{N}$, MSO query evaluation on an input BID instance of treewidth $\leqslant k$ has ra-linear data complexity.*

All probabilistic results are proven by rewriting to a formalism of relational instances with a circuit annotation, such that instance and circuit have a bounded-width joint decomposition. We compute a treelike provenance circuit for the instance using Theorem 4.2, combine it with the annotation circuit, and apply existing message passing techniques [20] to compute the probability of the circuit.

Counting. We turn to the problem of counting query results, and reduce it in ra-linear time to query evaluation on treelike instances, capturing a result of [2]:

Theorem 6.3 [2]. *For any fixed MSO query $q(\mathbf{x})$ with free first-order variables and $k \in \mathbb{N}$, the number of matching assignments to $\mathbf{x}$ on an input instance I of width $\leqslant k$ can be computed in ra-linear data complexity.*

7   Related Work

From the original results [9,12] on the linear-time data complexity of MSO evaluation on treelike structures, works such as [2] have investigated counting problems, including applications to probability computation (on graphs). A recent paper [5] also shows the linear-time data complexity of evaluating an MSO query on a treelike probabilistic network (analogous to a circuit). Such works, however, do not decouple the computation of a treelike *provenance* of the query and the *application* of probabilistic inference on this provenance, as we do. We also note results from another approach [22] on treelike structures, based on monadic Datalog (and not on MSO as the other works), that are limited to counting.

The *intensional* approach [26] to query evaluation on probabilistic databases is to compute a lineage of the query and evaluate its probability via general purpose methods; tree-like lineages allow for tractable probabilistic query evaluation

[18]. Many works in this field provide sufficient conditions for lineage tractability, only a few based on the data [24,25] but most based on the query [10,18]. For treelike instances, as we show, we can *always* compute treelike lineages, and we can do so for expressive queries (beyond UCQs considered in these works), or alternatively generalize Boolean lineages to connect them to more expressive semirings.

Our provenance study is inspired by the usual definitions of semiring provenance for the relational algebra and Datalog [17]. Another notion of provenance, for XQuery queries on trees, has been introduced in [13]. Both [17] and [13] provide *operational* definitions of provenance, which cannot be directly connected to tree automata. A different relevant work on provenance is [11], which introduces provenance circuits, but uses them for Datalog and only on *absorptive* semirings. Last, other works study provenance for *transducers* [6], but with no clear connections to semiring provenance or provenance for Boolean queries.

8   Conclusion

We have shown that two provenance constructions can be computed in linear time on trees and treelike instances: one for UCQs on arbitrary semirings, the other for arbitrary GSO queries as non-monotone Boolean expressions. A drawback of our results is their high combined complexity, as they rely on non-elementary encoding of the query to an automaton. One approach to fix this is monadic Datalog [14,22]; this requires defining and computing provenance in this setting.

Acknowledgments. This work was partly supported by a financial contribution from the Fondation Campus Paris-Saclay and the French ANR Aggreg project.

References

1. Amsterdamer, Y., Deutch, D., Tannen, V.: On the limitations of provenance for queries with difference. In: TaPP (2011)
2. Arnborg, S., Lagergren, J., Seese, D.: Easy problems for tree-decomposable graphs. J. Algorithms **12**(2) (1991)
3. Baget, J., Leclère, M., Mugnier, M.: Walking the decidability line for rules with existential variables. In: KR (2010)
4. Bodlaender, H.L.: A linear-time algorithm for finding tree-decompositions of small treewidth. SIAM J. Comput. **25**(6) (1996)
5. Bodlaender, H.L.: Probabilistic inference and monadic second order logic. In: Baeten, J.C.M., Ball, T., de Boer, F.S. (eds.) TCS 2012. LNCS, vol. 7604, pp. 43–56. Springer, Heidelberg (2012)
6. Bojańczyk, M.: Transducers with origin information. In: Esparza, J., Fraigniaud, P., Husfeldt, T., Koutsoupias, E. (eds.) ICALP 2014, Part II. LNCS, vol. 8573, pp. 26–37. Springer, Heidelberg (2014)
7. Cheney, J., Chiticariu, L., Tan, W.C.: Provenance in databases: Why, how, and where. Foundations and Trends in Databases **1**(4) (2009)

8. Cohen, S., Kimelfeld, B., Sagiv, Y.: Running tree automata on probabilistic XML. In: PODS (2009)
9. Courcelle, B.: The monadic second-order logic of graphs. I. Recognizable sets of finite graphs. Inf. Comput. **85**(1) (1990)
10. Dalvi, N., Suciu, D.: Efficient query evaluation on probabilistic databases. VLDBJ **16**(4) (2007)
11. Deutch, D., Milo, T., Roy, S., Tannen, V.: Circuits for Datalog provenance. In: ICDT (2014)
12. Flum, J., Frick, M., Grohe, M.: Query evaluation via tree-decompositions. J. ACM **49**(6) (2002)
13. Foster, J.N., Green, T.J., Tannen, V.: Annotated XML: queries and provenance. In: PODS (2008)
14. Gottlob, G., Pichler, R., Wei, F.: Monadic Datalog over finite structures of bounded treewidth. TOCL **12**(1) (2010)
15. Grädel, E.: Efficient evaluation methods for guarded logics and Datalog LITE. In: LPAR (2000)
16. Grädel, E., Hirsch, C., Otto, M.: Back and forth between guarded and modal logics. TOCL **3**(3) (2002)
17. Green, T.J., Karvounarakis, G., Tannen, V.: Provenance semirings. In: PODS (2007)
18. Jha, A.K., Suciu, D.: On the tractability of query compilation and bounded treewidth. In: ICDT (2012)
19. Kimelfeld, B., Senellart, P.: Probabilistic XML: models and complexity. In: Ma, Z., Yan, L. (eds.) Advances in Probabilistic Databases. STUDFUZZ, vol. 304, pp. 39–66. Springer, Heidelberg (2013)
20. Lauritzen, S.L., Spiegelhalter, D.J.: Local computations with probabilities on graphical structures and their application to expert systems. J. Royal Statistical Society, Series B (1988)
21. Meyer, A.R.: Weak monadic second order theory of succesor is not elementary-recursive. In: Logic Colloquium (1975)
22. Pichler, R., Rümmele, S., Woltran, S.: Counting and enumeration problems with bounded treewidth. Artificial Intelligence, and Reasoning, In Logic for Programming (2010)
23. Robertson, N., Seymour, P.D.: Graph minors. II. Algorithmic aspects of tree-width. J. Algorithms **7**(3) (1986)
24. Roy, S., Perduca, V., Tannen, V.: Faster query answering in probabilistic databases using read-once functions. In: ICDT (2011)
25. Sen, P., Deshpande, A., Getoor, L.: Read-once functions and query evaluation in probabilistic databases. PVLDB **3**(1–2) (2010)
26. Suciu, D., Olteanu, D., Ré, C., Koch, C.: Probabilistic Databases. Morgan & Claypool (2011)
27. Thatcher, J.W., Wright, J.B.: Generalized finite automata theory with an application to a decision problem of second-order logic. Math. systems theory **2**(1) (1968)

Language Emptiness of Continuous-Time Parametric Timed Automata

Nikola Beneš[1]($^{\boxtimes}$), Peter Bezděk[1], Kim G. Larsen[2], and Jiří Srba[2]

[1] Faculty of Informatics, Masaryk University Brno, Brno, Czech Republic
xbenes3@fi.muni.cz
[2] Department of Computer Science, Aalborg University, Aalborg, Denmark

Abstract. Parametric timed automata extend the standard timed automata with the possibility to use parameters in the clock guards. In general, if the parameters are real-valued, the problem of language emptiness of such automata is undecidable even for various restricted subclasses. We thus focus on the case where parameters are assumed to be integer-valued, while the time still remains continuous. On the one hand, we show that the problem remains undecidable for parametric timed automata with three clocks and one parameter. On the other hand, for the case with arbitrary many clocks where only one of these clocks is compared with (an arbitrary number of) parameters, we show that the parametric language emptiness is decidable. The undecidability result tightens the bounds of a previous result which assumed six parameters, while the decidability result extends the existing approaches that deal with discrete-time semantics only. To the best of our knowledge, this is the first positive result in the case of continuous-time and unbounded integer parameters, except for the rather simple case of single-clock automata.

1 Introduction

Timed automata [2] are a popular formalism used for modelling of real-time systems. In the classical definition, the clocks in guards are compared to fixed constants and one of the key problems, decidable in PSPACE [1], is the question of language emptiness. More than 20 years ago, Alur, Henzinger and Vardi [3] introduced a parametric variant of the language emptiness problem where clocks in timed automata can be additionally compared to a number of parameters. A clock is *nonparametric* if it is never compared with any of the parameters, otherwise the clock is *parametric*. The parametric language emptiness problem asks whether the parameters in the system can be replaced by constants so that the language of the resulting timed automaton becomes nonempty.

Nikola Beneš has been supported by the Czech Science Foundation grant project no. GA15-11089S.

Peter Bezděk has been supported by the Czech Science Foundation grant project no. GA15-08772S.

M.M. Halldórsson et al. (Eds.): ICALP 2015, Part II, LNCS 9135, pp. 69–81, 2015.
DOI: 10.1007/978-3-662-47666-6_6

Table 1. Decidability of the language (non)emptiness problems

	discrete time integer parameters	continuous time integer parameters	continuous time real parameters
n clocks, m parameters 1 parametric clock only	decidable [3]	**decidable**	undecidable [17]
3 clocks, 1 parameter	**undecidable**	**undecidable**	undecidable [17]
3 clocks, 6 parameters	undecidable [3]	undecidable [3]	undecidable [3]

Unfortunately, the parametric language emptiness problem is undecidable for timed automata with three parametric clocks [3]. Yet Alur, Henzinger and Vardi established a positive decidability result in the case of a single parametric clock. This decidability result was recently extended by Bundala and Ouaknine [10] to the case with two parametric clocks and an arbitrary number of nonparametric clocks. Both positive results are restricted to the discrete-time semantics with only integer delays. The problem of decidability of integer parametric language emptiness in the continuous-time semantics has been open for over 20 years. The parametric language emptiness problem has two variants, which we call *reachability* (existence of a parameter valuation s.t. the language is nonempty) and *safety* (existence of a parameter valuation s.t. the language is empty).

Our main contributions, summarised in Table 1, are: (i) undecidability of the reachability and safety problems (in discrete and continuous-time semantics) for three parametric clocks, no additional nonparametric clocks and one integer parameter and (ii) decidability of the reachability and safety problems in the continuous-time semantics for one parametric clock with an arbitrary number of integer parameters and an unlimited number of additional nonparametric clocks. For reachability the problem is further decidable in NEXPTIME.

Related work. Our undecidability result holds both for discrete and continuous time semantics and it uses only a single parameter with three parametric clocks, hence strengthening the result from [3] where six parameters were necessary for the reduction. In [10] the authors established NEXPTIME-completeness of the parametric reachability problem for the case of a single parametric clock but only for the discrete-time semantics. Parametric TCTL model checking of timed automata, in the discrete-time setting, was also studied in [9,19]. Our decision procedure for one parametric clock is, to the best of our knowledge, the first one that deals with continuous-time semantics without any restriction on the usage of parameters and without bounding the range of the parameters.

Reachability for parametric timed automata was shown decidable for certain (strict) subclasses of parametric timed automata, either by bounding the range of parameters [15] or by imposing syntactic restrictions on the use of parameters as in L/U automata [8,14]. The study of parametric timed automata in continuous time with parameters ranging over the rational or real numbers showed undecidability already for one parametric clock [17], or for two parametric clocks with exclusively strict guards [12]. We thus focus solely on integer-valued parameters in this paper.

Parametric reachability problems for interrupt timed automata were investigated by Bérard, Haddad, Jovanović and Lime [11] with a number of positive decidability results although their model is incomparable with the formalism of timed automata studied in this paper. Other approaches include the inverse method of [4] where the authors describe a procedure for deriving constrains on parameters in order to satisfy that timed automata remain time-abstract equivalent, however, the termination of the procedure is in general not guaranteed.

2 Definitions

We shall now introduce parametric timed automata, the studied problems and give an example of a parametric system for alarm sensor coordination.

Let $\mathbb{N}_0$ denote the set of nonnegative integers and $\mathbb{R}_{\geq 0}$ the set of nonnegative real numbers. Let $\mathcal{C}$ be a finite set of *clocks* and let $\mathcal{P}$ be a finite set of *parameters*. A *simple clock constraint* is an expression of the form $x \bowtie c$ where $x \in \mathcal{C}$, $c \in \mathbb{N}_0 \cup \mathcal{P}$ and $\bowtie \in \{<, \leq, =, \geq, >\}$. A *guard* is a conjunction of simple clock constraints, we denote the set of all guards by $\mathcal{G}$. A conjunction of simple clock constraints that contain only upper bounds on clocks, i.e. $\bowtie \in \{<, \leq\}$, is called an *invariant* and the set of all invariants is denoted by $\mathcal{I}$.

A *clock valuation* is a function $\nu : \mathcal{C} \to \mathbb{R}_{\geq 0}$ that assigns to each clock its nonnegative real-time age and *parameter valuation* is a function $\gamma : \mathcal{P} \to \mathbb{N}_0$ that assigns to each parameter its nonnegative integer value. Given a clock valuation ν, a parameter valuation γ and a guard (or invariant) $g \in \mathcal{G}$, we write $\nu, \gamma \models g$ if the guard expression g, after the substitution of all clocks $x \in \mathcal{C}$ with $\nu(x)$ and all parameters $p \in \mathcal{P}$ with $\gamma(p)$, is true. By ν_0 we denote the initial clock valuation where $\nu_0(x) = 0$ for all $x \in \mathcal{C}$. For a clock valuation ν and a delay $d \in \mathbb{R}_{\geq 0}$, we define the clock valuation $\nu + d$ by $(\nu + d)(x) = \nu(x) + d$ for all $x \in \mathcal{C}$.

Definition 1 (Parametric Timed Automaton). *A parametric timed automaton (PTA) over the set of clocks $\mathcal{C}$ and parameters $\mathcal{P}$ is a tuple $A = (\Sigma, L, \ell_0, F, I, \to)$ where Σ is a finite input alphabet, L is a finite set of locations, $\ell_0 \in L$ is the initial location, $F \subseteq L$ is the set of final (accepting) locations, $I : L \to \mathcal{I}$ is an invariant function assigning invariants to locations, and $\to \subseteq L \times \mathcal{G} \times \Sigma \times 2^{\mathcal{C}} \times L$ is the set of transitions, written as $\ell \xrightarrow{g,a,R} \ell'$ whenever $(\ell, g, a, R, \ell') \in \to$.*

For the rest of this section, let $A = (\Sigma, L, \ell_0, F, I, \to)$ be a fixed PTA. We say that a clock $x \in \mathcal{C}$ is a *parametric clock* in A if there is a simple clock constraint of the form $x \bowtie p$ with $p \in \mathcal{P}$ that appears in a guard or an invariant of A. Otherwise, if the clock x is never compared to any parameter, we call it a *nonparametric clock*.

A configuration of A is a pair (ℓ, ν) where $\ell \in L$ is the current location and ν is the current clock valuation. For every parameter valuation γ we define the corresponding timed transition system $T_\gamma(A)$ where states are all configurations (ℓ, ν) of A that satisfy the location invariants, i.e. $\nu, \gamma \models I(\ell)$, and the transition relation is defined as follows:

- $(\ell, \nu) \xrightarrow{d} (\ell, \nu + d)$ where $d \in \mathbb{R}_{\geq 0}$ if $\nu + d, \gamma \models I(\ell)$;
- $(\ell, \nu) \xrightarrow{a} (\ell', \nu')$ where $a \in \Sigma$ if there is a transition $\ell \xrightarrow{g,a,R} \ell'$ in A such that $\nu, \gamma \models g$ and $\nu', \gamma \models I(\ell')$ where for all $x \in C$ we define $\nu'(x) = 0$ if $x \in R$ and $\nu'(x) = \nu(x)$ otherwise.

A *timed language* of A under a parameter valuation γ, denoted by $L_\gamma(A)$, is the collection of all accepted *timed words* of the form $(a_0, d_0)(a_1, d_1) \ldots (a_n, d_n) \in (\Sigma \times \mathbb{R}_{\geq 0})^*$ such that in the transition system $T_\gamma(A)$ there is a computation $(\ell_0, \nu_0) \xrightarrow{d_0} (\ell'_0, \nu'_0) \xrightarrow{a_0} (\ell_1, \nu_1) \xrightarrow{d_1} \cdots \xrightarrow{a_{n-1}} (\ell_n, \nu_n) \xrightarrow{d_n} (\ell'_n, \nu'_n) \xrightarrow{a_n} (\ell_{n+1}, \nu_{n+1})$ where $\ell_{n+1} \in F$.

We can now define two problems for parametric timed automata, namely the reachability problem (reaching desirable locations) and safety problem (avoiding undesirable locations). Note that the problems are not completely dual, as the safety problem contains a hidden alternation of quantifiers.

Problem 1 (Reachability Problem for PTA). Given a PTA A, is there a parameter valuation γ such that $L_\gamma(A) \neq \emptyset$?

Problem 2 (Safety Problem for PTA). Given a PTA A, is there a parameter valuation γ such that $L_\gamma(A) = \emptyset$?

We shall now present a small case study of a wireless fire alarm system [13] modelled as a parametric timed automaton. In the alarm setup, a number of wireless sensors communicate with the alarm controller over a limited number of communication channels (in our simplified example we assume just a single channel). The wireless alarm system uses a variant of Time Division Multiple Access (TDMA) protocol in order to guarantee a safe communication of multiple sensors over a shared communication channel. In TDMA the data stream is divided into frames and each frame consists of a number of time slots allocated for exclusive use by the present wireless sensors. Each sensor is assigned a single slot in each frame where it can transmit on the shared channel.

We model each sensor as a timed automaton with two locations as shown in Figure 1a and 1b. The sensor in Figure 1a waits in its initial location until it receives a $wakeup_1$ message from the controller. After this, it takes strictly between 2 to 3 seconds to gather the current status of the sensor and transmit it as $result_1$ message back to the controller. Any subsequent wakeup signals during the transmission phase are ignored and after the transmission phase is finished, the sensor is ready to receive another wakeup signal. The sensor in Figure 1b has a more complex behaviour as transmitting the answer $result_2$ can take either strictly between 2 to 3 seconds, or 16 to 17 seconds.

The controller presented in Figure 1c is responsible for synchronising the two sensors and for assigning them their time slots so that no transmissions interfere. The parametric clock x of the controller determines the size of the time slots.

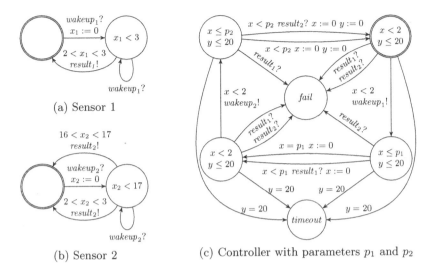

(a) Sensor 1

(b) Sensor 2

(c) Controller with parameters p_1 and p_2

Fig. 1. Wireless Fire Alarm System

First, it takes at most 2 seconds for the controller to wake up the first sensor after which it waits until the elapsed time reaches the value of the parameter p_1. If it receives the result of the reading of the first sensor in this time slot, it moves immediately into the next location where it performs the wakeup of the second sensor. If the first sensor does not deliver any result and the clock x reaches the value p_1, it also moves to the next location. Now a symmetric control is performed for the second sensor. If any of the two sensors transmit during the time the controller transmits the wakeup signals, we enter the location *fail*. The fail location is also reached if $result_2$ is received in the time slot of the first sensor and vice versa. The second clock y is used to simply measure the duration of the whole frame; whenever the duration of the frame reaches 20 seconds, the controller enters the *timeout* location.

We assume a standard handshake synchronisation of the controller and the two sensors running in parallel that results in a flat product timed automaton with two parameters p_1 and p_2. Note that x is the only parametric clock in our example. Now, our task is to find suitable values of the parameters that guide the duration of the time slots for the two sensors so that there is no behaviour of the protocol where it fails or timeouts. This question is equivalent to the safety problem on the constructed PTA where we mark *fail* and *timeout* as the accepting (undesirable) locations.

The obvious parameter valuation where $\gamma(p_1) = 5$ and $\gamma(p_2) = 19$ guarantees that the location *fail* is unreachable but it is not an acceptable solution as the duration of the frame becomes 24 and we reach *timeout*. However, there is another parameter valuation where $\gamma(p_1) = 5$ and $\gamma(p_2) = 9$ that guarantees that there is no possibility to fail or timeout. This is due to the fact that if the response time of the second sensor is too long, it skips one slot and the answer fits into an appropriate slot in the next frame.

In Section 4 we provide an algorithmic solution for finding such a parameter valuation that guarantees a given safety/reachability criterion. Note that as we are concerned with language (non)emptiness only, we employ two simplifications in the rest of the paper: First, we assume that the considered PTA have no invariants, as moving all invariants to guards preserves the language. Second, we assume that the alphabet is a singleton set as renaming all actions into a single action preserves language (non)emptiness.

3   Undecidability for Three Parametric Clocks

We shall now provide a reduction from the halting/boundedness problems of two counter Minsky machine to the reachability/safety problems on PTA. A *Minsky machine* with two nonnegative counters c_1 and c_2 is a sequence of labelled instructions $1 : inst_1$; $2 : inst_2$; $\ldots, n : inst_n$ where $inst_n = HALT$ and each $inst_i$, $1 \leq i < n$, is of one of the following forms (for $r \in \{1, 2\}$ and $1 \leq j, k \leq n$):

- (Increment) i: c_r++; goto j
- (Test and Decrement) i: if c_r=0 then goto k else (c_r--; goto j)

A configuration is a triple (i, v_1, v_2) where i is the current instruction and $v_1, v_2 \in \mathbb{N}_0$ are the values of the counters c_1 and c_2, respectively. A computation step between configurations is defined in the natural way. If starting from the initial configuration $(1, 0, 0)$ the machine reaches the instruction $HALT$ (note that the computation is deterministic) then we say it *halts*, otherwise it *loops*. The problem whether a given Minsky machine halts is undecidable [18]. The boundedness problem, i.e. the question whether there is a constant K such that $v_1 + v_2 \leq K$ for any configuration (i, v_1, v_2) reachable from $(1, 0, 0)$, is also undecidable [16].

The reduction from a two counter Minsky machine to PTA with a single parameter p and three parametric clocks x_1, x_2 and z is depicted in Figure 2. The reduction rules are shown only for the instructions handling the first counter. The rules for the second counter are symmetric. We also omit the transition labels as they are not relevant for the emptiness problem. The reduction preserves the property that whenever we are in a configuration (ℓ_i, ν) where $\nu(z) = 0$ then $\nu(x_1)$ and $\nu(x_2)$ represent the exact values of the counters c_1 resp. c_2, and the next instruction to be executed is the one with label i. Note also that there are no invariants used in the constructed automaton.

Lemma 1. *Let M be a Minsky machine. Let A be the PTA built according to the rules in Figures 2a and 2b (without the transitions for safety) and where ℓ_1 is the initial location and ℓ_n is the only accepting location. The Minsky machine M halts iff there is a parameter valuation γ such that $L_\gamma(A) \neq \emptyset$.*

Proof (Sketch). We only sketch a part of the proof to show the basic idea. We argue that from the configuration (ℓ_i, ν) where $\nu(z) = 0$ and where $\nu(x_1)$ and $\nu(x_2)$ represent the counter values, there is a unique way to move from ℓ_i to ℓ_j

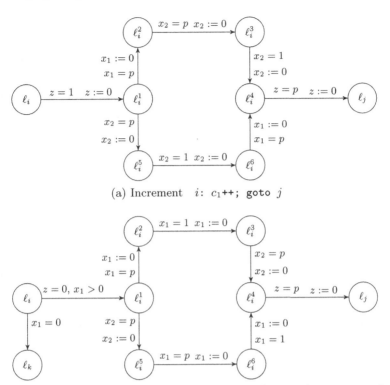

(a) Increment i: c_1++; goto j

(b) Test and decrement i: if c_1=0 then goto k else (c_1--; goto j)

(c) For safety, add this for every instruction i: c_1++; goto j

Fig. 2. Encoding of Minsky Machine as PTA with a single parameter p

(or possibly also to ℓ_k in the case of the test and decrement instruction) where again $\nu(z) = 0$ and the counter values are updated accordingly. As there are no invariants in the automaton, we can always delay long enough so that we get stuck in a given location, but this behaviour will not influence the language emptiness problem we are interested in.

Consider the automaton for the increment instruction from Figure 2a and assume we are in a configuration (ℓ_i, ν) where $\nu(z) = 0$, $\nu(x_1) = v_1$ and $\nu(x_2) = v_2$. First note that if $v_1 \geq p$ then there is no execution ending in ℓ_k due to the forced delay of one time unit on the transition from ℓ_i to ℓ_i^1 and the guard $x_1 = p$ tested in both the upper and lower branch in the automaton. Assume thus that $v_1 < p$. If $v_1 \geq v_2$ then we can perform the following execution with uniquely determined time delays: $(\ell_i, [x_1 \mapsto v_1, x_2 \mapsto v_2, z \mapsto 0]) \xrightarrow{1} (\ell_i^1, [x_1 \mapsto v_1+1, x_2 \mapsto v_2 + 1, z \mapsto 0]) \xrightarrow{p-v_1-1} (\ell_i^2, [x_1 \mapsto 0, x_2 \mapsto p - v_1 + v_2, z \mapsto p - v_1 - 1]) \xrightarrow{v_1-v_2}$

$(\ell_i^3, [x_1 \mapsto v_1 - v_2, x_2 \mapsto 0, z \mapsto p - v_2 - 1]) \xrightarrow{1} (\ell_i^4, [x_1 \mapsto v_1 - v_2 + 1, x_2 \mapsto 0, z \mapsto p - v_2]) \xrightarrow{v_2} (\ell_j, [x_1 \mapsto v_1 + 1, x_2 \mapsto v_2, z \mapsto 0])$. In this case where $v_1 \geq v_2$, executing the lower branch of the automaton will result in getting stuck in the location ℓ_i^6 as here necessarily $\nu(x_1) > p$. Clearly, there is a unique way of getting to ℓ_j in which the clock valuation of x_1 was incremented by one, hence faithfully simulating the increment instruction of the Minsky machine. The other cases and instructions are dealt with similarly, see [6].   □

Lemma 2. *Let M be a Minsky machine. Let A be the PTA built according to the rules in Figures 2a, 2b and 2c (including the transitions for safety) and where ℓ_1 is the initial location and ℓ_{acc} is the only accepting location. The Minsky machine M is bounded iff there is a parameter valuation γ such that $L_\gamma(A) = \emptyset$.*

Proof. If the computation of the Minsky machine is unbounded then clearly, for any parameter value of p, the Minsky machine will eventually try to make one of the counters larger or equal than p (using the increment instruction). Necessarily, we will then have $\nu(x_1) = p$ or $\nu(x_2) = p$ in the location ℓ_j where we end after performing the increment instruction i, implying that we can reach the accepting location ℓ_{acc} due to the transition added in Figure 2c and hence the language is nonempty. On the other hand, if the parameter p is large enough and the computation bounded (note that the boundedness condition $\exists K.\ v_1 + v_2 \leq K$ is equivalent to $\exists K.\ \max\{v_1, v_2\} \leq K$), we will not be able to enter the accepting location ℓ_{acc} and the language is empty.   □

We now conclude with the main theorem of this section, tightening the previously known undecidability result that used six parameters and three parametric clocks [3]. The theorem is valid for both the continuous-time and the discrete-time semantics due to the exact guards in all transitions of the constructed PTA that allow to take transitions only after integer delays.

Theorem 1. *The reachability and safety problems are undecidable for PTA with one integer parameter, three parametric clocks and no further nonparametric clocks in the continuous-time as well as the discrete-time semantics.*

4   Decidability for One Parametric Clock

In this section, we show that both the reachability and safety problems for PTA with a single parametric clock are decidable. Our general strategy is similar to that of [3], i.e. reducing the original PTA (which has continuous-time semantics in our case) into a so-called *parametric 0/1-timed automaton* with just a single clock. It is shown in [3] that the set of parameter valuations that ensure language nonemptiness of a given parametric 0/1-timed automaton with single clock is effectively computable. Moreover, in [10] the authors show that the reachability problem for parametric 0/1-timed automata is polynomial-time reducible to the halting problem of parametric bounded one-counter machines, which is in NP. As the parametric 0/1-timed automaton is going to be exponential in the size of the original PTA, this makes the reachability problem for PTA with a single parametric clock belong to the NEXPTIME complexity class.

A $0/1$-timed automaton is a timed automaton with discrete time, in which all the delays are explicitly encoded via two kinds of delay transitions: 0-transitions and 1-transitions. Formally, we enrich the syntax of a timed automaton with two transition relations $\xrightarrow{0}, \xrightarrow{1} \subseteq L \times L$ and modify the semantics so that $(\ell, \nu) \xrightarrow{0} (\ell', \nu)$ iff $\ell \xrightarrow{0} \ell'$ and $(\ell, \nu) \xrightarrow{1} (\ell', \nu + 1)$ iff $\ell \xrightarrow{1} \ell'$; other delays in the timed transition system are no longer possible.

Corner-Point Abstraction. As we are concerned with continuous time, our reduction to $0/1$-timed automata is more convoluted than that of [3], in which the nonparametric clocks were eliminated by moving their integer values into locations. In our setting, using region abstraction to eliminate nonparametric clocks will not allow us to correctly identify the $0/1$ delays. We thus choose to use *corner-point abstraction* [5] that is finer than the region-based one. In this abstraction, each region is associated with a set of its corner points. Note that the original definition only deals with timed automata that are bounded, while we want to be more general here. For this reason, we extend the original definition with extra corner points for unbounded regions.

We first define the region equivalence [2]. Let $M \in \mathbb{N}_0$ be the largest constant appearing in the constraints of a given timed automaton. Note that in the original definition the largest constant is considered for each clock independently. For the sake of readability, we consider M to be a common upper bound for each clock. Let ν, ν' be clock valuations. Let further $fr(t)$ be the fractional part of t and $\lfloor t \rfloor$ be the integral part of t. We define an equivalence relation $\equiv$ on clock valuations by $\nu \equiv \nu'$ if and only if the following three conditions are satisfied:

- for all $x \in \mathcal{C}$ either $\nu(x) \geq M$ and $\nu'(x) \geq M$ or $\lfloor \nu(x) \rfloor = \lfloor \nu'(x) \rfloor$;
- for all $x, y \in \mathcal{C}$ such that $\nu(x) \leq M$ and $\nu(y) \leq M$, $fr(\nu(x)) \leq fr(\nu(y))$ if and only if $fr(\nu'(x)) \leq fr(\nu'(y))$);
- for all $x \in \mathcal{C}$ such that $\nu(x) \leq M$, $fr(\nu(x)) = 0$ if and only if $fr(\nu'(x)) = 0$.

We define a *region* as an equivalence class of clock valuations induced by $\equiv$. A region r' is a time successor of a region r if for all $\nu \in r$ there exists $d \in \mathbb{R}_{>0}$ such that $\nu + d \in r'$ and for all $d', 0 \leq d' \leq d$, we have $\nu + d' \in r \cup r'$. As the time successor is unique if it exists, we use $succ(r)$ to denote the time successor of r. Moreover, if no time successor of r exists, we let $succ(r) = r$.

An $(M+1)$-*corner point* $\alpha : \mathcal{C} \to \mathbb{N}_0 \cap [0, M+1]$ is a function which assigns an integer value from the interval $[0, M + 1]$ to each clock. We define the successor of the M^{+1}-corner point α, denoted by $succ(\alpha)$, as follows:

$$\text{for each } x \in \mathcal{C}, \quad succ(\alpha)(x) = \begin{cases} \alpha(x) + 1 & \alpha(x) \leq M \\ M + 1 & \text{otherwise .} \end{cases}$$

For $R \subseteq \mathcal{C}$, we define the reset of the corner point α, denoted by $\alpha[R]$, as follows:

$$\text{for each } x \in \mathcal{C}, \quad \alpha[R](x) = \begin{cases} \alpha(x) & x \notin R \\ 0 & x \in R . \end{cases}$$

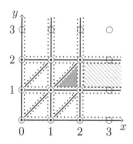
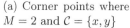
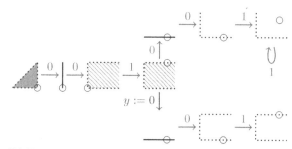

(a) Corner points where $M = 2$ and $\mathcal{C} = \{x, y\}$

(b) Fragment of an evolution of a region with a corner point (locations are omitted for simplicity)

Fig. 3. Corner point abstraction

We say α is a corner point of a region r whenever α is in the topological closure of r. The construction of the corner-point abstraction is illustrated in Figure 3. Notice the corner points in unbounded regions.

Construction of the Parametric 0/1-Timed Automaton. Now we show how to construct for a given PTA with one parametric clock an equivalent 0/1-PTA with just one clock. Let $A = (\Sigma, L, \ell_0, F, I, \rightarrow)$ be the original PTA over the set of clocks $\mathcal{C}$ and parameters $\mathcal{P}$. Let x_p denote the only parametric clock.

We first modify the automaton by adding a fresh clock z as follows: every transition $\ell \xrightarrow{g,a,R} \ell'$ is changed into $\ell \xrightarrow{g \wedge z < 1, a, R'} \ell'$ where $R' = R$ if $x_p \notin R$, and $R' = R \cup \{z\}$ otherwise. To every location ℓ we then add a new self-loop transition $\ell \xrightarrow{z=1,a,\{z\}} \ell$. Intuitively, the new clock z will always contain the fractional part of x_p. We call this new automaton A'. Clearly, this modification preserves the language (non)emptiness of the original automaton A.

In the second step, we use the corner-point abstraction of A' with respect to all clocks except for x_p to create the 0/1-timed automaton with a single clock. Let $\hat{\mathcal{C}} = (\mathcal{C} \cup \{z\}) \setminus \{x_p\}$ and let M be the largest constant appearing in the guards concerning the clocks in $\hat{\mathcal{C}}$. In the following, we consider regions and corner-points with respect to clocks in $\hat{\mathcal{C}}$ and the bound M. Let Reg denote the set of all such regions and let Cp denote the set of all corresponding corner-points, i.e. $Cp = (\mathbb{N}_0 \cap [0, M + 1])^{\hat{\mathcal{C}}}$.

We use the following auxiliary notation. Let $r \in Reg$ and $\alpha \in Cp$.

$$\iota(r, \alpha) = \begin{cases} LESS & \alpha(z) = 1 \text{ and } r \not\models z = 1 \\ MORE & \alpha(z) = 0 \text{ and } r \not\models z = 0 \\ EXACT & \text{otherwise} \end{cases}$$

The 0/1-timed automaton over the singleton set of clocks $\{\hat{x}_p\}$ is $\hat{A} = (\Sigma, L \times Reg \times Cp, (\ell_0, r_0, \alpha_0), F \times Reg \times Cp, I, \rightarrow)$ where r_0 is the initial region and $\alpha_0(x) = 0$ for all $x \in \hat{\mathcal{C}}$ is the initial corner-point. The transition relation is defined as follows:

- zero delay: $(\ell, r, \alpha) \xrightarrow{0} (\ell, r', \alpha)$ if $r' = succ(r)$ and α is a corner-point of both r and r';
- unit delay: $(\ell, r, \alpha) \xrightarrow{1} (\ell, r, \alpha')$ if $\alpha' = succ(\alpha)$ and both α and α' are corner-points of r;
- action: whenever $\ell \xrightarrow{g,a,R} \ell'$ in A' then let $g_1, \ldots, g_k$ be all the simple clock constraints appearing in g comparing clocks from $\hat{\mathcal{C}}$ and let $h_1, \ldots, h_n$ be the remaining simple clock constraints, i.e. those that consider x_p. For every (ℓ, r, α) that satisfies (1) $r \models g_1 \wedge \cdots \wedge g_k$ and (2) if $\iota(r, \alpha) \neq EXACT$ then no h_i contains equality $(=)$, we set $(\ell, r, \alpha) \xrightarrow{h_1 \wedge \cdots \wedge h_n, \hat{R}} (\ell', r[R \setminus \{x_\mathrm{p}\}], \alpha[R \setminus \{x_\mathrm{p}\}])$, where $\hat{R} = \{\hat{x}_\mathrm{p}\}$ if $x_\mathrm{p} \in R$ and $\hat{R} = \emptyset$ otherwise. The constraints h_i are created as follows: all x_p are changed into $\hat{x}_\mathrm{p}$; if $\iota(r, \alpha) = LESS$, all $<$ are changed into $\leq$ and all $\geq$ are changed into $>$; if $\iota(r, \alpha) = MORE$, all $\leq$ are changed into $<$ and all $>$ are changed into $\geq$.

Theorem 2. *The reachability and safety problems for parametric timed automata over integer parameters with one parametric clock in the continuous-time semantics are decidable. Moreover, the reachability problem is in NEXP-TIME.*

Proof (Idea). Due to space constraints, the complete proof can be found in [6]. As mentioned above, the modification from A to A' preserves the language (non)emptiness. The idea of the proof is to show that for every given parameter valuation, every run of A' has a corresponding run in $\hat{A}$ and vice versa. This shows that the reachability and safety problems for parametric timed automata with one parametric clock reduce to the reachability and safety problems for parametric 0/1-timed automata. These problems were shown decidable in [3]. The complexity argument is discussed in the beginning of this section. □

5   Conclusion

We have shown that for three parametric clocks with a single integer parameter, both the reachability and safety problems are undecidable in the discrete as well as the continuous semantics. This improves the previously known undecidability result by Alur, Henzinger and Vardi [3] where six parameters were needed. For the case with a single parametric clock with an unrestricted number of integer parameters and with any number of additional nonparametric clocks, we contributed to the solution of an open problem stated more than 20 years ago by proving a decidability result for reachability and safety problems in the continuous semantics, extending the previously known decidability result for the discrete-time semantics [3]. To achieve this result, we used the corner-point abstraction technique that had to be modified to handle also corner-points in unbounded regions, contrary to the use of the technique in [5]. Not surprisingly, the decidability of the problem in case of two parametric clocks in the continuous-time setting remains open, as it is the case also for a number of other problems over timed automata with two real-time clocks [7]. On the other hand, as demonstrated by our wireless fire alarm case study, the parameter synthesis problem for one parametric clock and an unlimited number of parameters is sufficiently expressive in order to describe nontrivial scheduling problems. As a next step, we will consider moving from corner-point regions into zones and provide an efficient implementation of the presented techniques.

Acknowledgments. We acknowledge a funding from the EU FP7 grant agreement nr. 318490 (SENSATION) and grant agreement nr. 601148 (CASSTING) and from the Sino-Danish Basic Research Center IDEA4CPS.

References

1. Alur, R., Courcoubetis, C., Dill, D.: Model-checking for real-time systems. In: LICS 1990. pp. 414–425. IEEE (1990)
2. Alur, R., Dill, D.: A theory of timed automata. Theoretical Computer Science **126**(2), 183–235 (1994)
3. Alur, R., Henzinger, T., Vardi, M.: Parametric real-time reasoning. In: Proceedings of 25th Annual Symposium on Theory of Computing (STOC 1993), pp. 592–601. ACM Press (1993)
4. André, É., Chatain, T., Fribourg, L., Encrenaz, E.: An inverse method for parametric timed automata. ENTCS **223**, 29–46 (2008)
5. Behrmann, Gerd, Fehnker, Ansgar, Hune, Thomas, Larsen, Kim Guldstrand, Pettersson, Paul, Romijn, Judi M.T., Vaandrager, Frits W.: Minimum-cost reachability for priced timed automata. In: Di Benedetto, Maria Domenica, Sangiovanni-Vincentelli, Alberto L. (eds.) HSCC 2001. LNCS, vol. 2034, pp. 147–161. Springer, Heidelberg (2001)
6. Beneš, N., Bezděk, P., Larsen, K.G., Srba, J.: Language emptiness of continuous-time parametric timed automata (2015). CoRR abs/1504.07838
7. Bouyer, P., Brihaye, T., Markey, N.: Improved undecidability results on weighted timed automata. Inform. Proc. Letters **98**(5), 188–194 (2006)
8. Bozzelli, L., La Torre, S.: Decision problems for lower/upper bound parametric timed automata. Formal Methods in Syst. Design **35**(2), 121–151 (2009)
9. Bruyère, Véronique, Raskin, Jean-François: Real-time model-checking: parameters everywhere. In: Pandya, Paritosh K., Radhakrishnan, Jaikumar (eds.) FSTTCS 2003. LNCS, vol. 2914, pp. 100–111. Springer, Heidelberg (2003)
10. Bundala, Daniel, Ouaknine, Joël: Advances in parametric real-time reasoning. In: Csuhaj-Varjú, Erzsébet, Dietzfelbinger, Martin, Ésik, Zoltán (eds.) MFCS 2014, Part I. LNCS, vol. 8634, pp. 123–134. Springer, Heidelberg (2014)
11. Bérard, Beatrice, Haddad, Serge, Jovanović, Aleksandra, Lime, Didier: Parametric interrupt timed automata. In: Abdulla, Parosh Aziz, Potapov, Igor (eds.) RP 2013. LNCS, vol. 8169, pp. 59–69. Springer, Heidelberg (2013)
12. Doyen, L.: Robust parametric reachability for timed automata. Information Processing Letters **102**(5), 208–213 (2007)
13. Feo-Arenis, Sergio, Westphal, Bernd, Dietsch, Daniel, Muñiz, Marco, Andisha, Ahmad Siyar: The wireless fire alarm system: ensuring conformance to industrial standards through formal verification. In: Jones, Cliff, Pihlajasaari, Pekka, Sun, Jun (eds.) FM 2014. LNCS, vol. 8442, pp. 658–672. Springer, Heidelberg (2014)

14. Hune, Thomas, Romijn, Judi M.T., Stoelinga, Mariëlle, Vaandrager, Frits W.:
    Linear parametric model checking of timed automata. In: Margaria, Tiziana, Yi,
    W. (eds.) TACAS 2001. LNCS, vol. 2031, pp. 189–203. Springer, Heidelberg (2001)
15. Jovanović, Aleksandra, Lime, Didier, Roux, Olivier H.: Integer parameter synthe-
    sis for timed automata. In: Piterman, Nir, Smolka, Scott A. (eds.) TACAS 2013
    (ETAPS 2013). LNCS, vol. 7795, pp. 401–415. Springer, Heidelberg (2013)
16. Kuzmin, E., Chalyy, D.: Decidability of boundedness problems for Minsky counter
    machines. Automatic Control and Computer Sciences 44(7), 387–397 (2010)
17. Miller, Joseph S.: Decidability and complexity results for timed automata and
    semi-linear hybrid automata. In: Lynch, Nancy A., Krogh, Bruce H. (eds.) HSCC
    2000. LNCS, vol. 1790, pp. 296–310. Springer, Heidelberg (2000)
18. Minsky, M.: Computation: Finite and Infinite Machines. Prentice (1967)
19. Wang, F.: Parametric timing analysis for real-time systems. Information and
    Computation 130(2), 131–150 (1996)

Analysis of Probabilistic Systems via Generating Functions and Padé Approximation

Michele Boreale$^{(\boxtimes)}$

Università di Firenze, Firenze, Italy
michele.boreale@unifi.it

Abstract. We investigate the use of generating functions in the analysis of discrete Markov chains. Generating functions are introduced as power series whose coefficients are certain hitting probabilities. Being able to compute such functions implies that the calculation of a number of quantities of interest, including absorption probabilities, expected hitting time and number of visits, and variances thereof, becomes straightforward. We show that it is often possible to recover this information, either exactly or within excellent approximation, via the construction of Padé approximations of the involved generating function. The presented algorithms are based on projective methods from linear algebra, which can be made to work with limited computational resources. In particular, only a black-box, on-the-fly access to the transition function is presupposed, and the necessity of storing the whole model is eliminated. A few numerical experiments conducted with this technique give encouraging results.

1 Introduction

Our goal is to understand if the concept of generating function [9] can play a useful role in the analysis of Markov chains. In the present paper, we focus on the reachability properties of time-homogeneous, finite Markov chains. The generating function of such a system is a power series in the variable z, $g(z) = \sum_{j \geq 0} a_j z^j$, whose coefficients, or *moments*, a_j are just the probabilities of hitting a state of interest exactly at time $j = 0, 1, 2, \cdots$. With $g(z)$, a whole host of information about the system is packed into a single mathematical object, including: the probability of the event itself - which is of course $g(1)$ - and various statistics, such as the expected hitting time and its variance. We will demonstrate that, by building a rational representation of $g(z)$, in a number of interesting situations it is possible to extract this information, either exactly or within excellent approximation, using limited computational resources. These limitations are mainly the fact that one can access the system's transition relation only in a black-box, on-the-fly[1] fashion, and can only store a small portion of its state space at time. We give a more detailed account of our approach and of our paper below.

Author's address: Michele Boreale, Università di Firenze, Dipartimento di Statistica, Informatica, Applicazioni (DiSIA) "G. Parenti", Viale Morgagni 65, I-50134 Firenze, Italy. E-mail: michele.boreale@unifi.it. Work partially supported by MIUR funded project CINA.

[1] That is, via a function that given a state returns the list of its successors together with their probabilities.

© Springer-Verlag Berlin Heidelberg 2015
M.M. Halldórsson et al. (Eds.): ICALP 2015, Part II, LNCS 9135, pp. 82–94, 2015.
DOI: 10.1007/978-3-662-47666-6_7

We first introduce and motivate the system's generating function $g(z)$, then establish some of its important properties, like its radius convergence and its rationality (Section 2). We then show (Section 3) that, via *Padé approximants* [3], an *exact* rational representation of $g(z)$ can be recovered from the knowledge of its first $2N$ moments, where N is the number of system's states. These moments can in principle be computed relying solely on a black-box, on-the-fly access to the transition relation, and do not require storing the entire model. Yet, limited resources imply that one is often forced to consider approximations. We argue (Section 4) that polynomial approximations, derived from truncating $g(z)$, may not be a good idea, and that rational ones should rather be preferred. This is especially true in the presence of clusters of states that are nearly uncoupled with other states in the chain. We then discuss a method to compute one such approximation effectively (Section 5). The basic idea here is to view the matrix P that represents the transition relation as a linear application on $\mathbb{R}^N$; then to take its projection $\hat{P}$ onto a small, m-dimensional subspace $\mathcal{K}_m$, with $m \ll N$. The generating function of the projected application, $\hat{g}(z)$, is a rational *Padé-type* approximation of the original $g(z)$. Accuracy is often very good already for small m: this somewhat surprising effectiveness is a consequence of the tendency of $\hat{P}$'s eigenvalues to be excellent approximations of P's ones. We show that the Arnoldi algorithm [2,8] can be used to effectively compute $\hat{g}(z)$, in a way that is compatible with an on-the-fly access to the transition relation and with limited computational resources. Notably, transitions need not be stored at all with this method. Error control and steady-state distributions are discussed in the full version [4]. We then present a few numerical experiments that have been conducted with a preliminary Matlab implementation of this idea (Section 6), and which give very encouraging results. For comparison, the results obtained on the same systems with a state-of-the-art probabilistic model checker are also reported. We conclude the paper with a discussion of future venues of research and related work (Section 7). All the proofs, some numerical examples and additional technical material can be found in the full version available online [4].

2   The System Generating Function $g(z)$

Consider a time-homogeneous Markov chain $\{X_j\}_{j \geq 0}$ over a finite set of states $S = \{1, ..., N\}$ with $N > 0$ and initial state $X_0 = 1$. To avoid uninteresting special cases, we will assume that all states of the chain are reachable from 1 (but need not assume the vice-versa, so the chain might well be reducible.) We want to study the event corresponding to reaching a (typically, 'bad') state $s_{bad} = N$, that is $Reach \stackrel{\Delta}{=} \{X_j = N$ for some $j \geq 0\}$ and denote by $p_{reach} \stackrel{\Delta}{=} \Pr(Reach)$ the probability of this event. Without loss of generality, we will assume that state N is absorbing that is $\Pr(X_{j+1} = N | X_j = N) = 1$. Later on in this section we will also consider another type of statistics, concerning the number of visits, where we will not assume N is absorbing. We will also be interested in statistics concerning the *hitting time* random variable, defined as: $T \stackrel{\Delta}{=} \inf\{j \geq 0 : X_j = N\}$. Let us define the j-th ($j \geq 0$) *moment* of the system as the probability of hitting state N for the first time exactly at time j, that is

$$a_j \overset{\triangle}{=} \Pr(X_j = N \text{ and } X_i \neq N \text{ for } i < j). \qquad (1)$$

Clearly, the probability of eventually reaching N is just the sum of the moments: $p_{reach} = \sum_{j \geq 0} a_j$. Our main object of study is defined below.

Definition 1 (Generating function). *The generating function of the system is the power series, in the complex variable z, $g(z) \overset{\triangle}{=} \sum_{j \geq 0} a_j z^j$.*

Note that the $g(z)/p_{reach}$ is just the probability generating function of the random variable T conditioned on the event *Reach*. As such, with $g(z)$ a whole host of information about T and its moments is packed into a single mathematical object. For instance, indicating with $g', g'', \ldots$ the derivatives of g, easy calculations show that (provided the mentioned quantities are all defined).

$$p_{reach} = g(1) \qquad E[T|Reach] = g'(1)/g(1) \qquad var[T|Reach] = (g''(1) + g'(1))/g(1) -$$
$$(g'(1)/g(1))^2 . \qquad (2)$$

More generally, information on higher moments of T can be extracted using the identity $E[X(X-1)\cdots(X-k)|Reach] = g^{(k+1)}(1)/g(1)$, although usually the first two moments are enough for a satisfactory analysis of the system. In practice, we will be able to extract this information only provided we are able to build an efficient representation of $g(z)$. As a first step towards this, we shall see in a moment that $g(z)$ can be represented as a rational function, that is, as the ratio of two polynomials in z. At this point it is convenient to introduce some notation.

Notation In the rest of the paper, some basic knowledge of linear algebra is presupposed. We let P denote the $N \times N$ stochastic matrix that defines the transition function of the chain. Departing from the usual convention, we will work with *column*-stochastic matrices, that is we let the element of row i and column j of P, denoted by p_{ij}, be $\Pr(X_{t+1} = i | X_t = j)$. In what follows, vectors are considered as column-vectors; in particular, e_i denotes the i-th canonical column vector of $\mathbb{R}^N$. So the vector $P^i e_1$ is just the probability distribution of the variable X_i of the chain. A vector is stochastic if its components are nonnegative and sum to 1. For a matrix or vector A, we let A^T denote its transpose. I_k will denote the $k \times k$ identity matrix; the index k will be omitted when clear from the context. A rational function in z of type $[h, k]$, for $k \geq 0$ and $h \geq 0$ or $h = -\infty$, is a ratio of two polynomials in z, $r(z)/t(z)$, such that $\deg(r) \leq h$ and $\deg(t) \leq k$. We will let z range over complex numbers and x on reals.

In the rest of the paper, we let $\tilde{e}_1 \overset{\triangle}{=} (P - I)e_1$. Note that $e_N^T P^j e_1$ is just the probability of being in state N at time j. Exploiting the fact that N is absorbing, it is easy to see that, for $j \geq 1$, $a_j = e_N^T P^j e_1 - e_N^T P^{j-1} e_1 = e_N^T P^{j-1} \tilde{e}_1$. Ignoring for a moment issues of convergence and singularity, we can then reason as follows. We first note that the following equality can be readily checked.

$$(I - zP)(I + zP^1 + z^2 P^2 + \cdots) = I$$

which implies that $(I - zP)^{-1} = (I + zP^1 + z^2 P^2 + \cdots)$. We then can write

$$g(z) = a_0 + \sum_{j \geq 1} a_j z^j \qquad\qquad = a_0 + \sum_{j \geq 1} e_N^T z^j P^{j-1} \tilde{e}_1$$

$$= a_0 + z \cdot e_N^T (\sum_{j \geq 0} z^j P^j) \tilde{e}_1 \quad = a_0 + z \cdot e_N^T (I - zP)^{-1} \tilde{e}_1$$

$$= a_0 + \frac{z \cdot e_N^T \mathrm{Adj}(I - zP) \tilde{e}_1}{\det(I - zP)} \tag{3}$$

where, in the last step, we have exploited Cramer's rule for the computation of the inverse (recall that Adj(A) denotes the *adjoint matrix* of a matrix A.) In the last expression, the denominator and numerator of the fraction are polynomials in z. This shows that $g(z)$ is a rational function. This informal reasoning can be made into a rigorous proof – a nontrivial point of which is related to the singularity of the matrix $(I - zP)$ at $z = 1$. Another important point is where the power series $g(z)$ is defined, that is, what is its radius of convergence. The the next theorem records these facts about $g(z)$.

Theorem 1 (convergence and rationality of g). *There is a real $R > 1$ such that the power series $g(z)$ in Definition 1 converges for all $|z| < R$. Moreover, there is a rational function $r(z)/t(z)$ such that for all such z's*

(a) $t(z) \neq 0$ (b) $\deg(r), \deg(t) \leq N - 1$ (c) $g(z) = r(z)/t(z)$. $\tag{4}$

The proof of the above theorem provides us also with an explicit expression for $g(z)$, that is (3). In what follows, $(I - zP)^{-1}$ will be used as an abbreviation for the matrix of rational expressions $\frac{\mathrm{Adj}(I-zP)}{\det(I-zP)}$, where it is understood that common factors are canceled out. Concerning this expression, note that $\det(I - zP) = z^N \det((1/z)I - P)$ is just the characteristic polynomial of P with coefficients reversed. That is, the relation between the characteristic polynomial and our $\det(I - zP)$ is as follows:

$$\det(zI - P) = \beta_N z^N + \beta_{N-1} z^{N-1} + \cdots + \beta_0 \qquad \text{and}$$
$$\det(I - zP) = \beta_N + \beta_{N-1} z + \cdots + \beta_0 z^N$$

(In passing, note that $\beta_N = 1$ and $\beta_0 = -\det(P)$.) In particular, from $\det(I - zP) = z^N \det((1/z)I - P)$ it is clear that the roots of the polynomial $\det(I - zP)$ are just the reciprocals of the nonzero roots of P: that is, the reciprocals of the nonzero eigenvalues of P. This fact can be exploited to give more precise information about R. We record these facts below.

Corollary 1. *There is $R > 1$ such that for $|z| < R$ and for $\tilde{e}_1 = (P - I)e_1$*

$$g(z) = a_0 + z \cdot e_N^T (I - zP)^{-1} \tilde{e}_1. \tag{5}$$

In particular, $g(z)$ has a radius of convergence either $R = |1/\lambda|$ for some eigenvalue $0 < |\lambda| < 1$ of P, or $R = +\infty$.

Let us now drop the assumption that N is absorbing. We are interested in counting the visits to state N, starting from $X_0 = 1$. To this purpose we introduce a different generating function: $f(z) \triangleq \sum_{j \geq 0} c_j z^j$, where $c_j \triangleq \Pr(X_j = N) = e_N^T P^j e_1$. By definition,

$f(1)$ is the expected number of visits of the chain to state N. Recall that $f(1) < +\infty$ iff N is transient. By paralleling the above development for $g(z)$, we can prove the following.

Theorem 2. *Let N be transient. The power series $f(z)$ has radius of convergence $R > 1$. Moreover, for $|z| < R$, one has $f(z) = e_N^T(I - zP)^{-1}e_1$. The expression on the right of the last equality gives rise to a rational function $r(z)/t(z)$ of type $[N - 1, N]$ such that $t(z) \neq 0$ for $|z| < R$.*

From now on, we will consider $g(z)$ only; statements and proofs for $f(z)$ can be obtained by obvious modifications. All the quantities of interest about the system, (2), will be easy to compute, provided we can recover the rational representation $r(z)/t(z)$ of $g(z)$ promised by Theorem 1. The expression provided by Corollary 1 can be useful for small values of N and provided one knows P explicitly. Here is a small example to illustrate.

Example 1. We consider a chain with $N \geq 3$ states $1, 2, ..., N$, where, for $1 \leq i \leq N - 3$ and a (small) $0 < \delta < 1$, there is a transition from i to 1 with probability $1 - \delta/i$, and to each of $i + 1, N - 1, N$ with probability $\delta/3i$; for $i = N - 2$, there is a transition from i to 1 with probability $1 - \delta/i$, and to each of $N - 1, N$ with probability $\delta/2i$; $N - 1$ and N are absorbing. For reasons that will become evident later on, we call this chain $Nasty(N, \delta)$.

The transition matrix P of $Nasty(6, \delta)$ is given on the right (recall that we work with column-stochastic matrices.) From symmetry considerations, it is clear that the probability of reaching either of the two absorbing states is $1/2$, thus $p_{reach} = 1/2$. Let us check this out via $g(z)$. With the help of a computer algebra system, we apply (5) and, taking into account that $a_0 = 0$, find:

$$P = \begin{bmatrix} 1-\delta & 1-\delta/2 & 1-\delta/3 & 1-\delta/4 & 0 & 0 \\ \delta/3 & 0 & 0 & 0 & 0 & 0 \\ 0 & \delta/6 & 0 & 0 & 0 & 0 \\ 0 & 0 & \delta/9 & 0 & 0 & 0 \\ \delta/3 & \delta/6 & \delta/9 & \delta/8 & 1 & 0 \\ \delta/3 & \delta/6 & \delta/9 & \delta/8 & 0 & 1 \end{bmatrix}$$

$$g(z) = \frac{1}{2} \frac{\delta^4 z^4 + 8\delta^3 z^3 + 72\delta^2 z^2 + 432\delta z}{(\delta^4 - 4\delta^3)z^4 + (12\delta^3 - 36\delta^2)z^3 + (108\delta^2 - 216\delta)z^2 + 648(\delta - 1)z + 648}.$$

When evaluating this at $z = 1$ we get $g(1) = p_{reach} = 1/2$. By differentiating the above expression of $g(z)$ and then evaluating the result at $z = 1$, we get $g'(1) = 2\left(\delta^3 + 9\delta^2 + 54\delta + 162\right)/\left(\delta^4 + 8\delta^3 + 72\delta^2 + 432\delta\right)$, which can be evaluated for instance at $\delta = 10^{-3}$ to compute $E[T|Reach] = g(1)^{-1}g'(1) \approx 1500.25$.

Computing an expression for $g(z)$ based on a direct application of Corollary 1 requires the explicit knowledge of the matrix P. Moreover, the computation relies on costly symbolic operations involving matrices whose entries are rational functions in z, rather than scalars. For these reasons, this method can only be practical for small values of N. The next section explains how to numerically calculate a rational representation of $g(z)$ out of the first $2N$ moments of the system, without having to know P explicitly.

3 Exact Reconstruction of $g(z)$

We first review Padé approximants and then explain how to employ them to exactly reconstruct $g(z)$. The exposition of Padé approximants in this section is standard. For an in depth treatment, see e.g. [3]. Let $f(z) = \sum_{i \geq 0} c_i z^i$ be a generic a power series in the complex variable z, with a nonzero radius of convergence. For any $n \geq 0$, we let the truncation of f at the n-th term be the polynomial $f_n(z) \triangleq \sum_{i=0}^{n} c_i z^i$. Let us indicate by $o(z^k)$ a generic power series divisible by z^k. Given two power series $f(z)$ and $d(z)$ we write $f(z) = d(z) \bmod z^{n+1}$ iff $f_n = d_n$, or, in other words, if $f(z) - d(z) = o(z^{n+1})$. Polynomials are of course considered as power series with only finitely many nonzero coefficients.

Definition 2 (Padé approximants). *Let $f(z) = \sum_{i \geq 0} c_i z^i$ be a power series in the complex variable z with a nonzero radius of convergence. Given integers $h, k \geq 0$, we say a pair of polynomials in z, say (r, t), is a $[h, k]$-Padé approximant of $f(z)$ if the following holds true, where $n = h + k$.*

$$(a)\ z \nmid t(z) \quad (b)\ \deg(r) \leq h,\ \deg(t) \leq k \quad (c)\ f(z)t(z) = r(z) \bmod z^{n+1}. \tag{6}$$

Seen as a *real* function $r(x)/t(x)$, a Padé approximant is a rational approximation of the function $f(x)$, up to and including the term of degree n of its Taylor of expansion. To see this, first note that equation (6)(c) is equivalent to saying that there exists a power series $k(x)$ such that $f(x)t(x) = r(x) + k(x)x^{n+1}$. As $t(0) \neq 0$, the last equation is equivalent to saying that, in a neighborhood of 0: $f(x) = r(x)/t(x) + x^{n+1}k(x)/t(x)$. This equation is equivalent to saying that the Taylor expansion of $r(x)/t(x)$ from $x = 0$, truncated at the n-th term, coincides with[2] $f_n(x)$, that is: $r(x)/t(x) = \sum_{i=0}^{n} c_i x^i + o(x^{n+1})$. In case f is itself a rational function, Padé approximants provide us with a method to actually find an *exact* representation of it, given only sufficiently many coefficients c_i, as we will see shortly. We first state a result about uniqueness of Padé approximants; its proof follows from easy manipulations on rational functions (or see [3, Th.1.1].)

Proposition 1. *Let (r, t) and (p, q) be two $[h, k]$-Padé approximants of f. Then they are the same as a function, in the sense that $r(z)/t(z) = p(z)/q(z)$.*

Given any power series $f(z)$, it is possible to compute a $[h, k]$-Padé approximant of it as follows. Here we assume for simplicity that $h \leq k$; the case $h > k$ does not interest us, and can be anyway treated with minor notational changes. Also, in view of condition (6)(a) of the definition of Padé approximant, without loss of generality we will restrict ourselves to the case where $t(0) = 1$, that is, the constant coefficient of t is always taken to be 1. Assume $n + 1$ ($n = h + k$) coefficients $c_0, c_1, ..., c_n$ of f are given. Arrange the coefficients from h through $h+k-1 = n-1$ to form a $k \times k$ matrix C as described next,

---

[2] To see this, observe that, for $0 \leq j \leq n$, by equating the j-th derivatives of the left and right hand side of (6)(c), one obtains that $f^{(j)}(x) = (r(x)/t(x))^{(j)} + o(x^{n-j+1})$, so that $c_j = f^{(j)}(0)/j! = (1/j!)(r(0)/t(0))^{(j)}$. Also note that the Taylor expansion of $r(x)/t(x)$ at $x = 0$ exists, as $t(0) \neq 0$.

where $c_l \overset{\triangle}{=} 0$ for indices $l < 0$. Let $r(z) = \alpha_h z^h + \cdots + \alpha_0$ and $t(z) = \beta_k z^k + \cdots + \beta_1 z + 1$ be two polynomials, for generic vectors of coefficients $\alpha = (\alpha_0, ..., \alpha_h)^T$ and $\beta = (\beta_1, ..., \beta_k)^T$. Assume (r, t) is a $[h, k]$-Padé approximant of f. Then we can equate coefficients of like powers on the left- and right-hand side of (6)(c).

$$C = \begin{bmatrix} c_h & c_{h-1} & \cdots & c_{h-k+1} \\ c_{h+1} & c_h & \cdots & c_{h-k+2} \\ & & & \vdots \\ c_{h+k-1} & c_{h+k-2} & \cdots & c_h \end{bmatrix}$$

In particular, coefficients from $h+1$ through $n = h+k$ are 0 on the right, thus coefficients on the left must satisfy the following, for $\gamma \overset{\triangle}{=} (-c_{h+1}, ..., -c_{h+k})^T$:

$$C\beta = \gamma. \tag{7}$$

On the other hand, assume (7), seen as a system of equations in the unknowns β, has a solution. Then by taking α given by:

$$\alpha = \tilde{C}\beta' \tag{8}$$

where $\beta' = (1, \beta_1, ..., \beta_h)^T$ and $\tilde{C}$ is the $(h+1) \times (h+1)$ lower-triangular matrix $\tilde{C}$ given on the right. We see that (6)(c) is satisfied by (r, t).

$$\tilde{C} = \begin{bmatrix} c_0 & & & \\ c_1 & c_0 & & \\ & & \vdots & \\ c_h & c_{h-1} & \cdots & c_1 & c_0 \end{bmatrix}.$$

Of course also (6)(a) and (6)(b) are satisfied. Therefore (r, t), as given by α and β, is a $[h, k]$-Padé approximant of f. In other words, we have shown the following.

Proposition 2. *A $[h, k]$-Padé approximant for f exists if and only if the system of equations (7) in the unknowns β has a solution. If it exists, the coefficients β and α for t and r are given, respectively, by a solution of (7) and by (8).*

Note that the procedure outlined above to reconstruct a $[h, k]$-Padé approximant takes $O(k^3 + h^2) = O(n^3)$ operations and $O(n^2)$ storage. Let us now come back to our Markov chain. Theorem 1 and Proposition 1 ensure that $g(z)$ coincides, as a function, with its $[N-1, N-1]$-Padé approximant. Using the above outlined method, one can reconstruct the rational form of $g(z)$, provided one knows (an upper bound on) N and the coefficients $a_0, ..., a_{2N-2}$. The latter can in principle be computed by a power iteration method: $a_0 = e_N^T e_1$ and $a_i = e_N^T P^{i-1} \tilde{e}_1$ for $i \geq 1$. For this, it is sufficient to obtain a black-box access to the function $u \mapsto Pu$, which is compatible with an on-the-fly implementation. A numerical example illustrating the method is reported in the full paper [4].

While dispensing with symbolic computations and the explicit knowledge of P, the method described in this section still suffers from drawbacks that confine its application to small-to-moderate values of N. Indeed, the solution of the linear system (7) has a time complexity of $O(N^3)$. Although this can be improved using known techniques from numerical linear algebra (see [4]), the real problem here is that the explicit computation of $2N$ moments a_j is in practice very costly and numerically instable, and should be avoided. Moreover, it is clear that a consistent gain in efficiency can only be obtained by accepting some degree of approximation in the computed solution.

4 Discussion: Approximating $g(z)$

Suppose that, possibly due to limited computational resources, we have access only to a limited number, say $m + 1$, of g's moments a_i. Or, more generally, we can access the transition relation of the Markov chain - the mapping $u \mapsto Pu$ - only a limited number m of times. Typically, we can only afford $m \ll N$. The resulting information may be not sufficient to recover an exact representation of $g(z)$. Then the problem becomes finding a a good approximating function $\hat{g}(z)$ of $g(z)$. "Good" here means at least that $g(z) - \hat{g}(z) = o(z^{m+1})$; but, even more than that, $\hat{g}(z)$ should approximate well $g(z)$ near $z = 1$, as we are mainly inter-

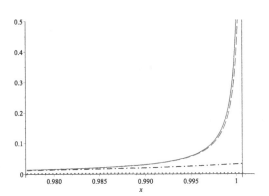

Fig. 1. Plots of $g_{10}(x)$ (dotted), of $g_{100}(x)$ (dash-dotted), of the $[1, 1]$-Padé approximant $\hat{g}(x) = 2x/(-5995x + 6000)$ (dashed) and of $g(x)$ (solid), for $Nasty(6, 10^{-3})$, near $x = 1$. Here $g(z)$ has a pole at $z = 1/\lambda$ with $\lambda \approx 0.9993$.

ested in evaluating $g(1), g'(1)$ and so on, see (2). The first, obvious attempt is to consider a polynomial approximation: $\hat{g}(z) = g_m(z) = \sum_{i=0}^{m} a_i z^i$. This is the truncation of $g(z)$ at the term of degree m.

Unfortunately, such a $\hat{g}(z)$ might be a very bad approximation of $g(z)$. The reason is that the rational representation $r(z)/t(z)$ of $g(z)$ may have a pole near[3] $z = 1$: that is, there can be a $z_0 \in \mathbb{C}$ such that $|z_0 - 1| \approx 0$ and $\lim_{z \to z_0} |r(z)/t(z)| = +\infty$. Then, as z approaches 1 from its convergence zone, $g(z)$ becomes extremely fast growing, and essentially impossible to approximate by means of a polynomial function, as polynomials have no finite poles.

As stated by Corollary 1, the pole of smallest modulus of $g(z)$, which determines its radius of convergence R, is of the form $z_0 = 1/\lambda$, for some subdominant eigenvalue λ of P, that is an eigenvalue with $|\lambda| < 1$. If 1 is "badly separated" from λ, that is if $|\lambda - 1| \approx 0$, the truncated sums $\sum_{i \leq m} a_i$ will converge very slowly to p_{reach}, as m grows. In this respect, a rational approximation $\hat{g}(z) = \frac{\hat{r}(z)}{\hat{t}(z)}$ can perform much better. Indeed, $\hat{t}(z)$ can be chosen so as to have a root near z_0. This in essence is what Padé approximation achieves. When building a $[h, k]$-Padé approximant with $h + k \leq m$, the same amount of information used to build the polynomial $g_m(z)$ above - the first $m + 1$ moments of $g(z)$ - is used to "guess" an approximation of z_0, that becomes a root of $\hat{t}(z)$ (this aspect will be further discussed in the next section, see Remark 1.) The benefit of rational over polynomial approximation is qualitatively illustrated by the plots in Fig. 1.

---

[3] Note that the rational function $r(z)/t(z)$, while coinciding with $g(z)$ within the disk $|z| < R$, will also be defined outside this disk.

It is well-known [5] that that the bad separation phenomenon (subdominant eigenvalues close to 1) occurs if there is a cluster of states that are strongly coupled with one another, but nearly uncoupled with other states in the chain, like $1, ..., n - 2$ in $Nasty(n, \delta)$ (see [4].) In the next section we explore an effective way of building rational approximations $\hat{g}(z) = \hat{r}(z)/\hat{t}(z)$.

5   Approximation of $g(z)$ via a Projection Method

The general idea of a projection method is as follows. Consider P as a linear map acting on the N-dimensional space $\mathbb{R}^N$. We identify a m-dimensional subspace, $\mathcal{K}_m$, and then consider the projection of P onto this space, say $\hat{P}$: this is our low-dimensional approximation of the original system. Here, $m \ll N$: practically m will be of the order of tens or hundreds. Formally, consider an integer $m \geq 1$ and the *Krylov subspace* of $\mathbb{R}^N$

$$\mathcal{K}_m(P, \tilde{e}_1) \triangleq \text{span}\{\tilde{e}_1, P\tilde{e}_1, P^2\tilde{e}_1, ..., P^{m-1}\tilde{e}_1\}$$

abbreviated as $\mathcal{K}_m$ in what follows. Now take any orthonormal basis of $\mathcal{K}_m$ and arrange the corresponding column vectors into a $N \times m$ matrix, $V_m = [v_1, ..., v_m]$. Note that orthonormality means that $V_m^T V_m = I_m$. We can consider the projection of P, seen as an application $\mathbb{R}^N \to \mathbb{R}^N$, onto $\mathcal{K}_m$. The representation of this application restricted to $\mathcal{K}_m$, in the basis V_m, is given by the $m \times m$ matrix

$$H_m = V_m^T P V_m . \tag{9}$$

(The matrix H_m will play the role played by $\hat{P}$ in the above informal description.) Note that if m is large enough then $\mathcal{K}_m$ will be a P-invariant subspace of $\mathbb{R}^N$, that is $P\mathcal{K}_m \subseteq \mathcal{K}_m$ [4].

Theorem 3. *Let $m \geq 1$. Consider the function, defined in a neighborhood of the origin*
$$\hat{g}(z) \triangleq a_0 + z \cdot (e_N^T V_m)(I_m - zH_m)^{-1}(V_m^T \tilde{e}_1) . \tag{10}$$
Then $\hat{g}(z)$ is a rational function of type $[m, m]$ and $g(z) - \hat{g}(z) = o(z^{m+1})$. Moreover, if 1 is not an eigenvalue of H_m, then $\hat{g}(z)$ is defined in a neighborhood of $z = 1$. Finally, if $\mathcal{K}_m$ is P-invariant, then $g(z) = \hat{g}(z)$.

Remark 1. Note from (10) that, while $\hat{g}(z)$ is a rational function of type $[m, m]$, it is not guaranteed that $g(z) - \hat{g}(z) = o(z^{2m+1})$. Thus $\hat{g}(z)$ is not, in general, a Padé approximant, but only a weaker *Padé-type* approximant. Comparing (5) and (10), we further see that how well $\hat{g}(z)$ approximates $g(z)$ depends on how well the polynomial $\det(I_m - zH_m)$ approximates the polynomial $\det(I - zP)$. We have already noted that the roots of these polynomials are the reciprocals of nonzero eigenvalues of H_m and P, respectively. It is known for general matrices P that, already for small values of m, H_m's eigenvalues - known as *Ritz values* in the literature - tend to be excellent approximations of the eigenvalues of P that are at the extreme of the spectrum, that is, those of either large or small modulus. The details and nature of such approximation are not yet fully understood,

---

[4] In particular, it is sufficient to take any $m \geq v$, where $v \leq N$ is the degree of the minimal polynomial of P, that is, the monic polynomial p of minimal degree such that $p(P) = 0_{N \times N}$: this is a consequence of the Cayley-Hamilton theorem.

Algorithm 1 Arnoldi based calculation of $\hat{g}(z)$

    Input: $m \geq 1$; a black-box mechanism for computing the function $u \mapsto Pu$
    Output: a triple (a_0, V_m, H_m)
1:    $a_0 = e_N^T e_1$
2:    $v_1 = \tilde{e}_1 / \|\tilde{e}_1\|_2$
3:    **for** $j = 1, 2, ..., m$ **do**
4:        **for** $i = 1, 2, ..., j$ **do** $h_{ij} = (Pv_j, v_i)$
5:        $w_j = Pv_j - \sum_{i=1}^{j} h_{ij} v_i$
6:        $h_{j+1,j} = \|w_j\|_2$
7:        **if** $(h_{j+1,j} = 0) \vee (j = m)$ **then break else** $v_{j+1} = w_j / h_{j+1,j}$
8:    $m = j$
9:    $V_m = [v_1, ..., v_m]$
10:    **return** (a_0, V_m, H_m)

---

but in the specialized literature there is abundant experimental and theoretical evidence about it, see e.g. [8]. In any case, this fact retrospectively sheds light on the somewhat surprising effectiveness of Padé approximation.

Formulae for the derivatives $\hat{g}^{(j)}(x) = \frac{d^j}{dx^j} \hat{g}(x)$ follow from (10) by the familiar rules of derivation, see [4]. We now show that the matrices V_m and H_m can be computed via an effective and numerically stable procedure known as the *Arnoldi process* [2,8]. Arnoldi does not require knowledge of the full matrix P, but only a black-box access to the matrix-vector multiplication function $u \mapsto Pu$, which makes it compatible with an on-the-fly approach. Algorithm 1 works incrementally and, for $j = 1, ..., m$, builds an orthonormal basis of $\mathcal{K}_j$, $V_j = [v_1, ..., v_j]$, and the corresponding projected version of P onto $\mathcal{K}_j$, H_j. The next vector v_{j+1} is built by orthonormalizing Pv_j against the available basis V_j (lines 4–7, which are essentially the Gram-Schmidt orthonormalization.) If this process results in the null vector ($h_{j+1,j} = 0$), then Pv_j is linearly dependent from vectors in V_j, thus the space $\mathcal{K}_j$ is P-invariant, and the main iteration stops.

    The algorithm makes use of the following variables, for $j = 1, ..., m$: the scalars $a_0, a_j \in \mathbb{R}$; vectors $v_j, w_j \in \mathbb{R}^N$; the matrix $H_m \in \mathbb{R}^{m \times m}$ whose nonzero elements are the reals $h_{l,l'}$ for $1 \leq l \leq l' + 1$ and $1 \leq l' \leq m$; the matrix $V_m \in \mathbb{R}^{N \times m}$. In line 4, $(\cdot, \cdot)$ denotes inner product. Of course, Pv_j needs to be computed only once per each j-iteration. Nesting of blocks is defined by indentation. The algorithm can take advantage of a sparse storage scheme. In what follows, we let W be the maximal number of nonzero elements in $P^j \tilde{e}_1$, for any $0 \leq j \leq m$, and B the maximal number of outgoing transitions from any state. Note that $W \cdot B$ is upper bounded by the overall number of transitions. Recall that a square matrix is in upper *Hessenberg* form if all its entries below the main subdiagonal are 0.

Theorem 4. *Let $m \geq 1$ and let (a_0, V_m, H_m) be the output returned by Algorithm 1. Then V_m is an orthonormal basis of $\mathcal{K}_m$ and (9) is satisfied. As a consequence, $\hat{g}(z)$ satisfies (10) given this choice of a_0, V_m, H_m. Moreover, H_m is in upper Hessenberg form and if $h_{m+1,m} = 0$ then $\hat{g}(z) = g(z)$. Assuming $u \mapsto Pu$ can be computed in $O(WB)$ operations* *and $O(W)$ storage, the algorithm takes $O(mWB)$ operations and $O(mW)$ storage to complete.*

A numerical example illustrating Algorithm 1 is reported in the full version [4]. Note that in principle we can calculate $\hat{g}(1)$, and more generally $\hat{g}(x)$ whenever defined for x, directly using the definition (10). However, it is computationally much better to proceed as follows: $\hat{g}(x) = a_0 + e_N^T V_m y$, where y is the (unique) solution of the system $(I_m - xH_m)y = V_m^T \tilde{e}_1 = e_1^{(m)} \cdot \|\tilde{e}_1\|_2$ (here $e_1^{(m)}$ is the first canonical vector of $\mathbb{R}^m$.) Since $(I_m - xH_m)$ is still quasi-triangular (upper Hessenberg), the system above can be solved with $O(m^2)$. The derivatives of g can be computed similarly, see [4]. In the end, via the Arnoldi algorithm 1 (cost $O(mWB)$), we have reduced the computation of all the important properties of the system to the resolution of small quasi-triangular systems, which cost approximately $O(m^2)$, for a small, "affordable" m. Computation of steady-state probabilities and of error control is also discussed in [4].

6   Experiments

We have put an on-the-fly implementation (in Matlab) of Algorithm 1 at work on a few simple probabilistic systems. With two exceptions, the chosen systems are relatively small, but, due to the presence of nearly uncoupled strongly connected components, they exhibit the bad separation phenomenon discussed in Section 4. In each case, the reachability probability of interest is easy to compute analytically, due to the symmetry of the system. The Matlab implementation, as well as the Matlab and PRISM specifications of the considered examples, are available at [4]. We give below an informal description of these systems.

The *Nasty(n, δ)* systems have been introduced in Example 1; here we have fixed $\delta = 10^{-3}$ and considered $n = 10^5, 10^6$. A *Queue(n)* system consists of n queueing processes, each of capacity four, running in parallel. At each time, either an enqueue (prob. 0.1) or a dequeue (prob. 0.9) request arrives, and is served by any process that is able to serve it. In case of global overflow, each process chooses either of two indeterminate error states, and remains there forever. This gives rise to 2^n possible overflow configurations, which are absorbing. The event of interest is that the system eventually reaches one specific such configuration; here the cases $n = 2, 3$ are considered. An *Ising(n)* system consists of n particles, each of which can take on, at each time, either the *up* or the *down* spin value, with a probability depending on how many *up*'s are in system and on a temperature parameter. The *all-up* and *all-down* configurations are absorbing, and the event of interest is that all-up is eventually reached, starting from the equilibrium configuration with $n/2$ particles down and $n/2$ up; here, the cases $n = 6, 8$ are considered. In *Chemical(n)*, a solution is initially composed by $n/2$ reactant pairs of type 1 and $n/2$ reactant pairs of type 2. A reactant pair of one type can get transformed into a pair of the other type, according to a chemical rule obeying the law of mass action – the probability that the reaction occurs depends on the concentration of the reactants. A solution consisting of reactants pairs all of the same type is absorbing. The probability of eventually reaching one specific such solution is sought for; here, the cases $n = 24, 26$ are considered.

Table 1. Results of some experiments. N = number of states; p_{reach} = exact reachability probability. $\hat{p}_{reach}$ = probability returned by the algorithms, truncated at the 4-th digit after decimal point; $\%error = 100 \times |p_{reach} - \hat{p}_{reach}|/p_{reach}$ is the relative error percentage; m = number of iterations; $time$ = time in seconds. For PRISM, default options have been employed except that for m (tweaking other options did not lead to significant improvements); model building time is included in $time$.

System			ALGORITHM 1				PRISM			
Name	N	p_{reach}	$\hat{p}_{reach}$	$\%error$	m	$time$	$\hat{p}_{reach}$	$\%error$	m	$time$
Nasty	10^5	0.5	0.5000	$< 10^{-11}$	5	0.08	0.4999	$< 10^{-4}$	27	5524
	10^6	0.5	0.5000	$< 10^{-10}$	5	0.06	-	-	-	-
Queue	49	0.25	0.2500	$< 10^{-5}$	15	0.92	0.2454	1.81	133176	0.34
	231	0.125	0.1248	0.11	37	11.21	0.0013	98.91	1983765	22.63
Ising	64	0.5	0.5000	$< 10^{-3}$	9	2.06	0.4982	0.34	26433	0.05
	256	0.5	0.4998	<0.10	18	20.58	0.0578	88.42	1257073	8.16
Chemical	25	0.5	0.4994	0.10	18	0.40	1.2×10^{-7}	99.99	2000030	1.55
	27	0.5	0.4937	1.25	20	0.45	7.8×10^{-9}	99.99	2000034	1.96

Table 1 displays the outcomes of these experiments. In all the considered cases, Algorithm 1 returned, in reasonable time, quite accurate results in terms of relative error. For comparison, results obtained with PRISM, a state-of-the-art probabilistic model checker [6], are also included. Results provided by PRISM were reasonably accurate only in three out of eight cases. In one case (*Nasty* with 10^6 states), PRISM was not able to build the model after about one hour. In five out of eight cases, the Matlab implementation of Algorithm 1 run anyway faster than PRISM.

7   Conclusion, Further and Related Work

We have demonstrated that, in the analysis of Markov chains, generating functions provide a bridge toward Padé approximation theory that is useful both at a conceptual and at a technical level. Direct extensions of the method to full temporal logics, such as LTL, seem worth studying, as well as extensions to richer models, like continuous Markov chains or Markov Decision Processes. Another potential field of application is the time-bounded analysis of infinite-state system.

Methods and tools based on a symbolic representations of the entire state-space, such as PRISM [6], can take great advantage of the presence of system regularities, as e.g. induced by massive interleaving: in those cases, an on-the-fly approach cannot be expected to match the performance of these tools. Nevertheless, as indicated by our small-scale experiment, the presented methodology turns out to be helpful in situations of bad eigenvalues separation, and/or when, for whatever reason, building the entire system's model turns out to be not feasible. In numerical linear algebra, numerous worksare devoted to the experimental evaluation of projective methods applied to Markov chains, see e.g. [7] and references therein. These works focus on the calculation of steady-state probabilities and no connection to generating functions is made. Further discussion on related work, including use of Padé approximation in Engineering [1], can be found in [4].

References

1. Antoulas, A.C.: Approximation of Large-scale Dynamical Systems. SIAM (2005)
2. Arnoldi, W.E.: The principle of minimized iterations in the solution of the matrix eigenvalue problem. Quarterly of Applied Mathematics **9**, 17–29 (1951)
3. Baker Jr., G.: Essentials of Padé Approximants. Academic Press (1975)
4. Boreale, M.: Full version of the present paper, Matlab and PRISM code. http://rap.dsi.unifi.it/~boreale/papers/GFviaKrylov.rar
5. Hartfiel, D.J., Meyer, C.D.: On the structure of stochastic matrices with a subdominant eigenvalue near 1. Linear Algebra Appl. **272**, 193–203 (1998)
6. Kwiatkowska, M., Norman, G., Parker, D.: PRISM 4.0: verification of probabilistic real-time systems. In: Gopalakrishnan, G., Qadeer, S. (eds.) CAV 2011. LNCS, vol. 6806, pp. 585–591. Springer, Heidelberg (2011)
7. Philippe, B., Saad, Y., Stewart, W.J.: Numerical Methods in Markov Chain Modelling. Operations Research **40**, 1156–1179 (1996)
8. Saad, Y.: Iterative methods for sparse linear systems. SIAM (2003)
9. Wilf, H.S.: Generatingfunctionology, 2/e. Academic Press (1994)

On Reducing Linearizability to State Reachability

Ahmed Bouajjani[1], Michael Emmi[2], Constantin Enea[1], and Jad Hamza[1]($\boxtimes$)

[1] LIAFA, Université Paris Diderot, Paris, France
[2] IMDEA Software Institute, Madrid, Spain
jhamza@liafa.univ-paris-diderot.fr

Abstract. Efficient implementations of atomic objects such as concurrent stacks and queues are especially susceptible to programming errors, and necessitate automatic verification. Unfortunately their correctness criteria — linearizability with respect to given ADT specifications — are hard to verify. Even on classes of implementations where the usual temporal safety properties like control-state reachability are decidable, linearizability is undecidable.

In this work we demonstrate that verifying linearizability for certain *fixed* ADT specifications is reducible to control-state reachability, despite being harder for *arbitrary* ADTs. We effectuate this reduction for several of the most popular atomic objects. This reduction yields the first decidability results for verification without bounding the number of concurrent threads. Furthermore, it enables the application of existing safety-verification tools to linearizability verification.

1 Introduction

Efficient implementations of atomic objects such as concurrent queues and stacks are difficult to get right. Their complexity arises from the conflicting design requirements of maximizing efficiency/concurrency with preserving the appearance of atomic behavior. Their correctness is captured by *observational refinement*, which assures that all behaviors of programs using these efficient implementations would also be possible were the atomic reference implementations used instead. Linearizability [12], being an equivalent property [5, 8], is the predominant proof technique: one shows that each concurrent execution has a linearization which is a valid sequential execution according to a specification, given by an abstract data type (ADT) or reference implementation.

Verifying automatically[1] that all executions of a given implementation are linearizable with respect to a given ADT is an undecidable problem [3], even on the typical classes of implementations for which the usual temporal safety properties are decidable, e.g., on finite-shared-memory programs where each thread is a finite-state machine. What makes linearization harder than typical temporal safety properties like control-state reachability is the existential quantification of a valid linearization per execution.

In this work we demonstrate that verifying linearizability for certain *fixed* ADTs is reducible to control-state reachability, despite being harder for *arbitrary* ADTs. We believe that fixing the ADT parameter of the verification problem is justified, since in

---

This work is supported in part by the VECOLIB project (ANR-14-CE28-0018).

[1] Without programmer annotation — see Section 6 for further discussion.

© Springer-Verlag Berlin Heidelberg 2015
M.M. Halldórsson et al. (Eds.): ICALP 2015, Part II, LNCS 9135, pp. 95–107, 2015.
DOI: 10.1007/978-3-662-47666-6_8

practice, there are few ADTs for which specialized concurrent implementations have been developed. We provide a methodology for carrying out this reduction, and instantiate it on four ADTs: the atomic queue, stack, register, and mutex.

Our reduction to control-state reachability holds on any class of implementations which is closed under intersection with regular languages[2] and which is *data independent* — informally, that implementations can perform only read and write operations on the data values passed as method arguments. From the ADT in question, our approach relies on expressing its violations as a finite union of regular languages.

In our methodology, we express the atomic object specifications using inductive rules to facilitate the incremental construction of valid executions. For instance in our atomic queue specification, one rule specifies that a dequeue operation returning empty can be inserted in any execution, so long as each preceding enqueue has a corresponding dequeue, also preceding the inserted empty-dequeue. This form of inductive rule enables a locality to the reasoning of linearizability violations.

Intuitively, first we prove that a sequential execution is invalid if and only if some subsequence could not have been produced by one of the rules. Under certain conditions this result extends to concurrent executions: an execution is not linearizable if and only if some projection of its operations cannot be linearized to a sequence produced by one of the rules. We thus correlate the finite set of inductive rules with a finite set of classes of non-linearizable concurrent executions. We then demonstrate that each of these classes of non-linearizable executions is regular, which characterizes the violations of a given ADT as a finite union of regular languages. The fact that these classes of non-linearizable executions can be encoded as regular languages is somewhat surprising since the number of data values, and thus alphabet symbols, is, a priori, unbounded. Our encoding thus relies on the aforementioned *data independence* property.

To complete the reduction to control-state reachability, we show that linearizability is equivalent to the emptiness of the language intersection between the implementation and finite union of regular violations. When the implementation is a finite-shared-memory program with finite-state threads, this reduces to the coverability problem for Petri nets, which is decidable, and EXPSPACE-complete.

To summarize, our contributions are:

– a generic reduction from linearizability to control-state reachability,
– its application to the atomic queue, stack, register, and mutex ADTs,
– the methodology enabling this reduction, which can be reused on other ADTs, and
– the first decidability results for linearizability without bounding the number of concurrent threads.

Besides yielding novel decidability results, our reduction paves the way for the application of existing safety-verification tools to linearizability verification.

Section 2 outlines basic definitions. Section 3 describes a methodology for inductive definitions of data structure specifications. In Section 4 we identify conditions under which linearizability can be reduced to control-state reachability, and demonstrate that

---

[2] We consider languages of well-formed method call and return actions, e.g., for which each return has a matching call.

typical atomic objects satisfy these conditions. Finally, we prove decidability of linearizability for finite-shared-memory programs with finite-state threads in Section 5. Proofs to technical results appear in the appendix.

2   Linearizability

We fix a (possibly infinite) set $\mathbb{D}$ of *data values*, and a finite set $\mathbb{M}$ of *methods*. We consider that methods have exactly one argument, or one return value. Return values are transformed into argument values for uniformity.[3] In order to differentiate methods taking an argument (e.g., the Enq method which inserts a value into a queue) from the other methods, we identify a subset $\mathbb{M}_{in} \subseteq \mathbb{M}$ of *input* methods which do take an argument. A *method event* is composed of a method $m \in \mathbb{M}$ and a data value $x \in \mathbb{D}$, and is denoted $m(x)$. We define the *concatenation* of method-event sequences $u \cdot v$ in the usual way, and ϵ denotes the empty sequence.

Definition 1. *A* sequential execution *is a sequence of method events,*

The projection $u_{|D}$ of a sequential execution u to a subset $D \subseteq \mathbb{D}$ of data values is obtained from u by erasing all method events with a data value not in D. The set of projections of u is denoted $\mathsf{proj}(u)$. We write $u \smallsetminus x$ for the projection $u_{|\mathbb{D} \setminus \{x\}}$.

Example 1. The projection $Enq(1)Enq(2)Deq(1)Enq(3)Deq(2)Deq(3) \smallsetminus 1$ is equal to $Enq(2)Enq(3)Deq(2)Deq(3)$.

We also fix an arbitrary infinite set $\mathbb{O}$ of operation (identifiers). A *call action* is composed of a method $m \in \mathbb{M}$, a data value $x \in \mathbb{D}$, an operation $o \in \mathbb{O}$, and is denoted $\mathtt{call}_o\ m(x)$. Similarly, a *return action* is denoted $\mathtt{ret}_o\ m(x)$. The operation o is used to match return actions to their call actions.

Definition 2. *A (concurrent) execution e is a sequence of call and return actions which satisfy a well-formedness property: every return has a call action before it in e, using the same tuple m, x, o, and an operation o can be used only twice in e, once in a call action, and once in a return action.*

Example 2. The sequence $\mathtt{call}_{o_1}\ Enq(7) \cdot \mathtt{call}_{o_2}\ Enq(4) \cdot \mathtt{ret}_{o_1}\ Enq(7) \cdot \mathtt{ret}_{o_2}\ Enq(4)$ is an execution, while $\mathtt{call}_{o_1}\ Enq(7) \cdot \mathtt{call}_{o_2}\ Enq(4) \cdot \mathtt{ret}_{o_1}\ Enq(7) \cdot \mathtt{ret}_{o_1}\ Enq(4)$ and $\mathtt{call}_{o_1}\ Enq(7) \cdot \mathtt{ret}_{o_1}\ Enq(7) \cdot \mathtt{ret}_{o_2}\ Enq(4)$ are not.

Definition 3. *An implementation $\mathcal{I}$ is a set of (concurrent) executions.*

Implementations represent libraries whose methods are called by external programs, giving rise to the following closure properties [5]. In the following, c denotes a call action, r denotes a return action, a denotes any action, and e, e' denote executions.

---

[3] Method return values are guessed nondeterministically, and validated at return points. This can be handled using the $\mathtt{assume}$ statements of typical formal specification languages, which only admit executions satisfying a given predicate. The argument value for methods without argument or return values, or with fixed argument/return values, is ignored.

- Programs can call library methods at any point in time:
  $e \cdot e' \in \mathcal{I}$ implies $e \cdot c \cdot e' \in \mathcal{I}$ so long as $e \cdot c \cdot e'$ is well formed.
- Calls can be made earlier:
  $e \cdot a \cdot c \cdot e' \in \mathcal{I}$ implies $e \cdot c \cdot a \cdot e' \in \mathcal{I}$.
- Returns been made later:
  $e \cdot r \cdot a \cdot e' \in \mathcal{I}$ implies $e \cdot a \cdot r \cdot e' \in \mathcal{I}$.

Intuitively, these properties hold because call and return actions are not visible to the other threads which are running in parallel.

For the remainder of this work, we consider only *completed* executions, where each call action has a corresponding return action. This simplification is sound when implementation methods can always make progress in isolation [11]: formally, for any execution e with pending operations, there exists an execution e' obtained by extending e only with the return actions of the pending operations of e. Intuitively this means that methods can always return without any help from outside threads, avoiding deadlock.

We simply reasoning on executions by abstracting them into *histories*.

Definition 4. *A history is a labeled partial order $(O, <, l)$ with $O \subseteq \mathbb{O}$ and $l : O \to \mathbb{M} \times \mathbb{D}$.*

The order $<$ is called the *happens-before relation*, and we say that o_1 *happens before* o_2 when $o_1 < o_2$. Since histories arise from executions, their happens-before relations are *interval orders* [5]: for distinct o_1, o_2, o_3, o_4, if $o_1 < o_2$ and $o_3 < o_4$ then either $o_1 < o_4$, or $o_3 < o_2$. Intuitively, this comes from the fact that concurrent threads share a notion of global time. $\mathbb{D}_h \subseteq \mathbb{D}$ denotes the set of data values appearing in h.

The *history of an execution e* is defined as $(O, <, l)$ where:

- O is the set of operations which appear in e,
- $o_1 < o_2$ iff the return action of o_1 is before the call action of o_2 in e,
- an operation o occurring in a call action $\texttt{call}_o \, m(x)$ is labeled by $m(x)$.

Example 3. The history of the execution $\texttt{call}_{o_1} \, Enq(7) \cdot \texttt{call}_{o_2} \, Enq(4) \cdot \texttt{ret}_{o_1} \, Enq(7) \cdot \texttt{ret}_{o_2} \, Enq(4)$ is $(\{o_1, o_2\}, <, l)$ with $l(o_1) = Enq(7)$, $l(o_2) = Enq(4)$, and with $<$ being the empty order relation, since o_1 and o_2 *overlap*.

Let $h = (O, <, l)$ be a history and u a sequential execution of length n. We say that h is *linearizable with respect to u*, denoted $h \sqsubseteq u$, if there is a bijection $f : O \to \{1, \ldots, n\}$ s.t.

- if $o_1 < o_2$ then $f(o_1) < f(o_2)$,
- the method event at position $f(o)$ in u is $l(o)$.

Definition 5. *A history h is* linearizable *with respect to a set $\mathcal{S}$ of sequential executions, denoted $h \sqsubseteq \mathcal{S}$, if there exists $u \in \mathcal{S}$ such that $h \sqsubseteq u$.*

A set of histories H is *linearizable* with respect to $\mathcal{S}$, denoted $H \sqsubseteq \mathcal{S}$ if $h \sqsubseteq \mathcal{S}$ for all $h \in H$. We extend these definitions to executions according to their histories.

A sequential execution u is said to be *differentiated* if, for all input methods $m \in \mathbb{M}_{in}$, and every $x \in \mathbb{D}$, there is at most one method event $m(x)$ in u. The subset of differentiated sequential executions of a set $\mathcal{S}$ is denoted by $\mathcal{S}_{\neq}$. The definition extends to (sets of) executions and histories. For instance, an execution is differentiated if for all input methods $m \in \mathbb{M}_{in}$ and every $x \in \mathbb{D}$, there is at most one call action $\mathtt{call}_o\ m(x)$.

Example 4. $\mathtt{call}_{o_1}\ Enq(7) \cdot \mathtt{call}_{o_2}\ Enq(7) \cdot \mathtt{ret}_{o_1}\ Enq(7) \cdot \mathtt{ret}_{o_2}\ Enq(7)$ is not differentiated, as there are two call actions with the same input method (Enq) and the same data value.

A *renaming* r is a function from $\mathbb{D}$ to $\mathbb{D}$. Given a sequential execution (resp., execution or history) u, we denote by $r(u)$ the sequential execution (resp., execution or history) obtained from u by replacing every data value x by $r(x)$.

Definition 6. *The set of sequential executions (resp., executions or histories) $\mathcal{S}$ is* data independent *if:*

- *for all $u \in \mathcal{S}$, there exists $u' \in \mathcal{S}_{\neq}$, and a renaming r such that $u = r(u')$,*
- *for all $u \in \mathcal{S}$ and for all renaming r, $r(u) \in \mathcal{S}$.*

When checking that a data-independent implementation $\mathcal{I}$ is linearizable with respect to a data-independent specification $\mathcal{S}$, it is enough to do so for differentiated executions [1]. Thus, in the remainder of the paper, we focus on characterizing linearizability for differentiated executions, rather than arbitrary ones.

Lemma 1 (Abdulla et al. [1]). *A data-independent implementation $\mathcal{I}$ is linearizable with respect to a data-independent specification $\mathcal{S}$, if and only if $\mathcal{I}_{\neq}$ is linearizable with respect to $\mathcal{S}_{\neq}$.*

3   Inductively-Defined Data Structures

A *data structure* $\mathcal{S}$ is given syntactically as an ordered sequence of rules $R_1, \ldots, R_n$, each of the form $u_1 \cdot u_2 \cdots u_k \in \mathcal{S} \wedge Guard(u_1, \ldots, u_k) \Rightarrow Expr(u_1, \ldots, u_k) \in \mathcal{S}$, where the variables u_i are interpreted over method-event sequences, and

- $Guard(u_1, \ldots, u_k)$ is a conjunction of conditions on $u_1, \ldots, u_k$ with atoms
  - $u_i \in M^*$ $(M \subseteq \mathbb{M})$
  - $\mathsf{matched}(m, u_i)$
- $Expr(u_1, \ldots, u_k)$ is an *expression* $E = a_1 \cdot a_2 \cdots a_l$ where
  - $u_1, \ldots, u_k$ appear in that order, exactly once, in E,
  - each a_i is either some u_j, a method m, or a Kleene closure m^* $(m \in \mathbb{M})$,
  - a method $m \in \mathbb{M}$ appears at most once in E.

We allow k to be 0 for base rules, such as $\epsilon \in \mathcal{S}$.

A condition $u_i \in M^*$ $(M \subseteq \mathbb{M})$ is satisfied when the methods used in u_i are all in M. The predicate $\mathsf{matched}(m, u_i)$ is satisfied when, for every method event $m(x)$ in u_i, there exists another method event in u_i with the same data value x.

Given a sequential execution $u = u_1 \cdot \ldots \cdot u_k$ and an expression $E = Expr(u_1, \ldots, u_k)$, we define $[\![E]\!]$ as the set of sequential executions which can be obtained from E by replacing the methods m by a method event $m(x)$ and the Kleene closures m^* by 0 or more method events $m(x)$. All method events must use the same data value $x \in \mathbb{D}$.

A rule $R \equiv u_1 \cdot u_2 \cdots u_k \in \mathcal{S} \wedge Guard(u_1, \ldots, u_k) \Rightarrow Expr(u_1, \ldots, u_k) \in \mathcal{S}$ is applied to a sequential execution w to obtain a new sequential execution w' from the set:

$$\bigcup_{\substack{w = w_1 \cdot w_2 \cdots w_k \wedge \\ Guard(w_1, \ldots, w_k)}} [\![Expr(w_1, \ldots, w_k)]\!]$$

We denote this $w \xrightarrow{R} w'$. The set of sequential executions $[\![\mathcal{S}]\!] = [\![R_1, \ldots, R_n]\!]$ is then defined as the set of sequential executions w which can be derived from the empty word:

$$\epsilon = w_0 \xrightarrow{R_{i_1}} w_1 \xrightarrow{R_{i_2}} w_2 \ldots \xrightarrow{R_{i_p}} w_p = w,$$

where $i_1, \ldots, i_p$ is a non-decreasing sequence of integers from $\{1 \ldots, n\}$. This means that the rules must be applied in order, and each rule can be applied 0 or several times.

Below we give inductive definitions for the atomic queue and stack data structures. Other data structures such as atomic registers and mutexes also have inductive definitions, as demonstrated in the appendix.

Example 5. The queue has a method Enq to add an element to the data structure, and a method Deq to remove the elements in a FIFO order. The method $DeqEmpty$ can only return when the queue is empty (its parameter is not used). The only input method is Enq. Formally, Queue is defined by the rules R_0, R_{Enq}, R_{EnqDeq} and $R_{DeqEmpty}$.

$$R_0 \equiv \epsilon \in \mathsf{Queue}$$
$$R_{Enq} \equiv u \in \mathsf{Queue} \wedge u \in Enq^* \Rightarrow u \cdot Enq \in \mathsf{Queue}$$
$$R_{EnqDeq} \equiv u \cdot v \in \mathsf{Queue} \wedge u \in Enq^* \wedge v \in \{Enq, Deq\}^* \Rightarrow Enq \cdot u \cdot Deq \cdot v \in \mathsf{Queue}$$
$$R_{DeqEmpty} \equiv u \cdot v \in \mathsf{Queue} \wedge \mathsf{matched}(Enq, u) \Rightarrow u \cdot DeqEmpty \cdot v \in \mathsf{Queue}$$

One derivation for Queue is:

$$\epsilon \in \mathsf{Queue} \xrightarrow{R_{EnqDeq}} Enq(1) \cdot Deq(1) \in \mathsf{Queue}$$
$$\xrightarrow{R_{EnqDeq}} Enq(2) \cdot Enq(1) \cdot Deq(2) \cdot Deq(1) \in \mathsf{Queue}$$
$$\xrightarrow{R_{EnqDeq}} Enq(3) \cdot Deq(3) \cdot Enq(2) \cdot Enq(1) \cdot Deq(2) \cdot Deq(1) \in \mathsf{Queue}$$
$$\xrightarrow{R_{DeqEmpty}} Enq(3) \cdot Deq(3) \cdot DeqEmpty \cdot Enq(2) \cdot Enq(1) \cdot Deq(2) \cdot Deq(1) \in \mathsf{Queue}$$

Similarly, Stack is composed of the rules $R_0, R_{PushPop}, R_{Push}, R_{PopEmpty}$.

$$R_0 \equiv \epsilon \in \mathsf{Stack}$$
$$R_{PushPop} \equiv u \cdot v \in \mathsf{Stack} \wedge \mathsf{matched}(Push, u) \wedge \mathsf{matched}(Push, v) \wedge u, v \in \{Push, Pop\}^*$$
$$\Rightarrow Push \cdot u \cdot Pop \cdot v \in \mathsf{Stack}$$
$$R_{Push} \equiv u \cdot v \in \mathsf{Stack} \wedge \mathsf{matched}(Push, u) \wedge u, v \in \{Push, Pop\}^* \Rightarrow u \cdot Push \cdot v \in \mathsf{Stack}$$
$$R_{PopEmpty} \equiv u \cdot v \in \mathsf{Stack} \wedge \mathsf{matched}(Push, u) \Rightarrow u \cdot PopEmpty \cdot v \in \mathsf{Stack}$$

We assume that the rules defining a data structure S satisfy a non-ambiguity property stating that the last step in deriving a sequential execution in $[\![S]\!]$ is unique and it can be effectively determined. Since we are interested in characterizing the linearizations of a history and its projections, this property is extended to permutations of projections of sequential executions which are admitted by S. Thus, we assume that the rules defining a data structure are *non-ambiguous*, that is:

- for all $u \in [\![S]\!]$, there exists a unique rule, denoted by $\mathtt{last}(u)$, that can be used as the last step to derive u, i.e., for every sequence of rules $R_{i_1}, \ldots, R_{i_n}$ leading to u, $R_{i_n} = \mathtt{last}(u)$. For $u \notin [\![S]\!]$, $\mathtt{last}(u)$ is also defined but can be arbitrary, as there is no derivation for u.
- if $\mathtt{last}(u) = R_i$, then for every permutation $u' \in [\![S]\!]$ of a projection of u, $\mathtt{last}(u') = R_j$ with $j \leq i$. If u' is a permutation of u, then $\mathtt{last}(u') = R_i$.

Given a (completed) history h, all the u such that $h \sqsubseteq u$ are permutations of one another. The last condition of non-ambiguity thus enables us to extend the function $\mathtt{last}$ to histories: $\mathtt{last}(h)$ is defined as $\mathtt{last}(u)$ where u is any sequential execution such that $h \sqsubseteq u$. We say that $\mathtt{last}(h)$ is the rule *corresponding* to h.

Example 6. For Queue, we define $\mathtt{last}$ for a sequential execution u as follows:

- if u contains a $DeqEmpty$ operation, $\mathtt{last}(u) = R_{DeqEmpty}$,
- else if u contains a Deq operation, $\mathtt{last}(u) = R_{EnqDeq}$,
- else if u contains only Enq's, $\mathtt{last}(u) = R_{Enq}$,
- else (if u is empty), $\mathtt{last}(u) = R_0$.

Since the conditions we use to define $\mathtt{last}$ are closed under permutations, we get that for any permutation u_2 of u, $\mathtt{last}(u) = \mathtt{last}(u_2)$, and $\mathtt{last}$ can be extended to histories. Therefore, the rules $R_0, R_{EnqDeq}, R_{DeqEmpty}$ are non-ambiguous.

4   Reducing Linearizability to State Reachability

Our end goal for this section is to show that for any data-independent implementation $\mathcal{I}$, and any specification S satisfying several conditions defined in the following, there exists a computable finite-state automaton $\mathcal{A}$ (over call and return actions) such that:

$$\mathcal{I} \sqsubseteq S \iff \mathcal{I} \cap \mathcal{A} = \emptyset$$

Then, given a model of $\mathcal{I}$, the linearizability of $\mathcal{I}$ is reduced to checking emptiness of the synchronized product between the model of $\mathcal{I}$ and $\mathcal{A}$. The automaton $\mathcal{A}$ represents (a subset of the) executions which are not linearizable with respect to S.

The first step in proving our result is to show that, under some conditions, we can partition the concurrent executions which are not linearizable with respect to S into a finite number of classes. Intuitively, each non-linearizable execution must correspond to a violation for one of the rules in the definition of S.

We identify a property, which we call *step-by-step linearizability*, which is sufficient to obtain this characterization. Intuitively, step-by-step linearizability enables us to build a linearization for an execution e incrementally, using linearizations of projections of e.

The second step is to show that, for each class of violations (i.e., with respect to a specific rule R_i), we can build a regular automaton $\mathcal{A}_i$ such that: a) when restricted to well-formed executions, $\mathcal{A}_i$ recognizes a subset of this class; b) each non-linearizable execution has a corresponding execution, obtained by data independence, accepted by $\mathcal{A}_i$. If such an automaton exists, we say that R_i is *co-regular* (formally defined later in this section).

We prove that, provided these two properties hold, we have the equivalence mentioned above, by defining $\mathcal{A}$ as the union of the $\mathcal{A}_i$'s built for each rule R_i.

4.1   Reduction to a Finite Number of Classes of Violations

Our goal here is to give a characterization of the sequential executions which belong to a data structure, as well as to give a characterization of the concurrent executions which are linearizable with respect to the data structure. This characterization enables us to classify the linearization violations into a finite number of classes.

Our characterization relies heavily on the fact that the data structures we consider are *closed under projection*, i.e., for all $u \in \mathcal{S}, D \subseteq \mathbb{D}$, we have $u_{|D} \in \mathcal{S}$. The reason for this is that the guards used in the inductive rules are closed under projection.

Lemma 2. *Any data structure $\mathcal{S}$ defined in our framework is closed under projection.*

A sequential execution u is said to *match* a rule R with conditions $Guard$ if there exist a data value x and sequential executions $u_1, \ldots, u_k$ such that u can be written as $[\![Expr(u_1, \ldots, u_k)]\!]$, where x is the data value used for the method events, and such that $Guard(u_1, \ldots, u_k)$ holds. We call x the *witness* of the decomposition. We denote by MR the set of sequential executions which match R, and we call it the *matching set* of R.

Example 7. MR_{EnqDeq} is the set of sequential executions of the form $Enq(x) \cdot u \cdot Deq(x) \cdot v$ for some $x \in \mathbb{D}$, and with $u \in Enq^*$.

Lemma 3. *Let $\mathcal{S} = R_1, \ldots, R_n$ be a data structure and u a differentiated sequential execution. Then,*

$$u \in \mathcal{S} \iff \mathsf{proj}(u) \subseteq \bigcup_{i \in \{1,\ldots,n\}} MR_i$$

This characterization enables us to get rid of the recursion, so that we only have to check non-recursive properties. We want a similar lemma to characterize $e \sqsubseteq \mathcal{S}$ for an execution e. This is where we introduce the notion of *step-by-step linearizability*, as the lemma will hold under this condition.

Definition 7. *A data structure $\mathcal{S} = R_1, \ldots, R_n$ is said be to* step-by-step linearizable *if for any differentiated execution e, if e is linearizable w.r.t. MR_i with witness x, we have:*

$$e \smallsetminus x \sqsubseteq [\![R_1, \ldots, R_i]\!] \implies e \sqsubseteq [\![R_1, \ldots, R_i]\!]$$

This notion applies to the usual data structures, as shown by the following lemma. The generic schema we use is the following: we let $u' \in [\![R_1, \ldots, R_i]\!]$ be a sequential execution such that $e \setminus x \sqsubseteq u'$ and build a graph G from u', whose acyclicity implies that $e \sqsubseteq [\![R_1, \ldots, R_i]\!]$. Then, we show that we can always choose u' so that G is acyclic.

Lemma 4. Queue, Stack, Register, *and* Mutex *are step-by-step linearizable.*

Intuitively, step-by-step linearizability will help us prove the right-to-left direction of Lemma 5 by allowing us to build a linearization for e incrementally, from the linearizations of projections of e.

Lemma 5. *Let S be a data structure with rules $R_1, \ldots, R_n$. Let e be a differentiated execution. If S is step-by-step linearizable, we have (for any j):*

$$e \sqsubseteq [\![R_1, \ldots, R_j]\!] \iff \text{proj}(e) \sqsubseteq \bigcup_{i \leq j} MR_i$$

Thanks to Lemma 5, if we're looking for an execution e which is not linearizable w.r.t. some data-structure S, we must prove that $\text{proj}(e) \not\sqsubseteq \bigcup_i MR_i$, i.e., we must find a projection $e' \in \text{proj}(e)$ which is not linearizable with respect to any MR_i ($e' \not\sqsubseteq \bigcup_i MR_i$).

This is challenging as it is difficult to check that an execution is not linearizable w.r.t. a union of sets simultaneously. Using non-ambiguity, we simplify this check by making it more modular, so that we only have to check one set MR_i at a time.

Lemma 6. *Let S be a data structure with rules $R_1, \ldots, R_n$. Let e be a differentiated execution. If S is step-by-step linearizable, we have:*

$$e \sqsubseteq S \iff \forall e' \in \text{proj}(e). \, e' \sqsubseteq MR \text{ where } R = \texttt{last}(e')$$

Lemma 6 gives us the finite kind of violations that we mentioned in the beginning of the section. More precisely, if we negate both sides of the equivalence, we have: $e \not\sqsubseteq S \iff \exists e' \in \text{proj}(e). \, e' \not\sqsubseteq MR$. This means that whenever an execution is not linearizable w.r.t. S, there can be only finitely reasons, namely there must exist a projection which is not linearizable w.r.t. the matching set of its corresponding rule.

4.2 Regularity of Each Class of Violations

Our goal is now to construct, for each R, an automaton $\mathcal{A}$ which recognizes (a subset of) the executions e, which have a projection e' such that $e' \not\sqsubseteq MR$. More precisely, we want the following property.

Definition 8. *A rule R is said to be* co-regular *if we can build an automaton $\mathcal{A}$ such that, for any data-independent implementation $\mathcal{I}$, we have:*

$$\mathcal{A} \cap \mathcal{I} \neq \emptyset \iff \exists e \in \mathcal{I}_{\neq}, e' \in \text{proj}(e). \, \texttt{last}(e') = R \wedge e' \not\sqsubseteq MR$$

A data structure S is co-regular *if all of its rules are co-regular.*

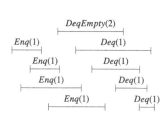 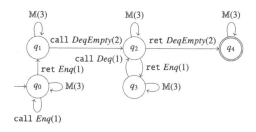

Fig. 1. A four-pair $R_{DeqEmpty}$ violation. The extended version of this paper demonstrates that this pattern with arbitrarily-many pairs is regular.

Fig. 2. An automaton recognizing $R_{DeqEmpty}$ violations, for which the queue is non-empty, with data value 1, for the span of $DeqEmpty$. We assume all call $Enq(1)$ actions occur initially without loss of generality due to implementations' closure properties.

Formally, the alphabet of $\mathcal{A}$ is $\{\texttt{call } m(x) \mid m \in \mathbb{M}, x \in D\} \cup \{\texttt{ret } m(x) \mid m \in\}$ $\mathbb{M}, x \in D$ for a finite subset $D \subseteq \mathbb{D}$. The automaton doesn't read operation identifiers, thus, when taking the intersection with $\mathcal{I}$, we ignore them.

Lemma 7. Queue, Stack, Register, *and* Mutex *are co-regular.*

Proof. To illustrate this lemma, we sketch the proof for the rule $R_{DeqEmpty}$ of Queue. The complete proof of the lemma can be found in the extended version of this paper.

We prove in the extended version that a history has a projection such that $\texttt{last}(h') = R_{DeqEmpty}$ and $h' \not\sqsubseteq MR_{DeqEmpty}$ if and only if it has a $DeqEmpty$ operation which is *covered* by other operations, as depicted in Fig. 1. The automaton $\mathcal{A}_{R_{DeqEmpty}}$ in Fig. 2 recognizes such violations.

Let $\mathcal{I}$ be any data-independent implementation. We show that

$$\mathcal{A}_{R_{DeqEmpty}} \cap \mathcal{I} \neq \emptyset \iff \exists e \in \mathcal{I}_{\neq}, e' \in \mathsf{proj}(e). \texttt{last}(e') = R_{DeqEmpty} \wedge e' \not\sqsubseteq MR_{DeqEm}$$

($\Rightarrow$) Let $e \in \mathcal{I}$ be an execution which is accepted by $\mathcal{A}_{R_{DeqEmpty}}$. By data independence, let $e_{\neq} \in \mathcal{I}$ and r a renaming such that $e = r(e_{\neq})$. Let $d_1, \ldots, d_m$ be the data values which are mapped to value 1 by r.

Let d be the data value which is mapped to value 2 by r. Let o the $DeqEmpty$ operation with data value d. By construction of the automaton we can prove that o is covered by $d_1, \ldots, d_m$, and conclude that h has a projection such that $\texttt{last}(h') = R_{DeqEmpty}$ and $h' \not\sqsubseteq MR_{DeqEmpty}$.

($\Leftarrow$) Let $e_{\neq} \in \mathcal{I}_{\neq}$ such that there is a projection e' such that $\texttt{last}(e') = R_{DeqEmpty}$ and $e' \not\sqsubseteq MR_{DeqEmpty}$. Let $d_1, \ldots, d_m$ be the data values given by the $R_{DeqEmpty}$-characterization in the full version of this paper, and let d be the data value corresponding to the $DeqEmpty$ operation.

Without loss of generality, we can always choose the cycle so that $Enq(d_i)$ doesn't happen before $Deq(d_{i-2})$ (if it does, drop d_{i-1}).

Let r be the renaming which maps $d_1, \ldots, d_m$ to 1, d to 2, and all other values to 3. Let $e = r(e_{\neq})$. The execution e can be recognized by automaton $\mathcal{A}_{R_{DeqEmpty}}$, and belongs to $\mathcal{I}$ by data independence.

When we have a data structure which is both step-by-step linearizable and co-regular, we can make a linear time reduction from the verification of linearizability with respect to $\mathcal{S}$ to a reachability problem, as illustrated in Theorem 1.

Theorem 1. *Let $\mathcal{S}$ be a step-by-step linearizable and co-regular data structure and let $\mathcal{I}$ be a data-independent implementation. There exists a regular automaton $\mathcal{A}$ such that:*

$$\mathcal{I} \sqsubseteq \mathcal{S} \iff \mathcal{I} \cap \mathcal{A} = \emptyset$$

5 Decidability and Complexity of Linearizability

Theorem 1 implies that the linearizability problem with respect to any step-by-step linearizable and co-regular specification is decidable for any data-independent implementation for which checking the emptiness of the intersection with finite-state automata is decidable. Here, we give a class $\mathcal{C}$ of data-independent implementations for which the latter problem, and thus linearizability, is decidable.

Each method of an implementation in $\mathcal{C}$ manipulates a finite number of local variables which store Boolean values, or data values from $\mathbb{D}$. Methods communicate through a finite number of shared variables that also store Boolean values, or data values from $\mathbb{D}$. Data values may be assigned, but never used in program predicates (e.g., in the conditions of if and while statements) so as to ensure data independence. This class captures typical implementations, or finite-state abstractions thereof, e.g., obtained via predicate abstraction.

Let $\mathcal{I}$ be an implementation from class $\mathcal{C}$. The automata $\mathcal{A}$ constructed in the proof of Lemma 7 use only data values 1, 2, and 3. Checking emptiness of $\mathcal{I} \cap \mathcal{A}$ is thus equivalent to checking emptiness of $\mathcal{I}_3 \cap \mathcal{A}$ with the three-valued implementation $\mathcal{I}_3 = \{e \in \mathcal{I} \mid e = e_{|\{1,2,3\}}\}$. The set $\mathcal{I}_3$ can be represented by a Petri net since bounding data values allows us to represent each thread with a finite-state machine. Intuitively, each token in the Petri net represents another thread. The number of threads can be unbounded since the number of tokens can. Places count the number of threads in each control location, which includes a local-variable valuation. Each shared variable also has one place per value to store its current valuation.

Emptiness of the intersection with regular automata reduces to the EXPSPACE-complete coverability problem for Petri nets. Limiting verification to a bounded number of threads lowers the complexity of coverability to PSPACE [7]. The hardness part of Theorem 2 comes from the hardness of state reachability in finite-state concurrent programs.

Theorem 2. *Verifying linearizability of an implementation in $\mathcal{C}$ with respect to a step-by-step linearizable and co-regular specification is PSPACE-complete for a fixed number of threads, and EXPSPACE-complete otherwise.*

6 Related Work

Several works investigate the theoretical limits of linearizability verification. Verifying a single execution against an arbitrary ADT specification is NP-complete [9]. Verifying all executions of a finite-state implementation against an arbitrary ADT specification (given as a regular language) is EXPSPACE-complete when program threads are bounded [2,10], and undecidable otherwise [3].

Existing automated methods for proving linearizability of an atomic object implementation are also based on reductions to safety verification [1,11,13]. Vafeiadis [13] considers implementations where operation's *linearization points* are fixed to particular source-code locations. Essentially, this approach instruments the implementation with ghost variables simulating the ADT specification at linearization points. This approach is incomplete since not all implementations have fixed linearization points. Aspect-oriented proofs [11] reduce linearizability to the verification of four simpler safety properties. However, this approach has only been applied to queues, and has not produced a fully automated and complete proof technique. Dodds et al. [6] prove linearizability of stack implementations with an automated proof assistant. Their approach does not lead to full automation however, e.g., by reduction to safety verification.

7 Conclusion

We have demonstrated a linear-time reduction from linearizability for fixed ADT specifications to control-state reachability, and the application of this reduction to atomic queues, stacks, registers, and mutexes. Besides yielding novel decidability results, our reduction enables the use of existing safety-verification tools for linearizability. While this work only applies the reduction to these four objects, our methodology also applies to other typical atomic objects including semaphores and sets. Although this methodology currently does not capture priority queues, which are not data independent, we believe our approach can be extended to include them. We leave this for future work.

References

1. Abdulla, P.A., Haziza, F., Holík, L., Jonsson, B., Rezine, A.: An integrated specification and verification technique for highly concurrent data structures. In: Piterman, N., Smolka, S.A. (eds.) TACAS 2013 (ETAPS 2013). LNCS, vol. 7795, pp. 324–338. Springer, Heidelberg (2013)
2. Alur, R., McMillan, K.L., Peled, D.: Model-checking of correctness conditions for concurrent objects. Inf. Comput. **160**(1–2) (2000)
3. Bouajjani, A., Emmi, M., Enea, C., Hamza, J.: Verifying concurrent programs against sequential specifications. In: Felleisen, M., Gardner, P. (eds.) ESOP 2013. LNCS, vol. 7792, pp. 290–309. Springer, Heidelberg (2013)
5. Bouajjani, A., Emmi, M., Enea, C., Hamza, J.: Tractable refinement checking for concurrent objects. In: POPL 2015. ACM (2015)
5. Bouajjani, A., Emmi, M., Enea, C., Hamza, J.: On reducing linearizability to state reachability. CoRR, abs/1502.06882 (2015). arxiv.org/abs/1502.06882
6. Dodds, M., Haas, A., Kirsch, C.M.: A scalable, correct time-stamped stack. In: POPL 2015. ACM (2015)
7. Esparza, J.: Decidability and complexity of petri net problems—an introduction. Lectures on Petri Nets I: Basic Models. Springer, Heidelberg (1998)
8. Filipovic, I., O'Hearn, P.W., Rinetzky, N., Yang, H.: Abstraction for concurrent objects. Theor. Comput. Sci. **411**(51–52) (2010)

9. Gibbons, P.B., Korach, E.: Testing shared memories. SIAM J. Comput. **26**(4) (1997)
10. Hamza, J.: On the complexity of linearizability. CoRR, abs/1410.5000 (2014). arxiv.org/abs/1410.5000
11. Henzinger, T.A., Sezgin, A., Vafeiadis, V.: Aspect-oriented linearizability proofs. In: D'Argenio, P.R., Melgratti, H. (eds.) CONCUR 2013 – Concurrency Theory. LNCS, vol. 8052, pp. 242–256. Springer, Heidelberg (2013)
12. Herlihy, M., Wing, J.M.: Linearizability: A correctness condition for concurrent objects. ACM Trans. Program. Lang. Syst. **12**(3) (1990)
13. Vafeiadis, V.: Automatically proving linearizability. In: Touili, T., Cook, B., Jackson, P. (eds.) CAV 2010. LNCS, vol. 6174, pp. 450–464. Springer, Heidelberg (2010)

The Complexity of Synthesis from Probabilistic Components

Krishnendu Chatterjee[1]([⊠]), Laurent Doyen[2], and Moshe Y. Vardi[3]

[1] IST Austria, Klosterneuburg, Austria
krish.chat@gmail.com
[2] CNRS and LSV, ENS Cachan, Cachan, France
[3] Rice University, Houston, USA

Abstract. The synthesis problem asks for the automatic construction of a system from its specification. In the traditional setting, the system is "constructed from scratch" rather than composed from reusable components. However, this is rare in practice, and almost every non-trivial software system relies heavily on the use of libraries of reusable components. Recently, Lustig and Vardi introduced *dataflow* and *controlflow* synthesis from libraries of reusable components. They proved that dataflow synthesis is undecidable, while controlflow synthesis is decidable. The problem of controlflow synthesis from libraries of *probabilistic components* was considered by Nain, Lustig and Vardi, and was shown to be decidable for qualitative analysis (that asks that the specification be satisfied with probability 1). Our main contribution for controlflow synthesis from probabilistic components is to establish better complexity bounds for the qualitative analysis problem, and to show that the more general quantitative problem is undecidable. For the qualitative analysis, we show that the problem (i) is EXPTIME-complete when the specification is given as a deterministic parity word automaton, improving the previously known 2EXPTIME upper bound; and (ii) belongs to UP ∩ coUP and is parity-games hard, when the specification is given directly as a parity condition on the components, improving the previously known EXPTIME upper bound.

1 Introduction

Synthesis from existing components. Reactive systems (hardware or software) are rarely built from scratch, but are mostly developed based on existing components. A component might be used in the design of multiple systems, e.g., function libraries, web APIs, and ASICs. The construction of systems from existing reusable components is an active research direction, with several important

This research was supported by Austrian Science Fund (FWF) Grant No P23499-N23, FWF NFN Grant No S11407-N23 (SHiNE), ERC Start grant (279307: Graph Games), EU FP7 Project Cassting, NSF grants CNS 1049862 and CCF-1139011, by NSF Expeditions in Computing project "ExCAPE: Expeditions in Computer Augmented Program Engineering", by BSF grant 9800096, and by gift from Intel.

M.M. Halldórsson et al. (Eds.): ICALP 2015, Part II, LNCS 9135, pp. 108–120, 2015.
DOI: 10.1007/978-3-662-47666-6_9

works, such as component-based construction [17], interface-based design [11]. The synthesis problem asks for the automated construction of a system given a logical specification. For example, in LTL (linear-time temporal logic) synthesis, the specification is given in LTL and the reactive system to be constructed is a finite-state transducer [16]. In the traditional LTL synthesis setting, the system is "constructed from scratch" rather than "composed" from existing components. Recently, Lustig and Vardi introduced the study of synthesis from reusable or existing components [13].

The model and types of composition. The precise mathematical model for the components and their composition is an important concern (and we refer the reader to [13,14] for a detailed discussion). As a basic model for a component, following [13], we abstract away the precise details of the component and model a component as a *transducer*, i.e., a finite-state machine with outputs. Transducers constitute a canonical model for reactive components, abstracting away internal architecture and focusing on modeling input/output behavior. In [13], two models of composition were studied, namely, *dataflow* composition, where the output of one component becomes an input to another component, and *controlflow* composition, where at every point of time the control resides within a single component. The synthesis problem for dataflow composition was shown to be undecidable, and the controlflow composition to be decidable [13].

Synthesis for probabilistic components. While [13] considered synthesis for non-probabilistic components which was extended to non-probabilistic recursive state components in [10], the study of synthesis for controlflow composition for probabilistic components was considered in [14]. Probabilistic components are transducers with a probabilistic transition function, that corresponds to modeling systems where there is probabilistic uncertainty about the effect of input actions. Thus the controlflow composition for probabilistic transducers aims at the construction of reliable systems from unreliable components. There is a rich literature about verification and analysis of such systems, cf. [3,9,12,18,19].

Qualitative and quantitative analysis. There are two probabilistic notions of correctness, namely, the *qualitative* criterion that requires the satisfaction of the specification with probability 1, and the more general *quantitative* criterion that requires the satisfaction of the specification with probability at least η, given $0 < \eta \leq 1$.

The synthesis questions and previous results. In the synthesis problem for controlflow composition, the input is a library $\mathcal{L}$ of probabilistic components, and we consider specifications given as parity conditions (that allow us to consider all ω-regular properties, which can express all commonly used specifications in verification). The *qualitative (resp., quantitative) realizability* and synthesis problems ask whether there exists a *finite* system S built from the components in $\mathcal{L}$, such that, regardless of the input provided by the external environment, the traces generated by the system S satisfy the specification with probability 1 (resp., probability at least η). Each component in the library can be instantiated an arbitrary number of times in the construction and there is no a-priori bound on the size of the system obtained. The way the specification is provided gives rise

Table 1. Computational complexity of synthesis from probabilistic components

	Qualitative		Quantitative	
	Our Results	Previous Results	Our Results	Previous Results
Embedded Parity	UP ∩ coUP (Parity-games hard)	EXPTIME	UP ∩ coUP (Parity-games hard)	Open
DPW Specifications	EXPTIME-c	2EXPTIME	Undecidable	Open

to two different problems: (i) *embedded parity realizability*, where the specification is given in the form of a parity index on the states of the components; and (ii) *DPW realizability*, where the specification is given as a separate deterministic parity word automaton (DPW). The results of [14] established the decidability of the qualitative realizability problem, namely, in EXPTIME for the embedded parity realizability problem and 2EXPTIME for the DPW realizability problem. The exact complexity of the qualitative problem and the decidability and complexity of the quantitative problem were left open, which we solve.

Our contributions. Our main contributions are (summarized in Table 1):

1. We show that both the qualitative and quantitative realizability problems for embedded parity lie in UP ∩ coUP, and even the qualitative problem is at least parity-games hard.
2. We show that the qualitative realizability problem for DPW specifications is EXPTIME-complete (an exponential improvement over the previous 2EXPTIME result). Finally, we show that the quantitative realizability problem for DPW specifications is undecidable.

Technical contributions. Our two main technical contributions are as follows. First, for the realizability of embedded parity specifications, while the most natural interpretation of the problem is as a partial-observation stochastic game (as also considered in [14]), we show that the problem can be reduced in polynomial time to a perfect-information stochastic game. Second, for the realizability of DPW specifications, we consider partial-observation stochastic games where the strategies correspond to a correct composition that defines, given an exit state of a component, to which component the control should be transferred. Since we aim at a finite-state system, we need to consider strategies with *finite memory*, and since the control flow is deterministic, we need to consider *pure* (non-randomized) strategies. Moreover, since the composition must be independent of the internal executions of the components, we need to consider strategies with *stuttering* invariance. We present polynomial-time reductions for stutter-invariant strategies to games with standard observation-based strategies. Our results establish optimal complexity results for qualitative analysis of partial-observation stochastic games with finite-memory stutter-invariant strategies, which are of independent interest. Finally, we present a polynomial reduction of the qualitative realizability for DPW specifications to partial-observation stochastic games with stutter-invariant strategies and obtain the EXPTIME-complete result. Detailed proofs are available in [6].

2   Definitions

Transducers. In this section we present the definitions of deterministic and probabilistic transducers, and strategies for them.

Deterministic transducers. A *deterministic transducer* is a tuple $B = \langle \Sigma_I, \Sigma_O, Q, q_0, \delta, L \rangle$, where: Σ_I is a finite input alphabet, Σ_O is a finite output alphabet, Q is a finite set of states, $q_0 \in Q$ is an initial state, $L : Q \to \Sigma_O$ is an output function labeling states with output letters, and $\delta : Q \times \Sigma_I \to Q$ is a transition function.

Probabilistic transducers. Let $\mathcal{D}(X)$ denote the set of all probability distributions on set X. A *probabilistic transducer* is a tuple $\mathcal{T} = \langle \Sigma_I, \Sigma_O, Q, q_0, \delta, F, L \rangle$, where: Σ_I is a finite input alphabet, Σ_O is a finite output alphabet, Q is a finite set of states, $q_0 \in Q$ is an initial state, $\delta : (Q \setminus F) \times \Sigma_I \to \mathcal{D}(Q)$ is a probabilistic transition function, $F \subseteq Q$ is a set of exit states, and $L : Q \to \Sigma_O$ is an output function labeling states with output letters. Note that there are no transitions out of an exit state. If F is empty, we say $\mathcal{T}$ is a probabilistic transducer without exits. Note that deterministic transducers can be viewed as a special case of probabilistic transducers.

Strategies for transducers, and probability measure. Given a probabilistic transducer $M = \langle \Sigma_I, \Sigma_O, Q, q_0, \delta, F, L \rangle$, a *strategy* for M is a function $f : Q^+ \to \mathcal{D}(\Sigma_I)$ that probabilistically chooses an input for each finite sequence of states. We denote by $\mathcal{F}$ the set of all strategies. A strategy is memoryless if the choice depends only on the last state in the sequence. A memoryless strategy can be written as a function $g : Q \to \mathcal{D}(\Sigma_I)$. A strategy is *pure* if the choice is deterministic. A pure strategy is a function $h : Q^+ \to \Sigma_I$, and a memoryless and pure strategy is a function $h : Q \to \Sigma_I$. A strategy f along with a probabilistic transducer M, with set of states Q, induces a probability distribution on Q^ω, denoted μ_f (see [6] for detailed definition).

Library of Components. A *library* is a finite set of probabilistic transducers that share the same input and output alphabets. Each transducer in the library is called a *component type*. Given a finite set of directions D, we say a library $\mathcal{L}$ has width D, if each component type in the library has exactly $|D|$ exit states. Since we can always add dummy unreachable exit states to any component, we assume, w.l.o.g., that all libraries have an associated width, usually denoted D. In the context of a particular component type, we often refer to elements of D as exits, and subsets of D as sets of exits.

Controlflow Composition from Libraries. We first informally describe the notion of controlflow composition of components from a library as defined in [14]. The components in the composition take turns interacting with the environment, and at each point in time, exactly one component is active. When the active component reaches an exit state, control is transferred to some other component. Thus, to define a controlflow composition, it suffices to name the components used and describe how control should be transferred between them. We use a deterministic transducer to define the transfer of control. Each library component can be used multiple times in a composition, and we treat these occurrences as distinct *component instances*. We emphasize that the composition can contain potentially arbitrarily many instances of each component type inside it. Thus, the size of the composition, a priori, is not bounded. Note that our notion of composition is *static*, where the components called are determined before run time, rather than *dynamic*, where the calls are determined during run time.

Let $\mathcal{L}$ be a library of width D. A *composer* over $\mathcal{L}$ is a deterministic transducer $C = \langle D, \mathcal{L}, \mathcal{M}, \mathsf{M}_0, \Delta, \lambda \rangle$. Here $\mathcal{M}$ is an arbitrary finite set of states. There is no bound on the size of $\mathcal{M}$. Each $\mathsf{M}_i \in \mathcal{M}$ is a component from $\mathcal{L}$ and $\lambda(\mathsf{M}_i) \in \mathcal{L}$ is the type of M_i. We use the following notational convention for component instances and names: the upright letter M always denotes component names (i.e., states of a composer) and the italicized letter M always denotes the corresponding component instances (i.e., elements of $\mathcal{L}$). Further, for notational convenience we often write M_i directly instead of $\lambda(\mathsf{M}_i)$. Note that while each M_i is distinct, the corresponding components M_i need not be distinct. Each composer defines a unique composition over components from $\mathcal{L}$. The current state of the composer corresponds to the component that is in control. The transition function Δ describes how to transfer control between components: $\Delta(\mathsf{M}, i) = \mathsf{M}'$ denotes that when the composition is in the ith final state of component M it moves to the start state of component M'. A composer can be viewed as an implicit representation of a composition. An explicit definition is presented in the full version of the paper. Note that the composition, denoted $\mathcal{T}_C$, is a probabilistic transducer without exits. When the composition $\mathcal{T}_C$ is in a state $\langle q, i \rangle$ corresponding to a non-exit state q of component M_i, it behaves like M_i. When the composition is in a state $\langle q_f, i \rangle$ corresponding to an exit state q_f of component M_i, the control is transferred to the start state of another component as determined by the transition function of the composer. Thus, at each point in time, only one component is active and interacting with the environment.

Parity objectives and values. An *index function* for a transducer is a function that assigns a natural number, called a priority index, to each state of the transducer. An index function α defines a parity objective Φ_α that is the subset of Q^ω consisting of the set of infinite sequence of states such that the largest priority that is visited infinitely often is even. Given a probabilistic transducer $\mathcal{T}$ and a parity objective Φ, the value of the probabilistic transducer for the objective, denoted as $\mathsf{val}(\mathcal{T}, \Phi)$, is $\inf_{f \in \mathcal{F}} \mu_f(\Phi)$, i.e., it is the minimal probability with which the parity objective is satisfied over all strategies in the transducer.

The Synthesis Questions. We consider two types of synthesis questions for controlflow composition. In the first problem (synthesis for embedded parity) the parity objective is specified directly on the state space of the library components, and in the second problem (synthesis from DPW specifications) the parity objective is specified by a separate deterministic parity automaton.

Synthesis for Embedded Parity. We first consider an index function that associates to each state of the components in the library a priority, and a specification defined as a parity condition over the sequence of visited states.

Exit Control Relation. Given a library $\mathcal{L}$ of width D, an *exit control relation* is a set $R \subseteq D \times \mathcal{L}$. We say that a composer $C = \langle D, \mathcal{L}, \mathcal{M}, \mathsf{M}_0, \Delta, \lambda \rangle$ is *compatible* with R, if the following holds: for all $\mathsf{M}, \mathsf{M}' \in \mathcal{M}$ and $i \in D$, if $\Delta(\mathsf{M}, i) = \mathsf{M}'$ then $\langle i, M' \rangle \in R$. Thus, each element of R can be viewed as a constraint on how the composer is allowed to connect components. An exit control relation is *non-blocking* if for every $i \in D$ there exists a component $M \in \mathcal{L}$ such that $\langle i, M \rangle \in R$ (i.e., every exit has at least one possible component for the next choice). For technical convenience we only consider non-blocking exit control relations.

Definition 1 (Embedded Parity Realizability and Synthesis). *Consider a library $\mathcal{L}$ of width D, an exit control relation R for $\mathcal{L}$, and an index function α for the components in $\mathcal{L}$ that defines the parity objective Φ_α. The* qualitative *(resp., quantitative) realizability problem for controlflow composition with embedded parity* is to decide whether there exists a composer C over $\mathcal{L}$, such that C is compatible with R, and $\mathsf{val}(\mathcal{T}_C, \Phi_\alpha) = 1$ *(resp., $\mathsf{val}(\mathcal{T}_C, \Phi_\alpha) \geq \eta$, given rational $\eta \in (0,1)$). A witness composer for the qualitative (resp., quantitative) problem is called an* almost-sure *(resp., η-optimal) composer. The corresponding embedded parity synthesis problems are to find such a composer C if it exists.*

Synthesis for DPW specifications. A *deterministic parity word automaton (DPW)* is a deterministic transducer where the labeling function is an index function that defines a parity objective. Given a DPW A, every word (infinite sequence of input letters) induces a run of the automaton, which is an infinite sequence of states, and the word is accepted if the run satisfies the parity objective. The language L_A of a DPW A is the set of words accepted by A. Let A be a deterministic parity automaton (DPW), M be a probabilistic transducer and $\mathcal{L}$ be a library of components. We say A is a *monitor* for M (resp. $\mathcal{L}$) if the input alphabet of A is the same as the output alphabet of M (resp. $\mathcal{L}$). Let A be a monitor for M and let L_A be the language accepted by A. The value of M for A, denoted as $\mathsf{val}(M, A)$, is $\inf_{f \in \mathcal{F}} \mu_f(\lambda^{-1}(L_A))$. The compatibility of the composer with an exit control relation can be encoded in the DPW (w.l.o.g., two distinct exit states do not have the same output).

Definition 2 (DPW realizability and synthesis). *Consider a library $\mathcal{L}$ and a DPW A that is a monitor for $\mathcal{L}$. The* qualitative *(resp., quantitative) realizability problem for controlflow composition with DPW specifications* is to decide whether there exists a composer C over $\mathcal{L}$, such that $\mathsf{val}(\mathcal{T}_C, A) = 1$ *(resp., $\mathsf{val}(\mathcal{T}_C, A) \geq \eta$, given rational $\eta \in (0,1)$). A witness composer for the qualitative (resp., quantitative) problem is called an* almost-sure *(resp., η-optimal) composer. The corresponding* DPW probabilistic synthesis problems *are to find such a composer C if it exists.*

Remark 1. The realizability problem for libraries with components can be viewed as a 2-player partial-observation stochastic parity game [14]. Informally, the game can be described as follows: the two players are the composer C and the environment E. The C player chooses components and the E player chooses sequence of inputs in the components chosen by C. However, C cannot see the inputs of E or even the length of the time inside a component. At the start C chooses a component M from the library $\mathcal{L}$. The turn passes to E, who chooses a sequence of inputs, inducing a probability distribution over paths in M from its start state to some exit x in D. The turn then passes to C, which must choose some component M' in $\mathcal{L}$ and pass the turn to E and so on. As C cannot see the moves made by E inside M, the choice of C cannot be based on the run in M, but only on the exit induced by the inputs selected by E and previous moves made by C. So C must choose the same next component M' for different runs that reach the same exit of M.

3   Realizability with Embedded Parity

We establish the results for the complexity of realizability with embedded parity. While the natural interpretation of the embedded parity problem is a partial-observation game, we show how the problem can be interpreted as a perfect-information stochastic game.

3.1   Perfect-Information Stochastic Parity Games

Perfect-Information Stochastic Games. A perfect-information stochastic game consists of a tuple $G = \langle S, S_1, S_2, A_1, A_2, \delta^G \rangle$, where S is a finite set of states partitioned into player-1 states (namely, S_1) and player-2 states (namely S_2), A_1 (resp., A_2) is the set of actions for player 1 (resp., player 2), and $\delta^G : (S_1 \times A_1) \cup (S_2 \times A_2) \to \mathcal{D}(S)$ is a probabilistic transition function that given a player-1 state and player-1 action, or a player-2 state and a player-2 action gives a probability distribution over the successor states. If the transition function is *deterministic* (that is the codomain of δ^G is S instead of $\mathcal{D}(S)$), then the game is a perfect-information deterministic game.

Plays and Strategies. A *play* is an infinite sequence of state-action pairs $\langle s_0 a_0 s_1 a_1 \ldots \rangle$ such that for all $j \geq 0$ we have that if $s_j \in S_i$ for $i \in \{1, 2\}$, then $a_j \in A_i$ and $\delta^G(s_j, a_j)(s_{j+1}) > 0$. A strategy is a recipe for a player to choose actions to extend finite prefixes of plays. Formally, a strategy π for player 1 is a function $\pi : S^* \cdot S_1 \to \mathcal{D}(A_1)$ that given a finite sequence of visited states gives a probability distribution over the actions (to be chosen next). A *pure* strategy chooses a deterministic action, i.e., is a function $\pi : S^* \cdot S_1 \to A_1$. A pure memoryless strategy is a pure strategy that does not depend on the finite prefix of the play but only on the current state, i.e., is a function $\pi : S_1 \to A_1$. The definitions for player-2 strategies τ are analogous. We denote by Π (resp., Π^{PM}) the set of all (resp., all pure memoryless) strategies for player 1, and analogously Γ (resp., Γ^{PM} for player 2). Given strategies $\pi \in \Pi$ and $\tau \in \Gamma$, and a starting state s, there is a unique probability measure over events (i.e., measurable subsets of S^ω), denoted by $\mathbb{P}_s^{\pi, \tau}(\cdot)$.

Finite-Memory Strategies. A pure player-1 strategy uses *finite-memory* if it can be encoded by a transducer $\langle \mathfrak{M}, m_0, \pi_u, \pi_n \rangle$ where $\mathfrak{M}$ is a finite set (the memory of the strategy), $m_0 \in \mathfrak{M}$ is the initial memory value, $\pi_u : \mathfrak{M} \times S \to \mathfrak{M}$ is the memory-update function, and $\pi_n : \mathfrak{M} \to A_1$ is the next-action function. Note that a finite-memory strategy is a deterministic transducer with input alphabet

S, output alphabet A_1, where π_u is the deterministic transition function, and π_n is the output labeling function. Formally, $\langle \mathfrak{M}, m_0, \pi_u, \pi_n \rangle$ defines the strategy π such that $\pi(\rho) = \pi_n(\widehat{\pi}_u(m_0, \rho))$ for all $\rho \in S^+$, where $\widehat{\pi}_u$ extends π_u to sequences of states as expected.

Parity Objectives, Almost-Sure, and Value Problem. Given a perfect-information stochastic game, a parity objective is defined by an index function α on the state space. Given a strategy π, the value of the strategy in a state s of the game G with parity objective Φ_α, denoted by $\mathsf{val}^G(\pi, \Phi_\alpha)(s)$, is the infimum of the probabilities among all player-2 strategies, i.e., $\mathsf{val}^G(\pi, \Phi_\alpha)(s) = \inf_{\tau \in \Gamma} \mathbb{P}_s^{\pi, \tau}(\Phi_\alpha)$. The value of the game is $\mathsf{val}^G(\Phi_\alpha)(s) = \sup_{\pi \in \Pi} \mathsf{val}^G(\pi, \Phi_\alpha)(s)$. A strategy π is almost-sure winning from s if $\mathsf{val}^G(\pi, \Phi_\alpha)(s) = 1$. Theorem 1 summarizes results about perfect-information games.

Theorem 1. *The following assertions hold* [1,4,7,8]: *(1) (Complexity). The quantitative decision problem (of whether* $\mathsf{val}^G(\Phi_\alpha) \geq \eta$, *given rational* $\eta \in (0, 1]$*) for perfect-information stochastic parity games lies in* UP $\cap$ coUP. *(2) (Memoryless determinacy). We have* $\mathsf{val}^G(\Phi_\alpha)(s) = \sup_{\pi \in \Pi^{PM}} \inf_{\tau \in \Gamma} \mathbb{P}_s^{\pi, \tau}(\Phi_\alpha) = \inf_{\tau \in \Gamma^{PM}} \sup_{\pi \in \Pi} \mathbb{P}_s^{\pi, \tau}(\Phi_\alpha)$ *(i.e., the quantification over the strategies can be restricted to* $\pi \in \Pi^{PM}$ *and* $\tau \in \Gamma^{PM}$ *).*

3.2 Complexity Results

The Upper-Bound Reduction. Consider a library $\mathcal{L}$ of width D, an exit control relation R for $\mathcal{L}$, and an index function α for $\mathcal{L}$ that defines the parity objective Φ_α. Let the number of components be $k + 1$, and let $M_i = \langle \Sigma_I, \Sigma_O, Q_i, q_0^i, \delta_i, F_i, L_i \rangle$ for $0 \leq i \leq k$, where $F_i = \{q_x^i : x \in D\}$. Let $[k] = \{0, 1, 2, \ldots, k\}$. We define a perfect-information stochastic game $G_{\mathcal{L}} = \langle S, S_1, S_2, A_1, A_2, \delta_{\mathcal{L}}^G \rangle$ with an index function α_G as follows: $S = \bigcup_{i=0}^k (Q_i \times \{i\}) \cup \{\bot\}$, $S_1 = \bigcup_{i=0}^k (F_i \times \{i\})$, $S_2 = S \setminus S_1$, $A_1 = [k]$, and $A_2 = \Sigma_I$. The state $\bot$ is a losing absorbing state (i.e., a state with self-loop as the only outgoing transition and assigned odd priority by the index function α_G), and the other transitions defined by the function $\delta_{\mathcal{L}}^G$ are as follows: (i) for $s = \langle q, i \rangle \in S_2$, and $\sigma \in A_2$, we have $\delta_{\mathcal{L}}^G(\langle q, i \rangle, \sigma)(\langle q', j \rangle) = \delta_i(q, \sigma)(q')$ if $i = j$, and 0 otherwise; and (ii) for $s = \langle q_x^i, i \rangle \in S_1$ and $j \in [k]$, we have that if $\langle x, M_j \rangle \in R$, then $\delta_{\mathcal{L}}^G(\langle q_x^i, i \rangle, j)(\langle q_0^j, j \rangle) = 1$, else $\delta_{\mathcal{L}}^G(\langle q_x^i, i \rangle, j)(\bot) = 1$. The intuitive description of the transitions is as follows: (1) given a player-2 state that is a non-exit state q in a component M_i, and an action for player 2 that is an input letter, the transition function $\delta_{\mathcal{L}}^G$ mimics the transition δ_i of M_i; and (2) given a player-1 state that is an exit state q_x^i in component i, and an action for player 1 that is the choice of a component j, if $\langle x, M_j \rangle$ is allowed by R, then the next state is the starting state of component j, and if the choice $\langle x, M_j \rangle$ is invalid (not allowed by R), then the next state is the losing absorbing state $\bot$. For all $\langle q, i \rangle \in S \setminus \{\bot\}$ define $\alpha_G(\langle q, i \rangle) = \alpha(q)$, and let Φ_{α_G} be the parity objective in $G_{\mathcal{L}}$.

Lemma 1. *Consider a library $\mathcal{L}$ of width D, an exit control relation R for $\mathcal{L}$, and an index function α for $\mathcal{L}$ that defines the parity objective Φ_α. Let $G_{\mathcal{L}}$ be* *the corresponding perfect-information stochastic game with parity objective Φ_{α_G}. There exists an almost-sure composer if and only if there exists an almost-sure winning strategy in $G_{\mathcal{L}}$ from $\langle q_0^0, 0 \rangle$, and there exists an η-optimal composer if and only if the value in $G_{\mathcal{L}}$ at $\langle q_0^0, 0 \rangle$ is at least η.*

Proof Sketch. There are two steps to establish correctness of the reduction. The first step is given a composer over $\mathcal{L}$ to construct a finite-memory strategy for player 1 in $G_{\mathcal{L}}$. Intuitively, this is simple as a composer represents a strategy for a partial-observation game (Remark 1), whereas in $G_{\mathcal{L}}$ we have perfect information. However, not every strategy in $G_{\mathcal{L}}$ can be converted to a composer. But we show that a pure memoryless strategy in $G_{\mathcal{L}}$ can be converted to a composer.

Valid Pure Memoryless Strategies in $G_{\mathcal{L}}$. A pure memoryless strategy π in $G_{\mathcal{L}}$ is *valid* if the following condition holds: for all states $\langle q_x^i, i \rangle \in S_1$ if $\pi(\langle q_x^i, i \rangle) = j$, then $\langle x, M_j \rangle \in R$, i.e., the choices of the pure memoryless strategies respect the exit control relation.

Valid Pure Memoryless Strategies to Composers. Given a valid pure memoryless strategy π in $G_{\mathcal{L}}$ we define a composer $C_\pi = \langle D, \mathcal{L}, \mathcal{M}, \mathsf{M}_0, \Delta, \lambda \rangle$ as follows: $\mathcal{M} = [k]$, $\mathsf{M}_0 = 0$, $\lambda(i) = M_i$, and for $0 \leq i \leq k$ and $x \in D$ we have that $\Delta(i, x) = j$ where $\pi(\langle q_x^i, i \rangle) = j$ for $q_x^i \in F_i$. In other words, for the composer there is a state for every component, and given a component and an exit state, the composer plays as the pure memoryless strategy. Since π is valid, the composer obtained from π is compatible with the relation R. Note that the composer mimics the pure memoryless strategy, and there is a one-to-one correspondence between strategies of player 2 in $G_{\mathcal{L}}$ and strategies of the environment in $\mathcal{T}_{C_\pi}$.

Theorem 2 (Complexity of Embedded Parity Realizability). *The qualitative and quantitative realizability problems for controlflow composition with embedded parity belong to UP $\cap$ coUP, and are at least as hard as the (almost-sure) decision problem for perfect-information deterministic parity games.*

4 Realizability with DPW Specifications

In this section we present three results. First, we present a new result for partial-observation stochastic parity games. Second, we show that the qualitative realizability problem for DPW specifications can be reduced to our solution for partial-observation stochastic games yielding an EXPTIME-complete result for the problem. Finally, we show that the quantitative realizability problem for DPW specifications is undecidable.

4.1 Partial-Observation Stochastic Parity Games

We consider partial-observation games with restrictions on strategies that correspond to the qualitative realizability problem, and present a new result to solve such games.

Partial-Observation Stochastic Games. In a stochastic game with partial observation, some states are not distinguishable for player 1. We say that they have the same observation for player 1. Formally, a partial-observation stochastic game consists of a stochastic game $G = \langle S, S_1, S_2, A_1, A_2, \delta^G \rangle$, a finite set $\mathcal{O}$ of observations, and a mapping $\mathsf{obs} : S \to \mathcal{O}$ that assigns to each state s of the game an observation $\mathsf{obs}(s)$ for player 1.

Observational Equivalence and Strategies. The observation mapping induces indistinguishability of play prefixes for player 1, and therefore we need to consider only the player-1 strategies that play in the same way after two indistinguishable play prefixes. We consider two classes of strategies depending on the indistinguishability of play prefixes for player 1 and they are as follows: (i) the play prefixes have the same observation sequence; and (ii) the play prefixes have the same sequence of distinct observations, that is they have the same observation sequence up to repetition (stuttering).

Classes of Strategies. The *observation sequence* of a sequence $\rho = s_0 s_1 \ldots s_n$ is the sequence $\mathsf{obs}(\rho) = \mathsf{obs}(s_0) \ldots \mathsf{obs}(s_n)$ of state observations; the *collapsed stuttering* of ρ is the sequence $\overline{\mathsf{obs}}(\rho) = o_0 o_1 o_2 \ldots$ of distinct observations defined as follows: $o_0 = \mathsf{obs}(s_0)$ and for all $i \geq 1$ we have $o_i = \mathsf{obs}(s_i)$ if $\mathsf{obs}(s_i) \neq \mathsf{obs}(s_{i-1})$, and $o_i = \epsilon$ otherwise (where ϵ is the empty sequence). We consider two types of strategies. A strategy π for player 1 is

- *observation-based* if for all sequences $\rho, \rho' \in S^+$ such that $\mathsf{last}(\rho) \in S_1$ and $\mathsf{last}(\rho') \in S_1$, if $\mathsf{obs}(\rho) = \mathsf{obs}(\rho')$ then $\pi(\rho) = \pi(\rho')$;
- *collapsed-stutter-invariant* if for all sequences $\rho, \rho' \in S^+$ such that $\mathsf{last}(\rho) \in S_1$ and $\mathsf{last}(\rho') \in S_1$, if $\overline{\mathsf{obs}}(\rho) = \overline{\mathsf{obs}}(\rho')$, then $\pi(\rho) = \pi(\rho')$.

We now present a polynomial-time reduction for deciding the existence of finite-memory almost-sure winning collapsed-stutter-invariant strategies to observation-based strategies, which is EXPTIME-complete [5].

Reduction of Collapsed-Stutter-Invariant Problem to Observation-Based Problem. There are two main ideas of the reduction. (1) First, whenever player 1 plays an action a, the action a is stored in the state space as long as the observation of the state remains the same. This allows to check that player 1 plays always the same action along a sequence of identical observations. (2) Second, whenever a transition is executed, player 2 is allowed to loop arbitrarily many times through the new state. This ensures that player 1 cannot rely on the number of times he sees an observation, thus that player 1 is collapsed-stutter-invariant. However, it should be forbidden for player 2 to loop forever in a state, which can be ensured by assigning priority 0 to the loop. Hence player 1 wins the parity objective if the loop is taken forever by player 2, and otherwise, visiting priority 0 infinitely often does not change the winner of the game.

The Formal Reduction. Given a partial-observation stochastic game $G = \langle S, S_1, S_2, A_1, A_2, \delta^G \rangle$ with observation mapping $\mathsf{obs} : S \to \mathcal{O}$, we construct a game $G' = \langle S', S_1', S_2', A_1, A_2', \delta^{G'} \rangle$ as follows:

- $S' = S \times (A_1 \cup \overline{A}_1 \cup \{0, \overline{0}\}) \cup \{\bot\}$ where $\overline{A}_1 = \{\overline{a} \mid a \in A_1\}$, assuming that $0 \notin A_1$. The states $\langle s, 0 \rangle$ are a copy of the state space of the original game, and in the states $\langle s, a \rangle$ with $s \in S_1$ and $a \in A_1$, player 1 is required to play action a; in the states $\langle s, \overline{0} \rangle$ and $\langle s, \overline{a} \rangle$, player 2 can stay for arbitrarily many steps. The state $\bot$ is absorbing and losing for player 1.

- $S_1' = S_1 \times (A_1 \cup \{0\}) \cup \{\bot\}$; $S_2' = S' \setminus S_1'$; and $A_2' = A_2 \cup \{\sharp\}$, assuming $\sharp \notin A_2$.
- The probabilistic transition function $\delta^{G'}$ is defined as follows: for all player-1 states $\langle s, x \rangle \in S_1'$ and actions $a \in A_1$:
  - if $x \in A_1 \setminus \{a\}$, then let $\delta^{G'}(\langle s, x \rangle, a))(\bot) = 1$, that is player 1 loses the game if he does not play the stored action;
  - if $x = a$ or $x = 0$, then for all $s' \in S$ let
    $\delta^{G'}(\langle s, x \rangle, a))(\langle s', \overline{a} \rangle) = \delta^G(s, a)(s')$ if $\mathsf{obs}(s') = \mathsf{obs}(s)$, and let
    $\delta^{G'}(\langle s, x \rangle, a))(\langle s', \overline{0} \rangle) = \delta^G(s, a)(s')$ if $\mathsf{obs}(s') \neq \mathsf{obs}(s)$; thus we store the action a as long as the state observation does not change;
  - All other probabilities $\delta^{G'}(\langle s, x \rangle, a))(\cdot)$ are set to 0, for example $\delta^{G'}(\langle s, 0 \rangle, a))(\langle s', y \rangle) = 0$ for all $y \neq \overline{a}$;

  and for all player-2 states $\langle s, x \rangle \in S_2'$, and actions $a \in A_2$:
  - if $x \in A_1 \cup \{0\}$, then for all $s' \in S$ let
    $\delta^{G'}(\langle s, x \rangle, a))(\langle s', \overline{x} \rangle) = \delta^G(s, a)(s')$ if $\mathsf{obs}(s') = \mathsf{obs}(s)$, and let
    $\delta^{G'}(\langle s, x \rangle, a))(\langle s', \overline{0} \rangle) = \delta^G(s, a)(s')$ if $\mathsf{obs}(s') \neq \mathsf{obs}(s)$; thus all actions are available to player 2 as in the original game, and the stored action x of player 1 is maintained if the state observation does not change;
  - if $x = \overline{b}$ for some $b \in A_1 \cup \{0\}$, then let
    $\delta^{G'}(\langle s, \overline{b} \rangle, \sharp))(\langle s, \overline{b} \rangle) = 1$, and
    $\delta^{G'}(\langle s, \overline{b} \rangle, a))(\langle s, b \rangle) = 1$ if $a \neq \sharp$; thus player 2 can decide to stay arbitrarily long in $\langle s, \overline{b} \rangle$ before going back to $\langle s, b \rangle$;
  - All other probabilities $\delta^{G'}(\langle s, x \rangle, a))(\cdot)$ and $\delta^{G'}(\langle s, x \rangle, \sharp))(\cdot)$ are set to 0.

The observation mapping obs' is defined according to the first component of the state: $\mathsf{obs}'(\langle s, x \rangle) = \mathsf{obs}(s)$. Given an index function α for G, define the index function α' for G' as follows: $\alpha'(\langle s, x \rangle) = \alpha(s)$ and $\alpha'(\langle s, \overline{x} \rangle) = 0$ for all $s \in S$ and $x \in A_1 \cup \{0\}$, and $\alpha'(\bot) = 1$. Hence, the state $\bot$ is losing for player 1, and the player-2 states $\langle s, \overline{x} \rangle$ are winning for player 1 if player 2 stays there forever.

Lemma 2. *Given a partial-observation stochastic game G with observation mapping obs and parity objective Φ_α defined by the index function α, a game G' with observation mapping obs' and parity objective $\Phi_{\alpha'}$ defined by the index function α' can be constructed in polynomial time such that the following statements are equivalent:*
- *there exists a finite-memory almost-sure winning collapsed-stutter-invariant strategy π for player 1 in G from s_0 for the parity objective Φ_α;*
- *there exists a finite-memory almost-sure winning observation-based strategy π' for player 1 in G' from $\langle s_0, \overline{0} \rangle$ for the parity objective $\Phi_{\alpha'}$.*

Theorem 3. *The qualitative problem of deciding the existence of a finite-memory almost-sure winning collapsed-stutter-invariant strategy in partial-observation stochastic games with parity objectives is EXPTIME-complete.*

4.2 Qualitative and Quantitative Realizability

We present a polynomial reduction for the qualitative realizability problem with DPW specifications to the existence of finite-memory collapsed-stutter-invariant almost-sure winning strategies, and thus show that the problem can be solved in EXPTIME. An EXPTIME lower bound is known for this problem [2].

Theorem 4. *The qualitative realizability problem for controlflow composition with DPW specifications is EXPTIME-complete.*

Finally, we establish undecidability of the quantitative realizability problem by a reduction from the quantitative decision problem for probabilistic automata (which is undecidable [15]).

Theorem 5. *The quantitative realizability problem for controlflow composition with DPW specifications is undecidable.*

References

1. Andersson, D., Miltersen, P.B.: The Complexity of Solving Stochastic Games on Graphs. In: Dong, Y., Du, D.-Z., Ibarra, O. (eds.) ISAAC 2009. LNCS, vol. 5878, pp. 112–121. Springer, Heidelberg (2009)
2. Avni, G., Kupferman, O.: Synthesis from Component Libraries with Costs. In: Baldan, P., Gorla, D. (eds.) CONCUR 2014. LNCS, vol. 8704, pp. 156–172. Springer, Heidelberg (2014)
3. Baier, C., Katoen, J.-P.: Principles of Model Checking (Representation and Mind Series). The MIT Press (2008)
4. Chatterjee, K.: Stochastic ω-regular Games. PhD thesis, UC Berkeley (2007)
5. Chatterjee, K., Doyen, L., Nain, S., Vardi, M.Y.: The Complexity of Partial-Observation Stochastic Parity Games with Finite-Memory Strategies. In: Muscholl, A. (ed.) FOSSACS 2014 (ETAPS). LNCS, vol. 8412, pp. 242–257. Springer, Heidelberg (2014)
6. Chatterjee, K., Doyen, L., Vardi. M. Y.: The complexity of synthesis from probabilistic components. CoRR, abs/1502.04844 (2015)
7. Chatterjee, K., Henzinger, T.A.: Reduction of stochastic parity to stochastic mean-payoff games. IPL **106**(1), 1–7 (2008)
8. Chatterjee, K., Jurdziński, M., Henzinger, T.A.: Quantitative stochastic parity games. In: SODA 2004, pp. 114–123 (2004)
9. Courcoubetis, C., Yannakakis, M.: The complexity of probabilistic verification. J. ACM **42**(4), 857–907 (1995)
10. De Crescenzo, I., La Torre, S.: Modular Synthesis with Open Components. In: Abdulla, P.A., Potapov, I. (eds.) RP 2013. LNCS, vol. 8169, pp. 96–108. Springer, Heidelberg (2013)
11. de Alfaro, L., Henzinger, T.A.: Interface Theories for Component-Based Design. In: Henzinger, T.A., Kirsch, C.M. (eds.) EMSOFT 2001. LNCS, vol. 2211, pp. 148–165. Springer, Heidelberg (2001)
12. Kwiatkowska, M., Norman, G., Parker, D.: PRISM 4.0: Verification of Probabilistic Real-Time Systems. In: Gopalakrishnan, G., Qadeer, S. (eds.) CAV 2011. LNCS, vol. 6806, pp. 585–591. Springer, Heidelberg (2011)

13. Lustig, Y., Vardi, M.Y.: Synthesis from Component Libraries. In: de Alfaro, L. (ed.) FOSSACS 2009. LNCS, vol. 5504, pp. 395–409. Springer, Heidelberg (2009)
14. Nain, S., Lustig, Y., Vardi, M.Y.: Synthesis from probabilistic components. LMCS 10(2) (2014)
15. Paz, A.: Introduction to probabilistic automata. Academic Press Inc. (1971)
16. Pnueli, A., Rosner, R.: On the synthesis of a reactive module. In: Proc. of POPL, pp. 179–190. ACM Press (1989)
17. Sifakis, J.: A framework for component-based construction extended abstract. In: SFEM 2005, pp. 293–300 (2005)
18. Vardi, M.Y.: Automatic verification of probabilistic concurrent finite-state systems. In: FOCS 1985, pp. 327–338 (1985)
19. Vardi, M.Y.: Probabilistic Linear-Time Model Checking: An Overview of the Automata-Theoretic Approach. In: Katoen, J.-P. (ed.) AMAST-ARTS 1999, ARTS 1999, and AMAST-WS 1999. LNCS, vol. 1601, pp. 265–276. Springer, Heidelberg (1999)

Edit Distance for Pushdown Automata

Krishnendu Chatterjee$^{(\boxtimes)}$, Thomas A. Henzinger, Rasmus Ibsen-Jensen, and Jan Otop

IST Austria, Wien-Umgebung, Austria
krish.chat@gmail.com

Abstract. The edit distance between two words w_1, w_2 is the minimal number of word operations (letter insertions, deletions, and substitutions) necessary to transform w_1 to w_2. The edit distance generalizes to languages $\mathcal{L}_1, \mathcal{L}_2$, where the edit distance is the minimal number k such that for every word from $\mathcal{L}_1$ there exists a word in $\mathcal{L}_2$ with edit distance at most k. We study the edit distance computation problem between pushdown automata and their subclasses. The problem of computing edit distance to pushdown automata is undecidable, and in practice, the interesting question is to compute the edit distance from a pushdown automaton (the implementation, a standard model for programs with recursion) to a regular language (the specification). In this work, we present a complete picture of decidability and complexity for deciding whether, for a given threshold k, the edit distance from a pushdown automaton to a finite automaton is at most k.

1 Introduction

Edit distance. The edit distance [13] between two words is a well-studied metric, which is the minimum number of edit operations (insertion, deletion, or substitution of one letter by another) that transforms one word to another. The edit distance between a word w to a language $\mathcal{L}$ is the minimal edit distance between w and words in $\mathcal{L}$. The edit distance between two languages $\mathcal{L}_1$ and $\mathcal{L}_2$ is the supremum over all words w in $\mathcal{L}_1$ of the edit distance between w and $\mathcal{L}_2$.

Significance of edit distance. The notion of *edit distance* provides a quantitative measure of "how far apart" are (a) two words, (b) words from a language, and (c) two languages. It forms the basis for quantitatively comparing sequences, a problem that arises in many different areas, such as error-correcting codes, natural language processing, and computational biology. The notion of edit distance between languages forms the foundations of a quantitative approach to verification. The traditional qualitative verification (model checking) question is the *language inclusion* problem: given an implementation (source language)

This research was funded in part by the European Research Council (ERC) under grant agreement 267989 (QUAREM), by the Austrian Science Fund (FWF) projects S11402-N23 (RiSE) and Z211-N23 (Wittgenstein Award), FWF Grant No P23499-N23, FWF NFN Grant No S11407-N23 (RiSE), ERC Start grant (279307: Graph Games), and MSR faculty fellows award.

© Springer-Verlag Berlin Heidelberg 2015
M.M. Halldórsson et al. (Eds.): ICALP 2015, Part II, LNCS 9135, pp. 121–133, 2015.
DOI: 10.1007/978-3-662-47666-6_10

defined by an automaton $\mathcal{A}_I$ and a specification (target language) defined by an automaton $\mathcal{A}_S$, decide whether the language $\mathcal{L}(\mathcal{A}_I)$ is included in the language $\mathcal{L}(\mathcal{A}_S)$ (i.e., $\mathcal{L}(\mathcal{A}_I) \subseteq \mathcal{L}(\mathcal{A}_S)$). The *threshold edit distance* (TED) problem is a generalization of the language inclusion problem, which for a given integer threshold $k \geq 0$ asks whether every word in the source language $\mathcal{L}(\mathcal{A}_I)$ has edit distance at most k to the target language $\mathcal{L}(\mathcal{A}_S)$ (with $k = 0$ we have the traditional language inclusion problem). For example, in simulation-based verification of an implementation against a specification, the measured trace may differ slightly from the specification due to inaccuracies in the implementation. Thus, a trace of the implementation may not be in the specification. However, instead of rejecting the implementation, one can quantify the distance between a measured trace and the specification. Among all implementations that violate a specification, the closer the implementation traces are to the specification, the better [5,7,10]. The edit distance problem is also the basis for *repairing* specifications [2,3].

Our models. In this work we consider the edit distance computation problem between two automata $\mathcal{A}_1$ and $\mathcal{A}_2$, where $\mathcal{A}_1$ and $\mathcal{A}_2$ can be (non)deterministic finite automata or pushdown automata. Pushdown automata are the standard models for programs with recursion, and regular languages are canonical to express the basic properties of systems that arise in verification. We denote by DPDA (resp., PDA) deterministic (resp., nondeterministic) pushdown automata, and DFA (resp., NFA) deterministic (resp., nondeterministic) finite automata. We consider source and target languages defined by DFA, NFA, DPDA, and PDA. We first present the known results and then our contributions.

Previous results. The main results for the classical language inclusion problem are as follows [11]: (i) if the target language is a DFA, then it can be solved in polynomial time; (ii) if either the target language is a PDA or both source and target languages are DPDA, then it is undecidable; (iii) if the target language is an NFA, then (a) if the source language is a DFA or NFA, then it is PSpace-complete, and (b) if the source language is a DPDA or PDA, then it is PSpace-hard and can be solved in ExpTime (to the best of our knowledge, there is a complexity gap where the upper bound is ExpTime and the lower bound is PSpace). The TED problem was studied for DFA and NFA, and it is PSpace-complete, when the source and target languages are given by DFA or NFA [2,3].

Our contributions. Our main contributions are as follows.

1. We show that the TED problem is ExpTime-complete, when the source language is given by a DPDA or a PDA, and the target language is given by a DFA or NFA. We present a hardness result which shows that the TED problem is ExpTime-hard for source languages given as DPDA and target languages given as DFA. We present a matching upper bound by showing that for source languages given as PDA and target languages given as NFA the problem can be solved in ExpTime. As a consequence of our lower bound we obtain that the language inclusion problem for source languages given by DPDA (or PDA) and target languages given by NFA is ExpTime-complete. Thus we present a complete picture of the complexity of the TED problem,

Table 1. Complexity of the language inclusion problem from C_1 to C_2. Our results are boldfaced.

	$C_2 = $ DFA	$C_2 = $ NFA	$C_2 = $ DPDA	$C_2 = $ PDA
$C_1 \in \{$DFA, NFA$\}$	PTime	PSpace-c	PTime	
$C_1 \in \{$DPDA, PDA$\}$		**ExpTime-c (Th. 2)**	undecidable	

Table 2. Complexity of FED(C_1, C_2). Our results are boldfaced. See Conjecture 14 for the open complexity problem of $C_1 \in \{$DPDA, PDA$\}$ and $C_2 = $ DFA.

	$C_2 = $ DFA	$C_2 = $ NFA	$C_2 = $ DPDA	$C_2 = $ PDA
$C_1 \in \{$DFA, NFA$\}$	coNP-c [3]	PSpace-c [3]	open (Conj. 18)	
$C_1 \in \{$DPDA, PDA$\}$	coNP-hard [3] **in ExpTime (Th. 8)**	**ExpTime-c (Th. 8)**	**undecidable (Prop. 15)**	

Table 3. Complexity of TED(C_1, C_2). Our results are boldfaced.

	$C_2 = $ DFA	$C_2 = $ NFA	$C_2 = $ DPDA	$C_2 = $ PDA
$C_1 \in \{$DFA, NFA$\}$	PSpace-c [2]		**undecidable (Prop. 17)**	
$C_1 \in \{$DPDA, PDA$\}$	**ExpTime-c (Th. 2 (1))**		undecidable	

and in addition we close a complexity gap in the classical language inclusion problem. In contrast, if the target language is given by a DPDA, then the TED problem is undecidable even for source languages given as DFA. Note that the interesting verification question is when the implementation (source language) is a DPDA (or PDA) and the specification (target language) is given as DFA (or NFA), for which we present decidability results with optimal complexity.

2. We also consider the *finite edit distance* (FED) problem, which asks whether there exists $k \geq 0$ such that the answer to the TED problem with threshold k is YES. For finite automata, it was shown in [2,3] that if the answer to the FED problem is YES, then a polynomial bound on k exists. In contrast, the edit distance can be exponential between DPDA and DFA. We present a matching exponential upper bound on k for the FED problem from PDA to NFA. Finally, we show that the FED problem is ExpTime-complete when the source language is given as a DPDA or PDA, and the target language as an NFA.

Our results are summarized in Tables 1, 2 and 3. Due to space constraints we omit some technical proofs, which are presented in the full version [6].

Related work. Algorithms for edit distance have been studied extensively for words [1,12,13,15–17]. The edit distance between regular languages was studied in [2,3], between timed automata in [8], and between straight line programs in [9,14]. A near-linear time algorithm to approximate the edit distance for a word to a DYCK language has been presented in [18].

2 Preliminaries

2.1 Words, Languages and Automata

Words. Given a finite alphabet Σ of letters, a *word* w is a finite sequence of letters. For a word w, we define $w[i]$ as the i-th letter of w and $|w|$ as its length. We denote the set of all words over Σ by Σ^*. We use ϵ to denote the empty word.

Pushdown Automata. A *(non-deterministic) pushdown automaton* (PDA) is a tuple $(\Sigma, \Gamma, Q, S, \delta, F)$, where Σ is the input alphabet, Γ is a finite stack alphabet, Q is a finite set of states, $S \subseteq Q$ is a set of initial states, $\delta \subseteq Q \times \Sigma \times (\Gamma \cup \{\bot\}) \times Q \times \Gamma^*$ is a finite transition relation and $F \subseteq Q$ is a set of final (accepting) states. A PDA $(\Sigma, \Gamma, Q, S, \delta, F)$ is a *deterministic pushdown automaton* (DPDA) if $|S| = 1$ and δ is a function from $Q \times \Sigma \times (\Gamma \cup \{\bot\})$ to $Q \times \Gamma^*$. We denote the class of all PDA (resp., DPDA) by PDA (resp., DPDA). We define the size of a PDA $\mathcal{A} = (\Sigma, \Gamma, Q, S, \delta, F)$, denoted by $|\mathcal{A}|$, as $|Q| + |\delta|$.

Runs of Pushdown Automata. Given a PDA $\mathcal{A}$ and a word $w = w[1] \ldots w[k]$ over Σ, a *run* π of $\mathcal{A}$ on w is a sequence of elements from $Q \times \Gamma^*$ of length $k + 1$ such that $\pi[0] \in S \times \{\epsilon\}$ and for every $i \in \{1, \ldots, k\}$ either (1) $\pi[i-1] = (q, \epsilon)$, $\pi[i] = (q', u')$ and $(q, w[i], \bot, q', u') \in \delta$, or (2) $\pi[i-1] = (q, ua)$, $\pi[i] = (q', uu')$ and $(q, w[i], a, q', u') \in \delta$. A run π of length $k+1$ is *accepting* if $\pi[k+1] \in F \times \{\epsilon\}$, i.e., the automaton is in an accepting state and the stack is empty. The *language recognized (or accepted)* by $\mathcal{A}$, denoted $\mathcal{L}(\mathcal{A})$, is the set of words that have an accepting run.

Context Free Grammar (CFG). A context free grammar in Chomsky normal form (CFG) is a tuple (Σ, V, s, P), where Σ is the alphabet, V is a set of *non-terminals*, $s \in V$ is a *start symbol* and P is a set of *production rules*. A production rule p has one of the following forms: (1) $p : v \rightarrow zu$, where $v, z, u \in V$; or (2) $p : v \rightarrow \alpha$, where $v \in V$ and $\alpha \in \Sigma$; or (3) $p : s \rightarrow \epsilon$.

Languages Generated by CFGs. Fix a CFG $G = (\Sigma, V, s, P)$. We define derivation $\rightarrow_G$ as a relation on $(\Sigma \cup V)^* \times (\Sigma \cup V)^*$ as follows: $w \rightarrow_G w'$ iff $w = w_1 v w_2$, with $v \in V$, and $w' = w_1 u w_2$ for some $u \in (\Sigma \cup V)^*$ such that $v \rightarrow u$ is a production from G. We define $\rightarrow_G^*$ as the transitive closure of $\rightarrow_G$. The *language generated by* G, denoted by $\mathcal{L}(G) = \{w \in \Sigma^* \mid s \rightarrow_G^* w\}$ is the set of words that can be derived from s. CFGs and PDAs are language-wise polynomially equivalent [11] (i.e., there is a polynomial-time procedure that, given a PDA, outputs a CFG of the same language and vice versa).

Finite Automata. A *non-deterministic finite automaton* (NFA) is a PDA with empty stack alphabet. We will omit Γ while referring to NFA, i.e., we will consider them as tuples $(\Sigma, Q, S, \delta, F)$. We denote the class of all NFA by NFA. Analogously to DPDA we define *deterministic finite automata* (DFA).

Language Inclusion. Let $\mathcal{C}_1, \mathcal{C}_2$ be subclasses of PDA. The *inclusion problem from $\mathcal{C}_1$ in $\mathcal{C}_2$* asks, given $\mathcal{A}_1 \in \mathcal{C}_1$, $\mathcal{A}_2 \in \mathcal{C}_2$, whether $\mathcal{L}(\mathcal{A}_1) \subseteq \mathcal{L}(\mathcal{A}_2)$.

Edit Distance between Words. Given two words w_1, w_2, the edit distance between w_1, w_2, denoted by $ed(w_1, w_2)$, is the minimal number of single letter operations: insertions, deletions, and substitutions, necessary to transform w_1 into w_2.

Edit Distance between Languages. Let $\mathcal{L}_1, \mathcal{L}_2$ be languages. We define the edit distance *from $\mathcal{L}_1$ to $\mathcal{L}_2$*, denoted $ed(\mathcal{L}_1, \mathcal{L}_2)$, as $\sup_{w_1 \in \mathcal{L}_1} \inf_{w_2 \in \mathcal{L}_2} ed(w_1, w_2)$. The edit distance between languages is not a distance function. In particular, it is not symmetric.

2.2 Problem Statement

In this section we define the problems of interest. Then, we recall the previous results and succinctly state our results.

Definition 1. *For $\mathcal{C}_1, \mathcal{C}_2 \in \{DFA, NFA, DPDA, PDA\}$ we define the following questions:*

1. *The threshold edit distance problem from $\mathcal{C}_1$ to $\mathcal{C}_2$ (denoted $TED(\mathcal{C}_1, \mathcal{C}_2)$): Given automata $\mathcal{A}_1 \in \mathcal{C}_1$, $\mathcal{A}_2 \in \mathcal{C}_2$ and an integer threshold $k \geq 0$, decide whether $ed(\mathcal{L}(\mathcal{A}_1), \mathcal{L}(\mathcal{A}_2)) \leq k$.*
2. *The finite edit distance problem from $\mathcal{C}_1$ to $\mathcal{C}_2$ (denoted $FED(\mathcal{C}_1, \mathcal{C}_2)$): Given automata $\mathcal{A}_1 \in \mathcal{C}_1$, $\mathcal{A}_2 \in \mathcal{C}_2$, decide whether $ed(\mathcal{L}(\mathcal{A}_1), \mathcal{L}(\mathcal{A}_2)) < \infty$.*
3. *Computation of edit distance from $\mathcal{C}_1$ to $\mathcal{C}_2$: Given automata $\mathcal{A}_1 \in \mathcal{C}_1$, $\mathcal{A}_2 \in \mathcal{C}_2$, compute $ed(\mathcal{L}(\mathcal{A}_1), \mathcal{L}(\mathcal{A}_2))$.*

We establish the complete complexity picture for the TED problem for all combinations of source and target languages given by DFA, NFA, DPDA and PDA:

1. TED for regular languages has been studied in [2], where PSpace-completeness of $TED(\mathcal{C}_1, \mathcal{C}_2)$ for $\mathcal{C}_1, \mathcal{C}_2 \in \{DFA, NFA\}$ has been established.
2. In Section 3, we study the TED problem for source languages given by pushdown automata and target languages given by finite automata. We establish ExpTime-completeness of $TED(\mathcal{C}_1, \mathcal{C}_2)$ for $\mathcal{C}_1 \in \{DPDA, PDA\}$ and $\mathcal{C}_2 \in \{DFA, NFA\}$.
3. In Section 5, we study the TED problem for target languages given by pushdown automata. We show that $TED(\mathcal{C}_1, \mathcal{C}_2)$ is undecidable for $\mathcal{C}_1 \in \{DFA, NFA, DPDA, PDA\}$ and $\mathcal{C}_2 \in \{DPDA, PDA\}$.

We study the FED problem for all combinations of source and target languages given by DFA, NFA, DPDA and PDA and obtain the following results:

1. FED for regular languages has been studied in [3]. It has been shown that for $\mathcal{C}_1 \in \{DFA, NFA\}$, the problem $FED(\mathcal{C}_1, DFA)$ is coNP-complete, while the problem $FED(\mathcal{C}_1, NFA)$ is PSpace-complete.

2. We show in Section 4 that for $\mathcal{C}_1 \in \{\mathsf{DPDA}, \mathsf{PDA}\}$, the problem $\mathsf{FED}(\mathcal{C}_1, \mathsf{NFA})$ is ExpTime-complete.
3. We show in Section 5 that (1) for $\mathcal{C}_1 \in \{\mathsf{DFA}, \mathsf{NFA}, \mathsf{DPDA}, \mathsf{PDA}\}$, the problem $\mathsf{FED}(\mathcal{C}_1, \mathsf{PDA})$ is undecidable, and (2) the problem $\mathsf{FED}(\mathsf{DPDA}, \mathsf{DPDA})$ is undecidable.

3   Threshold Edit Distance from Pushdown to Regular Languages

In this section we establish the complexity of the TED problem from pushdown to finite automata.

Theorem 2. *(1) For $\mathcal{C}_1 \in \{\mathsf{DPDA}, \mathsf{PDA}\}$ and $\mathcal{C}_2 \in \{\mathsf{DFA}, \mathsf{NFA}\}$, the $\mathsf{TED}(\mathcal{C}_1, \mathcal{C}_2)$ problem is ExpTime-complete. (2) For $\mathcal{C}_1 \in \{\mathsf{DPDA}, \mathsf{PDA}\}$, the language inclusion problem from $\mathcal{C}_1$ in NFA is ExpTime-complete.*

We establish the above theorem as follows: In Section 3.1, we present an exponential-time algorithm for $\mathsf{TED}(\mathsf{PDA}, \mathsf{NFA})$ (for the upper bound of (1)). Then, in Section 3.2 we show (2), in a slightly stronger form, and reduce it (that stronger problem) to $\mathsf{TED}(\mathsf{DPDA}, \mathsf{DFA})$, which shows the ExpTime-hardness part of (1).

3.1   Upper Bound

We present an ExpTime algorithm that, given (1) a PDA $\mathcal{A}_P$; (2) an NFA $\mathcal{A}_N$; and (3) a threshold t given in binary, decides whether the edit distance from $\mathcal{A}_P$ to $\mathcal{A}_N$ is above t. The algorithm extends a construction for NFA by Benedikt et al. [2].

Intuition. The construction uses the idea that for a given word w and an NFA $\mathcal{A}_N$ the following are equivalent: (i) $ed(w, \mathcal{A}_N) > t$, and (ii) for each accepting state s of $\mathcal{A}_N$ and for every word w', if $\mathcal{A}_N$ can reach s from some initial state upon reading w', then $ed(w, w') > t$. We construct a PDA $\mathcal{A}_I$ which simulates the PDA $\mathcal{A}_P$ and stores in its states all states of the NFA $\mathcal{A}_N$ reachable with at most t edits. More precisely, the PDA $\mathcal{A}_I$ remembers in its states, for every state s of the NFA $\mathcal{A}_N$, the minimal number of edit operations necessary to transform the currently read prefix w_p of the input word into a word w'_p, upon which $\mathcal{A}_N$ can reach s from some initial state. If for some state the number of edit operations exceeds t, then we associate with this state a special symbol $\#$ to denote this. Then, we show that a word w accepted by the PDA $\mathcal{A}_P$ has $ed(w, \mathcal{A}_N) > t$ iff the automaton $\mathcal{A}_I$ has a run on w that ends (1) in an accepting state of simulated $\mathcal{A}_P$, (2) with the simulated stack of $\mathcal{A}_P$ empty, and (3) the symbol $\#$ is associated with every accepting state of $\mathcal{A}_N$.

Lemma 3. *(1) Given (i) a PDA $\mathcal{A}_P$; (ii) an NFA $\mathcal{A}_N$; and (iii) a threshold t given in binary, the decision problem of whether $ed(\mathcal{A}_P, \mathcal{A}_N) \leq t$ can be reduced to the emptiness problem for a PDA of size $O(|\mathcal{A}_P| \cdot (t+2)^{|\mathcal{A}_N|})$. (2) $\mathsf{TED}(\mathsf{PDA}, \mathsf{NFA})$ is in ExpTime.*

3.2   Lower Bound

Our ExpTime-hardness proof of TED(DPDA, DFA) extends the idea from [2] that shows PSpace-hardness of the edit distance for DFA. The standard proof of PSpace-hardness of the universality problem for NFA [11] is by reduction to the halting problem of a fixed Turing machine M working on a bounded tape. The Turing machine M is the one that simulates other Turing machines (such a machine is called universal). The input to that problem is the initial configuration C_1 and the tape is bounded by its size $|C_1|$. In the reduction, the NFA recognizes the language of all words that do not encode valid computation of M starting from the initial configuration C_1, i.e., it checks the following four conditions: (1) the given word is a sequence of configurations, (2) the state of the Turing machine and the adjunct letters follow from transitions of M, (3) the first configuration is not C_1 and (4) the tape's cells are changed only by M, i.e., they do not change values spontaneously. While conditions (1), (2) and (3) can be checked by a DFA of polynomial size, condition (4) can be encoded by a polynomial-size NFA but not a polynomial-size DFA. However, to check (4) the automaton has to make only a single non-deterministic choice to pick a position in the encoding of the computation, which violates (4), i.e., the value at that position is different from the value $|C_1| + 1$ letters further, which corresponds to the same memory cell in the successive configuration, and the head of M does not change it. We can transform a non-deterministic automaton $\mathcal{A}_N$ checking (4) into a deterministic automaton $\mathcal{A}_D$ by encoding such a non-deterministic pick using an external letter. Since we need only one external symbol, we have $\mathcal{L}(\mathcal{A}_N) = \Sigma^*$ iff $ed(\Sigma^*, \mathcal{L}(\mathcal{A}_D)) = 1$. This suggests the following definition:

Definition 4. *An NFA* $\mathcal{A} = (\Sigma, Q, S, \delta, F)$ *is* nearly-deterministic *if* $|S| = 1$ *and* $\delta = \delta_1 \cup \delta_2$, *where* δ_1 *is a function and in every accepting run the automaton takes a transition from* δ_2 *exactly once.*

Lemma 5. *There exists a DPDA* $\mathcal{A}_P$ *such that the problem, given a nearly-deterministic NFA* $\mathcal{A}_N$, *decide whether* $\mathcal{L}(\mathcal{A}_P) \subseteq \mathcal{L}(\mathcal{A}_N)$, *is* ExpTime-hard.

Proof. Consider the *linear-space halting* problem for a (fixed) alternating Turing machine (ATM) M: given an input word w over an alphabet Σ, decide whether M halts on w with the tape bounded by $|w|$. There exists an ATM M_U, such that the linear-space halting problem for M_U is ExpTime-complete [4]. We show the ExpTime-hardness of the problem from the lemma statement by reduction from the linear-space halting problem for M_U.

We w.l.o.g. assume that existential and universal transitions of M_U alternate. Fix an input of length n. The main idea is to construct the language L of words that encode valid terminating computation trees of M_U on the given input. Observe that the language L depends on the given input. We encode a single configuration of M_U as a word of length $n + 1$ of the form $\Sigma^i q \Sigma^{n-i}$, where q is a state of M_U. Recall that a computation of an ATM is a tree, where every node of the tree is a configuration of M_U, and it is accepting if every leaf node is an accepting configuration. We encode computation trees T of M_U by traversing T preorder and executing the following: if the current node has only one successor, then write down the current configuration C, terminate it with $\#$ and move down to the successor node in T. Otherwise, if the current node has two successors s, t in the tree, then write down in order (1) the reversed current configuration C^R; and (2) the results of traversals on s and t, each surrounded by parentheses (and), i.e., $C^R (u^s) (u^t)$, where u^s (resp., u^t) is the result of the traversal of the subtree of T rooted at s (resp., t). Finally, if the current node is a leaf, write down the corresponding configuration and terminate with $\$$. For example, consider a computation with the initial configuration C_1, from which an existential transition leads to C_2, which in turn has a universal transition to C_3 and C_4. Such a computation tree is encoded as follows:

$$C_1 \# C_2^R (C_3 \ldots \$) (C_4 \ldots \$).$$

We define automata $\mathcal{A}_N$ and $\mathcal{A}_P$ over the alphabet $\Sigma \cup \{\#, \$, (,)\}$. The automaton $\mathcal{A}_N$ is a nearly deterministic NFA that recognizes only (but not all) words not encoding valid computation trees of M_U. More precisely, $\mathcal{A}_N$ accepts in four cases: (1) The word does not encode a tree (except that the parentheses may not match as the automaton cannot check that) of computation as presented above. (2) The initial configuration is different from the one given as the input. (3) The successive configurations, i.e., those that result from existential transitions or left-branch universal transitions (like C_2 to C_3), are valid. The right-branch universal transitions, which are preceded by the word ")(", are not checked by $\mathcal{A}_N$. For example, the consistency of the transition C_2 to C_4 is not checked by $\mathcal{A}_N$. Finally, (4) $\mathcal{A}_N$ accepts words in which at least one final configuration, which is a configuration followed by $\$$, is not final for M_U.

Next, we define $\mathcal{A}_P$ as a DPDA that accepts words in which parentheses match and right-branch universal transitions are consistent, e.g., it checks consistency of a transition from C_2 to C_4. The automaton $\mathcal{A}_P$ pushes configurations on even levels of the computation tree (e.g., C_2^R), which are reversed, on the stack and pops these configurations from the stack to compare them with the following configuration in the right subtree (e.g., C_4). In the example this means that, while the automaton processes the subword $(C_3 \ldots \$)$, it can use its stack to check consistency of universal transitions in that subword. We assumed that M_U does not have consecutive universal transitions. This means that, for example, $\mathcal{A}_P$ does not need to check the consistency of C_4 with its successive configuration. By construction, we have $L = \mathcal{L}(\mathcal{A}_P) \cap \mathcal{L}(\mathcal{A}_N)^c$ (recall that L is the language of encodings of computations of M_U on the given input) and M_U halts on the given input if and only if $\mathcal{L}(\mathcal{A}_P) \subseteq \mathcal{L}(\mathcal{A}_N)$ fails. Observe that $\mathcal{A}_P$ is fixed for all inputs, since it only depends on the fixed Turing machine M_U. $\square$

Now, the following lemma, which is (2) of Theorem 2, follows from Lemma 5.

Lemma 6. *The inclusion problem of* DPDA *in* NFA *is* ExpTime-*complete.*

Proof. The ExpTime upper bound is immediate (basically, an exponential determinization of the NFA, followed by complementation, product construction with the PDA, and the emptiness check of the product PDA in polynomial time in the size of the product). ExpTime-hardness of the problem follows from Lemma 5. □

Now, we show that the inclusion problem of DPDA in nearly-deterministic NFA, which is ExpTime-complete by Lemma 5, reduces to TED(DPDA, DFA). In the reduction, we transform a nearly-deterministic NFA $\mathcal{A}_N$ over the alphabet Σ into a DFA $\mathcal{A}_D$ by encoding a single non-deterministic choice by auxiliary letters. More precisely, for the transition relation $\delta = \delta_1 \cup \delta_2$ of $\mathcal{A}_N$, we transform every transition $(q, a, q') \in \delta_2$ into $(q, b^{(q,a,q')}, q')$, where $b^{(q,a,q')}$ is a fresh auxiliary letter. Now, consider a DPDA $\mathcal{A}_P$ over the alphabet Σ. As every word in $\mathcal{L}(\mathcal{A}_D)$ contains a single auxiliary letter $ed(\mathcal{L}(\mathcal{A}_P), \mathcal{L}(\mathcal{A}_D)) \geq 1$. Conversely, for every word $w \in \Sigma^*$ we have $ed(w, \mathcal{L}(\mathcal{A}_D)) \leq 1$ implies $w \in \mathcal{A}_N$. Therefore, $ed(\mathcal{L}(\mathcal{A}_P), \mathcal{L}(\mathcal{A}_D)) \leq 1$ if and only if $\mathcal{L}(\mathcal{A}_P) \subseteq \mathcal{L}(\mathcal{A}_N)$.

Lemma 7. TED(DPDA, DFA) *is* ExpTime-*hard.*

4   Finite Edit Distance from Pushdown to Regular Languages

In this section we study the complexity of the FED problem from pushdown automata to finite automata.

Theorem 8. *(1) For $\mathcal{C}_1 \in \{$DPDA, PDA$\}$ and $\mathcal{C}_2 \in \{$DFA, NFA$\}$ we have the following dichotomy: for all $\mathcal{A}_1 \in \mathcal{C}_1, \mathcal{A}_2 \in \mathcal{C}_2$ either $ed(\mathcal{L}(\mathcal{A}_1), \mathcal{L}(\mathcal{A}_2))$ is exponentially bounded in $|\mathcal{A}_1| + |\mathcal{A}_2|$ or $ed(\mathcal{L}(\mathcal{A}_1), \mathcal{L}(\mathcal{A}_2))$ is infinite. Conversely, for every n there exist a DPDA $\mathcal{A}_P$ and a DFA $\mathcal{A}_D$, both of the size $O(n)$, such that $ed(\mathcal{L}(\mathcal{A}_P), \mathcal{L}(\mathcal{A}_D))$ is finite and exponential in n (i.e., the dichotomy is asymptotically tight). (2) For $\mathcal{C}_1 \in \{$DPDA, PDA$\}$ the FED($\mathcal{C}_1$, NFA) problem is* ExpTime-*complete. (3) Given a PDA $\mathcal{A}_P$ and an NFA $\mathcal{A}_N$, we can compute the edit distance $ed(\mathcal{L}(\mathcal{A}_P), \mathcal{L}(\mathcal{A}_N))$ in time exponential in $|\mathcal{A}_P| + |\mathcal{A}_N|$.*

First, we show in Section 4.1 the exponential upper bound for (1), which together with Theorem 2, implies the ExpTime upper bound for (2). Next, in Section 4.2, we show that FED(DPDA, NFA) is ExpTime-hard. We also present the exponential lower bound for (1). Finally, (1), (2), and Theorem 2 imply (3) (by iteratively testing with increasing thresholds up to exponential bounds along with the decision procedure from Theorem 2).

4.1   Upper Bound

In this section we consider the problem of deciding whether the edit distance from a PDA to an NFA is finite. We start with a reduction of the problem. Given a language $\mathcal{L}$, we define $\overline{\mathcal{L}} = \{u : u \text{ is a prefix of some word from } \mathcal{L}\}$. We call an automaton $\mathcal{A}$ *safety* if every state of $\mathcal{A}$ is accepting. Note that an automaton is not necessarily total, i.e., some states might not have an outgoing transition for some input symbols, and thus a safety automaton does not necessarily accept all words. Note that for every NFA $\mathcal{A}_N$, the language $\mathcal{L}(\mathcal{A}_N)$ is the language of a safety NFA. We show that FED(PDA, NFA) reduces to FED from PDA to safety NFA.

Lemma 9. *Let $\mathcal{A}_P$ be a PDA and $\mathcal{A}_N$ an NFA. The following inequalities hold:*

$$ed(\mathcal{L}(\mathcal{A}_P), \mathcal{L}(\mathcal{A}_N)) \geq ed(\mathcal{L}(\mathcal{A}_P), \mathcal{L}(\mathcal{A}_N)) \geq ed(\mathcal{L}(\mathcal{A}_P), \mathcal{L}(\mathcal{A}_N)) - |\mathcal{A}_N|$$

The following definition and lemma can be seen as a reverse version of the pumping lemma for context free grammars (in that we ensure that the part which can not be pumped is small).

Compact G-decomposition. Given a CFG $G = (\Sigma, V, s, P)$, where $T = |V|$, and a word $w \in \mathcal{L}(G)$ we define *compact G-decomposition* of w as $w = (s_i u_i)_{i=1}^{k} s_{k+1}$, where s_i and u_i are subwords of w for all i, such that
  1. for all ℓ, the word $w(\ell) := (s_i u_i^\ell)_{i=1}^{k} s_{k+1}$ is in $\mathcal{L}(G)$; and
  2. $|w(0)| = \sum_{i=1}^{k+1} |s_i| \leq 2^T$ and $k \leq 2^{T+1} - 2$.

Lemma 10. *For every CFG $G = (\Sigma, V, s, P)$, every word $w \in \mathcal{L}(G)$ admits a compact G-decomposition.*

Intuition. The proof follows by repeated application of the principle behind the pumping lemma, until the part which is not pump-able is small.

Reachability Sets. Fix an NFA. Given a state q in the NFA and a word w, let Q_q^w be the set of states reachable upon reading w, starting in q. The set of states $R(w, q)$ is then the set of states reachable from Q_q^w upon reading any word. For a set Q' and word w, the set $R(w, Q')$ is $\bigcup_{q \in Q'} R(w, q)$.

We have the following **property of reachability sets**: Fix a word u, a number ℓ, an NFA and a set of states Q' of the NFA, where Q' is *closed under reachablity*, i.e., for all $q \in Q'$ and $a \in \Sigma$ we have $\delta(q, a) \subseteq Q'$. Let u' be a word with ℓ non-overlapping occurrences of u (e.g. u^ℓ). Consider any word w with edit distance strictly less than ℓ from u'. Any run on w, starting in some state of Q', reaches a state of $R(u, Q')$. This is because u must be a sub-word of w.

Lemma 11. *Let G be a CFG with a set of non-terminals of size T and let $\mathcal{A}_N$ be a safety NFA with state set Q of size n. The following conditions are equivalent:*

  (i) the edit distance $ed(\mathcal{L}(G), \mathcal{L}(\mathcal{A}_N))$ is infinite,
  (ii) the edit distance $ed(\mathcal{L}(G), \mathcal{L}(\mathcal{A}_N))$ exceeds $B := (2^{T+1} - 2) \cdot n + 2^T$, and
  (iii) there exists a word $w \in \mathcal{L}(G)$, with compact G-decomposition $w = (s_i u_i)_{i=1}^{k} s_{k+1}$, such that $R(u_k, R(u_{k-1}, R(u_{k-2}, \dots R(u_1, Q) \dots))) = \emptyset$.

Intuition behind the proof: Whenever we consider a word w, the compact G-representation of it is $w = (s_i u_i)_{i=1}^{k} s_{k+1}$. Let

$$R(w, j) = R(u_j, R(u_{j-1}, R(u_{j-2}, \dots R(u_1, Q) \dots)))$$

for all j and words w. Observe that **(i)** $\Rightarrow$ **(ii)** is trivial. Intuitively, the proof for **(ii)** $\Rightarrow$ **(iii)** is by contradiction: Consider a word w in $\mathcal{L}(G)$ with edit distance above B from $\mathcal{L}(\mathcal{A}_N)$. Assume towards contradiction that $R(w, k)$ is not empty. Then there is a word $w' = (s'_i u_i)^k_{i=1}$ in $\mathcal{L}(\mathcal{A}_N)$ where each s'_i has length at most n. But $ed(w, w') \le B$ by definition of compact G-representation (i.e. edit each s'_i to s_i separately), which is a contradiction. To show **(iii)** $\Rightarrow$ **(i)** we consider a word w where $R(w, k)$ is empty and we show that $w(\ell)$ (from compact G-representation) requires at least ℓ edits to $\mathcal{L}(\mathcal{A}_N)$. Inductively in j, there must be either at least ℓ edits on $(s_i u^\ell_i)^j_{i=1}$ or $R(w, j)$ has been reached, by the property of reachability sets. Since $R(w, k)$ is empty, there must be ℓ edits on $w(\ell)$.

The equivalence of (i) and (ii) of Lemma 11 gives a bound on the maximum finite edit distance. The following lemma follows from Lemmas 9 and 11, and Theorem 2 for testing given thresholds.

Lemma 12. *For all* $\mathcal{C}_1 \in \{\mathsf{DPDA}, \mathsf{PDA}\}, \mathcal{C}_2 \in \{\mathsf{DFA}, \mathsf{NFA}\}$ *the* $\mathsf{FED}(\mathcal{C}_1, \mathcal{C}_2)$ *problem is in* $\mathsf{ExpTime}$.

4.2 Lower Bound

We have shown the exponential upper bound on the edit distance if it is finite. It is easy to define a family of CFGs only accepting an exponential length word, using repeated doubling and thus the edit distance can be exponential between DPDA and DFA. We also show that the inclusion problem reduces to the finite edit distance problem $\mathsf{FED}(\mathsf{DPDA}, \mathsf{NFA})$ and get the following lemma.

Lemma 13. $\mathsf{FED}(\mathsf{DPDA}, \mathsf{NFA})$ *is* $\mathsf{ExpTime}$-*hard.*

We conjecture that, as for the case of language inclusion, for the finite edit distance problem the complexity of the DPDA/PDA to DFA problem matches the one for *NFA/DFA* to DFA.

Conjecture 14. $\mathsf{FED}(\mathsf{PDA}, \mathsf{DFA})$ is coNP-complete.

5 Edit Distance to PDA

Observe that the threshold distance problem from DFA to PDA with the fixed threshold 0 and a fixed DFA recognizing Σ^* coincides with the universality problem for PDA. Hence, the universality problem for PDA, which is undecidable, reduces to $\mathsf{TED}(\mathsf{DFA}, \mathsf{PDA})$. The universality problem for PDA reduces to $\mathsf{FED}(\mathsf{DFA}, \mathsf{PDA})$ as well by the same argument as in Lemma 13. Finally, we can reduce the inclusion problem from DPDA in DPDA, which is undecidable, to $\mathsf{TED}(\mathsf{DPDA}, \mathsf{DPDA})$ (resp., $\mathsf{FED}(\mathsf{DPDA}, \mathsf{DPDA})$). Again, we can use the same construction as in Lemma 13. In conclusion, we have the following proposition.

Proposition 15. *(1) For every class* $\mathcal{C} \in \{\mathsf{DFA}, \mathsf{NFA}, \mathsf{DPDA}, \mathsf{PDA}\}$, *the problems* $\mathsf{TED}(\mathcal{C}, \mathsf{PDA})$ *and* $\mathsf{FED}(\mathcal{C}, \mathsf{PDA})$ *are undecidable. (2) For every class* $\mathcal{C} \in \{\mathsf{DPDA}, \mathsf{PDA}\}$, *the problem* $\mathsf{FED}(\mathcal{C}, \mathsf{DPDA})$ *is undecidable.*

The results in (1) of Proposition 15 are obtained by reduction from the universality problem for PDA. However, the universality problem for DPDA is decidable. Still we show that TED(DFA, DPDA) is undecidable. The overall argument is similar to the one in Section 3.2. First, we define a pushdown counterpart of nearly-deterministic NFA. A PDA $\mathcal{A} = (\Sigma, \Gamma, Q, S, \delta, F)$ is *nearly-deterministic* if $|S| = 1$ and $\delta = \delta_1 \cup \delta_2$, where δ_1 is a function and for every accepting run, the automaton takes a transition from δ_2 exactly once.

By carefully reviewing the standard reduction of the halting problem for Turing machines to the universality problem for pushdown automata [11], we observe that the PDA that appear in the reduction are nearly-deterministic.

Lemma 16. *The problem, given a nearly-deterministic PDA $\mathcal{A}_P$, decide whether $\mathcal{L}(\mathcal{A}_P) = \Sigma^*$, is undecidable.*

Using the same construction as in Lemma 7 we show a reduction of the universality problem for nearly-deterministic PDA to TED(DFA, DPDA).

Proposition 17. *For every class $\mathcal{C} \in \{\mathsf{DFA}, \mathsf{NFA}, \mathsf{DPDA}, \mathsf{PDA}\}$, the problem TED($\mathcal{C}$, DPDA) is undecidable.*

We presented the complete decidability picture for the problems TED($\mathcal{C}_1, \mathcal{C}_2$), for $\mathcal{C}_1 \in \{\mathsf{DFA}, \mathsf{NFA}, \mathsf{DPDA}, \mathsf{PDA}\}$ and $\mathcal{C}_2 \in \{\mathsf{DPDA}, \mathsf{PDA}\}$. To complete the characterization of the problems FED($\mathcal{C}_1, \mathcal{C}_2$), with respect to their decidability, we still need to settle the decidability (and complexity) status of FED(DFA, DPDA). We leave it as an open problem, but conjecture that it is coNP-complete.

Conjecture 18. FED(DFA, DPDA) is coNP-complete.

References

1. Aho, A., Peterson, T.: A minimum distance error-correcting parser for context-free languages. SIAM J. of Computing **1**, 305–312 (1972)
2. Benedikt, M., Puppis, G., Riveros, C.: Regular repair of specifications. In: LICS 2011, pp. 335–344 (2011)
3. Benedikt, M., Puppis, G., Riveros, C.: Bounded repairability of word languages. J. Comput. Syst. Sci. **79**(8), 1302–1321 (2013)
4. Chandra, A.K., Kozen, D.C., Stockmeyer, L.J.: Alternation. J. ACM **28**(1), 114–133 (1981). http://doi.acm.org/10.1145/322234.322243
5. Chatterjee, K., Doyen, L., Henzinger, T.A.: Quantitative languages. ACM Trans. Comput. Log. 11(4) (2010)
6. Chatterjee, K., Henzinger, T.A., Ibsen-Jensen, R., Otop, J.: Edit distance for pushdown automata. CoRR abs/1504.08259 (2015). http://arxiv.org/abs/1504.08259
7. Chatterjee, K., Henzinger, T.A., Otop, J.: Nested weighted automata. CoRR abs/1504.06117 (2015). http://arxiv.org/abs/1504.06117 (to appear at LICS 2015)
8. Chatterjee, K., Ibsen-Jensen, R., Majumdar, R.: Edit distance for timed automata. In: HSCC 2014, pp. 303–312 (2014)

9. Gawrychowski, P.: Faster algorithm for computing the edit distance between SLP-compressed strings. In: Calderón-Benavides, L., González-Caro, C., Chávez, E., Ziviani, N. (eds.) SPIRE 2012. LNCS, vol. 7608, pp. 229–236. Springer, Heidelberg (2012)

10. Henzinger, T.A., Otop, J.: From model checking to model measuring. In: D'Argenio, P.R., Melgratti, H. (eds.) CONCUR 2013 – Concurrency Theory. LNCS, vol. 8052, pp. 273–287. Springer, Heidelberg (2013)

11. Hopcroft, J.E., Ullman, J.D.: Introduction to Automata Theory, Languages, and Computation. Adison-Wesley Publishing Company, Reading (1979)

12. Karp, R.: Mapping the genome: some combinatorial problems arising in molecular biology. In: STOC 93, pp. 278–285. ACM (1993)

13. Levenshtein, V.I.: Binary codes capable of correcting deletions, insertions, and reversals. Soviet physics doklady. 10, 707–710 (1966)

14. Lifshits, Y.: Processing compressed texts: a tractability border. In: Ma, B., Zhang, K. (eds.) CPM 2007. LNCS, vol. 4580, pp. 228–240. Springer, Heidelberg (2007)

15. Mohri, M.: Edit-distance of weighted automata: general definitions and algorithms. Intl. J. of Foundations of Comp. Sci. 14, 957–982 (2003)

16. Okuda, T., Tanaka, E., Kasai, T.: A method for the correction of garbled words based on the levenshtein metric. IEEE Trans. Comput. 25, 172–178 (1976)

17. Pighizzini, G.: How hard is computing the edit distance? Information and Computation 165, 1–13 (2001)

18. Saha, B.: The dyck language edit distance problem in near-linear time. In: FOCS 2014, pp. 611–620 (2014)

Solution Sets for Equations over Free Groups are EDT0L Languages

Laura Ciobanu[1], Volker Diekert[2]([⊠]), and Murray Elder[3]

[1] Institut de Mathématiques, Université de Neuchâtel, Neuchâtel, Switzerland
[2] Institut für Formale Methoden der Informatik, Universität Stuttgart,
Stuttgart, Germany
diekert@fmi.uni-stuttgart.de
[3] School of Mathematical and Physical Sciences, The University of Newcastle,
Callaghan, Australia

Dedicated to Manfred Kudlek (1940–2012)

Abstract. We show that, given a word equation over a finitely generated free group, the set of all solutions in reduced words forms an EDT0L language. In particular, it is an indexed language in the sense of Aho. The question of whether a description of solution sets in reduced words as an indexed language is possible has been open for some years [9,10], apparently without much hope that a positive answer could hold. Nevertheless, our answer goes far beyond: they are EDT0L, which is a proper subclass of indexed languages. We can additionally handle the existential theory of equations with rational constraints in free products $\star_{1 \leq i \leq s} F_i$, where each F_i is either a free or finite group, or a free monoid with involution. In all cases the result is the same: the set of all solutions in reduced words is EDT0L. This was known only for quadratic word equations by [8], which is a very restricted case. Our general result became possible due to the recent recompression technique of Jeż. In this paper we use a new method to integrate solutions of linear Diophantine equations into the process and obtain more general results than in the related paper [5]. For example, we improve the complexity from quadratic nondeterministic space in [5] to quasi-linear nondeterministic space here. This implies an improved complexity for deciding the existential theory of non-abelian free groups: NSPACE($n \log n$). The conjectured complexity is NP, however, we believe that our results are optimal with respect to space complexity, independent of the conjectured NP.

Research supported by the Australian Research Council FT110100178 and the University of Newcastle G1301377. The first author was supported by a Swiss National Science Foundation Professorship FN PP00P2-144681/1. The first and third authors were supported by a University of Neuchâtel Overhead grant in 2013.
Manfred Kudlek has the distinction of being the only person to have attended all ICALP conferences during his lifetime. He worked on Lindenmayer systems, visited Kyoto several times, and taught the second author that bikes are the best means of transport inside Kyoto.

M.M. Halldórsson et al. (Eds.): ICALP 2015, Part II, LNCS 9135, pp. 134–145, 2015.
DOI: 10.1007/978-3-662-47666-6_11

Introduction

The first algorithmic description of all solutions to a given equation over a free group is due to Razborov [17,18]. His description became known as a *Makanin-Razborov diagram*. This concept plays a major role in the positive solution of Tarski's conjectures about the elementary theory in free groups [12,21].

It was however unknown that there is an amazingly simple formal language description for the set of all solutions of an equation over free groups in reduced words: they are EDT0L. An EDT0L language L is given by a nondeterministic finite automaton (NFA), where transitions are labeled by endomorphisms in a free monoid which contains a symbol #. Such an NFA defines a rational language $\mathcal{R}$ of endomorphisms, and the condition on L is that $L = \{h(\#) \mid h \in \mathcal{R}\}$. The NFA we need for our result can be computed effectively in nondeterministic quasi-linear space, i.e., by some $\mathsf{NSPACE}(n \log n)$ algorithm. As a consequence, the automaton has singly exponential size $2^{\mathcal{O}(n \log n)}$ in the input size n.

A description of solution sets as EDT0L languages was known before only for quadratic word equations by [8]; the recent paper [5] did not aim at giving such a structural result. There is also a description of all solutions for a word equation by Plandowski in [14]. His description is given by some graph which can computed in singly exponential time, but without the aim to give any formal language characterization. Plandowski claimed in [14] that his method applies also to free groups with rational constraints, but he found a gap [15].

The technical results are as follows. Let $\mathrm{F}(A_+)$ be the free group over a finite generating set A_+ of (positive) letters. We let $A_\pm = A_+ \cup \{a^{-1} \mid a \in A_+\} \subseteq \mathrm{F}(A_+)$. We view $A_\pm$ as a finite alphabet (of *constants*) with the involution $\bar{a} = a^{-1}$. The involution is extended to the free monoid $A_\pm^*$ by $\overline{a_1 \cdots a_k} = \overline{a_k} \cdots \overline{a_1}$. We let $\pi : A_\pm^* \to \mathrm{F}(A_+)$ be the canonical morphism. As a set, we identify $\mathrm{F}(A_+)$ with the rational (i.e., regular) subset of reduced words inside $A_\pm^*$. A word is *reduced* if it does not contain any factor $a\bar{a}$ where $a \in A_\pm$. Thus, $w \in A_\pm^*$ is reduced if and only if $\pi(w) = w$. We emphasize that $\mathrm{F}(A_+)$ is realized as a subset of $A_\pm^*$. Let Ω be a set of *variables* with involution. An *equation* over $\mathrm{F}(A_+)$ is given as a pair (U, V), where $U, V \in (A_\pm \cup \Omega)^*$ are words over constants and variables. A *solution* of (U, V) is a mapping $\sigma : \Omega \to A_\pm^*$ which respects the involution such that $\pi\sigma(U) = \pi\sigma(V)$ holds in $\mathrm{F}(A_+)$. As usual, σ is extended to a morphism $\sigma : (A_\pm \cup \Omega)^* \to A_\pm^*$ by leaving constants invariant. Throughout we let # denote a special symbol, whose main purpose is to encode a tuple of words $(w_1, \ldots, w_k)$ as a single word $w_1 \# \cdots \# w_k$.

Theorem 1. *Let (U, V) be an equation over $\mathrm{F}(A_+)$ and $\{X_1, \ldots, X_k\}$ be any specified subset of variables. Then the solution set $\mathrm{Sol}(U, V)$ is EDT0L where $\mathrm{Sol}(U, V) = \{\sigma(X_1) \# \cdots \# \sigma(X_k) \mid \sigma \text{ solves } (U, V) \text{ in reduced words}\}$.*

Moreover, there is a nondeterministic algorithm which takes (U, V) as input and computes an NFA $\mathcal{A}$ such that $\mathrm{Sol}(U, V) = \{\varphi(\#) \mid \varphi \in L(\mathcal{A})\}$ in quasi-linear space.

The statement of Theorem 1 shifts the perspective on how to solve equations. Instead of solving an equation, we focus on an effective construction of some

NFA producing the EDT0L set. Once the NFA is constructed, the existence of a solution, or whether the number of solutions is zero, finite or infinite, become graph properties of the NFA.

Theorem 1 is a special case of a more general result involving the existential theory with rational constraints over free products. The generalization is done in several directions. First, we can replace $F(A_+)$ by any finitely generated free product $\mathbb{F} = \star_{1 \le i \le s} F_i$ where each F_i is either a free or finite group, or a free monoid with arbitrary involutions (including the identity). Thus, for example we may have $\mathbb{F} = \{a, b\}^* \star \mathbb{Z} \star \mathrm{PSL}(2, \mathbb{Z}) = \{a, b\}^* \star \mathbb{Z} \star (\mathbb{Z}/3\mathbb{Z}) \star (\mathbb{Z}/2\mathbb{Z})$ where $\bar{a} = a$ and $\bar{b} = b$. Second, we allow arbitrary rational constraints. We consider Boolean formulae Φ, where each atomic formula is either an equation or a *rational constraint*, written as $X \in L$, where $L \subseteq \mathbb{F}$ is a rational subset.

Theorem 2. *Let $\mathbb{F}$ be a free product as above, Φ a Boolean formula over equations and rational constraints, and $\{X_1, \ldots, X_k\}$ any subset of variables. Then $\mathrm{Sol}(\Phi) = \{\sigma(X_1)\# \cdots \#\sigma(X_k) \mid \sigma$ solves Φ in reduced words$\}$ is EDT0L.*

Moreover, there is an algorithm which takes Φ as input and produces an NFA $\mathcal{A}$ such that $\mathrm{Sol}(\Phi) = \{\varphi(\#) \mid \varphi \in L(\mathcal{A})\}$. The algorithm is nondeterministic and uses quasi-linear space in the input size $\|\Phi\|$.

For lack of space we present the main steps used to show Theorem 1, only. However, this covers the essential ideas to prove the more general result in Theorem 2 as well. All missing proofs and details are in our paper on the arXiv.

Preliminaries

The notion of a *rational set* is defined in any monoid, and a rational set can be specified by some NFA with arcs labeled by monoid elements, see [7]. Traditionally, rational sets in finitely generated free monoids are also called *regular*. If M is a monoid and $u, v \in M$, then we write $u \le v$ if u is a *factor* of v, which means we can write $v = xuy$ for some $x, y \in M$. We denote the neutral element in M by 1, thus, the empty word is also 1. The length of word w is denoted by $|w|$, and $|w|_a$ counts how often a letter a appears in w. An *involution* of a set A is a mapping $x \mapsto \bar{x}$ such that $\bar{\bar{x}} = x$ for all $x \in A$. For example, the identity map is an involution. A *morphism* between sets with involution is a mapping respecting the involution. A *monoid with involution* has to additionally satisfy $\overline{xy} = \bar{y}\,\bar{x}$. A *morphism* between monoids with involution is a homomorphism $\varphi : M \to N$ such that $\varphi(\bar{x}) = \overline{\varphi(x)}$. It is a Δ-*morphism* if $\varphi(x) = x$ for all $x \in \Delta$ where $\Delta \subseteq M$. In this article, whenever the term "morphism" is used it refers to a mapping which respects the underlying structure including the involution. All groups are monoids with involution given by $\bar{x} = x^{-1}$, and all group-homomorphisms are morphisms. Any involution on a set A extends to A^*: for a word $w = a_1 \cdots a_m$ we let $\bar{w} = \overline{a_m} \cdots \overline{a_1}$. If $\bar{a} = a$ for all $a \in A$ then $\bar{w}$ is simply the word w read from right-to-left. The monoid A^* is called a *free monoid with involution*.

The notion of an *EDT0L system* refers to **E**xtended, **D**eterministic, **T**able, **0** interaction, and **L**indenmayer. There is a vast literature on Lindenmayer systems, see [19], with various acronyms such as D0L, DT0L, ET0L, and HDT0L.

The subclass EDT0L is equal to HDT0L (see e.g. [20, Thm. 2.6]), and has received particular attention. We content ourselves to define EDT0L through a characterization (using rational control) due to Asveld [2]. The class of EDT0L languages is a proper subclass of indexed languages in the sense of [1], see [6]. For more background we refer to [20].

Definition 1. *Let A be an alphabet and $L \subseteq A^*$. We say that L is* EDT0L *if there is an alphabet C with $A \subseteq C$, a rational set of endomorphisms $\mathcal{R} \subseteq$ End(C^*), and a symbol $\# \in C$ such that $L = \{\varphi(\#) \mid \varphi \in \mathcal{R}\}$.*

Note that for a set $\mathcal{R}$ of endomorphisms of C^* we have $\{\varphi(\#) \mid \varphi \in \mathcal{R}\} \subseteq C^*$, in general. Our definition implies that $\mathcal{R}$ must guarantee that $\varphi(\#) \in A^*$ for all $\varphi \in R$. The set $\mathcal{R}$ is the *rational control*, and C is the *extended alphabet*.

Example 1. Let $A = \{a, b\}$ and $C = \{a, b, \#, \$\}$. We let H be the set of four endomorphisms f, g_a, g_b, h satisfing $f(\#) = \$\$$, $g_a(\$) = \$a$, $g_b(\$) = \$b$, and $h(\$) = 1$, and on all other letters the f, g_a, g_b, h behave like the identity. Consider the rational language $\mathcal{R} \subseteq H^*$ defined by $\mathcal{R} = h\{g_a, g_b\}^* f$ (where endomorphisms are applied right-to-left). A simple inspection shows that $\{\varphi(\#) \mid \varphi \in R\} = \{vv \mid v \in A^*\}$, which is not context-free.

Proof of Theorem 1

Preprocessing. We start by adding the special symbol $\#$ to the alphabet $A_\pm$, and define $A = A_\pm \cup \{\#\}$. We let $\overline{\#} = \#$; this will be the only self-involuting letter in this article. We must make sure that no solution uses $\#$ and every solution is in reduced words. We do so by introducing a finite monoid N with involution which plays the role of (a specific) rational constraint. Let $N = \{1, 0\} \cup A_\pm \times A_\pm$. We define a multiplication on N by $1 \cdot x = x \cdot 1 = x$, $0 \cdot x = x \cdot 0 = 0$, and

$$(a, b) \cdot (c, d) = \begin{cases} (a, d) & \text{if } b \neq \overline{c} \\ 0 & \text{otherwise.} \end{cases}$$

The monoid N has an involution by $\overline{1} = 1$, $\overline{0} = 0$, and $\overline{(a, b)} = (\overline{b}, \overline{a})$. Consider the morphism $\mu_0 : A^* \to N$ given by $\mu_0(\#) = 0$ and $\mu_0(a) = (a, a)$ for $a \in A_\pm$. It is clear that μ_0 respects the involution and $\mu_0(w) = 0$ if and only if either w contains $\#$ or w is not reduced. If, on the other hand, $1 \neq w \in A_\pm^*$ is reduced, then $\mu_0(w) = (a, b)$, where a is the first and b the last letter of w. Thus, if σ is a solution in reduced words, then for each variable $X \in \Omega$ there exists some element $\mu(X) \in N$ with $0 \neq \mu(X) = \overline{\mu(\overline{X})} \in N$ and $\mu(X) = \mu_0\sigma(X)$. Note that while rational constraints are not explicitly mentioned in Theorem 1, they play an essential role in ensuring that solutions are in reduced words.

Since EDT0L is closed under finite union, and since there are only finitely many choices for $\mu(X)$, we may assume that our input equation is specified together with a fixed morphism $\mu : \Omega \longrightarrow N$. A solution σ is now given by a mapping $\sigma : \Omega \to A^*$ satisfying three properties: $\pi\sigma(U) = \pi\sigma(V)$ (the equation holds in F(A_+)), $\overline{\sigma(X)} = \sigma(\overline{X})$, ($\sigma$ respects the involution), and $\mu(X) = \mu_0\sigma(X)$ for all $X \in \Omega$.

The next steps are standard, see [4]. With the help of additional variables we produce a system of equations (X_i, V_i), $1 \leq i \leq s$, such that each X_i is a variable and each V_i is a word of length 2. (The number s of equations is in $\mathcal{O}(|UV|)$ after this transformation.) Thus, we obtain a *triangular* system of equations. We may still assume that each variable X comes with a value $0 \neq \mu(X) = \overline{\mu(\overline{X})}$ and we extend μ to a morphism μ_{init} by $\mu_{\text{init}}(X) = \mu(X)$ for $X \in \Omega$ and $\mu_{\text{init}}(a) = \mu_0(a)$ for each $a \in A$. Next, by the following lemma, we switch to free monoids with involution. Lemma 1 is well-known and easy to see. Its geometric interpretation is the fact that the Cayley graph of a free group is a tree.

Lemma 1. *Let x, y, z be reduced words in $A_\pm^*$. Then $x = yz$ in the group $F(A_+)$ if and only if there are reduced words P, Q, R in $A_\pm^*$ such that $x = PR$, $y = PQ$, and $z = \overline{Q}R$ hold in the free monoid $A_\pm^*$.*

By Lemma 1 we content ourselves to prove the analogue of Theorem 1 for free monoids with involution, systems of equations $(U_i, V_i)_{1 \leq i \leq s}$, and a morphism $\mu_{\text{init}} : (A \cup \Omega)^* \to N$ such that $0 \neq \mu_{\text{init}}(X)$ for all $X \in \Omega$, where Ω is an enlarged set of variables. We return to a single equation (U', V') over $A \cup \Omega$ by letting $U' = U_1 \# \cdots \# U_s$ and $V' = V_1 \# \cdots \# V_s$. Notice that $\mu_{\text{init}}(X) \neq 0$ for all X and $|U_i|_\# = |V_i|_\# = 0$ for all i. A solution $\sigma : \Omega \to A_\pm^*$ must satisfy $\sigma(U') = \sigma(V')$, $\overline{\sigma(X)} = \sigma(\overline{X})$, and $\mu_{\text{init}}(X) = \mu_0 \sigma(X)$ for all $X \in \Omega$. The set of variables $\{X_1, \ldots, X_k\}$, specified in Theorem 1, is a subset of Ω, and the original solution set is a finite union of solution sets with respect to different choices for μ_{init}. In order to achieve our result we *protect* each variable X_i by defining a factor $\#X_i\#$ as follows. We assume $A_\pm \cup \Omega = \{x_1, \ldots, x_\ell\}$ with $x_i = X_i$ for $1 \leq i \leq k$ where $\{X_1 \ldots, X_k\}$ is the specified subset in the statement of Theorem 1. The word W_{init} over $(A \cup \Omega)^*$ is then defined as:

$$W_{\text{init}} = \#x_1\# \ldots \#x_\ell\#U'\#V'\#\overline{U'}\#\overline{V'}\#\overline{x_\ell}\# \ldots \#\overline{x_1}\#.$$

Observe that W_{init} is longer than (but still linear in) $|A| + |\Omega| + |UV|$. The number of $\#$'s in W_{init} is odd; and if $\sigma : (A \cup \Omega)^* \to A^*$ is a morphism with $\sigma(X)$ reduced for all $X \in \Omega$, then: $\pi\sigma(U) = \pi\sigma(V) \iff \sigma(U') = \sigma(V') \iff \sigma(W_{\text{init}}) = \sigma(\overline{W_{\text{init}}})$. Here (U, V) is the equation in Theorem 1 and (U', V') is the intermediate word equation over $A \cup \Omega$. Therefore Theorem 1 follows by showing that the following language is EDT0L:

$$\left\{ \sigma(X_1)\# \cdots \#\sigma(X_k) \,\middle|\, \sigma(W_{\text{init}}) = \sigma(\overline{W_{\text{init}}}) \wedge \mu_{\text{init}} = \mu_0 \sigma \wedge \forall X : \sigma(\overline{X}) = \overline{\sigma(X)} \right\}.$$

Partial Commutation and Extended Equations. Partial commutation is an important concept in our proof. It pops up where traditionally the unary case (solving a linear Diophantine equation) is used as a black box, as is done in [5]. At first glance it might seem like an unneccesary complication, but in fact the contrary holds. Using partial commutation allows us to encode all solutions completely in the edges of a graph, which we can construct in quasi-linear space, and is one of the major differences to [5]. As a (less important) side effect, results on linear Diophantine equations come for free as this is the special case $F(A_+) = \mathbb{Z}$: solving linear Diophantine equations becomes part of a more general process.

We fix $n = n_{\text{init}} = |W_{\text{init}}|$ and some $\kappa \in \mathbb{N}$ large enough, say $k = 100$. We let C be an alphabet with involution (of constants) such that $|C| = \kappa n$ and $A \subseteq C$. We define $\Sigma = C \cup \Omega$ and assume that $\#$ is the only self-involuting symbol of Σ. In the following $x, y, z, \ldots$ refer to words in Σ^* and $X, Y, Z, \ldots$ to variables in Ω. Throughout we let B, B' and $\mathcal{X}, \mathcal{X}'$ denote subsets which are closed under involution and satisfy $\mathcal{X}' \subseteq \mathcal{X} \subseteq \Omega$ and either $A \subseteq B \subseteq B' \subseteq C$ or $A \subseteq B' \subseteq B \subseteq C$. In particular, B and B' are always comparable.

We encode partial commutation by *types*. Let $\theta \subseteq (B \cup \mathcal{X}) \times B$ denote an irreflexive and antisymmetric relation. It is called a *type* if $(x, y) \in \theta$ implies:

- $(\overline{x}, \overline{y}) \in \theta$,
- $x, y \notin A$,
- $|\theta(x)| \leq 1$, where $\theta(x) = \{y \in B^* \mid (x, y) \in \theta\}$.

The type relation θ can be stored in quasi-linear space. Given θ and $\mu : B \cup \mathcal{X} \to N$ such that $\mu(xy) = \mu(yx)$ for all $(x, y) \in \theta$, we define a free partially commutative monoid with involution by $M(B, \mathcal{X}, \theta, \mu) = (B \cup \mathcal{X})^* / \{xy = yx \mid (x, y) \in \theta\}$ with a morphism $\mu : M(B, \mathcal{X}, \theta, \mu) \to N$. By $M(B, \theta, \mu)$ we denote the submonoid generated by B with the corresponding restrictions of θ and μ. Note that $M(A, \theta, \mu) = M(A, \emptyset, \mu_0)$ is the free monoid A^*.

If w is a factor of $W \in M(B, \mathcal{X}, \theta, \mu)$, then w is called a *proper factor* if $1 \neq w \neq W$ and $|w|_\# = 0$. The numbers $|u|$ and $|u|_a$ are well defined for every $u \in M(B, \mathcal{X}, \theta, \mu)$ since if two words represent the same monoid element then the number of occurences of each letter is the same. Typically we represent w, W by words $w, W \in (B \cup \mathcal{X})^*$, but their interpretation is always in $M(B, \mathcal{X}, \theta, \mu)$.

Definition 2. *We call* $W \in M(B, \mathcal{X}, \theta, \mu)$ *well-formed if* $|W| \leq \kappa n$, $|W|_\# = |W_{\text{init}}|_\#$, *and every proper factor* x *of* W *and every* $x \in B \cup \mathcal{X}$ *satisfies* $\mu(x) \neq 0$. *In addition, if* x *is a proper factor then* $\overline{x}$ *is also a proper factor and for each* $a \in A_\pm$ *there must be a factor* $\#a\#$ *in* W.

An extended equation is a tuple $V = (W, B, \mathcal{X}, \theta, \mu)$ *where* $W \in M(B, \mathcal{X}, \theta, \mu)$ *is well-formed. A* B-*solution of* V *is a* B-*morphism* $\sigma : M(B, \mathcal{X}, \theta, \mu) \to M(B, \theta, \mu)$ *such that* $\sigma(W) = \sigma(\overline{W})$ *and* $\sigma(X) \in y^*$ *whenever* $(X, y) \in \theta$. *A solution of* V *is a pair* (α, σ) *such that* $\alpha : M(B, \theta, \mu) \to A^*$ *is an* A-*morphism satisfying* $\mu_0 \alpha = \mu$ *and* σ *is a* B-*solution.*

W = equation, where the solution is a "palindrome" $\sigma(W) = \overline{\sigma(W)} \in A^*$.
B = alphabet of constants with $\# \in A \subseteq B = \overline{B} \subseteq C$.
$\mathcal{X}$ = variables appearing in W. Hence, $\mathcal{X} = \overline{\mathcal{X}} \subseteq \Omega$.
μ = morphism to control the constraint that the solution is reduced.
θ = partial commutation.

During the process of finding a solution, we change these parameters, and we describe the process in terms of a diagram (directed graph) of states and arcs between them.

The Directed Labeled Graph $\mathcal{G}$. We are now ready to define the directed labeled graph $\mathcal{G}$ which will be the core of the NFA defining the EDT0L language $\mathrm{Sol}(U,V) = \{\sigma(X_1)\# \cdots \#\sigma(X_k) \mid \sigma \text{ solves } (U,V) \text{ in reduced words}\}$.

Define the *vertex set* for $\mathcal{G}$ to be the set of all extended equations $V = (W, B, \mathcal{X}, \theta, \mu)$. The *initial vertices* are of the form $(W_{\mathrm{init}}, A, \Omega, \emptyset, \mu_{\mathrm{init}})$. Due to the different possible choices for μ_{init} there are exponentially many initial vertices. We define the set of *final vertices* by $\{(W, B, \emptyset, \emptyset, \mu) \mid W = \overline{W}\}$. By definition every final vertex trivially has a B-solution $\sigma = \mathrm{id}_B$. (Note that in a final vertex there are no variables.) The arcs in $\mathcal{G}$ are labeled and are of the form $(W, B, \mathcal{X}, \theta, \mu) \xrightarrow{h} (W', B', \mathcal{X}', \theta', \mu')$. Here $h : C^* \to C^*$ is an endomorphism given by a morphism $h : B' \to B^*$ such that h induces a well-defined morphism $h : M(B' \cup \mathcal{X}', \theta', \mu') \to M(B \cup \mathcal{X}, \theta, \mu)$. Note that the direction of the morphism is opposite to the direction of the arc. There will be further restrictions on arcs. For example, we will have $|h(b')| \leq 2$ for all b'. The main idea is as follows. Suppose $(W, B, \mathcal{X}, \theta, \mu) \xrightarrow{h} (W', B', \mathcal{X}', \theta', \mu')$ is an arc, $\alpha : M(B, \theta, \mu) \to M(A, \emptyset, \mu_0)$ is an A-morphism, and $(W', B', \mathcal{X}', \theta', \mu')$ has a B'-solution σ'; then there exists a solution (α, σ) of the vertex $(W, B, \mathcal{X}, \theta, \mu)$. Moreover, for the other direction if (α, σ) solves $V = (W, B, \mathcal{X}, \theta, \mu)$ and V is not final then we can follow an outgoing arc and recover (α, σ) from a solution at the target node. We will make this more precise below.

Compression Arcs. These arcs transform the sets of constants. Let $V = (W, B, \mathcal{X}, \theta, \mu)$ and $V' = (W', B', \mathcal{X}, \theta', \mu')$ be two vertices in $\mathcal{G}$. The compression arcs have the form $V \xrightarrow{h} V'$, where either $h = \mathrm{id}_{C^*}$ is the identity on C^* and we write $h = \varepsilon$ in this case, or h is defined by a mapping $c \mapsto h(c)$ where $c \in B'$. Recall that if a morphism h is defined by $h(c) = u$ for some letter c then, automatically, $h(\overline{c}) = \overline{u}$ and $h(x) = x$ for all $x \in \Sigma$ which are different from c and $\overline{c}$. We assume $0 \neq \mu'(c) = \mu(h(c)) \neq 1$ and $\mu(x) = \mu'(x)$ for all $x \in (B \cap B') \cup \mathcal{X}$ (if not explicitly stated otherwise).

We define compression arcs $(h(W'), B, \mathcal{X}, \theta, \mu) \xrightarrow{h} (W', B', \mathcal{X}, \theta', \mu')$ of the following three types.

1. **(Renaming.)** The morphism h is defined by $h(c) = a$ such that $B \subseteq B' = B \cup \{c, \overline{c}\}$, and $\theta \subseteq \theta'$. Thus, possibly, $\theta \subsetneq \theta'$.
2. **(Compression.)** We have $h(c) = u$ with $1 \neq |u| \leq 2$ and either $B = B'$ and $\theta' = \theta$ or $B \subsetneq B' = B \cup \{c, \overline{c}\}$ and $\theta = \theta' = \emptyset$.
3. **(Alphabet Reduction.)** We have $B' \subsetneq B$, $\theta' = \emptyset$, and h is induced by the inclusion $B' \subseteq B$ which leads to an arc label $h = \varepsilon = \mathrm{id}_{C^*}$.

For the proof of Theorem 1 it is enough to compress words of length at most 2 into a single letter. For Theorem 2 we need additionally arcs of type **2** where $u = a\overline{a}c$ with either $a = c$ (and $\overline{a} = \overline{c}$) or $(a\overline{a}, c\overline{c}) \in \theta$. In particular, the type relation has to be defined in slightly more complicated way. The purpose of arcs of type **3** is to remove letters in B that do not appear in the word W. This allows us to reduce the size of B and also to "kill" partial commutation.

Lemma 2. *Let* $(W, B, \mathcal{X}, \theta, \mu) \xrightarrow{h} (W', B', \mathcal{X}', \theta', \mu')$ *be a compression arc with* $W = h(W')$. *Let* $\alpha : M(B, \theta, \mu) \to M(A, \emptyset, \mu_0)$ *be an* A-*morphism at the vertex* $V = (h(W'), B, \mathcal{X}, \theta, \mu)$ *and let* σ' *be a* B'-*solution to* $V' = (W', B', \mathcal{X}', \theta', \mu')$. *Define a* B-*morphism* $\sigma : M(B, \mathcal{X}, \theta, \mu) \to M(B, \theta, \mu)$ *by* $\sigma(X) = h\sigma'(X)$. *Then* (α, σ) *is a solution at* V, $(\alpha h, \sigma')$ *is a solution at* V' *and* $\alpha\sigma(W) = \alpha h\sigma'(W')$.

Substitution Arcs. These arcs transform variables. Let $V = (W, B, \mathcal{X}, \theta, \mu)$ and $V' = (W', B, \mathcal{X}', \theta', \mu')$ be vertices in $\mathcal{G}$ and $X \in \mathcal{X}$. We assume that $\mathcal{X} = \mathcal{X}' \cup \{X, \overline{X}\}$ and $\mu(x) = \mu'(x)$, as well as $\theta(x) = \theta'(x)$ for all $x \in (B \cup \mathcal{X}) \setminus \{X, \overline{X}\}$. The set of constants is the same on both sides, but $\mathcal{X}'$ might have fewer variables. Substitution arcs are defined by a morphism $\tau : \{X\} \to BX \cup \{1\}$ such that we obtain a B-morphism $\tau : M(B, \mathcal{X}, \theta, \mu) \to M(B, \mathcal{X}', \theta', \mu')$. We let $\varepsilon = \mathrm{id}_{C^*}$ as before. We define substitution arcs $(W, B, \mathcal{X}, \theta, \mu) \xrightarrow{\varepsilon} (\tau(W), B, \mathcal{X}', \theta', \mu')$ if one of the following conditions apply.

4. **(Removing a Variable.)** Let $\mathcal{X}' = \mathcal{X} \setminus \{X, \overline{X}\}$. The B-morphism $\tau : M(B, \mathcal{X}, \theta, \mu) \to M(B, \mathcal{X}', \theta', \mu')$ is defined by $\tau(X) = 1$.
5. **(Variable Typing.)** The purpose of this arc is to introduce some type for variables without changing anything else, so $\mathcal{X}' = \mathcal{X}$ and $\mu' = \mu$. Suppose that $\theta(X) = \emptyset$ and $c \in B$ is a letter with $\mu(Xc) = \mu(cX)$ and such that $\theta' = \theta \cup \{(X, c), (\overline{X}, \overline{c})\}$. The B-morphism $\tau : M(B, \mathcal{X}, \theta, \mu) \to M(B, \mathcal{X}, \theta', \mu)$ is defined by the identity on $B \cup \mathcal{X}$. Note that the condition $\mu(Xc) = \mu(cX)$ implies that if $\mu : M(B, \mathcal{X}, \theta, \mu) \to N$ is well-defined, then $\mu : M(B, \mathcal{X}, \theta', \mu) \to N$ is well-defined, too. The other direction is trivial.
6. **(Substitution of a Variable.)** We have $(B, \mathcal{X}, \theta) = (B', \mathcal{X}', \theta')$. Let $a \in B$ be such that $\theta(X) \subseteq \{a\}$. (For $\theta(X) = \emptyset$ this is true for any $a \in B$.) We suppose that $\mu(X) = \mu(a)\mu'(X)$ (hence, automatically $\mu(\overline{X}) = \mu'(\overline{X})\mu(\overline{a})$) and that $\tau(X) = aX$ defines a morphism $\tau : M(B, \mathcal{X}, \theta, \mu) \to M(B, \mathcal{X}, \theta, \mu')$.

Lemma 3. *Let* $V = (W, B, \mathcal{X}, \theta, \mu) \xrightarrow{\varepsilon} (W', B, \mathcal{X}', \theta', \mu') = V'$ *with* $\varepsilon = \mathrm{id}_{C^*}$ *be a substitution arc with* $W' = \tau(W)$. *Let* $\alpha : M(B, \theta, \mu) \to M(A, \emptyset, \mu_0)$ *be an* A-*morphism at vertex* V *and* σ' *be a* B-*solution to* V'. *Define a* B-*morphism* $\sigma : M(B, \mathcal{X}, \theta, \mu) \to M(B, \theta, \mu)$ *by* $\sigma(X) = \sigma'\tau(X)$. *Then* (α, σ) *is a solution at* V *and* (α, σ') *is a solution at* V'. *Moreover,* $\alpha\sigma(W) = \alpha h\sigma'(W')$ *where* $h = \varepsilon$ *is viewed as the identity on* $\mathrm{id}_{M(B, \theta, \mu)}$.

Proof. Since σ' is a B-solution to V' we have $\sigma(W) = \sigma'(\tau(W)) = \sigma'(\overline{\tau(W)}) = \overline{\sigma'\tau(W)} = \overline{\sigma(W)}$. Hence, (α, σ) is a solution at V. Since $M(B, \theta, \mu) = M(B, \theta', \mu')$ (a possible change in μ or θ concerns variables, only), (α, σ') is a solution at V'. The assertion $\alpha\sigma(W) = \alpha h\sigma'(W')$ is trivial since $W' = \tau(W)$, $\sigma = \sigma'\tau$, and $h = \varepsilon$ induces the identity on $M(B, \theta, \mu)$. $\square$

Proposition 1. *Let* $V_0 \xrightarrow{h_1} V_1 \cdots \xrightarrow{h_t} V_t$ *be a path in* $\mathcal{G}$ *of length* t, *where* $V_0 = (W_{\mathrm{init}}, A, \Omega, \emptyset, \mu_{\mathrm{init}})$ *is an initial and* $V_t = (W', B, \emptyset, \emptyset, \mu)$ *is a final vertex. Then* V_0 *has a solution* (id_A, σ) *with* $\sigma(W_{\mathrm{init}}) = h_1 \cdots h_t(W')$. *Moreover, we have* $W' \in \#u_1\# \cdots \#u_k\#B^*$ *such that* $|u_i|_\# = 0$ *and we can write:*

$$h_1 \cdots h_t(u_1\# \cdots \#u_k) = \sigma(X_1)\# \cdots \#\sigma(X_k), \tag{1}$$

Proof. By definition of final vertices we have $\overline{W'} = W'$ and no variables occur in W'. Hence, id_{B^*} defines the (unique) B-solution of W'. By definition of the arcs, $h = h_1 \cdots h_t : M(B, \emptyset, \mu) \to A^* = M(A, \emptyset, \mu_{\text{init}})$ is an A-morphism which shows that (h, id_{B^*}) solves W'. There is only one A-morphism at V_0, namely id_{A^*}. Using Lemma 2 and Lemma 3 we see first that V_0 has some solution $(\mathrm{id}_{A^*}, \sigma)$ and second, that

$$\mathrm{id}_{A^*}\sigma(W_{\text{init}}) = \mathrm{id}_{A^*}h_1 \cdots h_t \mathrm{id}_{B^*}(W') = h_1 \cdots h_t(W'). \tag{2}$$

Finally, for $1 \leq j \leq t$ we have $h_j(\#) = \#$ and $|h_j(x)|_\# = 0$ for all other symbols. Hence the claim $h_1 \cdots h_t(u_1\# \cdots \#u_k) = \sigma(X_1)\# \cdots \#\sigma(X_k)$. $\qquad\square$

Compression[1]. Consider an initial vertex $V_0 = (W_{\text{init}}, A, \Omega, \emptyset, \mu_{\text{init}})$ with a solution (α, σ). We will show below that $\mathcal{G}$ contains a path $V_0 \xrightarrow{h_1} V_1 \cdots \xrightarrow{h_t} V_t$ to some final vertex $V_t = (W', B, \emptyset, \emptyset, \mu)$ such that $\sigma(W_{\text{init}}) = h_1 \cdots h_t(W')$, and so $\mathcal{G}$ contains all solutions to W_{init}. Let us show why then, indeed, we are almost done with Theorem 1. We augment the graph $\mathcal{G}$ by one more vertex which is just the symbol $\#$. Recall that $\{X_1, \ldots, X_k\}$ is the set of specified variables. Every final vertex $(W', B, \emptyset, \emptyset, \mu)$ has a unique factorization $W' = \#w'\#w''$ with $|w'|_\# = k$. Let us add arcs $(W', B, \emptyset, \emptyset, \mu) \xrightarrow{g_{w'}} \#$ where $g_{w'} : C^* \to C^*$ is the homomorphism (not necessarily respecting the involution) defined by $g_{w'}(\#) = w'$. If we define the NFA $\mathcal{A}$ as $\mathcal{G}$ with this augmentation and if we let $\#$ be the exclusive accepting vertex, then by Proposition 1 we obtain Theorem 1. The construction of $\mathcal{A}$ can easily be implemented by an $\mathsf{NSPACE}(n \log n)$ procedure in such a way that the NFA $\mathcal{A}$ becomes *trim*. This means that every vertex is on some path from an initial to a final vertex. Trimming is important to derive the complexity bounds announced in the abstract[2].

We show the existence of the path corresponding to the solution (α, σ) using an alternation between "block compression" and "pair compression", repeated until we reach a final vertex. The procedures use knowledge of the solution being aimed for. We proceed along arcs in $\mathcal{G}$ of the form $V = (W, B, \mathcal{X}, \theta, \mu) \xrightarrow{h} V' = (W', B', \mathcal{X}', \theta', \mu')$ thereby transforming a solution (α, σ) to V into a solution (α', σ') to V'. However, this is not allowed to be arbitrary: we must keep the invariant $\alpha\sigma(W) = \alpha'h\sigma'(W')$. For example, consider the alphabet reduction where $B' \subsetneq B$ and $W = W' \in (B' \cup \mathcal{X})^*$. In this case we have $h = \mathrm{id}_{C^*}$, which induces the inclusion $\varepsilon : M(B', \emptyset, \mu') \to M(B, \theta, \mu)$. If σ does not use letters outside B' there is no obstacle. In the other case, fortunately, we will need alphabet reduction only when the type relation is empty on both sides. Then we can define $\beta(b) = \alpha(b) \in A^*$ for $b \in B \setminus B'$ and $\beta(b) = b$ for $b \in B'$. We let $\sigma'(X) = \beta\sigma(X)$. This defines a B'-solution at V'. In some sense this is a huge "decompression".

---

[1] Compression became a main tool for solving word equations thanks to [16].

[2] The possibility to trim $\mathcal{A}$ in $\mathsf{NSPACE}(n \log n)$ uses the result of Immerman and Szelepcsényi that nondeterministic space complexity classes are closed under complementation. For a proof of the Immerman-Szelepcsényi Theorem, see e.g. [13].

A word $w \in \Sigma^*$ is a sequence of *positions*, say $1, 2, \ldots, |w|$, and each position is labeled by a letter from Σ. If $W = u_0 x_1 u_1 \cdots x_m u_m$, with $u_i \in C^*$ and $x_i \in \Omega$, then $\sigma(W) = u_0 \sigma(x_1) u_1 \cdots \sigma(x_m) u_m$ and the positions in $\sigma(W)$ corresponding to the u_i's are henceforth called *visible*.

Block compression. Let $V = (W, B, \mathcal{X}, \emptyset, \mu)$ be some current non-final vertex with an empty type relation and a solution (α, σ). We start a block compression only if $B \leq |W| \leq 29n$. Since $|C| = 100n$, there will be sufficiently many "fresh" letters in $C \setminus B$ at our disposal.

1. Follow substitution arcs to remove all variables with $|\sigma(X)| \leq 2$. If V became final, we are done and we stop. Otherwise, for each X we have $\sigma(X) = bw$ for some $b \in B$ and $w \in B^+$. Following a substitution arc we replace X by bX. Of course, we also replace $\overline{X}$ by $\overline{X}\,\overline{b}$, changing $\mu(X)$ to $\mu(X) = \overline{\mu(\overline{X})} = \mu(w)$ (from now on we will do this without comment). If $bX \leq W$ and $b'X \leq W$ are factors with $b, b' \in B$, then $\# \neq b = b'$ due to the previous substitution $X \mapsto bX$. For each $b \in B \setminus \{\#\}$ define sets $\Lambda_b \subseteq \mathbb{N}$ which contain those $\lambda \geq 2$ such that there is an occurrence of a factor $db^\lambda e$ in $\sigma(W)$ with $d \neq b \neq e$, where at least one of the b's is visible. We also let $\mathcal{X}_b = \{X \in \mathcal{X} \mid bX \leq W \wedge \sigma(X) \in bB^*\}$. Note that $\sum_b |\Lambda_b| + |\mathcal{X}_b| \leq |W|$.
2. Since W is well-formed we have $\Lambda_b = \Lambda_{\overline{b}}$. Fix some subset $B_+ \subseteq B$ such that for each $\# \neq b \in B$ we have $b \in B_+ \iff \overline{b} \neq B_+$. For each $b \in B_+$, where $\Lambda_b \neq \emptyset$, run the following *b-compression*:
3. *b*-**compression.** (This step removes all factors b^ℓ and $\overline{b}^\ell$, $\ell \geq 2$, from W.)
   (a) Introduce fresh letters $c_b, \overline{c_b}$ with $\mu(c_b) = \mu(b)$. In addition, for each $\lambda \in \Lambda_b$ introduce fresh letters $c_{\lambda,b}, \overline{c_{\lambda,b}}$ with $\mu(c_{\lambda,b}) = \mu(b)$. We abbreviate $c = c_b$, $\overline{c} = \overline{c_b}$, $c_\lambda = c_{\lambda,b}$, and $\overline{c_\lambda} = \overline{c_{\lambda,b}}$. We let $h(c_\lambda) = h(c) = b$ and we introduce a type by letting $\theta = \{(c_\lambda, c) \mid \lambda \in \Lambda_b\}$. Renaming arcs (type **1**) realize this transformation.
   So far we did not change W, but we enlarged the alphabet B to B', and introduced partial commutation between the fresh letters c_λ and c. The next steps change W and its solution.
   (b) Replace in $\sigma(W) \in B^*$ every factor $db^\lambda e$ (resp. $d\overline{b}^\lambda e$), where $d \neq b \neq e$ and $\lambda \in \Lambda_b$, by $dc^\lambda e$ (resp. $d\overline{c}^\lambda e$). This yields a new word $W' \in B'^*$, which was obtained via the renaming arc $h(c) = b$. Recall that for every $X \in \mathcal{X}_b$ we had $bX \leq W$ and for some positive ℓ we had $\sigma(X) = b^\ell w$ with $w \notin bB^*$. In the new word W' we have $cX \leq W'$ and for the new solution σ' we have $\sigma'(X) = c^\ell w'$ with $w' \notin cB'^*$. We rename $W', B', \alpha' = \alpha h, \sigma'$ as W, B, α, σ.
   (c) Enlarge θ by $\{(X, c) \mid X \in \mathcal{X}_b \wedge \sigma(X) \in c^*\}$ using a substitution arc.
   (d) The solution $\sigma(W)$ is still a word $\sigma(W) \in B^*$. Scan this word from left to right. Stop at each factor $dc^\lambda e$ with $d \neq c \neq e$ and $\lambda \in \Lambda_b$. If in this factor some position of the c's is visible then choose exactly one of these visible positions and replace that c by c_λ. If no c is visible, they are all inside some $\sigma(X)$; then choose any c and replace it by c_λ. Recall that c and

c_λ commute, hence $dc^\lambda e$ became $dc_\lambda c^{\lambda-1}e = dc^{\ell_1} c_\lambda c^{\ell_2} e \in M(B, \theta, \mu)$ for all $\ell_1 + \ell_2 = \lambda - 1$. In parallel we run the same steps for $\bar{c}$. The whole transformation can be realized by renaming arcs defined by $h(c_\lambda) = c$. There is a crucial observation: if $X \in \mathcal{X}_b$ and we had $\sigma(X) = c^\ell w$ with $w \notin cB^*$ before the transformation then now still $\sigma'(X) = c^\ell w'$. It is not clear which position has been occupied by c_λ, but due to commutation $c_\lambda \sigma'(X)$ is a factor in $\sigma'(W') \in M(B, \theta, \mu)$.

(e) Rename $W', B', \alpha' = \alpha h, \sigma'$ as W, B, α, σ. Perform the following loop 3(e)i – 3(e)iv until no c and no $X \in \mathcal{X}_b$ with $\sigma(X) \in c^*$ occurs in W.

   i. If $X \in \mathcal{X}_b$ and if the maximal ℓ is odd where $\sigma(X) \in c^\ell B^*$, then follow a substitution arc $X \mapsto cX$. Do the same for $\bar{c}$.

   ii. For all λ where there is some odd ℓ with $dc_\lambda c^\ell e \le \sigma'(W')$ follow a compression arc defined by $h(c_\lambda) = cc_\lambda$. This is possible since for each such factor $dc_\lambda c^\ell e$ either none of the positions in $c_\lambda c^\ell$ is visible or $c_\lambda c$ is visible. Thus, $dc_\lambda c^\ell e \le \sigma'(W')$ implies that ℓ is even.

   iii. Follow a compression arc defined by $h(c) = c^2$, after which $|W'|_c$ and $|W'|_{\bar{c}}$ are divided by 2. We obtain a new W'' with solution σ''.

   iv. Remove all X with $\sigma''(X) = 1$ by following a substitution arc (type **4**); rename all parameters back to $W, B, \mathcal{X}, \theta, \mu, \alpha, \sigma$.

(f) Let $B' = B \setminus \{c, \bar{c}\}$ and μ' be induced by μ. Observe that no letter c or $\bar{c}$ appears in $\sigma(W)$: they have all been consumed by c_λ. Thus, the type relation is empty again. Hence we can follow an alphabet reduction arc $(W, B, \mathcal{X}, \theta, \mu) \xrightarrow{\varepsilon} (W, B', \mathcal{X}, \emptyset, \mu')$. The new solution to $(W, B', \mathcal{X}', \emptyset, \mu')$ is the pair (α', σ) where $\alpha' = \alpha\varepsilon$ is defined by the restriction of α to $M(B', \emptyset, \mu')$.

Having performed b-compressions for all $b \in B_+$, we have increased the length of W. But it is not difficult to see that the total increase can be bounded in $\mathcal{O}(n)$. Actually, we have $|W| \le 31n$ at the end because we started with $|W| \le 29n$ and step 1 of block compression increases $|W|$ by at most $2n$. Now we use alphabet reduction in a final step of block compression in order to reduce the alphabet B such that $|B| \le |W|$. We end up at a vertex again named $V = (W, B, \mathcal{X}, \emptyset, \mu)$, which has a solution (α, σ). The new situation is that no proper factor b^2 appears in W anymore. The price is $|B| \le |W| \le 31n$.

We now run Jeż's procedure *pair compression*, which brings us back to $|B| \le |W| \le 29n$ and allows us to start another block compression. This keeps the length in $\mathcal{O}(n)$. Since our pair compression is very close to Jeż's presentation as published in [11] we content ourselves with the basic idea. For pair compression we begin with a partition $B \setminus \{\#\} = L \cup R$ such that $b \in L \iff \bar{b} \in R$. In general, there are many such partitions, but we choose with care a "good" partition, see below. Next, for all X, if $b \in R$ and $\sigma(X) \in bB^*$ then replace X by bX and $\overline{X}$ by $\overline{X}\bar{b}$. After that, no factor $ab \in LR$ is "crossing", i.e., is consisting of a visible and invisible letter, anymore. Moreover, $ab \in LR \iff \bar{b}\bar{a} \in LR$. Thus, we can follow compression arcs labeled by $h(c) = ab$, where c is a fresh letter. Let $s(i)$ denote the length of W after the i-th iteration of pair-compression. For at least one partition $B \setminus \{\#\} = L \cup R$, the "good" partition, it can be guaranteed that $s(i + 1) \in \frac{5s(i)}{6} + \mathcal{O}(n)$, see [11]. Together with $s(1) \in \mathcal{O}(n)$ this shows $s(i) \in \mathcal{O}(n)$ for all i. Another fact is crucial. We restricted ourselves to solutions in reduced words, which implies that whenever ab is a proper factor of $\sigma(W)$, then $b \neq \bar{a}$.

References

1. Aho, A.V.: Indexed grammars–an extension of context-free grammars. J. Assoc. Comput. Mach. **15**, 647–671 (1968)
2. Asveld, P.R.: Controlled iteration grammars and full hyper-AFL's. Information and Control **34**(3), 248–269 (1977)
3. Benois, M.: Parties rationelles du groupe libre. C. R. Acad. Sci. Paris, Sér. A **269**, 1188–1190 (1969)
4. Diekert, V., Gutiérrez, C., Hagenah, Ch.: The existential theory of equations with rational constraints in free groups is PSPACE-complete. Information and Computation, **202**, 105–140 (2005). Conference version in STACS 2001
5. Diekert, V., Jeż, A., Plandowski, W.: Finding all solutions of equations in free groups and monoids with involution. In: Hirsch, E.A., Kuznetsov, S.O., Pin, J.É., Vereshchagin, N.K. (eds.) CSR 2014. LNCS, vol. 8476, pp. 1–15. Springer, Heidelberg (2014)
6. Ehrenfeucht, A., Rozenberg, G.: On some context free languages that are not deterministic ET0L languages. RAIRO Theor. Inform. Appl. **11**, 273–291 (1977)
7. Eilenberg, S.: Automata, Languages, and Machines, vol. A. Acad Press (1974)
8. Ferté, J., Marin, N., Sénizergues, G.: Word-mappings of level 2. Theory Comput. Syst. **54**, 111–148 (2014)
9. Gilman, R.H.: Personal communication (2012)
10. Jain, S., Miasnikov, A., Stephan, F.: The complexity of verbal languages over groups. In: Proc. LICS 2012, pp. 405–414. IEEE Computer Society (2012)
11. Jeż, A.: Recompression: a simple and powerful technique for word equations. In: Proc. STACS. LIPIcs, 20:233–244 (2013). Journal version to appear in JACM
12. Kharlampovich, O., Myasnikov, A.: Elementary theory of free non-abelian groups. J. of Algebra **302**, 451–552 (2006)
13. Papadimitriou, C.H.: Computational Complexity. Addison Wesley (1994)
14. Plandowski, W.: An efficient algorithm for solving word equations. Proc. STOC 2006, pp. 467–476. ACM Press (2006)
15. Plandowski, W.: Personal communication (2014)
16. Plandowski, W., Rytter, W.: Application of Lempel-Ziv encodings to the solution of word equations. In: Larsen, K.G., Skyum, S., Winskel, G. (eds.) ICALP 1998. LNCS, vol. 1443, pp. 731–742. Springer, Heidelberg (1998)
17. Razborov, A.A.: On Systems of Equations in Free Groups. Ph.D thesis (1987)
18. Razborov, A.A.: On systems of equations in free groups. In: Combinatorial and Geometric Group Theory, pp. 269–283. Cambridge University Press (1994)
19. Rozenberg, G., Salomaa, A.: The Book of L. Springer (1986)
20. Rozenberg, G., et al. (Eds.): Handbook of Formal Languages, vol 1. Springer (1997)
21. Sela, Z.: Diophantine geometry over groups VIII: Stability. Annals of Math. **177**, 787–868 (2013)

Limited Set Quantifiers over Countable Linear Orderings

Thomas Colcombet and A.V. Sreejith[✉]

LIAFA, Université Paris-Diderot, Paris, France
{thomas.colcombet,sreejith}@liafa.univ-paris-diderot.fr

Abstract. In this paper, we study several sublogics of monadic second-order logic over countable linear orderings, such as first-order logic, first-order logic on cuts, weak monadic second-order logic, weak monadic second-order logic with cuts, as well as fragments of monadic second-order logic in which sets have to be well ordered or scattered. We give decidable algebraic characterizations of all these logics and compare their respective expressive power.

Keywords: Linear orderings · Algebraic characterization · Monadic second order logic

1 Introduction

Monadic second-order logic (*i.e.*, first-order logic extended with set quantifiers) is a concise and expressive logic that retains good decidability properties (though with a bad complexity). In particular, since the seminal works of Büchi [3], Rabin [11] and Shelah [13], it is known to be decidable over infinite linear orderings with countably many elements, such as $(\mathbb{Q}, <)$ [5,7]. A breakthrough result of Shelah (also in [13]) states that over general linear orderings (*i.e.*, not necessarily countable), or simply over $(\mathbb{R}, <)$, this logic is not decidable anymore. There is also a long line of research focusing on the analysis of the expressive power and decidability status of temporal logics, which, for most of them are equivalent in expressiveness to first-order logic (but much more tractable), and can be decided on some non-countable linear orderings.

Such studies are interesting for themselves, *i.e.*, for the techniques involved in their resolution and the understanding of the logics it requires for doing so. Such studies are also interesting since infinite linear orderings offer a natural model of continuous linear time.

Recently, another step in our understanding of monadic second-order logic over countable linear orderings has been made. An algebraic model, ◦-monoids, was proposed [4], yielding among other results the first known quantifier collapse of monadic second-order logic (to the one alternation fragment over set quantifiers), the resolution of a conjecture of Gurevich and Rabinovich [8] concerning

The research leading to these results has received funding from the European Union's Seventh Framework Programme (FP7/2007-2013) under grant agreement n° 259454.

© Springer-Verlag Berlin Heidelberg 2015
M.M. Halldórsson et al. (Eds.): ICALP 2015, Part II, LNCS 9135, pp. 146–158, 2015.
DOI: 10.1007/978-3-662-47666-6_12

the use of cuts "in the background" [6]. Algebraic recognizers give us a much deeper understanding of the expressive power of monadic second-order logic.

The next natural step is to follow the footprints of Schützenberger, who characterized algebraically first-order logic over finite words as languages that are recognized by aperiodic monoids [12] (in fact, the first-order logic terminology is in combination with McNaughton and Papert [10]) as these languages that are recognized by aperiodic monoids. Now that a suitable algebraic model is known for understanding monadic second-order logic, a similar study can be performed in this more general context. There exist already results of this kind, but these are so far restricted to the case of scattered linear orderings (*i.e.*, without any dense sub-ordering). In this context, first-order logic and first-order logic on cuts have been algebraically characterized [1], as well as weak monadic second-order logic [2]. Simple decision procedures are derived in all these situations.

In this paper, we perform a systematic analysis of sublogics of monadic second-order logic on countable linear orderings depending on the kind of sets over which set quantifiers range. If such sets are just singletons, we have exactly first-order logic (FO). If such sets are Dedekind cuts, we obtain first-order logic on cuts (FO[cut]). If finite sets only are allowed, this is weak monadic second-order logic (WMSO). If it is possible to quantify over both finite sets and cuts, we obtain weak monadic second-order on cuts (MSO[finite,cut]). We consider also MSO[ordinal] in which quantified sets need to be well-ordered. Finally MSO[scattered] corresponds to the case where quantified sets are required to be scattered. Our contribution is to compare the expressive power of all these logics (all are distinct but for MSO[finite,cut] which coincide with MSO[ordinal]), and characterize each of them by decidable algebraic means.

Structure of the Paper. In Section 2, we introduce linear orderings, words, and the logics we are interested in. In Section 3 we provide sufficient material concerning the algebraic framework of ∘-monoids, state the main characterization theorem, Theorem 2, and show the separation result, Theorem 3. Section 4 is devoted to the description of some ideas concerning one direction of the proof of Theorem 2. Section 5 concludes the paper.

2   Preliminaries

In this preliminary section, we introduce the notion of linear orderings (Section 2.1), (countable) words (Section 2.2) and the studied logics (Section 2.3).

2.1   Linear Orderings

A *linear ordering* $\alpha = (X, <)$ is a non-empty set X equipped with a total order $<$. A linear ordering α is *dense* if it contains at least two elements and for all $x < y \in \alpha$, there exists a z such that $x < z < y$. It is *scattered* if no subset of X induces a dense ordering. A *well ordering* is a linear ordering such that every non-empty subset has a minimal element. A subset of a linear ordering is well ordered (*resp.* scattered) if the linear ordering restricted to it is a well ordering (*resp.* scattered).

Given an element x, its *successor* (resp. *predecessor*) (if it exists) is the only $y > x$ (*resp.* $y < x$) such that there is no z such that $x < z < y$ (*resp.* $y < z < x$). A subset $I \subseteq \alpha$ of a linear ordering is *convex* if whenever $x, y \in I$ and $x < z < y$, $z \in I$. A *condensation* of a linear ordering is an equivalence relation $\sim$ such that all equivalence classes are convex. For a linear ordering α and a condensation $\sim$, we denote by $\alpha/\sim$, the *condensed linear ordering*: its elements are the equivalence classes for $\sim$, and the ordering is obtained by projection of the original ordering. Two convex subsets I, J of a linear ordering are *consecutive* if I and J are disjoint and their union is convex. Using the notations for elements, if $I < J$, then I is the predecessor of J, while J is the successor of I.

Given linear orderings $(\beta_i)_{i \in \alpha}$ (assumed disjoint up to isomorphism) indexed with a linear ordering α, their generalized sum $\sum_{i \in \alpha} \beta_i$ is the linear ordering over the (disjoint) union of the sets of the β_i's, with the order defined by $x < y$ if either $x \in \beta_i$ and $y \in \beta_j$ with $i < j$, or $x, y \in \beta_i$ for some i, and $x < y$ in β_i.

Given elements x, y, we denote by $[x, y)$ the set $\{z \mid x \leq z < y\}$, and similarly $[x, y], (x, y]$ and (x, y). We also denote as $(-\infty, x), (-\infty, x], (x, +\infty)$ and $[x, +\infty)$ the intervals that are unlimited to the left or to the right. Usually Dedekind cuts are defined as ordered pairs of sets (L, R) such that $L < R$. Here, we define a *Dedekind cut* (or simply a cut) as a left-closed subset X of a linear ordering, *i.e.*, for all $x < y$ with $y \in X$, then $x \in X$.

2.2   Infinite Words

Given a linear ordering α and a finite *alphabet* A, a *word* over A of *domain* α is a mapping $w : \alpha \to A$. The domain of a word is denoted $dom(w)$. In this work, all words are assumed of countable domain. The set of all *words of countable domain* is denoted by A°. A *language* is a subset of A°.

Given a convex set $X \subseteq dom(w)$ of word w, w_X denotes the word w *restricted* to X, *i.e.*, the word of domain X that coincides with w over X. A *factor* of a word w is any restriction of w to one of the convex subsets of its domain.

Given two words $u : \alpha \to A$ and $v : \beta \to A$ (where α and β are disjoint), we denote by uv the word over domain $\alpha + \beta$ such that each position $x \in \alpha$ (similarly $x \in \beta$) is labelled by $u(x)$ (by $v(x)$). The *generalized concatenation* of the words w_i (supposed of disjoint domain) indexed by a linear ordering α is

$$\prod_{i \in \alpha} w_i \; ,$$

and denotes the word of domain $\sum_i dom(w_i)$ which coincides with each w_i over $dom(w_i)$ for all $i \in \alpha$.

Some words will play an important role in the paper. The *empty word* ε, which is the only word of empty domain. The words denoted "$aaa\dots$" and "$\dots aaa$" are the words over the single letter a, and of respective domain $\omega = (\mathbb{N}, <)$ and

$\omega^* = (\mathbb{N}, >)$. Finally, *perfectshuffle*(A) for A, a non-empty finite set of letters, is a word of domain $(\mathbb{Q}, <)$ in which all non-empty intervals (x, y) contain at least once each letter of A. This word is unique up to isomorphism.

2.3   First-Order Logic, Monadic Second-Order Logic, and Between

We use logics for expressing properties of linear orderings or words. All of the several logics we study are all restrictions of monadic second-order logic (MSO). We very succinctly recall the basics of this logic here. The reader can refer to many other works on the subject, *e.g.*, [14]. We only consider word models.

Monadic second-order logic (MSO for short) is a logic with the following characteristics. It is possible to use *first-order variables* $x, y, z, \ldots$, ranging over positions of the word, and quantify over them thanks to $\exists x$ or $\forall y$. It is possible to use *monadic variables* $X, Y, \ldots$ (traditionally typeset in capital letters), that range over sets of positions of the word, and quantify over them using $\exists X, \forall Y$. Three atomic predicates can be used. The predicate $a(x)$, for a a letter, and x a position, holds if the letter carried at position x in the word is an a. The predicate $x < y$ for x, y, first-order variables denotes the order of the domain of the word. The membership predicate $x \in Y$ tests the membership of (the valuation of) a first-order variable x in (the valuation of) a monadic variable Y. All the Boolean connectives are also allowed. *First-order logic* (FO of short) is the fragment of this logic in which monadic variables, as well as quantifiers over them, are not allowed.

In this study, we are interested in the expressive power of logics weaker than MSO. There is a long tradition of such researches, initiated by the seminal work of Sch?tzenberger. For instance, it is classical to study first-order logic and its fragments when the quantifier alternation or the number of variables are restricted. In our case, our goal is to investigate the intricate relationship between the expressive power of the logic, and the infinite/dense nature of the linear orderings/words under study. The only parameter that we use for modifying the power of the logic is to change the range of monadic variables. By default, such variables range over any set of positions. We introduce now several *restricted set quantifiers* and the corresponding logics. Our simplest logic is first-order logic. The logic obtained by allowing monadic quantifiers restricted to Dedekind cuts is denoted *FO[cut]*. Another situation is when monadic second-order variables range over finite set, yielding *weak monadic second-order logic* (WMSO for short). We are also interested in the fragment in which it is possible to quantify both over finite sets and Dedekind cuts. We denote this logic *MSO[finite,cut]*. Then come logics in which monadic variables range over "infinite but small", sets of positions. We consider the case in which it is possible to quantify over well ordered sets, or scattered sets. We denote these logics *MSO[ordinal]* and *MSO[scattered]*.

We formally denote these restricted quantifiers as $\exists^V$ and $\forall^V$, where $V \subseteq \{\in, \notin\}^\circ$. A set belongs to the range of the quantifier $\exists^V$ or $\forall^V$ if its characteristic map (as a labelling of the domain by $\in, \notin$) is in V.

Given one of the above logics $\mathcal{L}$, a formula $\varphi \in \mathcal{L}$ and a countable word w we denote by $w \models \varphi$, the fact that the formula is true over w. We say that w is a *model* of φ. A language $L \subseteq A^\circ$ is *definable* in $\mathcal{L}$ if there exists a formula φ in $\mathcal{L}$ such that for all words $w \in A^\circ$, $w \in L$ if and only if $w \models \varphi$.

Remark 1. Some dependencies between these logics are simple to establish:

$$\begin{array}{ll} & \text{FO[cut]} \\ \text{FO} & \qquad\qquad \text{MSO[finite,cut]} - \text{MSO[ordinal]} - \text{MSO[scattered]} - \text{MSO} \\ & \text{WMSO} \end{array}$$

Indeed, FO[cut] is an extension of FO. Also WMSO extend FO since "being a singleton" is definable in WMSO. Similarly, MSO[finite,cut] is clearly an extension of both WMSO and FO[cut]. MSO[ordinal] can express finiteness, and represent cuts (as the left closure of a well ordered subset), and hence contains MSO[finite,cut]. In the same way, since being well ordered is expressible in MSO[scattered], MSO[scattered] contains MSO[ordinal]. Similarly, scatteredness being expressible in MSO, MSO[scattered] is a sublogic of MSO. In fact, all these logics are separated (Theorem 3), but for MSO[finite,cut] and MSO[ordinal] which happen to coincide (see Theorem 2).

The goal of this paper is to compare the expressive power of all these logics and be able to characterize them effectively.

3   The Algebraic Presentation: ∘-monoids

We now introduce the equivalent algebraic presentation of definable languages. We first describe the ∘-monoids in Section 3.1, and then the derived operations in Section 3.2, before presenting the theorems of characterization and separation in Section 3.3.

3.1   ∘-monoids, Syntactic ∘-monoids and Recognizability

As in the seminal work of Sch?tzenberger, we use algebraic acceptors for describing regular languages of countable words: ∘-monoids. A ∘-*monoid* is a set M equipped with an operation π, called the *product*, from M° to M, that satisfies $\pi(a) = a$ for all $a \in M$, and the *generalized associativity* property: for every words u_i over M° with i ranging over a countable linear ordering α,

$$\pi\left(\prod_{i \in \alpha} u_i\right) = \pi\left(\prod_{i \in \alpha} \pi(u_i)\right) .$$

Of course, an instance of ∘-monoids is the set of words over some alphabet A equipped with the generalized concatenation $\prod$, *i.e.*, $(A^\circ, \prod)$. It is even the *free ∘-monoid* generated by A. A ∘-*monoid morphism* from $\mathbf{M}$ to $\mathbf{N}$ (∘-monoids) is a map γ from M to N such that $\gamma(\prod_{i \in \alpha} a_i) = \pi(\prod_{i \in \alpha} \gamma(a_i))$.

Example 1. **Sing**$= (\{1, s, 0\}, \pi)$ where π is defined for all $u \in \{1, s, 0\}^\circ$ as:

$$
\pi(u) = \begin{cases} 1 & \text{if } u \in \{1\}^\circ, \\ s & \text{otherwise if } u \text{ contains no } 0, \text{ and exactly one } s, \\ 0 & \text{otherwise,} \end{cases}
$$

is a ∘-monoid (checking generalized associativity requires a case by case study).

By slightly modifying the example, we obtain the ∘-monoid **Fin** in which the second line in the definition of π is changed into "s if u contains no 0, and finitely many s's". The ∘-monoid **Ord** is when $\pi(u)$ evaluates to "s if u contains no 0, and a well ordered set of s's". Finally, the ∘-monoid **Scat** is when $\pi(u)$ evaluates to "s if u contains no 0, and a scattered set of s's". Once more, checking generalized associativity is by case analysis.

The element $\pi(\varepsilon)$ is called the *unit*, and it is customary to denote it 1 as done above. A *zero* (that does not necessarily exist) is an absorbing element, *i.e.*, an element such that $\pi(u0v) = 0$ whatever are u and v. It is denoted by convention 0 as in the above examples. An *idempotent* is an element e such that $\pi(ee) = e$.

A ∘-monoid can be used to recognize languages as follows. Consider a ∘-monoid $\mathbf{M} = (M, \pi)$, a map h from an alphabet A to M and a set $F \subseteq M$, then $(\mathbf{M}, h, F)$ *recognizes* the language $L = \{u \in A^\circ \mid \pi(h(u)) \in F\}$ (where h has been extended implicitly into a map from A° to M°). Said differently, L is the inverse image of F under the ∘-monoid morphism $\pi \circ h$. From [4], being recognizable by a ∘-monoid is equivalent to be definable in MSO.

Furthermore, when a language is recognizable by a finite ∘-monoid, then there is a minimal one called the *syntactic ∘-monoid*. It is minimal in the algebraic sense: all ∘-monoids that would recognize this language can be trimmed and quotiented yielding the syntactic one. We do not develop this aspect more in this short abstract.

Example 2. Coming back to the above examples, with $h(\in) = s$ and $h(\notin) = 1$, then $(\mathbf{Sing}, h, \{s\})$ recognizes the language L_{Sing} over the alphabet $\{\in, \notin\}$ of words that contain exactly one occurrence of $\in$. Similarly, $(\mathbf{Fin}, h, \{1, s\})$, $(\mathbf{Ord}, h, \{1, s\})$, and $(\mathbf{Scat}, h, \{1, s\})$ recognize the languages $L_{\text{Finite}}, L_{\text{Ord}}$ and L_{Scat} respectively, of words that contain "finitely many $\in$'s", "a well ordered set of $\in$'s", and "a scattered set of $\in$'s" respectively.

Let us note that these languages are the one used in the restricted quantifiers $\exists^V$ and $\forall^V$ for defining the logics (cuts are omitted for space considerations).

3.2 The Derived Operations

The product operation π is infinite, even in a finite $\circ$-monoid $\mathbf{M} = (M, \pi)$. Hence, π is *a priori* not representable in finite space (it has uncountably many possible inputs). This problems is resolved using derived operations.

The *operations derived* from π are the following:

- 1 is the unit constant $\pi(\varepsilon)$,
- $\cdot : M \times M \to M$ is defined for $a, b \in M$ as $a \cdot b = \pi(ab)$,
- $\omega : M \to M$ is defined for all $a \in M$ as $a^\omega = \pi(aaa \dots)$,
- $\omega* : M \to M$ is defined for all $a \in M$ as $a^{\omega*} = \pi(\dots aaa)$,
- $\eta : \mathcal{P}(M) \setminus \{\emptyset\} \to M$ is defined as $E^\eta = \pi(perfectshuffle(E))$ for $E \subseteq M$ non-empty.

Note that from the definitions, using generalized associativity, the unit element satisfies $1 \cdot 1 = 1^\omega = 1^{\omega*} = \{1\}^\eta = 1$, $a \cdot 1 = 1 \cdot a = a$, and $(E \cup \{1\})^\eta = E^\eta$ for all $a \in M$ and all non-empty $E \subseteq M$. Similarly, if there is a zero 0 then it satisfies $0 \cdot a = a \cdot 0 = 0^\omega = 0^{\omega*} = (E \cup \{0\})^\eta = 0$ for all $a \in M$ and $E \subseteq M$. This is why we usually do not mention these elements when describing derived operations.

Example 3. The derived operation of the above examples are entirely determined by the following table:

	$s \cdot s$	s^ω	$s^{\omega*}$	$\{s\}^\eta$
Sing	0	0	0	0
Fin	s	0	0	0
Ord	s	s	0	0
Scat	s	s	s	0

Though not essential in this short abstract, let us emphasize that the derived operations determine entirely the product π, as shown now.

Theorem 1. *There exists a set of equalities (A) involving the derived operations[1], such that:*

- *The operations derived from a $\circ$-monoid satisfy all the equations from (A).*
- *If $1, \cdot, \omega, \omega*, \eta$ are maps of correct type over a finite set M that satisfy the equalities of (A), then there exists one and only one product over M from which $1, \cdot, \omega, \omega*, \eta$ are derived.*

3.3 The Core Theorem

We state in this section our main results, Theorem 2 and 3. All $\circ$-monoids are assumed finite from now. We first refine our understanding of idempotents:

---

[1] These are variants of associativity, such as $x \cdot (y \cdot z) = (x \cdot y) \cdot z$, $1 \cdot x = x \cdot 1 = x$, $(a^\eta)^\omega = a^\omega$, and so on. A complete list is known [4], but of no use here.

- A *gap insensitive* idempotent e is an idempotent such that $e^\omega \cdot e^{\omega*} = e$.
- An *ordinal idempotent* e is an idempotent such that $e^\omega = e$. The name comes from the fact that in such a case, all words $u \in \{e\}^\circ$ that have a well ordered (*i.e.*, isomorphic to an ordinal) non-empty domain satisfy $\pi(u) = e$.
- Symmetrically, an *ordinal* idempotent* e is an idempotent such that $e^{\omega*} = e$.
- A *scattered idempotent* e is an idempotent which is at the same time an ordinal and an ordinal* idempotent. For such idempotents, all words $u \in \{e\}^\circ$ that have a scattered non-empty domain satisfy $\pi(u) = e$.
- A *shuffle idempotent* e is an idempotent such that $\{e\}^\eta = e$.
- A shuffle idempotent e is *shuffle simple* if for all $K \subseteq M$ such that $e \cdot a \cdot e = e$ for all $a \in K$, $(\{e\} \cup K)^\eta = e$.

Note that since in every $\circ$-monoid $(\{e\}^\eta)^\omega = (\{e\}^\eta)^{\omega*} = \{e\}^\eta$, every shuffle idempotent is a scattered idempotent. Note also that every scattered idempotent is by definition an ordinal idempotent and an ordinal* idempotent. Also, every scattered idempotent is obviously gap insensitive.

We define now the following properties of a $\circ$-monoid $\mathbf{M} = (M, \pi)$:

- *aperiodic* if for all $a \in \mathbf{M}$ there exists n such that $a^n = a^{n+1}$,
- i$\to$gi if all idempotents are gap insensitive,
- oi$\to$gi if all ordinal idempotents are gap insensitive,
- o*i$\to$gi if all ordinal* idempotents are gap insensitive,
- sc$\to$sh if all scattered idempotents are shuffle idempotent,
- sh$\to$ss if all shuffle idempotents are shuffle simple.

It is clear by definition that oi$\to$gi (as well as o*i$\to$gi) imply i$\to$gi. There is in fact another, slightly less direct, implication to mention:

Lemma 1. i$\to$gi *implies aperiodic.*

Proof. Let a be an element of a finite $\circ$-monoid M. There exists n such that a^n is idempotent. We compute $a^n = (a^n)^\omega \cdot (a^n)^{\omega*} = a \cdot (a^n)^\omega \cdot (a^n)^{\omega*} = a^{n+1}$. $\square$

We are now ready to state our core theorem.

Theorem 2. *Let* $\mathbf{M}$ *be the syntactic $\circ$-monoid of a language* $L \subseteq A^\circ$, *then:*

- *L is definable in FO iff* $\mathbf{M}$ *satisfies* i$\to$gi, sc$\to$sh *and* sh$\to$ss.
- *L is definable in FO[cut] iff* $\mathbf{M}$ *satisfies aperiodic,* sc$\to$sh *and* sh$\to$ss.
- *L is definable in WMSO iff* $\mathbf{M}$ *satisfies* oi$\to$gi, o*i$\to$gi, sc$\to$sh *and* sh$\to$ss.
- *L is definable in MSO[finite,cut] iff it is definable in MSO[ordinal] iff* $\mathbf{M}$ *satisfies* sc$\to$sh *and* sh$\to$ss.
- *L is definable in MSO[scattered] iff* $\mathbf{M}$ *satisfies* sh$\to$ss.

And as a consequence, these classes are decidable.

Example 4. Let us apply these characterizations to the ∘-monoids of Example 3:

	aper.	i→gi	oi→gi	o*i→gi	sc→sh	sh→ss	definable in
Sing	yes	yes	yes	yes	yes	yes	FO
Fin	yes	no	yes	yes	yes	yes	WMSO, FO[cut], not FO
Ord	yes	no	no	yes	yes	yes	FO[cut], not WMSO
Scat	yes	yes	yes	yes	no	yes	MSO[scattered], not MSO[ordinal]

Remark 2. One aspect of Theorem 2 is that MSO[finite,cut] and MSO[ordinal] are equivalent. If we apply this fact to the domain ω, then cuts can be eliminated easily, and MSO[finite,cut] coincide with WMSO. Still over ω, MSO[ordinal] obviously coincide with MSO. Hence Theorem 2 implies that WMSO and MSO coincide over ω (in fact, the same argument is valid over any well ordered countable word). This non-trivial fact is usually established using the deep result of determinization of McNaughton [9] (other proofs involve weak alternating automata or algebra).

Theorem 3. *There are languages separating all situations not covered by Theorem 2.*

Proof (sketch). In fact, two among the five separating languages were given in Example 4: $L_{\mathrm{Ord}} \in$FO[cut]\WMSO and $L_{\mathrm{Scat}} \in$MSO[scattered]\MSO[ordinal].

WMSO\FO[cut]$\neq \emptyset$: The witnessing language is "the domain is of even finite length". It is the classical example of non-aperiodicity over finite words, and it works as well in this case.

MSO[ordinal]\(FO[cut]∪WMSO) $\neq \emptyset$: For this, it is sufficient to take the disjoint union (for instance using disjoint alphabets) of a language in WMSO\FO[cut] and a language in FO[cut]\WMSO.

MSO\MSO[scattered]$\neq \emptyset$: Call a set X *perfectly dense* if all elements $x < y < z$ with $y \in X$ are such that (x,y) and (y,z) both intersect X. Said differently, all elements in X are limits from the left of elements from X, and symmetrically from the right. The language "there exists a set X of a-labelled positions which is perfectly dense" is obviously definable in MSO. Computing its syntactic ∘-monoid would yield four elements $1, a, b, 0$ with derived operations defined by $a \cdot a = a^{\omega} = a^{\omega*} = b \cdot b = b \cdot a = a \cdot b = b^{\omega} = b^{\omega*} = \{b\}^{\eta} = b$ and $\{a\}^{\eta} = \{a,b\}^{\eta} = 0$. The morphism sends a to a and b to b, and the accepting set is $\{0\}$. However, this language is not definable in MSO[scattered]: b is a shuffle idempotent which is not shuffle simple since $\{b\}^{\eta} = b = b \cdot a \cdot b$ and $\{a,b\}^{\eta} \neq b$.   □

4   From Logics to ∘-monoids

In this section, we show some of the results of the form "if a language $L \subseteq A^{\circ}$ is definable in logic $\mathcal{L}$, then its syntactic ∘-monoid satisfies property P" for suitable choices of $\mathcal{L}$ and P. The standard approach for such results is to use the technique of Ehrenfeucht-Fra?ss? games. We adopt a different presentation here, making use of our fine understanding of ∘-monoids.

Let us first recall that all the logics we work with differ by their use of restricted set quantifiers. These restricted quantifiers are parameterized by a language $V \subseteq \{\in, \notin\}^{\circ}$. The quantifier $\exists^V X$ signifies "there exists a set of positions X which, when written as a labelling of the linear ordering yields a word in V". We have seen the language L_{Sing}, L_{Finite}, L_{Ord}, L_{Scat} that correspond to the quantifiers over singletons, finite sets, well ordered sets, and scattered sets.

Thus, the core step in each of these proofs consists in showing that the operation of restricted set quantifier preserves the property we are interested in when done at the level of $\circ$-monoids. Essentially, this looks as follows: "assume that L_ϕ is recognized by a $\circ$-monoid that has property P' then $L_{\exists^V X \phi}$ also has property P". Thus, we start by describing how $\exists^V$ behaves.

Let us just mention here that the existential quantifier is the crux of the problem, and that the other constructions involved (atomic predicates and boolean connectives) have also to be treated, but do not involve interesting arguments. We also have to verify the closure of the properties we are interested in under quotient of $\circ$-monoids. This last step is usually not necessary, but, since we did not choose to present the properties as identities, it has to be done explicitly.

4.1   Restricted Quantifiers over $\circ$-monoids

Let us first recall how the existential set quantifier is implemented, from a language and algebraic theoretic point of view, and then refine this for restricted set quantifier.

Consider a language $L \in (A \times \{\in, \notin\})^{\circ}$. A word over this alphabet can be seen as a usual word over the alphabet A, enriched with the characteristic map of some set X: if a position belongs to X, then the second component is $\in$, otherwise it is $\notin$. The operation equivalent to existential set quantifier over such languages is $Proj(L)$ defined as:

$$Proj(L) = \{u_{|1} \in A^{\circ} \mid u \in L\} \ ,$$

where $u_{|1}$ denotes the word obtained by projecting each letter of u to its first component (similarly for $u_{|2}$). If furthermore L is recognized by some $\mathbf{M} = ((M, \pi), h, F)$, we define the new $\circ$-monoid $\mathcal{P}(\mathbf{M})$ to be $(\mathcal{P}(M), \pi)$, where

$$\text{for all } U \in (\mathcal{P}(M))^{\circ}, \quad \pi(U) = \{\pi(u) \mid u \in U\} \ ,$$

in which $u \in U$ holds if $dom(u) = dom(U)$ and for all $i \in dom(u)$, $u(i) \in U(i)$.

This construction is known to (1) produce a valid $\circ$-monoid, and (2) be such that $(\mathcal{P}(\mathbf{M}), h', F')$ recognizes $Proj(L)$ for $h'(a) = \{h(a, \in), h(a, \notin)\}$ and $F' = \{X \subseteq M \mid X \cap F \neq \emptyset\}$.

We present now a refinement of this construction, which furthermore restricts the range of the projection. Given a language $V \subseteq \{\in, \notin\}^{\circ}$ that represents the range of a restricted set quantifier, we define the *restricted projection* of L as:

$$Proj^V(L) = \{u_{|1} \in A^{\circ} \mid \text{ for some } u \in L \text{ such that } u_{|2} \in V\} \ .$$

This operation is the language theoretic counterpart to the logical restricted quantifier $\exists^V$. Let us assume furthermore that V is recognized by some $(\mathbf{V}, g, E)$. We assume (and this will always be the case) that $\mathbf{V}$ has a zero 0, and that $0 \notin E$. We define the new $\circ$-monoid $\mathcal{P}_\mathbf{V}(\mathbf{M})$ to be (N, π), where

$$\text{for all } U \in (\mathcal{P}(M \times V))^\circ, \quad \pi(U) = \{(\pi(u_{|1}), \pi(u_{|2})) \mid u \in U\} \setminus (M \times \{0\}) ,$$
$$\text{and} \quad N = \{\pi(U) \mid U \in \{\{(h(a, \in), g(\in)), (h(a, \not\in), g(\not\in))\} \mid a \in A\}^\circ\} .$$

We can recognize in this construction the above powerset construction, applied to the $\circ$-monoid $\mathbf{M} \times \mathbf{V}$, from which all occurrences of the zero of $\mathbf{V}$ are removed as well all all non-reachable elements.

Lemma 2. $\mathcal{P}_\mathbf{V}(\mathbf{M})$ *is a $\circ$-monoid.*
If L is recognized by $(\mathbf{M}, h, F)$, then $Proj^V(L)$ is recognized by $(\mathcal{P}_\mathbf{V}(\mathbf{M}), h', F')$ where $h'(a) = \{(h(a, \in), g(\in)), (h(a, \not\in), g(\not\in))\}$ and $F' = \{A \mid A \cap (F \times E) \neq \emptyset\}$.

4.2   Establishing Invariants

The core result in the translation from logics to $\circ$-monoids is the following.

Lemma 3. *Let $\mathbf{M}$ be a $\circ$-monoid.*

1. *If $\mathbf{M}$ satisfies i→gi then $\mathcal{P}_\mathbf{Sing}(\mathbf{M})$ satisfies i→gi.*
2. *If $\mathbf{M}$ satisfies aperiodic then $\mathcal{P}_\mathbf{Cut}(\mathbf{M})$ satisfies aperiodic[2].*
3. *If $\mathbf{M}$ satisfies oi→gi then $\mathcal{P}_\mathbf{Fin}(\mathbf{M})$ satisfies oi→gi (resp. o*i→gi).*
4. *If $\mathbf{M}$ satisfies sc→sh then $\mathcal{P}_\mathbf{Ord}(\mathbf{M})$ satisfies sc→sh.*
5. *If $\mathbf{M}$ satisfies sh→ss then $\mathcal{P}_\mathbf{Scat}(\mathbf{M})$ satisfies sh→ss.*

Let us give some ideas about its proof. Let $\mathbf{N}$ be $\mathcal{P}_\mathbf{V}(\mathbf{M})$ where $\mathbf{V}$ is one of **Sing, Fin, Ord** or **Scat** (unfortunately, **Cut** having a different structure, it has to be treated separately).

Lemma 4. *There exists a $\circ$-monoid morphism ρ from $\mathbf{N}$ to $\mathbf{M}$ such that for all $A \in N$, $(x, 1) \in A$ if and only if $x = \rho(A)$.*

Proof. Essentially, the point is to prove that for all $A \in N$, there is one and only one $\rho(A)$ such that $(\rho(A), 1) \in A$. The fact that this ρ is a $\circ$-monoid morphism is then straightforward. For proving it, it is sufficient to do it for the neutral element $\{(1, 1)\}$, the image of each letter 'a' which happens to be $\{(h(a), 1), (h(a), s)\}$, and then show the preservation of the property under $\cdot, \omega, \omega^*$ and η.   □

Let us show the simplest case of Lemma 3, the one for $\mathcal{P}_\mathbf{Sing}(\mathbf{M})$:

Lemma 5. *If a $\circ$-monoid $\mathbf{M}$ satisfies i→gi then $\mathcal{P}_\mathbf{Sing}(\mathbf{M})$ also does.*

---

[2] **Cut** is a $\circ$-monoid recognizing "cuts" that we omitted here for space reasons.

Proof. Let E be an idempotent in $\mathbf{N} = \mathcal{P}_{\mathbf{Sing}}(\mathbf{M})$. Our goal is to show that it is gap insensitive.

Let $(x, y) \in E$. Since $E = E \cdot E$, there exists $(x_1, y_1), (x_2, y_2) \in E$ such that $x_1 \cdot x_2 = x$ and $y_1 \cdot y_2 = y$. Since $y \neq 0$, at least one among y_1, y_2 is equal to 1. Without loss of generality, let us assume it is y_1. In this case, according to Lemma 4, $x_1 = \rho(E)$. In particular, since ρ is a morphism, this means that x_1 is an idempotent. Thus we can use the assumption that $\mathbf{M}$ satisfies i→gi on it, and get that $x_1^\omega \cdot x_1^{\omega*} = x_1$. It follows that the word

$$\overbrace{(x_1, 1)(x_1, 1) \ldots}^{\text{of domain } \omega} \overbrace{\ldots (x_1, 1)(x_1, 1)}^{\text{of domain } \omega*}(x_2, y_2)$$

has also value (x, y) under π (componentwise), and as a consequence $(x, y) \in E^\omega \cdot E^{\omega*}$. We have proved $E \subseteq E^\omega \cdot E^{\omega*}$.

Conversely, consider some $(x, y) \in E^\omega \cdot E^{\omega*}$. This means that there exists a word u of the form

$$\overbrace{(x_1, y_1)(x_2, y_2) \ldots}^{\text{of domain } \omega} \overbrace{\ldots (x_2', y_2')(x_1', y_1')}^{\text{of domain } \omega*}$$

which evaluates (componentwise) to (x, y), with (x_i, y_i) and $(x_i', y_i') \in E$ for all $i \in \mathbb{N}$. If all $y = 1$, then its clear. Otherwise, there is at most one among the y_i's and the y_i''s which is not equal to 1. Without loss of generality (by symmetry), we can assume that it is y_j. According to Lemma 4, $x_i = \rho(E)$ for all $i \neq j$ and $x_i' = \rho(E)$ for all i. Since ρ is a morphism, $\rho(E)$ is also an idempotent. Thus we can use the assumption that $\mathbf{M}$ satisfies i→gi. We obtain that $\rho(E)^\omega \cdot \rho(E)^{\omega*} = \rho(E)$. Thus, u evaluates to $(\rho(E), 1) \cdot (x_j, y_j) \cdot (\rho(E), 1) \in E^3 = E$. Hence $E^\omega \cdot E^{\omega*} \subseteq E$.

This terminates the proof that $\mathbf{N}$ satisfies i→gi. □

5   Conclusion

In this paper we have characterized algebraically and effectively several natural sublogics of MSO. Unfortunately the most involved arguments, namely the translation from algebra to logic, were not addressed in this short abstract. These can be found in the appendix.

References

1. Bès, A., Carton, O.: Algebraic characterization of FO for scattered linear orderings. In: CSL. LIPIcs, vol. 12, pp. 67–81. Schloss Dagstuhl - Leibniz-Zentrum fuer Informatik (2011)
2. Bès, A., Carton,O.: Algebraic characterization of WMSO for scattered linear orderings. Personal communication (2014)
3. Richard Büchi, J.: On a decision method in restricted second order arithmetic. In: Logic, Methodology and Philosophy of Science (Proc. 1960 Internat. Congr.), pp. 1–11. Stanford Univ. Press, Stanford (1962)

4. Carton, O., Colcombet, T., Puppis, G.: Regular languages of words over countable linear orderings. In: Aceto, L., Henzinger, M., Sgall, J. (eds.) ICALP 2011, Part II. LNCS, vol. 6756, pp. 125–136. Springer, Heidelberg (2011)

5. Chloé, R., Carton, O.: An algebraic theory for regular languages of finite and infinite words. Int. J. of Foundations of Comp. Sc. 16(04), 767–786 (2005)

6. Colcombet, T.: Monadic second-order logic and cuts in the backgrounds. In: CSR, page Invited Paper (2013)

7. Rosenstein, J.G.: Linear Orderings. Academic Press (1982)

8. Gurevich, Y., Rabinovich, A.M.: Definability and undefinability with real order at the background. J. Symb. Log. 65(2), 946–958 (2000)

9. McNaughton, R.: Testing and generating infinite sequences by a finite automaton. Information and Control 9, 521–530 (1966)

10. McNaughton, R., Papert, S.: Counter-free Automata. MIT Press (1971)

11. Rabin, M.O.: Decidability of second-order theories and automata on infinite trees. Trans. Amer. Math. Soc. 141, 1–35 (1969)

12. Schützenberger, M.-P.: On finite monoids having only trivial subgroups. Information and Control 8, 190–194 (1965)

13. Shelah, S.: The monadic theory of order. Ann. of Math. (2) 102(3), 379–419 (1975)

14. Thomas, W.: Languages, automata and logic. In: Rozenberg, G., Salomaa, A., (eds.) Handbook of Formal Languages, vol. 3, chapter 7, pp. 389–455. Springer (1997)

Reachability is in DynFO

Samir Datta[1], Raghav Kulkarni[2], Anish Mukherjee[1], Thomas Schwentick[3],
and Thomas Zeume[3]([✉])

[1] Chennai Mathematical Institute, Chennai, India
sdatta@cmi.ac.in
[2] Center for Quantum Technologies, Singapore, Singapore
kulraghav@gmail.com
[3] TU Dortmund University, Dortmund, Germany
{thomas.schwentick,thomas.zeume}@tu-dortmund.de

Abstract. We consider the *dynamic complexity* of some central graph problems such as Reachability and Matching and linear algebraic problems such as Rank and Inverse. As elementary change operations we allow insertion and deletion of edges of a graph and the modification of a single entry in a matrix, and we are interested in the complexity of maintaining a property or query. Our main results are as follows:

1. Reachability is in DynFO;
2. Rank of a matrix is in DynFO($+,\times$);
3. Maximum Matching (decision) is in non-uniform DynFO.

Here, DynFO allows updates of the auxiliary data structure defined in first-order logic, DynFO($+,\times$) additionally has arithmetics at initialization time and non-uniform DynFO allows arbitrary auxiliary data at initialization time. Alternatively, DynFO($+,\times$) and non-uniform DynFO allow updates by uniform and non-uniform families of poly-size, bounded-depth circuits, respectively.

The first result confirms a two decade old conjecture of Patnaik and Immerman [16]. The proofs rely mainly on elementary Linear Algebra. The second result can also be concluded from [7].

1 Introduction

Dynamic Complexity Theory studies dynamic problems from the point of view of Descriptive Complexity (see [13]). It has its roots in theoretical investigations of the view update problem for relational databases. In a nutshell, it investigates the logical complexity of updating the result of a query under deletion or insertion of tuples into a database.

As an example, the Reachability query asks whether in a directed graph there is a path from a distinguished node s to a node t. The correct result of this query (i.e., whether such a path exists in the current graph) can be maintained for *acyclic graphs* with the help of an auxiliary binary relation that is updated by a first-order formula after each insertion or deletion of an edge. In fact, one can simply maintain the transitive closure of the edge relation. In terms of Dynamic Complexity, we get that Acyclic Reachability is in DynFO. In this setting, a

© Springer-Verlag Berlin Heidelberg 2015
M.M. Halldórsson et al. (Eds.): ICALP 2015, Part II, LNCS 9135, pp. 159–170, 2015.
DOI: 10.1007/978-3-662-47666-6_13

sequence of change operations is applied to a graph with a fixed set of nodes whose edge set is initially empty [16].

Studying first-order logic as an update language in a dynamic setting is interesting for (at least) two reasons. In the context of relational databases, first-order logic is a natural update language as such updates can also be expressed in SQL. On the other hand, first-order logic also corresponds to circuit-based low level complexity classes; and therefore queries maintainable by first-order updates can be evaluated in a highly parallel fashion in dynamic contexts.

We also consider two extensions, DynFO($+,\times$) and non-uniform DynFO, whose programs can assume at initialization time multiplication and addition relations on the underlying universe of the graph, and arbitrarily pre-computed auxiliary relations, respectively. These two classes contain those problems that can be maintained by uniform and non-uniform families of poly-size, bounded-depth circuits, respectively.

The Reachability query is of particular interest here, as it is one of the simplest queries that can not be expressed (statically) in first-order logic, but rather requires recursion. Actually, it is in a sense prototypical due to its correspondence to transitive closure logic. The question whether the Reachability query can be maintained by first-order update formulas has been considered as one of the central open questions in Dynamic Complexity. It has been studied for several restricted graph classes and variants of DynFO [3,5,8,9,16,20]. In this paper, we confirm the conjecture of Patnaik and Immerman [16] that the Reachability query for general directed graphs is indeed in DynFO.

Theorem 1. *Directed Reachability is in* DynFO.

Our main tool is an update program (i.e., a collection of update formulas) for maintaining the rank of a matrix over finite fields $\mathbb{Z}_p$ against updates to individual entries of the matrix. The underlying algorithm works for matrix entries from arbitrary integer ranges, however, the corresponding DynFO update program assumes that only small numbers occur.[1]

Theorem 2. *Rank of a matrix is in* DynFO($+,\times$).

Theorem 1 follows from Theorem 2 by a simple reduction. Whether there is a path from s to t can be reduced to the question whether some (i,j)-entry of the inverse of a certain matrix has a non-zero value, which in turn can be reduced to a question about the rank of some matrix. This reduction (and similarly those mentioned below) is very restricted in the sense that a single change in the graph induces only a bounded number of changes in the matrix. We further use the observation that for domain independent queries as the Reachability query, DynFO is as powerful as DynFO($+,\times$). The combination of these ideas resolves the Patnaik-Immerman conjecture in a surprisingly elementary way.

By reductions to Reachability it further follows that Satisfiability of 2-CNF formulas and regular path queries for graph databases can be maintained in

---

[1] More precisely, it allows only integers whose absolute value is at most the possible number of rows and columns of the matrix.

DynFO. By another reduction to the matrix rank problem, we show that the existence of a perfect matching and the size of a maximum matching can be maintained in non-uniform DynFO.

Theorem 3. PERFECTMATCHING *and* MAXMATCHING *are in non-uniform* DynFO.

Related work Partial progress on the Patnaik-Immerman conjecture was achieved by Hesse [9], who showed that directed reachability can be maintained with first-order updates augmented with counting quantifiers, i.e., logical versions of uniform TC^0. More recently, Datta, Hesse and Kulkarni [3] studied the problem in the *non-uniform* setting and showed that it can in fact be maintained in non-uniform $AC^0[\oplus]$, i.e., non-uniform DynFO extended by parity quantifiers.

Dynamic algorithms for algebraic problems have been studied in [17–19]. The usefulness of matrix rank for graph problems in a logical framework has been demonstrated in [14]. Both [14,18] contain reductions from Reachability to matrix rank (different from ours). A dynamic algorithm for matrix rank, based on maintaining a reduced row echelon form, is presented in [7]. This algorithm can also be used to show that matrix rank is in $DynFO(+,\times)$. More details are discussed in Section 3.1.

In [18,19] a reduction from maximum matching to matrix rank has been used to construct a dynamic algorithm for maximum matching. While in this construction the inverse of the input matrix is maintained using Schwartz Zippel Lemma, we use the Isolation Lemma of Mulmuley, Vazirani and Vazirani's [15] to construct non-uniform dynamic circuits for maximum matching.

The question whether Reachability can be maintained by formulas from first-order logic has also been asked in the slightly different framework of First-Order Incremental Evaluation Systems (FOIES) [4]. It is possible to adapt our update programs to show that Reachability can be maintained by FOIES.

Organization After some preliminaries in Section 2, we describe in Section 3 dynamic algorithms for matrix rank and Reachability, independent of a particular dynamic formalism. In Section 4 we show how these algorithms can be implemented as DynFO programs. Section 5 contains open ends.

2   Preliminaries

We refer the reader to any standard text for an introduction to linear algebraic concepts (see, e.g., [2]). We briefly survey some relevant ones here. Apart from the concept of *vector space* use its *basis* i.e. a linearly independent set of vectors whose linear combination spans the entire vector space and its *dimension* i.e. the cardinality of any basis. We will use matrices as linear transformations. Thus an $n \times m$ matrix M over a field $\mathbb{F}$ yields a transformation $T_M : \mathbb{F}^m \to \mathbb{F}^n$ defined by $T_M : x \mapsto Mx$. We will abuse notation to write M for both the matrix and the transformation T_M. The *kernel* of M is the subspace of $\mathbb{F}^m$ consisting of vectors

x satisfying $Mx = \mathbf{0}$ where $\mathbf{0} \in \mathbb{F}^n$ is the vector of all zeroes. In this paper we mainly study the following algorithmic problems.

MATRIXRANK		REACH	
Given:	Integer matrix A	Given:	Directed graph G, nodes s, t
Output: rank(A) over $\mathbb{Q}$		Question: Is there a path from s to t in G?	

PERFECTMATCHING		MAXMATCHING	
Given:	Undirected graph G	Given:	Undirected graph G
Question: Is there a perfect matching in G?		Output: Maximum size of a matching in G	

For each natural number n, $[n]$ denotes $\{1, \ldots, n\}$.

3 Dynamic Algorithms for Rank, Reachability and Others

In this section, we present dynamic algorithms in an informal algorithmic framework. Their implementation as dynamic programs in the sense of Dynamic Complexity will be discussed in the next section. However, the reader will easily verify that these algorithms are highly parallelizable (in the sense of constant time parallel RAMs or the complexity class AC^0). We first describe how to maintain the rank of a matrix. Then we describe how to maintain an entry of the inverse of a matrix by a reduction to the rank of a matrix, and show that this immediately yields an algorithm for Reachability in directed graphs.

Theorem 3 uses the algorithm for rank in combination with the Isolation Lemma [15], more specifically its use described in [1], and a reduction from maximum matching to rank [11].

3.1 Maintaining the Rank of a Matrix

In this subsection we show that the rank of a matrix A can be maintained dynamically in a highly parallel fashion. We describe the algorithm for integer matrices, but it can be easily adapted for matrices with rational entries. At initialization time, the algorithm gets a number n of rows, a number m of columns, and a bound N for the absolute value of entries of the matrix A. Initially, all entries a_{ij} have value 0. Each change operation changes one entry of the matrix.

First, we argue that for maintaining the rank of A it suffices to maintain the rank of the matrix $(A \bmod p)$ for polynomially many primes of size $O(\max(n, \log N)^3)$. To this end recall that A has rank at least k if and only if A has a $k \times k$-submatrix A' whose determinant is non-zero. The value of this determinant is bounded by $n!N^n$, an integer with $O(n(\log n + \log N))$ many bits. Therefore, it is divisible by at most $O(n(\log n + \log N))$ many primes. By the Prime Number Theorem, there are $\sim \frac{\max(n, \log N)^3}{\log \max(n, \log N)^3}$ many primes in $[\max(n, \log N)^3]$. Hence for n large enough, the determinant of A' is non-zero if and only if there is a prime $p \in [\max(n, \log N)^3]$ such that the determinant of $(A' \bmod p)$ is non-zero. Hence the rank of A is at least k if and only if there is a prime p such that the rank of $(A \bmod p)$ is at least k. Thus in order to compute

$$
\text{Matrix } A \qquad\qquad \text{Basis } B \qquad\qquad A \cdot B
$$

$$
\begin{pmatrix} 0 & 1 & 0 & 1 & 0 \\ 0 & 1 & 0 & 1 & 0 \\ 0 & 1 & 0 & 1 & 0 \\ 1 & 0 & 0 & 1 & 0 \\ 1 & 1 & 0 & 0 & 0 \end{pmatrix} \bullet \begin{pmatrix} 0 & 1 & 0 & 0 & 0 \\ 0 & 1 & 0 & 0 & 1 \\ 1 & 0 & 0 & 0 & 0 \\ 0 & 1 & 0 & 1 & 0 \\ 0 & 0 & 1 & 0 & 0 \end{pmatrix} = \begin{pmatrix} 0 & 0 & 0 & 1 & 1 \\ 0 & 0 & 0 & 1 & 1 \\ 0 & 0 & 0 & 1 & 1 \\ 0 & 0 & 0 & 1 & 0 \\ 0 & 0 & 0 & 0 & 1 \end{pmatrix}
$$

Fig. 1. A basis A with an A-good basis B. The first three (column) vectors of B are in the kernel K. The principal components of the two other vectors are marked in red.

the rank of A it suffices to compute the rank of $(A \bmod p)$ in parallel for the primes in $[\max(n, \log N)^3]$, and to take maximum over all such ranks.

Now we show how to maintain the rank of a $n \times m$ matrix A over $\mathbb{Z}_p$. The idea is to maintain a basis of the column space that contains a basis of the kernel of A. The number of non-kernel vectors in the basis determines the rank of A.

By K we denote the kernel of A, i.e., the vector space of vectors v with $Av = 0$. For a vector v in $\mathbb{Z}_p^m$, we write $S(v)$ for the set of non-zero coordinates of Av, that is, the set of all i, for which $(Av)_i \neq 0$.

As auxiliary data structure, we maintain a basis B of $\mathbb{Z}_p^m$ with the following additional property, called A-good. A vector $v \in B$ is i-unique with respect to B and A, for some $i \in [n]$, if $i \in S(v)$ but $i \notin S(w)$, for every other $w \in B$. We omit A when it is clear from the context. A basis B of $\mathbb{Z}_p^m$ is A-good if every $v \in B - K$ is i-unique with respect to B and A, for some i. For $v \in B - K$ in an A-good basis B, the minimum i for which v is i-unique is called the *principal component* of v, denoted by $\mathrm{pc}(v)$. Figure 1 illustrates an A-good basis.

The following proposition shows that it suffices to maintain A-good bases in order to maintain matrix rank modulo p.

Proposition 1. *Let A be an $n \times m$ matrix over $\mathbb{Z}_p$ and B an A-good basis of $\mathbb{Z}_p^m$. Then $\mathrm{rank}(A) = n - |B \cap K|$.*

We now show how to maintain A-good bases modulo a prime p. Initially, the matrix A is all zero and every basis B of $\mathbb{Z}_p^m$ is A-good, as all its vectors are in K. Besides B, the algorithm also maintains the vector Av, for every $v \in B$, which is easy to do, as each change affects only one entry of A.

It is sufficient to describe how the basis can be adapted when one matrix entry a_{ij} of A is changed. We denote the new matrix by A', its entries by a'_{ij}, its kernel by K' and, for a vector v, the set of non-zero coordinates of $A'v$ by $S'(v)$. Clearly, for every vector v, Av and $A'v$ can only differ in the i-th coordinate as the only difference between A and A' is that $a_{ij} \neq a'_{ij}$. Therefore, if the A-good basis B is not A'-good, this can be only due to changes of the sets $S'(v)$ with respect to i. More specifically,

(a) there might be more than one vector $v \in B$ with $i \in S'(v)$, and
(b) there might be a vector $u \in B$ such that $\mathrm{pc}(u) = i$ but $i \notin S'(u)$.

Algorithm 1 Computation of B' from B.

---

(0) Copy all vectors from B to B'

(1) If $U \cup V \neq \emptyset$ then:
    (a) Choose $\hat{v}$ as follows:
        (i) If $V \neq \emptyset$, let $\hat{v}$ be the minimal element in V (with respect to the lexico-
        graphic order obtained from the order on V).
        (ii) If $V = \emptyset$ and $U \neq \emptyset$, let $\hat{v} \stackrel{\text{def}}{=} u$.
    (b) Make $\hat{v}$ i-unique by the following replacements in B':
        (i) Replace each element $w \in W$ by $w - (A'w)_i(A'\hat{v})_i^{-1}\hat{v}$.
        (ii) If $\hat{v} \in V$, replace each element $v \in V$, $v \neq \hat{v}$, by $v - (A'v)_i(A'\hat{v})_i^{-1}\hat{v}$.
        (iii) If $\hat{v} \in V$ and $U \neq \emptyset$, replace u by $\hat{u} \stackrel{\text{def}}{=} u - (A'u)_i(A'\hat{v})_i^{-1}\hat{v}$.
    (c) If u exists and $i \notin S'(u)$ (note: $U = \emptyset$) then let $\hat{u} \stackrel{\text{def}}{=} u$.

(2) If $\hat{u}$ has been defined (note: $i \notin S'(\hat{u})$) and $S'(\hat{u}) \neq \emptyset$ then:
    (a) Choose k minimal in $S'(\hat{u})$.
    (b) Make $\hat{u}$ k-unique by replacing every vector $v \in B'$ with $k \in S'(v)$ by $v - (A'v)_k(A'\hat{u})_k^{-1}\hat{u}$.

(3) Compute $A'v$, for every $v \in B'$ (with the help of the vectors Au, for $u \in B$)

When constructing an A'-good basis B' from the A-good basis B, those two issues have to be dealt with. To state the algorithm, the following definitions are useful. Let u denote the unique vector from B with $\text{pc}(u) = i$, if such a vector exists. The set of vectors $v \in B$ with $i \in S'(v)$ can be partioned into three sets U, V and W where

- $U = \{u\}$ if $i \in S'(u)$, otherwise $U = \emptyset$.
- V is the set of vectors $v \in B \cap K$ with $i \in S'(v)$; and
- W is the set of vectors $w \in B - K$, with $i \in S'(w)$ but $w \neq u$ (thus, in particular $\text{pc}(w) \neq i$).

For vectors $v \in V$, only i is a candidate for being the principal component since $S'(v) = \{i\}$ for such v because $Av = 0$ and the vectors Av and $A'v$ may only differ in the i-th component.

The idea for the construction of the basis B' is to apply modifications to B in two phases. In the first phase, when $U \cup V \neq \emptyset$, a vector $\hat{v} \in U \cup V$ is chosen as the new vector with principal component i. The i-uniqueness of $\hat{v}$ is ensured by replacing all other vectors x with $i \in S'(x)$ by $x - (A'x)_i(A'\hat{v})_i^{-1}\hat{v}$, where $(A'\hat{v})_i^{-1}$ denotes the inverse of the i-th entry of $A'\hat{v}$. The second phase assigns, when necessary, a new principal component k to the vector u or to its replacement from the first phase. Furthermore it ensures the k-uniqueness of this vector. The detailed construction of B' from B is spelled out in Algorithm 1.

Proposition 2. *Let A and A' be $n \times m$ matrices such that A' only differs from A in one entry $a'_{ij} \neq a_{ij}$. If B is an A-good basis of $\mathbb{Z}_p^m$ and B' is constructed according to Algorithm 1 then B' is an A'-good basis of $\mathbb{Z}_p^m$.*

An anonymous referee pointed out that the above stated algorithm for matrix rank (modulo p) is very similar to a dynamic algorithm for matrix rank presented as Algorithm 1 in [7] in a context where parallel complexity was not considered. Indeed, both algorithms essentially maintain Gaussian elimination, but the algorithm in [7] maintains a stronger normal form (reduced row echelon form) that differs by multiplication by a permutation matrix from our form. However, Algorithm 1 in [7], restricted to single entry changes and integers modulo p, can be turned into an AC^0 algorithm by observing that the sorting step 12 only requires moving two rows to the appropriate places.

3.2   Maintaining Reachability

Next, we give a dynamic algorithm for Reachability. To this end, we first show how to reduce Reachability to the test whether an entry of the inverse of an invertible matrix equals some small number. Testing such a property will in turn be reduced to matrix rank.

We remind the reader, that for Reachability the number n of nodes is fixed at initialization time and the edge set is initially empty. Afterwards in each step one edge can be deleted or inserted. For simplicity, we assume that two nodes s and t are fixed at initialization time and we are always interested in whether there is a path from s to t. To maintain Reachability for arbitrary pairs, the algorithm can be run in parallel, for each pair of nodes.

For a given directed graph $G = (V, E)$ with $|V| = n$, we define its adjacency matrix $A = A_G$, where $A_{u,v} = 1$ if $u \neq v$ and there is a directed edge $(u, v) \in E$, and otherwise $A_{u,v} = 0$.

The matrix $I - \frac{1}{n}A$ is strictly diagonally dominant, therefore it is invertible (see e.g. [12, Theorem6.1.10.]) and its inverse can be expressed by its Neumann series as $(I - \frac{1}{n}A)^{-1} = I + \sum_{i=1}^{\infty} (\frac{1}{n}A)^i$. The crucial observation is that the (s, t)-entry of the matrix on the right-hand side is non-zero if and only if there is a directed path from s to t. Therefore it suffices to maintain $(I - \frac{1}{n}A)^{-1}$ in order to maintain Reachability. To be able to work with integers, we consider the matrix $B \stackrel{\text{def}}{=} nI - A$ rather than $I - \frac{1}{n}A$. Clearly, the (s, t)-entry in B^{-1} is non-zero if and only if it is in $(I - \frac{1}{n}A)^{-1}$. Thus, for maintaining reachability it is sufficient to test whether the (s, t) entry of B^{-1} is non-zero.

More generally we show how to test whether the (i, j)-entry of the inverse B^{-1} of an invertible matrix B equals a number $a \leq n$ using matrix rank. A similar reduction has been used in [14, p.99]. Let b be the column vector with $b_j = 1$ and all other entries are 0. For every $l \leq n$, the lth entry of the vector $B^{-1}b$ is equal to the (l, j)-entry $(B^{-1})_{l,j}$ of B^{-1}. In particular, the unique solution of the equation $Bx = b$ has $(B^{-1})_{i,j}$ as ith entry. Now let B' be the matrix resulting from B by adding an additional row with 1 in the i-column and otherwise zero. Let further b' be b extended by another entry a. The equation $B'x = b'$ now corresponds to the equations $Bx = b$ and $x_i = a$ and, by the above, this system is feasible if and only if the (i, j)-entry of B^{-1} is equal to a. On the other hand, $B'x = b'$ is feasible if and only if $\text{rank}(B') = \text{rank}(B'|b')$, where $(B'|b')$ is the $(n + 1) \times (n + 1)$ matrix obtained by appending the column b' to B'. As B is invertible, $\text{rank}(B') = \text{rank}(B) = n$ and therefore, we get the following result.

Proposition 3. *Let B be an invertible matrix, $a \leq n$ a number, and B' and b' as just defined. Then, the (i,j)-entry of B^{-1} is equal to a if and only if $rank(B'|b') = n$.*

Thus, to maintain a small entry of the inverse of a matrix it suffices to maintain the rank of the matrix $B'|b'$ and to test, whether this rank is n (or, otherwise $n + 1$). As every change in B yields only one change in $B'|b'$, Algorithm 1 can be easily adapted for this purpose.

By choosing $a = 0$, the following corollary immediately follows from the observation made above, that the (s,t)-entry of the matrix $(nI - A)^{-1}$ is non-zero if and only if there is a directed path from s to t. It implies that also reachability can be maintained.

Corollary 1. *Let G be a directed graph with n vertices, A its adjacency matrix, $B = nI - A$, $a = 0$, and B' and b' as defined above (with s and t instead of i and j). Then, there is a path from node s to node t in G if and only if $rank(B'|b') = n + 1$.*

4   Matrix Rank and Reachability in DynFO

In this section we show Theorems 1 and 2. The proofs are based on the algorithms presented in Section 3. The proof for Theorem 3 is given in the full version of the paper. We first give the basic definitions for dynamic complexity, and show that, for domain-independent queries, DynFO programs with empty initialization are as powerful as DynFO programs with $(+, \times)$-initialization. Then we give proof sketches for the two theorems.

4.1   Dynamic Complexity

We basically adopt the original dynamic complexity setting from [16], although our notation is mainly from [20].

In a nutshell, inputs are represented as relational logical structures consisting of a universe, relations over this universe, and possibly some constant elements. For any change sequence, the universe is fixed from the beginning, but the relations in the initial structure are empty. This initially empty structure is then modified by a sequence of insertions and deletions of tuples. The goal of a dynamic program is to answer a given query after each prefix of a change sequence. To this end, the program can use some data structures, represented by auxiliary relations. Depending on the exact setting, these auxiliary relations might be initially empty or might contain some precomputed tuples.

We say that a dynamic program maintains a query q if it has a designated auxiliary relation that always coincides with the query result for the current database. An update program basically consists of two update formulas $\phi^R_{\text{INS}_S}(\boldsymbol{x}; \boldsymbol{y})$ and $\phi^R_{\text{DEL}_S}(\boldsymbol{x}; \boldsymbol{y})$ for each auxiliary relation R, and each input relation S; one for updates of R after insertions of S-tuples and one for deletions of S-tuples. Intuitively, when modifying the tuple $\boldsymbol{x}$ with the operation δ, then all tuples $\boldsymbol{y}$ satisfying $\phi_\delta^R(\boldsymbol{x}; \boldsymbol{y})$ will be contained in the updated relation R. Besides that, a program might have functions that define the initial values of the auxiliary relations.

Example 1. The transitive closure of an acyclic graph can be maintained by an update program with one binary auxiliary relation T which is intended to store the transitive closure [4, 16]. After inserting an edge (u, v) there is a path from x to y if, before the insertion, there has been a path from x to y or there have been paths from x to u and from v to y. Thus, T can be maintained for insertions by the formula $\phi_{\mathrm{INS}_E}^T(u, v; x, y) \stackrel{\mathrm{def}}{=} T(x, y) \vee (T(x, u) \wedge T(v, y))$. The formula for deletions is slightly more complicated.

Here, we concentrate on the following three dynamic complexity classes:

- DynFO is the class of all dynamic queries that can be maintained by dynamic programs with formulas from first-order logic starting from an empty database and empty auxiliary relations.
- DynFO$(+,\times)$ is defined as DynFO, but the programs have three particular auxiliary relations that are initialized as a linear order and the corresponding addition and multiplication relations. There might be further auxiliary relations, but they are initially empty.
- *Non-uniform* DynFO is defined as DynFO, but the auxiliary relations may be initialized by arbitrary functions.

4.2 DynFO and DynFO$(+,\times)$ Coincide for Domain Independent Queries

Next, we show that DynFO and DynFO$(+,\times)$ coincide for queries that are invariant under insertion and deletion of isolated elements. More precisely, a query q is *domain independent*, if $q(\mathcal{D}_1) = q(\mathcal{D}_2)$ for all databases $\mathcal{D}_1$ and $\mathcal{D}_2$ that coincide in all relations and constants (but possibly differ in the underlying domain). As an example, the Boolean Reachability query is domain independent, as its result is not affected by the presence of isolated nodes (besides s and t).

Theorem 4. *For every domain-independent query q,*
$$q \in \mathsf{DynFO}(+,\times) \ \text{if and only if} \ q \in \mathsf{DynFO}.$$

The proof idea is to simulate several computations. The DynFO program $\mathcal{P}'$ simulates several "runs" of the DynFO$(+,\times)$ program $\mathcal{P}$, one for each $m \leq \sqrt{n}$, where n is the domain size. Such a simulation starts as soon as m elements become "active" and serves to answer the query for the period when at least $(m - 1)^2$ but less than m^2 elements are active. More details about this construction can be found in the full paper.

Remark 1. Kousha Etessami already observed that arithmetic can be defined incrementally, so that at any point there are relations $<_{ad}$, $+_{ad}$ and $\times_{ad}$ that represent a linear order *on the activated domain*, and the ternary addition and multiplication relations [6].

4.3   Matrix Rank in DynFO$(+,\times)$

To the best of our knowledge, computational linear algebra problems like matrix rank and matrix inverse have not been studied before in dynamic complexity (with the notable exception of Boolean matrix multiplication in [10]). Therefore, there is no standard way of representing the matrix rank problem in the dynamic complexity framework. The key question is how to represent the numbers that appear in a matrix, as their size can grow arbitrarily compared to the dimensions of the matrix. We use a representation that does not allow matrices with large numbers but suffices for our applications in which matrix entries are not larger than the number of rows in the matrix.

More precisely, an input database for the matrix rank query MATRIXRANK consists of two ternary relations M_+, M_- and a linear order $<$. In the following, we identify the k-th element with respect to $<$ with the number k (and the minimal element represents 1). That the matrix has value $a > 0$ at position (i,j) is represented by a triple (i,j,a) in M_+. Likewise, $a_{ij} = a < 0$ is represented by a triple $(i,j,-a)$ in M_-. For each i,j, at most one triple (i,j,a) can be present in $M_+ \cup M_-$. If, for some i,j there is no triple (i,j,a) then $a_{i,j} = 0$. In this way, we can represent $n \times n$-matrices over $\{-n,\dots,n\}$ by databases with domain $\{1,\dots,n\}$. Non-square matrices can be represented as $n \times n$-matrices in a straightforward manner with the help of zero-rows or zero-columns.

Change operations might insert a triple (i,j,a) to M_+ or M_- (in case, no (i,j,b) is there), or delete a triple, but we do not allow change operations on $<$. That is, basically, single matrix entries can be set to 0 or from 0 to some other value. Initially, M_+ and M_- are empty, that is the matrix is the all-zero matrix, but $<$ is a complete linear order. The query MATRIXRANK maps a database $\mathcal{D}$ representing a matrix A in this way to the set $\{\mathrm{rank}(A)\}$, in case $\mathrm{rank}(A) > 0$ and to $\emptyset$ otherwise.

Theorem 5. MATRIXRANK *is in* DynFO$(+,\times)$.

There is a subtle technical point in the interpretation of the statement "MATRIXRANK is in DynFO$(+,\times)$". The database representing the input matrix A comes with a linear order $<_A$ and there is the linear order $<$ initially given to a DynFO$(+,\times)$ program. We require here that these orders are identical (as they are in our applications).

Proof. Algorithm 1 can be extended and translated into a dynamic program for MATRIXRANK in a straightforward manner. Let DOM be a given domain. In our setting, we have $N = n$, therefore it is sufficient to consider prime numbers $p \le n^2$. Such prime numbers and arithmetics in $\mathbb{Z}_p$ can be expressed with the help of pairs over DOM, via the bijection $(u_1, u_2) \mapsto (u_1 - 1) \times n + u_2$.   □

Remark 2. Due to the initial linear order, MATRIXRANK does not fit into the domain independence framework of Theorem 4. To maintain matrix rank in DynFO, we would need to build $<$ incrementally when entries are inserted to the matrix. However, when we use MATRIXRANK to maintain a domain independent query, Theorem 4 yields a DynFO upper bound.

4.4   Reachability in DynFO

As described at the beginning of this section, the reachability query REACH has a straightforward, and standard formalization in the dynamic complexity framework. Now we can sketch the proof of the main result of this paper.

Theorem 1 *(restated).* REACH *is in* DynFO.

Proof. It is straightforward to transform the approach of Proposition 3 into a dynamic program with arithmetic. The edge relation E can be viewed as an adjacency matrix A for the graph, and $nI - A$ and then $B'|b'$ can be easily defined in first-order logic. We note that each change of one pair in E only changes one entry in $B'|b'$. It thus suffices to maintain rank$(B'|b')$ by the program of Theorem 5 to maintain reachability from s to t. This shows REACH $\in$ DynFO$(+,\times)$. As the reachability query is domain independent, Theorem 4 yields the theorem. $\square$

Remark 3. In the DynFO-framework considered here, elements cannot be removed from the domain. Removal of nodes is allowed in the FOIES-framework of Dong, Su and Topor: when a node is not used in any edge, then it is removed from the domain. The proof above can be adapted to this framework.

By simple reductions we obtain the following further results.

Theorem 6. *(a) Regular path queries in directed labeled graphs can be maintained in* DynFO. *(b) Conjunctions of regular path queries in directed labeled graphs can be maintained in* DynFO. *(c) 2-SAT is in* DynFO.

5   Conclusion

The main technical contribution of the paper is that maintaining the rank of a matrix is in DynFO$(+,\times)$. From this we derive that Reachability can be maintained in DynFO improving on both the complexity and uniformity of previous results [3,9]. In the case of matching, we are able to prove only a non-uniform bound. As exemplified by regular path queries and 2-SAT, the fact that Reachability is in DynFO may help to show that many other queries and problems can be maintained in DynFO. However, the DynFO bound obtained here does not extend to all of NL, simply because DynFO is not known to be closed under even *unbounded* first order projection reductions. We believe that also the approach through Linear Algebra might yield further insights.

It is an interesting open question whether a Reachability witness can be maintained in DynFO and whether a Shortest Path witness can be maintained in DynTC^0. Some further reflections on our results can be found in the full version of the paper.

Acknowledgments. We would like to thank William Hesse for stimulating and illuminating discussions. We are grateful to the anonymous referee who brought [7] to our attention. Further we thank Nils Vortmeier for proofreading. The first and the third authors were partially funded by a grant from Infosys Foundation. The second author is supported by the Singapore National Research Foundation under NRF RF Award No. NRF-NRFF2013-13. The last two authors acknowledge the financial support by DFG grant SCHW 678/6-1.

References

1. Allender, E., Reinhardt, K., Zhou, S.: Isolation, matching, and counting uniform and nonuniform upper bounds. J. Comput. Syst. Sci. **59**(2), 164–181 (1999)
2. Artin, M.: Algebra. Featured Titles for Abstract Algebra. Pearson (2010)
3. Datta, S., Hesse, W., Kulkarni, R.: Dynamic Complexity of Directed Reachability and Other Problems. In: Esparza, J., Fraigniaud, P., Husfeldt, T., Koutsoupias, E. (eds.) ICALP 2014. LNCS, vol. 8572, pp. 356–367. Springer, Heidelberg (2014)
4. Dong, G., Jianwen, S.: Incremental and decremental evaluation of transitive closure by first-order queries. Information and Computation **120**(1), 101–106 (1995)
5. Dong, G., Jianwen, S.: Arity bounds in first-order incremental evaluation and definition of polynomial time database queries. J. Comput. Syst. Sci. **57**(3), 289–308 (1998)
6. Etessami, K.: Dynamic tree isomorphism via first-order updates. In: PODS, pp. 235–243 (1998)
7. Gudmund Skovbjerg Frandsen and Peter Frands Frandsen: Dynamic matrix rank. Theor. Comput. Sci. **410**(41), 4085–4093 (2009)
8. Grädel, E., Siebertz, S.: Dynamic definability. In: ICDT, pp. 236–248 (2012)
9. Hesse, W.: The dynamic complexity of transitive closure is in DynTC^0. Theor. Comput. Sci. **296**(3), 473–485 (2003)
10. Hesse, W., Immerman, N.: Complete problems for dynamic complexity classes. In: LICS, p. 313 (2002)
11. Hoang, T.M.: On the matching problem for special graph classes. In: IEEE Conference on Computational Complexity, pp. 139–150 (2010)
12. Horn, R.A., Johnson, C.R.: Matrix analysis. Cambridge University Press (2012)
13. Immerman, N.: Descriptive complexity. Graduate texts in computer science. Springer (1999)
14. Laubner, B.: The structure of graphs and new logics for the characterization of Polynomial Time. PhD thesis, Humboldt University of Berlin (2011)
15. Mulmuley, K., Vazirani, U.V., Vazirani, V.V.: Matching is as easy as matrix inversion. Combinatorica **7**(1), 105–113 (1987)
16. Patnaik, S., Immerman, N.: Dyn-FO: A parallel, dynamic complexityclass. Journal of Computer and System Sciences **55**(2), 199–209 (1997)
17. Reif, J.H., Tate, S.R.: On dynamic algorithms for algebraic problems. J. Algorithms **22**(2), 347–371 (1997)
18. Sankowski, P.: Dynamic transitive closure via dynamic matrix inverse (extended abstract). In: FOCS, pp. 509–517 (2004)
19. Sankowski, P.: Faster dynamic matchings and vertex connectivity. In: SODA, pp. 118–126 (2007)
20. Zeume, T., Schwentick, T.: On the quantifier-free dynamic complexity of reachability. Inf. Comput. **240**, 108–129 (2015)

Natural Homology

Jérémy Dubut[1,2], Éric Goubault[1], and Jean Goubault-Larrecq[2(✉)]

[1] LIX, Ecole Polytechnique, F-91128 Palaiseau Cedex, France
[2] LSV, ENS Cachan, F-94230 Cachan, France
`goubault@lsv.ens-cachan.fr`

Abstract. We propose a notion of homology for directed algebraic topology, based on so-called natural systems of abelian groups, and which we call natural homology. As we show, natural homology has many desirable properties: it is invariant under isomorphisms of directed spaces, it is invariant under refinement (subdivision), and it is computable on cubical complexes.

Keywords: Directed algebraic topology · Homology · Path space · Geometric semantics · Persistent homology · Natural system

1 Introduction

The purpose of this paper is to introduce a satisfactory notion of homology for directed algebraic topology. Let us clarify.

From a mathematical point of view, algebraic topology is a well-established, and rich domain. Its purpose is to classify shapes (topological spaces), disregarding differences in shapes that can be obtained from each other by continuous deformations (homotopy equivalence). A particularly useful notion there is *homology*, which is a sound abstraction of homotopy equivalence. Soundness means, notably, that if two spaces are not homologous, then one cannot deform one into the other continuously, in whichever way we attempt this. Homology is also computable [14,21] on finitely presented shapes (simplicial, resp. cubical sets), in sharp contrast with homotopy.

Directed algebraic topology is a variant of algebraic topology where the spaces also have a direction of time [12], and deformations must not only be continuous but also preserve the direction of time. Directed algebraic topology was born out of the so-called geometric semantics of concurrent processes (progress graphs [3], generally attributed to E. W. Dijkstra), and the higher-dimensional automaton model of true concurrency [18]. Imagine n concurrent processes, each with a local time $t_i \in [0,1]$. A configuration is a point in $[0,1]^n$, and a trajectory is a continuous *and monotonic* map from $[0,1]$ to $[0,1]^n$: monotonicity (a.k.a., directedness) reflects the fact that no process can go back in time. One can arguably consider as equivalent any two trajectories that are dihomotopic, namely that can be deformed into each other continuously, while respecting monotonicity at all times. This not only yields a geometric semantics for concurrency, but also one that is at the root of fast algorithms for state-space reduction, deadlock and

© Springer-Verlag Berlin Heidelberg 2015
M.M. Halldórsson et al. (Eds.): ICALP 2015, Part II, LNCS 9135, pp. 171–183, 2015.
DOI: 10.1007/978-3-662-47666-6_14

unreachable states detection, and verification of coordination properties, as in e.g. [6,8–10].

However intuitive the geometric semantics of processes may be, previous attempts at defining notions of homology suited to *directed* algebraic topology were somehow disappointing. We discuss them in Section 2. Our contribution is: 1. a new notion of directed homology, based on so-called natural systems of abelian groups, and which we call *natural homology*, and 2. the proofs of its basic properties, the most important probably being *invariance under refinement*— a central property of truly concurrent semantics [22]. We define and motivate natural systems (on pospaces) in Section 3, and the requirement for a notion of bisimilarity between them in Section 4. Finite cubical complexes provide finite presentations of pospaces, as explained in Section 5. We adapt the notion of natural homology to cubical complexes, and we prove the properties mentioned above in Section 6. We conclude in Section 7.

2   Related Work

Homology is a classical concept in (undirected) algebraic topology, and we shall only discuss the notions that various authors have proposed to fit the directed case. Non-abelian homology [17] may seem promising; so far, we are only aware of work by Krishnan in this direction [16]. All the other attempts we know [4,7,11,15] have the same weakness: they are not precise enough.

By this we mean, *first*, that directed homology should not be invariant under (undirected) homotopy. If it were, it would be blind to the essential feature of directed algebraic topology: that directions are important. In that case, we may as well use classical homology theories, which are already homotopy invariants. This is the bane of early directed homology theories [7].

Second, the Hurewicz theorems in classical algebraic topology state that the loss of information we must pay when replacing homotopy groups by homology groups is limited. One trivial consequence of these results is that a space X whose first homotopy group is non-trivial (different from 0) also has a non-trivial first homology group $H_1(X)$. Similarly, we would like any dihomotopically non-trivial shape to have non-trivial directed homology. This fails in any of the remaining proposals [4,11,15], as we explain now.

Consider the *matchbox* example, due to Fahrenberg [4], shown on the right. The exploded view is on the left, the finished product on the right. Note that this is not a cube: the bottom face and the interior are missing. The matchbox is meant to stand on its tip (vertex s), and time goes up, that is, a point is before another one if and only if its altitude is smaller.

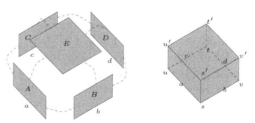

Fig. 1. Fahrenberg's matchbox example

Look at the edges a, b, c, d. The concatenation $a \star c$ of a and c is a directed path from s to t, and $b \star d$ is another one. A dihomotopy between these two would be a continuous map h from $[0, 1]$ to the space of directed paths from s to t such that $h(0) = a \star c$ and $h(1) = b \star d$. If h existed, then for some α, $h(\alpha)$ would be a directed path going through t', and then down to t: this is impossible, since $h(\alpha)$ must remain directed for all values of α. Indeed, $a \star c$ and $b \star d$ are not dihomotopic. In particular, the matchbox has non-trivial dihomotopy. This contrasts with the undirected view: the matchbox is contractible, and in particular all its classical homotopy and homology groups are trivial (equal to 0). We now examine why this example is not dealt properly by the existing proposals for directed homology.

Grandis [11] defined a notion of directed homology by enriching the classical homology groups of the space at hand with a partial order. The idea is that the generators of homology groups are holes, and the partial order serves to remember which holes come before which others. Since the classical homology groups of the matchbox are trivial, so is the ordered homology group that Grandis defines. However, the matchbox is dihomotopically non-trivial.

Kahl's homology graphs [15], which are defined, similarly, as homology groups with extra relations, suffer from the same problem.

Finally, the matchbox was produced by Fahrenberg to show that his own notion of directed homology [4] is unsatisfactory. The problem is common to any notion of directed homology that is based on homology *groups*, or even on cancellative monoids (an assumption used in [17]). Let e be the directed edge from t to t'. The directed paths $a \star c \star e$ and $b \star d \star e$ *are* dihomotopic, hence must be in the same equivalence class of (directed) homology. Cancellation of e then implies that $a \star c$ and $b \star d$ must be equivalent with respect to directed homology.

Our solution avoids cancellation by working with so-called *natural systems* of abelian groups, and builds upon several prior strands of research. The natural systems themselves arise from Baues and Wirsching's work on the cohomology of small categories [1]. The view of directed homotopy (resp., homology) as being based on classical homotopy (resp., homology) of spaces of traces can be traced to Raussen [19], and the carrier morphism we use near the end has its origin in work by Fajstrup [5]. The idea of using natural systems, indexed by so-called traces, to organize information on several topological objects together is also present in [19], although Raussen did not apply this to homology.

The notion of bisimulation of such natural systems, which we define to forget about irrelevant differences between them, is novel. We shall see later why this is needed. To make the argument short, a natural system of homology groups is an immense picture of homology groups, indexed by traces, but we do not care about the whole picture, rather about the patterns of change we see in groups as we extend the traces. This is a very similar concern as in persistent homology (of undirected spaces) [2]. However, the latter uses a very simple, linear ordering of the indices, and we must deal with a much more complex situation.

3   Natural Homology of Pospaces, and Natural Systems

Let X be a *pospace*, i.e., a topological space X with a partial order $\leq$ whose graph is closed in X^2. A fundamental example is I, the interval $[0, 1]$ with the usual ordering. A *path* from a to b in X is a continuous map $\pi \colon I \to X$ such that $\pi(0) = a$ and $\pi(1) = b$. It is a *dipath* (short for directed path) if and only if it is also monotonic. Following Raussen [19], a (directed) *trace* is the equivalence class $\langle \pi \rangle$ of a dipath π modulo reparametrization: a reparametrization is a monotonic continuous onto map φ from I to I, and π and π' are equivalent if and only if there are two reparametrizations φ, ψ such that $\pi \circ \varphi = \pi' \circ \psi$. Traces are dipaths, up to the speed at which we travel from time $t = 0$ to time $t = 1$, which is considered irrelevant. Given two (di)paths π from a to b and π' from b to c, the concatenation $\pi \star \pi'$ maps $t \in [0, 1/2]$ to $\pi(2t)$ and $t \in [1/2, 1]$ to $\pi'(2t - 1)$. This induces an associative operation $\star$ on the quotient space of traces.

A standard notion of classical algebraic topology is *homotopy*. Consider the n-cube I^n, and write ∂I^n for its boundary. Given a path-connected space Y, and fixing a so-called base point $y \in Y$, an n-loop in Y is a continuous map from I^n to Y that maps ∂I^n to y. In particular, for $n = 1$, a 1-loop is just a path from y to itself. A homotopy h between two n-loops λ, λ' is a continuous map from $I \times I^n$ to Y such that $h(0, \_) = \lambda$, $h(1, \_) = \lambda'$, and, for each α, $h(\alpha, \_)$ maps ∂I^n to the base point y. If such a homotopy exists, then one says that λ and λ' are *homotopic*—we can deform one continuously into the other. We let $\pi_n(Y)$ denote the set of equivalence classes of n-loops of Y modulo homotopy. It is useful to visualize the case $n = 1$, where loops modulo homotopy form a group $\pi_1(Y)$ under concatenation.

One can define *dihomotopy* and *dihomology* similarly, but one should be careful. For example, *directed* n-loops in a pospace are trivial. Instead, Raussen [19] proposes to consider $(n - 1)$-loops in the space $Y = Tr(X; a, b)$ of traces from a to b in X. For example, for $n = 1$, the points of Y are the traces from a to b, and any path between two such points $\langle \pi \rangle$ and $\langle \pi' \rangle$ is easily seen to be (up to reparametrization of π and π') a homotopy between π and π' that fixes the two endpoints a and b: a continuous map $h \colon I \times I \to X$ such that $h(0, \_) = \pi$, $h(1, \_) = \pi'$, $h(\_, 0) = a$, $h(\_, 1) = b$, and, for every value of the deformation parameter α, $h(\alpha, \_)$ is a dipath from a to b. The zeroth homology group $H_0(Y)$ of Y is of the form $\mathbb{Z}^k$ with k the number of equivalence classes of traces from a to b up to dihomotopy. In general, we may define the n-th directed homology group $\overrightarrow{H}_n(X; a, b)$ as the ordinary $(n - 1)$st singular homology group of $Tr(X; a, b)$.

While this looks like a perfect definition of directed homology, this is still unsatisfactory. Consider the following two pospaces. In each, time goes from left to right and from bottom to top, starting at $\mathbf{0}$ and ending at $\mathbf{1}$. The leftmost pospace is the geometric semantics of a PV-program [3] extended with global synchronization (written "•"), namely $(\mathsf{PaVa}\|\mathsf{PaVa}) \bullet (\mathsf{PaVaPbVb}\|\mathsf{PbVbPaVa})$.

The rightmost pospace is the geometric semantics of the PV-program $\mathsf{PaVaPaVa} \parallel \mathsf{PaVaPaVa}$. The four squares we carved out are those regions of space where the two processes would have acquired the lock a—which is impossible.

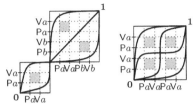

It is natural to compare pospaces X with two distinguished endpoints $\mathbf{0}$ and $\mathbf{1}$ by determining their dihomology groups $\overrightarrow{H}_n(X; \mathbf{0}, \mathbf{1})$. For $n = 1$, the two pospaces above both have exactly six traces up to dihomotopy, shown as thick lines: the two pospaces have the same dihomology group for $n = 1$, namely $\mathbb{Z}^6$. For $n \geq 2$, they also have the same dihomology groups $\overrightarrow{H}_n(X; \mathbf{0}, \mathbf{1})$, because the path-connected components of their trace spaces are contractible. Therefore, the $\overrightarrow{H}_n(\_; \mathbf{0}, \mathbf{1})$ construction does not distinguish the two pospaces, although they visibly have very different behaviors.

However, when we zoom in, and look at different pairs of endpoints, the situation changes. Consider the $(\mathsf{PaVa}\|\mathsf{PaVa}) \bullet (\mathsf{PaVaPbVb}\|\mathsf{PbVbPaVa})$ pospace again, but look at its dihomology group $\overrightarrow{H}_1(X; \mathbf{0}, t)$, where t is shown on the right: this is equal to $\mathbb{Z}^4$. However, no trace space of the other pospace $(\mathsf{PaVaPaVa}\|\mathsf{PaVaPaVa})$ has exactly four connected components, so $\mathbb{Z}^4$ cannot be a dihomology group of the latter. This detects an essential difference between the two pospaces.

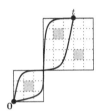

Given a trace $\langle \pi \rangle$, with π a dipath of X from a to b, we define $\overrightarrow{H}_n(X; \langle \pi \rangle) = \overrightarrow{H}_n(X; a, b)$. The family of groups $\overrightarrow{H}_n(X; \langle \pi \rangle)$, when $\langle \pi \rangle$ varies over traces, has extra structure: if α is a dipath from a' to a and β is a dipath from b to b', we obtain a continuous map from $Tr(X; a, b)$ to $Tr(X; a', b')$, which maps every trace $\langle \pi' \rangle$ to $\langle \alpha \star \pi' \star \beta \rangle$. We call $extensions$ the pairs $(\langle \alpha \rangle, \langle \beta \rangle)$. Applying the H_{n-1} functor to the map $\langle \pi' \rangle \mapsto \langle \alpha \star \pi' \star \beta \rangle$, we obtain a morphism of groups $\overrightarrow{H}_n(X; \langle \pi \rangle)$ to $\overrightarrow{H}_n(X; \langle \alpha \star \pi \star \beta \rangle)$, which we denote by $\langle \alpha \star \_ \star \beta \rangle$. This keeps track of how the homology picture formed by the traces from a to b inserts into the larger picture formed by the traces from the lower point a' to the higher point b'.

We are ready to give formal definitions. Let X be a pospace, and Fc_X be the small category whose objects are traces of X, and whose morphisms are extensions. This is the *factorization category* [1] of the small category whose objects are points of X, and whose morphisms are traces. A *natural system (of abelian groups)* is by definition a functor from the factorization category of a small category (e.g. Fc_X) to the category **Ab** of abelian groups.

Definition 1 (Natural homology). *The* natural homology *of X is the natural system $\overrightarrow{H}_n(X)$ that, as a functor, maps every trace $\langle \pi \rangle$ to $\overrightarrow{H}_n(X; \langle \pi \rangle)$, and every extension $(\langle \alpha \rangle, \langle \beta \rangle)$ to $\langle \alpha \star \_ \star \beta \rangle$.*

Figure 2 shows a few simple examples of natural homology systems $\overrightarrow{H}_1$. On top, we consider the pospace I itself. The middle diagram pictures the full subcat-

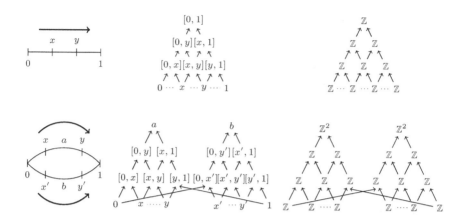

Fig. 2. Natural homology of two simple pospaces

egory of the (uncountable) category Fc_I whose objects are traces, which are identified to segments $[s, t]$ with $s, t \in \{0, 1, x, y\}$ and $s \leq t$ (x and y are shown on the left). $[s, s]$ simplifies to s. The rightmost diagram pictures the collection of dihomology groups above each object of Fc_I. The bottom row is similar, and applies to two copies of I glued at 0 and 1.

4   Bisimilarity of Natural Systems

The natural homology $\overrightarrow{H}_n(X)$ of a pospace X is very fine-grained: it not only records local homology groups $\overrightarrow{H}_n(X; \langle \pi \rangle)$, but also for which traces they occur. If we wish to compare the natural homology of two pospaces, the latter should be unimportant. Just as with persistent homology [2], it is the patterns of change, between groups $\overrightarrow{H}_n(X; \langle \pi \rangle)$ when $\langle \pi \rangle$ is changed into $\langle \alpha \star \pi \star \beta \rangle$ by extension, that count, not the values of the trace $\langle \pi \rangle$.

We introduce a notion of bisimulation of natural systems, and more generally of **Ab**-valued functors, that smoothes this out. Given two small categories X, Y and two functors $F : X \longrightarrow \mathbf{Ab}$ and $G : Y \longrightarrow \mathbf{Ab}$, we call *bisimulation* between F and G any set R of triples (x, η, y) with x an object of X, y an object of Y and η an isomorphism of groups from Fx to Gy such that:

1. for every object x of X, R contains some triple of the form (x, η, y), and similarly for every object y of Y;
2. for every triple $(x, \eta, y) \in R$ and every morphism $i : x \longrightarrow x'$ in X, there is a triple $(x', \eta', y') \in R$ (hence η' is an isomorphism) and a morphism $j : y \longrightarrow y'$ in Y such that $\eta' \circ Fi = Gj \circ \eta$, and symmetrically,

for every $(x, \eta, y) \in R$ and every morphism $j :$ $y \longrightarrow y'$ of Y there is a triple $(x', \eta', y') \in R$ and a morphism $i : x \longrightarrow x'$ such that $\eta' \circ Fi = Gj \circ \eta$.

$$
\begin{array}{ccc}
x & Fx \xrightarrow{\eta} Gy & y \\
i\downarrow & Fi\downarrow \qquad \downarrow Gj & \downarrow j \\
x' & Fx' \xrightarrow[\eta']{} Gy' & y'
\end{array}
$$

We say that F and G are *bisimilar* if and only if there is a bisimulation R between them. This is an equivalence relation.

A practical way of showing that two functors are bisimilar is by exhibiting an *open map* from one to the other. This arises from the theory of Joyal *et al.* [13]; we omit the details here. The open maps from a functor $F : E \longrightarrow \mathbf{Ab}$ to a functor $G : X \longrightarrow \mathbf{Ab}$ are the pairs (Φ, σ) where Φ is a fibration from E to X, and σ is a natural isomorphism from F to $G \circ \Phi$. We say that $\Phi : E \longrightarrow X$ is a *fibration* if and only if: (1) Φ is surjective on objects, i.e., for every object x of X there is an object e of E such that $\Phi(e) = x$, and (2) for every object e of E, every morphism $f : \Phi(e) \longrightarrow x'$ in X lifts to a morphism $h : e \longrightarrow e'$ in E such that $\Phi(h) = f$ (in particular, $\Phi(e') = x'$).

Proposition 1. *Two functors $F : X \longrightarrow \mathbf{Ab}$ and $G : Y \longrightarrow \mathbf{Ab}$ are bisimilar if and only if they are related by a span of open maps.*

We shall apply this to compare our natural homology of pospaces to a similar notion of natural homology of cubical complexes. We can think of the latter as a form of syntax for the latter. Their semantics is given by geometric realization, as we now explain.

5   Cubical Complexes and Their Geometric Realization

A cubical complex is a finite union of certain cubes of side-length 1 parallel to the axes in $\mathbb{R}^d$, whose vertices have integer coordinates [14]. Formally, let us define a (d-dimensional) *cubical complex* K as a finite set of *cubes* $(D, \boldsymbol{x})$, where $D \subseteq \{1, 2, \cdots, d\}$ and $\boldsymbol{x} \in \mathbb{Z}^d$, which is closed under taking past and future faces (to be defined shortly). The cardinality $|D|$ of D is the *dimension* of the cube $(D, \boldsymbol{x})$. Let $\mathbf{1}_k$ be the d-tuple whose kth component is 1, all others being 0. Each cube $(D, \boldsymbol{x})$ is *realized* as the geometric cube $\rho(D, \boldsymbol{x}) = I_1 \times I_2 \times \cdots \times I_d$ where $I_k = [x_k, x_k + 1]$ if $k \in D$, $I_k = [x_k, x_k]$ otherwise, matching the definition of [14].

When $|D| = n$, we write $D[i]$ for the ith element of D. For example, if $D = \{3, 4, 7\}$, then $D[1] = 3$, $D[2] = 4$, $D[3] = 7$. We also write $\partial_i D$ for D minus $D[i]$. Every n-dimensional cube $(D, \boldsymbol{x})$ has n *past faces* $\partial_i^0(D, \boldsymbol{x})$, defined as $(\partial_i D, x)$, and n *future faces* $\partial_i^1(D, \boldsymbol{x})$, defined as $(\partial_i D, x + \mathbf{1}_{D[i]})$, $1 \leq i \leq n$.

Together with these face operators, K exhibits the structure of a so-called *precubical set*, in the sense that the *precubical equations* $\partial_i^\alpha \partial_j^\beta = \partial_{j-1}^\beta \partial_i^\alpha$ ($1 \leq i < j$, $\alpha, \beta \in \{0, 1\}$) are satisfied. Precubical sets are a natural representation for truly concurrent processes, and occur as the main ingredient in the definition of *higher-dimensional automata* (HDA; see [18]). Cubical complexes are very particular precubical sets. Notably, they are non-looping in the sense of Fajstrup [5]. They are however enough for most purposes, including the definition of geometric semantics of finite PV-programs.

The *geometric realization* $\overrightarrow{Geom}(K)$ of a precubical set K is obtained, informally, by drawing it. For example, Fahrenberg's matchbox (Fig. 1) is really obtained by drawing a finite precubical set (a cubical complex, really) with 2-dimensional cubes A, B, C, D, and E, defined so that $\partial_1^0 A = \partial_1^0 B$ (the lower dashed connection in the exploded view), $\partial_2^0 A = a$, $\partial_2^0 B = b$, $\partial_1^0 a = \partial_1^0 b = s$, and so on. Formally, let $\overrightarrow{I}^n$ be the standard oriented cube $[0,1]^n$, with the pointwise ordering. Form the coproduct $A = \sum_{e \in K} \overrightarrow{I}^{n_e}$ where n_e is the dimension of e, i.e., the disjoint union of as many copies of $\overrightarrow{I}^n$ as there are n-dimensional cubes e, for $n \in \mathbb{N}$; the elements of A are pairs $(e, \boldsymbol{a})$ where e is an n-dimensional cube in K and $\boldsymbol{a} \in [0,1]^n$, for some n. For convenience, for $\boldsymbol{a} = (a_1, a_2, \cdots, a_n)$, we write $\delta_i^\alpha \boldsymbol{a}$ for $(a_1, a_2, \cdots, a_{i-1}, \alpha, a_i, \cdots, a_n)$. Finally, we glue all these cubes together, by defining $\overrightarrow{Geom}(K)$ as $A/\equiv$, where $\equiv$ is the smallest equivalence relation such that $(\partial_i^\alpha e, \boldsymbol{a}) \equiv (e, \delta_i^\alpha \boldsymbol{a})$. We shall write $[e, \boldsymbol{a}]$ for the point obtained as the equivalence class of $(e, \boldsymbol{a})$.

For a cubical complex K, the element $[(D, \boldsymbol{x}), \boldsymbol{a}]$ (with $D \subseteq \{1, 2, \cdots, d\}$, $|D| = n$, $\boldsymbol{x} \in \mathbb{Z}^d$, $\boldsymbol{a} \in [0,1]^n$) of $\overrightarrow{Geom}(K)$ defines a point $\epsilon([(D, \boldsymbol{x}), \boldsymbol{a}]) = \boldsymbol{x} + \sum_{i=1}^n a_i \mathbf{1}_{D[i]}$. One checks easily that ϵ is a pospace isomorphism of $\overrightarrow{Geom}(K)$ onto the union of the cubes $\rho(D, \boldsymbol{x})$, $(D, \boldsymbol{x}) \in K$. This observation is needed to relate the notions of geometric realization of precubical *sets* (as used, say, in [5]) and of cubical *complexes* (as used in [14]).

6   Discrete Natural Homology of Cubical Complexes

Paralleling the notion of trace in a pospace, for example as in [5], there is a notion of *discrete trace* in a precubical set K. Given $a, b \in K$, say that a is a *past boundary* of b if and only if $a = \partial_{i_0}^0 \partial_{i_1}^0 \cdots \partial_{i_k}^0 b$ for some $k \geq 0$, $i_0, i_1, \ldots, i_k$. For example, the edge a, the edge from s to s', and s, are past boundaries of A in the matchbox. *Future boundaries* are defined similarly, using the superscript 1 instead of 0: so the edge from u to u', the edge from s' to u', and u' itself, are future boundaries of A. We write $a \preceq b$ if and only if a is a past boundary of b or b is a future boundary of a. (Beware that this is not a transitive relation; we write $\preceq^*$ for its reflexive transitive closure.) A *discrete trace* from a to b in K is then a sequence $c_0 = a \preceq c_1 \preceq c_2 \preceq \cdots \preceq c_n = b$, $n \in \mathbb{N}$.

Abusing the Fc_X notation we used earlier for pospaces, let Fc_K be the small category whose objects are discrete traces. Its morphisms from a discrete trace from a to b to a discrete trace from a' to b' are the *discrete extensions*, namely pairs of discrete traces α from a' to a and β from b to b'. This is the factorization category of the small category whose objects are elements of K, and whose morphisms are discrete traces.

Note that we are not restricting a, b to be points, namely, of dimension 0; however, it is helpful to imagine, geometrically, that a full cube a stands for the point at its center. The construction is again due to Fajstrup [5]. Formally, for $a = (D, \boldsymbol{x})$, $n = |D|$, let $\hat{a}$ be the point $[a, \bullet]$ in $\overrightarrow{Geom}(K)$, where $\bullet = (\frac{1}{2}, \frac{1}{2}, \cdots, \frac{1}{2})$ is the center of the standard cube $\overrightarrow{I}^n$. Through the ϵ isomorphism, $\hat{a}$ is the point $\boldsymbol{x} + \sum_{i=1}^n \frac{1}{2} \mathbf{1}_{D[i]}$ in $\mathbb{R}^d$, the center of the cube $\rho(D, \boldsymbol{x})$.

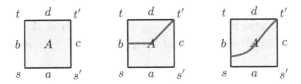

Fig. 3. From discrete traces to traces and vice versa

Every discrete trace α from a to b, say of the form $c_0 = a \preceq c_1 \preceq c_2 \preceq \cdots \preceq c_n = b$, defines a trace $\hat\alpha$ from $\hat a$ to $\hat b$, obtained by concatenating the n straight lines $\widehat{c_0 c_1}$, $\widehat{c_1 c_2}$, ..., $\widehat{c_{n-1} c_n}$. For a simple example, consider the cubical complex whose geometric realization is shown on Figure 3, left. There is a discrete trace α equal to $b \preceq A \preceq t'$, since $b = \partial_1^0 A$ is a past boundary of A and $t' = \partial_2^1 \partial_1^1 A$ is a future boundary of A. The corresponding trace $\hat\alpha$ is shown on the same figure, middle. Formally, if c_{i-1} is a past boundary $\partial_{i_1}^0 \partial_{i_2}^0 \cdots \partial_{i_k}^0 c_i$ of c_i, then $\hat c_{i-1} = [\partial_{i_1}^0 \partial_{i_2}^0 \cdots \partial_{i_k}^0 c_i, \bullet] = [c_i, \boldsymbol{a}]$ where $\boldsymbol{a} = \delta_{i_k}^0 \cdots \delta_{i_2}^0 \delta_{i_1}^0 \bullet$; define the dipath π by $\pi(t) = [c_i, (1-t)\boldsymbol{a} + t\bullet]$ for $t \in [0,1]$, and the trace $\widehat{c_{i-1} c_i}$ as $\langle \pi \rangle$. Similarly for future boundaries.

This allows us to transfer cubes a to points $\hat a \in \overrightarrow{Geom}(K)$, discrete traces α to traces $\hat\alpha$ in $\overrightarrow{Geom}(K)$, and also discrete extensions (α, β) to extensions $(\hat\alpha, \hat\beta)$. We can now mimic the natural homology of a pospace in the discrete setting of a cubical complex K: given a discrete trace γ from a to b, let $\overrightarrow{h}_n(K; \gamma)$ be the $(n-1)$st singular homology group of $Tr(\overrightarrow{Geom}(K); \hat a, \hat b)$. This defines another natural system $\overrightarrow{h}_n(K)$, this time from Fc_K instead of Fc_X, to **Ab**: the discrete traces γ are mapped to $\overrightarrow{h}_n(K; \gamma)$, and discrete extensions (α, β) are mapped to $H_{n-1}(\langle \hat\alpha \star \_ \star \hat\beta \rangle)$, mimicking the definition of $\overrightarrow{H}_n$.

For finite K, Raussen [20] shows that singular homology groups of trace spaces such as $Tr(\overrightarrow{Geom}(K); \hat a, \hat b)$ are computable, by computing a finite presentation of the trace spaces (a so-called prod-simplicial complex) from which we can compute homology using Smith normal form of matrices. As a consequence:

Proposition 2. *For a cubical complex K, for every $n \geq 1$, for all discrete trace γ of K, the nth discrete natural homology groups $\overrightarrow{h}_n(K; \gamma)$ are computable.*

By construction, the discrete natural homology group $\overrightarrow{h}_n(K; \gamma)$ is equal to the geometric homology group $\overrightarrow{H}_n(\overrightarrow{Geom}(K); \hat\gamma)$. However (for finite K) the discrete natural homology functor $\overrightarrow{h}_n(K)$ only lists those for the finitely many discrete traces, while $\overrightarrow{H}_n(\overrightarrow{Geom}(K))$ lists one group for each of the uncountably many traces in $\overrightarrow{Geom}(K)$. The discrete functor $\overrightarrow{h}_n(K)$ also has to cater for finitely many discrete extension morphisms, whereas $\overrightarrow{H}_n(\overrightarrow{Geom}(K))$ has to map uncountably many extension morphisms to group homomorphisms. This makes quite a difference—but not one up to bisimilarity:

Theorem 1 (Discrete Nat. Homology≡Geometric Nat. Homology).
For every cubical complex K, there is an open map from the natural system $\overrightarrow{H}_n(\overrightarrow{Geom}(K))$ to the discrete natural system $\overrightarrow{h}_n(K)$. In particular, they are bisimilar.

Before we describe the construction, notice that there is *no* open map in the other direction: remember that the open maps we consider have a fibration component, which must be surjective.

Proof. We need to define an open map $(\mathcal{C}, \sigma)$ from $\overrightarrow{H}_n(X)$, where $X = \overrightarrow{Geom}(K)$, to $\overrightarrow{h}_n(K)$. We start by building $\mathcal{C}$, which must be a fibration from Fc_X to Fc_K.

This is based on the notion of *carrier sequence* due to Fajstrup [5]. For a point s in $\overrightarrow{Geom}(K)$, there is a unique cube $e \in K$ of minimal dimension m such that s can be written as $[e, a]$, $a \in \overrightarrow{I}^m$. Write $\mathcal{C}(s)$ for this cube e, and call it the *carrier* of s. Every trace $\langle \pi \rangle$ in X gives rise to an ordered sequence of cubes $\mathcal{C}(\langle \pi \rangle)$ obtained as the carriers of $\pi(t)$, $t \in [0,1]$, and removing consecutive duplicates. This is formally defined in [5]. By a compactness argument $\mathcal{C}(\langle \pi \rangle)$ is a finite sequence, in fact a discrete trace, called the *carrier sequence* of $\langle \pi \rangle$. For example, the carrier sequence of the trace on the right of Figure 3 is $b \preceq A \preceq t'$.

We use this to define our functor $\mathcal{C}$, on objects by letting $\mathcal{C}(\langle \pi \rangle)$ be defined as above, and on morphisms by letting $\mathcal{C}(\langle \alpha \rangle, \langle \beta \rangle) = (\mathcal{C}(\langle \alpha \rangle), \mathcal{C}(\langle \beta \rangle))$ for every extension $(\langle \alpha \rangle, \langle \beta \rangle)$. This is surjective on objects since $\mathcal{C}(\hat{\gamma}) = \gamma$ for every discrete trace γ. We now claim that $\mathcal{C}$ is a fibration, and this amounts to show that: given any trace $\langle \pi \rangle$ of $\overrightarrow{Geom}(K)$, with carrier sequence $c_0 \preceq \cdots \preceq c_k$, if the latter extends to a discrete trace $c_{-p} \preceq \cdots \preceq c_{-1} \preceq c_0 \preceq \cdots \preceq c_k \preceq c_{k+1} \preceq \cdots \preceq c_{k+q}$ in K, then $\langle \pi \rangle$ extends to some trace $\langle \alpha \star \pi \star \beta \rangle$ such that $\mathcal{C}(\langle \alpha \star \pi \star \beta \rangle) = c_{-p} \preceq \cdots \preceq c_{-1} \preceq c_0 \preceq \cdots \preceq c_k \preceq c_{k+1} \preceq \cdots \preceq c_{k+q}$. By induction, the cases $(p, q) = (1, 0)$ and $(p, q) = (0, 1)$ suffice to establish the property. Some care has to be taken: the extension paths are *not* concatenations of simple straight lines joining the extra points $\hat{c}_j$, $j \geq k$ or $j \leq 0$. As the picture on the right shows (for $(p, q) = (0, 2)$), the dipath β does not—and cannot—go through $\hat{c}_3$.

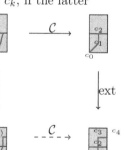

We now need to build a natural isomorphism $\sigma : \overrightarrow{H}_n(X) \longrightarrow \overrightarrow{h}_n(K) \circ \mathcal{C}$. In other words, we need to build group isomorphisms $\sigma_{\langle \pi \rangle} : \overrightarrow{H}_n(X; \langle \pi \rangle) \longrightarrow \overrightarrow{h}_n(K; \mathcal{C}(\langle \pi \rangle))$ that are natural, in the sense that, for every extension $(\langle \alpha \rangle, \langle \beta \rangle)$ of $\langle \pi \rangle$, and for $(\gamma, \delta) = \mathcal{C}(\langle \alpha \rangle, \langle \beta \rangle)$ the associated discrete extension, the following square commutes:

$$\overrightarrow{H}_n(X; \langle \pi \rangle) \xrightarrow{\quad \sigma_{\langle \pi \rangle} \quad} \overrightarrow{h}_n(K; \mathcal{C}(\langle \pi \rangle))$$

$$\langle \alpha \star \_ \star \beta \rangle \downarrow \qquad \qquad \downarrow \langle \hat{\gamma} \star \_ \star \hat{\delta} \rangle$$

$$\overrightarrow{H}_n(X; \langle \alpha \star \pi \star \beta \rangle) \xrightarrow{\quad \sigma_{\langle \alpha \star \pi \star \beta \rangle} \quad} \overrightarrow{h}_n(K; \mathcal{C}(\langle \alpha \star \pi \star \beta \rangle))$$

Let π be from s to t. Every cube I^k has a lattice structure whose meet $\wedge$ is pointwise min and whose join $\vee$ is pointwise max. Write s as $[\mathcal{C}(s), \boldsymbol{a}]$, and let $s_- = [\mathcal{C}(s), \boldsymbol{a} \wedge \bullet]$. Recall that $\bullet = (\frac{1}{2}, \cdots, \frac{1}{2})$, and that $\widehat{\mathcal{C}(s)} = [\mathcal{C}(s), \bullet]$. Similarly, let $\widehat{\mathcal{C}(t)} = [\mathcal{C}(t), \bullet]$, and we define $t_+ = [\mathcal{C}(t), \boldsymbol{b} \vee \bullet]$, where $t = [\mathcal{C}(t), \boldsymbol{b}]$. The situation is illustrated in the two gray boxes to the right. There are obvious dipaths $\eta_s, \lambda_s, \mu_t, \rho_t$ as displayed there, too. Those induce continuous maps between trace spaces by concatenation.

For example, there is a continuous map $\eta_s^* : Tr(X; s, t) \longrightarrow Tr(X; s_-, t)$ that sends each trace $\langle \pi' \rangle$ to $\langle \eta_s \star \pi' \rangle$. Similarly, $\lambda_s^*(\langle \pi' \rangle) = \langle \lambda_s \star \pi' \rangle$, and symmetrically, $^*\mu_t(\langle \pi' \rangle) = \langle \pi' \star \mu_t \rangle$, $^*\rho_t(\langle \pi' \rangle) = \langle \pi' \star \rho_t \rangle$. Each of these four maps is a homotopy equivalence (proof omitted), and therefore induce isomorphisms in homology. It remains to define $\sigma_{\langle \pi \rangle}$ as the composition $H_{n-1}(^*\mu_t)^{-1} \circ H_{n-1}(^*\rho_t) \circ H_{n-1}(\lambda_s^*)^{-1} \circ H_{n-1}(\eta_s^*)$ of those four isomorphisms. Naturality is, as usual, tedious but mechanical. $\qquad \square$

The potential problem mentioned at the beginning of Section 4 is then solved: the uncountable natural homology of $\overrightarrow{Geom}(K)$ is reduced, through bisimilarity, to the finite, discrete natural homology of K.

A *dihomeomorphism* is a continuous monotonic bijection between pospaces whose inverse is also continuous and monotonic.

Corollary 1 (Invariance under dihomeomorphism). *For any cubical complexes K, K' whose geometric realizations are dihomeomorphic, $\overrightarrow{H}_n(\overrightarrow{Geom}(K))$ and $\overrightarrow{H}_n(\overrightarrow{Geom}(K'))$ are isomorphic, and $\overrightarrow{h}_n(K)$ and $\overrightarrow{h}_n(K')$ are bisimilar.*

Of particular importance to the field of true concurrency is invariance under refinement [22]. In our case, this means that if we replace certain n-cubes in K by unions of 2^n smaller cubes with all dimensions halved, then the result should have the same natural homology. Indeed, such a process is called *subdivision* in the literature, and it is well-known that if K' is a subdivision of K, then $\overrightarrow{Geom}(K)$ and $\overrightarrow{Geom}(K')$ are dihomeomorphic. Hence:

Corollary 2 (Invariance under subdivision). *Let K be a cubical complex, and K' be a subdivision of K. Then $\overrightarrow{h}_n(K)$, and $\overrightarrow{h}_n(K')$ are bisimilar.*

7   Conclusion

We have defined a promising notion of homology for directed algebraic topology. We have shown that our natural systems of homology are computable on cubical complexes. We have also introduced a notion of bisimilarity with respect to which those natural systems should be compared. Importantly, natural homology is invariant under subdivision. We showed this as a special case of a more general result: that the natural homology of a cubical complex is bisimilar to that of its geometric realization.

As a litmus test, does our natural homology pass the criteria we set forth in Section 2? Look again at Fahrenberg's matchbox (Fig. 1). Its discrete natural homology would be too big to fit on a page, however its $\overrightarrow{H}_1$ at the trace $a \star c$ is equal to $\mathbb{Z}^2$. In particular, it has non-trivial natural homology, in the strong sense that its natural homology is not bisimilar to any natural system consisting only of copies of $\mathbb{Z}$ (e.g., the natural homology of a filled-out cube). This is the first proposal that distinguishes the matchbox from a trivial pospace.

References

1. Baues, H.-J., Wirsching, G.: Cohomology of small categories. Journal of Pure and Applied Algebra **38**(2–3), 187–211 (1985)
2. Carlsson, G.: Topology and data. AMS. Bulletin **46**(2), 255–308 (2009)
3. Coffman, E.G., Elphick, M.J., Shoshani, A.: System deadlocks. Computing Surveys **3**(2), 67–78 (1971)
4. Fahrenberg, U.: Directed homology. Electronic Notes in Theoretical Computer Science **100**, 111–125 (2004)
5. Fajstrup, L.: Dipaths and dihomotopies in a cubical complex. Advances in Applied Mathematics **35**(2), 188–206 (2005)
6. Fajstrup, L., Goubault, E., Haucourt, E., Mimram, S., Raussen, M.: Trace Spaces: An Efficient New Technique for State-Space Reduction. In: Seidl, H. (ed.) Programming Languages and Systems. LNCS, vol. 7211, pp. 274–294. Springer, Heidelberg (2012)
7. Goubault, E.: Géométrie du parallélisme. PhD thesis, Ecole Polytechnique (1995)
8. Goubault, E.: Geometry and concurrency: A user's guide. Mathematical Structures in Computer Science **10**(4), 411–425 (2000)
9. Goubault, E., Haucourt, E.: A Practical Application of Geometric Semantics to Static Analysis of Concurrent Programs. In: Abadi, M., de Alfaro, L. (eds.) CONCUR 2005. LNCS, vol. 3653, pp. 503–517. Springer, Heidelberg (2005)
10. Goubault, E., Heindel, T., Mimram, S.: A geometric view of partial order reduction. Electronic Notes in Theoretical Computer Science, 298 (2013)
11. Grandis, M.: Inequilogical spaces, directed homology and noncommutative geometry. Homology, Homotopy and Applications **6**, 413–437 (2004)
12. Grandis, M.: Directed Algebraic Topology. Models of non-reversible worlds. Cambridge University Press (2009)
13. Joyal, A., Nielsen, M., Winskel, G.: Bisimulation from open maps. Information and Computation **127**(2), 164–185 (1996)
14. T. Kaczynski, K. Mischaikow, and M. Mrozek. Computing homology. Homology, Homotopy and Applications 5(2), 233–256 (2003)
15. Kahl, T.: The homology graph of a HDA (2013). arXiv:1307.7994
16. Krishnan, S.: Flow-cut dualities for sheaves on graphs (2014). arXiv:1409.6712
17. Patchkoria, A.: On exactness of long sequences of homology semimodules. Journal of Homotopy and Related Structures **1**(1), 229–243 (2006)

18. Pratt, V.R.: Modeling Concurrency with Geometry. In: Wise, D.S. (ed.) 18th Ann. ACM Symp. Principles of Programming Languages, pp. 311–322 (1991)

19. Raussen, M.: Invariants of directed spaces. Applied Categorical Structures, 15 (2007)

20. Raussen, M.: Simplicial models for trace spaces II: General higher dimensional automata. Algebraic and Geometric Topology **12**(3), 1741–1762 (2012)

21. Sergeraert, F: The computability problem in algebraic topology. Advances in Mathematics, pp. 1–29 (1994)

22. van Glabeek, R.: Comparative concurrency semantics and refinement of actions. PhD thesis, Centrum voor Wiskunder en Informatica (1990)

Greatest Fixed Points of Probabilistic Min/Max Polynomial Equations, and Reachability for Branching Markov Decision Processes

Kousha Etessami[1]([⊠]), Alistair Stewart[1], and Mihalis Yannakakis[2]

[1] School of Informatics, University of Edinburgh, Edinburgh, UK
kousha@inf.ed.ac.uk, stewart.al@gmail.com
[2] Department of Computer Science, Columbia University, New York, USA
mihalis@cs.columbia.edu

Abstract. We give polynomial time algorithms for quantitative (and qualitative) *reachability* analysis for *Branching Markov Decision Processes* (BMDPs). Specifically, given a BMDP, and given an initial population, where the objective of the controller is to maximize (or minimize) the probability of eventually reaching a population that contains an object of a desired (or undesired) type, we give algorithms for approximating the supremum (infimum) reachability probability, within desired precision $\epsilon > 0$, in time polynomial in the encoding size of the BMDP and in $\log(1/\epsilon)$. We furthermore give P-time algorithms for computing ϵ-optimal strategies for both maximization and minimization of reachability probabilities. We also give P-time algorithms for all associated *qualitative* analysis problems, namely: deciding whether the optimal (supremum or infimum) reachability probabilities are 0 or 1. Prior to this paper, approximation of optimal reachability probabilities for BMDPs was not even known to be decidable.

Our algorithms exploit the following basic fact: we show that for any BMDP, its maximum (minimum) *non*-reachability probabilities are given by the *greatest fixed point* (GFP) solution $g^* \in [0,1]^n$ of a corresponding monotone max (min) Probabilistic Polynomial System of equations (max/min-PPS), $x = P(x)$, which are the Bellman optimality equations for a BMDP with non-reachability objectives. We show how to compute the GFP of max/min PPSs to desired precision in P-time.

1 Introduction

Multi-type branching processes (BPs) are infinite-state purely stochastic processes that model the stochastic evolution of a population of entities of distinct types. The BP specifies for every type a probability distribution for the offspring of entities of this type. Starting from an initial population, the process evolves from each generation to the next according to the probabilistic offspring

The full version of this paper is available at arxiv.org/abs/1502.05533. Research partially supported by the Royal Society and by NSF Grant CCF-1320654. Alistair Stewart's research supported by I. Diakonikolas's EPSRC grant EP/L021749/1.

M.M. Halldórsson et al. (Eds.): ICALP 2015, Part II, LNCS 9135, pp. 184–196, 2015.
DOI: 10.1007/978-3-662-47666-6_15

rules. Branching processes are a fundamental stochastic model with applications in many areas: physics, biology, population genetics, medicine etc. *Branching Markov Decision Processes* (BMDPs) provide a natural extension of BPs where the evolution is not purely stochastic but can be partially influenced or controlled: a controller can take actions which affect the probability distribution for the set of offspring of the entities of each type. The goal is to design a policy for choosing the actions in order to optimize a desired objective.

In recent years there has been great progress in resolving algorithmic problems for BMDPs with the objective of maximizing or minimizing the *extinction* probability, i.e., the probability that the population eventually becomes extinct. Polynomial time algorithms were developed for both maximizing and minimizing BMDPs for *qualitative* analysis, i.e. to determine whether the optimal extinction probability is 0, 1 or in-between [12], and for *quantitative* analysis, to compute optimal extinction probabilities to any desired precision [9]. However, key problems for optimizing BMDP *reachability* probability (probability that the population eventually includes an entity with a target type) have remained open.

Reachability objectives are very natural. Some types may be undesirable, in which case we want to avoid them to the extent possible. Or conversely, we may want to guide the process to reach certain desirable types. For example, branching processes have been used recently to model cancer tumor progression and multiple drug resistance of tumors due to multiple mutations ([1,15]). It could be fruitful to model the introduction of multiple drugs (each of which controls/influences cells with a different type of mutation) via a "controller" that controls the offspring of different types, thus extending the current models (and associated software tools) which are based on BPs only, to controlled models based on BMDPs. A natural question one could ask then is to compute the minimum probability of reaching a *bad* (malignant) cell type, and compute a drug introduction strategy that achieves (approximately) minimum probability. Doing this efficiently (in P-time) would avoid the combinatorial explosion of trying all possible combinations of drug therapies.

In this paper we provide the first polynomial time algorithms for quantitative (and also qualitative) *reachability* analysis for BMDPs. Specifically, we provide algorithms for ϵ-approximating the supremum probability, as well as the infimum probability, of reaching a given type (or a set of types) starting from an initial type (or an initial population of types), up to any desired additive error $\epsilon > 0$. We also give algorithms for computing ϵ-optimal strategies which achieve such ϵ-optimal values. The running time of these algorithms (in the standard Turing model of computation) is polynomial in both the encoding size of the BMDP and in $\log(\frac{1}{\epsilon})$. We also give P-time algorithms for the qualitative problems: we determine whether the supremum or infimum probability is 1 (or 0), and if so we actually compute an optimal strategy that achieves 1 (0, respectively).

In prior work [12], we studied optimization of extinction (a.k.a. termination) probabilities for BMDPs, and showed that optimal extinction probabilities are captured by the *least fixed point* (LFP) solution $q^* \in [0,1]^n$ of a corresponding system of monotone probabilistic max (min) polynomial equations called

maxPPSs (respectively minPPSs), which form the *Bellman optimality equations*
for termination of a BMDP. A maxPPS is a system of equations $x = P(x)$ over
a vector x of variables, where the right-hand-side of each equation is of the form
$\max_j\{p_j(x)\}$, where each $p_j(x)$ is a polynomial with non-negative coefficients
(including the constant term) that sum to at most 1 (such a polynomial is called
probabilistic). A minPPS is defined similarly. In [9], we introduced an algorithm,
called *Generalized Newton's Method* (GNM), for the solution of maxPPSs and
minPPSs, and showed that it computes the LFP of maxPPSs and minPPSs (and
hence also the optimal termination probabilities for BMDPs) to desired precision
in P-time. GNM is an iterative algorithm (like Newton's) which in each iteration
solves a suitable linear program (a different one for the max and min versions).

   In this paper we first model the reachability problem for a BMDP by an
appropriate system of equations: We show that the optimal *non-reachability*
probabilities for a given BMDP are captured by the *greatest fixed point* (GFP),
$g^* \in [0,1]^n$ of a corresponding maxPPS (or minPPS) system of Bellman equations. We then show that one can approximate the GFP solution $g^* \in [0,1]^n$ of
a maxPPS (or minPPS), $x = P(x)$, in time polynomial in both the encoding size
$|P|$ of the system of equations and in $\log(1/\epsilon)$, where $\epsilon > 0$ is the desired additive
error bound of the solution. (The model of computation is the standard Turing
machine model.) We also show that the qualitative analysis of determining the
coordinates of the GFP that are 0 and 1, can be done in P-time (and hence the
same holds for the optimal reachability probabilities of BMDPs).

   Our algorithms for computing the GFP of minPPS and maxPPS make use of
(a variant of) Generalized Newton Method adapted for the computation of GFP,
with a key important difference in the preprocessing step before applying GNM.
We first identify and remove only the variables that have value 1 in the GFP g^*
(we do not remove the variables with value 0, unlike the LFP case). We show
that for maxPPSs, once these variables are removed, the remaining system with
GFP $g^* < 1$ has a unique fixed point in $[0,1]^n$, hence the GFP is equal to the
LFP; applying GNM from the 0 initial vector converges quickly (in P-time, with
suitable rounding) to the GFP (by [9]). For minPPSs, even after the removal
of the variables x_i with $g_i^* = 1$, the remaining system may have multiple fixed
points, and we can have LFP < GFP. Nevertheless, we show that with the subtle
change in the preprocessing step, GNM, starting at the all-0 vector, remarkably
"skips over" the LFP and converges to the GFP solution g^*, in P-time.

   Comparing the properties of the LFP and GFP of max/minPPS, we note
that one difference for the qualitative problems is that for the GFP, both the
value=0 and the value=1 question depend only on the structure of the model
and not on its probabilities (the values of the coefficients), whereas in the LFP
case the value=1 question depends on the probabilities (see [12,13]).

   We also note some important differences regarding existence of optimal
strategies between extinction (termination) and reachability objectives for
BMDPs. We observe that, unlike optimization of termination probabilities for
BMDPs, for which there always exists a static deterministic optimal strategy
([12]), there need not exist any optimal strategy at all for maximizing reachabil-

ity probability in a BMDP, i.e. the supremum probability may not be attainable. If the supremum probability is 1 however, we show that there exists a strategy that achieves it (albeit, not necessarily a static one). For the min reachability objective there always exists an optimal deterministic and static strategy. In all cases, we show that we can compute in P-time an ϵ-optimal static (possibly randomized) policy, for both maximizing and minimizing reachability probability in a BMDP.

Related Work: BMDPs have been previously studied in both operations research (e.g., [14, 16]) and computer science (e.g., [6, 11, 12]). We have already mentioned the results in [9, 12] concerning the computation of the extinction probabilities of BMDPs and the computation of the LFP of max/minPPS. BPs are closely connected to stochastic context-free grammars, 1-exit Recursive Markov chains (1-RMC) [13], and the corresponding stateless probabilistic pushdown processes, pBPA [7]; their extinction or termination probabilities are interreducible, and they are all captured by the LFP of PPSs. The same is true for their controlled extensions, for example the extinction probability of BMDPs and the termination probabilities of 1-exit Recursive Markov Decision processes (1-RMDP) [12], are both captured by the LFP of maxPPS or minPPS. A different type of objective of optimizing the total expected reward for 1-RMDPs (and equivalently BMDPs) in a setting with positive rewards was studied in [11]; in this case the optimal values are rational and can be computed exactly in P-time.

The equivalence between BMDPs and 1-RMDPs however does not carry over to the reachability objective. The *qualitative* reachability problem for 1-RMDPs (equivalently BPA MDPs) and the extension to simple 2-person games 1-RSSGs (BPA games) were studied in [4] and [3] by Brazdil et al. It is shown in [4] that qualitative *almost-sure* reachability for 1-RMDPs can be decided in P-time (both for maximizing and minimizing 1-RMDPs). However, for maximizing reachability probability, almost-sure and limit-sure reachability are *not* the same: in other words, the supremum reachability probability can be 1, but it may not be achieved by any strategy for the 1-RMDP. By contrast, for BMDPs we show that if the supremum reachability probability is 1, then there is a strategy that achieves it. This is one illustration of the fact that the equivalence between 1-RMDP and BMDP does not hold for the reachability objective. The papers [3, 4] do not address the limit-sure reachability problem, and in fact even the decidability of limit-sure reachability for 1-RMDPs remains open.

Chen et. al. [5] studied model checking of branching processes with respect to properties expressed by deterministic parity tree automata and showed that the qualitative problem is in P (hence this holds in particular for reachability probability in BPs), and that the quantitative problem of comparing the probability with a rational is in PSPACE. Although not explicitly stated there, one can use Lemma 20 of [5] and our algorithm from [8] to show that the reachability probabilities of BPs can be approximated in P-time. Bonnet et. al. [2] studied a model of "probabilistic Basic Parallel Processes", which are syntactically close to Branching processes, except reproduction is asynchronous and the entity that reproduces in each step is chosen randomly (or by a scheduler/controller).

None of the previous results have direct bearing on the reachability problems for BMDPs.

Due to space limits, most proofs are omitted. See the full version [10].

2   Definitions and Background

We provide unified definitions of multi-type Branching processes (BPs), Branching MDPs (BMDPs), and Branching Simple Stochastic Games (BSSGs), by first defining BSSGs, and then specializing them to obtain BMDPs and BPs.

A *Branching Simple Stochastic Game* (BSSG), consists of a finite set $V = \{T_1, \ldots, T_n\}$ of types, a finite non-empty set $A_i \subseteq \Sigma$ of actions for each type (Σ is some finite action alphabet), and a finite set $R(T_i, a)$ of probabilistic rules associated with each pair (T_i, a), $i \in [n]$, where $a \in A_i$. Each rule $r \in R(T_i, a)$ is a triple (T_i, p_r, α_r), which we denote by $T_i \xrightarrow{p_r} \alpha_r$, where $\alpha_r \subseteq \mathbb{N}^n$ is a n-vector of natural numbers that denotes a finite multi-set over the set V, and where $p_r \in (0, 1]$ is the probability of the rule r, where $\sum_{r \in R(T_i, a)} p_r = 1$ for all $i \in [n]$ and $a \in A_i$. For BSSGs, the types are partitioned into two sets: $V = V_{\max} \cup V_{\min}$, $V_{\max} \cap V_{\min} = \emptyset$, where $V_{\max}$ contains those types "belonging" to player max, and $V_{\min}$ containing those belonging to player min. A *Branching Markov Decision Process* (BMDP) is a BSSG where one of the two sets $V_{\max}$ or $V_{\min}$ is empty. Intuitively, a BMDP (BSSG) describes the stochastic evolution of a population of entities of different types in the presence of a controller (or two players) that can influence the evolution. A *multi-type Branching Process* (BP), is a BSSG where all action sets A_i are singleton sets; hence in a BP players have no choices and thus don't exist: a BP defines a purely stochastic process.

A play (or trajectory) of a BSSG operates as follows: starting from an initial population (i.e., set of entities of given types) X_0 at time (generation) 0, a sequence of populations $X_1, X_2, \ldots$ is generated, where X_{k+1} is obtained from X_k as follows. Player max (min) selects for each entity e in set X_k that belongs to max (to min, respectively) an available action $a \in A_i$ for the type T_i of entity e; then for each such entity e in X_k a rule $r \in R(T_i, a)$ is chosen randomly and independently according to the rule probabilities p_r, where $a \in A_i$ is the action selected for that particular entity e. Every entity is then replaced by a set of entities with the types specified by the right-hand side multiset α_r of that chosen rule r. The process is repeated as long as the current population X_k is nonempty, and it is said to *terminate* (or become *extinct*) if there is some $k \geq 0$ such that $X_k = \emptyset$. When there are n types, we can view a population X_i as a vector $X_i \in \mathbb{N}^n$, specifying the number of objects of each type. We say that the process *reaches* a type T_j, if there is some $k \geq 0$ such that $(X_k)_j > 0$.

A player can base her decisions at each stage k on the entire past history, and may choose different actions for entities of the same type.[1] The decision

---

[1] We remark that, for optimizing termination and reachability probability, we could alternatively define the players' strategic role in BSSGs in various other ways, including asynchronous choice of (randomized) actions, as long as actions must eventually be chosen for all objects, without altering any of our results.

may be *randomized* (i.e. a probability distribution on the tuples of actions for the entities of the types controlled by the player) or *deterministic* (see the full version [10] for the formal definitions). Let Ψ_1, Ψ_2 be the set of all (randomized) strategies of the two players. We say that a strategy is *static* if for each type T_i controlled by that player the strategy always chooses the same action a_i, or the same probability distribution on actions, for all entities of type T_i in all histories.

We can consider different objectives by the players. Here we consider the *reachability* objective, where the goal of the two players, starting from a given population, is to maximize/minimize the probability of reaching a population which contains *at least* one entity of a given special type, T_{f^*}. It will follow from our results that a BSSG game with a reachability objective has a *value*.

Suppose that player 1 wants to maximize the probability of *not* reaching T_{f^*} and player 2 wants to minimize it. For strategies $\sigma \in \Psi_1$, $\tau \in \Psi_2$, and a given initial population $\mu \in \mathbb{N}^n$, with $(\mu)_{f^*} = 0$, we denote by $g^{*,\sigma,\tau}(\mu)$ the probability that $(X_d)_{f^*} = 0$ for all $d \geq 0$. The *value* of the non-reachability game for the initial population μ is $g^*(\mu) = \sup_{\sigma \in \Psi_1} \inf_{\tau \in \Psi_2} g^{*,\sigma,\tau}(\mu)$. We will show that determinacy holds for these games, i.e., $g^*(\mu) = \sup_{\sigma \in \Psi_1} \inf_{\tau \in \Psi_2} g^{*,\sigma,\tau}(\mu) = \inf_{\tau \in \Psi_2} \sup_{*, \sigma \in \Psi_1} g^{*,\sigma,\tau}(\mu)$. However, unlike the case for extinction probabilities ([12]), it does *not* hold that both players have optimal static strategies.

If μ has a single entity of type T_i, we will write g_i^* instead of $g^*(\mu)$. Given a BMDP (or BSSG), the goal is to compute the vector g^* of the g_i^*'s, i.e. the vector of non-reachability values of the different types. From the g_i^*'s, we can compute the value $g^*(\mu)$ for any initial population μ: $g^*(\mu) = \Pi_i(g_i^*)^{\mu_i}$.

We will associate a system of min/max probabilistic polynomial Bellman equations, $x = P(x)$, to each given BMDP or BSSG. A polynomial $p(x)$ is called *probabilistic* if all coefficients are nonnegative and sum to at most 1. A *probabilistic polynomial system (PPS)* is a system $x = P(x)$ where all $P_i(x)$ are probabilistic polynomials. A *max-min PPS* is a system $x = P(x)$ where each $P_i(x)$ is either: a Max-polynomial: $P_i(x) = \max\{q_{i,j}(x) : j \in \{1, ..., m_i\}\}$, or a Min-polynomial: $P_i(x) = \min\{q_{i,j}(x) : j \in \{1, ..., m_i\}\}$, where each $q_{i,j}(x)$ is a probabilistic polynomial, for every $j \in \{1, \ldots, m_i\}$. We shall call such a system a *maxPPS* (respectively, a *minPPS*) if for every $i \in \{1, \ldots, n\}$, $P_i(x)$ is a Max-polynomial (respectively, a Min-polynomial). We use *max/minPPS* to refer to a system of equations, $x = P(x)$, that is either a maxPPS or a minPPS.

For computational purposes we assume that all coefficients are rational, and that the polynomials are given in sparse form, i.e., by listing only the nonzero terms, with the coefficient and the nonzero exponents of each term given in binary. We let $|P|$ denote the total bit encoding length of a system $x = P(x)$ under this representation.

Any max-minPPS, $x = P(x)$, has a *least fixed point* (**LFP**) solution, $q^* \in [0,1]^n$, i.e., $q^* = P(q^*)$ and if $q = P(q)$ for some $q \in [0,1]^n$ then $q^* \leq q$ (coordinate-wise inequality). As observed in [12,13], q^* may in general contain irrational values, even in the case of pure PPSs. In this paper, we exploit the fact that every max-minPPS, $x = P(x)$, also has a *greatest fixed point* (**GFP**) solution, $g^* \in [0,1]^n$, i.e., such that $g^* = P(g^*)$ and if $q = P(q)$ for some

$q \in [0,1]^n$ then $q \leq g^*$. Again, g^* may contain irrational coordinates, so we in general want to approximate its coordinates.

We can consider a max-minPPS as a game between two players that control respectively the variables x_i where P_i is a max or a min polynomial. A (possibly randomized) policy σ for a player maps each of its variables x_i to a probability distribution $\sigma(i)$ over the indices $\{1, \ldots, m_i\}$ of the polynomials in P_i. A policy σ of the max player induces a minPPS $x = P_\sigma(x)$, where $(P_\sigma)_i(x) = \sum_{a \in A_i} \sigma(i)(a) \cdot q_{i,a}$. Let q_σ^* and g_σ^* denote the LFP and GFP of the min-PPS $x = P_\sigma(x)$. We say that σ is an *optimal* policy for the max player for the LFP (resp., the GFP) if $q_{\sigma^*}^* = q^*$ (resp., $g_{\sigma^*}^* = g^*$). The policy σ is ϵ-*optimal* for the LFP (resp. GFP) , if $\|q_{\sigma'}^* - q^*\|_\infty \leq \epsilon$ (resp., $\|g_{\sigma'}^* - g^*\|_\infty \leq \epsilon$). These concepts can be defined similarly for the min player and its policies.

It is convenient to put max-minPPSs in the following simple form.

Definition 1. *A max-minPPS, $x = P(x)$ in n variables is in* simple normal form (SNF) *if each $P_i(x)$, for all $i \in [n]$, is in one of the following three forms:*

Form L: $P(x)_i = a_{i,0} + \sum_{j=1}^n a_{i,j} x_j$, *where $a_{i,j} \geq 0$ for all j, &*
$\sum_{j=0}^n a_{i,j} \leq 1$.
Form Q: $P(x)_i = x_j x_k$ *for some j, k.*
Form M: $P(x)_i = \max\{x_j, x_k\}$ *or $P(x)_i = \min\{x_j, x_k\}$, for some j, k.*

We define SNF form for max/minPPSs analogously. Every max-minPPS, $x = P(x)$, can be transformed in P-time (as in [8,13]) to a suitably "*equivalent*" max-minPPS in SNF form (see the full version [10] for a formal statement and proof), where in particular both the LFP *and* GFP of the original system are projections of the LFP and GFP of the transformed systems. Thus *we may (and do) assume*, wlog, that all max/minPPSs are in SNF normal form.

The *dependency graph* of a max-minPPS $x = P(x)$ is a directed graph with one node for each variable x_i, and contains edge (x_i, x_j) iff x_j appears in $P_i(x)$.

For a max/minPPS, $x = P(x)$, with n variables (in SNF form), the *linearization* of $P(x)$ at a point $y \in \mathbb{R}^n$, is a system of max/min linear functions denoted by $P^y(x)$, which has the following form: if $P(x)_i$ has form L or M, then $P_i^y(x) = P_i(x)$, and if $P(x)_i$ has form Q, i.e., $P(x)_i = x_j x_k$ for some j, k, then $P_i^y(x) = y_j x_k + x_j y_k - y_j y_k$. We now recall and adapt from [9] the definition of distinct iteration operators for maxPPSs and minPPSs, both of which we shall refer to with the overloaded notation $I(x)$. These operators serve as the basis for *Generalized Newton's Method* (GNM) to be applied to maxPPSs and minPPSs, respectively. We need to slightly adapt the definition of operator $I(x)$, specifying the conditions on the GFP g^* under which the operator is well-defined:

Definition 2. *For a maxPPS, $x = P(x)$, with GFP g^*, with $0 \leq g^* < 1$, and for $0 \leq y \leq g^*$, define the operator $I(y)$ to be the unique optimal solution, $a \in \mathbb{R}^n$, to the following mathematical program:* Minimize: $\sum_i a_i$; Subject to: $P^y(a) \leq a$.

For a minPPS, $x = P(x)$, with GFP g^*, with $0 \leq g^* < 1$, and for $0 \leq y \leq g^*$, define the operator $I(y)$ to be the unique optimal solution $a \in \mathbb{R}^n$ to the following mathematical program: Maximize: $\sum_i a_i$; Subject to: $P^y(a) \geq a$.

These mathematical programs can be solved using Linear Programming. A priori, it is unclear whether the programs have a unique solution, i.e., whether the "definitions" of $I(x)$ for maxPPSs and minPPSs are well-defined. We show they are. We require *rounded* GNM, defined as follows ([9]).

GNM, with rounding parameter h: Starting at $x^{(0)} := \mathbf{0}$, For $k \geq 0$, compute $x^{(k+1)}$ from $x^{(k)}$ as follows: first calculate $I(x^{(k)})$, then for every coordinate i, set $x_i^{(k+1)}$ to be the maximum multiple of 2^{-h} which is $\leq \max\{0, I(x^{(k)})_i\}$.

3   Greatest Fixed Points Capture Non-reachability Values

For any given BSSG, $\mathcal{G}$, with a specified special type T_{f^*}, we will construct a max-minPPS, $x = P(x)$, and show that the vector g^* of *non*-reachability values for $(\mathcal{G}, T_{f^*})$ is precisely the *greatest fixed point* $g^* \in [0,1]^n$ of $x = P(x)$.

The system $x = P(x)$ has one variable x_i and one equation $x_i = P_i(x)$, for each type $T_i \neq T_{f^*}$. For each $i \neq f^*$, the min/max probabilistic polynomial $P_i(x)$ is constructed as follows. For all $j \in A_i$, let $R'(T_i, j) := \{r \in R(T_i, j) : (\alpha_r)_{f^*} = 0\}$ denote the set of rules for type T_i and action j that generate a multiset α_r not containing any element of type T_{f^*}. $P_i(x)$ contains one probabilistic polynomial $q_{i,j}(x)$ for each action $j \in A_i$, with $q_{i,j}(x) = \sum_{r \in R'(T_i,j)} p_r x^{\alpha_r}$. Note that we *do not* include, in the sum defining $q_{i,j}(x)$, any monomial $p_{r'} x^{\alpha_{r'}}$ associated with a rule r' which generates an object of the special type T_{f^*}. Then, if type T_i belongs to player max, who aims to *minimize* the probability of *not* reaching an object of type T_{f^*}, we define $P_i(x) \equiv \min_{j \in A_i} q_{i,j}(x)$. Likewise, if T_i belongs to min, whose aim is to *maximize* the probability of *not* reaching T_{f^*}, we define $P_i(x) \equiv \max_{j \in A_i} q_{i,j}(x)$. Note the swapped roles of max and min in the equations, versus the corresponding player's goal for the reachability objective. The following theorem is analogous to one in [12] for LFPs of max-minPPSs.

Theorem 1. *The value vector* $g^* \in [0,1]^n$ *of a BSSG is the GFP of the corresponding operator* $P(\cdot)$ *in* $[0,1]^n$. *Thus,* $g^* = P(g^*)$, *and* $\forall g' \in [0,1]^n$, $g' = P(g')$ *implies* $g' \leq g^*$. *Also, for any initial population* μ, *the non-reachability values satisfy* $g^*(\mu) = \sup_{\sigma \in \Psi_1} \inf_{\tau \in \Psi_2} g^{*,\sigma,\tau}(\mu) = \inf_{\tau \in \Psi_2} \sup_{\sigma \in \Psi_1} g^{*,\sigma,\tau}(\mu) = \Pi_i (g_i^*)^{\mu_i}$. *So, such games are* determined.

A direct corollary of the proof of Theorem 1 (see the full version [10]) is that the player maximizing non-reachability probability in a BSSG always has an optimal deterministic static strategy. The same is *not* true for the player trying to *minimize* this non-reachability probability (i.e. the player trying to maximize the reachability probability). We give two examples illustrating this (see [10] for details). The first example has types A, B, and C, start type A and target type B, only A is controlled; B is purely probabilistic. The rules are: $A \to AA$, $A \to B$, $B \xrightarrow{1/2} C$, $B \xrightarrow{1/2} \emptyset$. There is no randomized static optimal strategy for maximizing the reachability probability in this BMDP, although the supremum probability is 1. We show later however that for any BMDP, if the supremum reachability value is 1, then the player maximizing the reachability probability has a, not necessarily static, optimal strategy that achieves value 1. The second example shows that this is not the case if the value is strictly between 0 and 1. Consider the BMDP with types A, B, C, and D, start type A and target type D, with rules: $A \xrightarrow{2/3} BB$, $A \xrightarrow{1/3} \emptyset$, $B \to A$, $B \to C$, $C \xrightarrow{1/3} D$, $C \xrightarrow{2/3} \emptyset$. There is no optimal strategy for maximizing the reachability probability in this BMDP (i.e., the supremum, which is $1/2$, is not achievable by any strategy), see [10].

Qualitative = 1 non-reachability analysis for BSSGs & max-minPPSs. There are (easy) P-time algorithms to compute for a given max-minPPS the variables that have value 1 in the GFP, and thus also for deciding, for a given BSSG (or BMDP), whether $g_i^* = 1$ (i.e., whether the *non*-reachability value is 1). The easy algorithm boils down to AND-OR graph reachability.

Proposition 1. *There is a P-time algorithm that given a max-min-PPS, $x = P(x)$, with n variables, and with GFP $g^* \in [0,1]^n$, and given $i \in [n]$, decides whether $g_i^* = 1$, or $g_i^* < 1$; Moreover, when $g_i^* = 1$ the algorithm outputs a deterministic policy (i.e., deterministic static strategy for the BSSG) σ, for the max player which forces $g_i^* = 1$, Likewise, if $g_i^* < 1$, it outputs a deterministic static policy τ for the min player which forces $g_i^* < 1$.*

We consider detection of $g_i^* = 0$ for maxPPS and minPPS later; the minPPS case in particular is substantially more complicated.

4   maxPPSs

We first determine and remove the variables with value 1 in the GFP, after which we know $g^* < 1$. To analyze maxPPSs, we first perform a thorough structural analysis of PPSs (without max) and derive several properties that are useful in handling maxPPSs (and minPPSs). Building on these properties, we show:

Lemma 1. *For any maxPPS, $x = P(x)$, if GFP $g^* < 1$ then g^* is the unique fixed point of $x = P(x)$ in $[0,1]^n$. So $g^* = q^*$, where q^* is the LFP of $x = P(x)$.*

Thus, applying the algorithms from [9] for LFP computation of maxPPSs, yields:

Theorem 2. *Given a maxPPS, $x = P(x)$, with GFP g^*,*
1. *There is a P-time algorithm that determines, for $i \in [n]$, whether $g_i^* = 0$, and if $g_i^* > 0$ computes a deterministic static policy that achieves this.*
2. *Given any integer $j > 0$, there is an algorithm that computes a rational vector v with $\|g^* - v\|_\infty \leq 2^{-j}$, and also computes a deterministic static policy σ, such that $\|g^* - g_\sigma^*\| \leq 2^{-j}$, both in time polynomial in $|P|$ and j.*

Similar results follow for the maximization of nonreachability in BMDPs.

5   minPPSs

Theorem 3. *Given a minPPS, $x = P(x)$ with $g^* < 1$. If we use GNM with rounding parameter $h = j + 2 + 4|P|$, then after h iterations, we have $\|g^* - x^{(h)}\|_\infty \leq 2^{-j}$. This ϵ-approximates g^* in time polynomial in $|P|$ and $\log(\frac{1}{\epsilon})$.*

The minPPS case is much more involved. In order to prove this theorem, we need some structural lemmas about GFPs of minPPSs, and their relationship to static policies. There need not exist any policies σ with $g_\sigma^* = g^*$, so we need policies that can, in some sense, act as "surrogates" for it. We say that a PPS $x = P(x)$ is *linear degenerate* (LD) if every $P_i(x)$ is a convex combination of variables: $P_i(x) \equiv \sum_{j=1}^n p_{ij} x_j$ where $\sum_j p_{ij} = 1$. A PPS is *linear degenerate free (LDF)* if there is no bottom strongly connected component S of its dependency graph, whose induced subsystem $x_S = P_S(x_S)$ is linear degenerate. A policy σ for a max/minPPS, $x = P(x)$, is called linear degenerate free (LDF) if its associated PPS $x = P_\sigma(x)$ is an LDF PPS. It turns out there is an LDF policy σ^* whose associated LFP is the GFP of the minPPS, and we can get an ϵ-optimal policy by following σ^* with high probability and with low probability following some policy that can reach the target from anywhere.

Lemma 2. *If a minPPS $x = P(x)$ has $g^* < 1$ then:*

1. *There is an LDF policy σ with $g_\sigma^* < 1$,*
2. *$g^* \leq q_\tau^*$, for any LDF policy τ, and*
3. *There is an LDF policy σ^* whose associated LFP, $q_{\sigma^*}^*$, has $g^* = q_{\sigma^*}^*$.*

Note that the policy σ^* is not necessarily optimal because even though $g^* = q_{\sigma^*}^*$, there may be an i with $g_i^* = (q_{\sigma^*}^*)_i < (g_{\sigma^*}^*)_i = 1$. Next we show that Generalised Newton's Method (GNM) is well-defined. We use $\mathcal{N}_\sigma$ below to denote the standard Newton iteration operator applied to the PPS $x = P_\sigma(x)$ (see [10]).

Lemma 3. *Given a minPPS, $x = P(x)$, with GFP $g^* < 1$, and given y with $0 \leq y \leq g^*$, there exists an LDF policy σ with $P^y(\mathcal{N}_\sigma(y)) = \mathcal{N}_\sigma(y)$, the GNM operator $I(x)$ is defined at y, and for this policy σ, $I(y) = \mathcal{N}_\sigma(y)$.*

Using this, we can show a result for GFPs similar to one in [9] for LFPs:

Lemma 4. *Let $x = P(x)$ be a minPPS with GFP $g^* < 1$. For any $0 \leq x \leq g^*$ and $\lambda > 0$, $I(x) \leq g^*$, and if $g^* - x \leq \lambda(1 - g^*)$ then $g^* - I(x) \leq \frac{\lambda}{2}(1 - g^*)$.*

Theorem 3 follows by using Lemma 4 and Lemma 2(3.), and applying a similar inductive argument as in ([9], Section 3.5).

P-time detection of zeros in the GFP of a minPPS: $g_i^* \overset{?}{=} 0$.

We give a P-time algorithm for deciding whether the supremum reachability probability in a BMDP equals 1, in which case we show the supremum probability is achieved by a (memoryful but deterministic) strategy which we can compute in P-time (thus limit-sure and almost-sure reachability are the same). Let X be the set of all variables x_i in minPPS $x = P(x)$ in SNF form, with GFP $g^* < 1$.

1. Initialize $S := \{ x_i \in X \mid P_i(0) > 0,$ i.e., $P_i(x)$ contains a constant term $\}$.
2. Repeat the following until neither are applicable:
   (a) If a variable x_i is of form L and $P_i(x)$ has a term whose variable is already in S, then add x_i to S.

(b) If a variable x_i is of form Q or M and both variables in $P_i(x)$ are already in S, then add x_i to S.

3. Let $F := \{\ x_i \in X - S \mid P_i(1) < 1,\ \text{or}\ P_i(x)\ \text{has form Q}\ \}$.

4. Repeat the following until no more variables can be added:
   - If a variable $x_i \in X - S$ is of form L or M and P_i contains a term whose variable is in F, then add x_i to F.

5. If $X = S \cup F$, then terminate and output F.

6. Otherwise set $S := X - F$ and return to step 2.

Theorem 4. *Given a minPPS $x = P(x)$ with $g^* < 1$, this algorithm terminates and outputs precisely the variables x_i with $g_i^* = 0$, in time polynomial in $|P|$.*

Theorem 5. *There is a non-static deterministic optimal strategy for maximizing the probability of reaching a target type in a BMDP with probability 1, if the supremum probability of reaching the target is 1.*

We outline the non-static policy. The proof of Theorem 4 constructs a LDF policy σ with the property that $g_i^* = 0$ iff $(q_\sigma^*)_i = 0$. Let Z denote the set of variables with $g_i^* = 0 = (q_\sigma^*)_i$. From Proposition 1, we can also compute in P-time an LDF policy τ with $g_\tau^* < 1$. We combine σ and τ in the following non-static policy: We designate one member of our initial population with type in Z to be the queen. The rest of the population are workers. We use policy σ for the queen and τ for the workers. In following generations, if we have not reached an object of the target type, we choose one of the children in Z of the last generation's queen (which we show must exist) to be the new queen. Again, all other members of the population are workers.

Computing ϵ-optimal strategies for minPPSs in P-time:

We first use the following algorithm to find an LDF policy σ with $\|g^* - q_\sigma^*\|_\infty \leq \frac{1}{2}\epsilon$. We then use that policy to construct ϵ-optimal policies.

1. Compute, using GNM, a $0 \leq y \leq g^*$ with $\|g^* - y\|_\infty \leq 2^{-14|P|-3}\epsilon$;

2. Let $k := 0$, and let σ_0 be a policy that has $P_{\sigma_0}(y) = P(y)$ (i.e., σ_0 chooses the action with highest probability of reaching the target according to y).

3. Compute F_{σ_k}, the set of variables that, in the dependency graph of $x = P_{\sigma_k}(x)$, either are or depend on a variable x_i which either has form Q or else $P_i(1) < 1$ or $P_i(0) > 0$. Let D_{σ_k} be the complement of F_{σ_k}.

4. if D_{σ_k} is empty, we are done, and we output σ_k.

5. Find a variable[2] x_i of type M in D_{σ_k}, which has a choice x_j in F_{σ_k} (which isn't its current choice) such that $|y_i - y_j| \leq 2^{-14|P|-2}\epsilon$; Let policy σ_{k+1} choose x_j at x_i, & otherwise agree with σ_k. Let $k := k+1$; return to step 3.

Lemma 5. *The above algorithm terminates in P-time and outputs an LDF policy σ with $\|P_\sigma(y) - y\|_\infty \leq 2^{-14|P|-2}\epsilon$.*

---

[2] We show that such a variable x_i always exists whenever we reach this step.

We define a randomized static policy v as follows. With probability $2^{-28|P|-4}\epsilon$ we follow a (necessarily LDF) deterministic policy τ that satisfies $g_\tau^* < 1$. We can compute such a τ in P-time by Proposition 1. With the remaining probability $1 - 2^{-28|P|-4}\epsilon$, we follow the static deterministic policy σ that is output by the algorithm above. We can then show (see [10] for the involved proof):

Theorem 6. *The output policy σ of the algorithm satisfies $\|g^* - q_\sigma^*\|_\infty \leq \frac{1}{2}\epsilon$. Moreover, v satisfies $\|g^* - g_v^*\|_\infty \leq \epsilon$, i.e., it is ϵ-optimal.*

Theorem 7. *For a BMDP with minPPS $x = P(x)$, and minimum non-reachability probabilities given by the GFP $g^* < 1$, the following deterministic non-static non-memoryless strategy α is also ϵ-optimal starting with one object of any type:*
Use the policy σ output by the algorithm, until the population is at least $\frac{2^{4|P|+1}}{\epsilon}$ for the first time, thereafter use a deterministic static policy τ such that $g_\tau^* < 1$.

Corollary 1. *For maximizing BMDP reachability probability, we can compute in P-time a randomized static (or deterministic non-static) ϵ-optimal policy.*

References

1. Bozic, I., et al.: Evolutionary dynamics of cancer in response to targeted combination therapy. Elife **2**, e00747 (2013)
2. Bonnet, R., Kiefer, S., Lin, A.W.: Analysis of probabilistic basic parallel processes. In: Muscholl, A. (ed.) FOSSACS 2014 (ETAPS). LNCS, vol. 8412, pp. 43–57. Springer, Heidelberg (2014)
3. Brázdil, T., Brozek, V., Kucera, A., Obdrzálek, J.: Qualitative reachability in stochastic BPA games. Inf. Comput. **209**(8), 1160–1183 (2011)
4. Brázdil, T., Brozek, V., Forejt, V., Kucera, A.: Reachability in recursive Markov decision processes. Inf. Comput. **206**(5), 520–537 (2008)
5. Chen, T., Dräger, K., Kiefer, S.: Model checking stochastic branching processes. In: Rovan, B., Sassone, V., Widmayer, P. (eds.) MFCS 2012. LNCS, vol. 7464, pp. 271–282. Springer, Heidelberg (2012)
6. Esparza, J., Gawlitza, T., Kiefer, S., Seidl, H.: Approximative methods for monotone systems of min-max-polynomial equations. In: Aceto, L., Damgård, I., Goldberg, L.A., Halldórsson, M.M., Ingólfsdóttir, A., Walukiewicz, I. (eds.) ICALP 2008, Part I. LNCS, vol. 5125, pp. 698–710. Springer, Heidelberg (2008)
7. Esparza, J., Kučera, A., Mayr, R.: Model checking probabilistic pushdown automata. Logical Methods in Computer Science **2**(1), 1–31 (2006)
8. Etessami, K., Stewart, A., Yannakakis, M.: Polynomial-time algorithms for multitype branching processes and stochastic context-free grammars. In: Proc. 44th ACM Symposium on Theory of Computing (STOC) (2012)
9. Etessami, K., Stewart, A., Yannakakis, M.: Polynomial time algorithms for branching markov decision processes and probabilistic min(max) polynomial bellman equations. In: Czumaj, A., Mehlhorn, K., Pitts, A., Wattenhofer, R. (eds.) ICALP 2012, Part I. LNCS, vol. 7391, pp. 314–326. Springer, Heidelberg (2012)
10. Full preprint of this paper (2015). arXiv:1502.05533

11. Etessami, K., Wojtczak, D., Yannakakis, M.: Recursive stochastic games with positive rewards. In: Aceto, L., Damgård, I., Goldberg, L.A., Halldórsson, M.M., Ingólfsdóttir, A., Walukiewicz, I. (eds.) ICALP 2008, Part I. LNCS, vol. 5125, pp. 711–723. Springer, Heidelberg (2008)
12. Etessami, K., Yannakakis, M.: Recursive Markov decision processes and recursive stochastic games. Journal of the ACM (2015)
13. Etessami, K., Yannakakis, M.: Recursive Markov chains, stochastic grammars, and monotone systems of nonlinear equations. Journal of the ACM **56**(1) (2009)
14. Pliska, S.: Optimization of multitype branching processes. Management Sci., **23**(2), 117–124 (1976/1977)
15. Reiter, J.G., Bozic, I., Chatterjee, K., Nowak, M.A.: TTP: tool for tumor progression. In: Sharygina, N., Veith, H. (eds.) CAV 2013. LNCS, vol. 8044, pp. 101–106. Springer, Heidelberg (2013)
16. Rothblum, U., Whittle, P.: Growth optimality for branching Markov decision chains. Math. Oper. Res. **7**(4), 582–601 (1982)

Trading Bounds for Memory in Games with Counters

Nathanaël Fijalkow[1,2]([⊠]), Florian Horn[1], Denis Kuperberg[2,3],
and Michał Skrzypczak[1,2]

[1] LIAFA, Université Paris 7, Paris, France
nath@liafa.univ-paris-diderot.fr
[2] Institute of Informatics, University of Warsaw, Warsaw, Poland
[3] Onera/DTIM, Toulouse and IRIT, University of Toulouse, Toulouse, France

Abstract. We study two-player games with counters, where the objective of the first player is that the counter values remain bounded. We investigate the existence of a trade-off between the size of the memory and the bound achieved on the counters, which has been conjectured by Colcombet and Löding. We show that unfortunately this conjecture does not hold: there is no trade-off between bounds and memory, even for finite arenas. On the positive side, we prove the existence of a trade-off for the special case of thin tree arenas.

1 Introduction

This paper studies finite-memory determinacy for games with counters. The motivation for this investigation comes from the theory of regular cost functions, which we discuss now.

Regular Cost Functions. The theory of regular cost functions is a *quantitative* extension of the notion of regular languages, over various structures (words and trees). More precisely, it expresses *boundedness questions*. A typical example of a boundedness question is: given a regular language $L \subseteq \{a, b\}^*$, does there exist a bound N such that all words from L contain at most N occurences of a?

This line of work already has a long history: it started in the 80s, when Hashiguchi, and then later Leung, Simon and Kirsten solved the *star-height problem* by reducing it to boundedness questions [10,11,13,15]. Both the logics MSO+U and later cost MSO (as part of the theory of regular cost functions) emerged in this context [3,4,6,8], as quantitative extensions of the notion of regular languages that can express boundedness questions.

Consequently, developing the theory of regular cost functions comes in two flavours: the first is using it to reduce various problems to boundedness questions, and the second is obtaining decidability results for the boundedness problem for cost MSO over various structures.

The research leading to these results has received funding from the European Union's Seventh Framework Programme (FP7/2007-2013) under grant agreement 259454 (GALE).

© Springer-Verlag Berlin Heidelberg 2015
M.M. Halldórsson et al. (Eds.): ICALP 2015, Part II, LNCS 9135, pp. 197–208, 2015.
DOI: 10.1007/978-3-662-47666-6_16

For the first point, many problems have been reduced to boundedness questions. The first example is the star-height problem over words [11] and over trees [8], followed for instance by the boundedness question for fixed points of monadic formulae over finite and infinite words and trees [2]. The most important problem that has been reduced to a boundedness question is to decide the Mostowski hierarchy for infinite trees [7].

For the second point, it has been shown that over finite words and trees, a significant part of the theory of regular languages can successfully be extended to the theory of regular cost functions, yielding notions of regular expressions, automata, semigroups and logics that all have the same expressive power, and that extend the standard notions. In both cases, algorithms have been constructed to answer boundedness questions.

However, extending the theory of regular cost functions to infinite trees seems to be much harder, and the major open problem there is the decidability of cost MSO over infinite trees.

LoCo Conjecture. Colcombet and Löding pointed out that the only missing point to obtain the decidability of cost MSO is a finite-memory determinacy result for games with counters. More precisely, they conjectured that there exists a trade-off between the size of the memory and the bound achieved on the counters [5]. So far, this conjecture resisted both proofs and refutations, and one of the only non-trivial positive case known is due to Vanden Boom [16], which implied the decidability of the weak variant of cost MSO over infinite trees, later generalized to quasi-weak cost MSO in [1]. Unfortunately, Quasi-Weak cost MSO is strictly weaker than cost MSO, and this leaves open the question whether cost MSO is decidable.

Contributions. In this paper, we present two contributions:

- There is no trade-off, even for finite arenas, which disproves the conjecture,
- There is a trade-off for the special case of thin tree arenas.

Our first contribution does not imply the undecidability of cost MSO, it rather shows that proving the decidability will involve subtle combinatorial arguments that are yet to be understood.

Structure of This Document. The definitions are given in Section 2. We state the conjecture in Section 3. Section 4 disproves the conjecture. Section 5 proves that the conjecture holds for the special case of thin tree arenas.

2   Definitions

Arenas. The games we consider are played by two players, Eve and Adam, over potentially infinite graphs called arenas[1]. Formally, an arena $\mathcal{G}$ consists of a

---

[1] We refer to [9] for an introduction to games.

directed graph (V, E) whose vertex set is divided into vertices controlled by Eve
(V_E) and vertices controlled by Adam (V_A). A token is initially placed on a given
initial vertex v_0, and the player who controls this vertex pushes the token along
an edge, reaching a new vertex; the player who controls this new vertex takes
over, and this interaction goes on forever, describing an infinite path called a
play. Finite or infinite plays are paths in the graphs, seen as sequences of edges,
typically denoted π. In its most general form, a strategy for Eve is a mapping
$\sigma : E^* \cdot V_E \rightarrow E$, which given the history played so far and the current vertex
picks the next edge. We say that a play $\pi = e_0 e_1 e_2 \ldots$ is consistent with σ if
$e_{n+1} = \sigma(e_0 \cdots e_n \cdot v_n)$ for every n with $v_n \in V_E$.

Winning Conditions. A winning condition for an arena is a set of a plays for
Eve, which are called the winning plays for Eve (the other plays are winning
for Adam). A game consists of an arena $\mathcal{G}$ and a winning condition W for this
arena, it is usually denoted $(\mathcal{G}, W)$.

A strategy for Eve is winning for a condition, or ensures this condition, if all
plays consistent with the strategy belong to the condition. For a game $(\mathcal{G}, W)$,
we denote $\mathcal{W}_E(\mathcal{G}, W)$ the winning region of Eve, *i.e.* the set of vertices from
which Eve has a winning strategy.

Here we will consider the classical parity condition as well as *quantitative*
bounding conditions.

The parity condition is specified by a colouring function $\Omega : V \rightarrow \{0, \ldots, d\}$,
requiring that the maximum color seen infinitely often is even. The special case
where $\Omega : V \rightarrow \{1, 2\}$ corresponds to Büchi conditions, denoted Büchi(F) where
$F = \{v \in V \mid \Omega(v) = 2\}$, and $\Omega : V \rightarrow \{0, 1\}$ corresponds to CoBüchi conditions,
denoted CoBüchi(F) where $F = \{v \in V \mid \Omega(v) = 1\}$. We will also consider the
simpler conditions Safe(F) and Reach(F), for $F \subseteq V$: the first requires to avoid
F forever, and the second to visit a vertex from F at least once.

The bounding condition B is actually a family of winning conditions with an
integer parameter $B = \{B(N)\}_{N \in \mathbb{N}}$. We call it a quantitative condition because
it is monotone: if $N < N'$, all the plays in $B(N)$ also belong to $B(N')$.

The counter actions are specified by a function $c : E \rightarrow \{\varepsilon, i, r\}^k$, where k is
the number of counters: each counter can be incremented (i), reset (r), or left
unchanged (ε). The value of a play π, denoted $val(\pi)$, is the supremum of the
value of all counters along the play. It can be infinite if one counter is unbounded.
The condition $B(N)$ is defined as the set of plays whose value is less than N.

In this paper, we study the condition B-parity, where the winning condition
is the intersection of a bounding condition and a parity condition. The value of
a play that satisfies the parity condition is its value according to the bounding
condition. The value of a play which does not respect the parity condition is
∞. We often consider the special case of B-reachability conditions, denoted
B Until F. In such cases, we assume that the game stops when it reaches F.

Given an initial vertex v_0, the value $val(v_0)$ is:

$$\inf_\sigma \ \sup_\pi \ \{val(\pi) \mid \pi \text{ consistent with } \sigma \text{ starting from } v_0\} .$$

Finite-Memory Strategies. A *memory structure* $\mathcal{M}$ for the arena $\mathcal{G}$ consists of a set M of memory states, an initial memory state $m_0 \in M$ and an update function $\mu : M \times E \to M$. The update function takes as input the current memory state and the chosen edge to compute the next memory state, in a deterministic way. It can be extended to a function $\mu : E^* \cdot V \to M$ by defining $\mu^*(v) = m_0$ and $\mu^*(\pi \cdot (v, v')) = \mu(\mu^*(\pi \cdot v), (v, v'))$.

Given a memory structure $\mathcal{M}$, a strategy is induced by a next-move function $\sigma : V_E \times M \to E$, by $\sigma(\pi \cdot v) = \sigma(v, \mu^*(\pi \cdot v))$. Note that we denote both the next-move function and the induced strategy σ. A strategy with memory structure $\mathcal{M}$ has finite memory if M is a finite set. It is *memoryless*, or *positional* if M is a singleton: it only depends on the current vertex. Hence a memoryless strategy can be described as a function $\sigma : V_E \to E$.

An arena $\mathcal{G}$ and a memory structure $\mathcal{M}$ for $\mathcal{G}$ induce the expanded arena $\mathcal{G} \times \mathcal{M}$ where the current memory state is stored explicitly together with the current vertex: the vertex set is $V \times M$, the edge set is $E \times \mu$, defined by: $((v, m), (v', m')) \in E'$ if $(v, v') \in E$ and $\mu(m, (v, v')) = m'$. There is a natural one-to-one correspondence between strategies in $\mathcal{G} \times \mathcal{M}$ using $\mathcal{M}'$ as memory structure and strategies in $\mathcal{G}$ using $\mathcal{M} \times \mathcal{M}'$ as memory structure.

3   The Conjecture

In this section, we state the conjecture [5], and explain how positive cases of this conjecture imply the decidability of cost MSO.

3.1   Statement of the Conjecture

There exists mem $: \mathbb{N}^2 \to \mathbb{N}$ and $\alpha : \mathbb{N}^3 \to \mathbb{N}$ such that
for every B-parity game with k counters, $d + 1$ colors and initial vertex v_0,
there exists a strategy σ using mem(d, k) memory states, ensuring
$$B(\alpha(d, k, val(v_0))) \cap \text{Parity}(\Omega).$$

The function α is called a *trade-off function*: if there exists a strategy ensuring $B(N) \cap \text{Parity}(\Omega)$, then there exists a strategy with *small* memory that ensures $B(\alpha(d, k, N)) \cap \text{Parity}(\Omega)$. So, at the price of increasing the bound from N to $\alpha(d, k, N)$, one can use a strategy using a small memory structure.

To get a better understanding of this conjecture, we show three simple facts:

1. why reducing memory requires to increase the bound,
2. why the memory bound mem depends on the number of counters k,
3. why a weaker version of the conjecture holds, where mem depends on the value.

For the first point, we present a simple game, represented in Figure 1. It involves one counter and the condition B Until F. Starting from v_0, the game moves to v and sets the value of the counter to N. The objective of Eve is to take the edge to the right to F. However, this costs N increments, so if she wants the counter

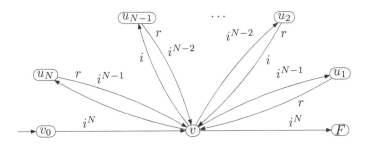

Fig. 1. A trade-off is necessary

value to remain smaller than N she has to set its value to 0 before taking this edge. She has N options: for $\ell \in \{1, \ldots, N\}$, the ℓth option consists in going to u_ℓ, involving the following actions:

- first, take $N - \ell$ increments,
- then, reset the counter,
- then, take $\ell - 1$ increments, setting the value to $\ell - 1$.

It follows that there is a strategy for Eve to ensure $B(N)$ Until F, which consists in going successively through u_N, u_{N-1}, and so on, until u_1, and finally to F. Hence to ensure that the bound is always smaller than N, Eve needs $N + 1$ memory states.

However, if we consider the bound $2N$ rather than N, then Eve has a very simple strategy, which consists in going directly to F, using no memory at all. This is a simple example of a trade-off: to ensure the bound N, Eve needs $N+1$ memory states, but to ensure the worse bound $2N$, she has a positional strategy.

For the second point, consider the following game with k counters (numbered cyclically) and only one vertex, controlled by Eve. There are k self-loops, each incrementing a counter and resetting the previous one. Eve has a simple strategy to ensure $B(1)$, which consists in cycling through the loops, and uses k memory states. Any strategy using less than k memory states ensures no bound at all, as one counter would be incremented infinitely many times but never reset. It follows that the memory bound mem in the conjecture has to depend on k.

For the third point, we give an easy result that shows the existence of finite memory strategies whose size depends on the value, even without losing anything on the bound.

Lemma 1. *For every B-parity game with k counters and initial vertex v_0, there exists a strategy σ ensuring $B(val(v_0)) \cap Parity(\Omega)$ with $(val(v_0) + 1)^k$ memory states.*

The proof consists in composing the B-parity game with a memory structure that keeps track of the counter values up to the value, reducing the B-condition to a safety condition.

3.2   The Interplay with Cost MSO

The aim of the conjecture stated above is the following: if true, it implies the decidability of cost MSO over infinite trees. More precisely, the technical difficulty to develop the theory of regular cost functions over infinite trees is to obtain effective constructions between variants of automata with counters, and this is what this conjecture is about.

In the qualitative case (without counters), to obtain the decidability of MSO over infinite trees, known as Rabin's theorem [14], one transforms MSO formulae into equivalent automata. The complementation construction is the technical cornerstone of this procedure. The *key* ingredient for this is games, and specifically positional determinacy for parity games. Similarly, other classical constructions, to simulate either two-way or alternating automata by non-deterministic ones, make a crucial use of positional determinacy for parity games.

In the quantitative case now, Colcombet and Löding [8] showed that to extend these constructions, one needs a similar result on parity games with counters, which is the conjecture we stated above.

So far, there is only one positive instance of this conjecture, which is the special case of B-Büchi games over chronological arenas[2].

Theorem 1 ([16]). *For every B-Büchi game played over a chronological arena with k counters and initial vertex v_0, Eve has a strategy ensuring $B(k^k \cdot val(v_0)^{2k}) \cap B\ddot{u}chi(F)$ with $2 \cdot k!$ memory states.*

This was the key ingredient in proving the decidability of weak cost MSO over infinite trees.

4   No Trade-off Over Finite Arenas

In this section, we show that the conjecture does not hold, even for finite arenas.

Theorem 2. *For all K, for all N, there exists a B-reachability game played over a finite arena $G_{K,N}$ with one counter and an initial vertex such that:*

  – *there exists a 3^K memory states strategy ensuring $B(K(K + 3))$ Until F,*
  – *no $K + 1$ memory states strategy ensure $B(N)$ Until F.*

We proceed in two steps. The first is an example giving a lower bound of 3, and the second is a nesting of this first example.

4.1   A First Lower Bound of 3

The game $\mathcal{G}_1$ giving a lower bound of 3 is represented in Figure 2. The condition is B Until F, with only one counter. In this game, Eve is torn between going to the right to reach F, which implies incrementing the counter, and going to the left, to reset the counter. The actions of Eve from the vertex u_n are:

---

[2] The definition of chronological arenas is given in Section 5.

- *increment*, and go one step to the right, to v_{n-1},
- *reset*, and go two steps to the left, to v_{n+2}.

The actions of Adam from the vertex v_n are:

- *play*, and go down to u_n,
- *skip*, and go to v_{n-1}.

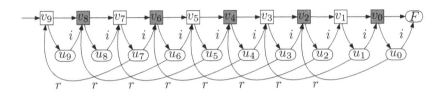

Fig. 2. Part of the game $\mathcal{G}_1$, where Eve needs 3 memory states. The colors on the vertices are used to construct a 3 memory states strategy.

Theorem 3. *In $\mathcal{G}_1$, from v_N:*

- *Eve has a 4 memory states strategy ensuring $B(3)$ Until F,*
- *Eve has a 3 memory states strategy ensuring $B(4)$ Until F,*
- *For all N, no 2 memory states strategy ensures $B(N)$ Until F.*

The first item follows from Lemma 1. However, to illustrate the properties of the game $\mathcal{G}_1$ we will provide a concrete strategy with 4 memory states that ensures $B(3)$ Until F. The memory states are i_1, i_2, i_3 and r, linearly ordered by $i_1 < i_2 < i_3 < r$. With the memory states i_1, i_2 and i_3, the strategy chooses to increment, and updates its memory state to the next memory state. With the memory state r, the strategy chooses to reset, and updates its memory state to i_1. This strategy satisfies a simple invariant: it always resets to the right of the previous reset, if any.

We show how to save one memory state, at the price of increasing the bound by one: we construct a 3 memory states strategy ensuring $B(4)$ Until F. The idea, as represented in Figure 2, is to color every second vertex and to use this information to track progress. The 3 memory states are called i, j and r. The update is as follows: the memory state is unchanged in uncoloured (white) states, and switches from i and j and from j to r on gray states. The strategy is as follows: in the two memory states i and j, Eve chooses to increment, and in r she chooses to reset. As for the previous strategy, this strategy ensures that it always resets to the right of the previous reset, if any.

4.2   General Lower Bound

We now push the example above further. A first approach is to modify $\mathcal{G}_1$ by increasing the length of the resets, going ℓ steps to the left rather than only 2.

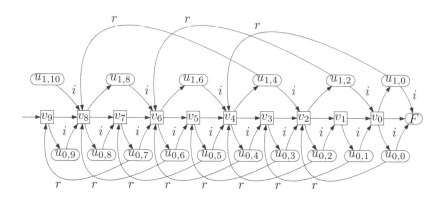

Fig. 3. The game with two levels.

However, this does not give a better lower bound: there exists a 3 memory states strategy in this modified game that ensures twice the value, following the same ideas as presented above.

We construct $\mathcal{G}_{K,N}$, a nesting of the game $\mathcal{G}_1$ with K levels. In Figure 3, we represented the interaction between two levels. Roughly speaking, the two levels are independent, so we play both games at the same time. Those two games use different timeline. For instance, in Figure 3, the bottom level is based on $(+1, -2)$ (an increment goes one step to the right, a reset two steps to the left), and the top level is based on $(+2, -4)$. This difference in timeline ensures that a strategy for Eve needs to take care somehow independently of each level, ensuring that the number of memory states depends on the number of levels.

Theorem 4. *In $\mathcal{G}_{K,N}$, from some initial vertex:*

- *Eve has a 3^K memory states strategy ensuring $B(K(K+3))$ Until F,*
- *No $K+1$ memory states strategy ensures $B(N)$ Until F.*

5 Existence of a Trade-off for Thin Tree Arenas

In this section, we prove that the conjecture holds for the special case of thin tree arenas[3].

Theorem 5. *There exist two functions $mem : \mathbb{N}^2 \to \mathbb{N}$ and $\alpha : \mathbb{N}^4 \to \mathbb{N}$ such that for every B-parity game played over a thin tree arena of width W with k counters, $d+1$ colors and initial vertex v_0, Eve has a strategy to ensure $B(k^k \cdot val(v_0)^{2k} \cdot \alpha(d, k, W+2, val(v_0))) \cap Parity(\Omega)$, with $W \cdot 3^k \cdot 2k! \cdot mem(d, k)$ memory states.*

The functions α and mem are defined inductively. Since the proof will work by induction on the number of colors, removing the least important color, we define four functions:

---

[3] The definitions of word and thin tree arenas are given in Subsection 5.1.

- α_0 and mem$_0$ when the set of colors is $\{0, \ldots, d\}$,
- α_1 and mem$_1$ when the set of colors is $\{1, \ldots, d\}$:

$$\alpha_0(d, k, W, N) = \alpha_1(d-1, k+1, 3W, W \cdot (N+1)^k),$$

$$\alpha_1(d, k, W, N) = \begin{cases} k^k \cdot N^{2k} & \text{if } d = 2, \\ \alpha_0(d, k, 2W, N) & \text{otherwise,} \end{cases}$$

$$\text{mem}_0(d, k) = 2 \cdot \text{mem}_1(d-1, k+1),$$

$$\text{mem}_1(d, k) = \begin{cases} 2 \cdot k! & \text{if } d = 1, \\ 2 \cdot \text{mem}_0(d, k) & \text{otherwise.} \end{cases}$$

Note that the functions α and mem depend on the width parameter; this is sufficient for the intended applications of the LoCo conjecture, as the width corresponds to the size of the automaton.

We focus in this extended abstract on the intermediate result for the special case of word arenas.

Theorem 6. *There exists two functions* mem : $\mathbb{N}^2 \to \mathbb{N}$ *and* α : $\mathbb{N}^4 \to \mathbb{N}$ *such that for every B-parity game played over a word arena of width W with k counters, $d+1$ colors and initial vertex v_0, Eve has a strategy to ensure $B(\alpha(d, W, k, val(v_0))) \cap Parity(\Omega)$, with mem$(d, k)$ memory states.*

5.1   Word and Thin Tree Arenas

A (non-labelled binary) *tree* is a subset $T \subseteq \{0, 1\}^*$ which is prefix-closed and non-empty. The elements of T are called *nodes*, and we use the natural terminology: for $n \in \{0, 1\}^*$ and $\ell \in \{0, 1\}$, the node $n \cdot \ell$ is a child of n, and a descendant of n if $\ell \in \{0, 1\}^*$.

A (finite or infinite) *branch* π is a word in $\{0, 1\}^*$ or $\{0, 1\}^\omega$. We say that π is a branch of the tree T if $\pi \subseteq T$ (or every prefix of π belongs to T when π is infinite) and π is maximal satisfying this property. A tree is called *thin* if it has only countably many branches. For example, the full binary tree $T = \{0, 1\}^*$ has uncountably many branches, therefore it is not thin.

Definition 1. *An arena is:*

- chronological *if there exists a function $r : V \to \mathbb{N}$ which increases by one on every edge: for all $(v, v') \in E$, $r(v') = r(v) + 1$.*
- a word arena of width W *if it is chronological, and for all $i \in \mathbb{N}$, the set $\{v \in V \mid r(v) = i\}$ has cardinal at most W.*
- a tree arena of width W *if there exists a function $R : V \to \{0, 1\}^*$ such that*
   1. for all $n \in \{0, 1\}^*$, the set $\{v \in V \mid R(v) = n\}$ has cardinal at most W.
   2. for all $(v, v') \in E$, we have $R(v') = R(v) \cdot \ell$ for some $\ell \in \{0, 1\}$.
   It is a thin tree arena if $R(V)$ is a thin tree.

The notions of word and tree arenas naturally appear in the study of automata over infinite words and trees. Indeed, the acceptance games of such automata, which are used to define their semantics, are played on word or tree arenas. Furthermore, the width corresponds to the size of the automaton.

5.2 Existence of a Trade-off for Word Arenas

We prove Theorem 6 by induction on the colors in the parity condition. Consider a B-parity game $\mathcal{G}$ with k counters and $d+1$ colors over a word arena of width W with initial vertex v_0. We examine two cases, depending whether the least important color (*i.e* the smallest) that appears is odd or even:

- if the set of colors is $\{1, \ldots, d\}$, then we construct a B-parity game $\mathcal{G}'$ using $\{2, \ldots, d\}$ as colors,
- if the set of colors is $\{0, \ldots, d\}$, then we construct a B-parity game $\mathcal{G}'$ using $\{1, \ldots, d\}$ as colors.

In both cases, we obtain from the induction hypothesis a winning strategy using few memory states in $\mathcal{G}'$, which we use to construct a winning strategy using few memory states in $\mathcal{G}$. The base case is given by Büchi conditions, and follows from Theorem 1.

Removing the Least Important Color: the Odd Case The first case we consider is when the least important color is 1. The technical core of the construction is motivated by the technique used in [16]. It consists in slicing the game horizontally, such that in each slice the strategy σ ensures to see a vertex of color greater than 1. This way, the combination of a memory structure of size 2 and a safety condition expresses that some color greater than 1 should be seen infinitely often, allowing us to remove the color 1.

Removing the Least Important Color: the Even Case The second case we consider is when the least important color is 0.

We explain the intuition for the case of CoBüchi conditions, *i.e* if there are only colors 0 and 1. Let $F = \{v \mid \Omega(v) = 1\}$. Define $X_0 = Y_0 = \emptyset$, and for $i \geq 1$:

$$\begin{cases} X_{i+1} = \mathcal{W}_E(\mathrm{Safe}(F) \text{ WeakUntil } Y_i) \\ Y_{i+1} = \mathcal{W}_E(\mathrm{Reach}(X_{i+1})) \end{cases}$$

The condition $\mathrm{Safe}(F)$ WeakUntil Y_i is satisfied by plays that do not visit F before Y_i: they may never reach Y_i, in which case they never reach F, or they reach Y_i, in which case they did not reach F before that.

We have $\bigcup_i Y_i = \mathcal{W}_E(\mathrm{CoBüchi}(F))$. A winning strategy based on these sets has two aims: in X_i it avoids F ("Safe" mode) and in Y_i it attracts to the next X_i ("Attractor" mode). The key property is that since the arena is a word arena of width W where Eve can bound the counters by N, she only needs to alternate between modes a number of times bounded by a function of N and W. In other words, the sequence $(Y_i)_{i \in \mathbb{N}}$ stabilizes after a number of steps bounded by a function of N and W. A remote variant of this bounded-alternation fact can be found in [12]. Hence the CoBüchi condition can be checked using a new counter and a Büchi condition, as follows.

There are two modes: "Safe" and "Attractor". The Büchi condition ensures that the "Safe" mode is visited infinitely often. In the "Safe" mode, only vertices of color 0 are accepted; visiting a vertex of color 1 leads to the "Attractor" mode and increments the new counter. At any time, she can reset the mode to "Safe". The counter is never reset, so to ensure that it is bounded, Eve must change modes finitely often. Furthermore, the Büchi condition ensures that the final mode is "Safe", implying that the CoBüchi condition is satisfied.

For the more general case of parity conditions, the same idea is used, but as soon as a vertex of color greater than 1 is visited, then the counter is reset.

Define $\mathcal{G}'$:

$$V' = \begin{cases} V'_E = V_E \times \{A, S\} \cup \overline{V} \\ V'_A = V_A \times \{A, S\} \, . \end{cases}$$

After each edge followed, Eve is left the choice to set the flag to S. The set of choice vertices is denoted $\overline{V}$. We define E' and the counter actions.

$$E' = \begin{cases} (v, A) \xrightarrow{c(v,v'),\varepsilon} \overline{v'} & \text{if } (v, v') \in E, \\ (v, S) \xrightarrow{c(v,v'),\varepsilon} (v', S) & \text{if } (v, v') \in E \text{ and } \Omega(v') = 0, \\ (v, S) \xrightarrow{c(v,v'),i} (v', A) & \text{if } (v, v') \in E \text{ and } \Omega(v') = 1, \\ (v, S) \xrightarrow{c(v,v'),r} (v', S) & \text{if } (v, v') \in E \text{ and } \Omega(v') > 1, \\ \overline{v} \xrightarrow{\varepsilon} (v, A) \text{ and } \overline{v} \xrightarrow{\varepsilon} (v, S) \end{cases}$$

Equip the arena $\mathcal{G}'$ with the colouring function Ω' defined by

$$\Omega'(v, m) = \begin{cases} 1 & \text{if } m = A, \\ 2 & \text{if } \Omega(v) = 0 \text{ and } m = S, \\ \Omega(v) & \text{otherwise.} \end{cases}$$

Remark that Ω' uses one less color than Ω, since no vertices have color 0 for Ω'.

Before stating the equivalence between $\mathcal{G}$ and $\mathcal{G}'$, we formalise the property mentioned above, that in word arenas Eve does not need to alternate an unbounded number of times between the modes "Safe" and "Attractor".

Lemma 2. *Let G be a word arena of width W, and a subset F of vertices such that every path in G contains finitely many vertices in F. Define the following sequence of subsets of vertices $X_0 = \emptyset$, and for $i \geq 0$*

$$\begin{cases} X_{2i+1} = \left\{ v \,\middle|\, \begin{array}{l} \text{all paths from } v \text{ contain no vertices in } F \\ \text{before the first vertex in } X_{2i}, \text{ if any} \end{array} \right\}, \\ X_{2i+2} = \{ v \,|\, \text{all paths from } v \text{ are finite or lead to } X_{2i+1} \} \, . \end{cases}$$

We have $X_0 \subseteq X_1 \subseteq X_2 \subseteq \cdots$, and X_{2W} covers the whole arena.

Denote $N = val(v_0)$.

Lemma 3.

1. *There exists a strategy σ' in $\mathcal{G}'$ that ensures $B(W \cdot (N+1)^k) \cap Parity(\Omega')$.*
2. *Let σ' be a strategy in $\mathcal{G}'$ ensuring $B(N') \cap Parity(\Omega')$ with K memory states, then there exists σ a strategy in $\mathcal{G}$ that ensures $B(N') \cap Parity(\Omega)$ with $2K$ memory states.*

Conclusion

We studied the existence of a trade-off between bounds and memory in games with counters, such as conjectured by Colcombet and Löding. We proved that there is no such trade-off in general, but that under some structural restrictions, as thin tree arenas, the conjecture holds.

We believe that the conjecture holds for all tree arenas, which would imply the decidability of cost MSO over infinite trees. A proof of this result would probably involve advanced combinatorial arguments, and require a deep understanding of the structure of tree arenas.

Acknowledgments. The unbounded number of fruitful discussions we had with Thomas Colcombet and Mikołaj Bojańczyk made this paper possible.

References

1. Blumensath, A., Colcombet, T., Kuperberg, D., Parys, P., Boom, M.V.: Two-way cost automata and cost logics over infinite trees. In: CSL-LICS, pp. 16–26 (2014)
2. Blumensath, A., Otto, M., Weyer, M.: Decidability results for the boundedness problem. Logical Methods in Computer Science **10**(3) (2014)
3. Bojańczyk, M.: A bounding quantifier. In: Marcinkowski, J., Tarlecki, A. (eds.) CSL 2004. LNCS, vol. 3210, pp. 41–55. Springer, Heidelberg (2004)
4. Bojańczyk, M., Colcombet, T.: Bounds in ω-regularity. In: LICS, pp. 285–296 (2006)
5. Colcombet, T.: Fonctions régulières de coût. Habilitation Thesis (2013)
6. Colcombet, T.: Regular cost functions, part I: logic and algebra over words. Logical Methods in Computer Science **9**(3) (2013)
7. Colcombet, T., Löding, C.: The non-deterministic mostowski hierarchy and distance-parity automata. In: Aceto, L., Damgård, I., Goldberg, L.A., Halldórsson, M.M., Ingólfsdóttir, A., Walukiewicz, I. (eds.) ICALP 2008, Part II. LNCS, vol. 5126, pp. 398–409. Springer, Heidelberg (2008)
8. Colcombet, T., Löding, C.: Regular cost functions over finite trees. In: LICS, pp. 70–79 (2010)
9. Farwer, B.: 1 omega-automata. In: Grädel, E., Thomas, W., Wilke, T. (eds.) Automata, Logics, and Infinite Games. LNCS, vol. 2500, pp. 3–21. Springer, Heidelberg (2002)
10. Hashiguchi, K.: Improved limitedness theorems on finite automata with distance functions. Theoretical Computer Science **72**(1), 27–38 (1990)
11. Kirsten, D.: Distance desert automata and the star height problem. ITA **39**(3), 455–509 (2005)
12. Kupferman, O., Vardi, M.Y.: Weak alternating automata are not that weak. In: 5th Israeli Symposium on Theory of Computing and Systems, pp. 147–158. IEEE Computer Society Press (1997)
13. Leung, H.: Limitedness theorem on finite automata with distance functions: An algebraic proof. Theoretical Computuer Science **81**(1), 137–145 (1991)
14. Rabin, M.O.: Decidability of second-order theories and automata on infinite trees. Transactions of the AMS **141**, 1–23 (1969)
15. Simon, I.: On semigroups of matrices over the tropical semiring. ITA **28**(3–4), 277–294 (1994)
16. Boom, M.V.: Weak cost monadic logic over infinite trees. In: Murlak, F., Sankowski, P. (eds.) MFCS 2011. LNCS, vol. 6907, pp. 580–591. Springer, Heidelberg (2011)

Decision Problems of Tree Transducers with Origin

Emmanuel Filiot[1], Sebastian Maneth[2], Pierre-Alain Reynier[3],
and Jean-Marc Talbot[3]([✉])

[1] Université Libre de Bruxelles, Brussels, Belgium
[2] University of Edinburgh, Edinburgh, Scotland
[3] Aix-Marseille Université and CNRS, Marseille, France
jean-marc.talbot@lif.univ-mrs.fr

Abstract. A tree transducer with origin translates an input tree into
a pair of output tree and origin info. The origin info maps each node
in the output tree to the unique input node that created it. In this
way, the implementation of the transducer becomes part of its semantics.
We show that the landscape of decidable properties changes drastically
when origin info is added. For instance, equivalence of nondeterministic
top-down and MSO transducers with origin is decidable. Both problems
are undecidable without origin. The equivalence of deterministic top-
down tree-to-string transducers is decidable with origin, while without
origin it is a long standing open problem. With origin, we can decide if
a deterministic macro tree transducer can be realized by a deterministic
top-down tree transducer; without origin this is an open problem.

Tree transducers were invented in the early 1970's as a formal model for
compilers and linguistics [23, 24]. They are being applied in many fields of computer science, such as syntax-directed translation [13], databases [15, 22], linguistics [4, 19], programming languages [21, 27], and security analysis [16]. The most
essential feature of tree transducers is their good balance between expressive
power and decidability.

Bojańczyk [3] introduces (string) transducers with origin. For "regular"
string-to-string transducers with origin he presents a machine independent characterization which admits Angluin-style learning and the decidability of natural
subclasses. These results indicate that classes of translations with origin are
mathematically better behaved than their origin-less counter parts.

We initiate a rigorous study of tree transducers with origin by investigating
the decidability of *equivalence*, *injectivity* and *query determinacy* on the following models: top-down tree-to-tree transducers [23, 24], top-down tree-to-string

The authors are grateful to Joost Engelfriet for his remarks for improvement and correction on an earlier version of this paper. This work has been carried out thanks to
the support of the ARCHIMEDE Labex (ANR-11-LABX-0033) and the A*MIDEX
project (ANR-11-IDEX-0001-02) funded by the "Investissements d'Avenir" French
Government program, managed by the French National Research Agency (ANR)
and by the PEPS project "Synthesis of Stream Processors" funded by CNRS.

© Springer-Verlag Berlin Heidelberg 2015
M.M. Halldórsson et al. (Eds.): ICALP 2015, Part II, LNCS 9135, pp. 209–221, 2015.
DOI: 10.1007/978-3-662-47666-6_17

Table 1. Decidability of equivalence

	top-down tree-to-tree		top-down tree-to-string		MSO tree-to-string	
	det	nd	det	nd	det	nd
	+ [12]	− [14]	?	− [14]	+ [10]	−
with origin	+	+	+	−	+	+

transducers [11], and MSO definable tree-to-string transducers (see, e.g., [10]). Unlike the string transducers of Bojańczyk [3], we will see that equivalent models of tree-to-string transducers do *not* remain equivalent in the presence of origin. This motivates the study of *subclass definability* problems (definability of a transduction from a class in a subclass) when considering the origin semantics.

Table 1 summarizes our results on equivalence; non-/deterministic are abbreviated by nd/det and decidable/undecidable by +/−. The "?" marks a long-standing open problem, already mentioned by Engelfriet [7]. The first change from − to + is the equivalence of nondeterministic top-down tree transducers. In the non-origin case this problem is already undecidable for restricted string-to-string transducers [14]. In the presence of origin it becomes decidable for tree transducers, because origin implies that any connected region of output nodes with the same origin is generated by one single rule. Hence, the problem reduces to letter-to-letter transducers [1]. What about nondeterministic top-down *tree-to-string* transducers (column four in Table 1)? Here output patterns cannot be treated as letters. By deferring output generation to a leaf they can simulate non-origin translations with undecidable equivalence [14]. Finally, we discuss column three. Here the origin information induces a structure on the output strings: recursive calls of origin-equivalent transducers must occur in similar "blocks", so that the same children of the current input node are visited in the same order (but possibly with differing numbers of recursive calls). This block structure allows to reason over single input paths, and to reduce the problem to deterministic tree-to-string transducers with monadic input. The latter can be reduced [20] to the famous HDT0L sequence equivalence problem.

Injectivity for deterministic transducers is undecidable for all origin-free models of Table 1. With origin, we prove undecidability in the tree-to-string case and decidability in the MSO and top-down tree cases. The latter is again due to the rigid structure implied by origins. We can track if two different inputs, over the *same* input nodes, produce the same output tree. We use the convenient framework of recognizable relations to show that the set of trees for which a transducer with origin produces the same output can be recognized by a tree automaton.

Motivation. Clearly, the more information we include in a transformation, the more properties become decidable. Consider invertability: on the one extreme, if all reads and writes are recorded (under ACID), then any computation becomes invertible. The question then arises, how much information needs to be included in order to be invertible. This problem has recently deserved much attention in the programming language community (see, e.g., [26]). Our work here was inspired by the very similar view/query determinacy problem. This problem asks for a given view and query, whether the query can be answered on the output of the view. It was shown decidable in [2] for views that are linear extended tree transducers, and queries that are deterministic MSO or top-down transducers. For views that include copying, the problem quickly becomes undecidable [2]. Our results show that such views *can* be supported, if origin is included. Consider for instance a view that regroups a list of publications into sublists of books, articles, etc. A tree transducer realizing this view *needs copying* (i.e., needs to process the original list multiple times). Without origin, we do not know a procedure that decides determinacy for such a view. With origin, we prove that determinacy is decidable. As expected: the world becomes safer with origin, but more restrictive (e.g., the query "is book X before article Y in the original list?" becomes determined when origin is added to the above view).

The tracking of origin information was studied in the programming language community, see [25]. As a technical tool it was used in [9] to characterize the MSO definable macro tree translations, and, in [17] to give a Myhill-Nerode theorem for deterministic top-down tree transducers. From a linguistic point of view, origin mappings on their own are subject of interest and are called "dependencies" or "links". Maletti [18] shows that dependencies (i.e., origins) give "surprising insights" into the structure of tree transformations: many separation results concerning expressive power can be obtained on the level of dependencies.

1 Preliminaries

For a nonnegative integer k we denote by $[k]$ the set $\{1, \ldots, k\}$. For an alphabet A, we denote by A^* the set of strings over A, and by ε the empty string. Let $w \in A^*$ be a string of length k. Its length is denoted by $|w| = k$, and its set of positions by $V(w) = [k]$. For $j \in [k]$, $w[j]$ denotes the j-th symbol of the string w. A *ranked alphabet* Σ is a finite set of symbols σ each with an associated natural k called its rank. We write $\sigma^{(k)}$ to denote that σ has rank k, and denote by $\Sigma^{(k)}$ the set of all symbols in Σ of rank k. The *set T_Σ of trees over* Σ is the smallest set T so that if $k \geq 0$, $t_1, \ldots, t_k \in T$, and $\sigma \in \Sigma^{(k)}$, then $\sigma(t_1, \ldots, t_k) \in T$. For the tree $\sigma()$ we simply write σ. The set $V(t)$ of nodes of tree $t \in T_\Sigma$ is the subset of $\mathbb{N}^*$ defined as $\{\varepsilon\} \cup \{iu \mid i \in [k], u \in V(t_i)\}$ if $t = \sigma(t_1, \ldots, t_k)$. Thus ε denotes the root node, and ui denotes the i-th child of a node u. For a tree t and $u \in V(t)$ we denote by $t[u]$ the label of node u in t, and by t/u the subtree of t rooted at u. For a tree t', we denote by $t[u \leftarrow t']$ the tree obtained from t by replacing the subtree rooted in position u by the tree t'. Given a tree $t \in T_\Sigma$ and $\Delta \subseteq \Sigma$, $V_\Delta(t)$ denotes the set of nodes $u \in V(t)$ such that $t[u] \in \Delta$.

Translations. Let Σ, Δ be two ranked alphabets. A *tree translation* (*from T_Σ to T_Δ*) is a relation $R \subseteq T_\Sigma \times T_\Delta$. Let A be an alphabet. A *tree-to-string translation* is a relation $R \subseteq T_\Sigma \times A^*$. The *domain* of a translation R, denoted $\mathsf{dom}(R)$, is defined as the projection of R on its first component. A translation R is *functional* if R is a function.

Origin Translations. Let s_1, s_2 be two structures (strings or trees). An *origin mapping* of s_2 in s_1 is a mapping $o : V(s_2) \to V(s_1)$. An *origin translation* is a set of pairs $(s_1, (s_2, o))$ such that o is an origin mapping of s_2 in s_1. Given $v \in V(s_2)$ and $u \in V(s_1)$, if $o(v) = u$ then we say that "v has origin u" or that "the origin of v is u".

2   Tree Translations with Origin

Top-down Transducers. A top-down tree transducer (TOP for short) is a rule-based finite-state machine that translates trees over a ranked alphabet Σ to trees over a ranked alphabet Δ. Rules are of the form $q(\sigma(x_1, \ldots, x_k)) \rightarrow \zeta$, where q is a state, $\sigma \in \Sigma^{(k)}$ a symbol of rank k, and ζ is a tree over Δ of which the leaves may also be labeled with symbols of the form $q'(x_i)$, for some state q' and $i \in [k]$. Applying this rule to a tree $s = \sigma(s_1, \ldots, s_k)$ produces, on the output, a tree t obtained by replacing in ζ all symbols $q'(x_i)$ by a tree over Δ, itself obtained as a result of evaluating s_i in state q'. The origin of all the nodes of ζ labeled in Δ is precisely the root node of s.

As an example, consider a TOP M_1 over $\Sigma = \{h^{(1)}, a^{(0)}\}$ and $\Delta = \{f^{(2)}, a^{(0)}\}$ with single state q and rules $q(h(x_1)) \rightarrow f(q(x_1), q(x_1))$ and $q(a) \rightarrow a$. It translates a monadic input tree of height n into a full binary tree of height n. Thus, the origin of any output node u is the input node $1^{|u|}$. As another example, consider M_2 with states q_0 and q and the rules $q_0(h(x_1)) \rightarrow f(q_0(x_1), q(x_1))$, $q(h(x_1)) \rightarrow h(q(x_1))$ and $q_0(a) \rightarrow a$, $q(a) \rightarrow a$. This transducer translates a monadic input tree of height n into a left-comb of monadic subtrees of decreasing height. Thus, $h(h(h(a)))$ is translated into $f(f(f(a, a), h(a)), h(h(a)))$. Again, the origin of node u is $1^{|u|}$.

Formally, a *top-down tree transducer* M is a tuple $(Q, \Sigma, \Delta, q_0, R)$ where Q is a finite set of states, $q_0 \in Q$ is the initial state, and R is a set of rules of the form $q(\sigma(x_1, \ldots, x_k)) \rightarrow \zeta$, where ζ is a tree over $\Delta \cup \{q'(x_i) \mid q' \in Q, i \in [k]\}$, where each symbol $q'(x_i)$ has rank 0. Every state $q \in Q$ realizes an origin translation $[\![q]\!]_o$ defined recursively as follows. Let $s = \sigma(s_1, \ldots, s_k)$ and ζ such that $q(\sigma(x_1, \ldots, x_k)) \rightarrow \zeta$ is a rule of M. Let $V = V(\zeta) \setminus V_\Delta(\zeta)$. For every $v \in V$, let $(s_{i_v}, (t_v, o_v)) \in [\![q_v]\!]_o$ where $q_v \in Q$ and $i_v \in [k]$ such that $\zeta[v] = q_v(x_{i_v})$. Then $(s, (t, o)) \in [\![q]\!]_o$ with

- $t = \zeta[v \leftarrow t_v \mid v \in V]$,
- $o(v') = \varepsilon$ for $v' \in V_\Delta(\zeta)$ and $o(vv') = i_v o_v(v')$ for $v \in V$ and $v' \in V(t_v)$.

The translation realized by q is defined as $[\![q]\!] = \{(s, t) \mid \exists o : (s, (t, o)) \in [\![q]\!]_o\}$. The *origin (tree) translation realized by* M is $[\![M]\!]_o = [\![q_0]\!]_o$, and the *(tree) translation realized by* M is $[\![M]\!] = [\![q_0]\!]$.

Note that TOPs are forced to produce at least one symbol when they read the leaf of an input tree. To inspect parts of the input tree without producing output, a TOP can be equipped with *regular look-ahead*, leading to the class of top-down tree transducers with regular look-ahead (TOP^R). This is done by changing the rules so that each left-hand side is of the form $q(\sigma(x_1, \ldots, x_k) : L)$ where L is a regular tree language. Such a rule can be applied only if the input tree $\sigma(s_1, \ldots, s_k)$ is in L. Alternatively, instead of L a state p of some given finite tree automaton can be used. A TOP^R M is *deterministic* if for any two rules with left-hand sides $q(\sigma(x_1, \ldots, x_k) : L_1)$ and $q(\sigma(x_1, \ldots, x_k) : L_2)$, we have $L_1 \cap L_2 = \emptyset$. Note that any TOP M can be transformed into a TOP^R M^R by adding universal look-ahead languages. M is deterministic if M^R is. The classes of deterministic top-down tree transducers without and with regular look-ahead are respectively denoted by DTOP and DTOP^R. Note that DTOP and DTOP^R realize only functional translations (and functional origin translations).

MSO Transducers. Deterministic MSO tree transducers (DMSOT for short) are logic-based tree transducers defined over monadic second-order logic (MSO). Any tree s over a ranked alphabet Σ of maximal rank k is seen as a logical structure of domain $V(s)$ over the node label predicates $\sigma(x)$, $\sigma \in \Sigma$, and successor predicates $i(x, y)$, $1 \leq i \leq k$, that relate a node x to its i-th child y. By MSO$[\Sigma]$ we denote all monadic second-order formulas over this signature, and write $s \models \phi$ whenever a tree s satisfies a formula $\phi \in$ MSO$[\Sigma]$. Let Δ be a ranked alphabet of maximal rank ℓ. To define the output tree $t \in T_\Delta$ of an input tree $s \in T_\Sigma$, a DMSOT uses MSO$[\Sigma]$ formulas with one or two free variables, interpreted over a fixed number of copies of s, to define the predicates of t over Δ. Formally, a DMSOT from T_Σ to T_Δ is a tuple $M = (C, \phi_{dom}, (\phi_n^c(x))_{c \in C}, (\phi_\delta^c(x))_{\delta \in \Delta, c \in C}, (\phi_i^{c,c'}(x, y))_{i \in [\ell], c, c' \in C})$ such that C is a finite set of copy indices, ϕ_{dom} is an MSO$[\Sigma]$-sentence which defines whether the input tree s is in the domain, ϕ_n^c, ϕ_δ^c are MSO$[\Sigma]$-formulas with one free variable x which respectively define the nodes $V(t)$ in the output tree t and their labels, and $\phi_i^{c,c'}$ are MSO$[\Sigma]$-formulas with two free variables x and y which define the edge relations between the nodes in $V(t)$.

Given a tree $s \in T_\Sigma$, $[\![M]\!]_o(s)$ is defined if $s \models \phi_{dom}$, and it is then equal to (t, o) where t is the structure whose domain is $D = \{(u, c) \mid s \models \phi_n^c(u)\}$ (each node (u, c) is denoted hereafter by u^c), and for all $u^c \in D$, u^c is labeled by $\delta \in \Delta$ if $s \models \phi_\delta^c(u)$, and a node $u_2^{c_2} \in D$ is the i-th child of a node $u_1^{c_1} \in D$ if $s \models \phi_i^{c_1, c_2}(u_1, u_2)$. The origin mapping o is defined by $o(u^c) = u$ (hence, for any input node u, there are at most $|C|$ nodes in the output with u as origin). Additionally, for M to be an MSO tree transducer, it is required that $t \in T_\Delta$.

For example, let $\Sigma = \{f^{(2)}, a^{(0)}, b^{(0)}\}$ and $\Delta = \{a^{(1)}, b^{(1)}, e^{(0)}\}$. The yield of a tree $s \in T_\Sigma$ is the monadic tree in T_Δ obtained from its leaves, in preorder. E.g., the yield of $f(f(a, b), b)$ is $a(b(b(e)))$. The yield translation is not in DTOP^R but in DMSOT. The preorder relation $\preceq$ on tree nodes obtained from the preorder traversal of the tree is known to be MSO$[\Sigma]$-definable. To realize the yield in DMSOT, we only need one copy ($C = 1$). The domain formula is true and all internal nodes are filtered out by $\phi_n^1(x) = \text{leaf}(x)$, where $\text{leaf}(x)$ holds true if x is a leaf. Labels are unchanged: $\phi_\sigma^1(x) = \sigma(x)$ for all $\sigma \in \Sigma$, and the first-child relation is defined by $\phi_1^{1,1}(x, y)$, which expresses that x and y are leaves, and that $x \preceq y \wedge \neg(\exists z. x \preceq z \wedge z \preceq y)$.

DMSOT can be extended with non-determinism, leading to the class of MSO tree transducers (MSOT). All formulas ϕ_{dom}, ϕ_δ^c and $\phi_i^{c,c'}$ can use a fixed additional finite set of free second-order variables $\overline{X}$. Once an assignment ν of each variable of $\overline{X}$ by a set of nodes of an input tree s is fixed, the previous formulas can be interpreted as before with respect to this assignment, thus defining an output pair (t_ν, o_ν) (if the domain formula holds true). The set of outputs associated with s is the set of all such pairs (t_ν, o_ν), for all assignments ν of $\overline{X}$. In the previous example, using a free variable X, one could also associate all scattered substrings (seen as monadic trees) of the yield. Only leaves in X are kept, by letting $\phi_n^1(x) = x \in X \wedge \text{leaf}(x)$, and $\phi_1^{1,1}(x, y)$ also requires that $x, y \in X$.

Origin-Equivalence Problem. Given two tree transducers M_1, M_2 which are either both in TOP^R or both in MSOT, decide whether $[\![M_1]\!]_o = [\![M_2]\!]_o$.

Theorem 1. *Origin-equivalence is decidable for MSO tree transducers and top-down tree transducers with regular look-ahead.*

Sketch of Proof. Given a tree transducer M, we simply write $\text{dom}(M)$ for $\text{dom}([\![M]\!])$. Inequality of the domains implies inequivalence. As a first step, for both classes of transducers, an equality test of the domains is performed. This is obviously decidable due to the effective regularity of these sets. Then, for TOP^R, by modifying the output alphabet of M_1 and M_2 and their rules, we show how to turn them into non-deleting [1] non-erasing [2] TOP without look-ahead, while preserving origin-equivalence. Let us notice then that the constraint that origins should be the same is strong: for all input trees $s \in \text{dom}(M_1)$, for all $(t, o) \in [\![M_1]\!]_o(s)$, M_2 must produce, in a successful execution, the symbols of t exactly at the same moment as M_1, and conversely for all $(t, o) \in [\![M_2]\!]_o(s)$. When considering non-erasing TOP, this property has a nice consequence: both M_1 and M_2 can be seen as symbol-to-symbol transducers, which means that each right-hand side of a rule contains exactly one node with a label from Δ. To be precise, the Δ-part of any right-hand side ζ of a rule of M_1 or M_2, with n leaves labeled by some $q(x_i)$, can be seen as a single symbol of rank n. For instance, if $\zeta = f(q_1(x_1), h(q_2(x_1)), q_3(x_2))$, then $f(., h(.), .)$ is seen as a single symbol of arity 3. M_1' and M_2', two non-deleting non-erasing TOP without look-ahead, can be built from M_1 and M_2 respectively such that $[\![M_1]\!]_o = [\![M_2]\!]_o$ iff $[\![M_1']\!] = [\![M_2']\!]$. Finally, it leads to solve an equivalence problem for (nondeterministic) non-deleting symbol-to-symbol top-down tree transducers which is known to be decidable [1].

For MSOT, we show that the equality set $E([\![M_1]\!]_o, [\![M_2]\!]_o)$ of $[\![M_1]\!]_o$ and $[\![M_2]\!]_o$, defined as the set of trees $s \in \text{dom}(M_1)$ such that $[\![M_1]\!]_o(s) = [\![M_2]\!]_o(s)$, is effectively regular. Without origins, even simple MSOT translations yield a non-regular set: *e.g.*, the translations $R_1 : f(s_1, s_2) \mapsto s_1$ and $R_2 : f(s_1, s_2) \mapsto s_2$ are both MSOT definable but their (origin-free) equality set $\{f(s, s) \mid s \in T_\Sigma\}$, is not regular. To show that the set $E([\![M_1]\!]_o, [\![M_2]\!]_o)$ is regular, we construct an MSO formula that defines it. This formula expresses for instance that any

---

[1] Non-deleting means that every x_i occurring in the left-hand side of a rule also occurs in its right-hand side.

[2] Non-erasing means that the right-hand side of each rule contains at least one symbol from Δ.

sequence of input nodes $u_1, \ldots, u_n$ that are connected with successor formulas $\phi_i^{c,c'}(x,y)$ by M_1 are also connected with successor formulas of M_2 with the same sequence of indices i, and conversely. It also expresses that the sequence of label formulas $\phi_\delta^c(x)$ of M_1 and M_2 that hold on $u_1, \ldots, u_n$ respectively, carry the same respective output symbols δ. As a consequence, to check origin-equivalence of M_1 and M_2, it suffices to check that $\mathsf{dom}(M_1) = \mathsf{dom}(M_2) = E(\llbracket M_1 \rrbracket_\mathrm{o}, \llbracket M_2 \rrbracket_\mathrm{o})$, which is decidable since all these sets are effectively regular. □

Origin-Injectivity Problem. Given a tree transducer M from T_Σ to T_Δ either in DMSOT or in DTOP$^\mathrm{R}$, decide whether the function $\llbracket M \rrbracket_\mathrm{o}$ is injective.

Theorem 2. *Origin-injectivity is decidable for deterministic MSO tree transducers and deterministic top-down tree transducers with regular look-ahead.*

Sketch of Proof. Let us denote by $R(\llbracket M \rrbracket_\mathrm{o})$ the set of pairs of trees $(s_1, s_2) \in \mathsf{dom}(M)^2$ such that $\llbracket M \rrbracket_\mathrm{o}(s_1) = \llbracket M \rrbracket_\mathrm{o}(s_2)$. Clearly, $\llbracket M \rrbracket_\mathrm{o}$ is injective iff $R(\llbracket M \rrbracket_\mathrm{o}) \cap (\neq_{T_\Sigma}) = \varnothing$, where $\neq_{T_\Sigma}$ is the difference relation over T_Σ. Take a pair $(s_1, s_2) \in R(\llbracket M \rrbracket_\mathrm{o})$. We can define its top- and left-most overlap $s_1 \otimes s_2$ that aligns the same nodes of s_1 and s_2 (recall that a node is a string over $\mathbb{N}$). Nodes in $V(s_1) \cap V(s_2)$ are labeled, in $s_1 \otimes s_2$, by the pair of their respective labels in s_1 and s_2. Nodes in $V(s_1) \backslash V(s_2)$ or $V(s_2) \backslash V(s_1)$ are labeled by pairs $(\sigma, \bot)$ or $(\bot, \sigma)$, for a special padding symbol $\bot$. Interestingly, $\llbracket M \rrbracket_\mathrm{o}(s_1) = \llbracket M \rrbracket_\mathrm{o}(s_2)$ if M produces the same output symbols when processing a node in $V(s_1) \cap V(s_2)$, and does not produce anything when processing a node in $V(s_2) \backslash V(s_1) \cup V(s_1) \backslash V(s_2)$.

When $M \in$ DTOP$^\mathrm{R}$, this last observation allows us to construct a DTOP$^\mathrm{R}$ M' reading trees $s_1 \otimes s_2 \in T_\Sigma \otimes T_\Sigma$, that simulates in parallel two executions of M on s_1 and s_2 respectively, and checks that M produces the same symbols at the same moment, for the common nodes of s_1 and s_2, and nothing elsewhere. Then, $\mathsf{dom}(M')$ equals the set of trees $s_1 \otimes s_2$ such that $(s_1, s_2) \in R(\llbracket M \rrbracket_\mathrm{o})$ and is regular, as DTOP$^\mathrm{R}$ have regular domains. In other words, the relation $R(\llbracket M \rrbracket_\mathrm{o})$ is recognizable [5]. It is easily shown that $\neq_{T_\Sigma}$ is, as well, recognizable. Since recognizable relations are closed under intersection, one gets decidability of injectivity for origin-translations of DTOP$^\mathrm{R}$.

When $M \in$ DMSOT, $R(\llbracket M \rrbracket_\mathrm{o})$ is also a recognizable relation. To prove that, we first transform M into two transducers $M_1, M_2 \in$ DMSOT that run on trees in $T_\Sigma \otimes T_\Sigma$. While processing trees $s_1 \otimes s_2$, M_i simulates M on s_i, so that $\llbracket M_i \rrbracket_\mathrm{o}(s_1 \otimes s_2) = \llbracket M \rrbracket_\mathrm{o}(s_i)$. Then, $\{s_1 \otimes s_2 \mid (s_1, s_2) \in R(\llbracket M \rrbracket_\mathrm{o})\} = E(\llbracket M_1 \rrbracket_\mathrm{o}, \llbracket M_2 \rrbracket_\mathrm{o})$, and the result follows since $E(\llbracket M_1 \rrbracket_\mathrm{o}, \llbracket M_2 \rrbracket_\mathrm{o})$ is regular for $M_1, M_2 \in$ DMSOT. □

Application to Query Determinacy. Let Q (resp. V) be a functional tree translation (resp. origin tree translation) from T_Σ to T_Δ, called the query (resp. the view). We say that Q is *determined* by V if for all trees $s_1, s_2 \in \mathsf{dom}(V)$, if $V(s_1) = V(s_2)$ then $Q(s_1) = Q(s_2)$. This generalizes the injectivity problem: the identity tree translation is determined by a view V iff V is injective.

Corollary 3. *Let Q (resp. V) be a tree translation (resp. an origin tree translation) defined by either a DTOP^R or a DMSOT. It is decidable whether Q is determined by V.*

Proof. Let Q_1, Q_2 be tree translations from $T_\Sigma \otimes T_\Sigma$ to T_Δ defined by $Q_i(s_1 \otimes s_2) = Q(s_i)$, for all $s_1, s_2 \in T_\Sigma$. Note that the Q_i are definable by DTOP^R (resp. DMSOT) if Q is. Let $r(V)$ be the set of trees $s_1 \otimes s_2$ such that $s_1, s_2 \in \text{dom}(V)$ and $V(s_1) = V(s_2)$. Q is determined by V iff $Q_1(s) = Q_2(s)$ for all s in $r(V)$, iff Q_1 and Q_2 are equivalent on $r(V)$. As seen before to solve the injectivity problem, we show that the pairs (s_1, s_2) such that $V(s_1) = V(s_2)$ is a recognizable relation, for transducers in DTOP^R or DMSOT. So, $r(V)$ is regular. The result follows as equivalence of DTOP^R or DMSOT is decidable on regular languages [20].   □

3   Tree-to-String Translations with Origin

A *top-down tree-to-string transducer* (yTOP for short) M is a tuple $(Q, \Sigma, \Delta, q_0, R)$ where Q, q_0, Σ are defined as for top-down tree transducers, Δ is an alphabet, and every rule is of the form $q(\sigma(x_1, \ldots, x_k)) \to \zeta$, where ζ is a *string* over Δ and the symbols $q'(x_i)$, with $q' \in Q$ and $i \in [k]$. The definition of $[\![q]\!]_o$ is as for top-down tree transducers, only that t and t_v are strings over Δ. In this way we obtain $[\![M]\!]_o$ and $[\![M]\!]$. yTOP generalize TOP, as right-hand sides of rules of TOP can be encoded as well-bracketed strings. The converse is false: even considering strings as monadic trees, a yTOP can for instance easily implement string reversal, which is impossible using TOP. As for TOP, we can equip this model with regular look-ahead, and consider deterministic machines. This defines the classes of deterministic top-down tree-to-string transducers (with regular look-ahead): yDTOP (yDTOP^R).

We give a yDTOP M implementing string reversal. It takes as input monadic trees over the alphabet $\Sigma = \{a^{(1)}, b^{(1)}, e^{(0)}\}$ and produces output strings over the alphabet $\Delta = \{a, b\}$. It has states q, q_a, q_b and rules $q(\sigma(x_1)) \to q(x_1)q_\sigma(x_1)$ and $q_\sigma(\sigma'(x_1)) \to q_\sigma(x_1)$ for $\sigma, \sigma' \in \Sigma^{(1)}$, and leaf rules $q(e) \to \varepsilon$ and $q_\sigma(e) \to \sigma$ for $\sigma \in \Sigma^{(1)}$. Clearly, for $s = a(a(b(e)))$ $[\![M]\!](s) = baa = w$. Note that the origin of each letter of w is the leaf of s (here, 111). Hence, the origin translation $[\![M]\!]_o$ is *not* MSO definable: there may be unboundedly many letters having such a leaf as origin (by contrast, the translation $[\![M]\!]$ is MSO definable, as tree-to-string MSO transducers are equivalent to yDTOP^R of linear size increase [8,9]). In Sec. 4 we show how to decide whether the origin translation of a yDTOP^R is MSO definable.

Undecidability Results.   By a construction similar to the above string reversal example, any string-to-string rational relation can be shown to be implementable as a yTOP such that the origin of every output symbol is the unique leaf of the input monadic tree. This "erasing" of origin info shows that origin-equivalence of yTOP is harder than equivalence of string-to-string rational relations known to be undecidable [14]. A similar technique can be used to show that origin-injectivity of yDTOP is also undecidable, by an encoding of the Post Correspondence Problem. This contrasts with the positive results presented in Sec. 2. There, origin-equivalence of MSOT relied on the regularity of the set $E([\![M_1]\!]_o, [\![M_2]\!]_o)$, and one can easily come up with two yDTOP M_1, M_2 such that this set is not regular. E.g., take the transducer $M_1 = M$ for string reversal of before, and M_2 the identity with rules $q(\sigma(x_1)) \to q_\sigma(x_1)q(x_1)$

and $q_\sigma(\sigma'(x_1)) \rightarrow q_\sigma(x_1)$ for $\sigma, \sigma' \in \Sigma^{(1)}$, and leaf rules as for M. Now $E(\llbracket M_1 \rrbracket_\mathrm{o}, \llbracket M_2 \rrbracket_\mathrm{o})$ is the set of palindromes on Δ^* (seen as monadic trees), which is not regular.

Equivalence of ydtop$^\mathrm{R}$. Though it is not possible to obtain decidability results by regular sets (or recognizable relations), we manage to prove, using more involved techniques, that origin-equivalence of yDTOP$^\mathrm{R}$ is decidable.

Theorem 4. *Origin-equivalence is decidable for deterministic top-down tree-to-string transducers with regular look-ahead.*

Sketch of Proof. Let M_1, M_2 be two yDTOP$^\mathrm{R}$. We first check whether M_1 and M_2 have the same domain D. If not we output "not equivalent". Otherwise, we build yDTOP transducers *without* look-ahead M_1', M_2', and a regular tree language D' such that $\llbracket M_1 \rrbracket_\mathrm{o} = \llbracket M_2 \rrbracket_\mathrm{o}$ iff M_1' and M_2' are origin-equivalent on D'.

Secondly, we transform the two yDTOP into end-marked leaf-producing yDTOP such that origin-equivalence is preserved: the leaf-producing property requires that transducers produce only at the leaves. The end-marked property means that every output string has a final end-marker. Both properties are obtained by modifying the input alphabet and the rules. For the last one, only the initial rules of the transducers are concerned. We still denote them $(M_i')_{i=1,2}$.

Last, we reduce the origin-equivalence problem of these yDTOP to the equivalence problem of monadic yDTOP, where 'monadic' means that every input symbol has rank 0 or 1. This is done by only considering partial output strings produced on root-to-leaf paths of the input tree. We give now some details on this last part. Let s be a tree, $w = a_1 \ldots a_n \in \Delta^*$ a string with $a_i \in \Delta$ for all i, and $o : V(w) \rightarrow V(s)$ an origin mapping. Let $U = \{u_1, \ldots, u_k\} \subseteq V(s)$ be a set of (distinct) nodes (in this order). We define $\Pi_{u_1, \ldots, u_k}(w, o)$ as the string in $(\Delta \times [k])^*$ obtained from w by erasing a_i if $o(i) \notin \{u_1, \ldots, u_k\}$, and changing a_i into (a_i, j) if $o(i) = u_j$. We give a key result which allows one to reduce our origin-equivalence problem of yDTOP to two instances of the equivalence problem of monadic yDTOP: for $s \in \mathrm{dom}(M_1') \cap \mathrm{dom}(M_2')$, $M_1'(s) = M_2'(s)$ iff the following two conditions are satisfied: (1) $\Pi_u(M_1'(s)) = \Pi_u(M_2'(s))$ for every $u \in V_{\Sigma^{(0)}}(s)$, (2) $\Pi_{u_1, u_2}(M_1'(s)) = \Pi_{u_1, u_2}(M_2'(s))$ for every $u_1 \neq u_2 \in V_{\Sigma^{(0)}}(s)$.

4 Subclass Definability Problems

Deterministic MSO tree-to-string transducers (DMSOTS for short) can be defined as a particular case of DMSOT transducers (their origin-equivalence is decidable by Theorem 1). While DMSOTS are equivalent to yDTOP$^\mathrm{R}$ of linear size increase [8, 9], this is not true in the presence of origin; there are such yDTOP$^\mathrm{R}$ for which no origin-equivalent DMSOTS exists (*e.g.* the string reversal example of Sec. 3). However, every DMSOTS effectively has an origin-equivalent yDTOP$^\mathrm{R}$ (obtained by following the respective constructions for origin-less transducers). Can we decide for a given yDTOP$^\mathrm{R}$ whether its origin translation is DMSOTS definable?

Theorem 5. *For a given* yDTOP^R *M, it is decidable whether or not there exists an origin-equivalent* DMSOTS. *If so, then such a* DMSOTS *can be constructed.*

Sketch of Proof. It relies on the notion of *bounded origin*: an origin translation τ is of bounded origin if there exists a number k such that for every $(s, (w, o)) \in \tau$ and $u \in V(s)$: $|\{v \in V(w) \mid o(v) = u\}| \leq k$, ie every input node can be the origin of only a bounded number of output positions. By their definition, origin translations of MSO transducers have bounded origin. The bounded origin property can be decided for a yDTOP^R M: we transform M into a yDTOP^R transducer M' that takes input trees of M, but with one node u marked. The transducer M' produces output only on the marked node. Thus, the length of its output equals the number of positions that have u as origin. Decidability follows from that of finiteness of ranges [6]. The proof then builds an origin-equivalent DMSOTS following the constructions in the literature.   □

Macro Tree Transducers. At last we consider a more powerful type of transducer: the macro tree transducer (MAC). For simplicity, we only look at total deterministic such transducers. A MAC extends a top-down tree transducer by *nesting* of recursive state calls. Thus, a state q is now of rank $m + 1$ and takes, besides the input tree, m arguments of type output trees. In the rules, these arguments are denoted by *parameters* $y_1, \ldots, y_m$. Thus, a rule is of the form $q(\sigma(x_1, \ldots, x_k), y_1, \ldots, y_m) \to \zeta$, where ζ is a tree over (nested) states, output symbols, and the parameters which may occur at leaves. As example, consider a MAC with initial rule $q_0(h(x_1)) \to q(x_1, a)$ and these rules: $q(h(x_1), y_1) \to q(x_1, q(x_1, y_1))$ and $q(a, y_1) \to b(y_1, y_1)$. For a monadic input tree $h(\ldots h(a) \ldots)$ of height $n + 1$, it produces a full binary tree of height 2^n. Thus, MACs can have *double-exponential* size increase; all models discussed so far have at most exponential size increase. A *total deterministic macro tree transducer* (MAC) is a tuple $M = (Q, \Sigma, \Delta, q_0, R)$ where Σ, Δ are as before, Q is a ranked alphabet with $Q^{(0)} = \emptyset$, $q_0 \in Q^{(1)}$, and R contains for every $q \in Q^{(m+1)}$, $m \geq 0$, $\sigma \in \Sigma^{(k)}$, and $k \geq 0$, a rule $q(\sigma(x_1, \ldots, x_k), y_1, \ldots, y_m) \to \zeta$, where ζ is a tree over $Q \cup \Delta \cup \{x_1, \ldots, x_k, y_1, \ldots, y_m\}$ such that x_i occurs in ζ at a node u if and only if u is the first child of a Q-labeled node (and symbols y_j are of rank zero). We denote ζ by $\mathsf{rhs}(q, \sigma)$. Every state $q \in Q^{(m+1)}$ of M induces a function $\llbracket q \rrbracket : T_\Sigma \times T_\Delta^m \to T_\Delta$. Let $s = \sigma(s_1, \ldots, s_k) \in T_\Sigma$ and $t_1, \ldots, t_m \in T_\Delta$. Then $\llbracket q \rrbracket(s, t_1, \ldots, t_m) = \llbracket \zeta \rrbracket$ where $\zeta = \mathsf{rhs}(q, \sigma)$ and $\llbracket \zeta \rrbracket$ is defined recursively as follows. If $\zeta = y_j$ then $\llbracket \zeta \rrbracket = t_j$. If $\zeta = d(\zeta_1, \ldots, \zeta_\ell)$ with $d \in \Delta^{(\ell)}$, then $\llbracket \zeta \rrbracket = d(\llbracket \zeta_1 \rrbracket, \ldots, \llbracket \zeta_\ell \rrbracket)$. If $\zeta = q'(x_i, \zeta_1, \ldots, \zeta_\ell)$ with $q' \in Q^{(\ell+1)}$ and $i \in [k]$, then $\llbracket \zeta \rrbracket = \llbracket q' \rrbracket(s_i, \llbracket \zeta_1 \rrbracket, \ldots, \llbracket \zeta_\ell \rrbracket)$.

Origin Semantics. We define the origin semantics of M using the MAC M^s and the *decorated version* $\mathsf{dec}(s)$ of an input tree s (see Definition 4.15 of [9]). Let s be an input tree of M. Then $\mathsf{dec}(s)$ is obtained from s by relabeling every node u by $\langle s[u], u \rangle$. For a state q and input symbol $\langle \sigma, u \rangle$ the MAC M^s applies the (q, σ)-rule of M, but with every output symbol d replaced by $\langle d, u \rangle$. The origin of an output node then simply is the second component of the label of that node. Intuitively, when a MAC applies a rule at input node u and generates output inside of parameter positions, then all these outputs have origin u. Note that such nodes may be duplicated later and appear unboundedly often (at arbitrary positions of the output tree). Let us see an example of an origin translation that cannot be defined by the previous models (but for which the non-origin translation can be defined): in fact we consider the identity on trees s over $\{f^{(2)}, a^{(0)}\}$. The MAC M has these rules for q_0: $q_0(a) \rightarrow a$ and $q_0(f(x_1, x_2)) \rightarrow f(q(x_1, a), q(x_2, a))$, and these rules for q: $q(f(x_1, x_2), y_1) \rightarrow f(q(x_1, y_1), q(x_2, y_1))$ and $q(a, y_1) \rightarrow y_1$. Thus, $[\![M]\!]_o(s) = (s, o)$ where $o(u) = u$ if u is an internal node, and $o(u) = \varepsilon$ if u is a leaf. Thus, all leaves have the root node as origin. Clearly, none of our previous models can realize such an origin translation.

Deciding whether the translation of a MAC can be defined by a DTOP^R is a difficult open problem: a MAC can use its parameters in complex ways, but still be definable by a DTOP^R. However, with origin, we are able to prove decidability.

Theorem 6. *For a given MAC M, it is decidable whether or not there exists an origin-equivalent DTOP^R. If so, then such a DTOP^R can be constructed.*

Sketch of Proof. First, if τ is the origin translation of a DTOP^R, and v, v' are nodes in $\tau(s)$ for some tree s such that v' is a descendant of v, then the origin of v' must be a descendant of the origin of v (see [17]). We call this property of an origin translation *order-preserving*. Consider now a MAC with order-preserving origin translation. Is it definable by a DTOP^R? To see this is not true, consider this MAC for the identity: $q_0(a) \rightarrow a$ and $q_0(f(x_1, x_2)) \rightarrow q(x_1, q_0(x_1), q_0(x_2))$, plus the rules $q(f(x_1, x_2), y_1, y_2) \rightarrow q(x_1, y_1, y_2)$ and $q(a, y_1, y_2) \rightarrow f(y_1, y_2)$. For the input tree $f(f(\cdots f(a, a) \cdots), a)$ that is a left-comb, the origin of each f-node is its left-most descendant leaf. Not the order is the problem, but, there are too many *connected* output nodes with the same origin. Intuitively, the connected output nodes of a DTOP^R can only span the size of a right-hand side. We say that an origin translation τ is *path-wise bounded-origin* if there exists a number k such that are at most k output nodes with the same origin on each path of the output tree. Both the order-preserving and path-wise bounded-origin properties can decided. For a MAC with these two properties, and nondeleting and nonerasing in its parameters (which can both be obtained by regular look-ahead), the depth of nested state calls on the same input node is bounded. An origin-equivalent DTOP^R can be constructed by introducing one state for each of these finitely many different nestings of state calls. □

Conclusions and Future Work. We have shown that several important decision problems for tree transducers become decidable in the presence of origin information. Some problems remain open, such as the decidability of equivalence of MACs with origin. In the future we would like to study other notions of origin such as unique identifiers instead of Dewey nodes, or, sets of nodes (one of which is guaranteed to be origin) instead of one node. A lot of work remains to be done on determinacy; for instance, we would like to show that a query determined by a view with origin can be rewritten into a tractable class of queries.

References

1. Andre, Y., Bossut, F.: On the equivalence problem for letter-to-letter top-down tree transducers. Theor. Comput. Sci. **205**(1–2), 207–229 (1998)
2. Benedikt, M., Engelfriet, J., Maneth, S.: Determinacy and rewriting of top-down and MSO tree transformations. In: Chatterjee, K., Sgall, J. (eds.) MFCS 2013. LNCS, vol. 8087, pp. 146–158. Springer, Heidelberg (2013)
3. Bojańczyk, M.: Transducers with origin information. In: Esparza, J., Fraigniaud, P., Husfeldt, T., Koutsoupias, E. (eds.) ICALP 2014, Part II. LNCS, vol. 8573, pp. 26–37. Springer, Heidelberg (2014)
4. Braune, F., Seemann, N., Quernheim, D., Maletti, A.: Shallow local multi-bottom-up tree transducers in statistical machine translation. In: ACL, pp. 811–821 (2013)
5. Comon, H., Dauchet, M., Gilleron, R., Jacquemard, F., Lugiez, D., Löding, C., Tison, S., Tommasi, M.: Tree automata techniques and applications (2007)
6. Drewes, F., Engelfriet, J.: Decidability of the finiteness of ranges of tree transductions. Inf. Comput. **145**(1), 1–50 (1998)
7. Engelfriet, J.: Some open questions and recent results on tree transducers and tree languages. In: Book, R. (ed.) Formal language theory; perspectives and open problems. Academic Press, New York (1980)
8. Engelfriet, J., Maneth, S.: Macro tree transducers, attribute grammars, and MSO definable tree translations. Inf. Comput. **154**(1), 34–91 (1999)
9. Engelfriet, J., Maneth, S.: Macro tree translations of linear size increase are MSO definable. SIAM J. Comput. **32**(4), 950–1006 (2003)
10. Engelfriet, J., Maneth, S.: The equivalence problem for deterministic MSO tree transducers is decidable. Inf. Process. Lett. **100**(5), 206–212 (2006)
11. Engelfriet, J., Rozenberg, G., Slutzki, G.: Tree transducers, L systems, and two-way machines. J. Comput. Syst. Sci. **20**(2), 150–202 (1980)
12. Ésik, Z.: Decidability results concerning tree transducers I. Acta Cybern. **5**(1), 1–20 (1980)
13. Fülöp, Z., Vogler, H.: Syntax-Directed Semantics - Formal Models Based on Tree Transducers. Springer, Monographs in Theoretical Computer Science. An EATCS Series (1998)
14. Griffiths, T.V.: The unsolvability of the equivalence problem for lambda-free non-deterministic generalized machines. J. ACM **15**(3), 409–413 (1968)
15. Hakuta, S., Maneth, S., Nakano, K., Iwasaki, H.: Xquery streaming by forest transducers. In: ICDE, pp. 952–963 (2014)
16. Küsters, R., Wilke, T.: Transducer-based analysis of cryptographic protocols. Inf. Comput. **205**(12), 1741–1776 (2007)
17. A. Lemay, S. Maneth, and J. Niehren. A learning algorithm for top-down XML transformations. In: PODS, pp. 285–296 (2010)
18. Maletti, A.: Tree transformations and dependencies. In: MOL, pp. 1–20 (2011)
19. Maletti, A., Graehl, J., Hopkins, M., Knight, K.: The power of extended top-down tree transducers. SIAM J. Comput. **39**(2), 410–430 (2009)
20. Maneth, S.: Equivalence problems for tree transducers (survey). In: AFL, pp. 74–93 (2014)
21. Matsuda, K., Inaba, K., Nakano, K.: Polynomial-time inverse computation for accumulative functions with multiple data traversals. In: PEPM, pp. 5–14 (2012)
22. Milo, T., Suciu, D., Vianu, V.: Typechecking for XML transformers. J. Comput. Syst. Sci. **66**(1), 66–97 (2003)

23. Rounds, W.C.: Mappings and grammars on trees. Math. Syst. Th. **4**(3), 257–287 (1970)
24. Thatcher, J.W.: Generalized sequential machine maps. JCSS **4**(4), 339–367 (1970)
25. van Deursen, A., Klint, P., Tip, F.: Origin tracking. J. Symb. Comput. **15**(5/6), 523–545 (1993)
26. Voigtländer, J., Hu, Z., Matsuda, K., Wang, M.: Enhancing semantic bidirectionalization via shape bidirectionalizer plug-ins. J. Funct. Program. **23**(5), 515–551 (2013)
27. Voigtländer, J., Kühnemann, A.: Composition of functions with accumulating parameters. J. Funct. Program. **14**(3), 317–363 (2004)

Incompleteness Theorems, Large Cardinals, and Automata over Infinite Words

Olivier Finkel[⊠]

Institut de Mathématiques de Jussieu - Paris Rive Gauche,
CNRS et Université Paris 7, Paris, France
Olivier.Finkel@math.univ-paris-diderot.fr

Abstract. We prove that there exist some 1-counter Büchi automata $\mathcal{A}_n$ for which some elementary properties are independent of theories like $T_n =:$ **ZFC** + "There exist (at least) n inaccessible cardinals", for integers $n \geq 1$. In particular, if T_n is consistent, then "$L(\mathcal{A}_n)$ is Borel", "$L(\mathcal{A}_n)$ is arithmetical", "$L(\mathcal{A}_n)$ is ω-regular", "$L(\mathcal{A}_n)$ is deterministic", and "$L(\mathcal{A}_n)$ is unambiguous" are provable from **ZFC** + "There exist (at least) $n + 1$ inaccessible cardinals" but not from **ZFC** + "There exist (at least) n inaccessible cardinals". We prove similar results for infinitary rational relations accepted by 2-tape Büchi automata.

Keywords: Automata and formal languages · Logic in computer science · Infinite words · 1-counter Büchi automaton · 2-tape Büchi automaton · Models of set theory · Incompleteness theorems · Large cardinals · Inaccessible cardinals · Independence from the axiomatic system "**ZFC** + there exist n inaccessible cardinals"

1 Introduction

The theory of automata reading infinite words, which is closely related to infinite games, is now a rich theory which is used for the specification and verification of non-terminating systems, see [GTW02,PP04].

As noticed in [Fin11], some connections between Automata Theory and Set Theory had arosen in the study of monadic theories of well orders, but this was related to automata reading much longer transfinite words than words of length ω or even than words of length a countable ordinal.

Then one usually thought that the finite or infinite computations appearing in Computer Science are "well defined" in the axiomatic framework of mathematics, and thus that a property on automata is either true or false and that one has not to take care of the different models of Set Theory (except perhaps for the Continuum Hypothesis **CH** which is known to be independent from **ZFC**). And the connections between Automata Theory and Set Theory seemed very far from the practical aspects of Computer Science.

In [Fin09] we recently proved a surprising result: the topological complexity of an ω-language accepted by a 1-counter Büchi automaton, or of an infinitary rational relation accepted by a 2-tape Büchi automaton, is not determined by the axiomatic system **ZFC**. In particular, there is a 1-counter Büchi automaton $\mathcal{A}$ (respectively, a 2-tape Büchi automaton $\mathcal{B}$) and two models $\mathbf{V}_1$ and $\mathbf{V}_2$ of **ZFC** such that the ω-language

© Springer-Verlag Berlin Heidelberg 2015
M.M. Halldórsson et al. (Eds.): ICALP 2015, Part II, LNCS 9135, pp. 222–233, 2015.
DOI: 10.1007/978-3-662-47666-6_18

$L(\mathcal{A})$ (respectively, the infinitary rational relation $L(\mathcal{B})$) is Borel in $\mathbf{V}_1$ but not in $\mathbf{V}_2$. We have proved in [Fin11] other independence results, showing that some basic cardinality questions on automata reading infinite words actually depend on the models of **ZFC** (see also [Fin10] for similar results for Büchi-recognizable languages of infinite pictures).

The next step in this research project was to determine which properties of automata actually depend on the models of **ZFC**, and to achieve a more complete investigation of these properties.

We obtain in this paper some more independence results which are more general and are related to the consistency of theories which are recursive extensions of the theory **ZFC** (while in the two papers [Fin09, Fin11] the independence results depended on the value of the ordinal $\omega_1^{\mathbf{L}}$ which plays the role of the first uncountable ordinal in the constructible universe **L**).

Recall that a large cardinal in a model of set theory is a cardinal which is in some sense much larger than the smaller ones. This may be seen as a generalization of the fact that ω is much larger than all *finite* cardinals. The inaccessible cardinals are the simplest such large cardinals. Notice that it cannot be proved in **ZFC** that there exists an inaccessible cardinal, but one usually believes that the existence of such cardinals is consistent with the axiomatic theory **ZFC**. The assumed existence of large cardinals have many consequences in Set Theory as well as in many other branches of Mathematics like Algebra, Topology or Analysis, see [Jec02].

We prove that there exist some 1-counter Büchi automata $\mathcal{A}_n$ for which some elementary properties are independent of theories like $T_n =:$ **ZFC** + "There exist (at least) n inaccessible cardinals", for integers $n \geq 1$. We first prove that "$L(\mathcal{A}_n)$ is Borel", "$L(\mathcal{A}_n)$ is arithmetical", "$L(\mathcal{A}_n)$ is ω-regular", "$L(\mathcal{A}_n)$ is deterministic", and "$L(\mathcal{A}_n)$ is unambiguous" are equivalent to the consistency of the theory T_n. This implies that, if T_n is consistent, all these statements are provable from **ZFC** + "There exist (at least) $n + 1$ inaccessible cardinals" but not from **ZFC** + "There exist (at least) n inaccessible cardinals". We prove similar results for infinitary rational relations accepted by 2-tape Büchi automata. Notice that the same reults can be proved for other large cardinals like hyperinaccessible or Mahlo cardinals, see [Jec02] for a precise definition of these cardinals.

The paper is organized as follows. We recall the notion of counter automata in Section 2. We expose some results of Set Theory in Section 3, and we prove our main results about 1-counter ω-languages in Section 4. We prove similar results for infinitary rational relations in Section 5. Concluding remarks are given in Section 6.

2   Counter Automata

We assume the reader to be familiar with the theory of formal (ω-)languages [Tho90, Sta97]. We recall the usual notations of formal language theory.

If Σ is a finite alphabet, a *non-empty finite word* over Σ is any sequence $x = a_1 \ldots a_k$, where $a_i \in \Sigma$ for $i = 1, \ldots, k$, and k is an integer ≥ 1. The *length* of x is k, denoted by $|x|$. The *empty word* has no letter and is denoted by λ; its length is 0. $\Sigma^\star$ is the *set of finite words* (including the empty word) over Σ.

The *first infinite ordinal* is ω. An ω-*word* over Σ is an ω-sequence $a_1 \ldots a_n \ldots$, where for all integers $i \geq 1$, $a_i \in \Sigma$. When $\sigma = a_1 \ldots a_n \ldots$ is an ω-word over Σ, we write $\sigma(n) = a_n$, $\sigma[n] = \sigma(1)\sigma(2) \ldots \sigma(n)$ for all $n \geq 1$ and $\sigma[0] = \lambda$.

The usual concatenation product of two finite words u and v is denoted $u.v$ (and sometimes just uv). This product is extended to the product of a finite word u and an ω-word v: the infinite word $u.v$ is then the ω-word such that:

$(u.v)(k) = u(k)$ if $k \leq |u|$, and $(u.v)(k) = v(k - |u|)$ if $k > |u|$.

The *set of ω-words* over the alphabet Σ is denoted by Σ^ω. An ω-*language* V over an alphabet Σ is a subset of Σ^ω, and its complement (in Σ^ω) is $\Sigma^\omega - V$, denoted V^-.

Let k be an integer ≥ 1. A k-counter machine has k *counters*, each of which containing a non-negative integer. The machine can test whether the content of a given counter is zero or not. And transitions depend on the letter read by the machine, the current state of the finite control, and the tests about the values of the counters. Notice that in this model some λ-transitions are allowed.

Formally a k-counter machine is a 4-tuple $\mathcal{M}=(K, \Sigma, \Delta, q_0)$, where K is a finite set of states, Σ is a finite input alphabet, $q_0 \in K$ is the initial state, and $\Delta \subseteq K \times (\Sigma \cup \{\lambda\}) \times \{0, 1\}^k \times K \times \{0, 1, -1\}^k$ is the transition relation. The k-counter machine $\mathcal{M}$ is said to be *real time* iff: $\Delta \subseteq K \times \Sigma \times \{0, 1\}^k \times K \times \{0, 1, -1\}^k$, i.e. iff there are no λ-transitions.

If the machine $\mathcal{M}$ is in state q and $c_i \in \mathbf{N}$ is the content of the ith counter $\mathcal{C}_i$ then the configuration (or global state) of $\mathcal{M}$ is the $(k + 1)$-tuple $(q, c_1, \ldots, c_k)$.

For $a \in \Sigma \cup \{\lambda\}$, $q, q' \in K$ and $(c_1, \ldots, c_k) \in \mathbf{N}^k$ such that $c_j = 0$ for $j \in E \subseteq \{1, \ldots, k\}$ and $c_j > 0$ for $j \notin E$, if $(q, a, i_1, \ldots, i_k, q', j_1, \ldots, j_k) \in \Delta$ where $i_j = 0$ for $j \in E$ and $i_j = 1$ for $j \notin E$, then we write:

$$a : (q, c_1, \ldots, c_k) \mapsto_\mathcal{M} (q', c_1 + j_1, \ldots, c_k + j_k).$$

Thus the transition relation must obviously satisfy:

if $(q, a, i_1, \ldots, i_k, q', j_1, \ldots, j_k) \in \Delta$ and $i_m = 0$ for some $m \in \{1, \ldots, k\}$ then $j_m = 0$ or $j_m = 1$ (but j_m may not be equal to -1).

Let $\sigma = a_1 a_2 \ldots a_n \ldots$ be an ω-word over Σ. An ω-sequence of configurations $r = (q_i, c_1^i, \ldots c_k^i)_{i \geq 1}$ is called a run of $\mathcal{M}$ on σ, iff:

(1) $(q_1, c_1^1, \ldots c_k^1) = (q_0, 0, \ldots, 0)$

(2) for each $i \geq 1$, there exists $b_i \in \Sigma \cup \{\lambda\}$ such that $b_i : (q_i, c_1^i, \ldots c_k^i) \mapsto_\mathcal{M}$ $(q_{i+1}, c_1^{i+1}, \ldots c_k^{i+1})$ and such that $a_1 a_2 \ldots a_n \ldots = b_1 b_2 \ldots b_n \ldots$

For every such run r, $\mathrm{In}(r)$ is the set of all states entered infinitely often during r.

Definition 1. *A Büchi k-counter automaton is a 5-tuple $\mathcal{M}=(K, \Sigma, \Delta, q_0, F)$, where $\mathcal{M}'=(K, \Sigma, \Delta, q_0)$ is a k-counter machine and $F \subseteq K$ is the set of accepting states. The ω-language accepted by $\mathcal{M}$ is:*

$L(\mathcal{M})= \{\sigma \in \Sigma^\omega \mid$ there exists a run r of $\mathcal{M}$ on σ such that $\mathrm{In}(r) \cap F \neq \emptyset\}$

The class of ω-languages accepted by Büchi k-counter automata is denoted **BCL**$(k)_\omega$. The class of ω-languages accepted by *real time* Büchi k-counter automata will be denoted **r-BCL**$(k)_\omega$.

We now recall the definition of classes of the arithmetical hierarchy of ω-languages, see [Sta97]. Let X be a finite alphabet. An ω-language $L \subseteq X^\omega$ belongs to the class Σ_n if and only if there exists a recursive relation $R_L \subseteq (\mathbb{N})^{n-1} \times X^\star$ such that:

$$L = \{\sigma \in X^\omega \mid \exists a_1 \ldots Q_n a_n \quad (a_1, \ldots, a_{n-1}, \sigma[a_n + 1]) \in R_L\},$$

where Q_i is one of the quantifiers $\forall$ or $\exists$ (not necessarily in an alternating order). An ω-language $L \subseteq X^\omega$ belongs to the class Π_n if and only if its complement $X^\omega - L$ belongs to the class Σ_n. The class Σ_1^1 is the class of *effective analytic sets* which are obtained by projection of arithmetical sets. An ω-language $L \subseteq X^\omega$ belongs to the class Σ_1^1 if and only if there exists a recursive relation $R_L \subseteq \mathbb{N} \times \{0, 1\}^\star \times X^\star$ such that: $L = \{\sigma \in X^\omega \mid \exists \tau (\tau \in \{0, 1\}^\omega \wedge \forall n \exists m((n, \tau[m], \sigma[m]) \in R_L))\}$.

Then an ω-language $L \subseteq X^\omega$ is in the class Σ_1^1 iff it is the projection of an ω-language over the alphabet $X \times \{0, 1\}$ which is in the class Π_2. The class Π_1^1 of *effective co-analytic sets* is simply the class of complements of effective analytic sets.

Recall that a Büchi Turing machine is just a Turing machine working on infinite inputs with a Büchi-like acceptance condition, and that the class of ω-languages accepted by Büchi Turing machines is the class Σ_1^1 of effective analytic sets [Sta97]. On the oher hand, one can construct, using a classical construction (see for instance [HMU01]), from a Büchi Turing machine $\mathcal{T}$, a 2-counter Büchi automaton $\mathcal{A}$ accepting the same ω-language. Thus one can state the following proposition.

Proposition 2. *An ω-language $L \subseteq X^\omega$ is in the class Σ_1^1 iff it is accepted by a non deterministic Büchi Turing machine, hence iff it is in the class* **BCL**$(2)_\omega$.

3   Some Results of Set Theory

We now recall some basic notions of set theory which will be useful in the sequel, and which are exposed in any textbook on set theory, like [Kun80, Jec02].

The usual axiomatic system **ZFC** is Zermelo-Fraenkel system **ZF** plus the axiom of choice **AC**. The axioms of **ZFC** express some natural facts that we consider to hold in the universe of sets. For instance a natural fact is that two sets x and y are equal iff they have the same elements. This is expressed by the *Axiom of Extensionality*:

$$\forall x \forall y \, [\, x = y \leftrightarrow \forall z (z \in x \leftrightarrow z \in y) \,].$$

Another natural axiom is the *Pairing Axiom* which states that for all sets x and y there exists a set $z = \{x, y\}$ whose elements are x and y:

$$\forall x \forall y \, [\, \exists z (\forall w (w \in z \leftrightarrow (w = x \vee w = y)))]$$

Similarly the *Powerset Axiom* states the existence of the set $\mathcal{P}(x)$ of subsets of a set x. Notice that these axioms are first-order sentences in the usual logical language of set theory whose only non logical symbol is the membership binary relation symbol $\in$. We refer the reader to any textbook on set theory for an exposition of the other axioms of **ZFC**.

A model $(\mathbf{V}, \in)$ of an arbitrary set of axioms $\mathbb{A}$ is a collection $\mathbf{V}$ of sets, equipped with the membership relation $\in$, where "$x \in y$" means that the set x is an element of the set y, which satisfies the axioms of $\mathbb{A}$. We often say " the model $\mathbf{V}$" instead of " the model $(\mathbf{V}, \in)$".

We say that two sets A and B have same cardinality iff there is a bijection from A onto B and we denote this by $A \approx B$. The relation $\approx$ is an equivalence relation. Using the axiom of choice **AC**, one can prove that any set A can be well-ordered so there is an ordinal γ such that $A \approx \gamma$. In set theory the cardinal of the set A is then formally defined as the smallest such ordinal γ.

The infinite cardinals are usually denoted by $\aleph_0, \aleph_1, \aleph_2, \ldots, \aleph_\alpha, \ldots$ The cardinal $\aleph_\alpha$ is also denoted by ω_α, when it is considered as an ordinal. The first infinite ordinal is ω and it is the smallest ordinal which is countably infinite so $\aleph_0 = \omega$ (which could be written ω_0). The first uncountable ordinal is ω_1, and formally $\aleph_1 = \omega_1$.

Let **ON** be the class of all ordinals. Recall that an ordinal α is said to be a successor ordinal iff there exists an ordinal β such that $\alpha = \beta + 1$; otherwise the ordinal α is said to be a limit ordinal and in this case $\alpha = \sup\{\beta \in \mathbf{ON} \mid \beta < \alpha\}$.

We recall now the notions of cofinality of an ordinal and of regular cardinal which may be found for instance in [Jec02]. Let α be a limit ordinal, the cofinality of α, denoted $cof(\alpha)$, is the least ordinal β such that there exists a strictly increasing sequence of ordinals $(\alpha_i)_{i<\beta}$, of length β, such that $\forall i < \beta$ $\quad \alpha_i < \alpha$ $\quad$ and $\sup_{i<\beta} \alpha_i = \alpha$. This definition is usually extended to 0 and to the successor ordinals: $cof(0) = 0$ and $cof(\alpha + 1) = 1$ for every ordinal α. The cofinality of a limit ordinal is always a limit ordinal satisfying: $\omega \leq cof(\alpha) \leq \alpha$. Moreover $cof(\alpha)$ is in fact a cardinal. A cardinal κ is said to be *regular* iff $cof(\kappa) = \kappa$. Otherwise $cof(\kappa) < \kappa$ and the cardinal κ is said to be *singular*.

A cardinal κ is said to be a *(strongly) inaccessible* cardinal iff $\kappa > \omega$, κ is regular, and for all cardinals $\lambda < \kappa$ it holds that $2^\lambda < \kappa$, where 2^λ is the cardinal of $\mathcal{P}(\lambda)$.

Recall that the class of sets in a model **V** of **ZF** may be stratified in a transfinite hierarchy, called the *Cumulative Hierarchy*, which is defined by $\mathbf{V} = \bigcup_{\alpha \in \mathbf{ON}} \mathbf{V}_\alpha$, where the sets $\mathbf{V}_\alpha$ are constructed by induction as follows:

(1). $\mathbf{V}_0 = \emptyset$
(2). $\mathbf{V}_{\alpha+1} = \mathcal{P}(\mathbf{V}_\alpha)$ is the set of subsets of $\mathbf{V}_\alpha$, and
(3). $\mathbf{V}_\alpha = \bigcup_{\beta<\alpha} \mathbf{V}_\beta$, for α a limit ordinal.

It is well known that if **V** is a model of **ZFC** and κ is an inaccessible cardinal in **V** then $\mathbf{V}_\kappa$ is also a model of **ZFC**. If there exist in **V** at least n inaccessible cardinals, where $n \geq 1$ is an integer, and if κ is the n-th inaccessible cardinal, then $\mathbf{V}_\kappa$ is also a model of **ZFC** + "There exist exactly $n - 1$ inaccessible cardinals" . This implies that one cannot prove in **ZFC** that there exists an inaccessible cardinal, because if κ is the first inaccessible cardinal in **V** then $\mathbf{V}_\kappa$ is a model of **ZFC** + "There exist no inaccessible cardinals".

We assume the reader to be familiar with basic notions of topology which may be found in [Mos80,LT94,Sta97,PP04]. There is a natural metric on the set Σ^ω of infinite words over a finite alphabet Σ containing at least two letters which is called the *prefix metric* and is defined as follows. For $u, v \in \Sigma^\omega$ and $u \neq v$ let $\delta(u, v) = 2^{-l_{\mathrm{pref}(u,v)}}$ where $l_{\mathrm{pref}(u,v)}$ is the first integer n such that the $(n + 1)^{st}$ letter of u is different from the $(n + 1)^{st}$ letter of v. This metric induces on Σ^ω the usual Cantor topology in which the *open subsets* of Σ^ω are of the form $W.\Sigma^\omega$, for $W \subseteq \Sigma^\star$.

Define now the *Borel Hierarchy* of subsets of Σ^ω:

Definition 3. *For a non-null countable ordinal α, the classes $\mathbf{\Sigma}_\alpha^0$ and $\mathbf{\Pi}_\alpha^0$ of the Borel Hierarchy on the topological space Σ^ω are defined as follows:*
$\mathbf{\Sigma}_1^0$ is the class of open subsets of Σ^ω, $\mathbf{\Pi}_1^0$ is the class of closed subsets of Σ^ω,

and for any countable ordinal $\alpha \geq 2$:

$\boldsymbol{\Sigma}_\alpha^0$ *is the class of countable unions of subsets of Σ^ω in* $\bigcup_{\gamma < \alpha} \boldsymbol{\Pi}_\gamma^0$.

$\boldsymbol{\Pi}_\alpha^0$ *is the class of countable intersections of subsets of Σ^ω in* $\bigcup_{\gamma < \alpha} \boldsymbol{\Sigma}_\gamma^0$.

The class of *Borel sets* is $\boldsymbol{\Delta}_1^1 := \bigcup_{\xi < \omega_1} \boldsymbol{\Sigma}_\xi^0 = \bigcup_{\xi < \omega_1} \boldsymbol{\Pi}_\xi^0$, where ω_1 is the first uncountable ordinal. The class of Borel subsets of Σ^ω is strictly included into the class $\boldsymbol{\Sigma}_1^1$ of *analytic sets* which are obtained by projection of Borel sets.

We now define completeness with regard to reduction by continuous functions. For a countable ordinal $\alpha \geq 1$, a set $F \subseteq \Sigma^\omega$ is said to be a $\boldsymbol{\Sigma}_\alpha^0$ (respectively, $\boldsymbol{\Pi}_\alpha^0$, $\boldsymbol{\Sigma}_1^1$)- *complete set* iff for any set $E \subseteq Y^\omega$ (with Y a finite alphabet): $E \in \boldsymbol{\Sigma}_\alpha^0$ (respectively, $E \in \boldsymbol{\Pi}_\alpha^0$, $E \in \boldsymbol{\Sigma}_1^1$) iff there exists a continuous function $f : Y^\omega \to \Sigma^\omega$ such that $E = f^{-1}(F)$.

4   Incompleteness Results for 1-counter ω-languages

We first recall that a (first-order) theory T in the language of set theory is a set of (first-order) sentences, called the axioms of the theory. If T is a theory and φ is a sentence then we write $T \vdash \varphi$ iff there is a formal proof of φ from T; this means that there is a finite sequence of sentences φ_j, $1 \leq j \leq n$, such that $\varphi_1 \vdash \varphi_2 \vdash \ldots \varphi_n$, where φ_n is the sentence φ and for each $j \in [1, n]$, either φ_j is in T or φ_j is a logical axiom or φ_j follows from $\varphi_1, \varphi_2, \ldots \varphi_{j-1}$ by usual rules of inference which can be defined purely syntactically. A theory is said to be consistent iff for no (first-order) sentence φ does $T \vdash \varphi$ and $T \vdash \neg\varphi$. If T is inconsistent, then for every sentence φ it holds that $T \vdash \varphi$. We shall denote Cons(T) the sentence "the theory T is consistent".

Recall that one can code in a recursive manner the sentences in the language of set theory by finite sequences over a finite alphabet, and then simply over the alphabet $\{0, 1\}$, by using a classical Gödel numbering of the sentences. We say that the theory T is recursive iff the set of codes of axioms in T is a recursive set of words over $\{0, 1\}$. In that case one can also code formal proofs from axioms of a recursive theory T and then Cons(T) is an arithmetical statement.

The theory **ZFC** is recursive and so are the theories $T_n =: $ **ZFC** + "There exist (at least) n inaccessible cardinals", for any integer $n \geq 1$.

We now recall Gödel's Second Incompleteness Theorem.

Theorem 4 (Gödel 1931). *Let T be a consistent recursive extension of* **ZF**. *Then $T \nvdash$ Cons(T).*

We now state the following lemmas.

Lemma 5. *Let T be a recursive theory in the language of set theory. Then there exists a Büchi Turing machine $\mathcal{M}_T$, reading words over a finite alphabet Σ, such that $L(\mathcal{M}_T) = \Sigma^\omega$ iff T is consistent and $L(\mathcal{M}_T) = \emptyset$ iff T is inconsistent. And there exists a Büchi Turing machine $\mathcal{M}'_T$, reading words over the finite alphabet Σ, such that $L(\mathcal{M}'_T) = \Sigma^\omega$ iff T is inconsistent and $L(\mathcal{M}'_T) = \emptyset$ iff T is consistent.*

Proof. We first describe informally the behaviour of the machine $\mathcal{M}_T$. The machine reads the input word but this does not affect the acceptance or non-acceptance of the word. Essentially the machine works as a program which enumerates all the formal proofs from T and enters each time in an accepting state iff the last sentence of the proof is not the sentence "$\exists x (x \neq x)$". If the theory T is consistent the machine will enter infinitely often in an accepting state q_f and thus the input ω-word will be accepted since the Büchi acceptance condition will be fulfilled. But if the theory is inconsistent then at some point of the computation the machine sees a proof whose last sentence is actually "$\exists x (x \neq x)$". In that case the machine enters in a rejecting state and stays forever in that state, and thus the input ω-word will be rejected.

The machine $\mathcal{M}'_T$ also works as a program which enumerates all the formal proofs from T. But this time it enters in an accepting state only when it sees a formal proof whose last sentence is actually "$\exists x (x \neq x)$", and then the machine $\mathcal{M}'_T$ stays in this accepting state forever. Thus the machine accepts all ω-words if the theory T is inconsistent and accepts not any ω-word if the theory T is consistent. □

Lemma 6. *Let T be a recursive theory in the language of set theory. Then there exists a Büchi Turing machine $\mathcal{M}_T$, reading words over a finite alphabet Σ, such that $L(\mathcal{M}_T) = \Sigma^\omega$ iff T is consistent and $L(\mathcal{M}_T)$ is Σ^1_1-complete iff T is inconsistent. And there exists a Büchi Turing machine $\mathcal{M}'_T$, reading words over the finite alphabet Σ, such that $L(\mathcal{M}'_T) = \Sigma^\omega$ iff T is inconsistent and $L(\mathcal{M}'_T)$ is Σ^1_1-complete iff T is consistent.*

Proof. This follows from the above Lemma 5, from the fact that there exists a Σ^1_1-complete ω-language accepted by a Büchi Turing machine (and even by a 1-counter Büchi automaton, see [Fin03]), and from the closure under finite union of the class of ω-languages accepted by *non-deterministic* Büchi Turing machines. □

We now state the following result.

Theorem 7. *Let T be a recursive theory in the language of set theory. Then there exists a real-time 1-counter Büchi automaton $\mathcal{A}_T$ reading words over a finite alphabet Γ such that $L(\mathcal{A}_T) = \Gamma^\omega$ iff T is consistent and $L(\mathcal{A}_T)$ is Σ^1_1-complete iff T is inconsistent. And there exists a real-time 1-counter Büchi automaton $\mathcal{A}'_T$ reading words over the finite alphabet Γ, such that $L(\mathcal{A}'_T) = \Gamma^\omega$ iff T is inconsistent and $L(\mathcal{A}'_T)$ is Σ^1_1-complete iff T is consistent.*

Proof. Let T be a recursive theory in the language of set theory, and $\mathcal{M}_T$ be the Büchi Turing machine, reading words over a finite alphabet Σ, which is given by Lemma 6. There exists a 2-counter Büchi automaton $\mathcal{C}_T$, such that $L(\mathcal{M}_T) = L(\mathcal{C}_T)$, and which can be effectively constructed from the machine $\mathcal{M}_T$.

We now use some constructions which were used in a previous paper [Fin06a] to study the topological properties of context-free ω-languages.

Let E be a new letter not in Σ, S be an integer ≥ 1, and $\theta_S : \Sigma^\omega \to (\Sigma \cup \{E\})^\omega$ be the function defined, for all $x \in \Sigma^\omega$, by:

$$\theta_S(x) = x(1).E^S.x(2).E^{S^2}.x(3).E^{S^3}.x(4)\ldots x(n).E^{S^n}.x(n+1).E^{S^{n+1}}\ldots$$

We proved in [Fin06a] that if $L \subseteq \Sigma^\omega$ is an ω-language in the class $\mathbf{BCL}(2)_\omega$ and $k = cardinal(\Sigma) + 2$, $S = (3k)^3$, then one can effectively construct from a Büchi 2-counter automaton $\mathcal{C}_T$ accepting L a real time Büchi 8-counter automaton $\mathcal{D}_T$ such that $L(\mathcal{D}_T) = \theta_S(L)$.

On the other hand, it is easy to see that $\theta_S(\Sigma^\omega)^- = (\Sigma \cup \{E\})^\omega - \theta_S(\Sigma^\omega)$ is accepted by a real time Büchi 1-counter automaton. The class $\mathbf{r\text{-}BCL}(8)_\omega$ is closed under finite union in an effective way and thus $\theta_S(L) \cup \theta_S(\Sigma^\omega)^-$ is accepted by a real time Büchi 8-counter automaton $\mathcal{E}_T$ which can be effectively constructed from $\mathcal{D}_T$.

Let now $K = 2 \times 3 \times 5 \times 7 \times 11 \times 13 \times 17 \times 19 = 9699690$ be the product of the eight first prime numbers. Let $\Gamma' = \Sigma \cup \{E\}$. An ω-word $x \in (\Gamma')^\omega$ is coded by the ω-word $\quad h_K(x) = A.C^K.x(1).B.C^{K^2}.A.C^{K^2}.x(2).B\ldots B.C^{K^n}.A.C^{K^n}.x(n).B\ldots$ over the alphabet $\Gamma'' = \Gamma' \cup \{A, B, C\}$, where A, B, C are letters not in Γ'. We proved in [Fin06a] that, from a real time Büchi 8-counter automaton $\mathcal{E}_T$ accepting $L(\mathcal{E}_T) \subseteq (\Gamma')^\omega$, one can effectively construct a Büchi 1-counter automaton $\mathcal{G}_T$ accepting the ω-language $h_K(L(\mathcal{E}_T)) \cup h_K((\Gamma')^\omega)^-$.

Consider now the mapping $\phi_K : (\Gamma' \cup \{A, B, C\})^\omega \to (\Gamma' \cup \{A, B, C, F\})^\omega$ which is defined by: for all $x \in (\Gamma' \cup \{A, B, C\})^\omega$,

$$\phi_K(x) = F^{K-1}.x(1).F^{K-1}.x(2)\ldots F^{K-1}.x(n).F^{K-1}.x(n+1).F^{K-1}\ldots$$

Then the ω-language $\phi_K(L(\mathcal{G}_T)) = \phi_K(h_K(L(\mathcal{E}_T)) \cup h_K((\Gamma')^\omega)^-)$ is accepted by a real time Büchi 1-counter automaton $\mathcal{H}_T$ which can be effectively constructed from the Büchi 1-counter automaton $\mathcal{G}_T$, [Fin06a]. And we set $\Gamma = \Gamma' \cup \{A, B, C, F\}$.

On the other hand, the ω-language $(\Gamma' \cup \{A, B, C, F\})^\omega - \phi_K((\Gamma' \cup \{A, B, C\})^\omega)$ is ω-regular and we can construct a (1-counter) Büchi automaton accepting it. Then one can effectively construct from $\mathcal{H}_T$ a real time Büchi 1-counter automaton $\mathcal{A}_T$ accepting the ω-language $\phi_K(h_K(L(\mathcal{E}_T)) \cup h_K((\Gamma')^\omega)^-) \cup \phi_K((\Gamma \cup \{A, B, C\})^\omega)^-$.

It suffices now to see that we have the two following cases:

If $L(\mathcal{M}_T) = L(\mathcal{C}_T) = \Sigma^\omega$, then we have successively the following equalities:
$L(\mathcal{E}_T) = (\Sigma \cup \{E\})^\omega = (\Gamma')^\omega$, $L(\mathcal{G}_T) = (\Gamma' \cup \{A, B, C\})^\omega$, $L(\mathcal{A}_T) = (\Gamma' \cup \{A, B, C, F\})^\omega = \Gamma^\omega$,

And if $L(\mathcal{M}_T) = L(\mathcal{C}_T)$ is Σ_1^1-complete, then $L(\mathcal{A}_T)$ is also Σ_1^1-complete. This follows from the fact that the mapping $\Psi : \Sigma^\omega \to (\Gamma' \cup \{A, B, C, F\})^\omega$ defined by $\Psi(x) = \phi_K(h_K(\theta_S(x)))$ is continuous and satisfies:
$$\forall x \in \Sigma^\omega \; [\, x \in L(\mathcal{M}_T) \Longleftrightarrow \Psi(x) \in L(\mathcal{A}_T) \,]$$

Finally the construction of the automaton $\mathcal{A}'_T$ is very similar except we start from the machine $\mathcal{M}'_T$ instead of the machine $\mathcal{M}_T$.   □

We now briefly recall a few definitions and facts about automata and ω-languages they accept.

An ω-language $L \subseteq \Gamma^\omega$ in $\mathbf{BCL}(1)_\omega$ is said to be unambiguous iff there exists a 1-counter Büchi automaton $\mathcal{A}$ such that $L = L(\mathcal{A})$ and every ω-word $x \in \Gamma^\omega$ has at most one accepting run by $\mathcal{A}$. In the other case the ω-language is said to be inherently ambiguous. An ω-language L accepted by a 1-counter Büchi automaton (respectively, a Büchi Turing machine) is said to have the maximum degree of ambiguity if for every 1-counter Büchi automaton (respectively, Büchi Turing machine) $\mathcal{A}$ such that $L = L(\mathcal{A})$ there exist $2^{\aleph_0}$ ω-words having $2^{\aleph_0}$ accepting runs by $\mathcal{A}$. Notice that this notion may depend on the accepting device which is used.

An ω-language accepted by a *deterministic* 1-counter Büchi (respectively, Muller) automaton is a Borel $\mathbf{\Pi}_2^0$-set (respectively, $\mathbf{\Delta}_3^0$-set); the Muller acceptance condition is stronger than the Büchi acceptance condition. The same result is true for any kinds of automata and in particular for Turing machines, see [Tho90,Sta97,PP04].

We now state the following result.

Theorem 8. *Let T be a recursive theory in the language of set theory. Then there exist two real-time 1-counter Büchi automata $\mathcal{A}_T$ and $\mathcal{A}'_T$, reading words over a finite alphabet Γ, such that* Cons(T) *is equivalent to each of the following items:*
(1) $L(\mathcal{A}_T) = \Gamma^\omega$; (2) $L(\mathcal{A}_T)$ is ω-regular; (3) $L(\mathcal{A}_T)$ is deterministic; (4) $L(\mathcal{A}_T)$ is Borel; (5) $L(\mathcal{A}_T)$ is in the Borel class $\mathbf{\Sigma}_\alpha^0$ (for a non-null countable ordinal α); (6) $L(\mathcal{A}_T)$ is in the Borel class $\mathbf{\Pi}_\alpha^0$ (for a non-null countable ordinal α); (7) $L(\mathcal{A}_T)$ is unambiguous; (8) $L(\mathcal{A}_T)$ is an arithmetical set; (9) $L(\mathcal{A}_T)$ is an hyperarithmetical set, i.e. an effective Δ_1^1-set; (10) $L(\mathcal{A}_T)$ is in the arithmetical class Σ_n (for $n \geq 1$); (11) $L(\mathcal{A}_T)$ is in the arithmetical class Π_n (for $n \geq 1$);
and also to each of the following items:
(1') $L(\mathcal{A}'_T) \neq \Gamma^\omega$; (2') $L(\mathcal{A}'_T)$ is not ω-regular; (3') $L(\mathcal{A}'_T)$ is not deterministic; (4') $L(\mathcal{A}'_T)$ is Σ_1^1-complete; (5') $L(\mathcal{A}'_T)$ is not Borel; (6') $L(\mathcal{A}'_T)$ is not in the Borel class $\mathbf{\Sigma}_\alpha^0$ (for a non-null countable ordinal α); (7') $L(\mathcal{A}'_T)$ is not in the Borel class $\mathbf{\Pi}_\alpha^0$ (for a non-null countable ordinal α); (8') $L(\mathcal{A}'_T)$ is inherently ambiguous; (9') $L(\mathcal{A}'_T)$ has the maximum degree of ambiguity (for acceptance by 1-counter automata or by Turing machines); (10') $L(\mathcal{A}'_T)$ is not an arithmetical set; (11') $L(\mathcal{A}'_T)$ is not an hyperarithmetical set; (12') $L(\mathcal{A}'_T)$ is not in the arithmetical class Σ_n (for $n \geq 1$); (13') $L(\mathcal{A}'_T)$ is not in the arithmetical class Π_n (for $n \geq 1$);

Proof. The real-time 1-counter Büchi automata $\mathcal{A}_T$ and $\mathcal{A}'_T$ are constructed in the proof of the preceding Theorem 7. It is straightforward to check that the ω-language Γ^ω is ω-regular, and even accepted by a deterministic Büchi automaton. Moreover it is in every Borel class and in every arithmetical class. It is also clearly unambiguous since it is deterministic. On the other hand a Σ_1^1-complete ω-language is not arithmetical, not hyperarithmetical, and not Borel. It cannot be ω-regular since ω-regular languages are Borel $\mathbf{\Delta}_3^0$-sets. Similarly it is not deterministic since it is not a $\mathbf{\Delta}_3^0$-set. Moreover any Σ_1^1-complete ω-language accepted by a 1-counter Büchi automaton (respectively, a Büchi Turing machine) has the maximum degree of ambiguity, see [Fin14]. $\square$

Recall that we denote T_n the theory **ZFC** + "There exist (at least) n inaccessible cardinals", for an integer $n \geq 0$. We can apply the preceding theorem to the theories T_n which are recursive, and get the real-time 1-counter Büchi automata $\mathcal{A}_{T_n}$ and $\mathcal{A}'_{T_n}$, which will be simply denoted $\mathcal{A}_n$ and $\mathcal{A}'_n$ in the sequel.

Theorem 9. *For every integer $n \geq 0$, there exist two real-time 1-counter Büchi automata $\mathcal{A}_n$ and $\mathcal{A}'_n$, reading words over a finite alphabet Γ, such that* Cons(T_n) *is equivalent to* each of the items (1)-(11) and (1')-(13') of the preceding theorem where $\mathcal{A}_T$ and $\mathcal{A}'_T$ are replaced by $\mathcal{A}_n$ and $\mathcal{A}'_n$. In particular, if **ZFC** + "There exist (at least) n inaccessible cardinals" is consistent, then each of the properties of $\mathcal{A}_n$ and $\mathcal{A}'_n$ given by these items (1)-(11) and (1')-(13') is provable from **ZFC** + "There exist (at least) $n+1$ inaccessible cardinals" but not from **ZFC** + "There exist (at least) n inaccessible cardinals".

Proof. The automata $\mathcal{A}_n$ and $\mathcal{A}'_n$ are given by the preceding theorem applied to the theories T_n. Recall that one can prove from **ZFC** + "There exist (at least) $n+1$ inaccessible cardinals" that if κ is the $n+1$-th inaccessible cardinal, then the set $\mathbf{V}_\kappa$ of the cumulative hierarchy is also a model of **ZFC** + "There exist n inaccessible cardinals". This implies that the theory **ZFC** + "There exist n inaccessible cardinals" is consistent and thus this implies also the properties of $\mathcal{A}_n$ and $\mathcal{A}'_n$ given by the items (1)-(11) and (1')-(13'). On the other hand if T_n is consistent, then these properties are not provable from T_n. Indeed T_n is then a consistent recursive extension of **ZFC** and thus by Gödel's Second Incompleteness Theorem we know that $T_n \nvdash \mathrm{Cons}(T_n)$.   □

5 Incompleteness Results for Infinitary Rational Relations

We now consider acceptance of binary relations over infinite words by 2-tape Büchi automata, firstly considered by Gire and Nivat in [GN84]. A 2-tape automaton is an automaton having two tapes and two reading heads, one for each tape, which can move asynchronously, and a finite control as in the case of a (1-tape) automaton. The automaton reads a pair of (infinite) words (u, v) where u is on the first tape and v is on the second tape, so that a 2-tape Büchi automaton $\mathcal{B}$ accepts an infinitary rational relation $L(\mathcal{B}) \subseteq \Sigma_1^\omega \times \Sigma_2^\omega$, where Σ_1 and Σ_2 are two finite alphabets. Notice that $L(\mathcal{B}) \subseteq \Sigma_1^\omega \times \Sigma_2^\omega$ may be seen as an ω-language over the product alphabet $\Sigma_1 \times \Sigma_2$.

We now use a coding we have defined in a previous paper [Fin06b] to study the topological complexity of infinitary rational relations. We first recall a coding of an ω-word over the finite alphabet $\Omega = \Sigma \cup \{A, B, C, E, F\}$, where 0 is assumed to be a letter of Σ, by an ω-word over the alphabet $\Omega' = \Omega \cup \{D\}$, where D is an additionnal letter not in Ω. For $x \in \Omega^\omega$ the ω-word $h(x)$ is defined by :

$$h(x) = D.0.x(1).D.0^2.x(2).D.0^3.x(3).D \ldots D.0^n.x(n).D.0^{n+1}.x(n+1).D \ldots$$

It is easy to see that the mapping h from Ω^ω into $(\Omega \cup \{D\})^\omega$ is injective. Let now α be the ω-word over the alphabet Ω' which is simply defined by:

$$\alpha = D.0.D.0^2.D.0^3.D.0^4.D \ldots D.0^n.D.0^{n+1}.D \ldots$$

The following result was proved in [Fin06b].

Proposition 10 ([Fin06b]). *Let $L \subseteq \Omega^\omega$ be in* **r-BCL**$(1)_\omega$ *and* $\mathcal{L} = h(L) \cup (h(\Omega^\omega))^-$. *Then* $R = \mathcal{L} \times \{\alpha\} \bigcup (\Omega')^\omega \times ((\Omega')^\omega - \{\alpha\})$ *is an infinitary rational relation. Moreover one can effectively construct from a real time 1-counter Büchi automaton $\mathcal{A}$ accepting L a 2-tape Büchi automaton $\mathcal{B}$ accepting the infinitary relation R.*

Using this Proposition 10 and Theorem 7 and a very similar reasoning as in the proofs of Theorems 8 and 9, we can now prove the following results.

Theorem 11. *For every integer* $n \geq 0$, *there exist two 2-tape Büchi automata* $\mathcal{B}_n$ *and* $\mathcal{B}'_n$, *reading words over a finite alphabet* $\Omega' \times \Omega'$, *such that* $\mathrm{Cons}(T_n)$ *is equivalent to each of the following items (1)-(11) and (1')-(13')*

(1) $L(\mathcal{B}_n) = (\Omega')^\omega \times (\Omega')^\omega$;   (2) $L(\mathcal{B}_n)$ is ω-regular;   (3) $L(\mathcal{B}_n)$ is deterministic; (4) $L(\mathcal{B}_n)$ is Borel;   (5) $L(\mathcal{B}_n)$ is in the Borel class $\mathbf{\Sigma}^0_\alpha$ (for a non-null countable ordinal α);   (6) $L(\mathcal{B}_n)$ is in the Borel class $\mathbf{\Pi}^0_\alpha$ (for a non-null countable ordinal α); (7) $L(\mathcal{B}_n)$ is unambiguous;   (8) $L(\mathcal{B}_n)$ is an arithmetical set;   (9) $L(\mathcal{B}_n)$ is an hyperarithmetical set, i.e. an effective Δ^1_1-set;   (10) $L(\mathcal{B}_n)$ is in the arithmetical class Σ_n (for $n \geq 1$);   (11) $L(\mathcal{B}_n)$ is in the arithmetical class Π_n (for $n \geq 1$);

(1') $L(\mathcal{B}'_n) \neq (\Omega')^\omega \times (\Omega')^\omega$;   (2') $L(\mathcal{B}'_n)$ is not ω-regular;   (3') $L(\mathcal{B}'_n)$ is not deterministic;   (4') $L(\mathcal{B}'_n)$ is Σ^1_1-complete;   (5') $L(\mathcal{B}'_n)$ is not Borel;   (6') $L(\mathcal{B}'_n)$ is not in the Borel class $\mathbf{\Sigma}^0_\alpha$ (for a non-null countable ordinal α);   (7') $L(\mathcal{B}'_n)$ is not in the Borel class $\mathbf{\Pi}^0_\alpha$ (for a non-null countable ordinal α);   (8') $L(\mathcal{B}'_n)$ is inherently ambiguous;   (9') $L(\mathcal{B}'_n)$ has the maximum degree of ambiguity (for acceptance by 2-tape automata or by Turing machines);   (10') $L(\mathcal{B}'_n)$ is not an arithmetical set; (11') $L(\mathcal{B}'_n)$ is not an hyperarithmetical set;   (12') $L(\mathcal{B}'_n)$ is not in the arithmetical class Σ_n (for $n \geq 1$);   (13') $L(\mathcal{B}'_n)$ is not in the arithmetical class Π_n (for $n \geq 1$);

In particular, if **ZFC** + "There exist (at least) n inaccessible cardinals" is consistent, then each of the properties of $\mathcal{B}_n$ and $\mathcal{B}'_n$ given by these items (1)-(11) and (1')-(13') is provable from **ZFC** + "There exist (at least) $n + 1$ inaccessible cardinals" but not from **ZFC** + "There exist (at least) n inaccessible cardinals".

6 Concluding Remarks

Using similar methods as above in this paper, we can construct, for a given theory T in the language of set theory and a given first-order sentence Φ in the language of set theory, a 1-counter Büchi automaton (or a 2-tape Büchi automaton) $\mathcal{A}_1$ (respectively, $\mathcal{A}_2$, $\mathcal{A}_3$) such that $L(\mathcal{A}_1)$ (respectively, $L(\mathcal{A}_2)$, $L(\mathcal{A}_3)$) is Borel (and deterministic, ω-regular, unambiguous, …) if and only if the sentence Φ is provable from T, (respectively, $\neg\Phi$ is provable from T, Φ is independent from T).

As an example recall that a famous open problem in Complexity Theory is the following question: " Is **P** equal to **NP**?" , see [HMU01]. Notice that "**P**= **NP**" can be expressed by a first-order sentence Ψ in the language of set theory. Thus one can construct a 2-tape Büchi automaton $\mathcal{A}_1$ (respectively, $\mathcal{A}_2$, $\mathcal{A}_3$) such that $L(\mathcal{A}_1)$ (respectively, $L(\mathcal{A}_2)$, $L(\mathcal{A}_3)$) is Borel if and only if the sentence Ψ is provable from T, (respectively, $\neg\Psi$ is provable from T, Ψ is independent from T). Since the "**P**= **NP**?" problem is one of the millennium problems for the solution of which one million dollars is offered by the Clay Institute, this is the sum one can win by proving that the infinitary rational relation $L(\mathcal{A}_1)$ (or $L(\mathcal{A}_2)$ or $L(\mathcal{A}_3)$) is Borel !

On the other hand, the results of this paper are true for other large cardinals than inaccessible ones. For instance we can replace inaccessible cardinals by hyperinaccessible, Mahlo, hyperMahlo, measurable, …(see [Jec02]) and still other ones and obtain similar results.

Finally we mention that in an extended version of this paper we prove similar independence results for timed automata reading timed words.

References

[Fin03] Finkel, O.: Borel hierarchy and omega context free languages. Theoretical Computer Science **290**(3), 1385–1405 (2003)

[Fin06a] Finkel, O.: Borel ranks and Wadge degrees of omega context free languages. Mathematical Structures in Computer Science **16**(5), 813–840 (2006)

[Fin06b] Finkel, O.: On the accepting power of 2-tape büchi automata. In: Durand, B., Thomas, W. (eds.) STACS 2006. LNCS, vol. 3884, pp. 301–312. Springer, Heidelberg (2006)

[Fin09] Finkel, O.: The complexity of infinite computations in models of set theory. Logical Methods in Computer Science **5**(4:4), 1–19 (2009)

[Fin10] Finkel, O.: Decision problems for recognizable languages of infinite pictures. In: Studies in Weak Arithmetics, Proceedings of the International Conference 28th Weak Arithmetic Days, June 17–19, vol. 196. Publications of the Center for the Study of Language and Information. Lecture Notes, pages 127–151. Stanford University (2010)

[Fin11] Finkel, O.: Some problems in automata theory which depend on the models of set theory. RAIRO - Theoretical Informatics and Applications **45**(4), 383–397 (2011)

[Fin14] Finkel, O.: Ambiguity of ω-languages of Turing machines. Logical Methods in Computer Science **10**(3:12), 1–18 (2014)

[GN84] Gire, F., Nivat, M.: Relations rationnelles infinitaires. Calcolo, pp. 91–125 (1984)

[GTW02] Grädel, E., Thomas, W., Wilke, W. (eds.): Automata, Logics, and Infinite Games: A Guide to Current Research, vol. 2500. LNCS. Springer, Heidelberg (2002)

[HMU01] Hopcroft, J.E., Motwani, R., Ullman, J.D.: Introduction to automata theory, languages, and computation. Addison-Wesley Series in Computer Science. Addison-Wesley Publishing Co., Reading (2001)

[Jec02] Jech, T.: Set theory, 3rd edn., Springer (2002)

[Kun80] Kunen, K.: Set theory. Studies in Logic and the Foundations of Mathematics, vol. 102. An introduction to independence proofs. North-Holland Publishing Co., Amsterdam (1980)

[LT94] Lescow, H., Thomas, W.: Logical specifications of infinite computations. In: de Bakker, J.W., de Roever, Willem-Paul, Rozenberg, Grzegorz (eds.) REX 1993. LNCS, vol. 803, pp. 583–621. Springer, Heidelberg (1994)

[Mos80] Moschovakis, Y.N.: Descriptive set theory. North-Holland Publishing Co., Amsterdam (1980)

[PP04] Perrin, D., Pin, J.-E.: Infinite words, automata, semigroups, logic and games. Pure and Applied Mathematics, vol. 141. Elsevier (2004)

[Sta97] Staiger, L.: ω-languages. In: Handbook of Formal Languages, vol. 3, pp. 339–387. Springer, Berlin (1997)

[Tho90] Thomas, W.: Automata on infinite objects. In: van Leeuwen, J. (ed.) Handbook of Theoretical Computer Science, volume B, Formal models and semantics, pp. 135–191. Elsevier (1990)

The Odds of Staying on Budget

Christoph Haase[1]([⊠]) and Stefan Kiefer[2]

[1] Laboratoire Spécification et Vérification (LSV),
CNRS and ENS de Cachan, Cachan Cedex, France
haase@lsv.ens-cachan.fr
[2] Department of Computer Science, University of Oxford, Oxford, UK

Abstract. Given Markov chains and Markov decision processes (MDPs) whose transitions are labelled with non-negative integer costs, we study the computational complexity of deciding whether the probability of paths whose accumulated cost satisfies a Boolean combination of inequalities exceeds a given threshold. For acyclic Markov chains, we show that this problem is PP-complete, whereas it is hard for the PosSLP problem and in PSPACE for general Markov chains. Moreover, for acyclic and general MDPs, we prove PSPACE- and EXP-completeness, respectively. Our results have direct implications on the complexity of computing reward quantiles in succinctly represented stochastic systems.

1 Introduction

Computing the shortest path from s to t in a directed graph is a ubiquitous problem in computer science, so shortest-path algorithms such as Dijkstra's algorithm are a staple for every computer scientist. These algorithms work in polynomial time even if the edges are weighted, so it is easy to answer questions like:

(I) *Is it possible to travel from Copenhagen to Kyoto in less than 15 hours?*

The shortest-path problem becomes more intricate as soon as uncertainties are taken into account. For example, additional information such as *"there might be congestion in Singapore, so the Singapore route will, with probability 10%, trigger a delay of 1 hour"* leads to questions of the following kind:

(II) *Is there a travel plan avoiding trips longer than 15 hours with probability at least 0.9?*

Markov decision processes (MDPs) are the established model to formalise problems such as (II). We consider MDPs where each transition is equipped with a non-negative "weight". The weight could be interpreted as time, distance, reward, or—as in this paper—as *cost*. For another example, imagine the plan of a research project whose workflow can be modelled by a directed weighted graph. In each project state the investigators can hire a programmer, travel to collaborators, acquire new equipment, etc., but each action costs money, and the result (i.e., the next project state) is probabilistic. The objective is to meet the goals of the project before exceeding its budget for the total accumulated cost.

C. Haase—Supported by Labex Digicosme, Univ. Paris-Saclay, project VERI-CONISS.

S. Kiefer—Supported by a Royal Society University Research Fellowship.

© Springer-Verlag Berlin Heidelberg 2015
M.M. Halldórsson et al. (Eds.): ICALP 2015, Part II, LNCS 9135, pp. 234–246, 2015.
DOI: 10.1007/978-3-662-47666-6_19

(III) *Is there a strategy to stay on budget with probability* ≥ 0.85?

MDP problems like (II) and (III) become even more challenging when each transition is equipped with both a cost and a *utility*. Such *cost-utility trade-offs* have recently been studied in [2].

   The problems (II) and (III) may become easier in *Markov chains*, which have no non-determinism, i.e., there are no actions. Referring to the project example above, the activities may be completely planned out, but their effects (i.e. cost and next state) may still be probabilistic, yielding problems of the kind:

(IV) *Will the budget be kept with probability* ≥ 0.85?

Closely related to the aforementioned decision problems is the following optimisation problem, referred to as the *quantile query* in [2,19]. A quantile query asked by a funding body, for instance, could be the following:

(V) *Given a probability threshold τ, compute the smallest budget that suffices with probability at least τ.*

Non-stochastic problems like (I) are well understood. The purpose of this paper is to investigate the complexity of MDP problems such as (II) and (III), of Markov-chain problems such as (IV), and of quantile queries like (V). More formally, the models we consider are Markov chains and MDPs with non-negative integer costs, and the main focus of this paper is on the *cost problem* for those models: Given a budget constraint φ represented as a Boolean combination of linear inequalities and a probability threshold τ, we study the complexity of determining whether the probability of paths reaching a designated target state with cost consistent with φ is at least τ.

   In order to highlight some issues, let us briefly discuss two approaches that do not, at least not in an obvious way, resolve the core challenges. First, one approach to answer the MDP problems could be to compute a strategy that minimises the *expected* total cost, which is a classical problem in the MDP literature, solvable in polynomial time using linear programming methods [13]. However, minimising the expectation may not be optimal: if you don't want to be late, it may be better to walk than to wait for the bus, even if the bus saves you time in average. The second approach with shortcomings is to phrase problems (II), (III) and (IV) as MDP or Markov-chain *reachability* problems, which are also known to be solvable in polynomial time. This, however, ignores the fact that numbers representing cost are commonly represented in their natural succinct *binary* encoding. Augmenting each state with possible accumulated costs leads to an exponential blow-up of the state space.

Our Contribution. The goal of this paper is to comprehensively investigate under which circumstances and to what extent the complexity of the cost problem and of quantile queries may be below EXP. We also significantly strengthen the NP lower bound derivable from [12]. We distinguish between acyclic and general control graphs. In short, we show the following for the cost problem: (1) for acyclic Markov chains it is PP-complete, (2) for general Markov chains it is hard for the PosSLP problem and in PSPACE, (3) for acyclic MDPs it is PSPACE-complete, and (4) for general MDPs it is EXP-complete. Due to space constraints, full details have been moved to a technical report [9].

Related Work. The motivation for this paper comes from the work on quantile queries in [2,19] mentioned above and on model checking so-called durational probabilistic systems [12] with a probabilistic timed extension of CTL. While the focus of [19] is mainly on "qualitative" problems where the probability threshold is either 0 or 1, an iterative linear-programming-based approach for solving quantile queries has been suggested in [2]. The authors report satisfying experimental results, the worst-case complexity however remains exponential time. Settling the complexity of quantile queries has been identified as one of the current challenges in the conclusion of [2].

Recently, there has been considerable interest in models of stochastic systems that extend weighted graphs or counter systems, see e.g. the survey [15]. Multi-dimensional percentile queries for various payoff functions are studied in [14]. The work by Bruyère et al. [6] has also been motivated by the fact that minimising the expected total cost is not always adequate. For instance, they consider the problem of computing a scheduler in an MDP with positive integer weights that ensures that both the expected and the maximum incurred cost remain below a given values. Other recent work also investigated MDPs with a single counter ranging over the non-negative integers, see e.g. [5]. However, in that work updates to the counter can be both positive and negative. For that reason, the analysis focuses on questions about reaching the counter value *zero*.

2   Preliminaries

We write $\mathbb{N} = \{0, 1, 2, \ldots\}$. For a countable set X we write $dist(X)$ for the set of *probability distributions* over X; i.e., $dist(X)$ consists of those functions $f : X \to [0, 1]$ such that $\sum_{x \in X} f(x) = 1$.

Markov Chains. A *Markov chain* is a triple $\mathcal{M} = (S, s_0, \delta)$, where S is a countable (finite or infinite) set of states, $s_0 \in S$ is an initial state, and $\delta : S \to dist(S)$ is a probabilistic transition function that maps a state to a probability distribution over the successor states. Given a Markov chain we also write $s \xrightarrow{p} t$ or $s \to t$ to indicate that $p = \delta(s)(t) > 0$. A *run* is an infinite sequence $s_0 s_1 \cdots \in \{s_0\} S^\omega$ with $s_i \to s_{i+1}$ for $i \in \mathbb{N}$. We write $Run(s_0 \cdots s_k)$ for the set of runs that start with $s_0 \cdots s_k$. To $\mathcal{M}$ we associate the standard probability space $(Run(s_0), \mathcal{F}, \mathcal{P})$ where $\mathcal{F}$ is the σ-field generated by all basic cylinders $Run(s_0 \cdots s_k)$ with $s_0 \cdots s_k \in \{s_0\} S^*$, and $\mathcal{P} : \mathcal{F} \to [0, 1]$ is the unique probability measure such that $\mathcal{P}(Run(s_0 \cdots s_k)) = \prod_{i=1}^{k} \delta(s_{i-1})(s_i)$.

Markov Decision Processes. A *Markov decision process (MDP)* is a tuple $\mathcal{D} = (S, s_0, A, En, \delta)$, where S is a countable set of states, $s_0 \in S$ is the initial state, A is a finite set of actions, $En : S \to 2^A \setminus \emptyset$ is an action enabledness function that assigns to each state s the set $En(s)$ of actions enabled in s, and

$\delta : S \times A \rightarrow dist(S)$ is a probabilistic transition function that maps a state s and an action $a \in En(s)$ enabled in s to a probability distribution over the successor states. A (deterministic, memoryless) *scheduler* for $\mathcal{D}$ is a function $\sigma : S \rightarrow A$ with $\sigma(s) \in En(s)$ for all $s \in S$. A scheduler σ induces a Markov chain $\mathcal{M}_\sigma = (S, s_0, \delta_\sigma)$ with $\delta_\sigma(s) = \delta(s, \sigma(s))$ for all $s \in S$. We write $\mathcal{P}_\sigma$ for the corresponding probability measure of $\mathcal{M}_\sigma$.

Cost Processes. A *cost process* is a tuple $\mathcal{C} = (Q, q_0, t, A, En, \Delta)$, where Q is a finite set of control states, $q_0 \in Q$ is the initial control state, t is the target control state, A is a finite set of actions, $En : Q \rightarrow 2^A \setminus \emptyset$ is an action enabledness function that assigns to each control state q the set $En(q)$ of actions enabled in q, and $\Delta : Q \times A \rightarrow dist(Q \times \mathbb{N})$ is a probabilistic transition function. Here, for $q, q' \in Q$, $a \in En(q)$ and $k \in \mathbb{N}$, the value $\Delta(q, a)(q', k) \in [0, 1]$ is the probability that, if action a is taken in control state q, the cost process transitions to control state q' and cost k is incurred. For the complexity results we define the *size* of $\mathcal{C}$ as the size of a succinct description, i.e., the costs are encoded in binary, the probabilities are encoded as fractions of integers in binary (so the probabilities are rational), and for each $q \in Q$ and $a \in En(q)$, the distribution $\Delta(q, a)$ is described by the list of triples (q', k, p) with $\Delta(q, a)(q', k) = p > 0$ (so we assume this list to be finite). Consider the directed graph $G = (Q, E)$ with

$$E := \{(q, q') \in (Q \setminus \{t\}) \times Q : \exists a \in En(q) \; \exists k \in \mathbb{N}. \; \Delta(q, a)(q', k) > 0\}.$$

We call $\mathcal{C}$ *acyclic* if G is acyclic (which can be determined in linear time).

A cost process $\mathcal{C}$ induces an MDP $\mathcal{D}_\mathcal{C} = (Q \times \mathbb{N}, (q_0, 0), A, En', \delta)$ with $En'(q, c) = En(q)$ for all $q \in Q$ and $c \in \mathbb{N}$, and $\delta((q, c), a)(q', c') = \Delta(q, a)(q', c' - c)$ for all $q, q' \in Q$ and $c, c' \in \mathbb{N}$ and $a \in A$. For a state $(q, c) \in Q \times \mathbb{N}$ in $\mathcal{D}_\mathcal{C}$ we view q as the current control state and c as the current cost, i.e., the cost accumulated thus far. We refer to $\mathcal{C}$ as a *cost chain* if $|En(q)| = 1$ holds for all $q \in Q$. In this case one can view $\mathcal{D}_\mathcal{C}$ as the Markov chain induced by the unique scheduler of $\mathcal{D}_\mathcal{C}$. For cost chains, actions are not relevant, so we describe cost chains just by the tuple $\mathcal{C} = (Q, q_0, t, \Delta)$.

Recall that we restrict schedulers to be deterministic and memoryless, as such schedulers will be sufficient for the objectives in this paper. Note, however, that our definition allows schedulers to depend on the current cost, i.e., we may have schedulers σ with $\sigma(q, c) \neq \sigma(q, c')$.

The Accumulated Cost K. In this paper we will be interested in the cost accumulated during a run before reaching the target state t. For this cost to be a well-defined random variable, we make two assumptions on the system: (i) We assume that $En(t) = \{a\}$ holds for some $a \in A$ and $\Delta(t, a)(t, 0) = 1$. Hence, runs that visit t will not leave t and accumulate only a finite cost. (ii) We assume that for all schedulers the target state t is almost surely reached, i.e., for all schedulers the probability of eventually visiting a state (t, c) with $c \in \mathbb{N}$ is equal to one. The latter condition can be verified by graph algorithms in time quadratic in the input size, e.g., by computing the *maximal end components* of the MDP obtained from $\mathcal{C}$ by ignoring the cost, see e.g. [3, Alg. 47].

Given a cost process $\mathcal{C}$ we define a random variable $K_\mathcal{C} : Run((q_0, 0)) \to \mathbb{N}$ such that $K_\mathcal{C}((q_0, 0)) (q_1, c_1) \cdots) = c$ if there exists $i \in \mathbb{N}$ with $(q_i, c_i) = (t, c)$. We often drop the subscript from $K_\mathcal{C}$ if the cost process $\mathcal{C}$ is clear from the context. We view $K(w)$ as the accumulated cost of a run w.

From the above-mentioned assumptions on t, it follows that for any scheduler the random variable K is almost surely defined. Dropping assumption (i) would allow the same run to visit states (t, c_1) and (t, c_2) for two different $c_1, c_2 \in \mathbb{N}$. There would still be reasonable ways to define a cost K, but no apparently best way. If assumption (ii) were dropped, we would have to deal with runs that do not visit the target state t. In that case one could study the random variable K as above *conditioned* under the event that t is visited. For Markov chains, [4, Sec. 3] describes a transformation that preserves the distribution of the conditional cost K, but t is almost surely reached in the transformed Markov chain. In this sense, our assumption (ii) is without loss of generality for cost chains. For general cost processes the transformations of [4] do not work. In fact, a scheduler that "optimises" K conditioned under reaching t might try to avoid reaching t once the accumulated cost has grown unfavourably. Hence, dropping assumption (ii) in favour of conditional costs would give our problems an aspect of multi-objective optimisation, which is not the focus of this paper.

The Cost Problem. Let x be a fixed variable. An *atomic cost formula* is an inequality of the form $x \leq B$ where $B \in \mathbb{N}$ is encoded in binary. A *cost formula* is an arbitrary Boolean combination of atomic cost formulas. A number $n \in \mathbb{N}$ *satisfies* a cost formula φ, in symbols $n \models \varphi$, if φ is true when x is replaced by n.

This paper mainly deals with the following decision problem: given a cost process $\mathcal{C}$, a cost formula φ, and a probability threshold $\tau \in [0, 1]$, the *cost problem* asks whether there exists a scheduler σ with $\mathcal{P}_\sigma(K_\mathcal{C} \models \varphi) \geq \tau$. The case of an atomic cost formula φ is an important special case. Clearly, for cost chains $\mathcal{C}$ the cost problem simply asks whether $\mathcal{P}(K_\mathcal{C} \models \varphi) \geq \tau$ holds. One can assume $\tau = 1/2$ without loss of generality, thanks to a simple construction, see [9]. Moreover, with an oracle for the cost problem at hand, one can use binary search over τ to approximate $\mathcal{P}_\sigma(K \models \varphi)$: i oracle queries suffice to approximate $\mathcal{P}_\sigma(K \models \varphi)$ within an absolute error of 2^{-i}.

By our definition, the MDP $\mathcal{D}_\mathcal{C}$ is in general infinite as there is no upper bound on the accumulated cost. However, when solving the cost problem, there is no need to keep track of costs above B, where B is the largest number appearing in φ. So one can solve the cost problem in so-called *pseudo-polynomial time* (i.e., polynomial in B, not in the size of the encoding of B) by computing an explicit representation of a restriction, say $\widehat{\mathcal{D}}_\mathcal{C}$, of $\mathcal{D}_\mathcal{C}$ to costs up to B, and then applying classical linear-programming techniques [13] to compute the optimal scheduler for the finite MDP $\widehat{\mathcal{D}}_\mathcal{C}$. Since we consider reachability objectives, the optimal scheduler is deterministic and memoryless. This shows that our restriction to deterministic memoryless schedulers is without loss of generality. In terms of our succinct representation we have:

Proposition 1. *The cost problem is in* EXP.

The subject of this paper is to investigate to what extent the EXP complexity is optimal.

3   Quantile Queries

In this section we consider the following function problem, referred to as *quantile query* in [2,19]. Given a cost chain $\mathcal{C}$ and a probability threshold τ, a quantile query asks for the smallest budget B such that $\mathcal{P}_\sigma(K_\mathcal{C} \leq B) \geq \tau$. We show that polynomially many oracle queries to the cost problem for atomic cost formulas "$x \leq B$" suffice to answer a quantile query. This can be done using binary search over the budget B. The following proposition provides a suitable general upper bound on this binary search, by exhibiting a concrete sufficient budget, computable in polynomial time:

Proposition 2. *Suppose $0 \leq \tau < 1$. Let p_{min} be the smallest non-zero probability and k_{max} be the largest cost in the description of the cost process. Then $\mathcal{P}_\sigma(K \leq B) \geq \tau$ holds for all schedulers σ, where*

$$B := k_{max} \cdot \left\lceil |Q| \cdot \left(- \ln(1 - \tau)/p_{min}^{|Q|} + 1 \right) \right\rceil .$$

The case $\tau = 1$ is covered by [19, Thm. 6], where it is shown that one can compute in polynomial time the smallest B with $\mathcal{P}_\sigma(K \leq B) = 1$ for all schedulers σ, if such B exists. We conclude that quantile queries are polynomial-time inter-reducible with the cost problem for atomic cost formulas.

4   Cost Chains

In this section we consider the cost problems for acyclic and general cost chains. Even in the general case we obtain PSPACE membership, avoiding the EXP upper bound from Prop. 1.

Acyclic Cost Chains. The complexity class PP [8] can be defined as the class of languages L that have a probabilistic polynomial-time bounded Turing machine M_L such that for all words x one has $x \in L$ if and only if M_L accepts x with probability at least $1/2$. The class PP includes NP [8], and Toda's theorem states that P^{PP} contains the polynomial-time hierarchy [17]. We show that the cost problem for acyclic cost chains is PP-complete.

Theorem 3. *The cost problem for acyclic cost chains is in PP. It is PP-hard under polynomial-time Turing reductions, even for atomic cost formulas.*

Proof (sketch). To show membership in PP, we construct a probabilistic Turing machine that simulates the acyclic cost chain, and keeps track of the currently accumulated cost on the tape. For the lower bound, it follows from [12, Prop. 4] that an instance of the KTH LARGEST SUBSET problem can be reduced to a cost problem for acyclic cost chains with atomic cost formulas. This problem is PP-hard under polynomial-time Turing reductions [10, Thm. 3].   □

PP-hardness strengthens the NP-hardness result from [12] substantially: by Toda's theorem it follows that any problem in the polynomial-time hierarchy can be solved by a deterministic polynomial-time bounded Turing machine that has oracle access to the cost problem for acyclic cost chains.

General Cost Chains. For the PP upper bound in Thm. 3, the absence of cycles in the control graph seems essential. Indeed, we can use cycles to show hardness for the PoSSLP problem, suggesting that the acyclic and the general case have different complexity. PoSSLP is a fundamental problem for numerical computation [1]. Given an arithmetic circuit with operators $+$, $-$, $*$, inputs 0 and 1, and a designated output gate, the PoSSLP problem asks whether the circuit outputs a positive integer. PoSSLP is in PSPACE; in fact, it lies in the 4th level of the *counting hierarchy (CH)* [1], an analogue to the polynomial-time hierarchy for classes like PP. We have the following theorem:

Theorem 4. *The cost problem for cost chains is in* PSPACE *and hard for* PoSSLP.

The remainder of this section is devoted to a proof sketch of this theorem. Showing membership in PSPACE requires non-trivial results. There is no agreed-upon definition of probabilistic PSPACE in the literature, but we can define it in analogy to PP as follows: *Probabilistic* PSPACE is the class of languages L that have a probabilistic polynomial-space bounded Turing machine M_L such that for all words x one has $x \in L$ if and only if M_L accepts x with probability at least $1/2$. The cost problem for cost chains is in this class, as can be shown by adapting the argument from the beginning of the proof sketch for Thm. 3, replacing PP with probabilistic PSPACE. It was first proved in [16] that probabilistic PSPACE equals PSPACE, hence the cost problem for cost chains is in PSPACE.

For the PoSSLP-hardness proof one can assume the following normal form, see the proof of [7, Thm.5.2]: there are only $+$ and $*$ operators, the corresponding gates alternate, and all gates except those on the bottom level have exactly two incoming edges, cf. the top of Fig. 1. We write $val(g)$ for the value output by gate g. Then PoSSLP asks: given an arithmetic circuit (in normal form) including gates g_1, g_2, is $val(g_1) \geq val(g_2)$?

As an intermediate step of independent interest, we show PoSSLP-hardness of a problem about deterministic finite automata (DFAs). Let Σ be a finite alphabet and call a function $f : \Sigma \to \mathbb{N}$ a *Parikh function*. The *Parikh image* of a word $w \in \Sigma^*$ is the Parikh function f such that $f(a)$ is the number of occurrences of a in w. We show:

Proposition 5. *Given an arithmetic circuit including gate g, one can compute in logarithmic space a Parikh function f (in binary encoding) and a DFA $\mathcal{A}$ such that $val(g)$ equals the number of accepting computations in $\mathcal{A}$ that are labelled with words that have Parikh image f.*

The construction is illustrated in Fig. 1. It is by induction on the levels of the arithmetic circuit. A gate labelled with "$+$" is simulated by *branching* into the inductively constructed gadgets corresponding to the gates this gate connects

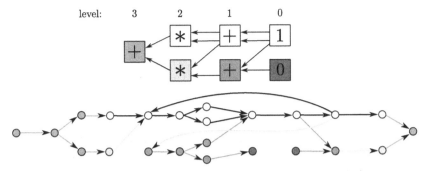

Fig. 1. Top: an arithmetic circuit in normal form. Bottom: a DFA (omitting input letters) corresponding to the construction of Prop. 5. Identical colours indicate a correspondence between gates and states.

to. Likewise, a gate labelled with "$*$" is simulated by *sequentially composing* the gadgets corresponding to the gates this gate connects to. It is the latter case that may introduce cycles in the structure of the DFA. Building on this construction, by encoding alphabet letters in natural numbers encoded in binary, we then show:

Proposition 6. *Given an arithmetic circuit including gate g on odd level ℓ, one can compute in logarithmic space a cost process $\mathcal{C}$ and $T \in \mathbb{N}$ with $\mathcal{P}(K_\mathcal{C} = T) = val(g)/m$, where $m = \exp_2(2^{(\ell-1)/2+1} - 1) \cdot \exp_d(2^{(\ell-1)/2+1} - 3)$.*

Towards the PosSLP lower bound from Thm. 4, given an arithmetic circuit including gates g_1, g_2, we use Prop. 6 to construct two cost chains $\mathcal{C}_1 = (Q, q_1, t, \Delta)$ and $\mathcal{C}_2 = (Q, q_2, t, \Delta)$ and $T_1, T_2 \in \mathbb{N}$ such that $\mathcal{P}(K_{\mathcal{C}_i} = T_i) = val(g_i)/m$ holds for $i \in \{1, 2\}$ and for $m \in \mathbb{N}$ as in Prop. 6. Then we compute a number $H \geq T_2$ such that $\mathcal{P}(K_{\mathcal{C}_2} > H) < 1/m$. The representation of m from Prop. 6 is of exponential size. However, using Prop. 2, H depends only logarithmically on $m + 1$. We combine $\mathcal{C}_1$ and $\mathcal{C}_2$ to a cost chain $\mathcal{C} = (Q \uplus \{q_0\}, q_0, t, \widetilde{\Delta})$, where $\widetilde{\Delta}$ extends Δ by $\widetilde{\Delta}(q_0)(q_1, H + 1) = 1/2$ and $\widetilde{\Delta}(q_0)(q_2, 0) = 1/2$. By this construction, the new cost chain $\mathcal{C}$ initially either incurs cost $H + 1$ and then emulates $\mathcal{C}_1$, or incurs cost 0 and then emulates $\mathcal{C}_2$. Those possibilities have probability $1/2$ each.

Finally, we compute a suitable cost formula φ such that we have $val(g_1) \geq val(g_2)$ if and only if $\mathcal{P}(K_\mathcal{C} \models \varphi) \geq 1/2$, completing the logspace reduction. We remark that the structure of the formula φ, in particular the number of inequalities, is fixed. Only the involved numbers depend on the concrete instance.

5   Cost Processes

Acyclic Cost Processes. We now prove that the cost problem for acyclic cost processes is PSPACE-complete. The challenging part is to show that PSPACE-hardness even holds for *atomic* cost formulas. For our lower bound, we reduce from a generalisation of the classical SUBSETSUM problem: Given a tuple $(k_1, \ldots, k_n, T)$ of natural numbers with n even, the QSUBSETSUM *problem* asks whether the following formula is true:

$$\exists x_1 \in \{0,1\} \ \forall x_2 \in \{0,1\} \ \cdots \ \exists x_{n-1} \in \{0,1\} \ \forall x_n \in \{0,1\} \ : \ \sum_{1 \leq i \leq n} x_i k_i = T$$

Here, the quantifiers $\exists$ and $\forall$ occur in strict alternation. It is shown in [18, Lem. 4] that QSUBSETSUM is PSPACE-complete. One can think of such a formula as a turn-based game, the QSUBSETSUM *game*, played between Player Odd and Player Even. If $i \in \{1, \ldots, n\}$ is odd (even), then turn i is Player Odd's (Player Even's) turn, respectively. In turn i the respective player decides to either *take k_i* by setting $x_i = 1$, or *not* to *take k_i* by setting $x_i = 0$. Player Odd's objective is to make the sum of the taken numbers equal T, and Player Even tries to prevent that. If Player Even is replaced by a random player, then Player Odd has a strategy to win with probability 1 if and only if the given instance is a "yes" instance for QSUBSETSUM. This gives a PSPACE-hardness proof for the cost problem with non-atomic cost formulas $\varphi \equiv (x = T)$. In order to strengthen the lower bound to atomic cost formulas $\varphi \equiv (x \leq B)$ we have to give Player Odd an incentive to take numbers k_i, although she is only interested in not exceeding the budget B. This challenge is addressed in our PSPACE-hardness proof.

The PSPACE-hardness result reflects the fact that the optimal strategy must take the current cost into account, not only the control state, even for atomic cost formulas. This may be somewhat counter-intuitive, as a good strategy should always "prefer small cost". But if there always existed a strategy depending *only* on the control state, one could guess this strategy in NP and invoke the PP-result of Sec. 4 in order to obtain an NP^{PP} algorithm, implying $\text{NP}^{\text{PP}} = \text{PSPACE}$ and hence a collapse of the counting hierarchy.

Indeed, for a concrete example, consider the acyclic cost process with $Q = \{q_0, q_1, t\}$, and $En(q_0) = \{a\}$ and $En(q_1) = \{a_1, a_2\}$, and $\Delta(q_0, a)(q_1, +1) = \frac{1}{2}$ and $\Delta(q_0, a)(q_1, +3) = \frac{1}{2}$ and $\Delta(q_1, a_1)(t, +3) = 1$ and $\Delta(q_1, a_2)(t, +6) = \frac{1}{2}$ and $\Delta(q_1, a_2)(t, +1) = \frac{1}{2}$. Consider the atomic cost formula $\varphi \equiv (x \leq 5)$. An optimal scheduler σ plays a_1 in $(q_1, 1)$ and a_2 in $(q_1, 3)$, because additional cost 3, incurred by a_1, is fine in the former but not in the latter configuration. For this scheduler σ we have $\mathcal{P}_\sigma(K \models \varphi) = \frac{3}{4}$.

Theorem 7. *The cost problem for acyclic cost processes is in* PSPACE. *It is* PSPACE-*hard, even for atomic cost formulas.*

Proof (sketch). To prove membership in PSPACE, we consider a procedure OPT that, given $(q, c) \in Q \times \mathbb{N}$ as input, computes the optimal (i.e., maximised over all schedulers) probability $p_{q,c}$ that starting from (q, c) one reaches (t, d) with $d \models \varphi$. The following procedure characterisation of $p_{q,c}$ for $q \neq t$ is crucial for OPT(q, c):

$$p_{q,c} = \max_{a \in En(q)} \sum_{q' \in Q} \sum_{k \in \mathbb{N}} \Delta(q, a)(q', k) \cdot p_{q',c+k}$$

So $\text{OPT}(q, c)$ loops over all $a \in En(q)$ and all $(q', k) \in Q \times \mathbb{N}$ with $\Delta(q, a)(q', k) > 0$ and recursively computes $p_{q', c+k}$. Since the cost process is acyclic, the height of the recursion stack is at most $|Q|$. The representation size of the probabilities that occur in that computation is polynomial. To see this, consider the product D of the denominators of the probabilities occurring in the description of Δ. The encoding size of D is polynomial. All probabilities occurring during the computation are integer multiples of $1/D$. Hence computing $\text{OPT}(q_0, 0)$ and comparing the result with τ gives a PSPACE procedure.

For the lower bound we reduce the QSUBSETSUM problem, defined above, to the cost problem for an atomic cost formula $x \leq B$. Given an instance $(k_1, \ldots, k_n, T)$ with n is even of the QSUBSETSUM problem, we construct an acyclic cost process $\mathcal{C} = (Q, q_0, t, A, En, \Delta)$ as follows. We take $Q = \{q_0, q_2, \ldots, q_{n-2}, q_n, t\}$. Those control states reflect pairs of subsequent turns that the QSUBSETSUM game can be in. The transition rules Δ will be set up so that probably the control states $q_0, q_2, \ldots, q_n, t$ will be visited in that order, with the (improbable) possibility of shortcuts to t. For even i with $0 \leq i \leq n-2$ we set $En(q_i) = \{a_0, a_1\}$. These actions correspond to Player Odd's possible decisions of not taking, respectively taking k_{i+1}. Player Even's response is modelled by the random choice of not taking, respectively taking k_{i+2} (with probability $1/2$ each). In the cost process, taking a number k_i corresponds to incurring cost k_i. We also add an additional cost ℓ in each transition.[1] Therefore we define our cost problem to have the atomic formula $x \leq B$ with $B := (n/2) \cdot \ell + T$. For some large number $M \in \mathbb{N}$, formally defined in [9], we set for all even $i \leq n-2$ and for $j \in \{0, 1\}$:

$$\Delta(q_i, a_j)(q_{i+2}, \ell + j \cdot k_{i+1}) = (1/2) \cdot (1 - (\ell + j \cdot k_{i+1})/M)$$
$$\Delta(q_i, a_j)(t, \ell + j \cdot k_{i+1}) = (1/2) \cdot (\ell + j \cdot k_{i+1})/M$$
$$\Delta(q_i, a_j)(q_{i+2}, \ell + j \cdot k_{i+1} + k_{i+2}) = (1/2) \cdot (1 - (\ell + j \cdot k_{i+1} + k_{i+2})/M)$$
$$\Delta(q_i, a_j)(t, \ell + j \cdot k_{i+1} + k_{i+2}) = (1/2) \cdot (\ell + j \cdot k_{i+1} + k_{i+2})/M$$

So with high probability the MDP transitions from q_i to q_{i+2}, and cost ℓ, $\ell + k_{i+1}$, $\ell + k_{i+2}$, $\ell + k_{i+1} + k_{i+2}$ is incurred, depending on the scheduler's (i.e., Player Odd's) actions and on the random (Player Even) outcome. But with a small probability, which is proportional to the incurred cost, the MDP transitions to t, which is a "win" for the scheduler as long as the accumulated cost is within budget B. We make sure that the scheduler loses if q_n is reached:

$$\Delta(q_n, a)(t, B+1) = 1 \qquad \text{with } En(q_n) = \{a\}$$

The MDP is designed so that the scheduler probably "loses" (i.e., exceeds the budget B); but whenever cost k is incurred, a winning opportunity with probability k/M arises. Since $1/M$ is small, the overall probability of winning is approximately C/M if total cost $C \leq B$ is incurred. In order to maximise this

---

[1] This is for technical reasons. Roughly speaking, this prevents the possibility of reaching the full budget B before an action in control state q_{n-2} is played.

chance, the scheduler wants to maximise the total cost without exceeding B, so the optimal scheduler will target B as total cost.

The values for ℓ, M and τ need to be chosen carefully, as the overall probability of winning is not exactly the sum of the probabilities of the individual winning opportunities. By the "union bound", this sum is only an upper bound, and one needs to show that the sum approximates the real probability closely enough.   □

General Cost Processes. We show the following theorem:

Theorem 8. *The cost problem is* EXP*-complete.*

The EXP upper bound was stated in Prop. 1. The lower bound is based on a reduction from *countdown games* [11].

The Cost-Utility Problem. MDPs with *two* non-negative and non-decreasing integer counters, viewed as cost and utility, respectively, were considered e.g. in [2]. Specifically, those works consider problems such as computing the minimal cost C such that the probability of gaining at least a given utility U is at least τ. Possibly the most fundamental of those problems is the following: the *cost-utility problem* asks, given an MDP with both cost and utility, and numbers $C, U \in \mathbb{N}$, whether one can, with probability 1, gain utility at least U using cost at most C. Using essentially the proof of Thm. 8 we show:

Corollary 9. *The cost-utility problem is* EXP*-complete.*

The Universal Cost Problem. We defined the cost problem so that it asks whether *there exists* a scheduler σ with $\mathcal{P}_\sigma(K_\mathcal{C} \models \varphi) \geq \tau$. A variant is the *universal cost problem*, which asks whether $\mathcal{P}_\sigma(K_\mathcal{C} \models \varphi) \geq \tau$ holds *for all* schedulers σ. Here the scheduler is viewed as an adversary which tries to prevent the satisfaction of φ. For cost chains the cost problem and the universal cost problem are equivalent. Thms. 7 and 8 hold analogously in the universal case:

Theorem 10. *The universal cost problem for acyclic cost processes is in* PSPACE*. It is* PSPACE*-hard, even for atomic cost formulas. The universal cost problem is* EXP*-complete.*

6   Conclusions and Open Problems

In this paper we have studied the complexity of analysing succinctly represented stochastic systems with a single non-negative and only increasing integer counter. We have improved the known complexity bounds significantly. Among other results, we have shown that the cost problem for Markov chains is in PSPACE and both hard for PP and the PosSLP problem. Can one prove PSPACE-hardness or membership in the counting hierarchy?

Regarding acyclic and general MDPs, we have proved PSPACE-completeness and EXP-completeness, respectively. Our results leave open the possibility that the cost problem for atomic cost formulas is not EXP-hard and even in PSPACE. The technique described in the proof sketch of Thm. 7 cannot be applied to general cost processes, because there we have to deal with paths of exponential length, which, informally speaking, have double-exponentially small probabilities. Proving hardness in an analogous way would thus require probability thresholds τ of exponential representation size.

Acknowledgments. The authors would like to thank Andreas Göbel for valuable hints, Christel Baier and Sascha Klüppelholz for thoughtful feedback on an earlier version of this paper, and anonymous referees for their helpful comments.

References

1. Allender, E., Bürgisser, P., Kjeldgaard-Pedersen, J., Bro, P.: Miltersen. On the complexity of numerical analysis. SIAM J. Comput. **38**(5), 1987–2006 (2009)
2. Baier, C., Dubslaff, C., Klüppelholz, S.: Trade-off analysis meets probabilistic model checking. In: Proc. CSL-LICS, pp. 1:1–1:10. ACM (2014)
3. Baier, C., Katoen, J.-P.: Principles of Model Checking. MIT Press (2008)
4. Baier, C., Klein, J., Klüppelholz, S., Märcker, S.: Computing conditional probabilities in markovian models efficiently. In: Ábrahám, E., Havelund, K. (eds.) TACAS 2014 (ETAPS). LNCS, vol. 8413, pp. 515–530. Springer, Heidelberg (2014)
5. Brázdil, T., Brožek, V., Etessami, K., Kučera, A., Wojtczak, D.: One-counter Markov decision processes. In: Proc. SODA, pp. 863–874. SIAM (2010)
6. Bruyère, V., Filiot, E., Randour, M., Raskin, J.-F.: Meet your expectations with guarantees: beyond worst-case synthesis in quantitative games. In: Proc. STACS, LIPIcs, vol. 25, pp. 199–213 (2014)
7. Etessami, K., Yannakakis, M.: Recursive Markov chains, stochastic grammars, and monotone systems of nonlinear equations. J. ACM **56**(1), 1:1–1:66 (2009)
8. Gill, J.: Computational complexity of probabilistic Turing machines. SIAM J. Comput. **6**(4), 675–695 (1977)
9. Haase, C., Kiefer, S.: The odds of staying on budget (2014). Technical Report at http://arxiv.org/abs/1409.8228
10. Haase, C., Kiefer, S.: The complexity of the Kth largest subset problem and related problems (2015). Technical Report at http://arxiv.org/abs/1501.06729
11. Jurdziński, M., Sproston, J., Laroussinie, F.: Model checking probabilistic timed automata with one or two clocks. Log. Meth. Comput. Sci. **4**(3), 12 (2008)
12. Laroussinie, F., Sproston, J.: Model checking durational probabilistic systems. In: Sassone, V. (ed.) FOSSACS 2005. LNCS, vol. 3441, pp. 140–154. Springer, Heidelberg (2005)
13. Puterman, M.L.: Markov Decision Processes: Discrete Stochastic Dynamic Programming. John Wiley and Sons (2008)
14. Randour, M., Raskin, J.-F., Sankur, O.: Percentile queries in multi-dimensional Markov decision processes. In: Proc. CAV, LNCS (2015)
15. Randour, M., Raskin, J.-F., Sankur, O.: Variations on the stochastic shortest path problem. In: D'Souza, D., Lal, A., Larsen, K.G. (eds.) VMCAI 2015. LNCS, vol. 8931, pp. 1–18. Springer, Heidelberg (2015)

16. Simon, J.: On the difference between one and many. In: Salomaa, A., Steinby, M. (eds.) ICALP 1977. LNCS, vol. 52, pp. 480–491. Springer, Heidelberg (1977)
17. Toda, S.: PP is as hard as the polynomial-time hierarchy. SIAM J. Comput. **20**(5), 865–877 (1991)
18. Travers, S.: The complexity of membership problems for circuits over sets of integers. Theor. Comput. Sci. **369**(1–3), 211–229 (2006)
19. Ummels, M., Baier, C.: Computing quantiles in Markov reward models. In: Pfenning, F. (ed.) FOSSACS 2013 (ETAPS 2013). LNCS, vol. 7794, pp. 353–368. Springer, Heidelberg (2013)

From Sequential Specifications to Eventual Consistency

Radha Jagadeesan and James Riely[✉]

DePaul University, Chicago, USA
jriely@gmail.com

Abstract. We address a fundamental issue of *interfaces* that arises in the context of cloud computing. We define what it means for a replicated and distributed implementation satisfy the standard sequential specification of the data structure. Several extant implementations of replicated data structures already satisfy the constraints of our definition. We describe how the algorithms discussed in a recent survey of convergent or commutative replicated datatypes [17] satisfy our definition. We show that our definition simplifies the programmer task significantly for a class of clients who conform to the CALM principle [10].

1 Introduction

An example serves to motivate the problem addressed in this paper. Consider an integer set interface with mutator methods add and remove and a single, boolean-valued accessor method get. We will assume that mutators do not return values (have return type Unit or void) and that accessors do not alter the state of the object. The sequential behavior of such a set can be defined as a set of strings such as $\mathsf{X}0$ $+0$ $\checkmark 0$ $\mathsf{X}1$ and $+0$ $+1$ $\checkmark 0$ $\checkmark 1$ -1 $\checkmark 0$ $\mathsf{X}1$, where $+k$ represents a call to add with argument k, $-k$ represents remove(k), $\checkmark k$ represents get(k) returning true and $\mathsf{X}k$ represents get(k) returning false. Since accessor methods do not alter the state of the object, the interface is closed under commutation of accessors: if $(s\ \checkmark 0\ \mathsf{X}1)$ is a valid traces in the interface, for some s, then so is $(s\ \mathsf{X}1\ \checkmark 0)$.

Consider the implementation of such a set as a cloud service that is implemented by replication of the data structure (eg. see [17]). In this distributed setting, we assume intra-node atomicity and sequencing of state transitions, whereas temporal relations between two computers that are distributed is only induced by the receipt of messages over the network. In this distributed context, there are two impediments to requiring the replicas to achieve consensus on a global total order [13] on the operations on the data structure. Firstly, the associated serialization bottleneck negatively affects performance and scalability (eg, see [6]). Secondly, the CAP theorem [8] imposes a tradeoff between consistency and partition-tolerance.

This has led to the emergence of alternative approaches based on *eventual consistency* and *optimistic replication* [16,19]. In such approaches, a replica may execute an operation without synchronizing with other replicas. The other replicas are updated asynchronously with the update operation. However, due to the vagaries of the network, even if every replica eventually receives and applies all updates, it could happen in possibly different orders. So, there has to be some mechanism to reconcile conflicting

© Springer-Verlag Berlin Heidelberg 2015
M.M. Halldórsson et al. (Eds.): ICALP 2015, Part II, LNCS 9135, pp. 247–259, 2015.
DOI: 10.1007/978-3-662-47666-6_20

updates (for illustrative examples, see [17, 18]). Thus, such approaches address the issue of efficiency (since any query to the state of the data structure at a replica is answered locally at the replica without any consensus overhead) and data remains available even in the presence of network partitions.

The literature on convergent or commutative replicated datatypes (CRDTs) (see [17] for a survey) provides a systematic attempt to design such datastructures. Consider the following diagram, in the style of [17].

$$\begin{array}{ccccccccc}
\longrightarrow & +0 & \Longrightarrow & \checkmark 0 & \longrightarrow & \chi 1 & \Longrightarrow\!\!\bullet\!\!\longrightarrow & \checkmark 0 & \longrightarrow & \checkmark 1 \\[2mm]
\longrightarrow & +1 & \Longrightarrow & \checkmark 1 & \longrightarrow & \chi 0 & \Longrightarrow\!\!\bullet\!\!\longrightarrow & \checkmark 0 & \longrightarrow & \checkmark 1
\end{array} \qquad (1)$$

In this sample execution, the mutators +0 and +1 are executed at distinct replicas. The actions in each replica are temporally ordered from left to right, as indicated by the horizontal arrows. We assume the local updates are atomic. After a local update, the replica forwards messages to the other replicas; in the diagram, the diagonal arrows between replicas indicate messages that propagate such local updates, with the interpretation that the operation is guaranteed to be finished at the recipient at the point the arrow appears on the recipients timeline. The accessors are executed locally and atomically at each replica. Of course, there is a consistent global state, testified by $\checkmark 0$ and $\checkmark 1$ at both replicas, after both messages have been delivered. Thus, the literature (eg. see [17] for a precise formalization) deems this implementation to be eventually consistent, since the states of all the replicas eventually converge at quiescent points, when all the messages have been delivered. This view is adequate for examples where we are interested only in the final state of the data structure.

Since eventual consistency only speaks about the quiescent points of the system, it does not address correctness of intermediate states in the evolution of the system. For example, all of the following implementation traces of a putative replicated set are deemed to be eventually consistent, even though we see very problematic behavior.

$$\begin{array}{ccc}
\begin{array}{l}
\rightarrow +0 \rightarrow\!\bullet\!\rightarrow \checkmark 1 \rightarrow \chi 1 \rightarrow \\
\rightarrow +1 \rule{3cm}{0.4pt}
\end{array}
&
\rightarrow \checkmark 0 \rightarrow +0 \rightarrow \checkmark 0
&
\begin{array}{l}
\rightarrow +0 \rightarrow\!\bullet\!\bullet\!\rightarrow \chi 0 \rightarrow +0 \rightarrow \checkmark 0 \\
\rightarrow -0 \rightarrow\!\bullet\!\rightarrow \checkmark 0 \rule{1cm}{0.4pt}\!\bullet\!\rightarrow \checkmark 0
\end{array} \\[4mm]
(2) & (3) & (4)
\end{array}$$

In figure (2), the accessor results regresses from $\checkmark 1$ to $\chi 1$ even though there is no remove invocation in the system; in figure (3), the initial accessor $\checkmark 0$ is not justified; in figure (4), the replicas conflict in their ordering of concurrent add/remove updates.

This problem is addressed by the seminal papers of [2,3]. [3] defines a notion of eventual consistency for transactions intuitively as compatibility with a serialization of them. In contrast, [2] views the interface of a replicated data structure as a *concurrent* specification that determines the valid result of an accessor from the context of a prior concurrent history. [1] extends this approach to allow for bounded rollbacks. In this style, the above examples are *declared* invalid; for example in figure (2), the result $\chi 1$ is deemed invalid in the context of its prior history.

In addition to capturing the properties of replicated implementations much more precisely than the traditional definitions of eventual consistency, this line of work has also lead to useful tools and techniques to aid the programmer: [3] provides general control flows sturctures that are guaranteed to yield eventually consistent implementations of transactions; [9] proves abstraction and composition theorems, applying it in particular to the replicated implementation of a graph data structure; [1] develops model checking techniques to reason about implementations relative to these specifications.

In these approaches, replicated data structures are specified directly, without any formal comparison to the sequential data structures that they are meant to approximate. This approach is (intentionally) agnostic to the design of the specifications themselves. For example, whereas the result ✗1 in figure (2) is not valid, the result ✗1 in figure (1) is valid. The justification for the different decisions about ✗1 in figures (1) and (2) is the traditional *sequential* specification of Set; namely, if there are no remove operations, ✓1 is acceptable iff there is a preceding +1.

In this paper, building on [3], we provide a definition of eventual consistency that develops precisely such a connection with the sequential specification. (As can be seen from figure (1), traditional criteria, such as linearizability [11] and quiescent consistency [5,12], do not apply here.) We show the utility of our definition by showing that clients satisfying the CALM principle (see [10] for a survey) can in fact abstract away completely from the distributed and replicated implementation and program against the sequential specification realized by the implementation.

Our work complements the research program of [1–3,9]. Our methods aim to provide a way to justify the interfaces described in this approach. In future work, we hope to use our methods to show that CALM clients of their interfaces can also be protected from details of distribution and replication. We also hope to adopt their methods to support a more general class of clients and to develop reasoning methods to show that implementations satisfy their specification.

An Informal Outline of our Approach. In a replicated data structure, a mutator m is *visible* to an event a if m executes at a's replica before a executes. We say that an implementation trace (such as those in the figures above) satisfies a sequential specification if for each event a, we can associate a string of events $t(a)$ that satisfies the following.

Mutator closed: $t(a)$ includes a as well as the mutator events that are visible to a

Validity: $t(a)$ is a valid sequential trace that ends in a

History consistency: For any events d and e, $t(d)$ and $t(e)$ agree on the ordering of mutator events that are visible to both d and e.

Figure (2) does not satisfy validity at the event ✗1 in the top replica since neither +0 +1 ✗1 nor +1 +0 ✗1 is a valid trace of a set. Figure (3) does not satisfy validity at the initial event ✓0 since the trace ✓0 is not a valid trace of a set. In Figure (4), to satisfy validity, we have to associate the trace −0 +0 at the ✓0 event in the top replica and the trace +0 −0 at the ✗0 event in the bottom replica, thus violating history-consistency.

For a positive example, in figure (1), the traces associated to the event ✗1 in the top replica is +0 and the trace associated to the event ✗0 in the bottom replica is +1. There is a choice for the trace associated with the events ✓1 in the top replica and ✓0 in the bottom replica. By consistency, they need to be the same, but they can both be chosen to be either +0 +1 or +0 +1.

Our definition is flexible enough to accommodate the data structures discussed in [17], a recent survey of the literature on CRDTs. For several of these data structures, the implementations unambiguously and categorically satisfy our definitions. There are particular subtleties that arise when we match some SET implementations (the OR-set and the 2P-set) against the sequential set specification that we discuss in the technical sections that follow.

Such data structures provide a particularly simple programming view for clients located at a replica. In a logically monotone execution, the arrival time of a concurrent mutator does not alter the evolution of the system (Our formalization of logically monotone executions is inspired by ideas in [14, 15].) We formalize a weaker monotonicity property: that there is *some* ordering of concurrent mutators that does not alter the evolution of a system. Under this weak monotonicity assumption (that is satisfied by all CALM executions), we prove abstraction [7] and composition [11] theorems. This is particularly relevant because it simplifies the programmer perspective for a large class of programs that includes those written in languages that realize the CALM principle, such as Bloom [4].

2 Bracketed Partial Orders and Labeled Visibility Relations

In this section, we define bracketed partial orders (BPOs). BPOs provide a formalization of diagrams such as those given in the introduction. BPOs are labelled partial orders, enriched with *replicas* and *bracketing*. Bracketing relates the remote execution of a mutator to the initial call of the mutator. Consider the following example.

$$
\begin{array}{c}
\rightarrow +0 \rightarrow +1 \\
\rightarrow \bullet \rightarrow \checkmark 1 \rightarrow X0 \rightarrow \bullet \rightarrow \checkmark 0
\end{array}
\tag{5}
$$

This is formalized as a BPO with seven events. There are two replicas: one for each horizontal line. The partial order is given by the arrows. Two events are labelled as mutators: +0 and +1. Three events are labelled as accessors: $\checkmark 1$, $X0$ and $\checkmark 0$. The remaining two events (shown without labels in the diagram) are bracketing events. In the formalism, bracketing events are labelled with the name of the preceding mutator event. Generally one is interested in the isomorphism class of labelled partial orders (the pomset), and therefore the event names themselves are uninteresting.

A BPO is *causal* if the order of mutator and bracketing events at each replica respects the partial order of the mutator events themselves. All of the figures in the introduction are causal. Figure (5) is not causal, however, since the mutator order is +0 +1 but the order at the bottom replica is +1 +0.

BPOs directly capture the notion of an *operation-based* CRDT (see [17]). *State-based* CRDTs can be considered a special case of causal BPOs that communicate multiple brackets with a single communication (modelled as an uninterrupted sequence of bracketed events at the receiving replica).

Let $\mathscr{A}$ and $\mathscr{M}$ be disjoint sets of *accessor labels* and *mutator labels*, respectively, and let $\mathscr{L} = \mathscr{A} \cup \mathscr{M}$ be a set of *labels*. We use metavariables s–v to range over various types of relations with labels in $\mathscr{L}$, which we generically refer to as "traces".

Example 2.1. In this paper we consider four implementations of an integer set datatype: the *G-set*, *U-set*, *OR-set*, and *2P-set*. See [17] for implementation details.

A G-set has mutator labels of the form $+k$, where k is an integer, and accessor labels of the form $\checkmark k$ and $\cancel{x}k$. A G-set is *grow only*; thus, once $\checkmark k$ has been observed for a particular k, it is impossible to subsequently observe $\cancel{x}k$. It is straightforward to specify the replicated implementation and, therefore, the corresponding BPO.

A U-set adds mutators of the form $-k$ to the labels of a G-set, denoting removal. A U-set requires that for every k, $+k$ may appear at most once in each execution—each $+k$ is *unique*. In addition, a $-k$ may only occur when $+k$ is visible. These requirements are imposed on the *client* of the U-set; it is not ensured by the U-set itself. The implementation is again straightforward. The client can guarantee uniqueness using various techniques; for example, take $k = 2^c \cdot 3^n$ where c is a globally unique client thread identifier and n is a monotone thread-local counter.

An OR-set (observed-remove set) has the same labels as a U-set, but does not require that $+k$ actions are unique. The implementation uses an underlying U-set and a map from the elements of the U-set to the elements of the OR-set. Consider the following BPO, from [17].

$$
\begin{array}{c}
\longrightarrow +0 \rightleftharpoons \rightarrow -0 \rightleftharpoons \rightarrow \bullet \longrightarrow \checkmark 0 \\
\longrightarrow +0 \rightleftarrows \\
\hspace{2cm} \searrow \bullet \hspace{0.5cm} \bullet \hspace{0.5cm} \bullet \longrightarrow \checkmark 0
\end{array}
\tag{6}
$$

This BPO is not a valid execution of a G-set (because of the -0) or a U-set (because of the two $+0$'s). However, this is a valid execution of an OR-set. The $+0$ in the middle replica is concurrent with the -0 of the top replica. Since they are working on top of an underlying U-set, the -0 only removes the $+0$ added by the top replica; the middle $+0$ is not affected and eventually prevails.

A 2P-set is implemented using two grow sets; one representing additions and one representing tombstones for removed elements, in the obvious way. Like a U-set, a 2P-set also constrains the behaviour of clients. A client must ensure that no element that is removed is subsequently re-added. The BPO in figure (6) is a valid 2P-set BPO if the events labeled $\checkmark 0$ are re-labeled to $\cancel{x}0$. In a OR-set, an add "wins" over a concurrent remove, whereas in a 2P-set, the remove wins. Thus these two examples represent different specializations of the set API. The OR-set resolves figure (6) to the sequential specification $+0 -0 +0 \checkmark 0 \checkmark 0$, whereas the 2P-set resolves it to $+0 +0 -0 \cancel{x}0 \cancel{x}0$.

The constraints on the clients of U-set and 2P-set are required for correct functioning *as a set*. The definition of *correctness* is given informally in [17]. The main contribution of this paper is to provide a formalization, which we do in Section 3. Under our definition, all executions of the G-set will be considered correct, and all causal executions of U-set will be considered correct, but only a subset of executions of the OR-set and 2P-set will be considered correct.   □

Definition 2.2. A *(replicated) bracketed partial order (BPO)* is a octuple $\langle E_A, E_M, E_B,$ $L, R, \lambda, \rho, \Rightarrow \rangle$ where R is a set of replicas, and the following hold.

(a) sets E_A, E_M and E_B are disjoint, $L \subseteq \mathcal{L}$, and $\langle E_A \cup E_M \cup E_B, \Rightarrow \rangle$ is a partial order,
(b) $\rho \in (E_A \cup E_M \cup E_B) \mapsto R$ and $\lambda \in (E_A \mapsto L \cap \mathcal{A}) \cup (E_M \mapsto L \cap \mathcal{M}) \cup (E_B \mapsto E_M)$,
(c) $\forall e \in E_B.\ \lambda(e) \Rightarrow e$ and $\rho(\lambda(e)) \neq \rho(e)$
(d) $\forall d, e \in E_B.$ if $\lambda(d) = \lambda(e)$ then either $d = e$ or $\rho(d) \neq \rho(e)$
(e) $\forall d, e \in E.$ if $\rho(d) = \rho(e)$ then either $d \Rightarrow e$ or $e \Rightarrow d$.

For a BPO s, we write $E_A(s)$ for the accessor events of s, $E_M(s)$ for the mutator events and $E_B(s)$ for the bracketing events. We also define $E_{AM}(s) \overset{\triangle}{=} E_A(s) \cup E_M(s)$. □

Condition (b) establishes the interpretation of the labelling function: The elements of E_A denote local events (accessors), the elements of E_M denote the origination of a global event (mutators), and the elements of E_B denote the remote reception of a global event (brackets). Events $m \in E_M$ and $b \in E_B$ are a *bracketed pair* when $\lambda(b) = m$. Condition (c) establishes that in a bracketed pair, the beginning must precede the end and occur at a separate replica. Condition (d) establishes that each mutator is bracketed at most once per replica. Thus, each mutator event has one "beginning" and as many as $|R| - 1$ "endings". Condition (e) establishes that events are totally ordered at each replica; concurrency within a replica can be handled via standard means.

Definition 2.3 (Causal). Let s be an BPO. Define $\text{remote}_s(e) \overset{\triangle}{=} \{b \in E_B(s) \mid \lambda_s(b) = e\}$. The BPO s is *causal* when $\forall d, e \in E_M(s)$. $\forall d' \in \text{remote}_s(d)$. $\forall e' \in \text{remote}_s(e)$. if $d \Rightarrow_s e$ and $\rho_s(d') = \rho_s(e')$ then $d' \Rightarrow_s e'$. □

BPOs have a clear operational intuition. We now provide an abstract view of BPOs which is sufficient to define correctness. The relations we need are weaker than labeled partial orders. In particular, we do not require transitivity. We refer to these potentially intransitive relations as *labeled visibility relations* (LVRs). For example, starting with the BPO given in figures (5) and (6), we derive the following LVRs.

In these diagrams, we use $\rightsquigarrow$ to represent an intransitive edge and $\rightarrow$ to represent a "transitive" edge. Thus, in the left diagram, the event $✗0$ sees $+1$ and $✓1$, but not $+0$, whereas $✓0$ sees all four prior events. Recall from figure (5) that the replica that generates $✗0$ sees $+1$ before $+0$, even though these are initiated in the reverse order. A causal BPO generates a transitive LVR, as in the right diagram above. Formally, LVRs are defined with a single visibility relation, which may or may not be transitive. We include replica identifiers to define liveness properties; we ignore them except when important.

Definition 2.4. Let $s = \langle E, L, R, \lambda, \rho, \rightsquigarrow \rangle$ be a sextuple such that E is a finite set of events, L is a set of labels, R is a set of replicas, $\lambda \in (E \mapsto L)$, $\rho \in (E \mapsto R)$ and $\rightsquigarrow \subseteq (E \times E)$. We say that s is a *labeled visibility relation (LVR)* if $\rightsquigarrow$ is reflexive and acyclic. We say that s is a *labeled partial order (LPO)* if $\rightsquigarrow$ is a partial order. We say that s is a *labeled total order (LTO)* if $\rightsquigarrow$ is a total order.

Given an LVR s, we write $E(s)$ for the event set of s, $L(s)$ for the label set, λ_s for the labeling function and $\rightsquigarrow_s$ for the visibility relation. Define $E_A(s) \overset{\triangle}{=} \{e \in E(s) \mid \lambda(e) \in \mathscr{A}\}$ and $E_M(s) \overset{\triangle}{=} \{e \in E(s) \mid \lambda(e) \in \mathscr{M}\}$. □

Below, we define the translation from BPOs to LVRs. For a BPO s, the relation $\xrightarrow{\text{local}}_s$ is the union of the local orders at each replica. Whenever $d \xrightarrow{\text{local}}_s e$, we have that $d \rightsquigarrow_s e$. For mutators m and accessors a, we have that $m \rightsquigarrow a$ if m has been received at a's replica. Otherwise, events d and e at different replicas are ordered when they are ordered

by $\Rightarrow_s$ and every mutator visible to d is also visible to e. The BPO $\xrightarrow{\ m\ \ a\ \ n\ } b$
translates to the LVR $m \rightsquigarrow a \rightsquigarrow n \rightsquigarrow b$, which we draw as $m \rightarrow a \rightarrow n \rightarrow b$. The BPO
$\xrightarrow{\ m\ \ a\ \ n\ } b$ translates to the LVR $m \rightarrow a \rightarrow n \rightsquigarrow b$.

For a BPO s, we have that $\forall m \in \mathsf{E_M}(s). \forall a, b \in \mathsf{E_A}(s).$ if $m \rightsquigarrow_s a \rightsquigarrow_s b$ then $m \rightsquigarrow_s b$.

Definition 2.5. For any sets $C \subseteq A$ and relation $\mathbf{R} \subseteq A \times A$, define $\mathbf{R} \setminus C \stackrel{\triangle}{=} \mathbf{R} \cap (C \times C)$.
Similarly, for $\mathbf{R} \subseteq A \times B$ and $C \subseteq A$, define $\mathbf{R} \setminus C \stackrel{\triangle}{=} \mathbf{R} \cap (C \times B)$.

Let s be a BPO. Define $(d \xrightarrow{\text{local}}_s e) \stackrel{\triangle}{=} (d \Rightarrow_s e)$ and $(\rho_s(d) = \rho_s(e))$. Recall Definition 2.3 of remote. Define $\mathsf{visM}_s(e) \stackrel{\triangle}{=} \{m \in \mathsf{E_M}(s) \mid m \xrightarrow{\text{local}}_s e$ or $\exists b \in \mathsf{remote}_s$
$(m). b \xrightarrow{\text{local}}_s e\}$. Then we define the LVR derived from s as follows: $\mathsf{lvr}(s) \stackrel{\triangle}{=} \langle \mathsf{E_{AM}}(s),$
$\mathsf{L}(s), \mathsf{R}(s), \rho_s, \lambda_s \setminus \mathsf{E_{AM}}(s), \rightsquigarrow \rangle$ where $\forall d, e \in \mathsf{E_{AM}}(s). d \rightsquigarrow e$ iff $d \in \mathsf{visM}_s(e)$ or $d \Rightarrow_s e$
and $\mathsf{visM}_s(d) \subseteq \mathsf{visM}_s(e)$. We write $\rightsquigarrow_s$ for the visibility relation of $\mathsf{lvr}(s)$. □

In a strongly distributed BPO, events at different replicas are only ordered via bracketed pairs; this disallows synchronization between replicas outside of the data structure formalized by the BPO.

Definition 2.6. A BPO is *strongly distributed* if $\forall d, e \in E_A \cup E_M \cup E_B$. if $\rho(d) \neq \rho(e)$
and $d \Rightarrow e$ then $\exists d' \in E_M, e' \in E_B. \lambda(e') = d'$ and $d \Rightarrow d' \Rightarrow e' \Rightarrow e$ □

Lemma 2.7. Let s be a strongly distributed BPO. Then the following three statements
are equivalent: (a) s is causal, (b) $(\rightsquigarrow_s)$ is transitive, and (c) $(\rightsquigarrow_s) = (\Rightarrow_s \setminus \mathsf{E_{AM}}(s))$. □

3 Eventual Consistency

Definitions of eventual consistency (EC) traditionally include both safety and liveness
properties. Liveness is purely a property of implementations. It can be expressed as a
simple closure property over sets of LVRs, which we call *eventual delivery*[1].

To define safety, we must first define specifications (Definition 3.1) and give some
basic vocabulary for permutations, order extensions and the like (Definition 3.2).

Specifications of sequential structures are typically given as sets of strings of labels.
To simplify the definitions, we use isomorphism closed sets of LTOs: the event set identifies a bijection between an implementation LVR and its specification as an LTO. Specification sets are closed with respect to renaming of events and arbitrary replacement
of the replica function (replicas don't matter in specifications). In addition, we ask that
specification sets be prefix closed, accessor enabled (an specification string can always
be extended by some accessor) and closed under reordering of adjacent accessors (if
there is no intervening mutator, then accessors commute).

Definition 3.1. Strings may be regarded as labeled total orders (LTOs) up to replica-insensitive isomorphism. LTOs s and t are *replica-insensitive isomorphic* if $\mathsf{L}(s) = \mathsf{L}(t)$
and there exists a bijection $\alpha : \mathsf{E}(s) \rightarrow \mathsf{E}(t)$ such that $\forall e \in \mathsf{E}(s). \lambda_s(e) = \lambda_t(\alpha(e))$ and
$\forall d, e \in \mathsf{E}(s). (d \rightsquigarrow_s e)$ iff $(\alpha(d) \rightsquigarrow_t \alpha(e))$.

---

[1] See Definition 3.2 of the *extension* of a partial order (notation $\subseteq$). A set S of LVRs satisfies *eventual delivery* if each mutator is eventually seen at every replica: $\forall s \in S. \forall m \in \mathsf{E_M}(s). \forall p \in \mathsf{R}(s). \exists t \in S. s \subseteq t$ and $\exists a \in \mathsf{E_A}(t). m \rightsquigarrow_t a$.

The following closure properties, defined on sets of strings, lift to isomorphism closed sets of LTOs. For strings $s, t \in \mathscr{L}^*$, let "st" denote concatenation. Let $T \subseteq \mathscr{L}^*$ be a set of strings. We say that T is *prefix closed* when $st \in T$ implies $s \in T$. We say that T is *accessor enabled* when $s \in T$ implies $\exists a \in \mathscr{A}. sa \in T$. We say that T is *accessor closed* when $\forall a, b \in \mathscr{A}. \{ta, tb\} \subseteq T$ implies $\{tab, tba\} \subseteq T$.

A *specification* is a set of total orders (LTOs) that is replica-insensitive isomorphism closed, prefix closed, accessor enabled and accessor closed. □

A specification, as given by Definition 3.1, is "sequential" because the orders are total.

Definition 3.2. We write $=_\pi$ for permutation equivalence; if $s \leq_\pi t$ then t may contain additional events that are not matched in s. If $s \sqsubseteq t$, then t is an *visibility-extension* of s, with the same events and greater visibility. (For an LPO this is an *order-extension*.) If $s \subseteq t$, then t is an *extension* of s, with both more events and greater visibility. $(s \setminus D)$ denotes the restriction of s to the events in D. Define $\{_s^{\mathsf{M}}} e \triangleq s \setminus (\{e\} \cup \{d \in \mathsf{E_M}(s) \mid d \rightsquigarrow e\})$ and $\{_s^{\mathsf{M}}} e \triangleq s \setminus (\{e\} \cup \{d \in \mathsf{E_M}(s) \mid e \not\rightsquigarrow d\})$. □

For trace s and $e \in \mathsf{E}(s)$, $\{_s^{\mathsf{M}}} e$ denotes the restriction of s to the mutator events visible to e, and $\{_s^{\mathsf{M}}} e$ denotes the restriction to the mutator events that are either visible to or "concurrent with" e. Both $\{_s^{\mathsf{M}}} e$ and $\{_s^{\mathsf{M}}} e$ include at most one accessor: e itself.

To establish eventual consistency of s with respect to T, we must exhibit a function t that maps each event in $\mathsf{E}(s)$ to a specification trace in T. The choice of t is constrained by two conditions.

Fix an event e and let $\mathsf{t}(e) = t$. The first condition requires that t include only events visible to or concurrent with e, and that t respect the order of those events in s. The requirement $\mathsf{E}(\{_s^{\mathsf{M}}} e) \subseteq \mathsf{E}(t)$ establishes that t includes e, as well as all of the mutators visible to e. The requirement $\mathsf{E}(t) \subseteq \mathsf{E}(\{_s^{\mathsf{M}}} e)$ establishes that t only includes mutators that are either visible to or concurrent with e. Finally, the requirement that $(s \setminus \mathsf{E}(t)) \sqsubseteq t$ establishes that t must respect the order of events in s.

Fix events d and e. The second condition requires that $\mathsf{t}(d)$ and $\mathsf{t}(e)$ agree on the order of mutator events in their intersection.

Definition 3.3. We say that t *refines* s at e if $\mathsf{E}(\{_s^{\mathsf{M}}} e) \subseteq \mathsf{E}(t) \subseteq \mathsf{E}(\{_s^{\mathsf{M}}} e)$ and $(s \setminus \mathsf{E}(t)) \sqsubseteq t$. We write $s \approx_{\mathsf{M}} t$ when $\forall m, n \in \mathsf{E_M}(s) \cap \mathsf{E_M}(t). m \rightsquigarrow_s n$ iff $m \rightsquigarrow_t n$.

An LVR s is *eventually consistent* (EC) with a specification T (notation $s \vDash_{\mathsf{ec}} T$) when there exists a map $\mathsf{t} : \mathsf{E}(s) \rightarrow T$ such that (a) $\forall e \in \mathsf{E}(s). \mathsf{t}(e)$ refines s at e, and (b) $\forall d, e \in \mathsf{E}(s). \mathsf{t}(d) \approx_{\mathsf{M}} \mathsf{t}(e)$.

Write $S \vDash_{\mathsf{ec}} T$ when $\forall s \in S. s \vDash_{\mathsf{ec}} T$. □

We call this "eventual consistency" because the definition ensures that at quiescent points the same accessors at all the replicas are mapped to the same sequential trace of visible mutator events. Given eventual delivery, then all replicas must eventually agree on the order of all mutators. In the case that specifications are mutator enabled, eventual consistency can be defined in terms of a global order on mutators (u in the proposition below) that all replicas must agree to.

Definition 3.4. A specification T is *mutator enabled* if $\forall s \in T. \forall m \in \mathscr{M}. sm \in T$. □

Proposition 3.5. Suppose T is mutator enabled specification. Then $s \models_{ec} T$ iff there exists a total order $u =_\pi s \setminus \mathcal{M}$ such that $\forall e \in \mathsf{E}(s). \exists t_e \in T. t_e$ refines s at e and $t_e \approx_M u$. □

Example 3.6. Any G-set execution s satisfies our definition. To see this, we follow the characterization from Proposition 3.5. Choose u to be any linearization of the mutators in s consistent with the execution. For any accessor (✓k or ✗k), choose t to be the subsequence of the prefix of u that contains *only* the adds that precede the accessor in s.

Any causal execution s of a U-set satisfies our definition. Again, we follow the characterization from Proposition 3.5. For any query, choose t to be the subsequence of the prefix of u that contains *only* the mutators (adds and removes) that precede the accessor in s. Causality, as assumed in [17], is necessary for the U-set to satisfy the specifcation. Without causality, the following execution $\longrightarrow +0 \rightarrow -0 \Longrightarrow \longrightarrow$ ✗0/✓0 is possible. In this execution, the initial remove does nothing to the state of the bottom replica's local copy of the set, leaving the two replicas out of sync.

We now turn to the OR-set and 2P-set. First a positive example. Consider figure (6) from Example 2.1. What are acceptable return values for the get actions? The top replica sees the actions +0 −0 +0 whereas the bottom replica sees +0 +0 −0. They see the same actions, but in different orders. In the 2P-set implementation, both gets return false (remove has priority over add). In the OR-set implementation, both gets return true (add has priority over remove). Both executions are EC. For the 2P-set, let u be +0 +0 −0. For the OR-set, let u be +0 −0 +0. An implementation which returns different values for the gets is not EC because there is no t that satisfies the requirements. Since the gets see the same mutators, the traces chosen by t must agree on their order.

As a negative example, consider the following OR-set execution.

$$\longrightarrow +0 \longrightarrow +1 \longrightarrow -1 \longrightarrow$$
$$\longrightarrow +1 \longrightarrow +0 \longrightarrow -0 \longrightarrow \checkmark 0 \rightarrow \checkmark 1 \longrightarrow$$

In an OR-set, removes only affect the adds that are visible. In this execution, the top −1 does not affect the bottom +1, and symmetrically, the bottom −0 does not affect the top +0; thus, the execution is possible. However, this execution is not EC with respect to any set trace: since the final mutators are both removes, at least one of ?0 and ?1 must return false in any sequential trace.

To guarantee EC executions of an OR-set, it is sufficient to require that every −k action be ordered before any concurrent +k of the same value. If the resulting enriched BPO is acyclic, then the OR-set execution is EC. The example above fails this test since we would have a cycle involving all of the mutators: +0 +1 −1 +1 +0 −0 +0.

The analysis of the 2P-set is symmetric. □

We end this section with the following simple fact about eventual consistency. The proof uses the fact that we allow events that are concurrent with e to be included in $t(e)$.

Lemma 3.7. If $v \models_{ec} T$ and $s \subseteq v$ then $s \models_{ec} T$. □

4 Results

We define a language of clients and define interaction between a client and data structure. We then define monotonicity and state the abstraction and composition results.

Clients. We consider a simple language for clients: parallel composition of sequential processes, which include method call, sequencing and conditional. Let tt and ff represent the boolean constants. Let k range over *values*, which include tt and ff. Let o range over *objects*, m over *mutator methods*, and a over *accessor methods*. Then *programs* (P), *configurations* (C) and *labels* (ℓ) are defined as follows.

$$P ::= \mathsf{stop} \mid o.m(k);P \mid \mathsf{if}\,o.a(k)\,\mathsf{then}\,P \mid \mathsf{if}\,o.a(k)\,\mathsf{then}\,P_1\,\mathsf{else}\,P_2$$
$$C ::= P_1 \parallel \cdots \parallel P_n$$
$$\ell ::= o.m(k) \mid o.a(k){:}\mathsf{tt} \mid o.a(k){:}\mathsf{ff}$$

For the most part, we elide occurrences of stop and explicit object references, writing $o.a(k);\mathsf{stop}$ as "$a(k)$". We also write $\mathsf{if}\,a(k)\,\mathsf{then}\,P\,\mathsf{else}\,P$ as "$a(k);P$". In our running example, we have been writing the label $\mathsf{add}(k)$ as "$+k$", $\mathsf{remove}(k)$ as "$-k$", $\mathsf{get}(k){:}\mathsf{tt}$ as "$✓k$" and $\mathsf{get}(k){:}\mathsf{ff}$ as "$✗k$".

Let $[\![\cdot]\!]$ be a semantic function mapping configurations to sets of LVRs. The definition is the obvious one. For example, let C be the configuration $\mathsf{add}(0);\mathsf{get}(1) \parallel \mathsf{add}$ $(1);\mathsf{get}(0);\mathsf{get}(1)$. Then $[\![C]\!]$ is a set of the following eight LVRs (up to isomorphism).

$$
\begin{array}{cccc}
+1 \xrightarrow{+0} ✗0 \xrightarrow{✗1} ✓1 &
+1 \xrightarrow{+0} ✓0 \xrightarrow{✗1} ✓1 &
+1 \xrightarrow{+0} ✗0 \xrightarrow{✓1} ✓1 &
+1 \xrightarrow{+0} ✓0 \xrightarrow{✓1} ✓1 \\[4pt]
+1 \xrightarrow{+0} ✗0 \xrightarrow{✗1} ✗1 &
+1 \xrightarrow{+0} ✓0 \xrightarrow{✗1} ✗1 &
+1 \xrightarrow{+0} ✗0 \xrightarrow{✓1} ✗1 &
+1 \xrightarrow{+0} ✓0 \xrightarrow{✓1} ✗1
\end{array}
\tag{7}
$$

Under what circumstances can such a client interact with a 2P-set or OR-set and expect that the observed behaviour if compatible with a sequential set? This question is addressed in our first result, known as *abstraction* : when is the actual implementation of a data structure a safe substitute for its "abstract" specification?

We must first define what it means for a client and a data structure to interact.

Interaction. From figure (7) it is clear that the data structure must be able to filter out executions of the client. The set datatype does not include any traces that are compatible with the four LPOs on the second line of figure (7).

From figure (7) it is equally clear that the data structure must be able to introduce visibility that is not found in the client. For example, to achieve the results on the first line, one must introduce visibility between the client programs, as follows.

$$
\begin{array}{cccc}
\begin{array}{l} +0 \longrightarrow ✗1 \\ +1 \to ✗0 \to ✓1 \end{array} &
\begin{array}{l} +0 \rightrightarrows ✗1 \\ +1 \rightrightarrows ✓0 \to ✓1 \end{array} &
\begin{array}{l} +0 \longrightarrow ✓1 \\ +1 \Rightarrow ✗0 \to ✓1 \end{array} &
\begin{array}{l} +0 \rightrightarrows ✓1 \\ +1 \rightrightarrows ✓0 \to ✓1 \end{array}
\end{array}
$$

It is safe for the data structure to add visibility (and therefore order) to the client; however, the reverse is not true. A client can only introduce order that is compatible with the data structure specification. Consider the sequential client $\mathsf{add}(0);\mathsf{get}(0);\mathsf{get}$ (0). If this client communicates to separate replicas in a G-set, the execution $+0\ ✓0\ ✗0$ is possible, via the BPO $\xrightarrow{} \overset{+0}{\underset{✗0}{\rightrightarrows}} \overset{✓0}{\rightarrow}$. To avoid such anomalies, it is sufficient to require that sequential clients alway move forward in the visibility relation. This can be achieved by restricting each client program to communicate with a single replica, or by other means. We include this requirement in our definition of composition, without specifying how it is fulfilled.

Definition 4.1. Let S be a set of LVRs. $[\![C]\!](S) \triangleq \{s \in S \mid \exists s' \in [\![C]\!].\, s' \subseteq s\}$ □

One reading of the asymmetry in this definition is that a data structure may introduce order, but not its clients. A more generous reading is that clients may require order that is compatible with the data structure (that the data structure *could* have), but may not introduce incompatible order.

Monotonicity and Abstraction. Even with this definition of the semantics, abstraction fails in general. Consider the client add(0); get(1) ‖ add(1); get(0). The BPO $\xrightarrow{} \overset{+0}{\underset{+1}{\to}} \overset{\chi 1}{\underset{\chi 0}{\to}} \xrightarrow{}$ has order agreeing with the client and is an EC execution of a set, but this behaviour is not observable by a client interacting with a sequential set. Abstraction holds for clients that ensure *monotone* access to the data structure.

A set V is monotone if whenever V contains a trace u with a mutator m that is concurrent with another event e, then V also contains a visibility extension v that orders m and e. Since v is an visibility extension of u, it must contain the same labels.

Definition 4.2. A set V of LVRs is *monotone* when $\forall u \in V. \forall m \in \mathsf{E_M}(u). \forall e \in \mathsf{E}(u)$. if $(m \not\leadsto_u e$ and $e \not\leadsto_u m)$ then $\exists v \in V. u \subseteq v$ and $(m \leadsto_u e$ or $e \leadsto_u m)$ □

Theorem 4.3. Let S be a set of LVRs and let T be a specification such that $S \vDash_{ec} T$. Let C be a client such that $[\![C]\!](S)$ is monotone. Then $\forall s \in [\![C]\!](S). \exists t \in [\![C]\!](T). s \subseteq t$. □

The theorem states that if there is an execution s in $[\![C]\!](S)$, then is a corresponding execution t in $[\![C]\!](T)$ that has exactly the same labels, and potentially more order. This says that any client behaviour possible with the implementation S is also possible using the sequential specification T.

Example 4.4. The G-set trace $\xrightarrow{} \overset{+0}{\underset{+1}{\to}} \overset{\chi 1}{\to} \xrightarrow{}$ can be allowed in a monotone subset of G-set executions, since we can order $\chi 1$ before $+1$ and still have a set execution; the events $+1$ and $+0$ can be ordered arbitrarily. The G-set trace $\xrightarrow{} \overset{+0}{\underset{+1}{\to}} \overset{\chi 1}{\underset{\chi 0}{\to}} \xrightarrow{}$, however, cannot be allowed in a monotone subset of G-set executions. In this case, if we order $+1$ before $\chi 1$, then the result is clearly not a set execution: 1 has been added, but is not reported present. If we choose the reverse order, we have $+0$ before $\chi 0$, and again the result fails to be a valid set execution. Example 4.7 below gives an example of a specific G-set client that satisfies monotonicity, under given assumptions. To design a general class of context-independent monotone clients for a given data structure, it is necessary to limit client programs, as done in languages in the CALM framework [10].

For example, in order to create a monotone subset of G-set traces, it is sound to restrict clients to disallow the two-armed if-then-else. The semantics of the one-armed if-then is blocking—the client must wait until the condition is true. The theorem establishes that such clients can safely us a G-set as though it were a sequential set.

The theorem provides guidance about how to design safe clients. In order to allow a two-armed conditional with the G-set, we must ensure that events occurring concurrently with a negative response cannot invalidate that response. One way to achieve this, following [10], is for the G-set to insert a barrier before returning a negative response. □

Composition of Data Structures. We now turn our attention to reasoning about compound data structures.

Definition 4.5. Given disjoint LTOs t_1 and t_2 (that is, $\mathsf{E}(t_1) \cap \mathsf{E}(t_2) = \emptyset$), let $t_1 \| t_2$ denote the set of their interleavings. This notion lifts to sets as follows: $(T_1 \| T_2) \triangleq \{t \in (t_1 \| t_2) \mid t_1 \in T_1 \text{ and } t_2 \in T_2 \text{ and } (\mathsf{E}(t_1) \cap \mathsf{E}(t_2) = \emptyset)\}$.

Given an LVR s and $L \subseteq \mathsf{L}(s)$, write $s \setminus L$ for the LVR that results by restricting s to events with labels in L. This notation lifts to sets in the obvious way: $S \setminus L \triangleq \bigcup_{s \in S} s \setminus L$. □

Theorem 4.6. Let $[\![C]\!](S)$ be a monotone set of LVRs. Let L_1 and L_2 be disjoint subsets of $\mathscr{L}$. For $i \in \{1,2\}$, let T_i be a specification with labels chosen from L_i. If $([\![C]\!](S) \setminus L_1) \vDash_{\mathsf{ec}} T_1$ and $([\![C]\!](S) \setminus L_2) \vDash_{\mathsf{ec}} T_2$ then $[\![C]\!](S) \vDash_{\mathsf{ec}} (T_1 \| T_2)$. □

Example 4.7. The following definitions implement a 2P-set p, using two G-sets, a for "added" and t for "tombstone": $p.\mathsf{add}(k) \triangleq a.\mathsf{add}(k)$, $p.\mathsf{remove}(k) \triangleq t.\mathsf{add}(k)$, and $p.\mathsf{get}(k) \triangleq a.\mathsf{get}(k) \wedge \neg t.\mathsf{get}(k)$. If we can establish the necessary monotonicity properties, then we can reason with the sequential specifications of a and t in proving p correct. An execution of a grow set g is monotone so long as for any $g.\textbf{✗}k$, there is no concurrent $g.{+}k$. We must show that both a and t are accessed monotonically, so long as p is accessed monotonically. An execution of a 2P-set p is monotone so long as (1) for any $p.\textbf{✓}k$, there is no concurrent $p.{-}k$, and (2) for any $p.\textbf{✗}k$, there is no concurrent $p.{+}k$.

The conditions for monotonicity of p are sufficient to establish monotonicity of a and t. There are two cases: (1) Suppose $p.\textbf{✓}k$. By monotonicity, we know there is no concurrent $p.{-}k$, therefore no concurrent $t.{+}k$. By definition of $p.\mathsf{get}$, we must have $a.\textbf{✓}k$ and $t.\textbf{✗}k$. Monotonicity imposes no constraints on $a.\textbf{✓}k$; to satisfy $t.\textbf{✗}k$, we must have no concurrent $t.{+}k$, but this is exactly guaranteed by monotonicity of p. (2) Suppose $p.\textbf{✗}k$. Then we know there is no concurrent $p.{+}k$, therefore no concurrent $a.{+}k$. By definition of $p.\mathsf{get}$, we must have either $a.\textbf{✗}k$ or $t.\textbf{✓}k$. The argument is as before. □

References

1. Bouajjani, A., Enea, C., Hamza, J.: Verifying eventual consistency of optimistic replication systems. In POPL 2014, pp. 285–296 (2014)
2. Burckhardt, S., Gotsman, A., Yang, H., Zawirski, M.: Replicated data types: specification, verification, optimality. In: POPL 2014, pp. 271–284 (2014)
3. Burckhardt, S., Leijen, D., Fähndrich, M., Sagiv, M.: Eventually consistent transactions. In: Seidl, H. (ed.) Programming Languages and Systems. LNCS, vol. 7211, pp. 67–86. Springer, Heidelberg (2012)
4. Conway, N., Marczak, W.R. et al.: Logic and lattices for distributed programming. In: ACM Symposium on Cloud Computing, pp. 1:1–1:14 (2012)
5. Derrick, J., Dongol, B., et al.: Quiescent consistency: defining and verifying relaxed linearizability. In: Formal, Methods, pp. 200–214 (2014)
6. Ellis, C.A., Gibbs, S.J.: Concurrency control in groupware systems. ACM SIGMOD Record **18**(2), 399–407 (1989)
7. Filipovic, I., O'Hearn, P.W., Rinetzky, N., Yang, H.: Abstraction for concurrent objects. Theoretical Comp. Sci. **411**, 4379–4398 (2010)

8. Gilbert, S., Lynch, N.: Brewer's conjecture and the feasibility of consistent, available, partition-tolerant web services. SIGACT News, pp. 51–59 (2002)
9. Gotsman, A., Yang, H.: Composite replicated data types. In: Vitek, J. (ed.) ESOP 2015. LNCS, vol. 9032, pp. 585–609. Springer, Heidelberg (2015)
10. Hellerstein, J.M.: The declarative imperative: Experiences and conjectures in distributed logic. SIGMOD Rec. **39**(1), 5–19 (2010)
11. Herlihy, M., Wing, J.M.: Linearizability: A correctness condition for concurrent objects. ACM TOPLAS **12**(3), 463–492 (1990)
12. Jagadeesan, R., Riely, J.: Between linearizability and quiescent consistency. In: Esparza, J., Fraigniaud, P., Husfeldt, T., Koutsoupias, E. (eds.) ICALP 2014, Part II. LNCS, vol. 8573, pp. 220–231. Springer, Heidelberg (2014)
13. Lamport, L.: Time, clocks, and the ordering of events in a distributed system. Commun. ACM **21**(7), 558–565 (1978)
14. Panangaden, P., Shanbhogue, V., Stark, E.W.: Stability and sequentiality in dataflow networks. In: ICALP 1990, pp. 308–321 (1990)
15. Panangaden, P., Stark, E.W.: Computations, residuals, and the power of indeterminacy. In: ICALP 1988, pp. 439–454 (1988)
16. Saito, Y., Shapiro, M.: Optimistic replication. Comput. Surv. **37**(1), 42–81 (2005)
17. Shapiro, M., Preguiça, N., Baquero, C., Zawirski, M.: A comprehensive study of Convergent and Commutative Replicated Data Types. TR 7506, Inria (2011)
18. Terry, D.B., Theimer, M.M. et al.: Managing update conflicts in bayou, a weakly connected replicated storage system. In: SOSP (1995)
19. Vogels, W.: Eventually consistent. Communications of the ACM **52**(1), 40–44 (2009)

Fixed-Dimensional Energy Games are in Pseudo-Polynomial Time

Marcin Jurdziński[1], Ranko Lazić[1], and Sylvain Schmitz[1,2]($\boxtimes$)

[1] DIMAP, Department of Computer Science, University of Warwick, Coventry, UK
[2] LSV, ENS Cachan & CNRS & INRIA, Cachan Cedex, France
schmitz@lsv.ens-cachan.fr

Abstract. We generalise the hyperplane separation technique (Chatterjee and Velner, 2013) from multi-dimensional mean-payoff to energy games, and achieve an algorithm for solving the latter whose running time is exponential only in the dimension, but not in the number of vertices of the game graph. This answers an open question whether energy games with arbitrary initial credit can be solved in pseudo-polynomial time for fixed dimensions 3 or larger (Chaloupka, 2013). It also improves the complexity of solving multi-dimensional energy games with given initial credit from non-elementary (Brázdil, Jančar, and Kučera, 2010) to 2EXPTIME, thus establishing their 2EXPTIME-completeness.

1 Introduction

Multi-Dimensional Energy Games are played turn-by-turn by two players on a finite *multi-weighted* game graph, whose edges are labelled with integer vectors modelling discrete energy consumption and refuelling. Player 1's objective is to keep the accumulated energy non-negative in every component along infinite plays. This setting is relevant to the synthesis of resource-sensitive controllers balancing the usage of various resources like fuel, time, money, or items in stock, and finding optimal trade-offs; see [3,4,10,11] for some examples. Maybe more importantly, energy games are the key ingredient in the study of several related resource-conscious games, notably multi-dimensional mean-payoff games [6] and games played on vector addition systems with states (VASS) [2,4,9].

The main open problem about these games has been to pinpoint the complexity of deciding whether Player 1 has a winning strategy when starting from a particular vertex and given an initial energy vector as part of the input. This particular *given initial credit* variant of energy games is also known as *Z-reachability* VASS games [4,5]. The problem is also equivalent via logarithmic-space reductions to deciding *single-sided* VASS games with a non-termination objective [2], and to deciding whether a given VASS (or, equivalently, a Petri net) simulates a given finite state system [1,9]. As shown by Brázdil, Jančar, and Kučera [4], all these problems can be solved in $(d-1)$EXPTIME where $d \geq 2$ is the number

Work funded in part by the ANR grant 11-BS02-001-01 REACHARD, the Leverhulme Trust Visiting Professorship VP1-2014-041, and the EPSRC grant EP/M011801/1.

© Springer-Verlag Berlin Heidelberg 2015
M.M. Halldórsson et al. (Eds.): ICALP 2015, Part II, LNCS 9135, pp. 260–272, 2015.
DOI: 10.1007/978-3-662-47666-6_21

of energy components, i.e. a TOWER of exponentials when d is part of the input. The best known lower bound for this problem is 2EXPTIME-hardness [9], leaving a substantial complexity gap. So far, the only tight complexity bounds are for the case $d = 2$: Chaloupka [5] shows the problem to be PTIME-complete when using unit updates, i.e. when the energy levels can only vary by -1, 0, or 1. However, quoting Chaloupka 'since the presented results about 2-dimensional VASS are relatively complicated, we suspect this [general] problem is difficult.'

When inspecting the upper bound proof of Brázdil et al. [4], it turns out that the main obstacle to closing the gap and proving 2EXPTIME-completeness lies in the complexity upper bounds for energy games with an *arbitrary initial credit*—which is actually the variant commonly assumed when talking about energy games. Given a multi-weighted game graph and an initial vertex v, we now wish to decide whether there exists an initial energy vector $\mathbf{b}$ such that Player 1 has a winning strategy starting from the pair $(v, \mathbf{b})$. As shown by Chatterjee, Doyen, Henzinger, and Raskin [6], this variant is simpler: it is coNP-complete. However, the parameterised complexity bounds in the literature [4,7] for this simpler problem involve an exponential dependency on the number $|V|$ of vertices in the input game graph, which translates into a tower of exponentials when solving the given initial credit variant.

Contributions. We show in this paper that the arbitrary initial credit problem for d-dimensional energy games can be solved in time $O(|V| \cdot \|E\|)^{O(d^4)}$ where $|V|$ is the number of vertices of the input multi-weighted game graph and $\|E\|$ the maximal value that labels its edges, i.e. in pseudo-polynomial time (see Thm. 3.3). We then deduce that the given initial credit problem for general multi-dimensional energy games is 2EXPTIME-complete, and also in pseudo-polynomial time when the dimension is fixed (see Thm. 3.5), thus closing the gap left open in [4,9]. Our parameterised bounds are of practical interest because typical instances of energy games would have small dimension but might have a large number of vertices. By the results of Chatterjee et al. [6], another consequence is that we can decide the existence of a *finite-memory* winning strategy for fixed-dimensional *mean-payoff* games in pseudo-polynomial time. The existence of a finite-memory winning strategy is the most relevant problem for controller synthesis, but until now, solving fixed-dimensional mean-payoff games in pseudo-polynomial time required infinite memory strategies [8].

Overview. We prove our upper bounds on the complexity of the arbitrary initial credit problem for d-dimensional energy games by reducing them to *bounding games*, where Player 1 additionally seeks to prevent arbitrarily high energy levels (Sec. 2.3). We further show these games to be equivalent to *first-cycle bounding games* in Sec. 5, where the total effect of the first simple cycle defined by the two players determines the winner. More precisely, first-cycle bounding games rely on a hierarchically-defined colouring of the game graph by *perfect half-spaces* (see Sec. 4), and the two players strive respectively to avoid or produce cycles in those perfect half-spaces.

First-cycle bounding games coloured with perfect half-spaces can be seen as generalising quite significantly both the 'local strategy' approach of Chaloupka

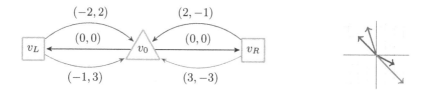

Fig. 1. A 2-dimensional multi-weighted game graph.

[5] for 2-dimensional energy games, and the 'separating hyperplane technique' of Chatterjee and Velner [8] for multi-dimensional mean-payoff games.

The reduction to first-cycle bounding games has several important corollaries: the *determinacy* of bounding games, and the existence of a *small hypercube property*, which in turn allow to derive the announced complexity bounds on energy games (see Sec. 3). In fact, we found with first-cycle bounding games a highly versatile tool, which we use extensively in our proofs on energy games.

We start by presenting the necessary background on energy and bounding games in Sec. 2. Some omitted material can be found in the full paper available from http://arxiv.org/abs/1502.06875.

2   Multi-Weighted Games

We define in this section the various games we consider in this work. We start by defining multi-weighted game graphs, which provide a finite representation for the infinite arenas over which our games are played. We then define energy games in Sec. 2.2, and their generalisation as bounding games in Sec. 2.3.

2.1   Multi-Weighted Game Graphs

We consider game graphs whose edges are labelled by vectors of integers. They are tuples of the form (V, E, d), where d is the dimension in $\mathbb{N}$, $V \overset{\text{def}}{=} V_1 \uplus V_2$ is a finite set of vertices, which is partitioned into Player 1 vertices (V_1) and Player 2 vertices (V_2), and E is a finite set of edges included in $V \times \mathbb{Z}^d \times V$, and such that every vertex has at least one outgoing edge; we call the labels in $\mathbb{Z}^d$ 'weights'.

Example 2.1. Figure 1 shows on its left-hand-side an example of a 2-dimensional multi-weighted game graph. Throughout this paper, Player 1 vertices are depicted as triangles and Player 2 vertices as squares.

Norms. For a vector **a**, we denote the maximum absolute value of its entries by $\|\mathbf{a}\| \overset{\text{def}}{=} \max_{1 \le i \le d} |\mathbf{a}(i)|$, and we call it the *norm* of **a**. By extension, for a set of edges E, we let $\|E\| \overset{\text{def}}{=} \max_{(v,\mathbf{u},v') \in E} \|\mathbf{u}\|$. We assume, without loss of generality, that $\|E\| > 0$ in our multi-weighted game graphs. Regarding complexity, we encode vectors of integers in binary, hence $\|E\|$ may be exponential in the size of the multi-weighted game graph.

Paths and Cycles. Given a multi-weighted game graph (V, E, d), a *configuration* is a pair $(v, \mathbf{a})$ with v in V and **a** in $\mathbb{Z}^d$. A *path* is a finite sequence of configurations

$\pi = (v_0, \mathbf{a}_0)(v_1, \mathbf{a}_1) \cdots (v_n, \mathbf{a}_n)$ in $(V \times \mathbb{Z}^d)^*$ such that for every $0 \leq j < n$ there exists an edge $(v_j, \mathbf{a}_{j+1} - \mathbf{a}_j, v_{j+1})$ in E (where addition is performed componentwise). The *total weight* of such a path π is $w(\pi) \overset{\text{def}}{=} \sum_{0 \leq j < n} \mathbf{a}_{j+1} - \mathbf{a}_j = \mathbf{a}_n - \mathbf{a}_0$. A *cycle* is a path $(v_0, \mathbf{a}_0)(v_1, \mathbf{a}_1) \cdots (v_n, \mathbf{a}_n)$ with $v_0 = v_n$. Such a cycle is *simple* if $v_j = v_k$ for some $0 \leq j < k \leq n$ implies $j = 0$ and $k = n$. We assume, without loss of generality, that every cycle contains at least one Player 1 vertex. We often identify simple cycles with their respective weights; the weights of the four simple cycles of the game graph in Fig. 1 are displayed on its right-hand-side.

Proposition 2.2. *In any game graph (V, E, d), the total weight of any simple cycle has norm at most $|V| \cdot \|E\|$.*

Plays and Strategies. Let v_0 be a vertex from V. A *play from v_0* is an infinite configuration sequence $\rho = (v_0, \mathbf{a}_0)(v_1, \mathbf{a}_1) \cdots$ such that $\mathbf{a}_0 = \mathbf{0}$ is the null vector and every finite prefix $\rho|_n \overset{\text{def}}{=} (v_0, \mathbf{a}_0) \cdots (v_n, \mathbf{a}_n)$ is a path. Note that, because $\mathbf{a}_0 = \mathbf{0}$, the total weight of this prefix is $w(\rho|_n) = \mathbf{a}_n$. We define the *norm* of a play ρ as the supremum of the norms of total weights of its prefixes: $\|\rho\| \overset{\text{def}}{=} \sup_n \|w(\rho|_n)\|$. A *strategy* for Player p, $p \in \{1, 2\}$, is a function σ_p taking as input a non-empty path $\pi \cdot (v, \mathbf{a})$ ending in a Player p vertex $v \in V_p$, and returning an edge $\sigma_p(\pi \cdot (v, \mathbf{a})) = (v, \mathbf{u}, v')$ from E. We employ the usual notions of plays *consistent* with strategies, and given some winning condition on plays, of *winning* strategies for a player.

Example 2.1 (continued). For instance, in the game graph depicted in Fig. 1, a strategy for Player 1 could be to move to v_L whenever the current energy level on the first coordinate is non-negative, and to v_R otherwise—note that this is an *infinite-memory* strategy—:

$$\sigma_1(\pi \cdot (v_0, \mathbf{a})) \overset{\text{def}}{=} \begin{cases} (v_0, (0,0), v_L) & \text{if } \mathbf{a}(1) \geq 0, \\ (v_0, (0,0), v_R) & \text{otherwise}, \end{cases} \tag{1}$$

and one for Player 2 could be to always select one particular edge in every vertex, regardless of the current energy vector—this is called a *counterless* strategy [4]—:

$$\sigma_2(\pi \cdot (v, \mathbf{a})) \overset{\text{def}}{=} \begin{cases} (v_L, (-2, 2), v_0) & \text{if } v = v_L \\ (v_R, (2, -1), v_0) & \text{otherwise}. \end{cases} \tag{2}$$

2.2 Multi-Dimensional Energy Games

Suppose (V, E, d) is a multi-weighted game graph, v_0 an initial vertex, and $\mathbf{b}$ is a vector from $\mathbb{N}^d$. A play ρ from v_0 is *winning* for Player 1 in the *energy game* $\Delta_{\mathbf{b}}(V, E, d)$ with *initial credit* $\mathbf{b}$ if, for all n, $\mathbf{b} + w(\rho|_n) \geq \mathbf{0}$, using the product ordering over $\mathbb{Z}^d$. Otherwise, Player 2 wins the play. An immediate property of energy games is *monotonicity*: if σ_1 is winning for Player 1 with some initial credit $\mathbf{b}$, and $\mathbf{b}' \geq \mathbf{b}$, then it is also winning for Player 1 with initial credit $\mathbf{b}'$.

Example 2.1 (continued). For example, one may observe that the strategy (1) for Player 1 is winning for the game graph of Fig. 1 with initial credit $(2, 2)$ (or larger). A geometric intuition comes from the directions of the total weights of simple cycles in Fig. 1: by choosing alternatively edges to v_L or v_R, Player 1 is able to balance the energy levels above the '$x + y = 0$' line.

2.3   Multi-Dimensional Bounding Games

A generalisation of energy games sometimes considered in the literature is to further impose a maximal *capacity* $\mathbf{c} \in \mathbb{N}^d$ (also called an upper bound) on the energy levels during the play [10,11]. Player 1 then wins a play ρ if $0 \leq \mathbf{b} + w(\rho|_n) \leq \mathbf{c}$ for all n.

In the spirit of the arbitrary initial credit variant of energy games, we also quantify $\mathbf{c}$ existentially. This defines the *bounding game* $\Gamma(V, E, d)$ over a multi-weighted game graph (V, E, d), where a play ρ is winning for Player 1 if its norm $\|\rho\|$ is finite, i.e. if the set $\{\|w(\rho|_n)\| : n \in \mathbb{N}\}$ of norms of total weights of all finite prefixes of ρ is bounded, and Player 2 wins otherwise, if it is unbounded. In other words, Player 1 strives to contain the current vector within some d-dimensional hypercube, while Player 2 attempts to escape.

Example 2.1 (continued). Note that Player 2 is now winning the bounding game defined by the game graph of Fig. 1 from any of the three vertices, for example using the strategy (2). Indeed, this strategy ensures that the only simple cycles that can be played have weights $(-2, 2)$ and $(2, -1)$. Because these vectors belong to an open half-plane, the total energy will drift deeper and deeper inside that open half-plane and its norm will grow unbounded.

Lossy Game Graphs. If Player 1 wins the bounding game, then there exists some initial credit for which she also wins the energy game. For a converse, let lossy(V, E, d) denote the *lossy* multi-weighted game graph obtained from (V, E, d) by inserting, at each Player 1 vertex and for each $1 \leq i \leq d$, a self-loop labelled by the negative unit vector $-\mathbf{e}_i$. In a bounding game played over a lossy game graph, it turns out that Player 1 can always bound the current vector from above by playing these unit decrements, hence she only has to ensure that the current vector remains bounded from below, i.e. she has to win the energy game for some initial credit. Formally (see the full paper for a proof):

Proposition 2.2. *From any vertex in any multi-weighted game graph (V, E, d):*
1. *Player 1 wins the energy game $\Delta_\mathbf{b}(V, E, d)$ for some $\mathbf{b} \in \mathbb{N}^d$ if and only if Player 1 wins the bounding game $\Gamma(\text{lossy}(V, E, d))$.*
2. *Player 2 wins the energy game $\Delta_\mathbf{b}(V, E, d)$ for all $\mathbf{b} \in \mathbb{N}^d$ if and only if Player 2 wins the bounding game $\Gamma(\text{lossy}(V, E, d))$.*

Our task in the following will be therefore to prove an upper bound on the time complexity required to solve bounding games.

Example 2.3. By Prop. 2.2, because she was winning the energy game of Fig. 1 with initial credit $(2, 2)$, Player 1 is now winning the bounding game played on the lossy version of the multi-weighted game graph of Fig. 1.

Example 2.4. As a rather different example, consider the multi-weighted game graph of Fig. 2. Although Player 2 does not control any vertex, and Player 1 controls the 'direction of divergence', Player 2 wins the associated bounding

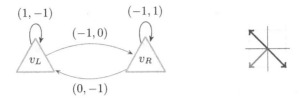

Fig. 2. A 2-dimensional game graph with only Player 1 vertices

game. Indeed, Player 1 can either eventually stay forever at one of the two vertices, or visit both vertices infinitely often. In any case, she loses.

3 Complexity Upper Bounds

Our main results are new parameterised complexity upper bounds for deciding whether Player 1 has a winning strategy in a given energy game. In turn, we rely for these results on a *small hypercube property* of bounding games, which we introduce next, and which will be a consequence of the study of first-cycle bounding games in Sec. 5.

3.1 Small Hypercube Property

In a bounding game, if Player 1 is winning, then by definition she has a winning strategy σ_1 such that for all plays ρ consistent with σ_1 there exists some bound B_ρ with $\|\rho\| \leq B_\rho$. We considerably strengthen this statement in Sec. 5 where we construct an explicit winning strategy, which yields an explicit *uniform* bound B for all consistent plays:

Lemma 3.1. *Let (V, E, d) be a multi-weighted game graph. If Player 1 wins the bounding game $\Gamma(V, E, d)$, then she has a winning strategy which ensures $\|\rho\| \leq (4|V| \cdot \|E\|)^{2(d+2)^3}$ for all consistent plays ρ.*

Note that our bound is polynomial in $|V|$ the number of vertices, unlike the bounds found in comparable statements by Brázdil et al. [4, Lem. 7] and Chatterjee et al. [7, Lem. 3], which incur an exponential dependence on $|V|$. This entails pseudo polynomial complexity bounds when d is fixed:

Corollary 3.2. *Bounding games on multi-weighted graphs (V, E, d) are solvable in time $(|V| \cdot \|E\|)^{O(d^4)}$.*

Proof. By Lem. 3.1, the bounding game is equivalent to a reachability game where Player 2 attempts to see the norm of the total weight exceed $B \stackrel{\text{def}}{=} (4|V| \cdot \|E\|)^{2(d+2)^3}$. This can be played within a finite arena of size $(2B+1)^d$ and solved in time linear in that size using the usual attractor computation algorithm. □

3.2 Energy Games with Arbitrary Initial Credit

The *arbitrary initial credit problem* for energy games takes as input a multi-weighted game graph and an initial vertex v_0 and asks whether there exists a vector **b** in $\mathbb{N}^d$ such that Player 1 wins $\Delta_{\mathbf{b}}(V, E, d)$ from v_0:

Theorem 3.3. *The arbitrary initial credit problem for energy games on multi-weighted game graphs* (V, E, d) *is solvable in time* $(|V| \cdot \|E\|)^{O(d^4)}$.

Proof. This follows from Prop. 2.2, and Cor. 3.2 applied to the game graph $\Gamma(\text{lossy}(V, E, d))$. □

3.3 Energy Games with Given Initial Credit

The *given initial credit problem* for energy games takes as input a multi-weighted game graph (V, E, d), an initial vertex v_0, and a credit $\mathbf{b}$ in $\mathbb{N}^d$ and asks whether Player 1 wins the energy game $\Delta_{\mathbf{b}}(V, E, d)$ from v_0. Thanks to Lem. 3.1, a proof of the upcoming Thm. 3.5 could be obtained using the work of Brázdil et al. [4], and more generally the techniques of Rackoff [12]. As usual in this work, we rather proceed by transferring that setting to that of bounding games (and thus to that of first-cycle bounding games). Our key lemma shows that any energy game with a given initial credit played over some multi-weighted game graph is equivalent to some bounding game played over a double-exponentially larger game graph:

Lemma 3.4. *Let* $\mathbf{b} \in \mathbb{N}^d$, (V, E, d) *be a multi-weighted game graph, and* $v \in V$. *One can construct in time* $O(|V'| \cdot |E| + d \cdot \log \|\mathbf{b}\|)$ *a multi-weighted game graph* (V', E', d) *and a vertex* $v_{\mathbf{b}}$ *in* V' *with* $|V'| \leq (4|V| \cdot \|E\|)^{2^d(d+3)^{3d}}$ *and* $\|E'\| = \|E\|$ *s.t., for all* $p \in \{1, 2\}$, *Player* p *wins the energy game* $\Delta_{\mathbf{b}}(V, E, d)$ *from* v *iff Player* p *wins the bounding game* $\Gamma(V', E', d)$ *from* $v_{\mathbf{b}}$.

By applying Cor. 3.2 to the game graph (V', E', d) and since $|E| \leq |V|^2 \cdot \|E\|^d$, we obtain a 2EXPTIME upper bound on the given initial credit problem, which is again pseudo-polynomial when d is fixed:

Theorem 3.5. *The given initial credit problem with credit* $\mathbf{b}$ *for energy games on multi-weighted game graphs* (V, E, d) *is solvable in time* $O(|V| \cdot \|E\|)^{2^{O(d \cdot \log d)}} + O(d \cdot \log \|\mathbf{b}\|)$.

This matches the 2EXPTIME lower bound from [9], and encompasses Chaloupka's PTIME upper bound in dimension $d = 2$ with unit updates, i.e. with $\|E\| = 1$. Because the given initial credit problem for energy games of fixed dimension $d \geq 4$ is EXPTIME-hard [9], the bound in terms of $\|E\|$ in Thm. 3.5 cannot be improved.

4 Perfect Half-Spaces

We recall in this section the definition of subsets of $\mathbb{Q}^d$ called *perfect half-spaces*. They will be used next in Sec. 5 to define a condition for Player 2 to win bounding games, which relies on Player 2's ability to force cycles inside perfect half-spaces. This can be understood as a generalisation of Chatterjee and Velner's approach for solving multi-dimensional *mean-payoff* games [8], which as we recall in the full paper relies on a similar ability to force cycles inside open half-spaces. We employ perfect half-spaces in Sec. 5 to colour the edges in *first-cycle bounding games*, which determine the winner using both the colours and the weight of the first cycle formed along a play.

4.1 Definitions from Linear Algebra

Given a subset $\mathbf{A}$ of $\mathbb{Q}^d$, we write $\mathsf{span}(\mathbf{A})$ (resp., $\mathsf{cone}(\mathbf{A})$) for the *vector space* (resp., the *cone*) *generated* by $\mathbf{A}$, i.e., the closure of $\mathbf{A}$ under addition and under multiplication by all (resp., nonnegative) rationals. A *k-perfect half-space* of $\mathbb{Q}^d$, where $k \in \{1, 2, \ldots, d\}$, is a (necessarily disjoint) union $H_d \cup \cdots \cup H_k$ such that:

- H_d is an open half-space of $\mathbb{Q}^d$;
- for all $j \in \{k, \ldots, d-1\}$, $H_j \subseteq \mathbb{Q}^d$ is an open half-space of the boundary of H_{j+1}.

Whenever we write a k-perfect half-space in form $H_d \cup \cdots \cup H_k$, we assume that each H_j is j-dimensional. We additionally define the $(d+1)$-perfect half-space as the empty set; a *partially-perfect half-space* is then a k-perfect half-space for some k in $\{1, \ldots, d+1\}$. A *perfect half-space* is a 1-perfect half-space.

4.2 Generated Perfect Half-Spaces

In order to pursue effective and parsimonious strategy constructions, we consider perfect half-spaces generated by particular sets of vectors, which will correspond to the total weights of simple cycles in multi-weighted game graphs. Given a norm M in $\mathbb{N}$, we say that an open half-space H is *M-generated* if its boundary equals $\mathsf{span}(\mathbf{B})$ for some set $\mathbf{B}$ of vectors of norm at most M. By extension, a partially-perfect half-space is *M-generated* if each of its open half-spaces is M-generated.

Proposition 4.1. *Any k-dimensional vector space of $\mathbb{Q}^d$ has at most $\mathcal{L}(k) \overset{\text{def}}{=} 2(2M+1)^{d(k-1)}$ open half-spaces that are M-generated.*

Example 4.2. In the game graph of Fig. 2, there are three 1-generated open half-spaces of interest: the half-plane $H_2 \overset{\text{def}}{=} \{(x,y) : x+y < 0\}$ with boundary $\mathsf{span}((-1,1),(1,-1))$ and containing $(-1,-1)$, and the two half-lines $H_1 \overset{\text{def}}{=} \{(x,y) : x+y = 0 \wedge x < 0\}$ and $H_1' \overset{\text{def}}{=} \{(x,y) : x+y = 0 \wedge x > 0\}$ with boundary $\mathsf{span}(\mathbf{0})$ and containing, respectively, $(-1,1)$ and $(1,-1)$. Those open half-spaces define two perfect half-spaces: $H_2 \cup H_1$ and $H_2 \cup H_1'$.

4.3 Hierarchy of Perfect Half-Spaces

Finally, we fix a ranked tree-like structure on all M-generated partially-perfect half-spaces, which provide a scaffolding on which we will build strategies in multi-dimensional bounding games. Observe that an M-generated partially-perfect half-space $H_d \cup \cdots \cup H_k$ for $k > 1$ can be extended using any of the M-generated open half-spaces H of the boundary of H_k; note that this boundary then equals $\mathsf{span}(H)$. In Example 4.2, H_2 can be extended using H_1 or H_1', and $\mathsf{span}(H_1) = \mathsf{span}(H_1') = \{(x,y) : x+y = 0\}$.

The set of M-generated perfect half-spaces can be totally ordered by positing a linear ordering $<$ between all M-generated open half-spaces. We write $\prec$ for the lexicographically induced linear ordering between all M-generated perfect half-spaces of $\mathbb{Q}^d$: if $\mathcal{H} = H_d \cup \cdots \cup H_1$ and $\mathcal{H}' = H_d' \cup \cdots \cup H_1'$, we define $\mathcal{H} \prec \mathcal{H}'$ to hold iff $H_j = H_j'$ for all $j \in \{k+1, \ldots, d\}$ and $H_k < H_k'$ for some $k \in \{1, 2, \ldots, d\}$.

5   First-Cycle Bounding Games

We define in this section *first-cycle bounding games*, which provide the key technical arguments for most of our results. Such games end as soon as a cycle is formed along a play, and the weight of this cycle determines the winner, along with a colouring information chosen by Player 2. In sections 5.2 and 5.3, we are going to show that first-cycle bounding games and infinite bounding games are equivalent, by translating winning strategies for each Player p, $p \in \{2, 1\}$, from first-cycle bounding games to bounding games. This yields in particular the small hypercube property of Lem. 3.1.

5.1   Definition

We define the *first-cycle bounding game* $G(V, E, d)$ on a multi-weighted game graph (V, E, d):

- at any Player-1 vertex, Player 2 chooses a $|V| \cdot \|E\|$-generated perfect half-space $\mathcal{H}$ of $\mathbb{Q}^d$, and then Player 1 chooses an outgoing edge, whose occurrence in the play becomes coloured by $\mathcal{H}$;
- at any Player-2 vertex, he chooses an outgoing edge;
- the game finishes as soon as a vertex is visited twice, which produces a simple cycle C with coloured Player-1 edges;
- Player 2 wins if $w(C)$, the total weight of the cycle, is in the largest partially-perfect half-space of $\mathbb{Q}^d$ that is contained in all the colours in C, i.e. the least common ancestor of all the colours in C; Player 1 wins otherwise.

Example 5.1. Player 2 wins the first-cycle bounding game played in Fig. 1 (but loses in its lossy version). For example, strategy (2) is winning for Player 2 if he colours the edges outgoing from v_0 by the perfect half-space $H'_2 \cup H_1$ where $H'_2 \overset{\text{def}}{=} \{(x, y) : x + y > 0\}$ and $H_1 \overset{\text{def}}{=} \{(x, y) : x + y = 0 \wedge x < 0\}$.

Example 5.2. Player 2 wins the first-cycle bounding game played in Fig. 2. Indeed, he can choose the colour $H_2 \cup H_1$ in v_L and the colour $H_2 \cup H'_1$ in v_R. Then Player 1 cannot avoid forming a simple cycle in either $H_2 \cup H_1$ (if cycling on v_L), in $H_2 \cup H'_1$ (if cycling on v_R), or in H_2 (if cycling between v_L and v_R).

Observe that first-cycle bounding games are finite perfect information games, and are thus *determined*: from any vertex, either Player 1 wins or Player 2 wins.

5.2   Winning Strategies for Player 2

Suppose σ is a strategy of Player 2 from a vertex v_0 in a first-cycle bounding game $G(V, E, d)$. Let $\widetilde{\sigma}$ be the following strategy of Player 2 in the infinite bounding game $\Gamma(V, E, d)$:

- at any Player-2 vertex, $\widetilde{\sigma}$ chooses the edge specified by σ;
- whenever a cycle is formed, $\widetilde{\sigma}$ cuts it out of its memory, and continues playing according to σ.

Lemma 5.3. *If σ is winning for Player 2 in $G(V, E, d)$ from some vertex v_0, then $\widetilde{\sigma}$ is winning for Player 2 in $\Gamma(V, E, d)$ from the same vertex v_0.*

Proof idea. Consider any infinite play $\tilde{\rho}$ consistent with $\tilde{\sigma}$, and let:

- ρ be obtained from $\tilde{\rho}$ by colouring all Player 1's edges with the $|V| \cdot \|E\|$-generated perfect half-spaces of $\mathbb{Q}^d$ as specified by σ;
- $C_1, C_2, \ldots$ be the cycle decomposition of ρ, and for each n, ρ_n be the simple path that remains after removing C_n;
- $\mathcal{H}_n$ be the largest partially-perfect half-space of $\mathbb{Q}^d$ that is contained in all the colours in C_n, for each n.

Since σ is winning for Player 2 in the first-cycle game, each cycle weight $w(C_n)$ belongs to the partially-perfect half-space $\mathcal{H}_n$. The bulk of the proof consists in extracting a 'direction of divergence' of the total energy, notwithstanding that the $\mathcal{H}_n$'s may keep varying.

In short, by distinguishing those n's for which the length of the simple path ρ_n is the smallest one that occurs infinitely often, we are able to show that the set of $\mathcal{H}_n$'s that occur infinitely often has a unique smallest element $\mathcal{H} = H_d \cup \cdots \cup H_k$ with respect to inclusion. Further linear-algebraic reasoning then shows that one of the component half-spaces $H_{k'}$ of $\mathcal{H}$ provides the desired direction of divergence: after some $N > 0$, all the sums of cycle weights $w(C_N) + w(C_{N+1}) + \cdots + w(C_n)$ belong to the topological closure $\overline{H_{k'}}$ and their distances from the boundary of $H_{k'}$ diverge. See the full paper for details. $\square$

5.3   Winning Strategies for Player 1

If there is no winning strategy for Player 2 in the first-cycle bounding game $G(V, E, d)$ from a vertex v_0, then by determinacy of first-cycle bounding games, there is a winning strategy σ for Player 1 in $G(V, E, d)$ from v_0.

Example 5.4. Consider the lossy version of the game graph in Fig. 1. Because Player 1 wins the energy game with initial credit $(2, 2)$, by Prop. 2.2 and Lem. 5.3, she wins the first-cycle bounding game. One winning strategy, whose moves depend only on the latest visited vertex (here only v_0) and colour $\mathcal{H}$ chosen by Player 2 in v_0, is as follows:

(i) if $(-2, 2)$ and $(-1, 3)$ are both outside $\mathcal{H}$, move to v_L, and
(ii) if $(2, -1)$ and $(3, -3)$ are both outside $\mathcal{H}$, move to v_R, and
(iii) otherwise perform the self-loop labelled $(-1, 0)$.

Observe that the first two cases (i) and (ii) are disjoint. Since there is no perfect half-space that contains $(-1, 0)$ and intersects both $\{(-2, 2), (-1, 3)\}$ and $\{(2, -1), (3, -3)\}$, this strategy is indeed winning for Player 1—the same would apply if she were to choose the other self-loop $(0, -1)$ instead.

The proof of our main result consists in constructing from σ a finite-memory winning strategy $\tilde{\sigma}$ for Player 1 in the infinite bounding game $\Gamma(V, E, d)$ from v_0, which ensures the small hypercube property stated in Lem. 3.1. Let us outline this construction. The memory of $\tilde{\sigma}$ consists of:

a simple path γ from the initial vertex v_0 to the current vertex v, in which Player 1's edges are coloured by $|V| \cdot \|E\|$-generated perfect half-spaces of $\mathbb{Q}^d$ (this can be represented concretely by a sequence of coloured edges from E);

a colour i.e. a $|V| \cdot \|E\|$-generated perfect half-space $\mathcal{H} = H_d \cup \cdots \cup H_1$ of $\mathbb{Q}^d$ (initially the $\prec$-minimal one);

counters $c(k, W)$ for every $k \in \{1, 2, \ldots, d\}$ and for every nonzero total weight W of a simple cycle, which are natural numbers (initially 0).

Strategy $\tilde{\sigma}$ copies its moves from strategy σ for the first-cycle bounding game, based on the coloured simple path and the colour it has in its memory. Whenever a cycle is formed it is removed from the simple path, and provided its weight W is nonzero, all the counters $c(k, W)$ are incremented.

Together with the current path, the counters provide the current energy level, which equals $w(\gamma) + \sum_W c(d, W) \cdot W$ throughout the play, where W ranges over all simple cycle weights. To keep the counters and thus the total energy bounded, $\tilde{\sigma}$ may perform one of the following operations after a counter increment:

- a k-*shift to* $H'_k > H_k$ changes the current colour $\mathcal{H}$ to the $\prec$-minimal perfect half-space of the form $H_d \cup \cdots \cup H_{k+1} \cup H'_k \cup \cdots \cup H'_1$, and resets to 0 all the counters $c(k', W)$ with $k' < k$;
- a k-*cancellation* changes the current colour $\mathcal{H}$ to the $\prec$-minimal perfect half-space of the form $H_d \cup \cdots \cup H_{k+1} \cup H'_k \cup \cdots \cup H'_1$. Simultaneously, given some simple cycle weights $W_1, \ldots, W_n$ and a positive integral solution $\mathbf{x}$ to $\sum_{i=1}^n \mathbf{x}(i) W_i = 0$, it subtracts $\mathbf{x} \cdot u(k)$ where $u(k) \overset{\text{def}}{=} (4|V| \cdot \|E\|)^{(2k-1)(d+2)^2}$ from all the tuples $(c(k', W_1), \ldots, c(k', W_n))$ with $k' \geq k$, and resets to 0 all the counters $c(k', W)$ with $k' < k$.

These operations allow to maintain two main invariants, from which the small hypercube property of Lem. 5.5 is derived. For all $1 \leq k \leq d$ and simple path weights W in the span of H_k:

- initially, after any $>k$-shift, and after any $\geq k$-cancellation, $c(k, W) < \mathcal{U}(k) \overset{\text{def}}{=} (4|V| \cdot \|E\|)^{2k(d+2)^2}$, the so-called k-*soft bound*;
- at all times, $c(k, W) < \mathcal{U}(k) + u(k)$, the so-called k-*hard bound*.

To ensure those invariants, strategy $\tilde{\sigma}$ further maintains that, whenever $c(k, W) \geq \mathcal{U}(k)$ and W is in $\text{span}(H_k)$, then W is in $\overline{H_k}$. When this new invariant cannot be preserved by any k-shift, then a version of the Farkas-Minkowski-Weyl Theorem implies that it can be enforced through a k-cancellation, in which a small positive integral solution can be found for the associated system of equations where $W_1, \ldots, W_n$ are the offending cycle weights.

This strategy shows a statement dual to Lem. 5.3, and thereby entails both the equivalence of infinite bounding games with first-cycle bounding games and the small hypercube property of Lem. 3.1 (see the full paper for a proof):

Lemma 5.5. *If σ is winning for Player 1 in $G(V, E, d)$ from some vertex v_0, then $\tilde{\sigma}$ is winning for Player 1 in $\Gamma(V, W, d)$ from v_0, and ensures energy levels of norm at most $(4|V| \cdot \|E\|)^{2(d+2)^3}$.*

6   Concluding Remarks

In this paper, we have shown in Thm. 3.3 and Thm. 3.5 that fixed-dimensional energy games can be solved in pseudo-polynomial time, regardless of whether the initial credit is arbitrary or fixed. For the variant with given initial credit, this closes a large complexity gap between the TOWER upper bounds of Brázdil, Jančar, and Kučera [4] and the lower bounds of Courtois and Schmitz [9], and also settles the complexity of simulation problems between VASS and finite state systems [9]:

Corollary 6.1. *The given initial credit problem for energy games is* 2EXPTIME-*complete, and* EXPTIME-*complete in fixed dimension* $d \geq 4$.

The main direction for extending these results is to consider a *parity* condition on top of the energy condition. Abdulla, Mayr, Sangnier, and Sproston [2] show that multi-dimensional energy parity games with given initial credit are decidable. They do not provide any complexity upper bounds—although one might be able to show TOWER upper bounds from the memory bounds on winning strategies shown by Chatterjee et al. [7, Lem. 3]—, leaving a large complexity gap with 2EXPTIME-hardness. This gap also impacts the complexity of *weak simulation* games between VASS and finite state systems [2].

Acknowledgments. The authors thank Dmitry Chistikov for his assistance in proving Lem. 5.3, the anonymous reviewers for their insightful comments, and Christoph Haase, Jérôme Leroux, and Claudine Picaronny for helpful discussions on linear algebra.

References

1. Abdulla, P.A., Atig, M.F., Hofman, P., Mayr, R., Kumar, K.N., Totzke, P.: Infinite-state energy games. In: CSL-LICS 2014. ACM (2014)
2. Abdulla, P.A., Mayr, R., Sangnier, A., Sproston, J.: Solving parity games on integer vectors. In: D'Argenio, P.R., Melgratti, H. (eds.) CONCUR 2013 – Concurrency Theory. LNCS, vol. 8052, pp. 106–120. Springer, Heidelberg (2013)
3. Brázdil, T., Chatterjee, K., Kučera, A., Novotný, P.: Efficient controller synthesis for consumption games with multiple resource types. In: Madhusudan, P., Seshia, S.A. (eds.) CAV 2012. LNCS, vol. 7358, pp. 23–38. Springer, Heidelberg (2012)
4. Brázdil, T., Jančar, P., Kučera, A.: Reachability games on extended vector addition systems with states. In: Abramsky, S., Gavoille, C., Kirchner, C., Meyer auf der Heide, F., Spirakis, P.G. (eds.) ICALP 2010. LNCS, vol. 6199, pp. 478–489. Springer, Heidelberg (2010)
5. Chaloupka, J.: Z-reachability problem for games on 2-dimensional vector addition systems with states is in P. Fund. Inform. **123**(1), 15–42 (2013)
6. Chatterjee, K., Doyen, L., Henzinger, T.A., Raskin, J.F.: Generalized mean-payoff and energy games. In: FSTTCS 2010. LIPIcs, vol. 8, pp. 505–516. LZI (2010)
7. Chatterjee, K., Randour, M., Raskin, J.F.: Strategy synthesis for multi-dimensional quantitative objectives. Acta Inf. **51**(3–4), 129–163 (2014)
8. Chatterjee, K., Velner, Y.: Hyperplane separation technique for multidimensional mean-payoff games. In: D'Argenio, P.R., Melgratti, H. (eds.) CONCUR 2013 – Concurrency Theory. LNCS, vol. 8052, pp. 500–515. Springer, Heidelberg (2013)
9. Courtois, J.-B., Schmitz, S.: Alternating vector addition systems with states. In: Csuhaj-Varjú, E., Dietzfelbinger, M., Ésik, Z. (eds.) MFCS 2014, Part I. LNCS, vol. 8634, pp. 220–231. Springer, Heidelberg (2014)

10. Fahrenberg, U., Juhl, L., Larsen, K.G., Srba, J.: Energy games in multiweighted automata. In: Cerone, A., Pihlajasaari, P. (eds.) ICTAC 2011. LNCS, vol. 6916, pp. 95–115. Springer, Heidelberg (2011)
11. Juhl, L., Guldstrand Larsen, K., Raskin, J.-F.: Optimal bounds for multiweighted and parametrised energy games. In: Liu, Z., Woodcock, J., Zhu, H. (eds.) Theories of Programming and Formal Methods. LNCS, vol. 8051, pp. 244–255. Springer, Heidelberg (2013)
12. Rackoff, C.: The covering and boundedness problems for vector addition systems. Theor. Comput. Sci. **6**(2), 223–231 (1978)

An Algebraic Geometric Approach to Nivat's Conjecture

Jarkko Kari$^{(\boxtimes)}$ and Michal Szabados

Department of Mathematics and Statistics,
University of Turku, 20014 Turku, Finland
{jkari,micsza}@utu.fi

Abstract. We study multidimensional configurations (infinite words) and subshifts of low pattern complexity using tools of algebraic geometry. We express the configuration as a multivariate formal power series over integers and investigate the setup when there is a non-trivial annihilating polynomial: a non-zero polynomial whose formal product with the power series is zero. Such annihilator exists, for example, if the number of distinct patterns of some finite shape D in the configuration is at most the size $|D|$ of the shape. This is our low pattern complexity assumption. We prove that the configuration must be a sum of periodic configurations over integers, possibly with unbounded values. As a specific application of the method we obtain an asymptotic version of the well-known Nivat's conjecture: we prove that any two-dimensional, non-periodic configuration can satisfy the low pattern complexity assumption with respect to only finitely many distinct rectangular shapes D.

1 Introduction

Consider configuration $c \in A^{\mathbb{Z}^d}$, a d-dimensional infinite array filled by symbols from finite alphabet A. Suppose that for some finite observation window $D \subseteq \mathbb{Z}^d$, the number of distinct patterns of shape D that exist in c is small, at most the cardinality $|D|$ of D. We investigate global regularities and structures in c that are enforced by such local complexity assumption.

Let us be more precise on the involved concepts. As usual, we denote by $c_{\boldsymbol{v}} \in A$ the symbol in c in position $\boldsymbol{v} \in \mathbb{Z}^d$. For $\boldsymbol{u} \in \mathbb{Z}^d$, we say that c is $\boldsymbol{u}$-periodic if $c_{\boldsymbol{v}} = c_{\boldsymbol{v}+\boldsymbol{u}}$ holds for all $\boldsymbol{v} \in \mathbb{Z}^d$, and c is periodic if it is $\boldsymbol{u}$-periodic for some $\boldsymbol{u} \neq 0$. For a finite domain $D \subseteq \mathbb{Z}^d$, the elements of A^D are D-patterns. For a fixed D, we denote by $c_{\boldsymbol{v}+D}$ the D-pattern in c in position $\boldsymbol{v}$, that is, the pattern $\boldsymbol{u} \mapsto c_{\boldsymbol{v}+\boldsymbol{u}}$ for all $\boldsymbol{u} \in D$. The number of distinct D-patterns in c is the D-pattern complexity $P_c(D)$ of c. Our assumption of low local complexity is

$$P_c(D) \leq |D|, \tag{1}$$

for some finite D.

© Springer-Verlag Berlin Heidelberg 2015
M.M. Halldórsson et al. (Eds.): ICALP 2015, Part II, LNCS 9135, pp. 273–285, 2015.
DOI: 10.1007/978-3-662-47666-6_22

Nivat's Conjecture

There are specific examples in the literature of open problems in this framework. *Nivat's conjecture* (proposed by M. Nivat in his keynote address in ICALP 1997 [Niv97]) claims that in the two-dimensional case $d = 2$, the low complexity assumption (1) for a rectangle D implies that c is periodic. The conjecture is a natural generalization of the one-dimensional Morse-Hedlund theorem that states that if a bi-infinite word contains at most n distinct subwords of length n then the word must be periodic [MH38]. In the two-dimensional setting for $m, n \in \mathbb{N}$ we denote by $P_c(m, n)$ the complexity $P_c(D)$ for the $m \times n$ rectangle D.

Conjecture 1 (Nivat's conjecture). If for some m, n we have $P_c(m, n) \leq mn$ then c is periodic.

The conjecture has recently raised wide interest, but it remains unsolved. In [EKM03] it was shown $P_c(m, n) \leq mn/144$ is enough to guarantee the periodicity of c. This bound was improved to $P_c(m, n) \leq mn/16$ in [QZ04], and recently to $P_c(m, n) \leq mn/2$ in [CK13b]. Also the cases of narrow rectangles have been investigated: it was shown in [ST02] and recently in [CK13a] that $P_c(2, n) \leq 2n$ and $P_c(3, n) \leq 3n$, respectively, imply that c is periodic.

The analogous conjecture in the higher dimensional setups $d \geq 3$ is false [ST00]. The following example recalls a simple counter example for $d = 3$.

Example 1. Fix $n \geq 3$, and consider the following $c \in \{0, 1\}^{\mathbb{Z}^3}$ consisting of two perpendicular lines of 1's on a 0-background, at distance n from each other: $c(i, 0, 0) = c(0, i, n) = 1$ for all $i \in \mathbb{Z}$, and $c(i, j, k) = 0$ otherwise. For D equal to the $n \times n \times n$ cube we have $P_c(D) = 2n^2 + 1$ since the D-patterns in c have at most a single 1-line piercing a face of the cube. Clearly c is not periodic although $P_c(D) = 2n^2 + 1 < n^3 = |D|$. Notice that c is a "sum" of two periodic components (the lines of 1's). Our results imply that any counter example must decompose into a sum of periodic components.   □

Periodic Tiling Problem

Another related open problem is the *periodic (cluster) tiling problem* by Lagarias and Wang [LW96]. A (cluster) tile is a finite $D \subset \mathbb{Z}^d$. Its co-tiler is any subset $C \subseteq \mathbb{Z}^d$ such that

$$D \oplus C = \mathbb{Z}^d. \tag{2}$$

The co-tiler can be interpreted as the set of positions where copies of D are placed so that they together cover the entire $\mathbb{Z}^d$ without overlaps. Note that the tile D does not need to be connected – hence the term "cluster tile" is sometimes used. The tiling is by translations of D only: the tiles may not be rotated.

It is natural to interpret any $C \subseteq \mathbb{Z}^d$ as the binary configuration $c \in \{0, 1\}^{\mathbb{Z}^d}$ with $c_v = 1$ if and only if $v \in C$. Then the tiling condition (2) states that C is a co-tiler for D if and only if the $(-D)$-patterns in the corresponding configuration c contain exactly a single 1 in the background of 0's. In fact, as co-tilers of D and $-D$ coincide [Sze98], this is equivalent to all D-patterns having a single 1.

We see that the set $\mathcal{C}$ of all co-tiler configurations for D is a *subshift of finite type* [LM95]. We also see that the low local complexity assumption (1) is satisfied. We even have $P_{\mathcal{C}}(D) \leq |D|$ where we denote by $P_{\mathcal{C}}(D)$ the number of distinct D-patterns found in the elements of the subshift $\mathcal{C}$.

Conjecture 2 (Periodic Tiling Problem). If tile D has a co-tiler then it has a periodic co-tiler.

This conjecture was first formulated in [LW96]. In the one-dimensional case it is easily seen true, but already for $d = 2$ it is open. Interestingly, it is known that if $|D|$ is a prime number then *every* co-tiler of D is periodic [Sze98] (see also our Example 2). The same is true if D is connected, that is, a polyomino [BN91].

Our Contributions

We approach these problems using tools of algebraic geometry. Assuming alphabet $A \subseteq \mathbb{Z}$, we express configuration c as a formal power series over d variables and with coefficients in A. The complexity assumption (1) implies that there is a non-trivial polynomial that annihilates the power series under formal multiplication (Lemma 1). This naturally leads to the study of the annihilator ideal of the power series, containing all the polynomials that annihilate it. Using Hilbert's Nullstellensatz we prove that the ideal contains polynomials of particularly simple form (Theorem 1). In particular, this implies that $c = c_1 + \cdots + c_m$ for some periodic $c_1, \ldots, c_m$ (Theorem 2). This decomposition result is already an interesting global structure on c, but to prove periodicity we would need $m = 1$.

We study the structure of the annihilator ideal in the two-dimensional setup, and prove that it is always a radical (Lemma 5). This leads to a stronger decomposition theorem (Theorem 3). In the case of Nivat's conjecture we then provide an asymptotic result (Theorem 4): for any non-periodic configuration c there are only finitely many pairs $m, n \in \mathbb{N}$ such that $P_c(m, n) \leq mn$.

Due to the page limit the proofs in the latter part of the paper are omitted.

2 Basic Concepts and Notation

For a domain R – which will usually be the whole numbers $\mathbb{Z}$ or complex numbers $\mathbb{C}$ – denote by $R[x_1, \ldots, x_d]$ the set of polynomials over R in d variables. We adopt the usual simplified notation: for a d-tuple of non-negative integers $\boldsymbol{v} = (v_1, \ldots, v_d)$ set $X^{\boldsymbol{v}} = x_1^{v_1} \ldots x_d^{v_d}$, then we write

$$R[X] = R[x_1, \ldots, x_d]$$

and a general polynomial $f \in R[X]$ can be expressed as $f = \sum a_{\boldsymbol{v}} X^{\boldsymbol{v}}$, where $a_{\boldsymbol{v}} \in R$ and the sum goes over finitely many d-tuples of non-negative integers $\boldsymbol{v}$. If we allow $\boldsymbol{v}$ to contain also negative integers we obtain *Laurent polynomials*, which are denoted by $R[X^{\pm 1}]$. Finally, by relaxing the requirement to have only finitely many $a_v \neq 0$ we get *formal power series*:

$$R[[X^{\pm 1}]] = \left\{ \sum a_v X^v \mid v \in \mathbb{Z}^d, \ a_v \in R \right\}.$$

Note that we allow negative exponents in formal power series.

Let d be a positive integer. Let us define a d-dimensional *configuration* to be any formal power series $c \in \mathbb{C}[[X^{\pm 1}]]$:

$$c = \sum_{v \in \mathbb{Z}^d} c_v X^v.$$

A configuration is *integral* if all coefficients c_v are integers, and it is *finitary* if there are only finitely many distinct coefficients c_v. In the case the coefficients are not given explicitly we denote the coefficient at position v by a subscript.

Classically in symbolic dynamics configurations are understood as elements of $A^{\mathbb{Z}^d}$. Because the actual names of the symbols in the alphabet A do not matter, they can be chosen to be integers. Then such a "classical" configuration can be identified with a finitary integral configuration by simply setting the coefficient c_v to be the symbol at position v.

The first advantage of using formal power series is that a multiplication by a Laurent polynomial is well defined and results again in formal power series. For example, $X^v c$ is a translation of c by the vector v. Another important example is that c is periodic if and only if there is a non-zero $v \in \mathbb{Z}^d$ such that $(X^v - 1)c = 0$. Here the right side is understood as a constant zero configuration.

For a polynomial $f(X) = \sum a_v X^v$ and a positive integer n define $f(X^n) = \sum a_v X^{nv}$. The following example, and the proof of Lemma 2, use the well known fact that for any integral polynomial f and prime number p, we have $f^p(X) \equiv f(X^p) \pmod{p}$.

Example 2. Our first example concerns the periodic tiling problem. We provide a short proof of the fact – originally proved in [Sze98] – that if the size $p = |D|$ of tile D is a prime number then all co-tilers C are periodic. When the tile D is represented as the Laurent polynomial $f(X) = \sum_{v \in D} X^v$ and the co-tiler C as the power series $c(X) = \sum_{v \in C} X^v$, the tiling condition (2) states that $f(X)c(X) = \sum_{v \in \mathbb{Z}^d} X^v$. Multiplying both sides by $f^{p-1}(X)$, we get

$$f^p(X)c(X) = \sum_{v \in \mathbb{Z}^d} p^{p-1} X^v \equiv 0 \pmod{p}.$$

On the other hand, since p is a prime, $f^p(X) \equiv f(X^p) \pmod{p}$ so that

$$f(X^p)c(X) \equiv 0 \pmod{p}.$$

Let $v \in D$ and $w \in C$ be arbitrary. We have

$$0 \equiv [f(X^p)c(X)]_{w+pv} = \sum_{u \in D} c(X)_{w+pv-pu} \pmod{p}.$$

The last sum is a sum of p numbers, each 0 or 1, among which there is at least one 1 (corresponding to $\boldsymbol{u} = \boldsymbol{v}$). The only way for the sum to be divisible by p is by having each summand equal to 1. We have that $\boldsymbol{w} + p(\boldsymbol{v} - \boldsymbol{u})$ is in C for all $\boldsymbol{u}, \boldsymbol{v} \in D$ and $\boldsymbol{w} \in C$, which means that C is $p(\boldsymbol{v} - \boldsymbol{u})$-periodic for all $\boldsymbol{u}, \boldsymbol{v} \in D$. □

The next lemma grants us that for low complexity configurations there exists at least one Laurent polynomial that annihilates the configuration by formal multiplication.

Lemma 1. *Let c be a configuration and $D \subset \mathbb{Z}^d$ a finite domain such that $P_c(D) \leq |D|$. Then there exists a non-zero Laurent polynomial $f \in \mathbb{C}[X^{\pm 1}]$ such that $fc = 0$.*

Proof. Denote $D = \{\boldsymbol{u}_1, \ldots, \boldsymbol{u}_n\}$ and consider the set

$$\{ (1, c_{\boldsymbol{u}_1 + \boldsymbol{v}}, \ldots, c_{\boldsymbol{u}_n + \boldsymbol{v}}) \mid \boldsymbol{v} \in \mathbb{Z}^d \}.$$

It is a set of complex vectors of dimension $n+1$, and because c has low complexity there is at most $n = |D|$ of them. Therefore there exists a common orthogonal vector $(a_0, \ldots, a_n)$. Let $g(X) = a_1 X^{-\boldsymbol{u}_1} + \cdots + a_n X^{-\boldsymbol{u}_n}$, then the coefficient of gc at position $\boldsymbol{v}$ is

$$(gc)_{\boldsymbol{v}} = a_1 c_{\boldsymbol{u}_1 + \boldsymbol{v}} + \cdots + a_n c_{\boldsymbol{u}_n + \boldsymbol{v}} = -a_0,$$

that is, gc is a constant configuration. Now it suffices to set $f = (X^{\boldsymbol{v}} - 1)g$ for arbitrary non-zero vector $\boldsymbol{v} \in \mathbb{Z}^d$. □

3   Annihilating Polynomials and Decomposition Theorem

Lemma 1 motivates the following definitions. Let c be a configuration. We say that a Laurent polynomial f *annihilates* (or is an *annihilator* of) the configuration if $fc = 0$. Define

$$\mathrm{Ann}(c) = \{ f \in \mathbb{C}[X] \mid fc = 0 \}.$$

It is the set of all annihilators of c. Clearly it is an ideal of $\mathbb{C}[X]$. The zero polynomial annihilates every configuration; let us call an annihilator *non-trivial* if it is non-zero. Note that the configuration is periodic if and only if $X^{\boldsymbol{v}} - 1 \in \mathrm{Ann}(c)$ for some non-zero $\boldsymbol{v} \in \mathbb{Z}^d$.

We defined $\mathrm{Ann}(c)$ to consist of complex polynomials, so that we can later use Hilbert's Nullstellensatz directly, as it requires polynomial ideals over algebraically closed field. We shall however occasionally work with integer coefficients and Laurent polynomials when it is more convenient.

Recall that in the case of Nivat's conjecture and Periodic tiling problem we study finitary integral configurations, which by Lemma 1 have a non-trivial annihilator. Moreover there is an integer annihilating polynomial – actually for integral configurations $\mathrm{Ann}(c)$ is always generated by integer polynomials.

If $Z = (z_1, \ldots, z_d) \in \mathbb{C}^d$ is a complex vector then it can be plugged into a polynomial. Plugging it into a monomial X^v results in $Z^v = z_1^{v_1} \cdots z_d^{v_d}$.

Lemma 2. *Let $c(X)$ be a finitary integral configuration and $f(X) \in \mathrm{Ann}(c)$ a non-zero integer polynomial. Then there exists an integer r such that for every positive integer n relatively prime to r we have $f(X^n) \in \mathrm{Ann}(c)$.*

Proof. Denote $f(X) = \sum a_v X^v$. First we prove the claim for the case when n is a large enough prime.

Let p be a prime, then we have $f^p(X) \equiv f(X^p) \pmod{p}$. Because f annihilates c, multiplying both sides by $c(X)$ results in

$$0 \equiv f(X^p)c(X) \pmod{p}.$$

The coefficients in $f(X^p)c(X)$ are bounded in absolute value by

$$s = c_{max} \sum |a_v|,$$

where c_{max} is the maximum absolute value of coefficients in c. Therefore if $p > s$ we have $f(X^p)c(X) = 0$.

For the general case, set $r = s!$. Now every n relatively prime to r is of the form $p_1 \cdots p_k$ where each p_i is a prime greater than s. Note that we can repeat the argument with the same bound s also for polynomials $f(X^m)$ for arbitrary m – the bound s depends only on c and the (multi)set of coefficients a_v, which is the same for all $f(X^m)$. Thus we have $f(X^{p_1 \cdots p_k}) \in \mathrm{Ann}(c)$. □

Lemma 3. *Let c be a finitary integral configuration and $f = \sum a_v X^v$ a nontrivial integer polynomial annihilator. Let $S = \{ v \in \mathbb{Z}^d \mid a_v \neq 0 \}$ and define*

$$g(X) = x_1 \cdots x_d \prod_{\substack{v \in S \\ v \neq v_0}} (X^{rv} - X^{rv_0})$$

where r is the integer from Lemma 2 and $v_0 \in S$ arbitrary. Then $g(Z) = 0$ for any common root $Z \in \mathbb{C}^d$ of $\mathrm{Ann}(c)$.

Proof. Fix Z. If any of its complex coordinates is zero then clearly $g(Z) = 0$. Assume therefore that all coordinates of Z are non-zero.

Let us define for $\alpha \in \mathbb{C}$

$$S_\alpha = \{ v \in S \mid Z^{rv} = \alpha \},$$
$$f_\alpha(X) = \sum_{v \in S_\alpha} a_v X^v.$$

Because S is finite, there are only finitely many non-empty sets $S_{\alpha_1}, \ldots, S_{\alpha_m}$ and they form a partitioning of S. In particular we have $f = f_{\alpha_1} + \cdots + f_{\alpha_m}$.

Numbers of the form $1 + ir$ are relatively prime to r for all non-negative integers i, therefore by Lemma 2, $f(X^{1+ir}) \in \text{Ann}(c)$. Plugging in Z we obtain $f(Z^{1+ir}) = 0$. Now compute:

$$f_\alpha(Z^{1+ir}) = \sum_{v \in S_\alpha} a_v Z^{(1+ir)v} = \sum_{v \in S_\alpha} a_v Z^v \alpha^i = f_\alpha(Z)\alpha^i$$

Summing over $\alpha = \alpha_1, \ldots, \alpha_m$ gives

$$0 = f(Z^{1+ir}) = f_{\alpha_1}(Z)\alpha_1^i + \cdots + f_{\alpha_m}(Z)\alpha_m^i.$$

Let us rewrite the last equation as a statement about orthogonality of two vectors in $\mathbb{C}^m$:

$$(f_{\alpha_1}(Z), \ldots, f_{\alpha_m}(Z)) \perp (\alpha_1^i, \ldots, \alpha_m^i)$$

By Vandermode determinant, for $i \in \{0, \ldots, m-1\}$ the vectors on the right side span the whole $\mathbb{C}^m$. Therefore the left side must be the zero vector, and especially for α such that $v_0 \in S_\alpha$ we have

$$0 = f_\alpha(Z) = \sum_{v \in S_\alpha} a_v Z^v.$$

Because Z does not have zero coordinates, each term on the right hand side is non-zero. But the sum is zero, therefore there are at least two vectors $v_0, v \in S_\alpha$. From the definition of S_α we have $Z^{rv} = Z^{rv_0} = \alpha$, so Z is a root of $X^{rv} - X^{rv_0}$. □

Theorem 1. *Let c be a finitary integral configuration with a non-trivial annihilator. Then there are non-zero $v_1, \ldots, v_m \in \mathbb{Z}^d$ such that the Laurent polynomial*

$$(X^{v_1} - 1) \cdots (X^{v_m} - 1)$$

annihilates c.

Proof. This is an easy corollary of Lemma 3. The polynomial $g(X)$ vanishes on all common roots of $\text{Ann}(c)$, therefore by Hilbert's Nullstellensatz there is n such that $g^n(X) \in \text{Ann}(c)$. Note that any monomial multiple of an annihilator is again an annihilator. Therefore also

$$\frac{g^n(X)}{x_1^n \cdots x_d^n X^{nrv_0(|S|-1)}}$$

is, and it is a Laurent polynomial of the desired form. □

Multiplying a configuration by $(X^v - 1)$ can be seen as a *"difference operator"* on the configuration. Theorem 1 then says, that there is a sequence of difference operators which annihilates the configuration. We can reverse the process: let us start by the zero configuration and step by step *"integrate"* until we obtain the original configuration. This idea gives the Decomposition theorem:

Theorem 2 (Decomposition Theorem). *Let c be a finitary integral config-uration with a non-trivial annihilator. Then there exist periodic integral config-urations $c_1, \ldots, c_m$ such that $c = c_1 + \cdots + c_m$.*

Example 3. Recall the 3D counter example in Example 1. It is the sum $c_1 + c_2$ where $c_1(i, 0, 0) = 1$ and $c_2(0, i, n) = 1$ for all $i \in \mathbb{Z}$, and all other entries are 0. Configurations c_1 and c_2 are $(1, 0, 0)$- and $(0, 1, 0)$-periodic, respectively, so that $(X^{(1,0,0)} - 1)(X^{(0,1,0)} - 1)$ annihilates $c = c_1 + c_2$. □

Example 4. The periodic configurations $c_1, \ldots, c_m$ in Theorem 2 may, for some configurations c, be necessarily non-finitary. Let $\alpha \in \mathbb{R}$ be irrational, and define three periodic two-dimensional configurations c_1, c_2 and c_3 by

$$c_1(i, j) = \lfloor i\alpha \rfloor, \qquad c_2(i, j) = \lfloor j\alpha \rfloor, \qquad c_3(i, j) = \lfloor (i + j)\alpha \rfloor.$$

Then $c = c_3 - c_1 - c_2$ is a finitary integral configuration (over alphabet $\{0, 1\}$), annihilated by the polynomial $(X^{(1,0)} - 1)(X^{(0,1)} - 1)(X^{(1,-1)} - 1)$, but it cannot be expressed as a sum of finitary periodic configurations. □

4   Structure of the Annihilator Ideal

In the rest of the paper we focus on two-dimensional configurations. We analyze $\mathrm{Ann}(c)$ using tools of algebraic geometry and provide a description of a polyno-mial ϕ which divides every annihilator. Moreover we show a theoretical result that $\mathrm{Ann}(c)$ is a radical ideal, which allows us to provide a stricter version of the Decomposition theorem for two-dimensional configurations.

The key ingredient needed for further analysis is the concept of a *line poly-nomial*. Let the *support* of a Laurent polynomial $f = \sum a_v X^v$ be defined as

$$\mathrm{supp}(f) = \{\, v \in \mathbb{Z}^d \mid a_v \neq 0 \,\}.$$

We say that f is a *line Laurent polynomial* if the support contains at least two points and all the points lie on a single line. Let us call a vector $v \in \mathbb{Z}^d$ *primitive* if its coordinates don't have a common non-trivial integer factor. Then every line Laurent polynomial can be expressed as

$$f(X) = X^{v'}(a_n X^{nv} + \cdots + a_1 X^v + a_0)$$

for some $a_i \in \mathbb{C}$, $n \geq 1$, $a_n \neq 0 \neq a_0$, $v', v \in \mathbb{Z}^d$, v primitive. Moreover, the vector v is determined uniquely up to the sign. We define the *direction* of a line Laurent polynomial to be the vector space $\langle v \rangle \subset \mathbb{Q}^d$.

To simplify the notation, we prefer to write $\mathbb{C}[x, y]$ in the place of $\mathbb{C}[x_1, x_2]$. We begin by a sequence of lemmas with a result from algebra. Recall that an ideal A is *prime* whenever $ab \in A$ implies $a \in A$ or $b \in A$. An ideal is *radical* if $a^n \in A$ implies $a \in A$.

Lemma 4.

1. *Prime ideals in $\mathbb{C}[x,y]$ are maximal ideals, principal ideals generated by irreducible polynomials, and the zero ideal.*
2. *Every radical ideal $A \leq \mathbb{C}[x,y]$ can be uniquely written as a finite intersection of prime ideals $P_1,\ldots,P_k$ where $P_i \not\subset P_j$ for $i \neq j$. Moreover*

$$A = \bigcap_{i=1}^{k} P_i = \prod_{i=1}^{k} P_i.$$

Lemma 5. *Let c be a two-dimensional, finitary and integral configuration with a non-trivial annihilator. Then $\mathrm{Ann}(c)$ is radical.*

Our proof of Lemma 5 relies on the decomposition of two-dimensional radical ideals into a product of primes from Lemma 4, which fails in higher dimensions. However, we conjecture that Lemma 5 is true for higher dimensions as well.

Lemma 6. *Let c be as in Lemma 5. Then there exist polynomials $\phi_1,\ldots,\phi_m$ and an ideal $H \leq \mathbb{C}[x,y]$ such that*

$$\mathrm{Ann}(c) = \phi_1 \cdots \phi_m H$$

where ϕ_i are line polynomials in pairwise distinct directions, and H is either an intersection of finitely many maximal ideals or $H = \mathbb{C}[x,y]$.

Moreover H is determined uniquely and ϕ_i are determined uniquely up to the order and multiplication by a constant.

Note that $H = \mathbb{C}[x,y]$ is not really a special case – it covers the case when H is the empty intersection. Let us denote the number m from Lemma 6 by $\mathrm{ord}(c)$. It is an important invariant of the configuration which provides information about its periodicity. A two-dimensional configuration is *doubly periodic* if there are two linearly independent vectors in which it is periodic. A configuration which is periodic but not doubly periodic is called *one-periodic*.

Theorem 3 (Strong Decomposition Theorem). *Let c, $m = \mathrm{ord}(c)$, and $\mathrm{Ann}(c) = \phi_1 \cdots \phi_m H$ be as in Lemma 6. Let $\phi = \phi_1 \cdots \phi_m$. Then there exist configurations $c_\phi, c_H, c_1,\ldots,c_m$ such that*

$$c = c_\phi + c_H$$
$$c_\phi = c_1 + \cdots + c_m,$$

where $\mathrm{Ann}(c_\phi) = \langle \phi \rangle$, $\mathrm{Ann}(c_H) = H$ and $\mathrm{Ann}(c_i) = \langle \phi_i \rangle$. Moreover c_ϕ and c_H are determined uniquely. Each c_i is one-periodic in the direction of ϕ_i, and c_H is doubly periodic.

Corollary 1. *Let c be as in Theorem 3. Then*

- *if $\mathrm{ord}(c) = 0$ the configuration is doubly periodic,*
- *if $\mathrm{ord}(c) = 1$ the configuration is one-periodic,*
- *if $\mathrm{ord}(c) \geq 2$ the configuration is non-periodic.*

5   Approaching Nivat's Conjecture

We already know that if a finitary integral configuration c satisfies the condition $P_c(m, n) \leq mn$ for some positive integers m, n, then it has an annihilating polynomial. The Nivat's conjecture claims that such a configuration is periodic, that is, $\text{ord}(c) \leq 1$. Our approach is the contrapositive: assume that c is a finitary integral configuration which is non-periodic, that is, $\text{ord}(c) \geq 2$. If c does not have an annihilating polynomial, we have $P_c(m, n) > mn$ for all m and n, and we are done. So we assume c has an annihilating polynomial so that the theory developed so far applies to c. We want to prove that c has high local complexity.

Assuming $\text{ord}(c) \geq 2$, let φ_1 and φ_2 be irreducible factors of ϕ_1 and ϕ_2. Any annihilator of c has a factor $f \in \text{Ann}(c)$ that can be written as $f = \varphi_1 \varphi_2 f'$ such that $c_1 = \varphi_2 f' c$ and $c_2 = \varphi_1 f' c$ are one-periodic configurations in different directions. Moreover, a block in c determines smaller blocks in $\varphi_2 f' c$ and $\varphi_1 f' c$ because the multiplication by a polynomial is a local operation on the configurations. We next estimate the number of distinct blocks in one-periodic configurations in order to lower bound the number of slightly bigger blocks in c.

Complexity of One-Periodic Configurations

Recall that for a finite domain $D \subset \mathbb{Z}^d$ we denote by c_{v+D} the pattern extracted from the position $v \in \mathbb{Z}^d$ in c. Let us define a *line of D-patterns in direction* $u \in \mathbb{Z}^d$ to be a set of the form

$$\mathcal{L} = \left\{ c_{v+ku+D} \mid k \in \mathbb{Z} \right\}$$

for some vector $v \in \mathbb{Z}^d$.

It is easy to characterize irreducible factors of line polynomials – every line polynomial can be decomposed as

$$f(X) = X^{v'}(a_n X^{nv} + \cdots + a_1 X^v + a_0)$$
$$= a_n X^{v'}(X^v - \lambda_1) \ldots (X^v - \lambda_n)$$

where $a_0 \neq 0 \neq a_n$, v is a primitive vector and $\lambda_1, \ldots, \lambda_n$ are complex roots of the polynomial $a_n t^n + \cdots + a_1 t + a_0$. A Laurent polynomial of the form $X^v - \lambda_i$ is irreducible. Therefore an irreducible polynomial factor of f either divides $X^{v'}$, 'or has to be up to a multiplicative constant of the form $X^{v''}(X^v - \lambda_i)$ for some $v'' \in \mathbb{Z}^d$.

The following two lemmas will be applied later on the one-periodic configurations $c_1 = \varphi_2 f' c$ and $c_2 = \varphi_1 f' c$, respectively. For a vector $v = (v_1, v_2) \in \mathbb{Z}^2$ let us denote the size of a minimal rectangle that contains it by $Box(v) := (|v_1|, |v_2|) \in \mathbb{Z}^2$.

Lemma 7. *Let c be a two-dimensional one-periodic configuration and $v', v \in \mathbb{Z}^2$, $0 \neq \lambda \in \mathbb{C}$ such that $\text{Ann}(c) = \langle X^{v'}(X^v - \lambda) \rangle$. Let $(m, n) = Box(v)$. Then for any non-negative integers M, N there are at least $Mn + mN + mn$ disjoint lines of blocks $(M + m) \times (N + n)$ in c in the direction of v.*

Lemma 8. *Let c be a two-dimensional one-periodic configuration and $u', u \in \mathbb{Z}^2$, $0 \neq \lambda \in \mathbb{C}$ such that $\mathrm{Ann}(c) = \langle X^{u'}(X^u - \lambda) \rangle$. Let $(m, n) = \mathrm{Box}(u)$ and $v \in \mathbb{Z}^2$ be a vector in a different direction than u.*

If $\mathcal{L}$ is any line of blocks $(M + m) \times (N + n)$ in direction v in c, then

$$|\mathcal{L}| > \frac{Mn + mN}{S},$$

where S is the positive area of the parallelogram specified by vectors u and v.

Putting Things Together

Applying Lemmas 7 and 8 on the configurations $c_1 = \varphi_2 f'c$ and $c_2 = \varphi_1 f'c$ provides the following lower bound for their common pre-image $c' = f'c$.

Lemma 9. *Let c' be a two-dimensional configuration such that*

$$\mathrm{Ann}(c') = \langle X^{v'}(X^{v_1} - \lambda_1)(X^{v_2} - \lambda_2) \rangle$$

where $\lambda_1, \lambda_2 \in \mathbb{C}$ are non-zero, $v', v_1, v_2 \in \mathbb{Z}^2$ and v_1, v_2 are primitive vectors. Denote $(m_i, n_i) = \mathrm{Box}(v_i)$. Then

$$P_{c'}(M + m_1 + m_2, N + n_1 + n_2) > \frac{(Mn_1 + m_1 N)(Mn_2 + m_2 N)}{m_1 n_2 + m_2 n_1}$$

for all non-negative integers M and N.

Let f be a Laurent polynomial in two variables and S its support. Let us extend the definition of the bounding box $\mathrm{Box}(\cdot)$ by setting

$$\mathrm{Box}(f) = (\max_{(a,b) \in S} a - \min_{(a,b) \in S} a, \max_{(a,b) \in S} b - \min_{(a,b) \in S} b).$$

Corollary 2. *Let c be a two-dimensional non-periodic finitary integral configuration and f its annihilator. Denote $(m, n) = \mathrm{Box}(f)$ and let $M \geq m, N \geq n$ be integers. Then:*

(a) $P_c(M, N) > (M - m)(N - n)$.
(b) If in the decomposition $\mathrm{Ann}(c) = \phi_1 \cdots \phi_{\mathrm{ord}(c)} H$ there are two ϕ_i, ϕ_j such that their directions are not horizontal or vertical, then $\exists \alpha > 1$:

$$P_c(M, N) > \alpha(M - m)(N - n).$$

(c) If $\mathrm{ord}(c) \geq 3$ then

$$P_c(M, N) > 2(M - m)(N - n).$$

The Main Result

Theorem 4. *Let c be a two-dimensional non-periodic configuration. Then $P_c(M, N) > MN$ holds for all but finitely many choices $M, N \in \mathbb{N}$.*

Corollary 3. *If c is a two-dimensional configuration such that $P_c(M, N) \leq MN$ holds for infinitely many pairs $M, N \in \mathbb{N}$, then c is periodic.*

The proof (details omitted) is structured as follows. Let c be non-periodic with a non-trivial annihilator, and let $\mathrm{Ann}(c) = \phi H$ be the decomposition of the annihilator as in Theorem 3, where $\phi = \phi_1 \cdots \phi_{\mathrm{ord}(c)}$. We consider different ranges of M and N.

Very Thin Blocks. Suppose N or M is so small that the support of ϕ does not fit inside the $M \times N$ rectangle. Then no annihilator of c fits inside the rectangle, and as in Lemma 1 we see that $P_c(M, N) > MN$, no matter how large the other dimension of the rectangle is.

Thin Blocks. Consider fixed N, large enough so that the support of ϕ fits inside a strip of height N. It can be shown that there exists M_0 such that for all $M > M_0$ we have $P_c(M, N) > MN$. Analogously for a fixed M.

Fat Blocks. We prove that there are constants M_0 and N_0 such that for $M > M_0$ and $N > N_0$ we have $P_c(M, N) > MN$. This follows directly from Corollary 2(c) and (b), respectively, in the cases when $\mathrm{ord}(c) \geq 3$, or when $\mathrm{ord}(c) = 2$ and ϕ_1 and ϕ_2 are not horizontal or vertical. The cases when $\mathrm{ord}(c) = 2$ and ϕ_1 is vertical (or the symmetric cases) require more careful analysis. In particular, we use the observation that it is enough to consider two letter configurations:

Lemma 10. *In any non-periodic configuration $c \in A^{\mathbb{Z}^2}$, letters can be merged to obtain a non-periodic configuration $c' \in \{0, 1\}^{\mathbb{Z}^2}$. Then $P_{c'}(D) \leq P_c(D)$ for all finite $D \subseteq \mathbb{Z}^d$. In particular, if Nivat's conjecture holds on binary configurations it holds in general.*

It is clear that the three ranges of M and N above cover everything so that $P_c(M, N) \leq MN$ can hold only for a finite number of $M, N \in \mathbb{N}$.

References

[BN91]   Beauquier, D., Nivat, M.: On translating one polyomino to tile the plane. In: Discrete & Computational Geometry **6** (1991)

[CK13a]   Cyr, V., Kra, B.: Complexity of short rectangles and periodicity. In: (submitted) (2013). arXiv: 1307.0098 [math.DS]

[CK13b]   Cyr, V., Kra, B.: Nonexpansive $\mathbb{Z}2$-subdynamics and Nivat's conjecture. Trans. Amer. Math. Soc. (2013). http://dx.doi.org/10.1090/S0002-9947-2015-06391-0

[EKM03]   Epifanio, C., Koskas, M., Mignosi, F.: On a conjecture on bidimensional words. In: Theor. Comput. Sci. 1–3(299) (2003)

[LW96]   Lagarias, J.C., Wang, Y.: Tiling the Line with Translates of One Tile. Inventiones Mathematicae **124**, 341–365 (1996)

[LM95]   Lind, D., Marcus, B.: An Introduction to Symbolic Dynamics and Coding. Cambridge University Press (1995)

[MH38]   Morse, M., Hedlund, G.A.: Symbolic Dynamics. American Journal of Mathematics **60**(4), 815–866 (1938)

[Niv97]   Nivat, M.: Invited talk at ICALP, Bologna (1997)

[QZ04]   Quas, A., Zamboni, L.Q.: Periodicity and local complexity. Theor. Comput. Sci. **319**(1-3), 229–240 (2004)

[ST00]   Sander, J.W., Tijdeman, R.: The complexity of functions on lattices. Theor. Comput. Sci. **246**(1-2), 195–225 (2000)

[ST02]   Sander, J.W., Tijdeman, R.: The rectangle complexity of functions on two-dimensional lattices. Theor. Comput. Sci. **270**(1-2), 857–863 (2002)

[Sze98]   Szegedy, M.: Algorithms to tile the infinite grid with finite clusters. In: FOCS, pp. 137–147. IEEE Computer Society (1998)

Nominal Kleene Coalgebra

Dexter Kozen[1], Konstantinos Mamouras[1], Daniela Petrişan[2],
and Alexandra Silva[2]([✉])

[1] Cornell University, Ithaca, USA
[2] Radboud University Nijmegen, Nijmegen, The Netherlands
alexandra@cs.ru.nl

Abstract. We develop the coalgebraic theory of nominal Kleene algebra, including an alternative language-theoretic semantics, a nominal
extension of the Brzozowski derivative, and a bisimulation-based decision procedure for the equational theory.

1 Introduction

Nominal Kleene algebra, introduced by Gabbay and Ciancia [12], is an algebraic
formalism intended for reasoning equationally about imperative programs with
statically scoped allocation and deallocation of resources. The system consists
of Kleene algebra, the algebra of regular expressions, augmented with a binding
operator ν that binds a named resource within a local scope.

Gabbay and Ciancia [12] proposed an axiomatization of the system consisting
of the axioms of Kleene algebra plus six equations capturing the behavior of the
binding operator ν and its interaction with the Kleene algebra operators. They
also defined a family of *nominal languages* consisting of certain sets of strings
over an infinite alphabet satisfying certain invariance properties and showed
soundness of the axioms over this class of interpretations. Their analysis revealed
some surprising subtleties arising from the non-compositionality of the sequential
composition and iteration operators.

In our previous work [15] we showed that the Gabbay-Ciancia axioms are
not complete for the semantic interpretation of [12], but we identified a slightly
wider class of language models over which they are sound and complete. The
proof of completeness of [15] consists of several stages of transformations to bring
expressions to a certain normal form. Although the construction is effective, one
of the transformations requires the intersection of several regular expressions, an
operation known to produce a double-exponential increase in size in the worst
case [13], thus the construction is unlikely to give a practical decision method.

In this paper, we investigate the coalgebraic theory of nominal Kleene algebra. The motivation for this investigation is to understand the structure of nominal Kleene algebra from a coalgebraic perspective with an eye toward a more
efficient decision procedure for the equational theory in the style of [4,5,24] for
Kleene algebra and Kleene algebra with tests.

The paper is organized as follows. In §3 we introduce a new class of language
models consisting of sets of equivalence classes of ν-*strings*. A ν-string is like a

© Springer-Verlag Berlin Heidelberg 2015
M.M. Halldórsson et al. (Eds.): ICALP 2015, Part II, LNCS 9135, pp. 286–298, 2015.
DOI: 10.1007/978-3-662-47666-6_23

string, except that it may contain binding operators. Two ν-strings are equivalent if they are provably so under the Gabbay-Ciancia axioms and associativity. The equivalence classes of ν-strings over a fixed set of variables form a nominal monoid. These language models are isomorphic to the free language models of [15], thus giving a new characterization of the free models, but more amenable to the development of the coalgebraic theory.

In §4 we introduce nominal versions of the semantic and syntactic Brzozowski derivatives. The derivatives are similar to their non-nominal counterparts, but extended to handle bound variables in such a way as to be invariant with respect to α-conversion. The semantic derivative is defined in terms of the new language model and characterizes the final coalgebra. We conclude the section with a result that relates the algebraic and coalgebraic structure and establishes the existence of minimal automata.

In the full version of this paper [16] we include all omitted proofs and extra examples. We also describe a data representation for the efficient calculation of the Antimirov derivative and give an exponential-space decision procedure.

Related Work. The notion of nominal sets goes back to work of Fraenkel and Mostowski in the early part of the twentieth century. The notion was first applied in computer science by Gabbay and Pitts [10] (see [22] for a survey).

Recently, there have been many studies involving nominal automata, automata on infinite alphabets, and regular expressions with binders that are closely related to the work presented here.

Montanari and Pistore [19–21] and Ferrari et al. [6] develop the theory of *history-dependent (HD) automata*, an operational model for process calculi such as the π-calculus. In these automata, there are mechanisms for explicit allocation and deallocation of names and for explicitly representing the history of allocated names.

A closely related model is the family of *finite memory automata* of Francez and Kaminski [8,9]. These are ordinary finite-state automata equipped with a finite set of *registers*. At any point in time, each register is either empty or contains a symbol from an infinite alphabet.

Bojanczyk, Klin, Lasota [3] undertake a comprehensive study of nominal automata and discuss the relationships between previous models. They consider nominal sets for arbitrary symmetries. They identify the important notion of *orbit-finiteness* as the appropriate analog of finiteness in the non-nominal case and show that their definitions are equivalent to previous definitions of finite memory automata [8,9]. Their paper does not consider the relationship with regular expressions.

Kurz, Suzuki, Tuosto [17,18] present a syntax of regular expressions with binders and consider its relationship with nominal automata. Their syntax includes operational mechanisms for the dynamic allocation and deallocation of fresh names and explicit permutations. Their semantics uses a name-independent combinatorial construct reminiscent of De Bruijn indices.

The most important distinguishing characteristic of our approach is that both the algebraic and coalgebraic structure are nominal. Our syntax, based on

Kleene algebra with ν-binders as introduced by Gabbay and Ciancia [12], and our final coalgebra semantics based on nominal sets of ν-strings, both carry a nominal coalgebraic structure given by the syntactic and semantic Brzozowski derivatives, and the interpretation map is the unique equivariant morphism to the final coalgebra.

2   Background

This section contains a severely abbreviated review of basic material on Kleene algebra, nominal sets, and the nominal extension of Kleene algebra (NKA) introduced by Gabbay and Ciancia [12], but prior familiarity with nominal sets, KA, and coalgebra will be helpful. For a more thorough introduction, the reader is referred to [11,22] for nominal sets, to [25] for Kleene (co)algebra, and to [12,15] for NKA.

Kleene Algebra (KA)   is the algebra of regular expressions. A *Kleene algebra* $(K, +, \cdot, {}^*, 0, 1)$ is an idempotent semiring with $*$ such that x^*y is the $\leq$-least z such that $y + xz \leq z$ and yx^* is the $\leq$-least z such that $y + zx \leq z$. Explicitly,

$$
\begin{array}{lll}
x + (y + z) = (x + y) + z & x(yz) = (xy)z & x + y = y + x \\
1x = x1 = x & x + 0 = x + x = x & x0 = 0x = 0 \\
x(y + z) = xy + xz & (x + y)z = xz + yz & 1 + xx^* \leq x^* \\
y + xz \leq z \ \Rightarrow\ x^*y \leq z & y + zx \leq z \ \Rightarrow\ yx^* \leq z & 1 + x^*x \leq x^*
\end{array}
$$

G-Sets.   A *group action* of a group G on a set X is a map $G \times X \to X$, written as juxtaposition, such that $\pi(\rho x) = (\pi\rho)x$ and $1x = x$ for $\pi, \rho \in G$ and $x \in X$. A *G-set* is a set X equipped with a group action $G \times X \to X$. The *orbit* of an element $x \in X$ is the set $\{\pi x \mid \pi \in G\} \subseteq X$. If X and Y are two G-sets, a function $f : X \to Y$ is called *equivariant* if $f \circ \pi = \pi \circ f$ for all $\pi \in G$.

The G-sets and equivariant functions form an elementary topos G-Set with group action on coproducts, products, and exponentials defined by

$$
\pi(\text{in } x) = \text{in}(\pi x) \quad \pi(x, y) = (\pi x, \pi y) \quad \pi() = () \quad \pi f = \pi \circ f \circ \pi^{-1}. \tag{1}
$$

In particular, for sets, $\pi A = \{\pi x \mid x \in A\}$. For $x \in X$ and $A \subseteq X$, define

$$
\text{fix}\, x = \{\pi \in G \mid \pi x = x\} \qquad\qquad \text{Fix}\, A = \bigcap_{x \in A} \text{fix}\, x.
$$

Note that $\text{Fix}\, A$ and $\text{fix}\, A$ are different: they are the subgroups of G that fix A pointwise and setwise, respectively.

Nominal Sets.   Fix a countably infinite set $\mathbb{A}$ of *atoms* and let $G_\mathbb{A}$ be the group of all finite permutations of $\mathbb{A}$ (permutations generated by transpositions $(a\ b)$). The set $\mathbb{A}$ is a $G_\mathbb{A}$-set under the group action $\pi a = \pi(a)$. If X is another $G_\mathbb{A}$-set, we say that $A \subseteq \mathbb{A}$ *supports* $x \in X$ if $\text{Fix}\, A \subseteq \text{fix}\, x$. An element $x \in X$ has *finite support* if there is a finite set $A \subseteq \mathbb{A}$ that supports x. If x has finite support, then there is a smallest set supporting x, called $\text{supp}\, x$. We write $a \# x$ and say a is *fresh for* x if $a \notin \text{supp}\, x$. A *nominal set* is a $G_\mathbb{A}$-set X of which every element has finite support. The nominal sets and equivariant functions form a full subcategory Nom of G-Set.

Expressions and ν-Strings.   NKA expressions are given by the grammar

$$e ::= a \in \mathbb{A} \mid e + e \mid ee \mid e^* \mid 0 \mid 1 \mid \nu a.e.$$

The scope of the binding νa in $\nu a.e$ is e. As a notational convention, we assign the binding operator νa lower precedence than product but higher precedence than sum; thus in products, scopes extend as far to the right as possible. For example, $\nu a.ab \; \nu b.ba$ should be read as $\nu a.(ab \; \nu b.(ba))$ and not $(\nu a.ab)(\nu b.ba)$. The set of NKA expressions over $\mathbb{A}$ is denoted $\mathsf{Exp}\,\mathbb{A}$.

The free variables $\mathsf{FV}(e)$ of an expression e are defined as usual, and the group $G_{\mathbb{A}}$ acts on $\mathsf{Exp}\,\mathbb{A}$ by permuting the variables in the obvious way. For example, $(a\,b)\nu a.b = \nu b.a$. The relation $\equiv_\alpha$ of α-equivalence on $\mathsf{Exp}\,\mathbb{A}$ is defined to be the least congruence containing the pairs $\{e \equiv_\alpha \pi e \mid \pi \in \mathsf{Fix}\,\mathsf{FV}(e)\}$. Let $[e]$ denote the $\equiv_\alpha$-congruence class of e.

Lemma 2.1. *The $\equiv_\alpha$-congruence classes of $\mathsf{Exp}\,\mathbb{A}$ form a nominal set with* $\mathsf{supp}\,[e] = \mathsf{FV}(e)$, *and the function FV is well defined and equivariant on $\equiv_\alpha$-classes.*

A ν-*string* is a string with νa binders; that is, it is an NKA expression with no occurrence of $+$, *, or 0 modulo multiplicative associativity, and no occurrence of 1 except to denote the null string, in which case we use ε instead.

$$x ::= a \in \Sigma \mid xx \mid \varepsilon \mid \nu a.x$$

The set of ν-strings over $\mathbb{A}$ is denoted $\mathbb{A}^\nu$.

NKA Axioms.   The axioms proposed by Gabbay and Ciancia [12] are:

$$\begin{array}{lll} \nu a.(d + e) = \nu a.d + \nu a.e & a\#e \Rightarrow \nu b.e = \nu a.(a\,b)e & \nu a.\nu b.e = \nu b.\nu a.e \\ a\#e \Rightarrow (\nu a.d)e = \nu a.de & a\#e \Rightarrow e(\nu a.d) = \nu a.ed & a\#e \Rightarrow \nu a.e = e. \end{array} \qquad (2)$$

Nominal ν-Monoids.   A *nominal ν-monoid over* $\mathbb{A}$ is a structure $(M, \cdot, 1, \mathbb{A}, \nu)$ with binding operation $\nu : \mathbb{A} \times M \to M$ such that
- $(M, \cdot, 1)$ is a monoid with group action $G_{\mathbb{A}} \times M \to M$ such that M is a nominal set;
- the operation ν satisfies the axioms (2);
- the monoid operations and ν are equivariant, or equivalently, every $\pi \in G_{\mathbb{A}}$ is an automorphism of M.

Nominal Kleene algebra (NKA).   A *nominal KA over* $\mathbb{A}$ is a structure $(K, +, \cdot, ^*, 0, 1, \mathbb{A}, \nu)$ with binding operation $\nu : \mathbb{A} \times K \to K$ such that
- $(K, +, \cdot, ^*, 0, 1)$ is a KA with group action $G_{\mathbb{A}} \times K \to K$ such that K is a nominal set;
- the operation ν satisfies the axioms (2);
- the KA operations and ν are equivariant in the sense that

$$\begin{array}{lll} \pi(x + y) = \pi x + \pi y & \pi(xy) = (\pi x)(\pi y) & \pi 0 = 0 \\ \pi(x^*) = (\pi x)^* & \pi(\nu a.x) = \nu(\pi a).(\pi x) & \pi 1 = 1, \end{array}$$

or equivalently, every $\pi \in G_{\mathbb{A}}$ is an automorphism of K.

3   A Nominal Language Model

Let M be a nominal ν-monoid over $\mathbb{A}$. Metasymbols $m, n, \ldots$ denote elements of M. Let $\wp M$ denote the powerset of M. On $\wp M$, define the KA operations and group action

$$A + B = A \cup B \quad AB = \{mn \mid m \in A,\ n \in B\} \quad A^* = \bigcup_k A^k \quad 0 = \varnothing$$
$$1 = \{\varepsilon\} \quad \nu a.A = \{\nu a.m \mid m \in A\} \quad \pi A = \{\pi m \mid m \in A\}. \tag{3}$$

We say that A is *uniformly finitely supported* if $\bigcup_{m \in A} \text{supp}\, m$ is finite. Let

$$\wp_{\mathrm{fs}} M = \{A \subseteq M \mid A \text{ is finitely supported}\}$$
$$\wp_{\mathrm{ufs}} M = \{A \subseteq M \mid A \text{ is uniformly finitely supported}\}.$$

Lemma 3.2 ([11, Theorem 2.29]). *For $A \subseteq M$, if A is uniformly finitely supported, then A is finitely supported and $\text{supp}\, A = \bigcup_{m \in A} \text{supp}\, m$.*

The converse is false in general. Both $\wp_{\mathrm{fs}} M$ and $\wp_{\mathrm{ufs}} M$ are closed under the operations (3).

Theorem 3.1. *The set $\wp_{\mathrm{ufs}} M$ with group action and KA operations (3) forms an NKA.*

3.1   Canonical Interpretation over $\mathbb{A}^\nu / \equiv$

For $x, y \in \mathbb{A}^\nu$, define $x \equiv y$ if x and y are provably equivalent using the axioms (2) (omitting the first, which is irrelevant as there is no occurrence of $+$ in ν-strings) and the axioms of equality and congruence. Let $[x]$ denote the $\equiv$-congruence class of x and $\mathbb{A}^\nu / \equiv$ the ν-monoid of all such congruence classes.

The *length* of $x \in \mathbb{A}^\nu$ is the number of occurrences of symbols of $\mathbb{A}$ in x, excluding binding occurrences νb. If $x \equiv y$, then x and y have the same length, and an occurrence of a symbol in x is free iff the corresponding occurrence in y is free. If both are free, then they are the same symbol. If both are bound, then they can be different symbols due to α-conversion. If two ν-strings are α-equivalent, then they are $\equiv$-equivalent.

Henceforth, let $M = \mathbb{A}^\nu / \equiv$. The map $L : \mathsf{Exp}\,\mathbb{A} \to \wp M$ is defined to be the unique homomorphism such that $L(a) = \{[a]\}$ for $a \in \mathbb{A}$. Explicitly,

$$L(e_1 + e_2) = L(e_1) \cup L(e_2) \qquad L(e_1 e_2) = \{mn \mid m \in L(e_1),\ n \in L(e_2)\}$$
$$L(e^*) = L(e)^* = \bigcup_k L(e)^k \qquad L(0) = \varnothing \qquad L(1) = \{\varepsilon\} \tag{4}$$
$$L(a) = \{[a]\},\ a \in \mathbb{A} \qquad L(\nu a.e) = \nu a.L(e) = \{\nu a.m \mid m \in L(e)\}.$$

The following lemma guarantees the existence of an equivariant homomorphism $L : \mathsf{Exp}\,\mathbb{A} / \equiv_\alpha \to \wp_{\mathrm{ufs}} M$.

Lemma 3.3. *The map L is well defined and equivariant on $\equiv_\alpha$-congruence classes and takes values in $\wp_{\mathrm{ufs}} M$.*

The following deconstruction lemma is important for our coalgebraic treatment of §4.

Lemma 3.4.
(i) If $ax \equiv by$, then $a = b$ and $x \equiv y$.
(ii) If $\nu a.ax \equiv \nu a.ay$, then $x \equiv y$.

Lemma 3.4(ii) is somewhat delicate. Note that $\nu a.x \equiv \nu a.y$ does not imply $x \equiv y$ in general: we have $\nu b.ab \not\equiv \nu b.ba$, but $\nu a.\nu b.ab \equiv \nu a.\nu b.ba$ by applying the permutation $(a\ b)$ and reversing the order of the bindings.

4   Coalgebraic Structure

We will presently define syntactic Brzozowski and Antimirov derivatives on NKA expressions over $\mathbb{A}$ and a corresponding semantic derivative on subsets of M. These constructs will be seen to comprise coalgebras for a Nom-endofunctor K defined by

$$KX = 2 \times X^{\mathbb{A}} \times [\mathbb{A}]X, \tag{5}$$

where the nominal set $X^{\mathbb{A}}$ consists of finitely supported functions $\mathbb{A} \to X$ and $[\mathbb{A}]X$ is the abstraction of the nominal set X; see [22] for a detailed account of the abstraction functor on Nom. We recall here that the nominal set $[\mathbb{A}]X$ is defined as the quotient of $\mathbb{A} \times X$ by the equivalence relation given by $(a, x) \sim (b, y)$ if and only if for any fresh c we have $(c\ a)x = (c\ b)y$. Furthermore, the abstraction functor $[\mathbb{A}](-)$ has a left adjoint $\mathbb{A}\#(-)$ defined on objects by

$$\mathbb{A}\#X = \{(a, x) \mid a\#x\}.$$

Hence a K-coalgebra is a tuple of the form $(X, \mathsf{obs}, \mathsf{cont}, \mathsf{cont}_\nu)$, where X is a nominal set and

$$\mathsf{obs} : X \to 2 \qquad \mathsf{cont} : X \to X^{\mathbb{A}} \qquad \mathsf{cont}_\nu : X \to [\mathbb{A}]X \tag{6}$$

are equivariant functions, called the *observation* and *continuation* maps, respectively. Using the cartesian closed structure on Nom and the adjunction $\mathbb{A}\#(-) \dashv [\mathbb{A}](-)$, the continuation maps are in one-to-one correspondence with maps defined on $\mathbb{A} \times X$ and $\mathbb{A}\#X$ respectively.

$$\frac{\mathsf{cont} : X \to X^{\mathbb{A}}}{\mathsf{cont}^\flat : \mathbb{A} \times X \to X} \qquad\qquad \frac{\mathsf{cont}_\nu : X \to [\mathbb{A}]X}{\mathsf{cont}_\nu^\flat : \mathbb{A}\#X \to X}$$

To simplify notation, we write

$$\mathsf{cont}_a : X \to X,\ a \in \mathbb{A} \qquad \mathsf{cont}_{\nu a} : \{s \in X \mid a\#s\} \to X,\ a \in \mathbb{A} \tag{7}$$

for the uncurried continuation maps obtained by fixing the first argument to $a \in \mathbb{A}$. Intuitively, cont_a tries to consume a free variable a and $\mathsf{cont}_{\nu a}$ tries to consume a bound variable a bound by νa. We will discuss the intuition behind these constructs more fully and justify the typing (6) in Example 4.1 below.

It follows from (1) that the equivariance of the structure map $(\mathsf{obs}, \mathsf{cont}, \mathsf{cont}_\nu)$ is equivalent to the properties

$$\mathsf{cont}_{\pi a} \circ \pi = \pi \circ \mathsf{cont}_a \qquad \mathsf{cont}_{\nu \pi a} \circ \pi = \pi \circ \mathsf{cont}_{\nu a} \qquad \mathsf{obs} \circ \pi = \mathsf{obs} \tag{8}$$

for all $\pi \in G_{\mathbb{A}}$.

Henceforth, the term *coalgebra* refers specifically to coalgebras for the Nom-functor K in (5).

4.1   Semantic Derivative

Let $M = \mathbb{A}^\nu/{\equiv}$. The semantic derivative is defined as a K-coalgebra with carrier the nominal set $\wp_{\mathrm{fs}} M$:

$$(\varepsilon, \delta, \delta_\nu) : \wp_{\mathrm{fs}} M \to 2 \times (\wp_{\mathrm{fs}} M)^\mathbb{A} \times [\mathbb{A}] \, \wp_{\mathrm{fs}} M$$

where

$$\varepsilon(A) = \begin{cases} 1, & \varepsilon \in A, \\ 0, & \varepsilon \notin A \end{cases} \qquad \begin{aligned} \delta_a(A) &= \{m \mid am \in A\}, \ a \in \mathbb{A} \\ \delta_{\nu a}(A) &= \{m \mid \nu a.am \in A\}, \ a \in \mathbb{A}. \end{aligned}$$

The maps δ_a and $\delta_{\nu a}$ are well defined by Lemma 3.4.

Example 4.1. The a in δ_a and $\delta_{\nu a}$ play very different roles. Intuitively, $\delta_a(A)$ tries to consume a free variable a at the front of strings in A. For example, for $b \neq a$,

- $\delta_a(\{aa, bb\}) = \{a\}$
- $\delta_a(\{\nu b.ab\}) = \{\nu b.b\}$
- $\delta_a(\{\nu a.ab\}) = \varnothing$ (since the first letter of $\nu a.ab$ is bound).

On the other hand, $\delta_{\nu a}(A)$ tries to consume a bound variable at the front of strings in A and change the remaining variables bound by the same binder to a. The bound variable need not be a, but it should be possible to change it to a by α-conversion. For example, for $b \neq a$,

1. $\delta_{\nu a}(\{\nu a.aa\}) = \delta_{\nu a}(\{\nu b.bb\}) = \{a\}$ (since $\nu b.bb = \nu a.aa$ in $\mathbb{A}^\nu/{\equiv}$)
2. $\delta_{\nu a}(\{\nu a.ab\}) = \{b\}$
3. $\delta_{\nu a}(\{\nu a.ba\}) = \varnothing$ (since the initial symbol b is not bound)
4. $\delta_{\nu a}(\{\nu b.ba\}) = \varnothing$ (since $\nu b.ba \neq \nu a.am$ for any $m \in \mathbb{A}^\nu/{\equiv}$)
5. $\delta_{\nu a}(\{(\nu a.aa)a\}) = \varnothing$ (since $(\nu a.aa)a \neq \nu a.am$ for any $m \in \mathbb{A}^\nu/{\equiv}$)
6. $\delta_{\nu a}(\{(\nu b.bb)b\}) = \{ab\}$ (since $(\nu b.bb)b = \nu a.aab$ in $\mathbb{A}^\nu/{\equiv}$).

Examples 4 and 5 do not arise in our coalgebraic semantics, since $\delta_{\nu a}$ may only be applied to A for which a is fresh due to the domain restriction in (7). If there are free occurrences of a, one cannot α-convert to obtain a string of the form $\nu a.am$, since those free occurrences would be captured.

4.2   Brzozowski Derivative

The syntactic Brzozowski derivative is defined inductively on the set of α-equivalence classes of NKA expressions $\mathsf{Exp}\,\mathbb{A}/{\equiv_\alpha}$. Like the semantic derivative, it can also be defined on a broader domain, but also will only make coalgebraic sense for the domain (6).

$$(\mathsf{E}, \mathsf{D}, \mathsf{D}_\nu) : \mathsf{Exp}\,\mathbb{A}/{\equiv_\alpha} \to 2 \times (\mathsf{Exp}\,\mathbb{A}/{\equiv_\alpha})^\mathbb{A} \times [\mathbb{A}](\mathsf{Exp}\,\mathbb{A}/{\equiv_\alpha})$$

The continuation maps D and D_ν can be further broken down as

$$\mathsf{D}_a : \mathsf{Exp}\,\mathbb{A}/{\equiv_\alpha} \to \mathsf{Exp}\,\mathbb{A}/{\equiv_\alpha} \qquad \mathsf{D}_{\nu a} : \{e \in \mathsf{Exp}\,\mathbb{A}/{\equiv_\alpha} \mid a \# e\} \to \mathsf{Exp}\,\mathbb{A}/{\equiv_\alpha}$$

for $a \in \mathbb{A}$. We first define these maps on $\mathsf{Exp}\,\mathbb{A}$, then argue that they are well defined on $\equiv_\alpha$-classes.

$$\mathsf{E}(e_1 + e_2) = \mathsf{E}(e_1) + \mathsf{E}(e_2) \qquad \mathsf{E}(e_1 e_2) = \mathsf{E}(e_1)\mathsf{E}(e_2) \qquad \mathsf{E}(a) = \mathsf{E}(0) = 0$$
$$\mathsf{E}(1) = \mathsf{E}(e^*) = 1 \qquad\qquad \mathsf{E}(\nu a.e) = \mathsf{E}(e)$$

$$\mathsf{D}_a(e_1 + e_2) = \mathsf{D}_a(e_1) + \mathsf{D}_a(e_2) \qquad \mathsf{D}_a(e_1 e_2) = \mathsf{D}_a(e_1)e_2 + \mathsf{E}(e_1)\mathsf{D}_a(e_2)$$
$$\mathsf{D}_a(e^*) = \mathsf{D}_a(e)e^* \qquad\qquad \mathsf{D}_a(0) = \mathsf{D}_a(1) = 0$$

$$\mathsf{D}_a(b) = \begin{cases} 1, & b = a \\ 0, & b \neq a \end{cases} \qquad\qquad \mathsf{D}_a(\nu b.e) = \begin{cases} 0, & b = a \\ \nu b.\mathsf{D}_a(e), & b \neq a \end{cases}$$

$$\mathsf{D}_{\nu a}(e_1 + e_2) = \mathsf{D}_{\nu a}(e_1) + \mathsf{D}_{\nu a}(e_2) \quad \mathsf{D}_{\nu a}(e_1 e_2) = \mathsf{D}_{\nu a}(e_1)e_2 + \mathsf{E}(e_1)\mathsf{D}_{\nu a}(e_2)$$
$$\mathsf{D}_{\nu a}(e^*) = \mathsf{D}_{\nu a}(e)e^* \qquad\qquad \mathsf{D}_{\nu a}(\nu b.e) = \nu b.\mathsf{D}_{\nu a}(e) + \mathsf{D}_a((a\ b)e), \ b \neq a$$
$$\mathsf{D}_{\nu a}(0) = \mathsf{D}_{\nu a}(1) = \mathsf{D}_{\nu a}(b) = 0$$

We can also define $\mathsf{D}_{\nu a}(\nu a.e) = \mathsf{D}_{\nu a}(\nu b.(a\ b)e)$ for an arbitrary b such that $b\#e$ and $b \neq a$, although strictly speaking this is not a function, since the choice of b is not determined. However, the choice of b does not matter, as we are considering expressions modulo α-equivalence.

Example 4.2. For $b \neq a$,
1. $\mathsf{D}_{\nu a}(\nu b.bb) = \nu b.\mathsf{D}_{\nu a}(bb) + \mathsf{D}_a((a\ b)bb) = 0 + a = a.$
2. $\mathsf{D}_{\nu a}(\nu a.aa) = \mathsf{D}_{\nu a}(\nu b.bb) = a.$
3. $\mathsf{D}_{\nu a}(\nu a.ab) = \mathsf{D}_{\nu a}(\nu c.cb) = \nu c.\mathsf{D}_{\nu a}(\varsigma b) + \mathsf{D}_a(ab) = 0 + b = b.$
4. $\mathsf{D}_{\nu a}(\nu b.ba) = \nu b.\mathsf{D}_{\nu a}(ba) + \mathsf{D}_a((a\ b)ba) = 0 + b = b.$

Example 4 will not arise in our coalgebraic semantics, since $\mathsf{D}_{\nu a}$ will only be applied to e for which a is fresh and the argument has a free variable a.

4.3   Final Coalgebra

The nominal coalgebra $(\wp_{\mathrm{fs}}\,M, \varepsilon, \delta, \delta_\nu)$ is final among coalgebras for the Nom-endofunctor K defined in (5). These are the coalgebras $(X, \mathsf{obs}, \mathsf{cont}, \mathsf{cont}_\nu)$ for which X is a nominal set and obs, cont and cont_ν are equivariant. Such a coalgebra can be viewed as an automaton with states X, transitions cont and cont_ν, and acceptance condition obs. The inputs to the automaton are elements of M. Starting from a state $s \in X$, an element $m \in M$ is *accepted* if $\mathsf{Accept}(s, m)$, where

$$\mathsf{Accept}(s, \varepsilon) = \mathsf{obs}(s) \qquad\qquad (9)$$
$$\mathsf{Accept}(s, am) = \mathsf{Accept}(\mathsf{cont}_a(s), m) \qquad\qquad (10)$$
$$\mathsf{Accept}(s, \nu a.m) = \mathsf{Accept}(\mathsf{cont}_{\nu a}(s), m), \ a\#s. \qquad\qquad (11)$$

Clause (11) requires some explanation. We must choose a representative element $\nu a.am$ of the $\equiv$-class such that a is fresh for s, so that $\mathsf{cont}_{\nu a}(s)$ will be defined. It is always possible to find such an a, since the $\equiv$-class is closed under α-conversion and s has finite support. However, the result is independent of the choice of a, as shown in part (ii) of the next lemma, so $\mathsf{Accept}(s, \nu a.am)$ is well defined.

Lemma 4.5.

(i) *The acceptance function is equivariant:*

$$\mathsf{Accept}(\pi s, \pi m) = \pi(\mathsf{Accept}(s, m)) = \mathsf{Accept}(s, m).$$

(ii) *If $b \# s$ and $c \# s$, then*

$$\mathsf{Accept}(s, \nu b.bm) = \mathsf{Accept}(s, \nu c.c(b\ c)m).$$

We do not explicitly require $c \# \nu b.bx$ in (ii); however, this is a consequence of (i) and the fact that if f is an equivariant function, then $\mathsf{supp}\, f(x) \subseteq \mathsf{supp}\, x$.

The unique coalgebra homomorphism from $(X, \mathsf{obs}, \mathsf{cont}, \mathsf{cont}_\nu)$ to the final coalgebra is just the automata-theoretic language semantics:

Theorem 4.2 (Final coalgebra). *The coalgebra $(\wp_{\mathsf{fs}}\, M, \varepsilon, \delta, \delta_\nu)$ is a final K-coalgebra. The unique coalgebra homomorphism $(X, \mathsf{obs}, \mathsf{cont}, \mathsf{cont}_\nu)$ to the final coalgebra is given by*

$$L_X : (X, \mathsf{obs}, \mathsf{cont}, \mathsf{cont}_\nu) \to (\wp_{\mathsf{fs}}\, M, \varepsilon, \delta, \delta_\nu) \quad L_X(s) = \{m \mid \mathsf{Accept}(s, m)\}.$$

Moreover, the coalgebra homomorphism $L_{\mathsf{Exp}\,\mathbb{A}} : \mathsf{Exp}\,\mathbb{A}/\equiv_\alpha \to \wp_{\mathsf{fs}}\, M$ coincides with the algebra homomorphism $L : \mathsf{Exp}\,\mathbb{A}/\equiv_\alpha \to \wp_{\mathsf{fs}}\, M$ defined in (4).

A more standard construction of the final coalgebra computed via the final sequence of the functor K [1] yields an equivalent presentation based on normal forms of ν-strings up to α-equivalence. However, this characterization is more cumbersome algebraically, as it requires explicit α-conversion to define sequential composition.

4.4   Automata Representation: Half of a Kleene Theorem

In this section we prove a theorem for NKA that relates the algebraic and coalgebraic structure. As noted in §4.3, a coalgebra can be regarded as an automaton acceptor with states X, transitions cont, and acceptance condition obs. The inputs to the automaton are elements of M. The state sets are nominal sets and may be formally infinite, but still may be essentially finite in a sense to be described next.

Following [3], we define the *size* of a coalgebra $(X, \mathsf{obs}, \mathsf{cont})$ to be the number of orbits of X under $G_\mathbb{A}$, where the *orbit* of $s \in X$ is the set $\{\pi s \mid \pi \in G_\mathbb{A}\}$. The orbit of s is the singleton $\{s\}$ if $\mathsf{supp}\, s = \varnothing$, otherwise it is infinite. The orbits partition X and determine an equivalence relation. The coalgebra is called *orbit-finite* if the total number of orbits is finite.

Lemma 4.6. *Let $(X, \mathsf{obs}, \mathsf{cont})$ be a coalgebra, $s \in X$, and $a \in \mathbb{A}$.*

(i) $\mathsf{supp}\,(\mathsf{cont}_{\nu a}(s)) \subseteq \{a\} \cup \mathsf{supp}\, s$.

(ii) *If $a \in \mathsf{supp}\, s$, then $\mathsf{supp}\,(\mathsf{cont}_a(s)) \subseteq \mathsf{supp}\, s$.*

(iii) *If $L(s)$ is uniformly finitely supported and $m \in L(s)$, then $\mathsf{supp}\, m \subseteq \mathsf{supp}\, s$.*

(iv) *If $a \# s$ and $L(s)$ is uniformly finitely supported, then $\mathsf{cont}_a(s)$ is a dead state (one for which $L(s) = \varnothing$).*

Theorem 4.3 (Half Kleene). *For every NKA expression e, there is a coalgebra X with designated start state s such that $L_X(s) = L(e)$. The coalgebra has an orbit-finite nondeterministic representation given by the Antimirov representation of the Brzozowski derivatives of e.*

It is interesting that the Antimirov derivatives give an orbit-finite representation, whereas the Brzozowski derivatives do not. More details, including a counterexample, can be found in the full version of this paper [16]. The orbit-finite representation underlies the decision procedure of the equational theory, which we also omit here for lack of space.

5 Conclusion and Open Problems

In this paper we have explored the coalgebraic theory of nominal Kleene algebra. We have introduced a new family of semantic models consisting of sets of nominal monoids and extended the coalgebraic structure of Kleene algebra to the nominal setting using these models. We have developed nominal versions of the Brzozowski and Antimirov derivatives that accommodate bound variables and are invariant with respect to α-conversion. We have proved a theorem relating the algebraic and coalgebraic structure, namely that every expression gives rise to an equivalent automaton. We have used this relationship to show that the equational theory can be decided in exponential space and described an efficient data representation that is amenable to implementation.

This work raises several intriguing questions. Foremost among them is the complexity of the equational theory. We have given a worst-case exponential-space decision procedure. On the other hand, the best lower bound we have is PSPACE-hardness, which follows from the PSPACE-completeness of the equivalence problem for regular expressions [26].

Despite the high complexity of the worst-case upper bound, much like the bisimulation-based algorithms for other KA-based systems [4,5,7,24], the situation may not be so bad in practice. To actually attain the worst-case bound seems to require highly pathological examples that would be unlikely to arise in practice. However, only implementation and experimentation can confirm or refute this view. This would be an interesting direction for future work.

Theorem 4.3 gives one direction of a Kleene theorem: expressions to automata. The converse is false, as the following example shows. Consider the nominal coalgebra with states and group action

- $s_0(a)$ for all $a \in \mathbb{A}$ with $\pi(s_0(a)) = s_0(\pi a)$,
- $s_1(a, b)$ for all $a, b \in \mathbb{A}$, $a \neq b$ with $\pi(s_1(a, b)) = s_1(\pi a, \pi b)$, and
- s_2 with $\pi s_2 = s_2$.

The transitions and observations are

$\mathrm{cont}_{\nu b}(s_0(a)) = s_1(a, b)$ $\mathrm{obs}(s_0(a)) = 1$

$\mathrm{cont}_a(s_1(a, b)) = s_0(b)$ $\mathrm{obs}(s_1(a, b)) = \mathrm{obs}(s_2) = 0$

for all $a, b \in \mathbb{A}$. All other transitions go to the dead state s_2. The set of ν-strings accepted from state $s_0(a)$ is

$$\{\varepsilon, \; \nu b.ba, \; \nu b.ba(\nu a.ab), \; \nu b.ba(\nu a.ab(\nu b.ba)), \; \nu b.ba(\nu a.ab(\nu b.ba(\nu a.ab))), \ldots\}$$

It can be shown using the normal form theorem of [15] that this set is not represented by any NKA expression, because it requires unbounded ν-depth.

Given that orbit-finite nominal automata are strictly more expressive than NKA expressions, several questions arise:

1. Can we characterize the subclass of orbit-finite nominal automata that are equivalent to NKA expressions? We conjecture that they are exactly those automata accepting sets of ν-strings of bounded ν-depth, although we are not sure how to characterize this class formally in a way that would lead to a converse of Theorem 4.3.
2. Can we extend the syntax of expressions to capture sets of unbounded ν-depth? The answer is yes: It is not difficult to show that orbit-finite nominal automata are equivalent to orbit-finite systems of right-linear equations. For example, the system corresponding to the automaton above would be

$$X_a = \varepsilon + \nu b.bY_{ab} \qquad\qquad Y_{ab} = aX_b.$$

   The set accepted by the automaton is the least solution of the system. This gives a full Kleene theorem, but of course we are now left with the open question of deriving proof rules for this new calculus and extending the completeness result of [15].
3. Can we prove a Kleene theorem for the nominal DFA and NFA models of Bojanczyk, Klin and Lasota [3], exposing the crucial difference that nondeterminism introduces in the nominal setting (nominal NFA are strictly more expressive that DFA)?
4. Can we use the coalgebraic setting to systematically develop a nominal Chomsky hierarchy and (semi-)decision procedures for different classes of languages?

The first two questions have an interesting interpretation in terms of the intended application of NKA, which was originally proposed in [12] as a framework for reasoning about *dynamic* allocation of resources. However, the ν-operators in NKA expressions are statically scoped, so *static* may be the more accurate adjective. The more expressive automata of [3,8,18,20] and of this paper may be the more appropriate vehicle for the study of dynamic allocation.

Acknowledgments. Thanks to Filippo Bonchi, Jamie Gabbay, Bart Jacobs, Tadeusz Litak, Damien Pous, and Ana Sokolova for many stimulating discussions and suggestions. This research was performed at Radboud University Nijmegen and supported by the Dutch Research Foundation (NWO), project numbers 639.021.334 and 612.001.113, and by the National Security Agency.

References

1. Adámek, J.: On final coalgebras of continuous functors. TCS **294**(12), 3–29 (2003)
2. Allauzen, C., Mohri, M.: A unified construction of the glushkov, follow, and antimirov automata. In: Královič, R., Urzyczyn, P. (eds.) MFCS 2006. LNCS, vol. 4162, pp. 110–121. Springer, Heidelberg (2006)
3. Bojanczyk, M., Klin, B., Lasota, S.: Automata theory in nominal sets. LMCS **10**(3) (2014)
4. Bonchi, F., Pous, D.: Checking NFA equivalence with bisimulations up to congruence. In: POPL 2013, pp. 457–468 (2013)
5. Braibant, T., Pous, D.: Deciding Kleene algebras in Coq. LMCS **8**(1:16), 1–42 (2012)
6. Ferrari, G.-L., Montanari, U., Tuosto, E., Victor, B., Yemane, K.: Modelling fusion calculus using HD-automata. In: Fiadeiro, J.L., Harman, N.A., Roggenbach, M., Rutten, J. (eds.) CALCO 2005. LNCS, vol. 3629, pp. 142–156. Springer, Heidelberg (2005)
7. Foster, N., Kozen, D., Milano, M., Silva, A., Thompson, L.: A coalgebraic decision procedure for NetKAT. In: POPL 2015, pp. 343–355 (2015)
8. Francez, N., Kaminski, M.: Finite-memory automata. TCS **134**(2), 329–363 (1994)
9. Francez, N., Kaminski, M.: An algebraic characterization of deterministic regular languages over infinite alphabets. TCS **306**(1–3), 155–175 (2003)
10. Gabbay, M., Pitts, A.M.: A new approach to abstract syntax involving binders. In: LICS 1999, pp. 214–224 (1999)
11. Gabbay, M.: Foundations of nominal techniques: logic and semantics of variables in abstract syntax. Bull. Symbolic Logic **17**(2), 161–229 (2011)
12. Gabbay, M.J., Ciancia, V.: Freshness and name-restriction in sets of traces with names. In: Hofmann, M. (ed.) FOSSACS 2011. LNCS, vol. 6604, pp. 365–380. Springer, Heidelberg (2011)
13. Gelade, W., Neven, F.: Succinctness of the complement and intersection of regular expressions. In: TACS 2008. Dagstuhl LIPIcs, vol. 1, pp. 325–336 (2008)
14. Kozen, D.: On the coalgebraic theory of Kleene algebra with tests. Tech. Rep., Cornell, March 2008. http://hdl.handle.net/1813/10173
15. Kozen, D., Mamouras, K., Silva, A.: Completeness and incompleteness in nominal Kleene algebra. Tech. Rep., Cornell, November 2014. http://hdl.handle.net/1813/38143
16. Kozen, D., Mamouras, K., Petrişan, D., Silva, A.: Nominal Kleene Coalgebra. Tech. Rep., Cornell, February 2015. http://hdl.handle.net/1813/39108
17. Kurz, A., Suzuki, T., Tuosto, E.: A characterisation of languages on infinite alphabets with nominal regular expressions. In: Baeten, J.C.M., Ball, T., de Boer, F.S. (eds.) TCS 2012. LNCS, vol. 7604, pp. 193–208. Springer, Heidelberg (2012)
18. Kurz, A., Suzuki, T., Tuosto, E.: On nominal regular languages with binders. In: Birkedal, L. (ed.) FOSSACS 2012. LNCS, vol. 7213, pp. 255–269. Springer, Heidelberg (2012)
19. Montanari, U., Pistore, M.: History dependent automata. Tech. Rep. TR-11-98, Computer Science, Università di Pisa (1998)
20. Montanari, U., Pistore, M.: History-dependent automata: an introduction. In: Bernardo, M., Bogliolo, A. (eds.) SFM-Moby 2005. LNCS, vol. 3465, pp. 1–28. Springer, Heidelberg (2005)
21. Pistore, M.: History Dependent Automata. PhD thesis, Università di Pisa (1999)

22. Pitts, A.M.: Nominal Sets: Names and Symmetry in Computer Science, Cambridge Tracts in Theoretical Computer Science 57. Cambridge University Press (2013)
23. Pous, D.: Symbolic algorithms for language equivalence and kleene algebra with tests. In: POPL 2015, pp. 357–368 (2015)
24. Silva, A.: Kleene Coalgebra. PhD thesis, Radboud University Nijmegen (2010)
25. Silva, A.: Position automata for Kleene algebra with tests. Scientific Annals of Computer Science **22**(2), 367–394 (2012)
26. Stockmeyer, L.J., Meyer, A.R.: Word problems requiring exponential time. In: STOC 1973, pp. 1–9 (1973)

On Determinisation of Good-for-Games Automata

Denis Kuperberg[1,2,3] and Michał Skrzypczak[3,4]([⊠])

[1] Onera/DTIM, Toulouse, France
[2] IRIT, University of Toulouse, Toulouse, France
[3] University of Warsaw, Warsaw, Poland
[4] LIAFA, University of Paris 7, Paris, France
mskrzypczak@mimuw.edu.pl

Abstract. In this work we study Good-For-Games (GFG) automata over ω-words: non-deterministic automata where the non-determinism can be resolved by a strategy depending only on the prefix of the ω-word read so far. These automata retain some advantages of determinism: they can be composed with games and trees in a sound way, and inclusion $L(\mathcal{A}) \supseteq L(\mathcal{B})$ can be reduced to a parity game over $\mathcal{A} \times \mathcal{B}$ if $\mathcal{A}$ is GFG. Therefore, they could be used to some advantage in verification, for instance as solutions to the synthesis problem.

The main results of this work answer the question whether parity GFG automata actually present an improvement in terms of state-complexity (the number of states) compared to the deterministic ones. We show that a frontier lies between the Büchi condition, where GFG automata can be determinised with only quadratic blow-up in state-complexity; and the co-Büchi condition, where GFG automata can be exponentially smaller than any deterministic automaton for the same language. We also study the complexity of deciding whether a given automaton is GFG.

1 Introduction

One of the classical problems of automata theory is synthesis — given a specification, decide if there exists a system that fulfils it and if there is, automatically construct one. The problem was solved positively by Büchi and Landweber [BL69] for the case of ω-regular specifications. There are two standard approaches to the problem: either by deterministic automata [McN66] or by tree automata [Rab72]. Henzinger and Piterman [HP06] have proposed a model of Good-For-Games (shortly GFG) automata that enjoy a weak form of non-determinism while still preserving soundness and completeness when solving the synthesis problem.

An automaton is Good-For-Games if there exists a strategy that resolves the non-deterministic choices, by taking into account only the prefix of the input

Research funded by ANR/DGA project Cx (ref. ANR-13-ASTR-0006); and by *fondation STAE* project BRIefcaSE. The second author has been supported by Poland's National Science Centre grant (decision DEC-2014-13/B/ST6/03595).

M.M. Halldórsson et al. (Eds.): ICALP 2015, Part II, LNCS 9135, pp. 299–310, 2015.
DOI: 10.1007/978-3-662-47666-6_24

ω-word read so far. The strategy is supposed to construct an accepting run of the automaton whenever an ω-word from the language is given. The motivation for this model in [HP06] was to simplify the transition structure of automata as solutions of the synthesis problem for Linear Temporal Logic. Experimental evaluation of GFG automata and their applications to stochastic problems were discussed in [KMBK14].

The notion of GFG automata was independently discovered in [Col09] under the name history-determinism, in the more general framework of regular cost functions. It turns out that deterministic cost automata have strictly smaller expressive power than non-deterministic ones and therefore history-determinism is used whenever a *sequential* model is needed.

In the survey [Col12] two important results about GFG automata over finite words are mentioned: first that every GFG automaton over finite words contains an equivalent deterministic subautomaton, second that it is decidable in PTIME if a given automaton over finite words is GFG. Additionally, a conjecture stating that every parity GFG automaton over ω-words contains an equivalent deterministic subautomaton is posed.

In [BKKS13], examples were given of Büchi and co-Büchi GFG automata which do not contain any equivalent deterministic subautomaton. Moreover, a link between GFG and tree automata was established: an automaton for a language L of ω-words is GFG if and only if its infinite tree version accepts the language of trees that have all their branches in L. However, the problem of the gap in the number of states between deterministic and GFG automata over ω-words was left open. Indeed, for all the available examples of GFG automata, there was an equivalent deterministic automaton of the same size.

We settle this question in the present paper. We show that for Büchi automata determinisation can be done with only a quadratic state-space blow-up. The picture is very different for co-Büchi automata (and all higher parity conditions), for which for every n we give an example of a GFG automaton with $2n + 1$ states that does not admit any equivalent deterministic automaton with less than $\frac{2^n}{2n+1}$ states.

The lower bound for determinising co-Büchi GFG automata shows that these automata can be exponentially more succinct than deterministic ones. Therefore, it indicates possibility of avoiding exponential blow-up by using GFG automata instead of deterministic automata in the problems of containment or synthesis. On the other hand, the quadratic determinisation construction for Büchi GFG automata shows that in this case GFG automata are close to deterministic ones. Therefore, the GFG model may be considered less relevant (with respect to succinctness) for Büchi condition than for general parity condition.

We emphasize the fact that although the model of GFG automata requires the existence of a strategy resolving the non-determinism, this strategy is not used in algorithms but only in proofs. Therefore, it is not a part of the size of the input in computations based on GFG automata. This is what allows an improvement on deterministic automata: we just rely on the existence of this strategy without having to explicit it.

In the present paper we additionally consider the problem of deciding whether a given parity automaton is GFG. The problem is decidable in EXPTIME (see [HP06]) but no efficient algorithm is known. In the special case where the automaton accepts all ω-words, we show that this is equivalent to solving a parity game, so it is in PTIME for any fixed parity condition, and in NP∩co–NP if the parity condition is a part of the input. The general case of deciding GFGness of parity automata is a priori more complicated. We show that it is in PTIME for co-Büchi automata, moreover the procedure involves building another automaton that could be GFG even if the input automaton is not. Therefore, this procedure could be used as a tool to produce co-Büchi GFG automata in some cases. The PTIME complexity in this case is surprising — although the required strategy can be of exponential size in the co-Büchi case, we can decide in polynomial time whether it exists. In the Büchi case we show that it is in NP to decide whether a given automaton is GFG. The problem of efficiently deciding GFGness of automata of higher parity indices remains open.

Structure of the Paper. In Section 2 we briefly introduce the basic notions used in our constructions. In Section 3 we provide the lower bound on the state-complexity of determinising co-Büchi GFG automata. Section 4 is devoted to the determinisation construction for Büchi GFG automata. In Section 5 we study the problem of deciding GFGness of a given automaton and in Section 6 we conclude. The technical details of the presented results are given in the long version, available online on the websites of the authors.

2   Definitions

By A we denote a finite *alphabet*, elements $a \in A$ are called *letters*. A^* is the set of finite words over A and A^ω is the set of ω-words over A. ϵ stands for the empty word. The successive letters of a word α are $\alpha(0), \alpha(1), \ldots$ The length of a finite word w is $|w|$. We use the standard notions of prefix and suffix of a word. By $u\alpha$ we denote the concatenation of a finite word u with a finite word or an ω-word α. If $K \subseteq A^\omega$ and $w \in A^*$ then we define $w^{-1}K \stackrel{\text{def}}{=} \{\alpha \in A^\omega \mid w\alpha \in K\}$.

In our constructions it is easier to work with an acceptance condition over transitions instead of states. Clearly, the translation from the state-based acceptance to the transition-based acceptance does not influence the number of states of a parity automaton. The opposite translation may increase the number of states by the factor corresponding to the acceptance condition but this translation is still polynomial (even linear for a fixed condition). Except that, the proposed definitions are standard.

2.1   Automata over ω-words

A non-deterministic parity automaton over ω-words (shortly *parity automaton*) is a tuple $\mathcal{A} = \langle A^{\mathcal{A}}, Q^{\mathcal{A}}, q_{\mathbf{I}}^{\mathcal{A}}, \Delta^{\mathcal{A}}, \Omega^{\mathcal{A}} \rangle$ that consists of: a finite set $A^{\mathcal{A}}$ called the *input alphabet*; a finite set $Q^{\mathcal{A}}$ of *states*; an *initial state* $q_{\mathbf{I}}^{\mathcal{A}} \in Q^{\mathcal{A}}$; a *transition*

302 D. Kuperberg and M. Skrzypczak

relation $\Delta^{\mathcal{A}} \subseteq Q^{\mathcal{A}} \times A^{\mathcal{A}} \times Q^{\mathcal{A}}$; and a *priority function* $\Omega^{\mathcal{A}} \colon \Delta^{\mathcal{A}} \to \mathbb{N}$. If the automaton $\mathcal{A}$ is known from the context then we skip the superscript $\mathcal{A}$.

Transitions $(q, a, q') \in \Delta$ are usually noted $q \xrightarrow{a} q'$. Similarly, if $w = a_0 a_1 \dots a_n$ and $q_i \xrightarrow{a_i} q_{i+1}$ is a transition of $\mathcal{A}$ for all $i \leq n$ then we write $q_0 \xrightarrow{w} q_{n+1}$ and call it a *path* in $\mathcal{A}$. We additionally require that for every $q \in Q$, $a \in A$ there is at least one transition in Δ of the form $q \xrightarrow{a} q'$ for some $q' \in Q$.

If $\Omega \colon \Delta \to \{i, i+1, \dots, j\}$ then we say that the *parity index* of $\mathcal{A}$ is (i, j). An automaton of parity index $(1, 2)$ is called a *Büchi* automaton and an automaton of parity index $(0, 1)$ is called a *co-Büchi* automaton. If $\mathcal{A}$ is a Büchi automaton then we additionally define $F \subseteq \Delta$ as $\Omega^{-1}(2)$ and call it the set of *accepting transitions*. Similarly, if $\mathcal{A}$ is a co-Büchi automaton then we define $R \subseteq \Delta$ as $\Omega^{-1}(1)$ and call it the set of *rejecting transitions*.

If Δ is such that for every $q \in Q$ and $a \in A$, there is a unique state $q' \in Q$ such that $q \xrightarrow{a} q'$ then $\mathcal{A}$ is a *deterministic* automaton. In this case, we might denote its transition relation by a function $\delta \colon Q \times A \to Q$ instead of Δ.

For an ω-word $\alpha \in A^{\omega}$, a *run* of $\mathcal{A}$ over α from a state $q \in Q$ is a function $\rho \colon \omega \to Q$ where for every $n \geq 0$, we have a transition of $\mathcal{A}$ $\rho(n) \xrightarrow{\alpha(n)} \rho(n+1)$ and $\rho(0) = q$. ρ is *accepting* over α if[1] $\limsup_{n \to \infty} \Omega\big((\rho(n), \alpha(n), \rho(n+1))\big)$ is even. In other words, the condition requires the highest priority that occurs infinitely often to be even. The priorities can be seen as *positive* (even) and *negative* (odd) events, ordered by their importance. The formula says that the most important event happening infinitely often has to be *positive*.

By the definition, if $\mathcal{A}$ is Büchi it means that the above sequence of transitions should contain infinitely many *accepting transitions*. Similarly, if $\mathcal{A}$ is co-Büchi then it should contain only finitely many *rejecting transitions*.

An automaton $\mathcal{A}$ *accepts* an ω-word α from $q \in Q$ if there exists an accepting run ρ of $\mathcal{A}$ from q over α. By $\mathrm{L}(\mathcal{A}, q)$ we denote the set of all ω-words that are accepted by $\mathcal{A}$ from q. The *language* of an automaton $\mathcal{A}$ is $\mathrm{L}(\mathcal{A}) \overset{\text{def}}{=} \mathrm{L}(\mathcal{A}, q_{\mathrm{I}})$.

An automaton $\mathcal{A}$ is *Good-For-Games* (GFG, for short) if there exists a function $\sigma \colon A^* \to Q$ that resolves the non-determinism of $\mathcal{A}$ depending only of the prefix of the input ω-word read so far: over every ω-word α, the function $n \mapsto \sigma\big(\alpha(0)\alpha(1) \dots \alpha(n-1)\big)$ is a run of $\mathcal{A}$ from q_{I} over α, and it is accepting over α whenever $\alpha \in \mathrm{L}(\mathcal{A})$. Clearly, every deterministic automaton is GFG.

3 Co-Büchi Case

In this section we provide the following result about the state-complexity of determinising co-Büchi GFG automata.

Theorem 1. *For every n there exists a co-Büchi GFG automaton $\mathcal{C}_n$ with $2n+1$ states such that any equivalent deterministic automaton has at least $\frac{2^n}{2n+1}$ states.*

---

[1] Note that whether a run ρ is accepting over α depends on the ω-word α.

All the automata C_n for $n \geq 1$ share the same alphabet consisting of four symbols $A \overset{\text{def}}{=} \{\iota, \sigma, \pi, \sharp\}$. The letters of the alphabet enable to manipulate on the set $\{0, 1, 2, \ldots, 2n-1\}$: ι, σ, π are three permutations of this set such that every permutation of this set can be obtained as a composition of these three (in fact ι is the identity permutation used for padding). The symbol $\sharp$ corresponds to the identity permutation on $\{1, \ldots, 2n-1\}$ but it is undefined on 0.

This way a finite word or an ω-word α over the alphabet A can be seen as a sequence of relations on the set $\{0, \ldots, 2n-1\}$ as depicted on Figure 1. We will represent these relations as a graph (denoted $\text{Graph}(\alpha)$). If α is finite let $D = \{0, 1, \ldots, |\alpha|\}$, otherwise $D = \omega$. The graph is a *plait* of width $2n$: the domain of $\text{Graph}(\alpha)$ is $\{0, 1, \ldots, 2n-1\} \times D$ and all the edges are of the form $(i, k) \to (\alpha(k)(i), k+1)$ for $i \in \{0, \ldots, 2n-1\}$ and $k, k+1 \in D$.

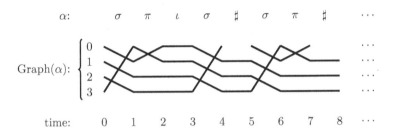

Fig. 1. The infinite sequence of relations on the set $\{0, \ldots, 3\}$ (i.e. $n = 2$) represented by an ω-word $\alpha \in A^\omega$

The language L_n contains an ω-word $\alpha \in A^\omega$ if and only if $\text{Graph}(\alpha)$ contains at least one infinite path.

The set of states of the automaton C_n is $Q = \{\bot, 0, 1, 2, \ldots, 2n-1\}$. The states $\{0, \ldots, 2n-1\}$ are deterministic: reading $a \in A$ in such a state q the automaton moves to the successive state according to the relation represented by a (or to $\bot$ if $a = \sharp$ and $q = 0$). The state $\bot$ is non-deterministic — the automaton can move from $\bot$ over any letter $a \in A$ to any state $q' \in \{0, \ldots, 2n-1\}$. Let the initial state of C_n be $\bot$ and the rejecting transitions be those of the form $\bot \overset{a}{\longrightarrow} q'$.

Note that every accepting run of C_n over an ω-word α indicates an infinite path in $\text{Graph}(\alpha)$. Therefore, we obtain the following fact.

Fact 2. $\text{L}(C_n) \subseteq L_n$.

Lemma 3. C_n is a GFG automaton recognising the language L_n.

Proof. It is enough to construct a function $\sigma \colon A^* \to Q$ that for every ω-word $\alpha \in L_n$ produces an accepting run of C_n over α — it will prove that $L_n \subseteq \text{L}(C_n)$ and that C_n is GFG. We will do it inductively with $\sigma(\epsilon) = \bot = q_{\mathbf{I}}^{C_n}$.

Let σ follow deterministically the transitions of C_n for all the states $q \neq \bot$. It remains to define $\sigma(wa)$ if $\sigma(w) = \bot$ and a successive letter a is given. Assume that $|wa| = k$.

For every $i \in \{0, 1, \ldots, 2n-1\}$ let p_i be the unique maximal path containing the node (i, k) in $\mathrm{Graph}(wa)$. Note that each of these paths p_i has a *starting position* — a node $(\bar{i}, k_i)$ on the path p_i with a minimal moment of time k_i. Clearly $k_i \leq k$. We say that p_i is *older* than $p_{i'}$ if $k_i < k_{i'}$ — in other words, p_i reaches further *to the left* than $p_{i'}$.

Let $\sigma(wa) = i$ such that p_i is the oldest among these paths (if there are two paths equally old, we move to that with smaller i).

Assume that $\alpha \in L_n$. We need to prove that σ produces an accepting run of $\mathcal{C}_n$ over α. Let $p_1, p_2, \ldots, p_m$ be the set of infinite paths in $\mathrm{Graph}(\alpha)$ (we know that $1 \leq m \leq 2n$). Assume that p_1 is an oldest among them and that it starts in a moment of time k_1. For every node (i, k_1) for $i = 0, \ldots, 2n-1$ that does not belong to any of these infinite paths, the unique maximal path containing (i, k_1) is finite. Therefore, for some $k' > k_1$, one of the paths $p_1, \ldots, p_m$ is the oldest among the paths intersecting the (k')th moment of time. So the function σ will use at most once a rejecting transition of $\mathcal{C}_n$ after reading the (k')th symbol of α and then it will follow one of the paths $p_1, \ldots, p_m$ and accept.   □

We now assume for the sake of contradiction that there exists a deterministic automaton $\mathcal{D}$ recognising L_n that has strictly less than $\frac{2^n}{2n+1}$ states. By Theorem 4 from [BKKS13] it means that we can use $\mathcal{D}$ as a memory structure for the automaton $\mathcal{C}_n$ to recognise L_n. Therefore, we focus on the product $\mathcal{C}_n \times \mathcal{D}$ with the acceptance condition taken from $\mathcal{C}_n$. What is important is that $\mathcal{C}_n \times \mathcal{D}$ has to follow the transitions of $\mathcal{C}_n$. We know that $\mathcal{C}_n \times \mathcal{D}$ is a deterministic co-Büchi automaton with strictly less than 2^n states and $\mathrm{L}(\mathcal{C}_n \times \mathcal{D}) = L_n$.

We will use the symbol ρ to denote finite and infinite runs of $\mathcal{C}_n \times \mathcal{D}$. For a given run ρ there are possibly many ω-words α that induce this run, since only the sequence of states is considered in ρ.

The rest of the argument aims at providing an ω-word α that belongs to L_n but is rejected by the product automaton $\mathcal{C}_n \times \mathcal{D}$. Intuitively, the construction of α requires to balance between the two aims: we need to infinitely often force the product automaton $\mathcal{C}_n \times \mathcal{D}$ to take a rejecting transition of $\mathcal{C}_n$ but at the same time to ensure that there is at least one infinite path in $\mathrm{Graph}(\alpha)$. The ω-word α, an infinite path in $\mathrm{Graph}(\alpha)$, and the rejecting run of $\mathcal{C}_n \times \mathcal{D}$ over α will be constructed as a limit of inductively constructed finite approximations. We will not control exactly the way $\mathcal{C}_n \times \mathcal{D}$ works in every position of our approximation, we will be interested only in some *checkpoints* controlled by partial runs.

Definition 4. *A partial run is a finite partial mapping* $\tau \colon \omega \rightharpoonup Q^{\mathcal{C}_n} \times Q^{\mathcal{D}}$ *such that* $\tau(0)$ *is defined and equal to* $(\bot, q_{\mathbf{I}}^{\mathcal{D}})$.

A partial run τ *is rejecting if all its states are of the form* $(\bot, m)$.

By $\tau \subseteq \rho$ *we denote the fact that a run* ρ *agrees with* τ *wherever* τ *is defined.*

The length *of* τ *is the maximal moment of time* k *such that* $\tau(k)$ *is defined.*

Note that the domain of a partial run τ does not have to be an initial segment of ω. The following definition is crucial.

Definition 5. *Let τ be a partial run of length k. We say that a value $i \in \{0, \ldots, 2n - 1\}$ is alive in τ if there exists an ω-word α such that for the run ρ of $\mathcal{C}_n \times \mathcal{D}$ over α we have $\tau \subseteq \rho$ and there exists a path $p: \{0, 1, \ldots, k\} \to \{0, 1, \ldots, 2n - 1\}$ in $\mathrm{Graph}(\alpha)$ that starts in the moment of time 0 and ends in the moment of time k with the value i (i.e. $p(k) = i$).*

Note that in the above definition we actually care only about the first k letters of α. However, it is cleaner to consider ω-words α here.

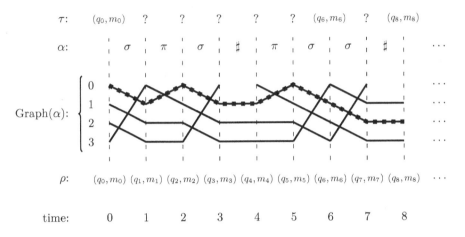

Fig. 2. An example of a partial run τ and an ω-word α that witnesses the fact that 2 is alive in τ. ρ is the run of $\mathcal{C}_n \times \mathcal{D}$ over α and the states of ρ and τ agree wherever defined. The dashed path is the path witnessing that 2 is alive in τ.

Figure 2 depicts a partial run and a witness that the value $i = 2$ is alive.

Our aim is to construct a sequence of partial rejecting runs of increasing lengths $\tau_0 \subset \tau_1 \subset \ldots$ such that for all $\ell \in \mathbb{N}$ there are at least n alive values in τ_ℓ. It will give a contradiction with our assumptions by the following lemma.

Lemma 6. *Assume that there exists a sequence of partial rejecting runs $\tau_0 \subset \tau_1 \subset \ldots$ of increasing lengths such that for all $\ell \in \mathbb{N}$ there exists an alive value in τ_ℓ. Then there exists an ω-word $\alpha \in L_n$ such that the run ρ of $\mathcal{C}_n \times \mathcal{D}$ over α is rejecting.*

Proof. Let k_ℓ be the length of τ_ℓ. Take any ℓ and assume that i_ℓ is a value that is alive in τ_ℓ. Observe that it is witnessed by:

- an ω-word α_ℓ,
- a run ρ_ℓ of $\mathcal{C}_n \times \mathcal{D}$ over α_ℓ, such that $\tau_\ell \subset \rho_\ell$,
- a path $p_\ell: \{0, \ldots, k_\ell\} \to \{0, \ldots, 2n - 1\}$ in $\mathrm{Graph}(\alpha_\ell)$ with $p_\ell(k_\ell) = i_\ell$.

Now we take a subsequence of $(\alpha_\ell, \rho_\ell, p_\ell)_{\ell \in \mathbb{N}}$ that is point-wise convergent to a triple

$$(\alpha, \rho, p) \in \left(A \times \left(Q^{\mathcal{C}_n} \times Q^{\mathcal{D}} \right) \times \{0, \ldots, 2n - 1\} \right)^\omega,$$

such that:

- ρ is the run of $\mathcal{C}_n \times \mathcal{D}$ over α,
- for infinitely many ℓ we have $\tau_\ell \subseteq \rho$,
- p encodes an infinite path in $\mathrm{Graph}(\alpha)$.

To formally construct (α, ρ, p) we can proceed similarly as in the proof of König's lemma. We fix $(\alpha(i), \rho(i), p(i))$ inductively for $i = 0, 1, \ldots$. At each moment we require that infinitely many $(\alpha_\ell, \rho_\ell, p_\ell)$ agree with (α, ρ, p) on the first i positions. Since for each i there are only finitely many choices of $(\alpha(i), \rho(i), p(i))$ so we can fix these values in such a way that still infinitely many $(\alpha_\ell, \rho_\ell, p_\ell)$ agree with them.

By the properties of (α, ρ, p) we know that ρ is rejecting as it contains infinitely many times a state of the form $(\bot, m)$. On the other hand, $\alpha \in L_n$ because p is a witness that $\mathrm{Graph}(\alpha)$ contains an infinite path. □

What remains is to construct the sequence τ_ℓ inductively. Our inductive assumption is that τ_ℓ is a partial rejecting run and the values $1, 3, 5, \ldots, 2n-1$ are alive in τ_ℓ (note that there is n such values). We put $\tau_0 = \left[0 \mapsto (\bot, q_\mathbf{I}^{\mathcal{D}})\right]$. Clearly τ_0 satisfies the inductive assumption (in fact all the values $i = 0, \ldots, 2n - 1$ are alive in τ_0).

Let k_ℓ be the length of τ_ℓ. We construct $\tau_{\ell+1}$ from τ_ℓ by applying some words to the last state $(\bot, m_\ell) = \tau_\ell(k_\ell)$ of τ_ℓ and observing the behaviour of $\mathcal{C}_n \times \mathcal{D}$.

Observe that there are $N = 2^n$ words $u_1, \ldots, u_N \in \{\iota, \sigma, \pi\}^*$ that encode distinct permutations P of $\{0, \ldots, 2n - 1\}$ such that for all $i \in \{0, \ldots, 2n - 1\}$, we have $\lfloor i/2 \rfloor = \lfloor P(i)/2 \rfloor$ i.e. such a permutation maps $\{2i, 2i + 1\}$ to itself.

We can assume that all the words $u_1, \ldots, u_N$ are of equal length by padding them with ι. Since there are strictly less than $N = 2^n$ states of $\mathcal{C}_n \times \mathcal{D}$, there are two distinct such words u, u' leading from $(\bot, m_\ell)$ to the same state (q'_ℓ, m'_ℓ) of $\mathcal{C}_n \times \mathcal{D}$. By the construction of $\mathcal{C}_n \times \mathcal{D}$ we know that $q'_\ell \in \{0, \ldots, 2n - 1\}$.

Assume that the permutations corresponding to u and u' differ on $2i + 1$, i.e. one of them maps $2i + 1$ to $2i$ and the other to $2i + 1$. Let X be the set of the values $\{u(1), u(3), \ldots, u(2n - 3), u(2n - 1), u'(2i + 1)\}$ (we write here $u(i')$ for the value assigned to i' by the permutation corresponding to u, the same for u'). By the above observations X contains exactly $n + 1$ elements.

Consider $w \in \{\iota, \sigma, \pi\}^*$ encoding a permutation that maps:

- q'_ℓ to 0,
- $X \setminus \{q'_\ell\}$ to $1, 3, 5, \ldots, 2n-1$ if $q'_\ell \in X$,
- X to $1, 3, \ldots, 2n-1$, and 2 if $q'_\ell \notin X$.

Since w as a permutation maps q'_ℓ to 0, we know that after reading $w\sharp$ from the state (q'_ℓ, m'_ℓ) the automaton $\mathcal{C}_n \times \mathcal{D}$ reaches a state of the form $(\bot, m_{\ell+1})$. For an illustration of these permutations, see Figure 3.

Fact 7. *Consider $\tau_{\ell+1}$ defined as $\tau_\ell \cup \left[k_\ell + |u| + |w| + 1 \mapsto (\bot, m_{\ell+1})\right]$. By the definition $\tau_\ell \subset \tau_{\ell+1}$, $\tau_{\ell+1}$ is rejecting, and all the values $1, 3, \ldots, 2n-1$ are alive in $\tau_{\ell+1}$ (it is witnessed by the fact that these values were alive in τ_ℓ and by the words $uw\sharp$ and $u'w\sharp$).*

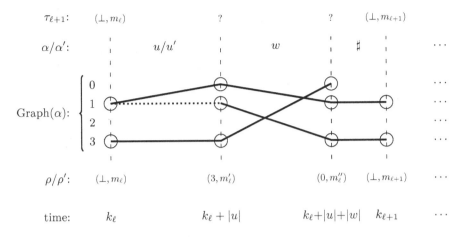

Fig. 3. The behaviour of $\mathcal{C}_n \times \mathcal{D}$ over $uw\sharp$ and $u'w\sharp$. The alive values are in circles, only edges between the alive values are drawn. The dashed edge corresponds to the action of the word u' on the value 1 (u and u' differ on this value). X is the set of values in circles at the moment of time $k_\ell + |u|$. $q'_\ell = 3$ is mapped to 0 by the permutation corresponding to w, the other elements of X are mapped to 1 and 3.

Therefore, we have constructed $\tau_{\ell+1}$ that satisfies the inductive invariant. This concludes the inductive construction of the sequence $(\tau_\ell)_{\ell \in \mathbb{N}}$. By Lemma 6 it finishes the proof of Theorem 1.

4 Büchi Case

In this section we discuss the quadratic upper bound for the state-complexity of determinising Büchi GFG automata, as expressed by the following theorem.

Theorem 8. *For every Büchi GFG automaton there exists an equivalent deterministic Büchi automaton with quadratic number of states.*

Here we provide some high level overview of the construction. A detailed description of it can be found in the full version of the paper.

The main part of the construction is an inductive normalisation of a given Büchi automaton $\mathcal{A}$. The normalisation is guided by the powerset automaton $\mathcal{D}$ having sets of states of $\mathcal{A}$ as its states. It turns out that if $\mathcal{A}$ is GFG then $L(\mathcal{A}) = L(\mathcal{D})$. During the normalisation we remove some irrelevant transitions of $\mathcal{A}$ and mark some existing transitions as accepting (while ensuring that we preserve the language $L(\mathcal{A})$ and the fact that $\mathcal{A}$ is GFG).

When reaching a fixed-point of the normalisation, we know that $\mathcal{A}$ is in certain formal sense *optimal*. This optimal $\mathcal{A}$ needs not be deterministic. However, we can prove that there is a function σ witnessing that $\mathcal{A}$ is GFG that uses $\mathcal{A}$ as a memory structure. Therefore, by combining $\mathcal{A}$ with σ, we can define a structure of a deterministic Büchi automaton for $L(\mathcal{A})$ over $\mathcal{A} \times \mathcal{A}$.

5   Recognising GFG Automata

We now investigate the algorithmic complexity of recognising whether a given automaton is GFG. We provide three results about general GFG-automata, Büchi GFG automata, and co-Büchi GFG-automata. Let us recall that in general, the problem of deciding if a given parity automaton is GFG was shown in [HP06] to belong to EXPTIME.

Equivalence with Parity Games. The following theorem shows that in general, the problem of GFGness of a given parity automaton is at least as hard as solving parity games. The later is known to be NP ∩ co–NP but there is no PTIME algorithm known.

Theorem 9. *Finding the winner of a parity game of index (i, j) is polynomially equivalent to deciding whether a given parity automaton of index (i, j) that accepts all ω-words is GFG.*

Indeed, we show that given a parity game $\mathcal{G}$ between the players $\exists$ and $\forall$, it is possible to build an automaton $\mathcal{A}$ accepting all ω-words, with the same parity index as $\mathcal{G}$, such that $\mathcal{A}$ is GFG if and only if $\exists$ wins $\mathcal{G}$. In the initial state of $\mathcal{A}$ the automaton is supposed to non-deterministically guess the next letter. If the guess is correct, we move to an accepting sink state, otherwise we move to a subautomaton mimicking the game $\mathcal{G}$, where moves of $\forall$ are represented by letters and moves of $\exists$ are represented by a choice of transition. This way $\mathcal{A}$ accepts all ω-words but no GFG strategy can guarantee to reach the accepting sink state. Therefore, $\mathcal{A}$ is GFG only if $\exists$ has a strategy to win the original game $\mathcal{G}$. A polynomial reduction from the problem of GFGness of an automaton accepting all ω-words to a parity game of the same index is an easy consequence of [HP06].

Recognising Büchi GFG Automata. The upper bounds given in Section 4 allow us to state the following theorem.

Theorem 10. *It is in NP to decide whether a given non-deterministic Büchi automaton $\mathcal{A}$ is GFG. Moreover, if $\mathcal{A}$ is GFG then we can construct an equivalent deterministic Büchi automaton in NP.*

Recognising co-Büchi GFG Automata.

Theorem 11. *Given a non-deterministic co-Büchi automaton, we can decide whether it is GFG in polynomial time.*

The cornerstone of the construction is a game called the *Joker Game*, defined relatively to a co-Büchi automaton $\mathcal{A}$. This is a perfect information two players game played between $\exists$ and $\forall$. The set of positions is $Q^{\mathcal{A}} \times Q^{\mathcal{A}}$, the initial position is $(q_{\mathbf{I}}^{\mathcal{A}}, q_{\mathbf{I}}^{\mathcal{A}})$, and at a round n starting in (p_n, q_n) the following choices are made by the players:

- $\forall$ chooses a letter $a_n \in A$,
- $\exists$ chooses a transition $p_n \xrightarrow{a_n} p_{n+1}$ of $\mathcal{A}$,
- $\forall$ chooses a transition $q_n \xrightarrow{a_n} q_{n+1}$ of $\mathcal{A}$ or plays JOKER and chooses a transition $p_n \xrightarrow{a_n} q_{n+1}$ of $\mathcal{A}$.

After that the game moves to the position (p_{n+1}, q_{n+1}). Player $\exists$ wins an infinite play if either:

- the run $(p_n)_n$ of $\mathcal{A}$ is accepting over $(a_n)_n$,
- $\forall$ played infinitely many times JOKER,
- or the run $(q_n)_n$ of $\mathcal{A}$ is not accepting over $(a_n)_n{}^2$.

Intuitively, the Joker Game forces $\exists$ to produce an accepting run of $\mathcal{A}$ over $(a_n)_n$ sequentially, whenever possible. However, since we cannot put the fact that $(a_n)_n \in L(\mathcal{A})$ into the acceptance condition (it would hide an exponential blow-up in the acceptance condition). Therefore, we ask $\forall$ to concurrently produce a run of $\mathcal{A}$ over $(a_n)_n$. If $\forall$ manages to produce an accepting run while $\exists$ fails to do so, it shows that $\mathcal{A}$ is not GFG. The other implication is problematic: the automaton $\mathcal{A}$ may not be GFG but $\exists$ may win the Joker Game by relying on the choices made by $\forall$.

We start by computing in polynomial time the winner of the Joker Game (a parity game of index $(0, 2)$) on $\mathcal{A}$. We show that if $\forall$ wins the Joker Game then $\mathcal{A}$ is not GFG. In the opposite case we are able to build a GFG automaton $\mathcal{B}$ of the same number of states as $\mathcal{A}$ that recognises the same language. Then, using again an appropriate game over $\mathcal{A} \times \mathcal{B}$ we can decide GFGness of $\mathcal{A}$ in polynomial time.

To build the automaton $\mathcal{B}$, we first compute a binary relation $\rightharpoonup$ on the states of $\mathcal{A}$. This relation is the winning region of yet another game, the *safety game*, which is the Joker Game where seeing a rejecting transition means immediate loss. By referring to the Joker Game we prove that for all q there is p such that $p \rightharpoonup p$ and $p \rightharpoonup q$.

This means that we can construct a deterministic safety automaton $\mathcal{D}$ with states p such that $p \rightharpoonup p$. Every ω-word that is accepted by $\mathcal{A}$ has a suffix accepted by $\mathcal{D}$ from some state p. It remains to add non-deterministic rejecting transitions to $\mathcal{D}$ in order to allow it to guess such a state p. For this, we compute an equivalence relation E on the states of $\mathcal{A}$ reflecting simultaneous reachability. We then use this relation to build $\mathcal{B}$ by connecting E-equivalent states of $\mathcal{D}$ using rejecting transitions. We finally show that the automaton $\mathcal{B}$ is GFG and recognises $L(\mathcal{A})$. The strategy witnessing GFGness of $\mathcal{B}$ uses the same intuition as the one in Lemma 3

6   Conclusion

The main result of this paper is a solution of the open problem asking what is the state-complexity of determinising parity GFG automata over ω-words. We prove

---

^{2} Formally, only the suffix of $(q_n)_n$ after the last JOKER played by $\forall$ is a run of $\mathcal{A}$ over the suffix of $(a_n)_n$.

that for co-Büchi GFG automata (and therefore all higher parity indices) the exponential blow-up cannot be avoided. For the remaining case of Büchi GFG automata we provide a construction of an equivalent deterministic automaton with quadratic number of states.

Using the tools developed to prove the above results, we are additionally able to study the complexity of the decision problem of verifying if a given parity automaton is GFG. We prove that for general parity automata the problem is at least as hard as solving parity games (for which no PTIME algorithm is known). Then we focus on the two subcases of Büchi and co-Büchi automata. In the case of Büchi automata we provide a very simple NP algorithm based on our determinisation construction. In the case of co-Büchi automata we have a bit more involved PTIME decision procedure. One of the advantages of the procedure is that, even if the automaton itself is not GFG, there could be cases when the procedure builds an equivalent GFG automaton with the same number of states. The possibilities of exploiting this fact are still to be studied.

Hopefully, the results presented in this paper will shed some light on possible efficient applications of GFG automata in the classical problems of verification.

For future research, in the Büchi case, both the exact time-complexity (between PTIME and NP) and state-complexity (between linear and quadratic) of the determinisation algorithm are still to be clarified.

The complexity of deciding GFGness for general parity automata is still open, with a lower bound of solving parity games and an EXPTIME upper bound.

References

BKKS13. Boker, U., Kuperberg, D., Kupferman, O., Skrzypczak, M.: Nondeterminism in the presence of a diverse or unknown future. In: Fomin, F.V., Freivalds, R., Kwiatkowska, M., Peleg, D. (eds.) ICALP 2013, Part II. LNCS, vol. 7966, pp. 89–100. Springer, Heidelberg (2013)

BL69. Büchi, J.R., Landweber, L.H.: Solving sequential conditions by finite-state strategies. Transactions of the American Mathematical Society **138**, 295–311 (1969)

Col09. Colcombet, T.: The theory of stabilisation monoids and regular cost functions. In: Albers, S., Marchetti-Spaccamela, A., Matias, Y., Nikoletseas, S., Thomas, W. (eds.) ICALP 2009, Part II. LNCS, vol. 5556, pp. 139–150. Springer, Heidelberg (2009)

Col12. Colcombet, T.: Forms of determinism for automata (invited talk). In: STACS, pp. 1–23 (2012)

HP06. Henzinger, T.A., Piterman, N.: Solving games without determinization. In: Ésik, Z. (ed.) CSL 2006. LNCS, vol. 4207, pp. 395–410. Springer, Heidelberg (2006)

KMBK14. Klein, J., Müller, D., Baier, C., Klüppelholz, S.: Are good-for-games automata good for probabilistic model checking? In: Dediu, A.-H., Martín-Vide, C., Sierra-Rodríguez, J.-L., Truthe, B. (eds.) LATA 2014. LNCS, vol. 8370, pp. 453–465. Springer, Heidelberg (2014)

McN66. McNaughton, R.: Testing and generating infinite sequences by a finite automaton. Information and Control **9**(5), 521–530 (1966)

Rab72. Rabin, M.O.: Automata on Infinite Objects and Church's Problem. American Mathematical Society, Boston (1972)

Owicki-Gries Reasoning for Weak Memory Models

Ori Lahav$^{(\boxtimes)}$ and Viktor Vafeiadis

Max Planck Institute for Software Systems (MPI-SWS), Kaiserslautern, Germany
orilahav@mpi-sws.org

Abstract. We show that even in the absence of auxiliary variables, the well-known Owicki-Gries method for verifying concurrent programs is unsound for weak memory models. By strengthening its non-interference check, however, we obtain OGRA, a program logic that is sound for reasoning about programs in the release-acquire fragment of the C11 memory model. We demonstrate the usefulness of this logic by applying it to several challenging examples, ranging from small litmus tests to an implementation of the RCU synchronization primitives.

1 Introduction

In 1976, Owicki and Gries [10] introduced a proof system for reasoning about concurrent programs, which formed the basis of rely/guarantee reasoning. Their system includes the usual Hoare logic rules for sequential programs, a rule for introducing auxiliary variables, and the following parallel composition rule:

$$\frac{\{P_1\}\,c_1\,\{Q_1\} \quad \{P_2\}\,c_2\,\{Q_2\} \quad \text{the two proofs are non-interfering}}{\{P_1 \wedge P_2\}\,c_1 \parallel c_2\,\{Q_1 \wedge Q_2\}}$$

This rule allows one to compose two verified programs into a verified concurrent program that assumes both preconditions and ensures both postconditions. The soundness of this rule requires that the two proofs are *non-interfering*, namely that every assertion R in the one proof is stable under any $\{P\}x := e$ (guarded) assignment in the other and vice versa; i.e., for every such pair, $R \wedge P \vdash R[e/x]$.

The Owicki-Gries system (OG) assumes a fairly simple but unrealistic concurrency model: *sequential consistency* (SC) [7]. This is essential: OG is complete for verifying concurrent programs under SC [12], and is therefore unsound under a weakly consistent memory semantics, such as TSO [9]. Auxiliary variables are instrumental in achieving completeness—without them, OG is blatantly incomplete; e.g., it cannot verify that $\{x = 0\}\,x \overset{\text{at}}{:=} x + 1 \parallel x \overset{\text{at}}{:=} x + 1\,\{x = 2\}$ (where "$\overset{\text{at}}{:=}$" denotes atomic assignment).

Nevertheless, many useful OG proofs do not use auxiliary variables, and one might wonder whether such proofs are sound under weak memory models. This is sadly not the case. Figure 1 presents an OG proof that a certain program cannot return $a = b = 0$ whereas under all known weak memory models it can in fact do so. Intuitively speaking, the proof is invalid under weak memory because the two threads may have different views

---

Due to space limits, supplementary material including full proofs and further examples is available at: http://plv.mpi-sws.org/ogra/.

© Springer-Verlag Berlin Heidelberg 2015
M.M. Halldórsson et al. (Eds.): ICALP 2015, Part II, LNCS 9135, pp. 311–323, 2015.
DOI: 10.1007/978-3-662-47666-6_25

$$\{x = 0 \wedge y = 0 \wedge a \neq 0\}$$

$$\{a \neq 0\} \;\big\|\; \{\top\}$$
$$x := 1; \;\big\|\; y := 1;$$
$$\{x \neq 0\} \;\big\|\; \{y \neq 0\}$$
$$a := y \;\big\|\; b := x$$
$$\{x \neq 0\} \;\big\|\; \{y \neq 0 \wedge (a \neq 0 \vee b = x)\}$$
$$\{a \neq 0 \vee b \neq 0\}$$

Non-interference checks are trivial. For example,

$$y \neq 0 \wedge (a \neq 0 \vee b = x) \wedge a \neq 0$$
$$\vdash y \neq 0 \wedge (a \neq 0 \vee b = 1)$$

and $y \neq 0 \wedge (a \neq 0 \vee b = x) \wedge x \neq 0$

$$\vdash y \neq 0 \wedge (y \neq 0 \vee b = x)$$

show stability of the last assertion of thread II
under $\{a \neq 0\}x := 1$ and $\{x \neq 0\}a := y$.

Fig. 1. OG proof that the "store buffering" program cannot return $a = b = 0$. This can also be proved in the restricted OG system with one (stable) global invariant [11]. Note that OG's "indivisible assignments" condition (3.1) is met: assignments mention at most one shared location.

of memory before executing each command. Thus, when the second thread terminates, the first thread may perform $a := y$ reading $y = 0$ and storing 0 in a, thereby invalidating the second thread's last assertion. We note that $y = 0$ was also readable by the second thread, albeit at an earlier point (before the $y := 1$ assignment). This is no accident, and this observation is essential for soundness of our proposed alternative.

In this paper we identify a stronger non-interference criterion that does not assume SC semantics. Thus, while considering the effect of an assignment $\{P\}x := y$ in thread I on the validity of an assertion R in thread II, one does not get to assume that R holds for the view of thread I while reading y. In fact, in some executions, the value read for y might even be inconsistent with R. Instead, the only allowed assumption is that some assertion that held not later than R in thread II was true while reading y. Thus our condition for checking stability of R under $\{P\}x := y$ is that $R \wedge P \vdash R[v/x]$ for every value v of y that is consistent with P and some non-later assertion of thread II.

We show that OG with our stronger non-interference criterion is sound under the release-acquire (RA) fragment of the C11 memory model [6], which exhibits a good balance between performance and mathematical sanity (see, e.g., [16,17]). Soundness under TSO follows, as TSO behaviors are all observable under RA (see [1]). Formalizing the aforementioned intuitions into a soundness proof for RA executions is far from trivial. Indeed, RA is defined axiomatically without an operational semantics and without the notion of a state. As a basis for the soundness proof, we introduce such a notion and study the properties of sequences of states observed by different threads.

We believe that the results of this paper may provide new insights for understanding weak memory models, as well as a *simple* and *useful* method for proving partial correctness of concurrent programs. We demonstrate the applicability of our logic (which we call OGRA) with several challenging examples, ranging from small litmus tests to an implementation of the read-copy-update (RCU) synchronization primitives [3]. We also provide support for fence instructions by implementing them as RMWs to an otherwise unused location and for a simple class of auxiliary variables, namely *ghost values*.

Related Work. Aiming to understand and verify high-performance realistic concurrent programs, program logics for weak memory models have recently received a lot of attention (see, e.g., [4,13,14,16,18]). Most of these logics concern the TSO memory model. Only two—RSL [18] and GPS [16]—can handle RA, but have a fairly complex foundation being based on separation logic. The most advanced of the two logics, GPS, has been used (with considerable ingenuity) to verify the RCU synchronization primitives [15], but simpler examples such as "read-read coherence" seem to be beyond its power (see Fig. 8). Finally, Cohen [2] studies an alternative memory model under which OG reasoning can be performed at the execution level.

2 Preliminaries

In this section, we present a simplified programming language, whose semantics adheres to that of the release-acquire fragment of C11's memory model [1]. We assume a finite set of locations $\mathsf{Loc} = \{v_1, \ldots, v_M\}$, a finite set Val of values with a distinguished value $0 \in \mathsf{Val}$, and any standard interpreted language for expressions containing at least all locations and values. We use x, y, z as metavariables for locations, v for values, e for expressions, and denote by $e(x_1, \ldots, x_n)$ an expression in which $x_1, \ldots, x_n$ are the only mentioned locations. The language's commands are given by the following grammar:

$$c ::= \mathbf{skip} \mid \mathbf{if}\ e(x)\ \mathbf{then}\ c\ \mathbf{else}\ c \mid \mathbf{while}\ e(x)\ \mathbf{do}\ c \mid c\,;c \mid c \parallel c \mid$$
$$x := v \mid x := e(y) \mid x \overset{y,z}{:=} e(y,z) \mid x \overset{at}{:=} e(x)$$

To keep the presentation simple, expressions in assignments are limited to mention at most two locations, and those in conditionals and loops mention one location. Assignments of expression mentioning two locations also specify the order in which these locations should be read (if one of them is local, this has no observable effect). The command $x \overset{at}{:=} e(x)$ is an atomic assignment corresponding to a primitive read-modify-write (RMW) instruction and, as such, mentions only one location.[1]

Now, as in the C11 formalization, the semantics of a program is defined to be its set of consistent executions [1]. An execution G is a triple $\langle A, L, E \rangle$ where:

- $A \subseteq \mathbb{N}$ is a finite set of *nodes*. We identify G with this set, e.g., when writing $a \in G$.
- L is a function assigning a *label* to each node, where a label is either $\langle \mathsf{S} \rangle$ ("Skip"), a triple of the form $\langle \mathsf{R}, x, v_r \rangle$ ("Read"), a triple of the form $\langle \mathsf{W}, x, v_w \rangle$ ("Write"), or a quadruple of the form $\langle \mathsf{U}, x, v_r, v_w \rangle$ ("Update"). For $\mathsf{T} \in \{\mathsf{S}, \mathsf{R}, \mathsf{W}, \mathsf{U}\}$, we denote by $G.\mathsf{T}$ the set of nodes $a \in A$ for which T is the first entry of $L(a)$, while $G.\mathsf{T}_x$ denotes the set of $a \in G.\mathsf{T}$ for which x is the second entry of $L(a)$. In addition, L induces the partial functions $G.loc : A \to \mathsf{Loc}$, $G.val_r : A \to \mathsf{Val}$, and $G.val_w : A \to \mathsf{Val}$ that respectively return (when applicable) the x, v_r and v_w components of a node.
- $E \subseteq (A \times A) \cup (A \times A \times \mathsf{Loc})$ is a set of *edges*, such that for every triple $\langle a, b, x \rangle \in E$ (reads-from edge) we have $a \in G.\mathsf{W}_x \cup G.\mathsf{U}_x$, $b \in G.\mathsf{S} \cup G.\mathsf{R}_x \cup G.\mathsf{U}_x$, and $G.val_w(a) = G.val_r(b)$ whenever $b \notin G.\mathsf{S}$.[2] The subset $E \cap (A \times A)$ is denoted by $G.po$ (*program order*), and $G.E_x$ denotes the set $\{\langle a, b \rangle \in A \times A \mid \langle a, b, x \rangle \in E\}$ (*x-reads-from*) for every $x \in \mathsf{Loc}$. Finally, $G.E_{all}$ denotes the set $G.po \cup \bigcup_{x \in \mathsf{Loc}} E_x$.

For all these notations, we often omit the "$G.$" prefix when it is clear from the context. Given an execution $G = \langle A, L, E \rangle$ and a set E' of edges we write $G \cup E'$ for the triple $\langle A, L, E \cup E' \rangle$ and $G \setminus E'$ for $\langle A, L, E \setminus E' \rangle$.

---

[1] Unlike usual OG [10], our assignments can mention more than one shared variable. In fact, our formal development does not differentiate between local and shared variables.

[2] Reads-from edges $\langle a, b, x \rangle$ with $b \in G.\mathsf{S}$ are used for defining visible states (see Definition 7).

$$\begin{aligned}
[\![\mathtt{skip}]\!] &= \mathcal{SG} \\
[\![\mathtt{if}\ e(x)\ \mathtt{then}\ c_1\ \mathtt{else}\ c_2]\!] &= \bigcup\{\mathcal{RG}(x, v); [\![c_i]\!] \mid v \in \mathsf{Val}, i \in \{1, 2\}, [\![e]\!](v) = 0\ \mathit{iff}\ i = 2\} \\
[\![\mathtt{while}\ e(x)\ \mathtt{do}\ c]\!] &= \bigcup_{n \geq 0}(\bigcup\{\mathcal{RG}(x, v) \mid v \in \mathsf{Val}, [\![e]\!](v) \neq 0\}; [\![c]\!])^n; \\
&\quad\ \bigcup\{\mathcal{RG}(x, v) \mid v \in \mathsf{Val}, [\![e]\!](v) = 0\} \\
[\![c_1; c_2]\!] &= [\![c_1]\!]; [\![c_2]\!] \\
[\![c_1 \parallel c_2]\!] &= \mathcal{SG}; ([\![c_1]\!] \parallel [\![c_2]\!]); \mathcal{SG} \\
[\![x := v]\!] &= \mathcal{WG}(x, v) \\
[\![x := e(y)]\!] &= \bigcup\{\mathcal{RG}(y, v); \mathcal{WG}(x, [\![e]\!](v)) \mid v \in \mathsf{Val}\} \\
[\![x \overset{y,z}{:=} e(y, z)]\!] &= \bigcup\{\mathcal{RG}(y, v_y); \mathcal{RG}(z, v_z); \mathcal{WG}(x, [\![e]\!](v_y, v_z)) \mid v_y, v_z \in \mathsf{Val}\} \\
[\![x \overset{\mathtt{at}}{:=} e(x)]\!] &= \bigcup\{\mathcal{UG}(x, v, [\![e]\!](v)) \mid v \in \mathsf{Val}\}
\end{aligned}$$

Fig. 2. Mapping of commands to sets of executions

Definition 1. A node a in an execution G is *initial* (*terminal*) in G if $\langle b, a \rangle \notin E_{all}$ ($\langle a, b \rangle \notin E_{all}$) for every $b \in G$. An edge $\langle a, b \rangle \in po$ is *initial* (*terminal*) in G if a is initial (b is terminal) in G.

Definition 2. Let $G = \langle A, L, E \rangle$ and $G' = \langle A', L', E' \rangle$ be two executions with disjoint sets of nodes.

- The execution $G \parallel G'$ is given by $\langle A \cup A', E \cup E', L \cup L' \rangle$.
- The execution $G; G'$ is given by $(G \parallel G') \cup (O \times I)$, where O is the set of terminal nodes of G, and I is the set of initial nodes of G'.
- Given $n \geq 0$, G^n is inductively defined by $G^0 = \langle \emptyset, \emptyset, \emptyset \rangle$ and $G^{n+1} = G^n; G$.

The above operations are extended to sets of executions in the obvious way (e.g., $\mathcal{G}; \mathcal{G}' = \{G; G' \mid G \in \mathcal{G}, G' \in \mathcal{G}', G; G' \text{ is defined}\}$).

Definition 3. Given $x \in \mathsf{Loc}$ and $v \in \mathsf{Val}$, an $\langle x, v \rangle$-*read gadget* is any execution of the form $\langle \{a\}, \{a \mapsto \langle \mathtt{R}, x, v \rangle\}, \emptyset \rangle$. $\langle x, v \rangle$-*write gadgets*, $\langle x, v_r, v_w \rangle$-*update gadgets* and *skip gadgets* are defined similarly. $\mathcal{RG}(x, v)$, $\mathcal{WG}(x, v)$, $\mathcal{UG}(x, v_r, v_w)$ and $\mathcal{SG}$ denote, respectively, the sets of all $\langle x, v \rangle$-read gadgets, all $\langle x, v \rangle$-write gadgets, all $\langle x, v_r, v_w \rangle$-update gadgets, and all skip gadgets.

Using these definitions, the mapping of commands to (sets of) executions is given in Fig. 2. Note that every execution $G \in [\![c]\!]$ for some command c satisfies $G.E_{all} = G.po$, and has a unique initial node that can reach any node, and a unique terminal node that can be reached from any node. We refer to such executions as *plain*. However, many of these executions are nonsensical as they can, for instance, read values never written in the program. We restrict our attention to *consistent* executions, as defined next.

Definition 4. A relation R is called a *modification order* for a location $x \in \mathsf{Loc}$ in an execution G if the following hold: (*i*) R is a total strict order on $\mathsf{W}_x \cup \mathsf{U}_x$; (*ii*) if $\langle a, b \rangle \in E_{all}^*$ then $\langle b, a \rangle \notin R$; (*iii*) if $\langle a, b \rangle \in E_{all}^+$ and $\langle c, b \rangle \in E_x$ then $\langle c, a \rangle \notin R$; and (*iv*) if $\langle a, b \rangle, \langle b, c \rangle \in R$ and $c \in \mathsf{U}$ then $\langle a, c \rangle \notin E_x$.

Definition 5. An execution $G = \langle A, L, E \rangle$ is called:

- *complete* if for every $b \in \mathsf{R} \cup \mathsf{U}$, we have $\langle a, b \rangle \in E_{loc(b)}$ for some $a \in \mathsf{W} \cup \mathsf{U}$.
- *coherent* if E_{all} is acyclic, and there is a modification order in G for each $x \in \mathsf{Loc}$.

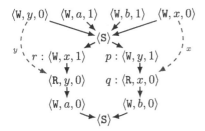

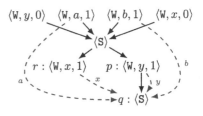

Fig. 3. Ignoring the dashed edges, this graph G is an initialized execution of the "store buffering" program (i.e., $G \in \mathcal{WG}(\top)$; $[\![c]\!]$, Def. 8). G is consistent as it can be extended with the set E' of the two dashed reads-from edges.

Fig. 4. Ignoring the dashed edges, we have the snapshot of $G \cup E'$ of Fig. 3 at $\langle p, q \rangle$ with respect to $\{r\}$. Adding the dashed edges results in a coherent execution; so the state $\{x \mapsto 1, y \mapsto 1, a \mapsto 1, b \mapsto 1\}$ is visible at $\langle p, q \rangle$ in $G \cup E'$.

- *consistent* if $G \cup E'$ is complete and coherent for some $E' \subseteq A \times A \times \mathsf{Loc}$.

To illustrate these definitions, Fig. 3 depicts a consistent non-SC execution of the "store buffering" program of Fig. 1 together with the implicit variable initializations.

While our notations are slightly different, the axiomatic semantics presented above corresponds to the semantics of C11 programs (see [1]) in which all locations are atomic, reads are acquire reads, writes are release writes, and updates are acquire-release RMWs. In addition, we do not allow reads from uninitialized locations. C11's "happens-before" relation corresponds to our E_{all}^+.

3   An Owicki-Gries Proof System for Release-Acquire

In this section, we present OGRA—our logic for reasoning about concurrent programs under release-acquire. As usual, the basic constructs are *Hoare triples* of the form $\{P\} c \{Q\}$, where P and Q are assertions and c is a command. To define validity of such a triple (in the absence of usual operational semantics), we formalize the notion of a visible state, taken to be a function from Loc to Val.

Definition 6. A *snapshot* of an execution $G = \langle A, L, E \rangle$ at an edge $\langle a, b \rangle \in po$ with respect to a set $B \subseteq A$ of nodes, denoted by $\mathcal{S}(G, \langle a, b \rangle, B)$, is the execution $\langle A' \uplus \{b\}, L|_{A'} \cup \{b \mapsto \langle S \rangle\}, E|_{A'} \cup \{\langle a, b \rangle\}\rangle$, where:
- $A' = \{a' \in A \setminus \{b\} \mid \exists c \in B \cup \{a\}. \langle a', c \rangle \in E_{all}^*\}$ and
- $E|_{A'} = E \cap ((A' \times A') \cup (A' \times A' \times \mathsf{Loc}))$.

Definition 7. Let G be an execution, and let $\langle a, b \rangle \in po$.
- A function $D : \mathsf{Loc} \to \mathbb{N}$ is called a $\langle G, \langle a, b \rangle \rangle$-*reader* of a state $\sigma : \mathsf{Loc} \to \mathsf{Val}$ if $D(x) \in \mathsf{W}_x \cup \mathsf{U}_x$ and $val_w(D(x)) = \sigma(x)$ for every $x \in \mathsf{Loc}$, and the execution $\mathcal{S}(G, \langle a, b \rangle, D[\mathsf{Loc}]) \cup \{\langle D(x), b, x \rangle \mid x \in \mathsf{Loc}\}$ is coherent.
- A state σ is called *visible* at $\langle a, b \rangle$ in G if there is a $\langle G, \langle a, b \rangle \rangle$-reader of σ.
- An assertion P *holds* at $\langle a, b \rangle$ in G if $\sigma \models P$ for every state σ visible at $\langle a, b \rangle$ in G.

In essence, the snapshot restricts the execution to the edge $\langle a, b \rangle$, all nodes in B, and all prior nodes and edges, and replaces the label of b by a skip. For a state to be visible at $\langle a, b \rangle$, additional reads-from edges should be added. For an example, see Fig. 4.

Definition 8. For a state σ, let $\mathcal{WG}(\sigma)$ be $\mathcal{WG}(v_1, \sigma(v_1)) \parallel \ldots \parallel \mathcal{WG}(v_M, \sigma(v_M))$, the set of all σ-initializations. Given an assertion P, $\mathcal{WG}(P) = \bigcup \{\mathcal{WG}(\sigma) \mid \sigma \models P\}$.

An execution G is called *initialized* if $G = (G_1; G_2) \cup E$ for some $G_1 \in \mathcal{WG}(\top)$, plain execution G_2, and set $E \subseteq A_1 \times A_2 \times \text{Loc}$ of edges. It can be shown that if G is coherent and initialized, then at least one state is visible at every program order edge.

Definition 9. A Hoare triple $\{P\} c \{Q\}$ is *valid* if Q holds at the terminal edge of $G \cup E'$ in $G \cup E'$ for every execution $G = \langle A, L, E \rangle$ in $\mathcal{WG}(P)$; $[\![c]\!]$; $\mathcal{SG}$ and set $E' \subseteq A \times A \times \text{Loc}$, such that $G \cup E'$ is a complete and coherent execution.

OG-style reasoning is often judged as non-compositional because it refers to non-interference of proof outlines that cannot be checked based solely on the two input Hoare triples. A straightforward remedy is to use a *rely/guarantee-style presentation* of OG, that permits compositional reasoning. In this case, the rely component, denoted by $\mathcal{R}$, consists of a set of assertions that are assumed to be stable under assignments performed by other threads. In turn, the guarantee component, denoted by $\mathcal{G}$, is a set of *guarded assignments*, that is assignments together with their immediate preconditions. Roughly speaking, a validity of an OG judgment $\mathcal{R}; \mathcal{G} \Vdash \{P\} c \{Q\}$ amounts to: "every terminating run of c starting from a state in P ends in a state in Q, and performs only assignments in $\mathcal{G}$, where each of which is performed while satisfying its guard; and moreover, the above holds in parallel to any run of a program c', provided that the assertions in $\mathcal{R}$ are stable under each of the assignments performed by c'."

Now, as demonstrated in the introduction, reasoning under RA requires a richer rely condition, as stability of an assertion in thread I under a guarded assignment of the form $\{P\} x := e(y)$ in thread II should be checked for all values readable for y in some non-later point of thread I. Similarly, stability under $\{P\} x \overset{y,z}{:=} e(y, z)$ should cover all values readable for y and z in two non-later points. Hence, we take $\mathcal{R}$ to consist of pairs of assertions, where the first component of each pair describes the current state and the second summarizes all non-later states. This leads us to the following definitions.

Definition 10. An *OG judgment* $\mathcal{R}; \mathcal{G} \Vdash \{P\} c \{Q\}$ extends a Hoare triple with two extra components:

- A finite set $\mathcal{R}$ of pairs of the form $R \nearrow C$, where R and C are assertions. We write $\mathcal{R}^R$ for $\bigvee \{R \mid R \nearrow \_ \in \mathcal{R}\}$ and $\mathcal{R}^C$ for $\bigwedge \{C \mid \_ \nearrow C \in \mathcal{R}\}$. We also write $\mathcal{R} \le \mathcal{R}'$ for such sets if for every $R \nearrow C \in \mathcal{R}$ there exists C' such that $R \nearrow C' \in \mathcal{R}'$ and $C \vdash C'$.
- A finite set $\mathcal{G}$ of *guarded assignments*, i.e., pairs of the form $\{R\} c$, where R is an assertion and c is an assignment command. We write $\mathcal{G} \le \mathcal{G}'$ for such sets if for every $\{R\} c \in \mathcal{G}$ there exists R' such that $\{R'\} c \in \mathcal{G}'$ and $R \vdash R'$.

Definition 11. A pair $R \nearrow C$ is *stable* under $\{P\} c$ if one of the following holds:

- c has the form $x := v$ and $R \wedge P \vdash R[v/x]$;
- c has the form $x := e(y)$ and $R \wedge P \vdash R[[\![e]\!](v_y)/x]$ for every $v_y \in \text{Val}$ such that $C \wedge P \nvdash y \ne v_y$ (i.e., for every $v_y \in \text{Val}$ such that $C \wedge P \wedge y = v_y$ is satisfiable);

$$(\text{CONSEQ}) \quad \frac{P' \vdash P \quad Q \vdash Q' \quad \mathcal{R} \leq \mathcal{R}' \quad \mathcal{G} \leq \mathcal{G}' \quad \mathcal{R}; \mathcal{G} \Vdash \{P\}c\{Q\}}{\mathcal{R}'; \mathcal{G}' \Vdash \{P'\}c\{Q'\}}$$

$$(\text{SEQ}) \quad \frac{\mathcal{R}_1; \mathcal{G}_1 \Vdash \{P\}c_1\{R\} \quad \mathcal{R}_2; \mathcal{G}_2 \Vdash \{R\}c_2\{Q\} \quad \mathcal{R}_1^R \vdash \mathcal{R}_2^C}{\mathcal{R}_1 \cup \mathcal{R}_2; \mathcal{G}_1 \cup \mathcal{G}_2 \Vdash \{P\}c_1; c_2\{Q\}}$$

$$(\text{SKIP}) \quad \frac{\{P \uparrow P\} \leq \mathcal{R}}{\mathcal{R}; \emptyset \Vdash \{P\}\texttt{skip}\{P\}}$$

$$(\text{PAR}) \quad \frac{\mathcal{R}_1; \mathcal{G}_1 \Vdash \{P_1\}c_1\{Q_1\} \quad \mathcal{R}_2; \mathcal{G}_2 \Vdash \{P_2\}c_2\{Q_2\}}{\mathcal{R}_1 \cup \mathcal{R}_2 \cup \{Q \uparrow (\mathcal{R}_1^R \vee \mathcal{R}_2^R \vee Q)\}; \mathcal{G}_1 \cup \mathcal{G}_2 \Vdash \{P_1 \wedge P_2\}c_1 \parallel c_2\{Q\}}$$
$$Q_1 \wedge Q_2 \vdash Q \quad \mathcal{R}_1; \mathcal{G}_1 \text{ and } \mathcal{R}_2; \mathcal{G}_2 \text{ are non-interfering}$$

$$(\text{ASSN}_0) \quad \frac{P \vdash Q[v/x] \quad \{P \uparrow P, Q \uparrow (P \vee Q)\} \leq \mathcal{R}}{\mathcal{R}; \{\{P\}x := v\} \Vdash \{P\}x := v\{Q\}}$$

$$(\text{ASSN}_1) \quad \frac{P \vdash Q[e(y)/x] \quad \{P \uparrow P, Q \uparrow (P \vee Q)\} \leq \mathcal{R}}{\mathcal{R}; \{\{P\}x := e(y)\} \Vdash \{P\}x := e(y)\{Q\}}$$

$$(\text{ASSN}_2) \quad \frac{P \vdash Q[e(y, z)/x] \quad \{P \uparrow P, Q \uparrow (P \vee Q)\} \leq \mathcal{R}}{\{(P \wedge (y = v)) \uparrow P \mid v \in \mathsf{Val}\} \leq \mathcal{R}}$$
$$\mathcal{R}; \{\{P\}x \overset{y,z}{:=} e(y, z)\} \Vdash \{P\}x \overset{y,z}{:=} e(y, z)\{Q\}$$

$$(\text{ASSN}_{at}) \quad \frac{P \vdash Q[e(x)/x] \quad \{P \uparrow P, Q \uparrow (P \vee Q)\} \leq \mathcal{R}}{\mathcal{R}; \{\{P\}x \overset{at}{:=} e(x)\} \Vdash \{P\}x \overset{at}{:=} e(x)\{Q\}}$$

$$(\text{ITE}) \quad \frac{\{P \uparrow P\} \leq \mathcal{R} \quad P \vdash \mathcal{R}^C \quad \mathcal{R}; \mathcal{G} \Vdash \{P \wedge (e(x) \neq 0)\}c_1\{Q\} \quad \mathcal{R}; \mathcal{G} \Vdash \{P \wedge (e(x) = 0)\}c_2\{Q\}}{\mathcal{R}; \mathcal{G} \Vdash \{P\}\texttt{if } e(x) \texttt{ then } c_1 \texttt{ else } c_2\{Q\}}$$

$$(\text{WHILE}) \quad \frac{P \uparrow\_ \in \mathcal{R} \quad \mathcal{R}^R \vdash \mathcal{R}^C \quad P \wedge (e(x) = 0) \vdash Q \quad \mathcal{R}; \mathcal{G} \Vdash \{P \wedge (e(x) \neq 0)\}c\{P\}}{\mathcal{R} \cup \{Q \uparrow (\mathcal{R}^R \vee Q)\}; \mathcal{G} \Vdash \{P\}\texttt{while } e(x) \texttt{ do } c\{Q\}}$$

Fig. 5. Owicki-Gries proof system for release-acquire.

- c has the form $x \overset{y,z}{:=} e(y, z)$ and $R \wedge P \vdash R[\llbracket e \rrbracket(v_y, v_z)/x]$ for every $v_y, v_z \in \mathsf{Val}$, such that $C \wedge P \nvdash y \neq v_y$ and $C \wedge P \nvdash z \neq v_z$; or
- c has the form $x \overset{at}{:=} e(x)$ and $R \wedge P \vdash R[e/x]$.

The proof system for deriving OGRA's judgments is given in Fig. 5. The rules are essentially those of Owicki and Gries [10] with minor adjustments due to our rely/guarantee style presentation and the more complex form of the $\mathcal{R}$ component. (To assist the reader, the supplementary material includes a similar presentation of usual OG.) Typically, we require the preconditions and postconditions to be included in $\mathcal{R}$, and make sure their second components keep track of (at least) all non-later assertions: for example, all the assignment rules require $\{P \uparrow P, Q \uparrow (P \vee Q)\} \leq \mathcal{R}$.

The rule for parallel composition (PAR) allows composing non-interfering judgments. Its precondition is the conjunction of the preconditions of the threads, while its postcondition, Q, is any stable assertion implied by the conjunction of the thread postconditions. (The asymmetry is because of the second components of the $\mathcal{R}$ entries: the states prior to the end of the parallel compositions are the union of those of each thread, and hence the stability of Q does not necessarily follow from that of Q_1 and Q_2.) Non-interference is checked for every rely condition of one thread and guarded assignment in the guarantee component of the other:

Definition 12. $\mathcal{R}_1; \mathcal{G}_1$ and $\mathcal{R}_2; \mathcal{G}_2$ are *non-interfering* if every $R \uparrow C \in \mathcal{R}_i$ is stable under every $\{P\}c \in \mathcal{G}_j$ for $i \neq j$.

$$
\begin{array}{ll}
\{\top\} & \quad \{x = 0\} \\
m := 42; & \quad \{x \neq 0 \rightarrow m = 42\} \\
\{m = 42\} & \quad \textbf{while } x = 0 \textbf{ do skip}; \\
x := 1 & \quad \{m = 42\} \\
\{\top\} & \quad a := m \\
 & \quad \{a = 42\}
\end{array}
$$

Fig. 6. Proof outline for a simple message passing idiom

$$
\begin{array}{ll}
 & \{f = 0\} \\
\{f \in \{0, 2\}\} & \quad \{f \in \{0, 1\}\} \\
x := 1; & \quad y := 1; \\
\{f \in \{0, 2\} \wedge x = 1\} & \quad \{f \in \{0, 1\} \wedge y = 1\} \\
f \overset{\text{at}}{:=} 10f + 1; & \quad f \overset{\text{at}}{:=} 10f + 2; \\
\{f \in \{1, 12, 21\} \wedge x = 1\} & \quad \{f \in \{2, 12, 21\} \wedge y = 1\} \\
a := y & \quad b := x \\
\begin{bmatrix} f \in \{1, 12, 21\} \wedge x = 1 \wedge \\ (f = 21 \rightarrow a = y) \end{bmatrix} & \quad \begin{bmatrix} f \in \{2, 12, 21\} \wedge y = 1 \wedge \\ (f = 12 \rightarrow b = x) \end{bmatrix} \\
\multicolumn{2}{c}{\{a = 1 \vee b = 1\}}
\end{array}
$$

Fig. 7. Proof outline for "store buffering" with fences

$$
\begin{array}{cccc}
\multicolumn{4}{c}{\{x = a = c = 0\}} \\
\begin{bmatrix} (x \neq 1 \wedge a \neq 1) \\ \nearrow x \neq 1 \end{bmatrix} & \begin{bmatrix} (x \neq 2 \wedge c \neq 2) \\ \nearrow x \neq 2 \end{bmatrix} & \{\top\} & \{\top\} \\
x := 1 & x := 2 & a := x; & c := x; \\
\{\top\} & \{\top\} & \{\top\} & \{\top\} \\
 & & b := x & d := x \\
 & & \{a = 1 \wedge b = 2 \rightarrow x = 2\} & \{c = 2 \wedge d = 1 \rightarrow x = 1\} \\
\multicolumn{4}{c}{\{a = 1 \wedge b = 2 \wedge c = 2 \rightarrow d \neq 1\}}
\end{array}
$$

Fig. 8. Proof outline for read-read coherence test (example CoRR2 in [8])

The consequence rule (CONSEQ) allows strengthening the precondition ($P' \vdash P$), weakening the postcondition ($Q \vdash Q'$), increasing the set of assertions required to be stable ($\mathcal{R} \leq \mathcal{R}'$), and increasing the set of allowed guarded assignments ($\mathcal{G} \leq \mathcal{G}'$).

The sequential composition rule (SEQ) collects the assertions and allowed assignments of both commands, and checks that $\mathcal{R}_1^R \vdash \mathcal{R}_2^C$. This ensures that stability of c_2's assertions would take into account all the states of c_1, that now become previous states.

The next interesting rule is ASSN$_2$ concerning assignments with expressions reading two variables. The rule requires that the value of the first variable being read (y) is stable assuming P also holds. This check is needed because of the way we interpret assertions as snapshot reads differs from the way that programs read the variables (one at a time): the stability check ensures that the difference is not observable. Note that the stability of y is trivial in case that there are no assignments to it in other threads.

Finally, the rules for conditionals and while-loops are standard: as with the SEQ rule, we require that the second component of $\mathcal{R}$ has taken into account all earlier states, and include the initial precondition in the set of stable assertions.

We can now state our main theorem, namely the soundness of OGRA.

Theorem 1. *If* $\mathcal{R}; \mathcal{G} \Vdash \{P\} c \{Q\}$ *is derivable, then* $\{P\} c \{Q\}$ *is valid.*

Before proving this theorem, we provide a few example derivations. The derivations are presented in a proof outline fashion. For each thread, the set $\mathcal{R}$ consists of all the assertions in its proof outline, with the second component being $\top$ (all values are possible) unless mentioned otherwise. The set $\mathcal{G}$ consists of all the assignments in the proof outline guarded by their immediate preconditions.

Our first example, shown in Fig. 6, is a simple message passing idiom. Thread I initializes a message m to 42 and then raises a flag x; thread II waits for x to have a non-zero value and then reads m, which should have value 42. To prove this, thread II assumes the invariant $x \neq 0 \rightarrow m = 42$ that holds initially and is stable.

Our next example, shown in Fig. 7, is a variant of the "store buffering" program (see Fig. 1) that uses fences to restore sequential consistency. Fence instructions are implemented as RMWs to a distinguished location f. The RA semantics enforces the corresponding update nodes to be linearly ordered by E^*_{all}, so this implementation imposes a synchronization between every pair of fences. These fences are stronger than C11's SC fences, as they restore full SC when placed between every pair of consecutive instructions. While any atomic assignment to f will have this effect, we choose commands that record the exact order in which the fences are linearized. By referring to this order in the proof, we can easily show that the outcome $a = b = 0$ is not possible.

Our third example, shown in Fig. 8, is a coherence test, demonstrating that threads cannot observe writes to the same location happen in different orders. The program consists of two independent writes to x and two readers: the goal is to prove that the first reader cannot read the one write and then the other, while the second reads them in the reverse order. The key to showing this are the assertions at the end of the reader threads saying that the value of x cannot change after both assignments have been observed. For these assertions to be stable, the writers correspondingly assert that the assignments to x happen before the corresponding reader observes x to have that value. Formally, the precondition of the $x := 1$ assignment is $(x \neq 1 \wedge a \neq 1)/x \neq 1$. This is stable under the $a := x$ assignment because 1 is not a readable value for x (we have: $x \neq 1 \nvdash x \neq v$ iff $v \neq 1$).

3.1 Soundness Proof

We present the main steps in the proof of Fig. 1. Annotations play a crucial role. An *annotation* is a function that assigns an assertion to every pair in $\mathbb{N} \times \mathbb{N}$. An annotation Θ is *valid* for an execution G if $\Theta(\langle a, b \rangle)$ holds at $\langle a, b \rangle$ in G for every $\langle a, b \rangle \in po$.

The proof consists of two parts. First, we show that derivability of a judgment $\mathcal{R}; \mathcal{G} \Vdash \{P\} c \{Q\}$ allows us to construct annotations of executions of c, that are *locally valid* and *stable*, as defined below. Then, we prove that such annotations, for complete and coherent executions, must also be valid. Theorem 1 is obtained as a corollary.

Definition 13. An annotation Θ is *locally valid* for an execution G if the following hold for every $\langle a, b \rangle \in po$, where $P = \bigwedge_{\langle a', a \rangle \in po} \Theta(\langle a', a \rangle)$ and $Q = \Theta(\langle a, b \rangle)$:

- If $L(a) = \langle S \rangle$ and a is not initial then $P \vdash Q$.
- If $L(a) = \langle R, x, v \rangle$ then $P \wedge (x = v) \vdash Q$.
- If $L(a) = \langle W, x, v \rangle$ then either $P \vdash Q[v/x]$, or there is a unique node a' such that $\langle a', a \rangle \in po$, and we have $a' \in R$ and $P \wedge (loc(a') = val_r(a')) \vdash Q[v/x]$.
- If $L(a) = \langle U, x, v_r, v_w \rangle$ then $P \wedge (x = v_r) \vdash Q[v_w/x]$.

Definition 14. Let G be an execution. An edge $\langle b_1, b_2 \rangle \in po$ is called G-*before* an edge $\langle a_1, a_2 \rangle \in po$ if either $\langle b_1, b_2 \rangle = \langle a_1, a_2 \rangle$ or $\langle b_2, a_1 \rangle \in po^*$.

Definition 15. Let G be an execution. A node $c \in G$ *interferes* with $\langle a, b \rangle \in po$ in G for an annotation Θ if the following hold:

- $\langle c, a \rangle \notin po^*$ and $\langle b, c \rangle \notin po^*$ (c is *parallel* to $\langle a, b \rangle$ in G).
- For all $c' \in R$ with $\langle c', c \rangle \in po$ and $\langle c', a \rangle \notin po^*$, we have $\Theta(\langle a', b' \rangle) \wedge \Theta(\langle c', c \rangle) \nvdash loc(c') \neq val_r(c')$ for some $\langle a', b' \rangle \in po$ such that $\langle a', b' \rangle$ is G-before $\langle a, b \rangle$ and $\langle b', c' \rangle \notin po^*$.

Definition 16. An annotation Θ is *stable* for an execution G if the following hold for every $\langle a, b \rangle \in po$ and node $c \in W \cup U$ that interferes with $\langle a, b \rangle$ in G for Θ, where $R = \Theta(\langle a, b \rangle)$ and $P = \bigwedge_{\langle c', c \rangle \in po} \Theta(\langle c', c \rangle)$:

- If $L(c) = \langle W, x, v \rangle$ then $P \wedge R \vdash R[v/x]$.
- If $L(c) = \langle U, x, v_r, v_w \rangle$ then $P \wedge (x = v_r) \wedge R \vdash R[v_w/x]$.

Definition 17. A Hoare triple $\{P\} c \{Q\}$ is *safe* if for every $G \in SG; [\![c]\!]; SG$, there is an annotation Θ that is locally valid and stable for G, and assigns some assertion P', such that $P \vdash P'$, to the initial edge of G, and some assertion Q', such that $Q' \vdash Q$, to its terminal edge.

Theorem 2. *If $R; G \Vdash \{P\} c \{Q\}$ is derivable for some R, G, then $\{P\} c \{Q\}$ is safe.*

Proof (Outline). Call a judgment $R; G \Vdash \{P\} c \{Q\}$ *good* if for every execution $G \in SG; [\![c]\!]; SG$, there exists an annotation Θ that satisfies the conditions given in Definition 17, as well as the following ones:

- R covers Θ for G, i.e., for every $\langle a_1, a_2 \rangle \in po$, there exist $P_1 / C_1, \ldots, P_n / C_n \in R$ such that $\bigwedge P_i \dashv\vdash \Theta(\langle a_1, a_2 \rangle)$ and $\Theta(\langle b_1, b_2 \rangle) \vdash \bigwedge C_i$ for every $\langle b_1, b_2 \rangle \in po$ that is G-before $\langle a_1, a_2 \rangle$ (in particular, for $\langle b_1, b_2 \rangle = \langle a_1, a_2 \rangle$).
- G covers Θ for G, i.e., for every $a_2 \in W \cup U$, there exist an edge $\langle a_1, a_2 \rangle \in po$ and an assertion P', such that $\Theta(\langle a_1, a_2 \rangle) \vdash P'$, and one of the following holds:
  - $L(a_2) = \langle W, x, v \rangle$ and $\{P'\} x := v \in G$.
  - $L(a_2) = \langle W, x, v \rangle, L(a_1) = \langle R, y, v_y \rangle$, and $\{P'\} x := e(y) \in G$ for some expression $e(y)$ such that $[\![e]\!](v_y) = v$.
  - $L(a_2) = \langle W, x, v \rangle, L(a_1) = \langle R, z, v_z \rangle, \Theta(\langle a_1, a_2 \rangle) \vdash y = v_y$ for some $v_y \in \text{Val}$, and $\{P'\} x \overset{y,z}{:=} e(y, z) \in G$ for some expression $e(y, z)$ such that $[\![e]\!](v_y, v_z) = v$.
  - $L(a_2) = \langle U, x, v_r, v_w \rangle$ and $\{P'\} x \overset{at}{:=} e(x) \in G$ for some expression $e(x)$ such that $[\![e]\!](v_r) = v_w$.

Next, by induction on the derivation, one shows that every derivable judgment $R; G \Vdash \{P\} c \{Q\}$ is good, and so $\{P\} c \{Q\}$ is safe. The non-interference condition is needed for showing that two annotations of executions G_1 and G_2 can be joined to a stable annotation of the parallel composition of G_1 and G_2. □

It remains to establish the link from safety of a Hoare triple to its validity.

Theorem 3. *Let G be a complete coherent initialized execution. If an annotation Θ is locally valid and stable for G, then it is valid for G.*

The proof (given in the full version of this paper) requires analyzing the relations between states that are visible on consecutive edges and parallel edges in the RA memory model. An alternative equivalent formulation of coherence, based on a new "write-before" relation, is particularly useful for this task.

$$\{r = 0\}$$

$$
\begin{array}{l|l|l}
\begin{array}{l}
\{\top\} \\
w := 1; \\
\{\top\} \\
\textbf{while } r \neq 1 \textbf{ do} \\
\quad \textbf{skip} \\
\{r = 1\}
\end{array}
&
\begin{array}{l}
\{r = 0\} \\
r := w; \\
\{r = 1 \rightarrow w = 1\} \\
r := w \left\{
\begin{array}{ll}
w = 1 & \text{for } 1 \\
r \neq 1 & \text{otherwise}
\end{array}
\right. \\
\{\top\}
\end{array}
&
\begin{array}{l}
\textbf{Main non-interference checks:} \\
\quad r = 1 \text{ under } \{r = 0\} r := w \\
\quad r = 1 \text{ under } \{w = 1\} r := w \\
\quad r = 1 \text{ under } \{r \neq 1\} r := w \\
r = 1 \rightarrow w = 1 \text{ under } \{\top\} w := 1 \\
\quad w = 1 \text{ under } \{\top\} w := 1 \\
\text{All the checks are trivial.}
\end{array}
\end{array}
$$

$$\{r = 1\}$$

Fig. 9. Simplified RCU example illustrating the use of the stronger assignment rule

$$
\begin{array}{l}
\{\top\} \\
x := 2; \;\Big\|\; y := 2; \\
y := 1 \quad\;\; x := 1 \\
\{x \neq 2 \vee y \neq 2\}
\end{array}
$$

Fig. 10. Auxiliary variables are necessary under SC

$$\{x = \langle 0, 0\rangle\}$$

$$
\begin{array}{l|l}
\{x \in \{\langle 0, 0\rangle, \langle 1, 2\rangle\}\} &
\{x \in \{\langle 0, 0\rangle, \langle 1, 1\rangle\}\} \\
\quad\quad\text{at} & \quad\quad\text{at} \\
x := \langle x_{\text{fst}} + 1, x_{\text{snd}} + 1\rangle &
x := \langle x_{\text{fst}} + 1, x_{\text{snd}} + 2\rangle \\
\{x \in \{\langle 1, 1\rangle, \langle 2, 3\rangle\}\} &
\{x \in \{\langle 1, 2\rangle, \langle 2, 3\rangle\}\}
\end{array}
$$

$$\{x = \langle 2, 3\rangle\}$$

Fig. 11. Verification of the parallel increment example

3.2 A Stronger Assignment Rule

Consider the program shown in Fig. 9, which contains an idiom found in the RCU implementation (verified in the supplementary material). Thread II reads w and writes its value to r twice, while thread I sets w to 1 and then waits for r to become 1. The challenge is to show that after thread I reads $r = 1$, the value of r does not change; i.e. that $r = 1$ is stable under the $r := w$ assignments. For the first $r := w$ assignment, this is easy because its precondition is inconsistent with $r = 1$. For the second assignment, however, there is not much we can do. Stability requires us to consider *any value* for w readable at some point by thread I. Our idea is to do a case split on the value that w reads. If w reads the value 1, then it writes $r := 1$, and so $r = 1$ is unaffected. If w reads a different value, then from the assignment's precondition, we can derive $r \neq 1$, which contradicts the $r = 1$ assertion.

To support such case splits, we provide the following stronger assignment rule. For simplicity, we consider only assignments of the form $x := e(y)$.

$$
(\text{ASSN}_1')\; \frac{P \vdash Q[e(y)/x] \quad \{P \nearrow P, Q \nearrow (P \vee Q)\} \leq \mathcal{R} \qquad \text{For every } v \in \text{Val}: \quad P \wedge (y = v) \vdash P_v \quad \{P_v \nearrow P\} \leq \mathcal{R}}{\mathcal{R}; \{\{P_v\} x := e(y) \mid v \in \text{Val}\} \Vdash \{P\} x := e(y)\, \{Q\}}
$$

The previous assignment rule is an instance of this rule by taking $P_v = P$ for all v.

4 Discussion and Further Research

While OGRA's non-interference condition appears to be restrictive, we note that it is unsound for weaker memory models, such as C11's relaxed accesses because it can prove, e.g., message passing, see Fig. 6. We also observe that OGRA's non-interference check coincides with the standard OG one for assignments of values ($x := v$) and atomic assignments ($x \overset{\text{at}}{:=} e(x)$). Moreover, the non-interference check is irrelevant for assignments to variables that do not occur in the proof outlines of other threads. Therefore, standard OG (without auxiliary variables) is sound under RA provided that all $x := e(y)$ and $x \overset{y,z}{:=} e(y, z)$ assignments write to variables that do not appear in the proof outlines of other threads. Fig. 6 and 7 provide two such cases in point. In addition, this entails, for instance, that the program in Fig. 10 cannot be verified in standard OG without auxiliary variables, as $x = 2 \wedge y = 2$ is a possible outcome for this program under RA.

OG's auxiliary variables, in general, are unsound under weak memory because they can be used to record the exact thread interleavings and establish completeness under SC [12]. A simple form of auxiliary state, which we call *ghost values*, however, *is* sound. The idea is as follows: given a program c, one may choose a domain G of "ghost" values, together with a function $\alpha : \text{G} \to \text{Val}$, and obtain a program c' by substituting each expression $e(x_1, \ldots, x_n)$ in c by an expression $e'(x_1, \ldots, x_n)$ such that $\alpha(\llbracket e' \rrbracket(g_1, \ldots, g_n)) = \llbracket e \rrbracket(\alpha(g_1), \ldots, \alpha(g_n))$ for all $g_1, \ldots, g_n \in \text{G}$. The validity of $\{P'\} c' \{Q'\}$ entails the validity of $\{P\} c \{Q\}$, provided that the following hold:

- If a state satisfies P then some corresponding ghost state satisfies P';
- If a state does not satisfy Q then any corresponding ghost state does not satisfy Q';

where a ghost state $\sigma' : \text{Loc} \to \text{G}$ *corresponds* to a state $\sigma : \text{Loc} \to \text{Val}$ iff $\alpha(\sigma'(x)) = \sigma(x)$ for every $x \in \text{Loc}$. This solution suffices, for instance, to reason about the parallel increment example, as shown in Fig. 11. There we took $\text{G} = \text{Val} \times \mathbb{N}$, with α being the first projection mapping. The second component tracks which of the assignments has already happened (0: none, 1: the first thread, 2: the second thread, otherwise: both). As a result, we obtain the validity of $\{x = 0\} x \overset{\text{at}}{:=} x + 1 \parallel x \overset{\text{at}}{:=} x + 1 \{x = 2\}$.

Analyzing soundness of other restricted forms of auxiliary variables is left for future work. Such extensions seem to be a prerequisite for obtaining a program logic that is both sound and complete under RA. Automation of proof search is another future goal. Our initial experiments show that, at least for the examples in this paper, HSF [5] is successful in automatically finding proofs in OGRA.

Acknowledgments. We would like to thank the ICALP'15 reviewers for their feedback. This work was supported by EC FET project ADVENT (308830).

References

1. Batty, M., Owens, S., Sarkar, S., Sewell, P., Weber, T.: Mathematizing C++ concurrency. In: POPL 2011, pp. 55–66. ACM (2011)
2. Cohen, E.: Coherent causal memory (2014). CoRR abs/1404.2187
3. Desnoyers, M., McKenney, P.E., Stern, A.S., Dagenais, M.R., Walpole, J.: User-level implementations of read-copy update. IEEE Trans. Parallel Distrib. Syst. **23**(2), 375–382 (2012)
4. Ferreira, R., Feng, X., Shao, Z.: Parameterized memory models and concurrent separation logic. In: Gordon, A.D. (ed.) ESOP 2010. LNCS, vol. 6012, pp. 267–286. Springer, Heidelberg (2010)

5. Grebenshchikov, S., Lopes, N.P., Popeea, C., Rybalchenko, A.: Synthesizing software verifiers from proof rules. In: PLDI 2012, pp. 405–416. ACM (2012)
6. ISO/IEC 14882:2011: Programming language C++ (2011)
7. Lamport, L.: How to make a multiprocessor computer that correctly executes multiprocess programs. IEEE Trans. Computers **28**(9), 690–691 (1979)
8. Maranget, L., Sarkar, S., Sewell, P.: A tutorial introduction to the ARM and POWER relaxed memory models (2012). http://www.cl.cam.ac.uk/~pes20/ppc-supplemental/test7.pdf
9. Owens, S., Sarkar, S., Sewell, P.: A better x86 memory model: x86-TSO. In: Berghofer, S., Nipkow, T., Urban, C., Wenzel, M. (eds.) TPHOLs 2009. LNCS, vol. 5674, pp. 391–407. Springer, Heidelberg (2009)
10. Owicki, S., Gries, D.: An axiomatic proof technique for parallel programs I. Acta Informatica **6**(4), 319–340 (1976)
11. Owicki, S., Gries, D.: Verifying properties of parallel programs: An axiomatic approach. Commun. ACM **19**(5), 279–285 (1976)
12. Owicki, S.S.: Axiomatic Proof Techniques for Parallel Programs. Ph.D. thesis, Cornell University, Ithaca, NY, USA (1975)
13. Ridge, T.: A rely-guarantee proof system for x86-TSO. In: Leavens, G.T., O'Hearn, P., Rajamani, S.K. (eds.) VSTTE 2010. LNCS, vol. 6217, pp. 55–70. Springer, Heidelberg (2010)
14. Sieczkowski, F., Svendsen, K., Birkedal, L., Pichon-Pharabod, J.: A separation logic for fictional sequential consistency. In: Vitek, J. (ed.) ESOP 2015. LNCS, vol. 9032, pp. 736–761. Springer, Heidelberg (2015)
15. Tassarotti, J., Dreyer, D., Vafeiadis, V.: Verifying read-copy-update in a logic for weak memory. In: PLDI 2015. ACM (2015)
16. Turon, A., Vafeiadis, V., Dreyer, D.: GPS: Navigating weak memory with ghosts, protocols, and separation. In: OOPSLA 2014, pp. 691–707. ACM (2014)
17. Vafeiadis, V., Balabonski, T., Chakraborty, S., Morisset, R., Nardelli, F.Z.: Common compiler optimisations are invalid in the C11 memory model and what we can do about it. In: POPL 2015, pp. 209–220. ACM (2015)
18. Vafeiadis, V., Narayan, C.: Relaxed separation logic: A program logic for C11 concurrency. In: OOPSLA 2013, pp. 867–884. ACM (2013)

On the Coverability Problem for Pushdown Vector Addition Systems in One Dimension

Jérôme Leroux[1], Grégoire Sutre[1], and Patrick Totzke[2]([✉])

[1] LaBRI, Univ. Bordeaux & CNRS, UMR 5800, Talence, France
[2] Department of Computer Science, University of Warwick, Coventry, UK
p.totzke@warwick.ac.uk

Abstract. Does the trace language of a given vector addition system (VAS) intersect with a given context-free language? This question lies at the heart of several verification questions involving recursive programs with integer parameters. In particular, it is equivalent to the coverability problem for VAS that operate on a pushdown stack. We show decidability in dimension one, based on an analysis of a new model called grammar-controlled vector addition systems.

1 Introduction

Pushdown systems are a well-known and natural formalization of recursive programs. Vector addition systems (VAS) are widely used to model concurrent systems and programs with integer variables. Pushdown vector addition systems (pushdown VAS) combine the two: They are VAS extended with a pushdown stack and allow to model, for instance, asynchronous programs [6] and, more generally, programs with recursion and integer variables.

Despite the model's relevance for automatic program verification, most classical model-checking problems are so far only partially solved. Termination and boundedness are decidable but their complexity is open [12]. Coverability and reachability are known to be TOWER-hard [9], but their decidability is open. In fact, reachability and the seemingly simpler coverability problem are essentially the same for pushdown VAS: there is a simple logarithmic-space reduction from reachability to coverability that only adds one extra dimension.

Contributions. Our main result is that coverability is decidable for 1-dimensional pushdown VAS. We work with a new grammar-based model called grammar-controlled vector addition systems (GVAS), which amounts to VAS restricted to firing sequences defined by a context-free grammar. In dimension one, this model corresponds to two-stack pushdown systems where one of the two stacks uses a single stack symbol. To prove our main result, we show that it is enough to check finitely many potential certificates of coverability. The latter are parse trees of the context-free grammar annotated with counter information from the 1-dimensional VAS. We truncate these annotated parse trees thanks to an

This work was partially supported by ANR project REACHARD (ANR-11-BS02-001).

M.M. Halldórsson et al. (Eds.): ICALP 2015, Part II, LNCS 9135, pp. 324–336, 2015.
DOI: 10.1007/978-3-662-47666-6_26

analysis of the asymptotic behavior of the summary function induced by the 1-dimensional GVAS. Asymptotically-linear summary functions are shown to be effectively Presburger-definable, which makes the above truncation effective.

Related Work. This paper continues a line of research that investigates the limitations of extending VAS while preserving the decidability of important verification questions, such as reachability, coverability and boundedness.

The coverability and boundedness problems for ordinary VAS are long known to be ExpSpace-complete [15,17] and reachability is decidable [8,11,16]. In recent years, several extensions of VAS have been considered with respect to decidability and complexity of reachability problems. For instance, Reinhardt [18] showed that reachability remains decidable for VAS in which one dimension can be tested for zero. *Branching VAS* introduce split-transitions and can be interpreted as bottom-up or top-down tree acceptors. *Alternating VAS* add a limited form of alternation where only one player is affected by the counters. Coverability and boundedness in these models are 2-ExpTime-complete [4,5], reachability is Tower-hard for branching and undecidable for alternating VAS [4,10].

Closer to this paper is the work of Bouajjani, Habermehl and Mayr [3], who study a model called BPA($\mathbb{Z}$). These are context-free grammars where nonterminals carry an integer parameter that can be evaluated and passed on when applying a production rule. They show how to compute a symbolic representation of the reachability set. Their formalism, like the 1-dimensional GVAS considered here, can model recursive programs with one integer variable. But while BPA($\mathbb{Z}$) allows arbitrary Presburger-definable operations on the variable, it cannot model return values.

Atig and Ganty [1] also study the context-free restriction of the reachability relation in vector addition systems. Instead of restricting the dimension of the VAS, they restrict the context-free language and show that reachability is decidable for the subclass of indexed context-free languages.

Outline. We first recall some background and notation for context-free grammars. Section 3 formally introduces grammar-controlled vector addition systems, their coverability problem and the required technology to solve it in dimension one. In Section 4, we show the existence of small certificates. These are subsequently proved to be recursive in two steps. Section 5 shows that, for so-called thin GVAS, the step relation is effectively Presburger-definable. Then, summary functions are shown to be computable by reduction to the thin case in Section 6.

2   Preliminaries

We let $\overline{\mathbb{R}} \stackrel{\text{def}}{=} \mathbb{R} \cup \{-\infty, +\infty\}$ denote the extended real number line and use the standard extensions of $+$ and $\leq$ to $\overline{\mathbb{R}}$. Recall that $(\overline{\mathbb{R}}, \leq)$ is a complete lattice. $\overline{\mathbb{Z}} \stackrel{\text{def}}{=} \mathbb{Z} \cup \{-\infty, +\infty\}$ and $\overline{\mathbb{N}} \stackrel{\text{def}}{=} \mathbb{N} \cup \{-\infty, +\infty\}$ denote the (complete) sublattices of extended integers and extended natural numbers, respectively.[1]

---

[1] Our extension of $\mathbb{N}$ contains $-\infty$ for technical reasons.

Words. Let A^* be the set of all finite *words* over the alphabet A. The *empty word* is denoted by ε. We write $|w|$ for the *length* of a word w in A^* and $w^k \stackrel{\text{def}}{=} ww \cdots w$ for its k-fold concatenation. The *prefix* partial order $\preceq$ over words is defined by $u \preceq v$ if $v = uw$ for some word w. We write $u \prec v$ if u is a proper prefix of v. A *language* is a subset $L \subseteq A^*$. A language L is said to be *prefix-closed* if $u \preceq v$ and $v \in L$ implies $u \in L$.

Trees. A *tree* T is a finite prefix-closed subset of $\mathbb{N}^*$ satisfying the property that if tj is in T then ti in T for all $i < j$. Elements of T are called *nodes*. Its *root* is the empty word ε. An *ancestor* of a node t is a prefix $s \preceq t$. A *child* of a node t in T is a node tj in T with j in $\mathbb{N}$. A node is called a *leaf* if it has no child, and is said to be *internal* otherwise. The *size* of a tree T is its cardinal $|T|$, its *height* is the maximal length $|t|$ for any of its nodes $t \in T$.

Context-Free Grammars. A *context-free grammar* is a triple $G = (V, A, R)$, where V and A are disjoint finite sets of *nonterminal* and *terminal* symbols, and $R \subseteq V \times (V \cup A)^*$ is a finite set of *production rules*. The *degree* of G is $\delta^G \stackrel{\text{def}}{=} \max\{|\alpha| \mid (X, \alpha) \in R\}$. We write
$$X \vdash \alpha_1 \mid \alpha_2 \mid \ldots \mid \alpha_k$$
to denote that $(X, \alpha_1), \ldots, (X, \alpha_k) \in R$. For all words $w, w' \in (V \cup A)^*$, the grammar admits a *derivation step* $w \Longrightarrow w'$ if there exist two words u, v in $(V \cup A)^*$ and a production rule (X, α) in R such that $w = uXv$ and $w' = u\alpha v$. Let $\stackrel{*}{\Longrightarrow}$ denote the reflexive and transitive closure of $\Longrightarrow$. The *language* of a word w in $(V \cup A)^*$ is the set $L_w^G \stackrel{\text{def}}{=} \{z \in A^* \mid w \stackrel{*}{\Longrightarrow} z\}$. A nonterminal X is said to be *derivable* from a word $w \in (V \cup A)^*$ if there exists $u, v \in (V \cup A)^*$ such that $w \stackrel{*}{\Longrightarrow} uXv$. A nonterminal $X \in V$ is called *productive* if $L_X^G \neq \emptyset$.

Parse Trees. A *parse tree* for a context-free grammar $G = (V, A, R)$ is a tree T equipped with a labeling function $sym : T \to (V \cup A \cup \{\varepsilon\})$ such that R contains the production rule $sym(t) \vdash sym(t0) \cdots sym(tk)$ for every internal node t with children $t0, \ldots, tk$. In addition, each leaf $t \neq \varepsilon$ with $sym(t) = \varepsilon$ is the only child of its parent. Notice that $sym(t) \in V$ for every internal node t. A parse tree is called *complete* when $sym(t) \in (A \cup \{\varepsilon\})$ for every leaf t. The *yield* of a parse tree (T, sym) is the word $sym(t_1) \cdots sym(t_\ell)$ where $t_1, \ldots, t_\ell$ are the leaves of T in lexicographic order (informally, from left to right). Observe that $S \stackrel{*}{\Longrightarrow} w$, where $S = sym(\varepsilon)$ is the label of the root and w is the yield. Conversely, a parse tree with root labeled by S and yield w can be associated to any derivation $S \stackrel{*}{\Longrightarrow} w$.

3   Grammar-Controlled Vector Addition Systems

We first recall the main concepts of vector addition systems. Fix $k \in \mathbb{N}$. A k-dimensional *vector addition system* (shortly, k-*VAS*) is a finite set $\boldsymbol{A} \subseteq \mathbb{Z}^k$ of *actions*. Its operational semantics is given by the binary *step* relations $\stackrel{\boldsymbol{a}}{\longrightarrow}$ over $\mathbb{N}^k$, where $\boldsymbol{a}$ ranges over $\boldsymbol{A}$, defined by $\boldsymbol{c} \stackrel{\boldsymbol{a}}{\longrightarrow} \boldsymbol{d}$ if $\boldsymbol{d} = \boldsymbol{c} + \boldsymbol{a}$. The step relations are extended to words and languages as expected: $\xrightarrow{\varepsilon}$ is the identity, $\xrightarrow{z\boldsymbol{a}} \overset{\text{def}}{=} \xrightarrow{\boldsymbol{a}} \circ \xrightarrow{z}$ for $z \in \boldsymbol{A}^*$ and $\boldsymbol{a} \in \boldsymbol{A}$, and $\xrightarrow{L} \overset{\text{def}}{=} \bigcup_{z \in L} \xrightarrow{z}$ for $L \subseteq \boldsymbol{A}^*$. For every word $z = \boldsymbol{a}_1 \cdots \boldsymbol{a}_k$ in $\boldsymbol{A}^*$, we let $\sum z$ denote the sum $\boldsymbol{a}_1 + \cdots + \boldsymbol{a}_k$. Notice that $\boldsymbol{c} \xrightarrow{z} \boldsymbol{d}$ implies $\boldsymbol{d} - \boldsymbol{c} = \sum z$, for every $\boldsymbol{c}, \boldsymbol{d} \in \mathbb{N}^k$.

The *VAS reachability problem* asks, given a k-VAS $\boldsymbol{A}$ and vectors $\boldsymbol{c}, \boldsymbol{d} \in \mathbb{N}^k$, whether $\boldsymbol{c} \xrightarrow{\boldsymbol{A}^*} \boldsymbol{d}$. This problem is known to be ExpSpace-hard [15], but no upper bound has been established yet. The *VAS coverability problem* asks, given a k-VAS $\boldsymbol{A}$ and vectors $\boldsymbol{c}, \boldsymbol{d} \in \mathbb{N}^k$, whether $\boldsymbol{c} \xrightarrow{\boldsymbol{A}^*} \boldsymbol{d}'$ for some vector $\boldsymbol{d}' \geq \boldsymbol{d}$. This problem is known to be ExpSpace-complete [15,17].

Definition 3.1 (GVAS). *A k-dimensional grammar-controlled vector addition system (shortly, k-GVAS) is a context-free grammar $G = (V, \boldsymbol{A}, R)$ with $\boldsymbol{A} \subseteq \mathbb{Z}^k$.*

We give the semantics of GVAS by extending the binary step relations of VAS to words over $V \cup \boldsymbol{A}$. Formally, for every word $w \in (V \cup \boldsymbol{A})^*$, we let $\xrightarrow{w} \overset{\text{def}}{=} \xrightarrow{L}$ where $L = L_w^G$ is the language of w. The *GVAS reachability problem* asks, given a k-GVAS $G = (V, \boldsymbol{A}, R)$, a nonterminal $S \in V$ and two vectors $\boldsymbol{c}, \boldsymbol{d} \in \mathbb{N}^k$, whether $\boldsymbol{c} \xrightarrow{S} \boldsymbol{d}$. The *GVAS coverability problem* asks, given the same input, whether $\boldsymbol{c} \xrightarrow{S} \boldsymbol{d}'$ for some vector $\boldsymbol{d}' \geq \boldsymbol{d}$. These problems can equivalently be rephrased in terms of VAS that have access to a pushdown stack, called *stack VAS* in [9] and *pushdown VAS* in [12]. Lazić [9] showed a Tower lower bound for these two problems, by simulating bounded Minsky machines. Their decidability remains open. As remarked in [9], GVAS reachability can be reduced to GVAS coverability. Indeed, a simple "budget" construction allows to reduce, in logarithmic space, the reachability problem for k-GVAS to the coverability problem for $(k + 1)$-GVAS. This induces a hierarchy of decision problems, consisting of, alternatingly, coverability and reachability for growing dimension. The decidability of all these problems is open. This motivates the study of the most simple case: the coverability problem in dimension one, which is the focus of this paper. Our main contribution is the following result.

Theorem 3.2. *The coverability problem is decidable for 1-GVAS.*

For the remainder of the paper, we restrict our attention to the dimension one, and shortly write GVAS instead of 1-GVAS. Every GVAS can be effectively normalized, by removing non-productive nonterminals, replacing terminals $a \in \mathbb{Z}$ by words over the alphabet $\{-1, 0, 1\}$, and enforcing, through zero padding (since $\xrightarrow{0}$ is the identity relation), that $|\alpha| \geq 2$ for some production rule $X \vdash \alpha$. So in order to simplify our proofs, we consider w.l.o.g. only GVAS of this simpler form.

Assumption. We restrict our attention to GVAS $G = (V, A, R)$ where every $X \in V$ is productive, where $A = \{-1, 0, 1\}$, and of degree $\delta^G \geq 2$.

We associate to a GVAS G and a word $w \in (V \cup A)^*$ the *displacement* $\Delta_w^G \in \overline{\mathbb{Z}}$ and the *summary* function $\sigma_w^G : \overline{\mathbb{N}} \to \overline{\mathbb{N}}$ defined by

$$\Delta_w^G \stackrel{\text{def}}{=} \sup\{\textstyle\sum z \mid z \in L_w^G\} \qquad \sigma_w^G(n) \stackrel{\text{def}}{=} \sup\{d \mid \exists c \le n : c \xrightarrow{w} d\}$$

Informally, Δ_w^G is the "best shift" achievable by a word in L_w^G, and $\sigma_w^G(n)$ gives the "largest" number that is reachable via some word in L_w^G starting from n or below. When no such number exists, $\sigma_w^G(n)$ is $-\infty$ (recall that $\sup \emptyset = -\infty$). Since all nonterminals are productive, the language L_w^G is not empty. Therefore, $\Delta_w^G > -\infty$ and $\sigma_w^G(n) > -\infty$ for some $n \in \mathbb{N}$.

Remark 3.3 (Monotonicity). For every $w \in (V \cup A)^*$ and $c, d, e \in \mathbb{N}$, $c \xrightarrow{w} d$ implies $c + e \xrightarrow{w} d + e$. Consequently, $\sigma_w^G(n + e) \ge \sigma_w^G(n) + e$ holds for every $w \in (V \cup A)^*$, $n \in \overline{\mathbb{N}}$ and $e \in \mathbb{N}$.

A straightforward application of Parikh's theorem shows that Δ_w^G is effectively computable from G and w. We will provide in Section 6 an effective characterization of σ_w^G when the displacement Δ_w^G is finite. In order to characterize functions σ_w^G where the displacement Δ_w^G is infinite, it will be useful to consider the *ratio* of w, defined as

$$\lambda_w^G \stackrel{\text{def}}{=} \liminf_{n \to +\infty} \frac{\sigma_w^G(n)}{n}$$

Notice that $\lambda_w^G \ge 1$. This fact follows from Theorem 3.3 and the observation that $\sigma_w^G(n) > -\infty$ for some $n \in \mathbb{N}$. From now on, we just write $L_w, \delta, \Delta_w, \sigma_w$ and λ_w when G is clear from the context.

Example 3.4. Multiplication by 2 can be expressed as a summary function using the GVAS with production rules $S \vdash -1\, S\, 1\, 1 \mid \varepsilon$. Indeed, for every c,

$$c \xrightarrow{S} d \iff \exists n \in \mathbb{N} : c \xrightarrow{(-1)^n (11)^n} d$$

$$\iff \exists n \le c : c \xrightarrow{(-1)^n} c - n \xrightarrow{(11)^n} c + n = d \iff c \le d \le 2c$$

Therefore, $\sigma_S(n) = 2n$ for every $n \in \mathbb{N}$. Observe that $\Delta_S = +\infty$ and $\lambda_S = 2$. □

Example 3.5. The Ackermann functions $A_m : \mathbb{N} \to \mathbb{N}$, for $m \in \mathbb{N}$, are defined by induction for every $n \in \mathbb{N}$ by:

$$A_m(n) \stackrel{\text{def}}{=} \begin{cases} n + 1 & \text{if } m = 0 \\ A_{m-1}^{n+1}(1) & \text{if } m > 0 \end{cases}$$

These functions are expressible as summary functions for the GVAS with nonterminals $X_0, \ldots, X_m$ and with production rules $X_0 \vdash 1$ and $X_i \vdash -1\, X_i\, X_{i-1} \mid 1 X_{i-1}$ for $1 \le i \le m$. It is routinely checked that $\sigma_{X_m}(n) = A_m(n)$ for every $n \in \mathbb{N}$. Notice also that $\lambda_{X_0} = 1$, $\lambda_{X_1} = 2$, and $\lambda_{X_m} = +\infty$ for every $m \ge 2$. □

Lemma 3.6. *For every two words $u, v \in (V \cup A)^*$, the following properties hold:*

1. $\Delta_{uv} = \Delta_u + \Delta_v$ *and* $\sigma_{uv} = \sigma_v \circ \sigma_u$.
2. *If* $u \overset{*}{\Longrightarrow} v$ *then* $\Delta_u \geq \Delta_v$, $\lambda_u \geq \lambda_v$, *and* $\sigma_u(n) \geq \sigma_v(n)$ *for all* $n \in \overline{\mathbb{N}}$.

An equivalent formulation of the coverability problem is the question whether $\sigma_S(c) \geq d$ holds, given a nonterminal $S \in V$ and two numbers $c, d \in \mathbb{N}$. We solve this problem by exhibiting small certificates for $\sigma_S(c) \geq d$, that take the form of (suitably truncated) annotated parse trees.

4   Small Coverability Certificates

To solve the coverability problem, we annotate parse trees in a way that is consistent with the summary functions. A *flow tree* for a GVAS G is a parse tree (T, sym) for G equipped with two functions $in, out : T \to \mathbb{N}$, assigning an *input* and an *output* value to each node, and satisfying, for every node $t \in T$, the following *flow conditions*:

1. If t is internal with children $t0, \ldots, tk$, then $in(t0) \leq in(t)$, $out(t) \leq out(tk)$, and $in(t(j+1)) \leq out(tj)$ for every $j = 0, \ldots, k-1$.
2. If t is a leaf then $out(t) \leq \sigma_{sym(t)}(in(t))$.

We shortly write $t : c\#d$ to mean that $(in(t), sym(t), out(t)) = (c, \#, d)$. A flow tree is called *complete* when the underlying parse tree is complete, i.e., when $sym(t) \in (A \cup \{\varepsilon\})$ for every leaf t. The following lemmas state useful properties of flow trees that can be shown using the flow conditions and the monotonicity of summary functions (see Theorem 3.3). A consequence is that $\sigma_S(c) \geq d$ holds if, and only if, there exists a complete flow tree with root $\varepsilon : cSd$.

Lemma 4.1. *It holds that* $\sigma_\#(c) \geq d$ *for every node* $t : c\#d$ *of a flow tree.*

Lemma 4.2. *Let* $S \in V$ *and* $c, d \in \mathbb{N}$. *If* $\sigma_S(c) \geq d$ *then there exists a complete flow tree with root* $\varepsilon : bSe$ *such that* $b \leq c$ *and* $e \geq d$.

We will need to compare flow trees. Let the *rank* of a flow tree (T, sym, in, out) be the pair $(|T|, \sum_{t \in T} in(t) + out(t))$. The lexicographic order lex over $\mathbb{N}^2$ is used to compare ranks of flow trees. A complete flow tree (T, sym, in, out) is called *optimal* if there exists no complete flow tree (T', sym', in', out') of strictly smaller rank such that $in'(\varepsilon) \leq in(\varepsilon)$, $sym'(\varepsilon) = sym(\varepsilon)$, and $out'(\varepsilon) \geq out(\varepsilon)$. Optimal flow trees enjoy the following important properties, stated formally below. Firstly, they are *tight*, meaning that the inequalities in the first flow condition are in fact equalities. Secondly, they are *balanced*, meaning that the input value of each node is never too large compared to its output value.

Lemma 4.3. *For every internal node* t *in an optimal complete flow tree, we have* $in(t0) = in(t)$, $in(t1) = out(t0)$, $\ldots$, $in(tk) = out(t(k-1))$, *and* $out(t) = out(tk)$, *where* $t0, \ldots, tk$ *are the children of* t.

Lemma 4.4. *For every node t in an optimal complete flow tree, it holds that* $in(t) \leq out(t) + \delta^{|V|}$.

Next, we show how to truncate flow trees while preserving enough information to decide that the *in* and *out* labelings satisfy the flow conditions. Our truncation is justified by the following lemma.

Lemma 4.5. *Let* $X \in V$ *and* $n \in \mathbb{N}$. *If* $\lambda_X = +\infty$ *and there is a derivation* $X \xrightarrow{*} uXv$ *such that* $\sigma_u(n) > n$, *then it holds that* $\sigma_X(n) = +\infty$.

Definition 4.6 (Certificates). *A* certificate *is a flow tree* (T, sym, in, out) *in which every leaf* t *with* $\lambda_{sym(t)} = +\infty$ *has a proper ancestor* $s \prec t$ *such that* $sym(s) = sym(t)$ *and* $in(s) < in(t)$.

Notice that every complete flow tree is a certificate. We now prove the existence of small certificates. Let $S \in V$ and $c, d \in \mathbb{N}$ such that $\sigma_S(c) \geq d$. We introduce the set $\mathcal{T}$ of all complete flow trees with root $\varepsilon : bSe$ satisfying $b \leq c$ and $e \geq d$. By Theorem 4.2, the set $\mathcal{T}$ is not empty. Let us pick (T, sym, in, out) in $\mathcal{T}$ among those of least rank. By definition, the root ε of T satisfies $in(\varepsilon) \leq c$ and $out(\varepsilon) = d$. Notice that the complete flow tree T is optimal. Let us introduce the set U of all nodes $t \in T$ such that every proper ancestor $s \prec t$ satisfies the following condition:

$$\text{For every ancestor } r \preceq s, \;\; sym(r) = sym(s) \implies in(r) \geq in(s) \qquad (1)$$

By definition, the set U is a nonempty and prefix-closed subset of T. The following fact derives from Theorem 4.1 and the property that T is a complete flow tree.

Fact 4.7. The tree U, equipped with the restrictions to U of the functions *sym*, *in* and *out*, is a certificate.

Our next step is to bound the height of U as well as the input and output values of its nodes. We will use the following properties, that are easily derived from the definition of U, the optimality of T, and Theorems 4.3 and 4.4.

Fact 4.8. Let r and s be nodes in U such that $r \prec s$.

1. If s is internal in U and $sym(r) = sym(s)$ then $out(s) < out(r)$, and
2. If s is a child of r then $out(s) \leq out(r) + (\delta - 1)\delta^{|V|}$.

Consider a leaf t in U. For each i in $\{0, \ldots, |t|\}$, let t_i denote the unique prefix $t_i \preceq t$ with length $|t_i| = i$, and let $(\#_i, d_i) = (sym(t_i), out(t_i))$. Note that $d_0 = out(\varepsilon) = d$. Fact 4.8 entails that for every i, j with $0 \leq i, j < |t|$,

$$d_{i+1} \leq d_i + \delta^{|V|+1} \qquad \text{and} \qquad (i < j \wedge \#_i = \#_j) \implies d_i > d_j \qquad (2)$$

Let $m_i = \max\{d_0, \ldots, d_i\}$ for all $i \in \{0, \ldots, |t|\}$. According to Equation (2), increasing pairs $m_i < m_{i+1}$ may occur in the sequence $m_0, \ldots, m_{|t|}$ only when

$\#_{i+1} \notin \{\#_0, \ldots, \#_i\}$ or $i + 1 = |t|$. So there are at most $|V|$ such increasing pairs. Moreover, for each increasing pair $m_i < m_{i+1}$, the increase $m_{i+1} - m_i$ is bounded by $\delta^{|V|+1}$. We derive that $d_i \leq m_{|t|} \leq d + |V| \cdot \delta^{|V|+1} < d + \delta^{2|V|+1}$ for all i with $0 \leq i \leq |t|$, since $\delta \geq 2$ by assumption. It follows from Equation (2) that each nonterminal in V appears at most $d + \delta^{2|V|+1}$ times in the sequence $(\#_i)_{0 \leq i < |t|}$. By the pigeonhole principle, we get that $|t| \leq |V| \cdot (d + \delta^{2|V|+1})$. We have thus shown that for every node $t \in U$,

$$|t| \leq d \cdot |V| + \delta^{3|V|+1} \qquad \text{and} \qquad in(t) + out(t) \leq 2d + \delta^{2|V|+3} \qquad (3)$$

This concludes the proof of the "only if" direction of the following proposition. The "if" direction follows from Lemma 4.1, since every certificate is a flow tree.

Proposition 4.9. *For every $S \in V$ and $c, d \in \mathbb{N}$, it holds that $\sigma_S(c) \geq d$ if, and only if, there exists a certificate with root $\varepsilon : bSd$ for some $b \leq c$ and whose nodes t satisfy Equation* (3).

The above proposition leads to a simple procedure to solve the coverability problem, as we only need to enumerate finitely many potential certificates. Checking whether an annotated parse tree is a certificate reduces to (a) the question whether a given nonterminal X has an infinite ratio, and (b) the coverability question $\sigma_X(c) \geq d$ for nonterminals X with finite ratio. Both questions will be shown to be decidable in Section 6 by reduction to the subclass of thin GVAS, which is the focus of the next section.

5 Semilinearity of the Step Relations for Thin GVAS

We turn to reachability relations in a particular subclass of GVAS called *thin*. A context-free grammar is said to be *thin*[2] if $\alpha \in A^* V A^*$ for every production rule $X \vdash \alpha$ such that X is derivable from α. Recall that *Presburger arithmetic* is the first-order theory of the natural numbers with addition. It is well-known that *semilinear sets* coincide with the sets definable in Presburger arithmetic [7].

Theorem 5.1. *For every nonterminal symbol S of a thin GVAS, the relation $\xrightarrow{S}$ is effectively definable in Presburger arithmetic.*

Our argument goes by a reduction to the reachability problem for 2-dimensional vector addition systems, and uses the following result.

Theorem 5.2 ([13]). *Let A be a 2-VAS and $\Pi \subseteq A^*$ be a regular language over its actions. The relation $\xrightarrow{\Pi}$ is effectively definable in the Presburger arithmetic.*

---

[2] Thinness entails that for any derivation $S \Rightarrow^* w$, the number of nonterminals in w is bounded by $\delta^{|V|}$. This entails that parse trees of thin GVAS are of bounded width. Thin GVAS are thus a subclass of the *finite-index* grammars of [1].

Let us call a GVAS $G = (V, A, R)$ *simple* if for every production rule $X \vdash \alpha$, either X is not derivable from α, or $\alpha \in AVA$. Clearly, every simple GVAS is thin. Conversely, every thin GVAS can be transformed into an equivalent simple GVAS by replacing production rules in $V \times A^*VA^*$ by finitely many new rules in $V \times AVA$. See the full paper [14] for details. Consequently, it suffices to show the claim of Theorem 5.1 for simple GVAS only.

We show by induction on $|V|$ that $\xrightarrow{S}$ is effectively definable in Presburger arithmetic for every simple thin GVAS $G = (V, A, R)$, and for every nonterminal $S \in V$. Naturally, if $|V|$ is empty the proof is immediate. Assume the induction is proved for a number $h \in \mathbb{N}$, and let us consider a simple thin GVAS $G = (V, A, R)$ with $|V| = h + 1$, and a nonterminal $S \in V$.

Notice that $A \overset{\text{def}}{=} \{-1, 0, 1\}^2$ is a vector addition system. We consider the finite, directed graph with set of nodes V that contains an $(a, -b)$-labeled edge from X to Y for every production rule $X \vdash aYb$ in R. To each nonterminal $X \in V$, we associate the regular language Π_X of words recognized by this finite graph starting from S and reaching X. By Theorem 5.2, $\xrightarrow{\Pi_X}$, the regular restriction of the reachability set of A, is effectively definable in Presburger arithmetic.

As a next ingredient, let Γ_X be the finite set of words $\alpha \in (V \cup A)^*$ such that $X \vdash \alpha$ is a production rule and X is not derivable from α. We observe that L_α^G is equal to the language of α in the simple grammar G', obtained from G by removing the nonterminal X and all production rules where X occurs. By induction, and since $\xrightarrow{a}$ are trivially Presburger-definable for terminals $a \in A$, we deduce that $\xrightarrow{\alpha}$ is effectively Presburger-definable as a composition of Presburger relations. Because Γ_X is finite, we deduce that $\xrightarrow{\Gamma_X} = \bigcup_{\alpha \in \Gamma_X} \xrightarrow{\alpha}$, is definable in the Presburger arithmetic as a finite disjunction of Presburger relations.

This following Lemma 5.3 concludes Theorem 5.1.

Lemma 5.3. *For for all $c, d \in \mathbb{N}$, $c \xrightarrow{S} d$ if, and only if, the following relation holds:*

$$\phi_S(c, d) \overset{\text{def}}{=} \bigvee_{X \in V} \exists c', d' \in \mathbb{N} \quad (c, d) \xrightarrow{\Pi_X} (c', d') \wedge c' \xrightarrow{\Gamma_X} d' \qquad (4)$$

Proof. Assume that $c \xrightarrow{S} d$. It means that there exists $w \in L_S$ such that $c \xrightarrow{w} d$. Since $w \in A^*$, we deduce that a sequence of derivation steps from S that produces w must necessarily derive at some point a nonterminal symbol X with a production rule $X \vdash \alpha$ such that $\alpha \in A^*$, and in particular $\alpha \in \Gamma_X$. By considering the first time a derivation step $X \Rightarrow$ with $\alpha \in \Gamma_X$ occurs, we deduce a sequence $X_0, \ldots, X_k$ of nonterminal symbols with $X_0 = S$, a sequence $r_1, \ldots, r_k$ of production rules $r_j \in R$ of the form $X_{j-1} \vdash a_j X_j b_j$ with $a_j, b_j \in A$, a production rule $r_{k+1} \in R$ of the form $X_k \vdash \alpha$ where $\alpha \in \Gamma_{X_k}$, and a word $w' \in L_\alpha$ such that $w = a_1 \ldots a_k w' b_k \ldots b_1$. Since $c \xrightarrow{w} d$, it follows that there exist $c', d' \in \mathbb{N}$ such that $c \xrightarrow{a_1 \ldots a_k} c' \xrightarrow{w'} d' \xrightarrow{b_k \ldots b_1} d$. Thus $(c, d) \xrightarrow{\pi} (c', d')$

with $\pi \overset{\text{def}}{=} (a_1, -b_1) \ldots (a_k, -b_k)$. It follows that $\phi_S(c, d)$ holds. Conversely, if $\phi_S(c, d)$ holds, by reversing the previous proof steps, if follows that $c \overset{S}{\longrightarrow} d$. A detailed proof is given in the full paper [14]. $\qquad \square$

6   Computation of Summaries for Bounded Ratios

In this section, we show that the summary function σ_X is effectively computable when the ratio λ_X is finite. In addition, the question whether λ_X is finite is shown to be decidable. These results are ultimately obtained by reduction to the thin GVAS case. We first consider nonterminals with finite displacements.

The next lemma follows from the observation that if the maximal displacement of a nonterminal is finite, then it can already be achieved by a short word.

Lemma 6.1. *Let $S \in V$ be a nonterminal with $\Delta_S < +\infty$. Then it holds that $\sigma_S(n) = n + \Delta_S$ for every $n \in \overline{\mathbb{N}}$ such that $n \geq \delta^{|V|}$.*

Proposition 6.2. *For every nonterminal $S \in V$ with $\Delta_S < +\infty$, the function σ_S is effectively computable.*

The following lemma will be useful in our reduction below.

Lemma 6.3. *Let $X \in V$ be a nonterminal. If there is a derivation $X \overset{*}{\Longrightarrow} uXv$ such that $\Delta_{uv} = +\infty$ then it holds that $\lambda_X = +\infty$.*

We will now show that summaries are computable for nonterminals with finite ratio. The main idea is to transform the given GVAS into an equivalent *thin* GVAS, by hard-coding the effect of nonterminals with finite displacement. This is effective due to Proposition 6.2. Computability of λ_X and σ_X then follows from Theorem 5.1. The following ad-hoc notion of equivalence is sufficient for this purpose. Crucially, it has no requirement for nonterminals with infinite ratio.

Two GVAS $G = (V, A, R)$ and $G' = (V', A', R')$ are called *equivalent* if firstly $V = V'$, secondly $\lambda_X^G = \lambda_X^{G'}$ for every nonterminal X, and thirdly $\sigma_X^G = \sigma_X^{G'}$ for every nonterminal X *with finite ratio*.

Unfoldings. For our first transformation, assume a nonterminal $X \in V$ with $\Delta_X^G < +\infty$. The *unfolding of X* is the GVAS $H = (V, A, R')$ where R' is obtained from R by removing all production rules $X \vdash \alpha$ and instead adding, for every $0 \leq i \leq \delta^{|V|}$ with $j = \sigma_X^G(i) > -\infty$, a rule $X \vdash (-1)^i (1)^j$.

Observe that the language L_X^H is finite, and that H can be computed from G and X because σ_X^G is computable by Proposition 6.2.

Fact 6.4. The unfolding of X is equivalent to G.

Expansions. Our second transformation completely inlines a given nonterminal with finite language. Given a nonterminal $Y \in V$ with L_Y^G finite, the *expansion of Y* is the GVAS $H = (V, A, R')$ where R' is obtained from R by replacing each production rule $X \vdash \alpha_0 Y \alpha_1 \cdots Y \alpha_k$, with Y not occurring in $\alpha_0 \cdots \alpha_k$, by the rules $X \vdash \alpha_0 z_1 \alpha_1 \cdots z_k \alpha_k$ where $z_1, \ldots, z_k \in L_Y^G$. Note that H can be computed from G and Y. Obviously, languages are preserved by this transformation, i.e., $L_w^G = L_w^H$ for every w in $(V \cup A)^*$. The following fact follows.

Fact 6.5. *The expansion of Y is equivalent to G.*

Abstractions. Our last transformation simplifies a given nonterminal with infinite ratio, in such a way that its ratio remains infinite. Given a nonterminal $X \in V$ with $\lambda_X^G = +\infty$, the *abstraction of X* is the GVAS $H = (V, A \cup \{1\}, R')$ where R' is obtained from R by removing all production rules $X \vdash \alpha$ and replacing them by the two rules $X \vdash 1X \mid \varepsilon$. Note that H can be computed from G and X.

Fact 6.6. *The abstraction of X is equivalent to G.*

We now show how to effectively transform a GVAS into an equivalent thin GVAS. As a first step, we hard-code the effect of nonterminals with finite displacement into the production rules, using unfoldings and expansions described above. By Fact 6.4 and 6.5, this results in an equivalent GVAS. Moreover, it now holds that every nonterminal Y occurring on the right handside α of some production rule $X \vdash \alpha$ has $\Delta_Y = +\infty$. Let (V, A, R) be the constructed GVAS and assume that it is not already thin. This means that there exists a production rule $X \vdash \alpha$ with $\alpha \notin A^*VA^*$ such that X is derivable from α. So $X \stackrel{*}{\Longrightarrow} uXv$ for some words u, v in $(V \cup A)^*$ such that uv contains some nonterminal Y. As Y occurs on the right handside of the initial production rule, it must have an infinite displacement. From Lemma 3.6 we thus get that also $\Delta_{uv} = +\infty$, and Lemma 6.3 lets us conclude that $\lambda_X = +\infty$. Therefore, by Fact 6.6, we may replace G by the abstraction of X. Observe that this strictly decreases the number of production rules violating the condition for the system to be thin and at the same time it preserves the property that $\Delta_Y = +\infty$ for every $Y \in V$ occurring in the right handside a production rule. By iterating this abstraction process, we obtain a thin GVAS that is equivalent to the GVAS that we started with. We have thus shown the following proposition. Its corollary follows from Theorem 5.1, and states the missing ingredients for the proof of the coverability problem.

Proposition 6.7. *For every GVAS G, there exists an effectively constructable thin GVAS that is equivalent to G.*

Corollary 6.8. *The question whether $\lambda_X < +\infty$ holds for a given GVAS G and a given nonterminal X, is decidable. Moreover, if $\lambda_X < +\infty$ then the function σ_X is effectively computable.*

Proof (of Theorem 3.2). Thanks to Proposition 4.9, it suffices to check finitely many candidate certificates, each consisting of a parse tree (T, sym) of bounded height and labeling functions $in, out : T \to \mathbb{N}$ with bounded values. It remains to show that it is possible to verify that a given candidate is in fact a certificate. For this, it needs to satisfy the two flow conditions from page 329 and moreover, every leaf t with $\lambda_{sym(t)} = +\infty$ must have some ancestor $s \prec t$ with $sym(s) = sym(t)$ and $in(s) < in(t)$.

The first flow condition can easily be verified locally. By Corollary 6.8, it is possible to check if $\lambda_{sym(t)} < +\infty$ for every leaf t and therefore verify the third condition. In order to verify the second flow condition, it suffices to check that $\sigma_{sym(t)}(in(t)) \geq out(t)$ holds for all leaves with finite ratio $\lambda_{sym(t)} < +\infty$. This is effective due to Corollary 6.8. Indeed, if none of the above checks fail then it follows from Lemma 4.5 that $\sigma_{sym(t)}(in(t)) \geq out(t)$ necessarily holds also for the remaining leaves t with $\lambda_{sym(t)} = +\infty$ (see the full paper [14] for details). This means that the candidate satisfies the second flow condition and therefore all requirements for a certificate. $\square$

7 Conclusion

The decidability of the coverability problem for pushdown VAS is a long-standing open question with applications for program verification. In this paper, we proved that coverability is decidable for 1-dimensional pushdown VAS. We reformulated the problem to the equivalent coverability problem for 1-dimensional grammar-controlled vector addition systems, and analyzed their behavior in terms of structural properties of derivation trees.

An NP lower complexity bound can be shown by reduction from the SUBSET SUM problem. A closer inspection of our approach allows to derive an EXPSPACE upper bound, using recent results by Blondin et al. [2] on 2-dimensional VAS reachability. The exact complexity is open, and so is the decidability of the problem for larger dimensions.

References

1. Atig, M.F., Ganty, P.: Approximating Petri net reachability along context-free traces. In: FSTTCS, pp. 152–163 (2011)
2. Blondin, M., Finkel, A., Göller, S., Haase, C., McKenzie, P.: Reachability in two-dimensional vector addition systems with states is PSPACE-complete. In: LICS (2015, to appear)
3. Bouajjani, A., Habermehl, P., Mayr, R.: Automatic verification of recursive procedures with one integer parameter. TCS **295**, 85–106 (2003)
4. Courtois, J.-B., Schmitz, S.: Alternating vector addition systems with states. In: Csuhaj-Varjú, E., Dietzfelbinger, M., Ésik, Z. (eds.) MFCS 2014, Part I. LNCS, vol. 8634, pp. 220–231. Springer, Heidelberg (2014)
5. Demri, S., Jurdzinski, M., Lachish, O., Lazic, R.: The covering and boundedness problems for branching vector addition systems. JCSS **79**(1), 23–38 (2013)

6. Ganty, P., Majumdar, R.: Algorithmic verification of asynchronous programs. ACM Trans. Progr. Lang. Syst. **34**(1), 6:1–6:48 (2012)
7. Ginsburg, S., Spanier, E.H.: Semigroups, Presburger formulas and languages. Pacific J. Math. **16**(2), 285–296 (1966)
8. Kosaraju, S.R.: Decidability of reachability in vector addition systems (preliminary version). In: STOC, pp. 267–281 (1982)
9. Lazic, R.: The reachability problem for vector addition systems with a stack is not elementary (2013). CoRR abs/1310.1767
10. Lazic, R., Schmitz, S.: Non-elementary complexities for branching VASS, MELL, and extensions. In: CSL/LICS (2014)
11. Leroux, J.: Vector addition system reachability problem: a short self-contained proof. In: POPL, pp. 307–316 (2011)
12. Leroux, J., Praveen, M., Sutre, G.: Hyper-ackermannian bounds for pushdown vector addition systems. In: CSL/LICS (2014)
13. Leroux, J., Sutre, G.: On flatness for 2-dimensional vector addition systems with states. In: Gardner, P., Yoshida, N. (eds.) CONCUR 2004. LNCS, vol. 3170, pp. 402–416. Springer, Heidelberg (2004)
14. Leroux, J., Sutre, G., Totzke, P.: On the coverability problem for pushdown vector addition systems in one dimension. CoRR abs/1503.04018, April 2015. http://arxiv.org/abs/http://arxiv.org/abs/1503.04018
15. Lipton, R.J.: The reachability problem requires exponential space. Tech. Rep. 63, Yale University, January 1976
16. Mayr, E.W.: An algorithm for the general Petri net reachability problem. In: STOC, pp. 238–246 (1981)
17. Rackoff, C.: The covering and boundedness problems for vector addition systems. TCS **6**(2), 223–231 (1978)
18. Reinhardt, K.: Reachability in Petri nets with inhibitor arcs. ENTCS **223**, 239–264 (2008)

Compressed Tree Canonization

Markus Lohrey[1]([✉]), Sebastian Maneth[2], and Fabian Peternek[2]

[1] Universität Siegen, Siegen, Germany
lohrey@eti.uni-siegen.de
[2] University of Edinburgh, Edinburgh, UK
{smaneth, f.peternek}@inf.ed.ac.uk

Abstract. Straight-line (linear) context-free tree (SLT) grammars have been used to compactly represent ordered trees. Equivalence of SLT grammars is decidable in polynomial time. Here we extend this result and show that isomorphism of unordered trees given as SLT grammars is decidable in polynomial time. The result generalizes to isomorphism of unrooted trees and bisimulation equivalence. For non-linear SLT grammars which can have double-exponential compression ratios, we prove that unordered isomorphism and bisimulation equivalence are PSPACE-hard and in EXPTIME.

1 Introduction

Deciding isomorphism between various mathematical objects is an important topic in theoretical computer science that has led to intriguing open problems like the precise complexity of the graph isomorphism problem. An example of an isomorphism problem, where the knowledge seems to be rather complete, is tree isomorphism. Aho, Hopcroft and Ullman [1, page 84] proved that isomorphism of unordered trees (rooted or unrooted) can be decided in linear time. An unordered tree is a tree, where the children of a node are not ordered. The precise complexity of tree isomorphism was finally settled by Lindell [13], Buss [5], and Jenner et al. [11]: tree isomorphism is LOGSPACE-complete if the trees are represented by pointer structures [11,13] and ALOGTIME-complete if the trees are represented by expressions [5,11]. All these results deal with trees that are given explicitly (either by an expression or a pointer structure). In this paper, we deal with the isomorphism problem for trees that are given in a succinct way. Several succinct encoding schemes for graphs exist in the literature. Galperin and Wigderson [8] considered graphs that are given by a Boolean circuit for the adjacency matrix. Subsequent work showed that the complexity of a problem undergoes an exponential jump when going from the standard input representation to the circuit representation; this phenomenon is known as upgrading, see [7] for more details and references. Concerning graph isomorphism, it was shown in [7] that its succinct version is PSPACE-hard, even for very restricted classes of Boolean circuits (DNFs and CNFs).

In this paper, we consider another succinct input representation that has turned out to be more amenable to efficient algorithms, and, in particular, does not show the upgrading phenomenon known for Boolean circuits: straight-line context-free grammars, i.e., context-free grammars that produce a single object. Such grammars have been intensively studied for strings and recently also for trees. Using a straight-line grammar,

© Springer-Verlag Berlin Heidelberg 2015
M.M. Halldórsson et al. (Eds.): ICALP 2015, Part II, LNCS 9135, pp. 337–349, 2015.
DOI: 10.1007/978-3-662-47666-6_27

repeated patterns in an string or tree can be abbreviated by a nonterminal which can be used in different contexts. For strings, this idea is known as grammar-based compression [6, 14], and it was extended to trees in [4]. In fact this approach can be also extended to general graphs by using hyperedge replacement graph grammars; the resulting formalism is known as hierarchical graph representation [12].

The main topic of this paper is the isomorphism problem for trees that are succinctly represented by straight-line context-free tree grammars (ST *grammars*). An example of such a grammar contains the productions $S \rightarrow A_0(a)$, $A_i(y) \rightarrow A_{i+1}(A_{i+1}(y))$ for $0 \leq i \leq n - 1$, and $A_n(y) \rightarrow f(y, y)$ (here y is called a parameter and in general several parameters may occur in a rule). This grammar produces a full binary tree of height 2^n and hence has $2^{2^n+1} - 1$ many nodes. Thus, an ST grammar may produce a tree, whose size is doubly exponential in the size of the grammar. The reason for this double exponential blow-up is copying: the parameter y occurs twice in the right-hand side of the production $A_n(y) \rightarrow f(y, y)$. If this is not allowed, i.e., if every parameter occurs at most once in every right-hand side, then the grammar is a *straight-line linear context-free tree grammar* (SLT *grammar*). The latter generalize dags (directed acyclic graphs) that allow to share repeated subtrees of a tree, whereas SLT grammars can also share repeated patterns that are not complete subtrees.

Several algorithmic problems are harder for trees represented by ST grammars than trees represented by SLT grammars. A good example is the membership problem for tree automata (PTIME-complete for SLT grammars and PSPACE-complete for ST grammars, see [14, Theorem 39]). A similar situation arises for the isomorphism problem: we prove that the isomorphism problem for (rooted or unrooted) unordered trees that are given by SLT grammars (resp., ST grammars) is PTIME-complete (resp., PSPACE-hard and in EXPTIME). Our polynomial time algorithm for SLT grammars constructs from a given SLT grammar G a new SLT grammar G' that produces a canonical representation (based on lexicographic ordering of depth-first left-to-right traversals) of the tree produced by G. For unrooted SLT-compressed trees, we first compute a compressed representation of the center node of a given SLT-compressed unrooted tree t. Then we compute an SLT grammar that produces the rooted version of t that is rooted in the center node. This is also the standard reduction of the unrooted isomorphism problem to the rooted isomorphism problem in the uncompressed setting, but it requires some work to carry out this reduction in polynomial time in the SLT-compressed setting.

Our techniques can be also used to show that checking bisimulation equivalence of trees that are represented by SLT grammars is PTIME-complete. This generalizes the well-known PTIME-completeness of bisimulation for dags [2]. In this context, it is interesting to note that bisimulation equivalence for graphs that are given by hierarchical graph representations is PSPACE-hard and in EXPTIME [3].

Full proofs can be found in the long version [13].

2 Preliminaries

For $k \geq 0$ let $[k] = \{1, \ldots, k\}$. Let Σ be an alphabet. By T_Σ we denote the set of all (ordered, rooted) trees over the alphabet Σ. It is defined recursively as the smallest set of strings such that if $t_1, \ldots, t_k \in T_\Sigma$ and $k \geq 0$ then also $\sigma(t_1, \ldots, t_k) \in T_\Sigma$. For the tree $a()$ we simply write a. The set $D(t)$ of *Dewey addresses* of a tree $t = \sigma(t_1, \ldots, t_k)$ is the subset of $\mathbb{N}^*$ defined recursively as $\{\varepsilon\} \cup \bigcup_{i \in [k]} i \cdot D(t_i)$. Thus ε is the root node of t and $u \cdot i$ is the i-th child of u. For $u \in D(t)$, we denote by $t[u] \in \Sigma$ the symbol at u, i.e., if $t = \sigma(t_1, \ldots, t_k)$, then $t[\varepsilon] = \sigma$ and $t[i \cdot u] = t_i[u]$. The *size* of t is $|t| = |D(t)|$.

A *ranked alphabet* N is a finite set of symbols each of which equipped with a non-negative integer, called its "rank". We write $N^{(k)}$ for the set of symbols in N that have rank k. For an alphabet Σ and a ranked alphabet N, we denote by $T_{N \cup \Sigma}$ the set of trees t over $N \cup \Sigma$ with the property that if $t[u] = A \in N^{(k)}$, then $u \cdot i \in D(t)$ if and only if $i \in [k]$. Thus, if a node is labeled by a ranked symbol, then the rank determines the number of children of the node. We fix a set $Y = \{y_1, y_2, \ldots\}$ of *parameters*, which are symbols of rank 0. For y_1 we also write y. We write $T_{\Sigma \cup N}(Y)$ for $T_{\Sigma \cup N \cup Y}$. For trees $t, t_1, \ldots, t_k \in T_{\Sigma \cup N}(Y)$ we denote by $t[y_j \leftarrow t_j \mid j \in [k]]$ the tree obtained from t by replacing in parallel every occurrence of y_j $(j \in [k])$ by t_j. A *context* is a tree in $T_{\Sigma \cup N}(\{y\})$ with exactly one occurrence of y. Let $\mathcal{C}_{\Sigma \cup N}$ be the set of all contexts and let $\mathcal{C}_\Sigma = \mathcal{C}_{\Sigma \cup N} \cap T_\Sigma(\{y\})$. For a context $t(y)$ and a tree t' we write $t[t']$ for $t[y \leftarrow t']$.

A *context-free tree grammar* is a tuple $G = (N, \Sigma, S, P)$ where N is a ranked alphabet of nonterminal symbols, Σ is an alphabet of terminal symbols with $\Sigma \cap N = \emptyset$, $S \in N^{(0)}$ is the start nonterminal, and P is a finite set of productions of the form $A(y_1, \ldots, y_k) \to t$ where $A \in N^{(k)}$, $k \geq 0$, and $t \in T_{N \cup \Sigma}(\{y_1, \ldots, y_k\})$. Occasionally, we consider context-free tree grammars without a start nonterminal. Two trees $\xi, \xi' \in T_{N \cup \Sigma}(Y)$ are in the one-step derivation relation $\Rightarrow_G$ induced by G, if ξ has a subtree $A(t_1, \ldots, t_k)$ with $A \in N^{(k)}$, $k \geq 0$ such that ξ' is obtained from ξ by replacing this subtree by $t[y_j \leftarrow t_j \mid j \in [k]]$, where $A(y_1, \ldots, y_k) \to t$ is a production in P. The tree language $L(G)$ produced by G is $\{t \in T_\Sigma \mid S \Rightarrow_G^* t\}$. The *size* of the grammar G is $|G| = \sum_{(A(y_1, \ldots, y_k) \to t) \in P} |t|$. The grammar $G = (N, \Sigma, S, P)$ is *deterministic* if for every $A \in N$ there is exactly one production of the form $A \to t$. The grammar G is *acyclic*, if there is a linear order $<$ on N such that $A < B$ whenever B occurs in a tree t with $(A \to t) \in P$. A deterministic and acyclic grammar is called *straight-line*. Note that $|L(G)| = 1$ for a straight-line grammar. We denote the unique tree t produced by the straight-line tree grammar G by $\text{val}(G)$. Moreover, for a tree $t \in T_{\Sigma \cup N}(Y)$ we denote with $\text{val}_G(t)$ the unique tree from $T_\Sigma(Y)$ such that $t \Rightarrow_G^* \text{val}_G(t)$. If G is clear from the context, we simply write $\text{val}(t)$ for $\text{val}_G(t)$. The grammar G is *linear* if for every production $(A \to t) \in P$ and every $y \in Y$, y occurs at most once in t.

For a straight-line context-free tree grammar (resp., straight-line linear context-free tree grammar) we say ST *grammar* (resp.. SLT *grammar.*) Occasionally, we also consider SLT grammars, where the start nonterminal belongs to $N^{(1)}$, i.e., has rank 1. For such a 1-SLT *grammar* G it holds that $\text{val}(G) \in \mathcal{C}_\Sigma$. Most of this paper is about SLT grammars, only at the very end of the paper we consider general ST grammars. SLT grammars generalize rooted node-labelled dags (directed acyclic graph), where the tree defined by such a dag is obtained by unfolding the dag starting from the root (formally, the nodes of the tree are the directed paths in the dag that start in the root). A dag can be viewed as an SLT grammar, where all nonterminals have rank 0 (the nodes of the dag correspond to the nonterminal of the SLT grammar). Dags are less succinct than SLT grammars (take the tree $f^N(a)$ for $N = 2^n$), which in turn are less succinct than general ST grammars (take a full binary tree of height 2^n).

In the literature, SLT grammars are usually defined over ranked terminal alphabets. The proof of the following result from [14] also works for an unranked alphabet Σ.

Lemma 1. *One can transform in polynomial time an SLT grammar into an equivalent SLT grammar, where every nonterminal has rank at most one and each production has one of the following four types (where $\sigma \in \Sigma$ and $A, B, C, A_1, \ldots, A_k \in N$):*

(1) $A \to \sigma(A_1, \ldots, A_k)$,   (3) $A(y) \to \sigma(A_1, \ldots, A_i, y, A_{i+1}, \ldots, A_k)$, or

(2) $A \to B(C)$,   (4) $A(y) \to B(C(y))$.

In the following, we will only deal with SLT grammars G having the property from Lemma 1. For $i \in [4]$, we denote with $G(i)$ the SLT grammar (without start nonterminal) consisting of all productions of G of type (i) from Lemma 1.

A *straight-line program* (SLP) can be seen as a 1-SLT grammar $G = (N, \Sigma, S, P)$ containing only productions of the form $A(y) \to B(C(y))$ and $A(y) \to \sigma(y)$ with $B, C \in N$ and $\sigma \in \Sigma$. Thus, G contains ordinary rules of a context-free string grammar in Chomsky normal form (but written as monadic trees). Intuitively, if $\text{val}(G) = a_1(\cdots a_n(y) \cdots)$ then G produces the string $a_1 \cdots a_n$ and we also write $\text{val}(G) = a_1 \cdots a_n$. For a string $w = a_1 \cdots a_n$ and two numbers $l, r \in [n]$ with $l \leq r$ we denote by $w[l, r]$ the substring $a_l a_{l+1} \cdots a_r$. The following result is well-known, see e.g. [14].

Lemma 2. *For a given SLP G and two binary encoded numbers $l, r \in [|\text{val}(G)|]$ with $l \leq r$ one can compute in polynomial time an SLP G' such that $\text{val}(G') = \text{val}(G)[l, r]$.*

3   Isomorphism of Rooted Unordered SLT-Compressed Trees

Let us fix an alphabet Σ. For $t \in T_\Sigma$ we denote with $\text{uo}(t)$ the unordered rooted version of t. It is the node-labeled directed graph (V, E, λ) where $V = D(t)$ is the set of nodes, $E = \{(u, u \cdot i) \mid i \in \mathbb{N}, u \in \mathbb{N}^*, u \cdot i \in D(t)\}$ is the edge relation, and λ is the node-labelling function with $\lambda(u) = t[u]$. For an SLT grammar G, we also write $\text{val}_{\text{uo}}(G)$ for $\text{uo}(\text{val}(G))$.

For reasons that will become clear in a moment we have to restrict in this section to *ranked* trees, i.e., trees $t \in T_\Sigma$ such that for all $u, v \in D(t)$, if $t[u] = t[v]$ then u and v have the same number of children (nodes with the same label have the same number of children). For the purpose of deciding the isomorphism problem for unordered SLT-represented trees this is not a real restriction. Denote for a tree $t \in T_\Sigma$ the ranked tree $\text{ranked}(t)$ such that $D(t) = D(\text{ranked}(t))$ and for every $u \in D(t)$ with $t[u] = \sigma$: if u has k children in t, then $\text{ranked}(t)[u] = \sigma_k$, where σ_k is a new symbol. Clearly, $\text{uo}(s)$ and $\text{uo}(t)$ are isomorphic if and only if $\text{uo}(\text{ranked}(s))$ and $\text{uo}(\text{ranked}(t))$ are isomorphic. Moreover, for an SLT grammar G we construct in polynomial time the SLT grammar $\text{ranked}(G)$ obtained from G by changing every production $A \to t$ into $A \to \text{ranked}(t)$, where ranked is extended to trees over Σ and nonterminals by defining $\text{ranked}(t)[u] = t[u]$ if $t[u]$ is a nonterminal. Then $\text{val}(\text{ranked}(G)) = \text{ranked}(\text{val}(G))$ holds. Hence, in the following we will only consider ranked trees, and all SLT grammars will produce ranked trees as well.

For a tree $t \in T_\Sigma$ we denote by $\mathsf{dflr}(t) \in \Sigma^*$ its depth-first left-to-right traversal string. It is defined as $\mathsf{dflr}(\sigma(t_1, \ldots, t_k)) = \sigma\,\mathsf{dflr}(t_1)\cdots\mathsf{dflr}(t_k)$ for $\sigma \in \Sigma$, $k \geq 0$, and $t_1, \ldots, t_k \in T_\Sigma$. Note that for ranked trees s and t it holds that: $\mathsf{dflr}(s) = \mathsf{dflr}(t)$ if and only if $s = t$. This is the reason for restricting to ranked trees: for unranked trees this equivalence fails. For instance, $\mathsf{dflr}((a(a(a)))) = a^3 = \mathsf{dflr}(a(a, a))$.

Let $<_\Sigma$ be an order on Σ; it induces the *length-lexicographical ordering* $<_\mathsf{lex}$ on Σ by $u <_\mathsf{lex} v$ iff (i) $|u| < |v|$ or (ii) $|u| = |v|$ and there exist $p, u', v' \in \Sigma^*$ and $a, b \in \Sigma$ with $a <_\Sigma b$, $u = pau'$, and $v = pbv'$. We extend $<_\mathsf{llex}$ to T_Σ by $s <_\mathsf{llex} t$ iff $\mathsf{dflr}(s) <_\mathsf{llex} \mathsf{dflr}(t)$.

Statement (1) in the following lemma was shown in [4] by computing from G, H in polynomial time SLPs G', H' with $\mathsf{val}(G') = \mathsf{dflr}(\mathsf{val}(G))$ and $\mathsf{val}(H') = \mathsf{dflr}(\mathsf{val}(H))$. Equivalence of SLPs can be decided in polynomial time (this result was independently shown by Plandowski, Hirshfeld, Jerrum, Moller, and Mehlhorn, Sundar, Uhrig, see [14] for references). For statement (2) one can do binary search to find the first position where the string $\mathsf{val}(G')$ and $\mathsf{val}(H')$ differ.

Lemma 3. *Let G, H be SLT grammars. It is decidable in polynomial time whether or not (1) $\mathsf{val}(G) <_\mathsf{llex} \mathsf{val}(H)$ and (2) whether or not $\mathsf{val}(G) = \mathsf{val}(H)$.*

For a tree $t \in T_\Sigma$ we define its *canon* $\mathsf{canon}(t)$ as the smallest tree s w.r.t. $<_\mathsf{llex}$ such that $\mathsf{uo}(s)$ is isomorphic to $\mathsf{uo}(t)$. In order to determine $\mathsf{canon}(t)$ for $t = \sigma(t_1, \ldots, t_k)$ let $c_i = \mathsf{canon}(t_i)$ for $i \in [k]$ and let $c_{i_1} \leq_\mathsf{llex} c_{i_2} \leq_\mathsf{llex} \cdots \leq_\mathsf{llex} c_{i_k}$ be the length-lexicographically ordered list of canons $c_1, \ldots, c_k$. Then $\mathsf{canon}(t) = \sigma(c_{i_1}, \ldots, c_{i_n})$. The following lemma can be easily shown by an induction on the tree structure:

Lemma 4. *Let $s, t \in T_\Sigma$. Then $\mathsf{uo}(s)$ is isomorphic to $\mathsf{uo}(t)$ iff $\mathsf{canon}(s) = \mathsf{canon}(t)$.*

In the following, we denote a tree $A_1(A_2(\cdots A_n(t)\cdots))$, where $A_1, A_2, \ldots, A_n$ are unary nonterminals with $A_1 A_2 \cdots A_n(t)$.

Theorem 5. *From a given SLT grammar G one can construct in polynomial time an SLT grammar G' such that $\mathsf{val}(G') = \mathsf{canon}(\mathsf{val}(G))$.*

Proof. Let $G = (N, \Sigma, S, P)$. We assume that G contains no distinct nonterminals $A_1, A_2 \in N^{(0)}$ such that $\mathsf{val}_G(A_1) = \mathsf{val}_G(A_2)$. This is justified because we can test $\mathsf{val}_G(A_1) = \mathsf{val}_G(A_2)$ in polynomial time by Lemma 3 (and replace A_2 by A_1 in G in such a case). We will add polynomially many new nonterminals to G and change the productions for nonterminals from $N^{(0)}$ such that for the resulting SLT grammar G': $\mathsf{val}_{G'}(Z) = \mathsf{canon}(\mathsf{val}_G(Z))$ for every $Z \in N^{(0)}$.

Consider a nonterminal $Z \in N^{(0)}$ and let M be the set of all nonterminals in G that can be reached from Z. By induction, we can assume that G already satisfies $\mathsf{val}_G(A) = \mathsf{canon}(\mathsf{val}_G(A))$ for every $A \in M^{(0)} \setminus \{Z\}$. We distinguish two cases.

Case (i). Z is of type (1) from Lemma 1, i.e., has a production $Z \to \sigma(A_1, \ldots, A_k)$. Using Lemma 3 we construct an ordering $i_1, \ldots, i_k$ of $[k]$ such that $\mathsf{val}_G(A_{i_1}) \leq_\mathsf{llex} \mathsf{val}_G(A_{i_2}) \leq_\mathsf{llex} \cdots \leq_\mathsf{llex} \mathsf{val}_G(A_{i_k})$. We obtain G' by replacing the production $Z \to \sigma(A_1, \ldots, A_k)$ by $Z \to \sigma(A_{i_1}, \ldots, A_{i_k})$ and get $\mathsf{val}_{G'}(Z) = \mathsf{canon}(\mathsf{val}_G(Z))$.

Case (ii). Z is of type (2), i.e., has a production $Z \to B(A)$. Let $\{S_1, \ldots, S_m\} = M^{(0)} \setminus \{Z\}$ be an ordering such that $\mathsf{val}_G(S_1) <_\mathsf{llex} \mathsf{val}_G(S_2) <_\mathsf{llex} \cdots <_\mathsf{llex} \mathsf{val}_G(S_m)$.

Note that A is one of these S_i. The sequence $S_1, S_2, \ldots, S_m$ partitions the set of all trees t in T_Σ into intervals $\mathcal{I}_0, \mathcal{I}_1, \ldots, \mathcal{I}_m$ with

- $\mathcal{I}_i = \{t \in T_\Sigma \mid \mathsf{val}_H(S_i) \leq_{\mathsf{llex}} t <_{\mathsf{llex}} \mathsf{val}_H(S_{i+1})\}$ for $1 \leq i \leq m - 1$,
- $\mathcal{I}_0 = \{t \in T_\Sigma \mid t <_{\mathsf{llex}} \mathsf{val}_H(S_1)\}$, and $\mathcal{I}_m = \{t \in T_\Sigma \mid \mathsf{val}_H(S_m) \leq_{\mathsf{llex}} t\}$.

Consider the maximal $G(4)$-derivation $B(A) \Rightarrow^*_{G(4)} B_1 B_2 \cdots B_N(A)$ starting from $B(A)$, where B_i is a type-(3) nonterminal. Clearly, the number N might be of exponential size, but the set $\{B_1, \ldots, B_N\}$ can be easily constructed. In order to construct an SLT for canon($\mathsf{val}_G(Z)$), it remains to reorder the arguments in right-hand sides of the type-(3) nonterminals B_i. The problem is of course that different occurrences of a type-(3) nonterminal in the sequence $B_1 B_2 \cdots B_N$ have to be reordered in a different way. But we will show that the sequence $B_1 B_2 \cdots B_N$ can be split into $m + 1$ blocks such that all occurrences of a type-(3) nonterminal in one of these blocks have to be reordered in the same way.

Let $t_k = \mathsf{val}_G(B_k B_{k+1} \cdots B_N(A))$ for $k \in [N]$ and $t_{N+1} = \mathsf{val}_G(A)$. Note that $t_1 = \mathsf{val}_G(Z) >_{\mathsf{llex}} \mathsf{val}_G(S_m)$ and that $t_{k+1} <_{\mathsf{llex}} t_k$ for all k. For $i \in [m]$ let k_i be the maximal position $k \leq N + 1$ such that $t_k \geq_{\mathsf{llex}} \mathsf{val}_G(S_i)$. Since $t_1 \geq_{\mathsf{llex}} \mathsf{val}_G(S_m) \geq_{\mathsf{llex}} \mathsf{val}_G(S_i)$ this position is well defined. Note that if $A = S_i$, then $k_i = k_{i-1} = \cdots = k_1 = N + 1$. For every $0 \leq i \leq m$, the interval $[k_{i+1} + 1, k_i]$ is the set of all k such that $\mathsf{val}_G(t_k) \in \mathcal{I}_i$. Here we set $k_{m+1} = 0$ and $k_0 = N+1$. Clearly, the interval $[k_{i+1}+1, k_i]$ might be empty. The positions $k_0, \ldots, k_m$ can be computed in polynomial time using binary search combined with Lemma 3. To apply the latter, note that for a given k we can compute in polynomial time an SLT grammar for the tree t_k using Lemma 2 for the SLP consisting of all type-(4) productions that are used to derive $B_1 B_2 \cdots B_N$.

We now factorize the string $B_1 B_2 \cdots B_N$ as $B_1 B_2 \cdots B_N = u_m u_{m-1} \cdots u_0$, where $u_m = B_1 \cdots B_{k_m - 1}$ and $u_i = B_{k_{i+1}} \cdots B_{k_i - 1}$ for $0 \leq i \leq m - 1$. By Lemma 2 we can compute in polynomial time an SLP G_i for the string u_i. For the further consideration, we view G_i as a 1-SLT grammar consisting only of type-(4) productions. Note that $\mathsf{val}(G_i)$ is a linear tree, where every node is labelled with a type-(3) nonterminal. We now add reordered versions of type-(3) productions to G_i. Consider a type-(3) production $(C(y) \to \sigma(A_1, \ldots, A_j, y, A_{j+1}, \ldots, A_k)) \in P$ where $C \in \{B_1, \ldots, B_N\}$. We add to G_i the type-(3) production $C(y) \to \sigma(A_{j_1}, \ldots, A_{j_\nu}, y, A_{j_{\nu+1}}, \ldots, A_{j_k})$, where $\{j_1, \ldots, j_k\} = [k]$ and $0 \leq \nu \leq k$ are chosen such that

(1) $\mathsf{val}_G(A_{j_1}) \leq_{\mathsf{llex}} \mathsf{val}_G(A_{j_2}) \leq_{\mathsf{llex}} \cdots \leq_{\mathsf{llex}} \mathsf{val}_G(A_{j_k})$ and
(2) $\mathsf{val}_G(A_{j_\nu}) \leq_{\mathsf{llex}} \mathsf{val}_G(S_i) <_{\mathsf{llex}} \mathsf{val}_G(A_{j_{\nu+1}})$.

Note that if $\nu = k$ then condition (2) states that $\mathsf{val}_G(A_{j_k}) \leq_{\mathsf{llex}} \mathsf{val}_G(S_i)$, and if $\nu = 0$ then it states that $\mathsf{val}_G(S_i) <_{\mathsf{llex}} \mathsf{val}_G(A_{j_1})$. Also note that condition (2) ensures that for every tree $t \in \mathcal{I}_i$: $\mathsf{val}_G(A_{j_\nu}) \leq_{\mathsf{llex}} t <_{\mathsf{llex}} \mathsf{val}_G(A_{j_{\nu+1}})$. Hence, $\mathsf{val}_G(\sigma(A_{j_1}, \ldots, A_{j_\nu}, t, A_{j_{\nu+1}}, \ldots, A_{j_k}))$ is a canon. The crucial observation now is that the above factorization $u_m u_{m-1} \cdots u_0$ of $B_1 B_2 \cdots B_N$ was defined in such a way that for every occurrence of a type-(3) nonterminal $C(y)$ in u_i, the parameter y will be substituted by a tree from $\mathcal{I}_i$ during the derivation from Z to $\mathsf{val}_G(Z)$. Hence, we reorder the arguments in the right-hand sides of nonterminal occurrences in u_i in the correct way to obtain a canon.

We now rename the nonterminals in the SLT grammars G_i (which are now of type-(3) and type-(4)) so that the nonterminal sets of $G, G_0, \ldots, G_m$ are pairwise disjoint.

Let $X_i(y)$ be the start nonterminal of G_i after the renaming. Then we add to the current SLT grammar G the union of all the G_i, and replace the production $Z \to B(A)$ by $Z \to X_m X_{m-1} \cdots X_0(A)$. The construction implies that $\mathsf{val}_{G'}(Z) = \mathsf{canon}(\mathsf{val}_G(Z))$ for the resulting grammar G'.

It remains to argue that the above construction can be carried out in polynomial time. All steps only need polynomial time in the size of the current SLT grammar. Hence, it suffices to show that the size of the SLT grammar is polynomially bounded. The algorithm is divided into $|N^{(0)}|$ many phases, where in each phase it enforces $\mathsf{val}_{G'}(Z) = \mathsf{canon}(\mathsf{val}_G(Z))$ for a single nonterminal Z. Consider a single phase, where $\mathsf{val}_{G'}(Z) = \mathsf{canon}(\mathsf{val}_G(Z))$ is enforced for a nonterminal Z. In this phase, we (i) change the production for Z and (ii) add new type-(3) and type-(4) productions to G (the union of the G_i above). But the number of these new productions is polynomially bounded in the size of the initial SLT grammar (the one before the first phase), because the nonterminals introduced in earlier phases are not relevant for the current phase. This implies that the additive size increase in each phase is bounded polynomially in the size of the initial grammar. □

Corollary 6. *The problem of deciding whether* $\mathsf{val}_{\mathsf{uo}}(G_1)$ *and* $\mathsf{val}_{\mathsf{uo}}(G_2)$ *are isomorphic for given* SLT *grammars* G_1 *and* G_2 *is* PTIME-*complete*

Proof. Membership in PTIME follows immediately from Lemma 3, Lemma 4, and Theorem 5. Moreover, PTIME-hardness already holds for dags, i.e., SLT grammars where all nonterminals have rank 0, as shown in [15]. □

4   Isomorphism of Unrooted Unordered SLT-Compressed Trees

In this section we show isomorphism for unrooted unordered trees represented by SLT grammars can be solved in polynomial time. An unrooted unordered tree t over Σ can be seen as a node-labeled (undirected) graph $t = (V, E, \lambda)$, where $E \subseteq V \times V$ is symmetric and $\lambda : V \to \Sigma$. Let $s = (V, E, \lambda)$ be a rooted unordered tree. The tree $\mathsf{ur}(s) = (V, E \cup E^{-1}, \lambda)$ is the unrooted version of s. An unrooted unordered tree t can be represented by an SLT grammar G by forgetting the order and root information present in G. Let $\mathsf{val}_{\mathsf{ur,uo}}(G) = \mathsf{ur}(\mathsf{uo}(\mathsf{val}(G)))$.

Let $t = (V, E, \lambda)$ be an unordered unrooted tree. For a node $v \in V$ we define the eccentricity $\mathsf{ecc}_t(v) = \max_{u \in V} \delta_t(u, v)$ and the diameter $\varnothing(t) = \max_{v \in V} \mathsf{ecc}_t(v)$, where $\delta_t(u, v)$ denotes the distance from u to v (i.e., the number of edges on the path from u to v in t). A node u of t is called *center node of* t if for all leaves v of t: $\delta_t(u, v) \le (\varnothing(s) + 1)/2$. Let $\mathsf{center}(t)$ be the set of all center nodes of t. One can compute the center nodes by deleting all leaves of the tree and iterating this step, until the current tree consists of at most two nodes. These are the center nodes of t. In particular, t has either one or two center nodes. Another characterization of center nodes that is important for our algorithm is via longest paths. Let $p = (v_0, v_1, \ldots, v_n)$ be a longest simple path in t, i.e., $n = \varnothing(t)$. Then the middle points $v_{\lfloor n/2 \rfloor}$ and $v_{\lceil n/2 \rceil}$ (which are identical if n is even) are the center nodes of t and are independent of the concrete longest path p.

Note that there are two center nodes if and only if $\varnothing(t)$ is odd. Since our constructions are simpler if a unique center node exists, we first make sure that $\varnothing(t)$ is even. Let $\#$ be a new symbol not in Σ. For an unrooted unordered tree t we denote by $\mathsf{even}(t)$ the tree where every pair of edge $(u, v), (v, u)$ is replaced by the edges $(u, v'), (v', v), (v, v'), (v', u)$, where v' is a new node labelled $\#$. Then for an SLT grammar $G = (N, \Sigma, P, S)$ we let $\mathsf{even}(G) = (N, \Sigma \cup \{\#\}, P', S)$ be the SLT grammar where P' is obtained from P by replacing every subtree $\sigma(t_1, \ldots, t_k)$ with $\sigma \in \Sigma$, $k \geq 1$, in a right-hand side by the subtree $\sigma(\#(t_1), \ldots, \#(t_k))$. Observe that (i) $\mathsf{val}_{\mathsf{ur,uo}}(\mathsf{even}(G)) = \mathsf{even}(\mathsf{val}_{\mathsf{ur,uo}}(G))$, (ii) $\varnothing(\mathsf{even}(t)) = 2 \cdot \varnothing(t)$ is even, i.e., $\mathsf{even}(t)$ has only one center node, and (iii) trees t and s are isomorphic if and only if $\mathsf{even}(t)$ and $\mathsf{even}(s)$ are isomorphic. Since $\mathsf{even}(G)$ can be constructed in polynomial time, we assume in the following that every SLT grammar produces a tree with a unique center node. For such a tree t we denote with $\mathsf{center}(t)$ its unique center node.

Let $u \in V$. The rooted version $\mathsf{root}(t, u)$ of t with root node u is $\mathsf{root}(t, u) = (V, E', \lambda)$, where $E' = \{(v, v') \in E \mid \delta_t(u, v) < \delta_t(u, v')\}$. Two unrooted unordered trees t_1, t_2 of even diameter are isomorphic iff $\mathsf{root}(t_1, \mathsf{center}(t_1))$ is isomorphic to $\mathsf{root}(t_2, \mathsf{center}(t_2))$. Thus, we can solve in polynomial time the isomorphism problem for unrooted unordered trees represented by SLT grammars G_1, G_2 by (i) computing for $i \in \{1, 2\}$ in polynomial time a compressed representation $\tilde{u}_i$ of $u_i = \mathsf{center}(\mathsf{val}_{\mathsf{ur,uo}}(G_i))$ (Section 4.1), (ii) computing for $i \in \{1, 2\}$ in polynomial time an SLT grammar G'_i such that $\mathsf{val}_{\mathsf{uo}}(G'_i) = \mathsf{root}(\mathsf{val}_{\mathsf{ur,uo}}(G_i), u_i)$ (Section 4.2) and (iii) testing in polynomial time if $\mathsf{val}_{\mathsf{uo}}(G'_1)$ is isomorphic to $\mathsf{val}_{\mathsf{uo}}(G'_2)$ (Corollary 6).

4.1 Finding Center Nodes

Let $G = (N, \Sigma, S, P)$ be an SLT grammar. A *G-compressed path* p is a string of pairs $p = (A_1, u_1) \cdots (A_n, u_n)$ such that for all $i \in [n]$, $A_i \in N$, $A_1 = S$, $u_i \in D(t_i)$ is a Dewey address in t_i where $(A_i \to t_i) \in P$, $t_i[u_i] = A_{i+1}$ for $i < n$, and $t_i[u_n] \in \Sigma$. If we omit the condition $t_i[u_n] \in \Sigma$, then p is a partial G-compressed path. Note that by definition, $n \leq |N|$. A partial G-compressed path uniquely represents one particular node in the derivation tree of G, and a G-compressed path represents a leaf of the derivation tree and hence a node of $\mathsf{val}(G)$. We denote this node by $\mathsf{val}_G(p)$. The concatenation $u_1, u_2, \ldots, u_n$ of the Dewey addresses is denoted by $u(p)$.

For a context $t(y) \in \mathcal{C}_\Sigma$ we define $\mathsf{ecc}(t) = \mathsf{ecc}_t(y)$ (recall that in a context there is a unique occurrence of the parameter y) and $\mathsf{rty}(t) = \delta_t(\varepsilon, y)$ (the distance from the root to the parameter y). For a tree $s \in T_\Sigma$ we denote with $h(s)$ its height. We extend these notions to contexts $t \in \mathcal{C}_{\Sigma \cup N}$ and trees $s \in T_{\Sigma \cup N}$ by $\mathsf{ecc}(t) = \mathsf{ecc}(\mathsf{val}_G(t))$, $\mathsf{rty}(t) = \mathsf{rty}(\mathsf{val}_G(t))$, and $h(s) = h(\mathsf{val}_G(s))$. Eccentricity, distance from root to y, and height can be computed in polynomial time for SLT-represented trees bottom-up. To do so, observe that for contexts $t(y), t'(y) \in \mathcal{C}_{\Sigma \cup N}$ and a tree $s \in T_{\Sigma \cup N}$: $\mathsf{rty}(t[t']) = \mathsf{rty}(t) + \mathsf{rty}(t')$, $\mathsf{ecc}(t[t']) = \max\{\mathsf{ecc}(t'), \mathsf{ecc}(t) + \mathsf{rty}(t')\}$, and $h(t[s]) = \max\{h(s), \mathsf{rty}(t) + h(s)\}$. Similarly, for $t(y) = \sigma(s_1, \ldots s_i, y, s_{i+1}, \ldots, s_k) \in \mathcal{C}_{\Sigma \cup N}$ and $s = \sigma(s_1, \ldots, s_k) \in T_{\Sigma \cup N}$: $\mathsf{rty}(t) = 1$, $\mathsf{ecc}(t) = 2 + \max\{h(s_i) \mid i \in [k]\}$, and $h(s) = 1 + \max\{h(s_i) \mid i \in [k]\}$.

Our search for the center node of an SLT-compressed tree is based on the following lemma. For a context $t(y) \in \mathcal{C}_\Sigma$, where u is the Dewey address of the parameter y, and a tree $s \in T_\Sigma$ we say that a node $v \in D(t[s])$ belongs to t if $v \in D(t) \setminus \{u\}$. Otherwise, we say that v belongs to s, which means that u is a prefix of v.

Lemma 7. *Let $t(y) \in C_\Sigma$ be a context and $s \in T_\Sigma$ a tree such that $\varnothing(t[s])$ is even. Let $c = \text{center}(t[s])$. Then c belongs to s if and only if $\text{ecc}(t) \leq h(s)$.*

Lemma 8. *For a given SLT grammar G such that $\text{val}_{ur,uo}(G)$ has even diameter, one can construct a G-compressed path for $\text{center}(\text{val}_{ur,uo}(G))$.*

Proof. Let $G = (N, \Sigma, S, P)$. Our algorithm for finding the center node for $\text{val}(G)$ stores at each point of time a single tuple (t_l, A, t_r, p), where $t_l \in C_{\Sigma \cup N}$ and $t_r \in T_{\Sigma \cup N} \cup \{\varepsilon\}$ are of polynomial size, $A \in N$, and p is a partial G-compressed path. It is started with the tuple $(y, S, \varepsilon, \varepsilon)$ The following invariants are preserved: If the current tuple is (t_l, A, t_r, p) is, then:

- If A has rank 0 then $t_r = \varepsilon$.
- $\text{val}(G) = \text{val}(t_l[A[t_r]])$ (here we set $t[\varepsilon] = t$).
- The tree $t_l[A[t_r]]$ can be derived from the start variable S.
- p is the partial G-compressed path to the distinguished A in $t_l[A[t_r]]$.
- $\text{center}(\text{val}_{ur,uo}(G))$ belongs to the subcontext $\text{val}(A)$ in $\text{val}(t_l)[\text{val}(A)[\text{val}(t_r)]]$.

For the tuple (t_l, A, t_r, p), the algorithm distinguishes on the right-hand side of A. If this right-hand side has the form $A(B)$ or $A(B(y))$, then, by comparing $\text{ecc}(t_l[B(y)])$ and $h(C[t_r])$ (we can compute these values in polynomial time, by constructing SLT grammars for $\text{val}(t_l[B(y)])$ and $\text{val}(C[t_r])$ and using the recursions for ecc, rty and h), we determine, whether the search for the center node has to continue in B or C, see Lemma 7. In the first case we continue with the tuple $(t_l, B, C[t_r], p \cdot (A, \varepsilon))$, and in the second case we continue with $(t_l[B(y)], C, t_r, p \cdot (A, 1))$.

Now assume that the right-hand side of A has the form $\sigma(A_1, \ldots, A_k)$. Let $t_i = t_l[\sigma(A_1, \ldots, A_{i-1}, y, A_{i+1}, \ldots, A_k)]$ for $i \in [k]$. Hence, $\text{val}(G) = \text{val}(t_i)[\text{val}(A_i)]$. By comparing $\text{ecc}(t_i)$ and $h(A_i)$ we want determine whether the center node belongs to $\text{val}(t_i)$ or $\text{val}(A_i)$, see Lemma 7. If the latter holds for some $i \in [k]$, we can continue the search in A_i, i.e., we continue with the tuple $(t_i, A_i, \varepsilon, p \cdot (A, i))$. On the other hand, assume that for all $i \in [k]$ the center point belongs to t_i. In particular, it does not belongs to any of the subtrees $\text{val}(A_i)$. But by the last invariant, we known that the center point belongs to the subcontext $\text{val}(A) = \sigma(\text{val}(A_1), \ldots, \text{val}(A_k))$. Hence, the σ-labelled node must be the center point and we can return its G-compressed path $p \cdot (A, \varepsilon)$. The case of a production $A(y) \to \sigma(A_1, \ldots A_{s-1}, y, A_{s+1}, \ldots, A_k)$ can be dealt with similarly. Note that $|t_l| + |t_r|$ stays bounded by the size of G. □

4.2 Re-Rooting of SLT Grammars

Let $G = (N, \Sigma, S, P)$ be an SLT grammar (as usual, having the normal form from Lemma 1) and p a G-compressed path. Let $s(p) \in T_{\Sigma \cup N}$ be the tree defined inductively as follows: Let $(A \to t) \in P$ and $u \in D(t)$. Then $s((A, u)) = t$. If $p = (A, t)p'$ with p' non-empty, then either *(i)* $u = \varepsilon$ and $t = B(C)$ or *(ii)* $u = i \in \mathbb{N}$ and $t[i] \in N^{(0)}$. In case *(i)* we set $s(p) = s(p')[C]$, in case *(ii)* we set $s(p) = t'[s(p')]$, where $t'(y)$ is obtained from t by replacing the i-th argument of the root by y. Note that $s(p') \in C_{\Sigma \cup N}(\{y\})$ if p' starts with a nonterminal of rank 1. Let $s = s(p)$; its size is bounded by the size of G. Note that $s[u(p)]$ is a terminal symbol (recall that $u(p)$ denotes the concatenation of the Dewey addresses in p). Assume that $s[u(p)] = \sigma \in \Sigma$. Let $\#$ be a fresh symbol and let s' be obtained from s by changing the label at $u(p)$ from σ to $\#$. Let $s' \Rightarrow_G^* s''$ be the shortest derivation such that $s''[\varepsilon] = \delta \in \Sigma$ (it consists of at most $|N|$ derivation steps). We denote the $\#$-labeled node in s'' by u. Finally, let t be obtained from s'' by changing the unique $\#$ into σ. We define the p-*expansion* of G, denoted $\mathsf{ex}_G(p)$, as the tuple (t, u, σ, δ). Note that $\mathsf{val}_G(p)$ is the unique $\#$-labelled node in $\mathsf{val}_G(s'')$. The p-expansion can be computed in polynomial time from G and p.

The p-expansion (t, u, σ, δ) has all information needed to construct a grammar G' representing the rooted version at p of $\mathsf{val}(G)$. If $u = \varepsilon$ then also $\mathsf{val}_G(p) = \varepsilon$. Since G is already rooted at ε nothing has to be done in this case and we return $G' = G$. If $u \neq \varepsilon$ then $\mathsf{val}_G(p) \neq \varepsilon$ and hence t contains two terminal nodes which uniquely represent the root node and the node $\mathsf{val}_G(p)$ of the tree $\mathsf{val}(G)$.

Let $s_1 \in T_\Sigma$ be a rooted ordered tree representing the unrooted unordered tree $\tilde{s}_1 = \mathsf{ur}(\mathsf{uo}(s_1))$. Let $u \neq \varepsilon$ be a node of s_1. Let $s_1[\varepsilon] = \delta \in \Sigma$ and $s_1[u] = \sigma \in \Sigma$. Since $u \neq \varepsilon$, we can write $s_1 = \delta(\zeta_1, \ldots, \zeta_{i-1}, t'[\sigma(\xi_1, \ldots, \xi_m)], \zeta_{i+1}, \ldots, \zeta_k)$, where t' is a context, and $u = iu'$, where u' is the Dewey address of the parameter y in t'. A rooted ordered tree s_2 that represents the rooted unordered tree $\tilde{s}_2 = \mathsf{root}(\tilde{s}_1, u)$ can be defined as $s_2 = \sigma(\xi_1, \ldots, \xi_m, \mathsf{rooty}(t')[\delta(\zeta_1, \ldots, \zeta_{i-1}, \zeta_{i+1}, \ldots, \zeta_k)])$, where rooty is a function mapping contexts to contexts defined recursively as follows ($f \in \Sigma$, $t_1, \ldots, t_{i-1}, t_{i+1}, \ldots, t_\ell \in T_\Sigma$, and $t(y), t'(y) \in \mathcal{C}_\Sigma$):

$$\mathsf{rooty}(y) = y \tag{1}$$
$$\mathsf{rooty}(f(t_1, \ldots, t_{i-1}, y, t_{i+1}, \ldots, t_\ell)) = f(t_1, \ldots, t_{i-1}, y, t_{i+1}, \ldots, t_\ell) \tag{2}$$
$$\mathsf{rooty}(t[t'(y)]) = \mathsf{rooty}(t')[\mathsf{rooty}(t(y))] \tag{3}$$

Intuitively, the mapping rooty unroots a context $t(y)$ towards its y-node u, i.e., it reverses the path from the root to u. Thus, for instance, $\mathsf{rooty}(f(a, y, b)) = f(a, y, b)$ and $\mathsf{rooty}(f(a, g(c, y, d), b)) = g(c, f(a, y, b), d)$.

Lemma 9. *From a given* SLT *grammar G and a G-compressed path p one can construct in polynomial time an* SLT *grammar G' such that $\mathsf{val}_{\mathsf{uo}}(G')$ is isomorphic to* $\mathsf{root}(\mathsf{val}_{\mathsf{ur},\mathsf{uo}}(G), \mathsf{val}_G(p))$.

Proof. Let $G = (N, \Sigma, S, P)$ and $\mathsf{ex}_G(p) = (t, u, \sigma, \delta)$. If $u = \varepsilon$ then define $G' = G$. If $u \neq \varepsilon$ then we can write $t = \delta(B_1, \ldots, B_{i-1}, t'[\sigma(\xi_1, \ldots, \xi_m)], B_{i+1}, \ldots, B_k)$, where $B_j \in N^{(0)}$, $\xi_j \in T_N$, t' is a context composed of nonterminals $A \in N^{(1)}$ and contexts $f(\zeta_1, \ldots, \zeta_{j-1}, y, \zeta_{j+1}, \ldots, \zeta_l)$ ($f \in \Sigma$, $\zeta_j \in T_N$), and $u = iu'$, where u' is the Dewey address of the parameter y in t'.

We define $G' = (N \uplus N', \Sigma, S, P')$ where $N' = \{A' \mid A \in N^{(1)}\}$. To define the production set P', we extend the definition of rooty to contexts from $\mathcal{C}_{\Sigma \cup N}$ by (i) allowing in the trees t_j from Equation (2) also nonterminals, and (ii) defining for every $B \in N^{(1)}$, $\mathsf{rooty}(B(y)) = B'(y)$. We now define the set of productions P' of P as follows: We put all productions from P except for the start production $(S \rightarrow s) \in P$ into P'. For the start variable S we add to P' the production

$$S \rightarrow \sigma(\xi_1, \ldots, \xi_m, \mathsf{rooty}(t')[\delta(B_1, \ldots, B_{i-1}, B_{i+1}, \ldots, B_k)]). \tag{4}$$

Moreover, let $A \in N^{(1)}$ and $(A(y) \to \zeta) \in P$. If this is a type-(3) production, then we add $A'(y) \to \zeta$ to P'. If $\zeta = B(C(y))$ then add $A'(y) \to C'(B'(y))$ to P'.

A simple induction shows that $\mathrm{val}_{G'}(A') = \mathrm{rooty}(\mathrm{val}_G(A))$ for every $A \in N^{(1)}$. This implies that $\mathrm{val}_{G'}(\mathrm{rooty}(c(y))) = \mathrm{rooty}(\mathrm{val}_G(c(y)))$ for every context $c(y)$ that is composed of contexts $f(\zeta_1, \ldots, \zeta_{j-1}, y, \zeta_{j+1}, \ldots, \zeta_l)$ ($\zeta_j \in T_N$) and nonterminals $A \in N^{(1)}$. In particular, $\mathrm{val}_{G'}(\mathrm{rooty}(t')) = \mathrm{rooty}(\mathrm{val}_G(t'(y)))$ for the context t'. This, and the form of the start production of G' (4) easily imply that $\mathrm{val}_{\mathrm{uo}}(G')$ is isomorphic to $\mathrm{root}(\mathrm{val}_{\mathrm{ur,uo}}(G), \mathrm{val}_G(p))$. $\qquad\Box$

Corollary 10. *The problem of deciding whether* $\mathrm{val}_{\mathrm{ur,uo}}(G_1)$ *and* $\mathrm{val}_{\mathrm{ur,uo}}(G_2)$ *are isomorphic for given* SLT *grammars* G_1 *and* G_2 *is* PTIME-*complete.*

Proof. The upper bound follows from Lemma 8, Lemma 9, and Corollary 6. Hardness for PTIME follows from the PTIME-hardness for dags [15] and the fact that isomorphism of rooted unordered trees can be reduced to isomorphism of unrooted unordered trees by labelling the roots with a fresh symbol. $\qquad\Box$

5   Further Results

Bisimulation on SLT-Compressed Trees. Fix a set Σ of node labels. Let $G = (V, E, \lambda)$ be a directed node-labelled graph, i.e., $E \subseteq V \times V$ and $\lambda : V \to \Sigma$. A binary relation $R \subseteq V \times V$ is a bisimulation on G, if for all $(u, v) \in R$ the following three conditions hold: (i) $\lambda(u) = \lambda(v)$, (ii) if $(u, u') \in E$ then there exists $v' \in V$ such that $(v, v') \in E$ and $(u', v') \in R$, and (iii) if $(v, v') \in E$ then there exists $u' \in V$ such that $(u, u') \in E$ and $(u', v') \in R$. Let the relation $\sim$ be the union of all bisimulations on G. It is the largest bisimulation and an equivalence relation. Two rooted unordered trees s, t with node labels from Σ and roots r_s, r_t are bisimulation equivalent if $r_s \sim r_t$ holds in the disjoint union of s and t. For instance, $f(a, a, a)$ and $f(a, a)$ are bisimulation equivalent but $f(g(a), g(b))$ and $f(g(a, b))$ are not. For a rooted unordered tree t we define the bisimulation canon $\mathrm{bcanon}(t)$ inductively: Let $t = f(t_1, \ldots, t_n)$ ($n \geq 0$) and let $b_i = \mathrm{bcanon}(t_i)$. Then $\mathrm{bcanon}(t) = f(s_1, \ldots, s_m)$, where (i) for every $i \in [m]$, s_i is isomorphic to one of the b_j, and (ii) for every $i \in [n]$ there is a unique $j \in [m]$ such that s_i and b_j are isomorphic as rooted unordered trees. In other words: Bottom-up, we eliminate repeated subtrees among the children of a node. For instance, $\mathrm{bcanon}(f(a, a, a)) = f(a) = \mathrm{bcanon}(f(a, a))$. Induction on the height of trees shows:

Lemma 11. *Let* s *and* t *be rooted unordered trees. Then* s *and* t *are bisimulation equivalent if and only if* $\mathrm{bcanon}(s)$ *and* $\mathrm{bcanon}(t)$ *are isomorphic.*

The proof of the following theorem is similar to those of Theorem 5.

Theorem 12. *From a given* SLT *grammar* G *one can compute a new* SLT *grammar* G' *such that* $\mathrm{val}_{\mathrm{uo}}(G')$ *is isomorphic to* $\mathrm{bcanon}(\mathrm{val}_{\mathrm{uo}}(G))$.

From Corollary 6, Lemma 11, and Theorem 12 we get:

Corollary 13. *For given* SLT *grammars* G_1 *and* G_2 *one can check in polynomial time, whether* $\mathsf{val_{uo}}(G_1)$ *and* $\mathsf{val_{uo}}(G_2)$ *are bisimulation equivalent.*

Non-linear ST Grammars. Recall from Section 2 that ST grammars are exponentially more succinct than SLT grammars. So, the following should not be surprising:

Lemma 14. *A given* ST *grammar can be transformed in exponential time into an equivalent* SLT *grammar.*

Using this lemma, the upper bounds in the following statement follow from Corollary 6, 10, and 13. For the lower bound, one can reduce from QBF using gadgets from [11].

Theorem 15. *The following questions are* PSPACE-*hard and in* EXPTIME *for given* ST *grammars* G_1 *and* G_2:

– *Are* $\mathsf{val_{uo}}(G_1)$ *and* $\mathsf{val_{uo}}(G_2)$ *isomorphic (resp., bisimulation equivalent)?*
– *Are* $\mathsf{val_{ur,uo}}(G_1)$ *and* $\mathsf{val_{ur,uo}}(G_2)$ *isomorphic?*

The precise complexity of these questions remains open. Since an ST grammar can be transformed into a hierarchical graph definition for a dag, we rediscover the following result from [3]: Bisimulation equivalence for dags given by hierarchical graph definitions is PSPACE-hard and in EXPTIME.

References

1. Aho, A., Hopcroft, J.E., Ullman, J.D.: The Design and Analysis of Computer Algorithms. Addison-Wesley, Reading (1974)
2. Balcázar, J., Gabarró, J., Sántha, M.: Deciding bisimilarity is P-complete. Formal Aspects of Computing **4**, 638–648 (1992)
3. Brenguier, R., Göller, S., Sankur, O.: A comparison of succinctly represented finite-state systems. In: Koutny, M., Ulidowski, I. (eds.) CONCUR 2012. LNCS, vol. 7454, pp. 147–161. Springer, Heidelberg (2012)
4. Busatto, G., Lohrey, M., Maneth, S.: Efficient memory representation of XML document trees. Inf. Syst. **33**(4–5), 456–474 (2008)
5. Buss, S.R.: Alogtime algorithms for tree isomorphism, comparison, and canonization. Kurt Gödel Colloquium **97**, 18–33 (1997)
6. Charikar, M., Lehman, E., Lehman, A., Liu, D., Panigrahy, R., Prabhakaran, M., Sahai, A., Shelat, A.: The smallest grammar problem. IEEE Trans. Inf. Theory **51**(7), 2554–2576 (2005)
7. Das, B., Scharpfenecker, P., Torán, J.: Succinct encodings of graph isomorphism. In: Dediu, A.-H., Martín-Vide, C., Sierra-Rodríguez, J.-L., Truthe, B. (eds.) LATA 2014. LNCS, vol. 8370, pp. 285–296. Springer, Heidelberg (2014)
8. Galperin, H., Wigderson, A.: Succinct representations of graphs. Inf. Contr. **56**, 183–198 (1983)
9. Jenner, B., Köbler, J., McKenzie, P., Torán, J.: Completeness results for graph isomorphism. J. Comput. Syst. Sci. **66**(3), 549–566 (2003)
10. Lengauer, T., Wagner, K.W.: The correlation between the complexities of the nonhierarchical and hierarchical versions of graph problems. J. Comput. Syst. Sci. **44**, 63–93 (1992)
11. Lindell, S.: A logspace algorithm for tree canonization (extended abstract). In: Proc. STOC 1992, pp. 400–404. ACM (1992)

12. Lohrey, M.: Algorithmics on SLP-compressed strings: a survey. Groups Complexity Cryptology **4**(2), 241–299 (2012)
13. Lohrey, M., Maneth, S., Peternek, F.: Compressed tree canonization (2015). arXiv.org http://arxiv.org/abs/1502.04625
14. Lohrey, M., Maneth, S., Schmidt-Schauß, M.: Parameter reduction and automata evaluation for grammar-compressed trees. J. Comput. Syst. Sci. **78**(5), 1651–1669 (2012)
15. Lohrey, M., Mathissen, C.: Isomorphism of regular trees and words. Inf. Comput. **224**, 71–105 (2013)

Parsimonious Types and Non-uniform Computation

Damiano Mazza[1]([✉]) and Kazushige Terui[2]

[1] CNRS, UMR 7030, LIPN, Université Paris 13, Sorbonne Paris Cité, Villetaneuse, France
Damiano.Mazza@lipn.univ-paris13.fr
[2] RIMS, Kyoto University, Kyoto, Japan
terui@kurims.kyoto-u.ac.jp

Abstract. We consider a non-uniform affine lambda-calculus, called parsimonious, and endow its terms with two type disciplines: simply-typed and with linear polymorphism. We show that the terms of string type into Boolean type characterize the class L/poly in the first case, and P/poly in the second. Moreover, we relate this characterization to that given by the second author in terms of Boolean proof nets, highlighting continuous affine approximations as the bridge between the two approaches to non-uniform computation.

1 Introduction

This paper is a contribution to the line of research known as *implicit computational complexity* (ICC), whose aim is to characterize complexity classes by means of logical systems and programming languages (which are intimately related via the Curry-Howard correspondence). Seminal work concerning this methodology includes [3,9,11]. The highlight of this paper consists in dealing with *non-uniform* complexity classes, which, as far as we know, no ICC work other than [13] have considered so far.

Computation is usually performed by a single machine or program which *uniformly* works on inputs of arbitrary length. On the other hand, some models of computation, such as Boolean circuits, only work on inputs of fixed length. Thus, to compute a function $f : \{0,1\}^* \longrightarrow \{0,1\}$, one needs to prepare a *family* $(C_n)_{n \in \mathbb{N}}$ of Boolean circuits, one for each input length n. The non-uniform perspective is important for hardware design and the study of small complexity classes. Two well-known non-uniform classes are P/poly, consisting of languages decided by families of polynomial size Boolean circuits, and L/poly, consisting of those decided by families of polynomial size branching programs (*i.e.*, decision trees with sharing). Such languages may well be non-recursive, but are still useful as they capture the combinatorial essence of P or L. For instance, a potential approach to separating P and NP is to show that NP is not included in P/poly.

We may call the above approach to non-uniform computation *family approach*. There is an alternative, which consists in using a uniform machine having access to arbitrary *advice*, *i.e.*, a fixed family $(w_n)_{n \in \mathbb{N}}$ of strings. Then, P/poly (resp. L/poly) is equivalently defined as the class of languages decided by a deterministic polytime (resp. logspace) Turing machine aided by a polynomial advice, *i.e.*, s.t. $|w_n|$ is polynomial in n. We may call this the *individual approach*.

Non-uniform complexity has been studied in the setting of proofs and functional programming. The second author showed in [19] that P/poly is precisely the class

© Springer-Verlag Berlin Heidelberg 2015
M.M. Halldórsson et al. (Eds.): ICALP 2015, Part II, LNCS 9135, pp. 350–361, 2015.
DOI: 10.1007/978-3-662-47666-6_28

of languages decided by families of polynomial size proof nets of multiplicative linear logic (*family approach*), whereas the first author introduced an infinitary affine λ-calculus in [13] exactly capturing the class P/poly (*individual approach*).

The present paper combines and extends these previous works. Our goal is twofold: to reconcile the two approaches, and to capture also L/poly.

The starting point is the idea of seeing the exponential modality of linear logic as some sort of limit, which is quite fruitful and is at the core of several recent developments [5, 16] as well as, albeit implicitly, older work on games semantics [1, 15] and intersection types [10]. Following this direction, in [12] the first author introduced an infinitary affine λ-calculus, in which the usual λ-calculus embeds, endowed with a topology such that finite affine terms form a dense subspace and reduction is continuous. This means that computation in the λ-calculus, which is non-linear, may be approximated arbitrarily well by computation on linear affine terms. In particular, since data type values (such as Booleans and strings) are only approximated by themselves, if a λ-term t is given in input a binary string w and t w $\rightarrow^l$ b (reduction in l steps) with b a Boolean, then there exists a finite affine term t_0 such that t_0w $\rightarrow^l$ b.

However, the "modulus of continuity" of reduction in the λ-calculus is ill-behaved: the size of t_0 may be exponential in l. The contribution of [13] was to find a subset of (infinitary) terms, called *parsimonious*, on which the "modulus of continuity" is polynomial. At this point, if one manages to bound l polynomially in the length of the input w, computation falls within P/poly: the polynomially-sized approximation t_0 of t may be given as advice, and normalization of the finite affine term t_0w may be done in deterministic polynomial time. In [13], *stratification* (originally due to Girard [9]) was used to obtain such a bound.

Here, instead, the bound will be enforced by types, which have the benefit of allowing us to modulate computational complexity via logical complexity. More specifically, in Sect. 2 we assign types to non-uniform parsimonious terms in two ways: with a simply-typed system called *non-uniform parsimonious logic* **nuPL**, and with its extension with *linear* polymorphism **nuPL$_{\forall \ell}$**. Let now nuPL (resp. nuPL$_{\forall \ell}$) be the class of languages decided by programs typable in **nuPL** (resp. **nuPL$_{\forall \ell}$**), and let APN (resp. APN^0) be the class of languages decided by polynomial size proof nets (resp. proof nets of *bounded height*). Our main result is

Theorem 1. nuPL $=$ APN^0 $=$ L/poly *and* nuPL$_{\forall \ell}$ $=$ APN $=$ P/poly.

The inclusion L/poly $\subseteq$ nuPL (resp. P/poly $\subseteq$ nuPL$_{\forall \ell}$) is shown by exhibiting a very natural encoding of branching programs (resp. Boolean circuits) as parsimonious programs (Sect. 3), while APN^0 $\subseteq$ L/poly follows from a simple observation on the geometry of interaction, *i.e.*, the persistency of paths in proof nets (Sect. 4). The inclusions nuPL $\subseteq$ APN^0 and nuPL$_{\forall \ell}$ $\subseteq$ P/poly are based on polynomial step normalization (Proposition 2) on top of continuity, as sketched above (Sect. 5). Finally, the equality APN $=$ P/poly was proved in [19].

Note that continuous approximation allows us to generate a family of proof nets from a single parsimonious term. It is thus continuity that reconciles the two approaches to non-uniformity (family and individual). From a more practical perspective, generation of proof nets from a single program is reminiscent of the work of Ghica [7], who exhibits a way to synthesize VLSI circuits from a functional program.

Our parsimonious framework has two aspects. On the logical side, parsimonious logic amounts to multiplicative affine logic with what we call *Milner's exponential modality*, enjoying monoidal functorial promotion and *Milner's law* $!A \cong A \otimes !A$. With respect to the usual exponential modality, Milner's exponential refuses *digging* $!A \multimap !!A$ and *contraction* $!A \multimap !A \otimes !A$. One side of Milner's law is an asymmetric form of contraction $!A \multimap A \otimes !A$, also known as *absorption*, whereas its dual $A \otimes !A \multimap !A$ has a differential flavor [4]. Indeed, parsimonious logic resembles an affine subsystem of differential linear logic, but we have not fully explored the connection.

On the computational side, the parsimonious λ-calculus may be seen as an affine λ-calculus with built-in streams. This is because Milner's law naturally makes $!A$ be the type of streams on A: absorption is "pop" and coabsorption is "push". Stream calculi abound in the literature, also in connection with (classical) logic [18] and ICC [6,17]. However, these are all orthogonal to the present work, both in terms of motivations (modeling streams is not our primary concern) and technically (our streams arise from a previously unremarked restriction of the exponential modality of linear logic).

We should also mention the companion paper [14], which focuses on the uniform version of the simply-typed parsimonious calculus presented here, showing that it exactly captures L, *i.e.*, uniform logspace, nicely complementing the results presented here.

2   The Non-uniform Parsimonious Lambda-Calculus

We introduce an alternative term syntax for the infinitary parsimonious calculus and its associated type system, improving on previous work [13] in two respects. First, we avoid use of *indices* when referring to instances of exponential variables (that are reminiscent of indices in AJM games [1]). We instead use more conventional tools like list and let constructs. Second, the calculus has a closer fit with the type system, *i.e.*, the type system is *syntax-directed*. This offers a better logical account of the programs, not to mention easier type inference.

In the following, $[k]$ stands for the set $\{0, \dots, k-1\}$ for every $k \in \mathbb{N}$.

Terms. We let $a, b, c, \dots$ (resp. $x, y, z, \dots$) range over a denumerably infinite set of *linear* (resp. *exponential*) variables. A *pattern* p is either $a \otimes b$ or a list $a_0 :: a_1 :: \cdots :: a_{n-1} :: x$ with $n \geq 0$ and $a_0, \dots, a_{n-1}$ distinct. If $n = 0$, the latter denotes the one element list x. We often use notation $p(x)$ to indicate the last exponential variable x. The *terms* are inductively generated by

$$t, u ::= \bot \mid a \mid x \mid \lambda a.t \mid tu \mid t \otimes u \mid t :: u \mid \text{let } p = t \text{ in } u \mid !_f(u_0, \dots, u_{k-1}),$$

where $k \geq 1$ and $f : \mathbb{N} \longrightarrow [k]$. The expression $!_f(u_0, \dots, u_{k-1})$ is called a *box*. We use **u** to range over boxes. When $k = 1$, f is obvious and we write $!(u_0)$, or even $!u_0$. Restricting to boxes of this form yields a *uniform* calculus, which is the object of [14].

The box **u** generates an infinite stream $u_{f(0)} :: u_{f(1)} :: u_{f(2)} :: \cdots$. We denote the n-th component of the stream by $\mathbf{u}(n)$, and the result of removing the first n elements by $\mathbf{u}^{+n}$, so that the stream can be expressed as $\mathbf{u}(0) :: \mathbf{u}(1) :: \cdots :: \mathbf{u}(n-1) :: \mathbf{u}^{+n}$. More precisely, $\mathbf{u}^{+n}$ denotes $!_{f+n}(u_0, \dots, u_{k-1})$, where $f^{+n}(i) := f(n+i)$.

Binders behave as expected; $\lambda a.u$ and (let $a \otimes b = t$ in u) bind linear variable a (and also b in the latter case) occurring in u, while (let $\mathsf{p} = t$ in u) with $\mathsf{p} = a_0 :: \cdots :: a_{n-1} :: x$ binds both linear variables $a_0, \ldots, a_{n-1}$ and an exponential variable x occurring in u. We adopt the standard α-equivalence and Barendregt's variable convention. The constant $\perp$ corresponds to *coweakening* in logic. It is only used for auxiliary purposes in this paper.

Informally, $t :: \mathbf{u}$ expresses the result of pushing an element t to the stream $\mathbf{u}$. It is convenient to think of pattern $a_0 :: \cdots :: a_{n-1} :: x$ as a "non-uniform" variable. When "substituting" a box $\mathbf{u}$ for it, the first n variables $a_0, \ldots, a_{n-1}$ are replaced by the first n components $\mathbf{u}(0), \ldots, \mathbf{u}(n-1)$ (with free variables renamed), while x takes $\mathbf{u}^{+n}$.

A *slice* of a term is obtained by removing all components but one from each box. A term is *parsimonious* if (i) all its slices are affine, *i.e.*, each variable (linear or not) occurs at most once; (ii) box subterms do not contain free linear variables; and (iii) all exponential variables belong to a box subterm. Thus each exponential variable corresponds to an "auxiliary door" of a unique box. The set of parsimonious terms in the above sense is denoted by nuPΛ.

Reduction. One-step reduction $t \xrightarrow{\sigma} u$ is defined relatively to a finite set σ of the form $\{b_1 :: x_1, \ldots, b_k :: x_k\}$. We denote the set $\{b_1, \ldots, b_k, x_1, \ldots, x_k\}$ by $\mathrm{fv}(\sigma)$. There are seven elementary rules, among which (beta), (com1) and (com2) are standard:

(beta) $\quad\quad (\lambda a.t)u \xrightarrow{\emptyset} t[u/a]$ $\quad\quad$ (merge) let $x = \mathbf{u}$ in $w \xrightarrow{\emptyset} w\{\mathbf{u}/x\}$

(com2) let $\mathsf{p} = C[t]$ in $u \xrightarrow{\emptyset} C[$let $\mathsf{p} = t$ in $u]$ $\quad$ (com1) $\quad\quad C[t]u \xrightarrow{\emptyset} C[tu]$

(cons) $\;$ let $a :: \mathsf{p} = t :: v$ in $w \xrightarrow{\emptyset}$ let $\mathsf{p} = v$ in $w[t/a]$

(dup) $\quad$ let $a :: \mathsf{p} = \mathbf{u}$ in $w \xrightarrow{\sigma}$ let $\mathsf{p} = \mathbf{u}^{+1}$ in $w[\mathbf{u}(0)'/a]$

(aux) $\;$ let $x = t :: v$ in $w[\mathbf{u}] \xrightarrow{\sigma}$ let $x = v$ in $w[\mathbf{u}(0)'[t/x] :: \mathbf{u}^{+1}]$

where $C[\bullet]$ stands for a context of the form (let $\mathsf{q} = v$ in $\bullet$). The term $\mathbf{u}(0)'$ is obtained from $\mathbf{u}(0)$ by replacing its free exponential variables $x_1, \ldots, x_m$ (except x in (aux)) with fresh linear variables $b_1, \ldots, b_m$. Thus the (dup) and (aux) rules introduce new free variables, which are recorded in $\sigma := \{b_1 :: x_1, \ldots, b_m :: x_m\}$ and are bound later. In the (aux) rule, the notation $w[\mathbf{u}]$ means that w contains a box $\mathbf{u}$ s.t. $x \in \mathrm{fv}(\mathbf{u})$. If no such box exists, the term t is erased, *i.e.*, the right hand side is let $x = v$ in w. This rule corresponds to a cut between a cocontraction and the auxiliary port of a box in differential linear logic. We include it for completeness but we never need it.

The substitution $w\{\mathbf{u}/x\}$ in the (merge) rule needs an explanation. Suppose that $\mathbf{u} = \;!_f(u_0, \ldots, u_{k-1})$ and that x occurs in a box $\mathbf{w} = \;!_g(w_0, \ldots, w_{l-1})$. Our intention is to replace the stream $\ldots \mathbf{w}(i) \ldots$ with $\ldots \mathbf{w}(i)[\mathbf{u}(i)/x] \ldots$ To achieve this, let $v_{ik+j} := w_i[u_j/x]$ for each $i \in [l]$ and $j \in [k]$. Then $w\{\mathbf{u}/x\}$ is the result of replacing the box $\mathbf{w}$ with $!_h(v_0, \ldots, v_{lk-1})$, where $h(m) = g(m)k + f(m)$.

The above rules are extended contextually. We have:

$$\frac{t \xrightarrow{\sigma} u}{C[t] \xrightarrow{\sigma} C[u]} \qquad\qquad \frac{w \xrightarrow{\sigma \cup \{b :: x\}} w'}{\text{let } \mathsf{p}(x) = v \text{ in } w \xrightarrow{\sigma} \text{let } \mathsf{p}(b :: x) = v \text{ in } w'}$$

where the left rule applies when $C[t]$ or $C[u]$ does not bind any variable in $\mathrm{fv}(\sigma)$.

$$\overline{\Gamma;\Delta, a : A \vdash a : A}\ \text{ax}$$

$$\overline{\Gamma;\Delta \vdash \bot : A}\ \text{coweak}$$

$$\frac{\Gamma;\Delta, a : A \vdash t : B}{\Gamma;\Delta \vdash \lambda a.t : A \multimap B}\ \multimap\text{I}$$

$$\frac{\Gamma;\Delta \vdash t : A \multimap B \quad \Gamma';\Delta' \vdash u : A}{\Gamma,\Gamma';\Delta,\Delta' \vdash tu : B}\ \multimap\text{E}$$

$$\frac{\Gamma;\Delta \vdash t : A \quad \Gamma';\Delta' \vdash u : B}{\Gamma,\Gamma';\Delta,\Delta' \vdash t \otimes u : A \otimes B}\ \otimes\text{I}$$

$$\frac{\Gamma;\Delta \vdash t : A \otimes B \quad \Gamma';\Delta', a : A, b : B \vdash u : C}{\Gamma,\Gamma';\Delta,\Delta' \vdash \text{let } a \otimes b = t \text{ in } u : C}\ \otimes\text{E}$$

$$\frac{\Gamma, \mathsf{p} : A;\Delta, a : A \vdash t : C}{\Gamma, (a :: \mathsf{p}) : A;\Delta \vdash t : C}\ \text{abs}$$

$$\frac{\Gamma;\Delta \vdash t : A \quad \Gamma';\Delta \vdash u : !A}{\Gamma,\Gamma';\Delta,\Delta' \vdash (t :: u) : !A}\ \text{coabs}$$

$$\frac{;\Delta \vdash u_0 : A \quad \cdots \quad ;\Delta \vdash u_{k-1} : A}{\Delta';\vdash !_f(u'_0, \ldots, u'_{k-1}) : !A}\ \text{!!}$$

$$\frac{\Gamma;\Delta \vdash t : !A \quad \Gamma', \mathsf{p} : A;\Delta' \vdash u : C}{\Gamma,\Gamma';\Delta,\Delta' \vdash \text{let } \mathsf{p} = t \text{ in } u : C}\ \text{!E}$$

$$\frac{\Gamma;\Delta \vdash t : A \quad \alpha \notin \text{fv}(\Gamma,\Delta)}{\Gamma;\Delta \vdash t : \forall \alpha.A}\ \forall\text{I}$$

$$\frac{\Gamma;\Delta \vdash t : \forall \alpha.A \quad B \text{ is !-free}}{\Gamma;\Delta \vdash t : A[B/\alpha]}\ \forall\text{E}$$

Fig. 1. The typing system $\mathbf{nuPL}_{\forall\ell}$. Removing the last two rules yields $\mathbf{nuPL}$.

Finally we write $t \to u$ if $t \xrightarrow{\emptyset} u$ holds. This is our notion of one-step reduction. *Reduction* is the reflexive-transitive closure of $\to$, denoted by $\to^*$.

For instance, if $t := \text{let } a :: \mathsf{p} = !u(x) \text{ in } a \otimes v$ and $t' := \text{let } \mathsf{p} = !u(x) \text{ in } u(b) \otimes v$, we have $t \xrightarrow{\{b::x\}} t'$, so $\text{let } x = w \text{ in } t \to \text{let } b :: x = w \text{ in } t'$. Thus the "uniform" variable x is replaced by the "non-uniform" $b :: x$.

Reduction may be shown to be confluent. However, termination is not guaranteed: take $\Delta := \lambda b.\text{let } a :: x = b \text{ in } a\,!x$ and let $\Omega := \Delta\,!\Delta$. These terms are parsimonious, and we have $\Omega \to \Omega$, like the namesake usual λ-term.

Types. We take as types the formulas of intuitionistic second order linear logic:

$$A, B ::= \alpha \mid A \multimap B \mid A \otimes B \mid !A \mid \forall \alpha.A$$

where α is a type variable. The set of types (resp. $\forall$-free types) is denoted by $\mathsf{Type}_{\forall\ell}$ (resp. Type).

We adopt a type system with dual contexts as in [2]. Typing judgments are of the form $\Gamma;\Delta \vdash t : A$, where Γ is a set of assignments of the form $\mathsf{p}(x) : C$, while Δ consists of assignments of the form $a : C$. Moreover, all variables occurring in Γ, Δ are distinct.

A term t is *typable in* $\mathbf{nuPL}_{\forall\ell}$ if the judgment $\Gamma;\Delta \vdash t : A$ may be derived according to the rules of Fig. 1. Likewise, t is *typable in* $\mathbf{nuPL}$ if $\Gamma;\Delta \vdash t : A$ is derived without any use of the quantifier $\forall$.

In the rule !!, if $\Delta = \{b_1 : B_1, \ldots, b_m : B_m\}$, then $\Delta' := \{y_1 : B_1, \ldots, y_m : B_m\}$ with $y_1, \ldots, y_m$ fresh, and $u'_i := u_i[y_1/b_1, \ldots, y_m/b_m]$. Notice that the rule $\forall\text{E}$ is applicable *only when B is !-free*. An induction on the last rule of derivations gives

Lemma 1. *Suppose that* $;\Delta \vdash t : A$. *Then:*

1. **parsimony:** $t \in \text{nuP}\Lambda$;
2. **typical ambiguity:** $;\Delta[B/\alpha] \vdash t : A[B/\alpha]$ *for any type B;*
3. **subject reduction:** $t \to t'$ *implies* $;\Delta \vdash t' : A$.

In the sequel, we will mainly deal with simple types in Type. Polymorphic types in Type$_{\forall\ell}$ are considered only when necessary. When working with Type, it is convenient to fix a propositional variable, which we denote by o. If $A, B \in$ Type, $A[B]$ stands for $A[B/o]$. We will even omit B, just writing $A[]$. This lack of information is harmless for composition: point 2 of Lemma 1 guarantees that terms of type $A[X] \multimap B$ and $B[Y] \multimap C$ may be composed to yield $A[X[Y]] \multimap C$. The only delicate point is iteration (see below), which requires *flat* terms, *i.e.*, of type $A \multimap A$ (identical domain and codomain).

Examples. We set $\lambda p.t := \lambda c.$let $p = c$ in t with c a fresh variable, so that we have

$$(\lambda a \otimes b.t(a,b)) \, u \otimes v \to^* t(u,v), \qquad (\lambda a :: x.t(a,x)) \, u :: v \to^* t(u,v).$$

With this notation, the two terms implementing Milner's law may be written as

$$\lambda a :: x.a \otimes !x : !A \multimap A \otimes !A, \qquad \lambda a \otimes b.\text{let } x = b \text{ in } a :: !x : A \otimes !A \multimap !A.$$

Fig. 2 shows some examples of data and functions represented in **nuPL**. As usual, a natural number n is expressed by a Church numeral n of type Nat. The type Nat supports iteration $\text{It}(n, step', base) := n \, !(step') \, base$, typed as:

$$\frac{; \Delta \vdash step : A \multimap A \quad \Gamma; \Sigma \vdash base : A}{\Delta', \Gamma; \Sigma, n : \text{Nat}[A] \vdash \text{It}(n, step', base) : A}$$

where Δ' and $step'$ are the results of systematically replacing linear variables by exponential ones. Note that the type of $step$ must be flat.

Since succ has a flat type Nat $\multimap$ Nat, it may be iterated to result in the addition function of type Nat$[] \multimap$ Nat $\multimap$ Nat. It is again flat with respect to the second argument, so a further iteration leads to the multiplication function of type Nat$[] \multimap$ Nat$[] \multimap$ Nat. Subtraction is defined similarly.

Notably, Church numerals are duplicable and storable as shown in Fig. 2. Using addition, multiplication, subtraction and duplication we may represent any polynomial with integer coefficients as a closed term of type Nat$[] \multimap$ Nat.

Moreover, all these unary constructions can be extended to binary ones. We define Str $:= !(o \multimap o) \multimap !(o \multimap o) \multimap o \multimap o$. The Church representation for the binary string $w = b_0 \cdots b_{n-1} \in \{0,1\}^n$ is given by

$$\text{w} := \lambda s_0 :: \cdots :: s_{n-1} :: x.\lambda s_0' :: \cdots :: s_{n-1}' :: y.\lambda d.c_0(\cdots c_{n-1}d \cdots) : \text{Str},$$

where $c_i = s_i$ or s_i' depending on the bit b_i being 0 or 1.

For the Boolean values, we adopt the multiplicative type Bool $:= o \otimes o \multimap o \otimes o$ used in [19]. Just as numerals, words and Booleans are duplicable and storable.

An advantage of multiplicative Booleans is that they support *flat* exclusive-or $\oplus$ in addition to flat negation $\neg$. On the other hand, the best we can do for conjunction (and disjunction) is $\wedge :$ Bool$[] \multimap$ Bool $\multimap$ Bool, *i.e.*, one of the arguments must be non-flat. By contrast, $\wedge$ admits a flat typing in **nuPL**$_{\forall\ell}$ by redefining Bool $:= \forall\alpha.\alpha \otimes \alpha \multimap \alpha \otimes \alpha$. As we will see in the next section, this results in greater expressiveness, showing in fact that a flat representation of $\wedge$ is impossible in **nuPL** unless L/poly = P/poly.

Let $t :$ Str$[] \multimap$ Bool be a closed term typable in **nuPL** (resp. in **nuPL**$_{\forall\ell}$). It defines a language $L(t) := \{w \in \{0,1\}^* : t\text{w} \to^* \text{tt}\}$. Let nuPL (resp. nuPL$_{\forall\ell}$) be the class of all such languages.

$$\text{Nat} := !(o \multimap o) \multimap o \multimap o, \qquad \text{Bool} := o \otimes o \multimap o \otimes o$$

n	$:= \lambda s_0 :: \cdots :: s_{n-1} :: x.\lambda d.s_0(\ldots s_{n-1} d \ldots)$	: Nat
succ	$:= \lambda n.\lambda s :: x.\lambda d.s(n!(x)d)$	: Nat $\multimap$ Nat
pred	$:= \lambda n.\lambda x.\lambda d.n((\lambda a.a) :: !x)d$	: Nat $\multimap$ Nat
dup	$:= \lambda n.\text{lt}(n, \lambda m_1 \otimes m_2.(\text{succ } m_1) \otimes (\text{succ } m_2), 0 \otimes 0)$	: Nat[] $\multimap$ Nat $\otimes$ Nat
store	$:= \lambda n.\text{lt}(n, \lambda x.!(\text{succ } x), !0)$	: Nat[] $\multimap$!Nat
tt	$:= \lambda c \otimes d.c \otimes d, \qquad$ ff $:= \lambda c \otimes d.d \otimes c$	: Bool
$\neg$	$:= \lambda b.\lambda c \otimes d.b(d \otimes c)$	: Bool $\multimap$ Bool
$\oplus$	$:= \lambda b_1.\lambda b_2.\lambda c.b_1(b_2 c)$	: Bool $\multimap$ Bool $\multimap$ Bool
$\wedge$	$:= \lambda b_1.\lambda b_2.\text{let } c \otimes d = b_1(b_2 \otimes \text{ff}) \text{ in } c$	: Bool[] $\multimap$ Bool $\multimap$ Bool

Fig. 2. Some data types and encodings

3   Expressiveness of Non-uniform Parsimonious Terms

In this section, we prove

Theorem 2. L/poly $\subseteq$ nuPL *and* P/poly $\subseteq$ nuPL$_{\forall \ell}$.

An *n-input branching program* is a triple $P = (G^P, v_s^P, v_t^P)$ where: G^P is a finite directed acyclic graph (dag); its nodes have out-degree 2 or 0 and are labelled in $[n]$; each node of out-degree 2 has one outgoing edge labelled by 0 and one by 1; and v_s^P, v_t^P are the *source* and *target* node, the former having in-degree 0, and the latter having out-degree 0. The *size* of P is the number of nodes of G^P.

A binary string $w = b_0 \cdots b_{n-1} \in \{0, 1\}^n$ and an n-input branching program P induce a directed forest $P(w)$, as follows: take G^P and, for each node v of out-degree 2 whose label is i, erase the edge labelled by $1 - b_i$. Thus, $P(w)$ is a dag of out-degree at most 1, *i.e.*, a forest, whose edges are directed towards the roots. We say that P accepts w if v_s^P is a leaf of the tree whose root is v_t^P; otherwise, it rejects.

A *family* of branching programs is a sequence $(P_n)_{n \in \mathbb{N}}$ s.t., for all $n \in \mathbb{N}$, P_n is an n-input branching program. It is of *polynomial size* if there exists a polynomial p such that, for all $n \in \mathbb{N}$, the size of P_n is bounded by $p(n)$. It is well known [20, Theorem 4.38] that L/poly is exactly the class of languages decided by families of branching programs of polynomial size. Therefore, to prove the first part of Theorem 2 it will be enough to encode polysize families of branching programs, as sketched below.

Encoding a forest. We have a very simple encoding of a forest G thanks to a flat encoding of the exclusive-or function. Suppose that the nodes of G are $\{0, \ldots, m - 1\}$ and think of a token traversing G. We express the state "the token is placed at node i" by the term @i $:= !_{\delta_i}(\text{ff}, \text{tt})$, where $\delta_i(j) = 1$ iff $i = j$. The term @i is of type

Fig. 3. A forest

!Bool and expresses the stream in which only the i-th component is tt and the rest is ff.

A forest is expressed by a term of type !Bool $\multimap$!Bool. For instance the forest in Fig. 3 is expressed by the term t below:

$$\lambda(c_0 :: c_1 :: c_2 :: c_3 :: c_4 :: c_5 :: x). (\text{ff} :: \text{ff} :: \text{ff} :: c_0 \oplus c_1 :: c_3 \oplus c_4 :: c_2 \oplus c_5 :: !x).$$

This term replaces the 0th, 1st and 2nd components of a given stream with ff since there are no edges coming into the nodes 0, 1, 2. The 3rd is the exclusive-or of the 0th and 1st, while the 4th (resp. 5th) is the exclusive-or of the 3rd and 4th (resp. 2nd and 5th). Thus the term actually represents a graph where self-loops are added to terminal nodes 4 and 5. As a consequence, we have $\mathsf{lt}(\mathsf{k}, t, @i) \to^* @j$ iff terminal node j is reachable from node i, as far as $k \geq 2$. It is clear that the same encoding works for arbitrary forests of arbitrary size.

Encoding a branching program. Actually the edges of a forest are to be chosen according to the input binary string $w = b_0 \cdots b_{n-1}$ ($b_i \in \{0, 1\}$) fed to a branching program P_n. Hence for instance the component $c_0 \oplus c_1$ above should be replaced by a term like $(b_i \wedge c_0) \oplus (\neg b_j \wedge c_1)$ where b_i, b_j are the Boolean values of type $\mathsf{Bool}[]$ depending on which the edges $0 \to 3$ and $1 \to 3$ are drawn; the former is drawn when $b_i = 1$, while the latter is drawn when $b_j = 0$. This raises no typing problem since conjunction $\wedge : \mathsf{Bool}[] \multimap \mathsf{Bool} \multimap \mathsf{Bool}$ is flat with respect to the second argument. We therefore obtain a term $t_n : !\mathsf{Bool}[] \multimap !\mathsf{Bool} \multimap !\mathsf{Bool}$ expressing a branching program $P_n = (G_n, 0, 1)$ (assuming that the source and target are respectively node 0 and 1). Converting an input string of type Str into a stream of type $!\mathsf{Bool}[]$ is easy.

Encoding a family of branching programs. This is not the end of the story. We cannot store the whole family $(P_n)_{n \in \mathbb{N}}$ in a term as it is, in spite of the infinite facility of our calculus. In fact, a box $!_f(u_0, \ldots, u_{n-1})$ is morally an infinite stream on the finite alphabet $u_0, \ldots, u_{n-1}$. In particular, streams containing items of unbounded size (such as the P_n) are not allowed. We are thus led to consider *advice*, an infinite stream containing finitely many instructions, according to which each P_n is "woven" step by step.

For instance, to compute the Boolean value associated with a single node k (rather than the whole state of type $!\mathsf{Bool}$) from the previous state, it is sufficient to consider four instructions: $\mathtt{iskip}, \mathtt{nskip}, \mathtt{pos}, \mathtt{neg}$. Let $\mathsf{Adv} := o^4 \multimap o$ and represent each instruction by $\lambda a_0 a_1 a_2 a_3 . a_i$ with $i \in [4]$. Then any advice can be represented by a box $!_f(\mathtt{iskip}, \mathtt{nskip}, \mathtt{pos}, \mathtt{neg}) : !\mathsf{Adv}[]$ with a suitable f.

We consider the following term which expects four inputs: advice (encoded by a term of type $!\mathsf{Adv}[]$), input string ($!\mathsf{Bool}[]$), previous state ($!\mathsf{Bool}$) and temporal value of node k (Bool), and returns the updated values of the same type.

$$\lambda a \! :: \! x. \, \lambda b \! :: \! y. \, \lambda c \! :: \! z. \, \lambda d. \text{ case } a \text{ of } \begin{array}{ll} \mathtt{iskip} \to !(x) \otimes !(y) \otimes c \! :: \! !(z) \otimes d; \\ \mathtt{nskip} \to !(x) \otimes b \! :: \! !(y) \otimes !(z) \otimes d; \\ \mathtt{pos} \quad \to !(x) \otimes !(y) \otimes !(z) \otimes ((b \wedge c) \oplus d); \\ \mathtt{neg} \quad \to !(x) \otimes !(y) \otimes !(z) \otimes ((\neg b \wedge c) \oplus d). \end{array}$$

The instruction $\mathtt{iskip}$ (resp. $\mathtt{nskip}$) skips the first bit b of the input (resp. c of the previous state). The behaviors of $\mathtt{pos}$ and $\mathtt{neg}$ are as expected. For instance, starting from initial input string $b_0 :: \cdots :: b_{n-1}$, previous state $c_0 :: \cdots :: c_{m-1}$ and temporal value ff, iteration of the above term four times with advice $\mathtt{iskip} :: \mathtt{pos} :: \mathtt{nskip} :: \mathtt{neg}$ yields value $(b_1 \wedge c_0) \oplus (\neg b_2 \wedge c_2)$ (together with the rest of input $b_3 :: \cdots :: b_{n-1}$ and the rest of nodes $c_3 :: \cdots :: c_{m-1}$).

Actually things are more complicated, since unused values c_i in the previous state are not just thrown away but to be preserved for later use. Each bit b_j of the input string is to be used several times. Most importantly, we need to compute not just one value d

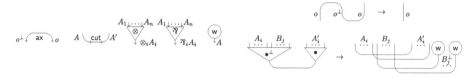

Fig. 4. Links (left) and cut-elimination steps (right). In the cut link, $A \perp A'$; $\bullet \in \{\otimes, \mathfrak{N}\}$.

(of type Bool) but a stream expressing the next state (of type !Bool). We can manage to do that by using more complicated instructions and types.

In the end, we obtain a term t_P : !Adv[] $\multimap$!Bool[] $\multimap$!Bool $\multimap$!Bool that "weaves" a forest $P_n(w)$ when an advice $\mathbf{a}_n$ of polynomial length $r(n)$ and an input string $\mathbf{w}$ of length n are provided. Actually the advice is given as a concatenation of all $(\mathbf{a}_i)_{i \in \mathbb{N}}$, from which a suitable advice for P_n is extracted by skipping the first $\sum_{i=1}^{n-1} r(i)$ components.

So far we gave a proof sketch of the first part of Theorem 2. For the second part, recall that $\mathbf{nuPL}_{\forall\ell}$ allows us to encode not only exclusive-or, but also conjunction and disjunction by terms of *flat* type Bool $\multimap$ Bool $\multimap$ Bool. Hence we may build a term similar to t_P above, which is able to "weave" a family of polynomial size Boolean circuits. This observation immediately leads to the second part of Theorem 2.

4 Boolean Nets and Logarithmic Space

We consider here multiplicative linear formulas of arbitrary arity, generated by

$$A, B ::= o \mid o^{\perp} \mid \otimes_{i \leq n} A_i \mid \mathfrak{N}_{i \leq n} A_i,$$

where $n \geq 1$ (the bound n will be omitted in the sequel unless necessary). Linear negation $(\cdot)^{\perp}$ is defined as usual via De Morgan laws (exchanging $\otimes$ and $\mathfrak{N}$). The *height* of a formula is its height as a tree. We will also need a more liberal notion of duality, which we denote by $A \perp B$, defined to be the smallest symmetric relation on formulas such that: $o \perp o^{\perp}$; if $A_1 \perp B_1, \ldots, A_n \perp B_n$, then $\otimes_{i \leq n} A_i \perp \mathfrak{N}_{i \leq n+k} B_i$ and $\mathfrak{N}_{i \leq n} A_i \perp \otimes_{i \leq n+k} B_i$, with $B_{n+1}, \ldots, B_{n+k}$ arbitrary. Of course $A \perp A^{\perp}$.

A *net* is formula-labelled directed graph built by composing the nodes of Fig. 4 (left), called *links*. Composition must respect the orientation, and the labeling must respect the constraints given in Fig. 4. Edges are allowed to have unconnected extremities. The edges incoming in (resp. outgoing from) a link are called *premises* (resp. *conclusions*) of the link. The number of premises of a $\otimes$ or $\mathfrak{N}$ link is called its *arity*. The premises of $\otimes$ and $\mathfrak{N}$ links are ordered, so we may speak of "the i-th premise". Each edge must be the conclusion of a link. The *size* of a net is the number of its links. Cut-elimination steps are defined in Fig. 4 (right).

Definition 1 (Boolean net). *An n-input Boolean net π is a net whose conclusions have type* Bool$[A_1]^{\perp}, \ldots,$ Bool$[A_n]^{\perp},$ Bool. *The height of π is the maximum height of A_i. A family of Boolean nets is sequence $(\pi_n)_{n \in \mathbb{N}}$ where each π_n is an n-input Boolean net. A family $(\pi_n)_{n \in \mathbb{N}}$ accepts (resp. rejects) a string $w \in \{0,1\}^*$ if the net obtained* *by cutting $\pi_{|w|}$ with $w_0, \ldots, w_{|w|-1}$ normalizes to* tt *(resp.* ff*). We denote by* APN^0 *the class of languages decided by families of Boolean nets of polynomial size and bounded height.*

A net π with n conclusions and k cut links has the shape given in the left hand side of Fig. 5, where $\rho_1, \ldots, \rho_n, \tau_1, \tau'_1, \ldots, \tau_k, \tau'_k$ are trees of $\otimes$, $\mathfrak{N}$ and w links, and ω consists solely of ax links. Moreover, since the leaves of ρ_i are labelled by atomic formulas, and since only ax and w

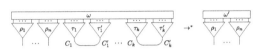

Fig. 5. A generic net π and its normal form π'

links may have atomic conclusions, the normal form π' of π (which exists and is unique) has the shape given in the right hand side of Fig. 5, where the conclusions of ω' are conclusions of ax or w links (ω' may contain irreducible cut links but this will not be important for us).

We will be interested in detecting *companion leaves* of ρ_i, ρ_j in π' (we allow $i = j$), *i.e.*, leaves which are conclusions of the same ax link of ω'. The geometry of interaction (GoI, [8]) gives us a way of doing this directly in π, without applying any cut-elimination step. The following is an immediate application of standard GoI definitions.

Proposition 1. *Let π be as in Fig. 5, of size s, and let h be the maximum height of $C_1, \ldots, C_k$. If e, e' are two leaves of ρ_i, ρ_j, deciding whether e, e' are companions in the normal form of π may be done in space $O(h \log s)$.*

Theorem 3. APN$^0 \subseteq$ L/poly.

Proof. Let $(\pi_n)_{n \in \mathbb{N}}$ be such a family, of size $p(n)$. Since p is a polynomial, the whole family may be encoded as advice. Deciding whether $w \in \{0, 1\}^n$ is accepted amounts to detecting whether a certain pair of leaves in π_n cut with the representations of the bits of w are companions in the normal form, which by Proposition 1 may be done in space $O(h \log(p(n))) = O(h \log n)$. But, by definition, the height h of cut formulas does not depend on n, so we conclude. $\qquad\square$

5   Approximations and Boolean Nets

We say that a term is *finite* if it contains only boxes of the form $!\bot$. Finite terms may be mapped to nets in a standard way, as follows. We first introduce the relation $A' \sqsubseteq A$ between classical multiplicative formulas and simple types, as the smallest such that: $o \sqsubseteq o$; if $A' \sqsubseteq A$ and $B' \sqsubseteq B$, then $(A')^{\perp} \mathfrak{N} B' \sqsubseteq A \multimap B$ and $A' \otimes B' \sqsubseteq A \otimes B$; if $A'_1, \ldots, A'_n \sqsubseteq A$, then $\otimes_{i \leq n} A'_i \sqsubseteq !A$. A straightforward induction on A gives

Lemma 2. $A', A'' \sqsubseteq A$ *implies* $(A')^{\perp} \perp A''$.

Let t be finite. We will map a type derivation of $\Gamma; \Delta \vdash t : A$ in **nuPL** to a net, which we abusively denote by $[\![t]\!]$, of conclusions Γ', Δ', A' such that $(\Gamma')^{\perp} \sqsubseteq !\Gamma$, $(\Delta')^{\perp} \sqsubseteq \Delta$ and $A' \sqsubseteq A$. The definition is by induction on the last rule of the derivation:

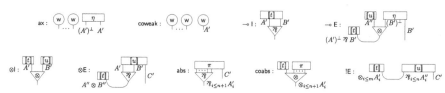

In the ax and coweak case, we choose Γ', Δ' and A' so that every ! is approximated in the minimal way, i.e., $n = 1$ in the definition of $\sqsubseteq$. In the cases ax and $\multimap$ E, η denotes the η-expansion net, defined as usual. In the cases $\multimap$ E, $\otimes$E and !E, Lemma 2 guarantees the soundness of the typing. For the rule !I, by finiteness the only possibility is that $k = 1$ and $u_0 = \bot$, so this case is treated as coweak. In the case abs (resp. coabs), the net π represents the net $[\![t]\!]$ (resp. $[\![u]\!]$) from which the $\mathfrak{P}$ (resp. $\otimes$) link corresponding to $p : A$ (resp. $!A$) has been removed. If no such link exists, it means that the conclusion came from a w (it cannot come from an atomic axiom), in which case we simply add a binary $\mathfrak{P}$ (resp. $\otimes$) link.

Note that the size of $[\![t]\!]$ is $O(s \cdot |t|)$, where s is the size of the types in t (this is because of the η nets). The following is standard:

Lemma 3. *Let t be finite and typable in* **nuPL.** *Then, $t \to t'$ implies $[\![t]\!] \to^* [\![t']\!]$.*

In what follows, for convenience we associate with each non-linear variable x a sequence $a_0, a_1, a_2, \ldots$ of pairwise distinct linear variables. We further suppose that if a (resp. b) is associated with x (resp. y), then $x \neq y$ implies $a \neq b$. The n-th *approximation* of $t \in$ nuPΛ is a finite term $\lfloor t \rfloor_n$ defined as follows: $\lfloor \bot \rfloor_n := \bot$; $\lfloor a \rfloor_n := a$; $\lfloor x \rfloor_n := \bot$; $\lfloor \lambda a.t \rfloor_n := \lambda a.\lfloor t \rfloor_n$; $\lfloor tu \rfloor_n := \lfloor t \rfloor_n \lfloor u \rfloor_n$; $\lfloor t \otimes u \rfloor_n := \lfloor t \rfloor_n \otimes \lfloor u \rfloor_n$; $\lfloor t :: u \rfloor_n := \lfloor t \rfloor_n :: \lfloor u \rfloor_n$; $\lfloor$ let $a \otimes b = u$ in $t \rfloor_n :=$ let $a \otimes b = \lfloor u \rfloor_n$ in $\lfloor t \rfloor_n$; $\lfloor$ let $p(x) = u$ in $t \rfloor_n :=$ let $p(a_0 :: \cdots :: a_{n-1} :: x) = \lfloor u \rfloor_n$ in $\lfloor t \rfloor_n$, where the a_i are associated with x; $\lfloor u \rfloor_n := \lfloor u(0)[\bar{a}_0/\bar{x}] \rfloor_n :: \cdots :: \lfloor u(n-1)[\bar{a}_{n-1}/\bar{x}] \rfloor_n :: !\bot$, where $\bar{x}$ are the free variables of the box and $\bar{a}_i$ are associated with $\bar{x}$.

Proposition 2. *Let t : Bool in* **nuPL$_{\forall \ell}$** *(or* **nuPL***). There exists a polynomial p depending solely on the types appearing in t and on its depth (the maximum number of nested boxes) such that, if $t \to^l$ b with $b \in \{tt, ff\}$, then $l \leq p(|t|)$. As a consequence, there is a polynomial q (with the same dependencies) s.t. $\lfloor t \rfloor_{q(|t|)} \to^* b$.*

Proof. The bound on the reduction length is proved by a careful reformulation of a standard cut-elimination argument. This immediately induces the approximation bound via continuity, as proved in [13, Lemmas 3 and 4]. □

Theorem 4. nuPL $\subseteq$ APN^0 *and* nuPL$_{\forall \ell}$ $\subseteq$ P/poly.

Proof. Given t : Str$[A] \multimap$ Bool and $n \in \mathbb{N}$, it is easy to obtain a term t' whose type is ; b_0 : Bool$[A \multimap A], \ldots, b_{n-1}[A \multimap A]$: Bool $\vdash t'(b_0, \ldots, b_{n-1})$: Bool. It takes n Booleans, converts them into a string of type Str$[A]$, and then passes it to t. If $w = b_0 \cdots b_{n-1} \in \{0, 1\}^n$, we have $u := t'(b_0, \ldots, b_{n-1}) \to^* tw \to^* b$. By Proposition 2, we have $\lfloor u \rfloor_{p(|u|)} \to^* b$ for p a polynomial not depending on w. But $|u| = O(n)$ by construction. Hence, by Lemma 3 the language decided by t may be decided by the family of Boolean nets $\pi_n := [\![\lfloor t' \rfloor_{q(n)}]\!]$ with q a polynomial, so the family is of polynomial size. Moreover, the conclusions of π_n are Bool$[B_n]^\perp, \ldots,$ Bool$[B_n]^\perp$, Bool, with

$B_n \sqsubseteq A \multimap A$, whose height does not depend on n (only the arity). Hence the family is also of bounded height. The second part is an immediate consequence of Proposition 2 as delineated in the introduction: the height of formulas plays no role, we normalize the underlying untyped term. $\square$

Acknowledgments. This work was partially supported by projects LOGOI ANR-2010-BLAN-0213-02, COQUAS ANR-12-JS02-006-01, ELICA ANR-14-CE25-0005 and JSPS KAKENHI 25330013.

References

1. Abramsky, S., Jagadeesan, R., Malacaria, P.: Full abstraction for PCF. Inf. Comput. **163**(2), 409–470 (2000)
2. Baillot, P., Terui, K.: Light types for polynomial time computation in lambda calculus. Inf. Comput. **207**(1), 41–62 (2009)
3. Bellantoni, S., Cook, S.A.: A new recursion-theoretic characterization of the polytime functions. Computational Complexity **2**, 97–110 (1992)
4. Ehrhard, T., Regnier, L.: Differential interaction nets. Electr. Notes Theor. Comput. Sci. **123**, 35–74 (2005)
5. Ehrhard, T., Regnier, L.: Uniformity and the taylor expansion of ordinary lambda-terms. Theor. Comput. Sci. **403**(2–3), 347–372 (2008)
6. Gaboardi, M., Péchoux, R.: Upper bounds on stream I/O using semantic interpretations. In: Grädel, E., Kahle, R. (eds.) CSL 2009. LNCS, vol. 5771, pp. 271–286. Springer, Heidelberg (2009)
7. Ghica, D.R.: Geometry of synthesis: a structured approach to VLSI design. In: Proceedings of POPL, pp. 363–375 (2007)
8. Girard, J.Y.: Geometry of interaction I: Interpretation of system F. Proccedings of Logic Colloquium **1988**, 221–260 (1989)
9. Girard, J.Y.: Light linear logic. Inf. Comput. **143**(2), 175–204 (1998)
10. Kfoury, A.J.: A linearization of the lambda-calculus and consequences. J. Log. Comput. **10**(3), 411–436 (2000)
11. Leivant, D., Marion, J.Y.: Lambda calculus characterizations of poly-time. Fundam. Inform. **19**(1/2) (1993)
12. Mazza, D.: An infinitary affine lambda-calculus isomorphic to the full lambda-calculus. In: Proceedings of LICS, pp. 471–480 (2012)
13. Mazza, D.: Non-uniform polytime computation in the infinitary affine lambda-calculus. In: Esparza, J., Fraigniaud, P., Husfeldt, T., Koutsoupias, E. (eds.) ICALP 2014, Part II. LNCS, vol. 8573, pp. 305–317. Springer, Heidelberg (2014)
14. Mazza, D.: Simple parsimonious types and logarithmic space (2015), available on the author's web page
15. Melliès, P.A.: Asynchronous games 2: The true concurrency of innocence. Theor. Comput. Sci. **358**(2–3), 200–228 (2006)
16. Melliès, P.-A., Tabareau, N., Tasson, C.: An explicit formula for the free exponential modality of linear logic. In: Albers, S., Marchetti-Spaccamela, A., Matias, Y., Nikoletseas, S., Thomas, W. (eds.) ICALP 2009, Part II. LNCS, vol. 5556, pp. 247–260. Springer, Heidelberg (2009)
17. Ramyaa, R., Leivant, D.: Ramified corecurrence and logspace. Electr. Notes Theor. Comput. Sci. **276**, 247–261 (2011)
18. Saurin, A.: Typing streams in the $\Lambda\mu$-calculus. ACM Trans. Comput. Log. **11**(4) (2010)
19. Terui, K.: Proof nets and boolean circuits. In: Proceedings of LICS, pp. 182–191 (2004)
20. Vollmer, H.: Introduction to circuit complexity - a uniform approach. Texts in theoretical computer science. Springer (1999)

Baire Category Quantifier
in Monadic Second Order Logic

Henryk Michalewski[1] and Matteo Mio[2]([✉])

[1] University of Warsaw, Warsaw, Poland
[2] CNRS/ENS-Lyon, Lyon, France
momatteo@gmail.com

Abstract. We consider Rabin's *Monadic Second Order* logic (MSO) of
the full binary tree extended with Harvey Friedman's "for almost all"
second-order quantifier ($\forall^*$) with semantics given in terms of Baire Category. In Theorem 1 we prove that the new quantifier can be eliminated
(MSO+$\forall^*$ = MSO). We then apply this result to prove in Theorem 2 that
the *finite–SAT* problem for the qualitative fragment of the probabilistic temporal logic pCTL* is decidable. This extends a previous result of
Brázdil, Forejt, Křetínský and Kučera valid for qualitative pCTL.

Keywords: Monadic second order logic · Baire category · pCTL*

1 Introduction

The main motivation of this paper is purely logical. We investigate the extension
of Rabin's *Monadic Second Order Theory of the Full Binary Tree* [14], henceforth simply shortened as MSO, with an additional "for almost all" second-order
quantifier $\forall^*$ whose set-theoretic semantics is defined as:

$$\forall^* X.\phi(X, \vec{Y}) \stackrel{\text{def}}{=} \left\{ \vec{Y} \mid \{X \mid \neg\phi(X, \vec{Y}) \text{ holds}\} \text{ is "topologically small"} \right\}$$

where topologically small is interpreted as *of Baire first category* (or *meager*)
in the standard topology on subsets of the full binary tree. Thus, for example,
the closed formula $\forall^* X.\phi(X)$ is valid if ϕ holds on all but a meager collection of
X's. To the best of our knowledge, the quantifier $\forall^*$ has been first introduced
and investigated, in the general context of First Order Logic, by H. Friedman in
unpublished manuscripts in 1978–79 (see [17] for an overview of this research).

Another extension of MSO with a *large cardinality* quantifier $\forall^{\aleph_0}$, defined by
replacing "is topologically small" with "has cardinality $\leq \aleph_0$" in the equation
defining $\forall^*$, has been recently investigated in [3] where it is proved that for every

H. Michalewski—Author supported by Polands National Science Centre grant no.
2014-13/B/ST6/03595.
M. Mio—Author supported by grant "Projet Émergent PMSO" of the École Normale
Supérieure de Lyon.

M.M. Halldórsson et al. (Eds.): ICALP 2015, Part II, LNCS 9135, pp. 362–374, 2015.
DOI: 10.1007/978-3-662-47666-6_29

MSO$+\forall^{\aleph_0}$ formula ψ there exists an equivalent MSO (without $\forall^{\aleph_0}$) formula $\hat{\psi}$ such that ψ and $\hat{\psi}$ denote the same set, that is the quantifier $\forall^{\aleph_0}$ can be expressed (i.e., *eliminated*) in ordinary MSO. This implies that the theory MSO$+\forall^{\aleph_0}$ is decidable. We prove a similar result for MSO$+\forall^*$.

Theorem 1. *For every* MSO$+\forall^*$ *formula ψ there exists an equivalent MSO (without $\forall^*$) formula $\hat{\psi}$ such that ψ and $\hat{\psi}$ denote the same set.*

Corollary 1. *The theory of* MSO$+\forall^*$ *is decidable.*

Our proof uses the fact, first proved in [11, Theorem 6.6], that every MSO–definable set of trees satisfies the *Baire property* (another, somewhat more elementary argument, can be deduced from Kolmogorov's theory of $\mathcal{R}$-sets, see [4, Theorem 3.8] and [9]). As a consequence, the Baire category of regular sets of trees can be determined using the well-known Banach–Mazur game (see, e.g., [12, §8]). Our main observation is that the game itself can be "implemented" via an alternating tree automaton (see Figure 4) which, while technically involved, is conceptually simple. For this reason our proof is radically different from that of [3] which is based on Shelah's *composition method* and does not involve automata constructions. An interesting topic for future research is to verify if, with techniques similar to those used in this work, the quantifier elimination theorem for MSO $+ \forall^{\aleph_0}$ of [3] can be proved using purely automata based methods. Further topics of future research and related work are discussed at the end of Section 3. The investigation of sets definable by formulas $\forall^*X.\phi(X, \vec{Y})$, with ϕ specifying Borel, analytic or $\mathcal{R}$–sets, instead of regular sets, has been an active area of research in descriptive set theory with contributions from R. Barua, J. P. Burgess, D. Miller, P. S. Novikov and R. Vaught among others. Our Theorem 1 can be considered an effective, automata–theoretic counterpart of set-theoretic results such as [20, Corollary1.10], [12, Theorem 29.22] and [4, Theorem 6.3].

An application to the theory of pCTL*. We apply our result on MSO$+\forall^*$ to the satisfiability problem of *probabilistic temporal logics* of programs (modeled as Markov chains) such as pCTL* [10] (see also [2, §10.4]). The work of Brázdil, Forejt, Křetínský and Kučera [6] is a main source of results in this area of research. A pCTL* formula can generally be satisfiable but only by infinite models, that is, by infinite Markov chains. Thus we distinguish between the *SAT problem* and the *finite–SAT problem* which, more restrictively, asks about the existence of finite models. In [6] the authors proved that both the SAT and the finite–SAT problems for the *qualitative fragment* of pCTL are decidable. We extend the second of these results to pCTL*.

Theorem 2. *The finite–SAT problem for qualitative pCTL* is decidable.*

Our proof method is based on a reduction of the finite–SAT problem to the satisfiability of MSO$+\forall^*$ which is decidable by Corollary 1. This proof technique is of general applicability and an equivalent of Theorem 2 can be proved even for more expressive probabilistic logics[1]. In contrast, the results of [6] provide much tighter algorithmic information, but the proof methods are specifically tailored to the logic pCTL and their applicability to other logics is not clear and requires a separate study.

The semantics of qualitative pCTL* is based on the (measure-theoretic) probabilistic concepts of *null set, set of positive measure* and *set of measure* 1. To reduce it to MSO+∀*, which can express the concept of Baire category, we apply a remarkable result of Ludwig Staiger ([16, Theorem 4]) which says that a regular set $L \subseteq \Sigma^\omega$ of infinite words is comeager if and only if $\mu(L) = 1$, where μ is the standard Lebesgue measure. This result is also used in [21] to develop a theory of fairness for concurrent systems. Using Staiger's theorem we prove that pCTL* with its standard probabilistic semantics and pCTL* with an alternative "Baire-categorical" semantics, where the state-formula $\mathbb{P}_{=1}\phi$ is interpreted as: "$s \models \mathbb{P}_{=1}\phi \Leftrightarrow$ the set of paths starting from s and satisfying the path–formula ϕ is comeager", agree on all finite models. This kind of observation is not new and has been already made (with respect to another logic) in the recent literature [1]. A proof of Theorem 2 is then obtained by combining these facts and by showing that the Baire–categorical semantics of pCTL* can be interpreted in MSO+∀*.

2   Background in Topology, Logic and Automata

Topology. Our exposition of topological and set–theoretical notions follows [12]. Given a topological space X, a set $A \subseteq X$ is *nowhere dense* if the interior of its closure is the empty set, that is $(\mathrm{int}(\mathrm{cl}(A)) = \emptyset$. A set $A \subseteq X$ is of *(Baire) first category* (or *meager*) if A can be expressed as countable union of nowhere dense sets. A set $A \subseteq X$ which is not meager is *of the second (Baire) category*. The complement of a meager set is called *comeager*. A set $B \subseteq X$ has the *Baire property* if $B = U \triangle M$, for some open set $U \subseteq X$ and meager set $M \subseteq X$, where $\triangle$ is the operation of symmetric difference $U \triangle M = (U \cup M) \setminus (U \cap M)$.

The set of natural numbers is denoted by ω. A topological space X is Polish if it is separable and completely metrizable. A main example of Polish space is the *Cantor space* $\{0, 1\}^\omega$ of infinite sequences of bits endowed with the product topology. The Cantor space is *zero-dimensional*, i.e., it has a basis of *clopen* (both open and closed) sets. We now describe the well-known Banach–Mazur game (see [12, 8.H] for a detailed overview) which characterizes Baire category.

Definition 1 (Banach–Mazur Game). *Let X be a zero–dimensional Polish space. For a given payoff set $A \subseteq X$ the infinite duration game $\mathbf{BM}(X, A)$ is played by Player I and Player II by sequentially choosing non-empty clopen sets*

$$
\begin{array}{lllll}
\textit{Player I} & U_0 & & U_2 & \cdots \\
\textit{Player II} & & U_1 & & U_3 \cdots
\end{array}
$$

---

[1] E.g., Theorem 2 can be proved for the, easily conceivable but to our knowledge never appeared in published work, probabilistic version of the non–probabilistic logic ECTL* (see, e.g., [18] for an introduction to non–probabilistic ECTL*).

with $U_{n+1} \subsetneq U_n$. *Player I wins if* $\bigcap_{n \in \omega} U_n \cap A \neq \emptyset$ *and Player II wins otherwise.*

Theorem 3. *Let X be a zero–dimensional Polish space. If $A \subseteq X$ has the Baire property, then* $\mathbf{BM}(X, A)$ *is determined and Player II wins iff A is meager.*

Monadic Second Order Logic. We assume familiarity of the reader with exposition of monadic second order logic (MSO). A standard reference is [19].

The set $\{L, R\}^*$ of finite words over the alphabet $\{L, R\}$ is called the *full binary tree* and each $w \in \{L, R\}^*$ is referred to as a *vertex*. The functions $\mathtt{Succ}_L$ ($w \mapsto w.L$) and $\mathtt{Succ}_R$ ($w \mapsto w.R$) are called *successor operations*. Given a finite alphabet Σ the function space $(\{L, R\}^* \to \Sigma)$ is denoted by $\mathcal{T}_\Sigma$ and an element $t \in \mathcal{T}_\Sigma$ is called a Σ-*labeled tree*, or just a Σ-*tree*. We identify $\{0, 1\}$-labeled trees, seen as characteristic functions, with sets of vertices. A tree $t \in \mathcal{T}_\Sigma$ is called *regular* if it has only finitely many subtrees up-to isomorphism. The space $\mathcal{T}_\Sigma$ has a natural topology homeomorphic to the Cantor space. A basis for the topology consists of clopen sets of Σ-trees extending a given finite prefix.

The language of MSO consists of *first order variables* x, y (ranging over vertices $w \in \{L, R\}^*$), second order variables X, Y (ranging over sets of vertices $t \in \mathcal{T}_{0,1}$), the set-theoretic membership relation $x \in X$ (interpreted as usual), the operations $\mathtt{Succ}_L$ and $\mathtt{Succ}_R$ (with interpretation given as above), the usual Boolean connectives ($\vee$, $\wedge$, $\neg$), first order quantifiers ($\forall x.\phi$, $\exists x.\phi$) and the second order quantifiers ($\forall X.\phi$, $\exists X.\phi$). In the rest of this paper we will consider MSO formulas whose free variables are all second order. This is not a significant restriction since it is well known (see, e.g., [19]) how to present MSO as a purely second order (i.e., without first order variables and quantifiers) theory. For a vector $\vec{Y} = (Y_1, \ldots, Y_n)$ and a variable X we write $\phi(X, \vec{Y})$ to denote that ϕ has precisely $n+1$ free variables $X, Y_1, \ldots, Y_n$. The set theoretic semantics of $\phi(X, \vec{Y})$ is the collection of $n + 1$-tuples of $\{0, 1\}$-trees $\langle t_0, t_1, \ldots, t_n \rangle$ satisfying the formula ϕ. Equivalently, $\phi(X, \vec{Y})$ defines a collection of Σ-trees with $\Sigma = \{0, 1\}^{n+1}$. A subset $A \subseteq \mathcal{T}_\Sigma$ is *regular* if it is definable by a MSO formula. Given a formula $\phi(X, \vec{Y})$ and a tuple $\vec{t} = \langle t_1, \ldots, t_n \rangle$ of $\{0, 1\}$-trees, the formula $\phi(X, \vec{t})$ with *parameters* $\vec{t}$ denotes the *section* $\{t_0 \mid \langle t_0, t_1, \ldots, t_n \rangle \in \phi(X, \vec{Y})\} \subseteq \mathcal{T}_{0,1}$. Note that $\phi(X, \vec{t})$ needs not be regular, but is regular when $\vec{t}$ is a regular tree.

Alternating Tree Automata. The importance of MSO stems from the fact that the theory is decidable. An approach to the proof of decidability taken in [13] is based on *alternating tree automata*. We include a brief exposition of alternating automata which follows the presentation in [13, Appendix C].

Definition 2 (Alternating automaton). *Given a finite set X, we denote with $\mathcal{DL}(X)$ the set of expressions e generated by the grammar $e ::= x \in X \mid e \wedge e \mid e \vee e$. An* alternating tree automaton *over a finite alphabet Σ is a tuple $\mathcal{A} = \langle \Sigma, Q, q_0, \delta, \mathcal{F} \rangle$ where Q is a finite set of states, $q_0 \in Q$ is the initial state, $\delta : Q \times \Sigma \to \mathcal{DL}(\{L, R\} \times Q)$ is the* alternating transition function*, $\mathcal{F} \subseteq \mathcal{P}(Q)$ is a set of subsets of Q called the* Muller condition*. The Muller condition $\mathcal{F}$ is* *called a* parity condition *if there exists a* parity assignment $\pi : Q \rightarrow \omega$ *such that:* $\mathcal{F} = \{F \subseteq Q \mid (\max_{q \in F} \pi(q))$ is even$\}$.

An alternating automaton $\mathcal{A}$ over the alphabet Σ defines, or "accepts", a set of Σ-trees. The *acceptance* of a tree $t \in \mathcal{T}_\Sigma$ is defined via a two-player ($\exists$ and $\forall$) game of infinite duration denoted as $\mathcal{A}(t)$. Game states of $\mathcal{A}(t)$ are of the form $\langle \vec{x}, q \rangle$ or $\langle \vec{x}, e \rangle$ with $\vec{x} \in \{L, R\}^*$, $q \in Q$ and $e \in \mathcal{DL}(\{L, R\} \times Q)$.

The game $\mathcal{A}(t)$ starts at state $\langle \epsilon, q_0 \rangle$. Game states of the form $\langle \vec{x}, q \rangle$, including the initial state, have only one successor state, to which the game progresses automatically. The successor state is $\langle \vec{x}, e \rangle$ with $e = \delta(q, a)$, where $a = t(\vec{x})$ is the labeling of the vertex $\vec{x}$ given by t. The dynamics of the game at states $\langle \vec{x}, e \rangle$ depends on the possibly nested shape of e. If $e = e_1 \vee e_2$, then Player $\exists$ moves either to $\langle \vec{x}, e_1 \rangle$ or $\langle \vec{x}, e_2 \rangle$. If $e = e_1 \wedge e_2$, then Player $\forall$ moves either to $\langle \vec{x}, e_1 \rangle$ or $\langle \vec{x}, e_2 \rangle$. If $e = (L, q)$ then the game progresses automatically to the state $\langle \vec{x}.L, q \rangle$. Lastly, if $e = (R, q)$ the game progresses automatically to the state $\langle \vec{x}.R, q \rangle$. Thus a play in the game $\mathcal{A}(t)$ is a sequence Π of game–states, that looks like: $\Pi = (\langle \epsilon, q_0 \rangle, \ldots, \langle L, q_1 \rangle, \ldots, \langle LR, q_2 \rangle, \ldots, \langle LRL, q_3 \rangle, \ldots, \langle LRLL, q_4 \rangle, \ldots)$, where the dots represent part of the play in game–states of the form $\langle \vec{x}, e \rangle$. Let $\infty(\Pi)$ be the set of automata states $q \in Q$ occurring infinitely often in configurations $\langle \vec{x}, q \rangle$ of Π. We then say that the play Π of $\mathcal{A}(t)$ is winning for $\exists$, if $\infty(\Pi) \in \mathcal{F}$. The play Π is winning for $\forall$ otherwise. The set (or "language") of Σ-trees defined by $\mathcal{A}$ is the collection $\{t \in \mathcal{T}_\Sigma \mid \exists$ has a winning strategy in the game $\mathcal{A}(t)\}$.

Definition 3. *An alternating automaton $\mathcal{A}$ is called* non-deterministic *if for all* $q \in Q$ *and* $a \in \Sigma$, *the expression* $\delta(q, a)$ *is n-ary disjunction* $e_1 \vee \cdots \vee e_n$ *where each disjunct e_i is a binary conjunction of the form* $\langle L, q_1 \rangle \wedge \langle R, q_2 \rangle$ *with* q_1, q_2 *not necessarily distinct.*

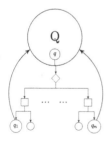

We will visualize alternating automata by diagrammatic pictures with the convention that $\Diamond$-shaped and $\Box$-shaped positions mark decisions of Player $\exists$ and Player $\forall$, respectively. For example, Figure 1 illustrates the shape of a non-deterministic automaton.

Fig. 1. A non–deterministic automaton $\mathcal{A}$ with states Q

The following theorem is of fundamental importance and states that alternating and nondeterministic automata have the same expressive power.

Theorem 4. *[13]   For every alternating automaton $\mathcal{A}$ there exists a non-deterministic parity automaton $\mathcal{B}$ defining the same set as $\mathcal{A}$.*

3   The Quantifier $\forall^*$ in MSO

In this section we introduce the extension of MSO with Friedman's "for almost all" quantifier interpreted using the concept of Baire category.

Definition 4 (MSO + $\forall^*$). *The syntax of* MSO + $\forall^*$ *extends that of* MSO *with the new second-order quantifier* $\forall^*$ *whose set-theoretic semantics is defined as:*

$$\forall^* X.\phi(X, \vec{Y}) \overset{\text{def}}{=} \left\{ \vec{t} \mid \neg\phi(X, \vec{t}) \subseteq \mathcal{T}_{0,1} \text{ is of first category} \right\}$$

The dual quantifier $\exists^* X.\phi$, *derivable as* $\exists^* X.\phi = \neg\forall^* X.(\neg\phi)$ *denotes the set* $\exists^* X.\phi(X, \vec{Y}) = \left\{ \vec{t} \mid \phi(X, \vec{t}) \subseteq \mathcal{T}_{0,1} \text{ is of second category} \right\}$.

The set denoted by $\forall^* X.\phi(X, \vec{Y})$ can be illustrated as in Figure 2, as the collection of trees $\vec{t}$ having a *large* (comeager) section $\phi(X, \vec{t})$. Informally, $\langle t_1, \ldots, t_n \rangle \in \forall^* X.\phi(X, \vec{Y})$ if, "for almost all" $t_0 \in \mathcal{T}_{0,1}$, the tuple $\langle t_0, t_1, \ldots, t_n \rangle$ satisfies ϕ.

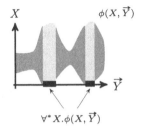

Fig. 2. Large section interpretation of Friedman's quantifier $\forall^*$. The large sections selected by quantifier $\forall^*$ are marked with lines.

Clearly, other kinds of "large section" quantifiers can be considered. Among others, the two quantifiers $\forall^{\aleph_0}$ and $\forall^{=1}$ obtained by replacing "is of first category" with "has cardinality $\leq \aleph_0$" and "has Lebesgue measure 0", are particularly natural since the σ-ideals of countable sets and of Lebesgue null sets are important set-theoretic notions of smallness. As already mentioned in the Introduction, the theory MSO+$\forall^{\aleph_0}$ has been first studied in [3]. Instead, to the best of our knowledge, the theories MSO+$\forall^*$ and MSO+$\forall^{=1}$ have never been investigated before.

Related and Future Work. The study of the system MSO+$\forall^{=1}$ appears to be an interesting topic of future research, especially in connection with investigations on probabilistic logics of programs. In this direction a relevant work is the recent paper of Carayol, Haddad and Serre [7] where the authors have developed a theory of nondeterministic tree automata with the usual acceptance condition on runs "every path must be accepting" replaced by "the set of accepting paths is of measure 1"[2]. Languages definable by such automata are called in [7] *qualitative tree languages*. Furthermore, Olivier Serre has explored in his habilitation thesis [15] and in a recent paper with Carayol [8], similar types of automata and languages obtained by replacing "measure 1" with "comeager" and "uncountable" in the acceptance condition.

The automata-based work of [7] and the approach based on extensions of MSO with large section quantifiers, followed in this paper and in [3], are to some extend similar, however the resulting theories diverge. Qualitative tree languages are not closed under complementation [7, Proposition 15], hence a comparison with a logic with negation such as MSO+$\forall^{=1}$ may be problematic.

---

[2] Other acceptance conditions based on measure are also investigated in [7].

4   Elimination of ∀* from MSO+∀*

This section is devoted to the proof of Theorem 1. The proof is by induction on the structure of the MSO + $\forall^*$ formula $\psi(\vec{Y})$ and consists in the construction of an alternating tree automaton $\mathcal{A}_\psi$ accepting the language defined by ψ. From the automaton $\mathcal{A}_\psi$, by Theorem 4 and by Rabin's theorem [14], one can effectively construct a purely MSO formula $\hat{\psi}(\vec{Y})$. The crucial step in the proof is associated with the case of $\psi(\vec{Y})$ having the form $\psi = \exists^* X . \phi(X, \vec{Y})$. By definition, $\exists^* X . \phi(X, \vec{Y})$ defines the set of n-tuples $\vec{t}$ of trees $\{\vec{t} \mid \phi(X, \vec{t})$ is not meager$\}$. The dual case of $\psi(\vec{Y}) = \forall^* X . \phi(X, \vec{Y})$ follows by complementation of automata.

By induction hypothesis, we can assume that the sub-formula ϕ is an ordinary MSO formula. So let $\mathcal{A}$ be a nondeterministic and parity automaton, schematically representable as in Figure 1, accepting the set of $(n{+}1)$-tuples of $\{0, 1\}$-trees defined by ϕ. Equivalently, the automaton $\mathcal{A}$ accepts Σ-trees with $\Sigma = \{0, 1\}^{n+1}$. We identify elements of Σ with sequences of bits $\langle b_0, b_1 \ldots b_n \rangle$ of length $n{+}1$. We now describe the well-known construction of automata $\mathcal{A}^\exists$ and $\mathcal{A}^\forall$ recognizing respectively the languages $\exists X . \phi(X, \vec{Y})$ and $\forall X . \phi(X, \vec{Y})$ over the restricted alphabet $\Sigma' = \{0, 1\}^n$. See [19] for a standard exposition.

Definition 5. *Let* $\mathcal{A} = (\Sigma, Q, q_0, \delta, \mathcal{F}_\pi)$ *be the nondeterministic parity automaton (with parity assignment* $\pi : Q \to \omega$*) accepting the language over the alphabet* $\Sigma = \{0, 1\}^{n+1}$ *defined by* ϕ*. The automaton* $\mathcal{A}^\exists$*, over the restricted alphabet* $\Sigma' = \{0, 1\}^n$*, is defined as* $\mathcal{A}^\exists = (\Sigma', Q, q_0, \delta^\exists, \mathcal{F}_\pi)$ *where* $\delta^\exists$ *is defined as*
$$\delta^\exists(q, \vec{b}) = e_0 \vee e_1 \iff \left(\delta(q, 0.\vec{b}) = e_0 \text{ and } \delta(q, 1.\vec{b}) = e_1 \right)$$
and $\vec{b} = \langle b_1, \ldots, b_n \rangle$*. Similarly,* $\mathcal{A}^\forall$ *is defined as:* $\mathcal{A}^\forall(\Sigma', Q, q_0, \delta^\forall, \mathcal{F}_\pi)$ *where*
$$\delta^\forall(q, \vec{b}) = e_0 \wedge e_1 \iff \left(\delta(q, 0.\vec{b}) = e_0 \text{ and } \delta(q, 1.\vec{b}) = e_1 \right).$$

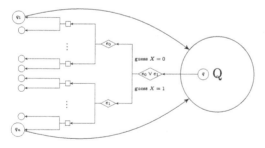

The three automata $\mathcal{A}$ (schematically depicted in Figure 1), $\mathcal{A}^\exists$ (Figure 3) and $\mathcal{A}^\forall$ have the same set of states Q and parity condition $\mathcal{F}_\pi$ with $\pi : Q \to \omega$. Note that $\mathcal{A}$ and $\mathcal{A}^\exists$ are non-deterministic while $\mathcal{A}^\forall$ is not. The following is a standard result.

Fig. 3. Automaton $\mathcal{A}^\exists$ constructed from $\mathcal{A}$

Proposition 1. *[13, 19] The automata* $\mathcal{A}^\exists$ *and* $\mathcal{A}^\forall$ *accept the languages defined by the formulas* $\exists X . \phi(X, \vec{Y})$ *and* $\forall X . \phi(X, \vec{Y})$*, respectively.*

We now introduce some convenient terminology. In the automaton $\mathcal{A}^\exists$, if the current state is q and the automata reads the letter $\vec{b}$ the transition $e_0 \vee e_1$ is reached. Here Player $\exists$ has two options:

Either choose the expression e_0, thus simulating the transition of $\mathcal{A}$ at q reading the letter $\langle 0, \vec{b} \rangle$. Then we say that "$\exists$ **chooses to label** X **with** 0",

Or choose the expression e_1, thus simulating the transition of $\mathcal{A}$ at q reading the letter $\langle 1, \vec{b} \rangle$. Then we say that "$\exists$ **chooses to label** X **with** 1".

Similarly, in the automaton $\mathcal{A}^\forall$, at the expression $e_0 \wedge e_1$, we say that "$\forall$ chooses to label X with 0" if Player $\forall$ moves to e_0 and that "$\forall$ chooses to label X with 1" if Player $\forall$ moves to e_1.

The alternating automaton $\mathcal{A}_\psi$ we are going to construct, recognizing the language of $\psi = \exists^* X.\phi(X, \vec{Y})$, is obtained by combining together the two automata $\mathcal{A}^\exists$ and $\mathcal{A}^\forall$. In what follows, to avoid confusion, we rename every state $q \in Q$ of the automaton $\mathcal{A}^\forall$ to q' so that $\mathcal{A}^\forall = (Q', q'_0, \delta^\forall, \mathcal{F}'_\pi)$, where π (and therefore $\mathcal{F}'_\pi$) and $\delta^\forall$ are defined over Q' as they were formerly defined over Q.

Before proceeding with the formal definition of $\mathcal{A}_\psi$ we describe informally the main ideas. The automaton $\mathcal{A}^\exists$ can be understood as a modified copy of $\mathcal{A}$ where player $\exists$ can choose (or "guess") labels of X, i.e., the $\{0,1\}$-labeled tree associated with the variable X in $\phi(X, \vec{Y})$. Similarly, in $\mathcal{A}^\forall$ choices related to X are made by Player $\forall$. In the automaton $\mathcal{A}_\psi$ we will implement the dynamics occurring in the Banach–Mazur game (see Definition 1) where Player I (as Player $\exists$) and Player II (as Player $\forall$) take turns in choosing how to label the tree associated with X. We will do so by defining $\mathcal{A}_\psi$ as an automaton consisting of two disjoint components $\mathcal{A}^\exists$ and $\mathcal{A}^\forall$ with special transitions allowing moving back and forth between these components. An appropriate Muller condition will be defined on $\mathcal{A}_\psi$ to enforce infinitely many alternations between components.

Definition 6. $\mathcal{A}_\psi = (\Sigma', Q \cup Q', q_0, \delta^\psi, \mathcal{F}_\psi)$ *where* $q_0 \in Q$ *is the initial state of* $\mathcal{A}^\exists$ *and, for* $\vec{q} = \langle q_1, \ldots, q_n \rangle \in \Sigma'$, *the transition function* δ^ψ *is defined as:*

$$\delta^\psi(q, \vec{b}) = e^\exists \vee e^\forall \iff \left(\delta^\exists(q, \vec{b}) = e^\exists \text{ and } \delta^\forall(q', \vec{b}) = e^\forall \right),$$

$$\delta^\psi(q', \vec{b}) = e^\exists \wedge e^\forall \iff \left(\delta^\exists(q, \vec{b}) = e^\exists \text{ and } \delta^\forall(q', \vec{b}) = e^\forall \right),$$

and $\mathcal{F}_\psi = \{ S \subseteq \mathcal{P}(Q \cup Q') \mid$ *the condition* $A \vee (B \wedge C)$ *holds*$\}$, *where* $A =$ "S *does not contain any* $q \in Q$", $B =$ "S *contains some* $q' \in Q'$" *and, lastly,* $C =$ "$proj(S) = \{q \in Q \mid q \in S \vee q' \in S\} \in \mathcal{F}_\pi$".

See Figure 4 for a graphical exposition of Definition 6. Some explanations are in order. The automaton $\mathcal{A}_\psi$ starts at the state q_0 which belongs to $\exists$-component. This is because the Banach–Mazur game starts with a move of Player I. When reading the letter $\vec{b}$ of the (root of the) input tree, the transition is of the form $\delta^\psi(q_0, \vec{b}) = e^\exists \vee e^\forall$ meaning that player $\exists$ has two options:

guess move: choose condition $e^\exists$, i.e., the same transition as in the automaton $\mathcal{A}^\exists$. Once transition $e^\exists$ is chosen, the first move associated with the expression $e^\exists$ will correspond to the choice of $\exists$ with regard to the labeling of X ($\exists$ should choose between 0 and 1). Note that the next state to be visited will be again in the $\mathcal{A}^\exists$ component because $e^\exists \in \mathcal{DL}(\{0,1\} \times Q)$.

skip move: choose condition $e^\forall$ as in the automaton $\mathcal{A}^\forall$. Once transition $e^\forall$ is chosen, the first move associated with the expression $e^\forall$ will correspond to the choice of $\forall$ with regard to the labeling of X ($\forall$ should choose between 0 and 1). Note that the next visited state is, this time, in the $\mathcal{A}^\forall$ component, because $e^\forall \in \mathcal{DL}(\{0,1\} \times Q')$.

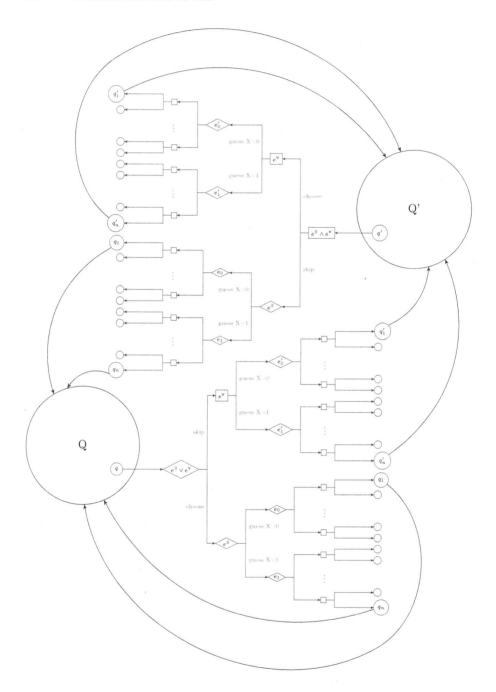

Fig. 4. Automaton $\mathcal{A}_\psi$ with states $Q \cup Q'$ and transitions depicted in this figure; $\diamond$-shaped positions mark decisionss of player $\exists$ and $\square$-shaped positions mark decisions of player $\forall$.

Guess and **skip** moves are marked in Figure 4. In a similar way, on states q' when reading the letter $\vec{b}$ the transition is of the form $\delta^\psi(q', \vec{b}) = e^\exists \wedge e^\forall$, hence Player $\forall$ can either make a **guess move** by picking $e^\forall$, "choose the labeling of X" and then remain in the $\forall$-component, or a **skip move** by picking $e^\exists$, thus allowing $\exists$ to "choose the labeling of X" and move to the $\exists$-component.

The Muller condition $\mathcal{F}_\psi$ captures the following aspects of the gameplay on the game tree $\mathcal{A}_\psi(\vec{t})$, where $\vec{t} = \langle t_1, \ldots, t_n \rangle$ with $t_i \in \mathcal{T}_{0,1}$:

A) a branch in $\mathcal{A}_\psi(\vec{t})$ is winning for $\exists$ if the $\exists$-component was visited only finitely many times. This intuitively means that, at some point, Player $\forall$ played "unfairly", never giving back the control to $\exists$.

B) if $\forall$ was "fair" and $\exists$ played unfairly, then $\exists$ loses and $\forall$ wins.

C) else, if both players played "fairly" alternating infinitely often between $\exists$ and $\forall$ components, then a branch in the game $\mathcal{A}_\psi(\vec{t})$ is winning for $\exists$ if and only if the sequence of visited states (ignoring the distinction between q and its copy q') is winning under the parity condition $\mathcal{F}_\pi$ (note that the parity assignment π is identical in all $\mathcal{A}$, $\mathcal{A}^\exists$ and $\mathcal{A}^\forall$).

Thus the automaton $\mathcal{A}_\psi$ implements the policy of infinite alternation between $\exists$ and $\forall$ in "guessing" the set X. Due to space limitations, a detailed proof of the fact that $\mathcal{A}_\psi$ accepts the set defined by $\psi = \exists^* X. \phi(X, \vec{Y})$ is not included and will appear in an extended version of this paper.

5   Applications to Probabilistic Logics

In this section we consider the *qualitative fragment* of the *probabilistic logic CTL** (pCTL*), as introduced in [10], which can express useful properties of Markov chains. We refer to the book [2, §10.4] for a detailed introduction.

Definition 7 (Markov Chain). *A Markov chain is a triple* (V, E, p), *where* (V, E) *is a directed graph and* $p: E \to [0, 1]$ *is a function assigning probabilities to each edge in such a way that the sum of the probabilities of edges leaving every vertex v is 1. A Markov chain is* finite *if V is finite and $p(e)$ is a rational number for all $e \in E$. A Markov chain is* simple *if every vertex has exactly two outgoing edges both labeled with probability* $\frac{1}{2}$.

Definition 8 (Syntax of Qualitative pCTL*). *Given a finite set of atomic predicates* $a_1, \ldots, a_n$, *qualitative pCTL* formulas are generated by the following two-sorted grammar:* state formulas $\Phi, \Psi ::= a_i \mid \neg \Phi \mid \Psi \vee \Phi \mid \phi \mid \mathbb{P}_{=1}\phi$, *and* path formulas $\phi, \psi ::= \Phi \mid \neg \phi \mid \phi \vee \psi \mid \circ \phi \mid \phi \, \mathcal{U} \, \psi$, *where* $\circ$ *and* $\mathcal{U}$ *are the usual* Next *and* Until *operators of linear time logic.*

Definition 9 (Semantics of Qualitative pCTL*). *Given a Markov chain* $M = (V, E, p)$ *and interpretations of the atomic predicates* $||a_i||_M \subseteq V$, *state formulas* Φ *are interpreted as sets* $||\Phi||_M \subseteq V$ *of vertices and path formulas* ϕ *are interpreted as sets* $||\phi||_M$ *of paths in the graph* (V, E). *The inductive definition is the same as that of CTL* (see [2, Definition 6.81]) with the addition of:* $v \in ||\mathbb{P}_{=1}\phi||_M \Leftrightarrow$ "$||\phi||_M$ *has measure 1 in the set of paths starting at v", where* *the measure on paths is defined in the standard way (see, e.g., [2, §10.4]) inferred from probabilities on the edges of M.*

Definition 10 (Finite Satisfiability). *We say that a formula Φ is finite satisfiable (finite–SAT) if there exists a finite Markov chain $M = (V, E, p)$ with interpretations $||a||_M$ of the atomic predicates such that $||\Phi||_M \neq \emptyset$.*

It has been observed in [5] that the finite–SAT problem can be restricted to finite–and–simple Markov chains. The observation follows from the fact, that one can transform any finite Markov chain into a finite and simple one, at the cost of introducing auxiliary "dummy states", by first simulating states with n outgoing edges by a sequence of binary choices and then simulating binary choices having arbitrary rational probabilities with a finite (cyclic) system of binary $\frac{1}{2}$-weighted choices. Furthermore, for every pCTL* formula Φ one can construct a formula $\hat{\Phi}$ (with an additional predicate for "dummy states") such that Φ is finite–SAT if and only if $\hat{\Phi}$ is satisfied by a finite and simple Markov chain. Thanks to this observation we can replace "finite" with "finite and simple" in Definition 10.

Definition 11 (Categorical Semantics of Qualitative pCTL*). *Given a Markov chain $M = (V, E, p)$, the categorical semantics of a given state formula Φ is a set $[\![\Phi]\!]_M \subseteq V$, defined as in the standard semantics $||\Phi||_M$ (Definition 9) on all connectives and on all path formulas except for $\mathbb{P}_{=1}$ which is instead defined as follows: $v \in [\![\mathbb{P}_{=1}\phi]\!]_M \Leftrightarrow$ "$||\phi||_M$ is comeager in the set of paths starting at v".*

Theorem 5. *Let $M = (V, E, p)$ be a finite and simple Markov chain and assume $||a_i||_M = [\![a_i]\!]_M$, for all atomic predicates a_i. Then $||\Phi||_M = [\![\Phi]\!]_M$ for every qualitative pCTL* formula Φ.*

Proof. The proof goes by induction on the structure of Φ with the only non-trivial case being $\Phi = \mathbb{P}_{=1}\phi$. Assume that ϕ is build from state formulas $\Psi_0, \ldots, \Psi_n$. By inductive hypothesis $||\Psi_i||_M = [\![\Psi_i]\!]_M$, for $0 \leq i \leq n$. It follows that $||\phi||_M = [\![\phi]\!]_M$. By standard arguments, $||\phi||_M$ denotes a regular set of paths in the graph (V, E). We only prove $||\Phi||_M \subseteq [\![\Phi]\!]_M$ as the case $[\![\Phi]\!]_M \subseteq ||\Phi||_M$ is similar. Assume, by contradiction, that $v \in ||\Phi||_M$ and $v \notin [\![\Phi]\!]_M$. This means that the regular set $||\phi||$ has measure 1 in the set of paths starting from v. By Staiger's theorem [16, Theorem 4] (see also [21, Theorem 9.8] for a convenient graph–theoretical formulation) this implies that $||\phi||_M$ is comeager in the set of paths starting from v. It then follows that $v \in [\![\Phi]\!]_M$ and thus we have the desired contradiction. $\square$

Theorem 2 in the Introduction, that is the decidability of the finite–SAT problem for qualitative pCTL*, follows from the theorem below along with the decidability of MSO+∀* (Corollary 1 of Theorem 1 in this paper).

Theorem 6. *To each qualitative pCTL* formula Φ with atomic predicates $a_1 \ldots a_n$ one can effectively associate a MSO+∀* formula $F_\Phi(x, X_{a_0}, \ldots, X_{a_n})$, such that "$\Phi$ is finite–SAT" $\Leftrightarrow$ "F_Φ is satisfiable."*

Proof. As discussed above, we can restrict attention to simple Markov chains M. The full binary tree $\{L, R\}^*$ can be viewed as a simple (infinite) Markov chain by labeling each edge with $\frac{1}{2}$. Furthermore, since pCTL* is invariant under *bisimulation* [2, Theorem 10.67], each Markov chain M (which is a binary graph since the probabilistic information is implicit) can be replaced by its unraveling $\{L, R\}^*$. Each interpretation $||a_i||_M$ of the atomic predicates can then by identified with a corresponding $t_i \in \mathcal{T}_{0,1}$. Hence simple (finite) Markov chains M with interpretations $[\![a_i]\!]_M$ of the atomic predicates can be identified with (regular) Σ-trees with $\Sigma = \{0, 1\}^n$. The proof goes by induction on the structure of state formulas Φ by defining formulas $F_\Phi(x, \vec{X}_{a_i})$ with the following property. An arbitrary Σ-tree $\vec{t} = \langle t_0, \ldots, t_n \rangle$ with vertex $w \in \{L, R\}^*$ satisfies F_Φ iff $w \in [\![F_\Phi]\!]_M$ with interpretation of the atomic predicates as $[\![a_i]\!] = t_i$. The construction of F_Φ follows the standard method (see, e.g., [18]) for interpreting CTL* into MSO. The only non-standard step is for $\Phi = \mathbb{P}_{=1}\phi$. The encoding in MSO + $\forall^*$ is not entirely trivial and will appear in an extended version of this paper. By Rabin's theorem [14], the formula F_Φ is satisfiable iff it satisfiable by a regular Σ-tree which can be interpreted as a finite-and-simple Markov chain M_t satisfying Φ. The desired result then follows by Theorem 5. $\square$

References

1. Baier, C., Bertrand, N., Bouyer, P., Brihaye, T., Größer, M.: Probabilistic and topological semantics for timed automata. In: Arvind, V., Prasad, S. (eds.) FSTTCS 2007. LNCS, vol. 4855, pp. 179–191. Springer, Heidelberg (2007)
2. Baier, C., Katoen, J.P.: Principles of Model Checking. The MIT Press (2008)
3. Bárány, V., Kaiser, Ł., Rabinovich, A.: Cardinality quantifiers in MLO over trees. In: Grädel, E., Kahle, R. (eds.) CSL 2009. LNCS, vol. 5771, pp. 117–131. Springer, Heidelberg (2009)
4. Barua, R.: R-sets and category. Trans. Amer. Math. Soc. **286**(1), 125–158 (1984)
5. Bertrand, N., Fearnley, J., Schewe, S.: Bounded satisfiability for PCTL. In: Proc. of CSL (2012)
6. Brázdil, T., Forejt, V., Křetínský, J., Kučera, A.: The satisfiability problem for probabilistic CTL. In: Proc. of LICS, pp. 391–402 (2008)
7. Carayol, A., Haddad, A., Serre, O.: Randomization in automata on infinite trees. ACM Transactions on Computational Logic **15**(3) (2014)
8. Carayol, A., Serre, O.: How good is a strategy in a game with nature? In: To appear in Proc. LICS (2015)
9. Gogacz, T., Michalewski, H., Mio, M., Skrzypczak, M.: Measure properties of game tree languages. In: Csuhaj-Varjú, E., Dietzfelbinger, M., Ésik, Z. (eds.) MFCS 2014, Part I. LNCS, vol. 8634, pp. 303–314. Springer, Heidelberg (2014)
10. Hansson, H., Jonsson, B.: A logic for reasoning about time and reliability. Formal Aspects of Computing **6**(5), 512–535 (1994)
11. Hjorth, G., Khoussainov, B., Montalban, A., Nies, A.: From automatic structures to Borel structures. In: Proc. of LICS, pp. 431–441 (2008)
12. Kechris, A.S.: Classical Descriptive Set Theory. Springer (1994)

13. Muller, D.E., Schupp, P.E.: Simulating alternating tree automata by nondeterministic automata: New results and new proofs of the theorems of Rabin. McNaughton and Safra. Theor. Comput. Sci. **141**(1&2), 69–107 (1995)
14. Rabin, M.O.: Decidability of second-order theories and automata on infinite trees. Transactions of American Mathematical Society **141**, 1–35 (1969)
15. Serre, O.: Playing with Trees and Logic. Habilitation Thesis, Université Paris Diderot (Paris 7) (2015)
16. Staiger, L.: Rich omega-words and monadic second-order arithmetic. In: Proc. of CSL, pp. 478–490 (1997)
17. Steinhorn, C.I.: Chapter XVI: Borel Structures and Measure and Category Logics. Perspectives in Mathematical Logic, vol. 8. Springer (1985)
18. Thomas, W.: On chain logic, path logic, and first-order logic over infinite trees. In: Proc. of LICS, pp. 245–256 (1987)
19. Thomas, W.: Languages, automata, and logic. In: Handbook of Formal Languages, pp. 389–455. Springer (1996)
20. Vaught, R.: Invariant sets in topology and logic. Fund. Math., 82, 269–294 (1974/1975), Collection of articles dedicated to Andrzej Mostowski on his sixtieth birthday, VII
21. Völzer, H., Varacca, D.: Defining fairness in reactive and concurrent systems. Journal of the ACM **59**(3) (2012)

Liveness of Parameterized Timed Networks

Benjamin Aminof[1], Sasha Rubin[2], Florian Zuleger[1], and Francesco Spegni[3]([✉])

[1] TU Vienna, Vienna, Austria
[2] Universitá degli Studi di Napoli "Federico II", Napoli, Italy
[3] Universitá Politecnica delle Marche, Ancona, Italy
spegni@dii.univpm.it

Abstract. We consider the model checking problem of infinite state systems given in the form of parameterized discrete timed networks with multiple clocks. We show that this problem is decidable with respect to specifications given by B- or S-automata. Such specifications are very expressive (they strictly subsume ω-regular specifications), and easily express complex liveness and safety properties. Our results are obtained by modeling the passage of time using symmetric broadcast, and by solving the model checking problem of parameterized systems of untimed processes communicating using k-wise rendezvous and symmetric broadcast. Our decidability proof makes use of automata theory, rational linear programming, and geometric reasoning for solving certain reachability questions in vector addition systems; we believe these proof techniques will be useful in solving related problems.

1 Introduction

Timed automata — finite state automata enriched by a finite number of dense- or discrete-valued clocks — can be used to model more realistic circuits and protocols than untimed systems [3,7]. A timed network consists of an arbitrary but fixed number of timed automata running in parallel [1,2]. In each computation step, either some fixed number of automata synchronize by a rendezvous-transition or time advances. We consider the *parameterized model-checking problem (PMCP)* for timed networks: Does a given specification (usually given by a suitable automaton) hold *for every system size*? Apart from a single result which deals with much weaker synchronization than rendezvous [13], no positive PMCP results for liveness specifications of timed automata are known.

System model: In this paper we prove the decidability of the PMCP for discrete timed networks with no controller and liveness specifications. To do this, we reduce the PMCP of these timed networks to the PMCP of *RB-systems* — systems of finite automata communicating via k-*wise rendezvous* and *symmetric*

Benjamin Aminof and Florian Zuleger were supported by the Austrian National Research Network S11403-N23 (RiSE) of the Austrian Science Fund (FWF) and by the Vienna Science and Technology Fund (WWTF) through grant ICT12-059. Sasha Rubin is a Marie Curie fellow of the Istituto Nazionale di Alta Matematica.

© Springer-Verlag Berlin Heidelberg 2015
M.M. Halldórsson et al. (Eds.): ICALP 2015, Part II, LNCS 9135, pp. 375–387, 2015.
DOI: 10.1007/978-3-662-47666-6_30

broadcast. This broadcast action is symmetric in the sense that there is no designated sender. In contrast, the standard broadcast action can distinguish between sender and receivers, and so the PMCP of liveness properties is undecidable even in the untimed setting [9].

Our Techniques and Results: Classical automata (e.g., nondeterministic Büchi word automata (NBW)) are not able to capture the behaviors of RB-systems. Thus, our decidability result uses nondeterministic BS-automata (and their fragments B- and S-automata) which strictly subsume NBW [6].

We show that the PMCP is decidable for controllerless discrete timed networks and (and systems communicating via k-wise rendezvous and symmetric broadcast) and specifications given by B-automata or S-automata (and in particular by NBW) or for negative specifications (i.e., the set of bad executions) given by BS-automata. We prove decidability by constructing a B-automaton that precisely characterizes the runs of a timed network from the point of view of a single process. Along the way, we also obtain an EXPSPACE upper bound for the PMCP of safety properties of discrete timed networks.

In order to build the B-automaton, an intricate analysis of the interaction between the transitions caused by the passage of time (modeled by broadcasts) which involve all processes, and those that are the result of rendezvousing processes, is needed. It is this interaction that makes the problem complicated. Thus, for example, results concerning pairwise rendezvous without broadcast [11] do not extend to our case. Our solution to this problem involves the introduction of the idea of a *rational relaxation* of a Vector Addition System, and geometric lemmas concerning paths in these relaxations. It is important to note that these vector addition systems can not capture the edges that correspond to the passage of time. However, they provide the much needed flexibility in capturing what happens in between time ticks *in the presence of* these ticks.

Related Work. Discrete timed networks with rendezvous and a controller were introduced in [1] where it was shown that safety is decidable using the technique of well-structured transition systems. Their result implies a non-elementary upper bound (which we improve to EXPSPACE) for the complexity of the PMCP of safety properties of timed networks without a controller. PMCP of liveness properties for continuous-time networks with a controller process is undecidable [2]. However, their proof heavily relies on time being dense and on the availability of a distinguished controller process. RB-systems with a controller were introduced in [12] where it is proved that under an additional strong restriction on the environment and process templates (called a shared simulation), such systems admit cutoffs that allow one to model check epistemic-temporal logic of the parameterised systems. The main difference between our work and theirs is: we do not have a controller, we make no additional restrictions, and we can model check specifications given by B- or S-automata. The authors in [13] proved that the PMCP is decidable for continuous timed networks synchronizing using conjunctive Boolean guards and MITL and TCTL specifications. Finally, there are many decidability and undecidability results in the untimed setting, e.g.,[4,5,8,9,14].

2   Definitions and Preliminaries

Labeled Transition Systems. A *(edge-)labeled transition system (LTS)* is a tuple $\langle S, I, R, \Sigma \rangle$, where S is the set of *states* (usually $S \subseteq \mathbb{N}$), $I \subseteq S$ are the *initial states*, $R \subseteq S \times \Sigma \times S$ is the *edge relation*, and Σ is the *edge-labels alphabet*. *Paths* are sequences of transitions, and runs are paths starting in initial states.

Automata. We use standard notation and results of automata, such as nondeterministic Büchi word automata (NBW) [15]. A *BS-word automaton (BSW)* ([6]) is a tuple $\langle \Sigma, Q, Q_0, \Gamma, \delta, \Phi \rangle$ where Σ is a finite *input alphabet*, Q is a set of *states*, $Q_0 \subseteq Q$ is a set of *initial states*, Γ is a set of *counter (names)*, $\delta \subseteq Q \times \Sigma \times \mathcal{C}^* \times Q$ is the *transition relation* where $\mathcal{C}$ is the set of *counter operations*, i.e. $c := 0, c := c + 1, c := d$ for $c, d \in \Gamma$, and Φ is the *acceptance condition* described below. A run ρ is defined like for nondeterministic automata over infinite words by ignoring the $\mathcal{C}^*$ component. Denote by $c(\rho, i)$ the ith value assumed by counter $c \in \Gamma$ along ρ. The acceptance condition Φ is a positive Boolean combination of the following conditions ($q \in Q, c \in \Gamma$): (i) q is visited infinitely often (Büchi-condition); (ii) $\limsup_i c(\rho, i) < \infty$ (B-condition); (iii) $\liminf_i c(\rho, i) = \infty$ (S-condition). An automaton that does not use B-conditions is called an *S-automaton (SW)*, and one that does not use S-conditions is called a *B-automaton (BW)*.

It is known that BSWs are relatively well behaved [6]: their emptiness problem is decidable; they are closed under union and intersection, but not complement; and BW (resp. SW) can be complemented to SW (resp. BW). Since BSWs are not closed under complement, we are forced, if we are to use the automata-theoretic approach for model checking (cf. [15]), to give the specification in terms of the undesired behaviours, or to consider specifications in terms of BWs or SWs (which both strictly extend ω-regular languages).

Rendezvous with Symmetric Broadcast (RB-System). Intuitively, RB-systems describe the parallel composition of $n \in \mathbb{N}$ copies of a process *template*. An RB-system evolves nondeterministically: either a k-wise rendezvous action is taken, i.e., k different processes instantaneously synchronize on a rendezvous action a, or the symmetric broadcast action is taken, i.e., all processes must take an edge labeled by b. Systems without the broadcast action are called R-systems.

In the rest of the paper, fix k (the number of processes participating in a rendezvous), a finite set Σ_{actn} of *rendezvous actions*, the *rendezvous alphabet* $\Sigma_{\mathsf{rdz}} = \cup_{\mathsf{a} \in \Sigma_{\mathsf{actn}}} \{\mathsf{a}_1, \ldots, \mathsf{a}_k\}$, and the *communication alphabet* Σ_{com} which is the union $\{((i_1, \mathsf{a}_1), \ldots, (i_k, \mathsf{a}_k)) \mid \mathsf{a} \in \Sigma_{\mathsf{actn}}, i_j \in \mathbb{N}, j \in [k]\} \cup \{\mathsf{b}\}$.

A *process template* (or *RB-template*) is a finite LTS $P = \langle S, I, R, \Sigma_{\mathsf{rdz}} \cup \{\mathsf{b}\} \rangle$ such that for every state $s \in S$ there is a transition $(s, \mathsf{b}, s') \in R$ for some $s' \in S$. We call edges labeled by b *broadcast edges*, and the rest *rendezvous edges*. For ease of exposition, we assume (with one notable exception, namely $P^{-\circ}$ defined in Section 3) that for every $\varsigma \in \Sigma_{\mathsf{rdz}}$ there is at most one edge in P labeled by

ς and we denote it by $\mathrm{edge}(\varsigma)$.[1] The *RB-system* $\mathcal{P}^n$ is defined, given a template P and $n \in \mathbb{N}$, is defined as the finite LTS $\langle Q^n, Q_0^n, \Delta^n, \Sigma_{\mathsf{com}} \rangle$[2] where:

1. Q^n is the set of functions (called *configurations*) of the form $f : [n] \to S$. We call $f(i)$ the *state of process i* in f. Note that we sometimes find it convenient to consider a more flexible naming of processes in which we let Q^n be the set of functions $f : X \to S$, where $X \subset \mathbb{N}$ is some set of size n.
2. The set of *initial configurations* $Q_0^n = \{f \in Q^n \mid f(i) \in I \text{ for all } i \in [n]\}$ consists of all configurations which map all processes to initial states of P.
3. The set of *global transitions* Δ^n are tuples $(f, \sigma, g) \in Q^n \times \Sigma_{\mathsf{com}} \times Q^n$ where one of the following two conditions hold:
   - $\sigma = \mathfrak{b}$, and for every $i \in [n]$ we have that $(f(i), \mathfrak{b}, g(i)) \in R$. This is called a *broadcast transition*.
   - $\sigma = ((i_1, \mathsf{a}_1), \ldots, (i_k, \mathsf{a}_k))$, where $\mathsf{a} \in \Sigma_{\mathsf{actn}}$ is the *action* taken, and $\{i_1, \ldots, i_k\} \subseteq [n]$ are k different processes. In this case, for every $1 \leq j \leq k$ we have that $(f(i_j), \mathsf{a}_j, g(i_j)) \in R$; and $f(i) = g(i)$ for every $i \notin \{i_1, \ldots, i_k\}$. This is called a *rendezvous transition*, and the processes in the set $\mathrm{prcs}(\sigma) := \{i_1, \ldots, i_k\}$ are called the *rendezvousing processes*.

We denote the *action taken* on a global transition $t = (f, \sigma, g)$ by $\mathsf{actn}(t)$. Thus, $\mathsf{actn}(t) := \mathsf{a}$ if $\sigma = ((i_1, \mathsf{a}_1), \ldots, (i_k, \mathsf{a}_k))$, and otherwise $\mathsf{actn}(t) := \mathfrak{b}$.

A process template P induces the *infinite RB-system* $\mathcal{P}$, i.e., the LTS $\mathcal{P} = \langle Q, Q_0, \Delta, \Sigma_{\mathsf{com}} \rangle$ where $Q = \cup_{n \in \mathbb{N}} Q^n$, $Q_0 = \cup_{n \in \mathbb{N}} Q_0^n$, $\Delta = \cup_{n \in \mathbb{N}} \Delta^n$.

Executions of an RB-System, and the Parameterized Model-Checking Problem. Given a global transition $t = (f, \sigma, g)$, and a process i, we say that i *moved* in t iff: $\sigma = \mathfrak{b}$, or $i \in \mathrm{prcs}(\sigma)$. We write $edge_i(t)$ for the edge of P taken by process i in the transition t, and $\perp$ if i did not move in t. Thus, if $\sigma = \mathfrak{b}$ then $edge_i(t) := (f(i), \mathfrak{b}, g(i))$; and if $\sigma = ((i_1, \mathsf{a}_1), \ldots, (i_k, \mathsf{a}_k))$ then $edge_i(t) := (f(i), \mathsf{a}_j, g(i))$ if $\sigma(j) = (i, \mathsf{a}_j)$ for some $j \in [k]$, and otherwise $edge_i(t) := \perp$. Take an RB-System $\mathcal{P}^n = \langle Q^n, Q_0^n, \Delta^n, \Sigma_{\mathsf{com}} \rangle$, a path $\pi = t_1 t_2 \ldots$ in $\mathcal{P}^n$, and a process i in $\mathcal{P}^n$. Define $\mathrm{proj}_\pi(i) := edge_i(t_{j_1}) edge_i(t_{j_2}) \ldots$, where $j_1 < j_2 < \ldots$ are all the indices j for which $edge_i(t_j) \neq \perp$. Intuitively, $\mathrm{proj}_\pi(i)$ is the path in P taken by process i during the path π. Define the set of *executions* $\mathrm{EXEC}_{\mathcal{P}}$ of $\mathcal{P}$ to be the set of the runs of $\mathcal{P}$ projected onto a single process. Note that, due to symmetry, we can assume w.l.o.g. that the runs are projected onto process 1. Formally, $\mathrm{EXEC}_{\mathcal{P}} = \{\mathrm{proj}_\pi(1) \mid \pi \text{ is a run of } \mathcal{P}\}$. We denote by $\mathrm{EXEC}_{\mathcal{P}}^{fin}$ (resp. $\mathrm{EXEC}_{\mathcal{P}}^{\infty}$) the finite (infinite) executions in $\mathrm{EXEC}_{\mathcal{P}}$.

For specifications $\mathcal{F}$ (e.g., LTL, NFWs) interpreted over infinite (resp. finite) words over the alphabet $S \times (\Sigma_{\mathsf{rdz}} \cup \{\mathfrak{b}\}) \times S$ of transitions,[3] the *Parameterized Model Checking Problem* (PMCP) for $\mathcal{F}$ is to decide, given a template P, and a specification $\varphi \in \mathcal{F}$, if all executions in $\mathrm{EXEC}_{\mathcal{P}}^{\infty}$ (resp. $\mathrm{EXEC}_{\mathcal{P}}^{fin}$) satisfy φ.

---

[1] This can always be assumed by increasing the size of the rendezvous alphabet.
[2] Even though Σ_{com} is infinite, Δ^n refers only to a finite subset of it.
[3] In this way we can also capture atomic propositions on edges or states since these atoms may be pushed into the rendezvous label.

Discrete Timed Networks. We refer the reader to [1] for a formal definition of timed networks. Here we describe the templates and informally describe the semantics. Fix a set C of *clocks*. A *timed network template* is a finite LTS $\langle Q, I, R, \Sigma_{\mathsf{rdz}} \rangle$. We associate to each letter $\mathsf{a}_i \in \Sigma_{\mathsf{rdz}}$ a *command* $r(\mathsf{a}_i) \subseteq C$ and a *guard* $p(\mathsf{a}_i)$. A guard p is a Boolean combination of predicates of the form $c \bowtie x$ where $c \in \mathbb{N}$ is a constant, $x \in C$ is a clock, and $\bowtie \in \{<, =\}$.

Intuitively, a discrete timed network consists of the parallel composition of $n \in \mathbb{N}$ template processes, each running a copy of the template. Each copy has a local state (q, t), where $q \in Q$ and $t : C \to \mathbb{N}$. A rendezvous action a is *enabled* if there are k processes in local states (q_i, t_i) $(i \in [k])$ and there are edges $(q_i, \mathsf{a}_i, q_i') \in R$ such that the clocks t_i satisfy the guards $p(\mathsf{a}_i)$. The rendezvous action is *taken* means that the k processes change state (to q_i') and each of the clocks in $r(\mathsf{a}_i)$ is reset to 0. The network evolves non-deterministically, in steps: either all clocks advance by one time unit (so every $t(c)$ increases by one)[4] or a rendezvous action $\mathsf{a} \in \Sigma_{\mathsf{rdz}}$ is taken. For a timed network template T let $\mathcal{T}^n$ denote the timed network composed of $n \in \mathbb{N}$ templates T and let $\mathcal{T}$ denote the union of the networks $\mathcal{T}^n$ for $n \in \mathbb{N}$.

Given a timed network template T one can build an equivalent RB-template P, i.e., $\mathrm{EXEC}_{\mathcal{P}} = \mathrm{EXEC}_{\mathcal{T}}$. The key insight is that the passage of time, that causes all clocks to advance by one time unit, is simulated by symmetric broadcast, and timed-guards are pushed into the template states. The RB-system $\mathcal{P}$ requires only a finite number of states since clock values bigger than the greatest constant appearing on the guards are collapsed to a single abstract value (cf. [1]).

Useful Lemmas. We state a few simple but useful lemmas. The first lemma states that, by partitioning processes of an RB-system into independent groups, a system with many processes can simulate in a single run multiple runs of smaller systems. If the simulated paths contain no broadcasts then the transitions of the simulated paths can be interleaved in any order. Otherwise, all simulated runs must have the same number of broadcasts, and the simulations of all the edges before the i'th broadcast on each simulated path must complete before taking the i'th broadcast on the simulating combined path.

Lemma 1 (RB-System Composition). *A system $\mathcal{P}^n$ can, using a single run, partition its processes into groups each simulating a run of a smaller system. All simulated paths must have the same number of broadcasts.*

Consider now an RB-system $\mathcal{P}^n$, and two configurations f, f' in it such that the number of processes in each state in f is equal to that in f', i.e., such that $|f^{-1}(s)| = |f'^{-1}(s)|$ for every $s \in S$. We call f, f' *twins*. A finite path π of length m for which $\mathsf{src}(\pi_1)$ and $\mathsf{dst}(\pi_m)$ are twins is called a *pseudo-cycle*. For example, for P in Figure 1, the following path in $\mathcal{P}^4$ is a pseudo-cycle that is not a cycle:
$$(p, q, q, r) \xrightarrow{((3,\mathsf{c}_1),(4,\mathsf{c}_2))} (p, q, r, p) \xrightarrow{((2,\mathsf{c}_1),(3,\mathsf{c}_2))} (p, r, p, p) \xrightarrow{((3,\mathsf{a}_1),(4,\mathsf{a}_2))} (p, r, q, q).$$

Lemma 2. *By renaming processes after each iteration, a pseudo-cycle π can be pumped to an infinite path which repeatedly goes through the actions on π.*

---

[4] Alternatively, as in [1], one can let time advance by any amount.

Fig. 1. R-template with $k = 2$ **Fig. 2.** RB-template with $k = 2$

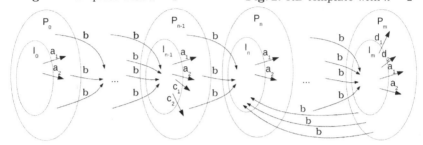

Fig. 3. A high level view of the reachability-unwinding lasso

3   The Reachability-Unwinding of a Process Template

Given template $P = \langle S, I, R, \Sigma_{\mathsf{rdz}} \cup \{\mathsf{b}\}\rangle$, our goal in this section is to construct a new process template $P^{\multimap} = \langle S^{\multimap}, I^{\multimap}, R^{\multimap}, \Sigma_{\mathsf{rdz}} \cup \{\mathsf{b}\}\rangle$, called the *reachability-unwinding* of P, see Figure 3. The template $P^{\multimap}$ will play a role in all our algorithms for solving the PMCP of RB-systems. Intuitively, $P^{\multimap}$ is obtained by alternating the following two operations: *(i)* taking a copy of P and removing from it all unreachable rendezvous edges; and *(ii)* unwinding on broadcast edges. This is repeated until a copy is created which is equal to a previous one, we then stop and close the unwinding back into the old copy, forming a high-level lasso structure.

Technically, it is more convenient to first calculate all the desired copies and then to arrange them in the lasso. Thus, we first calculate, for $0 \le i \le m$ (for an appropriate m), an R-template $P_i = \langle S_i, I_i, R_i, \Sigma_{\mathsf{rdz}}\rangle$ which is a copy of P with initial states redesignated and all broadcast edges, plus some rendezvous edges, removed. Second, we take $P_0, \dots, P_m$ and combine them, to create the single process template $P^{\multimap}$, by connecting the states in P_i with the initial states of P_{i+1} (P_n for $i = m$, where $n \le m$ is determined by the lasso structure) with broadcast edges, as naturally induced by P.

Construct the R-template $P_i = \langle S_i, I_i, R_i, \Sigma_{\mathsf{rdz}}\rangle$ (called the i'th *component* of $P^{\multimap}$) recursively: for $i = 0$, we let $I_0 := I$; and for $i > 0$ we let $I_i := \{s \in S \mid (h, \mathsf{b}, s) \in R$ for some $h \in S_{i-1}\}$ be the set of states reachable from S_{i-1} by a broadcast edge. The elements S_i and R_i are obtained using the following *saturation* algorithm, which is essentially a breadth-first search: start with $S_i := I_i$ and $R_i := \emptyset$; at each round of the algorithm, consider in turn each edge $e = (s, \mathsf{a}_h, t) \in R \setminus R_i$; if for every $l \in [k]$ there is some edge $(s', \mathsf{a}_l, t') \in R$ with $s' \in S_i$, then add e to R_i and add t (if not already there) to S_i. The algorithm ends when a fixed-point is reached. Observe a property of this algorithm: if $(s, \mathsf{a}_h, t) \in R_i$ then for all $l \in [k] \setminus \{h\}$ there exists $s', t' \in S_i$ such that $(s', \mathsf{a}_l, t') \in R_i$.

Now, P_i is completely determined by I_i (and P), and so there are at most $2^{|S|}$ possible values for it. Hence, for some $n \leq m < 2^{|S|}$ it must be that $P_n = P_{m+1}$. We stop calculating P_i's when this happens since for every $i \in \mathbb{N}_0$ it must be that $P_i = P_{n+((i-n) \bmod r)}$, where $r = m + 1 - n$. We call n the *prefix length* of $P^{-\circ}$ (usually denoted by ψ), call r the *period* of $P^{-\circ}$, and for $i \in \mathbb{N}_0$, call $n + ((i - n) \bmod r)$ the *associated component number* of i, and denote it by $\text{comp}(i)$.

We now construct from $P_0, \ldots, P_m$ the template $P^{-\circ} = \langle S^{-\circ}, I^{-\circ}, R^{-\circ}, \Sigma_{\mathsf{rdz}} \cup \{\mathsf{b}\} \rangle$, as follows: *(i)* $S^{-\circ} := \cup_{i=0}^{m} (S_i \times \{i\})$; *(ii)* $I^{-\circ} := I_0 \times \{0\}$ (recall that we also have $I_0 = I$); *(iii)* $R^{-\circ}$ contains the following transitions: the rendezvous transitions $\cup_{i=0}^{m} \{((s, i), \varsigma, (t, i)) \mid (s, \varsigma, t) \in R_i\}$, and the broadcast transitions $\cup_{i=0}^{m-1} \{((s, i), \mathsf{b}, (t, i + 1)) \mid (s, \mathsf{b}, t) \in R \text{ and } s \in S_i\}$ and $\{((s, m), \mathsf{b}, (t, n)) \mid (s, \mathsf{b}, t) \in R \text{ and } s \in S_m\}$.

We will abuse notation, and talk about the component P_i, referring sometimes to P_i as defined before (i.e., without the annotation with i), and sometimes to the part of $P^{-\circ}$ that was obtained by annotating the elements of P_i with i.

Observe that, by projecting out the component numbers (we will denote this projecting by superscript ⊚) from states in $P^{-\circ}$ (i.e., by replacing $(s, i) \in S^{-\circ}$ with $s \in S$), states and transitions in $P^{-\circ}$ induce states and transitions in P. Similarly, paths and runs in $\mathcal{P}^{-\circ}$ can be turned into paths and runs in $\mathcal{P}$. We claim that also the converse is true, i.e., that by adding component numbers, states and transitions in P can be lifted to ones in $P^{-\circ}$; and that by adding the correct (i.e., reflecting the number of previous broadcasts) component numbers to the states of the transitions of a run in $\mathcal{P}$, it too can be lifted to a run in $\mathcal{P}^{-\circ}$. However, a path in $\mathcal{P}$ that is not a run (i.e., that does not start at an initial configuration), may not always be lifted to a path in $\mathcal{P}^{-\circ}$ due to the removal of unreachable edges in the components making up $P^{-\circ}$.

The next lemma says that we may work with template $P^{-\circ}$ instead of P.

Lemma 3. *For every $n \in \mathbb{N}$, we have that $\text{runs}(\mathcal{P}^n) = \{\rho^{\circledcirc} \mid \rho \in \text{runs}((\mathcal{P}^{-\circ})^n)\}$.*

The following lemma says, intuitively, that for every component P_i there is a run of $\mathcal{P}^{-\circ}$ that "loads" arbitrarily many processes into every state of P_i.

Lemma 4. *For all $b, n \in \mathbb{N}$ there is a finite run π of $\mathcal{P}^{-\circ}$ with b broadcasts, s.t., $|f^{-1}(s)| \geq n$ for all states s in the component $P_{comp(b)}$, where $f = \mathsf{dst}(\pi)$.* □

The following lemma states that the set of finite executions of the RB-system $\mathcal{P}$ is equal to the set of finite *runs of the process template* $P^{-\circ}$ (modulo component numbers). This is very convenient since, whereas $\mathcal{P}$ is infinite, $P^{-\circ}$ is finite. Unfortunately, when it comes to infinite executions of $\mathcal{P}$ we only get that they are contained in (though in many cases not equal to) the set of infinite runs of $P^{-\circ}$. This last observation is also true for P: consider for example Figure 2 without the b edges, and an infinite repetition of the self loop.

Lemma 5. $\text{EXEC}_{\mathcal{P}}^{fin} = \{\pi^{\circledcirc} \mid \pi \in runs(P^{-\circ}), |\pi| \in \mathbb{N}\}$; and $\text{EXEC}_{\mathcal{P}}^{\infty} \subseteq \{\pi^{\circledcirc} \mid \pi \in runs(P^{-\circ}), |\pi| = \infty\}$

Solving PMCP for Regular Specifications. Given $P = \langle S, I, R, \Sigma_{\text{rdz}} \cup \{\flat\}\rangle$, let $\mathcal{A}_{P}^{fin}$ denote the reachability-unwinding $P^{-\circ}$ viewed as an automaton (NFW), with all states being accepting states, and transitions e are labeled $e^{\circledcirc}$ (i.e., they have the component number removed). Formally, $\mathcal{A}_{P}^{fin} = \langle R, S^{-\circ}, I^{-\circ}, R', S^{-\circ}\rangle$, so the input alphabet of $\mathcal{A}_{P}^{fin}$ is R (the transition relation of P), and $R' := \{(s, (s^{\circledcirc}, \sigma, t^{\circledcirc}), t) \mid (s, \sigma, t) \in R^{-\circ}\} \subseteq S^{-\circ} \times R \times S^{-\circ}$. Hence:

Theorem 1. *The PMCP of RB-systems (resp. discrete timed networks) for regular specifications is in* PSPACE *(resp.* EXPSPACE)

4   Solving PMCP of Liveness Specifications

In this section we show how to solve the PMCP for specifications concerning infinite executions. We begin with the following lemma showing that, if we want to use the automata theoretic approach, classical automata models (e.g. Büchi, Parity) are not up to the task.

Lemma 6. *There is a process template P such that* $\text{EXEC}_{\mathcal{P}}^{\infty}$ *is not ω-regular.*

Proof. Consider the process template given in Figure 2. It is not hard to see that in every infinite run of $\mathcal{P}^n$ there may be at most $n-1$ consecutive rendezvous transitions before a broadcast transition, resetting all processes to state 1, is taken. Overall, we have that $\text{EXEC}_{\mathcal{P}}^{\infty}$ is the set of words of the form $a_1^{n_1} a_2^{m_1} \flat$ $a_1^{n_2} a_2^{m_2} \flat \ldots$, where $m_i \in \{0, 1\}$ for every i, and $\limsup n_i < \infty$. This language is not ω-regular since the intersection of its complement with $\{a_1, \flat\}^{\omega}$ is not ω-regular (because it contains no ultimately periodic words).   □

In light of Lemma 6, we turn our attention to a stronger model, called BSW [6]. Thus, we solve the PMCP for liveness specifications as follows: given a process template P, we show how to build a BSW $\mathcal{A}_{P}^{\infty}$ accepting exactly the executions in $\text{EXEC}_{\mathcal{P}}^{\infty}$. Model checking of a specification given by a BSW $\mathcal{A}'$ accepting all undesired (i.e., bad) executions, is thus reduced to checking for the emptiness of the intersection of $\mathcal{A}_{P}^{\infty}$ and $\mathcal{A}'$.

Defining the Automaton $\mathcal{A}_{P}^{\infty}$. We now describe the structure of the BSW $\mathcal{A}_{P}^{\infty}$ (in fact we define a BW) accepting exactly the executions in $\text{EXEC}_{\mathcal{P}}^{\infty}$.

An important element in the construction is a classification of the edges in $P^{-\circ}$ into four types: blue, green, orange, and red. The red edges are those that appear at most finitely many times on any execution in $\text{EXEC}_{\mathcal{P}}^{\infty}$. An edge is blue if it appears infinitely many times on some execution in $\text{EXEC}_{\mathcal{P}}^{\infty}$ with finitely many broadcasts, but only finitely many times on every execution which has infinitely many broadcasts. An edge e is green if there is some run $\pi \in \text{EXEC}_{\mathcal{P}}^{\infty}$ with infinitely many broadcasts on which e appears unboundedly many times between broadcasts, i.e., if for every $n \in \mathbb{N}$ there are $i < j \in \mathbb{N}$ such that

$\pi_i \dots \pi_j$ contains n occurrences of e and no broadcast edges. An edge which is neither blue, green, nor red is orange. By definition, blue and green edges are not broadcast edges. Since the set $\mathrm{EXEC}_{\mathcal{P}}^{\infty}$ is infinite, it is not at all clear that the problem of determining the type of an edge is decidable. Indeed, this turns out to be a complicated question, and we dedicate Section 4.1 to show that one can decide the type of an edge.

The automaton $\mathcal{A}_{\mathcal{P}}^{\infty}$ is made up of three copies of $\mathcal{A}_{\mathcal{P}}^{fin}$ (called $\mathcal{A}_{\mathcal{P}}^{\infty 1}$, $\mathcal{A}_{\mathcal{P}}^{\infty 2}$, $\mathcal{A}_{\mathcal{P}}^{\infty 3}$), as follows: $\mathcal{A}_{\mathcal{P}}^{\infty 1}$ is an exact copy of $\mathcal{A}_{\mathcal{P}}^{fin}$; the copy $\mathcal{A}_{\mathcal{P}}^{\infty 2}$ has only the green and orange edges left; and $\mathcal{A}_{\mathcal{P}}^{\infty 3}$ has only the blue and green edges left (and in particular has no broadcast edges). Furthermore, for every edge (s, σ, s') in $\mathcal{A}_{\mathcal{P}}^{\infty 1}$ we add two new edges, both with the same source as the original edge, but one going to the copy of s' in the copy $\mathcal{A}_{\mathcal{P}}^{\infty 2}$, and one to the copy of s' in the copy $\mathcal{A}_{\mathcal{P}}^{\infty 3}$. The initial states of $\mathcal{A}_{\mathcal{P}}^{\infty}$ are the initial states of $\mathcal{A}_{\mathcal{P}}^{\infty 1}$. For the acceptance condition: every state in $\mathcal{A}_{\mathcal{P}}^{\infty 2}$ and $\mathcal{A}_{\mathcal{P}}^{\infty 3}$ is a Büchi-state, and there is a single counter $C \in \Gamma_B$ that is incremented whenever an orange rendezvous edge is taken in $\mathcal{A}_{\mathcal{P}}^{\infty 2}$ and reset if a broadcast edge is taken in $\mathcal{A}_{\mathcal{P}}^{\infty 2}$.

Formally, given a process template $P = \langle S, I, R, \Sigma_{rdz} \cup \{\mathfrak{b}\} \rangle$ and its unwinding $P^{-\circ} = \langle S^{-\circ}, I^{-\circ}, R^{-\circ}, \Sigma_{rdz} \cup \{\mathfrak{b}\} \rangle$ define $\mathcal{A}_{\mathcal{P}}^{\infty} = \langle \Sigma, Q, Q_0, \Gamma, \delta, \Phi \rangle$ as:

- The input alphabet Σ is the edge relation R of template P.
- The state set Q is $\{(i, s) \mid s \in S^{-\circ}, i \in \{1, 2, 3\}\}$.
- The initial state set Q_0 is $\{(1, s) \mid s \in I^{-\circ}\}$.
- There is one counter, $\Gamma = \{c\}$.
- The transition relation δ is $\delta_1 \cup \delta_2 \cup \delta_3$, where: δ_1 consists of all tuples $((1, s_1), (s_1^{\odot}, \sigma, s_2^{\odot}), \epsilon, (i, s_2))$ such that $(s_1, \sigma, s_2) \in R^{-\circ}, i \in \{1, 2, 3\}$; and δ_3 consists of all tuples $\{((3, s_1), (s_1^{\odot}, \sigma, s_2^{\odot}), \epsilon, (3, s_2))$ such that $(s_1, \sigma, s_2) \in R^{-\circ}$ is blue or green; and δ_2 consists of all tuples $((2, s_1), (s_1^{\odot}, \sigma, s_2^{\odot}), upd^{\sigma, \rho}, (2, s_2))$ such that $\rho := (s_1, \sigma, s_2) \in R^{-\circ}$ is green or orange, and $upd^{\sigma, \rho}$ is the single operation $c := 0$ if ρ is orange and $\mathsf{actn}(\sigma) = \mathfrak{b}$, and $upd^{\sigma, \rho}$ is the single operation $c := c + 1$ if ρ is orange and $\mathsf{actn}(\sigma) \neq \mathfrak{b}$, and $upd^{\sigma, \rho}$ is the empty sequence ϵ if ρ is green. Here ϵ is the empty sequence of operations (i.e., do nothing to the counter).
- The acceptance condition Φ states that $\limsup_i c(\rho, i) < \infty$ (i.e., counter c must be bounded) and some state $q \in Q \setminus \{(1, s) \mid s \in S^{-\circ}\}$ is visited infinitely often.

Lemma 7. *An edge (s_1, σ, s_2) of $P^{-\circ}$ is: (i) red iff it does not appear on any pseudo-cycle of $\mathcal{P}^{-\circ}$; (ii) blue iff it appears on a pseudo-cycle of $\mathcal{P}^{-\circ}$ with no broadcasts, but not on any that contain broadcasts; (iii) green iff it appears on a pseudo-cycle of $\mathcal{P}^{-\circ}$ with no broadcasts, that is part of a bigger pseudo-cycle with broadcasts; (iv) orange iff it appears on a pseudo-cycle C of $\mathcal{P}^{-\circ}$ that has broadcasts, but not on any without broadcasts.*

The following lemma states that we can assume that pseudo-cycles mentioned in Lemma 7 (that have broadcasts) are of a specific form.

Lemma 8. *An edge e appears on a pseudo-cycle D in $\mathcal{P}^{-\circ}$, which contains broadcasts, iff it appears on a pseudo cycle C of $\mathcal{P}^{-\circ}$ containing exactly r broadcast transitions and with all processes starting in the component P_n, where n, r are the prefix length and period of $P^{-\circ}$, respectively. Furthermore, C preserves any nested pseudo-cycles of D that contain no broadcasts.*

Theorem 2. *The language recognized by $\mathcal{A}_{\mathcal{P}}^{\infty}$ is exactly* EXEC$_{\mathcal{P}}^{\infty}$.

Proof (sketch). The fact that every word in EXEC$_{\mathcal{P}}^{\infty}$ is accepted by $\mathcal{A}_{\mathcal{P}}^{\infty}$ follows in a straightforward way from its construction. For the reverse direction, given $\alpha \in$ EXEC$_{\mathcal{P}}^{\infty}$ with an accepting run Ω in $\mathcal{A}_{\mathcal{P}}^{\infty}$, we need to construct a run π in $\mathcal{P}$ whose projection on process 1 is α. We consider the interesting case that α has infinitely many broadcasts (and thus finitely many red and blue edges). The challenging part is how to make process 1 trace the suffix β of α containing only green and orange edges. Since Ω is accepting, counter C_2 is bounded on Ω. Hence, there is a bound $\mathfrak{m}$ on the number of orange edges in β between any r broadcasts, where r is the period of $P^{-\circ}$.

For every green (resp. orange) edge e of P that appears on β, by Lemmas 7, 8, there is a pseudo-cycle C_e with r broadcasts on which e appears. Furthermore, if e is green it actually appears on an inner pseudo-cycle of C_e without broadcasts. Let E_{green} (resp. E_{orange}) be the set of green (resp. orange) edges that appear infinitely often on α. By taking exactly enough processes to assign them to one copy of C_e for every $e \in E_{\text{green}}$, and $\mathfrak{m}$ copies of C_e for every $e \in E_{\text{orange}}$, and composing them using Lemma 1 we can simulate all these copies of these pseudo-cycles in one pseudo-cycle D also with r broadcasts. By Lemma 2, we can pump this pseudo-cycle forever. Furthermore, between broadcasts we have freedom on how to interleave the simulations. We make process 1 trace β by making it successively swap places with the right process in the group simulating a copy of the cycle C_e where e is the next edge on β to be traced (just when the group is ready to use that edge). The key observation is that once a group is used by process 1 there are two options. If it is a group corresponding to a green edge then we can make the group (after 1 leaves it) traverse the inner pseudo-cycle (the one without broadcasts) thus making it ready to serve process 1 again. If the group corresponds to an orange edge e, then it will only be reusable when the whole pseudo-cycle C_e completes (since there is no inner pseudo-cycle to use), i.e., after r broadcasts. However, since there are $\mathfrak{m}$ groups for each such edge, and $\mathfrak{m}$ bounds from above the number of orange edges that need to be taken by process 1 between r broadcasts. □

As we show (Section 4.1, Theorem 4), the problem of determining the type (blue, green, orange, or red) of an edge in $P^{-\circ}$ is decidable, hence, we conclude this section by stating our main theorem (the proof is now immediate).

Theorem 3. *The PMCP (of RB-systems or discrete timed networks) for BW- or SW-specifications or complements of specifications given by BSW, is decidable.*

4.1   Deciding Edge Types

Theorem 4. *Given a process template* $P^{-\circ}$, *the problem of determining the type (blue, green, orange, red) of an edge* e *in* $P^{-\circ}$ *is decidable.*

A key observation for proving Theorem 4 is that by Lemma 7, the type of an edge can be decided by looking for witnessing pseudo-cycles C in $\mathcal{P}^{-\circ}$. Indeed, a witness can determine if an edge is green or not. If not, another witness can determine if it is orange or not, and the last witness can separate the blue from the red. We will show an algorithm that given an edge that is not green tells us if it is orange or not. The algorithm can be modified to check for the other types of witnesses without much difficulty.

By Lemma 8, we can assume that the pseudo-cycle C we are looking for has very specific structure. Our algorithm uses linear programming, in a novel and interesting way, to detect the existence of such a pseudo-cycle C.

Counter Representation. Given a process template $P = \langle S, I, R, \Sigma_{\mathsf{rdz}} \cup \{b\} \rangle$, let $d = |S|$, and fix once and for all some ordering $s_1, s_2, \ldots, s_d$ of the states in S. We associate with every configuration f in $\mathcal{P}$ a vector $f^\sharp := (|f^{-1}(s_1)|, \ldots, |f^{-1}(s_d)|) \in \mathbb{N}_0^d$, called the *counter representation* of f. We also associate with every transition $t = (f, \sigma, g)$ the vector $t^\sharp := g^\sharp - f^\sharp$ representing the change in the number of processes in each state. If t is a rendezvous transition then $g^\sharp - f^\sharp$ is completely determined by the action $\mathsf{a} \in \Sigma_{\mathsf{actn}}$ taken in σ. Indeed, if $\sigma = ((i_1, \mathsf{a}_1), \ldots, (i_k, \mathsf{a}_k))$ then $g^\sharp - f^\sharp = \mathsf{a}^\sharp$, where $\mathsf{a}^\sharp \in \mathbb{N}_0^d$ is the vector defined by letting $\mathsf{a}^\sharp(s) := |\{j \in [k] \mid \mathsf{dst}(\mathsf{edge}(\mathsf{a}_j)) = s\}| - |\{j \in [k] \mid \mathsf{src}(\mathsf{edge}(\mathsf{a}_j)) = s\}|$ for every $s \in S$.

Given $u \in \mathbb{Q}^d$, and a sequence of vectors $\varrho = \varrho_1 \ldots \varrho_m$ in $\mathbb{Q}^d$, the pair $\rho = (u, \varrho)$ is called a *path* from u to $v = u + \Sigma_{i=1}^m \varrho_i$. We write ρ_j for the vector $u + \Sigma_{i=1}^j \varrho_i$, for every $0 \le j \le m$. The path ρ is *legal* if $\rho_j \in \mathbb{Q}_{\ge 0}^d$ for every $0 \le j \le m$, i.e., if no coordinate goes negative at any point. Given a finite path $\pi_1 \ldots \pi_m$ in $\mathcal{P}$, we call the path $\pi^\sharp := (\mathsf{src}(\pi_1)^\sharp, \pi_1^\sharp \ldots \pi_m^\sharp)$ in $\mathbb{Q}^d$ its *counter representation*. Observe that $\pi^\sharp$ is always a legal path.

Rational Relaxation of VASs. Vector Addition Systems (VASs) or equivalently Petri nets are one of the most popular formal methods for the representation and the analysis of parallel processes [10]. Unfortunately, RB-systems **cannot** be modelled by VASs since a transition in a VAS only moves a constant number of processes, whereas a broadcast in an RB-system may move any number of processes. On the other hand, R-System can be modelled by VASs, and we do use this fact to analyze the behaviour of the counter representation between broadcasts. Moreover, we note that integer linear programming is a natural fit for describing paths and cycles in the counter representation. However, in order to apply linear programming to RB-systems we have to overcome two intertwined obstacles: *(i)* not every path in the counter representation induces a path in $\mathcal{P}$, and *(ii)* since we have no bound on the length of the pseudo-cycle C we cannot have variables describing each configuration on it, and we need to aggregate information. These obstacles are aggravated by the presence of broadcasts. Another difficulty of applying linear programming to RB-systems arises from the fact that the question of reachability in an RB-system with two (symmetric) broadcast actions and a controller is undecidable (which can be obtained by modifying a result in [9] concerning asymmetric broadcast).

The solution we propose to this problem, which we found to be surprisingly powerful, is to use linear programming but look for a solution in *rational* numbers and not in integers. Thus, we introduce the notion of the *rational relaxation* of a VAS, obtained by allowing any non-negative rational multiple of configurations and transitions of the original VAS. Since our linear programs use homogeneous systems of equations, multiplying a rational solution by a large enough number would yield another solution in integers. Thus the scaling property obtained a consequence of rational relaxation precludes the possibility of specifying a single controller! Thinking of the counter representation as vectors of rational numbers also allows us to use geometric reasoning to solve the two problems (i), (ii) described above. Essentially, by cutting transitions to smaller pieces (which cannot be done at will to integer vectors) and rearranging the pieces, we can transform a description of a path in an aggregated form, as it comes out of the linear program, into one which is legal and can be turned into a path in $\mathcal{P}$. We strongly believe that these techniques can be fruitfully used in other circumstances concerning counter-representations, and similar objects (such as vector addition systems and Petri nets).

Due to lack of space, the description of the linear programs we use, as well as the geometric machinery we develop will be published in an extended version.

References

1. Abdulla, P.A., Deneux, J., Mahata, P.: Multi-clock timed networks. In: Ganzinger, H. (ed.) LICS, pp. 345–354, July 2004
2. Abdulla, P.A., Jonsson, B.: Model checking of systems with manyidentical timed processes. TCS **290**(1), 241–264 (2003)
3. Alur, R.: Timed automata. In: Halbwachs, N., Peled, D.A. (eds.) CAV 1999. LNCS, vol. 1633, pp. 8–22. Springer, Heidelberg (1999)
4. Aminof, B., Jacobs, S., Khalimov, A., Rubin, S.: Parameterized model checking of token-passing systems. In: McMillan, K.L., Rival, X. (eds.) VMCAI 2014. LNCS, vol. 8318, pp. 262–281. Springer, Heidelberg (2014)
5. Aminof, B., Kotek, T., Rubin, S., Spegni, F., Veith, H.: Parameterized model checking of rendezvous systems. In: Baldan, P., Gorla, D. (eds.) CONCUR 2014. LNCS, vol. 8704, pp. 109–124. Springer, Heidelberg (2014)
6. Bojanczyk, M.: Beyond ω-regular languages. In: STACS 2010, pp. 11–16 (2010)
7. Chevallier, R., Encrenaz-Tiphene, E., Fribourg, L., Xu, W.: Timed verification of the generic architecture of a memory circuit using parametric timed automata. Formal Methods in System Design **34**(1), 59–81 (2009)
8. Delzanno, G., Sangnier, A., Zavattaro, G.: Parameterized verification of ad hoc networks. In: Gastin, P., Laroussinie, F. (eds.) CONCUR 2010. LNCS, vol. 6269, pp. 313–327. Springer, Heidelberg (2010)
9. Esparza, J., Finkel, A., Mayr, R.: On the verification of broadcast protocols. In: LICS, pp. 352–359 (1999)
10. Esparza, J., Nielsen, M.: Decidability issues for petri nets - a survey. Bulletin of the EATCS **52**, 244–262 (1994)

11. German, S.M., Sistla, A.P.: Reasoning about systems with many processes. JACM **39**(3), 675–735 (1992)
12. Kouvaros, P., Lomuscio, A.: A cutoff technique for the verification of parameterised interpreted systems with parameterised environments. In: IJCAI 2013 (2013)
13. Spalazzi, L., Spegni, F.: Parameterized model-checking of timed systems with conjunctive guards. In: Giannakopoulou, D., Kroening, D. (eds.) VSTTE 2014. LNCS, vol. 8471, pp. 235–251. Springer, Heidelberg (2014)
14. Suzuki, I.: Proving properties of a ring of finite-state machines. Inf. Process. Lett. **28**(4), 213–214 (1988)
15. Vardi, M.Y.: An automata-theoretic approach to linear temporallogic. In: Moller, F., Birtwistle, G. (eds.) Logics for Concurrency. LNCS, vol. 1043, pp. 238–266. Springer, Heidelberg (1996)

Symmetric Strategy Improvement

Sven Schewe[1]([✉]), Ashutosh Trivedi[2], and Thomas Varghese[1]

[1] University of Liverpool, Liverpool, UK
{Sven.Schewe,thomasmv}@liverpool.ac.uk
[2] Indian Institute of Technology Bombay, Mumbai, India
trivedi@cse.iitb.ac.in

Abstract. Symmetry is inherent in the definition of most of the two-player zero-sum games, including parity, mean-payoff, and discounted-payoff games. It is therefore quite surprising that no symmetric analysis techniques for these games exist. We develop a novel symmetric strategy improvement algorithm where, in each iteration, the strategies of both players are improved simultaneously. We show that symmetric strategy improvement defies Friedmann's traps, which shook the belief in the potential of classic strategy improvement to be polynomial.

1 Introduction

We study turn-based graph games between two players—Player Min and Player Max—who take turns to move a token along the vertices of a coloured finite graph so as to optimise their adversarial objectives. Various classes of graph games are characterised by the objective of the players, for instance in *parity games* the objective is to optimise the parity of the dominating colour occurring infinitely often, while in *discounted and mean-payoff games* the objective is the discounted and limit-average sum of the colours. Solving graph games is the central and most expensive step in many model checking [1,6,8,17,32], satisfiability checking [17, 27,30,32], and synthesis [23,28] algorithms. More efficient algorithms for solving graph games will therefore foster the development of performant model checkers and contribute to bringing synthesis techniques to practice.

Parity games enjoy a special status among graph games and the quest for performant algorithms [2–4,7,9,11,15–18,20–22,24–26,31,33,34] for solving them has therefore been an active field of research during the last decades. Traditional forward techniques ($\approx O(n^{\frac{1}{2}c})$ [15] for parity games with n positions and c colours), backward techniques ($\approx O(n^c)$ [9,21,33]), and their combination ($\approx O(n^{\frac{1}{3}c})$ [25]) provide good complexity bounds. These bounds are sharp, and techniques with good complexity bounds [15,25] frequently display their worst case complexity on practical examples. On the other hand, strategy improvement algorithms [3,11,20,24,26,31], a class of algorithms closely related to the Simplex for linear programming, are known to perform well in practice.

The work has been done while the second author was visiting the University of Liverpool supported by a Liverpool India Fellowship.

© Springer-Verlag Berlin Heidelberg 2015
M.M. Halldórsson et al. (Eds.): ICALP 2015, Part II, LNCS 9135, pp. 388–400, 2015.
DOI: 10.1007/978-3-662-47666-6_31

The standard strategy improvement algorithms are built around the existence of optimal positional strategies for both players. They start with an arbitrary positional strategy for a player and iteratively compute a better positional strategy in every step until the strategy cannot be further improved. Since there are only finitely many positional strategies in a finite graph, termination is guaranteed. The crucial step in a strategy improvement algorithm is to compute a better strategy from the current strategy. Given a current strategy σ of a player (say, Player Max), this step is performed by first computing the globally optimal counter strategy τ_σ^c of the opponent (Player Min) and then computing the value of each vertex of the game restricted to the strategies σ and τ_σ^c. For the games under discussion (parity, discounted, and mean-payoff) both of these computations are simple and tractable. This value dictates potentially locally profitable changes or switches $\mathsf{Prof}(\sigma)$ that Player Max can make vis-à-vis his previous strategy σ. For the correctness of the strategy improvement algorithm it is required that such locally profitable changes imply a global improvement. The strategy of Player Max can then be updated according to a *switching rule* (akin to pivoting rule of the Simplex) in order to give an improved strategy. This has led to the following template for classic strategy improvement algorithms.

---

1 determine an optimal counter strategy τ_σ^c for σ
2 evaluate the game for σ and τ_σ^c and determine profitable changes $\mathsf{Prof}(\sigma)$ for σ
3 update σ by applying changes from $\mathsf{Prof}(\sigma)$ to σ

A number of switching rules, including the ones inspired by Simplex pivoting rules, have been suggested for strategy improvement algorithms. The most widespread ones are to select changes for all game states where this is possible, choosing a combination of those with an optimal update guarantee, or to choose uniformly at random. For some classes of games, it is also possible to select an optimal combination of updates [26]. There have also been suggestions to use more advanced randomisation techniques with sub-exponential – $2^{O(\sqrt{n})}$ – bounds [3] and snare memory [11]. Unfortunately, all of these techniques have been shown to be exponential in the size of the game [12–14].

Classic strategy improvement algorithms treat the two players involved quite differently where at each iteration one player computes a globally optimal counter strategy, while the other player performs local updates. In contrast, a *symmetric strategy improvement* algorithm symmetrically improves the strategies of both players at the same time, and uses the finding to guide the strategy improvement. This suggests the following naïve symmetric approach.

---

1 determine $\tau' = \tau_\sigma^c$ determine $\sigma' = \sigma_\tau^c$
2 update σ to σ' update τ to τ'

This algorithm has earlier been suggested by Condon [5] where it was shown that a repeated application of this update can lead to cycles [5]. A problem with this naïve approach is that there is no guarantee that the primed strategies are generally better than the unprimed ones. With hindsight this is maybe not very surprising, as in particular no improvement in the evaluation of running the game with σ', τ' can be expected over running the game with σ, τ, as an improvement for one player is on the expense of the other. This observation led to the approach being abandoned.

The key contribution of this paper is the following more careful symmetric strategy improvement algorithm that guarantees improvements in each iteration similar to classic strategy improvement.

1 determine τ_σ^c	determine σ_τ^c
2 determine $\mathsf{Prof}(\sigma)$ for σ	determine $\mathsf{Prof}(\tau)$ for τ
3 update σ using $\mathsf{Prof}(\sigma) \cap \sigma_\tau^c$	update τ using $\mathsf{Prof}(\tau) \cap \tau_\sigma^c$

Observe that the main difference to classic strategy improvement approaches is that we exploit the strategy of the other player to inform the search for a good improvement step. In this algorithm we select only such updates to the two strategies that agree with the optimal counter strategy to the respective other's strategy. We believe that this will provide a gradually improving advice function that will lead to few iterations. We support this assumption by showing that this algorithm suffices to escape the traps Friedmann has laid to establish lower bounds for different types of strategy improvement algorithms [12–14].

2 Preliminaries

We focus on turn-based zero-sum games played between two players—Player Max and Player Min—over finite graphs. A game arena $\mathcal{A}$ is a tuple $(V_{\text{Max}}, V_{\text{Min}}, E, C, \phi)$ where $(V = V_{\text{Max}} \cup V_{\text{Min}}, E)$ is a finite directed graph with the set of vertices V partitioned into a set V_{Max} of vertices controlled by Player Max and a set V_{Min} of vertices controlled by Player Min, $E \subseteq V \times V$ is the set of edges, C is a set of colours, $\phi : V \to C$ is the colour mapping. We require that every vertex has at least one outgoing edge.

A turn-based game over $\mathcal{A}$ is played between players by moving a token along the edges of the arena. A play of such a game starts by placing a token on some initial vertex $v_0 \in V$. The player controlling this vertex then chooses a successor vertex v_1 such that $(v_0, v_1) \in E$ and the token is moved to this successor vertex. In the next turn the player controlling the vertex v_1 chooses the successor vertex v_2 with $(v_1, v_2) \in E$ and the token is moved accordingly. Both players move the token over the arena in this manner and thus form a play of the game. Formally, a play of a game over $\mathcal{A}$ is an infinite sequence of vertices $\langle v_0, v_1, \ldots \rangle \in V^\omega$ such that, for all $i \geq 0$, we have that $(v_i, v_{i+1}) \in E$. We write $\mathsf{Plays}_{\mathcal{A}}(v)$ for the set of plays over $\mathcal{A}$ starting from vertex $v \in V$ and $\mathsf{Plays}_{\mathcal{A}}$ for the set of plays of the game. We omit the subscript when the arena is clear from the context. We extend the colour mapping $\phi : V \to C$ from vertices to plays by defining the mapping $\phi : \mathsf{Plays} \to C^\omega$ as $\langle v_0, v_1, \ldots \rangle \mapsto \langle \phi(v_0), \phi(v_1), \ldots \rangle$.

A graph game $\mathcal{G}$ is a tuple $(\mathcal{A}, \eta, \prec)$, where $\mathcal{A}$ is an *arena*, $\eta : C^\omega \to \mathbb{D}$ is an evaluation function, and $\mathbb{D}$ is equipped with a preference order $\prec$. Parity, mean-payoff and discounted payoff games are graph games $(\mathcal{A}, \eta, \prec)$ played on game arenas $\mathcal{A} = (V_{\text{Max}}, V_{\text{Min}}, E, \mathbb{R}, \phi)$. For mean payoff games the evaluation function is $\eta : \langle c_0, c_1, \ldots \rangle \mapsto \liminf_{i \to \infty} \frac{1}{i} \sum_{j=0}^{i-1} c_j$, while for discounted payoff games with discount factor $\lambda \in [0, 1)$ it is $\eta : \langle c_0, c_1, \ldots \rangle \mapsto \sum_{i=0}^{\infty} \lambda^i c_i$ with $\prec$ as the natural order over the reals. For (max) parity games the evaluation function is $\eta : \langle c_0, c_1, \ldots \rangle \mapsto \limsup_{i \to \infty} c_i$ often used with a preference order $\prec_{\text{parity}}$ where higher even colours are preferred over smaller even ones, even colours are preferred over odd ones, and smaller odd colours are preferred over higher ones.

In the remainder, we will use parity games where every colour is unique, i.e., where ϕ is injective. The reason for this assumption is that we extended [19], which implements variants of [31], and the lower bounds from [12–14] refer to such games. All parity games can be translated into such games as discussed in [31]. For these games, we use a valuation function based on their progress measure. We define η as $\langle c_0, c_1, \ldots \rangle \mapsto (c, C, d)$, where $c = \limsup_{i \to \infty} c_i$ is the dominant colour of the colour sequence, $d = \min\{i \in \omega \mid c_i = c\}$ is the index of the first occurrence of c, and $C = \{c_i \mid i < d, c_i > c\}$ is the set of colours that occur before the first occurrence of c. The preference order is such that $(c', C', d') \prec (c, C, d)$ if either $c' \prec_{\text{parity}} c$, or $c = c'$ and following holds:

- $C \neq C'$ and $h = \max((C \setminus C') \cup (C' \setminus C))$ is even and belongs to C,
- $C \neq C'$ and $h = \max((C \setminus C') \cup (C' \setminus C))$ is odd and belongs to C',
- $C = C'$, c is even and $d < d'$, or $C = C'$, c is odd and $d > d'$.

A strategy of Player Max is a function $\sigma : V^* V_{\text{Max}} \to V$ such that $(v, \sigma(\pi v)) \in E$ for all $\pi \in V^*$ and $v \in V_{\text{Max}}$. Similarly, a strategy of Player Min is a function $\tau : V^* V_{\text{Min}} \to V$ such that $(v, \sigma(\pi v)) \in E$ for all $\pi \in V^*$ and $v \in V_{\text{Min}}$. We write Σ^∞ and T^∞ for the set of strategies of Player Max and Player Min, respectively. For a strategy pair $(\sigma, \tau) \in \Sigma^\infty \times T^\infty$ and an initial vertex $v \in V$ we denote the unique play starting from the vertex v by $\pi(v, \sigma, \tau)$ and we write $\text{val}_\mathcal{G}(v, \sigma, \tau)$ for the value of the vertex v under the strategy pair (σ, τ) defined as $\text{val}_\mathcal{G}(v, \sigma, \tau) \stackrel{\text{def}}{=} \eta(\phi(\pi(v, \sigma, \tau)))$. We also define the concept of the value of a strategy $\sigma \in \Sigma^\infty$ and $\tau \in T^\infty$ as $\text{val}_\mathcal{G}(v, \sigma) \stackrel{\text{def}}{=} \inf_{\tau \in T^\infty} \text{val}_\mathcal{G}(v, \sigma, \tau)$ and $\text{val}_\mathcal{G}(v, \tau) \stackrel{\text{def}}{=} \sup_{\sigma \in \Sigma^\infty} \text{val}_\mathcal{G}(v, \sigma, \tau)$. We also extend the valuation for vertices to a valuation for the whole game by defining $|V|$-dimensional vectors $\text{val}_\mathcal{G}(\sigma) : v \mapsto \text{val}_\mathcal{G}(v, \sigma)$ with the usual $|V|$-dimensional partial order $\sqsubseteq$, where $\text{val} \sqsubseteq \text{val}'$ iff $\text{val}(v) \preceq \text{val}'(v)$ for all $v \in V$.

We say that a strategy $\sigma \in \Sigma^\infty$ is memoryless or *positional* if for all $\pi, \pi' \in V^*$ and $v \in V_{\text{Max}}$ we have that $\sigma(\pi v) = \sigma(\pi' v)$. Thus, a positional strategy can be viewed as a function $\sigma : V_{\text{Max}} \to V$ such that for all $v \in V_{\text{Max}}$ we have that $(v, \sigma(v)) \in E$. The concept of positional strategies of Player Min is defined in an analogous manner. We write Σ and T for the set of positional strategies of Players Max and Min, respectively. We say that a game is positionally determined if:

- $\text{val}_\mathcal{G}(v, \sigma) = \min_{\tau \in T} \text{val}_\mathcal{G}(v, \sigma, \tau)$ holds for all $\sigma \in \Sigma$,
- $\text{val}_\mathcal{G}(v, \tau) = \max_{\sigma \in \Sigma} \text{val}_\mathcal{G}(v, \sigma, \tau)$ holds for all $\tau \in T$,
- **Existence of value**: for all $v \in V$ $\max_{\sigma \in \Sigma} \text{val}_\mathcal{G}(v, \sigma) = \min_{\tau \in T} \text{val}_\mathcal{G}(v, \tau)$ holds, and we use $\text{val}_\mathcal{G}(v)$ to denote this value, and

Fig. 1. Parity game arena with four vertices and unique colours

- **Existence of globally positional optimal strategies**: there is a pair τ_{min}, σ_{max} of strategies such that, for all $v \in V$, $\mathsf{val}_{\mathcal{G}}(v) = \mathsf{val}_{\mathcal{G}}(v, \sigma_{max}) = \mathsf{val}_{\mathcal{G}}(v, \tau_{min})$ holds. Observe that for all $\sigma \in \Sigma$ and $\tau \in T$ we have that $\mathsf{val}_{\mathcal{G}}(\sigma_{max}) \sqsupseteq \mathsf{val}_{\mathcal{G}}(\sigma)$ and $\mathsf{val}_{\mathcal{G}}(\tau_{min}) \sqsubseteq \mathsf{val}_{\mathcal{G}}(\tau)$.

Observe that the classes of games with positional strategies guarantee a positional optimal counter strategy for Player Min to all strategies $\sigma \in \Sigma$ of Player Max. We denote these strategies by τ_{σ}^{c}. Similarly, we denote the positional optimal counter strategy for Player Max to a strategy $\tau \in T$ by σ_{τ}^{c} of Player Min. While this counter strategy is not necessarily unique, we use the notational convention in all proofs that τ_{σ}^{c} is always the same counter strategy for $\sigma \in \Sigma$, and σ_{τ}^{c} is always the same counter strategy for $\tau \in T$.

Example 1. Consider the parity game arena shown in the Figure 1. We use circles for the vertices of Player Max and squares for Player Min. We label each vertex with its colour. Notice that a positional strategy can be depicted just by specifying an outgoing edge for all the vertices of a player. The positional strategies σ of Player Max is depicted in blue and the positional strategy τ of Player Min is depicted in red.

Classic Strategy Improvement Algorithm. For a strategy σ, an edge $(v, v') \in E$ with $v \in V_{Max}$ is a *profitable update* if $\sigma' \in \Sigma$ with $\sigma' : v \mapsto v'$ and $\sigma' : v'' \mapsto \sigma(v'')$ for all $v'' \neq v$ has a strictly greater evaluation than σ, i.e. $\mathsf{val}_{\mathcal{G}}(\sigma') \sqsupseteq \mathsf{val}_{\mathcal{G}}(\sigma)$. Let $\mathsf{Prof}(\sigma)$ be the set of profitable updates.

Example 2. Consider the strategies σ from the Example 1. Notice that strategy $\tau = \tau_{\sigma}^{c}$ is the optimal counter strategy to σ, i.e. $\mathsf{val}(\sigma) = \mathsf{val}(\sigma, \tau)$. It follows that $\mathsf{Prof}(\sigma) = \{(3, 4), (3, 0)\}$, because both the successor to the left and the successor to the right have a better valuation, $(3, \{4\}, 1)$ and $(0, \emptyset, 0)$, resp., than the successor on the selected self-loop, $(3, \emptyset, 0)$.

For a strategy σ and a functional (right-unique) subset $P \subseteq \mathsf{Prof}(\sigma)$ we define the strategy σ^{P} with $\sigma^{P} : v \mapsto v'$ if $(v, v') \in P$ and $\sigma^{P} : v \mapsto \sigma(v)$ if there is no $v' \in V$ with $(v, v') \in P$.

Algorithm 3 provides a generic template for strategy improvement algorithms. As we discussed in the introduction, the classic strategy improvement algorithms work well for classes of games that are positionally determined and have evaluation function are such that the set $\mathsf{Prof}(\sigma)$ of profitable updates is easy to identify, and reach an optimum exactly where there are no profitable updates. We next formalise these prerequisites for a class of games to be good for strategy improvement algorithm.

Algorithm 3. Classic strategy improvement algorithm

---
1 Let σ_0 be an arbitrary positional strategy. **Set** $i := 0$.
2 If $\mathsf{Prof}(\sigma_i) = \emptyset$ **return** σ_i
3 $\sigma_{i+1} := \sigma_i{}^P$ for some functional subset $P \subseteq \mathsf{Prof}(\sigma_i)$ s.t. $P \neq \emptyset$ if $\mathsf{Prof}(\sigma_i) \neq \emptyset$.
   Set $i := i + 1$. **go to** 2.

For a class of graph games, profitable updates are *combinable* if, for all strategies σ and all functional (right-unique) subsets $P \subseteq \mathsf{Prof}(\sigma)$ we have that $\mathsf{val}_{\mathcal{G}}(\sigma^P) \sqsupseteq \mathsf{val}_{\mathcal{G}}(\sigma)$. Moreover, we say that a class of graph games is *maximum identifying* if $\mathsf{Prof}(\sigma) = \emptyset \Leftrightarrow \mathsf{val}_{\mathcal{G}}(\sigma) = \mathsf{val}_{\mathcal{G}}$. We say that a class of games is *good for* max *strategy improvement* if they are positionally determined and have combinable and maximum identifying improvements.

Theorem 1. *Algorithm 3 returns an optimal strategy σ ($\mathsf{val}_{\mathcal{G}}(\sigma) = \mathsf{val}_{\mathcal{G}}$) of Player Max for all games that are good for* max *strategy improvement.*

As a remark, we can drop the combinability requirement while maintaining correctness when we restrict the updates to a single position, that is, when we require P to be singleton for every update. We call such strategy improvement algorithms *slow*, and a class of games *good for slow* max *strategy improvement* if it is maximum identifying and positionally determined.

Theorem 2. *Slow variants of Algorithm 3 returns an optimal strategy σ ($\mathsf{val}_{\mathcal{G}}(\sigma) = \mathsf{val}_{\mathcal{G}}$) of Player Max for all games that are positionally determined with maximum identifying improvement.*

Proof (of Theorems 1 and 2). The proof for both theorems is the same. The strategy improvement algorithm will produce a sequence $\sigma_0, \sigma_1, \sigma_2 \ldots$ of positional strategies with increasing quality $\mathsf{val}_{\mathcal{G}}(\sigma_0) \sqsubset \mathsf{val}_{\mathcal{G}}(\sigma_1) \sqsubset \mathsf{val}_{\mathcal{G}}(\sigma_2) \sqsubset \ldots$. As the set of positional strategies is finite, this chain must be finite. As the game is maximum identifying, the stopping condition provides optimality. $\square$

Various concepts and results extend naturally for analogous claims about Player Min. We call a class of game *good for strategy improvement* if it is good for max strategy improvement and good for min strategy improvement. Parity games, mean payoff games, and discounted payoff games are all good for strategy improvement (for both players). Moreover, the calculation of $\mathsf{Prof}(\sigma)$ is cheap in all of these instances, which makes them well suited for strategy improvement.

3 Symmetric Strategy Improvement Algorithm

We first extend the termination argument for classic strategy improvement techniques (Theorems 1 and 2) to symmetric strategy improvement given as Algorithm 4. In this section, we show the correctness of Algorithm 4.

Algorithm 4. Symmetric strategy improvement algorithm

1 Let σ_0 and τ_0 be arbitrary positional strategies. **set** $i := 0$.
2 Determine $\sigma_{\tau_i}^c$ and $\tau_{\sigma_i}^c$
3 $\sigma_i' := \sigma_i{}^P$ for $P \subseteq \mathsf{Prof}(\sigma_i) \cap \sigma_{\tau_i}^c$, s.t. $P \neq \emptyset$ if $\mathsf{Prof}(\sigma_i) \cap \sigma_{\tau_i}^c \neq \emptyset$.
4 $\tau_i' := \tau_i{}^P$ for $P \subseteq \mathsf{Prof}(\tau_i) \cap \tau_{\sigma_i}^c$, s.t. $P \neq \emptyset$ if $\mathsf{Prof}(\tau_i) \cap \tau_{\sigma_i}^c \neq \emptyset$.
5 **if** $\sigma_i' = \sigma_i$ and $\tau_i' = \tau_i$ **return** (σ_i, τ_i).
6 **set** $\sigma_{i+1} = \sigma_i'$; $\tau_{i+1} = \tau_i'$; $i := i + 1$. **go to** 2.

Lemma 1. *The symmetric strategy improvement algorithm terminates for all classes of games that are good for strategy improvement.*

Proof. We first observe that the algorithm yields a sequence $\sigma_0, \sigma_1, \sigma_2, \ldots$ of Player Max strategies for $\mathcal{G}$ with improving values $\mathsf{val}_\mathcal{G}(\sigma_0) \sqsubseteq \mathsf{val}_\mathcal{G}(\sigma_1) \sqsubseteq \mathsf{val}_\mathcal{G}(\sigma_2) \sqsubseteq \ldots$, where equality, $\mathsf{val}_\mathcal{G}(\sigma_i) \equiv \mathsf{val}_\mathcal{G}(\sigma_{i+i})$, implies $\sigma_i = \sigma_{i+1}$. Similarly, for the sequence $\tau_0, \tau_1, \tau_2, \ldots$ of Player Min strategies for $\mathcal{G}$, the values $\mathsf{val}_\mathcal{G}(\tau_0) \sqsupseteq \mathsf{val}_\mathcal{G}(\tau_1) \sqsupseteq \mathsf{val}_\mathcal{G}(\tau_2) \sqsupseteq \ldots$, improve (for Player Min), such that equality, $\mathsf{val}_\mathcal{G}(\tau_i) \equiv \mathsf{val}_\mathcal{G}(\tau_{i+i})$, implies $\tau_i = \tau_{i+1}$. As the number of values that can be taken is finite, eventually both values stabilise and the algorithm terminates. $\square$

What remains to be shown is that the symmetric strategy improvement algorithm cannot terminate with an incorrect result. In order to show this, we first prove the weaker claim that it is optimal in $\mathcal{G}(\sigma, \tau, \sigma_\tau^c, \tau_\sigma^c) = (V_{\max}, V_{\min}, E', \mathsf{val})$ such that $E' = \left\{ (v, \sigma(v)) \mid v \in V_{\max} \right\} \cup \left\{ (v, \tau(v)) \mid v \in V_{\min} \right\} \cup \left\{ (v, \sigma_\tau^c(v)) \mid v \in V_{\max} \right\} \cup \left\{ (v, \tau_\sigma^c(v)) \mid v \in V_{\min} \right\}$ is the subgame of $\mathcal{G}$ whose edges are those defined by the four positional strategies, when it terminates with the pair σ, τ.

Lemma 2. *When the symmetric strategy improvement algorithm terminates with the strategy pair σ, τ on games that are good for strategy improvement, then σ and τ are the optimal strategies for the respective players in $\mathcal{G}(\sigma, \tau, \sigma_\tau^c, \tau_\sigma^c)$.*

Proof. For $\mathcal{G}(\sigma, \tau, \sigma_\tau^c, \tau_\sigma^c)$, both update steps are not restricted: the changes Player Max can potentially select his updates from are the edges defined by σ_τ^c at the vertices $v \in V_{\max}$ where σ and σ_τ^c differ ($\sigma(v) \neq \sigma_\tau^c(v)$). Consequently, $\mathsf{Prof}(\sigma) = \mathsf{Prof}(\sigma) \cap \sigma_\tau^c$. Thus, $\sigma = \sigma'$ holds if, and only if, σ is the result of an update step when using classic strategy improvement in $\mathcal{G}(\sigma, \tau, \sigma_\tau^c, \tau_\sigma^c)$ when starting in σ. As game is maximum identifying, σ is the optimal Player Max strategy for $\mathcal{G}(\sigma, \tau, \sigma_\tau^c, \tau_\sigma^c)$. Likewise, the Player Min can potentially select every updates from τ_σ^c at vertices $v \in V_{\min}$, and we first get $\mathsf{Prof}(\tau) = \mathsf{Prof}(\tau) \cap \tau_\sigma^c$ with the same argument. As the game is minimum identifying, τ is the optimal Player Min strategy for $\mathcal{G}(\sigma, \tau, \sigma_\tau^c, \tau_\sigma^c)$. $\square$

We can now expand the optimality in the subgame $\mathcal{G}(\sigma, \tau, \sigma_\tau^c, \tau_\sigma^c)$ from Lemma 2 to global optimality the valuation of these strategies for $\mathcal{G}$.

Lemma 3. *When the symmetric strategy improvement algorithm terminates with the strategy pair σ, τ on a game $\mathcal{G}$ that is good for strategy improvement, then σ is an optimal Player Max strategy and τ an optimal Player Min strategy.*

Proof. Let σ, τ be the strategies returned by the symmetric strategy improvement algorithm for a game $\mathcal{G}$, and let $\mathcal{L} = \mathcal{G}(\sigma, \tau, \sigma_\tau^c, \tau_\sigma^c)$ denote the local game from Lemma 2 defined by them. Lemma 2 has established optimality in $\mathcal{L}$. Observing that the optimal responses in $\mathcal{G}$ to σ and τ, τ_σ^c and σ_τ^c, respectively, are available in $\mathcal{L}$, we first see that they are also optimal in $\mathcal{L}$. Thus, we have

- $\mathsf{val}_\mathcal{L}(\sigma) \equiv \mathsf{val}_\mathcal{L}(\sigma, \tau_\sigma^c) \equiv \mathsf{val}_\mathcal{G}(\sigma, \tau_\sigma^c)$ and
- $\mathsf{val}_\mathcal{L}(\tau) \equiv \mathsf{val}_\mathcal{L}(\sigma_\tau^c, \tau) \equiv \mathsf{val}_\mathcal{G}(\sigma_\tau^c, \tau)$.

Optimality in $\mathcal{L}$ then provides $\mathsf{val}_\mathcal{L}(\sigma) = \mathsf{val}_\mathcal{L}(\tau)$. Putting these three equations together, we get $\mathsf{val}_\mathcal{G}(\sigma, \tau_\sigma^c) \equiv \mathsf{val}_\mathcal{G}(\sigma_\tau^c, \tau)$.

Taking into account that τ_σ^c and σ_τ^c are the optimal responses to σ and τ, respectively, in $\mathcal{G}$, we expand this to $\mathsf{val}_\mathcal{G} \sqsupseteq \mathsf{val}_\mathcal{G}(\sigma) \equiv \mathsf{val}_\mathcal{G}(\sigma, \tau_\sigma^c) \equiv \mathsf{val}_\mathcal{G}(\sigma_\tau^c, \tau) \equiv \mathsf{val}_\mathcal{G}(\tau) \sqsupseteq \mathsf{val}_\mathcal{G}$ and get $\mathsf{val}_\mathcal{G} \equiv \mathsf{val}_\mathcal{G}(\sigma) \equiv \mathsf{val}_\mathcal{G}(\tau) \equiv \mathsf{val}_\mathcal{G}(\sigma, \tau)$.   □

The lemmas in this subsection yield the following results.

Theorem 3. *The symmetric strategy improvement algorithm is correct for games that are good for strategy improvement.*

Theorem 4. *The slow symmetric strategy improvement algorithm is correct for positionally determined games that are maximum and minimum identifying.*

We implemented our symmetric strategy improvement algorithm based on the progress measures introduced by Vöge and Jurdziński [31]. The first step is to determine the valuation for the optimal counter strategies to and the valuations for σ and τ.

Example 3. In our running example from Figure 1, we have discussed in the previous section that τ is the optimal counter strategy τ_σ^c and that $\mathsf{Prof}(\sigma) = \{(3,4),(3,0)\}$. In the optimal counter strategy σ_τ^c to τ, Player Max moves from 3 to 4, and we get $\mathsf{val}(1,\tau) = (1,\emptyset,0)$, $\mathsf{val}(4,\tau) = (4,\emptyset,0)$, $\mathsf{val}(3,\tau) = (4,\emptyset,1)$, and $\mathsf{val}(0,\tau) = (0,\emptyset,0)$. Consequently, $\mathsf{Prof}(\tau) = \{(4,1)\}$. For the update of σ, we select the intersection of $\mathsf{Prof}(\sigma)$ and σ_τ^c. In our example, this is the edge from 3 to 4 (depicted in green). To update τ, we select the intersection of $\mathsf{Prof}(\tau)$ and τ_σ^c. In our example, this intersection is empty, as the current strategy τ agrees with τ_σ^c.

A Minor Improvement on Stopping Criteria. We look at a minor but natural improvement over Algorithm 4. In Algorithm 4, we use termination on both sides as a condition to terminate the algorithm. We could alternatively check if *either* player has reached an optimum. Once this is the case, we can return the optimal strategy and an optimal counter strategy to it.

The correctness of this stopping condition is provided by Theorems 1 and 2. Checking this stopping condition is cheap: it suffices to check if $\mathsf{Prof}(\sigma_i)$ is empty—and to return $(\sigma_i, \tau_{\sigma_i}^c)$ in this case—and to check $\mathsf{Prof}(\tau_i)$ is empty—and then return $(\sigma_{\tau_i}^c, \tau_i)$.

Theorem 5. *The difference in the number of iterations of Algorithm 4 and the improved algorithm is at most linear in the number of states of $\mathcal{G}$.*

This holds because one simply converges against the optimal strategy: every change replcase a decision that deviates from it by a decision that concurs with it. While the improvement is minor, it implies that, for single player games, the number of updates required is at most linear in the number of states of $\mathcal{G}$. Consequently, exponential lower bounds for one player cases, e.g., for MDPs [10], do not apply for symmetric strategy improvement.

Friedmann's Traps.   A thorough discussion on how symmetric strategy improvement defies Friedmann's traps is provided in [29]. Broadly speaking, Friedmann's traps have two main ingredients. The most important one is a binary counter, which is counted up (hence providing an exponential bound) and a deceleration lane (a technical trick to orchestrate the timely counting).

This structure of Friedmann's traps has proven to be quite resistant against different update strategies. The traps for different update strategies differ mainly in the details of how the counter is incremented.

As we discuss in [29], symmetric strategy improvement defies the central counting strategy by setting the bits successively, starting with the most significant bit of the counter. Additionally, it also causes a malfunction of the deceleration lane. The deceleration lane ceases to work because, for all strategies of the opponent player, the optimal counter strategy is to take the same (the longest) path through the deceleration lane. Thus, the strategy for the deceleration lane will quickly converge to taking this path, and will never be reset in parts or full.

4   Experimental Results

We have implemented the symmetric strategy improvement algorithm for parity games and compared it with the standard strategy improvement algorithm with the popular locally optimising and other switching rules. To generate various examples we used the tools `steadygame` and `stratimprgen` that comes as a part of the parity game solver collection PGSolver [19]. We have compared the performance of our algorithm on parity games with 100 positions (see [29]) and found that the locally optimising policy outperforms other switching rules. We therefore compare our symmetric strategy improvement algorithm with the locally optimising strategy improvement below.

Since every iteration of both algorithms is rather similar—one iteration of our symmetric strategy improvement algorithm essentially runs two copies of an iteration of a classical strategy improvement algorithm—and tractable, the key data to compare these algorithms is the number of iterations taken.

Symmetric strategy improvement will often rule out improvements at individual positions: it disregards profitable changes of Player Max and Min if they do not comply with σ_τ^c and τ_σ^c, respectively. It is well known that considering fewer updates can lead to a significant increase in the number of updates on

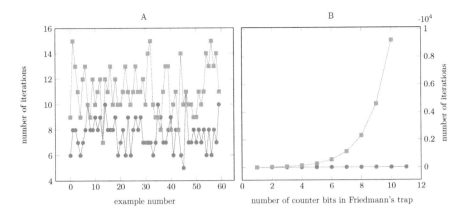

Fig. 2. The plots compare the performance of symmetric strategy improvement (data points in cyan circles) with classic strategy improvement using the locally optimising policy rule (data points in orange squares). The left plot refers to random examples generated using the `steadygame 1000 2 4 3 5 6` command. The right plot refers to Friedmann's trap [12] generated by the command `stratimprgen -pg switchallsubexp i`.

random examples and benchmarks. An algorithm based on the random-facet method [3,20], e.g., needs around a hundred iterations on the random examples with 100 positions we have drawn, simply because it updates only a single position at a time. The same holds for a random-edge policy where only a single position is updated. The figures for these two methods are given in [29].

It is therefore good news that symmetric strategy improvement does not display a similar weakness. It even uses less updates when compared to classic strategy improvement with the popular locally optimising and locally random policy rules. Note also that having less updates can lead to a faster evaluation of the update, because unchanged parts do not need to be re-evaluated [3].

Switch Rule	1	2	3	4	5	6	7	8	9	10
Cunningham	2	6	9	12	15	18	21	24	27	30
CunninghamSubexp	1	1	1	1	1	1	1	1	1	1
FearnleySubexp	4	7	11	13	17	21	25	29	33	37
FriedmannSubexp	4	9	13	15	19	23	27	31	35	39
RandomEdgeExpTest	1	2	2	2	2	2	2	2	2	2
RandomFacetSubexp	1	2	7	9	11	13	15	17	19	21
SwitchAllBestExp	4	5	8	11	12	13	15	17	18	19
SwitchAllBestSubExp	5	7	9	11	13	15	17	19	21	23
SwitchAllSubExp	3	5	7	9	10	11	12	13	14	15
SwitchAllExp	3	4	6	8	10	11	12	14	16	18
ZadehExp	-	6	10	14	18	21	25	28	32	35
ZadehSubexp	5	9	13	16	20	23	27	30	34	37

As shown in Figure 2, the symmetric strategy improvement algorithm performs better (on average) in comparison with the traditional strategy improvement algorithm with the locally optimising policy rule. It also avoids Friedmann's traps for the strategy improvement algorithm: the table above shows the performance of the symmetric strategy improvement algorithm for Friedmann's traps for other common switching rules. It is clear that our algorithm is not exponential for these classes of examples.

5   Discussion

We have introduced symmetric approaches to strategy improvement, where the players take inspiration from the respective other's strategy when improving theirs. This creates a rather moderate overhead, where each step is at most twice as expensive as a normal improvement step. For this moderate price, we have shown that we can break the traps Friedmann has introduced to establish exponential bounds for the different update policies in classic strategy improvement [12–14].

In hindsight, attacking a symmetric problem with a symmetric approach seems so natural, that it is quite surprising that it has not been attempted immediately. There are, however, good reasons for this, but one should also consent that the claim is not entirely true: the concurrent update to the respective optimal counter strategy has been considered quite early [12–14], but was dismissed, because it can lead to cycles [5].

The first reason is therefore that it was folklore that symmetric strategy improvement does not work. The second reason is that the argument for the techniques that we have developed in this paper would have been restricted to beauty until some of the appeal of classic strategy improvement was caught in Friedmann's traps. Friedmann himself, however, remained optimistic:

> We think that the strategy iteration still is a promising candidate for a polynomial time algorithm, however it may be necessary to alter more of it than just the improvement policy.

This is precisely, what the introduction of symmetry and co-improvement tries to do.

References

1. Alur, R., Henzinger, T.A., Kupferman, O.: Alternating-time temporal logic. Journal of the ACM **49**(5), 672–713 (2002)
2. Berwanger, D., Dawar, A., Hunter, P., Kreutzer, S.: DAG-width and parity games. In: Durand, B., Thomas, W. (eds.) STACS 2006. LNCS, vol. 3884, pp. 524–536. Springer, Heidelberg (2006)
3. Björklund, H., Vorobyov, S.: A combinatorial strongly subexponential strategy improvement algorithm for mean payoff games. Discrete Appl. Math. **155**(2), 210–229 (2007)
4. Browne, A., Clarke, E.M., Jha, S., Long, D.E., Marrero, W.: An improved algorithm for the evaluation of fixpoint expressions. TCS **178**(1–2), 237–255 (1997)
5. Condon, A.: On algorithms for simple stochastic games. In: Advances in Computational Complexity Theory, pp. 51–73. American Mathematical Society (1993)

6. de Alfaro, L., Henzinger, T.A., Majumdar, R.: From verification to control: dynamic programs for omega-regular objectives. In: Proc. of LICS, pp. 279–290 (2001)
7. Emerson, E.A., Jutla, C.S.: Tree automata, μ-calculus and determinacy. In: Proc. of FOCS, pp. 368–377. IEEE Computer Society Press, October 1991
8. Emerson, E.A., Jutla, C.S., Sistla, A.P.: On model-checking for fragments of μ-calculus. In: Courcoubetis, C. (ed.) CAV 1993. LNCS, vol. 697, pp. 385–396. Springer, Heidelberg (1993)
9. Emerson, E.A., Lei, C.: Efcient model checking in fragments of the propositional μ-calculus. In: Proc. of LICS, pp. 267–278. IEEE Computer Society Press (1986)
10. Fearnley, J.: Exponential lower bounds for policy iteration. In: Abramsky, S., Gavoille, C., Kirchner, C., Meyer auf der Heide, F., Spirakis, P.G. (eds.) ICALP 2010. LNCS, vol. 6199, pp. 551–562. Springer, Heidelberg (2010)
11. Fearnley, J.: Non-oblivious strategy improvement. In: Clarke, E.M., Voronkov, A. (eds.) LPAR-16 2010. LNCS, vol. 6355, pp. 212–230. Springer, Heidelberg (2010)
12. Friedmann, O.: An exponential lower bound for the latest deterministic strategy iteration algorithms. LMCS **7**(3) (2011)
13. Friedmann, O.: A Subexponential lower bound for Zadeh's pivoting rule for solving linear programs and games. In: Günlük, O., Woeginger, G.J. (eds.) IPCO 2011. LNCS, vol. 6655, pp. 192–206. Springer, Heidelberg (2011)
14. Friedmann, O.: A superpolynomial lower bound for strategy iteration based on snare memorization. Discrete Applied Mathematics **161**(10–11), 1317–1337 (2013)
15. Jurdziński, M.: Small progress measures for solving parity games. In: Reichel, H., Tison, S. (eds.) STACS 2000. LNCS, vol. 1770, pp. 290–301. Springer, Heidelberg (2000)
16. Jurdzinski, M., Paterson, M., Zwick, U.: A deterministic subexponential algorithm for solving parity games. SIAM J. Comput. **38**(4), 1519–1532 (2008)
17. Kozen, D.: Results on the propositional μ-calculus. TCS **27**, 333–354 (1983)
18. Lange, M.: Solving parity games by a reduction to SAT. In: Proc. of Int. Workshop on Games in Design and Verification (2005)
19. Lange, M., Friedmann, O.: The PGSolver collection of parity game solvers. Technical report, Institut für Informatik Ludwig-Maximilians-Universität (2010)
20. Ludwig, W.: A subexponential randomized algorithm for the simple stochastic game problem. Inf. Comput. **117**(1), 151–155 (1995)
21. McNaughton, R.: Infinite games played on finite graphs. Ann. Pure Appl. Logic **65**(2), 149–184 (1993)
22. Obdržálek, J.: Fast mu-calculus model checking when tree-width is bounded. In: Hunt Jr., W.A., Somenzi, F. (eds.) CAV 2003. LNCS, vol. 2725, pp. 80–92. Springer, Heidelberg (2003)
23. Piterman, N.: From nondeterministic Büchi and Streett automata to deterministic parity automata. In: Proc. of LICS, pp. 255–264. IEEE Computer Society (2006)
24. Puri, A.: Theory of hybrid systems and discrete event systems. PhD thesis, Computer Science Department, University of California, Berkeley (1995)
25. Schewe, S.: Solving parity games in big steps. In: Arvind, V., Prasad, S. (eds.) FSTTCS 2007. LNCS, vol. 4855, pp. 449–460. Springer, Heidelberg (2007)
26. Schewe, S.: An optimal strategy improvement algorithm for solving parity and payoff games. In: Kaminski, M., Martini, S. (eds.) CSL 2008. LNCS, vol. 5213, pp. 369–384. Springer, Heidelberg (2008)
27. Schewe, S., Finkbeiner, B.: Satisfiability and finite model property for the alternating-time μ-calculus. In: Ésik, Z. (ed.) CSL 2006. LNCS, vol. 4207, pp. 591–605. Springer, Heidelberg (2006)

28. Schewe, S., Finkbeiner, B.: Synthesis of asynchronous systems. In: Puebla, G. (ed.) LOPSTR 2006. LNCS, vol. 4407, pp. 127–142. Springer, Heidelberg (2007)
29. Schewe, S., Trivedi, A., Varghese, T.: Symmetric strategy improvement (2015). CoRR, abs/1501.06484
30. Vardi, M.Y.: Reasoning about the past with two-way automata. In: Larsen, K.G., Skyum, S., Winskel, G. (eds.) ICALP 1998. LNCS, vol. 1443, pp. 628–641. Springer, Heidelberg (1998)
31. Vöge, J., Jurdziński, K.: A discrete strategy improvement algorithm for solving parity games. In: Emerson, E.A., Sistla, A.P. (eds.) CAV 2000. LNCS, vol. 1855, pp. 202–215. Springer, Heidelberg (2000)
32. Wilke, T.: Alternating tree automata, parity games, and modal μ-calculus. Bull. Soc. Math. Belg. **8**(2), May 2001
33. Zielonka, W.: Infinite games on finitely coloured graphs with applications to automata on infinite trees. Theor. Comput. Sci. **200**(1–2), 135–183 (1998)
34. Zwick, U., Paterson, M.S.: The complexity of mean payoff games on graphs. Theoretical Computer Science **158**(1–2), 343–359 (1996)

Effect Algebras, Presheaves, Non-locality and Contextuality

Sam Staton[1](✉) and Sander Uijlen[2]

[1] University of Oxford, Oxford, England, UK
sam.staton@cs.ox.ac.uk
[2] Radboud University, Nijmegen, The Netherlands

Abstract. Non-locality and contextuality are among the most counter-intuitive aspects of quantum theory. They are difficult to study using classical logic and probability theory. In this paper we start with an effect algebraic approach to the study of non-locality and contextuality. We will see how different slices over the category of set valued functors on the natural numbers induce different settings in which non-locality and contextuality can be studied. This includes the Bell, Hardy and Kochen-Specker-type paradoxes. We link this to earlier sheaf theoretic approaches by defining a fully faithful embedding of the category of effect algebras in this presheaf category over the natural numbers.

1 Introduction

This paper is about generalized theories of probability that allow us to analyze the non-locality and contextuality paradoxes from quantum theory. Informally, the paradoxes have to do with the idea that it might not be possible to explain the outcomes of measurements in a classical way. We proceed by using now-standard techniques for local reasoning in computer science. Partial monoids play a crucial role in 'separation logic' which is a basic framework of locality especially relevant to memory locality (e.g. [3,4]). Presheaves on natural numbers have already been used to study local memory (e.g. [16]) and also to study contexts in abstract syntax (e.g. [7]).

The paper is in two parts. In the first we establish new relationships between two generalized theories of probability. In the second we analyze the paradoxes of contextuality using our theories of probability, and we use this to recover earlier formulations of them in different frameworks.

1.1 Generalized Probability Measures

Recall that a finite measurable space (X, Ω) comprises a finite set X and a sub-Boolean algebra Ω of the powerset $\Omega \subseteq \mathcal{P}(X)$, and recall:

Definition 1. *A probability distribution on a finite measurable space (X, Ω) is a function $p : \Omega \to [0,1]$ such that $p(X) = 1$ and if $A_1 \ldots A_n$ are disjoint sets in Ω, then $\sum_{i=1}^{n} p(A_i) = p(\bigcup_{i=1}^{n} A_i)$.*

© Springer-Verlag Berlin Heidelberg 2015
M.M. Halldórsson et al. (Eds.): ICALP 2015, Part II, LNCS 9135, pp. 401–413, 2015.
DOI: 10.1007/978-3-662-47666-6_32

We now analyze this definition to propose two general notions of probability measure. (NB. We will focus on finite probability spaces, because this is sufficient for our examples. We intend to return to infinite spaces in future work.)

Partial Monoids. Our first generalization involves partial monoids. Notice that the conditions on the probability distribution $p : \Omega \to [0,1]$ do not involve the space $\mathcal{P}(X)$. We only used the disjoint union structure of Ω. More generally, we can define a *pointed partial commutative monoid* (PPCM) to be a structure $(E, \oslash, 0, 1)$ where $\oslash : E \times E \to E$ is a commutative, associative partial binary operation with a unit 0. Then $(\Omega, \uplus, \emptyset, X)$ and the interval $([0,1], +, 0, 1)$ are PPCMs. A probability distribution is now the same thing as a PPCM homomorphism, $(\Omega, \uplus, \emptyset, X) \to ([0,1], +, 0, 1)$. Thus PPCMs are a candidate for a generalized probability theory. (This is a long-established position; see e.g. [6].)

Functors. Our second generalization goes as follows. Every finite Boolean algebra Ω is isomorphic to one of the form $\mathcal{P}(N)$ for a finite set N, called the atoms of Ω. Now, a probability distribution $p : \Omega \to [0,1]$ is equivalently given by a function $q : N \to [0,1]$ such that $\sum_{a \in N} q(a) = 1$. Let

$$D(N) = \{q : N \to [0,1] \mid \sum_{a \in N} q(a) = 1\} \tag{1}$$

be the set of all distributions on a finite set N. It is well-known that D extends to a functor $D : \mathbf{FinSet} \to \mathbf{Set}$. The Yoneda lemma gives a bijection between distributions in $D(N)$ and natural transformations $\mathbf{FinSet}(N, -) \to D$. Thus we are led to say that a generalized finite measurable space is a functor $F : \mathbf{FinSet} \to \mathbf{Set}$ (aka presheaf), and a probability distribution on F is a natural transformation $F \to D$. (This appears to be a new position.)

Relationship. Our main contribution in Section 2 and 3 is an adjunction between the two kinds of generalized measurable spaces: PPCMs, and presheaves $\mathbf{FinSet} \to \mathbf{Set}$. 'Effect algebras' are a special class of PPCMs [5,9]. We show that our adjunction restricts to a reflection from effect algebras into presheaves $\mathbf{FinSet} \to \mathbf{Set}$, which gives us a slogan that 'effect algebras are well-behaved generalized finite measurable spaces'.

1.2   Relating Non-locality and Contextuality Arguments

In the second part of the paper we investigate three paradoxes from quantum theory, attributed to Bell, Hardy and Kochen-Specker. We justify our use of effect algebras and presheaves by establishing relationships with earlier work by Abramsky and Brandenburger [1] and Hamilton, Isham and Butterfield [10]. For the purposes of introduction, we focus on the Bell paradox, and we focus on the mathematics. (Some physical intuitions are given in Section 4.)

The Bell paradox in terms of effect algebras and presheaves. As we show, the Bell scenario can be understood as a morphism of effect algebras $E \xrightarrow{t} [0,1]$, i.e., a generalized probability distribution. The paradox is that although this has a quantum realization, in that it factors through $Proj(\mathcal{H})$, the projections on a Hilbert space $\mathcal{H}$, it has no explanation in classical probability theory, in that there it does not factor through a given Boolean algebra Ω. Informally:

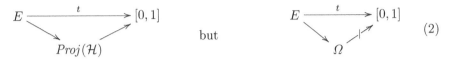

$$\tag{2}$$

Relationship with earlier sheaf-theoretic work on the Bell paradox. In [1], Abramsky and Brandenburger have studied Bell-type scenarios in terms of presheaves. We recover their results from our analysis in terms of generalized probability theory. Our first step is to notice that effect algebras essentially fully embed in the functor category $[\mathbf{FinSet} \to \mathbf{Set}]$. We step even closer by recalling the slice category construction. This is a standard technique of categorical logic for working relative to a particular object. As we explain in Section 4, the slice category $[\mathbf{FinSet} \to \mathbf{Set}]/\Omega$ is again a presheaf category. It is more-or-less the category used in [1]. Moreover, our non-factorization (2) transports to the slice category: Ω becomes terminal, and E is a subterminal object. Thus the non-factorization in diagram (2) can be phrased in the sheaf-theoretic language of Abramsky and Brandenburger: 'the family t has no global section'.

Other Paradoxes. Alongside the Bell paradox we study two other paradoxes:

– The Hardy paradox is similar to the Bell paradox, except that it uses possibility rather than probability. We analyze this by replacing the unit interval $([0,1], +, 0, 1)$ by the PPCM $(\{0,1\}, \vee, 0, 1)$ where $\vee$ is bitwise-or. Although this monoid is not an effect algebra, everything still works and we are able to recover the analysis of the Hardy paradox by Abramsky and Brandenburger.
– The Kochen-Specker paradox can be understood as saying that there is no PPCM morphism
$$Proj(\mathcal{H}) \to (\{0,1\}, \oslash, 0, 1) \tag{3}$$
with dim. $\mathcal{H} \geq 3$ and where $\oslash$ is like bitwise-or, except that $1 \oslash 1$ is undefined. Now, the slice category $[\mathbf{FinSet} \to \mathbf{Set}]/Proj(\mathcal{H})$ is again a presheaf category, and it is more-or-less the presheaf category used by Hamilton, Isham and Butterfield. The non-existence of a homomorphism (3) transports to this slice category: $Proj(\mathcal{H})$ becomes the terminal object, and $(\{0,1\}, \oslash, 0, 1)$ becomes the so-called 'spectral presheaf'. We are thus able to rephrase the non-existence of a homomorphism (3) in the same way as Hamilton, Isham and Butterfield [10]: 'the spectral presheaf does not have a global section'.

Summary. Motivated by techniques for locality in computer science, we have developed a framework for generalized probability theory based on effect algebras and presheaves. The relevance of the framework is demonstrated by the paradoxes of non-locality and contextuality, which arise as diagrams in one fundamental adjunction. Different analyses in the literature use different presheaf categories, but these all arise from our analysis by taking slice categories.

2   Pointed Partial Commutative Monoids

Definition 2. *A* pointed partial commutative monoid *(PPCM)* $(E, 0, 1, \oslash)$ *consists of a set E with a special element $0 \in E$, a chosen point $1 \in E$ and a partial function $\oslash : E \times E \to E$, such that for all $x, y, z \in E$ we have:*

1. *If $x \oslash y$ is defined, then $y \oslash x$ is also defined and $x \oslash y = y \oslash x$.*
2. *$x \oslash 0$ is always defined and $x = x \oslash 0$.*
3. *If $x \oslash y$ and $(x \oslash y) \oslash z$ are defined, then $y \oslash z$ and $x \oslash (y \oslash z)$ are defined and $(x \oslash y) \oslash z = x \oslash (y \oslash z)$.*

We write $x \perp y$ (say x is perpendicular to y), if $x \oslash y$ is defined. When we write $x \oslash y$, we tacitly assume $x \perp y$. We refer to $x \oslash y$ as the sum *of x and y.*

A morphism $f : E \to F$ of PPCMs is a map such that $f(0) = 0$, $f(1) = 1$ and $f(a \oslash b) = f(a) \oslash f(b)$ whenever $a \perp b$. This entails the category **PPCM**.

Definition 3. *An* effect algebra $(E, 0, \oslash, 1)$ *is a PPCM $(E, 0, \oslash, 1)$ such that*

1. *For every $x \in E$ there exists a unique $x^\perp$ such that $x \perp x^\perp$ and $x \oslash x^\perp = 1$.*
2. *$x \perp 1$ implies $x = 0$.*

We call $x^\perp$ the 'orthocomplement of x'. PPCM morphisms between effect algebras always preserve orthocomplements. We denote by **EA** the full subcategory of **PPCM** whose objects are effect algebras.

Example 4.   – *We will consider the set $2 = \{0, 1\}$ as a PPCM in two ways.*
  - *The initial PPCM $(2, \oslash, 0, 1)$ has $0 \oslash 0 = 0$ and $1 \oslash 0 = 0 \oslash 1 = 1$; this is an effect algebra.*
  - *The monoid $(2, \vee, 0, 1)$ with $0 \vee 0 = 0$ and $1 \vee 0 = 0 \vee 1 = 1 \vee 1 = 1$; this is not an effect algebra.*
- *Any Boolean algebra $(B, \vee, \wedge, 0, 1)$ is an effect algebra $(B, \oslash, 0, 1)$ where $x \perp y$ iff $x \wedge y = 0$, and then $x \oslash y \overset{def}{=} x \vee y$. A function between Boolean algebras is a Boolean algebra homomorphism iff it is a PPCM morphism.*
- *The projections on a Hilbert space form an effect algebra $(Proj(\mathcal{H}), +, 0, 1)$ where $p \perp q$ if their ranges are orthogonal.*
- *The unit interval $([0, 1], +, 0, 1)$ is an effect algebra when $x \perp y$ iff $x + y \leq 1$.*

3 Presheaves and Tests

In this section we consider a different notion of generalized probability space. Recall that for any finite set N we have a set $D(N)$ of distributions (Equation (1)). This construction is functorial in N. Consider the category $\mathbb{N}$, the skeleton of **FinSet**, whose objects are natural numbers considered as sets, $N = \{1, \ldots, n\}$, and whose morphisms are functions. Then $D : \mathbb{N} \to \textbf{Set}$, with $((D f)(q))(i) = \sum_{j \in f^{-1}(i)} q(j)$.

This leads us to a notion of generalized probability space via the Yoneda lemma. Write $\textbf{Set}^{\mathbb{N}}$ for the category of functors $\mathbb{N} \to \textbf{Set}$ (aka 'covariant presheaves') and natural transformations. The Yoneda lemma says $D(N) \cong \textbf{Set}^{\mathbb{N}}(\mathbb{N}(N, -), D)$. More generally we can thus understand natural transformations $F \to D$ as 'distributions' on a functor $F \in \textbf{Set}^{\mathbb{N}}$.

To make a connection between presheaves and PPCMs and effect algebras we recall the notion of test.

Definition 5. *Let E be a PPCM. An n-test in E is an n-tuple $(e_1, \ldots, e_n)$ of elements in E such that $e_1 \otimes \ldots \otimes e_n = 1$.*

The tests of a PPCM E form a presheaf $T(E) \in \textbf{Set}^{\mathbb{N}}$, where $T(E)(N)$ is the set of n-tests in E, and if $f : N \to M$ is a function then

$$T(E)(f)(e_1, \ldots, e_n) = (\bigotimes_{i \in f^{-1}(j)} e_i)_{j=1,\ldots,m}$$

This extends to a functor $T : \textbf{PPCM} \to \textbf{Set}^{\mathbb{N}}$. If $\psi : E \to A$ is a PPCM morphism, then we obtain the natural transformation $T(\psi)$ with components $T(\psi)_N(e_1, \ldots, e_n) = (\psi(e_1), \ldots, \psi(e_n))$. (See also [12, Def. 6.3].)

Example 6.   – $T(2, \oslash, 0, 1) \in \textbf{Set}^{\mathbb{N}}$ *is the inclusion:* $(T(2, \oslash, 0, 1))(N) = N$.
– $T(2, \vee, 0, 1) \in \textbf{Set}^{\mathbb{N}}$ *is the non-empty powerset functor:* $(T(2, \vee, 0, 1))(N) = \{S \subseteq N \mid S \neq \emptyset\}$.
– *Any finite Boolean algebra* $(B, \vee, \wedge, 0, 1)$ *is of the form* $\mathcal{P}(N)$ *for a finite set N; we have* $T(B, \oslash, 0, 1) = \mathbb{N}(N, -)$, *the representable functor.*
– *For the unit interval,* $T([0, 1], +, 0, 1) = D$, *the distribution functor.*

Our main result in this section is that the test functor essentially exhibits effect algebras as a full subcategory of $\textbf{Set}^{\mathbb{N}}$.

Theorem 7 *The induced function* $T_{A,B} : \textbf{PPCM}(A, B) \to \textbf{Set}^{\mathbb{N}}(TA, TB)$ *is a bijection when A is an effect algebra.*

Proof (summary). Since A is an effect algebra, every element $a \in A$ is part of a 2-test $(a, a^{\perp})$. It is then clear that $T_{A,B}$ is injective. Now suppose we have some natural transformation $\mu : T(A) \to T(B)$. The map $\psi_\mu : A \to B$ defined by $\psi_\mu(a) = x$, where $(x, x^{\perp}) = \mu_2(a, a^{\perp})$ has the property that $T(\psi_\mu) = \mu$.

Corollary 8. *The restriction to effect algebras,* $T : \textbf{EA} \to \textbf{Set}^{\mathbb{N}}$, *is full and faithful.*

We remark that a more abstract way to view the test functor is through the framework of nerves and realizations. For any natural number N the powerset $\mathcal{P}(N)$ is a Boolean algebra and hence an effect algebra. This extends to a functor $\mathcal{P} : \mathbb{N}^{\mathrm{op}} \to \mathbf{PPCM}$. The test functor T has a left adjoint, which is the left Kan extension of $\mathcal{P}$ along the Yoneda embedding. (This follows from Theorem 2 of [15, Ch. I.5]; $\mathbf{PPCM}$ is cocomplete by [2, Theorem 3.36].) Theorem 7 can be phrased 'the counit is an isomorphism at effect algebras', and Corollary 8 can be phrased 'finite Boolean algebras are dense in effect algebras'.

4   Non-Locality and Contextuality

In probability theory, questions of contextuality arise from the problem that the joint probability distribution for all outcomes of all measurements may not exist. We suppose a simple framework where Alice and Bob each have a measurement device with two settings. For simplicity we suppose that the device will emit 0 or 1, as the outcome of a measurement. We write $a_0{:}0$ for 'Alice measured 0 with setting a_0', $b_1{:}0$ for 'Bob measured 0 with setting b_1', and so on. To model this in classical probability theory we would consider a sample space S_A for Alice whose elements are functions $\{a_0, a_1\} \to \{0, 1\}$, i.e., assignments of outcomes to measurements. Similarly we have a sample space S_B for Bob. We would then consider a joint probability distribution on S_A and S_B.

In this model, we implicitly assume that Alice and Bob can not signal to each other. That is to say, for any joint distribution we can define marginal distributions each for Alice and Bob. However, the classical model does include an assumption: that Alice is able to record the outcome of the measurement in both settings. In reality, and in quantum physics, once Alice has recorded an outcome using one measurement setting, she cannot then know what the outcome would have been using the other measurement setting. Effect algebras provide a way to describe a kind of probability distribution that takes this measure-only-once phenomenon into account.

The non-locality 'paradox' is as follows: there are probability distributions in this effect algebraic sense (without signalling), which are physically realizable, but cannot be explained in a classical probability theory without signalling.

The main purpose of this section is not to study non-locality and contextuality in different systems, but rather to give a general framework to study them. We use this to recover earlier frameworks.

4.1   Bimorphisms, Joint Distributions, and Tables

It is convenient to first introduce a notion of bimorphism, which captures the notion of a probability distribution on joint measurements. Later we will see that bimorphisms are classified by a tensor product.

Definition 9. *Let A, B and C be pointed partial commutative monoids. A bimorphism $A, B \to C$ is a function $f : A \times B \to C$ such that for all $a, a_1, a_2 \in A$ and $b, b_1, b_2 \in B$ with $a_1 \perp a_2$ and $b_1 \perp b_2$ we have*

$$f(a, b_1 \oslash b_2) = f(a, b_1) \oslash f(a, b_2) \qquad f(a_1 \oslash a_2, b) = f(a_1, b) \oslash f(a_2, b)$$
$$f(a, 0) = f(0, b) = 0 \qquad f(1, 1) = 1$$

We now describe the scenario in the introduction to this section using bimorphisms. Let E_A be the effect algebra $\{0, a_0{:}0, a_0{:}1, a_1{:}0, a_1{:}1, 1\}$ with $0 \oslash x = x$ and $a_i{:}0 \oslash a_i{:}1 = 1$. This is the algebra for Alice's measurements. Similarly, let E_B be the algebra for Bob's measurements. A distribution on the joint measurements of Alice and Bob is a bimorphism $E_A, E_B \to [0, 1]$. We now give an elementary description of these bimorphisms. Each bimorphism $t : E_A, E_B \to [0, 1]$ restricts to a function

$$\tau : \{a_0{:}0, a_0{:}1, a_1{:}0, a_1{:}1\} \times \{b_0{:}0, b_0{:}1, b_1{:}0, b_1{:}1\} \to [0, 1]$$

which we call a probability table, and we characterize these:

Proposition 10. *A table* $\tau{:}\{a_0{:}0, a_0{:}1, a_1{:}0, a_1{:}1\} \times \{b_0{:}0, b_0{:}1, b_1{:}0, b_1{:}1\} \to [0, 1]$ *arises as the restriction of a bimorphism* $E_A, E_B \to [0, 1]$ *if and only if*

- it is a probability: $\sum_{o,o' \in \{0,1\}} \tau(a_i{:}o, b_j{:}o') = 1$, *for* $i, j \in \{0, 1\}$.
- it has marginalization, aka no signalling: *for all* $i, j \in \{0, 1\}$,

$$\tau(a_i{:}j, b_0{:}0) + \tau(a_i{:}j, b_0{:}1) = \tau(a_i{:}j, b_1{:}0) + \tau(a_i{:}j, b_1{:}1),$$
$$\tau(a_0{:}0, b_i{:}j) + \tau(a_0{:}1, b_i{:}j) = \tau(a_1{:}0, b_i{:}j) + \tau(a_1{:}1, b_i{:}j).$$

The standard *Bell table* is as below, and by Proposition 10 it extends to a bimorphism $E_A, E_B \to [0, 1]$. In this simple scenario we have two observers, each with two measurement settings, each with two outcomes, but it is straightforward to generalize to more elaborate Bell-like settings.

t	$a_0{:}0$	$a_0{:}1$	$a_1{:}0$	$a_1{:}1$
$b_0{:}0$	$\frac{1}{2}$	0	$\frac{3}{8}$	$\frac{1}{8}$
$b_0{:}1$	0	$\frac{1}{2}$	$\frac{1}{8}$	$\frac{3}{8}$
$b_1{:}0$	$\frac{3}{8}$	$\frac{1}{8}$	$\frac{1}{8}$	$\frac{3}{8}$
$b_1{:}1$	$\frac{1}{8}$	$\frac{3}{8}$	$\frac{3}{8}$	$\frac{1}{8}$

$$(4)$$

4.2   Realization and Bell's Paradox

Quantum realization. A table has a 'quantum realization' if there is a way to obtain it by performing quantum experiments. Recall that a quantum system is modelled by a Hilbert space $\mathcal{H}$, and a yes-no question such as "is the outcome of measuring a_0 equal to 1" is given by a projection on this Hilbert space. The projections form an effect algebra $Proj(\mathcal{H})$.

Definition 11. *A* quantum realization *for a distribution on joint measurements* $t : E, E' \to [0, 1]$ *is given by finite dimensional Hilbert spaces* $\mathcal{H}, \mathcal{H}'$, *two PPCM maps* $r : E \to Proj(\mathcal{H})$ *and* $r' : E' \to Proj(\mathcal{H}')$, *and a bimorphism* $p : Proj(\mathcal{H}), Proj(\mathcal{H}') \to [0, 1]$, *such that for all* $e \in E$ *and* $e' \in E'$ *we have* $p(r(e), r'(e')) = t(e, e')$.

The Bell table (4) has a quantum realization, with $\mathcal{H} = \mathcal{H}' = \mathbb{C}^2$.

Classical realization. Classically, every time Alice and Bob perform a measurement, nature determines an assignment of outcomes for all measurements, which determines the outcomes for Alice and Bob. In such a deterministic theory we can calculate a probability for things like $a_0{:}0 \wedge a_1{:}1 \wedge b_0{:}1 \wedge b_1{:}1$, in which case if Alice chose a_0 and Bob chose b_1, they would get the outcome 0 and 1, respectively. It can be shown (e.g., see [1]), that this is not the case for the standard Bell table.

Definition 12. *A classical realization for a distribution* $t : E, E' \rightarrow [0, 1]$ *is given by two Boolean algebras* B, B', *two effect algebra morphisms* $r : E \rightarrow B$, $r' : E' \rightarrow B'$ *and a bimorphism* $p : B, B' \rightarrow [0, 1]$ *such that for all* $e \in E$ *and* $e' \in E'$ *we have* $p(r(e), r'(e')) = t(e, e')$.

Consider the Boolean algebra, B_A, with atoms $\{a_1{:}i \wedge a_2{:}j \mid i, j \in \{0, 1\}\}$. Note that B_A is a free completion of the effect algebra E_A to a Boolean algebra, in that, under identification of $(a_1{:}0 \wedge a_2{:}0) \vee (a_1{:}0 \wedge a_2{:}1)$ with $a_1{:}0$, we have $E_A \subseteq B_A$ and every morphism $E_A \rightarrow B$, with B a Boolean algebra, must factor through B_A. Similarly, we have the algebra B_B for Bob.

Proposition 13. *The canonical maps* $r_A : E_A \rightarrow B_A$ *and* $r_B : E_B \rightarrow B_B$ *cannot be completed to a classical realization of Table 4. Therefore, Table 4 has no classical realization.*

4.3   Tensor Products

Definition 14. *The* tensor product *of two PPCMs* E, E' *is given by a PPCM* $E \otimes E'$ *and a bimorphism* $i : E, E' \rightarrow E \otimes E'$, *such that for every bimorphism* $f : E, E' \rightarrow F$ *there is a unique morphism* $g : E \otimes E' \rightarrow F$ *such that* $f = g \circ i$.

This gives a bijective correspondence between morphisms $E \otimes E' \rightarrow F$ and bimorphisms $E, E' \rightarrow F$. In fact, all tensor products of effect algebras exist (see e.g. [11]; but they can be trivial [8]). We return to the example of Alice and Bob.

Proposition 15.   – *The tensor product of Boolean algebras,* $B_A \otimes B_B$, *is the free Boolean algebra on the four elements* $\{a_1, a_2, b_1, b_2\}$, *where we identify, for example,* $a_1{:}1$ *with* a_1 *and* $a_1{:}0$ *with* $\neg a_1$.
  – *The tensor product of effect algebras* $E_A \otimes E_B$ *is the effect algebra generated by the 16 elements* $a_i{:}0 \wedge b_j{:}0$, $a_i{:}0 \wedge b_j{:}1$, $a_i{:}1 \wedge b_j{:}0$, $a_i{:}1 \wedge b_j{:}1$, *for* $i, j \in \{0, 1\}$, *such that each 4-tuple* $(a_i{:}0 \wedge b_j{:}0, a_i{:}0 \wedge b_j{:}1, a_i{:}1 \wedge b_j{:}0, a_i{:}1 \wedge b_j{:}1)$ *with* $i, j \in \{0, 1\}$ *is a 4-test. (For elements in such a 4-test we have that the effect algebra sum* $\oslash$ *is the Boolean join,* $\vee$. *Elements in different 4-tests are not perpendicular.)*

The statement of Bell's paradox can now be written in terms of homomorphisms, rather than bimorphisms:

Corollary 16. *Table 4,* $t : E_A \otimes E_B \rightarrow [0, 1]$, *does not factor through the embedding* $E_A \otimes E_B \rightarrow B_A \otimes B_B$.

4.4   Sheaf Theoretic Characterization

Since the test functor $T : \mathbf{EA} \rightarrow \mathbf{Set}^{\mathbb{N}}$ is full and faithful from effect algebras (Cor. 8), we can apply it to our effect algebra formulation of the Bell scenario, and arrive at a similar statement in terms of presheaves. Recall that $T([0,1]) = D$, the distributions functor, and so the Bell table yields a natural transformation $T(E_A \otimes E_B) \rightarrow D$. Recall that $T(B_A \otimes B_B) = \mathbb{N}(16, -)$, the representable functor, and so the non-existence of a classical realization (Cor. 16) amounts to the non-existence of a natural transformation as in the following diagram:

$$T(E_A \otimes E_B) \xrightarrow{\quad Tt \quad} D \qquad\qquad (5)$$
$$\underset{Ti}{\searrow} \quad \mathbb{N}(16, -) \nearrow$$

We can thus phrase Bell's paradox in the language of Grothendieck's sheaf theory. Since $i : (E_A \otimes E_B) \rightarrow (B_A \otimes B_B)$ is a subalgebra and T preserves monos, $T(E_A \otimes E_B)$ is a subpresheaf of $\mathbb{N}(16, -)$, aka a 'sieve' on 16. A map $T(E_A \otimes E_B) \rightarrow D$ out of a sieve is called a 'compatible family', and a map $\mathbb{N}(16, -) \rightarrow D$ amounts to a distribution in $D(16)$ (by the Yoneda lemma). Bell's paradox now states: "*the compatible family $T(t)$ has no amalgamation*".

4.5   Relationship with the Work of Abramsky and Brandenburger

Abramsky and Brandenburger [1] also phrase Bell's paradox in terms of a compatible family with no amalgamations. We now relate our statement with theirs.

Transferring the paradox to other categories. We can use adjunctions to transfer statements of non-factorization (such as Corollary 16) between different categories. Let $\mathcal{C}$ be a category and let $R : \mathbf{EA} \rightarrow \mathcal{C}$ be a functor with a left adjoint $L : \mathcal{C} \rightarrow \mathbf{EA}$. Let $j : X \rightarrow Y$ be a morphism in $\mathcal{C}$, and let $f : L(X) \rightarrow A$ be a morphism in $\mathbf{EA}$. Then f factors through $L(j)$ if and only if $f^{\sharp} : X \rightarrow R(A)$ factors through j, where $f^{\sharp}$ is the transpose of f.

$$L(X) \xrightarrow{\quad f \quad} A \qquad\qquad X \xrightarrow{\quad f^{\sharp} \quad} R(A)$$
$$\underset{L(j)}{\searrow} L(Y) \nearrow \qquad\qquad \underset{j}{\searrow} Y \nearrow$$

We use this technique to derive several equivalent statements of Bell's paradox. To start, the equivalence of the non factoring of the triangles (5) and (2) is immediate from the adjunction between the test functor and its left adjoint.

No global section. Recall that if X is an object of a category $\mathcal{C}$ then the objects of the *slice category* $\mathcal{C}/X$ are pairs (C, f) where $f : C \rightarrow X$. Morphisms are commuting triangles. The slice category $\mathcal{C}/X$ always has a terminal object, (X, id_X). The projection map $\Sigma_X : \mathcal{C}/X \rightarrow \mathcal{C}$, with $\Sigma_X(C, f) = C$, has a right adjoint $\Delta_X : \mathcal{C} \rightarrow \mathcal{C}/X$ with $\Delta_X(C) = (C \times X, \pi_2)$. First, notice that, using the adjunction $\Sigma_{\mathbb{N}(16,-)} \dashv \Delta_{\mathbb{N}(16,-)}$ we can rewrite diagram (5) in the slice category $(\mathbf{Set}^{\mathbb{N}})/\mathbb{N}(16,-)$ as:

$$(T(E_A \otimes E_B), Ti) \xrightarrow{\langle Tt, Ti \rangle} (D \times \mathbb{N}(16,-), \pi_2) \qquad (6)$$

$$\searrow \qquad \nearrow$$
$$(\mathbb{N}(16,-), \mathrm{id})$$

Since $(\mathbb{N}(16,-), \mathrm{id})$ is terminal, we can phrase Bell's paradox as *"the local section* $\langle Tt, Ti \rangle : (T(E_A \otimes E_B), Ti) \to (D \times \mathbb{N}(16,-), \pi_2)$ *has no global section"*.

Measurement Covers. The analysis of Abramsky and Brandenburger is based on a 'measurement cover', which corresponds to our effect algebra $E_A \otimes E_B$.

Fix a finite set X of measurements. In our Bell example, $X = \{a_0, a_1, b_0, b_1\}$. Also fix a finite set of O of outcomes. In our example, $O = \{0, 1\}$, so $O^X = 16$. Abramsky and Brandenburger work in the category of presheaves $\mathcal{P}(X)^{\mathrm{op}} \to \mathbf{Set}$ on the powerset $\mathcal{P}(X)$ (ordered by subset inclusion). They explain Bell-type paradoxes as statements that a certain compatible family for the presheaf $D(O^{(-)}) : \mathcal{P}(X)^{\mathrm{op}} \to \mathbf{Set}$ does not have a global section:

$$\mathcal{M} \xrightarrow{\hspace{3cm}} D(O^{(-)}) \qquad (7)$$
$$\searrow \qquad \nearrow$$
$$1$$

Here 1 is the terminal presheaf. The 'measurement cover' $\mathcal{M} \subseteq 1$ is defined by $\mathcal{M}(S) = \emptyset$ if $\{a_0, a_1\} \subseteq S$ or $\{b_0, b_1\} \subseteq S$, and $\mathcal{M}(S) = \{*\}$ otherwise. In general, $\mathcal{M}(S)$ is inhabited, i.e., non-empty, if the measurement context S is allowed in the Bell situation.

We now relate this diagram (7) with our diagram (2) by using an adjunction between $\mathbf{EA}$ and $\mathbf{Set}^{\mathcal{P}(X)^{\mathrm{op}}}$. We construct this adjunction as the following composite:

$$\mathbf{EA} \xrightarrow[\top]{T} \mathbf{Set}^{\mathbb{N}} \xrightarrow[\Sigma_{O^X}]{\Delta_{O^X}} \mathbf{Set}^{\mathbb{N}}/\mathbb{N}(O^X, -) \simeq \mathbf{Set}^{(\mathbb{N}^{\mathrm{op}}/(O^X))^{\mathrm{op}}} \xrightarrow[I_!]{I^*} \mathbf{Set}^{\mathcal{P}(X)^{\mathrm{op}}} \quad (8)$$

The first two adjunctions in this composite have already been discussed. The categorical equivalence $\mathbf{Set}^{\mathbb{N}}/\mathbb{N}(16,-) \simeq \mathbf{Set}^{(\mathbb{N}^{\mathrm{op}}/16)^{\mathrm{op}}}$ is an instance of a general fact about slices by representable presheaves (e.g. [13, Prop. A.1.1.7,Lem. C2.2.17]): in general, $\mathbf{Set}^{\mathbb{D}^{\mathrm{op}}}/\mathbb{D}(-, d) \simeq \mathbf{Set}^{(\mathbb{D}/d)^{\mathrm{op}}}$.

It remains to explain $I_! \dashv I^*$. The functor $I^* : \mathbf{Set}^{(\mathbb{N}^{\mathrm{op}}/16)^{\mathrm{op}}} \to \mathbf{Set}^{\mathcal{P}(X)^{\mathrm{op}}}$ is induced by precomposing with the functor $I : \mathcal{P}(X) \to \mathbb{N}^{\mathrm{op}}/O^X$ that takes a subset $U \subseteq X$ to the pair $(O^U, O^{i_U} : O^X \to O^U)$ where $i_U : U \to X$ is the set inclusion function. It has a left adjoint, $I_!$, for general reasons (e.g. [13, Prop. A.4.1.4]).

Corollary 17. *The right adjoint in (8) takes the effect algebra $[0, 1]$ to the presheaf $D(O^{(-)}) : \mathbf{Set}^{\mathcal{P}(X)^{\mathrm{op}}}$. The left adjoint in (8) takes the measurement cover $\mathcal{M} \subseteq 1$ to the effect algebra $E_A \otimes E_B \subseteq B_A \otimes B_B$.*

Thus the adjunction (8) *relates the effect algebra formulation of Bell's paradox* (2), *with the formulation of Abramsky and Brandenburger* (7).

4.6 Hardy Paradoxes

We now briefly consider a different kind of distribution. Not one where the entries are probabilities in the interval $[0, 1]$, but where they are *possibilities*, i.e., either 1 for "this outcome is possible" or 0 for "this outcome is not possible". The pointed monoid $(\{0, 1\}, \vee, 0, 1)$ is built from these two possibilities. For an effect algebra A, a possibility distribution is a morphism $A \to (\{0, 1\}, \vee, 0, 1)$, and a possibility distribution on joint measurements is a bimorphism $A, B \to (\{0, 1\}, \vee, 0, 1)$.

The PPCM morphism $s : ([0, 1], +, 0, 1) \to (\{0, 1\}, \vee, 0, 1)$ given by $s(0) = 0$, $s(x) = 1$ for $(x \neq 0)$ takes a probability distribution to its support, and by composing this with a probability distribution we get a possibility distribution.

The Hardy paradox concerns possibility, rather than probability. We can analyze it using PPCMs in a similar way to the way we analyzed the Bell paradox in Section 4.2. We can also relate our analysis with the analysis of Abramsky and Brandenburger [1], by embedding it in the presheaf category $\mathbf{Set}^{\mathbb{N}}$. Here the situation is slightly more subtle: we cannot use Corollary 8 since $(\{0, 1\}, \vee, 0, 1)$ is not an effect algebra, but we can still use Theorem 7, since it only appears on the right-hand-side of arrows.

4.7 Kochen-Specker Systems

A Kochen-Specker system is represented by a sub-effect algebra E of $Proj(\mathcal{H})$ such that there is no effect algebra morphism $E \to (\{0, 1\}, \otimes, 0, 1)$. This means we cannot assign a value 0 or 1 to every element of E in such a way that whenever $p_1, \ldots p_n \in E$ with $p_1 + \ldots p_n = 1$, exactly one of the p_i is assigned 1 and this assignment does not depend on the other $p_j, j \neq i$. (NB here we use partial join $\otimes$, with $1 \otimes 1$ undefined, whereas we used the total join $\vee$ in §4.6.)

We now view this in the presheaf category $\mathbf{Set}^{\mathbb{N}}$. Since there is no morphism $Proj(\mathcal{H}) \to (\{0, 1\}, \otimes, 0, 1)$, there is also no natural transformation $T(Proj(\mathcal{H})) \to T(\{0, 1\}, \otimes, 0, 1)$, by Corollary 8. We now explore this more explicitly.

The bounded operators on $\mathcal{H}$ form a C*-algebra, $B(\mathcal{H})$. An n-test in the effect algebra $Proj(\mathcal{H})$ can be identified with a unital *-homomorphism $\mathbb{C}^n \to B(\mathcal{H})$ from the commutative C*-algebra $\mathbb{C}^n$, by looking at the images of the characteristic functions on single points. So $T(Proj(\mathcal{H})) \cong \mathbf{C}^*(\mathbb{C}^-, B(\mathcal{H}))$. On the other hand, $T(\{0, 1\}, \otimes, 0, 1)(N) = N$.

There is another way to view this, via a restricted Gelfand duality. Let $\mathbf{CC}_f^*$ be the category of finite dimensional commutative C*-algebras. The functor $\mathbb{C}^- : \mathbb{N}^{\mathrm{op}} \to \mathbf{CC}_f^*$ is an equivalence of categories. Under this equivalence we have presheaves $T(Proj(\mathcal{H})), T(\{0, 1\}, \otimes, 0, 1) \in \mathbf{Set}^{\mathbf{CC}_f^{*\,\mathrm{op}}}$ with

$$T(Proj(\mathcal{H}))(A) = \mathbf{C}^*(A, B(\mathcal{H})) \qquad T(\{0, 1\}, \otimes, 0, 1)(A) = \mathrm{Spec}(A)$$

where $\mathrm{Spec}(A)$ is the Gelfand spectrum of A. Thus the Kochen-Specker paradox says that there is no natural transformation $\mathbf{C}^*(-, B(\mathcal{H})) \to \mathrm{Spec}$ in $\mathbf{Set}^{\mathbf{CC}_f^{*\,\mathrm{op}}}$.

We can use adjunctions to transport this statement to other categories. If a functor $R : \mathbf{Set}^{\mathbf{CC}_f^{*\,\mathrm{op}}} \to \mathcal{C}$ has a left adjoint $L : \mathcal{C} \to \mathbf{Set}^{\mathbf{CC}_f^{*\,\mathrm{op}}}$ and $L(X) = \mathbf{C}^*(-, B(\mathcal{H}))$ then the paradox says there is no morphism $X \to R(\mathrm{Spec})$ in $\mathcal{C}$.

In particular, we transport the paradox to the setting of Hamilton et al. [10], who were concerned with presheaves on the poset $C(B(\mathcal{H}))$ of commutative subalgebras of $B(\mathcal{H})$. We do this using the following composite adjunction:

$$\mathbf{Set}^{\mathbf{CC}_f^{*\,\mathrm{op}}} \underset{\Sigma_{\mathbf{C}^*(-, B(\mathcal{H}))}}{\overset{\Delta_{\mathbf{C}^*(-, B(\mathcal{H}))}}{\underset{\top}{\rightleftarrows}}} \mathbf{Set}^{\mathbf{CC}_f^{*\,\mathrm{op}}}/\mathbf{C}^*(-, B(\mathcal{H})) \;\simeq\; \mathbf{Set}^{(\mathbf{CC}_f^*\downarrow B(\mathcal{H}))^{\mathrm{op}}} \underset{J_!}{\overset{J^*}{\underset{\top}{\rightleftarrows}}} \mathbf{Set}^{C(B(\mathcal{H}))^{\mathrm{op}}}$$

The first adjunction between slice categories is as in Section 4.5. The middle equivalence is standard (e.g. [13, Prop. A.1.1.7]); here $(\mathbf{CC}_f^* \downarrow B(\mathcal{H}))$ is the category whose objects are pairs $(A, f : A \to B(\mathcal{H}))$ where A is a finite-dimensional commutative C*-algebra and f is a *-homomorphism. The adjunction $J_! \dashv J^*$ is induced by the evident embedding $J : C(B(\mathcal{H})) \to (\mathbf{CC}_f^* \downarrow B(\mathcal{H}))$.

The left adjoint of this composite takes the terminal presheaf on $C(B(\mathcal{H}))$ to the presheaf $\mathbf{C}^*(-, B(\mathcal{H}))$ on $\mathbf{CC}_f^*$. The right adjoint takes the spectral presheaf on $\mathbf{CC}_f^*$ to the spectral presheaf on $C(B(\mathcal{H}))$. Thus our statement of the paradox is equivalent to the statement of [10]: the spectral presheaf has no global section.

Summary. We have exhibited a crucial adjunction between two general approaches to finite probability theory: effect algebras and presheaves (Corollary 8). We have used this to analyze paradoxes of non-locality and contextuality (Section 4). There are simple algebraic statements of these paradoxes in terms of partial commutative monoids, but these transport across the adjunction to statements about presheaves on $\mathbb{N}$. By taking slice categories of the presheaf category, we recover earlier analyses of the paradoxes (e.g. Corollary 17).

Acknowledgments. We thank Robin Adams, Tobias Fritz, the Royal Society and the ERC.

References

1. Abramsky, S., Brandenburger, A.: The sheaf-theoretic structure of non-locality and contextuality. New J. Phys **13** (2011)
2. Adámek, J., Rosický, J.: Locally Presentable and Accessible Categories. CUP
3. Brotherston, J., Calcagno, C.: Classical BI: Its semantics and proof theory. Log. Meth. Comput. Sci. **6**(3) (2010)
4. Calcagno, C., O'Hearn, P.W., Yang: H.: Local action and abstract separation logic. In: Proc. LICS 2007, pp. 366–378 (2007)
5. Dvurečenskij, A., Pulmannová, S.: New trends in quantum structures. Kluwer
6. Engesser, K., Gabbay, D.M., Lehmann, D. (eds.): Handbook of Quantum Logic and Quantum Structures: Quantum Structures. Elsevier (2007)

7. Fiore, M.P., Plotkin, G.D., Turi, D.: Abstract syntax and variable binding. In: Proc. LICS 1999 (1999)
8. Foulis, D.J., Bennett, M.K.: Tensor products of orthoalgebras. Order **10**, 271–282
9. Foulis, D.J., Bennett, M.: Effect algebras and unsharp quantum logics. Found. Physics **24**(10), 1331–1352 (1994)
10. Hamilton, J., Isham, C.J., Butterfield, J.: Topos perspective on the Kochen-Specker theorem: III. Int. J. Theoret. Phys., 1413–1436 (2000)
11. Jacobs, B., Mandemaker, J.: Coreflections in algebraic quantum logic. Foundations of physics **42**(7), 932–958 (2012)
12. Jacobs, B.: Probabilities, distribution monads, and convex categories. Theor. Comput. Sci. **2**(28) (2011)
13. Johnstone, P.T.: Sketches of an elephant: a topos theory compendium. OUP (2002)
14. Kelly, G.M.: Basic concepts of enriched category theory. CUP (1980)
15. Mac Lane, S., Moerdijk, I.: Sheaves in geometry and logic. Springer-Verlag (1992)
16. Staton, S.: Instances of computational effects. In: Proc. LICS 2013 (2013)

On the Complexity of Intersecting Regular, Context-Free, and Tree Languages

Joseph Swernofsky[1] and Michael Wehar[2][✉]

[1] Independent Researcher, Los Altos, USA
joseph.swernofsky@gmail.com
[2] University at Buffalo, Buffalo, USA
mwehar@buffalo.edu

Abstract. We apply a construction of Cook (1971) to show that the intersection non-emptiness problem for one PDA (pushdown automaton) and a finite list of DFA's (deterministic finite automata) characterizes the complexity class P. In particular, we show that there exist constants c_1 and c_2 such that for every k, intersection non-emptiness for one PDA and k DFA's is solvable in $O(n^{c_1 k})$ time, but is not solvable in $O(n^{c_2 k})$ time. Then, for every k, we reduce intersection non-emptiness for one PDA and 2^k DFA's to non-emptiness for multi-stack pushdown automata with k-phase switches to obtain a tight time complexity lower bound. Further, we revisit a construction of Veanes (1997) to show that the intersection non-emptiness problem for tree automata also characterizes the complexity class P. We show that there exist constants c_1 and c_2 such that for every k, intersection non-emptiness for k tree automata is solvable in $O(n^{c_1 k})$ time, but is not solvable in $O(n^{c_2 k})$ time.

1 Introduction

To determine whether a mathematical object exists one could start by listing constraints for the proposed object to satisfy. Then, for each constraint one could build a verifier to computationally determine whether an input satisfies the constraint. If each constraint can be verified by an automaton, does that mean one could efficiently determine whether there exists an object that satisfies all of the constraints?

We will investigate problems where we are given an encoding of a finite list of automata and want to determine whether there exists a string that satisfies each automaton in the list. A problem of this form is referred to as an intersection non-emptiness problem because it is equivalent to determining whether the languages associated with the automata have a non-empty intersection.

Each of the intersection non-emptiness problems that we investigate will be viewed as an infinite family of problems indexed on the natural numbers by the number of machines k. For each problem in such a family, we will prove a time complexity lower bound. One may be tempted to view each family of problems as a single parameterized problem. Such an interpretation is fine as long as one realizes that we aren't simply proving a single parameterized complexity lower

© Springer-Verlag Berlin Heidelberg 2015
M.M. Halldórsson et al. (Eds.): ICALP 2015, Part II, LNCS 9135, pp. 414–426, 2015.
DOI: 10.1007/978-3-662-47666-6_33

bound. Rather, we are proving a lower bound for each of the infinitely many fixed levels of the parameterized problem.

In Section 2, we introduce some basic results that allow us to compare infinite families of problems. These results will be used to put all of our findings into a general framework to efficiently present our complexity lower bounds and shed light on the relationship between types of automata and complexity classes.

The intersection non-emptiness problem for DFA's, which we denote by $\mathrm{IE}_{\mathcal{D}}$, is a well known PSPACE-complete problem [7]. Consider fixing the number of machines in the input. Let $k\text{-}\mathrm{IE}_{\mathcal{D}}$ denote the restricted version of $\mathrm{IE}_{\mathcal{D}}$ such that only inputs with at most k machines are accepted. In [18], the second author proved a tight non-deterministic space complexity lower bound for the $k\text{-}\mathrm{IE}_{\mathcal{D}}$ problems. He showed that there exist c_1 and c_2 such that for every k, $k\text{-}\mathrm{IE}_{\mathcal{D}} \in \mathrm{NSPACE}(c_1 k \log(n))$ and $k\text{-}\mathrm{IE}_{\mathcal{D}} \notin \mathrm{NSPACE}(c_2 k \log(n))$ where space is measured relative to a fixed work tape alphabet. Therefore, we say that intersection non-emptiness for DFA's characterizes the complexity class NL.

First, we will investigate intersection non-emptiness for one PDA and a finite list of DFA's which we denote by $\mathrm{IE}_{1\mathcal{P}+\mathcal{D}}$. We will show that there exist constants c_1 and c_2 such that for every k, $k\text{-}\mathrm{IE}_{1\mathcal{P}+\mathcal{D}} \in \mathrm{DTIME}(n^{c_1 k})$ and $k\text{-}\mathrm{IE}_{1\mathcal{P}+\mathcal{D}} \notin \mathrm{DTIME}(n^{c_2 k})$. In order to show the lower bound, we will reduce the acceptance problem for $k \log(n)$-space bounded auxiliary pushdown automata to $k\text{-}\mathrm{IE}_{1\mathcal{P}+\mathcal{D}}$. Then, we will apply results from [4] to get that $k \log(n)$-space bounded auxiliary pushdown automata can be used to simulate $k \log(n)$-space bounded alternating Turing machines. Finally, we will apply results from [2] to get that $k \log(n)$-space bounded alternating Turing machines can be used to simulate n^k-time bounded deterministic Turing machines.

Next, we will investigate non-emptiness for multi-stack pushdown automata with k-phase switches which we denote by $k\text{-}\mathrm{MPDA}$. From [12], we know that $k\text{-}\mathrm{MPDA} \in \mathrm{DTIME}(n^{O(2^k)})$. This result was shown by reducing $k\text{-}\mathrm{MPDA}$ to non-emptiness for graph automata with bounded tree width and then further reducing to non-emptiness for tree automata. We will show that this upper bound is tight. In particular, we will show that there exist constants c_1 and c_2 such that for every k, $k\text{-}\mathrm{MPDA} \in \mathrm{DTIME}(n^{c_1 2^k})$ and $k\text{-}\mathrm{MPDA} \notin \mathrm{DTIME}(n^{c_2 2^k})$. In order to show the lower bound, we will reduce $2^k\text{-}\mathrm{IE}_{1\mathcal{P}+\mathcal{D}}$ to $k\text{-}\mathrm{MPDA}$ and then apply the lower bound for $2^k\text{-}\mathrm{IE}_{1\mathcal{P}+\mathcal{D}}$ from the preceding section. In addition, we will present a lower bound for the dual of the $k\text{-}\mathrm{MPDA}$ problem.

Finally, we will investigate intersection non-emptiness for tree automata which we denote by $\mathrm{IE}_{\mathcal{T}}$. In [16], it was shown that $\mathrm{IE}_{\mathcal{T}}$ is EXPTIME-complete. We will show that there exist constants c_1 and c_2 such that for every k, $k\text{-}\mathrm{IE}_{\mathcal{T}} \in \mathrm{DTIME}(n^{c_1 k})$ and $k\text{-}\mathrm{IE}_{\mathcal{T}} \notin \mathrm{DTIME}(n^{c_2 k})$. In order to show the lower bound, we will reduce the acceptance problem for $k \log(n)$-space bounded alternating Turing machines to $k\text{-}\mathrm{IE}_{\mathcal{T}}$. Then, we will again apply results from [2] to get that $k \log(n)$-space bounded alternating Turing machines can be used to simulate n^k-time bounded deterministic Turing machines.

2   Preliminaries

2.1   Complexity Classes

Each of the following complexity classes is associated with a machine class. In particular, a language X is in the complexity class if and only if there exists a machine M in the associated machine class such that M accepts X.

NL : Logarithmic space bounded non-deterministic Turing machines

AL : Logarithmic space bounded alternating Turing machines

AuxL : Logarithmic space bounded auxiliary pushdown automata

P : Polynomial time bounded deterministic Turing machines

Although $NL \subseteq P$, it is not known if $P = NL$. However, using machine simulations, it was proven that $P = AL$ in [2] and $P = AuxL$ in [4].

If one carefully looks at the simulations from [2], one will notice that there are universal constants c_1 and c_2 such that for every k, each n^k-time bounded deterministic Turing machine can be simulated by a $c_1 k \log(n)$-space bounded alternating Turing machine and each $k \log(n)$-space bounded alternating Turing machine can be simulated by a $n^{c_2 k}$-time bounded deterministic Turing machine. Hence, not only are P and AL equivalent, but the $\mathrm{DTIME}(n^k)$ classes that make up P and the $\mathrm{ASPACE}(k \log(n))$ classes that make up AL are in some sense level-by-level equivalent to each other. This example should motivate the notion of level-by-level equivalence that we introduce in Section 2.3.

2.2   Acceptance Problems

We will specify a Turing machine model and introduce acceptance problems for the machine classes associated with NL, AL, AuxL, and P.

By a Turing machine, we are referring to a machine with a single two-way read-only input tape and a single two-way read/write binary work tape. The condition on the work tape being binary is significant. In particular, for space complexity, constants matter when the alphabet is fixed. By an $f(n)$-time bounded Turing machine, we mean a Turing machine that runs for at most $f(n)$ steps on all inputs of length n. By an $f(n)$-space bounded Turing machine, we mean a Turing machine that uses at most $f(n)$ cells on the binary work tape for all inputs of length n. By uses at most $f(n)$ cells, we mean that the tape head never moves to the right of the $f(n)$th cell. Time and space bounded auxiliary pushdown automata can be defined similarly where the space bounds only apply to the auxiliary work tape. The space bounds do not apply to the stack.

The general form of an acceptance problem is as follows. Given an encoding of a machine M and an input x, does M accept x? For each k and each machine class that we discussed in Section 2.1, we can define an acceptance problem. Consider the following acceptance problems and their associated machine classes.

$$N^S_{k\,\log} \ : \ k\log(n)\text{-space bounded non-deterministic Turing machines}$$
$$A^S_{k\,\log} \ : \ k\log(n)\text{-space bounded alternating Turing machines}$$
$$Aux^S_{k\,\log} \ : \ k\log(n)\text{-space bounded auxiliary pushdown automata}$$
$$D^T_{n^k} \ : \ n^k\text{-time bounded deterministic Turing machines}$$

2.3   Level-By-Level Equivalence

For each k and each machine class from Section 2.1, we defined an acceptance problem. In other words, for each machine class, we defined an infinite family of acceptance problems. These infinite families of problems characterize their associated machine classes and their associated complexity classes.

Let's look at an example. Consider the family $\{D^T_{n^k}\}_{k\in\mathbb{N}}$. The proof of the time hierarchy theorem has two parts: universal simulation and diagonalization. From universal simulation of deterministic Turing machines, we get a constant c_1 such that for every k, $D^T_{n^k} \in \mathrm{DTIME}(n^{c_1 k})$. From diagonalization, we get a smaller constant c_2 such that for every k, $D^T_{n^k} \notin \mathrm{DTIME}(n^{c_2 k})$. Therefore, we say that this family characterizes the complexity class P.

The notion of an infinite family characterizing a complexity class leads us to the concept of LBL (level-by-level) reducibility. This concept will allow us to compare the complexity of infinite families of problems.[1]

Given two infinite families of problems $X := \{X_k\}_{k\in\mathbb{N}}$ and $Y := \{Y_k\}_{k\in\mathbb{N}}$, we say that X is (polynomial time) LBL-reducible to Y if there exists a constant c such that for every k, there exists an $O(n^c)$-time bounded reduction from X_k to Y_k where k is treated as a constant. If X is LBL-reducible to Y, then we write $X \leq_L Y$. If X is LBL-reducible to Y and Y is LBL-reducible to X, then we say that X and Y are LBL-equivalent and write $X \equiv_L Y$. Notice that $\leq_L$ is transitive and $\equiv_L$ is an equivalence relation.

To simplify how one shows that an infinite family X is LBL-reducible to $\{D^T_{n^k}\}_{k\in\mathbb{N}}$, we have the following proposition.

Proposition 1. *Let an infinite family X be given. If there exists c such that for every k, $X_k \in \mathrm{DTIME}(n^{ck})$, then X is LBL-reducible to $\{D^T_{n^k}\}_{k\in\mathbb{N}}$.*

To simplify how one shows a (near) tight time complexity lower bound for an infinite family X, we have the following proposition.

Proposition 2. *If an infinite family X is LBL-equivalent to $\{D^T_{n^k}\}_{k\in\mathbb{N}}$, then there exist c_1 and c_2 such that for every k, $X_k \in \mathrm{DTIME}(n^{c_1 k})$ and $X_k \notin \mathrm{DTIME}(n^{c_2 k})$.*

---

[1] An LBL-reduction is an infinite family of reductions. Such a family of reductions can be viewed as the non-uniform analogue of an fpt-reduction [5]. We introduce the distinct notion of an LBL-reduction to emphasize that our lower bounds will apply to each problem in the respective family of problems.

The simulations that we mentioned in Section 2.1 lead to two significant examples of LBL equivalence. In particular, from the simulations in [2], we have that $\{D^T_{n^k}\}_{k\in\mathbb{N}}$ is LBL-equivalent to $\{A^S_{k\log}\}_{k\in\mathbb{N}}$. Also, from the simulations in [4], we have that $\{D^T_{n^k}\}_{k\in\mathbb{N}}$ is LBL-equivalent to $\{Aux^S_{k\log}\}_{k\in\mathbb{N}}$. Now, we can apply Proposition 2 and the LBL equivalences to obtain (near) tight time complexity lower bounds. For example, consider the LBL equivalence between $\{D^T_{n^k}\}_{k\in\mathbb{N}}$ and $\{A^S_{k\log}\}_{k\in\mathbb{N}}$. By applying Proposition 2, we get that there exist c_1 and c_2 such that for every k, $A^S_{k\log} \in \mathrm{DTIME}(n^{c_1 k})$ and $A^S_{k\log} \notin \mathrm{DTIME}(n^{c_2 k})$.

We will further use the equivalences for deterministic time, alternating space, and auxiliary space to prove equivalences for intersection non-emptiness problems. Then, we will apply Proposition 2 to obtain (near) tight time complexity lower bounds for these problems.

3   One PDA and k DFA's

It is well known that the general intersection non-emptiness problem for DFA's is PSPACE-complete [7]. Further work has shown that variations of this problem are hard as well [9]. We consider the problem where in addition to a finite list of DFA's, we are also given a single pushdown automaton. Notice that it doesn't make sense to consider more than one PDA because the intersection non-emptiness problem for two pushdown automata is undecidable.

We will show that intersection non-emptiness for one PDA and k DFA's is equivalent to acceptance for n^k-time bounded deterministic Turing machines. In particular, we will show that $\{k\text{-IE}_{1\mathcal{P}+\mathcal{D}}\}_{k\in\mathbb{N}}$ is LBL-equivalent to $\{D^T_{n^k}\}_{k\in\mathbb{N}}$.

Using the product construction, one can solve each $k\text{-IE}_{1\mathcal{P}+\mathcal{D}}$ problem in $O(n^{ck})$ time for some constant c. Further, one can apply Proposition 1 to get the following result.

Proposition 3. $\{k\text{-IE}_{1\mathcal{P}+\mathcal{D}}\}_{k\in\mathbb{N}}$ is LBL-reducible to $\{D^T_{n^k}\}_{k\in\mathbb{N}}$.

In the following theorem, we reduce acceptance for space bounded auxiliary pushdown automata to intersection non-emptiness for one PDA and a finite list of DFA's. The reduction that we present is based on reductions from [6] and [7]. Our presentation is in the same format as that from the second author's previous work where he reduces acceptance for non-deterministic space bounded Turing machines to intersection non-emptiness for a finite list of DFA's [18].

Theorem 4. $\{Aux^S_{k\log}\}_{k\in\mathbb{N}}$ is LBL-reducible to $\{k\text{-IE}_{1\mathcal{P}+\mathcal{D}}\}_{k\in\mathbb{N}}$.

Proof. An auxiliary pushdown automaton has a stack, a two-way read-only input tape, and a single read/write work tape. We will restrict the read/write work tape to be binary and bound the amount of cells that the automaton can use in terms of the input length. In addition, we will only consider auxiliary pushdown automata where the stack alphabet is binary. Such restricted auxiliary PDA's are sufficient for carrying out the simulation in [4].

Let k be given. We will describe a reduction from $Aux^S_{k\log}$ to $k\text{-IE}_{1\mathcal{P}+\mathcal{D}}$. Let a $k\log(n)$-space bounded auxiliary pushdown automaton M of size n_M and an input string x of length n_x be given. Together, an encoding of M and x represent an arbitrary input for $Aux^S_{k\log}$. Let n denote the total size of M and x combined i.e. $n := n_M + n_x$.

Our task is to construct one PDA and k DFA's, denoted by PD and $\{D_i\}_{i\in[k]}$, each of size at most $O(n^c)$ for some fixed constant c such that M accepts x if and only if $L(PD) \cap \bigcap_{i\in[k]} L(D_i)$ is non-empty.

The automata will read in a string that represents a computation of M on x and verify that the computation is valid and accepting. The PDA PD will verify that the stack is managed correctly while the DFA's will verify that the work tape is managed correctly. In particular, the work tape of M will be split into k sections each consisting of $\log(n_x)$ sequential bits of memory. The ith DFA, D_i, will keep track of the ith section and verify that it is managed correctly. In addition, all of the DFA's will keep track of the tape head positions.

The following two concepts are essential to our construction.

A *section i configuration* of M is a tuple of the form:

(state, input position, work position, ith section of work tape).

A *forgetful configuration* of M is a tuple of the form:

(state, input position, work position, write bit, stack action, top bit).

The alphabet symbols are identified with forgetful configurations. The PDA PD only has two states. When it reads a forgetful configuration a, if a represents the top of the stack correctly, then PD loops in the initial/accepting state and pushes or pops based on the stack instruction that a represents. Otherwise, PD goes to the dead/rejecting state.

The states for the D_i's are identified with section i configurations. Each D_i has a single initial state. We identify this initial state with the section i configuration of M that represents the initial input and work positions, a blank ith section of the work tape, and the initial state of M. The final states of D_i represent accepting configurations of M.

Informally, the transitions are defined as follows. For each D_i, there is a transition from state r_1 to state r_2 with symbol a if a validly represents how the state and partial tapes for r_1 and r_2 could be manipulated in one step for the computation of M on input x. It's important to notice that in order to determine if there is a transition, the stack action and top bit of the stack must be taken into account.

We assert without proof that for every string y, y represents a valid accepting computation of M on x if and only if $y \in L(PD) \cap \bigcap_{i\in[k]} L(D_i)$. Therefore, M accepts x if and only if $L(PD) \cap \bigcap_{i\in[k]} L(D_i)$ is non-empty. By bounding the total number of section i configurations, one can show there exists a fixed two variable polynomial q such that each D_i has at most $q(n, k)$ states. Therefore, there is a constant d that does not depend on k such that each D_i has size at most $O(n^d)$ where k is treated as a constant. Further, we can compute each D_i's transition table by looping through every combination of a pair of states and an alphabet symbol, and marking the valid combinations. The number of possible combinations is a fixed polynomial blow-up from n^d. Therefore, we can compute the transition tables in $O(n^c)$ time for some slightly larger constant c that does not depend on k.

Since k was arbitrary, we have that for every k, there is an $O(n^c)$-time reduction from $Aux^S_{k\log}$ to $k\text{-IE}_{1\mathcal{P}+\mathcal{D}}$. □

In the preceding reduction, it was surprising that the PDA had a fixed number of states. Even more surprisingly, one could convert the automata constructed in the reduction to automata with a binary input alphabet. In doing so, the PDA can be made fixed. In other words, there is a fixed deterministic pushdown automaton for which the intersection non-emptiness problem is hard.

Corollary 5. $\{k\text{-IE}_{1\mathcal{P}+\mathcal{D}}\}_{k\in\mathbb{N}}$ and $\{D^T_{n^k}\}_{k\in\mathbb{N}}$ are LBL-equivalent.

Proof. From Section 2, we know that $\{Aux^S_{k\log}\}_{k\in\mathbb{N}} \equiv_L \{D^T_{n^k}\}_{k\in\mathbb{N}}$. Further, we have $\{Aux^S_{k\log}\}_{k\in\mathbb{N}} \leq_L \{k\text{-IE}_{1\mathcal{P}+\mathcal{D}}\}_{k\in\mathbb{N}} \leq_L \{D^T_{n^k}\}_{k\in\mathbb{N}}$ from Proposition 3 and Theorem 4. Combine to obtain the desired result. □

Corollary 6. $\exists c_1 \exists c_2 \forall k$ $k\text{-IE}_{1\mathcal{P}+\mathcal{D}} \in \text{DTIME}(n^{c_1 k})$ and $k\text{-IE}_{1\mathcal{P}+\mathcal{D}} \notin \text{DTIME}(n^{c_2 k})$.

Proof. Combine Corollary 5 with Proposition 2. □

4   MPDA's with k-Phase Switches

A two-stack pushdown automaton can simulate a Turing machine. Therefore, the non-emptiness problem for such machines is undecidable. However, we can restrict how and when the machines can access their stacks to obtain classes of machines whose non-emptiness problems are decidable [12]. In particular, we will discuss the k-phase switches restriction. This restriction forces a machine to designate a stack for popping. In other words, a restricted machine can push to any stack, but only pop from the designated stack. The k refers to how many times the machine can switch which stack is designated. We refer to a machine with such a restriction as a multi-stack pushdown automaton with k-phase switches. For background on such machines, we refer the reader to [14]. We also investigate what we refer to as the dual machines. These machines can pop from any stack, but can only push to the designated stack.

We will denote the non-emptiness problem for multi-stack pushdown automata with k-phase switches by $k\text{-MPDA}$. Similarly, we will denote the non-emptiness problem for the dual machines by $k\text{-co-MPDA}$. We will show that $\{k\text{-MPDA}\}_{k\in\mathbb{N}}$, $\{k\text{-co-MPDA}\}_{k\in\mathbb{N}}$, and $\{D^T_{n^{2^k}}\}_{k\in\mathbb{N}}$ are LBL-equivalent. As a result, we will obtain tight lower bounds for these non-emptiness problems.

Recently, the non-emptiness problem for a related class of infinite automata was shown to have a double exponential time lower bound [8]. In addition, the non-emptiness problem for ordered multi-stack pushdown automata was shown to have a double exponential time lower bound [1]. Our lower bound may be suggested by such sources, but we elegantly prove it using a novel reduction found in the proof of Theorem 8.

Proposition 7. $\{k\text{-MPDA}\}_{k\in\mathbb{N}}$ *and* $\{k\text{-co-MPDA}\}_{k\in\mathbb{N}}$ *are LBL-reducible to* $\{D^T_{n^{2^k}}\}_{k\in\mathbb{N}}$.

Sketch of proof. In [12], it was shown that k-MPDA and k-co-MPDA $\in$ DTIME$(n^{O(2^k)})$ by a reduction to non-emptiness for graph automata with bounded tree width and further to non-emptiness for tree automata. Then, one can apply a variation of Proposition 1 to get the desired result.   □

In the following theorem, we reduce intersection non-emptiness for one PDA and 2^k DFA's to non-emptiness for multi-stack pushdown automata with k-phase switches.

Theorem 8. $\{2^k\text{-IE}_{1\mathcal{P}+\mathcal{D}}\}_{k\in\mathbb{N}}$ *is LBL-reducible to* $\{k\text{-MPDA}\}_{k\in\mathbb{N}}$.

Sketch of proof. Let an input for 2^k-IE$_{1\mathcal{P}+\mathcal{D}}$ consisting of a PDA and 2^k DFA's be given. We will describe how to construct a multi-stack pushdown automaton M with k-phase switches whose language is non-empty if and only if the PDA and DFA's languages have a non-empty intersection.

The machine M will have k stacks. It will read its input and copy it onto all of the stacks besides the first stack. While it is reading the input, the first stack will be used to simulate the PDA on the input. Then, it will repeat the following procedure until each of the stacks have been designated once.

The procedure consists of popping from the designated stack and pushing what is being popped onto all of the other stacks. While it is popping, it is also simulating one DFA per copy of the input string or simulating one DFA in reverse per copy of the reversal of the input string. This will eventually create exponentially many copies of the input string followed by the reversal of the input string and lead to simulating each DFA or reversal on one of the copies.

If the PDA and all of the DFA's accept, then M will accept. Otherwise, M will reject. In total, we are able to simulate one PDA and $O(2^k)$ DFA's using only k-phase switches. Also, the size of M will be approximately the sum of the sizes of the PDA and DFA's.   □

A related reduction can be given for the dual machines. The proof of Theorem 9 has been omitted, but will be made available online for the interested reader.

Theorem 9. $\{2^k\text{-IE}_{1\mathcal{P}+\mathcal{D}}\}_{k\in\mathbb{N}}$ *is LBL-reducible to* $\{k\text{-co-MPDA}\}_{k\in\mathbb{N}}$.

Corollary 10. $\{k\text{-MPDA}\}_{k\in\mathbb{N}}$, $\{k\text{-co-MPDA}\}_{k\in\mathbb{N}}$, *and* $\{D^T_{n^{2^k}}\}_{k\in\mathbb{N}}$ *are LBL-equivalent.*

Sketch of proof. From Corollary 5, we have $\{2^k\text{-IE}_{1\mathcal{P}+\mathcal{D}}\}_{k\in\mathbb{N}} \equiv_L \{D^T_{n^{2^k}}\}_{k\in\mathbb{N}}$. Further, we have $\{2^k\text{-IE}_{1\mathcal{P}+\mathcal{D}}\}_{k\in\mathbb{N}} \leq_L \{k\text{-MPDA}\}_{k\in\mathbb{N}} \leq_L \{D^T_{n^{2^k}}\}_{k\in\mathbb{N}}$ from Proposition 7 and Theorem 8. Similarly, we have reductions for the dual machines. Combine to obtain the desired result.   □

Corollary 11. *There exist c_1 and c_2 such that for every k:*

$$i) \ k\text{-MPDA and } k\text{-co-MPDA} \in \text{DTIME}(n^{c_1 2^k})$$

$$ii) \ k\text{-MPDA and } k\text{-co-MPDA} \notin \text{DTIME}(n^{c_2 2^k}).$$

Sketch of proof.   Combine Corollary 10 and a variation of Proposition 2.   □

5 k Tree Automata

It is known that the general intersection non-emptiness problem for deterministic top-down tree automata is EXPTIME-complete [3]. We will show that intersection non-emptiness for k deterministic top-down tree automata is equivalent to acceptance for n^k-time bounded deterministic Turing machines. In particular, we will show that $\{k\text{-IE}_{\mathcal{T}}\}_{k \in \mathbb{N}}$ is LBL-equivalent to $\{D_{n^k}^T\}_{k \in \mathbb{N}}$. For background on decision problems for tree automata, we refer the reader to [3] and [13].

Using the product construction, one can solve each $k\text{-IE}_{\mathcal{T}}$ problem in $O(n^{ck})$ time for some constant c. Further, one can apply Proposition 1 to get the following result.

Proposition 12. $\{k\text{-IE}_{\mathcal{T}}\}_{k \in \mathbb{N}}$ *is LBL-reducible to* $\{D_{n^k}^T\}_{k \in \mathbb{N}}$.

In the following theorem, we reduce acceptance for alternating Turing machines to intersection non-emptiness for tree automata. The reduction that we present is similar to that found in [16] and briefly described in [3]. Our presentation is in the same format as Theorem 4.

Theorem 13. $\{A_{k \log}^S\}_{k \in \mathbb{N}}$ *is LBL-reducible to* $\{k\text{-IE}_{\mathcal{T}}\}_{k \in \mathbb{N}}$.

Proof.   An alternating Turing machine has existential states and universal states. Therefore, there are existential configurations and universal configurations. An existential configuration c leads to an accepting configuration if and only if there exists a valid transition out of c that leads to an accepting configuration. A universal configuration c leads to an accepting configuration if and only if every valid transition out of c leads to an accepting configuration. We will only consider alternating machines such that no universal configuration can have more than two valid outgoing transitions. We assert without proof that any alternating machine can be unraveled with intermediate universal states to satisfy this property in such a way that there is no more than a polynomial blow-up in the number of states.

Let k be given. We will describe a reduction from $A_{k \log}^S$ to $k\text{-IE}_{\mathcal{T}}$. Let a $k \log(n)$-space bounded alternating Turing machine M of size n_{M} and an input string x of length n_x be given. Together, an encoding of M and x represent an arbitrary input for $A_{k \log}^S$. Let n denote the total size of M and x combined i.e. $n := n_{\text{M}} + n_x$.

Our task is to construct k top-down deterministic tree automata, denoted by $\{T_i\}_{i \in [k]}$, each of size at most $O(n^c)$ for some fixed constant c such that M accepts x if and only if $\bigcap_{i \in [k]} L(T_i)$ is non-empty.

The tree automata will read in a labeled tree that represents a computation of M on x and verify that the computation is valid and accepting. The work tape of M will be split into k sections each consisting of $\log(n_x)$ sequential bits of memory. The ith tree automaton, T_i, will keep track of the ith section and verify that it is managed correctly. In addition, all of the tree automata will keep track of the tape head positions.

The following two concepts are essential to our construction.

A *section i configuration* of M is a tuple of the form:

$$\text{(state, input position, work position, ith section of work tape).}$$

A *forgetful configuration* of M is a tuple of the form:

$$\text{(state, input position, work position, write bit).}$$

The alphabet consists of symbols of arity 0, 1, and 2 such that each arity 0 symbol represents an accepting forgetful configuration, each arity 1 symbol represents an arbitrary forgetful configuration, and each arity 2 symbol represents a pair of forgetful configurations. We won't need any symbols of arity larger than 2 because each universal configuration has at most two outgoing transitions.

The states of T_i are identified with section i configurations. Each T_i has a single initial state. We identify this initial state with the section i configuration of M that represents the initial input and work positions, a blank ith section of the work tape, and the initial state of M.

We say that a section i configuration r extends a forgetful configuration a if r agrees with a on state, input position, and work position.

We say that a section i configuration r_1 transitions to a section i configuration r_2 on input x if either (a) the work position for r_1 is in the ith section and r_2 correctly represents how the tape positions and the ith section could change in one step of the computation on x, or (b) r_1 is not in the ith section and r_1 and r_2 agree on the ith section of the work tape.

For each T_i, we have the following transitions. Each arity 0 symbol a accepts on a state r if and only if r extends a and a represents an accepting state of M. Each arity 1 symbol a transitions from a state r_1 to a state r_2 if and only if (i) r_1 transitions to r_2 on input x (consistently with a's write bit), (ii) r_2 extends a, and (iii) if r_1 is a universal configuration and the work position of r_1 is in the ith section, then r_1 can only transition to r_2 on input x. Each arity 2 symbol (a_1, a_2) transitions from a state r to a pair of distinct states (r_1, r_2) if and only if r transitions to r_1 on input x, r transitions to r_2 on input x, r_1 extends a_1, and r_2 extends a_2.

We assert without proof that for every labeled tree y, y represents a valid accepting computation of M on x if and only if $y \in \bigcap_{i \in [k]} L(T_i)$. Therefore, M accepts x if and only if $\bigcap_{i \in [k]} L(T_i)$ is non-empty. By bounding the total number of section i configurations, one can show there exists a fixed two variable polynomial q such that each T_i has at most $q(n, k)$ states. Therefore, there is a constant d that does not depend on k such that each T_i has size at most $O(n^d)$ where k is treated as a constant. Further, we can compute each T_i's transition table by looping through every combination of a pair of states and an alphabet symbol, and marking the valid combinations. The number of possible combinations is a fixed polynomial blow-up from n^d. Therefore, we can compute the transition tables in $O(n^c)$ time for some slightly larger constant c that does not depend on k.

Since k was arbitrary, we have that for every k, there is an $O(n^c)$-time reduction from $A_{k\log}^S$ to k-IE$_{\mathcal{T}}$.   □

Corollary 14. $\{k\text{-IE}_{\mathcal{T}}\}_{k\in\mathbb{N}}$ and $\{D_{n^k}^T\}_{k\in\mathbb{N}}$ are LBL-equivalent.

Proof. From Section 2, we know that $\{A_{k\log}^S\}_{k\in\mathbb{N}} \equiv_L \{D_{n^k}^T\}_{k\in\mathbb{N}}$. Further, we have $\{A_{k\log}^S\}_{k\in\mathbb{N}} \leq_L \{k\text{-IE}_{\mathcal{T}}\}_{k\in\mathbb{N}} \leq_L \{D_{n^k}^T\}_{k\in\mathbb{N}}$ from Proposition 12 and Theorem 13. Combine to obtain the desired result.   □

Corollary 15. $\exists c_1 \exists c_2 \forall k \; k\text{-IE}_{\mathcal{T}} \in \text{DTIME}(n^{c_1 k})$ and $k\text{-IE}_{\mathcal{T}} \notin \text{DTIME}(n^{c_2 k})$.

Proof. Combine Corollary 14 with Proposition 2.   □

6   Conclusion

We introduced the notions of LBL reducibility and LBL equivalence for infinite families of problems.[2] We then used existing simulations to show that $\{D_{n^k}^T\}_{k\in\mathbb{N}}$, $\{A_{k\log}^S\}_{k\in\mathbb{N}}$, $\{Aux_{k\log}^S\}_{k\in\mathbb{N}}$, $\{k\text{-IE}_{1\mathcal{P}+\mathcal{D}}\}_{k\in\mathbb{N}}$, and $\{k\text{-IE}_{\mathcal{T}}\}_{k\in\mathbb{N}}$ are all polynomial time LBL-equivalent. Further, we applied that $\{D_{n^k}^T\}_{k\in\mathbb{N}}$ and $\{k\text{-IE}_{1\mathcal{P}+\mathcal{D}}\}_{k\in\mathbb{N}}$ are LBL-equivalent to show that $\{D_{n^{2^k}}^T\}_{k\in\mathbb{N}}$, $\{2^k\text{-IE}_{1\mathcal{P}+\mathcal{D}}\}_{k\in\mathbb{N}}$, $\{k\text{-MPDA}\}_{k\in\mathbb{N}}$, and $\{k\text{-}co\text{-MPDA}\}_{k\in\mathbb{N}}$ are all polynomial time LBL-equivalent. By combining these equivalences with Proposition 2, we get (near) tight time complexity lower bounds for all of these problems.

We claim that all of the polynomial time LBL-reductions that we presented can be carefully optimized to become log-space LBL-reductions. Formally, we say that a family X if log-space LBL-reducible to a family Y if there exists a constant c and a function f such that for every k, there exists a $(c+o(1))\log(n)$-space reduction from X_k to Y_k where k is treated as a constant.

The notion of log-space LBL equivalence can be used to express the P vs NL problem from structural complexity theory. Consider the machine classes consisting of polynomial time bounded deterministic Turing machines and log-space bounded non-deterministic Turing machines. These machine classes have associated acceptance problems $\{D_{n^k}^T\}_{k\in\mathbb{N}}$ and $\{N_{k\log}^S\}_{k\in\mathbb{N}}$, respectively. One can show that P = NL if and only if $\{D_{n^k}^T\}_{k\in\mathbb{N}}$ and $\{N_{k\log}^S\}_{k\in\mathbb{N}}$ are log-space LBL-equivalent.

Now, one might ask, "What's the relationship between P vs NL and intersection non-emptiness problems?" We know that $\{D_{n^k}^T\}_{k\in\mathbb{N}}$ and $\{k\text{-IE}_{1\mathcal{P}+\mathcal{D}}\}_{k\in\mathbb{N}}$ are log-space LBL-equivalent. In addition, from the second author's previous

---

[2] These concepts serve as the non-uniform analogues of fpt reducibility and fpt equivalence from the subject of parameterized complexity theory [5].

work [18], it can be shown that $\{N^S_{k\log}\}_{k\in\mathbb{N}}$ and $\{k\text{-IE}_\mathcal{D}\}_{k\in\mathbb{N}}$ are log-space LBL-equivalent. Therefore, we get that P = NL if and only if $\{k\text{-IE}_{1\mathcal{P}+\mathcal{D}}\}_{k\in\mathbb{N}}$ and $\{k\text{-IE}_\mathcal{D}\}_{k\in\mathbb{N}}$ are log-space LBL-equivalent. In other words, P = NL if and only if adding a PDA does not increase the difficulty of the intersection non-emptiness problem for DFA's.

We showed that intersection non-emptiness problems for DFA's, PDA's, and tree automata characterize complexity classes. There are many more types of automata and we suggest that one may be able to prove more characterizations using the notion of LBL reducibility. From this perspective, we intend to investigate decision problems for tree automata with auxiliary memory.

Acknowledgments. We greatly appreciate all of the help and suggestions that we received. We would especially like to thank Richard Lipton, Kenneth Regan, Atri Rudra, and all those from Carnegie Mellon University who were supportive of our research work while we were undergraduates.

References

1. Atig, M.F., Bollig, B., Habermehl, P.: Emptiness of multi-pushdown automata is 2ETIME-complete. In: Ito, M., Toyama, M. (eds.) DLT 2008. LNCS, vol. 5257, pp. 121–133. Springer, Heidelberg (2008)
2. Chandra, A.K., Kozen, D.C., Stockmeyer, L.J.: Alternation. J. ACM **28**(1), 114–133 (1981)
3. Comon, H., Dauchet, M., Gilleron, R., Löding, C., Jacquemard, F., Lugiez, D., Tison, S., Tommasi, M.: Tree automata techniques and applications, October 2007
4. Cook, S.A.: Characterizations of pushdown machines in terms of time-bounded computers. J. ACM **18**(1), 4–18 (1971)
5. Flum, J., Grohe, M.: Parameterized Complexity Theory. Springer-Verlag New York Inc., Secaucus (2006)
6. Karakostas, G., Lipton, R.J., Viglas, A.: On the complexity of intersecting finite state automata and NL versus NP. TCS **302**, 257–274 (2003)
7. Kozen, D.: Lower bounds for natural proof systems. In: Proc. 18th Symp. on the Foundations of Computer Science, pp. 254–266 (1977)
8. La Torre, S., Madhusudan, P., Parlato, G.: An infinite automaton characterization of double exponential time. In: Kaminski, M., Martini, S. (eds.) CSL 2008. LNCS, vol. 5213, pp. 33–48. Springer, Heidelberg (2008)
9. Lange, K.-J., Rossmanith, P.: The emptiness problem for intersections of regular languages. In: Havel, Ivan M., Koubek, Václav (eds.) MFCS 1992. LNCS, vol. 629, pp. 346–354. Springer, Heidelberg (1992)
10. Limaye, N., Mahajan, M.: Membership testing: removing extra stacks from multi-stack pushdown automata. In: Dediu, A.H., Ionescu, A.M., Martín-Vide, C. (eds.) LATA 2009. LNCS, vol. 5457, pp. 493–504. Springer, Heidelberg (2009)
11. Lipton, R.J.: On the intersection of finite automata. Gödel's Lost Letter and P=NP, August 2009
12. Madhusudan, P., Parlato, G.: The tree width of automata with auxiliary storage. POPL 2011 (2011)
13. Martens, W., Vansummeren, S.: Automata and logic on trees: Algorithms. ESSLLI 2007 (2007)

14. La Torre, S., Madhusudan, P., Parlato, G.: A robust class of context-sensitive languages. In: LICS 2007, pp. 161–170 (2007)
15. Valiant, L.G.: Decision Procedures for Families of Deterministic Pushdown Automata. Ph.D thesis, University of Warwick, August 1973
16. Veanes, M.: On computational complexity of basic decision problems of finite tree automata. UPMAIL Technical Report 133 (1997)
17. Wehar, M.: Intersection emptiness for finite automata. Honors thesis, Carnegie Mellon University (2012)
18. Wehar, M.: Hardness results for intersection non-emptiness. In: Esparza, J., Fraigniaud, P., Husfeldt, T., Koutsoupias, E. (eds.) ICALP 2014, Part II. LNCS, vol. 8573, pp. 354–362. Springer, Heidelberg (2014)

Containment of Monadic Datalog Programs via Bounded Clique-Width

Mikołaj Bojańczyk, Filip Murlak[✉], and Adam Witkowski

University of Warsaw, Warsaw, Poland
fmurlak@mimuw.edu.pl

Abstract. Containment of monadic datalog programs over data trees (labelled trees with an equivalence relation) is undecidable. Recently, decidability was shown for two incomparable fragments: downward programs, which never move up from visited tree nodes, and linear child-only programs, which have at most one intensional predicate per rule and do not use descendant relation. As different as the fragments are, the decidability proofs hinted at an analogy. As it turns out, the common denominator is admitting bounded clique-width counter-examples to containment. This observation immediately leads to stronger decidability results with more elegant proofs, via decidability of monadic second order logic over structures of bounded clique-width. An argument based on two-way alternating tree automata gives a tighter upper bound for linear child-only programs, closing the complexity gap: the problem is 2-ExpTime-complete. As a step towards these goals, complexity of containment over arbitrary structures of bounded clique-width is analysed: satisfiability and containment of monadic programs with stratified negation is in 3-ExpTime, and containment of a linear monadic program in a monadic program is in 2-ExpTime.

1 Introduction

One of the central questions of database theory is that of query containment: deciding if the answers to one query are always contained in the answers to another query, regardless of the content of the database. Being a generalization of satisfiability, containment is undecidable for queries expressed in first order logic (FO), but it is decidable for more restrictive classes of queries like unions of conjunctive queries (UCQs) [6], that is, queries expressible in the positive-existential fragment of FO. A way to go beyond FO without losing decidability, is to add recursion (equivalently, least fixed point operator) to unions of conjunctive queries: the resulting language is datalog. In general, containment of datalog programs is undecidable [19], but it becomes decidable under restriction to monadic datalog programs (equivalently, when the use of least fixed point operator is limited to unary formulae) [7].

Supported by Poland's National Science Centre grant UMO-2013/11/D/ST6/03075.

M.M. Halldórsson et al. (Eds.): ICALP 2015, Part II, LNCS 9135, pp. 427–439, 2015.
DOI: 10.1007/978-3-662-47666-6_34

In this work we are interested in a particular class of structures, called data trees, which are trees labelled with a finite alphabet with an additional equivalence relation over nodes (modelling data equality). The main motivation for studying data trees is that they are a convenient model for data organized in a hierarchical structure, for instance, XML documents. When the class of structures is restricted to data trees, containment is still decidable for UCQs [3], but it is undecidable for monadic datalog [1]. A line of research focused on XML applications investigates XPath, an XML query language, which in some variants allows recursion in the form of Kleene star [12,18]. The positive fragment of XPath can be translated to monadic datalog, but the converse translation is not possible due to XPath's limited abilities of testing data equality. In pure datalog setting, two natural fragments were recently shown decidable: downward programs, which never move up from visited tree nodes (cf. [12]), and linear child-only programs, which do not use descendant relation and do not branch (i.e., have at most one intensional predicate per rule) [17]. Relying on ad-hoc arguments, [17] sheds little light on the real reasons behind the decidability, and may give an impression that decidability of these two seemingly different fragments of datalog is pure coincidence. This work puts these two results in context and finds a common denominator, which leads to cleaner arguments, more general results, and—in some cases—tightened complexity bounds.

We show that the common feature of the two fragments is that they both admit counter-examples to containment of clique-width [10] linear in the size of the programs. This almost immediately gives decidability of containment, because monadic datalog is equivalent to monadic second order logic (MSO) [13], for which satisfiability is decidable over structures of bounded clique-width [9]. Unlike tree-width [14], clique-width has not been investigated in the context of datalog. The reason is that, for fixed k, a tree decomposition of width k can be computed in linear time for graphs of tree-width k [4], but for clique-width the best currently known polynomial time algorithm computes decompositions of width $2^{k+1} - 1$ for graphs of clique-width k [16]. Given that algorithms relying on decompositions are typically exponential in k, this results in a double exponential constant, which is impractical most of the time. For the purpose of our work, however, constructing a decomposition for a given structure is not an issue: we need to test if there *exists* a decomposition that yields some counter-example. A closer look at the MSO based approach gives a 3-EXPTIME upper bound; for linear programs we provide a more economic construction, which gives 2-EXPTIME upper bound (even for containment in arbitrary monadic programs).

This approach does not guarantee optimal complexity: for downward programs containment is 2-EXPTIME-complete, and EXPSPACE-complete under restriction to linear programs [17]. But in some cases it actually tightens the bounds: for linear child-only programs the complexity bounds were 2-EXPTIME-hard and in 3-EXPTIME, and our method gives a 2-EXPTIME algorithm, thus closing the complexity gap. Also, the classes of programs for which the algorithms work are broader; for instance, we can test containment in arbitrary monadic programs, not just downward, or linear child-only.

The paper is organized as follows. In Section 2 we recall basic definitions. In Section 3 we focus on datalog over (arbitrary) structures of bounded clique width. We show that a datalog program with stratified negation can be translated into a triple exponential tree automaton working over clique-width decompositions; this implies that satisfiability and containment of such programs over structures of bounded clique-width is in 3-ExpTime. For *linear* monadic programs without negation we provide a construction going via two-way alternating automata [7,21], which gives a 2-ExpTime upper bound for containment (even in arbitrary monadic programs). In Section 4 we apply these results to the problem of containment over data trees for downward programs and linear child-only programs. In Section 5 we conclude with a brief discussion of the obtained results.

2 Preliminaries

Finite structures and clique-width. Let $\tau = \{R_1, \ldots, R_\ell\}$ be a relational signature, i.e., a set of predicate symbols with arities $ar(R_i)$. A (finite) τ-structure $\mathbb{A}$ is a tuple $\langle A, R_1^{\mathbb{A}}, \ldots, R_\ell^{\mathbb{A}} \rangle$ consisting of universe A and relations $R_i^{\mathbb{A}} \subseteq A^{ar(R_i)}$ (interpretations of the predicates). A k-coloured τ-structure is a pair $(\mathbb{A}, \gamma)$, consisting of a τ-structure $\mathbb{A}$ and a mapping $\gamma : A \to \{1, \ldots, k\}$, assigning colours to elements of the universe of $\mathbb{A}$.

Clique width of structures is defined by means of an appropriate notion of decomposition, traditionally known as *k-expression (over τ)*. It is defined as a term over the following set of operations (function symbols) $\mathrm{Op}(\tau, k)$:

- $\mathrm{new}(i)$ for $1 \leq i \leq k$, nullary,
- $\rho(i, j)$ for $1 \leq i, j \leq k$, unary,
- $R(i_1, \ldots, i_r)$ for predicates $R \in \tau$ of arity r and $1 \leq i_1, \ldots, i_r \leq k$, unary,
- $\oplus$, binary.

With k-expression e we associate a k-coloured τ-structure $[\![e]\!]$:

- $[\![\mathrm{new}(i)]\!]$ is a structure with a single element, coloured i, and empty relations;
- $[\![\rho(i, j)(e)]\!]$ is obtained from $[\![e]\!]$ by recolouring all elements of colour i to j;
- $[\![R(i_1, \ldots, i_r)(e)]\!]$ is obtained from $[\![e]\!] = (\mathbb{A}, \gamma)$ by adding to $R^{[\![e]\!]}$ all tuples $(a_1, \ldots, a_r)$ such that $a_j \in A$ and $\gamma(a_j) = i_j$ for $1 \leq j \leq r$;
- $[\![e \oplus e']\!]$ is the disjoint union of $[\![e]\!]$ and $[\![e']\!]$.

A *k-expression for* $\mathbb{A}$ is any k-expression e such that $[\![e]\!] = (\mathbb{A}, \gamma)$ for some γ. The *clique-width* of $\mathbb{A}$ is the least k such that there exists a k-expression for $\mathbb{A}$.

Datalog. We assume some familiarity with datalog and only briefly recall its syntax and semantics; for more details see [2] or [5].

A *datalog program* $\mathcal{P}$ over a relational signature σ, split into *extensional* predicates σ_{ext} and *intensional* predicates σ_{int}, is a finite set of rules of the form *head* $\leftarrow$ *body*, where *head* is an atom over σ_{int} and *body* is a (possibly empty) conjunction of atoms over σ written as a comma-separated list. All variables in the body that are not used in the head are implicitly quantified existentially.

Program $\mathcal{P}$ is evaluated on σ_{ext}-structure $\mathbb{A}$ by generating all atoms over σ_{int} that can be inferred from $\mathbb{A}$ by applying the rules repeatedly, to the point of saturation. Each inferred atom can be witnessed by a *proof tree*: an atom inferred by a rule r from intensional atoms $A_1, A_2, \ldots, A_n$ (and some extensional atoms) is witnessed by a proof tree whose root has label r and n children which are the roots of the proof trees for atoms A_i (if r has no intensional predicates in its body then the root has no children). The program returns set $\mathcal{P}(\mathbb{A})$ consisting of those inferred atoms that match a distinguished *goal predicate* G.

In programs with *stratified negation* we assume that signature σ is partitioned into strata $\sigma_{\text{ext}} = \sigma_0, \sigma_1, \ldots, \sigma_{n-1}, \sigma_n = \{G\}$ for some $n \in \mathbb{N}$. For each $i > 0$, rules for predicates from stratum σ_i contain atoms over $\sigma_0 \cup \cdots \cup \sigma_i$ and negated atoms over predicates from $\sigma_0 \cup \cdots \cup \sigma_{i-1}$. The partition of σ induces a partition of $\mathcal{P}$ into $\mathcal{P}_1, \ldots, \mathcal{P}_n$. The evaluation is done stratum by stratum, that is $\mathcal{P}_i$ is run over atoms inferred by strata $\mathcal{P}_1, \ldots, \mathcal{P}_{i-1}$, including those coming directly from the structure.

In this paper we consider only *monadic* programs, i.e., programs whose intensional predicates are at most unary. A datalog program is *linear*, if the right-hand side of each rule contains at most one atom with an intensional predicate (proof trees for such programs are single branches, and we call them proof words).

For programs $\mathcal{P}, \mathcal{Q}$ with a common goal predicate G, we say that program $\mathcal{P}$ is *contained* in program $\mathcal{Q}$, written as $\mathcal{P} \subseteq \mathcal{Q}$, if

$$\mathcal{P}(\mathbb{A}) \subseteq \mathcal{Q}(\mathbb{A})$$

for each σ_{ext}-structure $\mathbb{A}$. Note that if goal predicate G is nullary, this means that if $G \in \mathcal{P}(t)$ then $G \in \mathcal{Q}(t)$; that is, if $\mathcal{P}$ says *true*, so does $\mathcal{Q}$.

Automata. A *ranked alphabet* Γ is a set of letters with arities. A *tree over ranked alphabet* Γ is an ordered tree labelled with elements of Γ such that the number of children of any given node is equal to the arity of its label. Trees over ranked alphabet Γ can be seen as terms over Γ, and vice versa. A term of the form $f(t_1, t_2, \ldots, t_n)$, $n = \text{ar}(f)$ corresponds to a tree whose root has label f, and children $v_1, \ldots, v_n$ where the subtree rooted at v_i corresponds to the term t_i. Thus, k-expressions over τ are trees over ranked alphabet $\text{Op}(\tau, k)$. An *unranked tree* is a tree without any constraints on the number of children.

A *two-way alternating tree automaton (2ATA)* $\mathcal{A} = \langle \Gamma, Q, q_I, \delta \rangle$ consists of a ranked alphabet Γ, a finite set of states Q, an initial state $q_I \in Q$, and a transition function

$$\delta \colon Q \times \Gamma \to \text{BC}^+(Q \times \mathbb{Z})$$

describing actions of automaton $\mathcal{A}$ in state q in a node with label f as a positive Boolean combination of atomic actions of the form (q, d), where $-1 \leq d \leq \text{ar} f$.

A *run* r of $\mathcal{A}$ over tree t is an unranked tree labelled with pairs (q, v), where q is a state of $\mathcal{A}$ and v is a node of t, satisfying the following condition: if a node of r with label (q, v) has children with labels $(q_1, v_1), \ldots, (q_n, v_n)$, and v has label f in t, then there exist $d_1, \ldots, d_n \in \mathbb{N}$ such that

- v_i is the d_i'th child of v in t for all i such that $d_i > 0$;
- $v_i = v$ for all i such that $d_i = 0$;
- v_i is the parent of v in t for all i such that $d_i = -1$; and
- Boolean combination $\delta(q, f)$ evaluates to *true* when atomic actions $(q_1, d_1), \ldots, (q_n, d_n)$ are substituted by *true*, and other atomic actions are substituted by *false*.

Tree t is *accepted* by automaton $\mathcal{A}$ if it admits a finite run. By $L(\mathcal{A})$ we denote the language *recognized* by $\mathcal{A}$; that is, the set of trees accepted by $\mathcal{A}$.

A *nondeterministic (one-way) tree automaton (NTA)* is an alternating two-way automaton such that each $\delta(q, f)$ is a disjunction of expressions of the from $(q_1, 1) \wedge (q_2, 2) \wedge \cdots \wedge (q_{\mathrm{ar}f}, \mathrm{ar}f)$.

3 Evaluating Monadic Datalog Over k-expressions

It is a part of the database theory folklore that every monadic datalog program can be translated to a formula of monadic second order logic (MSO) [13]. For concreteness, let us assume the syntax of MSO formulae over signature τ is

$$\varphi, \psi ::= \forall X\, \varphi \mid \exists X\, \varphi \mid \varphi \wedge \psi \mid \varphi \vee \psi \mid \varphi \to \psi \mid \varphi \leftrightarrow \psi \mid \neg \varphi \mid$$
$$\mid X \subseteq Y \mid X = \emptyset \mid singleton(X) \mid R(X_1, \ldots, X_r)$$

for $R \in \tau$, $r = \mathrm{ar}R$; the semantics is as usual. As is also well known, each MSO formula (over arbitrary structures) can be translated to an equivalent formula over k-expressions (see e.g. [15, Lemma 16]). Finally, each MSO formula over trees can be translated to an equivalent tree automaton [11,20]. Thus, evaluating a monadic Datalog program over a structure reduces to running an appropriate nondeterministic automaton over any k-expression for this structure. With a bit of care we can ensure that the automaton does not grow too fast.

Proposition 1. *Let k be a positive integer and $\mathcal{P}$ a monadic datalog program with stratified negation. One can construct (in time polynomial in the size of the output) a triple exponential NTA $\mathcal{A}_{\mathcal{P}}$ recognizing k-expressions e such that $[\![e]\!] = (\mathbb{A}, \gamma)$ and $\mathcal{P}(\mathbb{A}) \neq \emptyset$.*

Moreover, if $\mathcal{P}$ uses no negation, one can construct (in time polynomial in the size of the output) a double exponential NTA $\mathcal{A}_{\neg \mathcal{P}}$ recognizing k-expressions e such that $[\![e]\!] = (\mathbb{A}, \gamma)$ and $\mathcal{P}(\mathbb{A}) = \emptyset$.

Proof. Program $\mathcal{P}$ with p intensional predicates and at most q variables per rule can be translated to a linear-size MSO formula φ of the form

$$\forall X_1 \ldots \forall X_p\, \exists X_{p+1} \ldots \exists X_{p+q}\, \varphi_0(X_1, \ldots, X_{p+q})$$

where φ_0 is a quantifier-free formula over signature σ_{ext}, such that $\mathbb{A} \models \varphi$ if and only if $\mathcal{P}(\mathbb{A}) \neq \emptyset$ [13]. For programs with stratified negation we do not need to introduce arbitrary number of alternations. Without loss of generality we can assume that intensional predicates are split into positive, used only under even number of negations, and negative, used only under odd number of negations. One can obtain a linear-size formula of the form

$$\exists X_1 \ldots \exists X_\ell \forall X_{\ell+1} \ldots \forall X_m \exists X_{m+1} \ldots \exists X_n \, \varphi_0(X_1, \ldots, X_n)$$

where, roughly speaking, the first block of quantifiers introduces the negative predicates, the second deals with closure properties for the negative predicates and introduces the positive predicates, and the third deals with closure properties of the positive predicates.

The next step is to translate φ to a formula $\hat{\varphi}$ over the signature of k-expressions over σ_{ext}, such that for every σ_{ext}-structure $\mathbb{A}$ and every k-expression e for $\mathbb{A}$, $\mathbb{A} \models \varphi$ iff $e \models \hat{\varphi}$. We follow the translation from [15, Lemma 16]. It relies on the assumption that the universe of structure $\mathbb{A}$ is contained in the set of nodes of k-expression e: each node with label new(i) is identified with the element of $\mathbb{A}$ it represents. The translation does two things. It relativises the quantifiers to the set of leaves; that is, it replaces φ_0 with

$$\bigwedge_{i=1}^{\ell} \mathit{leaf}(X_i) \wedge \left(\bigwedge_{i=\ell+1}^{m} \mathit{leaf}(X_i) \rightarrow \bigwedge_{i=m+1}^{n} \mathit{leaf}(X_i) \wedge \varphi_0(X_1, \ldots, X_n) \right)$$

where $\mathit{leaf}(X_j)$ is an auxiliary formula saying that each element of X_j is a leaf. Then, it replaces each atomic formula $R(X_{j_1}, \ldots, X_{j_r})$ in φ_0 with formula

$$\psi_R(X_{j_1}, \ldots, X_{j_r})$$

saying that there is a node v with label $R(i_1, \ldots, i_r)$ such that some leaves $x_1 \in X_{j_1}, \ldots, x_r \in X_{j_r}$ are descendants of v and have colours $i_1, \ldots, i_r$ according to the current colouring in v. Note that the obtained formula $\hat{\varphi}$ only uses tree relations (child and labels) in the auxiliary formulae leaf and ψ_R for $R \in \sigma_{\text{ext}}$, which, incidentally, are not quantifier free. Instead of expressing these formulae in MSO, we shall keep them as primitives, to be translated directly to automata.

Translation of MSO formulae to automata [11,20] is done by induction over the structure of the formula: for each quantifier free subformula $\eta(X_{j_1}, \ldots, X_{j_\ell})$ of $\hat{\varphi}$ we construct a deterministic automaton over alphabet $\{0,1\}^n \times \text{Op}(\sigma_{\text{ext}}, k)$ that accepts tree t if and only if

$$(t', U_1, \ldots, U_\ell) \models \eta(X_{j_1}, \ldots, X_{j_\ell})$$

where tree t' is obtained from t by projecting the labels to $\text{Op}(\sigma_{\text{ext}}, k)$, and $U_1, \ldots, U_\ell$ are the sets of nodes whose label in t has 1 in coordinates $j_1, \ldots, j_\ell$, respectively. For subformulae $X_{j_1} \subseteq X_{j_2}$, $\mathit{singleton}(X_j)$, $X_j = \emptyset$, and $\mathit{leaf}(X_j)$, there are automata of constant-size state-space (though over exponential alphabet). For $\psi_R(X_{j_1}, \ldots, X_{j_r})$ the automaton has $2^{kr} + 1$ states: it works bottom-up maintaining sets $I_1, \ldots, I_r$ of colours assigned to elements of $X_{j_1}, \ldots, X_{j_r}$ by the colouring corresponding to the current node; it accepts if at any moment it finds a node with label $R(i_1, \ldots, i_r)$ for some $i_1 \in I_1, \ldots, i_r \in I_r$. The boolean connectives are realized by an appropriate product construction (negation is straightforward for deterministic automata). Altogether we end up with a product of linearly many deterministic automata of size at most single exponential; this gives a single exponential deterministic automaton $\mathcal{A}$ for the quantifier-free part of formula $\hat{\varphi}$. Automaton $\mathcal{A}_\mathcal{P}$ can be obtained from $\mathcal{A}$ by projecting out coordinates $m+1, \ldots, n$ of the labels, complementing, projecting out coordinates $\ell+1, \ldots, m$, complementing, and projecting out coordinates $1, \ldots, \ell$. Since each complementation involves exponential blow-up, the resulting automaton is triple exponential in the size of the program.

Automaton $\mathcal{A}_{\neg\mathcal{P}}$ is obtained from the negation of the first MSO formula in this proof; it requires only one complementation of a nondeterministic automaton, resulting in a double exponential bound.   □

As an immediate corollary we get that satisfiability in structures of bounded clique-width can be tested in 3-EXPTIME for monadic programs with stratified negation. The following is a special case of this.

Corollary 1. *Given monadic programs $\mathcal{P}, \mathcal{Q}$ and $k \in \mathbb{N}$, one can decide in 3-EXPTIME if $\mathcal{P}(\mathbb{A}) \subseteq \mathcal{Q}(\mathbb{A})$ for all structures $\mathbb{A}$ of clique-width at most k.*

If the "smaller" program is linear, we get better complexity. Our main technical contribution here is a direct translation from linear monadic datalog to 2ATA. Unlike in [7], were 2ATA worked on proof trees and essentially mimicked behaviour of datalog programs, we work on k-expressions, in which distant leaves may represent nodes that are in fact close together. The main idea is that each time the 2ATA sees $\oplus$, it guesses a way to cut the proof word into subwords to be realised in the left and right subtree (see appendix for the details).

Theorem 1. *For a linear monadic program $\mathcal{P}$ and $k \in \mathbb{N}$ one can construct (in time polynomial in the size of the output) a single exponential 2ATA $\mathcal{A}$ recognizing k-expressions e such that $[\![e]\!] = (\mathbb{A}, \gamma)$ and $\mathcal{P}(\mathbb{A}) \neq \emptyset$.*

We shall see later that these bounds are tight in the sense that obtaining a polynomial bound would violate lower bounds on the containment problem over data trees discussed in the following section.

The second, and last, step of the construction relies on the following theorem.

Theorem 2 ([7,21]). *For a given 2ATA $\mathcal{A}$ one can construct (in time polynomial in the size of the output) single exponential NTA $\mathcal{B}$ recognizing $L(\mathcal{A})$.*

Combining Theorem 1, Theorem 2, and the additional claim of Proposition 1, we obtain the following bounds.

Corollary 2. *Given a linear monadic program $\mathcal{P}$, a monadic program $\mathcal{Q}$, and $k \in \mathbb{N}$, one can decide in 2-EXPTIME if $\mathcal{P}(\mathbb{A}) \subseteq \mathcal{Q}(\mathbb{A})$ for all structures $\mathbb{A}$ of clique-width at most k.*

Proof. If the goal predicate G of $\mathcal{P}$ and $\mathcal{Q}$ is nullary, the claim follows immediately: one constructs an NTA $\mathcal{A}_\mathcal{P}$ recognizing the set of k-expressions e such that $[\![e]\!] = (\mathbb{A}, \gamma)$ and $G \in \mathcal{P}(\mathbb{A})$, and an NTA $\mathcal{A}_{\neg\mathcal{Q}}$ recognizing the set of k-expressions e such that $[\![e]\!] = (\mathbb{A}, \gamma)$ and $G \notin \mathcal{P}(\mathbb{A})$, and checks if $L(\mathcal{A}_\mathcal{P}) \cap L(\mathcal{A}_{\neg\mathcal{Q}}) \neq \emptyset$.

Assume that G is unary. Extend σ_{ext} with a fresh unary relation H. Let $\mathcal{P}_0$ be obtained from $\mathcal{P}$ by adding rule $G_0 \leftarrow G(x), H(x)$ for a fresh nullary predicate G_0 and changing the goal predicate to G_0; similarly construct $\mathcal{Q}_0$ from $\mathcal{Q}$. Now, it is enough to check if $L(\mathcal{A}_{\mathcal{P}_0}) \cap L(\mathcal{A}_{\neg \mathcal{Q}_0}) \cap L(\mathcal{B}) \neq \emptyset$, where $\mathcal{B}$ is an automaton recognizing the set of k-expressions e over signature $\sigma_{\text{ext}} \cup \{H\}$ such that $[\![e]\!] = (\mathbb{A}, \gamma)$ and $H^{\mathbb{A}}$ is a singleton.   □

4 Containment Over Data Trees

We now restrict the class of structures to *data trees*; that is, (finite) labelled unranked trees over $\Gamma \times \mathsf{DVal}$, where Γ is a finite alphabet and DVal is an infinite set of so-called *data values*. Datalog programs over data trees refer to relations: child $\downarrow$, descendant $\downarrow^+$, data value equality $\sim$, and label tests $a \in \Gamma$. That is, a data tree is seen as a relational structure over signature τ_{dt} consisting of binary relations $\{\downarrow, \downarrow^+, \sim\}$ and unary relations Γ.

We are interested in the problem of containment over data trees: we say that *program $\mathcal{P}$ is contained in program $\mathcal{Q}$ over data trees*, written as $\mathcal{P} \subseteq_{\text{dt}} \mathcal{Q}$, if

$$\mathcal{P}(t) \subseteq \mathcal{Q}(t)$$

for each data tree t. In this section by containment we always mean containment over data trees. Thus, the containment problem is: given programs $\mathcal{P}$, $\mathcal{Q}$ over data trees, decide if $\mathcal{P} \subseteq_{\text{dt}} \mathcal{Q}$.

We propose the following generic approach:

1. show that containment over all data trees is equivalent to containment over data trees of clique-width at most k;
2. test containment over k-expressions for data-trees.

The second step is easy for arbitrary monadic programs (Section 4.1). Hence, given that containment over data trees is undecidable for monadic programs [1], the first step can only be carried out for restricted fragments of monadic datalog. In what follows we shall consider two such fragments: downward programs (Section 4.2), and linear child-only programs (Section 4.3).

4.1 Containment Over Data Trees of Bounded Clique-width

In the light of the general results of the previous section, testing containment over data trees of clique-width at most k is almost straightforward: the only issue is that not all k-expressions over signature τ_{dt} yield data-trees. But those that do can be recognized by a tree automaton. Since the bound on the clique-width depends on the size of the program, the construction requires some care.

Lemma 1. *For all k, k-expressions for data trees form a regular language. The size of automaton recognizing those k-expressions is double exponential in k.*

Proof. To prove the claim, it suffices to give a double-exponential bottom-up automaton that given a k-expression for structure $\mathbb{A}$ verifies that:

- every element is reachable from root via directed $\downarrow$-path;
- the relation $\downarrow \cup \downarrow^{-1}$ is acyclic;
- $\downarrow^+$ is transitive closure of $\downarrow$;
- $\sim$ is reflexive, symmetric and transitive.

The construction is straightforward. The details are given in the appendix.   □

This suffices to solve containment over data trees of bounded clique-width.

Proposition 2. *Given monadic programs* $\mathcal{P}, \mathcal{Q}$ *and* $k \in \mathbb{N}$, *one can test in* 3-EXPTIME *whether* $\mathcal{P}(t) \subseteq \mathcal{Q}(t)$ *for every data tree* t *of clique-width at most* k. *If* $\mathcal{P}$ *is linear, the complexity drops to* 2-EXPTIME.

Proof. The proofs are just like for Corollary 1 and Corollary 2, except that the automaton recognizing counter-examples to containment has to be intersected with the automaton recognizing k-expressions that yield data trees.   □

4.2   Downward Programs

As we have explained, showing that containment over data trees of two programs is equivalent to containment over data trees of bounded clique-width immediately gives decidability of containment in 3-EXPTIME in general, and in 2-EXPTIME if the "smaller" problem is linear. In this section we apply this method to the class of downward programs. We do it mainly for illustrative purposes, as it is known that for downward programs containment is 2-EXPTIME-complete in general, and EXPSPACE-complete for linear programs [17]. But our method also gives broader results: it uses a relaxed definition of downward programs, and it works for testing containment in arbitrary monadic programs.

We begin with an observation that has been seminal to this work. Let $datacut(t)$ be the maximum over all $\downarrow$-edges (u, v) in t of the number of $\sim$-classes represented in both parts of t: the subtree rooted at v, and the rest of the tree.

Lemma 2. *Every data tree* t *has clique-width at most* $4 \cdot datacut(t) + 5$.

Proof. Let $k = 4 \cdot datacut(t) + 5$. We will use colours of the form

$$\{root, notroot\} \times \{0, 1, \ldots, datacut(d)\} \times \{old, new\}$$

plus an additional colour *temp*. We construct a k-expression inductively for subforests f of t, maintaining the following invariants for the induced colouring

1. all nodes have colours (x, y, old),
2. colours $(root, y, z)$ are reserved for the roots of f,
3. colours $(notroot, y, z)$ are used by other nodes,
4. node has colour $(x, 0, z)$ iff it carries a data value never used outside of f.

Colors (x, y, new) are used when combining two parts of the tree with $\oplus$, to avoid gluing colours together. Color $temp$ is used for recolouring.

For single-node tree s, we just use $(root, 1, old)$ or $(root, 0, old)$, in accordance with invariant 4.

Assume we have k-expressions for all immediate subtrees of subtree s, say $e_1, \ldots, e_\ell$. We use $\oplus$ to add them one by one, adding necessary $\sim$-edges. Each time we add another e_i, we first recolour (x, y, old) to (x, y, new) in e_i for all x, y. We add $\sim$-edges as required, and recolour the nodes (using $temp$) to restore invariants 1 and 4.

Next, we want to build a k-expression for s. We use $\oplus$ to add the root, coloured $(root, y, new)$, where y is chosen depending on $\sim$ relation between the root and the nodes in the immediate subtrees. We add $\downarrow$ edges between $(root, y, new)$ and $(root, y', old)$ and $\downarrow^+$ edges between $(root, y, new)$ and (x', y', old) for all x', y', as well as appropriate $\sim$-edges. Next, we recolour $(root, y', old)$ to $(notroot, y', old)$ for all y', and $(root, y, new)$ to $(root, y, old)$. If necessary, we do additional recolouring to restore invariant 4. $\qquad\square$

A datalog rule is essentially a conjunctive query, so one can associate with it a relational structure $\mathbb{A}_r$ in the usual way: the universe is the set $\mathrm{var}(r)$ of variables used in r, and relations are defined by extensional atoms of r. Recall that program $\mathcal{P}$ is *downward* if it had no nullary predicates and for each rule $r \in \mathcal{P}$, graph (V, E) with $V = \mathrm{var}(r)$ and $E = (\downarrow)^{\mathbb{A}_r} \cup (\downarrow^+)^{\mathbb{A}_r}$ was a tree in which the variable used in the head of r is the root [17]. Here we use a relaxed variant of this definition: we lift the restriction that the graph is a tree, but we keep the requirement that each node is reachable from the variable used in the head.

Theorem 3. *Let $\mathcal{P}$ be a downward program and $\mathcal{Q}$ an arbitrary monadic program. If $\mathcal{P} \not\subseteq \mathcal{Q}$, then there exists a witness with clique-width at most $4 \cdot \|\mathcal{P}\| + 5$.*

Proof. The theorem follows immediately from the following claim (with A set to the goal predicate G): for each tree t and each atom $A(v)$ inferred by $\mathcal{P}$ from t, there exists a data tree t' such that

- $datacut(t') \leq \|\mathcal{P}\|$,
- $A(root_{t'})$ can be inferred by $\mathcal{P}$ from t', where $root_{t'}$ is the root of t',
- there exists a homomorphism from t' to t that maps $root_{t'}$ to v (in particular, $G(root_{t'}) \notin \mathcal{Q}(t')$ unless $G(v) \in \mathcal{Q}(t)$).

We prove the claim by induction on the size of proof tree p witnessing $A(v)$. Let r be the rule in the root of p and let $h \colon \mathbb{A}_r \to t$ be the associated homomorphism (it maps the head variable to v). Let t_r be the data tree obtained from t_v (the subtree of t rooted at v) by interpreting $\sim$ as the least equivalence relation extending the image of $\sim^{\mathbb{A}_r}$ under h; it has at most $|r|$ non-singleton abstraction classes. Let $R_1(y_1), \ldots, R_m(y_m)$ be the intensional atoms in rule r. By the inductive hypothesis there exist appropriate $t_1, \ldots, t_m$ for $R_1(h(y_1)), \ldots, R_m(h(y_m))$. We obtain t' by taking disjoint union of t_r and $t_1, \ldots, t_m$ with the roots of $t_1, \ldots, t_m$ identified with $h(y_1), \ldots, h(y_m)$, and closing $\sim$ and $\downarrow^+$ under transitivity. $\qquad\square$

Corollary 3. *Containment of a downward program in a monadic program is in 3-ExpTime, and in 2-ExpTime if the downward program is also linear.*

4.3   Linear Child-Only Programs

In this section we show a more useful application of results from Section 3. Recall that by child-only programs we mean programs that use only $\downarrow$ and $\sim$ relation, while $\downarrow^+$ is forbidden. It is known that containment for linear child-only programs is in 3-ExpTime and 2-ExpTime-hard (for non-linear ones containment is undecidable) [17]. Here, we close this complexity gap and also extend the result to containment of linear child-only programs in arbitrary monadic ones. Similarly to the downward programs, we show that containment can be verified over data trees of bounded clique-width.

Theorem 4. *For a linear child-only program $\mathcal{P}$ and arbitrary monadic program $\mathcal{Q}$, if $\mathcal{P} \not\sqsubseteq \mathcal{Q}$, there exists a witness with clique-width at most $8\|\mathcal{P}\|^2 + 3\|\mathcal{P}\| + 2$.*

Because child-only programs can go up and down in the tree, it is not possible to bound *datacut* as we did for downward programs. It is possible, however, to bound the number of $\sim$-classes represented at the same time in an appropriately generalized subtree of a node v and in the rest of the tree. The proof of this fact is not very hard but it requires some technical definitions from [17] and is given in the appendix.

Just like for downward programs, the following is a straightforward application of Theorems 1 and 4 (and 2-ExpTime-hardness shown in [17]).

Corollary 4. *Containment of a linear child-only program in a monadic program is 2-ExpTime-complete.*

5   Conclusions

We have shown that over structures of bounded clique-width (arbitrary structures and data trees), containment of monadic datalog programs is decidable in 3-ExpTime, and containment of linear monadic programs in monadic programs is in 2-ExpTime. Consequently, decidability of containment for a fragment of monadic datalog reduces to showing that the fragment admits bounded clique-width counter-examples to containment. Graph decompositions have been used before for deciding properties of datalog programs: already in early 1990s Courcelle noticed a connection between runs of datalog programs and tree decompositions of structures, and concluded decidability of some properties of programs expressible in MSO over these decompositions [8]. Over data trees, however, tree-width is not useful: there is much less freedom in constructing models of a program, and neither $\downarrow^+$ nor $\sim$ are sparse relations, as is required by bounded tree-width. Clique-width seems to be exactly the notion that is needed.

We applied this method to generalize previously known decidability results for downward programs and linear child-only programs: we relaxed the definition of downward programs and, more importantly, we covered containment in arbitrary monadic programs. With a bit of extra effort one could further relax the notion of downward programs to allow disconnected rules. More interestingly, there is a relatively natural unifying fragment with decidable containment: linear child-only programs that use freely predicates defined by downward programs. The method seems flexible enough for further extensions. More generally, one could try to port the method to formalisms other then monadic datalog, e.g., XPath, both for data trees and data graphs.

References

1. Abiteboul, S., Bourhis, P., Muscholl, A., Wu, Z.: Recursive queries on trees and data trees. In: ICDT 2013, pp. 93–104 (2013)
2. Abiteboul, S., Hull, R., Vianu, V.: Foundations of Databases. Addison Wesley (1995)
3. Björklund, H., Martens, W., Schwentick, T.: Optimizing conjunctive queries over trees using schema information. In: Ochmański, E., Tyszkiewicz, J. (eds.) MFCS 2008. LNCS, vol. 5162, pp. 132–143. Springer, Heidelberg (2008)
4. Bodlaender, H.L.: A linear-time algorithm for finding tree-decompositions of small treewidth. SIAM J. Comput. 25(6), 1305–1317 (1996)
5. Ceri, S., Gottlob, G., Tanca, L.: Logic programming and databases. Springer-Verlag New York Inc. (1990)
6. Chandra, A.K., Merlin, P.M.: Optimal implementation of conjunctive queries in relational data bases. In: STOC 1977, pp. 77–90. ACM, New York (1977)
7. Cosmadakis, S.S., Gaifman, H., Kanellakis, P.C., Vardi, M.Y.: Decidable optimization problems for database logic programs (preliminary report). In: STOC 1988, pp. 477–490 (1988)
8. Courcelle, B.: Recursive queries and context-free graph grammars. Theor. Comput. Sci. 78(1), 217–244 (1991)
9. Courcelle, B., Makowsky, J.A., Rotics, U.: Linear time solvable optimization problems on graphs of bounded clique-width. Theory Comput. Syst. 33(2), 125–150 (2000)
10. Courcelle, B., Olariu, S.: Upper bounds to the clique width of graphs. Discrete Applied Mathematics 101(1–3), 77–114 (2000)
11. Doner, J.: Tree acceptors and some of their applications. J. Comput. Syst. Sci. 4(5), 406–451 (1970)
12. Figueira, D.: Satisfiability of downward XPath with data equality tests. In: PODS 2009, pp. 197–206 (2009)
13. Gottlob, G., Koch, C.: Monadic datalog and the expressive power of languages for web information extraction. J. ACM 51(1), 74–113 (2004)
14. Gottlob, G., Pichler, R., Wei, F.: Monadic datalog over finite structures of bounded treewidth. ACM Trans. Comput. Log. 12(1), 3 (2010)
15. Grohe, M., Turán, G.: Learnability and definability in trees and similar structures. Theory Comput. Syst. 37(1), 193–220 (2004)
16. Hlinený, P., Oum, S.: Finding branch-decompositions and rank-decompositions. SIAM J. Comput. 38(3), 1012–1032 (2008)
17. Mazowiecki, F., Murlak, F., Witkowski, A.: Monadic datalog and regular tree pattern queries. In: Csuhaj-Varjú, E., Dietzfelbinger, M., Ésik, Z. (eds.) MFCS 2014, Part I. LNCS, vol. 8634, pp. 426–437. Springer, Heidelberg (2014)

18. Neven, F., Schwentick, T.: On the complexity of XPath containment in the presence of disjunction, DTDs, and variables. Logical Methods in Computer Science **2**(3) (2006)

19. Shmueli, O.: Equivalence of datalog queries is undecidable. J. Log. Program. **15**(3), 231–241 (1993)

20. Thatcher, J.W., Wright, J.B.: Generalized finite automata theory with an application to a decision problem of second-order logic. Mathematical Systems Theory **2**(1), 57–81 (1968)

21. Vardi, M.Y.: Reasoning about the past with two-way automata. In: Larsen, K.G., Skyum, S., Winskel, G. (eds.) ICALP 1998. LNCS, vol. 1443, p. 628. Springer, Heidelberg (1998)

An Approach to Computing Downward Closures

Georg Zetzsche(✉)

Technische Universität Kaiserslautern, Fachbereich Informatik,
Concurrency Theory Group, Kaiserslautern, Germany
zetzsche@cs.uni-kl.de

Abstract. The downward closure of a word language is the set of all (not necessarily contiguous) subwords of its members. It is well-known that the downward closure of any language is regular. While the downward closure appears to be a powerful abstraction, algorithms for computing a finite automaton for the downward closure of a given language have been established only for few language classes.

This work presents a simple general method for computing downward closures. For language classes that are closed under rational transductions, it is shown that the computation of downward closures can be reduced to checking a certain unboundedness property.

This result is used to prove that downward closures are computable for (i) every language class with effectively semilinear Parikh images that are closed under rational transductions, (ii) matrix languages, and (iii) indexed languages (equivalently, languages accepted by higher-order pushdown automata of order 2).

1 Introduction

The *downward closure* $L{\downarrow}$ of a word language L is the set of all (not necessarily contiguous) subwords of its members. While it is well-known that the downward closure of any language is regular [15], it is not possible in general to compute them. However, if they are computable, downward closures are a powerful abstraction. Suppose L describes the behavior of a system that is observed through a lossy channel, meaning that on the way to the observer, arbitrary actions can get lost. Then, $L{\downarrow}$ is the set of words received by the observer [14]. Hence, given the downward closure as a finite automaton, we can decide whether two systems are equivalent under such observations, and even whether one system includes the behavior of another.

Further motivation for studying downward closures stems from a recent result of Czerwiński and Martens [8]. It implies that for language classes that are closed under rational transductions and have computable downward closures, separability by piecewise testable languages is decidable.

As an abstraction, compared to the Parikh image (which counts the number of occurrences of each letter), downward closures have the advantage of guaranteeing regularity for any language. Most applications of Parikh images, in contrast, require semilinearity, which fails for many interesting language classes.

© Springer-Verlag Berlin Heidelberg 2015
M.M. Halldórsson et al. (Eds.): ICALP 2015, Part II, LNCS 9135, pp. 440–451, 2015.
DOI: 10.1007/978-3-662-47666-6_35

An example of a class that lacks semilinearity of Parikh images and thus spurred interest in computing downward closures is that of the *indexed languages* [3] or, equivalently, those accepted by higher-order pushdown automata of order 2 [22].

It appears to be difficult to compute downward closures and there are few language classes for which computability has been established. Computability is known for *context-free languages* and *algebraic extensions* [7,21], *0L-systems* and *context-free FIFO rewriting systems* [1], *Petri net languages* [14], and *stacked counter automata* [29]. They are not computable for reachability sets of *lossy channel systems* [23] and *Church-Rosser languages* [13].

This work presents a new general method for the computation of downward closures. It relies on a fairly simple idea and reduces the computation to the so-called *simultaneous unboundedness problem (SUP)*. The latter asks, given a language $L \subseteq a_1^* \cdots a_n^*$, whether for each $k \in \mathbb{N}$, there is a word $a_1^{x_1} \cdots a_n^{x_n} \in L$ such that $x_1, \ldots, x_n \geq k$. This method yields new, sometimes greatly simplified, algorithms for each of the computability results above. It also opens up a range of other language classes to the computation of downward closures.

First, it implies computability for every language class that is closed under rational transductions and exhibits effectively semilinear Parikh images. This re-proves computability for *context-free languages* and *stacked counter automata* [29], but also applies to many other classes, such as the *multiple context-free languages* [26]. Second, the method yields the computability for *matrix grammars* [9,10], a powerful grammar model that generalizes Petri net and context-free languages. Third, it is applied to obtain computability of downward closures for the *indexed languages*.

Due to space restrictions, many proofs are only available in the full version of this work [28].

2   Basic Notions and Results

If X is an alphabet, X^* (X^+) denotes the set of (non-empty) words over X. The empty word is denoted by $\varepsilon \in X^*$. For a symbol $x \in X$ and a word $w \in X^*$, let $|w|_x$ be the number of occurrences of x in w. For words $u, v \in X^*$, we write $u \preceq v$ if $u = u_1 \cdots u_n$ and $v = v_0 u_1 v_1 \cdots u_n v_n$ for some $u_1, \ldots, u_n, v_0, \ldots, v_n \in X^*$. It is well-known that $\preceq$ is a well-quasi-order on X^* and that therefore the *downward closure* $L{\downarrow} = \{u \in X^* \mid \exists v \in L : u \preceq v\}$ is regular for any $L \subseteq X^*$ [15]. If X is an alphabet, $X^\oplus$ denotes the set of maps $\alpha \colon X \to \mathbb{N}$, which are called *multisets*. For $\alpha, \beta \in X^\oplus$, $k \in \mathbb{N}$ the multisets $\alpha + \beta$ and $k \cdot \alpha$ are defined in the obvious way. A set of the form $\{\mu_0 + x_1 \cdot \mu_1 + \cdots + x_n \cdot \mu_n \mid x_1, \ldots, x_n \geq 0\}$ for $\mu_0, \ldots, \mu_n \in X^\oplus$ is called *linear* and $\mu_1, \ldots, \mu_n$ are its *period elements*. A finite union of linear sets is called *semilinear*. The *Parikh map* is the map $\Psi \colon X^* \to X^\oplus$ defined by $\Psi(w)(x) = |w|_x$ for all $w \in X^*$ and $x \in X$. We lift Ψ to sets in the usual way: $\Psi(L) = \{\Psi(w) \mid w \in L\}$. If $\Psi(L) = \Psi(K)$, then L and K are *Parikh-equivalent*.

A *finite automaton* is a tuple (Q, X, E, q_0, F), where Q is a finite set of *states*, X is its input alphabet, $E \subseteq Q \times X^* \times Q$ is a finite set of *edges*, $q_0 \in Q$ is its *initial state*, and $F \subseteq Q$ is the set of its *final states*. If there is a path labeled

$w \in X^*$ from state p to q, we denote this fact by $p \xrightarrow{w} q$. The language accepted by A is denoted $\mathsf{L}(A)$.

A *(finite-state) transducer* is a tuple (Q, X, Y, E, q_0, F), where Q, X, q_0, F are defined as for automata and Y is its *output alphabet* and $E \subseteq Q \times X^* \times Y^* \times Q$ is the finite set of its *edges*. If there is a path from state p to q that reads the input word $u \in X^*$ and outputs the word $v \in Y^*$, we denote this fact by $p \xrightarrow{u,v} q$. In slight abuse of terminology, we sometimes specify transducers where an edge outputs a regular language instead of a word.

For alphabets X, Y, a *transduction* is a subset of $X^* \times Y^*$. If A is a transducer as above, then $\mathsf{T}(A)$ denotes its generated *transduction*, namely the set of all $(u, v) \in X^* \times Y^*$ such that $q_0 \xrightarrow{u,v} f$ for some $f \in F$. Transductions of the form $\mathsf{T}(A)$ are called *rational*. For a transduction $T \subseteq X^* \times Y^*$ and a language $L \subseteq X^*$, we write $TL = \{v \in Y^* \mid \exists u \in L \colon (u, v) \in T\}$. A class of languages $\mathcal{C}$ is called a *full trio* if it is effectively closed under rational transductions, i.e. if $TL \in \mathcal{C}$ for each $L \in \mathcal{C}$ and each rational transduction T.

Observe that for each full trio $\mathcal{C}$ and $L \in \mathcal{C}$, the language $L{\downarrow}$ is effectively contained in $\mathcal{C}$. By *computing downward closures* we mean finding a finite automaton for $L{\downarrow}$ when given a representation of L in $\mathcal{C}$. It will always be clear from the definition of $\mathcal{C}$ how to represent languages in $\mathcal{C}$.

The Simultaneous Unboundedness Problem. We come to the central decision problem in this work. Let $\mathcal{C}$ be a language class. The *simultaneous unboundedness problem* (SUP) *for $\mathcal{C}$* is the following decision problem:

Given A language $L \subseteq a_1^* \cdots a_n^*$ in $\mathcal{C}$ for some alphabet $\{a_1, \ldots, a_n\}$.
Question Does $L{\downarrow}$ equal $a_1^* \cdots a_n^*$?

The term "simultaneous unboundedness problem" reflects the fact that the equality $L{\downarrow} = a_1^* \cdots a_n^*$ holds if and only if for each $k \in \mathbb{N}$, there is a word $a_1^{x_1} \cdots a_n^{x_n} \in L$ such that $x_1, \ldots, x_n \geq k$.

After obtaining the results of this work, the author learned that Czerwiński and Martens considered a very similar decision problem [8]. Their *diagonal problem* asks, given a language $L \subseteq X^*$ whether for each $k \in \mathbb{N}$, there is a word $w \in L$ with $|w|_x \geq k$ for each $x \in X$. Czerwiński and Martens prove that for full trios with a decidable diagonal problem, it is decidable whether two given languages are separable by a piecewise testable language. In fact, their proof only requires decidability of the (ostensibly easier) SUP. Here, Theorem 1 implies that in each full trio, the diagonal problem is decidable if and only if the SUP is.

The following is the first main result of this work.

Theorem 1. *Let $\mathcal{C}$ be a full trio. Then downward closures are computable for $\mathcal{C}$ if and only if the SUP is decidable for $\mathcal{C}$.*

The proof of Theorem 1 uses the concept of simple regular expressions. Let X be an alphabet. An *atomic expression* is a rational expression of the form $(x \cup \varepsilon)$ with $x \in X$ or of the form $(x_1 \cup \cdots \cup x_n)^*$ with $x_1, \ldots, x_n \in X$. A *product* is a (possibly empty) concatenation $a_1 \cdots a_n$ of atomic expressions. A *simple* *regular expression (SRE)* is of the form $p_1 \cup \cdots \cup p_n$, where the p_i are products. Given an SRE r, we write $\mathsf{L}(r)$ for the language it describes. It was shown by Jullien [19] (and later rediscovered by Jonsson [2]) that SREs describe precisely the downward closed languages.

Proof (Theorem 1). Of course, if downward closures are computable for $\mathcal{C}$, then given a language $L \subseteq a_1^* \cdots a_n^*$ in $\mathcal{C}$, we can compute a finite automaton for $L{\downarrow}$ and check whether $L{\downarrow} = a_1^* \cdots a_n^*$. This proves the "only if" direction.

For the other direction, let us quickly observe that the emptiness problem can be reduced to the SUP. Indeed, if $L \subseteq X^*$ and T is the rational transduction $X^* \times \{a\}^*$, then $TL \subseteq a^*$ and $(TL){\downarrow} = a^*$ if and only if $L \neq \emptyset$.

Now, suppose the SUP is decidable for $\mathcal{C}$ and let $L \subseteq X^*$. Since we know that $L{\downarrow}$ is described by some SRE, we can enumerate SREs over X and are guaranteed that one of them will describe $L{\downarrow}$. Hence, it suffices to show that given an SRE r, it is decidable whether $\mathsf{L}(r) = L{\downarrow}$.

Since $\mathsf{L}(r)$ is a regular language, we can decide whether $L{\downarrow} \subseteq \mathsf{L}(r)$ by checking whether $L{\downarrow} \cap (X^* \setminus \mathsf{L}(r)) = \emptyset$. This can be done because we can compute a representation for $L{\downarrow} \cap (X^* \setminus \mathsf{L}(r))$ in $\mathcal{C}$ and check it for emptiness. It remains to be shown that it is decidable whether $\mathsf{L}(r) \subseteq L{\downarrow}$.

The set $\mathsf{L}(r)$ is a finite union of sets of the form $\{w_0\}{\downarrow}Y_1^*\{w_1\}{\downarrow} \cdots Y_n^*\{w_n\}{\downarrow}$ for some $Y_i \subseteq X$, $Y_i \neq \emptyset$, $1 \leq i \leq n$, and $w_i \in X^*$, $0 \leq i \leq n$. Therefore, it suffices to decide whether $\{w_0\}{\downarrow}Y_1^*\{w_1\}{\downarrow} \cdots Y_n^*\{w_n\}{\downarrow} \subseteq L{\downarrow}$. Since $L{\downarrow}$ is downward closed, this is equivalent to $w_0 Y_1^* w_1 \cdots Y_n^* w_n \subseteq L{\downarrow}$. For each $i \in \{1, \ldots, n\}$, we define the word $u_i = y_1 \cdots y_k$, where $Y_i = \{y_1, \ldots, y_k\}$. Observe that $w_0 Y_1^* w_1 \cdots Y_n^* w_n \subseteq L{\downarrow}$ holds if and only if for every $k \geq 0$, there are numbers $x_1, \ldots, x_n \geq k$ such that $w_0 u_1^{x_1} w_1 \cdots u_n^{x_n} w_n \in L{\downarrow}$. Moreover, if T is the rational transduction $T = \{(w_0 u_1^{x_1} w_1 \cdots u_n^{x_n} w_n, a_1^{x_1} \cdots a_n^{x_n}) \mid x_1, \ldots, x_n \geq 0\}$, then $T(L{\downarrow}) = \{a_1^{x_1} \cdots a_n^{x_n} \mid w_0 u_1^{x_1} w_1 \cdots u_n^{x_n} w_n \in L{\downarrow}\}$. Thus, the inclusion $w_0 Y_1^* w_1 \cdots Y_n^* w_n \subseteq L{\downarrow}$ is equivalent to $(T(L{\downarrow})){\downarrow} = a_1^* \cdots a_n^*$, which is an instance of the SUP, since we can compute a representation of $T(L{\downarrow})$ in $\mathcal{C}$. $\qquad\square$

Despite its simplicity, Theorem 1 has far-reaching consequences for the computability of downward closures. Let us record a few of them.

Corollary 1. *Suppose $\mathcal{C}$ and $\mathcal{D}$ are full trios such that given $L \in \mathcal{C}$, we can compute a Parikh-equivalent $K \in \mathcal{D}$. If downward closures are computable for $\mathcal{D}$, then they are computable for $\mathcal{C}$.*

Proof. We show that the SUP is decidable for $\mathcal{C}$. Given $L \in \mathcal{C}$, $L \subseteq a_1^* \cdots a_n^*$, we construct a Parikh-equivalent $K \in \mathcal{D}$. Observe that then $\Psi(K{\downarrow}) = \Psi(L{\downarrow})$. We compute a finite automaton A for $K{\downarrow}$ and then a semilinear representation of $\Psi(\mathsf{L}(A)) = \Psi(K{\downarrow}) = \Psi(L{\downarrow})$. Then $L{\downarrow} = a_1^* \cdots a_n^*$ if and only if some of the linear sets has for each $1 \leq i \leq n$ a period element containing a_i. Hence, the SUP is decidable for $\mathcal{C}$. $\qquad\square$

Note that if a language class has effectively semilinear Parikh images, then we can construct Parikh-equivalent regular languages. Therefore, the following is a special case of Corollary 1.

Corollary 2. *For each full trio with effectively semilinear Parikh images, downward closures are computable.*

Corollary 2, in turn, provides computability of downward closures for a variety of language classes. First, it re-proves the classical downward closure result for *context-free languages* [7,21] and thus *algebraic extensions* [21] (see [29] for a simple reduction of the latter to the former). Second, it yields a drastically simplified proof of the computability of downward closures for *stacked counter automata*, which was shown in [29] using the machinery of Parikh annotations. It should be noted, however, that the algorithm in [29] is easily seen to be primitive recursive, while this is not clear for the brute-force approach presented here.

Corollary 2 also implies computability of downward closures for *multiple context-free languages* [26], which have received considerable attention in computational linguistics. As shown in [26], the multiple context-free languages constitute a full trio and exhibit effectively semilinear Parikh images.

Our next application of Theorem 1 is an alternative proof of the computability of downward closures for *Petri net languages*, which was established by Habermehl, Meyer, and Wimmel [14]. Here, by Petri net language, we mean sequences of transition labels of runs from an initial to a final marking. Czerwiński and Martens [8] exhibit a simple reduction of the diagonal problem for Petri net languages to the place boundedness problem for Petri nets with one inhibitor arc, which was proven decidable by Bonnet, Finkel, Leroux, and Zeitoun [4]. Since the Petri net languages are well-known to be a full trio [18], Theorem 1 yields an alternative algorithm for downward closures of Petri net languages.

We can also use Corollary 1 to extend the computability of downward closures for Petri net languages to a larger class. *Matrix grammars* are a powerful formalism that is well-known in the area of regulated rewriting and generalizes numerous other grammar models [9,10]. They generate the *matrix languages*, a class which strictly includes both the context-free languages and the Petri net languages. It is well-known that the matrix languages are a full trio and given a matrix grammar, one can construct a Parikh-equivalent Petri net language [10]. Thus, the following is a consequence of Corollary 1.

Corollary 3. *Downward closures are computable for matrix languages.*

Finally, we apply Theorem 1 to the *indexed languages*. These were introduced by Aho [3] and are precisely those accepted by higher-order pushdown automata of order 2 [22]. Since indexed languages do not have semilinear Parikh images, downward closures are a promising alternative abstraction.

Theorem 2. *Downward closures are computable for indexed languages.*

The indexed languages constitute a full trio [3], and hence the remainder of this work is devoted to showing that their SUP is decidable. Note that since this class significantly extends the 0L-languages [11], Theorem 2 generalizes the computability result of Abdulla, Boasson, and Bouajjani for 0L-systems and context-free FIFO rewriting systems [1].

3   Indexed Languages

Let us define indexed grammars. The following definition is a slight variation[1] of the one from [17]. An *indexed grammar* is a tuple $G = (N, T, I, P, S)$, where N, T, and I are pairwise disjoint alphabets, called the *nonterminals*, *terminals*, and *index symbols*, respectively. P is the finite set of *productions* of the forms $A \to w$, $A \to Bf$, $Af \to w$, where $A, B \in N$, $f \in I$, and $w \in (N \cup T)^*$. We regard a word $Af_1 \cdots f_n$ with $f_1, \ldots, f_n \in I$ as a nonterminal to which a stack is attached. Here, f_1 is the topmost symbol and f_n is on the bottom. For $w \in (N \cup T)^*$ and $x \in I^*$, we denote by $[w, x]$ the word obtained by replacing each $A \in N$ in w by Ax. A word in $(NI^* \cup T)^*$ is called a *sentential form*. For $q, r \in (NI^* \cup T)^*$, we write $q \Rightarrow_G r$ if there are words $q_1, q_2 \in (NI^* \cup T)^*$, $A \in N$, $p \in (N \cup T)^*$ and $x, y \in I^*$ such that $q = q_1 Axq_2$, $r = q_1[p, y]q_2$, and one of the following is true: (i) $A \to p$ is in P, $p \in (N \cup T)^* \setminus T^*$, and $y = x$, (ii) $A \to p$ is in P, $p \in T^*$, and $y = x = \varepsilon$, (iii) $A \to pf$ is in P and $y = fx$, or (iv) $Af \to p$ is in P and $x = fy$. The language *generated by* G is $\mathsf{L}(G) = \{w \in T^* \mid S \Rightarrow_G^* w\}$, where $\Rightarrow_G^*$ denotes the reflexive transitive closure of $\Rightarrow_G$. *Derivation trees* are always unranked trees with labels in $NI^* \cup T \cup \{\varepsilon\}$ and a very straightforward analog to those of context-free grammars (see, for example, [27]). If t is a labeled tree, then its *yield*, denoted $\mathsf{yield}(t)$, is the word spelled by the labels of its leaves.

We will often assume that our indexed grammars are in *normal form*, which means that every production is in one of the following forms:

(i) $A \to Bf$,   (ii) $Af \to B$,   (iii) $A \to uBv$,   (iv) $A \to BC$,   (v) $A \to w$,

with $A, B, C \in N$, $f \in I$, and $u, v, w \in T^*$. Productions of these forms are called *push*, *pop*, *output*, *split*, and *terminal* productions, respectively. The normal form can be attained just like the Chomsky normal form of context-free grammars.

The SUP for Indexed Grammars. The SUP for indexed grammars does not seem to easily reduce to a decidable problem. In the case $L \subseteq a^*$, the SUP is just the finiteness problem, for which Hayashi presented a procedure using his pumping lemma [16]. However, neither Hayashi's nor any of the other pumping or shrinking lemmas [12,20,24,27] appears to yield decidability of the SUP. Therefore, this work employs a different approach: Given an indexed grammar G with $\mathsf{L}(G) \subseteq a_1^* \cdots a_n^*$, we apply a series of transformations, each preserving the simultaneous unboundedness (steps 1–3). These transformations leave us with an indexed grammar in which the number of nonterminals appearing in sentential forms is bounded. This allows us to construct an equivalent finite-index scattered context grammar (step 4), a type of grammars that is known to exhibit effectively semilinear Parikh images.

We begin with an analysis of the structure of index words that facilitate certain derivations. Let $G = (N, T, P, S)$ be an indexed grammar, $A \in N$ a nonterminal and $R \subseteq T^*$ a regular language. We write $\mathsf{IW}_G(A, R)$ for the set of

---

[1] We require that a nonterminal can only be replaced by a terminal word if it has no index attached to it. It is easy to see that this leads to the same languages [27].

index words that allow A the derivation of a word from R. Formally, we define $\mathsf{IW}_G(A, R) = \{x \in I^* \mid \exists y \in R\colon Ax \Rightarrow_G^* y\}$. The following lemma is essentially equivalent to the fact that the set of stack contents from which an alternating pushdown system can reach a final configuration is regular [5].

Lemma 1. *For an indexed grammar G, a nonterminal A, and a regular language R, the language $\mathsf{IW}_G(A, R)$ is effectively regular.*

Step 1: Productive Interval Grammars. We want to make sure that each nonterminal can only derive words in some fixed 'interval' $a_i^* \cdots a_j^*$. An *interval grammar* is an indexed grammar $G = (N, T, I, P, S)$ in normal form together with a map $\iota\colon N \to \mathbb{N} \times \mathbb{N}$, called *interval map*, such that for each $A \in N$ with $\iota(A) = (i, j)$, we have (i) $1 \leq i \leq j \leq n$, (ii) if $Ax \Rightarrow_G^* u$ for $x \in I^*$ and $u \in T^*$, then $u \in a_i^* \cdots a_j^*$, and (iii) if $S \Rightarrow_G^* u\, Ax\, v\, By\, w$ with $u, v, w \in (NI^* \cup T)^*$, $B \in N$, $x, y \in I^*$, and $\iota(B) = (k, \ell)$, then $j \leq k$.

We will also need our grammar to be 'productive', meaning that every derivable sentential form and every nonterminal in it contribute to the derived terminal words. A production is called *erasing* if its right-hand side is the empty word. A grammar is *non-erasing* if it contains no erasing productions. Moreover, a word $u \in (NI^* \cup T)^*$ is *productive* if there is some $v \in T^*$ with $u \Rightarrow_G^* v$. We call an indexed grammar G *productive* if (i) it is non-erasing and (ii) whenever $u \in (NI^* \cup T)^*$ is productive and $u \Rightarrow_G^* u'$ for $u' \in (NI^* \cup T)^*$, then u' is productive as well. The following proposition is shown in two steps. First, we construct an interval grammar and then use Lemma 1 to encode information about the current index word in each nonterminal. This information is then used, among other things, to prevent the application of productions that lead to nonproductive sentential forms. The proposition clearly implies that the SUP for indexed grammars can be reduced to the case of productive interval grammars.

Proposition 1. *For each indexed grammar G with $\mathsf{L}(G) \subseteq a_1^* \cdots a_n^*$, one can construct a productive interval grammar G' with $\mathsf{L}(G') = \mathsf{L}(G) \setminus \{\varepsilon\}$.*

Step 2: Partitioned Grammars. Our second step is based on the following observation. Roughly speaking, in an interval grammar, in order to generate an unbounded number of a_i's, there have to be derivation trees that contain either (i) an unbounded number of incomparable (with respect to the subtree ordering) a_i-subtrees (i.e. subtrees with yield in a_i^*) or (ii) a bounded number of such subtrees that themselves have arbitrarily large yields. In a partitioned grammar, we designate for each a_i, whether we allow arbitrarily many a_i-subtrees (each of which then only contains a single a_i) or we allow exactly one a_i-subtree (which is then permitted to be arbitrarily large).

Let us formalize this. A nonterminal A in an interval grammar is called *unary* if $\iota(A) = (i, i)$ for some $1 \leq i \leq n$. A *partitioned* grammar is an interval grammar $G = (N, T, I, P, S)$, with interval map $\iota\colon N \to \mathbb{N} \times \mathbb{N}$, together with a subset $D \subseteq T$ of *direct symbols* such that for each $a_i \in T$, the following holds: (i) If $a_i \in D$, then there is no $A \in N$ with $\iota(A) = (i, i)$, and (ii) if $a_i \notin D$ and t is a derivation tree of G, then all occurrences of a_i are contained in a single subtree whose root contains a unary nonterminal. In other words, direct symbols are never produced through unary nonterminals, but always directly through non-unary ones. If, on the other hand, a_i is not direct, then all occurrences of a_i stem from one occurrence of a suitable unary symbol. The next proposition clearly reduces the SUP for indexed grammars to the case of partitioned grammars.

Proposition 2. *Let G be a productive interval grammar with $\mathsf{L}(G) \subseteq a_1^* \cdots a_n^*$.* *Then, one can construct partitioned grammars $G_1, \ldots, G_m$ such that we have* $\mathsf{L}(G)\!\downarrow = a_1^* \cdots a_n^*$ *if and only if $\mathsf{L}(G_i)\!\downarrow = a_1^* \cdots a_n^*$ for some $1 \leq i \leq m$.*

Step 3: Breadth-bounded Grammars. The last step in our proof will be to solve the SUP in the case where we have a bound on the number of nonterminals in reachable sentential forms. The only obstacle to such a bound are the unary nonterminals corresponding to terminals $a_i \notin D$: All other nonterminals have $\iota(A) = (i, j)$ with $i < j$ and there can be at most $n - 1$ such symbols in a sentential form. However, for each $a_i \notin D$, there is at most one subtree with a corresponding unary nonterminal at its root. Our strategy is therefore to replace these problematic subtrees so as to bound the nonterminals: Instead of unfolding the subtree generated from $u \in NI^*$, we apply a transducer to u.

In order to guarantee that the replacement does not affect whether $\mathsf{L}(G)\!\downarrow$ equals $a_1^* \cdots a_n^*$, we employ a slight variant[2] of the equivalence that gives rise to the *cost functions* of Colcombet [6]. If $f \colon X \to \mathbb{N} \cup \{\infty\}$ is a partial function, we say that f is *unbounded on* $E \subseteq X$ if for each $k \in \mathbb{N}$, there is some $x \in E$ with $f(x) \geq k$ (in particular, $f(x)$ is defined). If $g \colon X \to \mathbb{N} \cup \{\infty\}$ is another partial function, we write $f \approx g$ if for each subset $E \subseteq X$, we have: f is unbounded on E if and only if g is unbounded on E. Note that if $h \colon Y \to X$ is a partial function and $f \approx g$, then $h \circ f \approx h \circ g$. Now, we compare the transducer and the original grammar on the basis of the following partial functions. Given an indexed grammar $G = (N, T, I, P, S)$ and a transducer A with $\mathsf{T}(A) \subseteq NI^* \times T^*$, we define the partial functions $f_G, f_A \colon NI^* \to \mathbb{N} \cup \{\infty\}$ by

$$f_G(u) = \sup\{|v| \mid v \in T^*,\ u \Rightarrow_G^* v\},$$
$$f_A(u) = \sup\{|v| \mid v \in T^*,\ (u, v) \in \mathsf{T}(A)\}.$$

Note that here, $\sup M$ is undefined if M is the empty set.

Proposition 3. *Given an indexed grammar G, one can construct a finite-state transducer A such that $f_A \approx f_G$.*

A simple argument (see the full version [28]) shows that it suffices to prove proposition 3 in the case that G is productive. Hence, we assume productivity of G. The construction of the transducer will involve deciding the *finiteness problem* for indexed languages, which asks, given G, whether $\mathsf{L}(G)$ is finite. Its decidability has been shown by Rounds [25] (and later again by Hayashi [16, Corollary5.1]).

---

[2] The difference is that we have an equivalence on partial instead of total functions.

Theorem 3 ([25]). *The finiteness problem for indexed languages is decidable.*

Let $R = \{Bw \mid B \in N,\ w \in I^*,\ w \in \mathsf{W}_G(B, T^*)\}$. Then f_G is clearly undefined on words outside of R. Therefore, it suffices to exhibit a finite-state transducer A with $f_A|_R \approx f_G$: The regularity of R means we can construct a transducer A' with $f_{A'} = f_A|_R$. In order to prove the relation $f_A|_R \approx f_G$, we employ the concept of shortcut trees.

Shortcut Trees. Note that since G is productive, the label ε does not occur in derivation trees for G. Let t be such a derivation tree. Let us inductively define the set of *shortcut trees* for t. Suppose t's root r has the label $\ell \in NI^* \cup T$. If $\ell \in N \cup T$, then the only shortcut tree for t consists of just one node with label ℓ. If $\ell = Bfv$, $B \in N$, $f \in I$, $v \in I^*$, then the shortcut trees for t are obtained as follows. We choose a set U of nodes in t such that

(i) each path from r to a leaf contains precisely one node in U,
(ii) the label of each $x \in U$ either equals Cv for some $C \in N$ or belongs to T,
(iii) the index word of each node on a path from r to any $x \in U$ has v as a suffix.

For each such choice of $U = \{x_1, \dots, x_n\}$, we take shortcut trees $t_1, \dots, t_n$ for the subtrees of $x_1, \dots, x_n$ and create a new shortcut tree for t by attaching $t_1, \dots, t_n$ to a fresh root node. The root node carries the label B. Note that every shortcut tree for t has height $|\ell| - 1$. We also call these *shortcut trees from* ℓ.

In other words, a shortcut tree is obtained by successively choosing a sentential form such that the topmost index symbol is removed, but the rest of the index is not touched. Note that if $\bar{t}$ is a shortcut tree for a derivation tree t, then $|\mathsf{yield}(\bar{t})| \leq |\mathsf{yield}(t)|$. On the other hand, every derivation tree has a shortcut tree with the same yield. Thus, if we define $\bar{f}_G \colon NI^* \to \mathbb{N} \cup \{\infty\}$ by

$$\bar{f}_G(u) = \sup\{|\mathsf{yield}(\bar{t})| \mid \bar{t} \text{ is a shortcut tree from } u\}$$

then we clearly have $\bar{f}_G \approx f_G$. Therefore, in order to prove $f_A|_R \approx f_G$, it suffices to show $f_A|_R \approx \bar{f}_G$. Let us describe the transducer A. For $B, C \in N$ and $g \in I$, consider the language $L_{B,g,C} = \{w \in (N \cup T)^* \mid Bg \Rightarrow'^*_G w,\ |w|_C \geq 1\}$. Here, $\Rightarrow'_G$ denotes the restricted derivation relation that forbids terminal productions. Then $L_{B,g,C}$ is the set of words $\mathsf{w}(\mathsf{root}(\bar{t}))$ for shortcut trees $\bar{t}$ of derivation trees from Bg (or, equivalently, Bgv with $v \in I^*$) such that C occurs in $\mathsf{w}(\mathsf{root}(\bar{t}))$. Here, $\mathsf{root}(\bar{t})$ denotes the root node of $\bar{t}$ and $\mathsf{w}(\mathsf{root}(\bar{t}))$ is the word consisting of the labels of the root's child nodes. Each $L_{B,g,C}$ belongs to the class of indexed languages, which is a full trio and has a decidable finiteness and emptiness problem. Hence, we can compute the following function, which will describe A's output. Pick an $a \in T$ and define for each $B, C \in N$ and $g \in I$:

$$\mathsf{Out}(B, g, C) = \begin{cases} \{a\}^* & \text{if } L_{B,g,C} \text{ is infinite,} \\ \{a\} & \text{if } L_{B,g,C} \text{ is finite and } L_{B,g,C} \cap (N \cup T)^{\geq 2} \neq \emptyset, \\ \{\varepsilon\} & \text{if } L_{B,g,C} \neq \emptyset \text{ and } L_{B,g,C} \subseteq N \cup T, \\ \emptyset & \text{if } L_{B,g,C} = \emptyset. \end{cases}$$

Note that for each $B, C \in N$ and $g \in I$, precisely one of the conditions on the right holds. The transducer A has states $\{q_0\} \cup N$ and edges $(q_0, B, \{\varepsilon\}, B)$ and $(B, g, \mathsf{Out}(B, g, C), C)$ for each $B, C \in N$ and $g \in I$. A's initial state is q_0 and its final states are all those $B \in N$ with $B \Rightarrow_G^* w$ for some $w \in T^*$. Hence, the runs of A on a word $Bw \in R$ correspond to paths (from root to leaf) in shortcut trees from Bw. Here, the productivity of the words in R guarantees that every run of A with input from R does in fact arise from a shortcut tree in this way.

Suppose A performs a run on input $Bw \in R$, $|w| = k$, and produces the outputs $a^{n_1}, \ldots, a^{n_k}$ in its k steps that read w. Then the definition of $\mathsf{Out}(\cdot, \cdot, \cdot)$ guarantees that there is a shortcut tree $\bar{t}$ such that the run corresponds to a path in which the i-th node has at least $n_i + 1$ children. In particular, $\bar{t}$ has at least $n_1 + \cdots + n_k$ leaves. Therefore, we have $f_A(Bw) \leq \bar{f}_G(Bw)$.

It remains to be shown that if $\bar{f}_G$ is unbounded on $E \subseteq R$, then f_A is unbounded on E. For this, we use a simple combinatorial fact. For a tree t, let $\delta(t)$ denote the maximal number of children of any node and let $\beta(t)$ denote the maximal number of branching nodes (i.e. those with at least two children) on any path from root to leaf.

Lemma 2. *In a set of trees, the number of leaves is unbounded if and only if δ is unbounded or β is unbounded.*

Suppose $\bar{f}_G$ is unbounded on $E \subseteq R$. Then there is a sequence of shortcut trees $t_1, t_2, \ldots$ from words in E such that $|\mathsf{yield}(t_1)|, |\mathsf{yield}(t_2)|, \ldots$ is unbounded. This means δ or β is unbounded on $t_1, t_2, \ldots$. Note that if t is a shortcut tree from $Bw \in R$, then the path in t with $\beta(t)$ branching nodes gives rise to a run of A on Bw that outputs at least $\beta(t)$ symbols. Hence, $f_A(Bw) \geq \beta(t)$. Thus, if β is unbounded on $t_1, t_2, \ldots$, then f_A is unbounded on E.

Suppose δ is unbounded on $t_1, t_2, \ldots$. Let x be an inner node of a shortcut tree $\bar{t}$. Then the subtree of x is also a shortcut tree, say of a derivation tree t from $Bgw \in R$ with $B \in N$, $g \in I$, $w \in I^*$. Moreover, x has a child node with a label $C \in N$ (otherwise, it would be a leaf of $\bar{t}$). We say that (B, g, C) is a *type* of x (note that a node may have multiple types). Since δ is unbounded on $t_1, t_2, \ldots$ and there are only finitely many possible types, we can pick a type (B, g, C) and a subsequence $t'_1, t'_2, \ldots$ such that each t'_k has an inner node x_k with at least k children and type (B, g, C). This means there are nodes of type (B, g, C) with arbitrarily large numbers of children and hence $L_{B,g,C}$ is infinite. We can therefore choose any t'_i and a run of A that corresponds to a path involving x_i. Since $L_{B,g,C}$ is infinite, this run outputs $\{a\}^*$ in the step corresponding to x_i. Moreover, this run reads a word in E and hence f_A is unbounded on E. This proves $f_A|_R \approx \bar{f}_G$ and thus Proposition 3.

A *breadth-bounded grammar* is an indexed grammar, together with a bound $k \in \mathbb{N}$, such that each of its reachable sentential forms contains at most k non-terminals. We can now show the following.

Proposition 4. *Let G be a partitioned grammar with $\mathsf{L}(G) \subseteq a_1^* \cdots a_n^*$. Then, one can construct a breadth-bounded grammar G' with $\mathsf{L}(G') \subseteq a_1^* \cdots a_n^*$ such that $\mathsf{L}(G){\downarrow} = a_1^* \cdots a_n^*$ if and only if $\mathsf{L}(G'){\downarrow} = a_1^* \cdots a_n^*$.*

The proof comprises two steps. First, we build a breadth-bounded grammar that, instead of unfolding the derivation trees below unary nonterminals, outputs their index words as terminal words, which results in a breadth-bounded grammar. Then, we apply our transducer from Proposition 3 to the resulting subwords. Since the breadth-bounded grammars generate a full trio, the proposition follows.

Step 4: Semilinearity. We have thus reduced the SUP for indexed grammars to the special case of breadth-bounded grammars. The last step in our proof is to prove the following. It clearly implies decidability of the SUP.

Proposition 5. *Breadth-bounded grammars exhibit effectively semilinear Parikh images.*

The basic idea of Proposition 5 is to use a decomposition of derivation trees into a bounded number of 'slices', which are edge sequences of either (i) only push and output productions ('positive slice') or (ii) only pop and output productions ('negative slice'). Furthermore, there is a relation between slices such that the index symbols that are pushed in a positive slice are popped precisely in those negative slices related to it. One can then mimic the grammar by simulating each positive slice in lockstep with all its related negative slices. This leads to a 'finite index scattered context grammar'. This type of grammars is well known to guarantee effectively semilinear Parikh images [9].

Acknowledgments. The author would like to thank Sylvain Schmitz, who pointed out to him that Jullien [19] was the first to characterize downward closed languages by SREs.

References

1. Abdulla, P.A., Boasson, L., Bouajjani, A.: Effective lossy queue languages. In: Proc. of ICALP 2001, pp. 639–651 (2001)
2. Abdulla, P.A., Collomb-Annichini, A., Bouajjani, A., Jonsson, B.: Using Forward Reachability Analysis for Verification of Lossy Channel Sys- tems. Form. Method. Syst. Des. **25**(1), 39–65 (2004)
3. Aho, A.V.: Indexed grammars-an extension of context-free grammars. J. ACM **15**(4), 647–671 (1968)
4. Bonnet, R., Finkel, A., Leroux, J., Zeitoun, M.: Model Checking Vector Addition Systems with one zero-test. In: LMCS 8.2:11 (2012)
5. Bouajjani, A., Esparza, J., Maler, O.: Reachability analysis of push- down automata: application to model-checking. In: Proc. of CONCUR 1997, pp. 135–150 (1997)
6. Colcombet, T.: Regular cost functions, Part I: logic and algebra over words. In: LMCS 9.3 (2013)
7. Courcelle, B.: On constructing obstruction sets of words. Bulletin of the EATCS **44**, 178–186 (1991)
8. Czerwiński, W., Martens, W.: A Note on Decidable Separability by Piece- wise Testable Languages (2014). arXiv:1410.1042 [cs.FL]

9. Dassow, J., Păun, G.: Regulated rewriting in formal language theory. Springer-Verlag, Berlin (1989)
10. Dassow, J., Păun, G., Salomaa, A.: Grammars with controlled derivations. In: Rozenberg, G., Salomaa, A. (eds.) Handbook of Formal Languages, vol. 2, pp. 101–154. Springer, Heidelberg (1997)
11. Ehrenfeucht, A., Rozenberg, G., Skyum, S.: A relationship between ET0L and EDT0L languages. Theor. Comput. Sci. **1**(4), 325–330 (1976)
12. Gilman, R.H.: A shrinking lemma for indexed languages. Theor. Comput. Sci. **163**(1-2), 277–281 (1996)
13. Gruber, H., Holzer, M., Kutrib, M.: The size of Higman-Haines sets. Theor. Comput. Sci. **387**(2), 167–176 (2007)
14. Habermehl, P., Meyer, R., Wimmel, H.: The downward-closure of petri net languages. In: Abramsky, S., Gavoille, C., Kirchner, C., Meyer auf der Heide, F., Spirakis, P.G. (eds.) ICALP 2010. LNCS, vol. 6199, pp. 466–477. Springer, Heidelberg (2010)
15. Haines, L.H.: On free monoids partially ordered by embedding. J. Combin. Theory **6**(1), 94–98 (1969)
16. Hayashi, T.: On Derivation Trees of Indexed Grammars-An Extension of the uvwxy-Theorem-. Publications of the Research Institute for Mathematical Sciences **9**(1), 61–92 (1973)
17. Hopcroft, J.E., Ullman, J.D.: Introduction to Automata Theory, Languages and Computation. Addison-Wesley, Reading (1979)
18. Jantzen, M.: On the hierarchy of Petri net languages. RAIRO Theor. Inf. Appl. **13**(1), 19–30 (1979)
19. Jullien, P.: Contribution à létude des types d'ordres dispersés. Université de Marseille, PhD thesis (1969)
20. Kartzow, A.: A pumping lemma for collapsible pushdown graphs of level 2. In: Proc. of CSL 2011, pp. 322–336 (2011)
21. van Leeuwen, J.: Effective constructions in well-partially-ordered free monoids. Discrete Math. **21**(3), 237–252 (1978)
22. Maslov, A.N.: Multilevel stack automata. Problems of Information Transmission **12**(1), 38–42 (1976)
23. Mayr, R.: Undecidable problems in unreliable computations. Theor. Comput. Sci. **297**(1-3), 337–354 (2003)
24. Parys, P.: A pumping lemma for pushdown graphs of any level. In: Proc. of STACS 2012, pp. 54–65 (2012)
25. Rounds, W.C.: Tree-oriented proofs of some theorems on context-free and indexed languages. In: Proc. of STOC 1970, pp. 109–116 (1970)
26. Seki, H., Matsumura, T., Fujii, M., Kasami, T.: On multiple context-free grammars. Theor. Comput. Sci. **88**(2), 191–229 (1991)
27. Smith, T.: On infinite words determined by indexed languages. In: Csuhaj-Varjú, E., Dietzfelbinger, M., Ésik, Z. (eds.) MFCS 2014, Part I. LNCS, vol. 8634, pp. 511–522. Springer, Heidelberg (2014)
28. Zetzsche, G.: An approach to computing downward closures (2015). arXiv:1503.01068 [cs.FL]
29. Zetzsche, G.: Computing downward closures for stacked counter au tomata. In: Proc. of STACS 2015, pp. 743–756 (2015)

How Much Lookahead is Needed to Win Infinite Games?

Felix Klein[(⊠)] and Martin Zimmermann

Reactive Systems Group, Saarland University, Saarbrücken, Germany
{klein,zimmermann}@react.uni-saarland.de

Abstract. Delay games are two-player games of infinite duration in which one player may delay her moves to obtain a lookahead on her opponent's moves. For ω-regular winning conditions it is known that such games can be solved in doubly-exponential time and that doubly-exponential lookahead is sufficient.

We improve upon both results by giving an exponential time algorithm and an exponential upper bound on the necessary lookahead. This is complemented by showing EXPTIME-hardness of the solution problem and tight exponential lower bounds on the lookahead. Both lower bounds already hold for safety conditions. Furthermore, solving delay games with reachability conditions is shown to be PSPACE-complete.

1 Introduction

Many of today's problems in computer science are no longer concerned with programs that transform data and then terminate, but with non-terminating reactive systems which have to interact with a possibly antagonistic environment for an unbounded amount of time. The framework of infinite two-player games is a powerful and flexible tool to verify and synthesize such systems. The seminal theorem of Büchi and Landweber [1] states that the winner of an infinite game on a finite arena with an ω-regular winning condition can be determined and a corresponding finite-state winning strategy can be constructed effectively.

Delay Games. In this work, we consider an extension of the classical framework: in a delay game, one player can postpone her moves for some time to obtain a lookahead on her opponent's moves. This allows her to win some games which she would loose without lookahead, e.g., if her first move depends on the third move of her opponent. Nevertheless, there are winning conditions that cannot be won with any finite lookahead, e.g., if her first move depends on every move of her opponent. Delay arises naturally if transmission of data in networks or components equipped with buffers are modeled.

From a more theoretical point of view, uniformization of relations by continuous functions [14,15] can be expressed and analyzed using delay games.

Partially supported by the DFG projects "TriCS" (ZI 1516/1-1) and "AVACS" (SFB/TR 14). The first author was supported by an IMPRS-CS PhD Scholarship.

M.M. Halldórsson et al. (Eds.): ICALP 2015, Part II, LNCS 9135, pp. 452–463, 2015.
DOI: 10.1007/978-3-662-47666-6_36

We consider games in which two players pick letters from alphabets Σ_I and Σ_O, respectively, thereby producing two infinite sequences α and β. Thus, a strategy for the second player induces a mapping $\tau\colon \Sigma_I^\omega \to \Sigma_O^\omega$. It is winning for her if $(\alpha, \tau(\alpha))$ is contained in the winning condition $L \subseteq \Sigma_I^\omega \times \Sigma_O^\omega$ for every α. If this is the case, we say that τ uniformizes L. In the classical setting, in which the players pick letters in alternation, the n-th letter of $\tau(\alpha)$ depends only on the first n letters of α. A strategy with bounded lookahead, i.e., only finitely many moves are postponed, induces a Lipschitz-continuous function τ (in the Cantor topology on Σ^ω) and a strategy with unbounded lookahead induces a continuous function (equivalently, a uniformly continuous function, as Σ^ω is compact).

Related Work. Hosch and Landweber proved that it is decidable whether a delay game with an ω-regular winning condition can be won with bounded lookahead [8]. Later, Holtmann, Kaiser, and Thomas revisited the problem and showed that if the delaying player wins such a game with unbounded lookahead, then she already wins it with doubly-exponential bounded lookahead, and gave a streamlined decidability proof yielding an algorithm with doubly-exponential running time [7]. Thus, the delaying player does not gain additional power from having unbounded lookahead, bounded lookahead is sufficient.

Going beyond ω-regularity by considering context-free conditions leads to undecidability and non-elementary lower bounds on the necessary lookahead, even for very weak fragments [5]. Nevertheless, there is another extension of the ω-regular conditions where one can prove the analogue of the Hosch-Landweber Theorem: it is decidable whether the delaying player wins a delay game with bounded lookahead, if the winning condition is definable in weak mondadic second order logic with the unbounding quantifier (WMSO+U) [16]. Furthermore, doubly-exponential lookahead is sufficient for such conditions, provided the delaying player wins with bounded lookahead at all. However, bounded lookahead is not always sufficient to win such games, i.e., the analogue of the Holtmann-Kaiser-Thomas Theorem does not hold for WMSO+U conditions. Finally, all delay games with Borel winning conditions are determined [11].

Stated in terms of uniformization, Hosch and Landweber proved decidability of the uniformization problem for ω-regular relations by Lipschitz-continuous functions and Holtmann et al. proved the equivalence of the existence of a continuous uniformization function and the existence of a Lipschitz-continuous uniformization function for ω-regular relations. Furthermore, uniformization of context-free relations is undecidable, even with respect to Lipschitz-continuous functions, but uniformization of WMSO+U relations by Lipschitz-continuous functions is decidable.

Furthermore, Carayol and Löding considered the case of finite words [3], and Löding and Winter [12] considered the case of finite trees, which are both decidable. However, the non-existence of MSO-definable choice functions on the infinite binary tree [2,6] implies that uniformization fails for such trees.

Although several extensions of ω-regular winning conditions for delay games have been considered, many problems remain open even for ω-regular conditions: there are no non-trivial lower bounds on the necessary lookahead and on the complexity of solving such games. Furthermore, only deterministic parity automata were used to specify winning conditions, and the necessary lookahead and the solution complexity is measured in their size. Thus, it is possible that considering weaker automata models like reachability or safety automata leads to smaller lookahead requirements and faster algorithms.

Our Contribution. We answer all these questions and improve upon both results of Holtmann et al. by determining the exact complexity of ω-regular delay games and by giving tight bounds on the necessary lookahead.

First, we present an exponential time algorithm for solving delay games with ω-regular winning conditions, an exponential improvement over the original doubly-exponential time algorithm. Both algorithms share some similarities: given a deterministic parity automaton $\mathcal{A}$ recognizing the winning condition of the game, a parity game is constructed that is won by the delaying player if and only if she wins the delay game with winning condition $L(\mathcal{A})$. Furthermore, both parity games are induced by equivalence relations that capture the behavior of $\mathcal{A}$. However, our parity game is of exponential size while the one of Holtmann et al. is doubly-exponential. Also, they need an intermediate game, the so-called block game, to prove the equivalence of the delay game and the parity game, while our equivalence proof is direct. Thus, our algorithm and its correctness proof are even simpler than the ones of Holtmann et al.

Second, we show that solving delay games is EXPTIME-complete by proving the first non-trivial lower bound on the complexity of ω-regular delay games. The lower bound is proved by a reduction from the acceptance problem for alternating polynomial space Turing machines [4], which results in delay games with safety conditions. Thus, solving delay games with safety conditions is already EXPTIME-hard. Our reduction is inspired by the EXPTIME-hardness proof for continuous simulation games [9], a simulation game on Büchi automata where Duplicator is able to postpone her moves to obtain a lookahead on Spoiler's moves. However, this reduction is from a two-player tiling problem while we directly reduce from alternating Turing machines.

Third, we determine the exact amount of lookahead necessary to win delay games with ω-regular conditions. From our algorithm we derive an exponential upper bound, which is again an exponential improvement. This upper bound is complemented by the first non-trivial lower bound on the necessary lookahead: there are reachability and safety conditions that are winning for the delaying player, but only with exponential lookahead, i.e., our upper bound is tight.

Fourth, we present the first results for fragments of ω-regular conditions. As already mentioned above, our lower bounds on complexity and necessary lookahead already hold for safety conditions, i.e., safety is already as hard as parity. Thus, the complexity of the problems manifests itself in the transition structure of the automaton, not in the acceptance condition. For reachability conditions, the situation is different: we show that solving delay games with reachability conditions is equivalent to universality of non-deterministic reachability automata and therefore PSPACE-complete.

Omitted proofs can be found in the full version [10].

2   Preliminaries

The non-negative integers are denoted by $\mathbb{N}$. An alphabet Σ is a non-empty finite set, Σ^* the set of finite words over Σ, Σ^n the set of words of length n, and Σ^ω the set of infinite words. The empty word is denoted by ε and the length of a finite word w by $|w|$. For $w \in \Sigma^* \cup \Sigma^\omega$ we write $w(n)$ for the n-th letter of w.

Automata. We use automata of the form $\mathcal{A} = (Q, \Sigma, q_I, \Delta, \varphi)$ where $\Delta: Q \times \Sigma \to 2^Q \setminus \{\emptyset\}$ is a a non-deterministic transition function and where the acceptance condition φ is either a set $F \subseteq Q$ of accepting states or a coloring $\Omega: Q \to \mathbb{N}$. An automaton is deterministic, if $|\Delta(q, a)| = 1$ for every q and a. In this case, we denote Δ by a function $\delta: Q \times \Sigma \to Q$. A state q of $\mathcal{A}$ is a sink, if $\Delta(q, a) = \{q\}$ for every $a \in \Sigma$. Finite and infinite runs are defined as usual. Given an automaton $\mathcal{A}$ over Σ with some set F of accepting states or with some coloring Ω, we consider the following acceptance modes:

Finite: $L_*(\mathcal{A}) \subseteq \Sigma^*$ denotes the set of finite words accepted by $\mathcal{A}$, i.e., the set of words that have a run ending in F.

Reachability: $L_\exists(\mathcal{A}) \subseteq \Sigma^\omega$ denotes the set of infinite words that have a run visiting an accepting state at least once. We have $L_\exists(\mathcal{A}) = L_*(\mathcal{A}) \cdot \Sigma^\omega$.

Safety: Dually, $L_\forall(\mathcal{A}) \subseteq \Sigma^\omega$ denotes the set of infinite words that have a run only visiting accepting states.

Parity: $L_p(\mathcal{A}) \subseteq \Sigma^\omega$ denotes the set of infinite words that have a run such that the maximal color visited infinitely often during this run is even.

Note that we require automata to be complete. For safety and parity acceptance this is no restriction, since we can always add a fresh rejecting sink and lead all missing transitions to this sink. However, incomplete automata with reachability acceptance are strictly stronger than complete ones, as incompleteness can be used to check safety properties. We impose this restriction since we are interested in pure reachability conditions.

Given a language $L \subseteq (\Sigma_I \times \Sigma_O)^\omega$ we denote by $\mathrm{pr}_I(L)$ its projection to the first component. Similarly, given an automaton $\mathcal{A}$ over $\Sigma_I \times \Sigma_O$, we denote by $\mathrm{pr}_I(\mathcal{A})$ the automaton obtained by projecting each letter to its first component.

Remark 1. Let $\mathrm{acc} \in \{*, \exists, \forall, p\}$, then $\mathrm{pr}_I(L_{\mathrm{acc}}(\mathcal{A})) = L_{\mathrm{acc}}(\mathrm{pr}_I(\mathcal{A}))$.

Games with Delay. A delay function is a mapping $f: \mathbb{N} \to \mathbb{N} \setminus \{0\}$, which is said to be constant, if $f(i) = 1$ for every $i > 0$. Given an ω-language $L \subseteq (\Sigma_I \times \Sigma_O)^\omega$ and a delay function f, the game $\Gamma_f(L)$ is played by two players, the input player "Player I" and the output player "Player O" in rounds $i = 0, 1, 2, \ldots$ as follows: in round i, Player I picks a word $u_i \in \Sigma_I^{f(i)}$, then Player O picks one letter $v_i \in \Sigma_O$. We refer to the sequence $(u_0, v_0), (u_1, v_1), (u_2, v_2), \ldots$ as a play of $\Gamma_f(L)$, which yields two infinite words $\alpha = u_0 u_1 u_2 \cdots$ and $\beta = v_0 v_1 v_2 \cdots$. Player O wins the play if the outcome $\binom{\alpha(0)}{\beta(0)}\binom{\alpha(1)}{\beta(1)}\binom{\alpha(2)}{\beta(2)} \cdots$ is in L, otherwise Player I wins.

Given a delay function f, a strategy for Player I is a mapping $\tau_I\colon \Sigma_O^* \to \Sigma_I^*$ where $|\tau_I(w)| = f(|w|)$, and a strategy for Player O is a mapping $\tau_O\colon \Sigma_I^+ \to \Sigma_O$. Consider a play $(u_0, v_0), (u_1, v_1), (u_2, v_2), \ldots$ of $\Gamma_f(L)$. Such a play is consistent with τ_I, if $u_i = \tau_I(v_0 \cdots v_{i-1})$ for every $i \in \mathbb{N}$. It is consistent with τ_O, if $v_i = \tau_O(u_0 \cdots u_i)$ for every $i \in \mathbb{N}$. A strategy τ for Player $p \in \{I, O\}$ is winning, if every play that is consistent with τ is winning for Player p. We say that a player wins $\Gamma_f(L)$, if she has a winning strategy.

Example 1. Consider L over $\{a, b, c\} \times \{b, c\}$ with $\binom{\alpha(0)}{\beta(0)}\binom{\alpha(1)}{\beta(1)}\binom{\alpha(2)}{\beta(2)} \cdots \in L$, if $\alpha(n) = a$ for every $n \in \mathbb{N}$ or if $\beta(0) = \alpha(n)$, where n is the smallest position with $\alpha(n) \neq a$. Intuitively, Player O wins, if the letter she picks in the first round is equal to the first letter other than a that Player I picks. Also, Player O wins, if there is no such letter. Note that L can be accepted by a safety automaton.

We claim that Player I wins $\Gamma_f(L)$ for every delay function f: Player I picks $a^{f(0)}$ in the first round and assume Player O picks b afterwards (the case where she picks c is dual). Then, Player I picks a word starting with c in the second round. The resulting play is winning for Player I no matter how it is continued. Thus, Player I has a winning strategy in $\Gamma_f(L)$.

Note that if a language L is recognizable by a (deterministic) parity automaton, then $\Gamma_f(L)$ is determined, as a delay game with parity condition can be expressed as an explicit parity game in a countable arena.

Also, note that universality of $\mathrm{pr}_I(L)$ is a necessary condition for Player O to win $\Gamma_f(L)$. Otherwise, Player I could pick a word from $\Sigma_I^\omega \setminus \mathrm{pr}_I(L)$, which is winning for him, no matter how Player O responds.

Proposition 1. *If Player O wins $\Gamma_f(L)$, then $\mathrm{pr}_I(L)$ is universal.*

3   Lower Bounds on the Lookahead

In this section, we prove lower bounds on the necessary lookahead for Player O to win delay games with reachability or safety conditions. Thus, the same bounds hold for more expressive conditions like Büchi, co-Büchi, and parity. They are complemented by an exponential upper bound for parity conditions in the next section. Note that both lower bounds already hold for deterministic automata.

Theorem 1. *For every $n > 1$ there is a language L_n such that*

- $L_n = L_\exists(\mathcal{A}_n)$ *for some deterministic automaton $\mathcal{A}_n$ with $|\mathcal{A}_n| \in \mathcal{O}(n)$,*
- *Player O wins $\Gamma_f(L_n)$ for some constant delay function f, but*
- *Player I wins $\Gamma_f(L_n)$ for every delay function f with $f(0) \leq 2^n$.*

Proof. Let $\Sigma_I = \Sigma_O = \{1, \ldots, n\}$. We say that w in Σ_I^* contains a bad j-pair, for $j \in \Sigma_I$, if there are two occurrences of j in w such that no $j' > j$ occurs in between. The automaton $\mathcal{B}_j$, depicted in Figure 1(a), accepts exactly the words with a bad j-pair. Now, consider the language L over Σ_I defined by

$$L = \bigcap_{1 \leq j \leq n} \{w \in \Sigma_I^* \mid w \text{ contains no bad } j\text{-pair}\}.$$

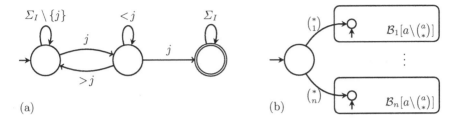

Fig. 1. (a) Automaton $\mathcal{B}_j$ for $j \in \Sigma_I$. (b) Construction of $\mathcal{A}_n$.

Straightforward inductions show that every $w \in L$ satisfies $|w| < 2^n$ and that there is a word $w_n \in L$ with $|w_n| = 2^n - 1$.

The winning condition L_n is defined as follows: $\binom{\alpha(0)}{\beta(0)}\binom{\alpha(1)}{\beta(1)}\binom{\alpha(2)}{\beta(2)} \cdots$ is in L_n if $\alpha(1)\alpha(2) \cdots$ contains a bad $\beta(0)$-pair, i.e., with her first move, Player O has to pick a j such that Player I has produced a bad j-pair. For technical reasons, the first letter picked by Player I is ignored. The construction of an automaton $\mathcal{A}_n$ recognizing L_n is sketched in Figure 1(b). Here, $\mathcal{B}_j[a \backslash \binom{a}{*})]$ denotes $\mathcal{B}_j$, where for each $a \in \Sigma_I$ every transition labeled by a is replaced by transitions labeled by $\binom{a}{b}$ for every $b \in \Sigma_O$. Clearly, we have $\mathcal{A}_n \in \mathcal{O}(n)$.

Player O wins $\Gamma_f(L_n)$ for every delay function with $f(0) > 2^n$. In the first round, Player I has to pick a word u_0 such that u_0 without its first letter is not in L. This allows Player O to find a bad j-pair for some j, i.e., she wins the play no matter how it is continued.

However, for f with $f(0) \le 2^n$, Player I has a winning strategy by picking the prefix of $1w_n$ of length $f(0)$ in the first round. Player O has to answer with some $j \in \Sigma_O$. In this situation, Player I can continue by finishing w_n and then playing some $j' \neq j$ ad infinitum, which ensures that the resulting sequence does not contain a bad j-pair. Thus, the play is winning for Player I. □

Using a similar construction, one can show exponential lower bounds for safety conditions as well.

Theorem 2. *For every $n > 1$ there is a language L'_n such that*

- *$L'_n = L_\forall(\mathcal{A}'_n)$ for some deterministic automaton $\mathcal{A}'_n$ with $|\mathcal{A}'_n| \in \mathcal{O}(n)$,*
- *Player O wins $\Gamma_f(L'_n)$ for some constant delay function f, but*
- *Player I wins $\Gamma_f(L'_n)$ for every delay function f with $f(0) \le 2^n$.*

The aforementioned constructions also work for constant-size alphabets, if we encode every $j \in \{1, \ldots, n\}$ in binary, resulting in automata $\mathcal{A}_n$ and $\mathcal{A}'_n$ whose sizes are in $\mathcal{O}(n \log n)$. It is open whether linear-sized automata and a constant-sized alphabet can be achieved simultaneously.

4 Computational Complexity of Delay Games

In this section, we determine the computational complexity of solving delay games. First, we consider the special case of reachability conditions and prove such games to be PSPACE-complete. Then, we show that games with safety conditions are EXPTIME-hard. The latter bound is complemented by an EXPTIME-algorithm for solving delay games with parity conditions. From this algorithm, we also deduce an exponential upper bound on the necessary lookahead for Player O, which matches the lower bounds given in the previous section.

4.1   Reachability Conditions

Recall that universality of the projection to the first component of the winning condition is a necessary condition for Player O for having a winning strategy in a delay game. Our first result in this section states that universality is also sufficient in the case of reachability conditions. Thus, solving delay games with reachability conditions is equivalent, via linear time reductions, to the universality problem for non-deterministic reachability automata, which is PSPACE-complete. Also, our proof yields an exponential upper bound on the necessary lookahead.

Theorem 3. *Let* $L = L_\exists(\mathcal{A})$, *where* $\mathcal{A}$ *is a non-deterministic reachability automaton. The following are equivalent:*

1. *Player* O *wins* $\Gamma_f(L)$ *for some delay function* f.
2. *Player* O *wins* $\Gamma_f(L)$ *for some constant delay function* f *with* $f(0) \leq 2^{|\mathcal{A}|}$.
3. $\mathrm{pr}_I(L)$ *is universal.*

Proof. To show the equivalence, it only remains to prove 3. $\Rightarrow$ 2.

We assume w.l.o.g. that the accepting states of $\mathcal{A}$ are sinks, which implies that $L_*(\mathrm{pr}_I(\mathcal{A}))$ is suffix-closed, i.e., $w \in L_*(\mathrm{pr}_I(\mathcal{A}))$ implies $ww' \in L_*(\mathrm{pr}_I(\mathcal{A}))$ for every $w' \in \Sigma_I^*$. Furthermore, let $\mathcal{A}^c$ be an automaton recognizing the complement of $L_*(\mathrm{pr}_I(\mathcal{A}))$, which is prefix-closed, as it is the complement of a suffix-closed language. We can choose $\mathcal{A}^c$ such that $|\mathcal{A}^c| \leq 2^{|\mathcal{A}|}$.

We claim that $L_*(\mathcal{A}^c)$ is finite. Assume it is infinite. Then, by König's Lemma there is an infinite word α whose prefixes are all in $L_*(\mathcal{A}^c)$. Due to universality, we have $\alpha \in L_\exists(\mathrm{pr}_I(\mathcal{A}))$, i.e., there is a prefix of α in $L_*(\mathrm{pr}_I(\mathcal{A}))$. Thus, the prefix is in $L_*(\mathrm{pr}_I(\mathcal{A}))$ and in the complement $L_*(\mathcal{A}^c)$ yielding the desired contradiction. An automaton with n states with a finite language accepts words of length at most $n - 1$. Thus, $w \in L_*(\mathrm{pr}_I(\mathcal{A}))$ for every $w \in \Sigma_I^*$ with $|w| \geq 2^{|\mathcal{A}|}$.

Using this, we show that Player O wins $\Gamma_f(L)$ if $f(0) = 2^{|\mathcal{A}|}$. Player I has to pick $f(0)$ letters with his first move, say $u_0 = \alpha(0) \cdots \alpha(f(0)-1)$. As $f(0)$ is large enough, we have $u_0 \in L_*(\mathrm{pr}_I(\mathcal{A}))$. Hence, there is a word $\beta(0) \cdots \beta(f(0) - 1) \in \Sigma_O^*$ such that $\binom{\alpha(0)}{\beta(0)} \cdots \binom{\alpha(f(0)-1)}{\beta(f(0)-1)} \in L_*(\mathcal{A})$. By picking $\beta(0), \ldots, \beta(f(0) - 1)$ in the first $f(0)$ rounds, Player O wins the play, no matter how it is continued. Hence, she has a winning strategy. $\square$

As universality for non-deterministic reachability automata is PSPACE-complete (see, e.g., [10]), we obtain the following consequence of Theorem 3.

Corollary 1. *The following problem is* PSPACE-*complete: Given a non-deterministic reachability automaton* $\mathcal{A}$, *does Player* O *win* $\Gamma_f(L_\exists(\mathcal{A}))$ *for some* f?

The upper bounds on complexity and necessary lookahead hold for non-deterministic automata while the lower bounds hold for deterministic ones [10].

4.2   Safety Conditions

Unsurprisingly, Example 1 shows that Theorem 3 does not hold for safety conditions: the projection $\mathrm{pr}_I(L)$ is universal, but Player O has no winning strategy for any delay function. It turns out that safety conditions are even harder than reachability conditions (unless PSpace equals ExpTime).

Theorem 4. *The following problem is* ExpTime-*hard: Given a deterministic safety automaton $\mathcal{A}$, does Player O win $\Gamma_f(L_\forall(\mathcal{A}))$ for some f?*

The proof proceeds by a reduction from the non-acceptance problem for alternating polynomial space Turing machines, which is sufficient due to APSpace = ExpTime [4] being closed under complement. Fix such a machine $\mathcal{M}$, an input x, and a polynomial p that bounds the space consumption of $\mathcal{M}$. We construct a safety automaton $\mathcal{A}$ of polynomial size in $|\mathcal{M}| + p(|x|)$ such that $\mathcal{M}$ rejects x if and only if Player O wins $\Gamma_f(L_\forall(\mathcal{A}))$ for some f. To this end, we give Player I control over the existential states while Player O controls the universal ones. Additionally, Player I is in charge of producing all configurations with his moves. He can copy configurations in order to wait for Player O's decisions of successors for universal transitions, which are delayed due to the lookahead.

More formally, the input alphabet Σ_I consists of the alphabet and the set of states of $\mathcal{M}$ and of two separators N and C while the output alphabet Σ_O consists of the transition relation of $\mathcal{M}$ and of two signals ✗ and ✓. Intuitively, Player I produces configurations of $\mathcal{M}$ of length $p(|x|)$ preceded by either C or N to denote whether the configuration is a *copy* of the previous one or a *new* one. Copying configurations is necessary to bridge the lookahead while waiting for Player O to determine the transition that is applied to a universal configuration. Player I could copy a configuration ad infinitum, but this will be losing for him, unless it is an accepting one. Player O chooses universal transitions at every separator[1] N by picking a transition of $\mathcal{M}$. At every other position, she has to pick a signal: ✗ allows her to claim an error in the configurations picked by Player O while ✓ means that she does not claim an error at the current position.

The automaton $\mathcal{A}$ checks that Player I produces only legal configurations, that he starts with the initial one, that Player O always picks a transition at the separators, and that the first error claimed by Player O is indeed an error. If it is one, then $\mathcal{A}$ goes to an accepting sink, otherwise to a rejecting sink. Finally, if Player I produces an accepting configuration without Player O correctly claiming an error in a preceding configuration, then $\mathcal{A}$ goes to a rejecting sink.

These properties can be checked by a deterministic safety automaton of polynomial size, as the configuations of $\mathcal{M}$ are of polynomial size. A detailed description of $\mathcal{A}$ and a proof that $\mathcal{M}$ rejects x if and only if Player O wins $\Gamma_f(L_\forall(\mathcal{A}))$ for some f can be found in the full version [10].

It is noteworthy that the ExpTime lower bound does not require the full exponential lookahead that might be necessary to win delay games with safety

---

[1] If the following configuration is existential or the separator is a C, then her choice is ignored.

conditions: Player O wins the game constructed above with constant lookahead that is smaller than $|\mathcal{A}|$, if she wins at all.

4.3   Parity Conditions

Now, we complement the ExpTime lower bound shown in the previous subsection with an exponential time algorithm for solving delay games with parity conditions. Thus, delay games with safety or parity conditions are ExpTime-complete. Also, we derive an exponential upper bound on the necessary lookahead from the algorithm. All results only hold for deterministic automata.

Theorem 5. *The following problem is in* ExpTime: *Given a deterministic parity automaton $\mathcal{A}$, does Player O win $\Gamma_f(L_p(\mathcal{A}))$ for some delay function f?*

We begin by constructing an exponentially-sized, delay-free parity game with the same number of colors as $\mathcal{A}$, which is won by Player O if and only if she wins $\Gamma_f(L_p(\mathcal{A}))$ for some delay function f.

Let $\mathcal{A} = (Q, \Sigma_I \times \Sigma_O, q_I, \delta, \Omega)$ with $\Omega\colon Q \to \mathbb{N}$. First, we adapt $\mathcal{A}$ to keep track of the maximal color visited during a run. To this end, we define the automaton $\mathcal{C} = (Q_\mathcal{C}, \Sigma_I \times \Sigma_O, q_I^\mathcal{C}, \delta_\mathcal{C}, \Omega_\mathcal{C})$ where $Q_\mathcal{C} = Q \times \Omega(Q)$, $q_I^\mathcal{C} = (q_I, \Omega(q_I))$,

$$\delta_\mathcal{C}((q,c), a) = (\delta(q,a), \max\{c, \Omega(\delta(q,a))\}),$$

and $\Omega_\mathcal{C}(q,c) = c$. We denote the size of $\mathcal{C}$ by n. Note that $\mathcal{C}$ does not recognize $L_p(\mathcal{A})$. However, we are only interested in runs on finite play infixes.

Remark 2. Let $w \in (\Sigma_I \times \Sigma_O)^*$ and let $(q_0, c_0)(q_1, c_1) \cdots (q_{|w|}, c_{|w|})$ be the run of $\mathcal{C}$ on w from some state $(q_0, c_0) \in \{(q, \Omega(q)) \mid q \in Q\}$. Then, $q_0 q_1 \cdots q_{|w|}$ is the run of $\mathcal{A}$ on w starting in q_0 and $c_{|w|} = \max\{\Omega(q_j) \mid 0 \le j \le |w|\}$.

In the following, we work with partial functions from $Q_\mathcal{C}$ to $2^{Q_\mathcal{C}}$, where we denote the domain of such a function r by $\mathrm{dom}(r)$. Intuitively, we use r to capture the information encoded in the lookahead provided by Player I. Assume Player I has picked $\alpha(0) \cdots \alpha(j)$ and Player O has picked $\beta(0) \cdots \beta(i)$ for $i < j$ such that the lookahead is $w = \alpha(i+1) \cdots \alpha(j)$. Then, we can determine the state q that $\mathcal{C}$ reaches after processing $\binom{\alpha(0)}{\beta(0)} \cdots \binom{\alpha(i)}{\beta(i)}$, but the automaton cannot process w, since Player O has not yet picked $\beta(i+1) \cdots \beta(j)$. However, we can determine the states Player O can enforce by picking an appropriate completion, which will be the ones contained in $r(q)$. Note that the function r depends on the lookahead w picked by Player I.

To formalize the functions capturing the lookahead picked by Player I, we define $\delta_\mathcal{P}\colon 2^{Q_\mathcal{C}} \times \Sigma_I \to 2^{Q_\mathcal{C}}$ via $\delta_\mathcal{P}(S, a) = \bigcup_{q \in S} \bigcup_{b \in \Sigma_O} \delta_\mathcal{C}(q, \binom{a}{b})$, i.e., $\delta_\mathcal{P}$ is the transition function of the powerset automaton of $\mathrm{pr}_I(\mathcal{C})$. As usual, we extend $\delta_\mathcal{P}$ to $\delta_\mathcal{P}^*\colon 2^{Q_\mathcal{C}} \times \Sigma_I^* \to 2^{Q_\mathcal{C}}$ via $\delta_\mathcal{P}^*(S, \varepsilon) = S$ and $\delta_\mathcal{P}^*(S, wa) = \delta_\mathcal{P}(\delta_\mathcal{P}^*(S, w), a)$.

Let $D \subseteq Q_\mathcal{C}$ be non-empty and let $w \in \Sigma_I^*$. We define the function r_w^D with domain D as follows: for every $(q, c) \in D$, we have

$$r_w^D(q, c) = \delta_\mathcal{P}^*(\{(q, \Omega(q))\}, w).$$

Note that we apply $\delta_{\mathcal{P}}$ to $\{(q, \Omega(q))\}$, i.e., the second argument is the color of q and not the color c from the argument to r_w^D. If $(q', c') \in r_w^D(q, c)$, then there is a word w' whose projection is w and such that the run of $\mathcal{A}$ on w' leads from q to q' and has maximal color c'. Thus, if Player I has picked the lookahead w, then Player O could pick an answer such that the combined word leads $\mathcal{A}$ from q to q' with minimal color c'.

We call w a witness for a partial function $r \colon Q_{\mathcal{C}} \to 2^{Q_{\mathcal{C}}}$, if we have $r = r_w^{\mathrm{dom}(r)}$. Thus, we obtain a language $W_r \subseteq \Sigma_I^*$ of witnesses for each such function r. We define $\mathfrak{R} = \{r \mid \mathrm{dom}(r) \neq \emptyset \text{ and } W_r \text{ is infinite}\}$.

Remark 3. Let $\mathfrak{R}$ be defined as above.

1. Let $r \in \mathfrak{R}$. Then, $r(q) \neq \emptyset$ for every $q \in \mathrm{dom}(r)$.
2. Let r be a partial function from $Q_{\mathcal{C}}$ to $2^{Q_{\mathcal{C}}}$. Then, W_r is recognized by a deterministic finite automaton with 2^{n^2} states.
3. Let $D \subseteq Q_{\mathcal{C}}$ be non-empty and let w be such that $|w| \geq 2^{n^2}$. Then, there exists some $r \in \mathfrak{R}$ with $\mathrm{dom}(r) = D$ and $w \in W_r$.

Now, we can define an abstract game $\mathcal{G}(\mathcal{A})$ which is played between Player I and Player O in rounds $i = 0, 1, 2, \ldots$ as follows: in each round, Player I picks a function from $\mathfrak{R}$ and Player O answers by a state of $\mathcal{C}$ subject to the following constraints. In the first round, Player I has to pick $r_0 \in \mathfrak{R}$ such that $\mathrm{dom}(r_0) = \{q_I^{\mathcal{C}}\}$ (C1) and Player O has to answer by picking a state $q_0 \in \mathrm{dom}(r_0)$, which implies $q_0 = q_I^{\mathcal{C}}$. Now, consider round $i > 0$: Player I has picked functions $r_0, r_1, \ldots, r_{i-1}$ and Player O has picked states $q_0, q_1, \ldots, q_{i-1}$. Now, Player I has to pick a function $r_i \in \mathfrak{R}$ such that $\mathrm{dom}(r_i) = r_{i-1}(q_{i-1})$ (C2). Then, Player O picks some state $q_i \in \mathrm{dom}(r_i)$.

Both players can always move: Player I can move, as $r_{i-1}(q_{i-1})$ is always non-empty (Remark 3.1) and thus the domain of some $r \in \mathfrak{R}$ (Remark 3.3), and Player O can move, as the domain of every $r \in \mathfrak{R}$ is non-empty by construction. The resulting play of $\mathcal{G}(\mathcal{A})$ is the sequence $r_0 q_0 r_1 q_1 r_2 q_2 \cdots$. It is won by Player O if the maximal color occurring infinitely often in $\Omega_{\mathcal{C}}(q_0) \Omega_{\mathcal{C}}(q_1) \Omega_{\mathcal{C}}(q_2) \cdots$ is even. Otherwise, Player I wins.

A strategy for Player I is a function τ_I' mapping the empty play prefix to a function r_0 satisfying (C1) and mapping a non-empty prefix $r_0 q_0 \cdots r_{i-1} q_{i-1}$ to a function r_i satisfying (C2). A strategy for Player O maps a play prefix $r_0 q_0 \cdots r_i$ to a state $q_i \in \mathrm{dom}(r_i)$. A play $r_0 q_0 r_1 q_1 r_2 q_2 \cdots$ is consistent with τ_I', if $r_i = \tau_I'(r_0 q_0 \cdots r_{i-1} q_{i-1})$ for every $i \in \mathbb{N}$ and it is consistent with τ_O', if $q_i = \tau_O'(r_0 q_0 \cdots r_i)$ for every $i \in \mathbb{N}$. A strategy τ' for Player $p \in \{I, O\}$ is winning, if every play that is consistent with τ' is winning for Player p. As usual, we say that a player wins $\mathcal{G}(\mathcal{A})$, if she has a winning strategy.

Lemma 1. *Player O wins $\Gamma_f(L_p(\mathcal{A}))$ for some delay function f if and only if Player O wins $\mathcal{G}(\mathcal{A})$.*

Now, we can prove Theorem 5. Due to Lemma 1, we just have to show that we can construct and solve an explicit version of $\mathcal{G}(\mathcal{A})$ in exponential time.

Proof. First, we argue that $\mathfrak{R}$ can be constructed in exponential time: to this end, one constructs for every partial function r from Q_C to 2^{Q_C} the automaton of Remark 3.2 recognizing W_r and tests it for recognizing an infinite language. There are exponentially many functions and each automaton is of exponential size, which yields the desired result. To conclude, we encode $\mathcal{G}(\mathcal{A})$ as a graph-based parity game of exponential size with the same number of colors as $\mathcal{A}$. Such a game can be solved in exponential time in the size of $\mathcal{A}$ [13].   □

The proof of Lemma 1 yields the exponential upper bound $2^{(|\mathcal{A}|k)^2+1}$ on the necessary lookahead, where k is the number of colors of $\mathcal{A}$. However, this can be improved by using a direct pumping argument.

Theorem 6. *Let* $L = L_p(\mathcal{A})$ *where* $\mathcal{A}$ *is a deterministic parity automaton with* k *colors. The following are equivalent:*

1. *Player O wins* $\Gamma_f(L)$ *for some delay function* f.
2. *Player O wins* $\Gamma_f(L)$ *for some constant delay function* f *with* $f(0) \le 2^{2|\mathcal{A}|k+2} + 2$.

5   Conclusion

We gave the first algorithm that solves ω-regular delay games in exponential time, which is an exponential improvement over the previously known algorithms. We complemented this by showing the problem to be EXPTIME-complete, even for safety conditions. Also, we determined the exact amount of lookahead that is necessary to win ω-regular delay games by proving tight exponential bounds, which already hold for safety and reachability conditions. Finally, we showed solving games with reachability conditions to be PSPACE-complete. To the best of our knowledge, all lower bounds are the first non-trivial ones for delay games.

Our lower bounds already hold for deterministic automata while our upper bounds (but the ones for reachability) only hold for deterministic automata. One can obviously obtain upper bounds for non-deterministic automata via determinization, but this incurs an exponential blowup, which might not be optimal. We leave the study of this problem for future work. Another open question concerns the influence of using different deterministic automata models that recognize the class of ω-regular conditions, e.g., Rabin, Streett, and Muller automata, on the necessary lookahead and the solution complexity, again measured in the size of the automata. Indeed, our construction used to prove Theorem 5 can be adapted to deal with these acceptance conditions, e.g., for conditions given by Muller automata, $\mathcal{C}$ keeps track of the vertices visited on a run and $\mathcal{G}(\mathcal{A})$ is a Muller game. This yields upper bounds, but it is open whether these are optimal.

Finally, we also considered winning conditions that are both reachability and safety conditions. Here, polynomial lookahead suffices and the problem is in Π_2^P, i.e., in the second level of the polynomial hierarchy. Again, both results only hold for deterministic automata and are presented in the full version [10]. In future work, we aim to find lower bounds.

Acknowledgments. We thank Bernd Finkbeiner for a fruitful discussion that lead to Theorem 1 and Theorem 3.

References

1. Büchi, J.R., Landweber, L.H.: Solving sequential conditions by finite-state strategies. Trans. Amer. Math. Soc. **138**, 295–311 (1969)
2. Carayol, A., Löding, C.: MSO on the infinite binary tree: choice and order. In: Duparc, J., Henzinger, T.A. (eds.) CSL 2007. LNCS, vol. 4646, pp. 161–176. Springer, Heidelberg (2007)
3. Carayol, A., Löding, C.: Uniformization in automata theory. In: Schroeder-Heister, P., Heinzmann, G., Hodges, W., Bour, P.E. (eds.) CLMPS. College Publications, London (2012) (to appear)
4. Chandra, A.K., Kozen, D., Stockmeyer, L.J.: Alternation. J. ACM **28**(1), 114–133 (1981)
5. Fridman, W., Löding, C., Zimmermann, M.: Degrees of lookahead in context-free infinite games. In: Bezem, M. (ed.) CSL 2011. LIPIcs, vol. 12, pp. 264–276. Schloss Dagstuhl - Leibniz-Zentrum für Informatik (2011)
6. Gurevich, Y., Shelah, S.: Rabin's uniformization problem. The Journal of Symbolic Logic **48**, 1105–1119 (1983)
7. Holtmann, M., Kaiser, L., Thomas, W.: Degrees of lookahead in regular infinite games. LMCS **8**(3) (2012)
8. Hosch, F.A., Landweber, L.H.: Finite delay solutions for sequential conditions. In: ICALP 1972, pp. 45–60 (1972)
9. Hutagalung, M., Lange, M., Lozes, É.: Buffered simulation games for Büchi automata. In: Ésik, Z., Fülöp, Z. (eds.) AFL 2014. EPTCS, vol. 151, pp. 286–300 (2014)
10. Klein, F., Zimmermann, M.: How much lookahead is needed to win infinite games? (2014). arXiv: 1412.3701
11. Klein, F., Zimmermann, M.: What are strategies in delay games? Borel determinacy for games with lookahead (2015). arXiv: 1504.02627
12. Löding, C., Winter, S.: Synthesis of deterministic top-down tree transducers from automatic tree relations. In: Peron, A., Piazza, C. (eds.) GandALF 2014. EPTCS, vol. 161, pp. 88–101 (2014)
13. Schewe, S.: Solving parity games in big steps. In: Arvind, V., Prasad, S. (eds.) FSTTCS 2007. LNCS, vol. 4855, pp. 449–460. Springer, Heidelberg (2007)
14. Thomas, W., Lescow, H.: Logical specifications of infinite computations. In: de Bakker, J.W., de Roever, W.-P., Rozenberg, G. (eds.) REX 1993. LNCS, vol. 803, pp. 583–621. Springer, Heidelberg (1994)
15. Trakhtenbrot, B., Barzdin, I.: Finite Automata; Behavior and Synthesis. Fundamental Studies in Computer Science, vol. 1. North-Holland Publishing Company, New York. American Elsevier (1973)
16. Zimmermann, M.: Delay games with WMSO+U winning conditions. In: CSR 2015 (2015, to appear). arXiv: 1412.3978

Track C: Foundations of Networked Computation: Models, Algorithms and Information Management

Symmetric Graph Properties
Have Independent Edges

Dimitris Achlioptas[1]($\boxtimes$) and Paris Siminelakis[2]

[1] Department of Computer Science, University of California, Santa Cruz, USA
optas@cs.ucsc.edu
[2] Department of Electrical Engineering, Stanford University, Stanford, USA
psimin@stanford.edu

Abstract. In the study of random structures we often face a trade-off between realism and tractability, the latter typically enabled by independence assumptions. In this work we initiate an effort to bridge this gap by developing tools that allow us to work with independence without assuming it. Let $\mathcal{G}_n$ be the set of all graphs on n vertices and let S be an arbitrary subset of $\mathcal{G}_n$, e.g., the set of all graphs with m edges. The study of random networks can be seen as the study of properties that are true for *most* elements of S, i.e., that are true with high probability for a uniformly random element of S. With this in mind, we pursue the following question: *What are general sufficient conditions for the uniform measure on a set of graphs $S \subseteq \mathcal{G}_n$ to be well-approximable by a product measure on the set of all possible edges?*

1 Introduction

Since their introduction in 1959 by Erdős and Rényi [6] and Gilbert [8], respectively, $G(n, m)$ and $G(n, p)$ random graphs have dominated the mathematical study of random networks [2,10]. Given n vertices, $G(n, m)$ selects uniformly among all graphs with m edges, whereas $G(n, p)$ includes each edge independently with probability p. A refinement of $G(n, m)$ are graphs chosen uniformly among all graphs with a given degree sequence, a distribution made tractable by the configuration model of Bollobás [2]. Due to their mathematical tractability these three models have become a cornerstone of Probabilistic Combinatorics and have found application in the Analysis of Algorithms, Coding Theory, Economics, Game Theory, and Statistical Physics.

At the foundation of this mathematical tractability lies symmetry: the probability of all edge sets of a given size is either the same, as in $G(n, p)$ and $G(n, m)$,

---

D. Achlioptas—Research supported by ERC Starting Grant StG-210743 and an Alfred P. Sloan Fellowship. Research co-financed by the European Union (European Social Fund ESF) and Greek national funds through the Operational Program "Education and Lifelong Learning" of the National Strategic Reference Framework (NSRF) - Research Funding Program: ARISTEIA II.
P. Siminelakis—Supported in part by an Onassis Foundation Scholarship.

M.M. Halldórsson et al. (Eds.): ICALP 2015, Part II, LNCS 9135, pp. 467–478, 2015.
DOI: 10.1007/978-3-662-47666-6_37

or merely a function of the potency of the vertices involved, as in the configuration model. This extreme symmetry bestows numerous otherworldly properties, including near-optimal expansion. Perhaps most importantly, it amounts to a complete lack of geometry, as manifest by the fact that the shortest path metric of such graphs suffers maximal distortion when embedded in Euclidean space [14]. In contrast, vertices of real networks are typically embedded in some low-dimensional geometry, either explicit (physical networks), or implicit (social and other latent semantics networks), with distance being a strong factor in determining the probability of edge formation.

While the shortcomings of the classical models have long been recognized, proposing more realistic models is not an easy task. The difficulty lies in achieving a balance between realism and mathematical tractability: it is only too easy to create network models that are both ad hoc and intractable. By now there are thousands of papers proposing different ways to generate graphs with desirable properties [9] the vast majority of which only provide heuristic arguments to support their claims. For a gentle introduction the reader is referred to the book of Newman [17] and for a more mathematical treatment to the books of Chung and Lu [4] and of Durrett [5].

In trying to replicate real networks one approach is to keep adding features, creating increasingly complicated models, in the hope of *matching* observed properties. Ultimately, though, the purpose of any good model is prediction. In that sense, the reason to study (random) graphs with certain properties is to understand what *other* graph properties are (typically) implied by the assumed properties. For instance, the reason we study the uniform measure on graphs with m edges, i.e., $G(n, m)$, is to understand "what properties are typically implied by the property of having m edges" (and we cast the answer as "properties that hold with high probability in a 'random' graph with m edges"). Notably, analyzing the uniform measure even for this simplest property is non-trivial. The reason is that it entails the single massive choice of an m-subset of edges, rather than m independent choices. In contrast, the independence of choices in $G(n, p)$ makes that distribution far more accessible, dramatically enabling analysis.

Connecting $G(n, m)$ and $G(n, p)$ is a classic result of random graph theory. The key observation is that to sample according to $G(n, p)$, since edges are independent and equally likely, we can first sample an integer $m \sim \mathrm{Bin}\left(\binom{n}{2}, p\right)$ and then sample a uniformly random graph with m edges, i.e., $G(n, m)$. Thus, for $p = p(m) = m/\binom{n}{2}$, the random graph $G \sim G(n, m)$ and the two random graphs $G^{\pm} \sim G(n, (1 \pm \epsilon)p)$ can be coupled so that, viewing each graph as a set of edges, with high probability,

$$G^- \subseteq G \subseteq G^+ \ . \tag{1}$$

The significance of this relationship between what we *wish* to study (uniform measure) and what we *can* study (product measure) can not be overestimated. It manifests most dramatically in the study of monotone properties: to study a monotone, say, increasing property in $G \sim G(n, m)$ it suffices to bound from above its probability in G^+ and from below in G^-. This connection has been thoroughly exploited to establish threshold functions for a host of monotone graph properties such as Connectivity, Hamiltonicity, and Subgraph Existence, making it the workhorse of random graph theory.

In this work we seek to extend the above relationship between the uniform measure and product measures to properties more delicate than having a given number of edges. In doing so we (i) provide a tool that can be used to revisit a number of questions in random graph theory from a more realistic angle and (ii) lay the foundation for designing random graph models eschewing independence assumptions. For example, our tool makes short work of the following set of questions (which germinated our work):

Given an arbitrary collection of n points on the plane what can be said about the set of all graphs that can be built on them using a given amount of wire, i.e., when connecting two points consumes wire equal to their distance? What does a uniformly random such graph look like? How does it change as a function of the available wire?

1.1   Our Contribution

A product measure on the set of all undirected simple graphs on n vertices, $\mathcal{G}_n$, is specified by a symmetric matrix $\mathbf{Q} \in [0, 1]^{n \times n}$ where $Q_{ii} = 0$ for $i \in [n]$. By analogy to $G(n, p)$ we denote by $G(n, \mathbf{Q})$ the measure in which every edge $\{i, j\}$ is included independently with probability $Q_{ij} = Q_{ji}$. Let $S \subseteq \mathcal{G}_n$ be arbitrary. Our main result is a sufficient condition for the uniform measure on S, denoted by $U(S)$, to be approximable by a product measure in the following sense.

Sandwichability. *The measure $U(S)$ is (ϵ, δ)-sandwichable if there exists an $n \times n$ symmetric matrix $\mathbf{Q}$ such that the distributions $G \sim U(S)$ and $G^{\pm} \sim G(n, (1 \pm \epsilon)\mathbf{Q})$ can be coupled so that $\Pr[G^- \subseteq G \subseteq G^+] \geq 1 - \delta$.*

Informally, the two conditions required for our theorem to hold are as follows:

Partition Symmetry. The set S should be symmetric with respect to *some* partition $\mathcal{P} = (P_1, \ldots, P_k)$ of the $\binom{n}{2}$ possible edges. More specifically, for a partition $\mathcal{P}$ define the *edge profile* of a graph G with respect to $\mathcal{P}$ to be the k-dimensional vector $\mathbf{m}(G) = (m_1(G), \ldots, m_k(G))$ where $m_i(G)$ counts the number of edges in G from part P_i. Partition symmetry amounts to the requirement that the characteristic function of S can depend on *how many* edges are included from each part but not on *which* edges. That is, if we let $\mathbf{m}(S) := \{\mathbf{m}(G) : G \in S\}$, then $\forall G \in \mathcal{G}_n, \mathbb{I}_S(G) = \mathbb{I}_{\mathbf{m}(S)}(\mathbf{m}(G))$. The $G(n, m)$ model is recovered by considering the trivial partition with $k = 1$ parts and $\mathbf{m}(S) = \{m\}$. Far more interestingly, in our motivating example edges are partitioned into equivalence classes according to their cost $\mathbf{c}$ (distance of endpoints) and the characteristic function allows graphs whose edge profile $\mathbf{m}(G)$ does not violate the total wire budget $C_B = \{\mathbf{v} \in \mathbb{N}^k : \mathbf{c}^\mathsf{T}\mathbf{v} \leq B\}$. We discuss the motivation for edge-partition symmetry at length in Section 2.

Convexity. Since membership in S depends solely on a graph's edge-profile, it follows that a uniformly random element of S can be selected as follows: (i) select an edge profile $\mathbf{v} = (v_1, \ldots, v_k) \in \mathbb{R}^k$ from the distribution on $\mathbf{m}(S)$ induced by $U(S)$, and then (ii) for each $i \in [k]$ independently select a uniformly random v_i-subset of P_i. In other words, the complexity of the uniform measure on S manifests entirely in the *induced* distribution on $\mathbf{m}(S) \in \mathbb{N}^k$ whose structure we need to capture.

Without any assumptions the set $\mathbf{m}(S)$ can be arbitrary, e.g., S can be the set of graphs having either $n^{1/2}$ or $n^{3/2}$ edges, rendering any approximation by a product measure hopeless. To impose some regularity we require the discrete set $\mathbf{m}(S)$ to be *convex* in the sense, that *it equals the set of integral points in its convex hull*. While convexity is not strictly necessary for our proof method to work (see Section 5), we feel that it provides a clean conceptual framework while still allowing very general properties to be expressed. These include all properties expressible as Linear Programs in the number of edges from each part, but also properties involving non-linear constraints, e.g., the absence of percolation. (Our original example, of course, amounts to a single linear inequality constraint.) Most importantly, since convex sets are closed under intersection, convex properties can be composed arbitrarily while remaining amenable to approximability by a product measure.

We state our results formally in Section 4. The general idea is this.

Theorem 1 (Informal). *If S is a convex symmetric set, then $U(S)$ is sandwichable by a product measure $G(n, \mathbf{Q}^*)$.*

The theorem is derived by following the *Principle of Maximum Entropy*, i.e., by proving that the induced measure on the set of edge-profiles $\mathbf{m}(S)$ concentrates around a unique vector $\mathbf{m}^*$, obtained by solving an entropy (concave function) maximization problem on the convex hull of $\mathbf{m}(S)$. The maximizer $\mathbf{m}^*$ can in many cases be computed explicitly, either analytically or numerically, and the product measure $\mathbf{Q}^*$ follows readily from it. Indeed, the maximizer $\mathbf{m}^*$ essentially characterizes the set S, as all quantitative requirements of our theorem are expressed only in terms of the number of vertices, n, the number of parts, k, and $\mathbf{m}^*$.

The proof relies on a new concentration inequality we develop for symmetric subsets of the binary cube which, as we shall see, is *sharp*. Besides enabling the study of monotone properties, our results allow one to obtain tight estimates of local graph features, such as the expectation and variance of subgraph counts.

2   Motivation

As stated, our goal is to enable the study of the uniform measure over sets of graphs. The first step in this direction is to identify a "language" for specifying sets of graphs that is expressive enough to be interesting but restricted enough to be tractable.

Arguably the most natural way to introduce structure on a set is to impose symmetry. Formally this is expressed as the *invariance* of the set's characteristic function under the action of a group of transformations. In this work, we explore the progress that can be made if we define an *arbitrary* partition of the edges and take the set of transformations to be the the the Cartesian product of all possible permutations of the edges (indices) within each part (symmetric group). While our work is only a first step towards a theory of extracting independence from symmetry, we argue that symmetry with respect to an edge partition is well-motivated for two reasons.

Existing Models. The first is that such symmetry, typically in a very rigid form, is already implicit in several random graph models besides $G(n, m)$. Among them are Stochastic Block Models (SBM), which assume the much stronger property of symmetry with respect to a *vertex* partition, and Stochastic Kronecker Graphs [13]. The fact that our notion of symmetry encompasses SBMs is particularly pertinent in light of the theory of Graph Limits [15], since inherent in the construction of the limiting object is an intermediate approximation of the sequence of graphs by a sequence of SBMs, via the (weak) Szemerédi Regularity Lemma [3,7]. Thus, any property that is encoded in the limiting object, typically subgraph densities, is expressible within our framework.

Enabling the Expression of Geometry. A strong driving force behind the development of recent random graph models has been the incorporation of geometry, an extremely natural backdrop for network formation. Typically this is done by embedding the vertices in some (low-dimensional) metric space and assigning probabilities to edges as a function of distance. Perhaps the *most significant feature* of our work is that it fully supports the expression of geometry but in a far more light-handed manner, i.e., without imposing any specific geometric requirement. This is achieved by (i) using edge-partitions to abstract away geometry as a symmetry rendering edges of the same length equivalent, while (ii) recognizing that there exist *macroscopic* constraints on the set of feasible graphs, e.g., the total edge length. Most obviously, in a physical network where edges (wire, roads) correspond to a resource (copper, concrete) there is a bound on how much can be invested to create the network while, more generally, cost (length) may represent a number of different notions that distinguish between edges.

3 Applications

A common assumption throughout the paper is the existence of a partition of the edges, expressing prior information about the setting at hand. Two prototypical examples are: *vertex-induced partitions*, as in the SBM, and *geometry induced partitions*, as in the d-dimensional lattice (torus). The applicability of our framework depends crucially on whether the partition is fine enough to express the desired property S. The typical pipeline is: (i) translate prior information in a partition of the edges, (ii) express the set of interest S as a specification on the edge-profile $\mathbf{m}$, (iii) solve the entropy-optimization problem and obtain the matrix $\mathbf{Q}^*$, and finally, (iv) perform all analyses and computations using the product measure $G(n, \mathbf{Q}^*)$, typically exploiting results from random graph theory and concentration of measure. Below are some examples.

Budgeted Graphs. Imagine that each possible edge e has multiple attributes that can be categorical (type of relation) or operational (throughput, latecency, cost, distance), compactly encoded as vector $\mathbf{X}_e \in \mathbb{R}^d$. We can form a partition $\mathcal{P}$ by grouping together edges that have identical attributes. Let $\boldsymbol{X} = [\boldsymbol{X}_1 \dots \boldsymbol{X}_k] \in \mathbb{R}^{d \times k}$ be the matrix where we have stacked the attribute vectors from each group and $\mathbf{b}$ be a vector of budgets. In this setting we might be interested in the *affine set* of graphs $S(\boldsymbol{X}, \boldsymbol{b}) = \{G \in \mathcal{G}_n | \boldsymbol{X} \cdot \mathbf{m}(G) \leq \boldsymbol{b}\}$, which can express a wide range of constraints. For such a set, besides generality of expression, the entropy optimization problem has a closed-form analytic solution in terms of the dual variables $\boldsymbol{\lambda} \in \mathbb{R}_+^d$. The probability of an edge (u, v) in part ℓ is given by:
$Q_{uv}^*(S) = [1 + \exp(\boldsymbol{X}_\ell^\mathsf{T} \boldsymbol{\lambda})]^{-1}$.

Navigability. In [11,12], Kleinberg gave sufficient conditions for greedy routing to discover paths of poly-logarithmic length between any two vertices in a graph. One of the most general settings where such navigability is possible is set-systems, a mathematical abstraction of the relevant geometric properties of grids, regular-trees and graphs of bounded doubling dimension. The essence of navigability lies in the requirement that for any vertex in the graph, the probability of having an edge to a vertex at distance in the range $[2^{i-1}, 2^i)$, i.e., at distance scale i, is approximately uniform for all $i \in [\log n]$. In our setting, we can partition the $\binom{n}{2}$ edges according to distance scale so that part P_i includes all possible edges between vertices at distance scale i. In [1] we prove that by considering a single linear constraint where the cost of edges in scale i is proportional to i, we recover Kleinberg's results on navigability in set-systems, *without* any independence assumptions regarding network formation, or coordination between the vertices (such as using the same probability distribution). Besides establishing the robustness of navigability, eschewing a specific mechanism for (navigable) network formation allows us to recast navigability as a property of networks brought about by economical (budget) and technological (cost) advancements.

Percolation Avoidance. To show that interesting non-linear constraints can also be accommodated we focus on the case of the Stochastic Block Model. Consider a social network consisting of q groups of sizes $(\rho_1, \dots, \rho_q) \cdot n$, where $\rho_i > 0$ for $i \in [q]$. As the partition of edges is naturally induced by the partition of vertices, for simplicity we adopt a double indexing scheme and instead of the edge-profile vector $\mathbf{m} \in \mathbb{R}^{\binom{q+1}{2}}$ we are going to use a symmetric edge-progile matrix $\mathbf{M} \in \mathbb{R}^{q \times q}$. Consider the property S_ϵ that a specific group s acts as the "connector", i.e., that the graph induced by the remaining groups should have no component of size greater than ϵn for some arbitrarily small $\epsilon > 0$. While the set S_ϵ is not symmetric with respect to this partition our result can still be useful, as follows.

Using a well known connection between multitype branching processes [16] and the existence of a giant component (percolation) in mean-field models, such as $G(n, p)$ and SBM, we can cast the existence of the giant component in terms of a condition on the number of edges between each block. Concretely, given the edge-profile matrix $\mathbf{M}$, for a given cluster $s \in [q]$ define the $(q - 1) \times (q - 1)$ matrix: $T(\mathbf{M})_{ij} := \frac{m_{ij}}{n^2 \rho_i}, \forall i, j \in [q] \setminus \{s\}$ that encapsulates the dynamics of a multi-type branching process. Let $\|\cdot\|_2$ denote the *operator norm* (maximum singular value). A classic result of branching processes asserts that if $\|T(\mathbf{M})\|_2 < 1$ no giant component exists. Thus, in our framework, the property $S_\epsilon = \{$ no giant component without vertices from $s\}$, can be accurately approximated under the specific partition $\mathcal{P}$ by the set of graphs S for which $\mathbf{M}(S) = \{\mathbf{M} : \|T(\mathbf{M})\|_2 < 1\}$, where additionally the set $\mathbf{M}(S)$ is convex as $\|T(M)\|_2$ is a convex function of $T(M)$ which is convex (linear) in $\mathbf{M}$.

4 Definitions and Results

We start with some notation. We will use lower case boldface letters to denote vectors and uppercase boldface letters to denote matrices. Further, we fix an arbitrary enumeration of the $N = \binom{n}{2}$ edges and sometimes represent the set of all graphs on n vertices as $H_N = \{0, 1\}^N$. We will refer to an element of $x \in H_N$ interchangeably as a graph and a string. Given a partition $\mathcal{P} = (P_1, \ldots, P_k)$ of $[N]$, we define $\Pi_N(\mathcal{P})$ to be the set of all permutations acting only within blocks of the partition.

Edge Block Symmetry. *Fix a partition $\mathcal{P}$ of $[N]$. A set $S \subseteq H_N$ is called $\mathcal{P}$-symmetric if it is invariant under the action of $\Pi_N(\mathcal{P})$. Equivalently, if $\mathbb{I}_S(x)$ is the indicator function of set S, then $\mathbb{I}_S(x) = \mathbb{I}_S(\pi(x))$ for all $x \in H_N$ and $\pi \in \Pi_N(\mathcal{P})$.*

The number of parts $k = |\mathcal{P}|$ gives a rough indication of the amount of symmetry present. For example, when $k = 1$ we have maximum symmetry as all edges are equivalent. In a stochastic block model (SBM) with ℓ vertex classes we have $k = \binom{\ell}{2}$. For a d-dimensional lattice, partitioning the $\binom{n}{2}$ edges by distance results in roughly $k = n^{1/d}$ parts. Finally, if $k = N$ there is no symmetry whatsoever. Our results accommodate partitions with as many as $O(n^{1-\epsilon})$ parts. This is way more than enough for most situations. For example, as we just saw, in d-dimensional lattices there are $O(n^{1/d})$ distances. Generically, if we have n points such that the nearest pair habe distance 1 and the farthest have distance D, fixing any $\delta > 0$ and binning together all edges of length $[(1+\delta)^i, (1+\delta)^{i+1})$ for $i \geq 0$, yields only $O(\delta^{-1} \log D)$ classes.

Recall that given a partition $\mathcal{P} = (P_1, \ldots, P_k)$ of H_N and a graph $x \in H_N$, the edge profile of x is $\mathbf{m}(x) := (m_1(x), \ldots, m_k(x))$, where $m_i(x)$ is the number of edges of x from P_i, and that the image of a $\mathcal{P}$-symmetric set S under $\mathbf{m}$ is denoted as $\mathbf{m}(S) \subseteq \mathbb{R}^k$. The edge-profile is crucial to the study of $\mathcal{P}$-symmetric sets due to the following intuitively obvious fact.

Proposition 1. *Any function $f : H_N \to \mathbb{R}$ invariant under $\Pi_N(\mathcal{P})$ depends only on the edge-profile $\mathbf{m}(\mathbf{x})$.*

Definition 1. *Let $p_i = |P_i|$ denote the number of edges in part i of partition $\mathcal{P}$.*

Edge Profile Entropy. *Given an edge profile $\mathbf{v} \in \mathbf{m}(S)$ define the entropy of $\mathbf{v}$ as* $\text{ENT}(\mathbf{v}) = \sum_{i=1}^{k} \log \binom{p_i}{v_i}$.

Using the edge-profile entropy we can express the induced distribution on $\mathbf{m}(S)$ as $\mathbb{P}(\mathbf{v}) = \frac{1}{|S|} e^{\mathrm{ENT}(\mathbf{v})}$. The crux of our argument is now this: the only genuine obstacle to S being approximable by a product measure is degeneracy, i.e., the existence of multiple, well-separated edge-profiles that maximize $\mathrm{ENT}(\mathbf{v})$. The reason we refer to this as degeneracy is that it typically encodes a hidden symmetry of S with respect to $\mathcal{P}$. For example, imagine that $\mathcal{P} = (P_1, P_2)$, where $|P_1| = |P_2| = p$, and that S contains all graphs with $p/2$ edges from P_1 and $p/3$ edges from P_2, or vice versa. Then, the presence of a single edge $e \in P_i$ in a uniformly random $G \in S$ boosts the probability of all other edges in P_i, rendering a product measure approximation impossible.

Note that since $\mathbf{m}(S)$ is a discrete set, it is non-trivial to quantify what it means for the maximizer of ENT to be "sufficiently unique". For example, what happens if there is a unique maximizer of $\mathrm{ENT}(\mathbf{v})$ strictly speaking, but sufficiently many near-maximizers to potentially receive, in aggregate, a majority of the measure? To strike a balance between conceptual clarity and generality we focus on the following.

Convexity. *Let* $\mathrm{Conv}(A)$ *denote the convex hull of a set* A. *Say that a* $\mathcal{P}$-*symmetric set* $S \subseteq \mathcal{G}_N$ *is convex iff the convex hull of* $\mathbf{m}(S)$ *contains no new integer points, i.e., if* $\mathrm{Conv}(\mathbf{m}(S)) \cap \mathbb{N}^k = \mathbf{m}(S)$.

Let $H_{\mathcal{P}}(\mathbf{v})$ be the approximation to $\mathrm{ENT}(\mathbf{v})$ that results by replacing each binomial term with its binary entropy approximation via the first term in Stirling's approximation.

Entropic Optimizer. *Let* $\mathbf{m}^* = \mathbf{m}^*(S) \in \mathbb{R}^k$ *be the solution to*
$$\max_{\mathbf{v} \in \mathrm{Conv}(\mathbf{m}(S))} H_{\mathcal{P}}(\mathbf{v}).$$

Defining the optimization over the convex hull of $\mathbf{m}(S)$ will allow us to study the set S by studying only the properties of the maximizer $\mathbf{m}^*$. Clearly, if a $\mathcal{P}$-symmetric set S has entropic optimizer $\mathbf{m}^* = (m_1^*, \ldots, m_k^*)$, the natural candidate product measure for each $i \in [k]$ assigns probability m_i^*/p_i to all edges in part P_i. The challenge is to relate this product measure to the uniform measure on S by proving concentration of the induced measure on $\mathbf{m}(S)$ around a point near $\mathbf{m}^*$. For that we need (i) the vector $\mathbf{m}^*$ to be "close" to a vector in $\mathbf{m}(S)$, and (ii) to control the decrease in entropy "away" from $\mathbf{m}^*$. To quantify this second notion we need the following parameters, expressing the geometry of convex sets.

Definition 2. *For a* $\mathcal{P}$-*symmetric convex set* S *define*

$$\text{Thickness:} \quad \mu = \mu(S) = \min_{i \in [k]} \min\{m_i^*, p_i - m_i^*\} \tag{2}$$

$$\text{Condition number:} \quad \lambda = \lambda(S) = \frac{5k \log n}{\mu(S)} \tag{3}$$

$$\text{Resolution:} \quad r = r(S) = \frac{\lambda + \sqrt{\lambda^2 + 4\lambda}}{2} > \lambda \tag{4}$$

The most important of the above three parameters is *thickness*. Its role is to quantify how close the optimizer $\mathbf{m}^*(S)$ comes, in any coordinate, to the natural boundary $\{0, p_1\} \times \ldots \times \{0, p_k\}$, where the entropy of a class becomes zero. As a result, thickness determines the *coordinate-wise concentration* around the optimum.

The *condition number* $\lambda(S)$, on the other hand, quantifies the robustness of S. To provide intuition, in order for the product measure approximation to be accurate for every class of edges (part of $\mathcal{P}$), fluctuations in the number of edges of order $\sqrt{m_i^*}$ need to be "absorbed" in the mean m_i^*. For this to happen with polynomially high probability for a single part, standard results imply we must have $m_i^* = \Omega(\log(n))$. We absorb the dependencies between parts by taking a union bound, thus multiplying by the number of parts, yielding the numerator in (3). Our results give strong probability bounds when $\lambda(S) \ll 1$, i.e., when in a typical graph in S the number of edges from each part P_i is $\Omega(k \log n)$ edges away from triviality, i.e., both from 0 and from $|P_i| = p_i$, a condition we expect to hold in all natural applications. We can now state our main result.

Theorem 2 (Main result). *Let $\mathcal{P}$ be any edge-partition and let S be any $\mathcal{P}$-symmetric convex set. For every $\epsilon > \sqrt{12\lambda(S)}$, the uniform measure on S is (ϵ, δ)-sandwichable, where $\delta = 2 \exp\left[-\mu(S)\left(\frac{\epsilon^2}{12} - \lambda(S)\right)\right]$.*

Remark 1. As a sanity check we see that as soon as $m \gg \log n$, Theorem 2 recovers the sandwichability of $G(n, m)$ by $G(n, p(m))$ as sharply as the Chernoff bound, up to the constant factor $\frac{1}{12}$ in the exponent.

Theorem 2 follows by analyzing the natural coupling between the uniform measure on S and the product measure corresponding to the entropic optimizer $\mathbf{m}^*$. Our main technical contribution is Theorem 3 below, a concentration inequality for $\mathbf{m}(S)$ when S is a convex symmetric set. The *resolution*, $r(S)$, defined in (4) above, reflects the narrowest *concentration interval* that can be proved by our theorem. When $\lambda(S) \ll 1$, as required for the theorem to be meaningfully applied, it scales optimally as $\sqrt{\lambda(S)}$.

Theorem 3. *Let $\mathcal{P}$ be any edge-partition, let S be any $\mathcal{P}$-symmetric convex set, and let $\mathbf{m}^*$ be the entropic optimizer of S. For all $\epsilon > r(S)$, if $G \sim U(S)$, then*

$$\mathbb{P}_S\left(|\mathbf{m}(G) - \mathbf{m}^*| \leq \epsilon \tilde{\mathbf{m}}^*\right) \geq 1 - \exp\left(-\mu(S)\left(\frac{\epsilon^2}{1+\epsilon} - \lambda(S)\right)\right) , \qquad (5)$$

where $\mathbf{x} \leq \mathbf{y}$ means that $x_i \leq y_i$ for all $i \in [k]$, and $\tilde{m}_i = \min\{m_i^*, p_i - m_i^*\}$.

The intuition driving concentration is that as *thickness* increases two phenomena occur: (i) vectors close to $\mathbf{m}^*$ capture a larger fraction of the measure, and (ii) the decay in entropy away from $\mathbf{m}^*$ becomes steeper. These joint forces compete against the probability mass captured by vectors "away" from the optimum. The point were they prevail corresponds to $\lambda(S) \ll 1$ or, equivalently, $\mu(S) \gg 5k \log(n)$. Assuming $\lambda(S) \ll 1$ the probability bounds we give scale as $n^{-\Omega(k\epsilon^2)}$. Without assumptions on S, and up to the constant 5 in (3), this is *sharp*, per Proposition 2 below.

5   Technical Overview

In this section, we present an overview of the technical work involved in proving Theorems 2 and 3. Most of the work lies in the concentration result, Theorem 3.

Concentration. The general idea is to identify a high-probability subset $\mathcal{L} \subseteq \mathbf{m}(S)$ by integrating the probability measure around the entropy-maximizing profile $\mathbf{m}^*$. Since ultimately our goal is to couple the uniform measure with a product measure, we need to establish concentration for the number of edges from each and every part, i.e., in every coordinate. There are two main issues: (i) we do not know $|S|$, and (ii) we must quantify the decrease in entropy as a function of the L_∞ distance from the maximizer $\mathbf{m}^*$. Our strategy to address these issues is:

Size of S. We bound $\log|S|$ from below by the contribution to $\log|S|$ of the entropic optimal edge-profile $\mathbf{m}^*$, thus upper-bounding the probability of every $\mathbf{v} \in \mathbf{m}(S)$ as

$$\log \mathbb{P}_S(\mathbf{v}) = \text{ENT}(\mathbf{v}) - \log(|S|) \leq \text{ENT}(\mathbf{v}) - \text{ENT}(\mathbf{m}^*) \ . \tag{6}$$

This is the crucial step that opens up the opportunity of relating the probability of a vector $\mathbf{v}$ to the distance $\|\mathbf{v} - \mathbf{m}^*\|_2$ through analytic properties of entropy. Key to this is the definition of $\mathbf{m}^*$ as the maximizer over $\text{Conv}(\mathbf{m}(S))$ instead of over $\mathbf{m}(S)$.

Proposition 2. *If $S = \mathcal{G}_n$ and $\mathcal{P}$ is any k-partition such that $|P_i| = \binom{n}{2}/k$ for all i, then $\log(|S|) - \text{ENT}(\mathbf{m}^*) = \Omega(k \log(n))$.*

Proposition 2 demonstrates that unless one utilizes specific geometric properties of the set S enabling integration around $\mathbf{m}^*$, instead of using a point-bound for $\log|S|$, a loss of $\Omega(k \log(n))$ is unavoidable. In other words, either one makes more assumptions on S besides symmetry and "convexity", or the claimed error term is optimal.

Distance bounds: To bound from below the rate at which entropy decays as a function of the component-wise distance from the maximizer $\mathbf{m}^*$, we first approximate $\text{ENT}(\mathbf{v})$ by $H_\mathcal{P}(\mathbf{v})$ (the binary entropy introduced earlier) to get a smooth function. Then, exploiting the separability, concavity and differentiability of binary entropy, we obtain component-wise distance bounds using a second-order Taylor approximation. At this step we also lose a cumulative factor of order $3k \log n$ stemming from Stirling approximations and the subtle point that the maximizer $\mathbf{m}^*$ might not be an integer point. The constant 3 can be improved, but in light of Proposition 2 this would be pointless and complicate the proof unnecessarily.

Union bound: Finally, we integrate the obtained bounds outside the set of interest by showing that even if all "bad" vectors where placed right at the boundary of the set, where the lower bound on the decay of entropy is smallest, the total probability mass would be exponentially small. The loss incurred at this step is of order $2k \log n$, since there are at most n^{2k} bad vectors.

Relaxing Conclusions. Our theorem seeks to provide concentration simultaneously for all parts. That motivates the definition of thickness parameter $\mu(S)$ as the minimum distance from the trivial boundary that any part has at the optimum $\mathbf{m}^*$. Quantifying everything in terms of $\mu(S)$ is a very conservative requirement. For instance, if we define the set S to have no edges in a particular part of the partition, then $\mu(S)$ is 0 and our conclusions become vacuous. Our proofs in reality generalize, to the case where we confine our attention only to a subset $I \subseteq [k]$ of blocks in the partition. In particular, if one defines I^* as the set of parts whose individual *thickness* parameter $\tilde{m}_i = \min\{m_i, p_i - m_i\}$ is greater than $5k \log n$, both theorems hold for the subset of edges $\cup_{i \in I^*} P_i$. In essence that means that for every part that is "well-conditioned", we can provide concentration of the number of edges and approximate monotone properties of only those parts by coupling them with product measures.

Relaxing Convexity. Besides partition symmetry, that comprises our main premise and starting point, the second main assumption made about the structure of S is *convexity*. In the proof convexity is used only to argue that: (i) the maximizer $\mathbf{m}^*$ will be close to some vector in $\mathbf{m}(S)$, and (ii) that the first order term in the Taylor approximation of the entropy around $\mathbf{m}^*$ is always negative. Since the optimization problem is defined on the convex hull of $\mathbf{m}(S)$, the convexity of $\mathrm{Conv}(\mathbf{m}(S))$ implies (ii), independently of whether $\mathbf{m}(S)$ is convex or not. We thus see that we can replace convexity of $\mathcal{P}$-symmetric sets with *approximate unimodality*.

Definition 3. *A $\mathcal{P}$-symmetric set S is called Δ-unimodal if the solution $\mathbf{m}^*$ to the entropy optimization problem defined in Section 2, satisfies:*

$$d_1(\mathbf{m}^*, S) := \min_{\mathbf{v} \in \mathbf{m}(S)} \|\mathbf{m}^* - \mathbf{v}\|_1 \leq \Delta \tag{7}$$

Convexity essentially implies that the set S is k-unimodal as we need to round each of the k coordinates of the solution to the optimization problem to the nearest integer. Under this assumption, all our results apply by only changing the condition number of the set to $\lambda(S) = \frac{(2\Delta + 3k) \log n}{\mu(S)}$. In this extended abstract, we opted to present our results by using the familiar notion of convexity to convey intuition on our results and postpone the presentation in full generality for the full version of the paper.

Coupling. To prove Theorem 2 using our concentration result, we argue as follows. Conditional on the edge-profile, we can couple the generation of edges in different parts independently, in each part the coupling being identical to that between $G(n, m)$ and $G(n, p)$. Then, using a union bound we can bound the probability that all couplings succeed, given an appropriate $\mathbf{v}$. Finally, using the concentration theorem we show that sampling an appropriate edge-profile $\mathbf{v}$ happens with high probability.

References

1. Achlioptas, D., Siminelakis, P.: Navigability is a robust property. CoRR, abs/1501.04931 (2015)
2. Bollobás, B.: Random graphs, vol. 73. Cambridge Studies in Advanced Mathematics, 2nd edn. Cambridge University Press, Cambridge (2001)
3. Borgs, C., Chayes, J.T., Cohn, H., Zhao, Y.: An $L\hat{}p$ theory of sparse graph convergence I: limits, sparse random graph models, and power law distributions. ArXiv e-prints, January 2014
4. Chung, F.R.K., Lu, L.: Complex graphs and networks, vol. 107. American mathematical society Providence (2006)
5. Durrett, R.: Random graph dynamics, vol. 20. Cambridge University Press (2007)
6. Erdős, P., Rényi, A.: On random graphs. Publicationes Mathematicae Debrecen **6**, 290–297 (1959)
7. Frieze, A., Kannan, R.: Quick approximation to matrices and applications. Combinatorica **19**(2), 175–220 (1999)
8. Gilbert, E.N.: Random graphs. The Annals of Mathematical Statistics, 1141–1144 (1959)
9. Goldenberg, A., Zheng, A.X., Fienberg, S.E., Airoldi, E.M.: A survey of statistical network models. Found. Trends Mach. Learn. **2**(2), 129–233 (2010)
10. Janson, S., Łuczak, T., Rucinski, A.: Random graphs. Wiley-Interscience Series in Discrete Mathematics and Optimization. Wiley-Interscience, New York (2000)
11. Kleinberg, J.M.: Navigation in a small world. Nature **406**(6798), 845 (2000)
12. Kleinberg, J.M.: Small-world phenomena and the dynamics of information. In: Dietterich, T.G., Becker, S., Ghahramani, Z. (eds.) NIPS, pp. 431–438. MIT Press (2001)
13. Leskovec, J., Chakrabarti, D., Kleinberg, J., Faloutsos, C., Ghahramani, Z.: Kronecker graphs: An approach to modeling networks. The Journal of Machine Learning Research **11**, 985–1042 (2010)
14. Linial, N., London, E., Rabinovich, Y.: The geometry of graphs and some of its algorithmic applications. Combinatorica **15**(2), 215–245 (1995)
15. Lovász, L.: Large networks and graph limits, vol. 60. American Mathematical Soc. (2012)
16. Mode, C.J.: Multitype branching processes: theory and applications. Modern analytic and computational methods in science and mathematics. American Elsevier Pub. Co. (1971)
17. Newman, M.: Networks: an introduction. Oxford University Press (2010)

Polylogarithmic-Time Leader Election
in Population Protocols

Dan Alistarh[1]([✉]) and Rati Gelashvili[2]

[1] Microsoft Research, Cambridge, UK
dan.alistarh@microsoft.com
[2] MIT, Cambridge, MA, USA
gelash@mit.edu

Abstract. Population protocols are networks of finite-state agents, interacting randomly, and updating their states using simple rules. Despite their extreme simplicity, these systems have been shown to cooperatively perform complex computational tasks, such as simulating register machines to compute standard arithmetic functions. The election of a unique *leader agent* is a key requirement in such computational constructions. Yet, the fastest currently known population protocol for electing a leader only has *linear* convergence time, and it has recently been shown that no population protocol using a *constant* number of states per node may overcome this linear bound.

In this paper, we give the first population protocol for leader election with *polylogarithmic* convergence time, using polylogarithmic memory states per node. The protocol structure is quite simple: each node has an associated value, and is either a *leader* (still in contention) or a *minion* (following some leader). A leader keeps incrementing its value and "defeats" other leaders in one-to-one interactions, and will drop from contention and become a minion if it meets a leader with higher value. Importantly, a leader also drops out if it meets a *minion* with higher absolute value. While these rules are quite simple, the proof that this algorithm achieves polylogarithmic convergence time is non-trivial. In particular, the argument combines careful use of concentration inequalities with anti-concentration bounds, showing that the leaders' values become spread apart as the execution progresses, which in turn implies that straggling leaders get quickly eliminated. We complement our analysis with empirical results, showing that our protocol converges extremely fast, even for large network sizes.

1 Introduction

Recently, there has been significant interest in modeling and analyzing interactions arising in biological or bio-chemical systems through an algorithmic lens. In particular, the population protocol model [AAD+06], which is the focus of this paper, consists of a set of n finite-state nodes interacting in pairs, where each interaction may update the states of both participants. The goal is to have all nodes converge on an output value, which represents the result of the

© Springer-Verlag Berlin Heidelberg 2015
M.M. Halldórsson et al. (Eds.): ICALP 2015, Part II, LNCS 9135, pp. 479–491, 2015.
DOI: 10.1007/978-3-662-47666-6_38

computation, usually a predicate on the initial state of the nodes. The set of interactions occurring at each step is assumed to be decided by an adversarial scheduler, which is usually subject to some fairness conditions. The standard scheduler when computing convergence bounds is the *probabilistic (uniform random) scheduler*, e.g., [AAE08b, PVV09, DV12], which picks the next pair to interact uniformly at random in each step. We adopt this probabilistic scheduler model in this paper. (Some references refer to this model as the *probabilistic* population model.) The fundamental measure of convergence is *parallel time*, defined as the number of scheduler steps until convergence, divided by n.[1]

The class of predicates computable by population protocols is now well-understood [AAD+06, AAE06, AAER07] to consist precisely of *semilinear predicates*, i.e. predicates definable in first-order Presburger arithmetic. The first such construction was given in [AAD+06], and later improved in terms of convergence time in [AAE06]. A parallel line of research studied the computability of deterministic functions in chemical reaction networks, which are also instances of population protocols [CDS14]. All three constructions fundamentally rely on the election of a single initial *leader* node, which co-ordinates phases of computation.

Reference [AAD+06] gives a simple protocol for electing a leader from a uniform population, based on the natural idea of having leaders eliminate each other directly through symmetry breaking. Unfortunately, this strategy takes at least linear parallel time in the number of nodes n: for instance, once this algorithm reaches *two* surviving leaders, it will require $\Omega(n^2)$ additional interactions for these two leaders to meet. Reference [AAE08a] proposes a significantly more complex protocol, conjectured to be sub-linear, and whose convergence is only studied experimentally. This reference posits the existence of a sublinear-time population protocol for leader election as a "pressing" open problem. In fact, the existence of a poly-logarithmic leader election protocol would imply that *any semilinear predicate* is computable in poly-logarithmic time by a *uniform population* [AAE06].

Recently, Doty and Soloveichik [DS15] showed that $\Omega(n^2)$ expected interactions are *necessary* for electing a leader in the classic probabilistic protocol model in which each node only has *constant* number of memory states (with respect to n). This negative result implies that computing semilinear predicates in leader-based frameworks is subject to the same lower bound. In turn, this motivates the question of whether faster computation is possible if the amount of memory per node is allowed to be a function of n.

Contribution. In this paper, we solve this problem by proposing a new population protocol for leader election, which converges in $O(\log^3 n)$ expected parallel time, using $O(\log^3 n)$ memory states per node. Our protocol, called *LM* for *Leader-Minion*, roughly works as follows. Throughout the execution, each node is either a *leader*, meaning that it can still win, or a *minion*, following some leader. Each node state is associated to some *absolute value*, which is a positive

---

[1] An alternative definition is when reactions occur in parallel according to a Poisson process [PVV09, DV12].

integer, and with a *sign*, positive if the node is still in contention, and negative if the node has become a minion.

If two leaders meet, the one with the larger absolute value survives, and increments its value, while the other drops out, becoming a minion, and adopting the other node's value, but with a negative sign. (If both leaders have the same value, they both increment it and continue.) If a leader meets a minion with *smaller* absolute value than its own, it increments its value, while the minion simply adopts the leader's value, but keeps the negative sign. Conversely, if a leader meets a minion with *larger* absolute value than its own, then the leader drops out of contention, adopting the minion's value, with negative sign. Finally, if two minions meet, they update their values to the maximum absolute value between them, but with a *negative* sign.

These rules ensure that, eventually, a single leader survives. While the protocol is relatively simple, the proof of poly-logarithmic time convergence is non-trivial. In particular, the efficiency of the algorithm hinges on the minion mechanism, which ensures that a leader with high absolute value can eliminate other contenders in the system, without having to directly interact with them.

Roughly, the argument is based on two technical insights. First, consider two leaders at a given time T, whose (positive) values are at least $\Theta(\log n)$ apart. Then, we show that, within $O(\log n)$ parallel time from T, the node holding the smaller value has become a minion, with constant probability. Intuitively, this holds since 1) this node will probably meet either the other leader or one of its minions within this time interval, and 2) it cannot increase its count fast enough to avoid defeat. For the second part of the argument, we show via anti-concentration that, after parallel time $\Theta(\log^2 n)$ in the execution, the values corresponding to an arbitrary pair of nodes will be separated by at least $\Omega(\log n)$.

We ensure that the values of nodes cannot grow beyond a certain threshold, and set the threshold in such a way that the total number of states is $\Theta(\log^3 n)$. We show that with high probability the leader will be elected before the values of the nodes reach the threshold. In the other case, remaining leaders with threshold values engage in a *backup* dynamics where minions are irrelevant and leaders defeat each other when they meet based on random binary indicators which are set using the randomness of the scheduler. This process is slower but deterministically correct, and only happens with very low probability, allowing to conclude that the algorithm converges to a single leader within $O(\log^3 n)$ parallel time, both with high probability and in expectation, using $O(\log^3 n)$ states.

In population protocols, in every interaction, one node is said to be the *initiator*, the other is the *responder*, and the state update rules can use this distinction. In our protocol, this would allow a leader (the initiator in the interaction) to defeat another leader with the same value (the responder), and could also simplify the backup dynamics of our algorithm. However, our algorithm has the nice property that the state update rules can be made completely symmetric with regards to the initiator and responder roles. (For this reason, *LM* works for $n > 2$ nodes, because to elect a leader among two nodes it is necessary to rely on the initiator-responder role distinction.)

Summing up, we give the first poly-logarithmic time protocol for electing a leader from a uniform population. We note that $\Omega(n \log n)$ interactions seem intuitively necessary for leader election, as this number is required to allow each node to interact at least once. However, this idea fails to cover all possible reaction strategies if nodes are allowed to have arbitrarily many states.

We complement our analysis with empirical data, suggesting that the convergence time of our protocol is close to logarithmic, and that in fact the asymptotic constants are small, both in the convergence bound, and in the upper bound on the number of states the protocol employs.

Related Work. We restrict our attention to work in the population model. The framework of population protocols was formally introduced in reference [AAD+06], to model interactions arising in biological, chemical, or sensor networks. It sparked research into its computational power [AAD+06, AAE06, AAER07], and into the time complexity of fundamental tasks such as majority [AAE08b, PVV09, DV12], and leader election [AAD+06, AAE08a].[2] References interested in *computability* consider an adversarial scheduler which is restricted to be *fair*, e.g., where each agent interacts with every other agent infinitely many times. For complexity bounds, the standard scheduler is *uniform*, scheduling each pair uniformly at random at each step, e.g., [AAE08b, PVV09, DV12]. This model is also known as the *probabilistic* population model.

To the best of our knowledge, no population protocol for electing a leader with sub-linear convergence time was known before our work. References [AAD+06, AAE06, CDS14] present leader-based frameworks for population computations, assuming the existence of such a node. The existence of such a sub-linear protocol is stated as an open problem in [AAD+06, AAE08a]. Reference [DH13] proposes a *leader-less* framework for population computation.

Recent work by Doty and Soloveichik [DS15] showed an $\Omega(n^2)$ lower bound on the number of interactions necessary for electing a leader in the classic probabilistic protocol model in which each node only has *constant* number of memory states with respect to the number of nodes n [AAER07]. The proof of this result is quite complex, and makes use of the limitation that the number of states remains constant even as the number of nodes n is taken to tend to infinity.

Thus, our algorithm provides a complexity separation between population protocols which may only use constant memory per node, and protocols where the number of states is allowed to be a function of n. We note that, historically, the classic population protocol model [AAD+06] only allowed a constant number of states per node, while later references relaxed this assumption.

A parallel line of research studied *self-stabilizing* population protocols, e.g., [AAFJ06, FJ06, SNY+10], that is, protocols which can converge to a correct solution from an *arbitrary* initial state. It is known that stable leader election is *impossible* from an arbitrary initial state [AAFJ06]. References [FJ06, SNY+10] circumvent this impossibility by relaxing the problem semantics. Our algorithm is not affected by this impossiblity result since it is not self-stabilizing.

---

[2] Leader election and majority are complementary tasks, and no complexity-preserving transformations exist, to our knowledge.

2   Preliminaries

Population Protocols. We assume a population consisting of n agents, or nodes, each executing as a deterministic state machine with states from a finite set Q, with a finite set of input symbols $X \subseteq Q$, a finite set of output symbols Y, a transition function $\delta : Q \times Q \rightarrow Q \times Q$, and an output function $\gamma : Q \rightarrow Y$. Initially, each agent starts with an input from the set X, and proceeds to update its state following interactions with other agents, according to the transition function δ. For simplicity of exposition, we assume that agents have identifiers from the set $V = \{1, 2, \ldots, n\}$, although these identifiers are not known to agents, and not used by the protocol.

The agents' interactions proceed according to a directed *interaction graph* G without self-loops, whose edges indicate possible agent interactions. Usually, the graph G is considered to be the complete directed graph on n vertices, a convention we also adopt in this paper.

The execution proceeds in *steps*, or *rounds*, where in each step a new edge (u, w) is chosen uniformly at random from the set of edges of G. Each of the two chosen agents updates its state according to function δ.

Parallel Time. The above setup considers sequential interactions; however, in general, interactions between pairs of distinct agents are independent, and are usually considered as occurring in parallel. In particular, it is customary to define one unit of *parallel time* as n consecutive steps of the protocol.

The Leader Election Problem. In the *leader election* problem, all agents start in the same initial state A, i.e. the only state in the input set $X = \{A\}$. The output set is $Y = \{Win, Lose\}$.

A population protocol solves leader election within ℓ steps with probability $1 - \phi$, if it holds with probability $1 - \phi$ that for any configuration $c : V \rightarrow Q$ reachable by the protocol after $\geq \ell$ steps, there exists a unique agent i such that, (1) for the agent i, $\gamma(c(i)) = Win$, and, (2) for any agent $j \neq i$, $\gamma(c(j)) = Lose$.

3   The Leader Election Algorithm

In this section, we describe the *LM* leader election algorithm. The algorithm has an integer parameter $m > 0$, which we set to $\Theta(\log^3 n)$. Each state corresponds to an integer value from the set $\{-m, -m+1, \ldots, -2, -1, 1, 2, m-1, m, m+1\}$. Respectively, there are $2m + 1$ different states. We will refer to states and values interchangeably. All nodes start in the same state corresponding to value 1.

The algorithm, specified in Figure 1, consists of a set of simple deterministic update rules for the node state. In the pseudocode, the node states before an interaction are denoted by x and y, while their new states are given by x' and y'. All nodes start with value 1 and continue to interact according to these simple rules. We prove that all nodes except one will converge to negative values, and

Parameters:
m, an integer > 0, set to $\Theta(\log^3 n)$
State Space:
$LeaderStates = \{1, 2, \ldots, m - 1, m, m + 1\}$,
$MinionStates = \{-1, -2, \ldots, -m + 1, -m\}$,
Input: States of two nodes, x and y
Output: Updated states x' and y'
Auxiliary Procedures:

$$is\text{-}contender(x) = \begin{cases} \text{true if } x \in LeaderStates; \\ \text{false otherwise.} \end{cases}$$

$$contend\text{-}priority(x, y) = \begin{cases} m & \text{if } \max(|x|, |y|) = m + 1; \\ \max(|x|, |y|) + 1 & \text{otherwise.} \end{cases}$$

$$minion\text{-}priority(x, y) = \begin{cases} -m & \text{if } \max(|x|, |y|) = m + 1; \\ -\max(|x|, |y|) & \text{otherwise.} \end{cases}$$

```
1  procedure update⟨x, y⟩
2      if is-contender(x) and |x| ≥ |y| then
3          x' ← contend-priority(x, y)
4      else x' ← minion-priority(x, y)
5      if is-contender(y) and |y| ≥ |x| then
6          y' ← contend-priority(x, y)
7      else y' ← minion-priority(x, y)
```

Fig. 1. The state update rules for the LM algorithm

that convergence is fast with high probability. This solves the leader election problem since we can define γ as mapping only positive states to *Win* (a leader).[3]

Since positive states translate to being a leader according to γ, we call a node a *contender* if it has a positive value, and a *minion* otherwise. We present the algorithm in detail below.

The state updates (i.e. the transition function δ) of the LM algorithm are completely symmetric, that is, the new state x' depends on x and y (lines 2-4) exactly as y' depends on y and x (lines 5-7).

If a node is a contender and has absolute value not less than the absolute value of the interaction partner, then the node remains a contender and updates its value using the *contend-priority* function (lines 3 and 6). The new value will be one larger than the previous value except when the previous value was $m + 1$, in which case the new value will be m.

If a node had a smaller absolute value than its interaction partner, or was a minion already, then the node will be a minion after the interaction. It will set its value using the *minion-priority* function, to either $-\max(|x|, |y|)$, or $-m$ if the maximum was $m + 1$ (lines 4 and 7).

Values $m + 1$ and m are treated the same way if the node is a minion (essentially corresponding to $-m$). These values serve as a binary tie-breaker among the contenders that reach the value m, as will become clear from the analysis.

---

[3] Alternatively, γ that maps states with values m and $m + 1$ to *WIN* would also work, but we will work with positive "leader" states for the simplicity of presentation.

4   Analysis

In this section, we provide a complete analysis of our leader election algorithm.

Notation. Throughout this proof, we denote the set of n nodes executing the protocol by V. We measure execution time in discrete steps (rounds), where each step corresponds to an interaction. The *configuration* at a given time t is a function $c : V \rightarrow Q$, where $c(v)$ is the state of the node v at time t. (We omit time t when clear from the context.) We call a node *contender* when the value associated with its state is positive, and a *minion* when the value is negative. As previously discussed, we assume $n > 2$. Also, for presentation purposes, consider n to be a power of two. We first prove that the algorithm never eliminates all contenders and that having a single contender means that a leader is elected.

Lemma 1. *There is always at least one contender in the system. After an execution reaches a configuration with only a single node v being a contender, then from this point, v will have $c(v) > 0$ (mapped to WIN by γ) in every reachable future configuration c, and there may never be another contender.*

Proof. By the structure of the algorithm, a node starts as a contender and may become a minion during an execution, but a minion may never become a contender. Moreover, an absolute value associated with the state of a minion node can only increase to an absolute value of an interaction partner. Finally, an absolute value may never decrease except from $m + 1$ to m.

Let us assume for contradiction that an execution reaches a configuration where all nodes are minions. Consider such a time point T_0 and let the maximum absolute value of the nodes at T_0 be u. Because the minions cannot increase the maximum absolute value in the system, there must have been a contender node v and a time $T_1 < T$ such that v had value u at time T_1. In order for this contender to have become a minion by time T_0, it must have interacted with another node with an absolute value strictly larger than u, after time T_1. However, the absolute value of a node never decreases except from $m+1$ to m, and despite the existence of an absolute value larger than u before time T_0, u is the largest absolute value at time T_0. The only way this can occur is if $u = m$ and the node v interacted with a node v' with value $m + 1$. But after such an interaction the node v' remains a contender with value m. In order for v' to become a minion by time T_0, it must have interacted with yet another node v'' of value $m + 1$ at some time T_2 between T_1 and T_0. But then this node v'' is left as a contender with value m, and the same reasoning applies to it. By infinite descent, we obtain a contradiction with the initial assumption that all nodes are minions.

Consequently, whenever there is a single contender in the system, it must have the largest absolute value. Otherwise, it could interact with a node with a larger absolute value and become a minion itself, contradicting the invariant that not all nodes can be minions at the same time.

Now we turn our attention to the convergence speed of the *LM* algorithm. Our goal is bound the number of rounds necessary to eliminate all except a single contender. In order for a contender to get eliminated, it must come across a larger value of another contender, the value possibly conducted through a chain of multiple minions via multiple interactions.

We first show by a rumor spreading argument that if the difference between the values of two contenders is large enough, then the contender with the smaller value will become a minion within the next $O(n \log n)$ rounds, with constant probability. Then using anti-concentration bounds we establish that for any two contenders, if no absolute value in the system reaches m, after $O(n \log^2 n)$ rounds the difference between their values is large enough with constant probability.

Lemma 2. *Consider two contender nodes with values u_1 and u_2, where $u_1 - u_2 \geq 4\xi \log n$ at time T for $\xi \geq 8$. Then, after $\xi n \log n$ rounds from T, the node that initially held the value u_2 will be a minion with probability at least $1/24$, independent of the history of previous interactions.*

Proof. Call a node that has an absolute value of at least u_1 an *up-to-date* node, and *out-of-date* otherwise. At time T, at least one node is up-to-date. Before an arbitrary round where we have x up-to-date nodes, the probability that an out-of-date node interacts with an up-to-date node, increasing the number of up-to-date nodes to $x + 1$, is $\frac{2x(n-x)}{n(n-1)}$. By a Coupon Collector argument, the expected number of rounds until every node is up-to-date is then $\sum_{x=1}^{n-1} \frac{n(n-1)}{2x(n-x)} \leq \frac{(n-1)}{2} \sum_{x=1}^{n-1} \left(\frac{1}{x} + \frac{1}{n-x} \right) \leq 2n \log n$.

By Markov's inequality, the probability that not all nodes are up-to-date after $\xi n \log n$ communication rounds is at most $2/\xi$. Let Y denote the number of up-to-date nodes at some given time after T. It follows that, after $\xi n \log n$ rounds, $\mathbb{E}[Y] \geq \frac{n(\xi-2)}{\xi}$. Let q be the probability of having at least $\frac{n}{3} + 1$ nodes after $\xi n \log n$ communication rounds. Then we have $qn + (1-q)(\frac{n}{3}+1) \geq \mathbb{E}[Y] \geq \frac{n(\xi-2)}{\xi}$, which implies that $q \geq \frac{1}{4}$ for $n > 2$ and $\xi \geq 8$.

Hence, with probability at least $1/4$, at least $n/3 + 1$ are nodes are up to date after $\xi n \log n$ rounds. By symmetry, the $n/3$ up-to-date nodes except the original node are uniformly random among the other $n - 1$ nodes. Therefore, any given node, in particular the node that had value u_2 at time T, has probability at least $1/4 \cdot 1/3 = 1/12$ to be up-to-date after $\xi n \log n$ rounds from T.

Let v_2 be the node that had value u_2 at time T. We now wish to bound the probability that v_2 is still a contender once it becomes up-to-date. The only way in which this can happen is if it increments its value at least $4\xi \log n$ times (so that its value can reach u_1) during the first $\xi n \log n$ rounds after T. We will show that the probability of this event is at most $1/24$.

In each round, the probability to select node v_2 is $2/n$ (selecting $n - 1$ out of $n(n - 1)/2$ possible pairs). Let us describe the number of times it is selected in $\xi n \log n$ rounds by considering a random variable $Z \sim \text{Bin}(\xi n \log n, 2/n)$. By a Chernoff Bound, the probability of being selected at least $4\xi \log n$ times in these rounds is at most $\Pr[Z \geq 4\xi \log n] \leq \exp(-2\xi \log n/3) \leq 1/n^{2\xi/3} \leq 1/24$.

The next Lemma shows that, after $\Theta(n \log^2 n)$ rounds, the difference between the values of any two given contenders is high, with reasonable probability.

Lemma 3. *Fix an arbitrary time T, and a constant $\xi \geq 1$. Consider any two contender nodes at time T, and time T_1 which is $32\xi^2 n \log^2 n$ rounds after T.*

If no absolute value of any node reaches m at any time until T_1, then, with probability at least $\frac{1}{24} - \frac{1}{n^{8\xi}}$, at time T_1, either at least one of the two nodes has become a minion, or the absolute value of the difference of the two nodes' values is at least $4\xi \log n$.

Proof. We will assume that no absolute value reaches m at any point until time T_1 and that the two nodes are still contenders at T_1. We should now prove that the difference of values is large enough.

Consider $32\xi^2 n \log^2 n$ rounds following time T. If a round involves an interaction with exactly one of the two fixed nodes we call it a *spreading* round. A round is spreading with probability $\frac{4(n-2)}{n(n-1)}$, which for $n > 2$ is at least $2/n$. So, we can describe the number of spreading rounds among the $32\xi^2 n \log^2 n$ rounds by a random variable $X \sim \mathrm{Bin}(32\xi^2 n \log^2 n, 2/n)$. Then, by Chernoff Bound, the probability of having at most $32\xi^2 \log^2 n$ spreading rounds is at most

$$\Pr\left[X \leq 32\xi^2 \log^2 n\right] \leq \exp\left(-\frac{64\xi^2 \log^2 n}{2^2 \cdot 2}\right) \leq 2^{-8\xi^2 \log^2 n} < \frac{1}{n^{8\xi}},$$

Let us from now on focus on the high probability event that there are at least $32\xi^2 \log^2 n$ spreading rounds between times T and T_1, and prove that the desired difference will be large enough with probability $\frac{1}{24}$. This implies the claim by Union Bound with the above event (note that for $n > 2$, $\frac{1}{n^{8\xi}} < \frac{1}{24}$ holds).

We assumed that both nodes remain contenders during the whole time, hence in each spreading round, a value of exactly one of them, with probability $1/2$ each, increases by one. Without loss of generality assume that at time T, the value of the first node was larger than or equal to the value of the second node. Let us now focus on the sum S of k independent uniformly distributed ± 1 Bernoulli trials x_i where $1 \leq i \leq k$, where each trial corresponds to a spreading round and outcome $+1$ means that the value of the first node increased, while -1 means that the value of the second node increased. In this terminology, we are done if we show that $\Pr[S \geq 4\xi \log n] \geq \frac{1}{24}$ for $k \geq 32\xi^2 \log^2 n$ trials.

However, we have that:

$$\Pr[S \geq 4\xi \log n] \geq \Pr[|S| \geq 4\xi \log n]/2 = \Pr[|S^2| \geq 16\xi^2 \log^2 n]/2 \qquad (1)$$

$$\geq \Pr[|S^2| \geq k/2]/2 = \Pr[|S^2| \geq \mathbb{E}[S^2]/2]/2 \qquad (2)$$

$$\geq \frac{1}{2^2 \cdot 2}\frac{\mathbb{E}[S^2]^2}{\mathbb{E}[S^4]} \geq 1/24 \qquad (3)$$

where (1) follows from the symmetry of the sum with regards to the sign. For (2) we have used that $k \geq 32\xi^2 \log^2 n$ and $\mathbb{E}[S^2] = k$. Finally, to get (3) we use the Paley-Zygmund inequality and the fact that $\mathbb{E}[S^4] = 3k(k-1) + k \leq 3k^2$. Evaluating $\mathbb{E}[S^2]$ and $\mathbb{E}[S^4]$ is simple by using the definition of S and the linearity of expectation. The expectation of each term then is either 0 or 1 and it suffices to count the number of terms with expectation 1, which are exactly the terms where each multiplier is raised to an even power.

Now we are ready to prove the bound on convergence speed.

Theorem 1. *There exists a constant α, such that for any constant $\beta \geq 3$ following holds: If we set $m = \alpha\beta \log^3 n = \Theta(\log^3 n)$, the algorithm elects a leader (i.e. reaches a configuration with a single contender) in at most $O(n \log^3 n)$ rounds (i.e. parallel time $O(\log^3 n)$) with probability at least $1 - 1/n^\beta$.*

Proof. Let us fix constants $0 < p < 1$ and $\xi \geq 8$ large enough such that

$$1/24 \cdot \left(1/24 - 1/n^{8\xi}\right) \geq p. \tag{4}$$

Let β be any constant ≥ 3 and take $\alpha = 16(33\xi^2)/p$. We set $m = \alpha\beta \log^3 n$ and consider the first $\alpha\beta n \log^3 n/4$ rounds of the algorithm's execution. For a fixed node, the probability that it interacts in each round is $2/n$. Let us describe the number of times a given node interacts within the first $\alpha\beta n \log^3 n/4$ rounds by a random variable $B \sim \mathrm{Bin}(\alpha\beta n \log^3 n/4, 2/n)$. By the Chernoff Bound, the probability of being selected more than m times during these rounds is at most:

$$\Pr\left[B \geq m\right] \leq \exp\left(-\alpha\beta \log^3 n/6\right) \leq 2^{-\frac{\alpha\beta}{6}\log^3 n} \leq 1/n^{\alpha\beta/6}.$$

Taking the Union Bound over all n nodes, with probability at least $1 - (n/n^{\alpha\beta/6})$, all nodes interact strictly less than m times during the first $\alpha\beta n \log^3 n/4$ rounds.

Next, let us focus on the high probability event above, meaning that all absolute values are strictly less than m during the first $\frac{\alpha\beta n \log^3 n}{4} = \frac{4\beta}{p}(33\xi^2)n \log^3 n$ rounds. For a fixed pair of nodes, this allows us to apply Lemma 3 followed by Lemma 2 (with parameter ξ) $\frac{4\beta(33\xi^2)n \log^3 n}{p(32\xi^2 n \log^2 n + \xi n \log n)} \geq \frac{4\beta \log n}{p}$ times. Each time, by Lemma 3, after $32\xi^2 n \log^2 n$ rounds with probability at least $1/24 - 1/n^{8\xi}$ the nodes get values at least $4\xi \log n$ apart. Then, after the next $\xi n \log n$ rounds, by Lemma 2, one of the nodes becomes a minion with probability at least $1/24$. Since Lemma 2 is independent from the interactions that precede it, by (4), each of the $\frac{4\beta \log n}{p}$ times if both nodes are contenders, we get probability at least p that one of the nodes becomes a minion. Consider a random variable $W \sim \mathrm{Bin}\left(4\beta \log n/p, p\right)$. By Chernoff bound the probability that both nodes in a given pair are still contenders after $\frac{\alpha\beta n \log^3 n}{4}$ rounds is at most:

$$\Pr\left[W \leq 0\right] = \Pr\left[W \leq 4\beta \log n\,(1-1)\right] \leq \exp\left(-\frac{4\beta \log n}{2}\right) \leq 2^{-2\beta \log n} < \frac{1}{n^{2\beta}},$$

By a Union Bound over all $< n^2$ pairs, for every pair of nodes, one of them is a minion after $\frac{\alpha\beta n \log^3 n}{4}$ communication rounds with probability at least $1 - \frac{n^2}{n^{2\beta}}$. Hence, with this probability, there will be only one contender.

Finally, combining with the conditioned event that none of the nodes interact m or more times gives that after the first $\frac{\alpha\beta n \log^3 n}{4} = O(n \log^3 n)$ rounds there must be a single contender with probability at least $1 - \frac{n^2}{n^{2\beta}} - \frac{n}{n^{\alpha\beta/6}} \geq 1 - \frac{1}{n^\beta}$ for $\beta \geq 3$. A single contender means that leader is elected by Lemma 1.

Finally, we can prove the expected convergence bound.

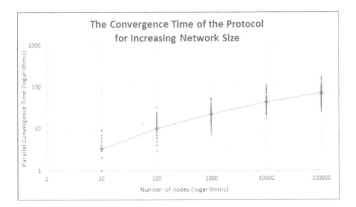

Fig. 2. The performance of the *LM* protocol. Both axes are logarithmic. The dots represent the results of individual experiments (100 for each network size), while the solid line represents the mean value for each network size.

Theorem 2. *There is a setting of parameter m of the algorithm such that $m = \Theta(\log^3 n)$, and the algorithm elects the leader in expected $O(n \log^3 n)$ rounds of communication (i.e. parallel time $O(\log^3 n)$).*

Proof. Let us prove that from any configuration, the algorithm elects a leader in expected $O(n \log^3 n)$ rounds. By Lemma 1, there is always a contender in the system and if there is only a single contender, then a leader is already elected. Now in a configuration with at least two contenders consider any two of them. If their values differ, then with probability at least $1/n^2$ these two contenders will interact in the next round and the one with the lower value will become a minion (after which it may never be a contender again). If the values are the same, then with probability at least $1/n$, one of these nodes will interact with one of the other nodes in the next round, leading to a configuration where the values of our two nodes differ[4], from where in the next round, independently, with probability at least $1/n^2$ these nodes meet and one of them again becomes a minion. Hence, unless a leader is already elected, in any case, in every two rounds, with probability at least $1/n^3$ the number of contenders decreases by 1.

Thus the expected number of rounds until the number of contenders decreases by 1 is at most $2n^3$. In any configuration there can be at most n contenders, thus the expected number of rounds until reaching a configuration with only a single contender is at most $2(n-1)n^3 \leq 2n^4$ from any configuration.

Now using Theorem 1 with $\beta = 4$ we get that with probability at least $1 - 1/n^4$ the algorithm converges after $O(n \log^3 n)$ rounds. Otherwise, with probability at most $1/n^4$ it ends up in some configuration from where it takes at most $2n^4$ expected rounds to elect a leader. The total expected number of rounds is therefore also $O(n \log^3 n) + O(1) = O(n \log^3 n)$, i.e. parallel time $O(\log^3 n)$.

---

[4] This is always true, even when the new value is not larger, for instance when the values were equal to $m + 1$, the new value of one of the nodes will be $m \neq m + 1$.

5   Experiments and Discussion

Empirical Data. We have also measured the convergence time of our protocol for different network sizes. (Figure 2 presents the results in the form of a log-log plot.) The protocol converges to a single leader quite fast, e.g., in less than 100 units of parallel time for a network of size 10^5. This suggests that the constants hidden in the asymptotic analysis are small. The shape of the curve confirms the poly-logarithmic behavior of the protocol.

Discussion. We have given the first population protocol to solve leader election in poly-logarithmic time, using a poly-logarithmic number of states per node. Together with [AAE06], the existence of our protocol implies that population protocols can compute any semi-linear predicate on their input in time $O(n \log^5 n)$, with high probability, as long as memory per node is poly-logarithmic.

Our result opens several avenues for future research. The first concerns *lower bounds*. We conjecture that the lower bound for leader election in population protocols is $\Omega(\log n)$, irrespective of the number of states. Further, empirical data suggests that the analysis of our algorithm can be tightened, cutting logarithmic factors. It would also be interesting to prove a tight a trade-off between the amount of memory available per node and the running time of the protocol.

Acknowledgments. Support is gratefully acknowledged from the National Science Foundation under grants CCF-1217921, CCF-1301926, and IIS-1447786, the Department of Energy under grant ER26116/DE-SC0008923, and the Oracle and Intel corporations."

References

[AAD+06] Angluin, D., Aspnes, J., Diamadi, Z., Fischer, M.J., Peralta, R.: Computation in networks of passively mobile finite-state sensors. Distributed computing **18**(4), 235–253 (2006)

[AAE06] Angluin, D., Aspnes, J., Eisenstat, D.: Stably computable predicates are semilinear. In: Proceedings of PODC 2006, pp. 292–299 (2006)

[AAE08a] Angluin, D., Aspnes, J., Eisenstat, D.: Fast computation by population protocols with a leader. Distributed Computing **21**(3), 183–199 (2008)

[AAE08b] Angluin, D., Aspnes, J., Eisenstat, D.: A simple population protocol for fast robust approximate majority. Distributed Computing **21**(2), 87–102 (2008)

[AAER07] Angluin, D., Aspnes, J., Eisenstat, D., Ruppert, E.: The computational power of population protocols. Distributed Computing **20**(4), 279–304 (2007)

[AAFJ06] Angluin, D., Aspnes, J., Fischer, M.J., Jiang, H.: Self-stabilizing population protocols. In: Anderson, J.H., Prencipe, G., Wattenhofer, R. (eds.) OPODIS 2005. LNCS, vol. 3974, pp. 103–117. Springer, Heidelberg (2006)

[CDS14] Chen, H.-L., Doty, D., Soloveichik, D.: Deterministic function computation with chemical reaction networks. Natural computing **13**(4), 517–534 (2014)

[DH13]   Doty, D., Hajiaghayi, M.: Leaderless deterministic chemical reaction networks. In: Soloveichik, D., Yurke, B. (eds.) DNA 2013. LNCS, vol. 8141, pp. 46–60. Springer, Heidelberg (2013)

[DS15]   Doty, D., Soloveichik, D.: Stable leader election in population protocols requires linear time (2015). ArXiv preprint. http://arxiv.org/abs/1502.04246

[DV12]   Draief, M., Vojnovic, M.: Convergence speed of binary interval consensus. SIAM Journal on Control and Optimization **50**(3), 1087–1109 (2012)

[FJ06]   Fischer, M., Jiang, H.: Self-stabilizing leader election in networks of finite-state anonymous agents. In: Shvartsman, M.M.A.A. (ed.) OPODIS 2006. LNCS, vol. 4305, pp. 395–409. Springer, Heidelberg (2006)

[PVV09]   Perron, E., Vasudevan, D., Vojnovic, M.: Using three states for binary consensus on complete graphs. In: IEEE INFOCOM 2009, pp. 2527–2535. IEEE (2009)

[SNY⁺10]   Sudo, Y., Nakamura, J., Yamauchi, Y., Ooshita, F., Kakugawa, H., Masuzawa, T.: Loosely-stabilizing leader election in population protocol model. In: Kutten, S., Žerovnik, J. (eds.) SIROCCO 2009. LNCS, vol. 5869, pp. 295–308. Springer, Heidelberg (2010)

Core Size and Densification
in Preferential Attachment Networks

Chen Avin[1], Zvi Lotker[1], Yinon Nahum[2(✉)], and David Peleg[2]

[1] Ben-Gurion University of the Negev, Beer Sheva, Israel
{avin,zvilo}@cse.bgu.ac.il
[2] The Weizmann Institute, Rehovot, Israel
{yinon.nahum,david.peleg}@weizmann.ac.il

Abstract. Consider a preferential attachment model for network evolution that allows both node and edge arrival events: at time t, with probability p_t a new node arrives and a new edge is added between the new node and an existing node, and with probability $1 - p_t$ a new edge is added between two existing nodes. In both cases existing nodes are chosen at random according to preferential attachment, i.e., with probability proportional to their degree. For $\delta \in (0, 1)$, the δ-*founders* of the network at time t is the minimal set of the first nodes to enter the network (i.e., founders) guaranteeing that the sum of degrees of nodes in the set is at least a δ fraction of the number of edges in the graph at time t. We show that for the common model where p_t is constant, i.e., when $p_t = p$ for every t and the network is sparse with linear number of edges, the size of the δ-founders set is concentrated around $\delta^{2/p} n_t$, and thus is linear in n_t, the number of nodes at time t. In contrast, we show that for $p_t = \min\{1, \frac{2a}{\ln t}\}$ and when the network is dense with super-linear number of edges, the size of the δ-founders set is sub-linear in n_t and concentrated around $\bar{\Theta}((n_t)^\eta)$, where $\eta = \delta^{1/a}$.

1 Introduction

Preferential Attachment is one of the prevalent mechanisms for network evolution, in the context of both social networks and other complex systems, and considerable efforts have been invested in studying its structure, properties and behavior. In particular, preferential attachment models were shown to generate networks with a power law degree distribution [9], a property that has been observed in many real life networks and is thought to be universal. Additional study of large scale social networks has revealed several other universal properties, for example the "small-world" phenomena, short average path lengths, navigability and high clustering coefficients.

Recently, two seemingly unrelated properties of social networks have been explored and analyzed. The first is a property of evolving networks known as *densification* [7], namely, the property that the network becomes denser over time. The second property concerns the *core-periphery* structure of the network:

Supported in part by the Israel Science Foundation (grant 1549/13).

M.M. Halldórsson et al. (Eds.): ICALP 2015, Part II, LNCS 9135, pp. 492–503, 2015.
DOI: 10.1007/978-3-662-47666-6_39

it is claimed that many networks exhibit a small and dense *core*, surrounded by a larger and sparser periphery [1].

Traditionally, most of the existing literature on preferential attachment networks and models deals with linear-sized networks, namely, networks whose number of edges is linear and whose average degree is constant over time. However, several recent empirical studies have suggested that the number of edges in a social network is sometimes *superlinear* [7], namely, the average degree grows over time, which leads to densification. Somewhat analogously, while much of the evidence and research concerning cores seems to indicate that core size is often linear, it was recently shown empirically that in certain settings, social networks have a dense *sub-linear* core [1]. These two seemingly unrelated discrepancies are the subject of the current paper, which focuses attention on the differences between linear and nonlinear sized preferential attachment social networks, and considers the effect of the number of edges in (or the density of) the network on the size of its core.

Once a universal property is observed experimentally, it is desirable to develop an evolutionary model in which the observed property naturally arises. For example, the preferential attachment model itself was proposed by Price [10] to explain observed power law distribution in citation networks and later by Barabasi et al. [2] in order to explain the observed heavy-tailed degree distributions in various other networks. Likewise, the model of Fraigniaud et al. [5] can be viewed as an evolutionary explanation for the navigability property observed by Milgram [8] and elaborated upon by Kleinberg [6]. This general approach is adopted in the current paper as well, namely, we attempt to explain the above discrepancies by analyzing the core size in a suitable evolutionary model. On the face of it, it is unclear why and how a dense sublinear core can emerge in an evolutionary model. In a nutshell, our main result shows that in the preferential attachment model, the core size depends "inversely" on the network density, in the sense that, whereas a sparse social network with linear number of edges (i.e., constant average degree) tend to have a linear-sized core, it turns out that denser social networks, which have a superlinear number of edges, tend to have a sublinear core.

While the notion of a core-periphery partition is intuitively clear, there are several different approaches to formalizing its definition [3,11,12]. All of these definitions share the common assumption, which is in fact inherently at the basis of the preferential attachment model, that the amount of power or influence of a vertex in a social network is proportional to its degree, that is, higher degree vertices are more powerful. Hence intuitively, the core should hold a sizable fraction of the degrees in the network while the periphery is much sparser and less connected. Two such examples are (i) to consider the core as the *"rich club"* [13] of the network, i.e., the set of nodes with the highest degrees in the network, and (ii) to define the core properties using an axiomatic approach [1], and leave open the possibility for several candidate sets.

Yet another characterization for the core can be motivated by a "historical" perspective on the evolution of a social network. By this viewpoint, the core should consist of the *"founders"* of the network, namely, the first vertices to join and establish it. Formally, the δ-founders set of a given network G consists of the γ vertices which were the first to join the network, where γ is the minimum integer such that the sum of the degrees of the vertices of the founders set is at least δ times the sum of the degrees of the vertices of the entire network.

The rest of the paper is organized as follows. Section 2 presents the model formally. Section 3 provides an overview of the results. The following sections present a detailed analysis of the model in its two cases: linear and super-linear number of edges. Due to lack of space we defer many of the technical details and proofs to the full version.

2   Model and Preliminaries

In this section we describe a slight generalization of the Preferential Attachment model $G(p_t)$ defined in [4] Ch 3.1. $G(p_t)$ has one parameter, p_t, which was assumed to be constant with t. Here we allow p_t to vary with t. Let us now introduce the model formally.

Consider a sequence $(G_t)_{t=1}^\infty$ of graphs, $G_t = (V_t, E_t)$, where V_t denotes the set of vertices in G_t and $n_t = |V_t|$. Let $d_t(w)$ denote the degree of node w in G_t. The initial graph G_1 consists of a single node v_1 and a single self loop (counted as degree 2), and for $t \geq 1$ the graph G_{t+1} is constructed from G_t at time t by performing either a *node event* with probability $p_t \in (0, 1]$, or an *edge event* with probability $1 - p_t$. In a node event, a new vertex v is added to the graph, along with a new edge (v, u) where $u \in V_t$ is chosen using *preferential attachment*, i.e., with probability proportional to its degree, i.e., with probability $p_u^t = \frac{d_t(u)}{\sum\limits_{w \in V_t} d_t(w)}$.

In an edge event, a new edge (v, u) is added, where both $u, v \in V_t$ are chosen by preferential attachment, independently of each other, i.e., (u, v) is chosen with probability $p_u^t \cdot p_v^t$. Note that each time step adds exactly one edge, so the number of edges in the graph G_t is t and the sum of degrees is $2t$.

Traditionally $G(p_t)$ was studied with constant p_t, i.e., $p_t = p$ for all t. In this paper we also study the case where $p_t = \min\{1, \frac{2a}{\ln t}\}$ for fixed $a > 0$.

The goal of the paper is to study the power and size of the *core* vs. the *periphery* of G_t. As mentioned earlier, intuitively the nodes in the core should be very powerful and well-connected within and to the network, but there is no single, clear definition for a core in the context of a social networks. Here we study a natural candidate set to be considered as the network core: the set of nodes who joined the network first, or the *founders* of the network. We conjecture that other possible definitions of the core will be highly correlated with the notion of founders, but we leave this direction to the full version of the paper.

Founders Definition. For G_t we name the vertices in the graph $v_1, v_2 \ldots$ according to their order of arrival, where $i < j$ if node v_i arrived before node v_j. Since at most one vertex joins the graph at each time step, this order is well defined. Intuitively, one may think of the nodes that arrived the network first as the founders of the society since they will be in the graph for the longest time and

Table 1. Summary of properties of the model $G(p_t)$ for constant p_t and $p_t = \min\{1, 2a/\ln t\}$. $\tilde{\Theta}$ hides polylogarithmic terms.

Parameters of $G(p_t)$	$p_t = p$	$p_t = \min\{1, \frac{2a}{\ln t}\}$
Number of edges	t	
Expected number of nodes $\mathbb{E}[n_t]$	$1 + p(t-1)$	$\frac{2at}{\ln t}(1 \pm o(1))$
Expected number of edges in terms of n_t	**Linear in n_t:** $p^{-1} n_t$	**super-linear in n_t:** $\frac{n_t \ln(n_t)}{2a}(1 \pm o(1))$
Expected δ-*founders* size $\mathbb{E}[\gamma_t^\delta]$	**Linear in n_t:** $\delta^{2/p} n_t (1 \pm o(1))$	**sub-linear in n_t:** $\tilde{\Theta}((n_t)^{\delta^{1/a}})$
Power law exponent β	$2 + \frac{p}{2-p}$	2

consequently, by the preferential attachment mechanism, one would expect them to be among the nodes of highest degrees in the graph. The degree of a vertex is well accepted as a measure of its power and influence, and similarly we use the sum of degrees of a group of vertices to capture its power. The power of the i founders at time t is denoted by $S_{i,t} = \sum_{j=1}^{i} d_t(v_j)$.

To consider the founders as a core, the founders set should have enough power, but what is enough? We use a parameter δ to quantify this and seek the minimal size set of founders that·has a δ fraction of the total power of the entire network (i.e, the sum of degrees of all nodes). This set is referred to as the δ-founders set. Formally:

Definition 1. *The δ-founders set of a graph G_t (for $\delta \in [0,1]$) is defined as $C_t^\delta = \{v_1, \ldots, v_{\gamma_t^\delta}\}$, where γ_t^δ is the minimum integer such that $S_{\gamma_t^\delta, t} \geq \delta \cdot 2t$.*

Recall that $\sum_{v \in V_t} d_t(v) = 2t$, thus the δ-founders set has at least δ fraction of the degrees. Our goal is to analyze $\gamma_t^\delta = |C_t^\delta|$ over time. Since γ_t^δ is increasing in δ, and $\gamma_t^0 = 0$, $\gamma_t^1 = n_t$, we are interested in γ_t^δ for $\delta \in (0,1)$. We note:

Lemma 1. *For every nonnegative integers t and i,*

$$\mathbb{P}[\gamma_t^\delta > i] = \mathbb{P}[S_{i,t} < 2\delta t] \qquad and \qquad \mathbb{P}[\gamma_t^\delta \leq i] = \mathbb{P}[S_{i,t} \geq 2\delta t].$$

Therefore, we can bound the size of the core C_t^δ by bounding $S_{i,t}$ for all i. Furthermore, We would like to bound γ_t^δ in terms of n_t.

3   Results Overview

The main contribution of the paper is in rigorously analyzing the $G(p_t)$ preferential attachment model when $p_t = \min\{1, \frac{2a}{\ln t}\}$. As it turns out, changing p_t from a constant to p_t s.t $p_t \xrightarrow[t \to \infty]{} 0$ leads to significant changes in the graph structure.

Moreover, the proofs for the concentration of the measures of interest (such as $S_{i,t}$ and γ_t^δ) are challenging.

Table 1 highlights the main changes in our properties and measures of interest for the two parameter settings of the model. For clarity, we present the expected values of the various measures, but the technical sections provide proofs for bounds that hold with high probability as well. These results offer a possible explanation for two seemingly unrelated phenomena, namely, *densefication* and *sub-linear core size*. We observe that if p_t is constant then the network does not exhibit densefication, namely, it has a linear number of edges and a constant average degree, which results also in a linear core size. In contrast, when $p_t = \min\{1, \frac{2a}{\ln t}\}$, the network becomes dense with average degrees increasing logarithmically, which results also in a sub-linear core size. These findings may suggest that these two phenomena are indeed related to each other.

It is important to note that one can also relate the above results to the degree distribution of the graph. It is known that for a constant $p_t = p > 0$ the degree distribution of $G(p)$ follows a power law (i.e., the fraction of nodes with degree k is proportional to $k^{-\beta}$) with exponent $2 < \beta \leq 3$, depending on p. In the full paper we show that for the case $p_t = \min\{1, \frac{2a}{\ln t}\}$ the degree distribution remains a power law, but for the special case where $\beta = 2$. Nevertheless, the bounds on the power law and β alone are insufficient to establish our conclusions with high probability, therefore the bounds on the power and size of the founders set have to be derived directly on $G(p_t)$ and not from the degree distribution.

Technically, our main analysis effort concentrated on proving our claims on the core sizes for the founders definition, namely bounding $S_{i,t}$ and γ, the size of the δ-founders set, with high probability. To do so, we first had to study the evolution process of the network and provide bounds for τ_i, the arrival time of node i. Once establishing the concentration of τ_i, it becomes possible to bound $S_{i,t}$ conditioned on τ_i. We believe that similar bounds can be derived for other definitions for the network core.

4   Expectation and Concentration of Degree Sums

Let $n_t = |V_t|$ denote the number of vertices at time t, and $\tau_i = \min\{t \mid n_t \geq i\}$ denote the arrival time of node v_i. Given τ_i, we analyze $\mathbb{E}[S_{i,t} \mid \tau_i]$, the expectation of $S_{i,t}$, for various times t. Formally we prove the following theorems.

Theorem 1 (Expectation of $S_{i,t}$). *For any positive integers i and t,*

$$\mathbb{E}[S_{i,t} \mid \tau_i] = 2t \prod_{j=\tau_i+1}^{t} \left(1 - \frac{p_{j-1}}{2j}\right), \tag{1}$$

where the empty product, corresponding to $t \leq \tau_i$, is defined to be equal to 1.

We also show the concentration of $S_{i,t}$ around $\mathbb{E}[S_{i,t} \mid \tau_i]$.

Theorem 2 (Concentration of $S_{i,t}$). *For any positive integers i and t, and real $\lambda \geq 0$,*

$$\mathbb{P}[S_{i,t} - \mathbb{E}\left[S_{i,t} \mid \tau_i\right] \geq \lambda \mid \tau_i] \leq \exp\left(-\frac{\lambda^2 \tau_i}{32t^2}\right), \tag{2}$$

$$\mathbb{P}[S_{i,t} - \mathbb{E}\left[S_{i,t} \mid \tau_i\right] \leq -\lambda \mid \tau_i] \leq \exp\left(-\frac{\lambda^2 \tau_i}{32t^2}\right), \tag{3}$$

We make use of the following technical lemma.

Lemma 2. *For integers t, k such that $t \geq k \geq 0$ and reals $r_k, r_{k+1}, \ldots, r_{t-1}$,*

$$k \prod_{j=k}^{t-1} \left(1 + \frac{1 - r_j}{j}\right) = t \prod_{j=k+1}^{t} \left(1 - \frac{r_{j-1}}{j}\right).$$

Proof of Thm. 1:   Note that for every $t \leq \tau_i$, we have $S_{i,t} = 2t = \mathbb{E}\left[S_{i,t} \mid \tau_i\right]$. Therefore, we prove Thm. 1 and Thm. 2 for times $t > \tau_i$. Then a node event will increase $S_{i,t+1}$ by 1 w.r.t $S_{i,t}$ if and only if the node $u \in V_t$, to which the new arriving v is attached, is one of the first i nodes, which happens with probability $\frac{S_{i,t}}{2t}$. Similarly, an edge event, where the edge (u, v) is added, will increase $S_{i,t+1}$ by 1 w.r.t $S_{i,t}$ if u is one of the first i nodes, which happens with probability $\frac{S_{i,t}}{2t}$ and by another 1, if v is one of the first i nodes, which happens, again, with probability $\frac{S_{i,t}}{2t}$. Hence, we have the following regression formula.

$$\mathbb{E}\left[S_{i,t+1} \mid \tau_i, S_{i,t}\right] = S_{i,t} + p_t \frac{S_{i,t}}{2t} + 2(1 - p_t)\frac{S_{i,t}}{2t} = \left(1 + \frac{2 - p_t}{2t}\right) S_{i,t}. \tag{4}$$

Taking expectations on both sides and proceeding with the recursion, we have

$$\mathbb{E}\left[S_{i,t} \mid \tau_i\right] = \begin{cases} 2t, & t \leq \tau_i, \\ 2\tau_i \cdot Z_{\tau_i, t}, & t \geq \tau_i, \end{cases} \tag{5}$$

where $Z_{j,t} = \prod_{\ell=j}^{t-1} \left(1 + \frac{2 - p_\ell}{2\ell}\right)$. Applying Lemma 2 with (k, r_j) set to $(\tau_i, p_j/2)$, Thm. 1 follows. $\square$

Proof of Thm. 2:   Again, it suffices to consider $t \geq \tau_i$. Let $X_{i,t} = \frac{S_{i,t}}{Z_{\tau_i, t}} \mid \tau_i$. We have that $(X_{i,t})_{t \geq \tau_i}$ is a martingale, since

$$\mathbb{E}\left[X_{i,t+1} \mid X_{i,t}, \ldots, X_{i,\tau_i}\right] = \mathbb{E}\left[X_{i,t+1} \mid X_{i,t}\right] = \mathbb{E}\left[X_{i,t+1} \mid \tau_i, S_{i,t}\right]$$
$$= \frac{\mathbb{E}\left[S_{i,t+1} \mid \tau_i, S_{i,t}\right]}{Z_{i,t+1}} = \frac{S_{i,t}}{Z_{\tau_i, t}} = X_{i,t},$$

where the third equality follows since given τ_i, $Z_{i,t+1}$ is a fixed number (and not a random variable), and the fourth equality follows by Eq. (4).

Since $|S_{i,t} - S_{i,t-1}| \leq 2$ and $S_{i,t} \leq 2t$, we have

$$
\begin{aligned}
Z_{\tau_i,t} \cdot |X_{i,t} - X_{i,t-1}| &= \left| S_{i,t} - \left(1 + \frac{2 - p_{t-1}}{2(t-1)} \right) S_{i,t-1} \right| \\
&\leq |S_{i,t} - S_{i,t-1}| + \frac{2 - p_{t-1}}{2(t-1)} S_{i,t-1} \leq 2 + (2 - p_{t-1}) \frac{S_{i,t-1}}{2(t-1)} \\
&\leq 4 - p_{t-1} \leq 4 .
\end{aligned}
$$

Hence, $|X_{i,t} - X_{i,t-1}| \leq 4/Z_{\tau_i,t}$. Applying Azuma's Inequality to the martingale $(X_{i,t})_{t \geq \tau_i}$ yields, for $\lambda \geq 0$,

$$
\mathbb{P}[X_{i,t} - \mathbb{E}[X_{i,t} \mid \tau_i] \geq \lambda \mid \tau_i] \leq \exp\left(-\lambda^2 / \left(32 \sum_{j=\tau_i+1}^{t} Z_{\tau_i,j}^{-2} \right) \right),
$$

$$
\mathbb{P}[X_{i,t} - \mathbb{E}[X_{i,t} \mid \tau_i] \leq -\lambda \mid \tau_i] \leq \exp\left(-\lambda^2 / \left(32 \sum_{j=\tau_i+1}^{t} Z_{\tau_i,j}^{-2} \right) \right).
$$

Substituting $\lambda/Z_{\tau_i,t}$ for λ, and noting that $Z_{a,c} = Z_{a,b} \cdot Z_{b,c}$ for integers $a \leq b \leq c$,

$$
\mathbb{P}[S_{i,t} - \mathbb{E}[S_{i,t} \mid \tau_i] \geq \lambda \mid \tau_i] \leq \exp\left(-\lambda^2 / \left(32 \sum_{j=\tau_i+1}^{t} Z_{j,t}^2 \right) \right),
$$

$$
\mathbb{P}[S_{i,t} - \mathbb{E}[S_{i,t} \mid \tau_i] \leq -\lambda \mid \tau_i] \leq \exp\left(-\lambda^2 / \left(32 \sum_{j=\tau_i+1}^{t} Z_{j,t}^2 \right) \right).
$$

Noting that $Z_{j,t} \leq \prod_{\ell=j}^{t-1} \left(1 + \ell^{-1} \right) = t/j$, and that

$$
\sum_{j=\tau_i+1}^{t} j^{-2} \leq \int_{\tau_i}^{t} j^{-2} dj \leq \int_{\tau_i}^{\infty} j^{-2} dj = \tau_i^{-1} ,
$$

Thm. 2 follows. $\square$

5   Concentration Inequalities for Arrival Times

The number of vertices at time t, denoted n_t, can be written as a sum of independent variables $n_t = 1 + \sum_{j=1}^{t-1} N_j$, where where N_j is an indicator which equals 1 if a new node arrived at time j, and 0 otherwise. Let $Q(t) = \mathbb{E}[n_t] = 1 + \sum_{j=1}^{t-1} p_t$ denote the expected number of vertices at time t. We then have the following bounds on τ_i, the expected arrival time of node i.

Lemma 3. *For any positive integers t_L, t_H, i satisfying $Q(t_L) \leq i \leq Q(t_H)+1$,*

$$
\mathbb{P}[\tau_i \leq t_L] \leq \exp\left(\frac{-2(i - Q(t_L))^2}{t_L - 1} \right) , \tag{6}
$$

$$
\mathbb{P}[\tau_i > t_H] \leq \exp\left(\frac{-2(Q(t_H) + 1 - i)^2}{t_H - 1} \right) , \tag{7}
$$

Proof. By definition of τ_i,

$$\mathbb{P}[\tau_i \le t_L] = \mathbb{P}[n_{t_L} \ge i] = \mathbb{P}\left[n_{t_L} - \mathbb{E}[n_{t_L}] \ge i - Q(t_L)\right]$$

$$= \mathbb{P}\left[\sum_{j=1}^{t_L-1} N_j - \sum_{j=1}^{t_L-1} p_j \ge i - Q(t_L)\right].$$

Recalling that $N_j \in \{0, 1\}$ and applying Hoeffding's Inequality, Ineq. (6) follows. Similarly for t_H,

$$\mathbb{P}[\tau_i > t_H] = \mathbb{P}[n_{t_H} < i] = \mathbb{P}[n_{t_H} \le i - 1] = \mathbb{P}\left[n_{t_H} - \mathbb{E}[n_{t_H}] \le i - Q(t_H) - 1\right],$$

yielding Ineq. (7). $\square$

By Lemma 3, we obtain the following bounds on γ_t^δ, the size of the core C_t^δ at time t.

Corollary 1. *For every positive integers t, i, t_L, t_H satisfying $Q(t_L) \le i \le Q(t_H) + 1$,*

$$\mathbb{P}[\gamma_t^\delta \le i] \le \exp\left(\frac{-2(i - Q(t_L))^2}{t_L - 1}\right) + \exp\left(\frac{-2(Q(t_H) + 1 - i)^2}{t_H - 1}\right)$$

$$+ \sum_{k=t_L+1}^{t_H} \left(\mathbb{P}[\tau_i = k] \cdot \mathbb{P}[S_{i,t} \ge 2\delta t \mid \tau_i = k]\right), \tag{8}$$

$$\mathbb{P}[\gamma_t^\delta > i] \le \exp\left(\frac{-2(i - Q(t_L))^2}{t_L - 1}\right) + \sum_{k=t_L+1}^{t} \left(\mathbb{P}[\tau_i = k] \cdot \mathbb{P}[S_{i,t} \ge 2\delta t \mid \tau_i = k]\right). \tag{9}$$

Proof. By Lemma 1,

$$\mathbb{P}[\gamma_t^\delta \le i] = \mathbb{P}[S_{i,t} \ge 2\delta t] = \sum_{k=i}^{\infty} \left(\mathbb{P}[\tau_i = k] \cdot \mathbb{P}[S_{i,t} \ge 2\delta t \mid \tau_i = k]\right)$$

$$\le \mathbb{P}[\tau_i \le t_L] + \sum_{k=t_L+1}^{t_H} \left(\mathbb{P}[\tau_i = k] \cdot \mathbb{P}[S_{i,t} \ge 2\delta t \mid \tau_i = k]\right) + \mathbb{P}[\tau_i > t_H].$$

Applying Lemma 3 to the last inequality, Ineq. (8) follows. Similarly,

$$\mathbb{P}[\gamma_t^\delta > i] \le \mathbb{P}[\tau_i \le t_L] + \sum_{k=t_L+1}^{\infty} \left(\mathbb{P}[\tau_i = k] \cdot \mathbb{P}[S_{i,t} < 2\delta t \mid \tau_i = k]\right).$$

Recalling that for $t \le \tau_i$ we have $S_{i,t} = 2t \ge 2\delta t$, the sum may be truncated once k reaches t. Applying Lemma 3, Ineq (9) follows.

6   Core Size in Linear-Sized Networks

In this section we assume p_t is constant, i.e., $p_t = p$ for every t. We now show that γ_t^δ, the size of the core C_t^δ at time t, is concentrated around ηpt, where $\eta = \delta^2/p$. Formally, we prove the following theorem.

Theorem 3. *For every nonnegative* $\alpha, \alpha_1, \alpha_2 = \Omega(1)$ *satisfying* $\alpha = \alpha_1 + \alpha_2$ *and* $\alpha = o(\sqrt{t})$,

$$
\begin{aligned}
(I) \;\; \mathbb{P}[\gamma_t^\delta \le \eta pt - \alpha\sqrt{\eta t}] \;&\le\; 2\exp\left(-2\alpha_1^2\left(1 - O\left(1/(\alpha_1\sqrt{t})\right)\right)\right) \\
&\quad + \exp\left(-\tfrac{\alpha_2^2\delta^2}{32}\left(1 - O\left(\alpha/\sqrt{t}\right)\right)\right), \\
(II) \;\; \mathbb{P}[\gamma_t^\delta > \eta pt + \alpha\sqrt{\eta t}] \;&\le\; \exp\left(-2\alpha_1^2\left(1 - O\left(\alpha_2/\sqrt{t}\right)\right)\right) \\
&\quad + \exp\left(-\tfrac{\alpha_2^2\delta^2}{32}\left(1 - O\left(\alpha_2/\sqrt{t}\right)\right)\right).
\end{aligned}
$$

Theorem 3 is a result of Corollary 1. Specifically, part (I) is obtained by applying the following lemma:

Lemma 4. *For any nonnegative* $\alpha, \alpha_1, \alpha_2 = \Omega(1)$ *satisfying* $\alpha = \alpha_1 + \alpha_2$ *and* $\alpha = o(\sqrt{t})$*, and for* $i = \lfloor \eta pt - \alpha\sqrt{\eta t}\rfloor$*,* $t_L = \lfloor \eta t - (\alpha + \alpha_1)\sqrt{\eta t}/p\rfloor$*,* $t_H = \lfloor \eta t - \alpha_2\sqrt{\eta t}/p\rfloor$ *and every* $k \in [t_L + 1, t_H]$*, the following hold:*

$$\mathbb{P}[\gamma_t^\delta \le i] = \mathbb{P}[\gamma_t^\delta \le \eta pt - \alpha\sqrt{\eta t}], \tag{10}$$

$$\frac{(Q(t_H) + 1 - i)^2}{t_H - 1} \ge \alpha_1^2, \tag{11}$$

$$\frac{(i - Q(t_L))^2}{t_L - 1} \ge \alpha_1^2\left(1 - O\left(\frac{1}{\alpha_1\sqrt{t}}\right)\right), \tag{12}$$

$$\mathbb{P}[S_{i,t} \ge 2\delta t \mid \tau_i = k] \le \exp\left(-\frac{\alpha_2^2\delta^2}{32}\left(1 - O\left(\frac{\alpha}{\sqrt{t}}\right)\right)\right). \tag{13}$$

Plugging Eq. (10) and Ineq. (11),(12),(13) into Ineq. (8) with these values of (i, t_L, t_H), Thm. 3(I) follows.

Part (II) is obtained by applying the following lemma:

Lemma 5. *For any nonnegative* $\alpha, \alpha_1, \alpha_2 = \Omega(1)$ *satisfying* $\alpha = \alpha_1 + \alpha_2$ *and* $\alpha = o(\sqrt{t})$*, and for* $i = \lfloor \eta pt + \alpha\sqrt{\eta t}\rfloor$*,* $t_L = \lceil \eta t + \alpha_2\sqrt{\eta t}/p\rceil$*, and every* $k \in [t_L + 1, t]$*, the following hold:*

$$\mathbb{P}[\gamma_t^\delta > i] \ge \mathbb{P}[\gamma_t^\delta > \eta pt + \alpha\sqrt{\eta t}], \tag{14}$$

$$\frac{(i - Q(t_L))^2}{t_L - 1} \ge \alpha_1^2\left(1 - O\left(\frac{\alpha_2}{\sqrt{t}}\right)\right), \tag{15}$$

$$\mathbb{P}[S_{i,t} < 2\delta t \mid \tau_i = k] \le \exp\left(-\frac{\alpha_2^2\delta^2}{32}\left(1 - O\left(\frac{\alpha_2}{\sqrt{t}}\right)\right)\right). \tag{16}$$

Plugging Ineq. (10),(12) and (13) into Ineq. (9) with these values of (i, t_L), Thm. 3(II) follows.

Given Thm. 3, and since $n_t = |V_t|$ is concentrated around $Q(t)$, it is easy to show that γ_t^δ is concentrated around $\eta \cdot n_t$ for $\eta = \delta^{2/p}$. Formally, letting

$$\mathbb{P}_- = \mathbb{P}[\gamma_t^\delta \le \eta n_t - \alpha \sqrt{\eta n_t/p}] \quad \text{and} \quad \mathbb{P}_+ = \mathbb{P}[\gamma_t^\delta > \eta n_t + \alpha \sqrt{\eta n_t/p}],$$

we prove the following.

Theorem 4. *For every nonnegative* $\alpha, \alpha_1, \alpha_2, \alpha_3 = \Omega(1)$ *satisfying* $\alpha = \alpha_1 + \alpha_2 + \alpha_3$ *and* $\alpha = o(\sqrt{t})$,

(I) $\mathbb{P}_- \le 2\exp\left(-2\alpha_1^2\left(1 - O\left(\frac{1}{\alpha_1\sqrt{t}}\right)\right)\right) + \exp\left(-\frac{\alpha_2^2\delta^2}{32}\left(1 - O\left(\frac{\alpha}{\sqrt{t}}\right)\right)\right)$

$\qquad\qquad + \exp\left(-\frac{2\alpha_3^2}{\eta}\left(1 - O\left(\frac{1}{\alpha_3\sqrt{t}}\right)\right)\right)$,

(II) $\mathbb{P}_+ \le \exp\left(-2\alpha_1^2\left(1 - O\left(\alpha_2/\sqrt{t}\right)\right)\right) + \exp\left(-\frac{\alpha_2^2\delta^2}{32}\left(1 - O\left(\alpha_2/\sqrt{t}\right)\right)\right)$

$\qquad\qquad + \exp\left(-\frac{2\alpha_3^2}{\eta}\left(1 - O\left(\frac{\alpha}{\sqrt{t}}\right)\right)\right)$.

Proof. Denoting $B = pt + \alpha_3\sqrt{t/\eta}$,

$$\mathbb{P}_- = \sum_{k=1}^t \mathbb{P}[n_t = k] \cdot \mathbb{P}[\gamma_t^\delta \le \eta n_t - \alpha\sqrt{\eta n_t/p} \mid n_t = k]$$

$$\le \sum_{k>B} \mathbb{P}[n_t = k] + \sum_{k\le B} \mathbb{P}[n_t = k] \cdot \mathbb{P}[\gamma_t^\delta \le \eta n_t - \alpha\sqrt{\eta n_t/p} \mid n_t = k]$$

$$\le \mathbb{P}[n_t > B] + \mathbb{P}[\gamma_t^\delta \le \eta B - \alpha\sqrt{\eta B/p}] .$$

Recalling that $\mathbb{E}[n_t] = \mathbb{E}[1 + \sum_{j=1}^{t-1} N_j] = 1 + p(t - 1)$, the first term can be bounded using Hoeffding's Inequality as follows.

$$\mathbb{P}[n_t > B] = \mathbb{P}\left[n_t - \mathbb{E}[n_t] > B - (1 + p(t-1))\right] = \mathbb{P}\left[n_t - \mathbb{E}[n_t] > \alpha_3\sqrt{\frac{t}{\eta}} - 1 + p\right] .$$

Recalling that $N_j \in \{0,1\}$, and applying Hoeffding's Inequality, we get

$$\mathbb{P}[n_t > B] \le \exp\left(-2\frac{\left(\alpha_3\sqrt{t/\eta} - 1 + p\right)^2}{t - 1}\right) = \exp\left(-\frac{2\alpha_3^2}{\eta}\left(1 - O\left(\frac{1}{\alpha_3\sqrt{t}}\right)\right)\right) .$$

The second term can be bounded by part (I) of Thm. 3 as follows.

$$\mathbb{P}[\gamma_t^\delta \le \eta B - \alpha\sqrt{\eta B/p}] = \mathbb{P}\left[\gamma_t^\delta \le \eta pt + \alpha_3\sqrt{\eta t} - \alpha\sqrt{\eta t + \alpha_3\sqrt{\eta t}/p}\right]$$

$$\le \mathbb{P}\left[\gamma_t^\delta \le \eta pt - (\alpha - \alpha_3)\sqrt{\eta t}\right] .$$

Applying part (I) of Thm. 3, part (I) of Thm. 4 follows. The proof of part (II) of Thm. 4 is similar. $\qquad\square$

7   Core Size in Superlinear-Sized Networks

In this section we assume $p_t = \min\{1, 2a/\ln t\}$ for some fixed real $a > 0$ and analyze γ_t^δ, the size of the core C_t^δ at time t. Specifically, fixing $\eta = \delta^{1/a}$ and letting $\sigma = \lfloor t^\eta \rfloor$, we show that γ_t^δ is concentrated around $Q(\sigma)$. Formally, we prove the following theorem.

Theorem 5. *For any nonnegative reals* $\alpha, \alpha_1, \alpha_2 = \Omega(1)$ *satisfying* $\alpha = \alpha_1 + \alpha_2$ *and* $\alpha = o(\sqrt{\sigma}/\ln \sigma)$, *we have*

(I) $\mathbb{P}[\gamma_t^\delta \leq Q(\sigma) - \alpha\sqrt{\sigma}] \leq 2\exp\left(-2\alpha_1^2\left(1 - O\left(\frac{\alpha_2^2}{\alpha_1\sqrt{\sigma}}\right)\right)\right)$
$\qquad\qquad\qquad\qquad + \exp\left(-\frac{\alpha_2^2\delta^2}{32}\left(1 - O\left(\frac{\alpha\ln\sigma}{\sqrt{\sigma}}\right)\right)\right),$

(II) $\mathbb{P}[\gamma_t^\delta > Q(\sigma) + \alpha\sqrt{\sigma}] \leq \exp\left(-2\alpha_1^2\left(1 - O\left(\frac{\alpha_2\ln\sigma}{\sqrt{\sigma}}\right)\right)\right)$
$\qquad\qquad\qquad\qquad + \exp\left(-\frac{\alpha_2^2\delta^2}{32}\left(1 - O\left(\frac{\alpha_2\ln\sigma}{\sqrt{\sigma}}\right)\right)\right).$

Note: *Since* $p_t = \min\{1, 2a/\ln t\}$, *then* $Q(\sigma)$ *is proportional to* $\frac{t^\eta}{\eta\ln t}$, *so we indeed get a concentration around* $Q(\sigma)$. *Also note that since* n_t *is concentrated around* $Q(t)$, *which is proportional to* $\frac{t}{\ln t}$, *we have that* $\mathbb{E}\left[\gamma_t^\delta\right] = \tilde{\Theta}(n_t^\eta)$.

Theorem 5 is a result of Corollary 1. Specifically, denoting $\Delta = \lfloor \alpha_2\sqrt{\sigma}/p_\sigma \rfloor$, part (I) is obtained by applying the following lemma:

Lemma 6. *For any nonnegative* $\alpha, \alpha_1, \alpha_2 = \Omega(1)$ *satisfying* $\alpha = \alpha_1 + \alpha_2$ *and* $\alpha = o(\sqrt{\sigma}/\ln\sigma)$, *and for* $i = \lfloor Q(\sigma) - \alpha\sqrt{\sigma} \rfloor$, $t_L = \sigma - \lceil(\alpha + \alpha_1)\sqrt{\sigma}/p_\sigma\rceil$, $t_H = \sigma - \Delta$ *and every* $k \in [t_L + 1, t_H]$, *the following hold:*

$$\mathbb{P}[\gamma_t^\delta \leq i] = \mathbb{P}[\gamma_t^\delta \leq Q(\sigma) - \alpha\sqrt{\sigma}] , \tag{17}$$

$$\frac{(Q(t_H) + 1 - i)^2}{t_H - 1} \geq \alpha_1^2\left(1 - O\left(\frac{\alpha_2^2}{\alpha_1\sqrt{\sigma}}\right)\right) , \tag{18}$$

$$\frac{(i - Q(t_L))^2}{t_L - 1} \geq \alpha_1^2\left(1 - O\left(\frac{1}{\alpha_1\sqrt{\sigma}}\right)\right) , \tag{19}$$

$$\mathbb{P}[S_{i,t} \geq 2\delta t \mid \tau_i = k] \leq \exp\left(-\frac{\alpha_2^2\delta^2}{32}\left(1 - O\left(\frac{\alpha\ln\sigma}{\sqrt{\sigma}}\right)\right)\right) . \tag{20}$$

Plugging Eq. (17) and Ineq. (18),(19),(20) into Ineq. (8) with these values of (i, t_L, t_H), Thm. 5(I) follows.

Part (II) is obtained by applying the following lemma:

Lemma 7. *For any nonnegative* $\alpha, \alpha_1, \alpha_2 = \Omega(1)$ *satisfying* $\alpha = \alpha_1 + \alpha_2$ *and* $\alpha = o(\sqrt{\sigma}/\ln\sigma)$, *and for* $i = \lfloor Q(\sigma) + \alpha\sqrt{\sigma} \rfloor$, $t_L = \sigma + \Delta$, *and every* $k \in [t_L + 1, t]$, *the following hold:*

$$\mathbb{P}[\gamma_t^\delta > i] \geq \mathbb{P}[\gamma_t^\delta > Q(\sigma) + \alpha\sqrt{\sigma}] , \tag{21}$$

$$\frac{(i - Q(t_L))^2}{t_L - 1} \geq \alpha_1^2\left(1 - O\left(\frac{\alpha_2\ln\sigma}{\sqrt{\sigma}}\right)\right) , \tag{22}$$

$$\mathbb{P}[S_{i,t} < 2\delta t \mid \tau_i = k] \leq \exp\left(-\frac{\alpha_2^2 \delta^2}{32}\left(1 - O\left(\frac{\alpha_2 \ln \sigma}{\sqrt{\sigma}}\right)\right)\right). \qquad (23)$$

Plugging Ineq. (17),(19) and (20) into Ineq. (9) with these values of (i, t_L), Thm. 5(II) follows.

References

1. Avin, C., Lotker, Z., Peleg, D., Pignolet, Y.-A., Turkel, I.: Core-periphery in networks: An axiomatic approach (2014). arXiv preprint arXiv:1411.2242
2. Barabási, A.-L., Albert, R.: Emergence of scaling in random networks. Science **286**, 509–512 (1999)
3. Borgatti, S.P., Everett, M.G.: Models of core/periphery structures. Social networks **21**(4), 375–395 (2000)
4. Chung, F.R.K., Lu, L.: Complex graphs and networks. AMS (2006)
5. Fraigniaud, P., Gavoille, C., Kosowski, A., Lebhar, E., Lotker, Z.: Universal augmentation schemes for network navigability: overcoming the sqrt(n)-barrier. In: Proc. 19th SPAA, pp. 1–7 (2007)
6. Kleinberg, J.: The small-world phenomenon: an algorithmic perspective. In: Proc. 32nd ACM Symp. on Theory of computing, pp. 163–170 (2000)
7. Leskovec, J., Kleinberg, J., Faloutsos, C.: Graph evolution: Densification and shrinking diameters. Trans. Knowledge Discovery from Data **1**, 2 (2007)
8. Milgram, S.: The small world problem. Psychology today **2**(1), 60–67 (1967)
9. Newman, M.: Networks: An Introduction. Oxford Univ. Press (2010)
10. de Price, D.S.: A general theory of bibliometric and other cumulative advantage processes. J. Amer. Soc. Inform. Sci. **27**(5), 292–306 (1976)
11. Rombach, M.P., Porter, M.A., Fowler, J.H., Mucha, P.J.: Core-periphery structure in networks. SIAM J. Applied Math. **74**(1), 167–190 (2014)
12. Zhang, X., Martin, T., Newman, M.E.J.: Identification of core-periphery structure in networks (2014). CoRR, abs/1409.4813
13. Zhou, S., Mondragón, R.J.: The rich-club phenomenon in the internet topology. IEEE Commun. Lett. **8**(3), 180–182 (2004)

Maintaining Near-Popular Matchings

Sayan Bhattacharya[1], Martin Hoefer[2(✉)], Chien-Chung Huang[3],
Telikepalli Kavitha[4], and Lisa Wagner[5]

[1] Institute of Mathematical Sciences, Chennai, India
bsayan@imsc.res.in
[2] MPI für Informatik and Saarland University, Saarbrücken, Germany
mhoefer@mpi-inf.mpg.de
[3] Department of Computer Science and Engineering, Chalmers University,
Gothenburg, Sweden
villars@mpi-inf.mpg.de
[4] Tata Institute of Fundamental Research, Mumbai, India
kavitha@tcs.tifr.res.in
[5] Department of Computer Science, RWTH Aachen University, Aachen, Germany
lwagner@rwth-aachen.de

Abstract. We study dynamic matching problems in graphs among
agents with preferences. Agents and/or edges of the graph arrive and
depart iteratively over time. The goal is to maintain matchings that are
favorable to the agent population and stable over time. More formally, we
strive to keep a small unpopularity factor by making only a small amortized number of changes to the matching per round. Our main result is
an algorithm to maintain matchings with unpopularity factor $(\Delta + k)$ by
making an amortized number of $O(\Delta + \Delta^2/k)$ changes per round, for any
$k > 0$. Here Δ denotes the maximum degree of any agent in any round.
We complement this result by a variety of lower bounds indicating that
matchings with smaller factor do not exist or cannot be maintained using
our algorithm.

As a byproduct, we obtain several additional results that might be of
independent interest. First, our algorithm implies existence of matchings
with small unpopularity factors in graphs with bounded degree. Second,
given any matching M and any value $\alpha \geq 1$, we provide an efficient
algorithm to compute a matching M' with unpopularity factor α over
M if it exists. Finally, our results show the absence of voting paths in
two-sided instances, even if we restrict to sequences of matchings with
larger unpopularity factors (below Δ).

1 Introduction

Matching arises as a fundamental task in many coordination, resource allocation, and network design problems. In many domains, matching and allocation
problems occur among agents with preferences, e.g., in job markets, when assigning residents to hospitals, or students to dormitory rooms, or when allocating

Supported by DFG Cluster of Excellence MMCI at Saarland University.

© Springer-Verlag Berlin Heidelberg 2015
M.M. Halldórsson et al. (Eds.): ICALP 2015, Part II, LNCS 9135, pp. 504–515, 2015.
DOI: 10.1007/978-3-662-47666-6_40

resources in distributed systems. There are a number of approaches for formal study of allocation under preferences, the most prominent being stable and popular matchings. Usually, there is a set of agents embedded into a graph, and each agent has a preference list over his neighbors. An edge is called a blocking pair if both agents strictly prefer each other to their current partners (if any). A matching without blocking pair is a *stable matching*. In a popular matching all agents get to vote between two matchings M and M'. They vote for M if it yields a partner which is strictly preferred to the one in M', or vice versa (they don't vote if neither of them is strictly preferred). The matching that receives more votes is more popular. For a *popular matching* there exists no other matching that is more popular.

Stable and (to a lesser extent) popular matchings have been studied intensively in algorithms, economics, operations research, and game theory, but mostly under the assumption that the set of agents and the set of possible matching edges remain static. In contrast, many application areas above are inherently dynamic. For example, in a large firm new jobs open up on a repeated basis, e.g., due to expansion into new markets, retirement of workers, or the end of fixed-term contracts. Similarly, new applicants from outside arrive, or internal employees seek to get promoted or move into a different department. The firm strives to fill its positions with employees in a way that is preferable to both firm and workers. The naive approach would be to compute, e.g., a stable or popular matching from scratch every time a change happens, but then employees might get assigned differently every time. Instead, the obvious goal is to maintain a stable or popular assignment at a small rate of change. Similar problems arise also in the context of dormitory room assignment or resource allocation in distributed systems. Perhaps surprisingly, these natural problems have not been studied in the literature so far.

Maintaining graph-theoretic solution concepts like matchings or shortest paths is an active research area in algorithms. In these works, the objective is to maintain matchings of maximum cardinality while making a small number of changes. These approaches are unsuitable for systems with agent preferences, which fundamentally change the nature and the characteristics of the problem.

More fundamentally, a central theme in algorithmic game theory is to study dynamics in games such as best response or no-regret learning. However, in the overwhelming majority of these works, the games themselves (agents, strategies, payoffs) are static over time, and the interest is to characterize the evolution of strategic interaction. In contrast, there are many games in which maintaining stability concepts at a small rate of change is a natural objective, such as in routing or scheduling problems. To the best of our knowledge, our paper is the first to study algorithms for maintaining equilibria in the prominent domain of matching and network design problems.

Model and Notation.   Before we state our results, let us formally introduce the model and notation. We consider a dynamic round-based matching scenario for a set V of agents. In each round t, there exists a graph $G^t = (V, E^t)$ with set E^t of possible matching edges among the agents. Initially, $E^0 = \emptyset$. For *edge* *dynamics*, in the beginning of each round $t \geq 1$ a single edge is added or deleted, i.e., E^t and E^{t+1} differ in exactly one edge. We denote this edge by e^t. Note that a particular edge e can be added and removed multiple times over time.

For *vertex dynamics*, in the beginning of each round $t \geq 1$ a single vertex arrives or departs along with all incident edges. We denote this vertex by v^t, where the same vertex can arrive and depart multiple times over time. Formally, in vertex dynamics all vertices exist throughout. We color them red and blue depending on whether they are currently present or not, respectively. Then, in the beginning of a round, if v^t arrives, it is colored red and all edges between v^t and red agents arrive. If v^t leaves, it is colored blue and all incident edges are removed. Thus, E^t and E^{t+1} differ by exactly a set of edges from v^t to red agents. Vertex dynamics also model the case when in each round the preference list of one vertex changes. Assume there is a separate vertex with the new preference list and consider two rounds in which the old vertex leaves and the new one arrives. Our asymptotic bounds will directly apply.

We consider several structures for the preferences. In the *roommates case* each agent $v \in V$ has a strict preference list $\succ_v$ over all other agents in V. In the *two-sided* case we have sets X and Y and $E^t \subseteq X \times Y$. In the *one-sided* case the elements in X do not have preferences, only agents in Y have preferences over elements in X. Each agent always prefers being matched over being unmatched.

Our goal is to maintain at small amortized cost a matching in each round that satisfies a preference criterion. Towards this end, we study several criteria in this paper. For matching M and agent v we denote by $M(v)$ the agent matched to v in M, where we let $v = M(v)$ when v is unmatched. In round t, an edge $e = (u, v) \in E^t \setminus M$ is called a *blocking pair* for matching $M \subseteq E^t$ if $u \succ_v M(v)$ and $v \succ_u M(u)$. M is a *stable matching* if it has no blocking pair.

For two matchings M and M', v is called a $(+)$-agent if $M'(v) \succ_v M(v)$. We call v a $(-)$-agent if $M(v) \succ_v M'(v)$ and (0)-agent if $M'(v) = M(v)$. We denote by V^+, V^- and V^0 the sets of $(+)$-, $(-)$- and (0)-agents, respectively. For $\alpha \geq 1$, we say M' is α-*more popular* than M if $|V^+| \geq \alpha \cdot |V^-|$. If $|V^+| = |V^-| = 0$, we say M' is *1-more popular* than M, and if $|V^+| > 0 = |V^-|$ then M' is ∞-*more popular* than M. In round t, the *unpopularity factor* $\rho(M) \in [1, \infty) \cup \{\infty\}$ of matching $M \subseteq E^t$ is the maximum α such that there is an α-more popular matching $M' \subseteq E^t$. M is a *c-unpopular matching* if it has unpopularity factor $\rho(M) \leq c$. A *1-unpopular matching* is called *popular matching*.

Our bounds depend on the *maximum degree* of any agent, where for one-sided instances this includes only the agents in Y. In round t, consider an agent v in G^t. We denote by $N^t(v)$ the set of current neighbors of v, by $d^t(v)$ the degree of v, by Δ^t the maximum degree of any agent. Finally, by $\Delta = \max_t \Delta^t$ we denote the maximum degree of any agent in any of the rounds. Observe that throughout the dynamics, we allow the same edge to arrive and depart multiple times. In addition, an agent v can have a much larger degree than Δ in $\bigcup_t E^t$.

Our Results. We maintain matchings when agents and/or edges of the graph arrive and depart iteratively over time. If every agent has degree at most Δ in every round, our algorithm maintains $O(\Delta)$-unpopular matchings by making an amortized number of $O(\Delta)$ changes to the matching per round. This result holds in one-sided, two-sided and roommates cases. It is almost tight with respect to the unpopularity factor, since there are instances where all matchings have unpopularity factor at least Δ. More formally, if there is one edge arriving or leaving per round, our algorithm yields a tradeoff. Given any number $k > 0$, the algorithm can maintain matchings with unpopularity factor $(\Delta + k)$ using an amortized number of $O(\Delta + \Delta^2/k)$ changes per round. If one vertex arrives or leaves per round, the algorithm needs $O(\Delta^2 + \Delta^3/k)$ changes per round.

The algorithm switches to a matching that is $\alpha > (\Delta + k)$-more popular whenever it exists, and we show that this strategy converges in every round. We can decide for a given matching M and value $\alpha \geq 1$ if there is a matching M' that yields an unpopularity factor at least α for M and compute M' if it exists. Our bounds imply the existence of matchings with small unpopularity factors in one-sided and roommates instances with bounded degree. These insights might be of independent interest.

For two-sided instances, stable and popular matchings exist, but we show that maintaining them requires an amortized number of $\Omega(n)$ changes to the matching per round, even when $\Delta = 2$. In addition, our algorithm cannot be used to maintain matchings with unpopularity factors below $\Delta - 1$. Iterative resolution of matchings with such unpopularity factors might not converge. In fact, we provide an instance and an initial matching from which every sequence of matchings with unpopularity factor greater than 1 leads into a cycle. In contrast to one-sided instances, this implies that two-sided instances might have no voting paths, even for complete and strict preferences. Furthermore, we show that cycling dynamics can evolve even when we restrict to resolution of matchings with higher unpopularity factors (up to Δ).

In summary, our results show that we can maintain a near-popular matching in a dynamic environment with relatively small changes, by pursuing a greedy improvement strategy. For the one-sided case, this achieves essentially the best unpopularity factor we can hope for. In the two-sided case, achieving a better factor with our strategy is bound to fail. Whether there are other strategies with better factors or smaller changes to maintain near-popular matchings is an interesting future direction.

Related Work. Stable matchings have been studied intensively and we refer to standard textbooks for an overview [10, 16, 19]. Perhaps closest to our paper are works on iterative resolution of blocking pairs. Knuth [15] provided a cyclic sequence of resolutions in a two-sided instance. Hence, even though stable matchings exist, iterative resolution of blocking pairs might not always lead there. Roth and Vande Vate [20] showed that there is always some sequence of polynomially many resolutions that leads to a stable matching. Ackermann et al [3] constructed instances where random sequences require exponential time with high probability. Although in the roommates case (for general graphs) stable matchings might not exist, Diamantoudi et al. [7] showed that there are always sequences of resolutions leading to a stable matching if it exists. Furthermore, the problem has been studied in constrained stable matching problems [11–13]. In contrast, our aim is to maintain matchings by making a small number of changes per round. However, we also show that, perhaps surprisingly, similar sequences do not exist for popular matchings in two-sided instances.

Stable matching turns out to be a very demanding concept that cannot be maintained at small cost. We obtain more positive results for near-popular matchings. The notion of popularity was introduced by Gärdenfors [8] in the two-sided case, who showed that every stable matching is popular when all preference lists are strict. When preference lists admit ties, it was shown by Biró, Irving, and Manlove [6] that the problem of computing an arbitrary popular matching in two-sided instances is NP-hard. They also provide an algorithm to decide if a matching is popular or not in the two-sided and roommates cases.

When agents on only one side have preferences, popular matchings might not exist. Abraham et al. [1] gave a characterization of instances that admit popular matchings; when preference lists are strict, they showed a linear-time algorithm to determine if a popular matching exists and if so, to compute one. Popular matchings in the one-sided case have been well-studied; closest to our paper is Abraham and Kavitha [2] that study the *voting paths* problem. Given an initial matching M_1, the problem is to find a voting path of least length, i.e., a sequence of matchings $M_1, M_2, \ldots, M_k$ of least length such that M_k is popular. In this sequence every M_i must be more popular than M_{i-1}. If a one-sided instance admits a popular matching, then from every M_1 there is always a voting path of length at most 2, and one of least length can be determined in linear time [2].

McCutchen [17] introduced the notion of *unpopularity factor* and showed that the problem of computing a least unpopular matching in one-sided instances is NP-hard. For a roommates instance, popular matchings might not exist. Huang and Kavitha [14] show that with strict preference lists, there is always a matching with unpopularity factor at most $O(\log n)$, and there exist instances where every matching has unpopularity factor $\Omega(\log n)$.

A prominent topic in algorithms is maintaining matchings in dynamic graphs that approximate the maximum cardinality matching. In graphs with n nodes and iterative arrival and departure of edges, Onak and Rubinfeld [18] design a randomized algorithm that maintains a matching which guarantees a large constant approximation factor and requires only $O(\log^2 n)$ amortized update time. Baswana, Gupta and Sen [4] provide a randomized 2-approximation in $O(\log n)$ amortized time. For deterministic algorithms, Gupta and Peng [9] gave a $(1+\varepsilon)$-approximation in $O(\sqrt{m}/\varepsilon^2)$ worst-case update time. Very recently, Bhattacharya et al. [5] showed a deterministic $(4 + \varepsilon)$-approximation in $O(m^{1/3}/\varepsilon^2)$ worst-case update time.

2   Maintaining $(\Delta + k)$-Unpopular Matchings

In this section, we present an algorithm that, given any number $k > 0$, maintains $(\Delta + k)$-unpopular matchings. Our approach applies in one-sided, two-sided and roommates instances. In the edge-dynamic case, it makes an amortized number of

---

Algorithm 1. DEFERREDRESOLUTION

---

1 **for** *every round* $t = 1, 2, \ldots$ **do**
2   Compute for matching M an α-more popular matching M' if it exists.
3   **while** M' *exists* **do**
4       $M \leftarrow M'$
5   Compute for matching M an α-more popular matching M' if it exists.

$O(\Delta + \Delta^2/k)$ changes to the matching per round. In every round, our algorithm DEFERREDRESOLUTION iteratively replaces the current matching with an α-more popular matching until no such matching exists (see Algorithm 1). We show in Section 2.1 that such matchings can be computed efficiently. In Section 2.2 we show that when $\alpha > \Delta$ the iterative replacement converges in every round and amortized over all rounds the number of changes made to the matching is at most $O(\Delta + \Delta^2/k)$ per round.

2.1 Finding an α-More Popular Matching

Let us first show that for any given matching M and any value α, we can decide in polynomial time if the unpopularity factor is $\rho(M) \geq \alpha$ and construct an α-more popular matching if it exists. While throughout this paper we assume agents to have strict preferences, this result holds even when the preferences have ties.

Theorem 1. *Let $G = (V, E)$ be a graph, and suppose for every agent $v \in V$ there is a preference order $\succeq_v$ (possibly with ties) over $N(v) \cup \{v\}$ such that $u \succeq_v v$ for all $u \in N(v)$. Then for every matching M in G and every value $\alpha \in \mathbb{R} \cup \{\infty\}$, we can decide in polynomial time if $\rho(M) \geq \alpha$ as well as compute an α-more-popular matching M' if it exists.*

Proof. The general structure of the algorithm is shown as Algorithm 2. The main idea is to construct an adjusted graph and find a maximum-weight matching, which allows to see if an α-more popular matching exists.

We first take a closer look at α. The case $\alpha \leq 1$ is trivial. If $\alpha > |V| - 1$, any α-more popular M' has no $(-)$-agent. So we are checking if $\rho(M) = \infty$ or, equivalently, if $\rho(M) \geq \alpha = |V|$. If $\rho(M) \in (1, |V| - 1]$, it is given as a ratio of two numbers $|V^+|$ and $|V^-|$, which are both integers in $\{1, \ldots, n\}$. Let $\mathbb{Q}_n$ be the set of rational numbers that can be expressed as a fraction of two integers in $\{1, \ldots, n\}$. Thus, when $\alpha \notin \mathbb{Q}_n$, we can equivalently test for $\rho(M) \geq \alpha^*$, where α^* is the smallest number of $\mathbb{Q}_n$ larger than α (see line 3 in the algorithm). Due to reasons mentioned below, we replace the test $\rho(M) \geq \alpha^*$ by testing $\rho(M) > \alpha'$, where α' is slightly smaller than α^*, but still larger than the next-smaller number of $\mathbb{Q}_n$. Formally, $\alpha' = \alpha^* - \epsilon$ with

$$\epsilon = \frac{1}{2} \cdot \min_{r, r' \in \mathbb{Q}_n} \{r - r' \mid r - r' > 0\}$$

Algorithm 2. Finding an α-more popular matching for M

---

1 **if** $\alpha \leq 1$ **then return** M
2 **else if** $\alpha > |V| - 1$ **then** set $\alpha^* \leftarrow |V|$
3 **else** set $\alpha^* \leftarrow \min\{r \in \mathbb{Q}_n \mid r \geq \alpha\}$
4 Set $\alpha' \leftarrow \alpha^* - \epsilon$
5 Construct $\tilde{G} = (\tilde{V}, \tilde{E})$ as union of two copies (V_1, E_1), (V_2, E_2) of G and edges E_3 between copies, and assign edge weights $w_2(e)$ to every edge $e \in \tilde{E}$
6 Compute a maximum-weight matching M^* in $\tilde{G}$
7 **if** $w_2(M^*) > |V|(2\alpha' + 1)$ **then return** $M^* \cap E_1$
8 **else return** $\emptyset$

---

half of the smallest strictly positive difference between any two numbers in $\mathbb{Q}_n$. Observe that $\rho(M) \geq \alpha$ if and only if $\rho(M) > \alpha'$.

For the test we construct M' via a maximum-weight matching in a graph structure $\tilde{G}$ indicating the gains and losses in popularity. $\tilde{G}$ contains two full copies of G. In addition, for each vertex v in G there is an edge connecting the two copies of v. More formally, $\tilde{G} = (\tilde{V}, \tilde{E})$, $\tilde{V} = V_1 \cup V_2$ and $\tilde{E} = E_1 \cup E_2 \cup E_3$. (V_1, E_1) and (V_2, E_2) constitute two copies of G. E_3 contains for each vertex v in G an edge (v_1, v_2) between its two copies $v_1 \in V_1$ and $v_2 \in V_2$. We define edge weights such that each maximum-weight matching M^* in $\tilde{G}$ is perfect. Then, we construct M' by restricting attention to V_1 and matching the same vertices as M^* within V_1. Vertices of V_1 matched to their copy remain unmatched in M'.

For clarity, we define the edge weights $w_2(e)$ in two steps. We first consider weights w_1 where, intuitively, $w_1(e)$ indicates whether the incident agents become $(+)$-, (0)-, or $(-)$-agents when e is added to M. The value of w_1 is used to charge the $(+)$-agents to the $(-)$-agents. Formally, let $e = (u_i, v_j) \in \tilde{E}$ and set

$$w_1(e) = \begin{cases} 2 & \text{if } v \succ_u M(u) \text{ and } u \succ_v M(v), \\ 1 & \text{if } v \succ_u M(u) \text{ and } u =_v M(v), \text{ or } v =_u M(u) \text{ and } u \succ_v M(v), \\ 0 & \text{if } v =_u M(u) \text{ and } u =_v M(v), \\ 1 - \alpha' & \text{if } v \succ_u M(u) \text{ and } M(v) \succ_v u, \text{ or } M(u) \succ_u v \text{ and } u \succ_v M(v), \\ -\alpha' & \text{if } v =_u M(u) \text{ and } M(v) \succ_v u, \text{ or } M(u) \succ_u v \text{ and } u =_v M(v), \\ -2\alpha' & \text{if } M(u) \succ_u v \text{ and } M(v) \succ_v u \end{cases}$$

We let $w_1(M) = \sum_{e \in M} w_1(e)$.

If there is an α^*-more popular matching M', there is a perfect matching $\tilde{M}$ in $\tilde{G}$ with total weight $w_1(\tilde{M}) > 0$. We simply install M' in both copies (V_1, E_1) and (V_2, E_2) and match single vertices to their copy. Then, for every $(+)$-agent in V^+ we add a weight of 2 on the incident edges of $\tilde{M}$. For every $(-)$-agent in V^- we subtract a weight of $2\alpha'$ on the incident edges of $\tilde{M}$. The contribution of (0)-agents in V^0 to the edge weight is 0. Thus, as $2|V^+| \geq 2\alpha^*|V^-| > 2\alpha'|V^-|$, we get $w_1(\tilde{M}) > 0$. In contrast, an arbitrary matching $\tilde{M}$ with $w_1(\tilde{M}) > 0$ might not be perfect and thus impossible to be transformed into a α^*-more popular matching in G. Towards this end, we change the weights to w_2 with

$w_2(e) = w_1(e) + 2\alpha' + 1$ for every $e \in \tilde{E}$. We show that there is an α^*-more popular matching M' if and only if a *maximum-weight matching* M^* for w_2 in $\tilde{G}$ has $w_2(M^*) > |V|(2\alpha' + 1)$. The key difference is that $w_2(e) > 0$ for all $e \in \tilde{E}$, and therefore under w_2 every maximum-weight matching is perfect.

More formally, if there is an α^*-more popular matching M', we construct $\tilde{M}$ as above and observe that $w_1(\tilde{M}) > 0$ if and only if $w_2(\tilde{M}) > |V|(2\alpha' + 1)$. For the other direction, we first claim that every maximum-weight matching M^* for w_2 is perfect. Assume first there is some maximum matching M^* where some vertex v remains single. By $M^*(V_1)$ we denote the part of M^* which only uses vertices in V_1. Similarly, $M^*(V_2)$ is the part of M^* which only uses vertices in V_2. W.l.o.g. we assume $w_2(M^*(V_1)) \geq w_2(M^*(V_2))$, and if $w_2(M^*(V_1)) = w_2(M^*(V_2))$ we assume the number of unmatched vertices in V_1 is larger or equal to the number of unmatched vertices in V_2. If $w_2(M^*(V_1)) > w_2(M^*(V_2))$, then M^* could be improved by matching V_2 in the same manner as V_1. Thus, $w_2(M^*(V_1)) = w_2(M^*(V_2))$, and there is at least one single vertex v_1 regarding M^* in V_1. If the corresponding copy $v_2 \in V_2$ is single as well, we can improve M^* by adding (v_1, v_2). If v_2 is matched, we can rearrange the matching on V_2 to mirror the one on V_1 without loss in total weight. Then (v_1, v_2) can be added. Hence, M^* has to be a perfect matching.

Suppose $w_2(M^*) > |V|(2\alpha' + 1)$, we construct an α^*-more popular matching as follows. As M^* has maximum-weight for w_2, by the observations above we can assume that $M^*(V_1)$ and $M^*(V_2)$ contain exactly the copies of the same edges of E. Since M^* is perfect, for each $v \in V$ both copies v_1, v_2 are matched. If they are matched via (v_1, v_2), we leave v single in M'. Otherwise, the non-single agents in M' are matched as their copies in $M^*(V_1)$. We claim that $w_2(M^*) > |V|(2\alpha' + 1)$ implies M' is α^*-more popular. First, note that $w_2(M^*) > |V|(2\alpha' + 1)$ implies $w_1(M^*) > 0$. Especially, this implies that $|V^+| > 0$. The preference of agent v for M' corresponds to the contribution of $v_1 \in V_1$ to $w_1(M^*)$, i.e., v_1 contributes $1, 0$, or $-\alpha'$ when $v \in V^+$, V^0, or V^-, respectively. By symmetry of M^* and of edge weights in E_3, the total contribution of vertices in V_1 to $w_1(M^*)$ is exactly $w_1(M^*)/2$. Hence, $w_1(M^*) > 0$ implies $|V^+| > \alpha'|V^-|$ for M'. Here the choice of $\alpha' = \alpha^* - \epsilon$ becomes crucial. By the choice of ϵ we know that the smallest value of $\mathbb{Q}_n$ larger than α' is α^*. Thus, $|V^+| > \alpha'|V^-|$ also implies $|V^+| \geq \alpha^*|V^-|$ which shows $|V^+| \geq \alpha|V^-|$. Hence, $w_2(M^*) > |V|(2\alpha' + 1)$ if and only if an α-more popular matching exists.

We can use the same approach for instances with one-sided preferences by simply defining the preferences of the other side to be indifferent between all potential matching partners as well as being single.   □

2.2   Convergence and Amortized Number of Changes

Given that we can decide and find α-more popular matchings efficiently, we now establish that for $\alpha > \Delta$ the iterative resolution does not lead into cycles and makes a small amortized number of changes per round.

Theorem 2. DEFERREDRESOLUTION *maintains a* $(\Delta+k)$-*unpopular matching by making an amortized number of* $O(\Delta + \Delta^2/k)$ *changes to the matching per round with edge dynamics, for any* $k > 0$.

Proof. Our proof is based on the following potential function

$$\Phi^t(M) = \sum_{v \in V} d^t(v) + 1 - rank(M(v)) \; ,$$

where $rank(M(v)) = i$ if in the preference list of v restricted to $N^t(v) \cup \{v\}$, partner $M(v)$ ranks at the ith position. Whenever DEFERREDRESOLUTION replaces a matching M in round t with any $(\Delta+k+\epsilon)$-*more popular* one M' with $\epsilon > 0$, we know that $|V^+| > (\Delta+k)|V^-|$.

Consider the symmetric difference $M' \oplus M = (M \cup M') \setminus (M \cap M')$. Observe that due to strictness of preference lists, we have $v \in V^0$ if and only if $M(v) = M'(v)$. In the two-sided or roommates case this also implies $M(v) \in V^0$. This implies that the number of changes between M and M' is at most $|M \oplus M'| \leq |V^+| + |V^-|$ (or in the one-sided case $|M \oplus M'| \leq 2(|V^+| + |V^-|)$).

First, suppose $|V^-| = 0$. In these steps, the potential strictly increases by at least $|V^+|$. Thus, on the average, for every unit of increase in the potential, the number of changes from M to M' is $O(1)$. Second, suppose $|V^-| \geq 1$. Then for every $v \in V^+$, the potential increases by at least 1. For every $v \in V^-$, it drops by at most Δ. Let $\delta = |V^+| - (\Delta+k)|V^-| > 0$. Thus,

$$\Phi^t(M') - \Phi^t(M) \geq |V^+| - \Delta|V^-| \geq \lceil \delta + k|V^-| \rceil$$

The average number of changes made per unit increase in the potential due to updates of the matching with $V^- > 0$ is at most

$$\frac{|M \oplus M'|}{\Phi^t(M') - \Phi^t(M)} = O\left(1 + \frac{\Delta}{k}\right) \; .$$

Finally, we bound the total increase in the potential function over time. Consider the rounds with additions and deletions of edges. If an edge is added in round t, the maximum potential value increases by at most 2 (or 1 in the one-sided case) and the current value of the potential does not decrease. If an edge is deleted, the maximum potential value decreases by at most 2 (or 1 in the one-sided case) and the current value of the potential decreases by at most 2Δ (or Δ in the one-sided case). Thus, in total we can increase the potential up to at most twice the number of edge additions. Also, each deletion creates the possibility to increase the potential by at most 2Δ in subsequent rounds. This implies an amortized potential increase of at most $O(\Delta)$ per round. Also, we get an average number of $O(1+\Delta/k)$ changes in the matching per unit of potential increase. Combining these insights yields the theorem. $\qquad\square$

We can strengthen the latter result in case we have only edge additions.

Corollary 1. DEFERREDRESOLUTION *maintains a* $(\Delta+k)$-*unpopular matching by making an amortized number of* $O(1+\Delta/k)$ *changes to the matching per round with edge dynamics without deletions, for any* $k > 0$.

Proof. In the previous proof we observed that rounds with edge additions generate an amortized potential increase of 1. Hence, we directly get the average number of $O(1 + \frac{\Delta}{k})$ changes in the matching per unit of potential increase also as amortized change per round. $\qquad\square$

The following corollary is due to the fact that we can simulate the addition or deletion of a single vertex by Δ additions or deletions of the incident edges. A similar reduction by Δ can be achieved without vertex deletions.

Corollary 2. DEFERREDRESOLUTION *maintains a* $(\Delta + k)$-*unpopular matching by making an amortized number of* $O(\Delta^2 + \Delta^3/k)$ *changes to the matching per round with vertex dynamics, for any* $k > 0$.

The above results apply in the roommates, two-sided, and one-sided cases. The bound on the unpopularity factor is almost tight, even in terms of existence in the one-sided case.

Proposition 1. *There exist one-sided instances with maximum degree* Δ *for every agent in* Y *such that every matching has unpopularity factor at least* Δ.

Proof. As an example establishing the lower bound consider a one-sided instance with $|X| = \Delta$ elements and $|Y| = \Delta + 1$ agents. We assume there is a global ordering $x_1, \ldots, x_\Delta$ over elements and $x_i \succ_y x_{i+1}$ for all agents $y \in Y$. If a matching M leaves an element in X unmatched, we can add any single edge and thereby create a matching with $|V^+| > 0$ and $|V^-| = 0$. By definition this new matching is now ∞-more popular, and the unpopularity factor becomes $\rho(M) = \infty$. For any matching M that matches all of X, we w.l.o.g. denote y_i as the agent with $M(y_i) = x_i$ for $i = 1, \ldots, \Delta$, and $y_{\Delta+1}$ the remaining unmatched agent. We show that M has unpopularity factor Δ by providing a matching M' that is Δ-more popular than M. Consider M' composed of edges (x_i, y_{i+1}) for $i = 1, \ldots, \Delta$ and y_1 unmatched. y_1 is a $(-)$-agent, all others are $(+)$-agents. $\qquad\square$

3   Two-Sided Matching and Lower Bounds

For the roommates case, the construction in [14] shows that there are instances in which every matching has unpopularity factor of $\Omega(\log \Delta)$. In contrast, in the two-sided case there always exists a stable matching, and every stable matching is a popular matching. However, we show that maintaining a stable or popular matching requires $\Omega(n)$ amortized changes per round, even in instances where we have only edge or vertex additions and every agent has degree at most 2.

Theorem 3. *There exist two-sided instances with* $\Delta = 2$ *such that maintaining a stable or popular matching requires* $\Omega(n)$ *amortized number of changes to the matching per round for (1) edge dynamics with only additions, (2) vertex dynamics with only additions in* X *and* Y, *(3) vertex dynamics with additions and deletions only in* X.

The case of vertex dynamics and only additions to X can be tackled using the standard DEFERREDACCEPTANCE algorithm of Gale-Shapley for stable matching.

Proposition 2. DEFERREDACCEPTANCE *maintains a stable matching by making an amortized number of $O(\Delta)$ changes to the matching per round with vertex dynamics and only additions to X.*

Hence, without any additional assumptions we can only expect to maintain α-unpopular matchings for $\alpha > 1$. Here we observe that our algorithm DEFERREDRESOLUTION cannot be used to maintain matchings with unpopularity factor significantly below Δ, even in the two-sided case. The problem is that the iterative resolution may be forced to cycle.

Theorem 4. *There is an instance with maximum degree Δ and an initial matching such that no sequence of iterative resolution of matchings with unpopularity factor $(\Delta - 1)$ leads to a α-unpopular matching, for any $\alpha < \Delta - 1$.*

It is easy to force DEFERREDRESOLUTION into the cycle. We first add the edges of one cycle matching, then the edges of the more popular cycle matching, and finally the edges of the third cycle matching. DEFERREDRESOLUTION will construct the first cycle matching and switch to the next one whenever it has arrived entirely.

The proof here uses a particular instance with degree $\Delta = 3$. Furthermore, it shows that even though two-sided instances always have popular matchings, there are instances and initial matchings such that no sequence of resolutions towards more popular matchings converges. The following corollary sharply contrasts the one-sided case, in which there always exist voting paths of length 2 whenever a popular matching exists.

Corollary 3. *There are two-sided matching instances and matchings from which there is no voting path to a popular matching.*

More generally, we can establish the following lower bound for any maximum degree $\Delta \geq 3$.

Theorem 5. *For every $\Delta \geq 3$ and $k = 3, \ldots, \Delta$ there is an instance with maximum degree Δ and an initial matching M such that any sequence of resolutions of matchings with unpopularity factor at least $k-1$ does not converge to a $(k-2)$-unpopular matching.*

We can again steer DEFERREDRESOLUTION into the cycle. We first let the edges (x_j, y_j) arrive that remain fixed throughout the cycle, for $j = \Delta - k + 1, \ldots, \Delta$. Then, we let the remaining incident edges arrive for these nodes. DEFERREDRESOLUTION will construct all edges (x_j, y_j) and keep them in the matching throughout. Then, we assume edges (x_j, y_j) arrive iteratively for $j = 1, \ldots, k-1$. DEFERREDRESOLUTION will include each of these edges into the matching. Subsequently, we consider the next matching from the cycle and let the edges arrive iteratively, and so on. DEFERREDRESOLUTION will switch to the next matching in the cycle whenever it has arrived entirely. It then infinitely runs through the cycle once all edges have arrived.

References

1. Abraham, D., Irving, R., Kavitha, T., Mehlhorn, K.: Popular matchings. SIAM J. Comput. **37**(4), 1030–1045 (2007)
2. Abraham, D., Kavitha, T.: Voting paths. SIAM J. Disc. Math. **24**(2), 520–537 (2010)
3. Ackermann, H., Goldberg, P., Mirrokni, V., Röglin, H., Vöcking, B.: Uncoordinated two-sided matching markets. SIAM J. Comput. **40**(1), 92–106 (2011)
4. Baswana, S., Gupta, M., Sen, S.: Fully dynamic maximal matching in $O(\log n)$ update time. In: Proc. 52nd Symp. Foundations of Computer Science (FOCS), pp. 383–392 (2011)
5. Bhattacharya, S., Henzinger, M., Italiano, G.: Deterministic fully dynamic data structures for vertex cover and matching. In: Proc. 25th Symp. Discrete Algorithms (SODA), pp. 785–804 (2015)
6. Biró, P., Irving, R.W., Manlove, D.F.: Popular matchings in the marriage and roommates problems. In: Calamoneri, T., Diaz, J. (eds.) CIAC 2010. LNCS, vol. 6078, pp. 97–108. Springer, Heidelberg (2010)
7. Diamantoudi, E., Miyagawa, E., Xue, L.: Random paths to stability in the roommates problem. Games Econom. Behav. **48**(1), 18–28 (2004)
8. Gärdenfors, P.: Match making: Assignments based on bilateral preferences. Behavioural Sciences **20**, 166–173 (1975)
9. Gupta, M., Peng, R.: Fully dynamic $(1+\varepsilon)$-approximate matchings. In: Proc. 54th Symp. Foundations of Computer Science (FOCS), pp. 548–557 (2013)
10. Gusfield, D., Irving, R.: The Stable Marriage Problem: Structure and Algorithms. MIT Press (1989)
11. Hoefer, M.: Local matching dynamics in social networks. Inf. Comput. **222**, 20–35 (2013)
12. Hoefer, M., Wagner, L.: Locally stable marriage with strict preferences. In: Fomin, F.V., Freivalds, R., Kwiatkowska, M., Peleg, D. (eds.) ICALP 2013, Part II. LNCS, vol. 7966, pp. 620–631. Springer, Heidelberg (2013)
13. Hoefer, M., Wagner, L.: Matching dynamics with constraints. In: Liu, T.-Y., Qi, Q., Ye, Y. (eds.) WINE 2014. LNCS, vol. 8877, pp. 161–174. Springer, Heidelberg (2014)
14. Huang, C., Kavitha, T.: Near-popular matchings in the roommates problem. SIAM J. Disc. Math. **27**(1), 43–62 (2013)
15. Knuth, D.: Marriages stables et leurs relations avec d'autres problemes combinatoires. Les Presses de l'Université de Montréal (1976)
16. Manlove, D.: Algorithmics of Matching Under Preferences. World Scientific (2013)
17. McCutchen, R.M.: The least-unpopularity-factor and least-unpopularity-margin criteria for matching problems with one-sided preferences. In: Laber, E.S., Bornstein, C., Nogueira, L.T., Faria, L. (eds.) LATIN 2008. LNCS, vol. 4957, pp. 593–604. Springer, Heidelberg (2008)
18. Onak, K., Rubinfeld, R.: Maintaining a large matching and a small vertex cover. In: Proc. 42nd Symp. Theory of Computing (STOC), pp. 457–464 (2010)
19. Roth, A., Sotomayor, M.O.: Two-sided Matching: A study in game-theoretic modeling and analysis. Cambridge University Press (1990)
20. Roth, A., Vate, J.V.: Random paths to stability in two-sided matching. Econometrica **58**(6), 1475–1480 (1990)

Ultra-Fast Load Balancing
on Scale-Free Networks

Karl Bringmann[1], Tobias Friedrich[2,3], Martin Hoefer[4],
Ralf Rothenberger[2,3]([✉]), and Thomas Sauerwald[5]

[1] Institute of Theoretical Computer Science, ETH Zurich, Switzerland
[2] Friedrich Schiller University Jena, Jena, Germany
[3] Hasso Plattner Institute, Potsdam, Germany
`ralf.rothenberger@hpi.de`
[4] Max Planck Institute for Informatics, Saarbrücken, Germany
[5] University of Cambridge, Cambridge, UK

Abstract. The performance of large distributed systems crucially depends on efficiently balancing their load. This has motivated a large amount of theoretical research how an imbalanced load vector can be smoothed with local algorithms. For technical reasons, the vast majority of previous work focuses on regular (or almost regular) graphs including symmetric topologies such as grids and hypercubes, and ignores the fact that large networks are often highly heterogenous.

We model large scale-free networks by Chung-Lu random graphs and analyze a simple local algorithm for iterative load balancing. On n-node graphs our distributed algorithm balances the load within $\mathcal{O}((\log \log n)^2)$ steps. It does not need to know the exponent $\beta \in (2,3)$ of the power-law degree distribution or the weights w_i of the graph model. To the best of our knowledge, this is the first result which shows that load-balancing can be done in double-logarithmic time on realistic graph classes.

1 Introduction

Load Balancing. Complex computational problems are typically solved on large parallel networks. An important prerequisite for their efficient usage is to balance the work load efficiently. Load balancing is also known to have applications to scheduling [17], routing [6], numerical computation such as solving partial differential equations [16,19], and finite element computations [13]. In the standard abstract formulation of load balancing, processors are represented by nodes of a graph, while links are represented by edges. The objective is to balance the load by allowing nodes to exchange loads with their neighbors via the incident edges. Particularly popular are decentralized, round-based iterative algorithms where a processor knows only its current load and that of the neighboring processors. We focus on *diffusive* load balancing strategies, where each processor decides how many jobs should be sent and balances its load with its neighbors in each round. As the degrees of the topologies of many networks follow heavy tailed statistics, our main interest lies on *scale-free* networks.

© Springer-Verlag Berlin Heidelberg 2015
M.M. Halldórsson et al. (Eds.): ICALP 2015, Part II, LNCS 9135, pp. 516–527, 2015.
DOI: 10.1007/978-3-662-47666-6_41

Diffusion. On networks with n nodes, our balancing model works as follows: At the beginning, each node i has some work load $x_i^{(0)}$. The goal is to obtain (a good approximation of) the balanced work load $\bar{x} := \sum_{i=1}^n x_i^{(0)}/n$ on all nodes. On heterogenous graphs with largely varying node degrees it is natural to consider a multiplicative quality measure: We want to find an algorithm which achieves $\max_i x_i^{(t)} = \mathcal{O}(\bar{x})$ at the earliest time t possible. Load-balancing is typically considered *fast* if this can be achieved in time logarithmic in the number of nodes. We aim at double-logarithmic time, which we call *ultra-fast* (following the common use of the superlative "ultra" for double-logarithmic bounds [4,10,18]).

The diffusion model was first studied by Cybenko [6] and, independently, Boillat [1]. The standard implementation is the *first order scheme* (FOS), where the load vector is multiplied with a diffusion matrix $\mathbf{P}$ in each step. For regular graphs with degree d, a common choice is $\mathbf{P}_{ij} = 1/(d+1)$ if $\{i,j\} \in E$. Already Cybenko [6] in 1989 shows for regular graphs a tight connection between the convergence rate of the diffusion algorithm and the absolute value of the second largest eigenvalue $\lambda_{\max}$ of the diffusion matrix $\mathbf{P}$. While FOS can be defined for non-regular graphs, its convergence is significantly affected by the loops which are induced by the degree discrepancies. Regardless of how the damping factor is chosen, FOS requires $\Omega(\log n)$ rounds on a broad class of non-regular graphs. For a proof and discussion of this statement we refer to the full version of this paper.

Scale-free Networks. Many real-world graphs have a power law degree distribution, meaning that the number of vertices with degree k is proportional to $k^{-\beta}$, where β is a constant intrinsic to the network. Such networks are synonymously called scale-free networks and have been widely studied. As a model for large scale-free networks we use the *Chung-Lu random graph model* with a power-law degree distribution with exponent $\beta \in (2,3)$. (See Section 2 for a formal definition.) This range of β's is typically studied as many scale-free networks (e.g. co-actors, protein interactions, internet, peer-to-peer [15]) have a power law exponent with $2 < \beta < 3$. It is known that the diameter of this graph model is $\Theta(\log n)$ while the average distance between two vertices is $\Theta(\log \log n)$ [3].

Results. Scale-free networks are omnipresent, but surprisingly few rigorous insights are known about their ability to efficiently balance load. Most results and developed techniques for theoretically studying load balancing only apply to regular (or almost-regular) graphs. In fact, we cannot hope for ultra-fast balancing on almost-regular graphs: Even for expander graphs of maximum degree d, there is a general lower bound of $\Omega(\log n/ \log d)$ iterations for *any* distributed load balancing algorithms (for a proof of this statement we refer to the full version of this paper). Our main result (cf. Theorem 2.1) shows that within $\mathcal{O}((\log \log n)^2)$ steps, our simple local balancing algorithm (cf. Algorithm 1) can balance the load on a scale-free graph with high probability. The algorithm assumes that the initial load is only distributed on nodes with degree $\Omega(\text{polylog } n)$ (cf. Theorem 2.2), which appears to be a natural assumption in typical load balancing applications. As the diameter of the graph is $\Theta(\log n)$, ultra-fast balancing is impossible if the initial load is allowed on arbitrary vertices. As standard FOS requires $\Omega(\log n)$ rounds, our algorithm uses a different, novel approach to overcome these restrictions.

Algorithm. The protocol proceeds in waves, and each wave (roughly) proceeds as follows. First, the remaining load is balanced within a core of high-degree nodes. These nodes are known to compose a structure very similar to a dense Erdős-Rényi random graph and thereby allow very fast balancing. Afterwards, the load is disseminated into the network from high- to low-degree nodes. Each node absorbs some load and forwards the remaining to lower-degree neighbors. If there are no such neighbors, the excess load is routed back to nodes it was received from. In this way, the load moves like a wave over the graph in decreasing order of degree and then swaps back into the core. We will show that each wave needs $\mathcal{O}(\log \log n)$ rounds. The algorithm keeps initiating waves until all load is absorbed, and we will show that only $\mathcal{O}(\log \log n)$ waves are necessary.

Techniques. There are a number of technical challenges in our analysis, mostly coming from the random graph model, and we have to develop new techniques to cope with them. For example, in scale-free random graphs there exist large sparse areas with many nodes of small degree that result in a high diameter. A challenge is to avoid that waves get lost by pushing too much load deep into these periphery areas. This is done by a partition of nodes into layers with significantly different degrees and waves that proceed only to neighboring layers. To derive the layer structure, we classify nodes based on their realized degrees. However, this degree might be different from the expected degree corresponding to the weights w_i of the network model, which is unknown to the algorithm. This implies that nodes might not play their intended role in the graph and the analysis (cf. Definition 4.2). This can lead to poor spread and the emersion of a few, large single loads during every wave. Here we show that several types of "wrong-degree" events causing this problem are sufficiently rare, or, more precisely, they tend to happen frequently only in parts of the graph that turn out not to be critical for the result. At the core, our analysis adjusts and applies fundamental probabilistic tools to derive concentration bounds, such as a variant of the method of bounded variances (cf. Theorem 4.1).

2   Model, Algorithms, and Formal Result

Chung-Lu Random Graph Model. We consider random graphs $G = (V, E)$ as defined by Chung and Lu [3]. Every vertex $i \in V = \{1, \ldots, n\}$ has a weight w_i with $w_i := \frac{\beta-2}{\beta-1} dn^{1/(\beta-1)} i^{-1/(\beta-1)}$ for $i = 1, 2 \ldots, n$. The probability for placing an edge $\{i, j\} \in E$ is then set to $\min\{w_i w_j / W, 1\}$ with $W := \sum_{i=1}^{n} w_i$. This creates a random graph where the expected degrees follow a power-law distribution with exponent $\beta \in (2, 3)$, the maximum expected node degree is $\frac{\beta-2}{\beta-1} dn^{1/(\beta-1)}$ and d influences the average expected node degree Chung and Lu [3]. The graph has a *core* of densely connected nodes which we define as

$$C := \left\{ i \in V \colon \deg_i \geqslant n^{1/2} - \sqrt{n^{1/2} \cdot (c+1) \ln n} \right\}.$$

Algorithm 1. Balance load in waves from core to all other nodes

---

repeat
    for *phase* $t \leftarrow 1$ **to** $\log \log n$ **do**
        for $\frac{32}{3-\beta}$ *rounds* **do**   // 1. diffusion on the core
            Nodes v with $\deg(v) \geqslant \omega_0$ perform diffusion with $\mathbf{P} = \mathbf{D}^{-1}\mathbf{A}$

        for L *rounds* **do**   // 2. downward propagation
            Every node absorbs at most m/nt^2 load.
            All remaining load is forwarded in equal shares to neighbors on
            the next lower layer.

        for L *rounds* **do**   // 3. upward propagation
            All nodes send their load back to the the next higher layer over
            the edges they received it from. The distribution of load
            amongst these edges can be arbitrary.
until *terminated*;

Distributing the Load in Waves. Our main algorithm is presented in Algorithm 1. It assumes that an initial total load of m resides exclusively on the core C of the network. The first rounds are spend on simple diffusion on the core with diffusion matrix $\mathbf{P} = \mathbf{D}^{-1}\mathbf{A}$, where $\mathbf{A}$ is the adjacency matrix and $\mathbf{D}$ is the degree matrix. Afterwards, the algorithm pushes the load to all other nodes in waves from the large to the small degree nodes and the other way around. To define the direction of the waves, the algorithm partitions the nodes into layers, where on layer k we have all nodes v of degree $\deg_v \in (\omega_k, \omega_{k-1}]$, where $\omega_0 = n^{1/2} - \sqrt{n^{1/2} \cdot (c+1)\ln n}$ and $\omega_{k+1} = \omega_k^{1-\varepsilon}$ for a constant

$$0 < \varepsilon < \min\left\{\frac{(3-\beta)}{(\beta-1)}, \frac{\beta-2}{3}, \frac{1}{2}\left(1 - \sqrt{\frac{3}{\beta+1}}\right)\right\}.$$

For every layer k we have $\omega_k > 2^{\frac{1}{\varepsilon(\beta-1)}}$. The last layer ℓ is the first, for which $\omega_\ell \leqslant 2^{\frac{1}{\varepsilon(\beta-1)}}$ holds. In this case, we define the interval simply to include all nodes with degree less than $\omega_{\ell-1}$. Note that in total we obtain at most $L := \frac{1}{\log(1/(1-\varepsilon))}\left(\log\log n + \log\frac{\varepsilon(\beta-1)}{2}\right)$ layers. To choose an appropriate ε, we have to know lower and upper bounds on β. These bounds are either known or can be chosen as constants arbitrarily close to 2 and 3. The algorithm therefore does not need to know the precise β. Our main result is then as follows.

Theorem 2.1. *Let $G = (V, E)$ be a Chung-Lu random graph as defined above. For any load vector $x^{(0)} \in \mathbb{R}^n_{\geqslant 0}$ with support only on the core C of the graph, there is a $\tau = \mathcal{O}((\log\log n)^2)$ such that for all steps $t \geqslant \tau$ of Algorithm 1, the resulting load vector $x^{(t)}$ fulfills $x_u^{(t)} = \mathcal{O}(\bar{x})$ for all $u \in V$ w. h. p.*[1]

---

[1] w. h. p. is short for "with high probability". We use w.h.p. to describe events that hold with probability $1 - n^{-c}$ for an arbitrary large constant c.

Reaching the Core. Algorithm 1 and Theorem 2.1 above require that the initial total load resides exclusively on the core C of the network. As the diameter of the network is $\Theta(\log n)$ [3], we cannot hope to achieve a double-logarithmic balancing time if all the initial load starts at an arbitrary small and remote vertex. However, we can allow initial load on all nodes with at least some polylogarithmic degree by adding an *initial phase* in which all nodes send all their load to an arbitrary neighbor on the next-highest layer. This initial local routing phase succeeds if all nodes with at least this polylogarithmic degree have at least one neighbor on the next-highest layer. The following theorem, the proof of which can be found in the full version of this paper, formalizes this result, while the rest of the paper proves Theorem 2.1.

Theorem 2.2. *Let* $G = (V, E)$ *be a Chung-Lu random graph as defined above. For any load vector* $x^{(0)} \in \mathbb{R}^n_{\geq 0}$ *with support only on nodes with degree* $\Omega((\log n)^{\max(3, 2/(3-\beta))})$, *the initial phase reaches after* $L = \Theta(\log \log n)$ *steps a load vector* $x^{(L)}$ *such that* $x_u^{(L)}$ *has support only on the core* C *w. h. p.*

3   Analysis of Load Balancing on the Core

We start our analysis of Algorithm 1 with its first step, the diffusion on the core. Recall the definition of the core C of the network and consider the core subgraph $\widetilde{G} = (\widetilde{V}, \widetilde{E})$ induced by C.

Lemma 3.1. *The core subgraph* $\widetilde{G}$ *of* G *fulfills*

$$|1 - \lambda_k(L)| \leqslant \Theta\left(\frac{\sqrt{(c+1)\ln(4n)}}{n^{(3-\beta)/4}} + \frac{(2c\ln n)^{1/4}}{n^{1/8}}\right)$$

for all eigenvalues $\lambda_k(L) > \lambda_{\min}(L)$ *of the normalized Laplacian* $L(\widetilde{G})$ *w. h. p.*

The proof of Lemma 3.1 is based on Theorem 2 from [12] and can be found in the full version of this paper. The following lemma states that after only a constant number of diffusion rounds in $\widetilde{G}$, the load of node $v \in C$ is more or less equal to $m \cdot w_v/W_0$.

Lemma 3.2. *After* $\frac{32}{3-\beta}$ *rounds of diffusion with* $\mathbf{P} = \mathbf{D}^{-1}\mathbf{A}$ *in the core subgraph* $\widetilde{G}$, *each node* $v \in C$ *has a load of at most* $\mathcal{O}\left(w_y/\sum_{x\in C} w_x\right)$ *w. h. p.*

The proof of Lemma 3.2 uses eq. 12.11 from [14] and can be found in the full version, too. An implication of the lemma is, that there is a constant $\varepsilon_0 > 0$ such that each node $v \in C$ has a load of at most $(1 + \varepsilon_0)\frac{w_v}{W_0}m$ after the first phase of Algorithm 1.

4   Analysis of Top-Down Propagation

We continue our analysis of Algorithm 1. This section studies the downward/upward propagation.

Many of the proofs in this section are based on the following variant of the method of bounded variances [8], which might be useful in its own respect.

Theorem 4.1. *Let* $X_1, \ldots, X_n$ *be independent random variables taking values in* $\{0, 1\}$, *and set* $\mu := \mathbf{E}\left[\sum_{i=1}^n X_i\right]$. *Let* $f := f(X_1, \ldots, X_n)$ *be a function satisfying*

$$|f| \leqslant M,$$

and consider an error event $\mathcal{B}$ *such that for every* $\mathbf{X}_n \in \overline{\mathcal{B}}$

$$|f(\mathbf{X}_n) - f(\mathbf{X}'_n)| \leqslant c$$

for every $\mathbf{X}'_n$ *that differs in only one position* X_i *from* $\mathbf{X}_n$, *and for some* $c > 0$. *Then for any* $0 \leqslant t \leqslant c\mu$ *we have*

$$\Pr\left[|f - \mathbf{E}[f]| > t + \tfrac{(2M)^2}{c}\Pr[\mathcal{B}]\right] \leqslant \tfrac{2M}{c}\Pr[\mathcal{B}] + 2\exp\left(-\tfrac{t^2}{16c^2\mu}\right).$$

The proof of this theorem closely follows the one of the method of bounded variances as can be found in [8]. For the sake of brevity, however, all proofs are omitted in this version of the paper and interested readers are referred to the full version.

First note that our algorithm deals with a random graph and therefore it might happen that some of the nodes' neighborhood look significantly different from what one would expect by looking at the expected values. We call these nodes *dead-ends* as they can not be utilized to effectively forward load. This definition will be made precise in Definition 4.2 below.

Only for the sake of analysis we assume that dead-ends do not push load to neighbors on the next lowest layer, but instead keep all of it. In reality the algorithm does *not* differentiate between nodes which are dead-ends and nodes which are no dead-ends. We also assume in this section that nodes do not consume any load during the top-down distribution.

The main goal of this section is twofold. We first show that no node which is not a dead-end, gets too much load. Then we show that the total load on all dead-ends from the core down to a layer with nodes of a certain constant degree is at most a constant fraction of the total load. The converse means that at least a constant fraction of load reaches the nodes of the last layer we are considering.

We define $V_k = \{v \mid w_v \in (\omega_k, \omega_{k-1}]\}$ as the set of nodes on layer k and $n_k = |V_k|$. Let $W_k = \sum_{v \in V_i} w_v$ be the total weight of nodes in layer k. Let $\gamma := \tfrac{1}{2}\left(d\tfrac{\beta-2}{\beta-1}\right)^{\beta-1}$. From the given weight sequence and the requirements $\omega_k > 2^{\frac{1}{\varepsilon(\beta-1)}}$ and $\omega_k < n^{1/(\beta-1)}$, we can easily derive the following bounds. For all $0 \leqslant k < \ell$ it holds that $\tfrac{\gamma}{2} \cdot n\omega_k^{1-\beta} \leqslant n_k \leqslant 4\gamma \cdot n\omega_k^{1-\beta}$. This implies $W_k \geqslant \tfrac{\gamma}{2} \cdot n\omega_k^{2-\beta}$. Let $\overline{d} = \tfrac{W}{n}$ the expected average degree.

For a node $v \in V_k$ we consider two partial degrees. Let D_v^h be the number of edges to nodes in the higher layer $k-1$, and D_v^ℓ is the number of edges to nodes in the lower layer $k + 1$. Note that D_v^h and D_v^ℓ are random variables, composed of sums of independent Bernoulli trials:

$$D_v^h = \sum_{u \in V_{k-1}} \text{Ber}\left(\frac{w_v \cdot w_u}{W}\right) \quad \text{and} \quad D_v^\ell = \sum_{u \in V_{k+1}} \text{Ber}\left(\frac{w_v \cdot w_u}{W}\right) .$$

In our proofs we will apply several well-known Chernoff bounds which use the fact that partial degrees are sums of independent Bernoulli trials.

We now define four properties which will be used throughout the analysis.

Definition 4.2. *A node $v \in V$ is a* dead-end *if one of the following holds:*

⟨D1⟩ **In-/Out-degree:** *A node $v \in V_k$ has this property if either $|D_v^h - \mathbf{E}[D_v^h]| \geqslant \mathbf{E}[D_v^h]^{2/3}$ or $|D_v^\ell - \mathbf{E}[D_v^\ell]| \geqslant \mathbf{E}[D_v^\ell]^{2/3}$.*

⟨D2⟩ **Wrong layer:** *A node $v \in V_k$ has this property if it has a degree that deviates by at least $w_v^{2/3}$ from its expected degree.*

⟨D3⟩ **Border:** *A node $v \in V_k$ has this property if it does not fulfill property ⟨D2⟩ and if it is of weight at least $\omega_{k-1} - \omega_{k-1}^{2/3}$ or at most $\omega_k + \omega_{k-1}^{2/3}$ and if it is assigned to the wrong layer.*

⟨D4⟩ **Induced Out-degree:** *A node $v \in V_k$ has this property if it fulfills none of the properties ⟨D1⟩ – ⟨D3⟩ and if it has at least $(\omega_k W_{k+1}/W)^{2/3}$ many lower-layer neighbors with properties ⟨D2⟩ or ⟨D3⟩.*

The next lemma shows that for a non-dead-end node $v \in V_k$ the received load x_v in phase k is almost proportional to the "layer-average load" $m \cdot w_v / W_k$. For dead-ends, the received load can be higher, but the probability to receive significantly higher load is small.

Lemma 4.3. *For $v_k \in V_k$ and the received load x_v in phase k the following holds. If v is not a dead-end,*

$$x_v \leqslant (1 + \varepsilon_k) \cdot m \cdot \frac{w_v}{W_k} ,$$

where for every layer k the error term ε_k is given by

$$(1 + \varepsilon_k) = (1 + \varepsilon_{k-1}) \cdot (1 + \mathcal{O}(\omega_k^{-1+\beta/3})) \cdot (1 + \mathcal{O}(\omega_k^{-(3-\beta)/6})) ,$$

so $\varepsilon_k \leqslant \varepsilon_{k+1}$ and $\varepsilon_k = \mathcal{O}(1)$.

Now we want to show that on each layer with sufficiently large constant weight at most a small fraction of the total load remains on dead-ends. To do so, we show that for each property ⟨D1⟩ – ⟨D4⟩ the nodes with these properties only contribute a small enough fraction to the total dead-end load of each layer. We begin by bounding the contribution of ⟨D1⟩-nodes to the total dead-end load.

Lemma 4.4. *If $\varepsilon \leqslant (3 - \beta)/(\beta - 1)$ and $\omega_k > \left(\frac{2\bar{d}}{\gamma}\left(\frac{1}{2e-1}\right)^3\right)^{2/(3-\beta)}$, the probability that a node $v \in V_k$ is a ⟨D1⟩-node is at most $2\exp(-c \cdot \omega_k^{(3-\beta)/6})$, for $c = \frac{1}{4}\left(\frac{\gamma}{2\bar{d}}\right)^{1/3}$.*

An implication of the former lemma is that there are no $\langle D1\rangle$-nodes on layers with weight at least polylog(n). Now that we have an understanding of which layers actually contain $\langle D1\rangle$-nodes, we can start to derive high probability upper bounds on the total load that is left on these nodes throughout the top-down phase.

Lemma 4.5. *If* $v \in V_k$ *is a* $\langle D1\rangle$*-node, then*

$$\mathbf{Pr}\left[x_v \geqslant \alpha \cdot \frac{m \cdot w_v}{W_k}\right] < \exp\left(-\Omega(\omega_k^{(3-\beta)/2} \cdot \min\{\alpha - 1, (\alpha - 1)^2\})\right) \ .$$

Now we use the tail bound from Lemma 4.5 and overestimate the load distribution of $\langle D1\rangle$-nodes with an exponential distribution. In particular, for each node $v \in V_k$ we introduce the variable X_v that measures the "$\langle D1\rangle$-load" of this node, i.e. the load that each $\langle D1\rangle$-node keeps. We can now show that for each node $v \in V_k$ the following random variable stochastically dominates the $\langle D1\rangle$-load X_v.

Definition 4.6. *For a node* $v \in V_k$ *let*

$$\widehat{X}_v = \begin{cases} 0 & \text{with prob. } 1 - \widehat{p}_v \\ \ell_v\left(1 + Exp(\lambda_v) + \mathbf{E}\left[D_v^h\right]^{-2/3}\right) & \text{with prob. } \widehat{p}_v \ , \end{cases}$$

where $\widehat{p}_v = 2\exp\left(-\frac{\mathbf{E}\left[D_v^h\right]^{1/3}}{4}\right)$ *is an upper bound for the probability that* v *is a* $\langle D1\rangle$*-node,* $\lambda_v = \frac{1}{4}\mathbf{E}\left[D_v^h\right]$ *and* $\ell_v = 2(1 + \varepsilon_k)m\frac{w_v}{W_k}$.

Note that our $\langle D1\rangle$-load overestimates the contribution of v to the total load left on $\langle D1\rangle$-nodes during the top-down phase. In particular, if v is not a $\langle D1\rangle$-node, then no $\langle D1\rangle$-load is left on v and consequently the contribution is 0. Otherwise, we use the tail bound from Lemma 4.5 as follows. We overestimate the load by assuming that at least twice the layer-average load is present on v. For the additional load, we can apply the tail bound under the condition $\alpha \geqslant 2$, which implies that this excess load is upper bounded by an exponentially distributed random variable with a parameter $\lambda_v = \frac{1}{4}\mathbf{E}\left[D_v^h\right]$.

We first obtain a high probability bound on the total load left on $\langle D1\rangle$-nodes in each layer k during the top-down phase.

Lemma 4.7. *For every constant* $c > 0$ *and any* k *the total load left on* $\langle D1\rangle$*-nodes in layer* k *is at most*

$$4(1 + \varepsilon_k)m\frac{\omega_{k-1}}{W_k}c\ln n + 40(1 + \varepsilon_k)m\frac{\omega_{k-1}}{W_k}n_k\exp\left(-\frac{1}{4}\left(\omega_k\frac{W_{k-1}}{W}\right)^{1/3}\right)$$

with probability at least $1 - n^{-c}$.

Now we take a closer look at nodes with property $\langle D2\rangle$. We can employ a Chernoff Bound to show that nodes with polylogarithmically large weights do not deviate by $w_v^{2/3}$ from their expected degree with high probability. This means that none of these nodes fulfills property $\langle D2 \rangle$ with high probability. In the following analysis we can therefore concentrate on nodes with weight at most polylog(n). This observation is crucial for the proof of Lemma 4.8.

Lemma 4.8. *For any k all nodes $v \in V_k$ with property $\langle D2 \rangle$ contribute at most*

$$\mathcal{O}\left(\left(1 + w_k^{-2/3}\right)^3 w_k^{\frac{4-\beta+\varepsilon(\beta-1)}{1-\varepsilon}} \cdot \exp\left(-w_k^{1/3}/4\right) m \right) + \mathcal{O}\left(\frac{\text{polylog}(n)}{\sqrt{n}} m \right)$$

to the total dead-end load of all layers with probability at most $1 - \frac{3}{n^C}$, for a constant $C > 1 + (\beta - 2)\left(1 + \frac{1}{1-\varepsilon}\right)$.

After successfully bounding the contribution of nodes with properties $\langle D1 \rangle$ and $\langle D2 \rangle$ to dead-end load, we will now turn to the border nodes with property $\langle D3 \rangle$. We already know that these nodes cannot deviate too much from their expected degrees, because they do not fulfill property $\langle D2 \rangle$ by definition. Therefore they can only be on one of two layers. We still have to differ between nodes in the upper half of a border and those in the lower half. The following lemma bounds the contribution of nodes in the upper half of a border.

Lemma 4.9. *For any k all nodes $v \in V_k$ with property $\langle D3 \rangle$ and $w_k \leqslant w_v \leqslant w_k + w_{k-1}^{2/3}$ contribute at most*

$$\Theta\left(w_k^{-\varepsilon(\beta-2)} + \frac{w_k^{\beta-2} w_{k+1}^{\beta-2} \cdot c \ln n}{n} \right) m$$

to the total dead-end load of layer $k+1$ w. h. p.

The following lemma about the contribution of nodes in the lower half of a border uses the smoothness of the weight distribution.

Lemma 4.10. *For any k all nodes $v \in V_{k+1}$ with property $\langle D3 \rangle$ and $w_k - w_k^{2/3} \leqslant w_v \leqslant w_k$ contribute at most*

$$(1 + \varepsilon_k)\left(\frac{6\left(d\frac{\beta-2}{\beta-1}\right)^{\beta-1}}{\frac{\gamma}{2}} w_k^{-1/3} + \frac{2 \cdot w_k^{\beta-1}}{\frac{\gamma}{2} n} + \frac{d \cdot w_k^{\beta-2} w_{k+1}^{\beta-2} \cdot c \ln n}{\left(\frac{\gamma}{2}\right)^2 n} \right) m$$

to the total dead-end load of layer k w. h. p.

At last we have to show that the dead-end load of nodes with property $\langle D4 \rangle$ is properly bounded. We already know, that each of these nodes obeys the upper bound from Lemma 4.3. Therefore it is sufficient to bound the number of these nodes. To bound the number of these nodes in V_k, we simply have to bound the total number of edges lost between nodes from V_k and nodes with properties $\langle D2 \rangle$ or $\langle D3 \rangle$ from V_{k+1}. This idea helps us to proof the following lemma.

Lemma 4.11. *Let* $\varepsilon < \min\left\{\frac{\beta-2}{3}, \frac{1}{2}\left(1 - \sqrt{\frac{3}{\beta+1}}\right)\right\}$. *Then the following statements hold:*

(1) For all $k > 0$ the total load of nodes $v \in V_k$ with property $\langle D4 \rangle$ is at most

$$\mathcal{O}\left(\omega_k^{\frac{2-\beta}{3(1-\varepsilon)}} + \omega_k^{\frac{(2\beta^2 - 11\beta + 14)(\beta-2)}{27(1-\varepsilon)}} + n^{\frac{3+1-\varepsilon+2(\beta-2)(1-\varepsilon)^2}{6} - 1}\right.$$
$$\left. + \exp\left(-\omega_k^{\frac{1-\varepsilon}{3}}/4\right)\omega_k^{\frac{1}{1-\varepsilon}+(\beta-2)} + \frac{\mathrm{polylog}(n)}{\sqrt{n}}\right)m \qquad w.\,h.\,p.$$

(2) For $k = 0$ there are no $\langle D4 \rangle$-nodes w. h. p.

Finally, we bound the total load left on dead-ends during the top-down phase.

Lemma 4.12. *For every constant c, there exists a constant c' such that if we run the top-down phase on layers with $\omega_i \geq c'$, then with probability at least $1 - 1/n^{-c}$ we obtain a total load of at most $m/2$ on all dead-ends on these layers.*

The last lemma implies that with high probability, for a suitably chosen *key layer* at most half of the load is left on dead-ends during the top-down phase on this and the above layers. In particular, our upper bound on the load of non-dead-end nodes in Lemma 4.3 implies that on this layer, every such node gets at most a load of $(1 + \varepsilon_k) \cdot m \cdot w_v/W_k$. On the other hand, a load of $m/2$ passes through this layer w.h.p. In the worst case all non-dead-ends get the maximum load of $(1 + \varepsilon_k) \cdot m \cdot w_v/W_k$. This results in at least $n \frac{\gamma}{4(1+\varepsilon_k)}\omega_k^{-\frac{\beta-1}{(1-\varepsilon)}}$ nodes which absorb m/n load each, causing a decrease of unassigned load by a constant fraction of at least $\frac{\gamma}{4(1+\varepsilon_k)}\omega_k^{-\frac{\beta-1}{(1-\varepsilon)}}$. Here, $\omega_k \geq c'$ where c' is as chosen in Lemma 4.12.

5 Analysis of Iterative Absorption

Algorithm 1 sends all unassigned load back to the top, balances it within the top layer, and restarts the top-down distribution step. Observe that all the arguments made for the analysis of the downward propagation can be applied for any value of m. The absorption of load during these iterations is adjusted according to the following scheme. We let each of the nodes absorb at most a load of $m/(n \cdot t^2)$ in round t. This scheme is executed for $t = \log \log n$ rounds and then repeated in chunks of $\log \log n$ rounds until all load is assigned. We will show that with high probability after a constant number of repetitions, all load is assigned. In addition, as $\sum_{t=1}^{\infty} 1/t^2 = \Pi^2/6$, each node receives a load of $(1 + \mathcal{O}(1)) \cdot m/n$.

In particular, our aim is to show that using this scheme we need only $\mathcal{O}(\log \log n)$ top-down distribution steps to reduce the total unassigned load in the system to $m' = m/\log^c n$, for any constant c. This is shown in the lemma below. Given this result, we run the protocol long enough such that c becomes a sufficiently large constant. We want to show that, if this is the case, each node on a layer with polylogarithmic degree gets a load of at most m/n, resulting in all remaining load being absorbed. As each non-dead-end on this layer gets a share of at most $w_v \frac{W_k}{W} m' = \frac{\text{polylog}(n)}{n} m' = m/n$ they fulfill the requirement. The same bound holds for $\langle D4 \rangle$-nodes by definition. As $\langle D1 \rangle$- and $\langle D2 \rangle$-nodes do not appear on layers of at least polylogarithmic degree, we can ignore them as well. All we need to care about now are $\langle D3 \rangle$-nodes. We can derive upper bounds on their load similar to the ones for non-dead-ends using results on the expected number of edges between these nodes and both their possible next-highest layers. This is a simple corollary from the proof of Theorem 2.2 which we defer to the full version of the paper. It now remains to show the following lemma, the proof of which can be found in the full version of this paper.

Lemma 5.1. *Using the repeated absorption scheme of Algorithm 1, for any constant c, only $\mathcal{O}(\log \log n)$ rounds suffice to reduce the unassigned load in the network to $m/\log^c n$.*

6   Discussion

To the best of our knowledge, we have presented the first double-logarithmic load balancing protocol for a realistic network model. Our algorithm reaches a balanced state in time *less than the diameter* of the graph, which is a common lower bound for other protocols (e.g. [9]). Note that our Theorem 2.1 can be interpreted outside of the intended domain: It reproves (without using the fact) that the giant component is of size $\Theta(n)$ (known from [3]) and that rumor spreading to most vertices can be done in $\mathcal{O}(\log \log n)$ (known from [10]).

Our algorithm works fully distributed, and nodes decide how many tokens should be sent or received based only on their current load (and those of its neighbors). We expect our wave algorithm to perform very robust against node and edge failures as it does *not* require global information on distances [9] or the computation of a balancing flow [7].

Our Theorem 2.2 allows initial load on nodes with degree $\Omega(\text{polylog } n)$. Future work includes a further relaxation of this assumption, for instance, by employing results about greedy local-search based algorithms to find high degree nodes [2,5]. Another interesting direction is to translate our load balancing protocol into an algorithm which samples a random node using the analogy between load and probability distributions. Such sampling algorithms are crucial for crawling large-scale networks such as online social networks like Facebook, where direct sampling is not supported [11].

References

1. Boillat, J.E.: Load balancing and poisson equation in a graph. Concurrency: Pract. Exper., **2**, 289–313 (1990)
2. Borgs, C., Brautbar, M., Chayes, J., Khanna, S., Lucier, B.: The power of local information in social networks. In: Goldberg, P.W. (ed.) WINE 2012. LNCS, vol. 7695, pp. 406–419. Springer, Heidelberg (2012)

3. Chung, F., Lu, L.: The average distances in random graphs with given expected degrees. Proceedings of the National Academy of Sciences **99**, 15879–15882 (2002)
4. Cohen, R., Havlin, S.: Scale-free networks are ultrasmall. Phys. Rev. Lett. **90**, 058701 (2003)
5. Cooper, C., Radzik, T., Siantos, Y.: A fast algorithm to find all high degree vertices in graphs with a power law degree sequence. In: Bonato, A., Janssen, J. (eds.) WAW 2012. LNCS, vol. 7323, pp. 165–178. Springer, Heidelberg (2012)
6. Cybenko, G.: Load balancing for distributed memory multiprocessors. J. Parallel and Distributed Comput. **7**, 279–301 (1989)
7. Diekmann, R., Frommer, A., Monien, B.: Efficient schemes for nearest neighbor load balancing. Parallel Computing **25**, 789–812 (1999)
8. Dubhashi, D., Panconesi, A.: Concentration of Measure for the Analysis of Randomized Algorithms. Cambridge University Press (2009)
9. Elsässer, R., Sauerwald, T.: Discrete load balancing is (almost) as easy as continuous load balancing. In: 29th Symp. Principles of Distributed Computing (PODC), pp. 346–354 (2010)
10. Fountoulakis, N., Panagiotou, K., Sauerwald, T.: Ultra-fast rumor spreading in social networks. In: 23rd Symp. Discrete Algorithms (SODA), pp. 1642–1660 (2012)
11. Gjoka, M., Kurant, M., Butts, C.T., Markopoulou, A.: Walking in Facebook: A case study of unbiased sampling of OSNs. In: 29th IEEE Conf. Computer Communications (INFOCOM), pp. 2498–2506 (2010)
12. Graham, F.C., Radcliffe, M.: On the spectra of general random graphs. Electr. J. Comb. **18** (2011)
13. Huebner, K.H., Dewhirst, D.L., Smith, D.E., Byrom, T.G.: The Finite Element Methods for Engineers. Wiley (2001)
14. Levin, D.A., Peres, Y., Wilmer, E.L.: Markov Chains and Mixing Times. AMS (2008)
15. Newman, M.E.J.: The structure and function of complex networks. SIAM Review **45**, 167–256 (2003)
16. Subramanian, R., Scherson, I.D.: An analysis of diffusive load-balancing. In: 6th Symp. Parallelism in Algorithms and Architectures (SPAA), pp. 220–225 (1994)
17. Surana, S., Godfrey, B., Lakshminarayanan, K., Karp, R., Stoica, I.: Load balancing in dynamic structured peer-to-peer systems. Performance Evaluation **63**, 217–240 (2006)
18. van der Hofstad, R.: Random graphs and complex networks (2011). www.win.tue.nl/rhofstad/NotesRGCN.pdf
19. Zhanga, D., Jianga, C., Li, S.: A fast adaptive load balancing method for parallel particle-based simulations. Simulation Modelling Practice and Theory **17**, 1032–1042 (2009)

Approximate Consensus in Highly Dynamic Networks: The Role of Averaging Algorithms

Bernadette Charron-Bost[1], Matthias Függer[2], and Thomas Nowak[3]($\boxtimes$)

[1] CNRS, École polytechnique, Palaiseau, France
charron@lix.polytechnique.fr
[2] Max-Planck-Institut für Informatik, Saarbrucken, Germany
mfuegger@mpi-inf.mpg.de
[3] ENS Paris, Paris, France
thomas.nowak@ens.fr

Abstract. We investigate the approximate consensus problem in highly dynamic networks in which topology may change continually and unpredictably. We prove that in both synchronous and partially synchronous networks, approximate consensus is solvable if and only if the communication graph in each round has a rooted spanning tree. Interestingly, the class of averaging algorithms, which have the benefit of being memoryless and requiring no process identifiers, entirely captures the solvability issue of approximate consensus in that the problem is solvable if and only if it can be solved using any averaging algorithm.

We develop a proof strategy which for each positive result consists in a reduction to the *nonsplit* networks. It dramatically improves the best known upper bound on the decision times of averaging algorithms and yields a quadratic time non-averaging algorithm for approximate consensus in non-anonymous networks. We also prove that a general upper bound on the decision times of averaging algorithms have to be exponential, shedding light on the price of anonymity.

Finally we apply our results to networked systems with a fixed topology and benign fault models to show that with n processes, up to $2n - 3$ of link faults per round can be tolerated for approximate consensus, increasing by a factor 2 the bound of Santoro and Widmayer for exact consensus.

1 Introduction

Recent years have seen considerable interest in the design of distributed algorithms for dynamic networked systems. Motivated by the emerging applications of the Internet and mobile sensor systems, the design of distributed algorithms for networks with a swarm of nodes and time-varying connectivity has been the subject of much recent work. The algorithms implemented in such dynamic networks ought to be decentralized, using local information, and resilient to mobility and link failures.

This work has been partially supported by a LIX-DGA contract.

© Springer-Verlag Berlin Heidelberg 2015
M.M. Halldórsson et al. (Eds.): ICALP 2015, Part II, LNCS 9135, pp. 528–539, 2015.
DOI: 10.1007/978-3-662-47666-6_42

A large number of distributed applications require to reach some kind of agreement in the network in finite time. For example, processes may attempt to agree on whether to commit or abort the results of a distributed database transaction; or sensors may try to agree on estimates of a certain variable; or vehicles may attempt to align their direction of motions with their neighbors. Another example is clock synchronization where processes attempt to maintain a common time scale. In the first example, an *exact consensus* is achieved on one of the outcomes (namely, commit or abort) as opposed to the other examples where processes are required to agree on values that are sufficiently close to each other. The latter type of agreement is referred to as *approximate consensus*.

For the exact consensus problem, one immediately faces impossibility results in truly dynamic networks in which some stabilization of the network during a sufficiently long period of time is not assumed (see, e.g., [21] and [19, Chapter5]). Because of its wide applicability, the approximate consensus problem appears as an interesting weakening of exact consensus to circumvent these impossibility results. The objective of the paper is exactly to study computability and complexity of approximate consensus in dynamic networks in which the topology may change continually and unpredictably.

Dynamic Networks. We consider a fixed set of processes that operate in rounds and communicate by broadcast. In the first part of this article, rounds are supposed to be synchronous in the sense that the messages received at some round have been sent at that round. Then we extend our results to partially synchronous rounds with a maximum allowable delay bound.

At each round, the communication graph is chosen arbitrarily among a set of directed graphs that determines the network model. Hence the communication graph can change continually and unpredictably from one round to the next. The local algorithm at each process applies a state-transition function to its current state and the messages received from its incoming neighbors in the current communication graph to obtain a new state.

While local algorithms can be arbitrary in principle, the basic idea is to keep them simple, so that coordination and agreement do not result from the local computational powers but from the flow of information across the network. In particular, we focus on *averaging algorithms* which repeatedly form convex combinations. These algorithms thus have the benefit of requiring little computational overhead, e.g., allowing for efficient implementations, even in hardware. One additional feature of averaging algorithms is to be memoryless in the sense that the next value of each process is entirely determined only from the values of its incoming neighbors in the current communication graph. More importantly, they work in anonymous networks, not requiring processes to have identifiers.

The network model we consider unifies a wide variety of dynamic networks. The most evident class of networks captured by this model are dynamic multiagent networks, in which communication links frequently go down while other links are established due to the mobility of the agents. The network model can also serve as an abstraction for static or dynamic wireless networks in which collisions and interferences make it difficult to predict which messages will be delivered in time. Finally, it can also be used to model traditional communication networks with a fixed communication graph and some transient link failures.

In our model, the number of processes n is fixed and known to each process. However, all of our results still hold when n is not the exact number of processes but only an upper bound. That allows us to extend the results to a completely dynamic network with a maximal number of processes that may join or leave.

Finally, for simplicity, we assume that all processes start computation at the same round. In fact, it is sufficient to assume that every process eventually participates to the computation either spontaneously (in other words, initiates the computation) or by receiving, possibly indirectly, a message from an initiator.

Contribution. We make the following contributions in this work:

(i) The main result is the exact characterization of the network models in which approximate consensus is solvable. We prove that the approximate consensus problem is solvable in a network model if and only if each communication graph in this model has a rooted spanning tree. This condition guarantees that the network has at least one coordinator in each round. The striking point is that coordinators may continually change over time without preventing nodes from converging to consensus. Accordingly, the network models in which approximate consensus is solvable are called *coordinated network models*. This result highlights the key role played by averaging algorithms in approximate consensus: the problem is solvable if and only if it can be solved using any averaging algorithm.

(ii) With averaging algorithms, we show that agreement with precision of ε can be reached in $O\left(n^{n+1} \log \frac{1}{\varepsilon}\right)$ rounds in a coordinated network model, which dramatically improves the previous bound in [8]. As a matter of fact, every general upper bound for the class of averaging algorithms has to be exponential in the size of the network as exemplified by the *butterfly network* model [11, 20]. Besides we derive a non-averaging algorithm that achieves agreement with precision of ε in $O\left(n^2 \log \frac{1}{\varepsilon}\right)$ rounds in non-anonymous networks.

(iii) We extend our computability and complexity results to the case of *partially synchronous rounds* in which communication delays may be non-null but are bounded by some positive integer Δ. We prove the same necessary and sufficient condition on network models for solvability of approximate consensus, and give an $O\left(n^{n\Delta+1} \log \frac{1}{\varepsilon}\right)$ upper bound on the number of rounds needed by averaging algorithms to achieve agreement with precision of ε.

(iv) Finally, as an application of the above results, we revisit approximate consensus in the context of communication faults. We prove a new result on the solvability of approximate consensus in a complete network model in the presence of benign communication faults, which shows that the number of link faults that can be tolerated increases by a factor 2 when solving approximate consensus instead of consensus.

Related Work. Agreement problems have been extensively studied in the framework of static communication graphs or with limited topology changes (see, e.g., [3,19]). In particular, the approximate consensus problem has been studied in numerous papers in the context of a complete graph and at most f faulty processes (see, e.g., [2,14,15]). In the case of benign failures, this yields communication graphs with a *fixed* core of at least $n - f$ processes that have outgoing links to all processes, and so play the role of steady coordinators of the network.

There is also a large body of previous work on general dynamic networks. However, in much of them, topology changes are restricted and the sequences of communication graphs are supposed to be "well-formed" in various senses. Such well-formedness properties are actually opposite to the idea of unpredictable changes. In [1], Angluin, Fischer, and Jiang study the *stabilizing consensus problem* in which nodes are required to agree exactly on some initial value, but without necessarily knowing when agreement is reached, and they assume that any two nodes can directly communicate infinitely often. In other words, they suppose the limit graph formed by the links that occur infinitely often to be complete. To solve the consensus problem, Biely, Robinson, and Schmid [5] assume that throughout every block of $4n - 4$ consecutive communication graphs there exists a stable set of roots. Coulouma and Goddard [12] weaken the latter stability condition to obtain a characterization of the sequences of communication graphs for which consensus is solvable. Kuhn, Lynch, and Oshman [17] study variations of the *counting* problem; they assume bidirectional links and a stability property, namely the *T-interval connectivity* which stipulates that there exists a stable spanning tree over every T consecutive communication graphs. All their computability results actually hold in the case of 1-interval connectivity which reduces to a property on the *set* of possible communication graphs, and the cases $T > 1$ are investigated just to improve complexity results. Thus they fully model unpredictable topology changes, at least for computability results on counting in a dynamic network.

The network model in [17], however, assumes a static set of nodes and communication graphs that are all bidirectional and connected. The same assumptions are made to study the time complexity of several variants of consensus [18] in dynamic networks. Concerning the computability issue, such strong assumptions make exact agreement trivially solvable: since communication graphs are continually strongly connected, nodes can collect the set of initial values and then make a decision on the value of some predefined function of this set.

The most closely related pieces of work are without a doubt those about asymptotic consensus and more specifically *consensus sets*: a consensus set is a set of stochastic matrices such that every infinite backward product of matrices from this set converges to a rank one matrix. Computations of averaging algorithms correspond to infinite products of stochastic matrices, and the property of asymptotic consensus is captured by convergence to a rank one matrix. Hence, when an upper bound on the number of nodes is known, the notion of network models in which approximate consensus is solvable reduces to the notion of consensus sets if we restrict ourselves to averaging algorithms. However the characterization of consensus sets in [6,13] is not included into our main computability result for approximate consensus since the fundamental assumption of a self loop at each node in communication graphs (a process can obviously communicate with itself) does not necessarily hold for the directed graphs associated to stochastic matrices in a consensus set. The characterization of compact consensus sets in [6,13] and our computability result of approximate consensus are thus incomparable.

In the same vein, some of our positive results can be shown equivalent to results about stochastic matrix products in the vast existing literature on asymptotic consensus. Notably Theorem 3 is similar to the central result in [7], but for that we develop a new proof strategy which consists in a reduction to *nonsplit network models*. The resulting proof is much simpler and direct as it requires neither star graphs [7] nor Sarymsakov graphs [23]. Moreover our proof yields a significantly better upper bound on the time complexity of averaging algorithms in coordinated network models, namely $O\left(n^{n+1} \log \frac{1}{\varepsilon}\right)$ instead of $O\left(n^{n^2} \log \frac{1}{\varepsilon}\right)$ in [8]. It also yields a non-averaging algorithm that achieves agreement with precision of ε in $O\left(n^2 \log \frac{1}{\varepsilon}\right)$ rounds in non-anonymous networks.

2   Approximate Consensus and Averaging Algorithms

We assume a distributed, round-based computational model in the spirit of the Heard-Of model [10]. A system consists of a set of processes $[n] = \{1, \ldots, n\}$. Computation proceeds in *rounds*: In a round, each process sends its state to its outgoing neighbors, receives values from its incoming neighbors, and finally updates its state. The value of the updated state is determined by a deterministic algorithm, i.e., a transition function that maps the values in the incoming messages to a new state value. Rounds are communication closed in the sense that no process receives values in round k that are sent in a round different from k.

Communications that occur in a round are modeled by a directed graph $G = ([n], E(G))$ with a self-loop at each node. The latter requirement is quite natural as a process can obviously communicate with itself instantaneously. Such a directed graph is called a *communication graph*. We denote by $\text{In}_p(G)$ the set of incoming neighbors of p and by $\text{Out}_p(G)$ the set of outgoing neighbors of p in G. Similarly $\text{In}_S(G)$ and $\text{Out}_S(G)$ denote the sets of the incoming and outgoing neighbors of the nodes in a non-empty set $S \subseteq [n]$. The cardinality of $\text{In}_p(G)$ is called the *in-degree* of process p in G.

A *communication pattern* is a sequence $(G(k))_{k \geqslant 1}$ of communication graphs. Here, $E(k)$, $\text{In}_p(k)$ and $\text{Out}_p(k)$ stand for $E(G(k))$, $\text{In}_p(G(k))$ and $\text{Out}_p(G(k))$, respectively.

Each process p has a *local state* s_p whose value at the end of round $k \geqslant 1$ is denoted by $s_p(k)$. Process p's initial state, i.e., its state at the beginning of round 1, is denoted by $s_p(0)$. The *global state* at the end of round k is the collection $s(k) = (s_p(k))_{p \in [n]}$. The *execution* of an algorithm from global initial state $s(0)$, with communication pattern $(G(k))_{k \geqslant 1}$ is the unique sequence $(s(k))_{k \geqslant 0}$ of global states defined as follows: for each round $k \geqslant 1$, process p sends $s_p(k-1)$ to all the processes in $\text{Out}_p(k)$, receives $s_q(k-1)$ from each process q in $\text{In}_p(k)$, and computes $s_p(k)$ from the incoming messages, according to the algorithm's transition function.

Consensus and Approximate Consensus. A crucial problem in distributed systems is to achieve agreement among local process states from arbitrary initial local states. It is a well-known fact that this goal is not easily achievable in the context of dynamic network changes [16, 21], and restrictions on communication patterns are required for that. A *network model* thus is a non-empty set $\mathcal{N}$ of communication graphs, those that may occur in communication patterns.

We now consider the above round-based algorithms in which the local state of process p contains two variables x_p and dec_p. Initially, the range of x_p is $[0, 1]$ and $dec_p = \bot$ (which informally means that p has not decided). Process p is allowed to set dec_p to the current value of x_p, and so to a value v different from $\bot$, only once; in that case we say that p *decides* v. An algorithm *achieves consensus with communication pattern* $(G(k))_{k \geqslant 1}$ if each execution from a global initial state as specified above and with the communication pattern $(G(k))_{k \geqslant 1}$ fulfills the following three conditions: *(i) Agreement:* The decision values of any two processes are equal. *(ii) Integrity:* The decision value of any process is an initial value. *(iii) Termination:* All processes eventually decide.

An algorithm *solves consensus* in a network model $\mathcal{N}$ if it achieves consensus with each communication pattern formed with graphs all in $\mathcal{N}$. Consensus is *solvable in* $\mathcal{N}$ if there exists an algorithm that solves consensus in $\mathcal{N}$. Observe that consensus is solvable in $n-1$ rounds if each communication graph is strongly connected. The following impossibility result due to Santoro and Widmayer [21], however, shows that network models in which consensus is solvable are highly constrained: consensus is not solvable in some "almost complete" graphs. Namely that consensus is not solvable in the network model comprising all communication graphs in which at least $n - 1$ processes have outgoing links to all other processes. The above theorem has been originally stated in the context of link faults in a complete communication graph but its scope can be trivially extended to dynamic communication networks.

To circumvent the impossibility of consensus even in such highly restricted network models, one may weaken Agreement into *ε-Agreement:* The decision values of any two processes are within an *a priori* specified $\varepsilon > 0$; and replace Integrity by *Validity:* All decided values are in the range of the initial values of processes.

An algorithm *achieves ε-consensus with communication pattern* $(G(k))_{k \geqslant 1}$ if each execution from a global initial state as specified above and with the communication pattern $(G(k))_{k \geqslant 1}$ fulfills Termination, Validity, and ε-Agreement. An algorithm *solves approximate consensus* in a network model $\mathcal{N}$ if for any $\varepsilon > 0$, it achieves ε-consensus with each communication pattern formed with graphs all in $\mathcal{N}$. Approximate consensus is *solvable in a network model* $\mathcal{N}$ if there exists an algorithm that solves approximate consensus in $\mathcal{N}$.

Averaging Algorithms. We focus on *averaging algorithms* which require little computational overhead and, more importantly, have the benefit of working in anonymous networks. The update rules for each variable x_p are of the form:

$$x_p(k) = \sum_{q \in \mathrm{In}_p(k)} w_{qp}(k) \, x_q(k-1), \tag{1}$$

where $w_{qp}(k)$ are positive reals and $\sum_{q \in \mathrm{In}_p(k)} w_{qp}(k) = 1$. In other words, at each round k, process p updates x_p to some weighted average of the values $x_q(k-1)$ it has just received. For convenience, we let $w_{qp}(k) = 0$ if $q \notin In_p(k)$.

An *averaging algorithm with parameter* $\varrho > 0$ is an averaging algorithm with the positive weights uniformly lower bounded by $\varrho \colon \forall k \geqslant 1,\, p, q \in [n] \colon$ $w_{qp}(k) \in \{0\} \cup [\varrho, 1]$. Since we strive for distributed implementations of averaging algorithms, $w_{qp}(k)$ is required to be locally computable. Finally note that the decision rule is not specified in the above definition: the decision time immediately follows from the number of rounds that is proven to be sufficient to reach ε-Agreement.

Some averaging algorithms with locally computable weights are of particular interest, such as the *equal neighbor averaging algorithm*, where at each round k process p chooses $w_{qp}(k) = 1/|\mathrm{In}_p(k)|$ for every q in $\mathrm{In}_p(k)$. It is clearly an averaging algorithm with parameter $\varrho = 1/n$.

3   Solvability and Complexity of Approximate Consensus

In this section, we characterize the network models in which approximate consensus is solvable. First we prove that every averaging algorithm solves approximate consensus in *nonsplit network models*, and extend this result to *coordinated network models* by a reduction to the nonsplit case. The latter result which is quite intuitive in the case of a fixed coordinator, actually holds when coordinators vary over time. Our proof of this known result (in the context of products of stochastic matrices) yields a new upper bound on the decision times of averaging algorithms and a quadratic time approximate consensus algorithm for non-anonymous coordinated networks. A classical partitioning argument combined with a characterization of rooted graphs [9] shows that the condition of rooted graphs is actually necessary to solve approximate consensus.

Nonsplit Network Model. A directed graph G is *nonsplit* if for all processes (p, q), it holds that $\mathrm{In}_p(G) \cap \mathrm{In}_q(G) \neq \emptyset$. A *nonsplit network model* is a network model in which each communication graph is nonsplit.

Intuitively, the occurrence of a nonsplit communication graph makes the variables x_p in an averaging algorithm to come closer together: any two processes p and q have at least one common incoming neighbor, leading to a common term in both p's and q's average. The convergence proof in [9] of infinite backward products of scrambling stochastic matrices, using the sub-multiplicativity of the Dobrushin's coefficient, formalizes this intuition and yields:

Theorem 1. *In a nonsplit network model of n processes, every averaging algorithm with parameter ϱ achieves ε-consensus in $\frac{1}{\varrho} \log \frac{1}{\varepsilon}$ rounds.*

Theorem 1 can be easily extended with respect to the granularity at which the assumption of nonsplit graphs holds. Let the *product* of two directed graphs G and H with the same set of nodes V be the directed graph $G \circ H$ with set of nodes V and a link from (p, q) if there exists $r \in V$ such that $(p, r) \in E(G)$

and $(r, q) \in E(H)$. For any positive integer K, we say a network model $\mathcal{N}$ is K-*nonsplit* if any product of K graphs from $\mathcal{N}$ is nonsplit.

Corollary 2. *In a K-nonsplit network model of n processes, every averaging algorithm with parameter ϱ achieves ε-consensus in $K\varrho^{-K} \log \frac{1}{\varepsilon} + K - 1$ rounds.*

Coordinated Network Model. A directed graph G is said to be p-*rooted*, for some node p, if for every node q, there exists a directed path from p to q. Then p is called a *root of G*.

While communication graphs remain p-rooted, process p can play the role of network coordinator: its particular position in the network allows p to impose its value on the network. Accordingly, a network model is said to be *coordinated* if each of its graphs is rooted. It is easy to grasp why in the case of a steady coordinator, processes converge to a common value and so achieve approximate consensus when running an averaging algorithm. We now show that the same still holds when coordinators change over time.

Theorem 3. *In a coordinated network model of n processes, every averaging algorithm with parameter ϱ achieves ε-consensus in $n\varrho^{-n} \log \frac{1}{\varepsilon} + n - 1$ rounds.*

The following lemma is the heart of our proof. Corollary 2 allows us to conclude.

Lemma 4. *Every coordinated network model with n processes is $(n-1)$-nonsplit.*

Proof. Let $H_1, \ldots, H_{n-1}$ be a sequence of $n - 1$ communication graphs, each of which is rooted. We recursively define the sets $S_p(k)$ by

$$S_p(0) = \{p\} \quad \text{and} \quad S_p(k) = \text{In}_{S_p(k-1)}(H_k) \text{ for } k \in \{1, \ldots, n - 1\}. \tag{2}$$

Then $S_p(k) = \text{In}_p(H_k \circ \cdots \circ H_1)$; because of the self-loops, $S_p(k) \subseteq S_p(k + 1)$ and none of the sets $S_p(k)$ is empty.

Now we have to show that for any $p, q \in [n]$,

$$S_p(n - 1) \cap S_q(n - 1) \neq \emptyset. \tag{3}$$

If $p = q$, then (3) trivially holds. Otherwise, assume by contradiction that (3) does not hold; for each $k \in \{0, \ldots, n - 1\}$, the sets $S_p(k)$ and $S_q(k)$ are disjoint. Consider the sequences $S_p(0) \subseteq \cdots \subseteq S_p(n - 1)$, $S_q(0) \subseteq \cdots \subseteq S_q(n - 1)$, and $S_p(0) \cup S_q(0) \subseteq \cdots \subseteq S_p(n - 1) \cup S_q(n - 1)$. Because $|S_p(0) \cup S_q(0)| \geq 2$ if $p \neq q$ and $|S_p(n - 1) \cup S_q(n - 1)| \leq n$, the latter sequence cannot be strictly increasing by the pigeonhole principle. Therefore $S_p(\ell) \cup S_q(\ell) = S_p(\ell + 1) \cup S_q(\ell + 1)$ for some $\ell \in \{0, \ldots, n - 2\}$. Since $S_p(\ell) \cap S_q(\ell) = \emptyset$ and $S_p(\ell + 1) \cap S_q(\ell + 1) = \emptyset$, it follows that $S_p(\ell) = S_p(\ell + 1)$ and $S_q(\ell) = S_q(\ell + 1)$. Hence both $S_p(\ell)$ and $S_q(\ell)$ have no incoming links in the graph $H_{\ell+1}$. This implies these sets both contain all the roots of $H_{\ell+1}$, a contradiction to the disjointness assumption.

The major difference with the previous proofs of this result [7,9] lies in the fact that we deal with "cumulative graphs" which are just nonsplit (scrambling matrices) instead of being star graphs (matrices with a positive column). In other words, we analyze the evolution of the lines and not of the columns of backward products of stochastic matrices. That allows for a drastic improvement of the decision time of averaging algorithms.

From [9], we derive that a directed graph G is rooted iff the acyclic condensation of G has a sole source. Combined with a simple partitioning argument, we show there exists no algorithm, whether or not it is an averaging algorithm, achieving approximate consensus in a network model with some non-rooted graphs. With our positive result in Theorem 3, this gives:

Theorem 5. *The approximate consensus problem is solvable in a synchronous network model $\mathcal{N}$ if and only if $\mathcal{N}$ is a coordinated model.*

Time Complexity of Approximate Consensus. Even with the improvement of Theorem 3, the upper bound on the decision times of averaging algorithms is exponential in the number n of processes. In particular, the equal neighbor averaging algorithm achieves ε-consensus in $O\left(n^{n+1}\log\frac{1}{\varepsilon}\right)$ rounds. The *butterfly network* model [11,20] is an example of a coordinated network model for which the equal neighbor averaging algorithm exhibits an exponentially large decision time. The example does not even require the network to be dynamic, using a time-constant network only. More precisely by spectral gap arguments, we show the following lower bound.

Theorem 6. *There is a coordinated model consisting of one graph such that, for any $\varepsilon > 0$, the equal neighbor averaging algorithm does not achieve ε-consensus by round k if $k = O\left(2^{n/3}\log\frac{1}{\varepsilon}\right)$.*

Another benefit of Lemma 4 is to provide an approximate consensus algorithm with a quadratic decision time. Indeed Lemma 4 corresponds to a *uniform translation* in the Heard-Of model [10] that transforms each block of $n-1$ consecutive coordinated rounds into one nonsplit macro-round. If each process applies an equal neighbor averaging procedure only at the end of each macro-round instead of applying it round by round, the resulting distributed algorithm (cf. Algorithm 1), which is no more an averaging algorithm and requires a unique identifier for each process, achieves ε-consensus in only $O\left(n^2\log\frac{1}{\varepsilon}\right)$ rounds, hinting at a price of of anonymity.

Algorithm 1. A quadratic time Approximate Consensus algorithm

---

Initially:
1: $x_p \in [0,1]$ and $V_p \leftarrow \{(p, x_p)\}$
In round $k \geqslant 1$ do:
2: send V_p to all processes in $\mathrm{Out}_p(k)$ and receive V_q from all processes q in $\mathrm{In}_p(k)$
3: $V_p \leftarrow \bigcup_{q \in \mathrm{In}_p(k)} V_q$
4: **if** $k \equiv 0 \mod n - 1$ **then**
5:      $x_p \leftarrow \sum_{(q, x_q) \in V_p} w_{qp}(k) x_q$
6:      $V_p \leftarrow \{(p, x_p)\}$
7: **end if**

---

4   Synchronism and Faults

Partially Synchronous Networks. Rounds so far have been supposed to be synchronous: messages are delivered in the same round in which they are sent. In [4,22], the latter condition is relaxed by allowing processes to receive *outdated* messages, and by bounding the number of rounds between the sending and the receipt of messages by some positive integer Δ. That results in the definitions of Δ-*partially synchronous rounds* and Δ-*bounded executions*. The communication graph at round k is understood to be the directed graph defined by the incoming messages at round k. In the case of averaging algorithms, $x_p(k) = \sum_{q \in \mathrm{In}_p(k)} w_{qp}(k)\, x_q\left(\kappa_q^p(k)\right)$, where $k - \Delta \leqslant \kappa_q^p(k) \leqslant k - 1$. Since process p has immediate access to x_p, we assume $\kappa_p^p(k) = k - 1$.

We now extend the results in the previous section to partially synchronous rounds. Our proof strategy is based on a reduction to the synchronous case: each process corresponds to a set of Δ virtual processes, and every Δ-bounded execution of an averaging algorithm with n processes coincides with a synchronous execution of an averaging algorithm with $n\Delta$ processes.

Unfortunately, Theorem 3 does not simply apply since the key property of a self-loop at each node is not preserved in this reduction. We overcome this difficulty by using Corollary 2 directly: First we prove that if all the graphs in the Δ-bounded execution are rooted, then each cumulative graph over $n\Delta$ consecutive rounds of the synchronous execution is nonsplit. To conclude, we observe that Corollary 2 holds even when some nodes have no self-loop. Again the reduction to nonsplit rounds allows for a much better upper bound on the decision time of the equal neighbor algorithm, namely $O\left(n^{n\Delta+1} \log \frac{1}{\varepsilon}\right)$ instead of $O\left(n^{(\Delta n)^2} \log \frac{1}{\varepsilon}\right)$ in [8,9].

We can now extend the characterization of the network models in which approximate consensus is solvable in Theorem 5 to computations with partially synchronous rounds.

Theorem 7. *The approximate consensus problem is solvable in a partially synchronous network model $\mathcal{N}$ if and only if $\mathcal{N}$ is a coordinated model.*

Communication Faults. Time varying communication graphs may result from benign communication faults (message losses) in a fixed network. In the light of Theorem 3, we revisit the problem of approximate consensus in the context of a complete network and communication faults.

Theorem 8. *Approximate consensus is solvable in a complete network with n processes if there are at most $2n - 3$ link faults per round.*

Proof. We actually prove that any directed graph with n nodes and at least $n^2 - 3n + 3$ links is rooted. Since $n^2 - 3n + 3 = (n^2 - n) - (2n - 3)$, the theorem immediately follows.

Assume that G is not a rooted graph. Then the condensation of G has two nodes without incoming link. We denote the corresponding two strongly connected components in G by S_1 and S_2, and their cardinalities by n_1 and n_2, respectively. Therefore the number of links in G that are not self-loops is at most equal to $n^2 - n - n_1(n - n_1) - n_2(n - n_2)$. Since $n^2 - n - n_1(n - n_1) - n_2(n - n_2) \leqslant n^2 - 3n + 2$ when $n_1, n_2 \in [n - 1]$, G has at most $n^2 - 3n + 2$ links.

Compared with the impossibility result established by Santoro and Widmayer [21] for exact consensus with $n - 1$ faults per round, the above theorem shows that the number of link faults that can be tolerated increases by a factor 2 when solving approximate consensus instead of consensus. Besides it is easy to construct a non-rooted communication graph with $n^2 - 2n + 2$ links which, combined with Theorem 5, shows that the bound in the above theorem is tight.

5   Discussion

The main goal of this paper has been to characterize the dynamic network models in which approximate consensus is solvable. As for exact consensus, approximate consensus does not require strong connectivity and it can be solved under the sole assumption of rooted communication graphs. However contrary to the condition of a stable set of roots and identifiers supposed in [5] for achieving consensus, approximate consensus can be solved even though roots arbitrarily change over time and processes are anonymous. In these respects, approximate consensus seems to be more suitable than consensus for handling real network dynamicity.

While anonymity of processes does not affect solvability, it could increase decision times in view of our quadratic time approximate consensus algorithm for a coordinated network with process identifiers, and the upper bound for averaging algorithms that we proved to be necessarily exponential.

Acknowledgments. We wish to thank Alex Olshevsky for helpful discussions on consensus sets, Martin Perner for many detailed comments, and Martin Biely for pointing out the full implications of the round translation in Lemma 4.

References

1. Angluin, D., Fischer, M.J., Jiang, H.: Stabilizing consensus in mobile networks. In: Gibbons, P.B., Abdelzaher, T., Aspnes, J., Rao, R. (eds.) DCOSS 2006. LNCS, vol. 4026, pp. 37–50. Springer, Heidelberg (2006)
2. Attiya, H., Lynch, N.A., Shavit, N.: Are wait-free algorithms fast? J. ACM **41**(4), 725–763 (1994)
3. Attiya, H., Welch, J.: Distributed Computing. Wiley, Hoboken (2005)
4. Bertsekas, D.P., Tsitsiklis, J.N.: Parallel and Distributed Computation: Numerical Methods. Athena Scientific, Belmont (1989)
5. Biely, M., Robinson, P., Schmid, U.: Agreement in directed dynamic networks. In: Even, G., Halldórsson, M.M. (eds.) SIROCCO 2012. LNCS, vol. 7355, pp. 73–84. Springer, Heidelberg (2012)

6. Blondel, V., Olshevshy, A.: How to decide consensus? A combinatorial necessary and sufficient condition and a proof that consensus is decidable but NP-hard. SIAM J. Control Optim. **52**(5), 2707–2726 (2014)
7. Cao, M., Morse, A.S., Anderson, B.D.O.: Reaching a consensus in a dynamically changing environment: a graphical approach. SIAM J. Control Optim. **47**(2), 575–600 (2008)
8. Cao, M., Morse, A.S., Anderson, B.D.O.: Reaching a consensus in a dynamically changing environment: convergence rates, measurement delays, and asynchronous events. SIAM J. Control Optim. **47**(2), 601–623 (2008)
9. Charron-Bost, B.: Orientation and connectivity based criteria for asymptotic consensus (2013). arXiv:1303.2043v1 [cs.DC]
10. Charron-Bost, B., Schiper, A.: The Heard-Of model: computing in distributed systems with benign faults. Distrib. Comput. **22**(1), 49–71 (2009)
11. Chung, F.R.: Spectral Graph Theory. AMS, Providence (1997)
12. Coulouma, É., Godard, E.: A characterization of dynamic networks where consensus is solvable. In: Moscibroda, T., Rescigno, A.A. (eds.) SIROCCO 2013. LNCS, vol. 8179, pp. 24–35. Springer, Heidelberg (2013)
13. Daubechies, I., Lagarias, J.C.: Sets of matrices all infinite products of which converge. Linear Algebra Appl. **161**, 227–263 (1992)
14. Dolev, D., Lynch, N.A., Pinter, S.S., Stark, E.W., Weihl, W.E.: Reaching approximate agreement in the presence of faults. J. ACM **33**(2), 499–516 (1986)
15. Fekete, A.D.: Asymptotically optimal algorithms for approximate agreement. Distrib. Comput. **4**(1), 9–29 (1990)
16. Fischer, M.J., Lynch, N.A., Paterson, M.S.: Impossibility of distributed consensus with one faulty process. J. ACM **32**(2), 374–382 (1985)
17. Kuhn, F., Lynch, N.A., Oshman, R.: Distributed computation in dynamic networks. In: 42nd ACM Symposium on Theory of Computing, pp. 513–522. ACM, New York City (2010)
18. Kuhn, F., Moses, Y., Oshman, R.: Coordinated consensus in dynamic networks. In: 30th Annual ACM Symposium on Principles of Distributed Computing, pp. 1–10. ACM, New York City (2011)
19. Lynch, N.A.: Distributed Algorithms. Morgan Kaufmann, San Francisco (1996)
20. Olshevsky, A., Tsitsiklis, J.N.: Degree fluctuations and the convergence time of consensus algorithms (2011), arXiv:1104.0454v1 [math.OC]
21. Santoro, N., Widmayer, P.: Time is not a healer. In: Monien, B., Cori, R. (eds.) 6th Symposium on Theoretical Aspects of Computer Science, LNCS, vol. 349, pp. 304–313. Springer, Heidelberg (1989)
22. Tsitsiklis, J.N.: Problems in Decentralized Decision Making and Computation. Ph.D. thesis, Massachusetts Institute of Technology (1984)
23. Xia, W., Cao, M.: Sarymsakov matrices and their application in coordinating multi-agent systems. In: 31st Chinese Control Conference, pp. 6321–6326. IEEE, New York City (2012)

The Range of Topological Effects
on Communication

Arkadev Chattopadhyay[1] and Atri Rudra[2]([⊠])

[1] Tata Institute of Fundamental Research, Mumbai, India
arkadev.c@tifr.res.in
[2] University at Buffalo, SUNY, Buffalo, NY, USA
atri@buffalo.edu

Abstract. We continue the study of communication cost of computing functions when inputs are distributed among k processors, each of which is located at one vertex of a network/graph called a terminal. Every other node of the network also has a processor, with no input. The communication is point-to-point and the cost is the total number of bits exchanged by the protocol, in the worst case, on all edges.

Chattopadhyay, Radhakrishnan and Rudra (FOCS'14) recently initiated the study of the effect of topology of the network on the total communication cost using tools from L_1 embeddings. Their techniques provided tight bounds for simple functions like Element-Distinctness (ED), which depend on the 1-median of the graph. This work addresses two other kinds of natural functions. We show that for a large class of natural functions like Set-Disjointness the communication cost is essentially n times the cost of the optimal Steiner tree connecting the terminals. Further, we show for natural composed functions like ED ∘ XOR and XOR ∘ ED, the naive protocols suggested by their definition is optimal for general networks. Interestingly, the bounds for these functions depend on more involved topological parameters that are a combination of Steiner tree and 1-median costs.

To obtain our results, we use some new tools in addition to ones used in Chattopadhyay et al. These include (i) viewing the communication constraints via a linear program; (ii) using tools from the theory of tree embeddings to prove topology sensitive direct sum results that handle the case of composed functions and (iii) representing the communication constraints of certain problems as a family of collection of multiway cuts, where each multiway cut simulates the hardness of computing the function on the star topology.

1 Introduction

We consider the following distributed computation problem $p \equiv (f, G, K, \Sigma)$: there is a set K of k processors that have to jointly compute a function $f : \Sigma^K \to \{0,1\}$. (Unless stated otherwise, we will assume that $\Sigma = \{0,1\}^n$.) Each of the k inputs to f is held by a distinct processor. Each processor is located on some node of a network (graph) $G \equiv (V, E)$. These nodes in V with an input are called *terminals* and the set of such nodes is denoted by K. The other nodes in

© Springer-Verlag Berlin Heidelberg 2015
M.M. Halldórsson et al. (Eds.): ICALP 2015, Part II, LNCS 9135, pp. 540–551, 2015.
DOI: 10.1007/978-3-662-47666-6_43

V have no input but have processors that also participate in the computation of f via the following communication process: there is some fixed a-priori protocol according to which, in each round of communication, nodes of the network send messages to their neighbors. The behavior of a node in any round is just a (randomized) function of inputs held by it and the sequence of bits it has received from its neighbors in the past. All communication is point-to-point in the sense that each edge of G is a private communication channel between its endpoints. In any round, if one of the endpoints of an edge is in a state where it expects to receive some communication from the other side, then silence from the other side is not allowed in a legal protocol. At the end of communication process, some pre-designated node of the network outputs the value of f on the input instance held by processors in K. We assume that protocols are randomized, using public coins that are accessible to all nodes of the network, and err with probability at most ϵ. The cost of a protocol on an input is the expected total number of bits communicated on all edges of the network. The main question we study in this work is how the cost of the best protocol on the worst input depends on the function f, the network G and the set of terminals K. This cost is denoted by $R_\epsilon(p)$ (and we use $R(p)$ to denote $R_{1/3}(p)$).

This communication model seems to be a natural abstraction of many distributed problems and was recently studied in its full generality by Chattopadhyay, Radhakrishnan and Rudra [9].[1] A noteworthy special case is when G is just a pair of nodes connected by an edge. This corresponds to the classical model of 2-party communication introduced by Yao [33] more than three decades ago. The study of the classical model has blossomed into the vibrant and rich field of communication complexity, which has deep connections to theoretical computer science in general and computational complexity in particular.

This point-to-point model had received early attention in the works of Tiwari [27], Dolev and Feder [11] and Duris and Rolim [13]. These early works seem to have entirely focused on deterministic and non-deterministic complexities. In particular, Tiwari [27] showed several interesting topology-sensitive bounds on the cost of deterministic protocols for simple functions. However, these bounds were for specific graphs like trees, grids, rings etc. More recently, there has been a resurgence of interest in the randomized complexity of functions in the point-to-point model. These have several motivations: BSP model of Valiant [28], models for MapReduce [18], parallel models to compute conjunctive queries [4], distributed models for learning [2], distributed streaming and functional monitoring [10], sensor networks [19] etc. Recently Drucker, Kuhn and Oshman [12] showed that some outstanding questions in this model (where one is interested in bounding the number of rounds of communication as opposed to bounding the total communication) have connections to well known hard problems on constant-depth circuits. Motivated by such diverse applications, a flurry of recent works [6,7,16,20,23,30–32] have proved strong lower bounds, developing very interesting techniques. All of these works, however, focus on the

---

[1] Related but different problems have been considered in distributed computing. Please see the full paper for more details.

star topology with k leaves, each having a terminal and a central non-terminal node. Note that every function on the star can be computed using $O(kn)$ bits of communication, by making the leaves simultaneously send each of their n-bit inputs to the center that outputs the answer. The aforementioned recent works show that this is an optimal protocol for various natural functions.

In contrast, on a general graph not all functions seem to admit $O(kn)$-bit protocols. Consider the naive protocol that makes all terminals send their inputs to a special node u. The speciality of u is the following: let the status of a node v in network G w.r.t. K, denoted by $\sigma_K(v)$, be given by $\sum_{w \in K} d_G(v, w)$, where $d_G(x, y)$ is the length of a shortest path in G between nodes x and y. Node u is special and called the *median* as it has a minimal status among all nodes, which we denote by $\sigma_K(G)$. Thus, the cost of the naive protocol is $\sigma_K(G) \cdot n$. For the star, the center is the median with status k. On the other hand, for the line, ring and grid, each having k nodes all of which are terminals, $\sigma_K(G)$ is $\Theta(k^2)$, $\Theta(k^2)$ and $\Theta(k^{3/2})$ respectively.

The work in [9] appears to be the first one to address the issue of randomized protocols over arbitrary G. It shows simple natural functions like Element-Distinctness[2], have $\Theta(\sigma_K(G))$ as the cost (up to a poly-log(k) factor) of the optimal randomized protocol computing them. In fact, all of the bounds in [9] are of the 1-median type (formally defined later). While these are essentially the strongest possible lower bounds[3], not all functions of interest have that high complexity. Consider the function Equality that outputs 1 precisely when all input strings at the nodes in K are the same. There is a randomized protocol of cost much less than $\sigma_K(G)$ for computing it: consider a minimal cost Steiner-tree with nodes in K as the terminals. Let the cost of this tree be denoted by $\mathrm{ST}(G, K)$. Root this tree at an arbitrary node. Each leaf node sends a hash (it turns out $O(1)$ bits of random hash suffices for our purposes[4]) of its string to its parent. Each internal node u collects all hashes that it receives from nodes in the sub-tree rooted at u, verifies if they are all equal to some string s. If so, it sends s to its parent and otherwise, it sends a special symbol to its parent indicating inequality. Thus, in cost $O(\mathrm{ST}(G, K))$, one can compute Equality with small error probability.[5]

For many scenarios in a distributed setting, the task to be performed is naturally layered in the following way. The set of terminal nodes is divided into t groups $K_1, \ldots, K_t$. Within a group of m terminals, the input needs to be pre-processed in a specified manner, expressed as a function $g : \left(\{0, 1\}^n\right)^m \to \{0, 1\}^n$. Finally the

---

[2] Given inputs $X^i \in \Sigma$ for every $i \in K$, the function ED : $\Sigma^K \to \{0, 1\}$ is defined as follows: ED $\left((X^i)_{i \in K}\right) = 1$ if and only if $X^i \neq X^j$ for every $i \neq j \in K$.

[3] Strictly speaking, the strongest lower bound is $\Omega(\sigma_K(G) \cdot n)$. Several functions, called linear 1-median type later, are shown to achieve this bound in [9].

[4] Observe that if two strings held at two terminals are not equal, each hash will detect inequality with probability at least $2/3$.

[5] In fact, we observe in the full paper that *any* function $f : \Sigma^K \to \{0, 1\}$ that depends on all of its input symbols needs $\Omega(\mathrm{ST}(G, K))$ amounts of communication (even for randomized protocols), which implies that the randomized protocol above for Equality is essentially optimal.

results of the computation of the groups need to be combined in a different way, given by another function $f : \left(\{0,1\}^n\right)^t \to \{0,1\}$. More precisely, we want to compute the composed function $f \circ g$. The canonical protocol will first compute in parallel all instances of the task g in groups using the optimal protocol for g and then use the optimal protocol for f on the outputs of g in each of K_i. However, this is not the optimal protocol for all f, g and network G. For example, consider the case when f is Equality and g is the bit-wise XOR function. As we show in the full paper, the optimal protocol for computing XOR has cost $\Theta\left(\mathrm{ST}\left(G, K\right) \cdot n\right)$. Hence, the naive protocol for EQ $\circ$ XOR will have cost $\Omega\left(\sum_{i=1}^{t}\left(\mathrm{ST}\left(G, K_i\right) \cdot n\right)\right)$. However, it is not hard to see that there is a protocol of cost $O\left(t \cdot \left(\mathrm{ST}\left(G, K\right)\right)\right)$. This cost can be much lower than the naive cost depending on the network.

Full version of the paper. All omitted material (including all the proofs) can be found in the full version of the paper [8].

2   Our Results

The first part of our work attempts to understand when the naive protocol cannot be improved upon for composed functions. Function composition is a widely used technique in computational complexity for building new functions out of more primitive ones [3,14,15,17,24]. Proving that the naive way of solving $f \circ g$ is essentially optimal, in many models remain open. In particular, even in the 2-party model of communication where the network is just an edge, this problem still remains unsolved (see [3]).To describe our results on composition, we need the following terminology: The cost of solving a problem $\left(f, G, K, \{0,1\}^n\right)$ will have a dependence on both n and the topology of G. We will deal with two kinds of dependence on n. If the cost depends linearly on n, we say f is of linear type. Otherwise, there is no dependence on n. (We typically ignore poly-log factors in this paper.) Call f a 1-*median* type function if its topology-sensitive complexity is $\sigma_K\left(G\right)$. We say f is of *Steiner tree* type, if its topology-sensitive complexity is $\mathrm{ST}\left(G, K\right)$. The protocol for a Steiner tree type problem f seems to move information around in a fundamentally different way from the one for a 1-median type problem g. It seems tempting to conjecture that there composition cannot be solved by any cheaper protocol than the naive ones. However, we are only able to prove this intuition for few natural instances in this work.

Consider the following composition: the first function is element distinctness function, denoted by ED, which was shown by [9] to be of 1-median type. The second is the bit-wise xor function (which we denote by XOR_n), which is shown to be of linear Steiner-tree type in the full paper. In particular, given a graph $G = (V, E)$ and t subsets $K_1, \ldots, K_t \subseteq V$, we define the composed function ED $\circ$ XOR_n as follows. Given $k_i \stackrel{def}{=} |K_i|$ n-bit vectors $X_1^i, \ldots, X_{k_i}^i \in \{0,1\}^n$ for every $i \in [t]$, define ED $\circ$ $\mathrm{XOR}_n\left(X_1^1, \ldots, X_{k_1}^1, \ldots, X_1^t, \ldots, X_{k_t}^t\right) =$ ED $\left(\mathrm{XOR}_n\left(X_1^1, \ldots, X_{k_1}^1\right), \ldots, \mathrm{XOR}_n\left(X_1^t, \ldots, X_{k_t}^t\right)\right)$. The naive algorithm mentioned earlier specializes for ED $\circ$ XOR_n as follows: compute the inner bit-wise

XOR's first[6] and then compute the ED on the intermediate values. This immediately leads to an upper bound of

$$
O\left(\sigma_{K_1,\ldots,K_t}(G) \cdot \log k + \sum_{i=1}^{t} \mathrm{ST}(G, K_i) \cdot \log k\right), \tag{1}
$$

where $\sigma_{K_1,\ldots,K_t}(G)$ is the minimum of $\sigma_{\bar{K}}(G)$ for every choice of $\bar{K}$ that has exactly one terminal from K_i for every $i \in [t]$. One of our results, stated below, shows that this upper bound is tight to within a poly-log factor:

Theorem 1

$$
R(\mathrm{ED} \circ \mathrm{XOR}_n, G, K, \{0,1\}^n) \geq \Omega\left(\frac{\sigma_{K_1,\ldots,K_t}(G)}{\log t} + \frac{\sum_{i=1}^{t} \mathrm{ST}(G, K_i)}{\log |V| \log \log |V|}\right).
$$

We prove the above result (and other similar results) by essentially proving a topology sensitive direct sum theorem (see Section 3.1 for more).

In the full paper, we further show that changing the order of composition to XOR ∘ ED also does not allow any cost savings over the naive protocol:

Theorem 2. *For every choice of $u_i \in K_i$ (where $k = \sum_{i=1}^{t} |K_i|$):*

$$
R(\mathrm{XOR}_1 \circ \mathrm{ED}, G, K, \{0,1\}^n) \geq \Omega\left(\mathrm{ST}(G, \{u_1, \ldots, u_t\}) + \frac{\sum_{i=1}^{t} \sigma_{K_i}(G)}{\log k}\right).
$$

The results discussed so far follow by appropriately reducing the problem on a general graph to a bunch of two-party lower bounds, one across each cut in the graph. This was the general idea in [9] as well but the reductions in this paper need to use different tools. However, the idea of two-party reduction seems to fail for the Set-Disjointness function, which is one of the centrally studied function in communication complexity. In our setting, the natural definition of Set-Disjointness (denoted by DISJ) is as follows: each of the k terminals in K have an n-bit string and the function tests if there is an index $i \in [n]$ such that all k strings have their ith bit set to 1. It is easy to check that this function can be computed with $O(\mathrm{ST}(G, K) \cdot n)$ bits of communication (in fact one can compute the bit-wise AND function with this much communication by successively computing the partial bit-wise AND as we go up the Steiner tree). Before our work, only a tight bound was known for the special case of G being a k-star (i.e. a lower bound of $\Omega(kn)$), due to the recent work of Braverman et al. [6]. In this work, we present a fairly general technique that ports a tight lower bound on a k-star to an almost tight lower bound for the general graph case. For the complexity of Set-Disjointness, this technique yields the following bound:

---

[6] In fact, we just need to compute the XOR of the hashes of the input, which with a linear hash is just the bit-wise XOR of $O(\log k)$-bits of hashes.

Theorem 3

$$R(\text{DISJ}, G, K, \{0,1\}^n) \geq \Omega\left(\frac{\text{ST}(G,K) \cdot n}{\log^2 k}\right).$$

Next, we present our key technical results and overviews of their proofs. We would like to point out that our proofs use many tools used in algorithm-design like (sub)tree embeddings, Borůvka's algorithm to compute an MST for a graph and integrality gaps of some well-known LPs, besides using L_1-embeddings of graph that was also used in [9]. We hope this work encourages further investigation of other algorithmic techniques to prove message-passing lower bounds.

3   Key Technical Results and Our Techniques

In the full paper, we present a simple re-formulation of the lower bound argument in [9] as a linear program (LP). The idea is simple: consider a problem $p = (f, G, K, \Sigma)$ and any cut C in the graph. Then the cut naturally gives rise to a two party problem for the induced function f_C across the cut C. Specifically, Alice gets the inputs from K on one side of the cut and Bob gets the rest of the inputs. Let Π be a protocol for p. Then by considering the messages exchanged by Π on the crossing edges of C induces a two party protocol for f_C. If x_e bits are transmitted on edge e, then $\sum_{e \text{ crosses } C} x_e \geq b(C)$, where $b(C)$ is a two-party communication complexity lower bound for f_C. Then $\sum_{e \in E(G)} x_e$ subject to the constraint $\sum_{e \text{ crosses } C} x_e \geq b(C)$ is a valid lower bound on $R(p)$. To make this idea work, we actually need to pick a hard distribution μ on Σ^K and use x_e to denote the expected communication over edge e under μ: see the full paper for more.

Using the scheme outlined above, we can prove our earlier claimed lower bound of $\Omega(\text{ST}(G,K) \cdot n)$ for the XOR_n problem. Further, this connection can also be used to recover the $\Omega(\sigma_K(G)/\log k)$ lower bound for the ED function from [9]– see the full paper. While LPs have been used to prove communication complexity lower bounds in the standard 2-party setting (see e.g. [25,26]), our use of LPs above seem to be novel for proving communication lower bounds. In the remainder of the section, we present two general results that we will use to prove our lower bounds for specific functions including those in Theorems 1, 2 and 3.

3.1   A Result on Two LPs

We now present a result that relates the objective values of two similar LPs. Both the LPs will involve the same underlying topology graph $G = (V, E)$.

We begin with the first LP, which we dub $\text{LP}^L(G)$ ($\ell \geq 1$ and $b^i(C) \geq 0$ for every $j \in [\ell]$ and cut C are integers):

$$\min \sum_{e \in E} x_e$$

subject to

$$\sum_{e \text{ crosses } C} x_e \geq \sum_{i=1}^{\ell} b^i(C) \qquad \text{for every cut } C$$

$$x_e \geq 0 \qquad \text{for every } e \in E.$$

In our results, we will use x_e to denote the expected communication on the edge e of an arbitrary protocol for a problem p over a distribution over the input. The constraint for each cut C will correspond to a two-party lower bound of $\sum_{i=1}^{\ell} b^i(C)$. Then the objective value of the above LP, which by abuse of notation we will also denote by $\text{LP}^L(G)$, will be a valid lower bound on $R(p)$.

Next we consider the second LP, which we dub $\text{LP}^U(G)$:

$$\min \sum_{i=1}^{\ell} \sum_{e \in E} x_{i,e}$$

subject to

$$\sum_{e \text{ crosses } C} x_{i,e} \geq b^i(C) \qquad \text{for every cut } C \text{ and } i \in [\ell]$$

$$x_{i,e} \geq 0 \qquad \text{for every } e \in E \text{ and } i \in [\ell].$$

In our results, we will connect the objective value of the above LP (which again with abuse of notation we denote by $\text{LP}^U(G)$) to the total communication of a trivial algorithm that solves problem p.

Our main aim is to show that for certain settings, the lower bound we get from $\text{LP}^L(G)$ is essentially the same as the upper bound we get from $\text{LP}^U(G)$.

Before we state our main technical result, we need to define the property we need on the values $b^i(C)$. We say that the values $b^i(C)$ satisfy the *sub-additive property* if for any three cuts C_1, C_2 and C_3 such that $C_1 \cup C_2 = C_3$,[7] we have that for every $i \in [\ell]$: $b^i(C_3) \leq b^i(C_1) + b^i(C_2)$. We remark that the two main families of functions that we consider in this paper lead to LPs that do satisfy the sub-additive property:

- Steiner Tree constraints. There are sets of terminals $T_i \subseteq V$ (for $i \in [\ell]$) and $b^i(C) = 1$ if C separates T_i and 0 otherwise.
- Multi-commodity flow constraints. We have a set of demands D_i (for $i \in [\ell]$) and $b^i(C)$ is the number of demand pairs in D_i that are separated by C.

We are now ready to state our first main technical result:

Theorem 4. *For any graph $G = (V, E)$ (and values $b^i(C)$ for any $i \in [\ell]$ and cut C with the sub-additive property), we have*

$$\text{LP}^U(G) \geq \text{LP}^L(G) \geq \Omega\left(\frac{1}{\log |V| \log \log |V|}\right) \cdot \text{LP}^U(G).$$

---

[7] This means that one side of the cut C_3 is the union of one side each of C_1 and C_2.

Theorem 4 is the main ingredient in proving the lower bound for a 1-median function composed with a Steiner tree function as given in Theorem 1. We can also use Theorem 4 to prove nearly tight lower bound for composing a Steiner tree type function XOR with a linear 1-median function IP as well as another 1-median function ED. However, it turns out for these functions, we can prove a better bound than Theorem 4. In particular, using techniques developed in [9], we can prove lower bounds given in Theorem 2 and the one stated below:

Corollary 1. *For every choice of $u_i \in K_i$ (where $k = \sum_{i=1}^{t} |K_i|$):*

$$R(\text{XOR} \circ \text{IP}_n, G, K, \{0,1\}^n) \geq \Omega \left(\text{ST}(G, \{u_1, \ldots, u_t\}) + \frac{\sum_{i=1}^{t} \sigma_{K_i}(G) \cdot n}{\log k} \right).$$

Proof Overview. We give an overview of our proof of Theorem 4 (specialized to the proof of Theorem 1). While the LP based lower bound argument for XOR_n in the full paper is fairly straightforward things get more interesting when we consider $\text{ED} \circ \text{XOR}_n$. It turns out that just embedding the hard distribution for ED from [9], one can prove a lower bound of just $\Omega \left(\frac{\sigma_{K_1, \ldots, K_t}(G)}{\log t} \right)$. The more interesting part is proving a lower bound of $\tilde{\Omega} \left(\sum_{i=1}^{t} \text{ST}(G, K_i) \right)$. It is not too hard to connect the *upper* bound of $\tilde{O} \left(\sum_{i=1}^{t} \text{ST}(G, K_i) \right)$ to the following LP, which we dub $LP_{\text{ST}}^U(G, K)$ (and is a specialization of $LP^U(G)$):

$$\min \sum_{i=1}^{t} \sum_{e \in E} x_{i,e}$$

subject to

$$\sum_{e \text{ crosses } C} x_{i,e} \geq 1 \qquad \text{for every cut } C \text{ that separates } K \text{ and } i \in [t]$$

$$x_{i,e} \geq 0 \qquad\qquad \text{for every } e \in E \text{ and } i \in [t].$$

Indeed the above LP is basically solving the sum of t independent linear programs: call them $LP_{\text{ST}}(G, K_i)$ for each $i \in [t]$. Hence, one can independently optimize each of these $LP_{\text{ST}}(G, K_i)$ and then just put them together to get an optimal solution for $LP_{\text{ST}}^U(G, K)$. This matches the claimed upper bounds since it is well-known that the objective value of $LP_{\text{ST}}(G, K_i)$ is $\Theta(\text{ST}(G, K_i))$ [29].

On the other hand, if one tries the approach we used to prove the lower bound for XOR_n, then one picks an appropriate hard distribution μ and shows that for every cut C the induced two-party problem has a high enough lower bound. In this case, it turns out that the corresponding two-party lower bound (ignoring constant factors) is the number of sets K_i separated by the cut. Then proceeding as in the argument for XOR_n if one sets y_e to be the expected (under μ) communication for any fixed protocol over any $e \in E$, then $(y_e)_{e \in E}$ is a feasible solution for the following LP, which we dub $LP_{\text{ST}}^L(G, K)$ (and is a specialization of $LP^L(G)$):

$$\min \sum_{e \in E} x_e$$

subject to

$$\sum_{e \text{ crosses } C} x_e \geq v(C, K) \qquad \text{for every cut } C$$

$$x_e \geq 0 \qquad \text{for every } e \in E,$$

where $v(C, K)$ is the number of subsets K_i that are separated by C. If we denote the objective value of the above LP by $LP_{\text{ST}}^L(G, K)$, then we have an overall lower bound of $\Omega(LP_{\text{ST}}^L(G, K))$. Thus, we would be done if we can show that $LP_{\text{ST}}^L(G, K)$ and $LP_{\text{ST}}^U(G, K)$ are close. It is fairly easy to see that $LP_{\text{ST}}^L(G, K) \leq LP_{\text{ST}}^U(G, K)$. However, to prove a tight lower bound, we need an approximate inequality in the other direction. We show this is true by the following two step process:

1. First we observe that if G is a tree T then $LP_{\text{ST}}^L(T, K) = LP_{\text{ST}}^U(T, K)$.
2. Then we use results from embedding graphs into sub-trees to show that there exists a subtree T of G such that $LP_{\text{ST}}^L(G, K) \approx LP_{\text{ST}}^L(T, K)$ and $LP_{\text{ST}}^U(G, K) \approx LP_{\text{ST}}^U(T, K)$, which with the first step completes our proof.

We would like to remark on three things. First, our proof can handle more general constraints than those imposed by the Steiner tree LP. In particular, we generalize the argument above to prove Theorem 4. Second, to the best of our knowledge this result relating the objective values of these two similar LPs seems to be new. However, we would like to point out that our proof follows (with minor modifications) a similar structure that has been used to prove other algorithmic results via tree embeddings (e.g. in [1]). Third, we find it interesting to observe that the upper bound on the gap between the two LP's is the key step in accomplishing a distributed direct-sum like result.

3.2   From Star to Steiner Trees

We define a multicut C of K to be a collection of non-empty pair-wise disjoint subsets $C_1, \ldots, C_r$ of K. Each such subset is called an explicit set of C and the (maybe empty) set $K \setminus \cup_{i=1}^r C_i$ is called its implicit set. We will call $f : \Sigma^K \to \{0, 1\}$ to be h-maximally hard on the star graph if the following holds for any multicut C. There exists a distribution μ_C^f such that the expected cost (under μ_C^f) of any protocol that correctly computes f on the following star graph is $\Omega(|C| \cdot h(|\Sigma|))$: each leaf of the star has all terminals from an explicit set from C, no two leaves have terminals from the same explicit set and the center contains terminals from the implicit set. The following is our second main technical result:

Theorem 5. *Let f be h-maximally hard on the star graph. Then*

$$R(f, G, K, \Sigma) \geq \Omega \left(\frac{\mathrm{ST}(G,K) \cdot h(|\Sigma|)}{\log^2 k} \right).$$

The above result easily implies the lower bound in Theorem 3.   Theorem 5 can also be used to prove a lower bound similar to Theorem 3 above for the Tribes function using the lower bound for Tribes on the star topology from [7].

Proof Overview. In all of the arguments so far, we reduce the lower bound problem on (G, K) to a bunch of two party lower bounds induced by cuts. However, we are not aware of any hard distribution such that one can prove a tight lower bound that reduces the set disjointness problem to a bunch of two-party lower bounds. In fact, the only non-trivial lower bound for set disjointness, in the point-to-point model, that we are aware of is the $\Omega(kn)$ lower bound for the k-star by Braverman et al. [6]. In particular, their proof does not seem to work by reducing the problem to two-party lower bounds. In this work, we are able to extend the set disjointness lower bound of [6] to Theorem 3.

We prove Theorem 3 by modifying the argument in [9] as follows. The idea in [9] is to construct a collection of cuts such that essentially every edge participates in $O(\log k)$ cuts and one can prove the appropriate two-party lower bound across each of the cuts in the collection so that when one sums up the contribution from each cut one gets the appropriate $\Omega(\sigma_K(G)/\log k)$ overall lower bound. (These collection of cuts were obtained via Bourgain's L_1 embedding [5,21]. As mentioned earlier, this trick does not seem to work for set disjointness and it is very much geared towards 1-median type functions). We modify this idea as follows: we construct a collection of *multi-cuts* such that (i) every edge in G appears in at most one multi-cut and (ii) one can use lower bounds on star graph to compute lower bounds for the induced function on each multi-cut, which can then be added up.

The main challenge in the above is to construct an appropriate collection of multi-cuts that satisfy properties (i) and (ii) above. The main idea is natural: we start with balls of radius 0 centered at each of the k terminals and then grow all the balls at the same rate. When two balls intersect, we combine the two balls and grows the larger ball appropriately. The multi-cut at any point of time is defined by the vertices in various balls. To argue the required properties, we observe that the algorithm above essentially simulates Borůvka's algorithm [22] on the *metric closure* of K with respect to the shortest path distances in G. In other words, we show that the sum of the contributions of the lower bounds from each multi-cut is related to the MST on the metric closure of K with respect to G, which is well-known to be closely related to $\mathrm{ST}(G, K)$ (see e.g. [29, Chap.2]). It turns out that for set disjointness, one has to define $O(\log k)$ different hard distributions (that depend on the structure of the multi-cuts above) and this is the reason why we lose a $O(\log k)$ factor in our lower bound. (We lose another $O(\log k)$ factor since we use lower bounds on the star topology.) To the best of our knowledge this is the first instance where the hard distribution actually depends on the graph structure– most of our results as well as those preceding ours use hard distributions that are *independent* of the graph structure. This argument generalizes easily to prove Theorem 5.

Acknowledgments. Thanks to Jaikumar Radhakrishnan for pointing out that the cost of minimum Steiner tree can bound the communication complexity of a class of functions. Many thanks to Anupam Gupta for answering our questions on tree embeddings and related discussions. We would like to thank the organizers of the 2014 Dagstuhl seminar on Algebra in Computational Complexity for inviting us to Dagstuhl, where some of the results in this paper were obtained. AC is supported by a Ramanujan Fellowship of the DST and AR is supported in part by NSF grant CCF-0844796.

References

1. Awerbuch, B., Azar, Y.: Buy-at-bulk network design. In: 38th Annual Symposium on Foundations of Computer Science, FOCS 1997, October 19–22, Miami Beach, Florida, USA, pp. 542–547 (1997). http://doi.ieeecomputersociety.org/10.1109/SFCS.1997.646143
2. Balcan, M.F., Blum, A., Fine, S., Mansour, Y.: Distributed learning, communication complexity and privacy. In: COLT, pp. 26.1–26.22 (2012)
3. Barak, B., Braverman, M., Chen, X., Rao, A.: How to compress interactive communication. SIAM J. Comput. **42**(3), 1327–1363 (2013). http://dx.doi.org/10.1137/100811969
4. Beame, P., Koutris, P., Suciu, D.: Communication steps for parallel query processing. In: PODS, pp. 273–284 (2013)
5. Bourgain, J.: On lipschitz embedding of finite metric spaces in hilbert space. Israel J. Math. **52**(1–2), 46–52 (1995)
6. Braverman, M., Ellen, F., Oshman, R., Pitassi, T., Vaikuntanathan, V.: A tight bound for set disjointness in the message-passing model. In: 54th Annual IEEE Symposium on Foundations of Computer Science (FOCS), pp. 668–677 (2013). http://doi.ieeecomputersociety.org/10.1109/FOCS.2013.77
7. Chattopadhyay, A., Mukhopadhyay, S.: Tribes is hard in the message-passing model. In: STACS, pp. 224–237 (2015)
8. Chattopadhyay, A., Rudra, A.: The Range of Topological Effects on Communication. ArXiv e-prints (April 2015)
9. Chattopadhyay, A., Radhakrishnan, J., Rudra, A.: Topology matters in communication. In: FOCS, pp. 631–640 (2014)
10. Cormode, G.: The continuous distributed monitoring model. SIGMOD Rec. **42**(1), 5–14 (2013). http://doi.acm.org/10.1145/2481528.2481530
11. Dolev, D., Feder, T.: Multiparty communication complexity. In: FOCS, pp. 428–433 (1989)
12. Drucker, A., Kuhn, F., Oshman, R.: On the power of the congested clique model. In: PODC, pp. 367–376 (2014)
13. Duris, P., Rolim, J.: Lower bounds on the multiparty communication complexity. J. Comput. Syst. Sci. **56**(1), 90–95 (1998)
14. Goldreich, O.: Three XOR-Lemmas — an exposition. In: Goldreich, O. (ed.) Studies in Complexity and Cryptography. LNCS, vol. 6650, pp. 248–272. Springer, Heidelberg (2011)

15. Goldreich, O., Nisan, N., Wigderson, A.: On yao's XOR-Lemma. In: Goldreich, O. (ed.) Studies in Complexity and Cryptography. LNCS, vol. 6650, pp. 273–301. Springer, Heidelberg (2011)

16. Huang, Z., Radunovic, B., Vojnovic, M., Zhang, Q.: Communication complexity of approximate maximum matching in distributed graph data. In: 32nd Symposium on Theoretical Aspects of Computer Science (STACS), pp. 460–473 (2015)

17. Karchmer, M., Raz, R., Wigderson, A.: Super-logarithmic depth lower bounds via the direct sum in communication complexity. Computational Complexity $5(3/4)$, 191–204 (1995)

18. Karloff, H.J., Suri, S., Vassilvitskii, S.: A model of computation for mapreduce. In: SODA, pp. 938–948 (2010)

19. Kowshik, H., Kumar, P.: Optimal function computation in directed and undirected graphs. IEEE Transcations on Information Theory $58(6)$, 3407–3418 (2012)

20. Li, Y., Sun, X., Wang, C., Woodruff, D.P.: On the communication complexity of linear algebraic problems in the message passing model. In: Kuhn, F. (ed.) DISC 2014. LNCS, vol. 8784, pp. 499–513. Springer, Heidelberg (2014). http://dx.doi.org/10.1007/978-3-662-45174-8_34

21. Linial, N., London, E., Rabinovich, Y.: The geometry of graphs and some of its algorithmic applications. Combinatorica $15(2)$, 215–245 (1995)

22. Nešetřil, J., Milková, E., Nešetřilová, H.: Otakar Borůvka on minimum spanning tree problem translation of both the 1926 papers, comments, history. Discrete Mathematics $233(13)$, 3–36 (2001). http://www.sciencedirect.com/science/article/pii/S0012365X00002247; czech and Slovak 2

23. Phillips, J.M., Verbin, E., Zhang, Q.: Lower bounds for number-in-hand multiparty communication complexity, made easy. In: Proceedings of the Twenty-Third Annual ACM-SIAM Symposium on Discrete Algorithms (SODA), pp. 486–501 (2012). http://portal.acm.org/citation.cfm?id=2095158&CFID=63838676&CFTOKEN=79617016

24. Raz, R., McKenzie, P.: Separation of the monotone NC hierarchy. Combinatorica $19(3)$, 403–435 (1999)

25. Sherstov, A.: Separating AC^0 from depth-2 majority circuits. SIAM J. Comput. $38(6)$, 2113–2129 (2009)

26. Shi, Y., Zhu, Y.: Quantum communication complexity of block-composed functions. Qunatum Computation and Information $9(5)$, 444–460 (2009)

27. Tiwari, P.: Lower bounds on communication complexity in distributed computer networks. J. ACM $34(4)$, 921–938 (1987)

28. Valiant, L.G.: A bridging model for parallel computation. Commun. ACM $33(8)$, 103–111 (1990)

29. Vazirani, V.V.: Approximation Algorithms. Springer-Verlag New York Inc., New York (2001)

30. Woodruff, D., Zhang, Q.: Tight bounds for distributed functional monitoring. In: STOC, pp. 941–960 (2012)

31. Woodruff, D.P., Zhang, Q.: When distributed computation is communication expensive. In: Afek, Y. (ed.) DISC 2013. LNCS, vol. 8205, pp. 16–30. Springer, Heidelberg (2013)

32. Woodruff, D., Zhang, Q.: An optimal lower bound for distinct elements in the message passing model. In: SODA, pp. 718–733 (2014)

33. Yao, A.C.C.: Some complexity questions related to distributed computing. In: 11th ACM Symposium on Theory of Computing (STOC), pp. 209–213 (1979)

Secretary Markets with Local Information

Ning Chen[1], Martin Hoefer[2]([⊠]), Marvin Künnemann[2,3], Chengyu Lin[4], and Peihan Miao[5]

[1] Nanyang Technological University, Singapore, Singapore
ningc@ntu.edu.sg
[2] MPI für Informatik, Saarbrücken, Germany
{mhoefer,marvin}@mpi-inf.mpg.de
[3] Saarbrücken Graduate School of Computer Science, Saarbrücken, Germany
[4] Chinese University of Hong Kong, Hong Kong, China
cylin@cse.cuhk.edu.hk
[5] University of California Berkeley, Berkeley, USA
peihan@berkeley.edu

Abstract. The secretary model is a popular framework for the analysis of online admission problems beyond the worst case. In many markets, however, decisions about admission have to be made in a decentralized fashion and under competition. We cope with this problem and design algorithms for secretary markets with limited information. In our basic model, there are m firms and each has a job to offer. n applicants arrive iteratively in random order. Upon arrival of an applicant, a value for each job is revealed. Each firm decides whether or not to offer its job to the current applicant without knowing the strategies, actions, or values of other firms. Applicants decide to accept their most preferred offer.

We consider the social welfare of the matching and design a decentralized randomized thresholding-based algorithm with ratio $O(\log n)$ that works in a very general sampling model. It can even be used by firms hiring several applicants based on a local matroid. In contrast, even in the basic model we show a lower bound of $\Omega(\log n/(\log \log n))$ for all thresholding-based algorithms. Moreover, we provide secretary algorithms with constant competitive ratios, e.g., when values of applicants for different firms are stochastically independent. In this case, we can show a constant ratio even when each firm offers several different jobs, and even with respect to its individually optimal assignment. We also analyze several variants with stochastic correlation among applicant values.

1 Introduction

The Voice is a popular reality television singing competition to find new singing talent contested by aspiring singers. The competition employs a panel of coaches; upon the arrival of a singer, every coach critiques the artist's performance and determines in real time if he/she wants the artist to be on his/her team. Among those who express "I want you", the artist selects a favorite coach. What strategy of picking artists should coaches adopt in order to select the best candidates?

M. Hoefer—Supported by DFG Cluster of Excellence MMCI at Saarland University.

M.M. Halldórsson et al. (Eds.): ICALP 2015, Part II, LNCS 9135, pp. 552–563, 2015.
DOI: 10.1007/978-3-662-47666-6_44

This problem is a reminiscent of the classic secretary problem: A firm interviews a set of candidates who arrive in an online fashion. When a candidate arrives, its value is revealed and the firm needs to make an immediate and irrevocable decision on whether to make an offer to the candidate, without knowing the values of future potential candidates. The objective is to maximize the (expected) value of the hired candidate. The secretary problem is well studied in social science and computer science. It is well known that the problem, in the worst case, does not admit an algorithm with any guaranteed competitive ratio. However, if candidates arrive in uniform random order, there is an online algorithm that achieves the optimal competitive ratio $1/e$ [7,21]. For a more detailed discussion on the secretary problem see, e.g., [1].

The scenario of The Voice is a generalization of the secretary problem from one firm to multiple firms and from one hire to multiple hires. Such a generalization yields several fundamental changes to the problem: Firms (i.e., coaches) are independent and compete with each other for candidates. Thus, each firm may determine on their own the strategy to adopt. Firms are decision makers; that is, there is no centralized authority and every firm can choose different strategies on its own (based on observed information). Each firm can only observe information revealed to itself, i.e., it has no knowledge on the values of other (firm, candidate)-pairs and selected strategies of other firms. Hence, adopting a best-response strategy in a game-theoretic sense might require learning other strategies and payoffs. Given the limited feedback this can be hard or even impossible. The same issues occur in many other decentralized markets, e.g., online dating and school admission, where entities behave individually and have to make decisions based on a very limited view on the market, the preferences, and the strategies used by potential competitors.

The objective of the present paper is to design and analyze strategies for all firms in such a decentralized, competitive enviroment to enable efficient allocations. Our algorithms are evaluated both globally and individually: On the one hand, we hope the outcomes achieve good social welfare (i.e., the total value obtained by all firms). Thus, we measure the competitive ratio compared to social welfare given by the optimal centralized online algorithm. On the other hand, considering that firms are self-interested entities, we hope that our algorithms generate a nearly optimal outcome for each individual firm. That is, although given the limited feedback it can be impossible to obtain best-response strategies, we nevertheless hope that (when applied in combination) our algorithms can approximate the outcome of a best response (in hindsight with full information) of every individual firm within a small factor.

We identify several settings that admit algorithms with small constant competitive ratio both globally and individually. This implies that even in decentralized markets with very limited feedback, there are algorithms to obtain a good social solution. For the general case, we provide a strategy to approximate social welfare within a logarithmic factor, and we show almost matching lower bounds on the competitive ratio for a very natural class of algorithms. Thus, in the general case centralized control seems to be necessary in order to achieve good social welfare.

Model and Preliminaries. We first outline our *basic model*, a decentralized online scenario for hiring a single applicant per firm with random arrival. There is a complete bipartite graph $G = (U, V, w)$ with sets $U = \{u_1, u_2, \ldots, u_m\}$ and $V = \{v_1, v_2, \ldots, v_n\}$ of firms and applicants, respectively. There is a *value* or *weight function*[1] $w : U \times V \to \mathbb{R}^+$. We assume that each firm can hire at most one applicant and there are more applicants than firms, i.e., $m \leq n$.

The weights describe an implicit preference of each individual to the other side. Each firm $u \in U$ prefers applicants according to the decreasing order of $w(u, \cdot)$ of the edges incident to u; similarly, each applicant $v \in V$ prefers firms according to the decreasing order of $w(\cdot, v)$ of the edges incident to v.[2]

Applicants in V arrive one by one in the market and reveal their edge weights to all firms. Upon the arrival of an applicant, each firm decides on whether to provide an offer for the applicant or not immediately; after collecting all job offers, the applicant then picks one that she prefers most, i.e., the one with the largest weight. Note that each firm can only see its own weights for the applicants and has no information about future applicants; in addition, all decisions cannot be revoked. In this paper, we consider the problem in the random permutation model, i.e., applicants arrive in a uniformly random order.

Our goal is to design decentralized algorithms when each firm makes its decision only based on its own previous information and there is no centralized authority that manages different firms altogether. There are two natural objectives to evaluate the performance of an algorithm, and due to online arrival some performance loss is unavoidable. The standard benchmark is *social welfare*, defined to be the total weight of assigned firm and applicant pairs. For an algorithm $\mathcal{A}$, we say the algorithm has a *competitive ratio* of α if for all instances, we have $\mathbb{E}\left[w(M^*)\right] / \mathbb{E}\left[w(M^{\mathcal{A}})\right] \leq \alpha$. Here the expectation is over random permutation, M^* is the maximum weight matching in G, and $M^{\mathcal{A}}$ is the matching returned when every firm runs algorithm $\mathcal{A}$. In addition, we would like to approximate the individual optimum assignment for each firm (i.e., the weight of its best candidate) and strive to obtain a constant competitive ratio for this benchmark.

Contribution and Techniques. As a natural first attempt, consider every firm running the classic secretary algorithm [7,21], which samples the first $r - 1$ applicants, records the best weight seen in the sample, and then offers to every applicant that exceeds this threshold. It turns out that such a strategy fails miserably in a decentralized market, even if each applicant has the same weight for all firms. For spatial reasons, the proof of this statement and many other formal arguments in this extended abstract are deferred to a full version.

Proposition 1. *For any constant $\beta < 1$, when setting $r = \lfloor \beta n \rfloor + 1$, then the classic secretary algorithm has a competitive ratio of $\Omega(n/\log n)$.*

---

[1] To avoid ties, we assume that no two edges have the same weight; this assumption is without loss of generality by using small perturbations or a fixed rule to break ties.

[2] In a more general preference model there are for each pair (u, v) different values obtained by u and v; we will not consider this general case in the present paper.

In contrast, we present in Section 2 a more careful approach based on sampling and thresholds that is $O(\log n)$-competitive. This algorithm can be applied in large generality (well beyond the basic model). In fact, we prove the guarantee in a scenario, where each firm u_i has a private matroid $\mathcal{S}_i$ and can accept any subset of applicants that forms an independent set in $\mathcal{S}_i$. Furthermore, our analysis extends to a general sampling model that encompasses the secretary model (random arrival, worst-case weights), prophet-inequality model (worst-case arrival, stochastic weights), as well as a variety of other mixtures of stochastic and worst-case assumptions [10]. Our main technique to handle decentralized thresholding is to bundle all stochastic decisions and treat correlations using linearity of expectation. The effects of applicant preferences and competition can then be analyzed in a pointwise fashion.

We contrast this result with an almost matching lower bound for thresholding-based algorithms in the basic model. A thresholding-based algorithm samples a number of applicants, determines a threshold, and then offers to every remaining applicant that has a weight above the threshold. Although such algorithms are nearly optimal in the centralized setting, every such algorithm must have a competitive ratio of at least $\Omega(\log n / \log \log n)$ in the decentralized setting. The lower bound carefully constructs a challenge to guess how many firms contribute to social welfare and to avoid overly high concentration of offers on a small number of valuable applicants.

In Section 3 we show that this challenge can be overcome if there is stochastic independence between the weights of an applicant to different firms. We study this property in a generalized model for decentralized k-secretary, where each firm u_i has k_i different jobs to offer. Upon arrival, an applicant reveals k_i weights for each firm u_i, one for each position. If each firm uses a variant of the optimal e-competitive algorithm for bipartite matching [16], independence between weights of different firms allows to show a constant competitive ratio. Moreover, each firm even manages to recover a constant fraction of the individual optimum matching and therefore almost plays a best response strategy.

Finally, in Section 4 we consider two additional variants with stochastically generated weights. In both variants we can show constant competitive ratios, and in one case firms can even hire their best applicant with constant probability.

Related Work. The secretary model is a classic approach to stopping problems and online admission [7]. The classic algorithm outlined in the previous section is e-competitive, which is the best possible ratio. In the algorithmic literature, recent work has addressed secretary models for packing problems with random arrival of elements. A prominent case is the matroid secretary problem [2], for which the first general algorithm was $O(\log k)$-competitive, where k is the rank of the matroid. The ratio was very recently reduced to $O(\log \log k)$ [8,20]. Constant-factor competitive algorithms have been obtained for numerous special cases [6, 11,14,18,24]. It remains a fascinating open problem if a general constant-factor competitive algorithm exists.

Another popular domain is bipartite matching in the secretary model, which has many applications in online revenue maximization via ad-auctions.

In Section 3 we use a variant of a recent optimal e-competitive algorithm [16], which tightened the ratio and improved it over previous algorithms [2,5,19]. The main idea has recently been extended to construct optimal secretary algorithms for packing linear problems [17], improving over previous approaches [4,23]. Algorithms based on primal-dual techniques are a popular approach, especially for budgeted online matching with different stochastic input assumptions [3,15,22].

Our analysis of the algorithm for the general case applies in a unifying sampling model recently proposed as a framework for online maximum independent set in graphs [10]. It encompasses many stochastic adversarial models for online optimization – the secretary model, the prophet inequality model, and various other mixtures of stochastic and worst-case adversaries.

Closer to our paper are studies of a secretary problem with k queues [9], or game-theoretic approaches with complete knowledge of opponent strategies [12, 13]. These scenarios, however, have significantly different assumptions on the firms and their feedback, and they do not target markets with both decentralized control and restricted feedback that we explore in this paper.

2   General Preferences

For general weights $w : U \times V \to \mathbb{R}^+$, Proposition 1 shows that the classic secretary algorithm may perform poorly in a decentralized market. A reasonable strategy has to be more careful in adopting a threshold to avoid extensive competition for few candidates. Inspired by Babaioff et al. [2], we overcome this obstacle with a randomized thresholding strategy, and analyze it in a very general distributed matroid scenario. We remark that our bounds apply even within a general sampling model [10] that encompasses the secretary model, prophet-inequality model, and many other approaches for stochastic online optimization.

For the combinatorial structure of the scenario, we consider that each firm u_i holds a possibly different matroid $\mathcal{S}_i$ over the set of applicants. Firm u_i can accept an applicant as long as the set of accepted applicants forms an independent set in $\mathcal{S}_i$. Special cases include hiring a single applicant or any subset of at most k_i many applicants. Each firm strives to maximize the sum of weights of hired applicants. The structure of $\mathcal{S}_i$ does not have to be known in advance. u_i only needs an oracle to test if a set of arrived applicants is an independent set in $\mathcal{S}_i$.

Algorithm 1 is executed in parallel by all firms u_i. We first sample a fraction of roughly $n/(c+1)$ applicants, where $c \geq 1$ is a global constant. Then we determine a random threshold based on the maximum weight seen by firm u_i in its sample. Firm u_i then greedily makes an offer only to those candidates whose values are above the threshold.

Theorem 1. *Algorithm 1 is $O(\log n)$-competitive.*

Proof. We denote by V_i^S the set of candidates in the sample and by V_i^I the other candidates. Note that by the choice of sample and the random arrival, we have that $\mathbf{Pr}\left[v_j \in V_i^S\right] = \frac{1}{c+1}$. More broadly, our subsequent arguments will require only the weaker bounds

---

Algorithm 1. Thresholding algorithm for u_i with matroids.

Draw a random number $k \sim Binom(n, 1/(c+1))$
Reject the first k applicants, denote this set by V_i^S
$m_i \leftarrow \arg\max_{v_j \in V_i^S} w(u_i, v_j)$
$X_i \leftarrow \text{Uniform}(-1, 0, 1, \ldots, \lceil \log n \rceil)$
$t_i \leftarrow w(u_i, m_i)/2^{X_i}$, $M_i \leftarrow \emptyset$
for all remaining v_j over time **do**

    if $w(u_i, v_j) \geq t_i$ and $M_i \cup \{v_j\}$ is independent set in $\mathcal{S}_i$ **then**

        make an offer to v_j

        if v_j accepts **then** $M_i \leftarrow M_i \cup \{v_j\}$

$$\mathbf{Pr}\left[v_j \in V_i^I\right] \geq \frac{1}{c+1} \quad \text{and} \quad \mathbf{Pr}\left[v_j \in V_i^S\right] \geq \frac{1}{c+1}. \tag{1}$$

These sampling inequalities obviously hold for every v_j, independently of $v_{j'} \in V_i^S$ or not for all other candidates $j' \neq j$.

Let $v_i^{\max} = \text{argmax}_j w(u_i, v_j)$ and $v_i^{2nd} = \text{argmax}_{j \neq v^{\max}} w(u_i, v_j)$ be a best and second best applicant for firm u_i, respectively (breaking ties arbitrarily). In addition, we denote by $w_i^{\max} = w(u_i, v_i^{\max})$ and $w_i^{2nd} = w(u_i, v_i^{2nd})$ their weights for firm u_i. For most of the analysis, we consider another weight function, the *capped weights* $\tilde{w}(u_i, v_j)$, based on thresholds t_i set by the algorithm as follows

$$\tilde{w}(u_i, v_j) = \begin{cases} w_i^{\max} & \text{if } v_j \in V_i^I, t_i = 2w_i^{2nd}, \text{ and } w(u_i, v_j) > 2w_i^{2nd}, \\ t_i & \text{if } v_j \in V_i^I \text{ and } w(u_i, v_j) \geq t_i, \\ 0 & \text{otherwise.} \end{cases}$$

Observe that the definition of $\tilde{w}$ relies on several random events, i.e., $v_j \in V_i^I$ and the choice of thresholds t_i. For any outcome of these events, however, we have that $\tilde{w}(u_i, v_j) \leq w(u_i, v_j)$ for all pairs (u_i, v_j), since if $t_i = 2w_i^{2nd}$ and $w(u_i, v_j) > 2w_i^{2nd}$, then $v_j = v_i^{\max}$. By the following lemma, in expectation over all the correlated random events, an optimal offline solution with respect to $\tilde{w}$ still gives an approximation to the optimal offline solution with respect to w.

Lemma 1. *Denote by $w(M)$ and $\tilde{w}(M)$ the weight and capped weight of a solution M. Let $\tilde{M}^*$ and M^* be optimal solutions for $\tilde{w}$ and w, respectively. Then*

$$\mathbb{E}\left[\tilde{w}(\tilde{M}^*)\right] \geq \Omega\left(\frac{1}{\log n}\right) \cdot w(M^*).$$

Proof. Let $(u_i, v_j) \in M^*$ be an arbitrary pair. First, assume that v_j maximizes $w(u_i, v_j)$, i.e., $v_j = v_i^{\max}$. By (1) with probability at least $1/(c+1)^2$, we have $v_j \in V_i^I$ and $v_i^{2nd} \in V_i^S$. For any such outcome, we have with probability $1/(\lceil \log(n) \rceil + 2)$ that either (1) $t_i = 2w_i^{2nd}$ and $\tilde{w}(u_i, v_j) = w_i^{\max}$ (if $w_i^{\max} \geq 2w_i^{2nd}$), or (2) $t_i = w_i^{2nd}$ and $t_i \leq w_i^{\max} < 2w_i^{2nd}$ (otherwise). This yields $\mathbb{E}[\tilde{w}(u_i, v_j)] \geq w(u_i, v_j)/(2(c+1)^2(\lceil \log(n) \rceil + 2))$.

Second, for any $v_j \neq v_i^{\max}$ with $w_i^{\max}/(2n) < w(u_i, v_j) \leq w_i^{\max}$, by (1) we know $v_i^{\max} \in V_i^S$ is an independent event which happens with probability at least $1/(c+1)$. Then, there is some $0 \leq k' \leq \lceil \log n \rceil + 1$, with $w(u_i, v_j) > w_i^{\max}/2^{k'} \geq w(u_i, v_j)/2$. With probability $1/(\lceil \log(n) \rceil + 2)$, we have that $X_i = k'$ and $\tilde{w}(u_i, v_j) = t_i \geq w(u_i, v_j)/2$. This yields $\mathbb{E}[\tilde{w}(u_i, v_j)] \geq w(u_i, v_j)/(2(c+1)^2(\lceil \log(n) \rceil + 2))$, since $v_j \in V_i^I$ with probability at least $1/(c+1)$ by (1).

Finally, we denote by $M^>$ the set of pairs $(u_i, v_j) \in M^*$ for which $w(u_i, v_j) > w_i^{\max}/(2n)$. The expected weight of the best assignment with respect to the threshold values is thus

$$\mathbb{E}\left[\tilde{w}(\tilde{M}^*)\right] \geq \sum_{(u_i,v_j)\in M^*} \mathbb{E}[\tilde{w}(u_i, v_j)] \geq \sum_{(u_i,v_j)\in M^>} \frac{w(u_i, v_j)}{2(c+1)^2(\lceil \log(n) \rceil + 2)}$$

$$= \frac{1}{2(c+1)^2(\lceil \log(n) \rceil + 2)} \cdot (w(M^*) - w(M^* \setminus M^>))$$

$$\geq \frac{1}{4(c+1)^2(\lceil \log(n) \rceil + 2)} \cdot w(M^*),$$

since $\sum_{(u_i,v_j)\in M^*\setminus M^>} w_i^{\max}/(2n) \leq \max_i w_i^{\max}/2 \leq w(M^*)/2$.   □

The previous lemma bounds the weight loss due to (i) all random choices inherent in the process of input generation and threshold selection and (ii) using the capped weights. The next lemma bounds the remaining loss due to adversarial arrival of elements in V_i^I, exploiting that $\tilde{w}$ equalizes equal-threshold firms.

Lemma 2. *Suppose subsets V_i^I and thresholds t_i are fixed arbitrarily and consider the resulting weight function $\tilde{w}$. Let $M^{\mathcal{A}}$ be the feasible solution resulting from Algorithm 1 using the thresholds t_i, for any arbitrary arrival order of applicants in $\bigcup V_i^I$. Then $w(M^{\mathcal{A}}) \geq \tilde{w}(\tilde{M}^*)/2$.*

Combining the insights, we see that that $w(M^*) \leq O(\log n) \cdot \mathbb{E}[w(M^{\mathcal{A}})]$, which proves the theorem.   □

Our general upper bound results from a thresholding-based algorithm. We constrast this result with a lower bound for thresholding-based algorithms when every firm wants to hire only a single applicant. It applies even when preferences of all firms over applicants are identical. More formally, an algorithm $\mathcal{A}$ is called *thresholding-based* if during its execution $\mathcal{A}$ rejects applicants for some number of rounds, then determines a threshold T and afterwards enters an *acceptance phase*. In the acceptance phase, it makes an offer to exactly those applicants whose weight exceeds threshold T. Note that the number of rejecting rounds in the beginning and the threshold T can be chosen arbitrarily at random.

The lower bound uses an *identical-firm* instance in which for each applicant v_j all firms have the same weight, i.e., there is $w(v_j) \geq 0$ such that $w(u_i, v_j) = w(v_j)$ for every firm u_i. It applies in the secretary model and the iid model. In the latter we draw the weight $w(v_j)$ for each v_j independently at random from a single distribution. The main difference is that M^* becomes a random variable.

Theorem 2. *Suppose every firm strives to hire a single applicant, and let $\mathcal{A}$ be any thresholding-based algorithm. If every firm adopts $\mathcal{A}$, there is an identical-firm instance I on which $\mathcal{A}$ has a competitive ratio of $\Omega(\log n / \log \log n)$. This lower bound applies in the iid model and the secretary model.*

Proof (Idea). For simplicity, we assume[3] that the thresholding-based algorithm $\mathcal{A}$ does not know the number of firms m. Assume that $n = \sum_{j=2}^{t} t^{2j}$ for some $t \in \mathbb{N}$. We construct a distribution $\mathcal{I}$ on a family of identical-firm instances by drawing the weight $w(v_{j'})$ of each applicant $v_{j'}$ according to $\Pr[w(v_{j'}) = t^{-j}] = t^{2j}/n$ for $j = 2, \ldots, t$. In the secretary model, we may assume that each applicant draws $w(v_{j'})$ at the moment it arrives in the random order, since the order is chosen independently of the weights. Since all applicant weights are identically distributed, we may even completely disregard the random arrival order.

We define classes $C_2, \ldots, C_t$, where each class C_j consists of all applicants $v_{j'}$ with value $w(v_{j'}) = t^{-j}$. Consider how $\mathcal{A}$ performs on $\mathcal{I}$ for some firm u_i. We can assume that $\mathcal{A}$ chooses a threshold among $\{t^{-2}, \ldots, t^{-t}\}$, since all other choices are equivalent concerning the set of applicants receiving an offer from u_i. Let p_j be the probability (over $\mathcal{I}$ and the random choices of $\mathcal{A}$) that threshold t^{-j} is chosen. Clearly, there is some $2 \leq k \leq t$ with $p_k \leq \frac{1}{t-1}$. By setting $m := \sum_{j=2}^{k} t^{2j}$, most firms should choose a threshold of t^{-k} to obtain a competetive solution, but by choice of k few firms do. Hence, the challenge for the firms is to guess m correctly and extract welfare from the right class of applicants.   □

3   Independent Preferences

In this section, we show improved results for decentralized matching in the secretary model when preferences are independent among firms. More formally, we assume firm u_i has a set U_i of k_i positions available. An adversary specifies a separate set $\mathcal{P}_i$ of n *applicant profiles* for each firm u_i. An applicant profile $p \in \mathcal{P}_i$ is a function $p : U_i \to \mathbb{R}^+$. In round t, when a new applicant v_t arrives, we pick one remaining profile $p_{it} \in \mathcal{P}_i$ for each $u_i \in U$ independently and uniformly at random. The weight for position $u_{ij} \in U_i$ is then given by $w(u_{ij}, v_t) = p_{it}(u_{ij})$. We pick profiles from $\mathcal{P}_i$ uniformly at random without replacement. Special cases of this model are, e.g., when all weights for all positions are independently sampled from a certain distribution, or for each firm u_i the weights of all applicants are sampled independently from a different distribution for each position.

In contrast to the previous section, we assume that each applicant has k_i weight values for each firm u_i. A straightforward $O(\log n)$-competitive algorithm is to run Algorithm 1 separately for each position of each firm. In contrast, when $n \geq \sum_{i=1}^{m} k_i$ and $k_i \leq \alpha n$ for all $i \in [m]$ and some constant $\alpha \in (0, 1)$, we can achieve a constant competitive ratio using Algorithm 2. This algorithm resembles an optimal algorithm for secretary matching with a single firm [16]. Each firm rejects a number of applicants and enters an acceptance phase. In this phase, it maintains two virtual solutions: (1) an individual virtual optimum $M_{i,t}^*$ with

---

[3] This assumption can be dropped by introducing firms with neglibly small preferences.

Algorithm 2. Matching algorithm for firm u_i for independent weights

---

Reject the first $r_i - 1$ applicants
$M_i, M_i' \leftarrow \emptyset$
for *applicant v_t arriving in round $t = r_i, \ldots, n$* **do**
    Let $M_{i,t}^*$ be optimum matching for firm u_i and applicants $\{v_1, \ldots, v_t\}$
    if *v_t is matched to position u_{ij} in M_i^* and u_{ij} unmatched in M_i'* **then**
        Make an offer for position u_{ij} to v_t
        $M_i' \leftarrow M_i' \cup \{(u_{ij}, v_t)\}$
        if *v_t accepts* **then**
            $M_i \leftarrow M_i \cup \{(u_{ij}, v_t)\}$

---

respect to applicants arrived up to and including round t, and (2) a virtual solution M_i' where all applicants are assumed to accept the offers of u_i. If the newly arrived applicant v_t is matched in $M_{i,t}^*$, it is offered the same position unless this position is already filled in M_i'.

Theorem 3. *Algorithm 2 achieves a constant competitive ratio.*

Proof. Fix a firm u_i. The matching M_i' is constructed by assuming that u_i is the only firm in the market, i.e., every applicant accepts the offer of firm u_i. Consider the individual optimum $M_{i,n}^*$ in hindsight. Then, by repeating the analysis of [16], the expected value of M_i' is

$$\mathbb{E}\left[w(M_i')\right] \geq \frac{r_i - 1}{n} \ln\left(\frac{n}{r_i - 1}\right) \cdot w(M_{i,n}^*) = f(r_i) \cdot w(M_{i,n}^*) \ ,$$

where we denote the ratio by $f(r_i)$. Recall $k_i \leq \alpha n$. Set r_i in the interval $(\beta n, \gamma n)$ for some appropriate constants $\beta, \gamma \in (0, 1)$ such that $\beta > \alpha$. This ensures that $f(r_i)$ becomes a constant.

Let us now analyze the performance of the algorithm in the presence of competition. Consider applicant v_t in round t and the following events: (1) $P(u_i, v_t)$ is the event that u_i sends an offer to v_t, and (2) $A(u_i, v_t)$ is the event that u_i sends an offer to v_t and he accepts it. u_i's decision to offer depends only on $M_{i,t}^*$ and M_i', but not on the acceptance decisions of earlier applicants. v_t for sure accepts an offer from u_i if u_i offers and no other firm offers. Offers from other firms $u_{i'}$ occur only if $u_{i'}$ is matched in $M_{i',t}^*$. More formally, $A(u_i, v_t)$ occurs (at least) if $P(u_i, v_t)$ and none of the $P(u_{i'}, v_t)$ occur. Since the profiles for different firms are combined independently

$$\mathbf{Pr}\left[A(u_i, v_t) \mid P(u_i, v_t)\right] \geq \prod_{i \neq i'}(1 - \mathbf{Pr}\left[P(u_{i'}, v_t)\right])$$

Consider the probability that v_t is matched in $M_{i',t}^*$. Since the order of profiles for $u_{i'}$ is independent of the order for u_i, we can imagine again choosing t

profiles at random. Of those a random profile is chosen to be one of v_t. The t profiles determine $M^*_{i',t}$, which matches $\min(t, k_{i'})$ profiles. Since the last profile is determined at random, the probability that v_t is matched in $M^*_{i',t}$ is at most $\min(1, k_{i'}/t)$. As $t \geq r_{i'} \geq \beta n$, we have

$$\mathbf{Pr}\left[P(u_{i'}, v_t)\right] \leq \begin{cases} 0 & \text{if } t \leq r_{i'} - 1, \\ k_{i'}/(\beta n) & \text{otherwise.} \end{cases}$$

Thus,

$$\mathbf{Pr}\left[A(u_i, v_t) \mid P(u_i, v_t)\right] \geq \prod_{i \neq i'}(1 - \mathbf{Pr}\left[P(u_{i'}, v_t)\right]) \geq \exp\left(\sum_{i=1}^{m} \ln\left(1 - \frac{k_i}{\beta n}\right)\right)$$

$$\geq \exp\left(-\sum_{i=1}^{m} \frac{1}{1 - (\alpha/\beta)} \cdot \frac{k_i}{\beta n}\right) \geq \exp\left(-\frac{1}{\beta - \alpha}\right).$$

The third inequality follows from $k_i \leq \alpha n$ by $(1 - k_i/(\beta n)) \geq 1 - \alpha/\beta$. Furthermore, it holds that $\ln(1 - x) \geq -\frac{x}{1-x}$ for all $x \in (0, 1)$. The last inequality is due to $n \geq \sum_j k_j$.

Consequently, $\mathbb{E}\left[w(M_i)\right]$ recovers at least a constant fraction of $\mathbb{E}\left[w(M'_i)\right]$, which represents a constant factor approximation to the individual optimum $M^*_{i,n}$ for i in hindsight. By linearity of expectation, the algorithm achieves a constant competitive ratio for the expected weight of the optimum matching. □

4   Correlated Preferences

In this section, we treat the basic model where every firm strives to hire one applicant. We consider stochastic input generation which allows correlations on the weights incident to an applicant. Specifically, assume that each applicant v_i has a parameter q_i, measuring his built-in quality, and the weights of edges incident to v_i are generated independently from a distribution D_i with mean q_i. Note that the lower bound for the classical e-competitive algorithm for the secretary problem (Proposition 1) applies to this general setting. As a natural candidate, we consider in particular normal distributions and assume that $D_i \sim N(q_i, \sigma^2)$ where q_i is the quality of applicant v_i and σ is a fixed constant.

We analyze correlations in two regimes: When the random noise is small and the preference lists of each firm are unlikely to differ significantly and when large variance has substantial effects on the preferences.

Small Variance. We consider the case of highly correlated preferences of an applicant to all firms with possibly small fluctuations around an applicant's quality. Consider the list-based approach of Algorithm 3 that first samples a linear number $r = \Theta(n)$ of applicants and afterwards maintains a list of the top m candidates observed so far. The key observation we exploit is that Algorithm 3,

---

Algorithm 3. List-based algorithm for firm u

Initialize list $L_u = (\ell_{u,1}, \ldots, \ell_{u,m})$, initialized with $(-\infty, \ldots, -\infty)$
(maintain L_u to contain the top m weights u observed so far, where
$\ell_{u,1} \geq \cdots \geq \ell_{u,m})$
Reject the first $(r-1)$ applicants, denote the set by R
for *applicant v_t arriving in round $t = r, \ldots, n$* **do**
  if $w(u, v_t) \geq \ell_{u,m}$ **then**
   Update L_u: Push w_{u,v_t} into L_u and pop $\ell_{u,m}$ out.
   if *popped out $\ell_{u,m} = -\infty$ or corresponds to an applicant in R* **then**
    Make an offer to v_t, stop if v_t accepts

---

in contrast to the classical algorithm for the secretary problem, can cope well with competition, provided that applicants have a global quality that all firms roughly agree on. In particular, each of the top m applicants will be matched to her best firm with constant probability.

Without loss of generality, let $q_1 \geq \cdots \geq q_n$. Formally, define the parameter $\delta_{min} := \min_{i \neq j} |q_i - q_j|$, and $\psi = \frac{\delta_{min}}{\sigma}$.

Theorem 4. *Let $\psi = \omega(n)$ and $r = \Theta(n)$. Algorithm 3 achieves a constant competitive ratio with high probability, i.e., with probability approaching 1 over all possible weights, we have $\mathbb{E}[w(M^*)] \leq c \cdot \mathbb{E}[w(M^{\mathcal{A}})]$ for some constant $c > 0$.*

Large Variance. If weights are perturbed by high-variance normal distributions, this yields a natural situation in which the classic algorithm for the secretary problem achieves a constant competitive ratio. Let $\delta_{max} := \max_{i \neq j} |q_i - q_j|$ denote the largest difference in applicants' qualifications and define the parameter $\varphi := \frac{\delta_{max}}{\sigma}$.

Theorem 5. *The classic secretary algorithm achieves a constant competitive ratio when $\varphi = O(\frac{1}{n^2})$ and $r = \Theta(n)$. Each firm hires its best applicant with constant probability.*

References

1. Babaioff, M., Immorlica, N., Kempe, D., Kleinberg, R.: Online auctions and generalized secretary problems. SIGecom Exchanges **7**(2) (2008)
2. Babaioff, M., Immorlica, N., Kleinberg, R.: Matroids, secretary problems, and online mechanisms. In: Proc. 18th Symp. Discrete Algorithms (SODA), pp. 434–443 (2007)
3. Devanur, N., Hayes, T.: The adwords problem: Online keyword matching with budgeted bidders under random permutations. In: Proc. 10th Conf. Electronic Commerce (EC), pp. 71–78 (2009)
4. Devanur, N., Jain, K., Sivan, B., Wilkens, C.: Near optimal online algorithms and fast approximation algorithms for resource allocation problems. In: Proc. 12th Conf. Electronic Commerce (EC), pp. 29–38 (2011)
5. Dimitrov, N., Plaxton, G.: Competitive weighted matching in transversal matroids. Algorithmica **62**(1–2), 333–348 (2012)

6. Dinitz, M., Kortsarz, G.: Matroid secretary for regular and decomposable matroids. SIAM J. Comput. **43**(5), 1807–1830 (2014)
7. Dynkin, E.: The optimum choice of the instant for stopping a Markov process. Sov. Math. Dokl. **4**, 627–629 (1963)
8. Feldman, M., Svensson, O., Zenklusen, R.: A simple O(log log(rank))-competitive algorithm for the matroid secretary problem. In: Proc. 26th Symp. Discrete Algorithms (SODA), pp. 1189–1201 (2015)
9. Feldman, M., Tennenholtz, M.: Interviewing secretaries in parallel. In: Proc. 13th Conf. Electronic Commerce (EC), pp. 550–567 (2012)
10. Göbel, O., Hoefer, M., Kesselheim, T., Schleiden, T., Vöcking, B.: Online independent set beyond the worst-case: secretaries, prophets, and periods. In: Esparza, J., Fraigniaud, P., Husfeldt, T., Koutsoupias, E. (eds.) ICALP 2014, Part II. LNCS, vol. 8573, pp. 508–519. Springer, Heidelberg (2014)
11. Im, S., Wang, Y.: Secretary problems: Laminar matroid and interval scheduling. In: Proc. 22nd Symp. Discrete Algorithms (SODA), pp. 1265–1274 (2011)
12. Immorlica, N., Kalai, A., Lucier, B., Moitra, A., Postlewaite, A., Tennenholtz, M.: Dueling algorithms. In: Proc. 43rd Symp. Theory of Computing (STOC), pp. 215–224 (2011)
13. Immorlica, N., Kleinberg, R.D., Mahdian, M.: Secretary problems with competing employers. In: Spirakis, P.G., Mavronicolas, M., Kontogiannis, S.C. (eds.) WINE 2006. LNCS, vol. 4286, pp. 389–400. Springer, Heidelberg (2006)
14. Jaillet, P., Soto, J.A., Zenklusen, R.: Advances on matroid secretary problems: free order model and laminar case. In: Goemans, M., Correa, J. (eds.) IPCO 2013. LNCS, vol. 7801, pp. 254–265. Springer, Heidelberg (2013)
15. Karande, C., Mehta, A., Tripathi, P.: Online bipartite matching with unknown distributions. In: Proc. 43rd Symp. Theory of Computing (STOC), pp. 587–596 (2011)
16. Kesselheim, T., Radke, K., Tönnis, A., Vöcking, B.: An optimal online algorithm for weighted bipartite matching and extensions to combinatorial auctions. In: Bodlaender, H.L., Italiano, G.F. (eds.) ESA 2013. LNCS, vol. 8125, pp. 589–600. Springer, Heidelberg (2013)
17. Kesselheim, T., Radke, K., Tönnis, A., Vöcking, B.: Primal beats dual on online packing LPs in the random-order model. In: Proc. 46th Symp. Theory of Computing (STOC), pp. 303–312 (2014)
18. Kleinberg, R.: A multiple-choice secretary algorithm with applications to online auctions. In: Proc. 16th Symp. Discrete Algorithms (SODA), pp. 630–631 (2005)
19. Korula, N., Pál, M.: Algorithms for secretary problems on graphs and hypergraphs. In: Albers, S., Marchetti-Spaccamela, A., Matias, Y., Nikoletseas, S., Thomas, W. (eds.) ICALP 2009, Part II. LNCS, vol. 5556, pp. 508–520. Springer, Heidelberg (2009)
20. Lachish, O.: O(log log rank) competitive ratio for the matroid secretary problem. In: Proc. 55th Symp. Foundations of Computer Science (FOCS), pp. 326–335 (2014)
21. Lindley, D.: Dynamic programming and decision theory. Applied Statistics **10**, 39–51 (1961)
22. Mehta, A., Saberi, A., Vazirani, U., Vazirani, V.: Adwords and generalized online matching. J. ACM **54**(5) (2007)
23. Molinaro, M., Ravi, R.: Geometry of online packing linear programs. Math. Oper. Res. **39**(1), 46–59 (2014)
24. Soto, J.: Matroid secretary problem in the random-assignment model. SIAM J. Comput. **42**(1), 178–211 (2013)

A Simple and Optimal Ancestry Labeling Scheme for Trees

Søren Dahlgaard, Mathias Bæk Tejs Knudsen, and Noy Rotbart$^{(\boxtimes)}$

Department of Computer Science, University of Copenhagen,
Universitetsparken 5, 2100 Copenhagen, Denmark
{soerend,knudsen,noyro}@di.ku.dk

Abstract. We present a $\lg n + 2 \lg \lg n + 3$ ancestry labeling scheme for trees. The problem was first presented by Kannan et al. [STOC 88'] along with a simple $2 \lg n$ solution. Motivated by applications to XML files, the label size was improved incrementally over the course of more than 20 years by a series of papers. The last, due to Fraigniaud and Korman [STOC 10'], presented an asymptotically optimal $\lg n + 4 \lg \lg n + O(1)$ labeling scheme using non-trivial tree-decomposition techniques. By providing a framework generalizing interval based labeling schemes, we obtain a simple, yet asymptotically optimal solution to the problem. Furthermore, our labeling scheme is attained by a small modification of the original $2 \lg n$ solution.

1 Introduction

The concept of *labeling schemes*, introduced by Kannan, Naor and Rudich [16], is a method to assign bit strings, or labels, to the vertices of a graph such that a query between vertices can be inferred directly from the assigned labels, without using a centralized data structure. A labeling scheme for a family of graphs $\mathcal{F}$ consists of an *encoder* and a *decoder*. Given a graph $G \in \mathcal{F}$, the encoder assigns labels to each node in G, and the decoder can infer the query given only a set of labels. The main quality measure for a labeling scheme is the size of the largest label size it assigns to a node of any graph of the entire family. One of the most well studied questions in the context of labeling schemes is the *ancestry problem*. An ancestry labeling scheme for the family of rooted trees of n nodes $\mathcal{F}$ assigns labels to a tree $T \in \mathcal{F}$ such that given the labels $\ell(u), \ell(v)$ of any two nodes $u, v \in T$, one can determine whether u is an *ancestor* of v in T.

Improving the label size for this question is highly motivated by XML search engines. An XML document can be viewed as a tree and queries over such

S. Dahlgaard—Research partly supported by Mikkel Thorup's Advanced Grant from the Danish Council for Independent Research under the Sapere Aude research career programme.

M. Bæk Tejs Knudsen—Research partly supported by Mikkel Thorup's Advanced Grant from the Danish Council for Independent Research under the Sapere Aude research career programme and the FNU project AlgoDisc - Discrete Mathematics, Algorithms, and Data Structures.

© Springer-Verlag Berlin Heidelberg 2015
M.M. Halldórsson et al. (Eds.): ICALP 2015, Part II, LNCS 9135, pp. 564–574, 2015.
DOI: 10.1007/978-3-662-47666-6_45

documents amount to testing ancestry relations between nodes of these trees [2,10,11]. Search engines process queries using an index structure summarizing the ancestor relations. It is imperative to the performance of such engines, that as much as possible of this index can reside in main memory. The big size of web data thus implies that even a small reduction in label size may significantly improve the memory cost and performance. A more detailed explanation can be found in [1].

One solution to this problem, dating back to at least '74 by Tarjan [19] and used in the seminal paper in '92 by Kannan et al. [16] is the following: Given an n node tree T rooted in r, perform a DFS traversal and for each $u \in T$ let dfs(u) be the index of u in the traversal. Assign the label of u as the pair $\ell(u) = (\text{dfs}(u), \text{dfs}(v))$, where v is the descendant of u of largest index in the DFS traversal. Given two labels $\ell(u) = (\text{dfs}(u), \text{dfs}(v))$ and $\ell(w)$, u is an ancestor of w iff dfs(u) $\leq$ dfs(w) $\leq$ dfs(v). In other words the label $\ell(u)$ represents an interval, and ancestry is determined by interval containment. Since every DFS index is a number in $\{1, \ldots, n\}$ The size of a label assigned by this labeling scheme is at most $2 \lg n$.[1]

This labeling scheme, denoted CLASSIC from hereon, was the first in a long line of research to minimize the label size to $1.5 \lg n$ [3], $\lg n + O(\lg n / \lg \lg n)$ [20][2], $\lg n + O(\sqrt{\lg n})$ [1,8] and finally $\lg n + 4 \lg \lg n + O(1)$ [14], essentially matching a lower bound of $\lg n + \lg \lg n - O(1)$ [4]. Additional results for this labeling scheme are a $\lg n + O(\lg \delta)$ labeling scheme for trees of depth at most δ [13] and investigation on a dynamic variant with some pre-knowledge on the structure of the constructed tree [10]. The asymptotically optimal labeling scheme [14] also implies the existence of a *universal poset* of size $O(n^k \lg^{4k} n)$.

Labeling schemes for other functions were considered. Among which are adjacency [7,9,16], routing [12,20], nearest common ancestor [5,6] connectivity [18], distance [15], and flow [17].

1.1 Our Contribution

We present a *simple* ancestry labeling scheme of size $\lg n + 2 \lg \lg n + 3$. Similarly to [1,14] our labeling scheme is based on assigning an interval to each node of the tree. Our labeling scheme can be seen as an extension of the CLASSIC scheme described above, with the key difference that rather than storing the exact size of the interval, we store only an approximation thereof. In order to store only this approximation, our scheme assigns intervals larger than needed, forcing us to use a range larger than $1, \ldots, n$. Our main technical contribution is to minimize the label size by balancing the approximation ratio with the size of the range required to accommodate the resulting labels. While it is a challenge to prove the mathematical properties of our labeling scheme, describing it can be done in a few lines of pseudocode.

---

[1] Throughout this paper we use $\lg n$ to denote the base 2 logarithm $\lg_2 n$.
[2] The paper discussed proves this bound on routing, which can be used to determine ancestry.

The simplicity of our labeling scheme contrast the labeling scheme of [14] , which relies among other things, on a highly nontrivial tree decomposition that must precede it's encoding. As a concrete example, our encoder can be implemented using a single DFS traversal, and results in a label size of $\lg n + 2\lg\lg n + O(1)$ bits compared to $\lg n + 4\lg\lg n + O(1)$. However, for trees of constant depth, our scheme has size $\lg n + \lg\lg n + O(1)$, which is worse than the bound achieved in [13].

The paper is structured as follows: In Section 2 we present a general framework for describing ancestry labeling schemes based on the notion of *left-including intervals*, which we introduce in Definition 1. Lemmas 1 and 2 show the correctness of all labeling schemes which construct labels by assigning *left-including intervals* to nodes. We illustrate the usefulness of this framework by using it to describe CLASSIC in Section 3. Finally, In Section 4 we use the framework to construct our new approximation-based ancestry labeling scheme.

The proofs of Lemmas 1 to 3 are deferred to the full version.

1.2   Preliminaries

We use the notation $[n] = \{0, 1, \ldots, n-1\}$ and denote the concatenation of two bit strings a and b by $a \circ b$. Let $T = (V, E)$ be a tree rooted in r, where $u, v \in V$. The node u is an *ancestor* of v if u lies on the unique path from the root to v, and we call v a *descendant* of u iff u is an ancestor of v. We denote the subtree rooted in u as T_u, i.e. the tree consisting of all descendants of u, and stress that a node is both an ancestor and descendant of itself. The encoding and decoding running time are described under the word-RAM model.

We denote the *interval* assigned to a node u by $I(u) = [a(u), b(u)]$, where $a(u)$ and $b(u)$ denote the lower and upper part of the interval, respectively. We also define $\bar{a}(u)$ and $\bar{b}(u)$ to be the maximum value of $a(v)$ respectively $b(v)$, where v is a descendant of u (note that this includes u itself). We will use the following notion:

Definition 1. *Let T be a rooted tree and I an interval assignment defined on $V(T)$. We say that the interval assignment I is* left-including *if for each $u, v \in T$ it holds that u is an ancestor of v iff $a(v) \in I(u)$.*

In contrast to Definition 1, the literature surveyed [3, 8, 14, 16] considers intervals where u is an ancestor of v iff $I(v) \subseteq I(u)$, i.e. the interval of a descendant node is fully contained in the interval of the ancestor. This distinction is amongst the unused leverage points which we will use to arrive at our new labeling scheme.

2   A Framework for Interval Based Labeling Schemes

In this section we introduce a framework for assigning intervals to tree nodes. We will see in Sections 3 and 4 how this framework can be used to describe ancestry labeling schemes. The framework relies heavily on the values defined in Section 1.2, namely $a(u), b(u), \bar{a}(u), \bar{b}(u)$. An illustration of these values is found

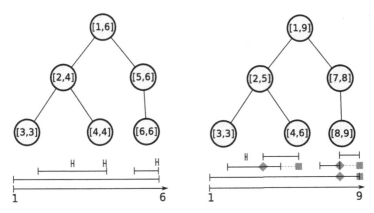

Fig. 1. Two examples of left-including interval assignments to a tree. Left: a left-including assignment as used for CLASSIC in the introduction corresponding to $b(u) = \overline{a}(u)$. Right: a different left-including assignment for the same tree. For internal nodes where $\overline{a}(u)$ and $\overline{b}(u)$ do not coincide with $b(u)$, we have marked these by a gray diamond and square respectively.

in Figure 1 below. The interval $[\overline{a}(u), \overline{b}(u)]$ can be seen as a slack interval from which $b(u)$ can be chosen. This will prove useful in Section 4.

The following lemmas contain necessary and sufficient conditions for interval assignments satisfying the left inclusion property.

Lemma 1. *Let T be a rooted tree and I a left-including interval assignment defined on $V(T)$. Then the following is true:*

1. *For each $u \in T$, $b(u) \geq \overline{a}(u)$.*
2. *For each $u \in T$ and $v \in T_u \setminus \{u\}$ a descendant of u, $a(v) > a(u)$.*
3. *For each $u \in T$, $[a(u), \overline{b}(u)] = \bigcup_{v \in T_u} I(v) = \bigcup_{v \in T_u} [a(v), b(v)]$*
4. *For any two distinct nodes $u, v \in T$ such that u is not an ancestor of v and v is not an ancestor of u the intervals $[a(u), \overline{b}(u)]$ and $[a(v), \overline{b}(v)]$ are disjoint.*

Lemma 2. *Let T be a rooted tree and I an interval assignment defined on $V(T)$. If the following conditions are satisfied, then I is a left-including interval assignment.*

i *For each $u \in T$, $b(u) \geq \overline{a}(u)$.*
ii *For each $u \in T$ and $v \in T$ a child of u, $a(v) > a(u)$.*
iii *For any two siblings $u, v \in T$ the intervals $[a(u), \overline{b}(u)]$ and $[a(v), \overline{b}(v)]$ are disjoint.*

2.1 The Framework

We now consider a general approach for creating left-including interval assignments. For a node $u \in T$ and a positive integer t we define the procedure ASSIGN(u, t) that assigns intervals to T_u recursively and in particular, assigns $a(u) = t$. For pseudocode of the procedure see Algorithm 1.

Algorithm 1. Assigning intervals to all nodes in the subtree T_u rooted at u ensuring $a(u) = t$.

1: **procedure** ASSIGN(u, t)
2:     $(a(u), \bar{a}(u), b(u), \bar{b}(u)) \leftarrow (t, t, t, t)$
3:     **for** $v \in children(u)$ **do**
4:         ASSIGN($v, \bar{b}(u) + 1$)
5:         $(\bar{a}(u), \bar{b}(u)) \leftarrow (\bar{a}(v), \bar{b}(v))$
6:     Assign $b(u)$ such that $b(u) \geq \bar{a}(u)$.
7:     $\bar{b}(u) \leftarrow \max\{b(u), \bar{b}(u)\}$

Algorithm 1 provides a general framework for assigning intervals using a depth-first traversal. We can use it to design an actual interval assignment by specifying: (1) the way we choose $b(u)$, and (2) the order in which the children are traversed. These specifications correspond to Algorithm 1 and Algorithm 1, respectively, and determine entirely the way the intervals are assigned. It may seem counter-intuitive to pick $b(u) > \bar{a}(u)$, but we will show that doing so in a systematic way, we are able to describe the interval using fewer bits by limiting the choices for $b(u)$. In the remainder of this paper, we will see how these two decisions impact also the label size, and produce our claimed labeling scheme.

We now show that any ordering of the children and any way of choosing $b(u)$ satisfying $b(u) \geq \bar{a}(u)$ generates a left-including interval assignment.

Lemma 3. *Let T be a tree rooted in r. After running Algorithm 1 with* ASSIGN($r, 0$) *the values of $\bar{a}(u), \bar{b}(u)$ are correct, i.e. for all $u \in T$:*

$$\bar{a}(u) = \max_{v \in T_u}\{a(v)\}, \quad \bar{b}(u) = \max_{v \in T_u}\{b(v)\}.$$

The following Lemma is useful for showing several properties in the framework.

Lemma 4. *Let u be a node in a tree T with children $v_1 \ldots v_k$. After running Algorithm 1 with parameters* ASSIGN($r, 0$) *where $v_1 \ldots v_k$ are processed in that order, the following properties hold:*

1. $\bar{b}(u) - a(u) + 1 = \left(\sum_{i=1}^{k} \bar{b}(v_i) - a(v_i) + 1\right) + 1$.
2. $\bar{a}(u) - a(u) + 1 = \bar{a}(v_k) - a(v_k) + \left(\sum_{i=1}^{k-1} \bar{b}(v_i) - a(v_i) + 1\right) + 1$.

Proof. By the definition of ASSIGN we see that for all $i = 1, \ldots, k-1$, $a(v_{i+1}) = 1 + \bar{b}(v_i)$. Furthermore $a(v_1) = a(u) + 1$ and $\bar{b}(v_k) = \bar{b}(u)$. Hence:

$$\bar{b}(u) - a(u) + 1 = \bar{b}(v_k) - a(v_1) + 2$$

$$= \left(\sum_{i=2}^{k} \bar{b}(v_i) - \bar{b}(v_{i-1})\right) + \bar{b}(v_1) - a(v_1) + 2$$

$$= \left(\sum_{i=1}^{k} \bar{b}(v_i) - a(v_i) + 1\right) + 1.$$

The second equality follows by the same line of argument.   □

Theorem 1. *Let T be a tree rooted in r. After running Algorithm 1 with parameters* ASSIGN$(r, 0)$ *the set of intervals produced are left-including.*

Proof. Consider any node $u \in T$ and a call ASSIGN(u, t). We will prove each of the conditions of Lemma 2, which implies the theorem.

   i This condition is trivially satisfied by Algorithm 1.
  iii First, observe that any interval assigned to a node w by a call to ASSIGN(v, t) has $a(w) \geq t$, and by i it has $\bar{b}(w) \geq b(w) \geq a(w)$. Let $v_1, \ldots, v_k$ be the children in the order of the for loop in Algorithm 1. By Algorithms 1 to 1 we have $a(v_1) = a(u) + 1, a(v_2) = \bar{b}(v_1) + 1, a(v_3) = \bar{b}(v_2) + 1, \ldots, a(v_k) = \bar{b}(v_{k-1}) + 1$, thus the condition is satisfied.
  ii The first child v of u has $a(v) = t + 1 = a(u) + 1$. By the same line of argument as in *iii* we see that all other children w of u must have $a(w) > a(v) = a(u) + 1$.   □

3   The Classic Ancestry Labeling Scheme

To get acquainted with the framework of Section 2, we use it to redefine CLASSIC, the labeling scheme introduced in Section 1.

Let T be a tree rooted in r. We first modify the function ASSIGN to create ASSIGN-CLASSIC such that the intervals $I(u) = [a(u), b(u)]$ correspond to the intervals of the CLASSIC algorithm described in the introduction. To do this we set $b(u) = \bar{a}(u)$ in Algorithm 1 and traverse the children in any order in Algorithm 1. We note that there is a clear distinction between an algorithm such as ASSIGN-CLASSIC and an encoder. This distinction will be more clear in Section 4. We will need the following lemma to describe the encoder.

Lemma 5. *After* ASSIGN-CLASSIC(u, t) *is called the following invariant is true:*

$$\bar{b}(u) - a(u) + 1 = |T_u|$$

Proof. We prove the claim by induction on $|T_u|$. When $|T_u| = 1$ u is a leaf and hence $\bar{b}(u) = a(u) = t$ and the claim holds.

Let $|T_u| = m > 1$ and assume that the claim holds for all nodes with subtree size $< m$. Let $v_1, \ldots, v_k$ be the children of u. By Lemma 4 and the induction hypothesis we have:

$$\bar{b}(u) - a(u) + 1 = \left(\sum_{i=1}^{k} \bar{b}(v_i) - a(v_i) + 1 \right) + 1$$

$$= \left(\sum_{i=1}^{k} |T_{v_i}| \right) + 1 = |T_u|.$$

This completes the induction.   □

Description of the Encoder: Let T be an n-node tree rooted in r. We first invoke a call to ASSIGN-CLASSIC$(r, 0)$. By Lemma 5 we have $\bar{b}(r) - a(r) + 1 = n$ and this implies $0 \leq a(u), b(u) \leq n - 1$ for every $u \in T$. Let x_u and y_u be the encoding of $a(u)$ and $b(u)$ using exactly[3] $\lceil \lg n \rceil$ bits respectively. We set the label of u to be the concatenation of the two bitstrings, i.e. $\ell(u) = x_u \circ y_u$.

Description of the Decoder: Let $\ell(u)$ and $\ell(v)$ be the labels of the nodes u and v in a tree T. By the definition of the encoder, the labels have the same size and it is $2z$ for some integer $z \geq 1$. Let $\ell(u) = x_u \circ y_u$ where x_u and y_u are the first and last z bits of $\ell(u)$ respectively. Let a_u and b_u the integers from $[2^z]$ corresponding to the bit strings x_u and y_u respectively. We define a_v and b_v analogously. The decoder responds `True`, i.e. that u is the ancestor of v, iff $a_v \in [a_u, b_u]$.

The correctness of the labeling scheme follows from Theorem 1 and the description of the decoder.

4   An Approximation-Based Approach

In this section we present the main result of this paper:

Theorem 2. *There exist an ancestry labeling scheme of size* $\lceil \lg n \rceil + 2 \lceil \lg \lg n \rceil + 3$.

To prove this theorem, we use the framework introduced in Section 2. The barrier in reducing the size of the CLASSIC labeling scheme is that the number of different intervals that can be assigned to a node is $\Theta(n^2)$. It is impossible to encode so many different intervals without using at least $2 \lg n - O(1)$ bits. The challenge is therefore to find a smaller set of intervals $I(u) = [a(u), b(u)]$ to assign to the nodes. First, note that Lemma 1 points 2 and 4 imply that any two nodes u, v must have $a(u) \neq a(v)$. By considering the n node tree T rooted in r where r has $n - 1$ children, we also see that there must be at least $n - 1$ different values of $b(u)$ (by Lemma 1 point 4). One might think that this implies the need for $\Omega(n^2)$ different intervals. This is, however, not the case. We consider a family of intervals, such that $a(u) = O(n)$ and the size of each interval, $b(u) - a(u) + 1$, comes from a much smaller set, S. Since there are $O(n |S|)$ such intervals we are able to encode them using $\lg n + \lg |S| + O(1)$ bits.

We now present a modification of ASSIGN called ASSIGN-NEW. Calling ASSIGN-NEW$(r, 0)$ on an n-node tree T with root r will result in each $a(u) \in [2n]$ and $b(u) \in S$, where S is given by:

$$S = \left\{ \left\lfloor (1 + \varepsilon)^k \right\rfloor \mid k \in \left[4 \lceil \lg n \rceil^2 \right] \right\}, \tag{1}$$

where ε is the unique solution to the equation $\lg(1 + \varepsilon) = (\lceil \lg n \rceil)^{-1}$. First, we examine some properties of S:

---

[3] This can be accomplished by padding with zeros if necessary.

Lemma 6. *Let S be defined as in (1). For every $m \in \{1, 2, \ldots, 2n\}$ there exists $s \in S$ such that:*

$$m \le s < m(1 + \varepsilon) \ .$$

Furthermore, $s = \lfloor (1 + \varepsilon)^k \rfloor$ for some $k \in \left[4 \lceil \lg n \rceil^2 \right]$, and both s and k can be computed in $O(1)$ time.

Proof. Fix $m \in \{1, 2, \ldots, 2n\}$. Let k be the largest integer such that $(1 + \varepsilon)^{k-1} < m$. Equivalently, k is the largest integer such that:

$$k - 1 < \frac{\lg m}{\lg (1 + \varepsilon)} = (\lg m) \cdot \lceil \lg n \rceil \ .$$

In other words we choose k as $\lceil (\lg m) \cdot \lceil \lg n \rceil \rceil$ and note that k is computed in $O(1)$ time. Since $\lg m \le \lg(2n) \le 2 \lg n$:

$$k \le \lceil 2(\lg n) \cdot \lceil \lg n \rceil \rceil \le 2 \lceil \lg n \rceil^2 < 4 \lceil \lg n \rceil^2 \ .$$

By setting $s = \lfloor (1 + \varepsilon)^k \rfloor$ we have $s \in S$. By the definition of k we see that $(1 + \varepsilon)^k \ge m$ and thus also $s \ge m$. Similarly:

$$m(1 + \varepsilon) > (1 + \varepsilon)^{k-1} \cdot (1 + \varepsilon) = (1 + \varepsilon)^k \ge s.$$

This proves that $s \in S$ satisfies the desired requirement. Furthermore s can be computed in $O(1)$ time by noting that:

$$s = \lfloor (1 + \varepsilon)^k \rfloor = \left\lfloor 2^{\lg(1+\varepsilon)k} \right\rfloor = \left\lfloor 2^{\lceil \lg n \rceil^{-1} \cdot k} \right\rfloor .$$

$\square$

We now define ASSIGN-NEW by modifying ASSIGN in the following two ways. First, we specify the order in which the children are traversed in Algorithm 1. This is done in non-decreasing order of their subtree size, i.e. we iterate $v_1, \ldots, v_k$, where $|T_{v_1}| \le \ldots \le |T_{v_k}|$. Second, we choose $b(u)$ in Algorithm 1 as the smallest value, such that $b(u) \ge \bar{a}(u)$ and $b(u) - a(u) + 1 \in S$. This is done by using Lemma 6 with $m = \bar{a}(u) - a(u) + 1$ and setting $b(u) = a(u) + s - 1$. In order to do this we must have $m \le 2n$. To do this, we show the following lemma corresponding to Lemma 5 in Section 3.

Lemma 7. *After ASSIGN-NEW(u, t) is called the following invariants holds:*

$$\bar{a}(u) - a(u) + 1 \le |T_u| (1 + \varepsilon)^{\lfloor \lg |T_u| \rfloor} \tag{2}$$

$$\bar{b}(u) - a(u) + 1 \le |T_u| (1 + \varepsilon)^{\lfloor \lg |T_u| \rfloor + 1} \tag{3}$$

Proof. We prove the claim by induction on $|T_u|$. When $|T_u| = 1$, u is a leaf, so $\bar{b}(u) = \bar{a}(u) = a(u) = t$ and the claim holds.

Now let $|T_u| = m > 1$ and assume that the claim holds for all nodes with subtree size $< m$. Let $v_1, \ldots, v_k$ be the children of u such that $|T_{v_1}| \le \ldots \le |T_{v_k}|$.

First, we show that (2) holds. By Lemma 4 we have the following expression for $\overline{a}(u) - a(u) + 1$:

$$\overline{a}(u) - a(u) + 1 = (\overline{a}(v_k) - a(v_k) + 1) + \left(\sum_{i=1}^{k-1} \overline{b}(v_i) - a(v_i) + 1 \right) + 1. \quad (4)$$

It follows from the induction hypothesis that:

$$\overline{a}(v_k) - a(v_k) + 1 \leq |T_{v_k}| \, (1 + \varepsilon)^{\lfloor \lg |T_{v_k}| \rfloor} \leq |T_{v_k}| \, (1 + \varepsilon)^{\lfloor \lg |T_u| \rfloor}. \quad (5)$$

Furthermore, by the ordering of the children, we have $\lg |T_{v_i}| \leq \lg |T_u| - 1$ for every $i = 1, \ldots, k - 1$. Hence:

$$\overline{b}(v_i) - a(v_i) + 1 \leq |T_{v_i}| \, (1 + \varepsilon)^{\lfloor \lg |T_{v_i}| \rfloor + 1} \leq |T_{v_k}| \, (1 + \varepsilon)^{\lfloor \lg |T_u| \rfloor}. \quad (6)$$

Inserting (5) and (6) into (4) proves invariant (2).

Since $\overline{b}(u) = \max \{\overline{b}(v_k), b(u)\}$ we only need to upper bound $\overline{b}(v_k) - a(u) + 1$ and $b(u) - a(u) + 1$. First we note that since $b(u)$ is chosen smallest possible such that $b(u) \geq \overline{a}(u)$ and $b(u) - a(u) + 1 \in S$, it is guaranteed by Lemma 6 that:

$$b(u) - a(u) + 1 < (1 + \varepsilon) \, (\overline{a}(u) - a(u) + 1) \leq |T_u| \, (1 + \varepsilon)^{\lfloor \lg |T_u| \rfloor + 1}$$

Hence we just need to upper bound $\overline{b}(v_k) - a(u) + 1$. First we note that just as in (4):

$$\overline{b}(v_k) - a(u) + 1 = \left(\sum_{i=1}^{k} \overline{b}(v_i) - a(v_i) + 1 \right) + 1 \quad (7)$$

By the induction hypothesis, for every $i = 1, \ldots, k$:

$$\overline{b}(v_i) - a(v_i) + 1 \leq |T_{v_i}| \, (1 + \varepsilon)^{\lfloor \lg |T_{v_i}| \rfloor + 1} \leq |T_{v_i}| \, (1 + \varepsilon)^{\lfloor \lg |T_u| \rfloor + 1} \quad (8)$$

Inserting (8) into (7) gives the desired:

$$\overline{b}(v_k) - a(u) + 1 \leq 1 + \sum_{i=1}^{k} |T_{v_i}| \, (1 + \varepsilon)^{\lfloor \lg |T_u| \rfloor + 1} \leq |T_u| \, (1 + \varepsilon)^{\lfloor \lg |T_u| \rfloor + 1}.$$

This completes the induction. □

By Lemma 7 we see that for a tree T with n nodes and $u \in T$:

$$\overline{a}(u) - a(u) + 1 \leq |T_u| \, (1 + \varepsilon)^{\lfloor \lg |T_u| \rfloor} \leq n \cdot 2^{\lfloor \lg n \rfloor \lg(1+\varepsilon)} \leq n \cdot 2^1 = 2n.$$

In particular, for any $u \in T$ we see that $a(u) \leq 2n$, and by Lemma 6 the function ASSIGN-NEW is well-defined.

We are now ready to describe the labeling scheme:

Description of the Encoder: Given an n-node tree T rooted in r, the encoding algorithm works by first invoking a call to APPROX-NEW$(r, 0)$. Recall that by Lemma 6 we find $b(u)$ such that $b(u) - a(u) + 1 = \lfloor (1 + \varepsilon)^k \rfloor$ as well as the value of k in $O(1)$ time. For a node u, denote the value of k by $k(u)$ and let x_u and y_u be the bit strings representing $a(u)$ and $k(u)$ respectively, consisting of exactly $\lceil \lg(2n) \rceil$ and $\left\lceil \lg(4 \lceil \lg n \rceil^2) \right\rceil$ bits (padding with zeroes if necessary). This is possible since $a(u) \in [2n]$ and $k(u) \in \left[4 \lceil \lg n \rceil^2 \right]$.

For each node $u \in T$ we assign the label $\ell(u) = x_u \circ y_u$. Since

$$\lceil \lg(2n) \rceil = 1 + \lceil \lg n \rceil, \quad \left\lceil \lg(4 \lceil \lg n \rceil^2) \right\rceil = 2 + \lceil 2 \lg(\lceil \lg n \rceil) \rceil = 2 + \lceil 2 \lg \lg n \rceil,$$

the label size of this scheme is $\lceil \lg n \rceil + \lceil 2 \lg \lg n \rceil + 3$.

Description of the Decoder: Let $\ell(u)$ and $\ell(v)$ be the labels of the nodes u and v in a tree T. By the definition of the encoder the labels have the same size and it is $s = z + \lceil 2 \lg z \rceil + 3$ for some integer $z \geq 1$. By using that $s - \lceil 2 \lg s \rceil - 3 = z - O(1)$ we can compute z in $O(1)$ time. We know that the number of nodes n in T satisfies $\lceil \lg n \rceil = z$. We can therefore define ε to be the unique solution to $\lg(1 + \varepsilon) = \lceil \lg n \rceil^{-1} = z^{-1}$. Let x_u and y_u be the first $z + 1$ bits and last $\lceil 2 \lg z \rceil + 2$ bits of $\ell(u)$ respectively. We let a_u and k_u be the integers in $[2^{z+1}]$ and $[4z^2]$ corresponding to the bit strings x_u and y_u respectively. We define s_u as $\lfloor (1 + \varepsilon)^{k_u} \rfloor$ and $b_u = s_u + a_u - 1$. We define a_v, b_v, k_v, s_v analogously. The decoder responds **True**, i.e. that u is the ancestor of v, iff $a_v \in [a_u, b_u]$.

Theorem 2 is now achieved by using the labeling scheme described above. Correctness follows from Theorem 1.

References

1. Abiteboul, S., Alstrup, S., Kaplan, H., Milo, T., Rauhe, T.: Compact labeling scheme for ancestor queries. SIAM J. Comput. 35(6), 1295–1309 (2006)
2. Abiteboul, S., Buneman, P., Suciu, D.: Data on the Web: From Relations to Semistructured Data and XML. Morgan Kaufmann (1999)
3. Abiteboul, S., Kaplan, H., Milo, T.: Compact labeling schemes for ancestor queries. In: Proc. 25th ACM-SIAM Symposium on Discrete Algorithms, pp. 547–556 (2001)
4. Alstrup, S., Bille, P., Rauhe, T.: Labeling schemes for small distances in trees. SIAM J. Discret. Math. 19(2), 448–462 (2005)
5. Alstrup, S., Gavoille, C., Kaplan, H., Rauhe, T.: Nearest common ancestors: A survey and a new distributed algorithm. pp. 258–264. ACM Press (2002)
6. Alstrup, S., Halvorsen, E.B., Larsen, K.G.: Near-optimal labeling schemes for nearest common ancestors. In: SODA, pp. 972–982 (2014)
7. Alstrup, S., Kaplan, H., Thorup, M., Zwick, U.: Adjacency labeling schemes and induced-universal graphs. CoRR, abs/1404.3391 (2014) (to appear in STOC 2015)
8. Alstrup, S., Rauhe, T.: Improved labeling scheme for ancestor queries. In: Proc. 13th ACM-SIAM Symposium on Discrete Algorithms, pp, 947–953 (2002)
9. Alstrup, S., Rauhe, T.: Small induced-universal graphs and compact implicit graph representations. In: FOCS 2002, pp. 53–62 (2002)

10. Cohen, E., Kaplan, H., Milo, T.: Labeling dynamic xml trees. SIAM Journal on Computing **39**(5), 2048–2074 (2010)
11. Deutsch, A., Fernández, M.F., Florescu, D., Levy, A.Y., Suciu, D.: A query language for XML. Computer Networks **31**(11–16), 1155–1169 (1999)
12. Fraigniaud, P., Gavoille, C.: Routing in Trees. In: Orejas, F., Spirakis, P.G., van Leeuwen, J. (eds.) ICALP 2001. LNCS, vol. 2076, pp. 757–772. Springer, Heidelberg (2001)
13. Fraigniaud, P., Korman, A.: Compact ancestry labeling schemes for xml trees. In: Proc. 21st ACM-SIAM Symp. on Discrete Algorithms (SODA) (2010)
14. Fraigniaud, P., Korman, A.: An optimal ancestry scheme and small universal posets. In: STOC 2010, pp. 611–620 (2010)
15. Gavoille, C., Peleg, D., Pérennesc, S., Razb, R.: Distance labeling in graphs. Journal of Algorithms **53**, 85–112 (2004)
16. Kannan, S., Naor, M., Rudich, S.: Implicit representation of graphs. SIAM Journal on Discrete Mathematics, 334–343 (1992)
17. Katz, M., Katz, N.A., Korman, A., Peleg, D.: Labeling schemes for flow and connectivity. SIAM Journal on Computing **34**(1), 23–40 (2004)
18. Korman, A.: Labeling schemes for vertex connectivity. ACM Transactions on Algorithms **6**(2) (2010)
19. Robert Endre Tarjan: Finding dominators in directed graphs. SIAM J. Comput. **3**(1), 62–89 (1974)
20. Thorup, M., Zwick, U.: Compact routing schemes. In: SPAA 2001, pp. 1–10 (2001)

Interactive Communication with Unknown Noise Rate

Varsha Dani[1], Mahnush Movahedi[1]([✉]), Jared Saia[1], and Maxwell Young[2]

[1] University of New Mexico, Albuquerque, USA
{movahedi,saia}@cs.unm.edu
[2] Drexel University, Philadelphia, USA
myoung@cs.drexel.edu

Abstract. Alice and Bob want to run a protocol over an noisy channel, where a certain number of bits are flipped adversarially. Several results take a protocol requiring L bits of noise-free communication and make it robust over such a channel. In a recent breakthrough result, Haeupler described an algorithm that sends a number of bits that is conjectured to be near optimal in such a model. However, his algorithm critically requires *a priori* knowledge of the number of bits that will be flipped by the adversary.

We describe an algorithm requiring no such knowledge. If an adversary flips T bits, our algorithm sends $L + O\left((T + \sqrt{LT + L})\log(LT + L)\right)$ bits in expectation and succeeds with high probability in L. It does so without any *a priori* knowledge of T. Assuming a conjectured lower bound by Haeupler, our result is optimal up to logarithmic factors.

Our algorithm critically relies on the assumption of a private channel. We show that privacy is necessary when the amount of noise is unknown.

1 Introduction

How can two parties run a protocol over a noisy channel? Interactive communication seeks to solve this problem while minimizing the total number of bits sent. Recently, Haeupler [12] gave an algorithm for this problem that is conjectured to be optimal. However, as in previous work [1,2,4,6,9–11,17], his algorithm critically relies on the assumption that the algorithm knows the noise rate in advance, *i.e.*, the algorithm knows in advance the number of bits that will be flipped by the adversary.

In this paper, we remove this assumption. To do so, we add a new assumption of privacy. In particular, in our model, an adversary can flip an unknown number of bits, at arbitrary times, but he never learns the value of any bits sent over the channel. This assumption is necessary: with a public channel and unknown noise rate, the adversary can run a man-in-the-middle attack to mislead either party (see Theorem 3, Section 4).

This research was supported in part by NSF grants CNS-1318294 and CCF-1420911. An extended version of the paper is available at http://arxiv.org/abs/1504.06316

M.M. Halldórsson et al. (Eds.): ICALP 2015, Part II, LNCS 9135, pp. 575–587, 2015.
DOI: 10.1007/978-3-662-47666-6_46

Problem Overview. We assume that Alice and Bob are connected by a noisy binary channel. Our goal is to build an algorithm that takes as input some distributed protocol π that works over a noise-free channel and outputs a distributed protocol π' that works over the noisy channel.

We assume an adversary chooses π, and which bits to flip in the noisy channel. The adversary knows our algorithm for transforming π to π'. However, he neither knows the private random bits of Alice and Bob, nor the bits sent over the channel, except when it is possible to infer these from knowledge of π and our algorithm.

We let T be the number of bits flipped by the adversary, and L be the length of π. As in previous work, we assume that Alice and Bob know L.

Our Results. Our main result is summarized in the following theorem.

Theorem 1. *Algorithm 1 (Section 2) tolerates an unknown number of adversarial errors, T, succeeds with high probability in the transcript length[1], L, and if successful, sends $L + O\left((T + \sqrt{LT + L})\log(LT + L)\right)$ bits in expectation.*

The number of bits sent by our algorithm is within logarithmic factors of optimal, assuming a conjecture from [12]. Details are in the full version.

In the full version, we also show a relationship between the overhead incurred when $T = 0$ and when $T \geq 1$ for a very general class of algorithms, and derive some near-optimal results under this class.

Challenges. Can we adapt prior results by guessing the noise rate? Underestimation threatens correctness if the actual number of bit flips exceeds the algorithm's tolerance. Conversely, overestimation leads to sending more bits than necessary. Thus, we need a protocol that adapts to the adversary's actions.

One idea is to adapt the amount of communication redundancy based on the number of errors detected thus far. However, this presents a new challenge because the parties may have different views of the number of errors. They will need to synchronize their adaptions over the noisy channel.

Another technical challenge is termination. The length of the simulated protocol is necessarily unknown, so the parties will likely not terminate at the same time. After one party has terminated, it is a challenge for the other party to detect this fact based on bits received over the noisy channel.

1.1   Related Work

For L bits to be transmitted from Alice to Bob, Shannon [19] proposes an error correcting code of size $O(L)$ that yields correct communication over a *noisy* channel with probability $1 - e^{-\Omega(L)}$. At first glance, this may appear to solve our problem. But consider an *interactive* protocol with communication complexity L, where Alice sends one bit, then Bob sends back one bit, and so forth where the value of each bit sent *depends on the previous bits received*. Two problems arise. First, using block codewords is not efficient; to achieve a small error probability,

---

[1] Specifically with probability at least $1 - 1/L^c$, for some constant $c > 0$.

"dummy" bits may be added to each bit prior to encoding, but this results in a superlinear blowup in overhead. Second, due to the interactivity, an error that occurs in the past can ruin all computation that comes after it. Thus, error correcting codes fall short when dealing with interactive protocols.

The seminal work of Schulman [17,18] overcame these obstacles by describing a deterministic method for simulating interactive protocols on noisy channels with only a constant-factor increase in the total communication complexity. This work spurred vigorous interest in the area (see [3] for an excellent survey).

Schulman's scheme tolerates an adversarial noise rate of $1/240$. It critically depends on the notion of a *tree code* for which an exponential-time construction was originally provided. This exponential construction time motivated work on more efficient constructions [4,13,16]. There were also efforts to create alternative codes [9,15]. Recently, elegant computationally-efficient schemes that tolerate a constant adversarial noise rate have been demonstrated [1,10]. Additionally, a large number of powerful results have improved the tolerable adversarial noise rate [2,5,6,8,11].

The closest prior work to ours is that of Haeupler [12]. His work assumes a fixed and known adversarial noise rate ϵ, the fraction of bits flipped by the adversary. Communication efficiency is measured by *communication rate* which is L divided by the total number of bits sent. Haeupler [12] describes an algorithm that achieves a communication rate of $1 - O(\sqrt{\epsilon \log \log(1/\epsilon)})$, which he conjectures to be optimal. We compare our work to his in Section 4.

Feinerman, Haeupler and Korman [7] recently studied the interesting related problem of spreading a rumor of a single bit in a noisy network. In their framework, in each synchronous round, each agent can deliver a single bit to a random anonymous agent. This bit is flipped independently at random with probability $1/2 - \epsilon$ for some fixed $\epsilon > 0$. Their algorithm ensures with high probability that in $O(\log n/\epsilon^2)$ rounds and with $O(n \log n/\epsilon^2))$ messages, all nodes learn the correct rumor. They also present a majority-consensus algorithm with the same resource costs, and prove these resource costs are optimal for both problems.

2   Our Algorithm

In this section, we first provide a formal definition of our model. We then summarize the main components of our algorithm along with the intuition behind our design.

2.1   Formal Model

Our algorithm takes as input a protocol π which is a sequence of L bits, each of which is transmitted either from Alice to Bob or from Bob to Alice. As in previous work, we also assume that Alice and Bob both know L. We let Alice be the party who sends the first bit in π.

Channel Steps. We assume communication over the channel is synchronous and individual computation is instantaneous. We define a *channel step* as the amount of time that it takes to send one bit over the channel.

Silence on the Channel. When neither Alice nor Bob sends in a channel step, we say that the channel is silent. In any contiguous sequence of silent channel steps, the bit received on the channel in the first step is set by the adversary for free. By default, the bit received in subsequent steps of the sequence remains the same, unless the adversary pays for one bit flip in order to change it. In short, the adversary pays a cost of one bit flip each time it wants to change the value of the bit received in any contiguous sequence of silent steps.

2.2 Overview, Notation and Definitions

Our algorithm is presented as Algorithm 1. The overall idea of the algorithm is simple: the parties run the original protocol π for b steps as if there was no noise. Then, they verify whether an error has occurred by checking the fingerprints. Based on the result of this verification procedure, the computation of π either moves forward b steps or is rewound b steps. We now define some notation.

Fingerprints. To verify communication, our algorithm uses randomized hash functions as described in the following well-known theorem.

Theorem 2. *(Naor and Naor [14]) For any positive integer $\mathcal{L}$ and any probability p, there exists a hash function $\mathcal{F}$ that given a uniformly random bit string S as the seed, maps any string of length at most $\mathcal{L}$ bits to a bit string hash value H, such that the collision probability of any two strings is at most p, and the length of S and H are $|S| = \Theta(\log(\mathcal{L}/p))$ and $|H| = \Theta(\log(1/p))$ bits.*

Transcripts. We define Alice's *transcript*, $\mathcal{T_A}$, as the sequence of possible bits of π that Alice has either sent or received up to the current time. Similarly, we let $\mathcal{T_B}$ denote Bob's transcript. For both Alice or Bob, we define a *verified* transcript to be the longest prefix of a transcript for which a verified fingerprint has been received. We denote the verified transcript for Alice as $\mathcal{T_A^*}$, and for Bob as $\mathcal{T_B^*}$.

Blocks. We define a *block* as b contiguous bits of protocol π that are run during Line 8 of Alice's protocol. The value b denotes the *size* of the block which is set by Alice, and decreases based on the number of blocks in which she detects a corruption.

Rounds. We define a *round* as one iteration of the repeat loop in Alice's protocol. This loop starts at Line 2 and ends at Line 12. A round consists of executing b channel steps of π (a block), as well as sending and receiving some fingerprints required for verification purposes.

Other Notation. For a transcript $\mathcal{T}$ and integer x, we define $\mathcal{T}[0,\ldots,x]$ to be the first x bits of $\mathcal{T}$ if $\mathcal{T}$ is of length at least x. Otherwise, we define $\mathcal{T}[0,\ldots,x]$ as *null*. For two strings x and y, we define $x \odot y$ to be the concatenation of x and

y. We define the function **getH** based on Theorem 2. When given a probability p, **getH** returns the tuple $(\mathbf{h}, \ell_h, \ell_s)$, where $\mathbf{h}$ denotes the hash function from Theorem 2; ℓ_h is $|H|$, the size of the output of the hash function; and ℓ_s is $|S|$, the size of the seed. The probability used in **getH** is $\frac{1}{(Lx)^c}$ for some $c \geq 2$.

2.3 Algorithm Design

At the start of each round, Alice chooses a random seed $\mathcal{S}_\mathcal{A}$ and sends 1) $\mathcal{S}_\mathcal{A}$; 2) $\mathcal{H}_\mathcal{A}$ which is the fingerprint of $\mathcal{T}_\mathcal{A}$; and 3) $|\mathcal{T}_\mathcal{A}|$ which is size of $\mathcal{T}_\mathcal{A}$. This allows Bob to determine which block of π they should compute. Next, Alice and Bob continue the computation of π for b steps as if the channel was noiseless. Finally, Bob sends $\mathcal{H}_\mathcal{B}$, which is the hash of $\mathcal{T}_\mathcal{B}$ using the same seed $\mathcal{S}_\mathcal{A}$ previously received from Alice. Alice listens for the hash value from Bob and checks if it matches her own transcript. If she receives a correct hash value, she proceeds to the next round. If she does not, she rewinds the transcript $\mathcal{T}_\mathcal{A}$ by b steps, decreases the value of b, and begins a new round.

Adaptive Block Size. The block size b decreases each time Alice fails to receive a correct fingerprint from Bob. We can use our lower-bound to calculate how to decrease the b during the protocol. To prevent progress in computation of π in a round, it suffices to corrupt a single bit in the round. Drawing intuition from our lower bound, we want our algorithm to send $L + \tilde{\Theta}(\sqrt{LT})$ bits per block. Thus, we initially set $b = \sqrt{L}$, and after t errors, we set $b = \sqrt{L/t}$.

Adapting the block size is critical to achieving a good communication rate, but it raises a new challenge: How can Alice and Bob agree on the current block size? Our solution is to make Alice authoritative on the current block size. Therefore, while Alice knows the start and the end of each block, Bob may not.

To overcome this problem, Alice includes the block size at the start of each block. If Bob detects an error, he enters into a "listening loop". During this time, Bob continually checks all received bits and exits the loop if a valid fingerprint is received. Alice, having not received a correct fingerprint from Bob, will rewind all the bits added to her transcript in the current round and reduce the current block size. Intuitively, this ensures that 1) Alice is never more than one block ahead of Bob; and 2) Bob is always allowed to "catch up" to Alice.

Adapting Fingerprint Size. The execution time of our algorithm depends on the unknown value T. But a fixed-sized fingerprint will fail if T is very large. For example, consider the case where the fingerprint size is always $\Theta(\log L)$ with collision probability of $\frac{1}{L^c}$ for some constant c. The adversary can choose $T = \omega(L^c)$ and flip one bit in each block. Since Alice and Bob must send $\Omega(T)$ blocks, the probability of failure due to hash collision is no longer negligible.

To solve this problem, we adaptively increase the fingerprint size based on the channel step, x. In particular, we use a family of hash functions with $|S| = |H| = \Theta(\log(Lx))$ and probability of failure of $p = \frac{1}{(Lx)^c}$, for some constant $c \geq 2$. This ensures the probability that there is any fingerprint collision is $o(1)$.

Algorithm 1. Interactive Communication

ALICE'S PROTOCOL

Data: $\pi \leftarrow L$-step protocol to be simulated augmented by $2\sqrt{L}$ extra steps of sending random bits; x is the channel step number;

1 **Initialization:**
$\mathcal{T_A} \leftarrow null;\ \mathcal{T_A^*} \leftarrow null;$
$i \leftarrow 1;\ b \leftarrow \sqrt{L};$

2 **repeat**

3     $(\mathbf{h}, \ell_h, \ell_s) \leftarrow \mathbf{getH}\left(\frac{1}{(Lx)^c}\right);$

4     $S_\mathcal{A} \leftarrow \ell_s$ uniformly random bits;

5     $\mathcal{H_A} \leftarrow \mathbf{h}(\mathcal{T_A} \odot |\mathcal{T_A}| \odot b, S_\mathcal{A});$

6   Send $(|\mathcal{T_A}|, b, S_\mathcal{A}, \mathcal{H_A});$

7     $\mathcal{T_A^*} \leftarrow \mathcal{T_A}$;

8   Resume π
      for b steps; record in $\mathcal{T_A};$

9     $\mathcal{H_B'} \leftarrow$ the next ℓ_h bits received;

10   **if** $\mathcal{H_B'} \neq \mathbf{h}(\mathcal{T_A}, S_\mathcal{A})$ **then**

11        $\mathcal{T_A} \leftarrow \mathcal{T_A^*}$;

12    $i \leftarrow i + 1;$

13    $b \leftarrow \max\left(\log L, \sqrt{\frac{L}{i}}\right);$

14 **until** $|\mathcal{T_A}| \geq L + \sqrt{L};$

15 **Terminate and Output:** the outcome of π based on transcript $\mathcal{T_A^*};$

BOB'S PROTOCOL

Data: $\pi \leftarrow L$-step protocol to be simulated augmented by $2\sqrt{L}$ extra steps of listening; x is the channel step number;

1 **Initialization:**
$\mathcal{T_B} \leftarrow null;\ \mathcal{T_B^*} \leftarrow null;$
$(\mathbf{h}, \ell_h, \ell_s) \leftarrow \mathbf{getH}\left(\frac{1}{L^c}\right);$ Listen for $\ell_h + \ell_s + 2\log L - 1$ channel steps ;

2 **while** $TRUE$ **do**

3   **repeat**

4   Listen for 1 step;

5   Find x' such that $x' + \ell_h + \ell_s + 2\log L = x$ and $(\mathbf{h}, \ell_h, \ell_s) = \mathbf{getH}\left(\frac{1}{(Lx')^c}\right);$

6         $(\mathbf{h}, \ell_h, \ell_s) \leftarrow \mathbf{getH}\left(\frac{1}{(Lx')^c}\right);$

7   Set $(t_\mathcal{A}', b', S_\mathcal{A}', \mathcal{H_A'})$ based on last $\ell_h + \ell_s + 2\log L$ bits received;

8   **if** $|\mathcal{T_B^*}| \geq L$ and the last $10\log x$ received bits are the same **then**

9   **Terminate and Output:** the outcome of π based on $\mathcal{T_B^*};$

10   **until** $\mathcal{H_A'} = \mathbf{h}(\mathcal{T_B}[0,\dots,t_\mathcal{A}] \odot t_\mathcal{A}' \odot b', S_\mathcal{A}');$

11        $\mathcal{T_B^*} \leftarrow \mathcal{T_B}[0,\dots,t_\mathcal{A}']$;

12        $\mathcal{T_B} \leftarrow \mathcal{T_B^*};$

13   Resume π for b' steps; record in $\mathcal{T_B};$

14        $\mathcal{H_B} \leftarrow \mathbf{h}(\mathcal{T_B}, S_\mathcal{A}');$

15   Send $\mathcal{H_B};$

    end

Handling Termination. In previous work, since ϵ and L' are known, both parties know when to terminate, and can do so at the same time. However, since we know neither parameter, termination is now more challenging.

In our algorithm, π is padded with $2\sqrt{L}$ additional bits at the end. Each of these bits is set independently and uniformly at random by Alice. Alice terminates when her transcript is of length equal to $L + \sqrt{L}$. Bob terminates when 1) his verified transcript is of length L; and 2) he has received a sequence of $10 \log x$ consecutive bits that are all the same. These conditions ensure that 1) Bob is very unlikely to terminate before Alice; and 2) Bob terminates soon after Alice, unless the adversary pays a significant cost to delay this.

3   Proof of Theorem 1

3.1   Probabilities of Bad Events

Before proceeding to our proof, we bound the probability of two *bad* events.

- Event 1 - Hash Collision. Either Alice or Bob incorrectly validates a fingerprint and updates their verified transcript to include bits not in π.
- Event 2 - Consecutive Bit Match. In the last $2\sqrt{L}$ padded random bits of π sent by Alice, Bob receives $10 \log x$ consecutive bits that are all the same.

Lemma 1. *The probability of Event 1 is $O(1/L^c)$ for any constant $c > 1$.*

Proof. Alice chooses a fresh random seed S each time she sends a fingerprint and Bob's fingerprints use the received seed. The adversary never learns these random bits. Assume the adversary flips some bits in the block and also possibly flips some bits in the fingerprint sent. Then, by Theorem 2, the probability of a hash collision is at most $2/(Lx)^c$, where x is the channel step in which Alice set the hash function. Let ξ be the event that there is ever a collision for any of the hash functions. Then $Pr(\xi) \leq \frac{2}{L^c} \sum_{x=1}^{\infty} \frac{1}{x^c}$ which is $O(1/L^c)$ when $c > 1$.   □

Lemma 2. *The probability of Event 2 is $O(1/L^9)$.*

Proof. Let ξ be the event that Bob receives $10 \log x$ bits that are all the same after $|T_B^*| \geq L$, but before Alice terminates. The bits sent by Alice during this time are random and will remain random, even if the adversary flips them. Thus, for a fixed channel step x, the probability that the last $10 \log x$ random bits received by Bob are all the same is $1/x^{10}$. Thus, by a union bound, $Pr(\xi) \leq \sum_{x=L}^{\infty} 1/x^{10} = O(1/L^9)$.   □

3.2   Remaining Proof

We now complete the proof of correctness.

Lemma 3. *In Line 5 of Bob's algorithm, there is only one possible value for* $(\mathbf{h}, \ell_h, \ell_s)$.

Proof. We give a proof by contradiction. We assume that there are two values x' and x'' that meet the two conditions of Line 5. Thus, $x' = x - (\ell'_h + \ell'_s + 2\log L)$ and $x'' = x - (\ell''_h + \ell''_s + 2\log L)$. Without loss of generality, let $x' < x''$. Thus, $\ell'_h + \ell'_s \leq \ell''_h + \ell''_s$ since they will be computed based on $\mathbf{getH}\left(\frac{1}{(Lx')^c}\right)$ and $\mathbf{getH}\left(\frac{1}{(Lx'')^c}\right)$. But then, $x'' - x' = \ell'_h + \ell'_s - \ell''_h + \ell''_s \leq 0$ or $x' \geq x''$ which is a contradiction.   □

Lemma 4. $|\mathcal{T}_A|$ *never decreases in any round.*

Proof. In each round, Alice adds b bits to $\mathcal{T}_A$. At the end of the round, she either keeps these b bits if there is a fingerprint match or rewinds b bits if not.   □

For transcripts T and T', we denote $T \preccurlyeq T'$ if and only if T is a prefix of T'.

Lemma 5. *Our algorithm has the following properties:*

1. *When Alice is in Line 2 through Line 6 of her protocol, Bob is listening on the channel.*
2. *When Alice is in Line 2 through Line 6 of her protocol and also at the end of each round, $\mathcal{T}_A \preccurlyeq \mathcal{T}_B$.*
3. *It is always the case that $|\mathcal{T}_A| - b \leq |\mathcal{T}_B^*| \leq |\mathcal{T}_A|$.*
4. *Alice terminates before Bob.*

Proof. We prove by induction on the number of times that Alice executes Line 6.
Base Case: If $i = 0$, then $\mathcal{T}_A = \mathcal{T}_B = \mathcal{T}_B^* = null$, and Bob is listening at the start of his protocol, so all of the lemma's statements hold.
Inductive Step: In Line 6 of round $i-1$, Alice sends a fingerprint. By the inductive hypothesis, Bob is listening at this point. We do a case analysis:

- Case 1: Bob does not verify the fingerprint Alice sends. In this case, Bob will continue listening in his repeat loop for the entire time that Alice completes round $i - 1$ and continues to Line 6 of round i. Bob will not update $\mathcal{T}_B$ and $\mathcal{T}_B^*$ in round $i - 1$ and up to Line 6 of round i. After Line 6 of round $i - 1$, Alice continues computation of π for b steps, but the value $\mathcal{H}'_B$ received in Line 9 will not match her own transcript. Thus, she will undo the updates to her transcript. Therefore, all of the lemma statements hold from Line 6 of round $i - 1$ up to Line 6 of round i.

- Case 2: Bob verifies the fingerprint Alice sends. Thus, $\mathcal{T}_{\mathcal{B}}[0, \ldots, t'_a] = \mathcal{T}_{\mathcal{A}}$ and $b' = b$ at the end of Line 6 of round $i - 1$. Next, Bob exits his repeat loop and sets $\mathcal{T}_{\mathcal{B}}$ and $\mathcal{T}_{\mathcal{B}}^*$ both to $\mathcal{T}_{\mathcal{B}}[0, \ldots, t'_a] = \mathcal{T}_{\mathcal{A}}$. Then, Alice and Bob compute b steps of π. Finally, Bob sends his fingerprint to Alice.

  If Bob's fingerprint matches Alice's transcript, she will update her transcript by b steps. This means at the end of round $i - 1$, $\mathcal{T}_{\mathcal{A}} = \mathcal{T}_{\mathcal{B}}$ and they are at most b steps ahead of $\mathcal{T}_{\mathcal{B}}^*$. If Bob's fingerprint does not match Alice's transcript, she sets her transcript to $\mathcal{T}_{\mathcal{A}}^*$ which is equal to $\mathcal{T}_{\mathcal{A}}$ at the beginning of round $i - 1$ which is itself equal to $\mathcal{T}_{\mathcal{B}}^*$. Thus, $\mathcal{T}_{\mathcal{A}} \preceq \mathcal{T}_{\mathcal{B}}$ at the end of round $i - 1$ up to Line 6 of round i. In both cases it holds that $|T_A| - b \le |T_B^*| \le |T_A|$ from Line 6 of round $i - 1$ up to Line 6 of round i.

We now show that Alice terminates before Bob. Alice terminates if $|T_A| \ge L + \sqrt{L}$. Bob terminates when both $|T_B^*| \ge L$ and he receives $10 \log x$ consecutive bits that are the same. When $|T_B^*| \ge L$, since $|T_A| - b \le |T_B^*| \le |T_A|$ and $b \le \sqrt{L}$, we know $L \le |T_A| \le L + \sqrt{L}$. This means that either Alice terminates in the round that $|T_B^*| \ge L$, or she sends extra random bits to Bob for one more round. Since Event 2 does not occur, we know that Bob will not terminate until after Alice has finished sending these random bits.   □

In the following lemmas, we say that a round is *successful* if $\mathcal{T}_{\mathcal{A}}$ increases by b bits at the end of the round, *i.e.*, a successful round simulates one block of π.

Lemma 6. *In a round, if no bit flip occurs in lines 2 through 6 of Alice's algorithm, then when Alice completes Line 6, Bob will exit his listening loop (repeat loop) and $\mathcal{T}_{\mathcal{A}} = \mathcal{T}_{\mathcal{B}}$.*

Proof. Based on Lemma 5, when Alice sends her message in Line 6, Bob is in his listening loop, and he receives values $\mathcal{H}_{\mathcal{A}}$, $t_a = |\mathcal{T}_{\mathcal{A}}|$ and b. If no bit flip occurs before or at Line 6 of Alice's algorithm, these values are received correctly by Bob. Also, by Lemma 5, $\mathcal{T}_{\mathcal{A}}$ is a prefix of $\mathcal{T}_{\mathcal{B}}$ at this point. Since there are no bit flips through Line 6 of Alice's algorithm, Bob verifies Alice's fingerprint and exits the listening loop. When Bob updates his transcript in Line 12, he sets it to $\mathcal{T}_{\mathcal{A}}$. So, after Alice completes Line 6, both parties have the same transcript.   □

Lemma 7. *If no bit flip occurs in a round, then the round is successful.*

Proof. By Lemma 6, when Alice completes Line 6 in a round with no bit flips, Bob exits his listening loop, and $\mathcal{T}_{\mathcal{A}} = \mathcal{T}_{\mathcal{B}}$. Thus, both parties run π for b bits. Since there are no bit flips during the computation of π or during the transmission of Bob's fingerprint, at the end of the round, $\mathcal{T}_{\mathcal{A}}$ and $\mathcal{T}_{\mathcal{B}}$ will increase by b bits.   □

The proof of the following lemma is technical and is deferred to the full version.

Lemma 8. *The size of the fingerprints that Alice or Bob send in round i is $\Theta(\log Li)$.*

Lemma 9. *Alice terminates after at most* $L + O\left((T + \sqrt{LT + L})\log(LT + L)\right)$ *channel steps and Bob terminates in* $O\left((T + \sqrt{LT + L})\log(LT + L)\right)$ *channel steps after Alice.*

Proof. There are T total bit flips, so bit flips occur in at most T rounds. Thus, by Lemma 7, at most T rounds are not successful. The value of b decreases only in rounds that are not successful, so the value of b never gets smaller than $\sqrt{\frac{L}{T+1}}$.

By definition of a successful round, $\mathcal{T}_A$ increases by b steps in such a round. The number of successful rounds until $|\mathcal{T}_A| \geq L + \sqrt{L}$ is at most $\sqrt{LT + L} + \sqrt{T+1}$. Thus, Alice terminates after at most $r_A = T + \sqrt{LT + L} + \sqrt{T+1}$ rounds.

We first calculate the number of fingerprint bits sent by Alice and Bob in these rounds. Let x_i denote the number of channel steps that have elapsed when Alice or Bob is sending the fingerprint in round i. Note that any block is of size at most $\sqrt{L}$ and at least $\log L$. We let the size of the fingerprint be equal to $c_1 \log L x_i$ for some constant $c_1 > 0$. Then, $i(\log L + c \log x_i) \leq x_i \leq i(\sqrt{L} + c \log x_i)$, and thus $c_1 \log L x_i = \Theta(\log L i)$ (see Lemma 8). So, the total number of fingerprint bits sent is no more than

$$\sum_{i=1}^{r_A} c_1 \log L x_i = \Theta(\sum_{i=1}^{r_A} \log L i) = \Theta((T + \sqrt{LT + L})\log(LT + L)).$$

We now count the number of channel steps (called x_A) until Alice terminates. This will be $L + \sqrt{L}$ plus the number of bits sent in rounds that were not successful plus the number of bits sent for fingerprints, *i.e.*,

$$x_A = L + \sqrt{L} + \sum_{i=1}^{T} \sqrt{\frac{L}{i}} + \Theta\left((T + \sqrt{LT + L})\log(LT + L)\right)$$

$$= L + \Theta((T + \sqrt{LT + L})\log(LT + L)).$$

By Lemma 5, $|\mathcal{T}_A| - b \leq |\mathcal{T}_B^*|$ at each channel step. Thus, in the round that Alice terminates, $|\mathcal{T}_B^*| \geq L$ since b is at most $\sqrt{L}$. Bob terminates when, in some step x, he has seen $10 \log x$ consecutive bits that are all the same. Let x_B be the time when Bob terminates. After Alice terminates, the adversary must flip at least one bit every $10 \log x_B$ channel steps. So,

$$\frac{x_B - x_A}{10 \log x_B} \leq T. \tag{1}$$

Note that,

$$L + c_5\left((T + \sqrt{LT + L})\log(LT + L)\right) \leq x_A \leq L + c_4\left((T + \sqrt{LT + L})\log(LT + L)\right).$$

Let $x_B = x_A + \alpha(x_A - L)$ for some $\alpha \geq 1$. By substituting x_A and x_B in the left side of Formula 1,

$$\frac{x_B - x_A}{10 \log x_B} = \frac{\alpha(x_A - L)}{10 \log((1+\alpha)x_A - \alpha L)}$$

$$\geq \frac{\alpha c_4 \big((T + \sqrt{LT + L}) \log(LT + L)\big)}{10 \log(L + (1 + \alpha)c_5((T + \sqrt{LT + L}) \log(LT + L))))}$$

$$\geq \frac{\alpha c_4}{c_6 \log \alpha} T$$

This is a contradiction if $c_4 \alpha > c_6 \log \alpha$ or $\frac{\alpha}{\log \alpha} > \frac{c_6}{c_4}$. This means that α must be smaller than a constant depending only on c_6 and c_4. $\qquad \square$

Lemma 10. *If a party terminates, its output will be correct.*

Proof. Conditioned on both Event 1 and Event 2 not happening, all bits added to $\mathcal{T}_{\mathcal{A}}^*$ and $\mathcal{T}_{\mathcal{B}}^*$ have already been verified to be correct. Upon termination, both $|\mathcal{T}_{\mathcal{A}}^*|$ and $|\mathcal{T}_{\mathcal{B}}^*|$ are of size at least L by the termination conditions. Thus, upon termination both Alice and Bob will have the first L bits of π which is all that is needed to have the correct output. $\qquad \square$

The proof of Theorem 1 now follows from Lemmas 1, 2, 9, and 10.

4   Some Additional Remarks

Need for Private Channels The following theorem justifies our assumption of private channels. The proof is in the extended version of this paper.

Theorem 3. *Consider any algorithm for interactive communication over a public channel that works with unknown T and always terminates in the noise-free case. Any such algorithm succeeds with probability at most $1/2$.*

Communication Rate Comparison. In Haeupler's algorithm [12], the noise rate ϵ is known in advance and is used to design an algorithm with a communication rate of $1 - O(\sqrt{\epsilon \log \log 1/\epsilon})$. Let L' be the length of π'. Then in his algorithm, $L' = O(L)$, and so the adversary is restricted to flipping $\epsilon L' = O(L)$ bits. Thus, in his model, T and L' are always $O(L)$. In our model, the values of T and L' are not known in advance, and so both T and L' may be asymptotically larger than L.

How do our results compare with [12]? As noted above, a direct comparison is only possible when $T = O(L)$. Restating our algorithm in terms of ϵ, we have the following theorem whose proof is extended version of this paper.

Theorem 4. *If the adversary flips $O(L)$ bits and the noise rate is ϵ then our algorithm guarantees a communication rate of $1 - O\left(\frac{1}{\sqrt{L}} + \sqrt{\epsilon}\right) \log L$.*

Our algorithm has the property that a noise rate greater than $1/\log L$ gives communication rate no less than a noise rate of $1/\log L$. In particular, it does not help the adversary to flip more than 1 bit per block. The above communication rate is meaningful when $\epsilon \le 1/\log^2 L$. An optimized version of our algorithm, not included here for ease of presentation[2], achieves, for $T < L$, a communication rate of $1 - O\left(\frac{1}{\sqrt{L}} + \sqrt{\epsilon \log L}\right)$. This communication rate is always positive, and hence meaningful.

A Note on Fingerprint Size. A natural question is whether more powerful probabilistic techniques than union bound could enable us to use smaller fingerprints as done in [12]. The variability of block sizes poses a challenge to this approach since Alice and Bob must either agree on the current block size, or be able to recover from a disagreement by having Bob stay in the listening loop so he can receive Alice's message. If their transcripts diverge by more than a constant number of blocks, it may be difficult to make such a recovery, and therefore it seems challenging to modify our algorithm to use smaller fingerprints. However, it is a direction for further investigation.

5 Conclusion

We have described the first algorithm for interactive communication that tolerates an unknown but finite amount of noise. Against an adversary that flips T bits, our algorithm sends $L + O\left((T + \sqrt{LT + L})\log(LT + L)\right)$ bits in expectation where L is the transcript length of the computation. We prove this is optimal up to logarithmic factors, assuming a conjectured lower bound by Haeupler. Our algorithm critically relies on the assumption of a private channel, an assumption that we show is necessary in order to tolerate an unknown noise rate.

Several open problems remain including the following. First, can we adapt our results to interactive communication that involves more than two parties? Second, can we more efficiently handle an unknown amount of stochastic noise? Finally, for any algorithm, what are the optimal tradeoffs between the overhead incurred when $T = 0$ and the overhead incurred for $T > 0$?

Acknowledgments. We are grateful to Bernhard Haeupler, Tom Hayes, Mahdi Zamani, and the anonymous reviewers for their useful discussions and comments.

References

1. Brakerski, Z., Kalai, Y.T.: Efficient interactive coding against adversarial noise. In: Foundations of Computer Science (FOCS), pp. 160–166 (2012)
2. Brakerski, Z., Naor, M.: Fast algorithms for interactive coding. In: Symposium on Discrete Algorithms (SODA), pp. 443–456 (2013)
3. Braverman, M.: Coding for interactive computation: progress and challenges. In: Communication, Control, and Computing (Allerton), pp. 1914–1921, October 2012

---

[2] The basic idea is to start the block size at $\sqrt{L \log L}$ instead of $\sqrt{L}$.

4. Braverman, M.: Towards deterministic tree code constructions. In: Innovations in Theoretical Computer Science Conference (ITCS), pp. 161–167 (2012)
5. Braverman, M., Efremenko, K.: List and unique coding for interactive communication in the presence of adversarial noise. In: Foundations of Computer Science (FOCS), pp. 236–245 (2014)
6. Braverman, M., Rao, A.: Towards coding for maximum errors in interactive communication. In: Symposium on Theory of Computing (STOC), pp. 159–166 (2011)
7. Feinerman, O., Haeupler, B., Korman, A.: Breathe before speaking: efficient information dissemination despite noisy, limited and anonymous communication. In: Principles of Distributed Computing (PODC), pp. 114–123. ACM (2014)
8. Franklin, M., Gelles, R., Ostrovsky, R., Schulman, L.: Optimal Coding for Streaming Authentication and Interactive Communication. IEEE Transactions on Information Theory $61(1)$, 133–145 (2015)
9. Gelles, R., Moitra, A., Sahai, A.: Efficient and explicit coding for interactive communication. In: Foundations of Computer Science (FOCS), pp. 768–777, October 2011
10. Ghaffari, M., Haeupler, B.: Optimal Error Rates for Interactive Coding II: Efficiency and List Decoding (2013) http://arxiv.org/abs/1312.1763
11. Ghaffari, M., Haeupler, B., Sudan, M.: Optimal Error Rates for Interactive Coding I: Adaptivity and Other Settings. In: Symposium on Theory of Computing (STOC), pp. 794–803 (2014)
12. Haeupler, B.: Interactive channel capacity revisited. In: Foundations of Computer Science (FOCS), pp. 226–235. IEEE (2014)
13. Moore, C., Schulman, L.J.: Tree Codes and a Conjecture on Exponential Sums. In: Innovations in Theoretical Computer Science (ITCS), pp. 145–154 (2014)
14. Naor, J., Naor, M.: Small-bias probability spaces: Efficient constructions and applications. SIAM Journal on Computing (SICOMP) $22(4)$, 838–856
15. Ostrovsky, R., Rabani, Y., Schulman, L.J.: Error-Correcting Codes for Automatic Control. IEEE Transactions on Information Theory $55(7)$, 2931–2941 (2009)
16. Peczarski, M.: An Improvement of the Tree Code Construction. Information Processing Letters $99(3)$, 92–95 (2006)
17. Schulman, L.: Communication on noisy channels: a coding theorem for computation. In: Foundations of Computer Science (FOCS), pp. 724–733, October 1992
18. Schulman, L.J.: Deterministic coding for interactive communication. In: Symposium on Theory of Computing (STOC), pp. 747–756 (1993)
19. Shannon, C.E.: A Mathematical Theory of Communication. Bell System Technical Journal $27(3)$, 379–423 (1948)

Fixed Parameter Approximations for k-Center Problems in Low Highway Dimension Graphs

Andreas Emil Feldmann$^{(\boxtimes)}$

Department of Combinatorics and Optimization, University of Waterloo,
Waterloo, Canada
andreas.feldmann@uwaterloo.ca

Abstract. We consider the k-*Center* problem and some generalizations. For k-Center a set of k *center vertices* needs to be found in a graph G with edge lengths, such that the distance from any vertex of G to its nearest center is minimized. This problem naturally occurs in transportation networks, and therefore we model the inputs as graphs with bounded *highway dimension*, as proposed by Abraham et al. [ICALP 2011].

We show both approximation and fixed-parameter hardness results, and how to overcome them using *fixed-parameter approximations*. In particular, we prove that for any $\varepsilon > 0$ computing a $(2-\varepsilon)$-approximation is W[2]-hard for parameter k, and NP-hard for graphs with highway dimension $O(\log^2 n)$. The latter does not rule out fixed-parameter $(2-\varepsilon)$-approximations for the highway dimension parameter h, but implies that such an algorithm must have at least doubly exponential running time in h if it exists, unless the ETH fails. On the positive side, we show how to get below the approximation factor of 2 by combining the parameters k and h: we develop a fixed-parameter $3/2$-approximation with running time $2^{O(kh \log h)} \cdot n^{O(1)}$.

We also provide similar fixed-parameter approximations for the *weighted k-Center* and $(k, \mathcal{F})$-*Partition* problems, which generalize k-Center.

1 Introduction

In this paper we consider the k-*Center* problem and some of its generalizations. For the problem, k locations need to be found in a network, so that every node in the network is close to a location. More formally, the input is specified by an integer $k \in \mathbb{N}$ and a graph $G = (V, E)$ with positive edge lengths. A *feasible solution* to the problem is a set $C \subseteq V$ of *centers* such that $|C| \leq k$. The aim is to minimize the maximum distance between any vertex and its closest center. That is, let $\mathrm{dist}_G(u, v)$ denote the shortest-path distance between two vertices $u, v \in V$ of G according to the edge lengths, and $B_v(r) = \{u \in V \mid \mathrm{dist}_G(u, v) \leq r\}$ be the *ball of radius r around v*. We need to minimize the *cost* of the solution C, which is the smallest value ρ for which $\bigcup_{v \in C} B_v(\rho) = V$. We say that a center $v \in C$ *covers* a vertex $u \in V$ if $u \in B_v(\rho)$. Hence we can see the problem as finding k centers covering all vertices of G with balls of minimum radius.

I would like to thank Jochen Könemann for reading a draft of this paper. This work was supported by ERC Starting Grant PARAMTIGHT (No. 280152).

M.M. Halldórsson et al. (Eds.): ICALP 2015, Part II, LNCS 9135, pp. 588–600, 2015.
DOI: 10.1007/978-3-662-47666-6_47

The k-Center problem naturally arises in transportation networks, where for instance it models the need to find locations for manufacturing plants, hospitals, police stations, or warehouses under a budget constraint. Unfortunately it is NP-hard to solve the problem in general [23], and the same holds true in various models for transportation networks, such as planar graphs [22] and metrics using Euclidean (L_2) or Manhattan (L_1) distance measures [13]. A more recent model for transportation networks uses the *highway dimension*, which was introduced as a graph parameter by Abraham et al. [1]. The intuition behind its definition comes from the empirical observation [6,7] that in a road network, starting from any point A and travelling to a sufficiently far point B along the quickest route, one is bound to cross one of relatively few "access points". There are several formal definitions for the highway dimension that differ slightly [1,3,2]. All of them however imply the existence of *locally sparse shortest path covers*. Therefore, in this paper we consider this as a generalization of the original highway dimension definitions (in fact the definition given in [2] is equivalent to this).

Definition 1. *Given a graph $G = (V, E)$ with edge lengths and a scale $r \in \mathbb{R}^+$, let $\mathcal{P}_{(r,2r]} \subseteq 2^V$ contain all vertex sets given by shortest paths in G of length more than r and at most $2r$. A shortest path cover $\mathrm{SPC}(r) \subseteq V$ is a hitting set for the set system $\mathcal{P}_{(r,2r]}$, i.e. $P \cap \mathrm{SPC}(r) \neq \emptyset$ for each $P \in \mathcal{P}_{(r,2r]}$. We call the vertices in $\mathrm{SPC}(r)$ hubs. A hub set $\mathrm{SPC}(r)$ is called locally h-sparse, if for every vertex $v \in V$ the ball $B_v(2r)$ of radius $2r$ around v contains at most h vertices from $\mathrm{SPC}(r)$. The highway dimension of G is the smallest integer h such that there is a locally h-sparse shortest path cover $\mathrm{SPC}(r)$ for every scale $r \in \mathbb{R}^+$ in G.*

Abraham et al. [1] introduced the highway dimension in order to explain the fast running times of various shortest-path heuristics. However they also note that "conceivably, better algorithms for other [optimization] problems can be developed and analysed under the small highway dimension assumption". In this paper we investigate the k-Center problem and focus on graphs with low highway dimension as a model for transportation networks. One advantage of using such graphs is that they do not only capture road networks but also networks with transportation links given by air-traffic or railroads. For instance, introducing connections due to airplane traffic will render a network non-planar, while it can still be argued to have low highway dimension. This is because longer flight connections tend to be serviced by bigger but sparser airports, which act as hubs. This can for instance be of interest in applications where warehouses need to be placed to store and redistribute goods of globally operating enterprises. Unfortunately however, we show in this paper that the k-Center problem also remains NP-hard on graphs with low highway dimension.

Two popular and well-studied ways of coping with NP-hard problems is to devise *approximation* [23] and *fixed-parameter* [12] algorithms. For the former we demand polynomial running times but allow the computed solution to deviate from the optimum cost. That is, we compute a *c-approximation*, which is a feasible solution with a cost that is at most c times worse than the best possible for the given instance. A problem that allows a polynomial-time c-approximation for any input is *c-approximable*, and *c* is called the *approximation factor* of the corresponding algorithm. The rational behind fixed-parameter algorithms is that some *parameter* p of the input is small and we can therefore afford running times that are super-polynomial in p, where however we demand optimum solutions. That is, we compute a solution with optimum cost in time $f(p) \cdot n^{O(1)}$ for some function $f(\cdot)$ that is independent of the input size n. A problem that has a fixed-parameter algorithm for a parameter p is called *fixed-parameter tractable* (FPT) for p. What however, if a problem is neither approximable nor FPT? In this case it may be possible to overcome the complexity by combining these two paradigms. In particular, the objective becomes to develop *fixed-parameter c-approximation* (*c*-FPA) algorithms that compute a *c*-approximation in time $f(p) \cdot n^{O(1)}$ for a parameter p.

The idea of combining the two paradigms of approximation and fixed-parameter tractability has been suggested before. However only few results are known for this setting (cf. [21]). In this paper we show that for the *k*-Center problem it is possible to overcome lower bounds for its approximability and its fixed-parameter tractability using fixed-parameter approximations. For many different input classes, such as planar graphs [22], and L_2- and L_∞-metrics [13], the *k*-Center problem is 2-approximable but not $(2 - \varepsilon)$-approximable for any $\varepsilon > 0$, unless P=NP. We show that, unless P=W[2], for general graphs there is no $(2 - \varepsilon)$-FPA algorithm for the parameter k. Additionally, we prove that, unless P=NP, *k*-Center is not $(2 - \varepsilon)$-approximable on graphs with highway dimension $O(\log^2 n)$. This does not rule out $(2 - \varepsilon)$-FPA algorithms for the highway dimension parameter, and we leave this as an open problem. However the result implies that if such an algorithm exists then its running time must be enormous. In particular, unless the exponential time hypothesis (ETH) fails, there can be no $(2 - \varepsilon)$-FPA algorithm with doubly exponential $2^{2^{o(\sqrt{h})}} \cdot n^{O(1)}$ running time in the highway dimension h.

In face of these hardness results, it seems tough to beat the approximation factor of 2 for *k*-Center, even when considering fixed-parameter approximations for either the parameter k or the highway dimension. Our main result however is that we can obtain a significantly better approximation factor for *k*-Center when combining these two parameters. Such an algorithm is useful when aiming for high quality solutions, for instance in a setting where only few warehouses should be built in a transportation network, since warehouses are expensive or stored goods should not be too dispersed for logistical reasons.

It is known [2] that locally $O(h \log h)$-sparse shortest path covers can be computed for graphs of highway dimension h in polynomial time. In the following theorem summarizing our main result, the first given running time assumes this approximation. In general it is NP-hard to compute the highway dimension [14], but it is unknown whether this problem is FPT. If this is the case and the running time is sufficiently small, this can be used as an oracle in our algorithm.

Theorem 2. *For any graph G with n vertices and of highway dimension h, there is an algorithm that computes a 3/2-approximation to the k-Center problem in time $2^{O(kh \log h)} \cdot n^{O(1)}$. If locally h-sparse shortest path covers are given by an oracle the running time is $9^{kh} \cdot n^{O(1)}$.*

We leave open whether approximation factors better than $3/2$ can be obtained for the combined parameter (k, h). Even if we also leave open whether $(2 - \varepsilon)$-FPA algorithms exist for the parameter h alone, we are able to prove that the techniques we use to obtain Theorem 2 cannot omit using both k and h as parameters. To obtain a $(2 - \varepsilon)$-FPA algorithm with running time $f(h) \cdot n^{O(1)}$ for any function $f(\cdot)$ independent of n, a lot more information of the input would need to be exploited than the algorithm of Theorem 2 does. To explain this, we now turn to the used techniques.

1.1 Used Techniques

A crucial observation for our algorithm is that at any scale r, a graph of low highway dimension is structured in the following way (Figure 1). We will prove that the vertices are either at distance at most r from some hub, or they lie in clusters of diameter at most r that are at distance more than $2r$ from each other. Hence, for the cost ρ of the optimum solution, at scale $r = \rho/2$ a center that resides in a cluster cannot cover any vertices of some other cluster. In this sense the clusters are "independent" of each other. At the same time we

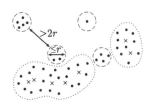

Fig. 1. Clusters (dashed circles) are far from hubs (crosses). They have small diameter and are far from each other.

are able to bound the number of hubs of scale $\rho/2$ in terms of k and the highway dimension. Intuitively, this is comparable to graphs with small vertex cover, since the vertices that are not part of the vertex cover form an independent set. In this sense the highway dimension is a generalization of the vertex cover number (this is in fact the reason why computing the highway dimension is NP-hard [14]).

At the same time the k-Center problem is a generalization of the Dominating Set problem. This problem is W[2]-hard [12], which, as we will show, is also why k-Center is W[2]-hard to approximate for parameter k. However Dominating Set is FPT using the vertex cover number as the parameter [5]. This is one of the reasons why combining the two parameters k and h yields a $3/2$-FPA algorithm for k-Center. In fact the similarity seems so striking at first that one is tempted to either reduce the problem of finding a $3/2$-approximation for k-Center on low highway dimension graphs to solving Dominating Set on a graph of low vertex cover number, or at least use the known techniques for the latter problem to solve the former. However it is unclear how this can be made to work. Instead we devise a more involved algorithm that is driven by the intuition that the two problems are similar. The intuition works on two levels. The first part of the algorithm will determine some of the approximate centers by exploiting the fact that a center in a cluster cannot cover vertices of other clusters. In the second part of our algorithm we will actually reduce the problem of finding the remaining approximate centers to Dominating Set in a graph with small vertex cover number. Proving that the found centers form a feasible approximate solution needs a non-trivial proof.

The algorithm will guess the cost ρ of the optimum solution in order to exploit the structure of the graph given by the locally h-sparse shortest path cover for scale $r = \rho/2$. In particular, the shortest path covers of other scales do not need to be locally sparse in order for the algorithm to succeed. We will show that there are graphs for which k-Center is not $(2-\varepsilon)$-approximable, unless P=NP, and for which the shortest path cover for scale $\rho/2$ is locally 25-sparse. Hence our techniques, which only consider the shortest path cover of scale $\rho/2$, cannot yield a $(2 - \varepsilon)$-FPA algorithm for parameter h. The catch is though that the reduction produces graphs which do not have locally sparse shortest path covers for scales significantly larger than $\rho/2$. Hence a $(2 - \varepsilon)$-FPA algorithm for parameter h might still exist. However such an algorithm would have to take larger scales into account than just $\rho/2$, and as argued above, it would have to have at least doubly exponential running time in h.

Proving that no $(2 - \varepsilon)$-FPA algorithm for parameter k exists for k-Center, unless P=W[2], is straightforward given the original reduction of Hsu and Nemhauser [19] from the W[2]-hard Dominating Set problem. For parameter h however we develop some more advanced techniques. For the reduction we show how to construct a graph of low highway dimension given a metric of low *doubling dimension*, so that distances between vertices are preserved by a $(1 + \varepsilon)$ factor. The doubling dimension [16] is a parameter that captures the bounded volume growth of metrics, such as given by Euclidean and Manhattan distances. Formally, a metric has doubling dimension d if for every $r \in \mathbb{R}^+$ and vertex v, the ball $B_v(2r)$ of radius $2r$ is the union of at most 2^d balls of radius r. Since k-Center is not $(2-\varepsilon)$-approximable in L_∞-metrics [13], unless P=NP, and these have doubling dimension 2, we are able to conclude that the hardness translates to graphs of highway dimension $O(\log^2 n)$.

1.2   Generalizations

In addition to k-Center, we can obtain similar results for two generalizations of the problem by appropriately modifying our techniques.[1] For the *weighted k-Center* problem, the vertices have integer weights and the objective is to choose centers of total weight at most k to cover all vertices with balls of minimum radius (see [23] for a formal definition). This problem is 3-approximable [18,23] and no better approximation factor is known. However we are able to modify our techniques to obtain a 2-FPA algorithm for the combined parameter (k, h).

An alternative way to define the k-Center problem is in terms of finding a *star cover* of size k in a metric, where the cost of the solution is the longest of any star edge in the solution. More generally, in their seminal work Hochbaum and Shmoys [18] defined the $(k, \mathcal{F})$-*Partition* problem (see [18] for a formal definition). Here a family of (unweighted) graphs $\mathcal{F}$ is given and the aim is to partition the vertices of a metric into k sets and connect the vertices of each set by a graph from the family $\mathcal{F}$. The solution cost is measured by the "bottleneck", which is the longest distance between any two vertices of the metric that are connected by an edge in a graph from the family $\mathcal{F}$. The case when $\mathcal{F}$ contains only stars

---

[1] The details of the generalizations, together with all missing proofs of this paper, are deferred to the full version.

is exactly the k-Center problem, given the shortest-path metric as input. The $(k, \mathcal{F})$-Partition problem is $2d$-approximable [18], where d is the largest diameter of any graph in $\mathcal{F}$. We show that a 3δ-FPA algorithm for the combined parameter (k, h) exists, where δ is the largest radius of any graph in $\mathcal{F}$. Hence for graph families in which $3\delta < 2d$ this improves on the general algorithm by Hochbaum and Shmoys [18]. This is for example the case when $\mathcal{F}$ contains "stars of paths", i.e. stars for which each edge is replaced by a path of length δ. The diameter of such a graph is 2δ, while the radius is δ, and hence $3\delta < 2d = 4\delta$.

1.3 Related Work

Given its applicability to various problems in transportation networks, but also in other contexts such as image processing and data-compression, the k-Center problem has been extensively studied in the past. We only mention closely related results here, that were not mentioned before. For planar and map graphs the k-Center problem is FPT [11] for the combined parameter (k, ρ). Note though that k and ρ are somewhat opposing parameters in the sense that typically if k is small then ρ will be large, and vice versa. It would therefore be interesting to know if there are $(2 - \varepsilon)$-FPA algorithms for k-Center on planar or map graphs that do not use ρ as a parameter. For metrics with Euclidean or Manhattan distances, $(1 + \varepsilon)$-FPA algorithms for the combined parameter (k, ε, D) can be obtained [4,17], where D is the dimension of the geometric space.

Generally, fixed-parameter approximations have not been intensively studied so far. A survey is given by Marx [21]. Some newer developments include $(1 + \varepsilon)$-FPA algorithms [20] for problems such as Max Cut, Edge Dominating Set, or Graph Balancing, for parameters such as treewidth and cliquewidth combined with ε. In terms of lower bounds, Bonnet et al. [9] make a connection between the linear PCP conjecture and fixed-parameter inapproximability for problems such as Independent Set.

Abraham et al. [1] introduce the highway dimension, and study it in various papers [1,3,2]. Their main interest is in explaining the good performance of various shortest-path heuristics assuming low highway dimension. In [2] they show that a locally $O(h \log h)$-sparse shortest path cover can be computed in polynomial time for any scale if the highway dimension of the input graph is h, and each shortest path is unique. We will assume that the latter is always the case, since we can slightly perturb the edge lengths. Feldmann et al. [14] consider computing approximations for various other problems that naturally arise in transportation networks. They show that quasi-polynomial time approximation schemes can be obtained for problems such as Travelling Salesman, Steiner Tree, or Facility Location, if the highway dimension is constant. This is done by probabilistically embedding a low highway dimension graph into a bounded treewidth graph while introducing arbitrarily small distortions of distances. Known algorithms to compute optimum solutions on low treewidth graphs then imply the approximation schemes. It is interesting to note that this approach does not work for the k-Center problem since, in contrast to the above mentioned problems, its objective function is not linear in the edge lengths. The only other theoretical result mentioning the highway dimension that we are aware of is by Bauer et al. [8], who show that for any graph G there exist edge lengths such that the highway dimension is $\Omega(\mathrm{pw}(G)/\log n)$, where $\mathrm{pw}(G)$ is the pathwidth of G.

2   Highway Dimension and Vertex Covers

We observe that the vertices of a low highway dimension graph are highly structured for any scale r: the vertices that are far from any hub of a shortest path cover for scale r are clustered into sets of small diameter and large inter-cluster distance (Figure 1). This observation was already made in [14]. We first formally define the clusters and then prove that they have the claimed properties. For a set $S \subseteq V$ let $\mathrm{dist}_G(u, S) = \min_{v \in S} \mathrm{dist}_G(u, v)$ be the distance from u to the closest vertex in S.

Definition 3. *Fix $r \in \mathbb{R}^+$ and $\mathrm{SPC}(r) \subseteq V$ in a graph $G = (V, E)$. We call an inclusion-wise maximal set $T \subseteq \{v \in V \mid \mathrm{dist}_G(v, \mathrm{SPC}(r)) > r\}$ with $\mathrm{dist}_G(u, v) \leq r$ for all $u, v \in T$ a cluster, and we denote the set of all clusters by $\mathcal{T}$. The non-cluster vertices are those which are not contained in any cluster of $\mathcal{T}$.*

Note that the set $\mathcal{T}$ is specific for the scale r and the hub set $\mathrm{SPC}(r)$. The following lemma characterizes the structure of the clusters and non-cluster vertices. Here we let $\mathrm{dist}_G(S, S') = \min_{v \in S} \mathrm{dist}_G(v, S')$ be the minimum distance between vertices of two sets S and S'.

Lemma 4. *Let $\mathcal{T}$ be the cluster set for a scale r and a shortest path cover $\mathrm{SPC}(r)$. For each non-cluster vertex v there is a hub in $\mathrm{SPC}(r)$ with $\mathrm{dist}_G(u, v) \leq r$. The diameter of any cluster $T \in \mathcal{T}$ is at most r, and $\mathrm{dist}_G(T, T') > 2r$ for any distinct pair of clusters $T, T' \in \mathcal{T}$.*

Intuitively, Lemma 4 means that a low highway dimension graph has similarities to a graph with small vertex cover. For this reason, as part of our algorithm we compute the optimum dominating set of a graph with a small vertex cover. To show how this can be done, we begin by formally defining the relevant notions. Let $G = (V, E)$ be an (unweighted) graph. A *dominating set* $D \subseteq V$ is a set of vertices such that every vertex of V is adjacent to some vertex of D. The *Dominating Set* problem is to find a dominating set of minimum size. A *vertex cover* $W \subseteq V$ is a set of vertices such that every edge of E is incident to some vertex of W. The *vertex cover number* is the size of the smallest vertex cover. The reason why the Dominating Set problem is FPT for this parameter essentially is that the vertex cover number is an upper bound on the pathwidth [15]. For the following lemma, we simplify an algorithm due to Alber et al. [5] to solve Dominating Set on bounded treewidth graphs. This improves the running time. Note that the algorithm assumes the vertex cover to be given explicitly.

Lemma 5. *Given a graph G and a vertex cover W of G, the Dominating Set problem can be solved in time $O(3^l \cdot n^2)$, where $l = |W|$.*

3 The Fixed-Parameter Approximation Algorithm

If the optimum cost for k-Center is ρ and there is a locally s-sparse shortest path cover $\text{SPC}(r)$ for scale $r = \rho/2$, then $|\text{SPC}(r)| \leq ks$. This is because there are k balls of radius ρ covering the whole graph, and by Definition 1 there are at most s hubs in each ball. If the input graph has highway dimension h and there is an oracle that gives locally h-sparse shortest path covers for each scale, then we can set $s = h$. Otherwise, Abraham et al. [2] show how to compute $O(\log h)$-approximations of shortest path covers in polynomial time, if shortest paths have unique lengths. The latter can be assumed by perturbing the edge lengths and therefore we can set $s = O(h \log h)$.

The hubs of $\text{SPC}(r)$ hit all shortest paths of length in $(r, 2r]$ in the same way a vertex cover hits all single edges, which can be thought of as unit length paths in an unweighted graph. Furthermore, we know that the clusters are all more than $2r = \rho$ apart by Lemma 4. Hence any center of the optimum solution to k-Center that lies in a cluster cannot cover any vertex of another cluster. Compared to the Dominating set problem, this is analogous to the vertices of a graph that do not belong to the vertex cover and therefore form an independent set: a vertex of an optimum dominating set that is part of the independent set cannot dominate another vertex of the independent set.

The first part of our algorithm is driven by this intuition. After guessing the optimum cost ρ and computing $\text{SPC}(\rho/2)$ together with its cluster set $\mathcal{T}$, we will see how the algorithm computes three approximate center sets. For the first set C_1 the algorithm guesses a subset of the hubs of $\text{SPC}(\rho/2)$ that are close to the optimum center set. This can be done in exponential time in ks because there are at most that many hubs. The independence property of the clusters then makes it easy to determine a second set C_2 of approximate centers, each of which lies in a cluster that must contain an optimum center. To determine the third set of centers C_3, the similarity of our problem to Dominating Set on graphs with small vertex cover number becomes more concrete in our algorithm: we reduce finding C_3 to Dominating Set in such a way that the constructed graph has a vertex cover number that can be bounded in the number of hubs of $\text{SPC}(\rho/2)$.

More concretely, consider an input graph $G = (V, E)$ with an optimum k-Center solution C^* with cost ρ. For an index $i \in \{1, 2, 3\}$ we denote by $R_i^* = \bigcup_{v \in C_i^*} B_v(\rho)$ and $R_i = \bigcup_{v \in C_i} B_v(\frac{3}{2}\rho)$ the regions covered by some set of optimum centers $C_i^* \subseteq C^*$ (with balls of radius ρ) and approximate centers $C_i \subseteq V$ (with balls of radius $\frac{3}{2}\rho$), respectively. The algorithm tries every scale r in order to guess the correct value for which $r = \rho/2$. After computing $\text{SPC}(r)$ together with its cluster set $\mathcal{T}$, the algorithm first checks that the number of hubs is not too large, in order to keep the running time low. In particular, since we know that $|\text{SPC}(\rho/2)| \leq ks$, we can dismiss any shortest path cover containing more hubs. Assume that $r = \rho/2$ was found. The next step is to guess a minimal set H of hubs in $\text{SPC}(\rho/2)$, such that the balls of radius $\rho/2$ around hubs in H cover all optimum non-cluster centers. That is, if $C_1^* \subseteq C^*$ denotes the set of optimum centers which each are at distance at most $\rho/2$ from some hub in $\text{SPC}(\rho/2)$, then $C_1^* \subseteq \bigcup_{v \in H} B_v(\rho/2)$ and H is minimal with this property.

We choose this set of hubs H as the first set of centers C_1 for our approximate solution. Note that due to the minimality of H, $|C_1| \leq |C_1^*|$. Also $R_1^* \subseteq R_1$ since for any center in C_1^* there is a center at distance at most $\rho/2$ in C_1.

The next step is to compute a set of centers so that all clusters of the cluster set $\mathcal{T}$ of $\mathrm{SPC}(\rho/2)$ are covered. Some of the clusters are already covered by the first set of centers C_1, and thus in this step we want to cover all clusters in $\mathcal{U} = \{T \in \mathcal{T} \mid T \setminus R_1 \neq \emptyset\}$. By the definition of C_1^*, any remaining optimum center in $C^* \setminus C_1^*$ must lie in a cluster. Furthermore, the distance between clusters of $\mathrm{SPC}(\rho/2)$ is more than ρ by Lemma 4, so that a center of $C^* \setminus C_1^*$ in a cluster T cannot cover any vertices of another cluster $T' \neq T$. Hence if we guessed H correctly we can be sure that each cluster $T \in \mathcal{U}$ must contain a center of $C^* \setminus C_1^*$. For each such cluster we pick an arbitrary vertex $v \in T$ and declare it a center of the second set C_2 for our approximate solution. Thus if the optimum set of centers for $\mathcal{U}$ is $C_2^* = \{v \in C^* \mid \exists T \in \mathcal{U} : v \in T\}$, we have $|C_2| \leq |C_2^*|$. Moreover, since the diameter of each cluster is at most $\rho/2$ by Lemma 4, we get $R_2^* \subseteq R_2$.

At this time we know that all clusters in $\mathcal{T}$ are covered by the region $R_1 \cup R_2$. Hence if any uncovered vertices remain in $V \setminus (R_1 \cup R_2)$ for our current approximate solution, they must be non-cluster vertices. By our definition of C_1^* and C_2^*, all remaining optimum centers $C_3^* = C^* \setminus (C_1^* \cup C_2^*)$ lie in clusters of $\mathcal{T} \setminus \mathcal{U}$. Since $R_1^* \subseteq R_1$ and $R_2^* \subseteq R_2$, any remaining vertex of $V \setminus (R_1 \cup R_2)$ must be in the region R_3^* covered by centers in C_3^*. Next we show how to compute a set C_3 such that the region R_3 includes all remaining vertices of the graph and $|C_3| \leq |C_3^*|$. Note that the latter means that the number of centers in $C_1 \cup C_2 \cup C_3$ is at most k, since C_1^*, C_2^*, and C_3^* are disjoint.

To control the size of C_3 we will compute the smallest number of centers that cover parts of R_3^* with balls of radius ρ. In particular, we will guess the set of hubs $H' \subseteq \mathrm{SPC}(\rho/2) \setminus H$ that lie in the region R_3^* (note that we exclude hubs of H from this set). We then compute a center set C_3 of minimum size such that $H' \subseteq \bigcup_{v \in C_3} B_v(\rho)$. For this we reduce the problem of computing centers covering H' to the Dominating Set problem in a graph of fixed vertex cover number. The reduction follows the lines of the bipartite reduction from Set Cover to Dominating Set [10]. More concretely, let $W = \bigcup_{T \in \mathcal{T} \setminus \mathcal{U}} T$ denote the vertices in the clusters that include C_3^*. For each vertex $w \in W$ we encode the set of vertices of H' that it could cover with a ball of radius ρ, in an instance G' of Dominating Set as follows. The vertices of G' include all vertices of W and H', and we introduce an edge between a vertex $w \in W$ and $u \in H'$ if the distance between w and u is at most ρ. Hence a vertex $w \in W$ will dominate exactly the vertices of H' in the ball $B_w(\rho)$. Since we are only interested in covering vertices of H', additionally we introduce *apex vertices* a and a' to G', together with an edge between every $w \in W$ and a, and an edge between a and a'. This way all vertices of W can be dominated by only a, and one of the apexes must be part of any dominating set in G'. Note that by our choice of the edges, $H' \cup \{a\}$ is a vertex cover for G' and so its vertex cover number is at most $ks + 1$, as required.

If the optimum dominating set of G' is D, we let $C_3 = D \setminus \{a, a'\}$. The following lemma shows that the size of C_3 is as required, and the centers of C_3 cover all hubs of H' with balls of radius ρ.

Lemma 6. *Assuming that the algorithm guessed the correct scale $r = \rho/2$ and the correct sets H and H', the set $C_3 = D \setminus \{a, a'\}$ is of size at most $|C_3^*|$ and $H' \subseteq \bigcup_{v \in C_3} B_v(\rho)$.*

It remains to show that the three computed center sets C_1, C_2, and C_3 cover all vertices of G, which we do in the following lemma.

Lemma 7. *Assuming that the algorithm guessed the correct scale $r = \rho/2$ and the correct sets H and H', the approximate center sets C_1, C_2, and C_3 cover all of G, i.e. $R_1 \cup R_2 \cup R_3 = V$.*

Proof. The proof is by contradiction. Assume there is a $v \in V \setminus (R_1 \cup R_2 \cup R_3)$ that is not covered by the computed approximate center sets. The idea is to identify a hub $y \in \mathrm{SPC}(\rho/2)$ on the shortest path between v and an optimum center $w \in C^*$ covering v. We will show that this hub y must however be in H' and therefore v is in fact in R_3, since v also turns out to be close to y.

To show the existence of y, we begin by arguing that the closest hub $x \in \mathrm{SPC}(\rho/2)$ to v is neither in H nor in H'. We know that each cluster of $\mathcal{T}$ is in $R_1 \cup R_2$, so that $v \notin R_1 \cup R_2$ must be a non-cluster vertex. Thus by Lemma 4, $\mathrm{dist}_G(v, x) \le \rho/2$. The region R_1 in particular contains all vertices that are at distance at most $\rho/2$ from any hub in $H = C_1$. Since $v \notin R_1$ and $\mathrm{dist}_G(v, x) \le \rho/2$, this means that $x \notin H$. From $v \notin R_3$ we can also conclude that $x \notin H'$ as follows. By Lemma 6, C_3 covers all hubs of H' with balls of radius ρ. Hence if $x \in H'$ then v is at distance at most $\frac{3}{2}\rho$ from a center of C_3, i.e. $v \in R_3$.

From $x \notin H \cup H'$ we can conclude the existence of y as follows. Consider an optimum center $w \in C^*$ that covers v, i.e. $v \in B_w(\rho)$. Recall that $R_1^* \subseteq R_1$ and $R_2^* \subseteq R_2$. Since $v \notin R_1 \cup R_2$, this means that w is neither in C_1^* nor in C_2^* so that $w \in C_3^*$. By definition of H', any hub at distance at most ρ from a center in C_3^* is in H', unless it is in H. Hence, as $x \notin H \cup H'$, the distance between x and w must be more than ρ. Since $\mathrm{dist}_G(v, x) \le \rho/2$, we get $\mathrm{dist}_G(v, w) > \rho/2$. We also know that $\mathrm{dist}_G(v, w) \le \rho$, because w covers v. Hence the shortest path cover $\mathrm{SPC}(\rho/2)$ must contain the hub y, which lies on the shortest path between v and w. In particular, $\mathrm{dist}_G(v, y) \le \rho$ and $\mathrm{dist}_G(y, w) \le \rho$. Analogous to the argument used for x above, R_1 contains all vertices at distance at most ρ from H, so that $y \notin H$ since $v \notin R_1$. However, then the distance bound for y and w yields $y \in H'$, as $w \in C_3^*$.

Since C_1^* contains all non-cluster centers but $w \notin C_1^*$, by Lemma 4 we get $\mathrm{dist}_G(y, w) > \rho/2$, which implies $\mathrm{dist}_G(v, y) < \rho/2$. But then v is contained in the ball $B_y(\rho/2)$, which we know is part of the third region R_3 since $y \in H'$. This contradicts the assumption that v was not covered by the approximate center set. $\square$

Note that the proof of Lemma 7 does not imply that $R_3^* \subseteq R_3$, as was the case for R_1 and R_2. It suffices though to establish the correctness of the algorithm. The runtime analysis concluding the proof of Theorem 2 is deferred to the full version of the paper.

4   Hardness Results

We begin by observing that the original reduction of Hsu and Nemhauser [19] for k-Center also implies that there are no $(2 - \varepsilon)$-FPA algorithms.

Theorem 8. *It is W[2]-hard for parameter k to compute a $(2-\varepsilon)$-approximation to the k-Center problem, for any $\varepsilon > 0$.*

We now turn to proving that $(2 - \varepsilon)$-approximations are hard to compute on graphs with low highway dimension. For this we introduce a general reduction from low doubling metrics to low highway dimension graphs in the next lemma. Here the *aspect ratio* of a metric (X, dist_X) is the maximum distance between any two vertices of X divided by the minimum distance.

Lemma 9. *Given any metric (X, dist_X) with constant doubling dimension d and aspect ratio α, for any $\varepsilon > 0$ there is a graph $G = (X, E)$ of highway dimension $O((\log(\alpha)/\varepsilon)^d)$ on the same vertex set such that for all $u, v \in X$, $\mathrm{dist}_X(u, v) \leq \mathrm{dist}_G(u, v) \leq (1 + \varepsilon) \mathrm{dist}_X(u, v)$. Furthermore, G can be computed in polynomial time from the metric.*

Feder and Greene [13] show that, for any $\varepsilon > 0$, it is NP-hard to compute a $(2 - \varepsilon)$-approximation for the k-Center problem in two-dimensional L_∞ metrics. Furthermore all edges of the instance they construct in the reduction have unit length, and thus the aspect ratio is at most n. The doubling dimension of any such metric is 2, since a ball of radius $2r$ (a "square" of side-length $4r$) can be covered by 4 balls of radius r ("squares" of side-length $2r$). By the reduction given in Lemma 9 we thus get the following result.

Corollary 10. *For any constant $\varepsilon > 0$ it is NP-hard to compute a $(2 - \varepsilon)$-approximation for the k-Center problem on graphs of highway dimension $O(\log^2 n)$.*

The challenge remains is to push the highway dimension bound of this inapproximability result down to a constant. This would mean that no $(2 - \varepsilon)$-FPA algorithm for k-Center exists if the parameter is the highway dimension h, unless P=NP. However, assuming the exponential time hypothesis, any $(2-\varepsilon)$-FPA algorithm for parameter h must have doubly exponential running time. In particular, by Corollary 10 any algorithm with running time $2^{2^{o(\sqrt{h})}} \cdot n^{O(1)}$ would solve an NP-hard problem in subexponential $2^{o(n)}$ time when $h \in O(\log^2 n)$. Thus if a $(2 - \varepsilon)$-approximation algorithm for k-Center with parameter h exists, it is fair to assume that its running time depending on h must be extremely large.

The following lemma gives further evidence that obtaining a $(2 - \varepsilon)$-FPA algorithm for parameter h is hard. As argued below, it excludes the existence of such algorithms that only use shortest path covers of constant scales.

Lemma 11. *For any $\varepsilon > 0$ it is NP-hard to compute a $(2 - \varepsilon)$-approximation for the k-Center problem on graphs for which on any scale $r > 0$ there is a locally $(3 \cdot 2^{2r-1} + 1)$-sparse shortest path cover $\mathrm{SPC}(r)$. Moreover, this is true for instances where the optimum cost ρ is at most 4.*

Consider a $(2-\varepsilon)$-FPA algorithm for k-Center, which only takes shortest path covers of constant scales into account, where the parameter is their sparseness. That is, the algorithm computes a $(2 - \varepsilon)$-approximation using hub sets SPC(r) only for values $r \leq R$ for some $R \in O(1)$, and the parameter is a value s such that for every $r \leq R$, SPC(r) is locally s-sparse. By Lemma 11 such an algorithm would imply that P=NP. Moreover this is true even if $R \in O(\rho)$. Hence if it is possible to beat the inapproximability barrier of 2 using the local sparseness as a parameter, then such an algorithm would have to take large (non-constant) scales into account. Note that the running time of our 3/2-FPA algorithm can in fact be bounded in terms of the local sparseness of SPC($\rho/2$) instead of the highway dimension. Therefore, by Theorem 8 and Lemma 11, our algorithm necessarily needs to combine the parameter h with k in order to achieve its approximation guarantee.

References

1. Abraham, I., Fiat, A., Goldberg, A.V., Werneck, R.F.: Highway dimension, shortest paths, and provably efficient algorithms. In: SODA, pp. 782–793 (2010)
2. Abraham, I., Delling, D., Fiat, A., Goldberg, A.V., Werneck, R.F.: VC-Dimension and shortest path algorithms. In: Aceto, L., Henzinger, M., Sgall, J. (eds.) ICALP 2011, Part I. LNCS, vol. 6755, pp. 690–699. Springer, Heidelberg (2011)
3. Abraham, I., Delling, D., Fiat, A., Goldberg, A.V., Werneck, R.F.: Highway dimension, shortest paths, and provably efficient shortest path algorithms. Techical Report (2013)
4. Agarwal, P.K., Procopiuc, C.M.: Exact and approximation algorithms for clustering. Algorithmica **33**(2), 201–226 (2002)
5. Alber, J., Bodlaender, H.L., Fernau, H., Kloks, T., Niedermeier, R.: Fixed parameter algorithms for dominating set and related problems on planar graphs. Algorithmica **33**(4), 461–493 (2002)
6. Bast, H., Funke, S., Matijevic, D., Sanders, P., Schultes, D.: In transit to constant time shortest-path queries in road networks. In: ALENEX (2007)
7. Bast, H., Funke, S., Matijevic, D.: Ultrafast shortest-path queries via transit nodes. 9th DIMACS Implementation. Challenge **74**, 175–192 (2009)
8. Bauer, R., Columbus, T., Rutter, I., Wagner, D.: Search-Space size in contraction hierarchies. In: Fomin, F.V., Freivalds, R., Kwiatkowska, M., Peleg, D. (eds.) ICALP 2013, Part I. LNCS, vol. 7965, pp. 93–104. Springer, Heidelberg (2013)
9. Bonnet, E., Escoffier, B., Kim, E.J., Paschos, V.T.: On subexponential and FPT-time inapproximability. In: Gutin, G., Szeider, S. (eds.) IPEC 2013. LNCS, vol. 8246, pp. 54–65. Springer, Heidelberg (2013)
10. Chlebík, M., Chlebíková, J.: Approximation hardness of dominating set problems in bounded degree graphs. Information and Computation **206**(11), 1264–1275 (2008)
11. Demaine, E.D., Fomin, F.V., Hajiaghayi, M., Thilikos, D.M.: Fixed-parameter algorithms for (k, r)-center in planar graphs and map graphs. TALG **1**(1), 33–47 (2005)
12. Downey, R.G., Fellows, M.R.: Fundamentals of parameterized complexity, vol. 4. Springer (2013)
13. Feder, T., Greene, D.: Optimal algorithms for approximate clustering. In: STOC, pp. 434–444 (1988)

14. Feldmann, A.E., Fung, W.S., Könemann, J., Post, I.: A $(1 + \varepsilon)$-embedding of low highway dimension graphs into bounded treewidth graphs. In: ICALP (2015)
15. Fellows, M.: Towards fully multivariate algorithmics: some new results and directions in parameter ecology. In: Fiala, J., Kratochvíl, J., Miller, M. (eds.) IWOCA 2009. LNCS, vol. 5874, pp. 2–10. Springer, Heidelberg (2009)
16. Gupta, A., Krauthgamer, R., Lee, J.R.: Bounded geometries, fractals, and low-distortion embeddings. In: FOCS, pp. 534–543 (2003)
17. Har-Peled, S., Mazumdar, S.: On coresets for k-means and k-median clustering. In: STOC, pp. 291–300 (2004)
18. Hochbaum, D.S., Shmoys, D.B.: A unified approach to approximation algorithms for bottleneck problems. JACM **33**(3), 533–550 (1986)
19. Hsu, W.-L., Nemhauser, G.L.: Easy and hard bottleneck location problems. Discrete Applied Mathematics **1**(3), 209–215 (1979)
20. Lampis, M.: Parameterized approximation schemes using graph widths. In: Esparza, J., Fraigniaud, P., Husfeldt, T., Koutsoupias, E. (eds.) ICALP 2014. LNCS, vol. 8572, pp. 775–786. Springer, Heidelberg (2014)
21. Marx, D.: Parameterized complexity and approximation algorithms. The Computer Journal **51**(1), 60–78 (2008)
22. Plesník, J.: On the computational complexity of centers locating in a graph. Aplikace Matematiky **25**(6), 445–452 (1980)
23. Vazirani, V.V.: Approximation Algorithms. Springer-Verlag New York Inc. (2001); ISBN 3-540-65367-8

A Unified Framework for Strong Price of Anarchy in Clustering Games

Michal Feldman and Ophir Friedler[✉]

Blavatnik School of Computer Science, Tel Aviv University, Tel Aviv, Israel
michal.feldman@cs.tau.ac.il, ophirf@mail.tau.ac.il

Abstract. We devise a unified framework for quantifying the inefficiency of equilibria in clustering games on networks. This class of games has two properties exhibited by many real-life social and economic settings: (a) an agent's utility is affected only by the behavior of her direct neighbors rather than that of the entire society, and (b) an agent's utility does not depend on the actual strategies chosen by agents, but rather by whether or not other agents selected the same strategy. Our framework is sufficiently general to account for unilateral versus coordinated deviations by coalitions of different sizes, different types of relationships between agents, and different structures of strategy spaces. Many settings that have been recently studied are special cases of clustering games on networks. Using our framework: (1) We recover previous results for special cases and provide extended and improved results in a unified way. (2) We identify new settings that fall into the class of clustering games on networks and establish price of anarchy and strong price of anarchy bounds for them.

1 Introduction

Suppose that mobile phone providers offer a significant discount for calls between their subscribers. In such a case, users selecting a provider would benefit the most by subscribing to the provider of the friends with whom they talk most. Alternatively, consider radio stations selecting radio frequencies on which to broadcast. Because nearby stations that select the same frequency incur interference, each station would favor a frequency that is used the least by its nearby stations. Finally, in some opinion formation settings, an agent forming an opinion aims to have the same opinion as similar agents and an opinion that is different from dissimilar agents.

At a first glance the different settings described above seem different in their nature. In the first example, people want to make similar choices to others, while in the second example they wish to differentiate themselves from others, and in the third example their choice depends on the type of relationship they have with

This work was partially supported by the European Research Council under the European Union's Seventh Framework Programme (FP7/2007-2013) / ERC grant agreement number 337122.

M.M. Halldórsson et al. (Eds.): ICALP 2015, Part II, LNCS 9135, pp. 601–613, 2015.
DOI: 10.1007/978-3-662-47666-6_48

others. Indeed, different models for these real-life settings have been devised and studied in the algorithmic game theory literature [2,13–15,18]. For example, in coordination games on graphs [2], each agent has her own set of feasible options, and an agent benefits by sharing her option with as many friends as possible. In Max-Cut games [15], all agents choose one of two options, and they all seek to distinguish themselves from others. In 2-NAE-SAT games [13], CNF clauses are satisfied when their literals have different values. Agents correspond to literals that select truth values, where agents derive utility from the satisfied clauses of which they are part of. In this case, an agent may wish to share her truth value with some (e.g. in a clause where she is negated, and the other literal is not), while distinguishing herself from others.

Inefficiency of Equilibria. It is well known that settings in which individual agents follow their self interests may exhibit economic inefficiencies. A great deal of the work in the algorithmic game theory literature has attempted to quantify the inefficiencies that may arise as a result of selfish behavior. The most common measure that has been used is the *price of anarchy* (PoA) [17], which is defined as the ratio of the welfare obtained in the worst Nash equilibrium (NE) and the unconstrained optimal welfare. But because of some weaknesses of the NE solution concept, researchers have turned to additional solution concepts, such as *strong equilibrium* (SE), and *q-strong equilibrium* (q-SE). Whereas an NE is an outcome that is resilient against unilateral deviations, a q-SE is resilient against coordinated deviations of coalitions of size at most q. Therefore, a 1-SE is a NE, and an n-SE is resilient against *any* coordinated coalitional deviation, i.e., an n-SE is an SE. In settings in which agents can coordinate their actions (in particular their deviations), the SE concept is more suitable. One can then study the *strong price of anarchy* (SPoA) [1], which is the worst case ratio of an SE and the social optimum, or the q-SPoA, in which the SE is replaced by the q-SE.

The inefficiency of equilibria, as measured by the PoA (and in some cases also the SPoA and q-SPoA), has been studied in all of the above-mentioned games [2,13–15,18]. Because these models correspond to different settings, each work developed its own tools and techniques to compute the (q-)strong price of anarchy.

Clustering. The starting point of this paper is the observation that all the above-mentioned settings are special cases of a very well known setting — that of *clustering* [5]. In this setting, a network is given, with each edge labeled as either a *coordination* or an *anti-coordination* edge. The objective is to find a clustering (i.e., a partition of the nodes into clusters) that maximizes the number of edges (or the total weight of edges in the weighted case) that are *satisfied*, where a coordination edge is satisfied if its adjacent nodes are in the same cluster, and an anti-coordination edge is satisfied if its adjacent nodes are in different clusters.

The above description represents the traditional perspective on the problem, but one can also consider its game-theoretic variant. In the game-theoretic setting each node corresponds to a strategic agent whose strategy space is a feasible subset of all clusters. Each strategic agent chooses a cluster, and given the selected clusters of all the agents (i.e. an *outcome* of the game), the payoff of an agent is the number (or total weight, in the weighted case) of the satisfied edges to which she is incident. We refer to the class of games that arises from this setting as *clustering games on networks*.

Let us recall some of the classes mentioned above, and restate them as special cases of clustering games on networks. In coordination games on graphs [2], each agent has a specified subset of clusters she can join, and all edges are coordination edges. In Max-k-Cut games [15] all the edges are anti-coordination edges, and all agents can join any of the k existing clusters. In 2-NAE-SAT games [13], each agent can join one of two clusters, and edges can be either coordination or anti-coordination edges.

As mentioned above, each case has its own analysis. In light of the observation that all of these settings are special cases of clustering games on networks, the objective of the persent paper is the following:

> *Construct a unified recipe for quantifying the degradation of social welfare (i.e., PoA, SPoA, q-SPoA) in various settings that fall into the class of clustering games on networks.*

The class of clustering games on networks is a rich class that lies at the intersection of two important classes of games: hedonic [11] and graphical games [16]. In hedonic games the utility of each player is fully determined by the set of players that selected the same strategy. They serve as a fundamental model in the study of coalition formation, which has implications for many social, economic, and political scenarios [4,6,8–11]. In graphical games (see [16,20] and references therein) the agents are nodes in a graph, and the payoff of each agent depends strictly on the strategies of her neighbors. Such games attempt to capture the local nature of interactions in a network.

Our Contribution Is Four-Fold:

1. We provide a unified framework for computing the PoA, SPoA, and q-SPoA in clustering games on graphs.
2. We use our framework to recover previous results on special cases.
3. We use our framework to establish new PoA, SPoA, and q-SPoA bounds on previously studied games.
4. We identify new settings that fall into the class of clustering games on networks and establish PoA, SPoA, and q-SPoA bounds for them.

A summary of our results appears in Fig. 1. The different rows correspond to different games, whose description is given in the three left columns. For each class of games, we provide results on the PoA, SPoA, and q-SPoA. For q-SPoA, we find it convenient to present our results using the *coordination factor* $z(q) = \frac{q-1}{n-1}$, which is a real number in $[0,1]$ that equals 0 in the case of PoA (i.e., $q = 1$) and 1 in the case of SPoA (i.e., $q = n$). Because of lack of space we elaborate below on two types of games that appear in the table. For the other games, we refer the reader to the corresponding sections.

Symmetric Coordination Games on Graphs (SCGGs). We introduce the class of SCGGs, in which all the relationships are coordination, and all players have the same k strategies. We use our framework to establish tight bounds of $1/k$ for the PoA, and $\frac{k}{2k-1}$ for the SPoA, and provide lower bounds on the q-SPoA.

Max-Cut games. It is part of folklore that the PoA of Max-Cut games is exactly 1/2. Recently, [13] showed that the SPoA is exactly 2/3, and that the q-SPoA remains roughly 1/2 even for $q = O(n^{\frac{1}{2}-\epsilon})$ for any $\epsilon > 0$. Using our framework, we provide more refined results on the q-SPoA, specifically, we show that the q-SPoA is at least $\frac{2}{4-z(q)}$ for every q. A direct corollary of this result is that when the coordination factor is at least some constant $\alpha \in [0,1]$ (i.e., q is roughly αn), the q-SPoA strictly improves to $\frac{2}{4-\alpha}$ (in particular, recovering the PoA and SPoA results, which correspond to the two extremes, $\alpha = 0$ and $\alpha = 1$, respectively). On the negative side, we construct an instance in which the q-SPoA remains roughly 1/2 even for $q = O(n^{1-\epsilon})$ (and even slightly more relaxed values of q), thus further tightening the previous negative results. Referred to Fig. 1 for a complete list of our results.

Class	Case Description			Result		
Name	$+/-$	# of Str.	Sym	PoA	SPoA	q-SPoA
Max-Cut	$-$	2	$\checkmark$	1/2	2/3	$\frac{2}{4-z(q)}$ ⋆
2-NAE-SAT	$+/-$	2	$\checkmark$	1/2	2/3	$\frac{2}{4-z(q)}$ ⋆
Max-k-Cut	$-$	k	$\checkmark$	$\frac{k-1}{k}$	$\frac{k-1}{k-\frac{1}{2(k-1)}}$ ⋆	$\frac{k-1}{k-\frac{1}{2(k-1)}\cdot z(q)}$ ⋆
SCGGs	$+$	k	$\checkmark$	$1/k$ ⋆	$\frac{k}{2k-1}$ ⋆	$\frac{2+(k-2)\cdot z(q)}{2k-z(q)}$ ⋆
CGGs	$+$	k	$\times$	0	1/2	$\frac{z(q)}{2}$
SCGs	$+/-$	k	$\checkmark$	$1/k$	$\frac{1}{2-\frac{1}{k(k-1)}}$ ⋆	$\frac{2+(k-2)\cdot z(q)}{2k-\frac{1}{k-1}\cdot z(q)}$ ⋆
CGs	$+/-$	k	$\times$	0	1/2	$\frac{z(q)}{2}$ ⋆

Fig. 1. In the first column: SCGs = Symmetric Clustering Games on Networks, SCGGs = Symmetric Coordination Games on Graphs, CGGs = Coordination games on graphs, CGs = Clustering Games on Networks. Each class is fully described using three attributes: (a) the "$+/-$" column states the type of relationships on the edges ($+$ for coordination, $-$ for anti-coordination); (b) the "# of Str." column states how many strategies exist in the game; And (c) the "Sym" (i.e., symmetry) column states whether all players have the same strategies or not. The *coordination factor* $z(q)$ is $\frac{q-1}{n-1}$. New results are marked with a '⋆', the rest are recovered by our framework. PoA results are tight. SPoA results for Max-Cut, 2-NAE-SAT, SCGGs, CGGs and CGs are tight. The q-SPoA results still have a gap between lower and upper bounds.

Existence of Equilibria. While there exist clustering games on networks that do not possess any SE (see, e.g., [2]), for some special cases, it has been shown that an SE always exists if the size of the strategy space is 2 [2,13]. Theorem 1 extends this result to every clustering game with two strategies.

Our Techniques. In our analysis, we utilize equilibrium properties to order the agents of a coalition so that each agent does not benefit when deviating together with the agents following her. We obtain a lower bound on the total welfare of a coalition, with respect to a different outcome (typically an optimal one), and utilize the potential function to translate it into an expression which we break into combinatorial objects such as cuts – edges between disjoint sets of agents (e.g. the coalition and its complement), and interiors – edges between agents in the same set (e.g. edges between agents in the coalition). The obtained expression for the lower bound is generic enough to encompass all clustering games on networks, yet expressive enough to give each special case its unique treatment. For symmetric games, rather than providing a lower bound with respect to a single (optimal) outcome, we generalize the method from [13,14] and consider all the optimal outcomes that can be obtained by permuting the agents strategies. We combine all these lower bounds to achieve an improved lower bound which is quantified for each special case separately.

Related Work. PoA analysis was initiated by [17], and continued in a long line of studies in the algorithmic game theory literature. PoA with respect to SE and q-SE was first considered in [1]. The notion of *coalitional smoothness* was introduced by [3]. Specifically, if a game is (λ, μ)-coalitionally smooth, the SPoA is at least $\frac{\lambda}{1+\mu}$, and the same bound extends to more general solution concepts (see [3] and references therein for details on strong correlated and strong coarse correlated equilibria). In clustering games on networks, half the social welfare is a *potential function* [19], therefore a result from [3] implies that clustering games on networks are $(1/2, 0)$-coalitionally smooth, which implies that the SPoA is at least $1/2$ (which is tight for clustering games on networks). We improve the lower bound for SPoA in most of the special cases, and provide q-SPoA bounds, but our technique does not extend to the more general solution concepts.

Many clustering models have been studied from a game theoretic perspective. In [12] two variants are considered, and in both the weights of the edges are derived from a metric. In one the utilities are computed differently, and in the other the number of clusters is not pre-defined. Moreover, only NE is studied. In [7], all edges are coordination edges and the utility of each player is the utility of clustering games on networks, divided by the number of players that selected the same strategy, therefore each player derives her utility not only from her neighbours but from everyone in her cluster. Finally, clustering, i.e. the partitioning of objects with respect to similarity measures, is an area of research with an explosive amount of studies (e.g. [5]).

2   Model and Preliminaries

A Clustering Game (CG) is a tuple $\langle G = (V, E), (w_e)_{e \in E}, (b_e \in \{0, 1\})_{e \in E}, (\Sigma_i)_{i \in V} \rangle$ where G is an undirected graph with no self-loops. Each node corresponds to a player, and Σ_i is the (finite) strategy space of player i. Each edge e has a weight $w_e \in \mathbb{R}_{\geq 0}$, and a type $b_e \in \{0, 1\}$, where 0 implies that e is an *anti-coordination* edge and 1 implies that e is a *coordination* edge. Let $|V| = n$, $|E| = m$. Each element $\sigma = (\sigma_1, \ldots, \sigma_n) \in \times_i \Sigma_i$ is an *outcome* of the game. For each edge $e = \{i, j\}$, if e is a coordination edge, then it is *satisfied* if and only if $\sigma_i = \sigma_j$; if e is an anti-coordination edge, then it is satisfied if and only if $\sigma_i \neq \sigma_j$. Let 1_e^σ equal 1 if the edge e is satisfied in outcome σ and 0 otherwise. The utility of player i is the weight of satisfied edges she is incident to, i.e., $u_i(\sigma) = \sum_{e:i\in e} w_e \cdot 1_e^\sigma$. A CG is *symmetric* if $\Sigma_i = \{1, \ldots, k\}$ for every player i. In this case we abuse notation and denote the strategy space by k.

We identify several games from the literature as special cases of clustering games on networks and introduce *symmetric coordination games on graphs* as another special case. Due to space limitations, the proof of the following proposition as well as most proofs are deferred to the full version.

Proposition 1. *The following games are special cases of clustering games.*

1. *Max-Cut games [13] are CGs of the form:* $\langle G, (w_e)_{e\in E}, (0)_{e\in E}, 2\rangle$
2. *2-NAE-SAT games [13] are CGs of the form:* $\langle G, (w_e)_{e\in E}, (b_e \in \{0,1\})_{e\in E}, 2\rangle$
3. *Max-k-Cut games [14] are CGs of the form:* $\langle G, (w_e)_{e\in E}, (0)_{e\in E}, k\rangle$
4. *Coordination games on graphs [2] are CGs of the form:* $\langle G, (w_e)_{e\in E},$ $(1)_{e\in E}, (\Sigma_i)_{i\in V}\rangle$
5. *Symmetric coordination games on graphs are CGs of the form:* $\langle G = (V, E),$ $(w_e)_{e\in E}, (1)_{e\in E}, k\rangle$.

Given an outcome σ, $(\sigma_i^*, \sigma_{-i})$ denotes the outcome where σ_i is replaced by $\sigma_i^* \in \Sigma_i$, e.g., $\sigma = (\sigma_i, \sigma_{-i})$. Given outcomes σ, σ^* and a set of players $A \subseteq V$, $(\sigma_A^*, \sigma_{-A})$ denotes the outcome where all players in A play by σ^*, and all players in A^c play by σ, where $A^c = V \setminus A$ (and so $\sigma = (\sigma_A, \sigma_{-A})$).

Definition 1. *A Nash equilibrium (NE)[1] is an outcome σ such that no player can strictly increase her utility by deviating unilaterally, i.e., if for every player i and strategy $a \in \Sigma_i$, $u_i(\sigma_i, \sigma_{-i}) \geq u_i(a, \sigma_{-i})$. σ is a q-strong equilibrium (q-SE) if there is no coalition A of size at most q and outcome σ_A^* of A's members such that $u_i(\sigma_A^*, \sigma_{-A}) > u_i(\sigma_A, \sigma_{-A})$ for all $i \in A$. For $q = n$, the outcome σ is called a strong equilibrium (SE).*

Throughout this paper the quality of an outcome σ is measured by its *social welfare*, i.e., the sum: $SW(\sigma) = \sum_{i\in V} u_i(\sigma)$. In addition, for every set of players $A \subseteq V$, denote the total welfare of the players in A by $SW_A(\sigma) = \sum_{i\in A} u_i(\sigma)$.

Theorem 1. *For every clustering game with two strategies, any optimal outcome is an SE.*

The degradation of social welfare is commonly quantified as follows: Consider a solution concept (e.g. a NE), and quantify the ratio between the social welfare of the worst solution and that of an optimal outcome. When the solution concept is a q-SE the ratio is called the *q-strong price of anarchy*, which is the measure of interest in this paper. Note that for $q = 1$ the ratio is the *price of anarchy* (PoA), and for $q = n$ it is the *strong price of anarchy* (SPoA) which was defined in [1]. Formally, given a class of games Γ, let $\mathcal{G} \in \Gamma$, and let $q\text{-}SE(\mathcal{G})$ be the set of q-SE in game $\mathcal{G}$, and let σ^* be an optimal outcome of $\mathcal{G}$ (i.e. $\sigma^* \in \arg\max_\sigma SW(\sigma)$):

---

[1] In this paper we restrict attention to pure-strategy equilibrium (i.e., no randomization is used in an agent's behavior).

$$q\text{-SPoA} = \min_{\mathcal{G} \in \Gamma} \frac{\min\{SW(\sigma) : \sigma \in q\text{-}SE(\mathcal{G})\}}{SW(\sigma^*)} \tag{1}$$

By definition, the q-SPoA ranges from 0 to 1, where 1 corresponds to full efficiency. In this terminology lower bounds are positive results on the efficiency and upper bounds are negative results. Note that in order to lower bound the q-SPoA, it is enough to bound the term $SW(\sigma)/SW(\sigma^*)$ for every q-SE σ.

2.1   Welfare Guarantees in Equilibrium

In this section we establish the lemmas required to analyse the efficiency loss quantified by the q-SPoA in various special cases of clustering games on networks.

For two m-vectors $v = (v_1, \ldots v_m)$, $u = (u_1, \ldots u_m)$, let $v \cdot u = (v_1 \cdot u_1, \ldots, v_m \cdot u_m)$, and denote by $\langle v \rangle$ the inner product with the edge weights, i.e., $\langle v \rangle = \sum_{e \in E} w_e \cdot v_e$. Observe that the operator $\langle \cdot \rangle$ is linear, i.e., given vectors $v^1, \ldots, v^k$, it holds that: $\sum_{i=1}^k \langle v^i \rangle = \sum_{i=1}^k \sum_{e \in E} w_e \cdot v_e^i = \sum_{e \in E} w_e \cdot \sum_{i=1}^k v_e^i = \langle \sum_{i=1}^k v^i \rangle$.

Edge Partition. Let $1^\sigma = (1_e^\sigma)_{e \in E}$. Consider a pair of outcomes σ and σ^*. Let $\mathcal{B}$ be the characteristic vector of edges that are satisfied *both* by σ and σ^*, i.e., $\mathcal{B} = 1^\sigma \cdot 1^{\sigma^*}$. Let $\mathcal{E}$ be the characteristic vector of edges that are satisfied in σ but not satisfied in σ^*, i.e., $\mathcal{E} = 1^\sigma \cdot \overline{1^{\sigma^*}}$. Let $\mathcal{O}$ be the characteristic vector of edges that are satisfied in σ^* but not satisfied in σ, i.e., $\mathcal{O} = \overline{1^\sigma} \cdot 1^{\sigma^*}$. The choice of these notations will become clear later, regardless, observe that $1^\sigma = \mathcal{B} + \mathcal{E}$ and $1^{\sigma^*} = \mathcal{B} + \mathcal{O}$. Sometimes we abuse notation and let a characteristic vector be the set of edges it represents (e.g. $e \in 1^\sigma$ if and only if e is satisfied in σ).

Consider a set of players A. Let 1_e^A equal 1 if $e \cap A \neq \emptyset$ and 0 otherwise. Let $1^A = (1_e^A)_{e \in E}$. The *interior of A* is the set $\{e \in E : e \subseteq A\}$. Observe that $\mathcal{I}^A = \overline{1^{A^c}}$ is the characteristic vector of the interior of A, since $e \subseteq A$ if and only if $e \cap A^c = \emptyset$. Moreover, for two disjoint sets of players $A, B \subseteq V$, the vector $\delta^{A,B} = 1^A \cdot 1^B$ is the characteristic vector of the A–B cut, since an edge is in the cut if and only if it intersects with A and B.

Lemma 1. *For every set of players $A \subseteq V$ it holds that:*
$SW_A(\sigma) = \langle (2 \cdot \mathcal{I}^A + \delta^{A,A^c}) \cdot 1^\sigma \rangle$. *As a result, if $A = V$ then $SW(\sigma) = 2\langle 1^\sigma \rangle$*

Clustering games admit a *potential function*. A function Φ is a potential if it encodes the difference in utility of a player when deviating from one strategy to another, i.e., $\forall i, \forall a, b \in \Sigma_i, \forall \sigma : u_i(a, \sigma_{-i}) - u_i(b, \sigma_{-i}) = \Phi(a, \sigma_{-i}) - \Phi(b, \sigma_{-i})$.

Theorem 2. *The function $\Phi(\sigma) = \langle 1^\sigma \rangle$ is a potential for every clustering game.*

A direct corollary of Theorem 2 and Lemma 1 is that for every outcome σ, $\Phi(\sigma) = \frac{1}{2} SW(\sigma)$, therefore the problem of bounding the q-SPoA reduces to bounding the ratio $\Phi(\sigma)/\Phi(\sigma^*)$ for every q-SE σ and optimal outcome σ^*. By rearranging the terms of the potential function property, we get:

$$u_i(a, \sigma_{-i}) = \Phi(a, \sigma_{-i}) - \Phi(b, \sigma_{-i}) + u_i(b, \sigma_{-i}) \tag{2}$$

The Renaming Procedure. Let σ be a q-SE, σ^* an optimal outcome, and $K \subseteq V$ a set of players of size at most q. If all players in K deviate together from σ to σ^*, then by the definition of a q-SE, there is some player $i \in K$ so that $u_i(\sigma) \geq u_i(\sigma^*_K, \sigma_{-K})$. Rename the players so that this is player 1. Similarly, there is some player i in $K' = K \setminus \{1\}$ so that $u_i(\sigma) \geq u_i(\sigma^*_{K'}, \sigma_{-K'})$. Rename the players in K' so that this is player 2. Iterate this argument to rename all players in K and conclude that for each $i \in K$ it holds that $u_i(\sigma) \geq u_i(\sigma^*_{\{i...|K|\}}, \sigma_{-\{i...|K|\}})$. To simplify notation, for each $i \in K$ let $p_{K,i}^{\sigma,\sigma^*} = (\sigma^*_{\{i...|K|\}}, \sigma_{-\{i...|K|\}})$. When clear in the context, we omit the outcomes and denote it by $p_{K,i}$. Note that $p_{K,1} = (\sigma^*_K, \sigma_{-K})$ and $p_{K,|K|+1} = \sigma$. By this notation, we get $u_i(\sigma) \geq u_i(p_{K,i})$ for all $i \in K$. Therefore, it holds that $SW_K(\sigma) = \sum_{i \in K} u_i(\sigma) \geq \sum_{i \in K} u_i(p_{K,i})$. For each $i \in K$ we apply (2) to $u_i(p_{K,i})$ and $u_i(p_{K,i+1})$ and we get that $\sum_{i \in K} u_i(p_{K,i}) = \sum_{i \in K} (\Phi(p_{K,i}) - \Phi(p_{K,i+1}) + u_i(p_{K,i+1}))$. Observe that the sum on the potential function telescopes. We conclude that

$$SW_K(\sigma) \geq \Phi(\sigma^*_K, \sigma_{-K}) - \Phi(\sigma) + \sum_{i \in K} u_i(p_{K,i+1}) \tag{3}$$

Given an ordering o on a set of players K, for every edge $e = \{i,j\}$ from the interior of K, if $o(i) < o(j)$ then let $[K]_e^{\sigma,\sigma^*} = 1_e^{(\sigma_i, \sigma^*_{-i})}$. For each edge $e \not\subseteq K$, let $[K]_e^{\sigma,\sigma^*} = 0$. Let $[K]^{\sigma,\sigma^*} = ([K]_e^{\sigma,\sigma^*})_e$. The essence of $[K]_e^{\sigma,\sigma^*}$ is that the *first* player in the edge e plays according to her strategy in σ, and the *second* player plays according to her strategy in σ^*. Therefore, when considering edges from $[K]^{\sigma,\sigma^*}$, we are guaranteed that *exactly* one player changes color (from σ to σ^*) on each edge in the interior of K. We prove in the full version that the right-most sum of (3) can be expressed as follows:

$$\sum_{i \in K} u_i(p_{K,i+1}) = \langle \delta^{K,K^c} \cdot 1^\sigma \rangle + \langle \mathcal{I}^K \cdot \left(1^\sigma + [K]^{\sigma,\sigma^*} \right) \rangle \tag{4}$$

The object $[K]^{\sigma,\sigma^*}$ formally describes the result of the renaming procedure, i.e., it encodes the fact that we are considering edges from the interior of K such that *one* player plays according to σ, and *the other* plays according to σ^*. Equation (3) together with (4) provides a welfare guarantee for a set of players of size q. To provide a welfare guarantee for a larger set of players A, $|A| > q$, apply (3) for each subset $K \subseteq A$, and sum all inequalities. The resulting equation is:

$$SW_A(\sigma) \geq \frac{q-1}{|A|-1} \langle \mathcal{I}^A \cdot (\mathcal{B} + \mathcal{O}) \rangle +$$

$$\binom{|A|-1}{q-1}^{-1} \sum_{K \subseteq A, |K|=q} \langle \left(\mathcal{I}^K \cdot [K]^{\sigma,\sigma^*} + \delta^{K,K^c} \cdot 1^{(\sigma^*_K, \sigma_{-K})} \right) \cdot (\mathcal{B} + \mathcal{O} + \mathcal{E}) \rangle \tag{5}$$

Let Π be the set of all permutations $\pi : \{1, \ldots, k\} \to \{1, \ldots, k\}$. Given a permutation π, let σ^*_π be the outcome in which every player i plays $\pi(\sigma^*_i)$, i.e., $\sigma^*_\pi = (\pi(\sigma^*_i))_{i \in V}$. For every permutation π, $1^{\sigma^*} = 1^{\sigma^*_\pi}$. Therefore, $\mathcal{B}$, $\mathcal{O}$ and $\mathcal{E}$

are *permutation invariant*, i.e., for every π, $\mathcal{B} = 1^\sigma \cdot 1^{\sigma^*_\pi}$, and similarly for $\mathcal{O}$ and $\mathcal{E}$. Permutation invariance was previously considered for Max-k-Cut games [14]. Let $D_\pi(\sigma, \sigma^*)$ be the set of players with *different* strategies in σ and σ^*_π, i.e., $D_\pi(\sigma, \sigma^*) = \{i : (\sigma^*_\pi)_i \neq \sigma_i\}$. When clear in the context we omit the outcomes and write D_π. Let $q_\pi = \min\{q, |D_\pi|\}$. Lemma 2 establishes our general inequality for symmetric clustering games.

Lemma 2. *For every q-SE σ, and optimal outcome σ^*, it holds that:*

$$\sum_{\pi \in \Pi} SW_{D_\pi}(\sigma) \geq \sum_{\pi \in \Pi} \left(\langle \left(\frac{q_\pi - 1}{|D_\pi| - 1} \mathcal{I}^{D_\pi} + \delta^{D_\pi, D^c_\pi} \right) \cdot (\mathcal{B} + \mathcal{O}) \rangle + \right.$$
$$\left. \binom{|D_\pi| - 1}{q_\pi - 1}^{-1} \sum_{K \subseteq D_\pi, |K| = q_\pi} \langle \left(\mathcal{I}^K \cdot [K]^{\sigma, \sigma^*_\pi} + \delta^{K, D_\pi \setminus K} \cdot 1^{((\sigma^*_\pi)_K, \sigma_{-K})} \right) \cdot (\mathcal{B} + \mathcal{O} + \mathcal{E}) \rangle \right)$$

Lemma 3. *Let σ and σ^* be two outcomes. Then it holds that:* $\sum_{\pi \in \Pi} SW_{D_\pi}(\sigma) = (k - 1)(k - 1)! \, SW(\sigma)$.

Game Specific Analysis. For each special case of clustering games, given a set K, the values of $[K]^{\sigma, \sigma^*_\pi}_e$ and $1^{((\sigma^*_\pi)_K, \sigma_{-K})}_e$ in the right-hand side of Lemma 2 may be interpreted differently by making arguments that are specific to the special case. While nothing is assumed on the order of the players in any set K, since K is a subset of D_π, it is guaranteed that for each edge $e \in \mathcal{I}^K$, the outcome that is considered in $[K]^{\sigma, \sigma^*_\pi}_e$ is such that one player plays her strategy in σ, while the other player plays her strategy in σ^*_π *which is different* from her strategy in σ. Similarly, in the outcome $((\sigma^*_\pi)_K, \sigma_{-K})$, all the players in K play their strategies in σ^*_π *which are different* from their strategies in σ.

3 Strong Price of Anarchy Bounds

In this section we show how to utilize the proposed framework to establish bounds on the q-strong price of anarchy. Define the *coordination factor* to be $z(q) = \frac{q-1}{n-1}$, with $z(1) = 0$ and $z(n) = 1$, which correspond to the PoA and SPoA, respectively. Intuitively, it measures the amount of coordination that is assumed to exist in the regarded solution concept. Our first result generalizes the lower bound in [2].

Theorem 3. *The q-SPoA of every clustering game is at least $\frac{z(q)}{2}$*

Proof. Let σ be a q-SE and σ^* an optimal outcome. Substitute V for A in (5) and omit all terms with $[K]^{\sigma, \sigma^*}_e$ and $1^{(\sigma^*_K, \sigma_{-K})}$ (since they are non-negative). Observe that $1^V = (1, \ldots, 1)$ and recall that $\langle 1^{\sigma^*} \rangle = \frac{1}{2} SW(\sigma^*)$. Therefore,

$$SW(\sigma) = SW_V(\sigma) \geq \frac{q-1}{n-1} \langle \mathcal{B} + \mathcal{O} \rangle = z(q) \cdot \langle 1^{\sigma^*} \rangle = z(q) \cdot \frac{1}{2} \cdot SW(\sigma^*)$$

$\square$

A tight upper bound of $1/2$ for $q = n$, and an upper bound of $\frac{z(q)}{2-z(q)}$ for every q was shown in [2] for coordination games on graphs. For $q = n$, the same lower bound is also achieved by Theorem 10 in [3]. Indeed, clustering games on networks are utility-maximization potential games with only positive externalities, and $\Phi(\sigma) = \frac{1}{2}SW(\sigma)$. Therefore, the game is $(1/2, 0)$-*coalitionally smooth*[2] which implies that the SPoA is at least $1/2$. This bound extends to more general solution concepts (see [3] for details regarding mixed, correlated and coarse-correlated strong equilibria).

3.1   Symmetric Coordination Games on Graphs (SCGGs)

When all edges are coordination edges, and all players have the same strategy space $\{1, \ldots, k\}$, the class that is obtained is SCGGs with k strategies.

Theorem 4. *The SPoA of SCGGs with k strategies is at least $\frac{k}{2k-1}$.*

Proof. (sketch.) Let σ be an SE and σ^* an optimal outcome. Consider Lemma 2 in the context of SCGGs. Since all edges are coordination edges, any edge e that is satisfied in σ or σ_π^* is certainly unsatisfied when exactly one of its adjacent nodes changes strategy, i.e. $[D_\pi]_e^{\sigma, \sigma^*} = 0$. Therefore, for $q = n$ we get:

$$\sum_{\pi \in \Pi} SW_{D_\pi}(\sigma) \geq \sum_{\pi \in \Pi} \left\langle \left(\mathcal{I}^{D_\pi} + \delta^{D_\pi, D_\pi^c} \right) \cdot (\mathcal{B} + \mathcal{O}) \right\rangle \tag{6}$$

By Lemma 3, in the left-hand side we get $(k-1)(k-1)!\,SW(\sigma)$. In the right-hand side, since $\mathcal{B}$ and $\mathcal{O}$ are permutation invariant, we get four expressions:

$$\left\langle \left(\sum_\pi \mathcal{I}^{D_\pi} \right) \cdot \mathcal{B} \right\rangle + \left\langle \left(\sum_\pi \delta^{D_\pi, D_\pi^c} \right) \cdot \mathcal{B} \right\rangle + \left\langle \left(\sum_\pi \mathcal{I}^{D_\pi} \right) \cdot \mathcal{O} \right\rangle + \left\langle \left(\sum_\pi \delta^{D_\pi, D_\pi^c} \right) \cdot \mathcal{O} \right\rangle \tag{7}$$

We use simple combinatorial arguments to compute the expressions above.

First expression: $e = \{i, j\} \in \mathcal{B}$ if and only if $\sigma_i = \sigma_j$ and $\sigma_i^* = \sigma_j^*$. In such a case, for every permutation π we get that $e \in \mathcal{I}^{D_\pi}$ if and only if $\pi(\sigma_i^*) \neq \sigma_i$. There are $(k-1)$ options to fix $\pi(\sigma_i^*)$, and for each such option there are $(k-1)!$ options to set the other $(k-1)$ values of π (as there are no additional restrictions). Therefore, $\left\langle \left(\sum_\pi \mathcal{I}^{D_\pi} \right) \cdot \mathcal{B} \right\rangle = (k-1)(k-1)!\,\langle \mathcal{B} \rangle$. Second expression: Since e cannot be in the D_π–D_π^c cut, it holds that $\left\langle \left(\sum_\pi \delta^{D_\pi, D_\pi^c} \right) \cdot \mathcal{B} \right\rangle = 0$. Third expression: $e = \{i, j\} \in \mathcal{O}$ if and only if $\sigma_i \neq \sigma_j$ and $\sigma_i^* = \sigma_j^*$. In such a case $e \in \mathcal{I}^{D_\pi}$ if and only if $\{\pi(\sigma_i^*)\} \cap \{\sigma_i, \sigma_j\} = \emptyset$. There are $(k-2)$ options to fix $\pi(\sigma_i^*)$, and for each such option there are $(k-1)!$ options to set the other $(k-1)$ values of π (as there are no additional restrictions). Therefore, $\left\langle \left(\sum_\pi \mathcal{I}^{D_\pi} \right) \cdot \mathcal{O} \right\rangle = (k-2)(k-1)!\,\langle \mathcal{O} \rangle$. Fourth expression: $e \in \delta^{D_\pi, D_\pi^c}$ in exactly two disjoint events: $\pi(\sigma_i^*) = \sigma_i$ or $\pi(\sigma_j^*) = \sigma_j$, therefore, $\left\langle \left(\sum_\pi \delta^{D_\pi, D_\pi^c} \right) \cdot \mathcal{O} \right\rangle = 2(k-1)!\,\langle \mathcal{O} \rangle$.

In total, we conclude that (7) is at least $k!\,\Phi(\sigma^*) - (k-1)!\,\Phi(\sigma)$. Recall that $SW(\sigma) = 2 \cdot \Phi(\sigma)$. Divide both sides by $(k-1)!$ and reorganize terms to get $(2k-1) \cdot \Phi(\sigma) \geq k \cdot \Phi(\sigma^*)$, as desired. $\square$

---

[2] The reader is referred to [3] for the exact definition.

The following proposition shows that Theorem 4 is tight.

Proposition 2. *The symmetric coordination game on a line graph with $2k$ nodes and k strategies for each player has a SPoA of $\frac{k}{2k-1}$.*

Theorem 4 extends to q-SPoA case as follows:

Theorem 5. *The q-SPoA of SCGGs with k strategies is at least $\frac{2+z(q)\cdot(k-2)}{2k-z(q)}$.*

Proposition 3 shows that the PoA (i.e., for $q = 1$) bound is tight.

Proposition 3. *There exists a SCGG with k strategies, with a PoA $= 1/k$.*

3.2   Symmetric Anti-coordination Games on Graphs

When all edges are anti-coordination edges, and all players have the same strategy space $\{1, \ldots, k\}$, the class that is obtained is Max-k-Cut [14]. Previous work [15,18] showed that the PoA exactly $(k-1)/k$. For $k = 2$, [13] showed that the SPoA is exactly $2/3$. In [14] an upper bound of $\frac{2k-2}{2k-1}$ was established[3]. Theorem 6 uses the framework to establish lower bounds on the q-SPoA for any k and q.

Theorem 6. *The q-SPoA of Max-k-Cut games is at least: $\frac{k-1}{k-\frac{1}{2(k-1)}\cdot z(q)}$.*

3.3   Symmetric Clustering Games on Networks (SCGs)

In this class, edges are either coordination or anti-coordination edges, and all players have k strategies. For $k = 2$, the class coincides with 2-NAE-SAT, for which a tight SPoA bound of $2/3$ was shown [13]. We establish a bound on the q-SPoA for every q and k:

Theorem 7. *The q-SPoA of SCGs with k strategies is at least $\frac{2+z(q)\cdot(k-2)}{2k-z(q)\cdot\frac{1}{k-1}}$.*

A special case of Theorem 7 is that the SPoA of SCGs is at least $\frac{1}{2-\frac{1}{k(k-1)}}$, for which a simplified proof is provided in the full version. For the class of 2-NAE-SAT ($k = 2$), it was shown in [13] that for $q = O(n^{\frac{1}{2}-\epsilon})$ (for any $\epsilon > 0$), the q-SPoA is $1/2$. Theorem 8 shows this is true even for $q = O(n^{1-\epsilon})$.

Theorem 8. *For any $\epsilon > 0$ and $q = O(n^{1-\epsilon})$, the q-SPoA of Max-Cut is $1/2$.*

On the other hand, the positive result that is implied by Theorem 7, is that for any $\alpha \in (0,1]$ and $z(q) \geq \alpha$, (roughly, when $q \geq \alpha n$), the q-SPoA is $\frac{2}{4-\alpha}$.

Corollary 1. *The q-SPoA of 2-NAE-SAT games is at least $\frac{2}{4-z(q)}$.*

---

[3] In [14], the authors also presented a matching lower bound of $\frac{2k-2}{2k-1}$. However, one of the derivations in the analysis contained an error. This was verified by personal communication with the authors. Their corrected analysis leads to a lower bound that matches the lower bound, established in this paper in Theorem 6 for $z(q) = 1$.

4   Future Directions

Our results and analysis suggest interesting directions for future research. First, the existence of q-SE in clustering games on networks is only partially understood. For example, it is conjectured in [14] that every Max-k-Cut game admits an SE. This is clearly an interesting open problem, as well as the more general problem of whether all symmetric clustering games on networks possess an SE.

A full characterization of q-SE existence for clustering games on networks is an interesting open problem.

Second, our analysis leaves a gap in the q-SPoA for Max-k-Cut games. We note that the analysis of this gap gives rise to a combinatorial problem that is of independent interest.

Third, our proof techniques provide q-SPoA results only with respect to pure equilibria. It is desired to extend this analysis to handle other solution concepts such as mixed, correlated and coarse correlated equilibria. Finally, it would be interesting to explore ways in which our analysis can shed light on coalitional dynamics in clustering games on networks.

References

1. Andelman, N., Feldman, M., Mansour, Y.: Strong price of anarchy. In: SODA, pp. 189–198 (2007)
2. Apt, K.R., Rahn, M., Schäfer, G., Simon, S.: Coordination games on graphs (Extended Abstract). In: Liu, T.-Y., Qi, Q., Ye, Y. (eds.) WINE 2014. LNCS, vol. 8877, pp. 441–446. Springer, Heidelberg (2014)
3. Bachrach, Y., Syrgkanis, V., Tardos, É., Vojnović, M.: Strong price of anarchy, utility games and coalitional dynamics. In: Lavi, R. (ed.) SAGT 2014. LNCS, vol. 8768, pp. 218–230. Springer, Heidelberg (2014)
4. Banerjee, S., Konishi, H., Sönmez, T.: Core in a simple coalition formation game. Social Choice and Welfare 18(1), 135–153 (2001)
5. Bansal, N., Blum, A., Chawla, S.: Correlation clustering. Machine Learning 56(1–3), 89–113 (2004)
6. Barberà, S., Gerber, A.: On coalition formation: durable coalition structures. Mathematical Social Sciences 45(2), 185–203 (2003)
7. Bilò, V., Fanelli, A., Flammini, M., Monaco, G., Moscardelli, L.: Nash stability in fractional hedonic games. In: Liu, T.-Y., Qi, Q., Ye, Y. (eds.) WINE 2014. LNCS, vol. 8877, pp. 486–491. Springer, Heidelberg (2014)
8. Bloch, F., Diamantoudi, E.: Noncooperative formation of coalitions in hedonic games. International Journal of Game Theory 40(2), 263–280 (2011)
9. Bogomolnaia, A., Jackson, M.O.: The stability of hedonic coalition structures. Games and Economic Behavior 38(2), 201–230 (2002)
10. Diamantoudi, E., Xue, L.: Farsighted stability in hedonic games. Social Choice and Welfare 21(1), 39–61 (2003)
11. Dreze, J.H., Greenberg, J.: Hedonic coalitions: Optimality and stability. Econometrica: Journal of the Econometric Society, 987–1003 (1980)
12. Feldman, M., Lewin-Eytan, L., Naor, J.: Hedonic clustering games. In: SPAA, pp. 267–276. ACM (2012)

13. Gourvès, L., Monnot, J.: On strong equilibria in the max cut game. In: Leonardi, S. (ed.) WINE 2009. LNCS, vol. 5929, pp. 608–615. Springer, Heidelberg (2009)
14. Gourvès, L., Monnot, J.: The max k-cut game and its strong equilibria. In: Kratochvíl, J., Li, A., Fiala, J., Kolman, P. (eds.) TAMC 2010. LNCS, vol. 6108, pp. 234–246. Springer, Heidelberg (2010)
15. Hoefer, M.: Cost sharing and clustering under distributed competition. Ph.D thesis, University of Konstanz (2007)
16. Kearns, M., Littman, M.L., Singh, S.: Graphical models for game theory. In: UAI, pp. 253–260. Morgan Kaufmann Publishers Inc. (2001)
17. Koutsoupias, E., Papadimitriou, C.: Worst-case equilibria. In: Meinel, C., Tison, S. (eds.) STACS 1999. LNCS, vol. 1563, p. 404. Springer, Heidelberg (1999)
18. Kun, J., Powers, B., Reyzin, L.: Anti-coordination games and stable graph colorings. In: Vöcking, B. (ed.) SAGT 2013. LNCS, vol. 8146, pp. 122–133. Springer, Heidelberg (2013)
19. Monderer, D., Shapley, L.: Potential games. Games and economic behavior **14**(1), 124–143 (1996)
20. Nisan, N., Roughgarden, T., Tardos, É., Vazirani, V.V.: Algorithmic game theory, chapter 7. Cambridge University Press (2007)

On the Diameter of Hyperbolic Random Graphs

Tobias Friedrich[1,2] and Anton Krohmer[1,2](✉)

[1] Friedrich-Schiller-Universität Jena, Jena, Germany
[2] Hasso Plattner Institute, Potsdam, Germany
anton.krohmer@hpi.de

Abstract. Large real-world networks are typically scale-free. Recent research has shown that such graphs are described best in a geometric space. More precisely, the internet can be mapped to a hyperbolic space such that geometric greedy routing performs close to optimal (Boguná, Papadopoulos, and Krioukov. Nature Communications, 1:62, 2010). This observation pushed the interest in hyperbolic networks as a natural model for scale-free networks. Hyperbolic random graphs follow a power-law degree distribution with controllable exponent β and show high clustering (Gugelmann, Panagiotou, and Peter. ICALP, pp. 573–585, 2012).

For understanding the structure of the resulting graphs and for analyzing the behavior of network algorithms, the next question is bounding the size of the diameter. The only known bound is $\mathcal{O}((\log n)^{32/((3-\beta)(5-\beta))})$ (Kiwi and Mitsche. ANALCO, pp. 26–39, 2015). We present two much simpler proofs for an improved upper bound of $\mathcal{O}((\log n)^{2/(3-\beta)})$ and a lower bound of $\Omega(\log n)$. If the average degree is bounded from above by some constant, we show that the latter bound is tight by proving an upper bound of $\mathcal{O}(\log n)$ for the diameter.

1 Introduction

Large real-world networks are almost always sparse and non-regular. Their degree distribution typically follows a *power law*, which is synonymously used for being *scale-free*. Since the 1960's, large networks have been studied in detail and hundreds of models were suggested. In the past few years, a new line of research emerged, which showed that scale-free networks can be modeled more realistically when incorporating *geometry*.

Euclidean Random Graphs. It is not new to study graphs in a geometric space. In fact, graphs with *Euclidean geometry* have been studied intensively for more than a decade. The standard Euclidean model are random geometric graphs which result from placing n nodes independently and uniformly at random on an Euclidean space, and creating edges between pairs of nodes if and only if their distance is at most some fixed threshold r. These graphs have been studied in relation to subjects such as cluster analysis, statistical physics, hypothesis testing, and wireless sensor networks [23]. The resulting graphs are more or less regular and hence do not show a scale-free behavior with power-law degree distribution as observed in large real-world graphs.

© Springer-Verlag Berlin Heidelberg 2015
M.M. Halldórsson et al. (Eds.): ICALP 2015, Part II, LNCS 9135, pp. 614–625, 2015.
DOI: 10.1007/978-3-662-47666-6_49

Table 1. Known diameter bounds for various random graphs. In all cases the diameter depends on the choice of the model parameters. Here we consider a constant average degree. For scale-free networks, we also assume a power law exponent $2 < \beta < 3$.[1]

Random Graph Model	Diameter	
Sparse Erdős-Rényi [5]	$\Theta(\log n)$ [24]	
d-dim. Euclidean [23]	$\Theta(n^{1/d})$ [15]	
Watts-Strogatz [26]	$\Theta(\log n)$ [6]	
Kleinberg [18]	$\Theta(\log n)$ [21]	
Chung-Lu [8]	$\Theta(\log n)$ [8]	⎫
Pref. Attachment [1]	$\Theta(\log \log n)$ [10]	⎬ power-law graphs
Hyperbolic [19]	$\mathcal{O}((\log n)^{\frac{32}{(3-\beta)(5-\beta)}})$ [17]	⎭

Hyperbolic Random Graphs. For modeling scale-free graphs, it is natural to apply a non-Euclidean geometry with negative curvature. Krioukov et al. [19] introduced a new graph model based on *hyperbolic geometry*. Similar to euclidean random graphs, nodes are uniformly distributed in a hyperbolic space and two nodes are connected if their hyperbolic distance is small. The resulting graphs have many properties observed in large real-world networks. This was impressively demonstrated by Boguná et al. [4]: They computed a maximum likelihood fit of the internet graph in the hyperbolic space and showed that greedy routing in this hyperbolic space finds nearly optimal shortest paths in the internet graph. The quality of this embedding is an indication that hyperbolic geometry naturally appears in large scale-free graphs.

Known Properties. A number of properties of hyperbolic random graphs have been studied. Gugelmann et al. [16] compute exact asymptotic expressions for the expected number of vertices of degree k and prove a constant lower bound for the clustering coefficient. They confirm that the clustering is non-vanishing and that the degree sequence follows a power-law distribution with controllable exponent β. For $2 < \beta < 3$, the hyperbolic random graph has a giant component of size $\Omega(n)$ [2,3], similar to other scale-free networks like Chung-Lu [8]. Other studied properties include the clique number [14], bootstrap percolation [7]; as well as algorithms for efficient generation of hyperbolic random graphs [25] and efficient embedding of real networks in the hyperbolic plane [22].

Diameter. The diameter, the length of the longest shortest path, is a fundamental property of a network. It also sets a worst-case lower bound on the number of steps required for all communication processes on the graph. In contrast to the average distance, it is determined by a single—atypical—long path. Due to this sensitivity to small changes, it is notoriously hard to analyze. Even subtle

---

[1] Note that the table therefore refers to a non-standard Preferential Attachment version with adjustable power law exponent $2 < \beta < 3$ (normally, $\beta = 3$).

changes to the graph model can make an exponential difference in the diameter, as can be seen when comparing Chung-Lu (CL) random graphs [8] and Preferential Attachment (PA) graphs [1] in the considered range of the power law exponent $2 < \beta < 3$: On the one hand, we can embed a CL graph in the PA graph and they behave effectively the same [13]; on the other hand, the diameter of CL graphs is $\Theta(\log n)$ [8] while for PA graphs it is $\Theta(\log \log n)$ [10]. Table 1 provides an overview over existing results. It was open so far how the diameter of hyperbolic random graphs compares to the aforementioned bounds for other scale-free graph models. The only known result for their diameter is $\mathcal{O}((\log n)^{\frac{32}{(3-\beta)(5-\beta)}})$ by Kiwi and Mitsche [17].

Our Contribution. We improve upon the result of Kiwi and Mitsche [17] in the three directions, as described by the following theorems. First, we present a much simpler proof which also shows polylogarithmic upper bound for the diameter, but with a better (i.e. smaller) exponent.

Theorem 1. *Let $2 < \beta < 3$. The diameter of the giant component in the hyperbolic random graph $\mathcal{G}(n, \alpha, C)$ is $\mathcal{O}((\log n)^{\frac{2}{3-\beta}})$ with probability $1 - \mathcal{O}(n^{-3/2})$.*

The proof of Theorem 1 is presented in Section 3. It serves as an introduction to the more involved proof of a logarithmic upper bound for the diameter presented in Section 4. There we show with more advanced techniques that for small average degrees the following theorem holds.

Theorem 2. *Let $2 < \beta < 3$, and C be a large enough constant. Then, the diameter of the giant component in the hyperbolic random graph $\mathcal{G}(n, \alpha, C)$ is $\mathcal{O}(\log n)$ with probability $1 - \mathcal{O}(n^{-3/2})$.*

The logarithmic upper bound is best possible. In particular, we show that Theorem 2 is tight by presenting the following matching lower bound.

Theorem 3. *Let $2 < \beta < 3$. Then, the diameter of the giant component in the hyperbolic random graph $\mathcal{G}(n, \alpha, C)$ is $\Omega(\log n)$ with probability $1 - n^{-\Omega(1)}$.*

Due to space constraints, the proof of Theorem 3 can be found in the long version. We point out that although we prove all diameter bounds on the giant component, our proofs will make apparent that the giant component is in fact the component with the largest diameter in the graph.

Used Techniques. Our formal analysis of the diameter has to deal with a number of technical challenges. First, in contrast to proving a bound on the average distance, it is not possible to average over all path lengths. In fact, it is not even sufficient to exclude a certain kind of path with probability $1 - \mathcal{O}(n^{-c})$; as this has to hold for all possible $\Omega(n!)$ paths. This makes a union bound inapplicable. We solve this by introducing *upwards paths* (cf. Definition 12), which are in a sense "almost" shortest paths, and of which there are only two per node. We prove deterministically that their length asymptotically bounds the diameter. Then, we bound the length of a single upwards path by a multiplicative drift argument known from evolutionary computation [20]; and show that the length of conjunctions of upwards paths follows an Erlang distribution.

A second major challenge is the fact that a probabilistic analysis of shortest paths (and likewise, upwards paths) typically uncovers the probability space in a consecutive fashion. Revealing the positions of nodes on the path successively introduces strong stochastic dependencies that are difficult to handle with probabilistic tail bounds [11]. Instead of studying the stochastic dependence structure in detail, we use the geometry and model the hyperbolic random graph as a Poisson point process. This allows us to analyze different areas in the graph independently, which in turn supports our stochastic analysis of shortest paths.

2   Notation and Preliminaries

In this section, we briefly introduce hyperbolic random graphs. Although this paper is self-contained, we recommend to a reader who is unfamiliar with the notion of hyperbolic random graphs the more thorough investigations [16,19].

Let $\mathbb{H}_2$ be the hyperbolic plane. Following [19], we use the *native* representation; in which a point $v \in \mathbb{H}_2$ is represented by polar coordinates (r_v, φ_v); and r_v is the hyperbolic distance of v to the origin.[2]

To construct a hyperbolic random graph $G(n, \alpha, C)$, consider now a circle D_n with radius $R = 2 \ln n + C$ that is centered at the origin of $\mathbb{H}_2$. Inside D_n, n points are distributed independently as follows. For each point v, draw φ_v uniformly at random from $[0, 2\pi)$, and draw r_v according to the probability density function

$$\rho(r) := \frac{\alpha \sinh(\alpha r)}{\cosh(\alpha R) - 1} \approx \alpha e^{\alpha(r-R)}.$$

Next, connect two points u, v if their hyperbolic distance is at most R, i.e. if

$$d(u, v) := \cosh^{-1}(\cosh(r_u) \cosh(r_v) - \sinh(r_u) \sinh(r_v) \cos(\Delta\varphi_{u,v})) \leqslant R. \quad (1)$$

By $\Delta\varphi_{u,v}$ we describe the small relative angle between two nodes u, v, i.e. $\Delta\varphi_{u,v} := \cos^{-1}(\cos(\varphi_u - \varphi_v)) \leqslant \pi$.

This results in a graph whose degree distribution follows a power law with exponent $\beta = 2\alpha + 1$, if $\alpha \geqslant \frac{1}{2}$, and $\beta = 2$ otherwise [16]. Since most real-world networks have been shown to have a power law exponent $2 < \beta < 3$, we assume throughout the paper that $\frac{1}{2} < \alpha < 1$. Gugelmann et al. [16] proved that the average degree in this model is then $\delta = (1 + o(1)) \frac{2\alpha^2 e^{-C/2}}{\pi(\alpha - 1/2)^2}$.

We now present a handful of Lemmas useful for analyzing the hyperbolic random graph. Most of them are taken from [16]. We begin by an upper bound for the angular distance between two connected nodes. Consider two nodes with radial coordinates r, y. Denote by $\theta_r(y)$ the maximal radial distance such that these two nodes are connected. By equation (1),

$$\theta_r(y) = \arccos\left(\frac{\cosh(y)\cosh(r) - \cosh(R)}{\sinh(y)\sinh(r)}\right). \quad (2)$$

This terse expression is closely approximated by the following Lemma.

---

[2] Note that this seemingly trivial fact does not hold for conventional models (e.g. Poincaré halfplane) for the hyperbolic plane.

Lemma 4 ([16]). *Let* $0 \leqslant r \leqslant R$ *and* $y \geqslant R - r$. *Then,*

$$\theta_r(y) = \theta_y(r) = 2e^{\frac{R-r-y}{2}}(1 \pm \Theta(e^{R-r-y})).$$

For most computations on hyperbolic random graphs, we need expressions for the probability that a sampled point falls into a certain area. To this end, Gugelmann et al. [16] define the *probability measure* of a set $S \subseteq D_n$ as

$$\mu(S) := \int_S f(y)\,dy,$$

where $f(r)$ is the probability mass of a point $p = (r, \varphi)$ given by $f(r) := \frac{\rho(r)}{2\pi} = \frac{\alpha \sinh(\alpha r)}{2\pi(\cosh(\alpha R)-1)}$. We further define the *ball* with radius x around a point (r, φ) as

$$B_{r,\varphi}(x) := \{(r', \varphi') \mid d((r', \varphi'), (r, \varphi)) \leqslant x\}.$$

We write $B_r(x)$ for $B_{r,0}(x)$. Note that $D_n = B_0(R)$. Using these definitions, we can formulate the following Lemma.

Lemma 5 ([16]). *For any* $0 \leqslant r \leqslant R$ *we have*

$$\mu(B_0(r)) = e^{-\alpha(R-r)}(1 + o(1)) \tag{3}$$

$$\mu(B_r(R) \cap B_0(R)) = \frac{2\alpha e^{-r/2}}{\pi(\alpha - 1/2)} \cdot (1 \pm \mathcal{O}(e^{-(\alpha-1/2)r} + e^{-r})) \tag{4}$$

Since we often argue over sequences of nodes on a path, we say that a node v is *between* two nodes u, w, if $\Delta\varphi_{u,v} + \Delta\varphi_{v,w} = \Delta\varphi_{u,w}$. Recall that $\Delta\varphi_{u,v} \leqslant \pi$ describes the small angle between u and v. E.g., if $u = (r_1, 0), v = (r_2, \frac{\pi}{2}), w = (r_3, \pi)$, then v lies between u and w. However, w does not lie between u and v as $\Delta\varphi_{u,v} = \pi/2$ but $\Delta\varphi_{u,w} + \Delta\varphi_{w,v} = \frac{3}{4}\pi$.

Finally, we define the area $B_I := B_0(R - \frac{\log R}{1-\alpha} - c)$ as the *inner band*, and $B_O := D_n \setminus B_I$ as the *outer band*, where $c \in \mathbb{R}$ is a large enough constant.

The Poisson Point Process. We often want to argue about the probability that an area $S \subseteq D_n$ contains one or more nodes. To this end, we usually apply the simple formula

$$\Pr[\exists v \in S] = 1 - (1 - \mu(S))^n \geqslant 1 - \exp(-n \cdot \mu(S)). \tag{5}$$

Unfortunately, this formula significantly complicates once the positions of some nodes are already known. This introduces conditions on $\Pr[\exists v \in S]$ which can be hard to grasp analytically. To circumvent this problem, we use a Poisson point process $\mathcal{P}_n$ [23] which describes a different way of distributing nodes inside D_n. It is fully characterized by the following two properties:

- If two areas S, S' are disjoint, then the number of nodes that fall within S and S' are independent random variables.
- The expected number of points that fall within S is $\int_S n\mu(S)$.

One can show that these properties imply that the number of nodes inside S follows a Poisson distribution with mean $n\mu(S)$. In particular, we obtain that the number of nodes $|\mathcal{P}_n|$ inside D_n is distributed as $\text{Po}(n)$, i.e. $\mathbb{E}[|\mathcal{P}_n|] = n$, and

$$\Pr(|\mathcal{P}_n| = n) = \frac{e^{-n} n^n}{n!} = \Theta(n^{-1/2}).$$

Let the random variable $\mathcal{G}(\mathcal{P}_n, n, \alpha, C)$ denote the resulting graph when using the Poisson point process to distribute nodes inside D_n. Since it holds

$$\Pr[\mathcal{G}(\mathcal{P}_n, n, \alpha, C) = G \mid |\mathcal{P}_n| = n] = \Pr[\mathcal{G}(n, \alpha, C) = G],$$

we have that every property p with $\Pr[p(\mathcal{G}(\mathcal{P}_n, n, \alpha, C))] \leqslant \mathcal{O}(n^{-c})$ holds for the hyperbolic random graphs with probability $\Pr[p(\mathcal{G}(n, \alpha, C))] \leqslant \mathcal{O}(n^{\frac{1}{2}-c})$.

We explicitly state whenever we use the Poisson point process $\mathcal{G}(\mathcal{P}_n, n, \alpha, C)$ instead of the normal hyperbolic random graph $\mathcal{G}(n, \alpha, C)$. In particular, we can use a matching expression for equation (5): $\Pr[\exists v \in S] = 1 - \exp(-n \cdot \mu(S))$.

3   Polylogarithmic Upper Bound

As an introduction to the main proof, we first show a simple polylogarithmic upper bound on the diameter of the hyperbolic random graph. We start by investigating nodes in the inner band B_I and show that they are connected by a path of at most $\mathcal{O}(\log \log n)$ nodes. We prove this by partitioning D_n into R layers of constant thickness 1. Then, a node in layer i has radial coordinate $\in (R - i, R - i + 1]$. We denote the layer i by $L_i := B_0(R - i + 1) \setminus B_0(R - i)$.

Lemma 6. *Let $1 \leqslant i, j \leqslant R/2$, and consider two nodes $v \in L_i, w \in L_j$. Then,*

$$\frac{2}{e} e^{\frac{i+j-R}{2}} (1 - \Theta(e^{i+j-R})) \leqslant \theta_{r_u}(r_v) \leqslant 2 e^{\frac{i+j-R}{2}} (1 + \Theta(e^{i+j-R})),$$

Furthermore, we have $\mu(L_j \cap B_R(v)) = \Theta(e^{-\alpha j + \frac{i+j-R}{2}})$, and, if $(i+j)/R < 1 - \varepsilon$ for some constant $\varepsilon > 0$, we have for large n

$$\frac{1}{e} e^{-\alpha j + \frac{i+j-R}{2}} \leqslant \mu(L_j \cap B_R(v)) \leqslant 4 e^{-\alpha j + \frac{i+j-R}{2}}.$$

Proof. The statements follow directly from Lemmas 4 and 5 and the fact that we have $R - i < r_v \leqslant R - i + 1$ for a node $v \in L_i$.   □

Using Lemma 6, we can now prove that a node $v \in B_I$ has a path of length $\mathcal{O}(\log \log n)$ that leads to $B_0(R/2)$. Recall that the inner band was defined as $B_I := B_0(R - \frac{\log R}{1-\alpha} - c)$, where c is a large enough constant.

Lemma 7. *Consider a node v in layer i. With probability $1 - \mathcal{O}(n^{-3})$ it holds*

1. *if $i \in [\frac{\log R}{1-\alpha} + c, \frac{2 \log R}{1-\alpha} + c]$, then v has a neighbor in layer L_{i+1}, and*
2. *if $i \in [\frac{2 \log R}{1-\alpha} + c, R/2]$, then v has a neighbor in layer L_j for $j = \frac{\alpha}{2\alpha-1} i$.*

Proof. The probability that node $v \in L_i$ does not contain a neighbor in L_{i+1} is

$$(1 - \Theta(e^{-\alpha(i+1)+i+\frac{1-R}{2}}))^n \leqslant \exp(-\Theta(1) \cdot e^{\log R + c(1-\alpha)}).$$

Since $R = 2 \log n + C$ and c is a large enough constant, this proves part (1) of the claim. An analogous argument shows part (2).   □

Lemma 7 shows that there exists a path of length $\mathcal{O}(\log \log n)$ from each node $v \in B_I$ to some node $u \in B_0(R - \frac{2 \log R}{1-\alpha} - c)$. Similarly, from u there exists a path of length $\mathcal{O}(\log \log n)$ to $B_0(R/2)$ with high probability. Since we know that the nodes in $B_0(R/2)$ form a clique by the triangle inequality, we therefore obtain that all nodes in B_I form a connected component with diameter $\mathcal{O}(\log \log n)$.

Corollary 8. *Let $\frac{1}{2} < \alpha < 1$. With probability $1 - \mathcal{O}(n^{-3})$, all nodes $u, v \in B_I$ in the hyperbolic random graph are connected by a path of length $\mathcal{O}(\log \log n)$.*

3.1   Outer Band

By Corollary 8, we obtain that the diameter of the graph induced by nodes in B_I is at most $\mathcal{O}(\log \log n)$. In this section, we show that each component in B_O has a polylogarithmic diameter. Then, one can easily conclude that the overall diameter of the giant component is polylogarithmic, since all nodes in $B_0(R/2)$ belong to the giant component [3]. We begin by presenting one of the crucial Lemmas in this paper that will often be reused.

Lemma 9. *Let $u, v, w \in V$ be nodes such that v lies between u and w, and let $\{u, w\} \in E$. If $r_v \leqslant r_u$ and $r_v \leqslant r_w$, then v is connected to both u and w. If $r_v \leqslant r_u$ but $r_v \geqslant r_w$, then v is at least connected to w.*

Proof. By [3, Lemma 5.28], we know that if two nodes $(r_1, \varphi_1), (r_2, \varphi_2)$ are connected, then so are $(r_1', \varphi_1), (r_2', \varphi_2)$ where $r_1 \leqslant r_1'$ and $r_2' \leqslant r_2$. Since the distance between nodes is monotone in the relative angle $\Delta\varphi$, this proves the first part of the claim. The second part can be proven by an analogous argument.   □

For convenience, we say that an edge $\{u, w\}$ *passes under* v if one of the requirements of Lemma 9 is fulfilled. Using this, we are ready to show Theorem 1. In this argument, we investigate the angular distance a path can at most traverse until it passes under a node in B_I. By Lemma 9, we then have with high probability a short path to the center $B_0(R/2)$ of the graph.

(Proof of Theorem 1). Partition the hyperbolic disc into n disjoint sectors of equal angle $\Theta(1/n)$. The probability that k consecutive sectors contain no node in B_I is

$$(1 - \Theta(k/n) \cdot \mu(B_0(R - \tfrac{\log R}{1-\alpha} - c)))^n \leqslant \exp(-\Theta(1) \cdot k \cdot e^{-\alpha \log R/(1-\alpha)})$$
$$= \exp(-\Theta(1) \cdot k \cdot (\log n)^{-\frac{\alpha}{1-\alpha}}).$$

Hence, we know that with probability $1 - \mathcal{O}(n^{-3})$, there are no $k :=$ $\Theta((\log n)^{\frac{1}{1-\alpha}})$ such consecutive sectors. By a Chernoff bound, the number of nodes in k such consecutive sectors is $\Theta((\log n)^{\frac{1}{1-\alpha}})$ with probability $1 - \mathcal{O}(n^{-3})$. Applying a union bound, we get that with probability $1 - \mathcal{O}(n^{-2})$, every sequence of k consecutive sectors contains at least one node in B_I and at most $\Theta(k)$ nodes in total. Consider now a node $v \in B_O$ that belongs to the giant component. Then, there must exist a path from v to some node $u \in B_I$. By Lemma 9, this path can visit at most k sectors—and therefore use at most $\Theta(k)$ nodes—before reaching u. From u, there is a path of length $\mathcal{O}(\log \log n)$ to the center $B_0(R/2)$ of the hyperbolic disc by Corollary 8. Since this holds for all nodes, and the center forms a clique, the diameter is therefore $\mathcal{O}((\log n)^{\frac{1}{1-\alpha}}) = \mathcal{O}((\log n)^{\frac{2}{3-\beta}})$.   □

This bound slightly improves upon the results in [17] who show an upper bound of $\mathcal{O}((\log n)^{\frac{8}{(1-\alpha)(2-\alpha)}})$. As we will see in Theorem 3, however, the lower bound on the diameter is only $\Omega(\log n)$. We bridge this gap in the remaining part of the paper by analyzing the behavior in the outer band more carefully.

4   Logarithmic Upper Bound

In this section, we show that the diameter of the hyperbolic random graph is actually $\mathcal{O}(\log n)$, as long as the average degree is a small enough constant. We proceed by the following proof strategy. Consider a node $v \in B_O$. We investigate the *upwards path* from this node, which is intuitively constructed as follows: Each node on an upwards path has the smallest radial coordinate among all neighbors of the preceding node.

We first show that the diameter is asymptotically bounded by the longest upwards path in the graph. Afterwards, we prove that an upwards path is at most of length $\mathcal{O}(\log n)$ with high probability by investigating a random walk whose hitting time dominates the length of the upwards path. A simple union bound over all nodes will conclude the proof.

We start by stating a bound that shows that if v is between two nodes u, w that are connected by an edge, then v is either connected to u or v, or one of these nodes has a radial coordinate at least 1 smaller than v. Due to space constraints, this and all following proofs can be found in the long version.

Lemma 10. *Let u, v, w be nodes in the outer band such that v lies between u and w. Furthermore, let $\{u, w\} \in E$, but $\{u, v\}, \{v, w\} \notin E$. Then, for large n, at least one of the following holds: $r_u \leqslant r_v - 1$ or $r_w \leqslant r_v - 1$.*

Similarly to Lemma 9, we say that an edge $\{u, w\}$ *passes over* v, if the requirements of Lemma 10 are fulfilled. Before we introduce the formal definition of an upwards path, we define the notion of a straight path.

Definition 11. *Let $\pi = [v_1, \ldots, v_k]$ be a path in the hyperbolic random graph where $\forall i, v_i \in B_O$. We say that π is straight, if $\forall i \in \{2, \ldots, k-1\}$ the node v_i lies between v_{i-1} and v_{i+1}.*

The definition of a straight path captures the intuitive notion that the path does not "jump back and forth". Next, we define an upwards path, which is a special case of a straight path.

Definition 12. *Let $v \in B_O$ be a node in the hyperbolic random graph and define $\tilde{\varphi}_u := (\pi + \varphi_u - \varphi_v) \bmod 2\pi$. Furthermore, we define the neighbors to the right of u as*

$$\tilde{\Gamma}(u) := \Gamma(u) \cap \{w \in B_O \mid \tilde{\varphi}_w \geqslant \tilde{\varphi}_u\}. \tag{6}$$

Then we say that $\pi_v = [v = v_0, v_1, \ldots, v_k]$ is an upwards path from left to right *from v if $\forall i \in \{0, \ldots, k-1\}$: $v_{i+1} = \text{argmax}_{u \in \tilde{\Gamma}(v_i)}\{r_u\}$, and there is no longer upwards path $\pi'_v \supsetneq \pi_v$.*

Analogously, we define an upwards path from right to left *by replacing $\tilde{\varphi}_w \geqslant \tilde{\varphi}_u$ by $\tilde{\varphi}_w \leqslant \tilde{\varphi}_u$ in equation (6).*

Observe that there are two upwards paths from each node: One from right to left, and the other from left to right. An upwards path also only uses nodes in B_O. The exclusion of B_I can only increase the diameter of a component in the outer band.

The next Lemma shows that the length of the longest upwards path asymptotically bounds the length of all straight shortest paths in the outer band.

Lemma 13. *Assume that for all nodes $v \in B_O$, the upwards paths in both directions are of length $|\pi_v| \leqslant f(n)$. Let $\pi = [u_1, u_2, \ldots, u_k]$ be a straight shortest path. Then, $|\pi| \leqslant 2 \cdot f(n) + 1 = \mathcal{O}(f(n))$.*

We proceed by arguing that all upwards paths in the outer band are of length $\mathcal{O}(\log n)$ at most. In fact, we show a stronger statement by deriving an exponential tail bound on the length of an upwards path. To this end, we model an upwards path as a random walk. Consider for some node v all neighbors to the right of v. Among those, the neighbor in the largest layer (or equivalently, the smallest radial coordinate) is the neighbor on which any upwards path from left to right will continue. We formulate a probability that the upwards path jumps into a certain layer and analyze the probability that after T steps, the random walk modeled by this process is absorbed, i.e. we reach a node that has no further neighbors in this direction.

Let the random variables $[u = V_0, V_1, \ldots]$ describe the upwards path from u, and let $X_i := \ell$ if V_i is in layer L_ℓ, and $X_i := 0$ if the upwards path consists of $< i$ nodes. Without loss of generality, the upwards path is from left to right. Then, we have

$$\Pr[X_{i+1} = m \mid X_1, \ldots, X_i] \leqslant \tfrac{1}{2} \Pr[\exists w \in L_m \text{ such that } d(V_i, w) \leqslant R]$$
$$\cdot \Pr[\nexists m' > m \text{ such that } \exists w' \in L_{m'} \text{ with } d(V_i, w') \leqslant R] \quad (7)$$

Note that in $Pr[X_{i+1} = m \mid X_1, \ldots, X_i]$ we implicitly condition on the fact that some preceding node $V_{i'}$ with $i' < i$ on the upwards path was *not* connected to V_{i+1}; which technically excludes some subset of $B_R(V_i)$. We fix this issue by considering the Poisson point process and exposing the randomness as follows. First, we assume that there are no preconditions on $\Pr[X_{i+1} = m]$ (i.e. the upwards path begins in V_i). Then, the above formula is exact. We now expose all neighbors of V_i, and obtain $w \in L_m$ as the neighbor in the uppermost layer. Now we expose the conditions. This is possible, since in the Poisson point model, each area disjoint from other areas can be treated independently. The exposing of the conditions can only delete nodes. In this process, w might be deleted, lowering the probability of $\Pr[X_{i+1} = m]$.

Therefore, our stated formula is indeed an upper bound.

Lemma 14. *Let the random variables $[u = V_0, V_1, \ldots]$ describe the upwards path from u, and let $\forall i, X_i := \ell$ if V_i is in layer L_ℓ, and $X_i := 0$ if the upwards path consists of $< i$ nodes. Then, if C is large enough, we have $\mathbb{E}[X_{i+1}] \leqslant 0.99 \cdot X_i$.*

Lemma 14 shows that X_i has a multiplicative drift towards 0. Let $T := \min\{i \mid X_i = 0\}$ be the random variable describing the length of an upwards path. We now bound T by a multiplicative drift theorem as presented by Lehre and Witt [20, Theorem 7] and originally developed by Doerr and Goldberg [9, Theorem 1] for the analysis of evolutionary algorithms. For the sake of completeness, we restate their result.

Theorem 15 (from [9, 20]). *Let $(X_t)_{t \geqslant 0}$ be a stochastic process over some state space $\{0\} \cup [x_{\min}, x_{\max}]$, where $x_{\min} > 0$. Suppose that there exists some $0 < \delta < 1$ such that $\mathbb{E}[X_t - X_{t+1} \mid X_0, \ldots, X_t] \geqslant \delta X_t$. Then for the first hitting time $T := \min\{t \mid X_t = 0\}$ it holds*

$$\Pr[T \geqslant \tfrac{1}{\delta}(\ln(X_0/x_{\min}) + r) \mid X_0] \leqslant e^{-r} \text{ for all } r > 0. \qquad \bullet$$

In our case, $X_0 \leqslant \frac{\log R}{1-\alpha} + c$ and $x_{\min} = 1$. Using Lemma 14 this shows that

$$\Pr[T \geqslant 101 \cdot (\log \log \log n + r)] \leqslant e^{-r}. \qquad (8)$$

Hence, with probability $1 - \mathcal{O}(n^{-3})$ the random walk process described by X_i terminates after $\mathcal{O}(\log n)$ steps. By a union bound we have that all upwards paths in G are of length $\mathcal{O}(\log n)$ with probability $1 - \mathcal{O}(n^{-2})$. To show that all shortest paths are of length $\mathcal{O}(\log n)$, however, we need a slightly stronger statement, namely that the sum of $\mathcal{O}(\log \log n)$ upwards paths is at most of length $\mathcal{O}(\log n)$.

Lemma 16. *Let $(T_i)_{i=1\ldots X}$ be distributed according to equation (8), where $X = c \log \log n$. Then, with probability $1 - \mathcal{O}(n^{-3})$, $\sum_{i=1}^X T_i \leqslant \mathcal{O}(\log n)$.*

To conclude our result on the diameter, it is left to investigate shortest paths in B_O that are not straight. The general proof strategy for those paths is as follows. First, we show that such a path has edges that switch directions. Such an edge must pass over all preceding (or all following) nodes, as will become apparent in the next Lemma.

Lemma 17. *Consider a shortest path $\pi = [u_1, \ldots, u_k]$ that is not straight. In particular, π then has one or more sequence of nodes u_{i-1}, u_i, u_{i+1} such that u_i is not between u_{i-1} and u_{i+1}. Then, for all such positions i it holds that either*

1. $\forall j < i: u_j$ *is between u_i and u_{i+1}, or*
2. $\forall j > i: u_j$ *is between u_i and u_{i-1}.*

By Lemma 10, we know that can only be $\mathcal{O}(\log \log n)$ changes of directions; and Lemma 16 lets us conclude that the total path length is still $\mathcal{O}(\log n)$. This can be used to show that *all* shortest paths are of length $\mathcal{O}(\log n)$.

Lemma 18. *Let* $\pi = [u_1, u_2, \ldots, u_k]$ *be a shortest path where* $\forall i, u_i \in B_O$. *Then, with probability* $1 - \mathcal{O}(n^{-3/2})$, $|\pi| = \mathcal{O}(\log n)$.

Lemma 18 in conjunction with Lemma 7 then proves Theorem 2, i.e. that the diameter of the hyperbolic random graph is $\mathcal{O}(\log n)$ if the average degree is a small enough constant.

5   Conclusion

We derive a new polylogarithmic upper bound on the diameter of hyperbolic random graphs; and show that it is $\mathcal{O}(\log n)$ if the average degree is small. We further prove a matching lower bound. This immediately yields lower bounds for any broadcasting protocol that has to reach all nodes. Processes such as bootstrap percolation or rumor spreading therefore must run at least $\Omega(\log n)$ steps until they inform all nodes in the giant component.

Our work focuses on power law exponents $2 < \beta < 3$, but we believe that our proof can be extended to bound the diameter for $\beta > 3$ by $\Theta(\log n)$. For other scale-free models it was also interesting to study the phase transition at $\beta = 2$ and $\beta = 3$. Another natural open question is the average distance (also known as average diameter) between two random nodes. We conjecture that the average distance is $\Theta(\log \log n)$, but leave this open for future work.

Acknowledgements. We thank Konstantinos Panagiotou for many useful discussions, and suggesting a layer-based proof as in Lemma 6.

References

1. Barabási, A.-L., Albert, R.: Emergence of scaling in random networks. Science **286**, 509–512 (1999)
2. Bode, M., Fountoulakis, N., Müller, T.: On the giant component of random hyperbolic graphs. In: 7th European Conference on Combinatorics, Graph Theory and Applications (EuroComb), pp. 425–429 (2013)
3. Bode, M., Fountoulakis, N., Müller, T.: The probability that the hyperbolic random graph is connected (2014). www.math.uu.nl/Muell001/Papers/BFM.pdf
4. Boguná, M., Papadopoulos, F., Krioukov, D.: Sustaining the internet with hyperbolic mapping. Nature Communications **1**, 62 (2010)
5. Bollobás, B.: Random graphs. Springer (1998)
6. Bollobás, B., Chung, F.R.K.: The diameter of a cycle plus a random matching. SIAM Journal of Discrete Mathematics **1**, 328–333 (1988)
7. Candellero, E., Fountoulakis, N.: Bootstrap percolation and the geometry of complex networks (2014). arxiv1412.1301
8. Chung, F., Lu, L.: The average distances in random graphs with given expected degrees. Proceedings of the National Academy of Sciences **99**, 15879–15882 (2002)
9. Doerr, B., Goldberg, L.A.: Drift analysis with tail bounds. In: Schaefer, R., Cotta, C., Kołodziej, J., Rudolph, G. (eds.) PPSN XI. LNCS, vol. 6238, pp. 174–183. Springer, Heidelberg (2010)

10. Dommers, S., van der Hofstad, R., Hooghiemstra, G.: Diameters in preferential attachment models. Journal of Statistical Physics **139**, 72–107 (2010)
11. Dubhashi, D.P., Panconesi, A.: Concentration of measure for the analysis of randomized algorithms. Cambridge University Press (2009)
12. Evans, M., Hastings, N., Peacock, B.: Statistical Distributions, chapter 12, 3rd edn., pp. 71–73. Wiley-Interscience (2000)
13. Fountoulakis, N., Panagiotou, K., Sauerwald, T.: Ultra-fast rumor spreading in models of real-world networks (2015). Unpublished draft
14. Friedrich, T., Krohmer, A.: Cliques in hyperbolic random graphs. In: 34th IEEE Conference on Computer Communications (INFOCOM) (2015). https://hpi.de/fileadmin/user_upload/fachgebiete/friedrich/publications/2015/cliques2015.pdf
15. Friedrich, T., Sauerwald, T., Stauffer, A.: Diameter and broadcast time of random geometric graphs in arbitrary dimensions. Algorithmica **67**, 65–88 (2013)
16. Gugelmann, L., Panagiotou, K., Peter, U.: Random hyperbolic graphs: degree sequence and clustering. In: Czumaj, A., Mehlhorn, K., Pitts, A., Wattenhofer, R. (eds.) ICALP 2012, Part II. LNCS, vol. 7392, pp. 573–585. Springer, Heidelberg (2012)
17. Kiwi, M., Mitsche, D.: A bound for the diameter of random hyperbolic graphs. In: 12th Workshop on Analytic Algorithmics and Combinatorics (ANALCO), pp. 26–39 (2015)
18. Kleinberg, J.: Navigation in a small world. Nature **406**, 845 (2000)
19. Krioukov, D., Papadopoulos, F., Kitsak, M., Vahdat, A., Boguñá, M.: Hyperbolic geometry of complex networks. Physical Review E **82**, 036106 (2010)
20. Lehre, P.K., Witt, C.: General drift analysis with tail bounds (2013). arxiv1307.2559
21. Martel, C.U., Nguyen, V.: Analyzing Kleinberg's (and other) small-world models. In: 23rd Annual ACM Symposium on Principles of Distributed Computing (PODC), pp. 179–188 (2004)
22. Papadopoulos, F., Psomas, C., Krioukov, D.: Network mapping by replaying hyperbolic growth. IEEE/ACM Transactions on Networking, 198–211 (2014)
23. Penrose, M.: Random Geometric Graphs. Oxford scholarship online. Oxford University Press (2003)
24. Riordan, O., Wormald, N.: The diameter of sparse random graphs. Combinatorics, Probability and Computing **19**, 835–926 (2010)
25. von Looz, M., Staudt, C.L., Meyerhenke, H., Prutkin, R.: Fast generation of dynamic complex networks with underlying hyperbolic geometry (2015). arxiv1501.03545
26. Watts, D.J., Strogatz, S.H.: Collective dynamics of 'small-world' networks. Nature **393**, 440–442 (1998)

Tight Bounds for Cost-Sharing
in Weighted Congestion Games

Martin Gairing[1]([✉]), Konstantinos Kollias[2], and Grammateia Kotsialou[1]

[1] University of Liverpool, Liverpool, Merseyside L69 3BX, UK
m.gairing@liverpool.ac.uk
[2] Stanford University, Stanford, CA 94305, USA

Abstract. This work studies the price of anarchy and the price of stability of cost-sharing methods in weighted congestion games. We require that our cost-sharing method and our set of cost functions satisfy certain natural conditions and we present general tight price of anarchy bounds, which are robust and apply to general equilibrium concepts. We then turn to the price of stability and prove an upper bound for the Shapley value cost-sharing method, which holds for general sets of cost functions and which is tight in special cases of interest, such as bounded degree polynomials. Also for bounded degree polynomials, we close the paper with a somehow surprising result, showing that a slight deviation from the Shapley value has a huge impact on the price of stability. In fact, for this case, the price of stability becomes as bad as the price of anarchy.

1 Introduction

The class of weighted congestion games [16,17] encapsulates a large collection of important applications in the study of the inefficiencies induced by strategic behavior in large systems. The applications that fall within this framework involve a set of players who place demands on a set of resources. As an example, one of the most prominent such applications is *selfish routing* in a telecommunications or traffic network [3,8,20]. When more total demand is placed on a resource, the resource becomes scarcer, and the quality of service experienced by its users degrades. More specifically, in weighted congestion games there is a set of players N and a set of resources E. Each player $i \in N$ has a weight $w_i > 0$ and she gets to select the subset of the resources that she will use. The possible subsets she can pick are given in her set of possible strategies, $\mathcal{P}_i$. Once players make their decisions, each resource $e \in E$ generates a joint cost $f_e \cdot c_e(f_e)$, where f_e is the total weight of the users of e and c_e is the cost function of e. The joint cost of a resource is covered by the set of players S_e using e, i.e., $\sum_{i \in S_e} \chi_{ie} = f_e \cdot c_e(f_e)$, where χ_{ie} is the *cost share* of player i on resource e.

The way the cost shares χ_{ie} are calculated is given by the *cost-sharing method* used in the game. A cost-sharing method determines the cost-shares of the players on a resource, given the joint cost that each subset of them generates, i.e., the

This work was supported by EPSRC grants EP/J019399/1 and EP/L011018/1.

M.M. Halldórsson et al. (Eds.): ICALP 2015, Part II, LNCS 9135, pp. 626–637, 2015.
DOI: 10.1007/978-3-662-47666-6_50

cost shares are functions of the state on that resource alone. In most applications of interest, it is important that the cost-sharing method possesses this *locality property*, since we expect the system designer's control method to scale well with the size of the system and to behave well as resources are dynamically added to or removed from the system. Altering our cost-sharing method of choice changes the individual player costs. Given that our candidate outcomes are expected to be game-theoretic equilibrium solutions, this modification of the individual player costs also changes the possible outcomes players can reach. The *price of anarchy* (POA) and the *price of stability* (POS) measure the performance of a cost-sharing method by comparing the worst and best equilibrium, respectively, to the optimal solution, and taking the worst-case ratio over all instances.

Certain examples of cost-sharing methods include *proportional sharing* (PS) and the *Shapley value* (SV). In PS, the cost share of a player is proportional to her weight, i.e., $\chi_{ie} = w_i \cdot c_e(f_e)$, while the SV of a player on a resource e is her average marginal cost increase over a uniform ordering of the players in S_e. Other than different POA and POS values, different cost-sharing methods also possess different equilibrium existence properties. The *pure Nash equilibrium* (PNE) is the most widely accepted solution concept in such games. In a PNE, no player can improve her cost with a unilateral deviation to another strategy. In a *mixed Nash equilibrium* (MNE) players randomize over strategies and no player can improve her expected cost by picking a different distribution over strategies. By Nash's famous theorem, a MNE is guaranteed to exist in every weighted congestion game. However, existence of a PNE is not guaranteed for some cost-sharing methods. As examples, PS does not guarantee equilibrium existence (see [12] for a characterization), while the SV does. In [11], it is shown that only the class of *generalized weighted Shapley values* (see Section 2 for a definition) guarantees the existence of a PNE in such games.

As a metric that is worst-case by nature, the POA of a method that does not always induce a PNE must be measured with respect to more general concepts, such as the MNE, which are guaranteed to exist. Luckily, POA upper bounds are typically *robust* [18] which means they apply to MNE and even more general classes (such as correlated and coarse-correlated equilibria). On the other hand, the motivations behind the study of the POS assume a PNE will exist, hence the POS is more meaningful when the method does guarantee a PNE.

1.1 Our Contributions

In this work we make two main contributions, one with respect to the POA and one with respect to the POS.

General POA bounds: On the POA side, we present *tight* bounds for general classes of allowable cost functions and for general cost-sharing methods, i.e., we parameterize the POA by (i) the set of allowable cost functions (which changes depending on the application under consideration) and (ii) the cost-sharing method. To obtain our tight bounds we make use of the following natural assumptions, which we explain in more detail in Section 2:

1. Every cost function in the game is continuous, nondecreasing, and convex.
2. Cost-sharing is consistent when player sets generate costs in the same way.
3. The cost share of a player on a resource is a convex function of her weight.

We now briefly discuss these assumptions. Assumption 1 is standard in congestion-type settings. For example, linear cost functions have obvious applications in many network models, as do queueing delay functions, while higher degree polynomials (such as quartic) have been proposed as realistic models of road traffic [22]. Assumption 2 asks that the cost-sharing method only looks at how players generate costs and does not discriminate between them in any other way. Assumption 3 asks that the curvature of the cost shares is consistent, i.e., given Assumption 1, that the share of a player on a resource is a convex function of her weight (otherwise, we would get that the share of the player increases in a slower than convex way but the total cost of the constant weight players increases in a convex way, which we view as unfair). We note that our upper bounds are robust and apply to general equilibrium concepts that are guaranteed to exist for all cost-sharing methods.

SV based POS bounds: Studying the POS is most well-motivated in settings where a trusted mediator or some other authority can place the players in an initial configuration and they will not be willing to deviate from it. For this reason, the POS is a very interesting concept, especially for games possessing a PNE. Hence, we focus on cost sharing methods which always induce games with a PNE. For SV cost sharing, we prove an upper bound on the POS which holds for all sets of cost functions that satisfy Assumption 1. We show that for the interesting case of polynomials of bounded degree d, this upper bound is $d + 1$, which is asymptotically tight and always very close to the lower bound in [7].

Moreover, we show that this linear dependence on the maximum degree d is very fragile. To do so, we consider a parameterized class of weighted Shapley values, where players with larger weight get an advantage or disadvantage, which is determined by a single parameter γ. When $\gamma = 0$ this recovers the SV. For all other values $\gamma \neq 0$, we show that the POS is very close and for $\gamma > 0$ even matches the upper bound on the *price of anarchy* in [10]. In other words, for this case the POS and the POA coincide, which we found very surprising, in particular because the upper bound in [10] even applies to general cost-sharing methods. We note that these weighted Shapley values are the only cost-sharing methods that guarantee existence of a PNE and satisfy Assumption 2 [10,11].

1.2   Related Work and Comparison to Previous Results

POA. The POA was proposed in [15]. Most work on the inefficiency of equilibria for weighted congestion games has focused on PS. Tight bounds for the case of linear cost functions have been obtained in [3,8]. The case of bounded degree polynomials was resolved in [1] and subsequent work [4,9,18] concluded the study of PS. In particular, [18] formalized the *smoothness* framework which shows how robust POA bounds (i.e., POA bounds that apply to general equilibrium concepts) are obtained.

Further cost-sharing methods have been considered in [10,14]. Here, [14] provides tight bounds for the SV in games with convex cost functions, while [10] proves that the SV is optimal among the methods that guarantee the existence of a PNE and that PS is near-optimal in general, for games with polynomial cost functions. The authors also show tight bounds on the *marginal contribution method* (which charges a player the increase her presence causes to the joint cost) in games with polynomial costs. Optimality of the SV in closely related settings has also been discussed in [13,19].

POS. The term price of stability was introduced in [2] for the network cost-sharing game, which was further studied in [5,6,14] for weighted players and various cost-sharing methods. With respect to congestion games, results on the POS are only known for polynomial *unweighted* games, for which [7] provides exact bounds. Work in [13,19] studies the POS of the Shapley value in related settings.

Comparison to previous work. Our POA results greatly generalize the work on cost-sharing methods for weighted congestion games and give a recipe for tight bounds in a large array of applications. Prior to our work only a handful of cost-sharing methods have been tightly analyzed. Our results facilitate the better design of such systems, beyond the optimality criteria considered in [10]. For example, the SV has the drawback that it can't be computed efficiently, while PS (on top of not always inducing a PNE) might have equilibria that are hard to compute. In cases where existence and efficient computation of a PNE is considered important, the designer might opt for a different cost-sharing method, such as a priority based one (that fixes a weight-dependent ordering of the players and charges them the marginal increase they cause to the joint cost in this order) which has polynomial time computable shares and equilibria. Our results show how the inefficiency of equilibria is quantified for all such possible choices, to help evaluate the tradeoffs between different options. Our work closely parallels the work on network cost-sharing games in [14], which provides tight bounds for general cost-sharing methods.

Our POS upper bound is the first for weighted congestion games that applies to any class of convex costs. The work in [19] presents SV POS bounds in a more general setting with non-anonymous but submodular cost functions. In a similar vein, [13] presents tight POS bounds on the SV in games with non-anonymous costs, by allowing any cost function and parameterizing by the number of players in the game, i.e., they show that for the set of all cost functions the POS of the Shapley value is $\Theta(n \log n)$ and for the set of supermodular cost functions it becomes n, where n is the number of players. These upper bounds apply to our games as well, however we adopt a slightly different approach. We allow an infinite number of players for our bounds to hold and parameterize by the set of possible cost functions, to capture the POS of different applications. For example, for polynomials of degree at most d, we show that the POS is at most $d + 1$, even when $n \to \infty$. Observe that for unweighted games PS and SV are identical.

Thus, the lower bound in [7], which approaches $d+1$, also applies to our setting, showing that our bound for polynomials is asymptotically tight.

Our lower bounds on the POS for the parameterized class of weighted Shapley values build on the corresponding lower bounds on the POA in [10]. Our construction matches these bounds by ensuring that the instance possesses a unique Nash equilibrium. Together with our upper bound this shows an interesting contrast: For the special case of SV the POS is exponentially better than the POA, but as soon as we give some weight dependent priorities to the players, the POA and the POS essentially coincide.

2   Preliminaries

In this section we present our model in more detail. We write $N = \{1, 2, \ldots, n\}$ for the *players* and $E = \{1, 2, \ldots, m\}$ for the *resources*. Each player $i \in N$ has a positive weight w_i and a *strategy set* $\mathcal{P}_i$, each element of which is a subset of the resources, i.e., $\mathcal{P}_i \subseteq 2^E$. We write $P = (P_1, \ldots, P_n)$ for an *outcome*, with $P_i \in \mathcal{P}_i$ the *strategy* of player i. Let $S_e(P) = \{i : e \in P_i\}$ be the set of users of e and $f_e(P) = \sum_{i \in S_e(P)} w_i$ be the *total weight* on e. The *joint cost* on e is $C_e(f_e(P)) = f_e(P) \cdot c_e(f_e(P))$, with c_e a function that is drawn from a given set of allowable functions $\mathcal{C}$. We write $\chi_{ie}(P)$ for the *cost share* of player i on resource e. These shares are such that $\sum_{i \in S_e(P)} \chi_{ie}(P) = C_e(f_e(P))$. We make the following assumptions on the *cost-sharing method* and the set of allowable cost functions:

(1) Every function that can appear in the game is continuous, nondecreasing, and convex. We also make the mild technical assumption that $\mathcal{C}$ is closed under dilation, i.e., that if $c(x) \in \mathcal{C}$, then also $c(a \cdot x) \in \mathcal{C}$ for $a > 0$. We note that without loss of generality, every $\mathcal{C}$ is also closed under scaling, i.e., if $c(x) \in \mathcal{C}$, then also $a \cdot c(x) \in \mathcal{C}$ for $a > 0$ (this is given by simple scaling and replication arguments).

(2) Given a player set S and a cost function c, suppose we alter the players (e.g., change their weights or identities) and the cost function c in a manner such that the cost generated by every subset of S on c remains unchanged. (For example suppose we initially have two players with weights 1 and 2 and cost function $c(x) = x^2$ and we modify them so that the weights are now 2 and 4 and the cost function is now $c(x) = x^2/4$.) Our second assumption states that the cost shares of the players will remain the same. In effect, we ask that the cost-sharing method only charges players based on how they contribute to the joint cost. We also assume, without loss of generality, that if the costs of all subsets of $\hat{S} = \{1, 2, \ldots, k\}$ are scaled versions of those corresponding to $S = \{1, 2, \ldots, k\}$, then the cost shares are also simply scaled by the same factor (again this is given by simple scaling and replication arguments).

(3) Since each cost share is a function of the cost function and player set on a resource, we will also write $\xi_c(i, S)$ for the share of a player i when the cost function is c and the player set is S. This means that the cost share of player i on resource e can be written both as $\chi_{ie}(P)$ and as $\xi_{c_e}(i, S_e(P))$. Our third assumption now states that the expression $\xi_{c_e}(i, S_e(P))$ is a continuous, nondecreasing, and convex function of the weight of player i. This is something to expect from a reasonable cost-sharing method, given that the joint cost on the resource is a continuous nondecreasing convex function of the weight of player i.

The *pure Nash equilibrium* (PNE) condition on an outcome P states that for every player i it must be the case that:

$$\sum_{e \in P_i} \chi_{ie}(P) \leq \sum_{e \in P_i'} \chi_{ie}(P_i', P_{-i}), \text{ for every } P_i' \in \mathcal{P}_i. \tag{1}$$

The social cost in the game will be the sum of the player costs, i.e.,

$$C(P) = \sum_{i \in N} \sum_{e \in P_i} \chi_{ie}(P) = \sum_{i \in N} \sum_{e \in P_i} \xi_{c_e}(i, S_e(P)) = \sum_{e \in E} f_e(P) \cdot c_e(f_e(P)). \tag{2}$$

Let $\mathcal{P}$ be the set of outcomes and $\mathcal{P}^N$ be the set of PNE outcomes of the game. Then the *price of anarchy* (POA) is defined as $POA = \frac{\max_{P \in \mathcal{P}^N} C(P)}{\min_{P \in \mathcal{P}} C(P)}$, and the *price of stability* (POS) is defined as $POS = \frac{\min_{P \in \mathcal{P}^N} C(P)}{\min_{P \in \mathcal{P}} C(P)}$. The POA and POS for a class of games are defined as the largest such ratios among all games in the class.

Weighted Shapley Values. The weighted Shapley value defines how the cost $C_e(\cdot)$ of resource e is partitioned among the set of players S_e using e. Given an ordering π of the players in S_e, the marginal cost increase by players $i \in S_e$ is $C(f_i^\pi + w_i) - C(f_i^\pi)$, where f_i^π is the total weight of players preceding i in the ordering. For a given distribution Π over orderings, the cost share of player i is $E_{\pi \sim \Pi}[C(f_i^\pi + w_i) - C(f_i^\pi)]$. For the weighted Shapley value, the distribution over orderings is given by a sampling parameter λ_i for each player i. The last player of the ordering is picked proportional to the sampling parameter λ_i. This process is then repeated iteratively for the remaining players.

As in [10], we study a *parameterized class of weighted Shapley values* defined by a parameter γ. For this class $\lambda_i = w_i^\gamma$ for all players i. For $\gamma = 0$ this reduces to the (normal) *Shapley value* (SV), where we have a uniform distribution over orderings.

3   Tight POA Bounds for General Cost-Sharing Methods

We first generalize the (λ, μ)-smoothness framework of [18] to accommodate any cost-sharing method and set of possible cost functions. Suppose we identify positive parameters λ and $\mu < 1$ such that for every cost function in our allowable set $c \in \mathcal{C}$, and every pair of sets of players T and T^*, we get

$$\sum_{i \in T^*} \xi_c(i, T \cup \{i\}) \leq \lambda \cdot w_{T^*} \cdot c(w_{T^*}) + \mu \cdot w_T \cdot c(w_T), \tag{3}$$

where $w_S = \sum_{i \in S} w_i$ for any set of players S. Then, for P a PNE and P^* the optimal solution, we would get

$$C(P) \overset{(2)}{=} \sum_{i \in N} \sum_{e \in P_i} \xi_{c_e}(i, S_e(P)) \overset{(1)}{\leq} \sum_{i \in N} \sum_{e \in P_i^*} \xi_{c_e}(i, S_e(P) \cup \{i\})$$

$$= \sum_{e \in E} \sum_{i \in S_e(P^*)} \xi_{c_e}(i, S_e(P) \cup \{i\})$$

$$\overset{(3)}{\leq} \sum_{e \in E} \lambda \cdot w_{S_e(P^*)} c_e(w_{S_e(P^*)}) + \mu \cdot w_{S_e(P)} \cdot c_e(w_{S_e(P)})$$

$$\overset{(2)}{=} \lambda \cdot C(P^*) + \mu \cdot C(P). \tag{4}$$

Rearranging (4) yields a $\lambda/(1 - \mu)$ upper bound on the POA. The same bound can be easily shown to apply to MNE and more general concepts (correlated and coarse correlated equilibria), though we omit the details (see, e.g., [18] for more). We then get the following lemma.

Lemma 1. *Consider the following optimization program with variables λ, μ.*

Minimize $\quad \frac{\lambda}{1-\mu}$ $\hfill (5)$

Subject to $\mu \leq 1$ $\hfill (6)$

$\quad \sum_{i \in T^*} \xi_c(i, T \cup \{i\}) \leq \lambda \cdot w_{T^*} \cdot c(w_{T^*}) + \mu \cdot w_T \cdot c(w_T), \forall c, T, T^* (7)$

Every feasible solution yields a $\lambda/(1 - \mu)$ upper bound on the POA of the cost sharing method given by $\xi_c(i, S)$ and the set of cost functions $\mathcal{C}$.

The upper bound holds for any cost-sharing method and set of cost functions. We now proceed to show that the optimal solution to Program (5)-(7) gives a tight upper bound when our assumptions described in Section 2 hold.

Theorem 1. *Let (λ^*, μ^*) be the optimal point of Program (5)-(7). The POA of the cost-sharing method given by $\xi_c(i, S)$ and the set of cost functions $\mathcal{C}$ is precisely $\lambda^*/(1 - \mu^*)$.*

Proof. First define $\zeta_c(y, x)$ for $y, x > 0$ as

$$\zeta_c(y, x) = \max_{T^*: w_{T^*} = y, T: w_T = x} \sum_{i \in T^*} \xi_c(i, T \cup \{i\}). \tag{8}$$

With this definition, we can rewrite Program (5)-(7) as

Minimize $\quad \frac{\lambda}{1-\mu}$ $\hfill (9)$

Subject to $\quad \mu \leq 1$ $\hfill (10)$

$\quad \zeta_c(y, x) \leq \lambda \cdot y \cdot c(y) + \mu \cdot x \cdot c(x), \ \forall c \in \mathcal{C}, x, y \hfill (11)$

Observe that for every constraint, we can scale the weights of the players by a factor a, dilate the cost function by a factor $1/a$, and scale the cost function by an arbitrary factor and keep the constraint intact (by Assumption 2). This suggests we can assume that every constraint has $y = 1$ and $c(1) = 1$. Then we rewrite Program (9)-(11) as

$$\text{Minimize} \quad \frac{\lambda}{1-\mu} \tag{12}$$

$$\text{Subject to} \quad \mu \leq 1 \tag{13}$$

$$\zeta_c(1, x) \leq \lambda + \mu \cdot x \cdot c(x), \ \forall c \in \mathcal{C}, x \tag{14}$$

The Lagrangian dual of Program (12)-(14) is

$$\text{Minimize} \quad \frac{\lambda}{1-\mu} + \sum_{c \in \mathcal{C}, x > 0} z_{cx} \cdot (\zeta_c(1, x) - \lambda - \mu \cdot x \cdot c(x)) + z_\mu \cdot (\mu - 1) \tag{15}$$

$$\text{Subject to} \quad z_{cx}, z_\mu \geq 0 \tag{16}$$

Our primal is a semi-infinite program with an objective that is continuous, differentiable, and convex in the feasible region, and with linear constraints. We get that strong duality holds (see also [21,23] for a detailed treatment of strong duality in this setting). We first treat the case when the optimal value of the primal is finite and is given by point (λ^*, μ^*). Before concluding our proof we will explain how to deal with the case when the primal is infinite or infeasible. The KKT conditions yield for the optimal $\lambda^*, \mu^*, z_{cx}^*$:

$$\frac{1}{1-\mu^*} = \sum_{c \in \mathcal{C}, x > 0} z_{cx}^* \tag{17}$$

$$\frac{\lambda^*}{(1-\mu^*)^2} = \sum_{c \in \mathcal{C}, x > 0} z_{cx}^* \cdot x \cdot c(x) \tag{18}$$

Calling $\eta_{cx} = z_{cx}^* / \sum_{c \in \mathcal{C}, x > 0} z_{cx}^*$ and dividing (18) with (17) we get

$$\frac{\lambda^*}{1-\mu^*} = \sum_{c \in \mathcal{C}, x > 0} \eta_{cx} \cdot x \cdot c(x). \tag{19}$$

By (19) and the fact that all constraints for which $z_{cx}^* > 0$ are tight (by complementary slackness), we get

$$\sum_{c \in \mathcal{C}, x > 0} \eta_{cx} \cdot \zeta_c(1, x) = \sum_{c \in \mathcal{C}, x > 0} \eta_{cx} \cdot x \cdot c(x). \tag{20}$$

Lower bound construction. Let $\mathcal{T} = \{(c, x) : z_{cx}^* > 0\}$. The construction starts off with a single player i, who has weight 1 and, in the PNE, uses a single resource e_i by herself. The cost function of resource e_i is an arbitrary function from $\mathcal{C}$ such that $c_{e_i}(1) \neq 0$ (it is easy to see that such a function exists, since $\mathcal{C}$ is closed under dilation, unless all function are 0, which is a trivial case) scaled so that $c_{e_i}(1) = \sum_{(c,x) \in \mathcal{T}} \eta_{cx} \cdot \zeta_c(1, x)$. The other option of player i is to use a set of resources, one for each $(c, x) \in \mathcal{T}$ with cost functions $\eta_{cx} \cdot c(\cdot)$. The resource corresponding to each (c, x) is used in the PNE by a player set that is equivalent to the T that maximizes the expression in (8) for the corresponding c, x. We now prove that player i does not gain by deviating to her alternative strategy. The key point is that due to convexity of the cost shares (Assumption 3), the worst case T^* in definition (8) will always be a single player. Then we can see that the cost share of i on each (c, x) resource in her potential deviation will be $\eta_{cx} \cdot \zeta_c(1, x)$. It then follows that she is indifferent between her two strategies. Note that the PNE cost of i is $\sum_{(c,x) \in T} \eta_{cx} \cdot \zeta_c(1, x)$, which by (19) and (20) is equal to $\lambda^*/(1 - \mu^*)$. Also note that if player i could use her alternative strategy by herself, her cost would be 1.

We now make the following observation which allows us to complete the lower bound construction: Focus on the players and resources of the previous paragraph. Suppose we scale the weight of player i, as well as the weights of the users of the resources in her alternative strategy by the same factor $a > 0$. Then, suppose we dilate the cost functions of all these resources (the one used by i in the PNE and the ones in her alternative strategy) by a factor $1/a$ so that the costs generated by the players go back to the values they had in the previous paragraph. Finally, suppose we scale the cost functions by an arbitrary factor $b > 0$. We observe that the fact that i has no incentive to deviate is preserved (by Assumption 2) and the ratio of PNE cost versus alternative cost for i remains the same, i.e., $\lambda^*/(1 - \mu^*)$. This suggests that for every player generated by our construction so far in the PNE, we can repeat these steps by looking at her weight and PNE cost and appropriately constructing her alternative strategy and the users therein. After repeating this construction for a large number of layers $M \to \infty$, we complete the instance by creating a single resource for each of the players in the final layer. The cost functions of these resources are arbitrary nonzero functions from $\mathcal{C}$ scaled and dilated so that each one of these players is indifferent between her PNE strategy and using the newly constructed resource.

Consider the outcome that has all players play their alternative strategies and not the ones they use in the PNE. Evey player other than the ones in the final layer would have a cost $\lambda^*/(1 - \mu^*)$ smaller, as we argued above. We can now see that, by (20), the cost of every player in the PNE is the same as that of the players in the resources of her alternative strategy. This means the cost across levels of our construction is identical and the final layer is negligible, since $M \to \infty$. This proves that the cost of the PNE is $\lambda^*/(1 - \mu^*)$ times larger than the outcome that has all players play their alternative strategies, which gives the tight lower bound.

Note on case with primal infeasibility. Recall that during our analysis we assumed that the primal program (12)-(14) had a finite optimal solution. Now suppose the program is either infeasible or $\mu = 1$, which means the minimizer yields an infinite value. This implies that, if we set μ arbitrarily close to 1, then there exists some $c \in \mathcal{C}$, such that, for any arbitrarily large λ, there exists $x > 0$ such that $\zeta_c(1, x) > \lambda + \mu \cdot x \cdot c(x)$. We can rewrite this last expression as $\zeta_c(1, x)/(x \cdot c(x)) > \mu + \lambda/(x \cdot c(x))$, which shows we have c, x values such that $\zeta_c(1, x)$ is arbitrarily close to $x \cdot c(x)$ or larger (since μ is arbitrarily close to 1). We can then replace λ with λ' such that the constraint becomes tight. It is not hard to see that these facts give properties parallel to (19) and (20) by setting $\eta_{cx} = 1$ for our c, x and every other such variable to 0. Then our lower bound construction goes through for this arbitrarily large $\lambda'/(1-\mu)$, which shows we can construct a lower bound with as high POA as desired.   □

4   Shapley Value POS

In this section we study the POS for a class of weighted Shapley values, where the sampling parameter of each player i is defined by $\lambda_i = w_i^\gamma$ for some γ.

We start with an upper bound on the POS for the case that $\gamma = 0$, i.e., for the *Shapley value* (SV) cost-sharing method. For the SV, existence of a PNE has been shown in [14] with the help of the following potential function, which is defined for an arbitrary ordering of the players:

$$\Phi(P) = \sum_{e \in E} \Phi_e(P) = \sum_{e \in E} \sum_{i \in S_e(P)} \xi_{c_e}(i, \{j : j \leq i, j \in S_e(P)\}). \tag{21}$$

We first prove the following lemma which is the main tool for proving our upper bound on the POS.

Lemma 2. *Suppose we are given an outcome of the game and suppose we substitute any given player i with two players who have weight $w_i/2$ each and who use the exact same resources as i. Then the value of the potential function will be at most the same as before the substitution.*

Proof. First rename the players so that the substituted player i has the highest index. Assign indices i' and i'' to the new players, with $i'' > i' > i$. On every resource e that is used by these players, the potential decreases by $\xi_{c_e}(i, S_e(P))$, while it increases by $\xi_{c_e}(i', S_e(P) \cup \{i'\} \setminus \{i\}) + \xi_{c_e}(i'', S_e(P) \cup \{i', i''\} \setminus \{i\})$. Hence, it suffices to show that

$$\xi_{c_e}(i', S_e(P) \cup \{i'\} \setminus \{i\}) + \xi_{c_e}(i'', S_e(P) \cup \{i', i''\} \setminus \{i\}) \leq \xi_{c_e}(i, S_e(P)). \tag{22}$$

For simplicity, in what follows call $\xi = \xi_{c_e}(i, S_e(P)), \xi' = \xi_{c_e}(i', S_e(P) \cup \{i'\} \setminus \{i\})$, and $\xi'' = \xi_{c_e}(i'', S_e(P) \cup \{i', i''\} \setminus \{i\})$. Consider every ordering π of the players in $S_e(P) \setminus \{i\}$ and every possible point in the ordering where a new player can be placed. If we assume that player is i and we average all possible joint cost jumps i can cause (by definition of the SV) we get ξ. Similarly, with i', we get ξ'. If we repeat the same thought process for i'', we are not getting ξ'', since the position of i' in the ordering is unspecified. However, we get a value that is larger than ξ'' if we always place i' right before i''. Call this larger value $\hat{\xi}''$. Observe that if we take every ordering π of $S_e(P) \setminus \{i\}$ and in every possible position, we place first i' and then i'' and we take the average of the combined joint cost jump that they cause, we will be getting $\xi' + \hat{\xi}''$, which, as we explained is at least $\xi' + \xi''$. Now note that this combined jump of the two players will also be the jump that i would cause in that particular position (since $w_{i'} + w_{i''} = w_i$), which means $\xi' + \hat{\xi}'' = \xi$, which in turn gives $\xi \geq \xi' + \xi''$ and completes the proof.   □

By repeatedly applying Lemma 2, we can break the total weight on each resource in players of infinitesimal size and the value of the potential will not increase. This suggests:

$$\Phi_e(P) \geq \int_0^{f_e(P)} c_e(x)dx. \tag{23}$$

Now call P^* the optimal outcome and $P = \arg\min_{P'} \Phi(P')$ the minimizer of the potential function, which is, by definition, also a PNE. We get:

$$C(P^*) \overset{(21)}{\geq} \Phi(P^*) \overset{\text{Def.}P}{\geq} \Phi(P) \overset{(23)}{\geq} \sum_{e \in E} \int_0^{f_e(P)} c_e(x)dx$$

$$= \frac{\sum_{e \in E} \int_0^{f_e(P)} c_e(x)dx}{\sum_{e \in E} f_e(P) \cdot c_e(f_e(P))} \cdot C(P) \geq \min_{e \in E} \frac{\int_0^{f_e(P)} c_e(x)dx}{f_e(P) \cdot c_e(f_e(P))} \cdot C(P).$$

Rearranging yields the following theorem.

Theorem 2. *The POS of the SV with C the set of allowable cost functions is at most $\max_{c \in C, x > 0} \frac{x \cdot c(x)}{\int_0^x c(x')dx'}$.*

Corollary 1. *For polynomials with non-negative coefficients and degree at most d, the POS of the SV is at most $d + 1$, which asymptotically matches the lower bound of [7] for unweighted games.*

In the remainder of this section, we show that this linear dependence on the maximum degree d of the polynomial cost functions is very fragile. More precisely, for all values $\gamma \neq 0$, we show an exponential (in d) lower bound which matches the corresponding lower bound on the POA in [10]. Our bound for $\gamma > 0$ even matches the upper bound on the POA [10], which holds for the weighted Shapley value in general. Our constructions modify the corresponding instances in [10], making sure that they have a unique Nash equilibrium. Due to page restrictions we defer the proof to our full version.

Theorem 3. *For polynomial cost functions with non-negative coefficients and maximum degree d, the POS for the class of weighted Shapley values with sampling parameters $\lambda_i = w_i^\gamma$ is at least*

(a) $(2^{\frac{1}{d+1}} - 1)^{-(d+1)}$, for all $\gamma > 0$, and
(b) $(d + 1)^{d+1}$, for all $\gamma < 0$.

References

1. Aland, S., Dumrauf, D., Gairing, M., Monien, B., Schoppmann, F.: Exact price of anarchy for polynomial congestion games. SIAM Journal on Computing 40(5), 1211–1233 (2011)
2. Anshelevich, E., Dasgupta, A., Kleinberg, J., Tardos, E., Wexler, T., Roughgarden, T.: The price of stability for network design with fair cost allocation. SIAM Journal on Computing 38(4), 1602–1623 (2008)

3. Awerbuch, B., Azar, Y., Epstein, A.: The price of routing unsplittable flow. In: Proceedings of STOC, pp. 57–66. ACM (2005)
4. Bhawalkar, K., Gairing, M., Roughgarden, T.: Weighted congestion games: Price of anarchy, universal worst-case examples, and tightness. ACM Transactions on Economics and Computation $2(4)$, 14 (2014)
5. Chen, H.L., Roughgarden, T.: Network design with weighted players. Theory of Computing Systems $45(2)$, 302–324 (2009)
6. Chen, H.L., Roughgarden, T., Valiant, G.: Designing network protocols for good equilibria. SIAM Journal on Computing $39(5)$, 1799–1832 (2010)
7. Christodoulou, G., Gairing, M.: Price of stability in polynomial congestion games. In: Fomin, F.V., Freivalds, R., Kwiatkowska, M., Peleg, D. (eds.) ICALP 2013, Part II. LNCS, vol. 7966, pp. 496–507. Springer, Heidelberg (2013)
8. Christodoulou, G., Koutsoupias, E.: The price of anarchy of finite congestion games. In: Proceedings of STOC, pp. 67–73. ACM (2005)
9. Gairing, M., Schoppmann, F.: Total latency in singleton congestion games. In: Deng, X., Graham, F.C. (eds.) WINE 2007. LNCS, vol. 4858, pp. 381–387. Springer, Heidelberg (2007)
10. Gkatzelis, V., Kollias, K., Roughgarden, T.: Optimal cost-sharing in weighted congestion games. In: Liu, T.-Y., Qi, Q., Ye, Y. (eds.) WINE 2014. LNCS, vol. 8877, pp. 72–88. Springer, Heidelberg (2014)
11. Gopalakrishnan, R., Marden, J.R., Wierman, A.: Potential games are necessary to ensure pure Nash equilibria in cost sharing games. Mathematics of Operations Research (2014)
12. Harks, T., Klimm, M.: On the existence of pure Nash equilibria in weighted congestion games. Mathematics of Operations Research $37(3)$, 419–436 (2012)
13. Klimm, M., Schmand, D.: Sharing non-anonymous costs of multiple resources optimally. In: Paschos, V.T., Widmayer, P. (eds.) CIAC 2015. LNCS, vol. 9079, pp. 274–287. Springer, Heidelberg (2015). arXiv preprint arXiv:1412.4456
14. Kollias, K., Roughgarden, T.: Restoring pure equilibria to weighted congestion games. In: Aceto, L., Henzinger, M., Sgall, J. (eds.) ICALP 2011, Part II. LNCS, vol. 6756, pp. 539–551. Springer, Heidelberg (2011)
15. Koutsoupias, E., Papadimitriou, C.: Worst-case equilibria. Computer Science Review $3(2)$, 65–69 (2009)
16. Monderer, D., Shapley, L.S.: Potential games. Games and Economic Behavior $14(1)$, 124–143 (1996)
17. Rosenthal, R.W.: A class of games possessing pure-strategy Nash equilibria. International Journal of Game Theory $2(1)$, 65–67 (1973)
18. Roughgarden, T.: Intrinsic robustness of the price of anarchy. In: Proceedings of STOC, pp. 513–522. ACM (2009)
19. Roughgarden, T., Schrijvers, O.: Network cost-sharing without anonymity. In: Lavi, R. (ed.) SAGT 2014. LNCS, vol. 8768, pp. 134–145. Springer, Heidelberg (2014)
20. Roughgarden, T., Tardos, É.: How bad is selfish routing? Journal of the ACM (JACM) $49(2)$, 236–259 (2002)
21. Shapiro, A.: On duality theory of convex semi-infinite programming. Optimization $54(6)$, 535–543 (2005)
22. Sheffi, Y.: Urban transportation networks: equilibrium analysis with mathematical programming methods. Prentice-Hall (1985)
23. Wu, S.Y., Fang, S.C.: Solving convex programs with infinitely many linear constraints by a relaxed cutting plane method. Computers & Mathematics with Applications $38(3)$, 23–33 (1999)

Distributed Broadcast Revisited: Towards Universal Optimality

Mohsen Ghaffari$^{(\boxtimes)}$

MIT, Cambridge, USA
ghaffari@mit.edu

Abstract. This paper revisits the classical problem of multi-message broadcast: given an undirected network G, the objective is to deliver k messages, initially placed arbitrarily in G, to all nodes. Per round, one message can be sent along each edge. The standard textbook result is an $O(D + k)$ round algorithm, where D is the diameter of G. This bound is *existentially* optimal, which means there exists a graph G' with diameter D over which any algorithm needs $\Omega(D + k)$ rounds.

In this paper, we seek the stronger notion of optimality—called *universal optimality* by Garay, Kutten, and Peleg [FOCS'93]—which is with respect to the best possible for graph G itself. We present a distributed construction that produces a k-message broadcast schedule with length roughly within an $\tilde{O}(\log n)$ factor of the best possible for G, after $\tilde{O}(D + k)$ pre-computation rounds.

Our approach is conceptually inspired by that of Censor-Hillel, Ghaffari, and Kuhn [SODA'14, PODC'14] of finding many essentially-disjoint trees and using them to parallelize the flow of information. One key aspect that our result improves is that our trees have sufficiently low diameter to admit a nearly-optimal broadcast schedule, whereas the trees obtained by the algorithms of Censor-Hillel et al. could have arbitrarily large diameter, even up to $\Theta(n)$.

1 Introduction and Related Work

Broadcasting messages is a basic primitive used in the vast majority of global[1] distributed algorithms. This paper presents an algorithm that moves towards *universal optimality* for this decades-old problem, in contrast to the standard *existential optimality*.

Model and Problem: We use the standard message passing model CONGEST [10] where the network is represented by an undirected graph $G = (V, E)$, $n := |V|$ and initially each node only knows its neighbors. Communications occur in synchronous rounds, where per round, one B-bit message can be sent along each edge — typically one assumes $B = O(\log n)$. In the *multi-message broadcast problem*, k many B-bit messages are each initially placed in an arbitrary location of the network G, possibly multiple ones in the same node, and the objective is to deliver them to all nodes.

---

[1] Global problems are those in which the solution in a node can depend on the information residing in far away parts of the network.

© Springer-Verlag Berlin Heidelberg 2015
M.M. Halldórsson et al. (Eds.): ICALP 2015, Part II, LNCS 9135, pp. 638–649, 2015.
DOI: 10.1007/978-3-662-47666-6_51

1.1   The Standard Algorithm, and Existential vs. Universal Optimality

The standard k-message broadcast algorithms work in $O(D + k)$ rounds, where D denotes the network diameter. The solution that appears in most textbooks is via communication on a BFS tree, e.g., Peleg [10, Section 4.3.2] writes:

> "A natural approach ... is to collect the items at the root of the tree and then broadcast them one by one. Clearly since broadcasting the items can be done in a pipelined fashion (releasing one item in each round), the broadcast phase can be completed within $(k + depth(T))$ time. ... the collection phase can be performed by an upcast operation with a similar complexity. Hence, the entire problem can be solved in time $O(k + depth(T))$."

This $O(k + depth(T))$ time is $O(k + D)$ rounds in our terminology as for the BFS tree T, $depth(T) = O(D)$. An even more natural approach that achieves the same bound is *flooding*—in each round, each node sends to all its neighbors a message that it has not sent before, if it has one. Analyzing multi-message flooding is non completely trivial. An elegant, but surprisingly[2] less-known, analysis is given by Topkis [13]. See Appendix A of the full-version for a streamlined description.

Existential vs. Universal Optimality: In the theoretical distributed computing community, we often call this $O(k + D)$ time algorithm *optimal*, without paying close attention to the meaning of the implied optimality. Interestingly, Peleg [10, Chapter4,Excercise1] asks (implicitly) to explain why the $\Omega(D)$ term is a *universal* lower bound while the $\Omega(k)$ term is only *existential*. The simple explanation is that broadcasting even a single message takes $\Omega(D)$, regardless of its initial placement. At the same time, on many graphs, one can solve the problem much faster than $O(k + D)$. To give an example, and to take it to the extreme, in the complete graph $D = 1$ but it would be quite embarrassing if $O(k)$ time was the best that we could do for broadcasting k messages; one can easily achieve an $O(k/n)$ time.

To the best of our knowledge, the distinction between the two notions was first emphasized by Garay, Kutten, and Peleg in 1993[6]. Referring to an $O(n)$ time solution for leader election, they write:

> "This solution is optimal in the sense that there exist networks for which this is the best possible. This type of optimality may be thought of as 'existential' optimality. Namely, there are points in the class of input instances under consideration, for which the algorithm is optimal. A stronger type of optimality—which we may analogously call 'universal' optimality—is when the proposed algorithm solves the problem optimally on every instance."

Although the motivation for universal optimality is clear and strong, achieving it is not straightforward and thus, to the best of our knowledge, there is no known distributed algorithm that (non-trivially) achieves universal optimality. In fact,

---

[2] The author believes that this analysis should make its way to the textbooks and courses on distributed algorithms.

even for the MST problem that Garay, Kutten, and Peleg [6] chose to seek such a universal optimality, currently only an *existential* (near) optimality is known[3]. Furthermore, this is even after more than 20 years which includes developments such as the $\tilde{O}(D + \sqrt{n})$ upper bound of Kutten and Peleg [7], the $\tilde{O}(\mu(G, w) + \sqrt{n})$ upper bound of Elkin [4] where $\mu(G, w)$ denotes the *MST-radius*[4], and the (existential) $\tilde{\Omega}(D + \sqrt{n})$ lower bounds of Peleg and Rubinovich[11] and Das Sarma et al. [3].

1.2 Main Result

We present a distributed algorithm that, with high probability[5], produces a k-message broadcast schedule of length $O((\mathsf{OPT} \cdot \log n + \log^2 n) \log \log n)$ rounds, after $\tilde{O}(D + k)$ pre-computation rounds. Here, we use OPT to denote the minimum k-message broadcast schedule length, which depends on graph G and the initial placement of the messages in G.

Remark: We note that the length of the pre-computation phase is certainly a weakness of our result, as our aim is to move *"towards universal optimality"*. See the brief discussion at the end of this section in this regard.

1.3 Our Approach in a Nutshell

A key concept in our approach is what we call *shallow-tree packing*, which is an adaptation of the *tree packing* concept Censor-Hillel et al. [1,2] used for *connectivity decomposition*.

Shallow-Tree Packing: Imagine an optimal broadcast schedule that delivers the k messages to all nodes in OPT rounds. If we trace how each message spreads to all the nodes, for each message $i \in [k]$, we get a spanning subgraph H_i of G that has diameter at most OPT, i.e., the spanning subgraph defined by edge-set having all the edges that message i was transmitted over them. Out of these, we can carve out spanning trees T_i of depth OPT. Clearly, each edge is used by at most OPT of the trees, at most one per round. Let $\mathcal{T}$ be the set of all spanning trees of G with depth at most OPT. For each tree $T \in \mathcal{T}$, define $\lambda(T)$ to be the number of messages $i \in [k]$ for which $T_i = T$, divided by OPT. Thus, we know that in G, there exists a weighted collection $\mathcal{F} = (\lambda(T), T)$ of spanning

---

[3] It is presumable that, to achieve universal optimality for MST, it might be helpful to first understand universal optimality for multi-message broadcast. This is because, the $\tilde{O}(D + \sqrt{n})$-round MST algorithm of Kutten and Peleg [7] uses an k-message broadcast subroutine, with $k = \tilde{O}(\sqrt{n})$, which currently is the standard existentially optimal $O(k + D)$ scheme, thus taking $\tilde{O}(D + \sqrt{n})$ rounds.

[4] The MST-radius is defines as, roughly speaking, the maximum radius around each edge e up to which one has to look to ensure that this edge e is not the heaviest edge in a cycle. See [4] for the formal definition.

[5] We use the phase "with high probability" (w.h.p.) to indicate that an event happens with probability at least $1 - 1/n$.

trees of depth $\alpha = \text{OPT}$ such that

$$\forall e \in G, \sum_{T \in \mathcal{F}, e \in T} \lambda(T) \le 1, \ and \ \sum_{T \in \mathcal{F}} \lambda(T) = \beta = \frac{k}{\text{OPT}}.$$

This tree collection $\mathcal{F}$ is what we call a *shallow-tree packing* with depth $\alpha = \text{OPT}$ and size $\beta = k/\text{OPT}$. Intuitively, the reader can think of this as a *fractional* variant of the cleaner *integral* version where we have $\beta = k/\text{OPT}$ *edge-disjoint* spanning trees, each of depth $\alpha = \text{OPT}$.

A Rough Sketch of Our Algorithm: Our algorithm tries to mimic the optimal broadcast schedule—which we only know exists but we do not know what it looks like—by constructing a shallow-tree packing with depth $O(\text{OPT} \log n)$ and size $O(k/(\text{OPT} \log n))$.

The high-level outline of this construction is via an instantiation of the *Lagrangian relaxation* (see e.g., [12,14]): Roughly speaking, we start with an empty initial collection, and we add shallow-trees to it, one by one. In each iteration, we assign to each edge a cost defined by an exponential in the hitherto load, where load means the total weight of trees crossing the edge. We then compute a tree of depth $O(\text{OPT} \log n)$ with cost within $O(\log n)$ factor of the optimal, and add this tree to the collection (with a small weight). We show that, after a number of these iterations, we arrive at the claimed shallow-tree packing. The Lagrangian relaxation analysis we provide for this can be viewed as considerably simplifying the counterpart in [1], thanks to some small changes we have in the outline of the iterative packing method. To distributedly compute the shallow tree of nearly minimum cost in each iteration, we present an algorithm which has an outline close to the MST algorithm of Gallager, Humblet, Spira [5], but with many vital changes both in the algorithm and in the analysis.

Once the shallow-tree packing is constructed, we use it to produce the near-optimal broadcast schedule. Roughly speaking, we almost evenly distribute the messages among the trees, and each tree broadcasts its own share of messages. While this part would have been essentially trivial if we had a packing of edge-disjoint trees, some complications arise due to the fractionality of our packing. Some additional algorithmic ideas are used to overcome this challenge. The analysis of this part includes a novel technique for tightly bounding the information dissemination time in probabilistic scenarios, which we believe can be of interest well-beyond this paper. To make the technique more accessible, the full-version of the paper explains this analysis also in the context of a very simple toy problem.

On the Length of the Pre-computation Phase: While the length of the pre-computation weakens our result, we believe that there is still important merit in the result, because: (1) The pre-computation is only performed once, regardless of the number of the times one needs to solve k-message broadcast, and even for different initial message placements. Hence its cost amortizes over multiple iterations. (2) Even the pre-computation phase is not significantly longer than the standard time bound. (3) Most importantly, the author believes that the general framework set forth here and particularly our characterization of distributed broadcast algorithms as special combinatorial graph structures, i.e., the *shallow-tree packings*, are an important step towards universal optimality. Furthermore, the technical method we use for computing these structures arguably has the right overall outline. However, it is conceivable that one might be able to improve this pre-computation phase, while using the same outline.

2   Distributed Shallow-Tree Packing

In this section, we explain how we compute the shallow-tree packing, which is our pre-computation phase. In the next section, we explain how to use this structure to produce the near-optimal broadcast schedule. The main result of this section is captured by the following theorem:

Theorem 1. *There is a distributed algorithm that, with high probability, produces a shallow-tree packing of depth $O(\mathsf{OPT}\log n)$ and size $O(k/(\mathsf{OPT}\log n))$, in $\tilde{O}(D+k)$ rounds.*

Throughout this section, we assume that a 2-approximation of OPT is known. This assumption can be removed using standard techniques, simply by repeating our algorithm for $O(\log n)$ iterations, once for each 2-approximation guess $\overline{\mathsf{OPT}} = 2^j$, and picking the smallest guess $\overline{\mathsf{OPT}}$ for which the construction was *successful*. That is, the first guess that lead to a successful construction of a shallow-tree packing with depth $O(\overline{\mathsf{OPT}}\log n)$ and size $O(k/(\overline{\mathsf{OPT}}\log n))$. Since overall this is a standard technique, we do not explain the details any further. Also, note that these $O(\log n)$ iterations increase the round complexity of our pre-computation by an $O(\log n)$ factor, which gets absorbed in the $\tilde{O}()$ notation. Furthermore, at the end only this one successful shallow-tree packing is kept, which means this part only effects the pre-cumputation.

2.1   Packing via Lagrangian Relaxations

The algorithm we present here is to some extent similar to an algorithm in [1], which itself is rooted in a long line of research on Lagrangian relaxations (see e.g. the prominent works of [12,14]). Our algorithm, however, has small but vital changes, which allow us to extend the result to our approximate setting, and furthermore, interestingly, they allow us to significantly simplify the analysis, in contrast to the counterpart in [1, AppendixF].

Lemma 1. *Suppose that given costs for the edges, each in range $[n^{-\Theta(1)}, n^{\Theta(1)}]$, we can distributedly compute a depth $O(d\log n)$ tree with cost within $O(\log n)$ factor of the min-cost depth d tree, in $O(d\log^2 n)$ rounds. Then, there is a distributed algorithm that produces a shallow-tree packing with depth $O(\mathsf{OPT}\log n)$ and size $O(k/(\mathsf{OPT}\log n))$, in $\tilde{O}(k/\mathsf{OPT})$ iterations, each taking $\tilde{O}(\mathsf{OPT})$ rounds.*

We note that any broadcast schedule needs at least $\Omega(D)$ rounds, which means that $\mathsf{OPT} \geq \Omega(D)$. This will be used throughout our calculations.

Let us first present the outline of the algorithm: We maintain a weighted collection $\mathcal{F}$ of trees of depth $O(\mathsf{OPT} \log n)$, where each tree $\tau \in \mathcal{F}$ has weight $w_\tau \in [0,1]$. Initially, the collection is empty. We iteratively add trees to our collection, one by one, as follows: In each iteration, we define an edge cost $c_e = e^{\alpha \mathsf{load}_e}$, where $\alpha = 10 \log n$ and $\mathsf{load}_e = \sum_{\tau \in \mathcal{F}, e \in \tau} w_\tau$. We then find a tree τ^* of depth at most $\mathsf{OPT} \cdot O(\log n)$ such that its cost is within an $O(\log n)$ factor of the min-cost depth OPT tree. This part will be done using the distributed algorithm that we describe later in Section 2.2. We add τ^* to our collection with weight $\frac{\delta}{\alpha}$, where $\delta > 0$ is a small constant, e.g., $\delta = 0.01$. If there is an edge that has $\mathsf{load}_e \geq 1$, or if $\Theta(\frac{k}{\mathsf{OPT}})$ iterations have passed, we stop.

We get the following result about the total weight of the collection. We only provide a proof sketch; the full proof appears in the full-version of this paper.

Lemma 2. *At termination, the total weight in the collection is at least* $\frac{k}{\mathsf{OPT} \cdot O(\log n)}$.

Proof (Proof Skecth). The claim is trivial if termination is because of finishing $\Theta(\frac{k}{\mathsf{OPT}})$ iterations. Suppose that the algorithm terminated because there was an edge with load greater than 1. Consider the potential function $\phi = \sum_{e \in E} e^{\alpha \mathsf{load}_e}$. Initially, this potential is equal to $m \leq n^2$ as each edge has load 0. When we stop, there is at least one edge with load 1, which means that the potential at that point is at least $e^\alpha \geq n^{10}$. Also notice that the potential is non-decreasing over time because the loads are non-decreasing. Using some calculations, we can show that per iteration, the potential increases by at most a $(1 + \frac{\delta \cdot O(\log n) \cdot \mathsf{OPT}}{k})$ factor. Thus, we conclude that there are at least $\Theta(\frac{k}{\delta \cdot \mathsf{OPT}})$ iterations, and therefore, the total weight at the end is at least $\frac{k}{\mathsf{OPT} \cdot O(\log n)}$, as claimed.

2.2 Distributed Approximation of Min-Cost Shallow-Tree

The general outline of our distributed min-cost shallow-tree approximation algorithm is inspired by that of the MST algorithm of Boruvka from 1926 [9], which has the same outline as the algorithm of Gallager, Humblet, and Spira[5]. That is, our algorithm also works in $O(\log n)$ iterations, each time reducing the number of connected components by a constant factor. However, there are significant changes both in the algorithm and in the analysis, which allow us to upper bound the depth of the produced tree to $O(d \log n)$, while sacrificing only an $O(\log n)$ factor in the cost, in comparison to the min-cost depth d tree.

Theorem 2. *There is a distributed algorithm that in $O(d \log^2 n)$ rounds computes a depth $O(d \log n)$ tree with cost within $O(\log n)$ factor of the min-cost depth d tree.*

General Outline: The algorithm works in $O(\log n)$ iterations, each taking $O(d \log n)$ rounds. During these iterations, we maintain a spanning forest, which changes from each iteration to the next. Initially, the forest has no edges, thus each node forms its own connected component of the forest, i.e., its own cluster. In each iteration, some components merge with each other, hence forming the new clustering, i.e., the new forest. Throughout, we have a center for each cluster, and the centers of the next iteration are a subset of the centers of the current iteration. We maintain the invariant that at the end of iteration i, each forest component has radius at most id (computed as farthest distance from its center). Therefore, the final depth of the tree after $O(\log n)$ iterations is at most $O(d \log n)$, as promised. This linear depth growth is a part that makes our algorithm significantly different than what happens in the aforementioned MST algorithms, as there the radius can grow much faster, up to exponentially fast, i.e., by a multiplicative constant factor per iteration.

Let C^* be the cost of the min-cost depth d tree. Per iteration, we spend a cost of $O(C^*)$ to *buy* more edges that are added to the forest, in such a way that the number of components goes down by a constant factor in expectation (and while maintaining the radius invariant). This will let us prove that after $O(\log n)$ iterations, w.h.p, we have a single spanning tree. Therefore, the total cost is at most $O(C^* \log n)$, as promised.

Merges in Each Iteration: Per iteration, for each center v, we find a path P_v^* with at most d hops that connects v to one of the other centers $u \neq v$, and such that P_v^* has a 2-factor minimal cost, considering all the d-hop paths connecting v to other centers. Then, this path gets suggested to be added to the forest. For each center, we toss a fair coin. Each center v that draws a head will "buy" and add to its component all the paths coming from centers u that draw a tail and suggested a path towards v. Then, v stays as a component center of this bigger component and those centers u loose their centrality role. A center that draw a tail but the end point of its suggested path also draw a tail stays as a center for the next iteration. Since each of the previous components had radius $(i-1)d$, by induction hypothesis, and we added only a path of length d from center u to one of the centers v of the previous components, the new component has radius at most id, thus proving the induction about the radius of the clusters.

Note that the added paths might produce cycles. To clean up these cycles, we run a BFS from each cluster center that decided to stay alive for this iteration, along the already-bought edges (those of the paths bought in this iteration or in the past iterations), and the BFS grows for only id rounds. Each (non-center) node joins the cluster related to the first BFS that reaches it, and only forwards the BFS token of this center (along its already-bought edges). This way, the cluster of each center has radius at most id. Note that each node v will still be in a cluster, because the center of cluster that absorbed the old cluster of v is at most id hops away from v, considering the already-bought edges.

Lemma 3. *After $O(\log n)$ iterations, with high probability, we have a single connected component, i.e., a spanning tree.*

Proof. Consider a single iteration. For each component, the probability that it merges with at least one other component is at least $1/4$, which happens when this component draws a tail and the receptive end of the path draws a head.

Hence, we get that in expectation, the total number of the new components is at most $3/4$ fraction of the old number of components. Therefore, after say $20 \log n$ iterations, the expected number of connected components is at most $(\frac{3}{4})^{20 \log n} \leq \frac{1}{n^{10}}$. Hence, Markov's inequality tells us that the probability that we have 2 or more components left is at most $\frac{2}{n^{10}}$. That is, w.h.p., the number of connected components is at most one, i.e., we have arrived at a tree.

What remains is to explain how we compute the approximately min-cost suggested paths, from each cluster center to another cluster along a path of at most d-hops, and why all these paths in one iteration together have cost at most $O(C^*)$. We first explain the path finding part, and then present our cost analysis.

Path Computations: To recap the setting, we are given a set S of centers, and we want to find for each center $v \in S$ a d-hop path P_v^* to a center $u \neq v$ such that the cost of P_v^* is within 2-factor of the minimum cost such path for v. Furthermore, we want to do this for all nodes $v \in S$, altogether in time $\tilde{O}(d)$. If we had a single center, we could run a d-round truncated version of Bellman-Ford which would find the best path for the center exactly. However, doing this for many centers at the same time is nontrivial, due the congestion. Intuitively, the challenge is rooted in the fact that the edges have non-uniform costs, and thus, a path with larger hop-count can have significantly lower cost to alternative paths with smaller number of hops. To overcome this, we use a clever *rounding* trick of Nanongkai [8] presented in the context of single-source shortest path approximation[6]. Roughly speaking, this trick allows us to translate the problem approximately to one in the setting with uniform edge costs.

Consider a threshold $W = 2^j$, for $j \in [-\Theta(\log n), \Theta(\log n)]$. Roughly speaking, we want for each $v \in S$ to know if it has a d-hop path of cost at most W to another node in S or not. We round up the cost of each edge to the next multiplicative factor of $\varepsilon W/d$, where $\varepsilon = 0.01$. That is, for each edge e, define $cost'(e) = \frac{W\varepsilon}{d} \lceil \frac{cost(e)}{\varepsilon W/d} \rceil$. Thus, $cost(e) \leq cost'(e) \leq cost(e) + W\varepsilon/d$. Hence, over a path made of at most d hops, this rounding can only increase the cost by most $W\varepsilon$, which is much smaller than W. Hence, if for example we had a d-hop path with length at most $W(1 - \varepsilon)$, it still is below our threshold W.

What did we gain by doing this rounding? First, consider a single center and imagine a *BFS* growing synchronously from all sides at a speed of $\frac{W\varepsilon}{d}$ units of cost per round. Any node at distance (according to $cost'$) at most W from the BFS source would be hit by the BFS after at most $\frac{W}{W\varepsilon/d} = d/\varepsilon$ such rounds.

Even nicer, we can run many BFSs each growing from one of the centers, in parallel and at the same speed. At some point some BFS tokens might simultaneously reach the same node which gives rise to the question of which one should be forwarded. Since we only care to find one nearly min-cost path for each center, to control the congestion, when many BFS tokens reach the same point at the same time, we only keep two of the tokens and discard the rest. We keep two tokens so that, for any center node that will be reached later, at least

---

[6] Nanongkai[8] mentions that this idea has been rediscovered a few times before, in different contexts, which is a testament to its elegance and naturalness.

one token is different from this center's. Hence, we can run the BFS growth in $2d/\varepsilon$ rounds, still growing synchronously but at the speed of speed of $\frac{W\varepsilon}{d}$ units of cost per two rounds. As a result, for each center v, if there is a center u at d-hop distance at most $(1 - \varepsilon)W$ from v, then at the end of the BFS growth, center v will receive the token of u, or potentially a different center u' at d-hop distance at most W from v.

The above description was for a given threshold $W = 2^j$. We simply run it for all possible values $j \in [-\Theta(\log n), \Theta(\log n)]$, one by one, and we let each node v remember the first center u that its BFS token reaches v. This way, each center node v will know a different center u with a 2-factor-minimal d-hop path to v. The path is not known by v itself, but v can reconstruct the path by sending a token in the reverse direction of the path, towards u. This token traversal does not create congestion either as at most two tokens will be sent in the reverse direction, per each value of j.

Note that the heads and tails outcomes of the coin tosses of the centers can already be put in these tokens, and we only spread BFS tokens out of head-centers and only tail-centers accept a token. At the end, the tail tokens send the tokens in the reverse direction. Thus, all the edges that are *bought* in this iteration get identified distributedly.

Lemma 4. *Per iteration, we spend a cost of at most $O(C^*)$.*

Proof. We prove that, the total cost of all the suggested paths is at most $O(C^*)$. For this, we turn our attention to the d-depth spanning tree τ^* which has (the minimum) cost C^*. Note that the algorithm does not know this tree, and it only appears in our analysis. Recall that for each center v, we used the notation P_v^* to denote a 2-factor minimal-cost d-hop path connecting v to a different center, which gets suggested by v. We show that there exist alternate d-hop paths P_v', one for each v, connecting v to a different center, such that $\sum_v cost(P_v') = O(C^*)$. This would finish our cost analysis proof, simply because for each v, we have $cost(P_v^*) \leq 2cost(P_v')$.

To define the paths P_v', we present a simple (and in our opinion cute) distributed algorithm, which we imagine it to be run on the tree τ^*. We emphasize that in reality there is no such algorithm being run, and we are only imagining such an algorithm for the purpose of analysis.

Pick an arbitrary root for τ^* and orient τ^* outwards, making each node know its parent, and its depth in τ^*, and the maximum depth of τ^*. Then, we start a synchronous upcast in τ^* starting from the deepest level and moving upwards, one hop at a time, from all sides. Each center node v starts this upcast with a *pebble* tagged with its name and we gradually send up these pebbles. We maintain that each node will send to its parent at most one pebble. When the upcast reaches a node w, this node w might have received a number of pebbles from its children, at most one from each of them, and might have one of its own as well, if it is a center. Node w matches a maximal even number of the pebbles pairwise, except for at most one which remains when the number of pebbles is odd. This matching defines a unique τ^*-path P' for the owner center of each pebble, the unique τ^*-path connecting the pebble owner to the origin of the other pebble in the pair. Furthermore, this way, each edge of τ^* appears in at most two of the paths, once for each end of the matching. If w has an odd number of pebbles, it sends the one remaining up to its parent. The same process is repeated until the upcast reaches the root. If the root has no pebble left once it performs its step of matching, we are done as we have found paths for centers with total cost at most $2C^*$. If there is one pebble left at the root, grab an arbitrary one of the other centers and assign to this pebble owner the unique τ^*-path to that other center. This increases the total cost of paths P' to at most $3C^*$.

3   Broadcast on the Shallow-Tree Packing

We now can assume that we are given a shallow-tree packing of depth $O(\mathsf{OPT} \cdot \log n)$ and size $k/O(\mathsf{OPT} \cdot \log n)$, as computed in Section 2, and use it to produce a broadcast schedule of length $O(\mathsf{OPT} \log n \log \log n + \log^2 n)$ rounds.

Theorem 3. *Given the shallow-tree packing of Section 2, we can broadcast k messages in $O((\mathsf{OPT} \log n + \log^2 n) \log \log n)$ rounds.*

Due to the space limitations, this algorithm and its analysis are deferred to the full-version of the paper. Here, we describe the general outline of the approach and present a much simpler algorithm that produces a broadcast schedule of length $O(\mathsf{OPT} \log^2 n)$.

3.1   General Outline for Broadcast Using Shallow-Tree Packing

Consider the given shallow-tree packing, which is a collection $\mathcal{F} = (\lambda(T), T)$ of depth-$O(\mathsf{OPT} \log n)$ spanning trees such that

$$\forall e \in G, \sum_{T \in \mathcal{F}, e \in T} \lambda(T) \leq 1, \text{ and } \sum_{T \in \mathcal{F}} \lambda(T) = \frac{k}{\mathsf{OPT} \cdot O(\log n)}.$$

Our general broadcast method is as follows: for each message $i \in [k]$, we assign it to one of the trees, chosen randomly with probabilities proportional to the weights $\lambda(T)$. This choice will be made by the node initially holding the message, which then tags the message with this tree. Each message $i \in [k]$ is assigned to tree T with probability $\frac{\lambda(T)}{\sum_{T \in \mathcal{F}} \lambda(T)}$. Since $\sum_{T \in \mathcal{T}} \lambda(T) = \frac{k}{\mathsf{OPT} \cdot O(\log n)}$ and for each tree $\lambda(T) \leq 1$, we can infer that each tree in expectation receives at most $O(\log n) \cdot \mathsf{OPT}$ messages, and due to the independence of the choices of the messages, a Chernoff bound and then a union bound over all trees tell us that, w.h.p., each tree receives at most $O(\log n) \cdot \mathsf{OPT}$ messages.

What is left at this point is to broadcast the messages assigned to each tree, inside it. If we were in the hypothetical/ideal scenario where trees are truly non-overlapping (i.e., edge-disjoint), we could perform the broadcasts in different trees with no interference from each other, and the broadcasts in tree T would be done in $O(k_T + D_T)$ where k_T is the number of messages assigned to T and D_T is its depth. Hence, the broadcast in each tree would be done in at most $O(\mathsf{OPT} \cdot \log n)$ rounds. This would mean all broadcasts finish in this time, as the broadcasts in different trees happen in parallel. Hence, that would give us a broadcast schedule with length at most an $O(\log n \cdot \mathsf{OPT})$.

However, this outline misses a critical point, namely that the trees are not edge-disjoint; they are only fractionally disjoint which means that they can overlap but this overlap is controlled, i.e., $\forall e \in G, \sum_{T \in \mathcal{T}, e \in T} \lambda(T) \leq 1$. We next describe in more detail the challenge caused by this fractionality and the simpler method of overcoming it while sacrificing an $O(\log n)$ factor.

3.2   The Challenge, and Our Simpler $O(\log^2 n) \cdot \mathsf{OPT}$ Schedule

The Challenge: Here, we try to provide an intuitive description of the challenge. Consider an edge e, and suppose that there are many trees in the collection that include e. Per round, only one of the trees can use the edge. The natural methods for giving the edge to the trees would be by various methods of time-sharing, proportional to the weight of the trees. For instance, per round, we could pick the tree that will use the edge randomly, with probabilities proportional to $\lambda(T)$. The total weight of the trees going through an edge is at most 1, but also typically at least a constant. That is, we can not count on it being considerably smaller than 1. Hence, with these methods of time sharing, roughly speaking, a tree with weight $\lambda(T)$ will get to use the edge only once per about $\Theta(1/\lambda(T))$ rounds. Notice that a similar difficulty occurs with other simple methods of time-sharing as well. Such a delay would effectively make the time to even broadcast a single message in the tree grow by about a $\Theta(1/\lambda(T))$ factor, i.e., to $\Omega(\mathsf{OPT} \cdot \log n/\lambda(T))$, which can be larger than $\omega(\mathsf{OPT} \cdot \log n)$ because $\lambda(T) \in [0, 1]$.

Our Simpler Broadcast Schedule: One saving grace which allows us to find an $O(\log^2 n) \cdot \mathsf{OPT}$ length broadcast schedule without too much of work is that, the construction we presented in Section 2 is not too far from integrality[7]. More concretely, we in fact have $O(k/\mathsf{OPT})$ trees, one per iteration of the Lagrangian relaxation, and each of these trees has $\lambda(T) = \delta/\alpha = \Theta(1/\log n)$. Hence, each edge is in at most $O(\log n)$ trees. In the random distribution of the messages among these trees, each gets $O(\mathsf{OPT} + \log n)$ messages, with high probability.

Now to broadcast messages, divide the time into phases, each having $O(\log n)$ rounds: For each edge, assign each of the rounds of the phase to one of the trees that goes through this edge. Hence, each tree gets to use each of its edges at least once per phase. Therefore, the messages that are being broadcast in each tree proceed with a speed of one hop per phase, at least. This means that after $O(k_T + D_T) = O(\mathsf{OPT} + \log n + \mathsf{OPT} \cdot \log n) = O(\mathsf{OPT} \cdot \log n)$ phases, which is equal to $O(\mathsf{OPT} \log^2 n)$ rounds, all the messages of the tree are delivered to all its nodes. As this holds for each of our trees, this method finishes the broadcast in $O(\mathsf{OPT} \log^2 n)$ rounds.

---

[7] We note that already having this property ready by our construction is not vital. One can obtain similar guarantees for any other construction, using *randomized rounding*.

Acknowledgments. The author thanks Christoph Lenzen for helpful conversations about the concept of shallow-tree packings. The author is also grateful to Bernhard Haeupler for discussions regarding correctness of the max-flow min-cut based analysis technique that we present for probabilistic message dissemination.

References

1. Censor-Hillel, K., Ghaffari, M., Kuhn, F.: Distributed connectivity decomposition. In: The Proc. of the Int'l Symp. on Princ. of Dist. Comp. (PODC), pp. 156–165 (2014)
2. Censor-Hillel, K., Ghaffari, M., Kuhn, F.: A new perspective on vertex connectivity. In: Pro. of ACM-SIAM Symp. on Disc. Alg. (SODA), pp. 546–561 (2014)
3. Das Sarma, A., Holzer, S., Kor, L., Korman, A., Nanongkai, D., Pandurangan, G., Peleg, D., Wattenhofer, R.: Distributed verification and hardness of distributed approximation. In: Proc. of the Symp. on Theory of Comp. (STOC), pp. 363–372 (2011)
4. Elkin, M.: A faster distributed protocol for constructing a minimum spanning tree. In: Pro. of ACM-SIAM Symp. on Disc. Alg. (SODA), pp. 359–368 (2004)
5. Gallager, R.G., Humblet, P.A., Spira, P.M.: A distributed algorithm for minimum-weight spanning trees. ACM Trans. on Prog. Lang. and Sys. 5(1), 66–77 (1983)
6. Garay, J., Kutten, S., Peleg, D.: A sub-linear time distributed algorithm for minimum-weight spanning trees. In: Proc. of the Symp. on Found. of Comp. Sci. (FOCS), pp. 659–668 (1993)
7. Kutten, S., Peleg, D.: Fast distributed construction of k-dominating sets and applications. In: The Proc. of the Int'l Symp. on Princ. of Dist. Comp. (PODC), pp. 238–251 (1995)
8. Nanongkai, D.: Distributed approximation algorithms for weighted shortest paths. In: Proc. of the Symp. on Theory of Comp. (STOC), pp. 565–573 (2014)
9. Nesetril, J., Milkova, E., Nesetrilova, H.: Otakar boruvka on minimum spanning tree problem translation of both the 1926 papers, comments, history. Discrete Mathematics **233**(1), 3–36 (2001)
10. Peleg, D.: Distributed Computing: A Locality-sensitive Approach. Society for Industrial and Applied Mathematics, Philadelphia (2000)
11. Peleg, D., Rubinovich, V.: A near-tight lower bound on the time complexity of distributed MST construction. In: Proc. of the Symp. on Found. of Comp. Sci. (FOCS) (1999)
12. Plotkin, S.A., Shmoys, D.B., Tardos, É.: Fast approximation algorithms for fractional packing and covering problems. Mathematics of Operations Research **20**(2), 257–301 (1995)
13. Topkis, D.M.: Concurrent broadcast for information dissemination. IEEE Transactions on Software Engineering **10**, 1107–1112 (1985)
14. Young, N.E.: Sequential and parallel algorithms for mixed packing and covering. In: Proc. of the Symp. on Found. of Comp. Sci. (FOCS), pp. 538–546. IEEE (2001)

Selling Two Goods Optimally

Yiannis Giannakopoulos$^{(\boxtimes)}$ and Elias Koutsoupias

University of Oxford, Oxford, UK
ygiannak@cs.ox.ac.uk

Abstract. We provide sufficient conditions for revenue maximization in a two-good monopoly where the buyer's valuations for the items come from independent (but not necessarily identical) distributions over bounded intervals. Under certain distributional assumptions, we give exact, closed-form formulas for the prices and allocation rules of the optimal selling mechanism. As a side result we give the first example of an optimal mechanism in an i.i.d. setting over a support of the form $[0, b]$ which is *not* deterministic. Since our framework is based on duality techniques, we were also able to demonstrate how slightly relaxed versions of it can still be used to design mechanisms that have very good approximation ratios with respect to the optimal revenue, through a "convexification" process.

1 Introduction

The problem of designing auctions that maximize the seller's revenue in settings with many heterogeneous goods has attracted a large amount of interest in the last years, both from the Computer Science as well as the Economics community (see e.g. [2,3,5,6,8,11,15]). Here the seller faces a buyer whose true values for the m items come from a probability distribution over $\mathbb{R}_+^m$ and, based only on this incomplete prior knowledge, he wishes to design a selling mechanism that will maximize his expected revenue. For the purposes of this paper, the prior distribution is a product one, meaning that the item valuations are independent. The buyer is additive, in the sense that her happiness from receiving any subset of items is the sum of her values of the individual items in that bundle. The buyer is also selfish and completely rational, thus willing to lie about her true values if this is to improve her own happiness. So, the seller should also make sure to give the right incentives to the buyer in order to avoid manipulation of the protocol by misreporting.

The special case of a single item has been very well understood since the seminal work of Myerson [13]. However, when one moves to settings with multiple goods, the problem becomes notoriously difficult and novel approaches are necessary. Despite the significant effort of the researchers in the field, essentially only specialized, partial results are known: there are exact solutions for

The research leading to these results has received funding from the European Research Council under the European Union's Seventh Framework Programme (FP7/2007-2013)/ERC grant agreement no. 321171.

© Springer-Verlag Berlin Heidelberg 2015
M.M. Halldórsson et al. (Eds.): ICALP 2015, Part II, LNCS 9135, pp. 650–662, 2015.
DOI: 10.1007/978-3-662-47666-6_52

two items in the case of identical uniform distributions over unit-length intervals [11,15], exponential over $[0, \infty)$ [3] or identical Pareto distributions with tail index parameters $\alpha \geq 1/2$ [6]. For more items optimal results are only known for uniform valuations over the unit interval [5], and due to the difficulty of exact solutions most of the work focuses in showing approximation guarantees for simple selling mechanisms [1,4,6,9]. This difficulty is further supported by the complexity #P-hardness results of Daskalakis et al. [2]. It is important to point out that even for two items *we know of no general and simple, closed-form conditions framework under which optimality can be extracted when given as input the item distributions, in the case when these are not necessarily identical.* This is our goal in the current paper.

Our Contribution. We introduce general but simple and clear, closed-form distributional conditions that can guarantee optimality and immediately give the form of the revenue-maximizing selling mechanism (its payment and allocation rules), for the setting of two goods with valuations distributed over bounded intervals (Theorem 1). For simplicity and a clearer exposition we study distributions supported over the real unit interval $[0,1]$. By scaling, the results generalize immediately to intervals that start at 0, but more work would be needed to generalize them to arbitrary intervals. We use the closed forms to get optimal solutions for a wide class of distributions satisfying certain simple analytic assumptions (Theorem 2 and Sect. 4). As useful examples, we provide exact solutions for families of monomial ($\propto x^c$) and exponential ($\propto e^{-\lambda x}$) distributions (Corollaries 1 and 2 and Sect. 4), and also near-optimal results for power-law ($\propto (x+1)^{-\alpha}$) distributions (Sect. 5). This last approximation is an application of a more general result (Theorem 3) involving the relaxation of some of the conditions for optimality in the main Theorem 1; the "solution" one gets in this new setting might not always correspond to a feasible selling mechanism, however it still provides an upper bound on the optimal revenue as well as hints as to how to design a well-performing mechanism, by "convexifying" it into a feasible mechanism (Sect. 5).

Particularly for the family of monomial distributions it turns out that the optimal mechanism is a very simple deterministic mechanism that offers to the seller a menu of size [7] just 4: fixed prices for each one of the two items and for their bundle, as well as the option of not buying any of them. For the rest of the distributions randomization is essential for optimality, as is generally expected in such problems of multidimensional revenue maximization (see e.g. [3,8,15]). For example, this is the case for two i.i.d. exponential distributions over the unit interval $[0,1]$, which gives the first such example where determinism is suboptimal even for regularly[1] i.i.d. items. A point worth noting here is the striking difference between this result and previous results [3,4] about i.i.d. exponential distributions which have as support the entire $\mathbb{R}_+$: the optimal selling mechanism there is the deterministic one that just offers the full bundle of both items.

---

[1] A probability distribution F is called *regular* if $t - \frac{1-F(t)}{f(t)}$ is increasing.

Although the conditions that the probability distributions must satisfy are quite general, they leave out a large class of distributions. For example, they do not apply to power-law distributions with parameter $\alpha > 2$. In other words, this work goes some way towards the complete solution for arbitrary distributions for two items, but the general problem is still open. In this paper, we opted towards simple conditions rather than full generality, but we believe that extensions of our method can generalize significantly the range of distributions; we expect that a proper "ironing" procedure will enable our technique to resolve the general problem for two items.

Techniques. The main result of the paper (Theorem 1) is proven by utilizing the *duality* framework of [5] for revenue maximization, and in particular using complementarity: the optimality of the proposed selling mechanism is shown by verifying the existence of a dual solution with which they satisfy together the required complementary slackness conditions of the duality formulation. Constructing these dual solutions explicitly seems to be a very challenging task and in fact there might not even be a concise way to do it, especially in closed-form. So instead we just prove the existence of such a dual solution, using a *max-flow min-cut* argument as main tool (Lemma 3). This is, in a way, an abstraction of a technique followed in [5] for the case of uniform distributions which was based on Hall's theorem for bipartite matchings. Since here we are dealing with general and non-identical distributions, this kind of refinement is essential and non-trivial, and in fact forms the most technical part of the paper. Our approach has a strong geometric flavor, enabled by introducing the notion of the *deficiency* of a two-dimensional body (Definition 1, Lemma 2), which is inspired by classic matching theory [10,14].

All omitted proofs and further discussion can be found in the full version of our paper.

1.1   Model and Notation

We study a two-good monopoly setting in which a seller deals with a buyer who has values $x_1, x_2 \in I$ for the items, where $I = [0,1]$. The seller has only an incomplete knowledge of the buyer's preference, in the form of two independent distributions (with densities) f_1, f_2 over I from which x_1 and x_2 are drawn, respectively. The cdf of f_j will be denoted by F_j. We will also use standard vector notation $\mathbf{x} = (x_1, x_2)$. For any item $j \in \{1,2\}$, index $-j$ will refer the complementary item, that is $3 - j$, and as it's standard in game theory $\mathbf{x}_{-j} = x_{-j}$ will denote the remaining of vector $\mathbf{x}$ if the j-th coordinate is removed, so $\mathbf{x} = (x_j, x_{-j})$ for any $j = 1, 2$.

The seller's goal is to design a selling mechanism that will maximize his revenue. Without loss[2] we can focus on direct-revelation mechanisms: the bidder will be asked to submit bids b_1, b_2 and the mechanism consists simply of an allocation rule $a_1, a_2 : I^2 \to I$ and a payment function $p : I^2 \to \mathbb{R}_+$ such that

---

[2] This is due to the celebrated Revelation Principle [13].

$a_j(b_1, b_2)$ is the probability of item j being sold to the buyer (notice how we allow for randomized mechanisms, i.e. lotteries) and $p(b_1, b_2)$ is the payment that the buyer expects to pay; it is easier to consider the expected payment for all allocations, rather than individual payments that depend on the allocation of items. The reason why the bids b_j are denoted differently than the original values x_j for the items is that, since the bidder is a rational and selfish agent, she might lie and misreport $b_j \neq x_j$ if this is to increase her personal gain given by the quasi-linear *utility* function

$$u(\mathbf{b}; \mathbf{x}) \equiv a_1(\mathbf{b})x_1 + a_2(\mathbf{b})x_2 - p(\mathbf{b}), \tag{1}$$

the expected happiness she'll receive by the mechanism minus her payment. Thus, we will demand our selling mechanisms to satisfy the following standard conditions:

- *Incentive Compatibility (IC)*, also known as truthfulness, saying that the player would have no incentive to misreport and manipulate the mechanism, i.e. her utility is maximized by truth-telling: $u(\mathbf{b}; \mathbf{x}) \leq u(\mathbf{x}; \mathbf{x})$
- *Individual Rationality (IR)*, saying that the buyer cannot harm herself just by truthfully participating in the mechanism: $u(\mathbf{x}; \mathbf{x}) \geq 0$.

It turns out the critical IC property comes without loss[3] for our revenue-maximization objective, so for now on we will only consider truthful mechanisms, meaning we can also relax the notation $u(\mathbf{b}; \mathbf{x})$ to just $u(\mathbf{x})$.

There is a very elegant and helpful analytic characterization of truthfulness, going back to Rochet [16] (for a proof see e.g. [6]), which states that the player's utility function must be *convex* and that the allocation probabilities are simply given by the utility's derivatives, i.e. $\partial u(\mathbf{x})/\partial x_j = a_j(\mathbf{x})$. Taking this into consideration and rearranging (1) with respect to the payment, we define

$$\mathcal{R}_{f_1, f_2}(u) \equiv \int_0^1 \int_0^1 \left(\frac{\partial u(\mathbf{x})}{\partial x_1} x_1 + \frac{\partial u(\mathbf{x})}{\partial x_2} x_2 - u(\mathbf{x}) \right) f_1(x_1) f_2(x_2) \, dx_1 \, dx_2$$

for every absolutely continuous function $u : I^2 \longrightarrow \mathbb{R}_+$. If u is convex with partial derivatives in $[0, 1]$ then u is a valid utility function and $\mathcal{R}_{f_1, f_2}(u)$ *is the expected revenue of the seller under the mechanism induced by* u. Let $\text{REV}(f_1, f_2)$ denote the best possible such revenue, i.e. the supremum of $\mathcal{R}_{f_1, f_2}(u)$ when u ranges over the space of all feasible utility functions over I^2. So the problem we want to deal with in this paper is exactly that of $\sup_u \mathcal{R}_{f_1, f_2}(u)$.

We now present the conditions on the probability distributions which enable our technique to provide a closed-form of the optimal auction.

Assumption 0. *We assume that the density functions* f_1, f_2 *are absolutely continuous over* I *and almost everywhere[4] (a.e.) differentiable. Furthermore, we assume that they are bounded from below, except for small values of* x_i; *in particular, we assume that there exists some small* ϵ *such that* $f_i(x_i) > \epsilon$, *for every* $x_i > \epsilon$.

---

[3] Also due to the Revelation Principle.

[4] Everywhere except a set of zero Lebesgue measure.

Assumption 1. *The probability distributions f_1, f_2 are such that functions $h_{f_1,f_2}(\mathbf{x}) - f_2(1)f_1(x_1)$ and $h_{f_1,f_2}(\mathbf{x}) - f_1(1)f_2(x_2)$ are nonnegative, where*

$$h_{f_1,f_2}(\mathbf{x}) \equiv 3f_1(x_1)f_2(x_2) + x_1 f_1'(x_1)f_2(x_2) + x_2 f_2'(x_2)f_1(x_1). \qquad (2)$$

Function h_{f_1,f_2} will also be assumed to be absolutely continuous.

We will drop the subscript f_1, f_2 in the above notations whenever it is clear which distributions we are referring to. Assumption 1 is a slightly stronger condition than $h(\mathbf{x}) \geq 0$ which is a common regularity assumption in the economics literature for multidimensional auctions with m items: $(m+1)f(\mathbf{x}) + \nabla f(\mathbf{x}) \cdot \mathbf{x} \geq 0$, where f is the joint distribution for the item valuations (see e.g. [11,12,15]). In fact, Manelli and Vincent [11] make the even stronger assumption that for each item j, $x_j f_j(x_j)$ is an increasing function. Even more recently, that assumption has also been deployed by Wang and Tang [17] in a two-item setting as one of their sufficient conditions for the existence of optimal auctions with small-sized menus.

Strengthening the regularity condition $h(\mathbf{x}) \geq 0$ to that of Assumption 1 is essentially only used as a technical tool within the proof of Lemma 2, and as a matter of fact we don't really need it to hold in the entire unit box I^2 but just in a critical sub-region $D_{1,2}$ which corresponds to the valuation subspace where both items are sold with probability 1 (see Fig. 1 and Sect. 2.1). The same is true for the second leg of Assumption 0, which is used in the proof of Lemma 3. As mentioned earlier in the Introduction, we introduce these technical conditions in order to simplify our exposition and enforce the clarity of the techniques, but we believe that a proper "ironing" [13] process can probably bypass these restrictions and generalize our results.

2   Sufficient Conditions for Optimality

This section is dedicated to proving the main result of the paper:

Theorem 1. *If there exist decreasing, concave functions $s_1, s_2 : I \to I$, with $s_1'(t), s_2'(t) > -1$ for all $t \in I$, such that for almost every $x_1, x_2 \in I$*

$$\frac{s_1(x_2)f_1(s_1(x_2))}{1 - F_1(s_1(x_2))} = 2 + \frac{x_2 f_2'(x_2)}{f_2(x_2)} \quad and \quad \frac{s_2(x_1)f_2(s_2(x_1))}{1 - F_2(s_2(x_1))} = 2 + \frac{x_1 f_1'(x_1)}{f_1(x_1)}, \quad (3)$$

then there exists a constant $p \in [0, 2]$ such that

$$\int_D h(\mathbf{x}) \, dx_1 \, dx_2 = f_1(1) + f_2(1) \qquad (4)$$

where D is the region of I^2 enclosed by curves[5] $x_1 + x_2 = p, x_1 = s_1(x_2)$ and $x_2 = s_2(x_1)$ and including point $(1,1)$, i.e. $D = \{\mathbf{x} \in I \mid x_1 + x_2 \geq p \lor x_1 \geq$

---

[5] See Fig. 1.

$s_1(x_2) \lor x_2 \geq s_2(x_1)\}$, *and the optimal selling mechanism is given by the utility function*

$$u(\mathbf{x}) = \max\{0, x_1 - s_1(x_2), x_2 - s_2(x_1), x_1 + x_2 - p\}. \tag{5}$$

In particular, if $p \leq \min\{s_1(0), s_2(0)\}$, then the optimal mechanism is the deterministic full-bundling with price p.

2.1   Partitioning of the Valuation Space

Due to the fact that the derivatives of functions s_j in Theorem 1 are above -1, each curve $x_1 = s_1(x_2)$ and $x_2 = s_2(x_1)$ can intersect the full-bundle line $x_1 + x_2 = p$ at most at a single point. So let $x_2^* = x_2^*(p), x_1^* = x_1^*(p)$ be the

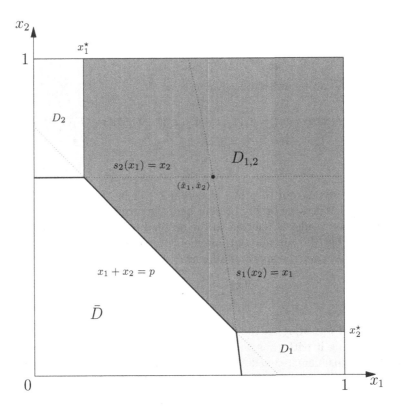

Fig. 1. The valuation space partitioning of the optimal selling mechanism for two independent items, one following a uniform distribution and the other an exponential with parameter $\lambda = 1$. Here $s_1(t) = (2-t)/(3-t)$, $s_2(t) = 2 - W(2e) = 0.625$ and $p = 0.787$. In region D_1 (light grey) item 1 is sold deterministically and item 2 with a probability of $-s_1'(x_2)$, in D_2 (light grey) only item 2 is sold and region $D_{1,2}$ (dark grey) is where the full bundle is sold deterministically, for a price of p.

coordinates of these intersections, respectively, i.e. $s_1(x_2^*) = p - x_2^*$ and $s_2(x_1^*) = p - x_1^*$. If such an intersection does not exist, just define $x_2^* = 0$ or $x_1^* = 0$.

The construction and the optimal mechanism given in Theorem 1 gives then rise to the following partitioning of the valuation space I^2 (see Fig. 1):

- Region $\bar{D} = I^2 \setminus D$ where no item is allocated
- Region $D_1 = \{ \mathbf{x} \in I^2 \mid x_1 \geq s_1(x_2) \wedge x_2 \leq x_2^* \}$ where item 1 is sold with probability 1 and item 2 with probability $s_1'(x_2)$ for a price of $s_1(x_2) - x_2 s_1'(x_2)$
- Region $D_2 = \{ \mathbf{x} \in I^2 \mid x_2 \geq s_2(x_1) \wedge x_1 \leq x_1^* \}$ where item 2 is sold with probability 1 and item 1 with probability $s_2'(x_1)$ for a price of $s_2(x_1) - x_1 s_2'(x_1)$
- Region $D_{1,2} = D \setminus D_1 \cup D_2 = \{ \mathbf{x} \in I^2 \mid x_1 + x_2 \geq p \wedge x_1 \geq x_1^* \wedge x_2 \geq x_2^* \}$ where both items are sold deterministically in a full bundle of price p.

Under this decomposition:

$$\int_{D_1} h(\mathbf{x}) \, dx_1 \, dx_2 = \int_0^{x_2^*} \int_{s_1(x_2)}^1 h(\mathbf{x}) \, dx_1 \, dx_2 = f_1(1) F_2(x_2^*)$$

so expression (4) can be written equivalently as

$$\int_{D_{1,2}} h(\mathbf{x}) \, dx_1 \, dx_2 = f_1(1)(1 - F_2(x_2^*)) + f_2(1)(1 - F_1(x_1^*)). \tag{6}$$

2.2 Duality

The major underlying tool to prove Theorem 1 will be the duality framework of [5]. For completeness we briefly present here the formulation and key aspects, and the interested reader is referred to the original text for further details.

Remember that the revenue optimization problem we want to solve here is to maximize $\mathcal{R}(u)$ over the space of all convex functions $u : I^2 \longrightarrow \mathbb{R}_+$ with

$$0 \leq \frac{\partial u(\mathbf{x})}{\partial x_j} \leq 1, \qquad j = 1, 2, \tag{7}$$

for a.e. $\mathbf{x} \in I^2$. First we relax this problem by dropping the convexity assumption and replacing it with (absolute) continuity. We also drop the lower bound in (7). Then this new relaxed program is dual to the following: minimize $\int_0^1 \int_0^1 z_1(\mathbf{x}) + z_2(\mathbf{x}) \, d\mathbf{x}$ where the new dual variables $z_1, z_2 : I^2 \longrightarrow \mathbb{R}_+$ are such that z_j is (absolutely) continuous with respect to its j-coordinate and the following conditions are satisfied for all $x_1, x_2 \in I$:

$$z_j(0, x_{-j}) = 0, \qquad j = 1, 2, \tag{8}$$

$$z_j(1, x_{-j}) \geq f_j(1) f_{-j}(x_{-j}), \qquad j = 1, 2, \tag{9}$$

$$\frac{\partial z_1(\mathbf{x})}{\partial x_2} + \frac{\partial z_2(\mathbf{x})}{\partial x_2} \leq 3 f_1(x_1) f_2(x_2) + x_1 f_1'(x_1) f_2(x_2) + x_2 f_1(x_1) f_2'(x_2). \tag{10}$$

We will refer to the first optimization problem, where u ranges over the relaxed space of continuous, nonnegative functions with derivatives at most 1, as the *primal program* and to the second as the *dual*. Intuitively, every dual solution z_j must start at zero and grow all the way up to $f_j(1)f_{-j}(x_{-j})$ while travelling in interval I, in a way that the sum of the rate of growth of both z_1 and z_2 is never faster than the right hand side of (10). In [5] is proven that indeed these two programs satisfy both weak duality, i.e. for any feasible u, z_1, z_2 we have

$$\mathcal{R}(u) \le \int_0^1 \int_0^1 z_1(\mathbf{x}) + z_2(\mathbf{x})\, d\mathbf{x}$$

as well as complementary slackness, in the form of the even stronger following form of ε-complementarity:

Lemma 1 (Complementarity). *If u, z_1, z_2 are feasible primal and dual solutions, respectively, $\varepsilon > 0$ and the following complementarity constraints hold for a.e. $\mathbf{x} \in I^2$,*

$$u(\mathbf{x})\left(h(\mathbf{x}) - \frac{\partial z_1(\mathbf{x})}{\partial x_1} - \frac{\partial z_2(\mathbf{x})}{\partial x_2}\right) \le \varepsilon f_1(x_1)f_2(x_2), \tag{11}$$

$$u(1, x_{-j})\left(z_j(1, x_{-j}) - f_j(1)f_{-j}(x_{-j})\right) \le \varepsilon f_j(1)f_{-j}(x_{-j}), \quad j = 1, 2, \tag{12}$$

$$z_j(\mathbf{x})\left(1 - \frac{\partial u(\mathbf{x})}{\partial x_j}\right) \le \varepsilon f_1(x_1)f_2(x_2), \quad j = 1, 2, \tag{13}$$

where h is defined in (2), then the values of the primal and dual programs differ by at most 7ε. In particular, if the conditions are satisfied with $\varepsilon = 0$, both solutions are optimal.

Our approach into proving Theorem 1 will be to show the existence of a pair of dual solutions z_1, z_2 with respect to which the utility function u given by the theorem indeed satisfies complementarity. Notice here the existential character of our technique: our duality approach offers the advantage to use the proof of just the existence of such duals, without having to explicitly describe them and compute their objective value in order to prove optimality, i.e. that the primal and dual objectives are indeed equal. Also notice that the utility function u given by Theorem 1 is convex by construction, so in case someone shows optimality for u in the relaxed setting, then u must also be optimal among all feasible mechanisms.

2.3   Deficiency

The following notion will be the tool that gives a very useful geometric interpretation to the rest of the proof of Theorem 1 and it will be critical into proving Lemma 3.

Definition 1. *For any body $S \subseteq I^2$ define its* deficiency *(with respect to distributions f_1, f_2) to be*

$$\delta(S) \equiv \int_S h(\mathbf{x})\,d\mathbf{x} - f_2(1) \int_{S_1} f_1(x_1)\,dx_1 - f_1(1) \int_{S_2} f_2(x_2)\,dx_2,$$

where S_1, S_2 denote S's projections to the x_1 and x_2 axis, respectively.

Lemma 2. *If the requirements of Theorem 1 hold, then no body $S \subseteq D_{1,2}$ has positive deficiency.*

2.4   Dual Solution and Optimality

The following lemma will complete the proof of Theorem 1. It is the most technical part of this paper, and utilizes a max-flow min-cut argument in order to prove the existence of a feasible dual pair z_1, z_2 that satisfies the complementarity conditions with respect to the utility function given by Theorem 1, thus establishing optimality. It is inspired by the bipartite matching approach in [5] where Hall's theorem is used in order to prove existence, in the special case of uniformly distributed items. Here we need to abstract and generalize our approach in order to incorporate general distributions in the most smooth way possible. The proof has a strong geometric flavor, which is achieved by utilizing the notion of deficiency that was introduced in Sect. 2.3 and using Lemma 2.

Lemma 3. *Assume that the conditions of Theorem 1 hold. Then for arbitrary small $\varepsilon > 0$, there exist feasible dual solutions z_1, z_2 which are ε-complementary to the (primal) u given by (5). Therefore, the mechanism induced by u is optimal.*

3   The Case of Identical Items

In this section we focus in the case of identically distributed valuations, i.e. $f_1(t) = f_2(t) \equiv f(t)$ for all $t \in I$, and we provide clear and simple conditions under which the critical property (3) of Theorem 1 hold.

First notice that in this case the regularity Assumption 1 gives $3 + \frac{x_1 f'(x_1)}{f(x_1)} + \frac{x_2 f'(x_2)}{f(x_2)} \geq 0$ a.e. in I^2 (since f is positive) and thus $\frac{t f'(t)}{f(t)} \geq -\frac{3}{2}$ for a.e. $t \in I$. An equivalent way of writing this is that $t^{3/2} f(t)$ is increasing, which interestingly is the complementary case of that studied by Hart and Nisan [6] for two i.i.d. items: they show that when $t^{3/2} f(t)$ is decreasing, then deterministically selling in a full bundle is optimal.

Theorem 2. *Assume that $G(t) = t f(t)/(1 - F(t))$ and $H(t) = t f'(t)/f(t)$ give rise to well defined, differentiable functions over I, G being strictly increasing and convex, H decreasing and concave, with $G + H$ increasing and $G(1) \geq 2 + H(0)$. Then the requirements of Theorem 1 are satisfied. In particular*

$$s(t) = G^{-1}(2 + H(t))$$

and, if

$$\int_0^1 \int_0^1 h(\mathbf{x})\, d\mathbf{x} - \int_0^p \int_0^{p-x_2} h(\mathbf{x})\, d\mathbf{x} - 2f(1) \tag{14}$$

is nonpositive for $p = s(0)$ then the optimal selling mechanism is the one offering deterministically the full bundle for a price of p being the root of (14) in $[0, s(0)]$, otherwise the optimal mechanism is the one defined by the utility function

$$u(\mathbf{x}) = \max\{0, x_1 - s(x_2), x_2 - s(x_1), x_1 + x_2 - p\}$$

with $p = x^* + s(x^*)$, where $x^* \in [0, s(0)]$ is the constant we get by solving

$$\int_{x^*}^{s(x^*)} \int_{s(x^*)+x^*-x_2}^1 h(\mathbf{x})\, d\mathbf{x} + \int_{s(x^*)}^1 \int_{x^*}^1 h(\mathbf{x})\, d\mathbf{x} = 2f(1)(1 - F(x^*)). \tag{15}$$

Corollary 1 (Monomial Distributions). *The optimal selling mechanism for two i.i.d. items with valuations from the family of distributions with densities $f(t) = (c+1)t^c$, $c \geq 0$, is deterministic. In particular, it offers each item for a price of $s = {}^{c+1}\sqrt{\frac{c+2}{2c+3}}$ and the full bundle for a price of $p = s + x^*$, where x^* is the solution to (15).*

Notice that for $c = 0$ the setting of Corollary 1 reduces to a two uniformly distributed goods setting, and gives the well-known results of $s = 2/3$ and $p = (4 - \sqrt{2})/3$ (see e.g. [11]). For the linear distribution $f(t) = 2t$, where $c = 1$, we get $s = \sqrt{3/5}$ and $p \approx 1.091$.

Corollary 2 (Exponential Distributions). *The optimal selling mechanism for two i.i.d. items with valuations exponentially distributed over I, i.e. having densities $f(t) = \lambda e^{-\lambda t}/(1 - e^{-\lambda})$, with $0 < \lambda \leq 1$, is the one having $s(t) = \frac{1}{\lambda}\left[2 - \lambda t - W\left(e^{2-\lambda-\lambda t}(2 - \lambda t)\right)\right]$ and a price of $p = x^* + s(x^*)$ for the full bundle, where x^* is the solution to (15). Here W is Lambert's product logarithm function[6].*

For example, for $\lambda = 1$ we get $s(t) = 2 - t - W\left(e^{1-t}(2 - t)\right)$ and $p \approx 0.714$. Interestingly, to our knowledge this is the first example for an i.i.d. setting with valuations coming from a regular, continuous distribution over an interval $[0, b]$, where an optimal selling mechanism is *not* deterministic. Also notice how this case of exponential i.i.d. items on a bounded interval is different from the one on $[0, \infty)$: by [3,4] we know that at the unbounded case the optimal selling mechanism for two exponential i.i.d. items is simply the deterministic full-bundling, but in our case of the bounded I this is not the case any more.

4   Non-identical Items

An interesting aspect of the technique of Theorem 2 is that it can readily be used also for non identically distributed valuations. One just has to define

---

[6] Function W can be defined as the solution to $W(t)e^{W(t)} = t$.

$G_j(t) \equiv tf_j(t)/(1 - F_j(t))$ and $H_j(t) = tf_j'(t)/f_j(t)$ for both items $j = 1, 2$ and check again whether G_1, G_2 are strictly increasing and convex and H_1, H_2 non-negative, decreasing and concave. Then, we can get $s_j(t) = G_j^{-1}(2 + H_{-j}(t))$ and check if $s_j(1) > -1$ and the price p of the full bundle can be given by (4). Again, a quick check of whether full bundling is optimal is to see if for $p = \min \{s_1(0), s_2(0)\}$ expression $\int_0^1 \int_0^1 h(\mathbf{x}) \, d\mathbf{x} - \int_0^p \int_0^{p-x_2} h(\mathbf{x}) \, d\mathbf{x} - f_1(1) - f_2(1)$ is nonpositive.

Example 1. Consider two independent items, one having uniform valuation $f_1(t) = 1$ and one exponential $f_2(t) = e^{-t}/(1 - e^{-1})$. Then we get that $s_1(t) = (2 - t)/(3 - t)$, $s_2(t) = 2 - W(2e) \approx 0.625$ and $p \approx 0.787$. The optimal selling mechanism offers either only item 2 for a price of $s_2 \approx 0.625$, or item 1 deterministically and item 2 with a probability $s_1'(x_2)$ for a price of $s_1(x_2) - x_2 s_1'(x_2)$, or the full bundle for a price of $p \approx 0.787$. You can see the allocation space of this mechanism in Fig. 1.

5   Approximate Solutions

In the previous sections we developed tools that, under certain assumptions, can give a complete closed-form description of the optimal selling mechanism. However, remember that the initial primal-dual formulation upon which our analysis was based, assumes a relaxed optimization problem. Namely, we dropped the convexity assumption of the utility function u. In the results of the previous sections this comes for free: the optimal solution to the relaxed program turns out to be convex anyways, as a result of the requirements of Theorem 1. But what happens if that was not the case? The following tool shows that even in that case our results are still applicable and very useful into both finding good upper bounds on the optimal revenue (Theorem 3) as well as designing almost-optimal mechanisms that have provably very good performance guarantees (Sect. 5.1).

Theorem 3. *Assume that all the requirements of Theorem 2 are satisfied, except from the concavity of function H. Then, the function u given by that theorem might not be convex any more and thus not a valid utility function, but it generates an* upper bound *to the optimal revenue, i.e.* $\mathrm{REV}(f, f) \leq \mathcal{R}_{f,f}(u)$.

Example 2 (Power-Law Distributions). A class of important distributions that falls into the description of Theorem 3 are the power-law distributions with factors $\alpha \leq 2$. More specifically, these are the distributions having densities $f(t) = c/(t+1)^\alpha$, with the normalization factor c selected so that $\int_0^1 f(t) \, dt = 1$, i.e. $c = (\alpha - 1)/(1 - 2^{1-\alpha})$. It is not difficult to verify that these distributions satisfy Assumption 1. For example, for $\alpha = 2$ one gets $f(x) = 2/(x + 1)^2$, the *equal revenue* distribution shifted in the unit interval. For this we can compute via (3) that $s(t) = \frac{1}{2}\sqrt{5 + 2t + t^2} - \frac{1}{2}(1 + t)$ and $p \approx 0.665$, which gives an upper bound of $R_{f,f}(u) \approx 0.383$ to the optimal revenue $\mathrm{REV}(f, f)$.

5.1   Convexification

The approximation results described in Theorem 3 can be used not only for giving upper bounds on the optimal revenue, but also as a *design* technique for good selling mechanisms. Since the only deviation from a feasible utility function is the fact that function s is not concave (and thus u is not convex), why don't we try to "convexify" u, by replacing s by a concave function $\tilde{s}$? If $\tilde{s}$ is "close enough" to the original s, by the previous discussion this would also result in good approximation ratios for the new, feasible selling mechanism.

Let's demonstrate this by an example, using the equal revenue distribution $f(t) = 2/(t+1)^2$ of the previous example. We need to replace s with a concave $\tilde{s}$ in the interval $[0, x^*]$. So let's choose $\tilde{s}$ to be the concave hull of s, i.e. the minimum concave function that dominates s. Since s is convex, this is simply the line that connects the two ends of the graph of s in $[0, x^*]$, that is, the line

$$\tilde{s}(t) = \frac{s(0) - s(x^*)}{x^*}(x^* - t) + s(x^*).$$

A calculation shows that this new *valid* mechanism has an expected revenue which is within a factor of just $1 + 3 \times 10^{-9}$ of the upper bound given by s using Theorem 3, rendering it essentially optimal.

Acknowledgments. We thank Anna Karlin, Amos Fiat, Costis Daskalakis and Ian Kash for insightful discussions. We also thank the anonymous reviewers for their useful comments.

References

1. Babaioff, M., Immorlica, N., Lucier, B., Weinberg, S.M.: A simple and approximately optimal mechanism for an additive buyer. In FOCS 2014 (2014)
2. Daskalakis, C., Deckelbaum, A., Tzamos, C.: The complexity of optimal mechanism design. In: SODA 2013, pp. 1302–1318 (2013)
3. Daskalakis, C., Deckelbaum, A., Tzamos, C.: Mechanism design via optimal transport. In: EC 2013, pp. 269–286 (2013)
4. Giannakopoulos, Y.: Bounding the optimal revenue of selling multiple goods. CoRR, abs/1404.2832 (2014)
5. Giannakopoulos, Y., Koutsoupias, E.: Duality and optimality of auctions for uniform distributions. In: EC 2014, pp. 259–276 (2014)
6. Hart, S., Nisan, N.: Approximate revenue maximization with multiple items. In: EC 2012 (2012)
7. Hart, S., Nisan, N.: The menu-size complexity of auctions. In: EC 2013, pp. 565–566 (2013)
8. Hart, S., Reny, P.J.: Maximal revenue with multiple goods: nonmonotonicity and other observations. Technical report, The Center for the Study of Rationality, Hebrew University, Jerusalem (2012)
9. Li, X., Yao, A.C.-C.: On revenue maximization for selling multiple independently distributed items. Proc. Natl. Acad. Sci. **110**(28), 11232–11237 (2013)
10. Lovász, L., Plummer, M.D.: Matching theory, North-Holland (1986)

11. Manelli, A.M., Vincent, D.R.: Bundling as an optimal selling mechanism for a multiple-good monopolist. J. Econ. Theory **127**(1), 1–35 (2006)
12. McAfee, R.P., McMillan, J.: Multidimensional incentive compatibility and mechanism design. J. Econ. Theory **46**(2), 335–354 (1988)
13. Myerson, R.B.: Optimal auction design. Mathematics of Operations Research **6**(1), 58–73 (1981)
14. Ore, O.: Graphs and matching theorems. Duke Mathematical Journal **22**(4), 625–639 (1955)
15. Pavlov, G.: Optimal mechanism for selling two goods. The BE Journal of Theoretical Economics **11**(1) (2011)
16. Rochet, J.-C.: The taxation principle and multi-time hamilton-jacobi equations. Journal of Mathematical Economics **14**(2), 113–128 (1985)
17. Wang, Z., Tang, P.: Optimal mechanisms with simple menus. In: EC 2014 (2014)

Adaptively Secure Coin-Flipping, Revisited

Shafi Goldwasser[1,3], Yael Tauman Kalai[2], and Sunoo Park[3]([✉])

[1] The Weizmann Institute of Science, Rehovot, Israel
[2] Microsoft Research, Cambridge, USA
[3] MIT, Cambridge, USA
sunoo@csail.mit.edu

Abstract. The question of how much bias a coalition of faulty players can introduce into distributed sampling protocols in the full information model was first studied by Ben-Or and Linial in 1985. They focused on the problem of *collective coin-flipping*, in which a set of n players wish to use their private randomness to generate a common random bit b in the presence of $t(n)$ faulty players, such that the probability that $b = 0$ (and 1) are at least ε for some constant $\varepsilon > 0$. They showed that the majority function can tolerate $t = \Theta(\sqrt{n})$ corruptions even in the presence of adaptive adversaries and conjectured that this is optimal in the adaptive setting. Shortly thereafter, Lichtenstein, Linial, and Saks proved that the conjecture holds for protocols where each player sends a single bit. Their result has been the main progress on the conjecture for the last 30 years.

In this work we revisit this question, and ask: what about protocols where players can send longer messages? Can increased communication enable tolerance of a larger fraction of corrupt players?

We introduce a model of *strong adaptive* corruptions, in which an adversary sees all messages sent by honest parties in any given round, and based on the message content, decides whether to corrupt a party (and alter its message) or not. This is in contrast to the (classical) adaptive adversary, who corrupts parties based on prior communication history, and cannot alter messages already sent. Such strongly adaptive corruptions seem to be a realistic concern in settings where malicious parties can alter (or sabotage the delivery) of honest messages depending on their content, yet existing adversarial models do not take this into account.

We prove that any one-round coin-flipping protocol, *regardless of message length*, can be secure against at most $\widetilde{O}(\sqrt{n})$ strong adaptive corruptions. Thus, increased message length does not help in this setting.

We then shed light on the connection between adaptive and strongly adaptive adversaries, by proving that for any symmetric one-round coin-flipping protocol secure against t adaptive corruptions, there is a symmetric one-round coin-flipping protocol secure against t strongly adaptive corruptions. Going back to the standard adaptive model, we can now prove that any symmetric one-round protocol with arbitrarily long messages can tolerate at most $\widetilde{O}(\sqrt{n})$ adaptive corruptions.

At the heart of our results there is a new technique for converting any one-round secure protocol with arbitrarily long messages into a secure one where each player sends only $\mathsf{polylog}(n)$ bits. This technique may be of independent interest.

© Springer-Verlag Berlin Heidelberg 2015
M.M. Halldórsson et al. (Eds.): ICALP 2015, Part II, LNCS 9135, pp. 663–674, 2015.
DOI: 10.1007/978-3-662-47666-6_53

1 Introduction

A collective coin-flipping protocol is one where a set of n players use private randomness to generate a common random bit b. Several protocol models have been studied in the literature. In this work, we focus on the model of *full information* [1] where all parties communicate via a single broadcast channel.

The challenge is that $t = t(n)$ of the parties may be corrupted and aim to bias the protocol outcome (i.e. the "coin") in a particular direction. We focus on *Byzantine faults*, where once a party is corrupted, the adversary completely controls the party and can send any message on its behalf. Two types of Byzantine adversaries have been considered in the literature: *static* adversaries and *adaptive* adversaries. A static adversary is one that chooses which t players to corrupt *before the protocol begins*. An adaptive adversary is one who may choose which t players to corrupt adaptively, as the protocol progresses.

Collective coin-flipping in the case of static adversaries is well understood (see section 1.2). In this work, our focus is on the setting of adaptive adversaries, which has received considerably less attention. A collective coin-flipping protocol is said to be secure against t adaptive (resp. static) corruptions if for any adaptive adversary corrupting t parties, there is a constant $\varepsilon > 0$ such that the probability that the protocol outputs 0 (and the probability that the protocol outputs 1) is at least ε, where the probability is taken over the randomness of the players and the adversary.

The question we study is: *What is the maximum number of adaptive corruptions that a secure coin-flipping protocol can tolerate?* On the positive side, it has been shown by Ben-Or and Linial [1] in 1985 that the majority protocol (where each party sends a random bit, and the output is equal to the majority of the bits sent), is resilient to $\Theta(\sqrt{n})$ adaptive corruptions. Ben-Or and Linial conjectured that this is in fact optimal.

Conjecture 1 ([1]). Majority is the optimal coin-flipping protocol against adaptive adversaries. In particular, any coin-flipping protocol is resilient to at most $O(\sqrt{n})$ adaptive corruptions.

Shortly thereafter, Lichtenstein, Linial, and Saks [6] proved the conjecture for a restricted class of protocols: namely, those in which each player sends only a single bit. Their result has been the main progress on the conjecture of [1] during the last 30 years.

1.1 Our Contribution

We first define a new adversarial model of *strong adaptive* corruptions. Informally, an adversary is strongly adaptive if he can corrupt players depending on the content of their messages. More precisely, in each round, he can see all the messages that honest players "would" send, and then decide which of them to corrupt. This is in contrast to a (traditionally defined) adaptive adversary who can, at any point in the protocol, corrupt any player who has not yet spoken based on the history of communication, but cannot alter the message of a player who has already spoken. Thus, strong adaptive adversaries are more powerful than adaptive adversaries.

We believe that the notion of *strong adaptive security* gives rise to a natural and interesting new adversarial model[1] in which to study multi-party protocols in general. Indeed, it is a realistic concern in many settings that malicious parties may decide to stop or alter messages sent by honest players *depending on message content*, and it is a shortcoming that existing adversarial models fail to take such behavior into account.

Our main result is that the conjecture of [1] holds (up to polylogarithmic factors) for any one-round coin-flipping protocol in the presence of *strong adaptive* corruptions.

Theorem. Any secure one-round coin-flipping protocol Π can tolerate at most $t = \widetilde{O}(\sqrt{n})$ strong adaptive corruptions.

This is shown by a generic reduction of communication in the protocol: first, we prove that any strongly adaptively secure protocol Π can be converted to one where players send messages of no more than polylogarithmic length, while preserving the number of corruptions that can be tolerated. Then, we show that any protocol with messages of polylogarithmic length can be converted to one where each player sends only a single bit, at the cost of a polylogarithmic factor in the number of corruptions. Finally, we reach the *single-bit* setting in which the bound of Lichtenstein et al. [6] can be applied to obtain the theorem. We believe that our technique of converting any protocol into one with short messages is of independent interest and will find other applications.

Furthermore, we prove that strongly adaptively secure protocols are a more general class of protocols than symmetric adaptively secure protocols. A symmetric protocol Π is a one that is oblivious to the order of its inputs: that is, where for any permutation $\pi : [n] \to [n]$ of the players, it holds that the protocol outcome $\Pi(r_1, \ldots, r_n) = \Pi(r_{\pi(1)}, \ldots, r_{\pi(n)})$ is the same.

Theorem. For any symmetric one-round coin-flipping protocol Π secure against $t = t(n)$ adaptive corruptions, there is a symmetric one-round coin-flipping protocol Π' secure against $\Omega(t)$ strong adaptive corruptions.

Curiously, this proof makes a novel use of the Minimax Theorem [7,8] from game theory, in order to take any symmetric, adaptively secure protocol and convert it to a new protocol which is strongly adaptively secure. This technique views the protocol as a zero-sum game between two players $\mathcal{A}_0$ and $\mathcal{A}_1$, where $\mathcal{A}_0$

---

[1] We consider our strong adaptive adversarial notion to be closely tied to the notion of a *rushing* adversary in the setting of static corruptions. The intuitive idea of a rushing adversary is that the adversary *sees all possible information in each round, before making his move*. We remark that a notion of "rushing adaptive adversary" has been previously proposed in the literature, but such an adversary is weaker than our strong adaptive adversary. We argue that our strong adaptive adversary better captures the idea that the adversary *sees all possibly relevant information in each round, before making his move*, since in the adaptive setting, the adversary's strategy must decide not only what messages to send, but also *which players to corrupt*.

wins if the protocol outcome is 0 and $\mathcal{A}_1$ wins if the outcome is 1. We analyze the "minimax strategy" in which the players try to minimize their maximum loss, in order to deduce the strong adaptive security of the new protocol. Whereas some prior works have made use of game theory in the analysis of (two-party) protocols, this is the first use of these game-theoretic concepts in the *construction* of distributed multiparty protocols.

Finally, using the above results as stepping stones, we return to the classical conjecture of [1], in the model of adaptive adversaries, and show that the conjecture holds (up to polylogarithmic factors) for any *symmetric* one-round protocol with arbitrarily long messages.

Theorem. Any secure symmetric one-round coin-flipping protocol Π can tolerate at most $t = \widetilde{O}(\sqrt{n})$ adaptive corruptions.

1.2   Related Work

The full-information model (also known as the *perfect information model*) was introduced by Ben-Or and Linial [1] to study the problem of collective coin-flipping when no secret communication is possible between honest players.

In the Static Setting. Protocols for collective coin-flipping in the presence of static corruptions have been constructed in a series of works that variously focus on improving the fault-tolerance, round complexity, and/or bias of the output bit. Feige [3] gave a protocol that is $(\delta^{1.65}/2)$-secure[2] in the presence of $t = (1 + \delta) \cdot n/2$ static corruptions for any constant $0 < \delta < 1$. Russell, Saks, and Zuckerman [9] then showed that any protocol that is secure in the presence of linearly many corruptions must either have at least $(1/2 - o(1)) \cdot \log^*(n)$ rounds, or communicate many bits per round.

Interestingly, nearly all proposed *multi-round* protocols for collective coin-flipping first run a *leader election* protocol in which one of the n players is selected as a "leader", who then outputs a bit that is taken as the protocol outcome. We remark that this approach is inherently unsuitable for adaptive adversaries, which can always corrupt the leader after he is elected, and thereby surely control the protocol outcome.

In the Adaptive Setting. The study of coin-flipping protocols has been predominantly in the static setting. The problem of adaptively secure coin-flipping was introduced by Ben-Or and Linial [1] and further examined by Lichtenstein, Linial, and Saks [6] as described in the previous section. In addition, Dodis [2] proved that through "black-box" reductions from *non-adaptive* coin-flipping, it is not possible to tolerate significantly more corruptions than the majority protocol. The definition of "black-box" used in [2] is rather restricted: it only considers sequential composition of non-adaptive coin-flipping protocols, followed by a (non-interactive) function computation on the coin-flips thus obtained.

---

[2] A coin-flipping protocol is ε-secure against t static corruptions if for any static adversary that corrupts up to t parties, the probability that the protocol outputs 0 is at least ε.

In the point-to-Point Channels Setting. An adversarial model bearing some resemblance to our strong adaptive adversary model was introduced and analyzed by Hirt and Zikas [5] in the *pairwise communication channels* model, rather than the full-information model. In their model, the adversary can corrupt a party P based on some of the messages that P sends within a round, then the adversary controls the rest of P's messages in that round (and for future rounds). Unlike in our strong adaptive model, the adversary of [5] cannot "see inside all players' heads" and overwrite arbitrary honest messages based on their content before they are sent.

A brief survey of the literature on coin-flipping in the setting of computationally bounded players is given in the full version [4].

2 Preliminaries

We consider coin-flipping protocols in the *full-information model* (also known as the *perfect information model*), where n computationally unbounded players communicate via a single broadcast channel. The network is synchronized between rounds, but is asynchronized within each round (that is, there is no guarantee on message ordering within a round, and an adversary can see the messages of all honest players in a round before deciding his own messages).

In this work, we focus on *one-round* protocols, and we consider protocols that terminate (and produce an output) with probability 1. In particular, we focus on *coin-flipping* protocols, which are defined as follows.

Definition 1 (Coin-flipping protocol). *A coin-flipping protocol $\Pi = \{\Pi_n\}_{n \in \mathbb{N}}$ is a family of protocols where each Π_n is a n-player protocol which outputs a bit in $\{0, 1\}$.*

Notation. We write $\overset{s}{\approx}$ for statistical indistinguishability of distributions. We denote by $\Pr^{\Pi}(b)$ the probability that an honest execution of Π will lead to the outcome $b \in \{0, 1\}$. We denote by $\Pr^{\Pi, \mathcal{A}}(b)$ the probability that an execution of Π in the presence of an adversary $\mathcal{A}$ will lead to the outcome $b \in \{0, 1\}$. The probability is over the random coins of the honest players and the adversary.

For one-round protocols, we write $\Pi_n(r_1, \ldots, r_n)$ to denote the outcome of the protocol Π_n when each player i sends message r_i. (The vector $(r_1, \ldots, r_n)$ is a protocol *transcript*.)

2.1 Properties of Protocols

Definition 2 (Symmetric protocol). *A protocol Π is symmetric if the outcome of a protocol execution is the same no matter how the messages within each round are permuted. In particular, a one-round protocol Π is symmetric if for all $n \in \mathbb{N}$ and any permutation $\pi \in [n] \to [n]$,*

$$\Pi_n(r_1, \ldots, r_n) = \Pi_n(r_{\pi(1)}, \ldots, r_{\pi(n)}).$$

We remark, for completeness, that in the multi-round case, the outcome of a symmetric protocol should be unchanged even if different permutations are applied in different rounds.

Definition 3 (Single-bit/multi-bit protocol). *A protocol is* single-bit *if each player sends at most one bit over the course of the protocol execution. Similarly, a protocol is* m-bit *if each player sends at most m bits over the course of the protocol execution. More generally, a protocol which is not single-bit is called* multi-bit.

Definition 4 (Public-coin protocol). *A protocol is* public-coin *if each honest player broadcasts all of the randomness he generates (i.e. his "local coin-flips"), and does not send any other messages.*

2.2 Adversarial Models in the Literature

The type of adversary that has been by far the most extensively studied in the coin-flipping literature is the static adversary, which chooses a subset of players to corrupt *before* the protocol execution begins, and controls the behavior of the corrupt players arbitrarily throughout the protocol execution.

A stronger type of adversary is the *adaptive* adversary, which may choose players to corrupt at any point during protocol execution, and controls the behavior of the corrupt players arbitrarily from the moment of corruption until protocol termination.

Definition 5 (Adaptive adversary). *Within each round, the adversary chooses players one-by-one to send their messages; and he can perform corruptions at any point during this process.*

2.3 Security of Coin-Flipping Protocols

The security of a coin-flipping protocol is usually measured by the extent to which an adversary can, by corrupting a subset of parties, bias the protocol outcome towards his desired bit.

Definition 6 (ε-security). *A coin-flipping protocol Π is* ε-secure *against $t = t(n)$ adaptive (or static or strong adaptive) corruptions if for all $n \in \mathbb{N}$, it holds that for any adaptive (resp. static or strong adaptive) adversary $\mathcal{A}$ that corrupts at most $t = t(n)$ players,*

$$\min \left(\Pr^{\Pi_n, \mathcal{A}}(0), \Pr^{\Pi_n, \mathcal{A}}(1) \right) \geq \varepsilon.$$

We remark that this definition of ε-security is sometimes referred to as ε-*control* or ε-*resilience* in other works. We next define a *secure* protocol to be one with "minimal" security properties (that is, one where the adversary does not almost always get the outcome he wants).

Definition 7 (Security). *A coin-flipping protocol is* secure *against $t = t(n)$ corruptions if it is ε-secure against t corruptions for some constant $0 < \varepsilon < 1$.*

In this work, we investigate the maximum proportion of adaptive corruptions that can be tolerated by *any* secure protocol.

3 Our Results

3.1 Strongly Adaptive Adversaries

In this work, we propose a new, stronger adversarial model than those that have been studied thus far (see section 2.2), in which the adversary can see all honest players' messages within any given round, and *subsequently* decide which players to corrupt. That is, he can see all the messages that the honest players "would have sent" in a round, and then selectively intercept and alter these messages.

Definition 8 (Strong adaptive adversary). *Within each round, the adversary sees all the messages that honest players would have sent, then gets to choose which (if any) of those messages to corrupt (i.e. replace with messages of his choice).*

This notion is an essential tool underlying the proof techniques in our work. Moreover, we believe that the notion of *strong adaptive security* gives rise to a natural and interesting new adversarial model in which to study multi-party protocols, which is of independent interest beyond the scope of this work.

3.2 Corruption Tolerance in Secure Coin-Flipping Protocols

Our main contributions consist of the following three results. These can be viewed as partial progress towards proving the 30-year-old conjecture of [1].

Theorem 1. *Any one-round coin-flipping protocol Π can be secure against at most $t = \widetilde{O}(\sqrt{n})$ strong adaptive corruptions.*

Theorem 2. *For any symmetric one-round coin-flipping protocol Π secure against $t = t(n)$ adaptive corruptions, there is a symmetric one-round coin-flipping protocol Π' secure against $\Omega(t)$ strong adaptive corruptions.*

Corollary 1. *Any symmetric one-round coin-flipping protocol Π can be secure against at most $t = \widetilde{O}(\sqrt{n})$ adaptive corruptions.*

In the next sections, we proceed to give detailed proofs of the theorems.

3.3 Proof of Theorem 1

We begin by recalling the result of Lichtenstein et al. [6] which proves that the maximum number of adaptive corruptions for any secure *single-bit* coin-flipping protocol is $O(\sqrt{n})$. Note that the *majority protocol* is the one-round protocol in which each player broadcasts a random bit, and the majority of broadcasted bits is taken to be the protocol outcome.

Theorem 3 ([6]). *Any coin-flipping protocol in which each player broadcasts at most one bit can be secure against at most $t = O(\sqrt{n})$ corruptions. Moreover, the majority protocol achieves this bound.*

Next, we establish some definitions and supporting lemmas.

Definition 9 (Distance between message-vectors). *For vectors* $r, r' \in \mathcal{M}^n$, *let* $\mathsf{dist}(r, r')$ *be equal to the number of coordinates* $i \in [n]$ *for which* $r_i \neq r'_i$.

Definition 10 (Robust sets). *Let* Π *be a one-round coin-flipping protocol in which each player sends a message from a message space* $\mathcal{M}$. *For any* $n \in \mathbb{N}$ *and* $b \in \{0, 1\}$, *define the set* $\mathsf{Robust}^{\Pi_n}(b, t)$ *as follows:*

$$\mathsf{Robust}^{\Pi_n}(b, t) = \{r \in \mathcal{M}^n \;:\; \forall r' \in \mathcal{M}^n \text{ s.t. } \mathsf{dist}(r, r') \leq t, \; \Pi_n(r) = \Pi_n(r') = b\}.$$

Lemma 1. *Let* Π *be a one-round coin-flipping protocol in which each player sends a random message from a message space* $\mathcal{M}$. Π *is secure against* $t = t(n)$ *strong adaptive corruptions if and only if there exists a constant* $0 < \varepsilon < 1$ *such that for all* $n \in \mathbb{N}$ *and each* $b \in \{0, 1\}$,

$$\Pr_{r \leftarrow \mathcal{M}} \left[r \in \mathsf{Robust}^{\Pi_n}(b, t) \right] \geq \varepsilon.$$

Proof. Given in the full version [4].

Since players are computationally unbounded and we consider one-round protocols, we may without loss of generality consider public-coin protocols[3]: for any one-round protocol Π in the full-information model, there is a protocol Π' with an identical output distribution (in the presence of any adversary), in which honest players send random messages in $\{0, 1\}^k$ for some $k = \mathsf{poly}(n)$.

The following lemma serves as a stepping-stone to our final theorem.

Lemma 2. *For any one-round multi-bit coin-flipping protocol* Π *secure against* $t = t(n)$ *strong adaptive corruptions, and any constant* $\delta > 0$, *there is a one-round* ℓ-*bit coin-flipping protocol* Π' *that is secure against* t *strong adaptive corruptions, where* $\ell = O(\log^{1+\delta}(n))$.

Proof. Without loss of generality, we consider only public-coin protocols, and assume that each player sends a message of the same length (say, $k = k(n)$ bits). Let $\delta > 0$ be any constant, let $\ell = O(\log^{1+\delta}(n))$, and let $\ell' = 2^\ell$.

For an $\ell' \times n$ matrix of messages $M \in (\{0, 1\}^k)^{\ell' \times n}$, we define the protocol Π^M as follows: each player P_i broadcasts a random integer $a_i \leftarrow [\ell']$, and the protocol outcome is defined by

$$\Pi_n^M(a_1, \ldots, a_n) = \Pi_n(M_{(a_1, 1)}, \ldots, M_{(a_n, n)}),$$

where $M_{(i,j)}$ denotes the message at the ith row and j^{th} column of the matrix M. For notational convenience, define $M(a_1, \ldots, a_n) = (M_{(a_1, 1)}, \ldots, M_{(a_n, n)})$. Notice that by construction of the protocol Π^M, it holds that for any message-vector $a \in [\ell']^n$,

$$M(a) \in \mathsf{Robust}^{\Pi_n}(b, t) \implies a \in \mathsf{Robust}^{\Pi_n^M}(b, t). \tag{1}$$

---

[3] This is without loss of generality: each player can simply send his random coin tosses, and security holds since we are in the full-information model.

Suppose each entry of the matrix M is a uniformly random message in $\{0,1\}^k$. Note that the length of each player's message in Π^M is $\log(\ell') = \ell$. We want to show that Π^M is a secure coin-flipping protocol against t strong adaptive corruptions, for some M. By Lemma 1, it is sufficient to show that there exists $M \in (\{0,1\}^k)^{\ell' \times n}$ such that for all $b \in \{0,1\}$, $\Pr_{\boldsymbol{a}\leftarrow[\ell']^n}\left[\boldsymbol{a} \in \mathsf{Robust}^{\Pi_n^M}(b,t)\right] \geq \varepsilon$, where $0 < \varepsilon < 1$ is constant. By implication (1), it actually suffices to prove:

$$\exists M \in (\{0,1\}^k)^{\ell' \times n} \text{ s.t. } \forall b \in \{0,1\}, \quad \Pr_{\boldsymbol{a}\leftarrow[\ell']^n}\left[M(\boldsymbol{a}) \in \mathsf{Robust}^{\Pi_n}(b,t)\right] \geq \varepsilon. \quad (2)$$

Suppose the matrix M is chosen uniformly at random. Let $\boldsymbol{a}_1, \ldots \boldsymbol{a}_n$ be sampled independently and uniformly from $[\ell']^n$. Since, the number of matrix rows $\ell' = 2^{O(\log^{1+\delta}(n))}$ is super-polynomial, it is overwhelmingly likely that $\boldsymbol{a}_1, \ldots \boldsymbol{a}_n$ will be composed of distinct elements in $[\ell']$. That is, to be precise,

$$\Pr_{\boldsymbol{a}_1,\ldots,\boldsymbol{a}_n}\left[\forall (i,j) \neq (i',j') \in [n] \times [n], \ (\boldsymbol{a}_i)_j \neq (\boldsymbol{a}_{i'})_{j'}\right] \geq 1 - \mathsf{negl}(n).$$

If $\boldsymbol{a}_1, \ldots, \boldsymbol{a}_n$ are indeed composed of distinct elements, the message-vectors $M(\boldsymbol{a}_1), \ldots, M(\boldsymbol{a}_n)$ are independent random elements in $(\{0,1\}^k)^n$. Thus,

$$(M(\boldsymbol{a}_1), \ldots, M(\boldsymbol{a}_n)) \overset{s}{\approx} (\boldsymbol{r}_1, \ldots, \boldsymbol{r}_n), \quad (3)$$

when M is a random matrix in $(\{0,1\}^k)^{\ell' \times n}$, the (short) message-vectors $\boldsymbol{a}_1, \ldots, \boldsymbol{a}_n$ are random in $[\ell']^n$, and the (long) message-vectors $\boldsymbol{r}_1, \ldots, \boldsymbol{r}_n$ are random in $(\{0,1\}^k)^n$.

Since Π is a secure coin-flipping protocol, there is a constant $0 < \varepsilon' < 1$ such that for all $n \in \mathbb{N}$ and $b \in \{0,1\}$ and $i \in [n]$,

$$\Pr_{\boldsymbol{r}_i}\left[\boldsymbol{r}_i \in \mathsf{Robust}^{\Pi_n}(b,t)\right] \geq \varepsilon'.$$

The rest of the proof follows from a series of Chernoff bounds. Details are given in the full version [4].

Having reduced the length of players' messages to $\mathsf{polylog}(n)$ in Lemma 2, we now prove the following lemma which reduces the required communication even further, so that each player sends only one bit. This comes at the cost of a polylogarithmic factor reduction in the number of corruptions. Then, finally, we bring together Lemmas 2 and 3 to prove the theorem.

Lemma 3. *For any one-round ℓ-bit coin-flipping protocol Π secure against $t = t(n)$ strong adaptive corruptions, there is a one-round single-bit coin-flipping protocol Π' that is secure against t/ℓ strong adaptive corruptions.*

Proof (sketch). Let Π be any one-round ℓ-bit coin-flipping protocol secure against $t = t(n)$ strong adaptive corruptions. We construct a new protocol Π' in which there are a factor of ℓ more players, and each player sends only a single bit. To compute the outcome, the messages of the first ℓ players in Π' are concatenated and interpreted as the message of the first player in Π, and the rest of the players' messages are constructed analogously. Due to the strong adaptive security of the original protocol Π, any adversary attacking the new protocol $\Pi'_{n\cdot\ell}$ that corrupts up to $t(n)$ bits can be perfectly simulated by an adversary attacking the original protocol Π', so $\Pi'_{n\cdot\ell}$ is secure against up to $t(n)$ corruptions. We remark that this argument does not hold for *adaptive* adversaries, because seeing the players' messages bit-by-bit in the new protocol Π' may give an adversary more power to attack $\Pi'_{n\cdot\ell}$ than Π_n, given the same number t of corruptions in each. The full proof is given in the full version [4].

Theorem 1. *Any one-round coin-flipping protocol Π can be secure against at most $t = \widetilde{O}(\sqrt{n})$ strong adaptive corruptions.*

Proof. Suppose, for contradiction, that there exists a one-round coin-flipping protocol Π which is secure against t corruptions, where $t = \omega(\sqrt{n} \cdot \mathsf{polylog}(n))$. Then, by Lemma 2, there is an ℓ-bit one-round coin-flipping protocol Π' that is secure against t strong adaptive corruptions, where $\ell = \mathsf{polylog}(n)$. By applying Lemma 3 to the protocol Π', we deduce that there is a single-bit one-round coin-flipping protocol Π'' which is secure against $t/\ell = \widetilde{\Omega}(t)$ strong adaptive corruptions. Since a *strongly adaptive* adversary can perfectly simulate any strategy of an *adaptive* adversary, it follows that Π'' is secure against $\widetilde{\Omega}(t)$ adaptive corruptions. Since Π'' is single-bit, this contradicts Theorem 3.

3.4   Proof of Theorem 2

In this section, we show that for any symmetric one-round coin-flipping protocol secure against t *adaptive* corruptions, there is a one-round coin-flipping protocol secure against $\Omega(t)$ corruptions by *strong adaptive* adversaries. That is, one-round strong adaptively secure protocols are a more general class than one-round symmetric, adaptively secure protocols.

The Minimax Theorem – a classic tool in game theory – will be an important tool in our proof. Due to space constraints, we refer to the full version [4] for the statement of the Minimax Theorem and supporting game-theoretic definitions.

Theorem 2. *For any symmetric one-round coin-flipping protocol Π secure against $t = t(n)$ adaptive corruptions, there is a symmetric one-round coin-flipping protocol Π' secure against $s = t/2$ strong adaptive corruptions.*

Proof. Let Π be a symmetric one-round coin-flipping protocol secure against $t = t(n)$ adaptive corruptions, and define $s(n) = t(n)/2$. We define a new protocol $\Pi' = \{\Pi'_n\}_{n\in\mathbb{N}}$ as follows:

$$\Pi'_n(r_1, \ldots, r_n) = \min_{r'_1, \ldots, r'_s} \max_{r''_1, \ldots, r''_s} \Pi_{n+2s}(r_1, \ldots, r_n, r'_1, \ldots, r'_s, r''_1, \ldots, r''_s),$$

where $s = s(n)$ and honest players in Π'_n send messages exactly as in Π_{n+2s}. Observe that Π_{n+2s} is secure against $t(n + 2s(n)) > t(n)$ corruptions. We show that Π'_n is secure against $s(n) = t(n)/2$ strong adaptive corruptions.

CASE 1. Suppose that the adversary aims to bias the outcome towards 0. By the security of Π_{n+2s}, there is a constant $0 < \varepsilon < 1$ such that $\mathrm{Pr}^{\Pi_{n+2s}, \mathcal{A}}(1) \geq \varepsilon$ for any adaptive adversary $\mathcal{A}$ that corrupts up to $t = 2s$ players. Without loss of generality (since the protocol is symmetric), suppose that the adversary corrupts the last $2s$ players in Π_{n+2s}.

We say that the honest players' messages $r_1, \ldots, r_n$ "fix" the outcome of Π_{n+2s} to be 1 if for any possibly malicious messages $\hat{r}_1, \ldots, \hat{r}_{2s}$, it holds that $\Pi_{n+2s}(r_1, \ldots, r_n, \hat{r}_1, \ldots, \hat{r}_{2s}) = 1$. Then, with probability at least ε, the honest players' messages $r_1, \ldots, r_n$ "fix" the outcome of Π_{n+2s} to be 1. (To see this: suppose not. Then there would exist an adversary which could set the corrupt messages $\hat{r}_1, \ldots, \hat{r}_{2s}$ so that the protocol outcome is 0 with probability $1 - \varepsilon$. But this cannot be, since we already established that $\mathrm{Pr}^{\Pi_{n+2s}, \mathcal{A}}(1) \geq \varepsilon$.)

Define $R_1 \stackrel{\text{def}}{=} \{(r_1, \ldots, r_n) : \forall \hat{r}_1, \ldots, \hat{r}_{2s}, \ \Pi_{n+2s}(r_1, \ldots, r_n, \hat{r}_1, \ldots, \hat{r}_{2s}) = 1\}$ to be the set of those honest message-vectors that fix the output of Π_{n+2s} to 1.

Take any $(r_1, \ldots, r_n) \in R_1$. We now show that the outcome of Π'_n when the honest players send messages $r_1, \ldots, r_n$ is equal to 1, even in the presence of a strong adaptive adversary $\mathcal{A}'$ that corrupts up to s players and aims to bias the outcome towards 0. Without loss of generality, suppose that $\mathcal{A}'$ corrupts the first s players in Π'_n, and replaces their honest messages $r_1, \ldots, r_s$ with some maliciously chosen messages $\hat{r}_1, \ldots, \hat{r}_s$. In this case, the outcome of Π'_n is

$$
\begin{aligned}
\Pi'_n(\hat{r}_1, &\ldots, \hat{r}_s, r_{s+1}, \ldots, r_n) \\
&= \min_{r'_1, \ldots, r'_s} \max_{r''_1, \ldots, r''_s} \Pi_{n+2s}(\hat{r}_1, \ldots, \hat{r}_s, r_{s+1}, \ldots, r_n, r'_1, \ldots, r'_s, r''_1, \ldots, r''_s) \\
&\geq \min_{r'_1, \ldots, r'_s} \Pi_{n+2s}(\hat{r}_1, \ldots, \hat{r}_s, r_{s+1}, \ldots, r_n, r'_1, \ldots, r'_s, r_1, \ldots, r_s) \\
&= \min_{r'_1, \ldots, r'_s} \Pi_{n+2s}(r_1, \ldots, r_n, \hat{r}_1, \ldots, \hat{r}_s, r'_1, \ldots, r'_s) \qquad \text{(by symmetry)} \\
&= 1,
\end{aligned}
$$

since we started with $(r_1, \ldots, r_n) \in R_1$.

We already established that the probability that the honest players' messages fall in R_1 is at least ε. Thus we deduce that with probability at least ε, the outcome of the new protocol Π'_n is equal to 1, even in the presence of a strong adaptive adversary corrupting s players and aiming to bias towards 0.

CASE 2. Suppose instead that the adversary $\mathcal{A}'$ aims to bias the outcome towards 1. We apply the Minimax Theorem to a zero-sum game where player 1 chooses the messages $r'_1, \ldots, r'_s$ and player 2 chooses the messages $r''_1, \ldots, r''_s$, and player 1 "wins" if the protocol outcome is 0, and player 2 wins otherwise. The rest of Case 2 is similar to Case 1; see the full version [4] for details.

4   Conclusion

We believe that this work paves the way to a number of little-explored research directions. We highlight some interesting questions for future work:

- To study the extent to which *communication can be reduced in protocols in general*, and to extend our communication-reduction techniques to the settings of multi-round protocols and/or adaptive security.
- To apply the *strong adaptive security notion* in the context of other types of protocols and settings, and to design protocols secure in the presence of strong adaptive adversaries.
- To consider whether adaptively secure *asymmetric* coin-flipping protocols can be converted to adaptively secure *symmetric* protocols, in general. This is not known even for the one-round case, and the question is moreover of interest since there are known one-round protocols which are not symmetric.
- To extend this work to prove (or disprove) the long-open conjecture of Lichtenstein et al. [6] that *any* adaptively secure coin-flipping protocol can tolerate at most $O(\sqrt{n})$ corruptions.

References

1. Ben-Or, M., Linial, N.: Collective coin flipping, robust voting schemes and minima of banzhaf values. In: FOCS, pp. 408–416. IEEE Computer Society (1985)
2. Dodis, Y.: Impossibility of black-box reduction from non-adaptively to adaptively secure coin-flipping. In: Electronic Colloquium on Computational Complexity (ECCC) 7.39 (2000)
3. Feige, U.: Noncryptographic selection protocols. In: FOCS, pp. 142–153. IEEE Computer Society (1999)
4. Goldwasser, S., Kalai, Y.T., Park, S.: Adaptively Secure Coin-Flipping, Revisited. (2015). arXiv: 1503.01588 [cs]
5. Hirt, M., Zikas, V.: Adaptively secure broadcast. In: Gilbert, H. (ed.) EUROCRYPT 2010. LNCS, vol. 6110, pp. 466–485. Springer, Heidelberg (2010)
6. Lichtenstein, D., Linial, N., Saks, M.E.: Some extremal problems arising form discrete control processes. Combinatorica 9(3), 269–287 (1989)
7. Nash, J.F.: Equilibrium points in n-person games. In: Proceedings of the National Academy of Sciences 36(1), 1950, pp. 48–49. doi:10.1073/pnas.36.1.48. eprint: http://www.pnas.org/content/36/1/48.full.pdf+html. http://www.pnas.org/content/36/1/48.short
8. Von Neumann, J., Morgenstern, O.: Theory of Games and Economic Behavior. Princeton University Press. ISBN: 0691119937 (1944)
9. Russell, A., Saks, M.E., Zuckerman, D.: Lower Bounds for Leader Election and Collective Coin-Flipping in the Perfect Information Model. SIAM J. Comput. 31(6), 1645–1662 (2002). doi:10.1137/S0097539700376007. http://dx.doi.org/10.1137/S0097539700376007

Optimal Competitiveness for the Rectilinear Steiner Arborescence Problem

Erez Kantor[1][(✉)] and Shay Kutten[2]

[1] MIT CSAIL, Cambridge, MA, USA
erezk@csail.mit.edu
[2] Technion, 32000 Haifa, Israel
kutten@ie.technion.ac.il

Abstract. We present optimal online algorithms for two related known problems involving Steiner Arborescence, improving both the lower and the upper bounds. One of them is the well studied continuous problem of the *Rectilinear Steiner Arborescence* (RSA). We improve the lower bound and the upper bound on the competitive ratio for RSA from $O(\log N)$ and $\Omega(\sqrt{\log N})$ to $\Theta(\frac{\log N}{\log \log N})$, where N is the number of Steiner points. This separates the competitive ratios of RSA and the Symetric-RSA (SRSA), two problems for which the bounds of Berman and Coulston is STOC 1997 were identical. The second problem is one of the Multimedia Content Distribution problems presented by Papadimitriou et al. in several papers and Charikar et al. SODA 1998. It can be viewed as the discrete counterparts (or a network counterpart) of RSA. For this second problem we present tight bounds also in terms of the network size, in addition to presenting tight bounds in terms of the number of Steiner points (the latter are similar to those we derived for RSA).

1 Introduction

Steiner trees, in general, have many applications, see e.g. [12] for a rather early survey that already included hundreds of items. In particular, Steiner Arborescences[1] are useful for describing the evolution of processes in time. Intuitively, directed edges represent the passing of time. Since there is no way to go back in time in such processes, all the directed edges are directed away from the initial state of the problem (the root), resulting in an arborescence. Various examples are given in the literature such as processes in constructing a Very Large Scale Integrated electronic circuits (VLSI), optimization problems computed in iterations (where it was not feasible to return to results of earlier iterations), dynamic programming, and problems involving DNA, see, e.g. [3,4,6,13]. Papadimitriou at al. [19,20] and Charikar et al. [5] presented the discrete version, in the context

---

E. Kantor– in a part by NSF Awards 0939370-CCF, CCF-1217506 and CCF-AF-0937274 and AFOSR FA9550-13-1-0042.

S. Kutten–Supported in part by the ISF, Israeli ministry of science and by the Technion Gordon Center.

[1] A Steiner arborescence is a Steiner tree directed away from the root.

M.M. Halldórsson et al. (Eds.): ICALP 2015, Part II, LNCS 9135, pp. 675–687, 2015.
DOI: 10.1007/978-3-662-47666-6_54

of Multimedia Content Delivery (MCD) to model locating and moving caches for titles on a path graph. The formal definition of (one of the known versions) of this problem, Directed-MCD, appears in Section 2.

We present new tight lower and upper bounds for two known interrelated problems involving Steiner Arborescences: *Rectilinear Steiner Arborescence (RSA)* and Directed-MCD (DMCD). We also deal indirectly with a third known arborescence problem: the *Symmetric*-RSA (SRSA) problem by separating its competitive ratio from that of RSA. That is, when the competitive ratios of RSA and SRSA were discussed originally by Berman and Coulston [4], the same lower and upper bounds were presented for both problems.

The *RSA* Problem: This is a rather heavily studied problem, described also e.g. in [4,9,17,18,22]. A rectilinear line segment in the plane is either horizontal or vertical. A rectilinear path contains only rectilinear line segments. This path is also *y-monotone* (respectively, *x-monotone*) if during the traversal, the y (resp., x) coordinates of the successive points are never decreasing. The input is a set of *requests* $\mathcal{R} = \{r_1 = (x_1, y_1), ..., r_N = (x_N, y_N)\}$ called Steiner terminals (or points) in the positive quadrant of the plane. A feasible solution to the problem is a set of rectilinear segments connecting all the N terminals to the origin $r_0 = (0,0)$, where the path from the origin to each terminal is both x-monotone and y-monotone (rectilinear shortest path). The goal is to find a feasible solution in which the sum of lengths of all the segments is the minimum possible. The above mentioned third problem, SRSA was defined in the same way, except that the above paths were not required to be x-monotone (only y-monotone).

Directed-MCD defined in Section 2 is very related to RSA. Informally, one difference is that it is discrete (Steiner points arrive only at discrete points) whiling RSA is continuous. In addition, in DMCD each "X coordinates" represents a network nodes. Hence, the number of X coordinates is bounded from above by the network size. This resemblance turned out to be very useful for us, both for solving RSA and for solving DMCD.

The *online* Version of *RSA* [4]: the given requests (terminals) are presented to the algorithm with nondecreasing y-coordinates. After receiving the i'th request $r_i = (x_i, y_i)$ (for $i = 1, ..., N$), the on-line RSA algorithm must extend the existing arborescence solution to incorporate r_i. There are two additional constraints: (1) a line, once drawn (added to the solution), cannot be deleted, and (2) a segment added when handling a request r_i, can only be drawn in the region between y_{i-1} (the y-coordinates of the previous request r_{i-1}) and upwards (grater y-coordinates). If an algorithm obeys constraint (1) but not constraint (2), then we term it a *pseudo online* algorithm. Note that quite a few algorithms known as "online", or as "greedy offline" fit this definition of "pseudo online".

Additional Related Works. Online algorithms for RSA and SRSA were presented by Berman and Coulston [4]. The online algorithms in [4] were $O(\log N)$ competitive (where N was the number of the Steiner points) both for RSA and SRSA. Berman and Coulston also presented $\Omega(\sqrt{\log N})$ lower bounds for both continuous problems. Note that the upper bounds for both problems were equal, and were the squares of the lower bounds. A similar gap for MCD arose from results of Halperinet al. [11], who gave a similar competitive ratio of $O(\log N)$, while Charikaret al. [5] presented a lower bound of $\Omega(\sqrt{\log n})$ for various variants of MCD, where n was the network size. Their upper bound was again the square of the lower bound. Berman and Coulston also conjectured that to close these gaps, both the upper bound and the lower bound for both problems could be improved. This conjecture was disproved in the cases of SRSA and of MCD on undirected line networks [15]. The latter paper closed the gap by presenting an optimal competitive ratio of $O(\sqrt{\log N})$ for SRSA and $O(\min\{\sqrt{n}, \sqrt{\log N}\})$ for MCD on the undirected line network with n nodes. They left the conjecture of Berman and Coulston open for RSA and for MCD on directed line networks. In the current paper, we prove this conjecture (for RSA and for Directed-MCD), thus separating RSA and SRSA in terms of their competitive ratios.

Charikar et al. [5] also studied the the offline case for MCD, for which they gave a constant approximation. The offline version of RSA is heavily studied. It was attributed to [18] who gave an exponential integer programming solution and to [9] who gave an exponential time dynamic programming algorithm. An exact and polynomial algorithm was proposed in [24], which seemed surprising, since many Steiner problems are NP-Hard. Indeed, difficulties in that solution were noted by Rao et al. [22], who also presented an approximation algorithm. Efficient algorithms are claimed in [7] for VLSI applications. However, the problem was proven NP-Hard in [23]. (The rectilinear Steiner tree problem was proven NP-Hard in [10]). Heuristics that are fast "in practice" were presented in [8]. A PTAS was presented by [17].

An optimal logarithmic competitive ratio for MCD on *general undirected* networks was presented in [2]. They also present a constant off-line approximation for MCD on grid networks.

On the Relation Between this Paper and [15]. An additional contribution of the current paper is the further development of the approach of developing (fully) online algorithms in two stages: (a) develop a pseudo online algorithm; and (b) convert the pseudo online into an online algorithm. As opposed to the problem studied in [15] where a pseudo online algorithm was known, here the main technical difficulty was to develop such an algorithm. From [15] we also borrowed an interesting twist on the rather common idea to translate between instances of a discrete and a continuous problems: we translate in *both* directions, the discrete solutions helps in optimizing the continuous one *and vice versa*.

Our Contributions. We improve both the upper and the lower bounds of RSA to show that the competitive ratio is $\Theta(\frac{\log N}{\log \log N})$. This proves the conjecture for RSA of Berman and Coulston [4] and also separates the competitive ratios of RSA and SRSA. We also provide tight upper and lower bound for Directed-MCD, the network version of RSA (both in terms of n and of N). The main technical innovation is the specific pseudo online algorithm we developed here, in order to convert it later to an online algorithm. The previously known offline algorithms for RSA and for DMCD where *not* pseudo online, so we could not use them. In addition to the usefulness of the new algorithm in generating the online algorithm, this pseudo online algorithm may be interesting in itself: It is $O(1)$-competitive for DMCD and for RSA (via the transformation) for a *different* (but rather common) online model (where each request must be served before the next one arrives, but no time passes between requests).

Paper Structure. Definitions are given in Section 2. The pseudo online algorithm SQUARE for DMCD is presented and analyzed in Section 3. In Section 4, we transform SQUARE to a (fully) online algorithm D-LINE^{on} for DMCD. Then, Section 5 describes the transformation of the online DMCD algorithm D-LINE^{on} to become an optimal online algorithm for RSA, as well as a transformation back from RSA to DMCD to make the DMCD online algorithm also optimal in terms of n (not just N). These last two transformations are taken from [15]. Finally, a lower bound is given in Section 6.

Because of space considerations, some of the proofs are omitted. However, all the proofs are given in the full version [16]. Moreover, the best way to understand the algorithms in this paper may be from a geometric point of view. Hence, in [16], we added multiple drawings to illustrate both the algorithms and the proofs.

2   Preliminaries

The Network×Time Grid (Papadimitriou et. al, [20]). A *directed line network* $L(n) = (V_n, E_n)$ is a network whose node set is $V_n = \{1, ..., n\}$ and its edge set is $E_n = \{(i, i+1) \mid i = 1, ..., n-1\}$. Given a directed line network $L(n) = (V_n, E_n)$, construct "time-line" graph $\mathcal{L}(n) = (\mathcal{V}_n, \mathcal{E}_n)$, intuitively, by "layering" multiple replicas of $L(n)$, one per time unit, where in addition, each node in each replica is connected to the same node in the next replica. Formally, the node set $\mathcal{V}_n$ contains a *node replica* (sometimes called just a *replica*) (v, t) of every $v \in V_n$, coresponding to each time step $t \in \mathbb{N}$. That is, $\mathcal{V}_n = \{(v, t) \mid v \in V_n, t \in \mathbb{N}\}$. The set of directed edges $\mathcal{E}_n = \mathcal{H}_n \cup \mathcal{A}_n$ contains *horizontal directed edges* $\mathcal{H}_n = \{((u, t), (v, t)) \mid (u, v) \in E_n, t \in \mathbb{N}\}$, connecting network nodes in every time step (round), and directed *vertical edges*, called *arcs*, $\mathcal{A}_n = \{((v, t), (v, t+1)) \mid v \in V_n, t \in \mathbb{N}\}$, connecting different copies of V_n. When n is clear from the context, we may write just X rather than X_n, for every $X \in \{V, E, \mathcal{V}, \mathcal{H}, \mathcal{A}\}$. Notice that $\mathcal{L}(n)$ can be viewed geometrically as a grid of n by ∞ whose grid points are the replicas. We consider the time as if it proceeds upward. We use such geometric presentations also in the text, to help clarifying the description.

The DMCD Problem. We are given a directed line network $L(n)$, an *origin* node $v_0 \in V$, and a set of *requests* $\mathcal{R} \subseteq \mathcal{V}$. A feasible solution is a subset of directed edges $\mathcal{F} \subseteq \mathcal{E}$ such that for every request $r \in \mathcal{R}$, there exists a path in $\mathcal{F}$ from the origin $(v_0, 0)$ to r. Intuitively a directed horizontal edge $((u, t), (v, t))$ is for delivering a copy of a multimedia title from node u to node v at time t.

A directed vertical edge (arc) $((v, t), (v, t+1))$ is for storing a copy of the title at node v from time t to time $t+1$. For convenience, the endpoints $\mathcal{V}_\mathcal{F}$ of edges in $\mathcal{F}$ are also considered parts of the solution. For a given algorithm A, let $\mathcal{F}_A$ be the solution of A, and let $cost(A, \mathcal{R})$, (the cost of algorithm A), be $|\mathcal{F}_A|$. (We assume that each storage cost and each delivery cost is 1.) The goal is to find a minimum cost feasible solution. Let OPT be the set of edges in some optimal solution whose cost is $|\text{OPT}|$.

Online *DMCD*. In the online versions of the problem, the algorithm receives as input a sequence of events. One type of events is a request in the (ordered) set $\mathcal{R}$ of requests $\mathcal{R} = \{r_1, r_2, ..., r_N\}$, where the requests times are in a non-decreasing order, i.e., $t_1 \leq t_2 \leq ... \leq t_N$ (as in RSA). A second type of events is a time event (this event does not exists in RSA), where we assume a clock that tells the algorithm that no additional requests for time t are about to arrive (or that there are no requests for some time t at all). The algorithm then still has the opportunity to complete its calculation for time t (e.g., add arcs from some replica (v, t) to $(v, t + 1)$). Then time $t + 1$ arrives.

When handling an event ev, the algorithm only knows the following: (a) all the previous requests $r_1, ..., r_i$; (b) time t; and (c) the solution arborescence $\mathcal{F}_{ev}$ it constructed so far (originally containing only the origin). In each event, the algorithm may need to make decisions of two types, before seeing future events:

(1.DMCD) If the event is the arrival of a request $r_i = (v_i, t_i)$, then from which *current* (time t_i) cache (a point already in the solution arborescence $\mathcal{F}_{ev}$ when r_i arrives) to serve r_i by adding *horizontal* directed edges to $\mathcal{F}_{ev}$.

(2.DMCD) If this is the time event for time t, then at which nodes to store a copy for time $t + 1$, for future use: select some replica (or replicas) (v, t) already in the solution $\mathcal{F}_{ev}$ and add to $\mathcal{F}_{ev}$ an edge directed from (v, t) to $(v, t + 1)$.

Note that at time t, the online algorithm cannot add nor delete any edge with an endpoint that corresponds to previous times. Similarly to e.g. [2,5,19–21], at least one copy must remain in the network at all times.

General Definitions and Notations. Consider an interval $J = \{v, v + 1, ..., v + \rho\} \subseteq V$ and two integers $s, t \in \mathbb{N}$, s.t. $s \leq t$. Let $J[s, t]$ be the *"rectangle subgraph"* of $\mathcal{L}(n)$ corresponding to vertex set J and time interval $[s, t]$. This rectangle consists of the replicas and edges of the nodes of J corresponding to every time in the interval $[s, t]$. For a given subsets $\mathcal{V}' \subseteq \mathcal{V}$, $\mathcal{H}' \subseteq \mathcal{H}$ and $\mathcal{A}' \subseteq \mathcal{A}$, denote by (1) $\mathcal{V}'[s, t]$ replicas of $\mathcal{V}'$ corresponding to times $s, ..., t$. Define similarly (2) $\mathcal{H}'[s, t]$ for horizontal edges of $\mathcal{H}'$; and (3) $\mathcal{A}'[s, t]$ arcs of $\mathcal{A}'$. (When $s = t$, we may write $\mathcal{X}[t] = \mathcal{X}[s, t]$, for $\mathcal{X} \in \{J, \mathcal{V}', \mathcal{H}'\}$.) Consider also two nodes $v, u \in V$ s.t. $u \leq v$. Let $\mathcal{P}_\mathcal{H}[(u, t), (v, t)]$ be the set of horizontal directed edges of the path from (u, t) to (v, t). Let $\mathcal{P}_\mathcal{A}[(v, s), (v, t)]$ be the set of arcs of the path from (v, s) to (v, t). Let $dist_\infty^\rightarrow((u, s), (v, t))$ be the "directed" distance from (u, s) to (v, t) in L_∞ norm. Formally, $dist_\infty^\rightarrow((u, s), (v, t)) = \max\{t - s, v - u\}$, if $s \leq t$ and $u \leq v$ and $dist_\infty^\rightarrow((u, s), (v, t)) = \infty$, otherwise.

3   Algorithm SQUARE, a Pseudo Online Algorithm

This section describes a pseudo online algorithm named SQUARE for the DMCD problem. Developing SQUARE was the main technical difficulty of this paper. Consider a requests set $\mathcal{R} = \{r_0 = (0,0), r_1 = (v_1, t_1), ..., r_N = (v_N, t_N)\}$ such that $0 \leq t_1 \leq t_2 \leq ... \leq t_N$. When Algorithm SQUARE starts, the solution includes just $r_0 = (0,0)$. Then, SQUARE handles, first, request r_1, then, request r_2, etc... In handling a request r_i, the algorithm may add some edges to the solution. (It never deletes any edge from the solution.) After handling r_i, the solution is an arborescence rooted at r_0 that spans the request replicas $r_1, ..., r_i$. Denote by SQUARE(i) the solution of SQUARE after handling the i'th request. For a given replica $r = (v, t) \in \mathcal{V}$ and a positive integer ρ, let

$$\mathcal{S}[r, \rho] = [v - \rho, v] \times [t - \rho, t]$$

denotes the rectangle subgraph (of the layered graph) whose top right corner is r induced by the set of replicas that contains every replica q such that (1) there is a directed path in the layer graph from q to r; and (2) the distance from q to r in L_∞ is at most ρ. For each request $r_i \in \mathcal{R}$, for $i = 1, ..., N$, SQUARE performs the following.

(SQ1) Add the vertical path from $(0, t_{i-1})$ to $(0, t_i)$.

(SQ2) Let replica $q_i^{\text{close}} = (u_i^{\text{close}}, s_i^{\text{close}})$ be such that q_i^{close} is already in the solution SQUARE($i - 1$) and (1) the distance in L_∞ norm from q_i^{close} to r_i is minimum (over the replicas already in the solution); and (2) over those replicas choose the latest, that is, $s_i^{\text{close}} = \max\{t \leq t_i \mid (u_i^{\text{close}}, t) \in \text{SQUARE}(i - 1)\}$. Define the *radius* of r_i as $\rho^{\text{SQ}}(i) = dist_\infty(q_i^{\text{close}}, r_i) = \max\{|v_i - u_i^{\text{close}}|, |t_i - s_i^{\text{close}}|\}$. Call q_i^{close} the *closest* replica of the i'th request.

(SQ3) Choose a replica $q_i^{\text{serve}} = (u_i^{\text{serve}}, s_i^{\text{serve}}) \in \mathcal{S}[r_i, 5 \cdot \rho^{\text{SQ}}(i)]$ such that q_i^{serve} is already in the solution SQUARE($i - 1$) and u_i^{serve} is the leftmost node (over the nodes corresponding to replicas of $\mathcal{S}[r_i, 5 \cdot \rho^{\text{SQ}}(i)]$ that are already in the solution). Call q_i^{serve} the *serving replica* of the i'th request.

(SQ4) Deliver a copy from q_i^{serve} to r_i via $(u_i^{\text{serve}}, t_i)$. This is done by storing a copy in node u_i^{serve} from time s_i^{serve} to time t_i, and then delivering a copy from $(u_i^{\text{serve}}, t_i)$ to (v_i, t_i) .

(SQ5) Store a copy in u_i^{serve} from time t_i to time $t_i + 4 \cdot \rho^{\text{SQ}}(i)$.

Intuitively, steps SQ1–SQ4 utilize previous replicas in the solution, while step SQ5 prepares the contribution of r_i to serve later requests. Note that SQUARE is not an online algorithm, since in step SQ4, it may add to the solution some arcs corresponding to previous times. Such an action cannot be preformed by an online algorithm. Denote by $\mathcal{F}^{\text{SQ}} = \mathcal{H}^{\text{SQ}} \cup \mathcal{A}^{\text{SQ}}$ the feasible solution SQUARE(N) of SQUARE. Let BASE(i) $= \{(u, t_i) \mid u_i^{\text{serve}} \leq u \leq v_i\}$ and let BASE $= \cup_{i=1}^{N}$ BASE(i) (notice that BASE $\subseteq \mathcal{F}^{\text{SQ}}$ because of step SQ4). Similarly, let TAIL(i) $= \{(u_i^{\text{serve}}, t) \mid t_i \leq t \leq t_i + 4\rho^{\text{SQ}}(i)\}$ be the nodes of the path $\mathcal{P}_A[(u_i^{\text{serve}}, t_i), (u_i^{\text{serve}}, t_i + 4 \cdot \rho^{\text{SQ}}(i))]$ (added to the solution in step SQ5) and let $\text{TAIL} = \cup_{i=1}^N \text{TAIL}(i)$. Note that $\mathcal{F}^{\text{SQ}}$ is indeed an arborescence rooted at $(0,0)$.

Analysis of SQUARE. First, bound the cost of SQUARE as a function of the radii (defined in SQ2).

Observation 1 $cost(\text{SQUARE}, \mathcal{R}) \leq 14 \sum_{i=1}^N \rho^{\text{SQ}}(i)$.

(For lack of space, some of the proofs are omitted. Still, Observation 1 is obvious from the description of SQUARE.) It is left to bound from below the cost of the optimal solution as a function of the radii.

Quarter Balls. Our analysis is based on the following notion. A *quarter-ball*, or a Q-BALL, of *radius* $\rho \in \mathbb{N}$ centered at a replica $q = (v, t) \in \mathcal{V}$ contains every replica from which there exists a path of length ρ to q [2]. For every request $r_i \in \mathcal{R}$, denote by $Q\text{-BALL}^{\text{SQ}}(r_i, \rho^{\text{SQ}}(i))$ [3] (also $Q\text{-BALL}^{\text{SQ}}(i)$ for short) the quarter-ball centered at r_i with radius $\rho^{\text{SQ}}(i)$.

Intuitively, for every request $r_i \in \mathcal{R}'$ (where $\mathcal{R}'$ obey the observation's condition below), OPT's solution starts outside of $Q\text{-BALL}^{\text{SQ}}(i)$, and must reach r_i with a cost of $\rho^{\text{SQ}}(i)$ at least.

Observation 2 *Consider some subset $\mathcal{R}' \subseteq \mathcal{R}$ of requests. If the Q-balls, $Q\text{-BALL}^{\text{SQ}}(i)$ and $Q\text{-BALL}^{\text{SQ}}(j)$, of every two requests $r_i, r_j \in \mathcal{R}'$ are edges disjoint, then $|\text{OPT}| \geq \sum_{r_i \in \mathcal{R}'} \rho^{\text{SQ}}(i)$.*

Covered and Uncovered Requests. Consider some request $r_i = (v_i, t_i)$ and its serving replica $q_i^{\text{serve}} = (u_i^{\text{serve}}, s_i^{\text{serve}})$ (see step SQ3). We say that r_i is *covered*, if $v_i - u_i^{\text{serve}} \geq \rho^{\text{SQ}}(i)$ (see SQ2 and SQ3). Intuitively, this means the solution $\mathcal{F}^{\text{SQ}}$ is augmented by the whole top of the square $\text{SQUARE}[r_i, \rho^{\text{SQ}}(i)]$. Otherwise, we say that r_i is *uncovered*. Let $\text{COVER} = \{i \mid r_i \text{ is a covered request}\}$ and let $\text{UNCOVER} = \{i \mid r_i \text{ is an uncovered request}\}$. Given Observation 2, the following lemma implies that

$$|\text{OPT}| \geq \sum_{i \in \text{COVER}} \rho^{\text{SQ}}(i). \tag{1}$$

Lemma 1. *Consider two **covered** Requests r_i and r_j. The quarter balls $Q\text{-BALL}^{\text{SQ}}(i)$ and $Q\text{-BALL}^{\text{SQ}}(j)$ are edge disjoint.*

The lemma follows easily from geometric considerations, see figures 5–6 and the proof in [16]. By observations 1, 2, and Inequality (1), we have:

---

[2] This is, actually, the definition of the geometric place "ball". We term them "quarter ball" to emphasize that we deal with directed edges. That is, it is not possible to reach (v, t) from above nor from the right.

[3] Note that $Q\text{-BALL}^{\text{SQ}}(r_i, \rho^{\text{SQ}}(i))$ is different from $\mathcal{S}[r_i, \rho^{\text{SQ}}(i)]$, since the first ball considers distances in L_2 norm and the last considers distances in L_∞ norm.

Observation 3 SQUARE*'s cost for covered requests is no more than* $14 \cdot$ OPT.

It is left to bound the cost of SQUARE for the uncovered requests.

Overview of the Analysis of the Cost of Uncovered Requests. Unfortunately, unlike the case of covered requests, balls of two *uncovered* requests may not be disjoint. Still, we managed to have a somewhat similar argument that we now sketch. The formal analysis appears in [16]. Below, we partition the balls of uncovered requests into disjoint subsets. Each has a representative request, a *root*. We show that the Q-BALL of roots *are* edge disjoint. This implies by Observation 1 and Observation 2 that the cost SQUARE pays for the roots is smaller than 14 times the total cost of an optimal solution. Finally, we show that the cost of SQUARE for all the requests in each subset is at most twice the cost of SQUARE for the root of the subset. Hence, the total cost of SQUARE for the uncovered requests is also just a constant times the total cost of the optimum.

To construct the above partition, we define the following relation: ball Q-BALL$^{SQ}(j)$ is the *child* of Q-BALL$^{SQ}(i)$ (for two *uncovered* requests r_i and r_j) intuitively, if the Q-BALL$^{SQ}(i)$ is the first ball (of a request later then r_j) such that Q-BALL$^{SQ}(i)$ and Q-BALL$^{SQ}(j)$ are *not* edge disjoint. Clearly, this parent-child relation induces a forest on the Q-BALLs of uncovered requests. The following observation follows immediately from the definition of a root.

Observation 4 *The quarter balls of every two root requests are edge disjoint.*

The above observation together with Observation 2, implies the following.

Observation 5 *The cost of* SQUARE *for the roots is* $14 \cdot |$OPT$|$ *at most.*

It is left to bound the cost that SQUARE pays for the balls in each tree (in the forest of Q-BALLs) as a constant function of the cost it pays for the tree root. Specifically, we show that the sum of the radii of the Q-BALLs in the tree (including that of the root) is at most twice the radius of the root. This implies the claim for the costs by Observation 1 and Observation 2. To show that, given any non leaf ball Q-BALL$^{SQ}(i)$ (not just a root), we first analyze only Q-BALL$^{SQ}(i)$'s "latest child" Q-BALL$^{SQ}(j)$. That is, $j = \max_k \{Q$-BALL$^{SQ}(k)$ is a child of Q-BALL$^{SQ}(i)\}$. We show that the radius of the latest child is, at most, a quarter of the radius of Q-BALL$^{SQ}(i)$. Second, we show that the *sum* of the radii of the rest of the children (all but the latest child) is, at most, a quarter of the radius of Q-BALL$^{SQ}(i)$ too. Hence, the radius of a parent ball is at least twice as the sum of its children radii. This implies that the sum of the radii of all the Q-BALLs in a tree is at most twice the radius of the root.

The hardest technical part here is in the following lemma that, intuitively, states that "a lot of time" (proportional to the request's radius) passes between the time one child ball ends and the time the next child ball starts, see Fig. 1.

Lemma 2. *Consider some uncovered request* r_i *which has at least two children. Let* Q-BALL$^{SQ}(j)$, Q-BALL$^{SQ}(k)$ *some two children of* Q-BALL$^{SQ}(i)$, *such that* $k < j$. *Then,* $t_j - \rho^{SQ}(j) \geq t_k + 4\rho^{SQ}(k)$.

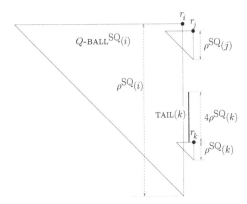

Fig. 1. Geometric look on a parent Q-BALL$^{SQ}(i)$ (note that a Q-BALL is a triangle) and its children Q-BALL$^{SQ}(j)$ and Q-BALL$^{SQ}(k)$

Intuitively, the radius of a parent Q-BALL is covered by the radii of its children Q-BALLs, plus the tails (see step SQ5) between them. Restating the lemma, the time of the earliest replica in Q-BALL$^{SQ}(j)$ is not before the time of the latest replica in TAIL(k). Intuitively, recall that the tail length of a request is much grater than the radius of the request's Q-BALL. Hence, the fact that the radius of a latest child is at most a quarter of the radius of its parent, together with Lemma 2, imply that the sum of the childrens radii is less than half of the radius of the parent Q-BALL . The full proof of Lemma 2 (appears in [16]) uses geometric considerations. Outlining the proof, we first establish an additional lemma. Given any two requests r_j and r_ℓ such that $j > \ell$, the following lemma formalizes the following: Suppose that the node v_j of request r_j is "close in space (or in the network)" to the node v_ℓ of another request r_ℓ. Then, the whole Q-BALL of r_j is "far in time" (and later) from r_j.

Lemma 3. *Suppose that, $j > \ell$ and $v_j - \rho^{SQ}(j) + 1 \leq u_\ell^{serve} \leq v_j$. Then, the time of the earliest replica in Q-BALL$^{SQ}(j)$ is not before the time of the latest replica in TAIL(ℓ), i.e., $t_j - \rho^{SQ}(j) \geq t_\ell + 4\rho^{SQ}(\ell)$.*

Intuitively, Lemma 3 follows thanks to the tail left in step SQ5 of SQUARE, as well as to the action taken in SQ3 for moving u^{serve} further left of u^{close}. In the proof of Lemma 2, we show that in the case that two requests r_k and r_j are siblings, either **(1)** they satisfy the conditions of Lemma 3, or **(2)** there exists some request r_ℓ such that $k < \ell < j$ such that r_ℓ and r_j satisfy the conditions of Lemma 3. Moreover, the time of the last replica in TAIL(ℓ) is even later then the time of the last replica in TAIL(k). In both cases, we apply Lemma 3 to show that the time of the earliest replica in Q-BALL$^{SQ}(j)$ is not before the time of the latest replica in TAIL(k) as needed for the lemma.

   To summarize, we show (1) For *covered* requests the cost of SQUARE is $O(1)$ of $|$OPT$|$; see Observation 3. (2) For *uncovered* requests, we prove in [16] (as overviewed above) two facts: (2.a) the Q-BALLs of the root requests are edges disjoint, and hence by Observation 5, the sum of their radii is $O(1)$ of $|$OPT$|$ too.

(2.b) On the other hand, the sum of root's radii is at least half of the sum of the radii of all the uncovered requests. This establishes Theorem 6.

Theorem 6. *Algorithm* SQUARE *is $O(1)$-competitive for* DMCD *under the pseudo online model.*

4   Algorithm D-LINE^{on} - The "real" Online Algorithm

In this section, we transform the pseudo online algorithm SQUARE of Section 3 into a (fully) online algorithm D-LINE^{on} for DMCD. The full details as well as the formal proof of this transformation appears in [16][4]. Let us nevertheless give some intuition here.

The reason Algorithm SQUARE is *not* online, is one of the the actions it takes at step SQ4. There, it stores a copy at the serving replica u_i^{serve} for request r_i from time s_i^{serve} to time t_i. This requires "going back in time" in the case that the time $s_i^{\text{serve}} < t_i$. A (full) online algorithm cannot perform such an action. Intuitively, Algorithm D-LINE^{on} "simulates" the impossible action by (1) storing additional copies (beyond those stored by SQUARE); and (2) shifting the delivery to request r_i (step SQ4 of SQUARE) from an early time to time t_i of r_i. It may happen that the serving node u_i^{serve} of r_i does not have a copy (in SQUARE) at t_i. In that case, Algorithm D-LINE^{on} also (3) delivers first a copy to $(u_i^{\text{serve}}, t_i)$ from some node w on the left of u_i^{serve}. Simulation step (1) above (that we term the storage phase) is the one responsible for ensuring that such a node w exists, and is "not too far" from u_i^{serve}.

For the storage phase, Algorithm D-LINE^{on} covers the network by "intervals" of various lengthes (pathes that are subgraphs of the network graph). There are overlaps in this cover, so that each node is covered by intervals of various lengthes. Let the length of some interval I be $length(I)$. Intuitively, given an interval I and a time t, if SQUARE kept a copy in a node of interval I "recently" ("recent" is proportional to $length(I)$), then D-LINE^{on} makes sure that a copy is kept at the left most node of this interval, or "nearby" (in some node in the interval just left to I).

Theorem 7. D-LINE^{on} *is $O(\frac{\log n}{\log \log n})$-competitive for* DMCD *problem.*

5   Optimal Algorithm for RSA and for DMCD

Algorithm D-LINE^{on} in Section 4 solves DMCD. To solve also RSA, we transform Algorithm D-LINE^{on} to an algorithm RSA^{on} that solves RSA. First, let us view the reasons why the solution for DMCD (Section 4) does not yet solve

---

[4] We comment that it bears similarities to the transformation of the pseudo online algorithm Triangle to a (full) online algorithm for *undirected* MCD in [15]. The transformation here is harder, since there the algorithm sometimes delivered a copy to a node v from some node on v's right, which we had to avoid here (since the network is directed to the right).

RSA. In DMCD, the X coordinate of every request (in the set $\mathcal{R}$) is taken from a known set of size n (the network nodes $\{1, 2, ..., n\}$). On the other hand, in RSA, the X coordinate of a *point* is arbitrary. (A lesser obstacle is that the Y coordinate is a real number, rather than an integer.) The main idea is to make successive guesses of the number of Steinr points and of the largest X coordinate and solve under is proven wrong (e.g. a point with a larger X coordinate arrives) then readjust the guess for future request. Fortunately, the transformation is exactly the same as the one used in [14,15] to transform the algorithm for undirected MCD to solve $SRSA$.

Theorem 8. *Algorithm* RSA^{on} *is optimal and is* $O(\frac{\log N}{\log \log N})$-*competitive.*

5.1   Optimizing DMCD for a Small Number of Requests

Algorithm D-LINE^{on} was optimal only as the function of the network size. Recall that our solution for RSA was optimal as a function of the number of requests. We obtain this property for the solution of DMCD too, by transforming our RSA algorithm back to solve DMCD, and obtain the promised competitiveness, $O(\min\{\frac{\log N}{\log \log N}, \frac{\log n}{\log \log n}\})$, see [16].

6   Lower Bound for RSA

In this section, we prove the following theorem, establishing a tight lower bound for RSA and for DMCD on directed line networks. Interestingly, this lower bound is not far from the one proven by Alon and Azar for *undirected* Euclidian Steiner trees [1]. Unfortunately, the lower bound of [1] does not apply to our case since their construct uses edges directed in what would be the wrong direction in our case (from a high Y value to a low one).

Theorem 9. *The competitive ratio of any deterministic online algorithm for* DMCD *in directed line networks is* $\Omega(\frac{\log n}{\log \log n})$, *implying also an* $\Omega(\frac{\log N}{\log \log N})$ *lower bound for* RSA.

Proof: We first outline the proof. Informally, given a deterministic online algorithm ONALG$_{\text{MCD}}$, we construct an adversarial input sequence. Initially, the request set includes the set DIAG $= \{(k, k) \mid 0 \leq k \leq n\}$. That is, at each time step t, the request (t, t) is made. In addition, if the algorithm leaves "many copies" then the lower bound is easy. Otherwise, the algorithm leaves "too few copies" from some time $t - 1$ until time t. For each such time, the adversary makes another request at $(t - k, t)$ for some k defined later. The idea is that the adversary can serve this additional request from the diagonal copy at $(t-k, t-k)$ paying the cost of k. On the other hand, the algorithm is not allowed at time t to decide to serve from $(t - k, t - k)$. It must serve from a copy it did leave. Since the algorithm left only "few" copies to serve time t the replica, $(t, t - k)$ can be chosen at least at distance $k(\log n)$ from any copy the algorithm did leave. Hence, the algorithm's cost for such a time t is $\Omega(\log n)$ times greater than that of the adversary. The full proof appears in [16].   ∎

References

1. Alon, N., Azar, Y.: On-line Steine trees in the euclidean plane. Discrete & Computational Geometry **10**, 113–121 (1993)
2. Bar-Yehuda, R., Kantor, E., Kutten, S., Rawitz, D.: Growing half-balls: minimizing storage and communication costs in CDNs. In: Czumaj, A., Mehlhorn, K., Pitts, A., Wattenhofer, R. (eds.) ICALP 2012, Part II. LNCS, vol. 7392, pp. 416–427. Springer, Heidelberg (2012)
3. Bein, W., Golin, M., Larmore, L., Zhang, Y.: The Knuth-Yao quadrangle-inequality speedup is a consequence of total monotonicity. ACM Transactions on Algorithms, **6**(1) (2009)
4. Berman, P., Coulston, C.: On-line algorrithms for Steiner tree problems. In: STOC, pp. 344–353 (1997)
5. Charikar, M., Halperin, D., Motwani, R.: The dynamic servers problem. In: 9th Annual Symposium on Discrete Algorithms (SODA), pp. 410–419 (1998)
6. Cheng, X., Dasgupta, B., Lu, B.: Polynomial time approximation scheme for symmetric rectilinear Steiner arborescence problem. J. Global Optim., **21**(4) (2001)
7. Cho, J.D.: A min-cost flow based min-cost rectilinear Steiner distance-preserving tree construction. In: ISPD, pp. 82–87 (1997)
8. Cong, J., Kahng, A.B., Leung, K.S.: Efficient algorithms for the minimum shortest path Steiner arborescence problem with applications to VLSI physical design. IEEE Trans. on CAD of Integrated Circuits and Systems **17**(1), 24–39 (1998)
9. Ladeira de Matos, R.R.: A rectilinear arborescence problem. Dissertation, University of Alabama (1979)
10. Garey, M.R., Johnson, D.S.: The rectilinear Steiner tree problem is NP-complete. SIAM J. Appl. Math. **32**(4), 826–834 (1977)
11. Halperin, D., Latombe, J.C., Motwani, R.: Dynamic maintenance of kinematic structures. In: Laumond, J.P., Overmars, M. (eds.) Algorithmic Foundations of Robotics, pp. 155–170. A.K. Peters Publishing (1997)
12. Hwang, F.K., Richards, D.S.: Steiner tree problems. Networks **22**(1), 55–897 (1992)
13. Kahng, A., Robins, G.: On optimal interconnects for VLSI. Kluwer Academic Publishers (1995)
14. Kantor, E., Kutten, S.: Optimal competitiveness for symmetric rectilinear Steiner arborescence and related problems (2013). CoRR, abs/1307.3080
15. Kantor, E., Kutten, S.: Optimal competitiveness for symmetric rectilinear steiner arborescence and related problems. In: Esparza, J., Fraigniaud, P., Husfeldt, T., Koutsoupias, E. (eds.) ICALP 2014, Part II. LNCS, vol. 8573, pp. 520–531. Springer, Heidelberg (2014)
16. Kantor, E., Kutten, S.: Optimal competitiveness for the rectilinear steiner arborescence problem (2015). CoRR, arxiv.org/abs/1504.08265
17. Lu, B., Ruan, L.: Polynomial time approximation scheme for rectilinear Steiner arborescence problem. Combinatorial Optimization **4**(3), 357–363 (2000)
18. Nastansky, L., Selkow, S.M., Stewart, N.F.: Cost minimum trees in directed acyclic graphs. Z. Oper. Res. **18**, 59–67 (1974)
19. Papadimitriou, C.H., Ramanathan, S., Rangan, P.V.: Information caching for delivery of personalized video programs for home entertainment channels. In: IEEE International Conf. on Multimedia Computing and Systems, pp. 214–223 (1994)
20. Papadimitriou, C.H., Ramanathan, S., Rangan, P.V.: Optimal information delivery. In: Staples, J., Katoh, N., Eades, P., Moffat, A. (eds.) ISAAC 1995. LNCS, vol. 1004, pp. 181–187. Springer, Heidelberg (1995)

21. Papadimitriou, C.H., Ramanathan, S., Rangan, P.V., Sampathkumar, S.: Multimedia information caching for personalized video-on demand. Computer Communications **18**(3), 204–216 (1995)
22. Rao, S., Sadayappan, P., Hwang, F., Shor, P.: The Rectilinear Steiner Arborescence problem. Algorithmica, pp. 277–288 (1992)
23. Shi, W., Su, C.: The rectilinear Steiner arborescence problem is NP-complete. In: SODA, pp. 780–787 (2000)
24. Trubin, V.A.: Subclass of the Steiner problems on a plane with rectilinear metric. Cybernetics and Systems Analysis **21**(3), 320–324 (1985)

Normalization Phenomena in Asynchronous Networks

Amin Karbasi[1], Johannes Lengler[2]([⊠]), and Angelika Steger[2]

[1] School of Engineering and Applied Science, Yale University, New Haven, CT, USA
amin.karbasi@yale.edu
[2] Department of Computer Science, ETH Zurich, Zurich, Switzerland
{johannes.lengler,angelika.steger}@inf.ethz.ch

Abstract. In this work we study a diffusion process in a network that consists of two types of vertices: *inhibitory* vertices (those obstructing the diffusion) and *excitatory* vertices (those facilitating the diffusion). We consider a continuous time model in which every edge of the network draws its transmission time randomly. For such an asynchronous diffusion process it has been recently proven that in Erdős-Rényi random graphs a *normalization phenomenon* arises: whenever the diffusion starts from a large enough (but still tiny) set of active vertices, it only percolates to a certain level that depends only on the activation threshold and the ratio of inhibitory to excitatory vertices. In this paper we extend this result to all networks in which the percolation process exhibits an explosive behaviour. This includes in particular inhomogeneous random networks, as given by Chung-Lu graphs with degree parameter $\beta \in (2,3)$.

1 Introduction

One of the main goals in studying complex networks (e.g., social, neural networks) is to better understand the interconnection between the elements of such systems, and as a result being able to reason about their accumulated behavior. Of particular interest is to make a connection between the network's *structure* and *function*: once we have quantified the configuration of a network, how can we turn the results into predictions on the overall system behaviour?

A natural setting in which network structure plays a central role is the *diffusion of innovation* where a new product is introduced to the market (e.g., a new search engine), a new drug is suggested to doctors, or a new political movement has gained power in an unstable society. Once an innovation appears, people may have different reactions: some embrace it and try to promote the innovation, and others may refute it and try to obstruct the circulation of innovation. As a result, depending on the nature of the innovation (how much intrinsically people like or dislike the new product, idea, etc) and the structure of the network, the diffusion may die out quickly or spread explosively through the population. In order to understand to what extent an innovation is accepted it is important to understand the dynamics of the diffusion within the underlying network.

M.M. Halldórsson et al. (Eds.): ICALP 2015, Part II, LNCS 9135, pp. 688–700, 2015.
DOI: 10.1007/978-3-662-47666-6_55

We make the following assumptions: after trying an innovation, an individual's reaction (i.e., like or dislike of the innovation) is parametrized by a binary random variable that takes -1 with probability τ (indicating that she dislikes the innovation) and $+1$ with probability $1 - \tau$ (indicating that she likes the innovation). We assume that each individual's reaction is *independent* of that of its neighbors. However, an individual's tendency to become *active*, meaning that she tries the new innovation, depends on input from neighbors who have already tried the innovation and either promote or bash it. More precisely, we adopt the so-called *linear threshold model* [16]: an individual switches her state from inactive to active if the difference between its active supporting and its active inhibiting neighbors is above a threshold k. Given an initial set of early active individuals (those who were first exposed to the innovation and tried it), the diffusion process unfolds according to the aforementioned model. What we are interested in is to understand under what conditions an initially active set will spread through a non-trivial portion of the population.

Related Work. Classically, such diffusion processes have been studied without inhibition under the name of *bootstrap percolation* [9], [11]. With hindsight we note that in such processes the exact timing of the transmission of information from an active vertex to its neighbors plays no role and we can thus assume that all transmission times are exactly one. The activation then takes place in rounds. Such process were first studied by Chalupa et al. [9] on a 2-dimensional lattice in the context of magnetic disordered systems. Since then it has been the subject of intense research and found numerous applications for modeling complex systems' behaviors such as social influence [16], infectious disease propagation [18], jamming transitions [20], and neuronal activity [8], to name a few. The main focus of the previous work has been to understand the final number of active vertices. In the case of the 2-dimensional lattice, Holroyd [15] determined a sharp asymptotic threshold: as the size of the initial active set goes above a threshold, the process percolates to almost all vertices. This result was then generalized to 3-dimensional [4] and recently to any d-dimensional grid [3]. Bootstrap percolation has also been studied on a variety of graphs such as trees [5], random regular graphs [6], Erdős-Rényi graphs [17], and power-law graphs [2]. All this work exhibits an all-or-nothing phenomenon: either the initial set is small and the diffusion stops quickly or it is large enough and percolates to almost all vertices.

In [12] the authors introduced the bootstrap percolation process with inhibitory vertices as a model for the spread of activity through a population of neurons. That paper analyzed the bootstrap percolation process for Erdős-Rényi random graphs. The authors showed that on the one hand, if activation takes place in rounds inhibition makes the network highly susceptible to small changes in the starting set, but that on the other hand it does not so in the *asynchronous case* (where the spread of activity requires an exponentially distributed, random transmission time). However, the proofs relied substantially on the symmetry of Erdős-Rényi random graphs. The aim of the paper at hand is to extend the results from [12] (in the asynchronous setting) to a much larger class of graphs that do not exhibit a uniform degree distribution as the Erdős-Rényi random graphs do. This graph class contains in particular power-law distributed random graphs as provided by the so-called Chung-Lu model.

It has been empirically observed that many real networks follow a power-law degree distribution, including internet topology [13], World Wide Web [1], and Facebook social network [21]. That is, for most real-world graphs, the fraction of vertices with degree d scale like $d^{-\beta}$ where β is usually between 2 and 3. As a result, power-law graphs represent real-world networks more realistically than Erdős-Rényi graphs [7]. There are many generative models that construct such power-law graphs. In the present work we use the model of Chung and Lu [10]. It is known that for a graph of size n and in the absence of any inhibition, $\theta(n) := n^{(\beta-2)/(\beta-1)}$ is a coarse threshold for bootstrap percolation in such scale-free graphs [2].

Main Results. We say that a process looks *explosive* to an individual if in a short time a large number of her neighbors become active (for a formal definition, see Section 3). In [12] it was shown (without using this terminology) that asynchronous bootstrap percolation is explosive for Erdős-Rényi random graphs. In Theorem 2, we show that explosive percolation is more prevalent and happens also for *power-law graphs*, if we restrict ourselves to vertices of large degree.

In Theorem 1 we prove that an explosive process is automatically *normalizing*: of all individuals to which the process looks explosive, a certain fraction will turn active that can be accurately estimated as a function of τ and k that is *independent* of the structure of the graph. In contrast to all-or-nothing phenomena where the size of the final active set can either be very small or very large, our result provides a middle ground for asynchronous percolation processes. Note that to some vertices the process does not look explosive for a rather trivial reason: they may not have many neighbors at all. For these low degree vertices we can still give a heuristic prediction on how many of them will become active (cf. Section 5.1).

To further support our theoretical results, we perform experiments on real-world networks such as the Epinions social network. We observe that already for a small number of vertices our theoretical estimates for the final set of active vertices matches the numbers provided by the experimental data.

2   Formal Definitions and Notation

Let $G = (V, E)$ be a finite graph, and let $A \subset V$. Let $k \geq 2$ be an integer, and let $\tau \in [0, 1]$. Then the (k, τ)-*bootstrap percolation process* on G with starting set A is defined as follows. We first split the set of vertices randomly into two subsets V^+ and V^-. Each vertex is independently assigned to V^- with probability τ, and to V^+ otherwise. We call the vertices in V^+ *excitatory*, and the vertices in V^- *inhibitory*. We start the percolation process at time $t = 0$. At this time, all vertices in A turn *active*, and all other vertices are *inactive*. Whenever a vertex becomes active, it sends out signals to all its neighbors. Each signal takes a random *transmission time* to travel to its target; all transmission times are independently drawn from the exponential distribution Exp(1) with expectation 1. For every vertex v and time t, let $S^+(v,t)$ and $S^-(v,t)$ be the number of signals that have arrived at v until time t and that originated from excitatory and inhibitory vertices, respectively. Let $S(v,t) := S^+(v,t) - S^-(v,t)$. Whenever there is an inactive vertex v and a time $t > 0$ such that $S(v,t) \geq k$ then v immediately turns active, and sends out signals to all its neighbors. Active vertices do not turn inactive again. In particular, during the process each active vertex sends out exactly one signal to each of its neighbors. For $t \geq 0$, we denote by $a(t)$ the number of active neighbors at time t, and by a^* the number of active neighbors at termination, i.e., $a^* = \lim_{t \to \infty} a(t)$. As further notation, let $\Gamma_s^+(v)$ and $\Gamma_s^-(v)$ be the number of active excitatory and inhibitory neighbors of v among the first s active neighbors, respectively. (We only consider these random variables when there are at least s active neighbors.) Note that $\Gamma_s^+(v,t) + \Gamma_s^-(v,t) = s$. Finally, let $X_i(v) \in \{\pm 1\}$ be the random variable that describes whether the i-th signal arriving at v is excitatory or inhibitory.

Chung-Lu Model. To generate power law graphs of size n, we use the model of Chung and Lu [10]. In this model, each vertex i is assigned a positive weight w_i. The probability p_{ij} of having an edge between vertices i and j is $\min\{1, w_i w_j / z\}$ where $z = \sum_{i=1}^n w_i$. Note that $G(n,p)$ can be viewed as a special case where for all vertices i we set $w_i = pn$. To generate a power-law graph with exponent $2 < \beta < 3$ and (constant) average degree $\bar{d}$, we set $w_i = \bar{d}\frac{\beta-2}{\beta-1}(\frac{n}{i})^{1/(\beta-1)}$.

Basic Notation. For a sequence of events $\mathcal{E} = \mathcal{E}(n)$ we say that $\mathcal{E}$ holds *with high probability* (w.h.p.) if $\Pr[\mathcal{E}(n)] \to 1$ for $n \to \infty$. For any $x, y, z \in \mathbb{R}$, we use $x \in y \pm z$ to abbreviate the inequalities $y - z \leq x \leq y + z$.

3   Results

The main result of our paper is that the normalization phenomenon occurs for a large class of graphs, i.e., there is a universal constant $\alpha = \alpha(\tau, k)$ that does not depend on the structure of the graph such that bootstrap percolation activates an α-fraction of all vertices to which the process appears *fast*. We start with a precise definition of what we mean by "fast".

Definition 1. *Let $G = (V, E)$ be a graph[1] with n vertices, and let $A, S \subset V$. Furthermore, let $C \in \mathbb{N}$ and $\eta, \delta > 0$. We say that percolation on G with starting set A is* locally (C, η, δ)-explosive *for the set S if the following holds with probability at least $1 - \delta$. For all but at most $\eta|S|$ vertices $v \in S$ there are times $t = t(v)$ such that at time t the vertex v has no active neighbor and at time $t + \eta$ it has at least C active neighbors.*

In the above definition, the probability is taken with respect to the random choices of inhibitory/excitatory signs and random delays. Also note that the

---
[1] All results and proofs carry over immediately to directed graphs as well. In this case, explosiveness needs to be defined with respect to the in-neighbors of G.

time t may well depend on the run of the diffusion and how it unfolds. If t can be chosen independently of v (but possibly still depending on the run), then we call the process *globally* (C, η, δ)-*explosive for* S. In this case, we call the time t the *start of the explosion.*[2]

Intuitively, a locally (C, η, δ)-explosive process looks fast from the point of view of individual vertices $v \in S$: the time that is needed to go from 0 active neighbors to C active neighbors is at most η. A *globally* (C, η, δ)-explosive process also looks fast from a global point of view, since the process needs only time η to go from a situation where almost no vertex in S has any active neighbors to a point where almost all of them have many neighbors.

We have defined (C, η, δ)-explosive processes for all values of C and η, but we will use them only in the case that C is large and $\eta > 0$ is very small. It turns out that for many standard graph models, percolation processes are locally (C, η, δ)-explosive for a large enough set S. For example, for the Erdős-Rényi random graph model $G(n, p)$ (with $p \gg 1/n$) it was observed in [12] that this is the case for $S = V$ and for all constants $C, \eta > 0$ (see also Theorem 3 in Section 4.1 of the present paper). In Theorem 2 below we prove the same for power law graphs, and in Section 5 we perform experiments indicating that percolation is also explosive on real-world networks.

The following theorem tells us that every process that is locally (C, η, δ)-explosive for some set S is also *normalizing* for that set, i.e., the number of active vertices in S at the end of the process is roughly $\alpha|S|$, where the constant α given by

$$\alpha = \min\left\{ \left(\frac{1-\tau}{\tau}\right)^k, 1 \right\}. \tag{1}$$

Note crucially that α does not depend on the structure of the graph or on the size of S.

Theorem 1. *For every $\varepsilon > 0$ there exist positive constants C_S, C, η, $\delta > 0$ such that the following holds. For a graph $G = (V, E)$ with n vertices, and sets $A, S \subset V$ such that $|S| \geq C_S$ and $|A \cap S| \leq \eta|S|$, if G with starting set A is locally (C, η, δ)-explosive for S, then with probability at least $1 - \varepsilon$ the percolation process will terminate with $(\alpha \pm \varepsilon)|S|$ active vertices in S. In particular, for $S = V$ the final active set has size $a^* = (\alpha \pm \varepsilon)n$.*

Theorem 1 has vast implications since many bootstrap percolation processes are (C, η, δ)-explosive – mostly even globally explosive. In the latter case, our proofs imply that we truly have an explosive behaviour in the set S: for every $\varepsilon, \varepsilon' > 0$ there is a constant C and a time t (that may depend on the process at hand) such that at time t there are at most $\varepsilon|S|$ active vertices in S, while at time $t + \varepsilon'$ there are at least $(\alpha - \varepsilon)|S|$ active vertices in S. Since we know that the final number of active vertices is at most $(\alpha + \varepsilon)|S|$, we may informally restate this fact as follows. For sufficiently large C and sufficiently small η, in a (C, η, δ)-explosive process all but an arbitrarily small fraction of the activations in S will happen in an arbitrarily short time interval.

---

[2] By slight abuse of notation, as t is not unique.

Recall that the C'-core of a graph is defined as the largest subgraph in which all vertices have degree at least C'. In the following, we show that for every $C, \eta > 0$, *scale free networks* are locally (C, η, δ)-explosive if we choose S to be the C'-core for some constant $C' = C'(C, \eta, \delta)$.

Theorem 2. *Let $G = G_n = (V_n, E_n)$ be a Chung-Lu power law graph with exponent $\beta \in (2,3)$. Moreover, let $a = a_n$ be such that $a_n \in \omega(n^{(\beta-2)/(\beta-1)})$ and $a_n \in o(n)$, and let $A = A_n \subset V_n$ be a random set of size a. Then for all constants $C, \eta, \delta > 0$ there exists $C' > 0$ such that w.h.p. G with starting set A is globally (C, η, δ)-explosive for the C'-core of G. In particular, by Theorem 1 for every $\varepsilon > 0$ w.h.p. the fraction of active vertices in the C'-core is $\alpha \pm \varepsilon$ for sufficiently large C', where α is given by Equation (1).*

It should not be surprising to see that for normalization in power-law graphs we need to restrict ourselves to large degree vertices. Most generative models for power-law graphs (including the Chung-Lu model) contain linearly many vertices with degrees strictly less than k. Typically, there are even linearly many isolated vertices. It is clear that none of these vertices can be activated by a bootstrap percolation process unless they are in the initial active set. In Section 5, we develop a heuristic estimate on the fraction of low degree vertices that finally become active. The fraction of active vertices can be as small as 0 (for vertices with degree less than k), and approaches α as the degree grows. Note, however, that the exact fraction for low degree vertices that become active depends on the degree distribution of the graph, whereas for high degree vertices this fraction is a universal constant that is independent of the graph structure.

4   Proofs

This section contains the proofs of Theorem 1 and Theorem 2. Due to space restrictions, we only give rough sketches. Full proofs can be found in the appendix. The proof of Theorem 1 takes up ideas from [12], where it was proven that bootstrap percolation with inhibition is normalizing on Erdős-Rényi graphs.

Proof of Theorem 1

We start with some basic facts about the percolation process. Fix some vertex $v \in V$, and recall that $\Gamma_s^+(v)$ and $\Gamma_s^-(v)$ are the number of excitatory and inhibitory vertices among the first s active neighbors of v, respectively. When the process starts, we do not need to decide right away for the signs of all the vertices in V. Rather, we can postpone the decision until a vertex becomes active. So whenever a neighbor of v turns active, it flips a coin to decide on its sign, and this coin flip is independent of any other coin flips. Hence, the number of inhibitory vertices among its first s active neighbors is binomially distributed $\text{Bin}(s, \tau)$. By the Chernoff bounds, if s is sufficiently large then $\Gamma_s^+(v)$ and $\Gamma_s^-(v)$ are concentrated around their expectations $(1 - \tau)s$ and τs, respectively.

We will link the probability that a vertex $v \in S$ becomes active with a random walk on $\mathbb{Z}$, using the following fact about random walks (see [14], Problem 5.3.1.).

Lemma 1. *Let $X_1, X_2, \ldots, X_n$ be a sequence of independent random variables, each of which is equal to 1 with probability $p \in [0, 1]$ and -1 otherwise. Consider the biased random walk $Z_i = X_1 + X_2 + \cdots + X_i$. Then there exists for every $\varepsilon > 0$ and $k \in \mathbf{N}$ a constant $C_0 = C_0(\varepsilon, k)$ such that the following is true:*

$$\Pr[\exists i \leq C_0 \ s.t. \ Z_i = k] \in (1 \pm \varepsilon) \cdot \min \left\{ 1, p^k / (1 - p)^k \right\}.$$

Recall that $X_i(v)$ is 1 if the i-th signal arriving in v is excitatory, and -1 otherwise, and let $Z_i(v) := X_1(v) + X_2(v) + \cdots + X_i(v)$. We know that the vertex v becomes active with the arrival of the first signal that causes $Z_i(v)$ to become k, if such a signal exists. We will show that $Z_i(v)$ follows (essentially) a one-dimensional random walk with bias τ.

There are two problems which complicate the analysis: the first being that the processes $(Z_i(v))_{i \in \mathbb{N}}$ and $(Z_i(u))_{i \in \mathbb{N}}$ are not independent for different vertices u and v, and the second being that for a fixed vertex v, the variables $X_i(v)$ and $X_j(v)$ are not independent for $i \neq j$, meaning that $(Z_i(v))_{i \in \mathbb{N}}$ is not a true random walk.

We overcome these problems as follows. Fix some large constant $\tilde{C} > 0$. Since the process is explosive, for the typical vertex v there exists a time at which v has at least $\tilde{C}$ active neighbors, but has not yet received any signals (if all $\tilde{C}$ neighbors become active in a sufficiently small time interval, then the signals are very unlikely to arrive within this interval). If $\tilde{C}$ is sufficiently large, then the fraction of positive signals among all signals on their way will be roughly $1 - \tau$. Since the transmission delays are distributed with an exponential distribution, which is *memoryless*, all the signals on their way are equally likely to arrive first. In particular, $X_1(v)$ is positive with probability roughly $1 - \tau$, and this holds *independent* of the sign of the first incoming signal of other vertices. After the first signal has arrived, the fraction of positive signals on their way will still be roughly $1 - \tau$, since removing a single signal has only a very small impact on this fraction. Thus $X_2(v)$ is also positive with probability $\approx 1 - \tau$, and the same holds for the first few incoming signals. Therefore, for small i the random variable $Z_i(v)$ resembles a one-dimensional random walk as in Lemma 1. As i grows larger, $Z_i(v)$ fails to follow a random walk, but then $Z_i(v)$ is typically already very negative. Thus we can show directly that most likely $Z_i(v)$ never becomes positive again. The details are rather involved, and we omit them due to space restrictions.

We remark that if we would assign positive or negative labels to the *edges* instead of the vertices, then similar, but substantially simpler arguments apply. E.g., the one-dimensional random walks for different vertices are independent of each other, so it is not necessary to condition on the history of the process.

4.1   Proof of Theorem 2

In this section we prove Theorem 2, which states that bootstrap percolation on a power law random graph G is locally (C, η, δ)-explosive for the C'-core of G. We remark that the condition $\mathfrak{a} \gg n^{(\beta-2)/(\beta-1)}$ is necessary since the threshold for bootstrap percolation without inhibition in Chung-Lu graphs is $n^{(\beta-2)/(\beta-1)}$, see [2, Theorem 2.3]. We want to apply Theorem 1. The main idea to see that bootstrap percolation is explosive is to observe that the C-core contains an Erdős-Rényi random graph $G(n', p)$, where n' and p depend on C. In order to prove Theorem 2, we will use the following statement about bootstrap percolation in Erdős-Rényi random graphs $G(n, p)$ with n vertices, where each edge is present independently of each other with probability p. For convenience, let

$$\Lambda := \left(\frac{(k-1)!}{(1-\tau)^k np^k} \right)^{1/(k-1)}. \tag{2}$$

In [12, Theorem 2] it was shown that the threshold for bootstrap percolation is $(1 - 1/k)\Lambda$, i.e., w.h.p. a random set of size $a_0 = (1 + \varepsilon)(1 - 1/k)\Lambda$ will activate almost all of the graph, while for a random set of size $a_0 = (1 - \varepsilon)(1 - 1/k)\Lambda$ the bootstrap percolation process dies out with $a^* \leq 2a_0$. Moreover, it was shown that the bootstrap percolation process on $G(n, p)$ is globally explosive and normalizing. More precisely, we have the following.

Theorem 3. *For every $\varepsilon > 0$ there is a constant $D > 0$ such that the following holds. Assume $D/n \leq p \ll n^{-1/k}$. Let G contain an Erdős-Rényi random graph $G(n, p)$. Let $x \geq D$, and let A be a random set of size $|A| = x\Lambda$. Then with probability at least $1 - O(1/\log x)$ the bootstrap percolation process on G with starting set A activates at least $(\alpha - \varepsilon)n$ vertices in time $O(x^{-1/(2k)} + (pn)^{-1})$.*

Apart from some technical differences, Theorem 3 differs from the statement proven in [12] in an important aspect: there the statement was only proven for the case $G = G(n, p)$, while we only require the graph to *contain* a $G(n, p)$. Note that this a non-trivial extension since additional edges may obstruct percolation due to inhibition. For this reason the proof becomes more subtle at some points even though the main line of argument remains similar to the proof in [12].

We are now ready to prove Theorem 2. The key observation is that the vertices of weight at least w induce a graph that contains an Erdős-Rényi graph as a subgraph, and thus by Theorem 3 percolation is explosive. However, we cannot immediately use $w = C'$ since then the size of the initial set would be below threshold, so we need to iterate the argument for several values of w. For convenience, let the *w-weight core* $G_{\geq w}$ of a Chung-Lu power law graph be the subgraph induced by all vertices of weight at least w. Since the weight corresponds to the expected degree, this notion is closely related to the w-core.

Proof (of Theorem 2). Let $C, \eta > 0$. Further let $\varepsilon > 0$, and let $D > 0$ be the constant given by Theorem 3. For sake of exposition, we assume that the i-th weight is given by $(n/i)^{1/(\beta-1)}$, i.e., that the average degree is $\bar{d} = (\beta-1)/(\beta-2)$.

Other values of $\bar{d}$ will change the calculations below only by constant factors. Since $2 < \beta < 3$, we may choose $\gamma > 0$ such that $(\beta - 1)/2 < \gamma < 1$. Let $\gamma' := (\beta-1)(k-1)/(2k+1-\beta)$. It is easy to check that $0 < \gamma' < (\beta-1)/2 < \gamma < 1$. For all $i \in \mathbb{N}$, let $w_i := n^{\gamma^i/(\beta-1)}$. Then the w_i-weight core G_i has size n_i, where n_i is given by the equation $(n/n_i)^{1/(\beta-1)} = w_i$. We easily deduce $n_i = n^{1-\gamma^i}$. Any two vertices in G_i have weight at least w_i, so they are connected with probability at least $p_i := \min\{1, w_i^2/n\} = \min\{1, n^{-1+2\gamma^i/(\beta-1)}\}$.

Let θ_i be the threshold for percolation in an Erdős-Rényi random graph $G(n_i, p_i)$. By Theorem 3 we know that there is an absolute constant $\tilde{C}$ (depending only on k and τ) such that $\theta_i = \tilde{C}^{-1} \cdot (n_i p_i^k)^{-1/(k-1)} = \tilde{C}^{-1} \cdot (n^{1-\gamma^i/\gamma'})$. Therefore, $n_{i-1}/\theta_i = \tilde{C} \cdot ((n/n_{i-1})^{(-1+\gamma/\gamma')})$ for all i. In particular, for $\tilde{D} := ((\alpha - \varepsilon)\tilde{C}/D)^{\gamma'/(\gamma-\gamma')}$ note that $(\alpha - \varepsilon)n_{i-1} > D\theta_i$ whenever $n_{i-1} \leq \tilde{D}n$. Let i_0 be the some index such that $n_{i_0-1} \leq \tilde{D}n$. Note that equivalently i_0 satisfies $n^{-\gamma^{i_0-1}} < \tilde{D}$.

We show inductively that we have explosive percolation on G_i for all $1 \leq i \leq i_0$. For $i = 1$, the number of vertices in G_1 that are initially active is $n^{-1+(\beta-2)(\beta-1)}n_1 = n^{\gamma-(\beta-1)/2} = n^{\Omega(1)} = n_1^{\Omega(1)}$. Any two vertices in G_1 are connected with probability $p_1 = 1$. We apply Theorem 1 (with some probability $p_1' = n_1^{-1/k-\varepsilon} < p_1$ to ensure the condition $p_1' \ll n_1^{-1/k}$), and obtain that an $(\alpha - \varepsilon)$-fraction of G_1 is activated after time $o(1)$. For convenience, set the time $t = 0$ to be the time *after* this phase, since we may safely ignore an additional time of $o(1)$.

For the inductive step, assume that an $(\alpha - \varepsilon)$-fraction of G_{i-1} is activated at some time $t_{i-1} < \eta$. Obviously G_{i-1} is a subgraph of G_i, so there are at least $(\alpha - \varepsilon)n_{i-1}$ active vertices in G_i. Since $(\alpha - \varepsilon)n_{i-1} > D\theta_i$ for $i - 1 < i_0$, we may apply Theorem 3 to deduce that bootstrap percolation activates with probability at least $1 - q_i$ at least an $(\alpha-\varepsilon)$-fraction of G_{i-1} until time $t_{i-1}+\Delta_i$, where $q_i = O(1/\log(n_{i-1}/\theta_i))$ and $\Delta_i = O((n_{i-1}/\theta_i)^{-1/(2k)} + (p_i n_i)^{-1})$. Since $n_{i-1}/\theta_i = \Theta(n^{\gamma^i(\gamma-\gamma')/\gamma'})$ and $p_i n_i = \Omega(n^{\gamma^i(3-\beta)/(\beta-1)})$, let $c := \min\{(\gamma - \gamma')/(2k\gamma'), (3 - \beta)/(\beta - 1)\} > 0$. Then

$$t_{i_0} = \sum_{i=1}^{i_0} \Delta_i = O\left(\sum_{i=1}^{i_0} n^{-c\gamma^i}\right) = O\left(n^{-c\gamma^{i_0}}\right) = O((n_{i_0}/n)^c).$$

In particular, by choosing i_0 such that n_{i_0}/n is a sufficiently small constant, we can achieve $t_{i_0} \leq \eta$. Similarly, since $q_i = O((\gamma^i \log n)^{-1})$, the accumulated error probability is $\sum_{i=1}^{i_0} q_i = O(q_{i_0}) = O(1/\log(n/n_{i_0}))$, and again by choosing i_0 such that n_{i_0}/n is a sufficiently small constant, we can achieve $\sum_{i=1}^{i_0} q_i \leq \delta$.

This proves that with probability at least $1-\delta$ bootstrap percolation activates an $(\alpha - \varepsilon)$-fraction of the w_{i_0}-weight core of G. Finally observe that w_{i_0} is a constant if n_{i_0}/n is constant. Now choose $C' := c\max\{2C/(\alpha-\varepsilon), w_{i_0}\}$ for some $c > 1$. Observe that for suitable c at least a $(1 - \eta)$-fraction of the C'-core is also in the w_{i_0}-weight core, and each of those vertices has at time η an expected number of at least $(1-\eta)C'(\alpha-\varepsilon) \geq cC$ active neighbors in the w_{i_0}-weight core. Thus the bootstrap percolation process is (C, η, δ)-explosive for the C'-core.

5   Experiments

Note that Theorem 2 explains what happens to the set of high degree vertices. However, many interesting graphs have a quite large number of low-degree vertices. To this end, we first develop a heuristic estimate for what happens in the remainder of the graph.

5.1   Heuristic Estimation

The proof of Theorem 1 shows that if a process is explosive for a set S then the vertices in S follow closely a random walk with bias $1 - \tau$ (cf. Lemma 1). For our estimation we will assume that this is in fact true for all vertices in G. So consider a random walk with bias $1 - \tau$. Let $p_{i,k,j}$ denote the probability of reaching k within i steps if we start from j. Clearly we have $p_{0,k,k} = 1$ and $p_{0,k,j} = 0$ for all $j < k$. For all $i \geq 1$ we can recursively compute $p_{i,k,j}$ as follows:

$$
p_{i,k,j} = \begin{cases} 1; & j = k \\ (1 - \tau) \cdot p_{i-1,k,j+1} + \tau \cdot p_{i-1,k,j-1} & \text{otherwise} \end{cases}
$$

Assume we know that (after percolation has terminated) a $\hat{p}$-fraction of all edges starts at an active vertex. Then we can assume that the probability that a vertex in $V \setminus A$ with degree i has been activated is given by $\sum_{j=k}^{i} \Pr[\text{Bin}(i, \hat{p}) = j] \cdot p_{j,k,0}$. Using this idea we can now set up an approximation for a given graph $G = (V, E)$ with n vertices and m edges. Let n_i denote the number of vertices with degree i, for $0 \leq i \leq n$, and let $p_{\text{boot}} = a_0/n$ be the probability that a vertex belongs to the initial active set A. Then $\hat{p}$ satisfies the following equation:

$$
\hat{p} \cdot m = \sum_{i=0}^{n} \left((p_{\text{boot}} i n_i) + (1 - p_{\text{boot}}) \cdot \sum_{j=k}^{i} \Pr[\text{Bin}(i, \hat{p}) = j] \cdot p_{j,k,0} \cdot i \cdot n_i \right). \quad (3)
$$

By Equation (3), we can numerically compute $\hat{p}$. Afterwards, we can estimate the number a^* of vertices that will turn active:

$$
a^* \approx \sum_{i=0}^{n} \left(p_{\text{boot}} \cdot n_i + (1 - p_{\text{boot}}) \cdot \sum_{j=k}^{i} \Pr[\text{Bin}(i, \hat{p}) = j] \cdot p_{j,k,0} \cdot n_i \right). \quad (4)
$$

Note that the i-th summand in (4) predicts the number of vertices of degree i, so we also obtain an estimate for the number of active vertices of a given degree.

5.2   Simulations

To test Equation (4), we compare it with simulations. We use the Epinion social network [19] that describes the trust relationship between its members. The size of the network is 75879, but we only consider its largest connected component with size $n = 75877$. We use $k = 4$ as activation parameter, and we start with a random active set A of size $a_0 = 2000$. In all the experiments, we simulate the asynchronous percolation process for different values of τ ranging from 0.1 to 0.9 with the step size of 0.1. For each value of τ we run the diffusion process 20

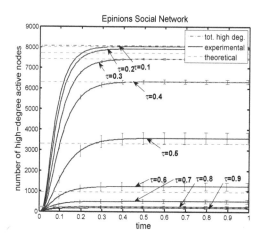

(a) Percolation process (high-degree vertices)

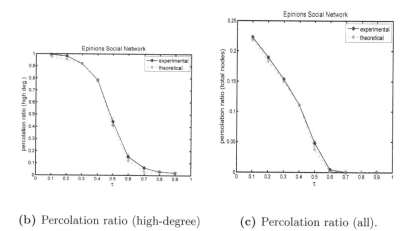

(b) Percolation ratio (high-degree)   (c) Percolation ratio (all).

Fig. 1. Asynchronous bootstrap percolation in the Epinions social network. *(a)* shows the evolution of the process for different values of τ among high degree vertices, i.e., those with degrees at least 16. *Red dashed line:* number of high-degree vertices in $V \setminus A$; *blue lines:* growth of number of active high-degree vertices in $V \setminus A$ over time; *green dashed lines:* prediction from Theorem 1 (Equation (1)) for the C'-core if C' is sufficiently large. *(b)* shows the active fraction (blue lines) among all high-degree vertices in $V \setminus A$ after termination, and the prediction from Equation (1) (green lines), cf. above. *(c)* shows the corresponding fraction among all vertices in $V \setminus A$, and the prediction from equation (4).

times and report the average size of the final active vertices along with standard deviations. We plot the fraction of active vertices within $V \setminus A$ (or within $\{v \in V \setminus A \mid \deg(v) \geq 16\}$ for Figures 1a and 1b).

Acknowledgements. We thank Qi Zhang for her help in preparing Figure 1.

References

1. Adamic, L.A., Huberman, B.A.: Power-law distribution of the world wide web. Science **287**(5461), 2115–2115 (2000)
2. Amini, H., Fountoulakis, N.: Bootstrap Percolation in Power-Law Random Graphs. Journal of Statistical Physics **155**(1), 72–92 (2014)
3. Balogh, J., Bollobás, B., Duminil-Copin, H., Morris, R.: The sharp threshold for bootstrap percolation in all dimensions. Transactions of the American Mathematical Society (2012)
4. Balogh, J., Bollobás, B., Morris, R.: Bootstrap percolation in three dimensions. The Annals of Probability, 1329–1380 (2009)
5. Balogh, J., Peres, Y., Pete, G.: Bootstrap percolation on infinite trees and non-amenable groups. Combinatorics, Probability & Computing **15**(5), 715–730 (2006)
6. Balogh, J., Pittel, B.G.: Bootstrap percolation on the random regular graph. Random Structures & Algorithms **30**(1–2), 257–286 (2007)
7. Barabási, A.-L., Albert, R.: Emergence of scaling in random networks. Science **286**(5439), 509–512 (1999)
8. Breskin, I., Soriano, J., Moses, E., Tlusty, T.: Percolation in living neural networks. Physical Review Letters (2006)
9. Chalupa, J., Leath, P., Reich, G.: Bootstrap percolation on a bethe lattice. Journal of Physics C: Solid State Physics **12**(1), L31 (1979)
10. Chung, F., Lu, L.: The average distances in random graphs with given expected degrees. Proceedings of the National Academy of Sciences **99**(25), 15879–15882 (2002)
11. Dorogovtsev, S.N., Goltsev, A.V., Mendes, J.F.: Critical phenomena in complex networks. Reviews of Modern Physics **80**(4), 1275 (2008)
12. Einarsson, H., Lengler, J., Panagiotou, K., Mousset, F., Steger, A.: Bootstrap percolation with inhibition (2014). arXiv preprint arXiv:1410.3291
13. Faloutsos, M., Faloutsos, P., and Faloutsos, C. On power-law relationships of the internet topology. In: ACM SIGCOMM Computer Communication Review, vol. 29, pp. 251–262. ACM (1999)
14. Grimmett, G., Stirzaker, D.: One Thousand Exercises in Probability. OUP Oxford (2001)
15. Holroyd, A.E.: Sharp metastability threshold for two-dimensional bootstrap percolation. Probability Theory and Related Fields **125**(2), 195–224 (2003)
16. Kempe, D., Kleinberg, J., Tardos, É.: Maximizing the spread of influence through a social network. In: ACM SIGKDD Knowledge Discovery and Data Mining, pp. 137–146. ACM (2003)
17. Lelarge, M.: Diffusion and cascading behavior in random networks. Games and Economic Behavior **75**(2), 752–775 (2012)
18. Moore, C., Newman, M.E.: Epidemics and percolation in small-world networks. Physical Review E **61**(5), 5678 (2000)

19. Richardson, M., Agrawal, R., Domingos, P.: Trust management for the semantic web. In: Fensel, D., Sycara, K., Mylopoulos, J. (eds.) ISWC 2003. LNCS, vol. 2870, pp. 351–368. Springer, Heidelberg (2003)
20. Toninelli, C., Biroli, G., Fisher, D.S.: Jamming percolation and glass transitions in lattice models. Physical review letters **96**(3), 035702 (2006)
21. Ugander, J., Karrer, B., Backstrom, L., Marlow, C.: The anatomy of the facebook social graph (2011). arXiv preprint arXiv:1111.4503

Broadcast from Minicast Secure Against General Adversaries

Pavel Raykov[1,2(✉)]

[1] School of Electrical Engineering, Tel-Aviv University, Tel-Aviv, Israel
[2] ITMO University, 49 Kronverkskiy av., Saint-Petersburg 197101, Russia
pavelraykov@post.tau.ac.il

Abstract. Byzantine broadcast is a distributed primitive that allows a specific party to consistently distribute a message among n parties in the presence of potential misbehavior of up to t of the parties. The celebrated result of [PSL80] shows that broadcast is achievable from point-to-point channels if and only if $t < n/3$.

The following two generalizations have been proposed to the original broadcast problem. In [FM98] the authors considered a *general adversary* characterized by the sets of parties that can be corrupted. It was shown that broadcast is achievable from point-to-point channels if and only if no three possible corrupted sets can cover the whole party set. In [CFF+05] the notion of point-to-point channels has been extended to the b-minicast channels allowing to locally broadcast among any subset of b parties. It has been shown that broadcast secure against adversaries corrupting up to t parties is achievable from b-minicast if and only if $t < \frac{b-1}{b+1}n$.

In this paper we combine both generalizations by considering the problem of achieving broadcast from b-minicast channels secure against general adversaries. Our main result is a condition on the possible corrupted sets such that broadcast is achievable from b-minicast if and only if this condition holds.

1 Introduction

1.1 Byzantine Broadcast

The Byzantine broadcast problem (aka Byzantine generals) is formulated as follows [PSL80]: A specific party (the sender) wants to distribute a message among n parties in such a way that all correct parties obtain the same message, even when some of the parties are malicious. The malicious misbehavior is modeled by a central adversary corrupting up to t parties and taking full control of their actions. Corrupted parties are called *Byzantine* and the remaining parties are

---

The unabridged version of this paper appears in [Ray15].

P. Raykov – Supported by ISF grant 1155/11, Israel Ministry of Science and Technology (grant 3-9094), GIF grant 1152/2011, and the Check Point Institute for Information Security.

© Springer-Verlag Berlin Heidelberg 2015
M.M. Halldórsson et al. (Eds.): ICALP 2015, Part II, LNCS 9135, pp. 701–712, 2015.
DOI: 10.1007/978-3-662-47666-6_56

called *correct*. Broadcast requires that all correct parties agree on the same value v, and if the sender is correct, then v is the value proposed by the sender. Broadcast is one of the most fundamental primitives in distributed computing. It is used to implement various protocols like voting, bidding, collective contract signing, etc. Basically, this list can be continued with all protocols for secure multi-party computation (MPC) as defined in [Yao82, GMW87].

There exist various implementations of Byzantine broadcast from synchronous point-to-point communication channels with different security guarantees. In the model without trusted setup, perfectly-secure Byzantine broadcast is achievable if and only if $t < n/3$ [PSL80, BGP92, CW92]. In the model with trusted setup, information-theoretically or cryptographically secure Byzantine broadcast is achievable for any $t < n$ [DS83, PW96].

1.2 Extending the Broadcast Problem

We consider the following two extensions of the broadcast problem.

b-**Minicast Communication Model.** The classical Byzantine broadcast problem [PSL80] assumes that the parties communicate with point-to-point channels only. Fitzi and Maurer [FM00] considered a model where the parties can additionally access partial broadcast channels among any set of b parties. We call such a partial broadcast channel b-*minicast*. Then, one can interpret the point-to-point communication model as the 2-minicast model. Hence, the result of [PSL80] states that broadcast is achievable from 2-minicast if and only if the number of corrupted parties $t < n/3$. In [FM00] it is shown that broadcast is achievable from 3-minicast if and only if $t < n/2$. Later this was generalized for the arbitrary b-minicast model—in [CFF+05] it is proved that broadcast is achievable from b-minicast if and only if $t < \frac{b-1}{b+1}n$.

It has also been studied how many 3-minicast channels need to be available in order to achieve broadcast [JMS12]. A general MPC protocol in the asynchronous model using 3-minicasts has been considered in [BBCK14].

General Adversaries. Originally, the broadcast problem has been stated for a *threshold* adversary that can corrupt any set of parties A such that $|A| \leq t$ for some threshold t. Fitzi and Maurer [FM98] considered a model with a *general* adversary that can corrupt a set of parties A such that $A \in \mathcal{A}$ for some family of possible corrupted sets $\mathcal{A}$. In [FM98] it has been shown that broadcast is achievable from point-to-point channels if and only if the adversary cannot corrupt three sets of parties from $\mathcal{A}$ that cover the whole party set.

A MPC protocol secure against general adversaries is given in [HM97]. More efficient MPC protocols secure against general adversaries are studied in [Mau02, HMZ08, HT13, LO14].

One of the most prominent approaches to construction of protocols secure against general adversaries is the "player emulation" technique of [HM97]. Its main idea is a generation of a new set of "virtual parties" $\mathcal{V}$ that are emulated by the original parties $\mathcal{P}$. Then the problem of constructing a protocol among $\mathcal{P}$ is reduced to a protocol construction among $\mathcal{V}$. As an example application of the player emulation technique, consider the protocols by [RVS+04, CDI+13]. These protocols construct broadcast from 3-minicast and tolerate any general adversary who cannot corrupt two sets of parties that cover the whole party set. In [RVS+04, CDI+13], triples of actual parties from $\mathcal{P}$ are used to emulate virtual parties, where the emulation protocol is implemented with the help of 3-minicast.

1.3 Contributions

We consider the combination of the two described extensions for the broadcast problem. That is, we study which general adversaries can be tolerated while constructing broadcast from b-minicast channels. We completely resolve this question by **(1)** giving a condition on a general adversary that can be tolerated while implementing broadcast from b-minicast and **(2)** showing that this condition is tight. Our results improve the previous work on generalized adversaries in the minicast communication model [FM00, RVS+04, CDI+13]. For example, consider a setting with 4 parties P_1, P_2, P_3, P_4 in the 3-minicast model. In Table 1 we illustrate for which general adversaries our results are new.

To show **(1)** we construct a protocol that realizes broadcast from minicast and is secure against general adversaries. The protocol we give does not employ the player emulation technique and is inspired by the original protocol of [CFF+05] that is secure against threshold adversaries. To show **(2)** we reduce a protocol secure against a general adversary to a protocol that is secure against a threshold adversary such that the former is impossible according to [CFF+05].

2 Model and Definitions

Parties. We consider a setting with n parties (that are also called players) $\mathcal{P} = \{P_1, \ldots, P_n\}$. We employ the following notion of party set partition:

Definition 1. *A list $\mathcal{S} = (S_0, \ldots, S_{k-1})$ is a k-partition of $\mathcal{P}$ if $\bigcup_{i=0}^{k-1} S_i = \mathcal{P}$ and all S_i, S_j are pair-wise disjoint. Furthermore, if all S_i are non-empty we call such a partition* proper.

We introduce additional notation to denote the set of parties from $\mathcal{P}$ without two sets S_i and S_j from the partition $\mathcal{S}$: Let $\mathcal{S}_{\downarrow i,j} := \mathcal{P} \setminus (S_{i \bmod k} \cup S_{j \bmod k})$.

Adversary. We assume that some of the parties can be corrupted by a central adversary making them deviate from the prescribed protocol in any desired manner. Before the execution of the protocol the adversary must specify the set of parties $A \subseteq \mathcal{P}$ to corrupt. The choice of the adversary is limited by means of a family of possible corrupted sets $\mathcal{A} \subseteq 2^{\mathcal{P}}$, i.e., the corrupted set A can be chosen only from $\mathcal{A}$. We assume that $\mathcal{A}$ is monotone, i.e., for $\forall S, S'$ $(S \in \mathcal{A}) \wedge (S' \subseteq S) \Rightarrow S' \in \mathcal{A}$. The set $\mathcal{A}$ is called an *adversarial structure*. We consider *perfect security* which captures the fact that the protocol never fails even if the adversary has unbounded computing power.

Table 1. The overview of tolerable general adversaries in the 3-minicast model. In the first column each maximal corrupted set is shown with an oval. In the third column we give a reference to the first work to show whether one can construct broadcast from 3-minicasts in the setting where any of the possible corrupted sets can be controlled by the adversary.

Possible corrupted sets	Broadcast possible?	Literature
	No	[FM00]
	No	**This work**
	Yes	**This work**
	Yes	[RVS+04, CDI+13]
	Yes	[FM00]

The parties that are corrupted are also called *Byzantine* or *malicious*, while the remaining uncorrupted parties are called *correct* or *honest*.

Communication. In the classical setting [PSL80], it is assumed that the parties are connected with a synchronous authenticated point-to-point network. Synchronous means that all parties share a common clock and that the message delay in the network is bounded by a constant. In this paper we consider an extended model where messages can be consistently delivered to more than one recipient via a b-minicast channel.

Definition 2 (b-minicast model). *In the b-minicast model, in every subset $Q \subseteq \mathcal{P}$ of at most b parties each party $P_i \in Q$ has access to a b-minicast channel that takes an input v from some domain $\mathcal{D}$ from P_i and outputs v to all parties* *in Q. Each of the b-minicast channels is also synchronous, i.e., the message delivery delay to all recipients is bounded by a constant.*

Note that the classical setting with point-to-point channels can be seen as an instantiation of the b-minicast model for $b = 2$.

Broadcast Protocols. A broadcast protocol allows parties to simulate a global broadcast channel with the help of the communication means available in the presence of an adversary. Formally, this is defined as follows.

Definition 3 (Broadcast). *A protocol among the parties $\mathcal{P}$ where some party $P_s \in \mathcal{P}$ (called the sender) holds an input $v \in \mathcal{D}$ and every party $P_i \in \mathcal{P}$ outputs a value $y_i \in \mathcal{D}$ achieves broadcast if the following holds:*

VALIDITY: *If the sender P_s is correct, then every correct party $P_i \in \mathcal{P}$ outputs the sender's value $y_i = v$.*

CONSISTENCY: *All correct parties in $\mathcal{P}$ output the same value.*

3   The Main Result

We now characterize which adversary structures $\mathcal{A}$ can be tolerated while implementing broadcast from b-minicast channels. Our condition is inspired by the impossibility proof of [CFF+05]. There it is shown that no protocol can realize broadcast among $b + 1$ parties from b-minicast where the adversary can corrupt any number of parties. The impossibility proof of [CFF+05] is built by considering a chain of parties $P_1, \ldots, P_{b+1}$ where any pair of parties $P_i, P_{i+1 \bmod b+1}$ $(i = 1, \ldots, b + 1)$ can be honest, while the remaining parties are corrupted. We generalize this to a chain of party *sets*:

Definition 4. *An adversarial structure $\mathcal{A}$ is said to contain a k-chain if there exists a proper k-partition $\mathcal{S} = (S_0, \ldots, S_{k-1})$ of $\mathcal{P}$ such that $\forall i \in [0, k - 1]$ $\mathcal{S}_{\downarrow i, i+1} \in \mathcal{A}$. A structure is called k-chain-free if it does not have a k-chain.*

Our main result can be formulated as follows.

Theorem 1. *In the b-minicast communication model, broadcast tolerating adversary structure $\mathcal{A}$ is achievable if and only if $\mathcal{A}$ is $(b + 1)$-chain-free.*

The proof of the theorem is split into two parts. In Section 4 we give a protocol that realizes broadcast from b-minicast channels and tolerates any $(b+1)$-chain-free $\mathcal{A}$. In Section 5 we show that no protocol can implement broadcast in this model while tolerating some $\mathcal{A}$ that has a $(b + 1)$-chain. Some of the proofs are omitted and appear only in the full version of this paper [Ray15].

4   The Feasibility Proof

In this section we construct a broadcast protocol that uses b-minicast channels and tolerates any adversarial structure $\mathcal{A}$ which is $(b + 1)$-chain-free. Our construction consists of three steps and is based on [CFF+05]. First, we introduce a new distributed primitive called proxcast and show how to realize proxcast from b-minicast while tolerating arbitrary corruptions. Second, we consider a broadcast primitive with hybrid security, i.e., depending on which set of parties the adversary corrupts, the primitive satisfies only one of the broadcast security guarantees. We then show how to implement such a hybrid broadcast primitive from proxcast. Finally, given broadcast with hybrid security, we implement broadcast secure against any $\mathcal{A}$ which is $(b+1)$-chain-free.

4.1   Proxcast

Proxcast is a relaxed version of the binary broadcast primitive in that it has a weakened consistency property. As a result of a proxcast invocation, each of the parties P_i outputs a level ℓ_i from some range $[0, \ell - 1]$ indicating whether 0 was likely to be proxcast (lower levels) or 1 (higher levels). It is guaranteed that if the sender is correct, then any correct P_i outputs $\ell_i = 0$ if 0 is proxcast, and $\ell_i = \ell - 1$ if 1 is proxcast. The consistency property of proxcast guarantees that levels of correct parties are close, however, they may be different. Formally, we define proxcast as follows.

Definition 5 (ℓ-proxcast). *Let $\ell \in \mathbb{N}$. A protocol among $\mathcal{P}$ where the sender $P_s \in \mathcal{P}$ holds an input $v \in \{0, 1\}$ and every party $P_i \in \mathcal{P}$ finally outputs a level $\ell_i \in [0, \ell - 1]$, achieves ℓ-proxcast if the following holds:*

VALIDITY: *If P_s is correct then all correct P_i output $\ell_i = (\ell - 1)v$.*

CONSISTENCY: *There exists $k \in [0, \ell - 2]$ such that each correct P_i outputs level $\ell_i \in \{k, k + 1\}$.*

Our construction of b-proxcast from b-minicasts is a simplification of the proxcast protocol of [CFF+05]. We let the sender P_s b-minicast the value he holds among all subsets of b parties. Then each P_i computes the level ℓ_i to be the minimum number of parties with whom P_i sees only zeros.

Protocol Proxcast$_b(\mathcal{P}, P_s, v)$

1. If $|\mathcal{P}| \le b$ then broadcast v using b-minicast. Let y_i denote the output of P_i. Each P_i decides on $\ell_i := (b - 1)y_i$.
2. Otherwise:
   2.1 Let $\mathcal{B} = \{S \subseteq \mathcal{P} \mid (P_s \in S) \land (|S| = b)\}$.
       $\forall S \in \mathcal{B}$: P_s b-minicasts v among S. Let y_S be the output of each $P_i \in S$.
   2.2 $\forall P_i \in \mathcal{P} \setminus \{P_s\}$: For each $T \subseteq \mathcal{P} \setminus \{P_s, P_i\}$ with $|T| \le b - 2$, let $V_i^T := \{y_S \mid (S \in \mathcal{B}) \land (P_i \in S) \land (T \subseteq S)\}$. Output ℓ_i to be the minimum $|T|$ such that $V_i^T = \{0\}$ (if no such T exists output $\ell_i := b - 1$).
       The sender P_s: Output $\ell_s := v(b - 1)$.

Lemma 1. *In the b-minicast model, the protocol Proxcast$_b$ perfectly securely achieves b-proxcast in the presence of any adversary.*

4.2   Broadcast with Hybrid Security

In [FHHW03] the authors proposed a more fine-grained security definition of broadcast in the setting with a threshold adversary. The new primitive is called a two-threshold broadcast. In the two-threshold version of broadcast, the consistency and validity properties of broadcast are guaranteed to be achieved in the presence of up to t_c and t_v malicious parties, respectively. It is shown in [FHHW03] that one can implement two-threshold broadcast from point-to-point channels if and only if $t = 0$ or $t + 2T < n$ (where $t = \min(t_c, t_v)$ and $T = \max(t_c, t_v)$).

We consider an extended notion of the two-threshold broadcast with respect to general adversaries. We call the resulting primitive hybrid broadcast.[1]

Definition 6 (Hybrid broadcast). *A protocol among $\mathcal{P}$ where the sender $P_s \in \mathcal{P}$ holds an input $v \in \mathcal{D}$ and every party $P_i \in \mathcal{P}$ finally outputs a value $y_i \in \mathcal{D}$, achieves hybrid broadcast with respect to a pair of adversarial structures $(\mathcal{A}_c, \mathcal{A}_v)$ if the following holds:*

VALIDITY: *If the sender P_s is honest and a set $A \in \mathcal{A}_v$ of parties is corrupted then all honest $P_i \in \mathcal{P} \setminus A$ output sender's value $y_i = v$.*

CONSISTENCY: *If a set $A \in \mathcal{A}_c$ of parties is corrupted then all honest $P_i \in \mathcal{P} \setminus A$ output the same value.*

We construct a hybrid broadcast protocol in two steps. First, we introduce additional operations on adversary structures and prove properties of them. Then, we present the protocol.

Additional Tools. We start by defining two operators del and proj that transform adversarial structures. The first operator del chooses only the sets $A \in \mathcal{A}$ that do not contain a specific party P_i. The second operator proj selects only the sets $A \in \mathcal{A}$ that can be corrupted together with a specific party P_i. Formally, $\mathsf{del}(\mathcal{A}, P_i) := \{A \in \mathcal{A} \mid P_i \notin A\}$ and $\mathsf{proj}(\mathcal{A}, P_i) := \{A \in \mathsf{del}(\mathcal{A}, P_i) \mid A \cup \{P_i\} \in \mathcal{A}\}$. We prove that $\mathsf{del}(\mathcal{A}, P_i)$ and $\mathsf{proj}(\mathcal{A}, P_i)$ are adversarial structures over $\mathcal{P} \setminus \{P_i\}$.

Lemma 2. *If $\mathcal{A}$ is an adversarial structure over $\mathcal{P}$, then $\mathsf{del}(\mathcal{A}, P_i)$ and $\mathsf{proj}(\mathcal{A}, P_i)$ are adversarial structures over $\mathcal{P} \setminus \{P_i\}$.*

---

[1] Note that our extension for non-threshold adversaries is more general than the one given in [FHHW03], because we allow any $\mathcal{A}_c$ and $\mathcal{A}_v$, and not only $\mathcal{A}_c \subseteq \mathcal{A}_v$ or $\mathcal{A}_v \subseteq \mathcal{A}_c$ as [FHHW03]. This difference stems from the fact that we treat the broadcast security properties (validity and consistency) equally, while [FHHW03] considers the case where the broadcast security properties can be "degraded" if the adversarial power grows. That is, in [FHHW03], if the adversary corrupts A from some $\mathcal{A}_{small}$ then the broadcast protocol must satisfy both validity and consistency, while if the adversary corrupts A from some $\mathcal{A}_{big} \supseteq \mathcal{A}_{small}$, then only validity (or only consistency) is required to be satisfied.

We now give a condition on adversarial structures $(\mathcal{A}_c, \mathcal{A}_v)$, such that if this condition holds then there exists a protocol that constructs hybrid broadcast from b-minicasts and tolerates $(\mathcal{A}_c, \mathcal{A}_v)$. Similarly to the k-chain-free condition, it is inspired by the chain of parties considered in the impossibility proof of [CFF+05].

Definition 7. *A pair of structures $(\mathcal{A}_c, \mathcal{A}_v)$ is said to be k-chain-free-compatible if for any proper k-partition $\mathcal{S}$ of $\mathcal{P}$, there exists $i \in [0, k-3]$ such that $\mathcal{S}_{\downarrow i, i+1} \notin \mathcal{A}_c$ or $\mathcal{S}_{\downarrow k-2, k-1} \notin \mathcal{A}_v$ or $\mathcal{S}_{\downarrow k-1, 0} \notin \mathcal{A}_v$.*

We show that k-chain-free-compatible structures satisfy the following useful properties:

Lemma 3. *Let $(\mathcal{A}_c, \mathcal{A}_v)$ be k-chain-free-compatible and $P_i \in \mathcal{P}$. Then the pair $(\mathsf{proj}(\mathcal{A}_c, P_i), \mathsf{del}(\mathcal{A}_v, P_i))$ is also k-chain-free-compatible.*

Lemma 4. *If $\mathcal{A}$ is k-chain-free, then $(\mathcal{A}, \mathcal{A})$ is k-chain-free-compatible.*

The protocol. The protocol $\mathtt{HybridBC}$ works recursively as follows. First, the sender P_s proxcasts the value v he holds to everyone in $\mathcal{P}$. Then, each of the receivers in $\mathcal{P} \setminus \{P_s\}$ invokes $\mathtt{HybridBC}$ again to hybridly broadcast his level among the receivers. Now the view of every receiver P_i consists of $n-1$ levels of the others. Then, P_i partitions the receiver set $\mathcal{P} \setminus \{P_s\}$ into b subsets according to the level that is hybridly broadcast by the parties in the set. By analyzing the properties of this partition, each P_i takes his final decision.

Protocol $\mathtt{HybridBC}(\mathcal{P}, P_s, v, \mathcal{A}_c, \mathcal{A}_v)$

1. If $|\mathcal{P}| \le b$ then broadcast v using b-minicast.
2. Otherwise:
   2.1 Parties in $\mathcal{P}$ invoke $\mathtt{Proxcast}_b(\mathcal{P}, P_s, v)$. Let ℓ_i denote the output of P_i.
   2.2 Let $\mathcal{A}'_c := \mathsf{proj}(\mathcal{A}_c, P_s)$, $\mathcal{A}'_v := \mathsf{del}(\mathcal{A}_v, P_s)$ and $\mathcal{P}' := \mathcal{P} \setminus \{P_s\}$.
   2.3 $\forall P_j \in \mathcal{P}'$: Parties in $\mathcal{P}'$ invoke $\mathtt{HybridBC}(\mathcal{P}', P_j, \ell_j, \mathcal{A}'_c, \mathcal{A}'_v)$.[2]
   $\forall P_j \in \mathcal{P}'$: Let ℓ^i_j denote the output of P_i.
   2.4 $\forall P_i \in \mathcal{P}'$: For all $\ell \in [0, b-1]$ let $L^i_\ell := \{P_j \in \mathcal{P}' \mid \ell^i_j = \ell\}$. Let $L^i_b = \{P_s\}$. Let $\mathcal{L}^i = (L^i_0, \ldots, L^i_b)$ be a $(b+1)$-partition of $\mathcal{P}$.
   2.5 $\forall P_i \in \mathcal{P}'$: Output 0 if

   $$(\mathcal{L}^i_{\downarrow b, 0} \in \mathcal{A}_v) \wedge \left(\bigwedge_{k=0}^{\ell_i} L^i_k \ne \emptyset \right) \wedge \left(\bigwedge_{k=0}^{\ell_i-1} \mathcal{L}^i_{\downarrow k, k+1} \in \mathcal{A}_c \right).[3]$$

   Otherwise, output 1.
   2.6 The sender P_s outputs v.

---

[2] The protocol $\mathtt{HybridBC}$ works for binary values only. This invocation is translated into $\log b$ parallel invocations of $\mathtt{HybridBC}$ to broadcast ℓ_j bit by bit.

[3] We assume that $\bigwedge_{k=0}^{-1}(\ldots)$ is always true.

Lemma 5. *In the b-minicast model, the protocol* HybridBC *perfectly securely achieves binary hybrid broadcast if* $(\mathcal{A}_c, \mathcal{A}_v)$ *is* $(b+1)$-*chain-free-compatible.*

Proof. The proof proceeds by induction over the size of the party set $|\mathcal{P}| = n$. If $|\mathcal{P}| \le b$ then broadcast is directly achieved with the help of b-minicast. Now we prove the induction step. We assume that the protocol HybridBC achieves hybrid broadcast for any set of $n - 1$ parties.

Consider now any $(b+1)$-chain-free-compatible $(\mathcal{A}_c, \mathcal{A}_v)$ over the set of n parties $\mathcal{P}$. Due to Lemma 3, the pair $(\mathcal{A}'_c, \mathcal{A}'_v)$ computed at Step 2.2 is $(b+1)$-chain-free-compatible. Hence, we can assume that each recursive invocation of the protocol HybridBC at Step 2.3 achieves hybrid broadcast.

Now we prove each of the hybrid broadcast security properties:

VALIDITY: We assume that the sender P_s is correct, and we show that all correct parties in $\mathcal{P}$ output v. Because P_s always outputs v, we are left to show only that all correct receivers in $\mathcal{P}' = \mathcal{P} \backslash \{P_s\}$ output v. Assume that the adversary corrupts some $A \in \mathcal{A}_v$. Let $H = \mathcal{P}' \backslash A$ denote the set of the remaining honest receivers. We consider two cases:

- **($v = 0$)** Because P_s is correct, $\mathtt{Proxcast}_b$ guarantees that every $P_i \in H$ has $\ell_i = 0$. Now consider any correct P_x. We have that $H \subseteq L_0^x$. Because $\mathcal{A}_v$ is monotone and $\mathcal{P}' \backslash H \in \mathcal{A}_v$, we get that $\mathcal{P}' \backslash L_0^x \in \mathcal{A}_v$. Hence, $\mathcal{L}_{\downarrow b,0}^x \in \mathcal{A}_v$. We also have that $L_0^x \neq \emptyset$ since $P_x \in L_0^x$. Consequently, each correct $P_x \in H$ verifies that his $\mathcal{L}_{\downarrow b,0}^x \in \mathcal{A}_v$, $L_0^x \neq \emptyset$ and decides on 0 at Step 2.5.

- **($v = 1$)** Because P_s is correct, $\mathtt{Proxcast}_b$ guarantees that every $P_i \in H$ has $\ell_i = b - 1$. Now consider any correct P_x. We have that $H \subseteq L_{b-1}^x$. Because $\mathcal{A}_v$ is monotone and $\mathcal{P}' \backslash H \in \mathcal{A}_v$, we get that $\mathcal{P}' \backslash L_{b-1}^x \in \mathcal{A}_v$. Hence, $\mathcal{L}_{\downarrow b-1,b}^x \in \mathcal{A}_v$. Assume for the sake of contradiction that P_x decides on 0 instead of 1. This means that $\mathcal{L}_{\downarrow b,0}^x \in \mathcal{A}_v$, $\bigwedge_{k=0}^{b-1} (L_k^x \neq \emptyset)$ and $\bigwedge_{k=0}^{b-2} (\mathcal{L}_{\downarrow k,k+1}^x \in \mathcal{A}_c)$. Together with $\mathcal{L}_{\downarrow b-1,b}^x \in \mathcal{A}_v$ this implies that $\mathcal{L}^x$ is a proper $(b+1)$-partition of $\mathcal{P}$, showing that $(\mathcal{A}_c, \mathcal{A}_v)$ is not $(b+1)$-chain-free-compatible, a contradiction.

CONSISTENCY: If P_s is correct, then consistency holds because of the validity property. Assume now the adversary corrupts some $A \in \mathcal{A}_c$ such that $P_s \in A$. Let $H := \mathcal{P}' \backslash A$ denote the set of correct receivers. Because HybridBC satisfies the consistency property if $A \backslash \{P_s\} \in \mathcal{A}'_c$ is corrupted, we have that receivers in H compute the same set $\mathcal{L}$, i.e., for all $P_i, P_j \in H$ holds $\mathcal{L}^i = \mathcal{L}^j$.

Consider a party $P_i \in H$ with the smallest ℓ_i. The properties of $\mathtt{Proxcast}_b$ guarantee that any $P_j \in H$ has $\ell_j \in \{\ell_i, \ell_i + 1\}$. If $\ell_j = \ell_i$, then P_j decides on the same value as P_i at Step 2.5 because P_j uses the same $\mathcal{L}$. Assume now that $\ell_j = \ell_i + 1$. Consider two possible cases:

(P_i decides on 0) If P_i decides on 0, then

$$\left(\mathcal{L}_{\downarrow b,0} \in \mathcal{A}_v\right) \wedge \left(\bigwedge_{k=0}^{\ell_i} L_k^i \neq \emptyset\right) \wedge \left(\bigwedge_{k=0}^{\ell_i-1} \mathcal{L}_{\downarrow k,k+1} \in \mathcal{A}_c\right)$$

is true. Because any correct $P_x \in H$ has $\ell_x \in \{\ell_i, \ell_i + 1\}$, we have that $H \subseteq L_{\ell_i} \cup L_{\ell_i+1}$. Because $\mathcal{A}_c$ is monotone, we get that $\mathcal{P}' \setminus (L_{\ell_i} \cup L_{\ell_i+1}) \in \mathcal{A}_c$. Hence, $\mathcal{L}_{\downarrow \ell_i, \ell_i+1} \in \mathcal{A}_c$. Since $P_j \in L_{\ell_i+1}$, we have that $L_{\ell_i+1} \neq \emptyset$. Consequently, P_j verifies that

$$(\mathcal{L}_{\downarrow b,0} \in \mathcal{A}_v) \wedge \left(\bigwedge_{k=0}^{\ell_i+1} L_k^i \neq \emptyset \right) \wedge \left(\bigwedge_{k=0}^{\ell_i} \mathcal{L}_{\downarrow k, k+1} \in \mathcal{A}_c \right)$$

is true and decides on 0.

(P_i decides on 1) If P_i decides on 1, then the following formula is false:

$$(\mathcal{L}_{\downarrow b,0} \in \mathcal{A}_v) \wedge \left(\bigwedge_{k=0}^{\ell_i} L_k^i \neq \emptyset \right) \wedge \left(\bigwedge_{k=0}^{\ell_i-1} \mathcal{L}_{\downarrow k, k+1} \in \mathcal{A}_c \right).$$

If the above formula is false, then so is the one below:

$$(\mathcal{L}_{\downarrow b,0} \in \mathcal{A}_v) \wedge \left(\bigwedge_{k=0}^{\ell_i+1} L_k^i \neq \emptyset \right) \wedge \left(\bigwedge_{k=0}^{\ell_i} \mathcal{L}_{\downarrow k, k+1} \in \mathcal{A}_c \right).$$

Hence, P_j also decides on 1. □

4.3   The Broadcast Protocol

In order to achieve broadcast secure against an adversarial structure $\mathcal{A}$ which is $(b+1)$-chain-free, we let the parties invoke the hybridly secure broadcast protocol with $\mathcal{A}_c$ and $\mathcal{A}_v$ set to $\mathcal{A}$.

Protocol Broadcast$(\mathcal{P}, P_s, v, \mathcal{A})$
1. Parties $\mathcal{P}$ invoke HybridBC$(\mathcal{P}, P_s, v, \mathcal{A}, \mathcal{A})$. Let y_i denote the output each P_i receives.
2. $\forall P_i \in \mathcal{P}$: decide on y_i.

Note that the protocol Broadcast achieves broadcast for binary domains only. In order to achieve broadcast for arbitrary input domains efficiently one can use broadcast amplification protocols of [HMR14, HR14].

Lemma 6. *In the b-minicast model, the protocol Broadcast perfectly securely achieves broadcast if $\mathcal{A}$ is $(b+1)$-chain-free.*

5   The Impossibility Proof

We employ Lemma 2 of [CFF+05]:

Lemma 7. *In the b-minicast communication model, broadcast among $b+1$ parties $\{Q_0, \ldots, Q_b\}$ is not achievable if any pair Q_i, $Q_{(i+1) \bmod (b+1)}$ can be honest while the adversary corrupts the remaining parties.*

Now we proceed to the main impossibility statement (based on [CFF+05, Theorem2]).

Lemma 8. *In the b-minicast communication model, there is no secure broadcast protocol among $\mathcal{P}$ that tolerates an adversarial structure $\mathcal{A}$ which is not $(b+1)$-chain-free.*

Proof. For the sake of contradiction, assume that there exists a broadcast protocol π tolerating $\mathcal{A}$ which is not $(b+1)$-chain-free. Because $\mathcal{A}$ is not $(b+1)$-chain-free, there exists a proper $(b+1)$-partition $\boldsymbol{S} = (S_0, \ldots, S_b)$ of $\mathcal{P}$ such that all $\boldsymbol{S}_{\downarrow i,i+1} \in \mathcal{A}$. Using π, we now construct a protocol π' among $b+1$ parties $\{Q_0, \ldots, Q_b\}$ for achieving broadcast from b-minicast. The protocol π' lets each Q_i simulate parties in S_i. If a party Q_i is corrupted, then the simulated parties in S_i can behave arbitrarily. If a party Q_i is honest, then the simulated parties in S_i follow their protocol specification, i.e., behave correctly. Because π is secure against corruption of any set $\boldsymbol{S}_{\downarrow i,i+1}$, the protocol π' is secure whenever any pair $Q_i, Q_{i+1 \bmod b+1}$ is honest and the remaining parties are corrupted. This contradicts Lemma 7. □

6 Conclusions

We showed that broadcast secure against any adversarial structure $\mathcal{A}$ is achievable from b-minicast channels if and only if $\mathcal{A}$ is $(b+1)$-chain-free. This result is a generalization of [PSL80,FM98,FM00,CFF+05,RVS+04,CDI+13]. An interesting open question is to continue this line of research and study broadcast achieveability in communication models where only some subset of b-minicast channels is available.

Acknowledgments. We would like to thank Martin Hirt, Sandro Coretti and anonymous referees for their valuable comments about the paper.

References

[BBCK14] Backes, M., Bendun, F., Choudhury, A., Kate, A.: Asynchronous MPC with a strict honest majority using non-equivocation. In: PODC (2014)

[BGP92] Berman, P., Garay, J.A., Perry, K.J.: Bit optimal distributed consensus. In: Computer Science Research (1992)

[CDI+13] Cohen, G., Damgård, I.B., Ishai, Y., Kölker, J., Miltersen, P.B., Raz, R., Rothblum, R.D.: Efficient multiparty protocols via log-depth threshold formulae. In: Canetti, R., Garay, J.A. (eds.) CRYPTO 2013, Part II. LNCS, vol. 8043, pp. 185–202. Springer, Heidelberg (2013)

[CFF+05] Considine, J., Fitzi, M., Franklin, M., Levin, L.A., Maurer, U., Metcalf, D.: Byzantine agreement given partial broadcast. Journal of Cryptology (2005)

[CW92] Coan, B.A., Welch, J.L.: Modular construction of a byzantine agreement protocol with optimal message bit complexity. Inf. and Comp. (1992)

[DS83] Dolev, D., Strong, H.R.: Authenticated algorithms for Byzantine agreement. SIAM Journal on Computing (1983)

[FHHW03] Fitzi, M., Hirt, M., Holenstein, T., Wullschleger, J.: Two-threshold broadcast and detectable multi-party computation. In: Biham, E. (ed.) EUROCRYPT 2003. LNCS, vol. 2656, pp. 51–67. Springer, Heidelberg (2003)

[FM98] Fitzi, M., Maurer, U.M.: Efficient byzantine agreement secure against general adversaries. In: Kutten, S. (ed.) DISC 1998. LNCS, vol. 1499, pp. 134–148. Springer, Heidelberg (1998)

[FM00] Fitzi, M., Maurer, U.: From partial consistency to global broadcast. In: STOC (2000)

[GMW87] Goldreich, O., Micali, S., Wigderson, A.: How to play any mental game. In: STOC (1987)

[HM97] Hirt, M., Maurer, U.M.: Complete characterization of adversaries tolerable in secure multi-party computation. In: PODC (1997)

[HMR14] Hirt, M., Maurer, U., Raykov, P.: Broadcast amplification. In: Lindell, Y. (ed.) TCC 2014. LNCS, vol. 8349, pp. 419–439. Springer, Heidelberg (2014)

[HMZ08] Hirt, M., Maurer, U.M., Zikas, V.: MPC vs. SFE : unconditional and computational security. In: Pieprzyk, J. (ed.) ASIACRYPT 2008. LNCS, vol. 5350, pp. 1–18. Springer, Heidelberg (2008)

[HR14] Hirt, M., Raykov, P.: Multi-valued byzantine broadcast: the $t < n$ case. In: Sarkar, P., Iwata, T. (eds.) ASIACRYPT 2014, Part II. LNCS, vol. 8874, pp. 448–465. Springer, Heidelberg (2014)

[HT13] Hirt, M., Tschudi, D.: Efficient general-adversary multi-party computation. In: Sako, K., Sarkar, P. (eds.) ASIACRYPT 2013, Part II. LNCS, vol. 8270, pp. 181–200. Springer, Heidelberg (2013)

[JMS12] Jaffe, A., Moscibroda, T., Sen, S.: On the price of equivocation in byzantine agreement. In: PODC (2012)

[LO14] Lampkins, J., Ostrovsky, R.: Communication-efficient MPC for general adversary structures. In: Abdalla, M., De Prisco, R. (eds.) SCN 2014. LNCS, vol. 8642, pp. 155–174. Springer, Heidelberg (2014)

[Mau02] Maurer, U.M.: Secure multi-party computation made simple. In: Cimato, S., Galdi, C., Persiano, G. (eds.) SCN 2002. LNCS, vol. 2576, pp. 14–28. Springer, Heidelberg (2003)

[PSL80] Pease, M.C., Shostak, R.E., Lamport, L.: Reaching agreement in the presence of faults. Journal of the ACM (1980)

[PW96] Pfitzmann, B., Waidner, M.: Information-theoretic pseudosignatures and Byzantine agreement for $t \geq n/3$. Technical report, IBM Research (1996)

[Ray15] Raykov, P.: Broadcast from minicast secure against general adversaries (2015). Cryptology ePrint Archive, Report 2015/352 http://eprint.iacr.org/

[RVS+04] Ravikant, D.V.S., Muthuramakrishnan, V., Srikanth, V., Srinathan, K., Pandu Rangan, C.: On byzantine agreement over (2,3)-uniform hypergraphs. In: Guerraoui, R. (ed.) DISC 2004. LNCS, vol. 3274, pp. 450–464. Springer, Heidelberg (2004)

[Yao82] Yao, A.C.: Protocols for secure computations. In: FOCS (1982)

Author Index

Abramsky, Samson II-31
Achlioptas, Dimitris II-467
Agrawal, Shweta I-1
Ailon, Nir I-14
Aisenberg, James II-44
Albers, Susanne I-26
Alistarh, Dan II-479
Amanatidis, Georgios I-39
Amarilli, Antoine II-56
Aminof, Benjamin II-375
Anshelevich, Elliot I-52
Aronov, Boris I-65
Avigdor-Elgrabli, Noa I-78
Avin, Chen II-492
Azar, Yossi I-91

Beame, Paul I-103
Behsaz, Babak I-116
Bei, Xiaohui I-129
Beigi, Salman I-143
Beneš, Nikola II-69
Berkholz, Christoph I-155
Bernstein, Aaron I-167
Beyersdorff, Olaf I-180
Bezděk, Peter II-69
Bhangale, Amey I-193
Bhattacharya, Sayan I-206, II-504
Bienvenu, Laurent I-219
Björklund, Andreas I-231, I-243
Bodirsky, Manuel I-256
Bojańczyk, Mikołaj II-427
Bonet, Maria Luisa II-44
Boreale, Michele II-82
Bouajjani, Ahmed II-95
Bourhis, Pierre II-56
Bringmann, Karl II-516
Bun, Mark I-268
Burton, Benjamin A. I-281
Buss, Sam II-44

Canonne, Clément L. I-294
Cao, Yixin I-306
Charron-Bost, Bernadette II-528

Chatterjee, Krishnendu II-108, II-121
Chattopadhyay, Arkadev II-540
Chekuri, Chandra I-318
Chen, Ning I-129, II-552
Chew, Leroy I-180
Ciobanu, Laura II-134
Cohen, Aloni I-331
Cohen, Gil I-343
Cohen, Ilan Reuven I-91
Colcombet, Thomas II-146
Coudron, Matthew I-355
Crăciun, Adrian II-44
Cseh, Ágnes I-367
Curticapean, Radu I-380
Czyzowicz, Jurek I-393

Dahlgaard, Søren II-564
Dani, Varsha II-575
Datta, Samir II-159
Dell, Holger I-231
Desfontaines, Damien I-219
Diekert, Volker II-134
Disser, Yann I-406
Doron, Dean I-419
Doyen, Laurent II-108
Dubut, Jérémy II-171
Dvořák, Zdeněk I-432

Elder, Murray II-134
Emmi, Michael II-95
Enea, Constantin II-95
Erlebach, Thomas I-444
Etesami, Omid I-143
Etessami, Kousha II-184

Faonio, Antonio I-456
Feldman, Michal II-601
Feldmann, Andreas Emil I-469, II-588
Fijalkow, Nathanaël II-197
Filiot, Emmanuel II-209
Finkel, Olivier II-222
Fomin, Fedor V. I-481, I-494
Fontes, Lila I-506

Frascaria, Dario   I-26
Friedler, Ophir   II-601
Friedrich, Tobias   II-516, II-614
Friggstad, Zachary   I-116
Függer, Matthias   II-528
Fulla, Peter   I-517
Fung, Wai Shing   I-469

Gairing, Martin   II-626
Galanis, Andreas   I-529
Ganguly, Sumit   I-542
Garg, Jugal   I-554
Gąsieniec, Leszek   I-393
Gaspers, Serge   I-567
Gawrychowski, Paweł   I-580, I-593
Gelashvili, Rati   II-479
Georgiadis, Loukas   I-605
Ghaffari, Mohsen   II-638
Gharibian, Sevag   I-617
Giannakopoulos, Yiannis   II-650
Giannopoulou, Archontia C.   I-629
Göbel, Andreas   I-642
Gohari, Amin   I-143
Goldberg, Leslie Ann   I-529, I-642, I-654
Goldreich, Oded   I-666
Goldwasser, Shafi   II-663
Golovnev, Alexander   I-481, I-1046
Goubault, Éric   II-171
Goubault-Larrecq, Jean   II-171
Grohe, Martin   I-155
Große, Ulrike   I-678
Gudmundsson, Joachim   I-678
Gupta, Shalmoli   I-318
Gur, Tom   I-666
Gysel, Rob   I-654

Haase, Christoph   II-234
Hamza, Jad   II-95
Hansen, Thomas Dueholm   I-689
Hemenway, Brett   I-701
Henzinger, Monika   I-206, I-713, I-725
Henzinger, Thomas A.   II-121
Hoefer, Martin   II-504, II-516, II-552
Hoffmann, Michael   I-444
Holmgren, Justin   I-331
Horn, Florian   II-197
Huang, Chien-Chung   I-367, II-504
Huang, Lingxiao   I-910
Husfeldt, Thore   I-231

Ibsen-Jensen, Rasmus   II-121
Im, Sungjin   I-78, I-737
Ishai, Yuval   I-1
Istrate, Gabriel   II-44
Italiano, Giuseppe F.   I-206, I-605

Jagadeesan, Radha   II-31, II-247
Jahanjou, Hamid   I-749
Jain, Rahul   I-506
Jansen, Bart M.P.   I-629
Jerrum, Mark   I-529
Jin, Yifei   I-898
Jurdziński, Marcin   II-260

Kalai, Yael Tauman   II-663
Kamat, Vikram   I-243
Kammer, Frank   I-444
Kaniewski, Jedrzej   I-761
Kannan, Sampath   I-773
Kantor, Erez   II-675
Kaplan, Haim   I-689
Kar, Koushik   I-52
Karbasi, Amin   II-688
Kari, Jarkko   II-273
Kari, Lila   I-1022
Karpinski, Marek   I-785
Kaski, Petteri   I-494
Katz, Matthew J.   I-65
Kavitha, Telikepalli   I-367, II-504
Kawarabayashi, Ken-ichi   II-3
Kawase, Yasushi   I-797
Kayal, Neeraj   I-810
Kerenidis, Iordanis   I-506
Khot, Subhash   I-822
Khurana, Dakshita   I-1
Kiefer, Stefan   II-234
Klein, Felix   II-452
Klimm, Max   I-406
Knauer, Christian   I-678
Knudsen, Mathias Bæk Tejs   II-564
Kobayashi, Yusuke   I-797
Koiran, Pascal   I-810
Kollias, Konstantinos   II-626
Komarath, Balagopal   I-834
Könemann, Jochen   I-469
Kopecki, Steffen   I-1022
Kopparty, Swastik   I-193
Kosowski, Adrian   I-393
Kotsialou, Grammateia   II-626

Koutsoupias, Elias   II-650
Kowalik, Łukasz   I-243
Kozen, Dexter   II-286
Kozik, Marcin   I-846
Kranakis, Evangelos   I-393
Kreutzer, Stephan   II-3
Krinninger, Sebastian   I-713, I-725
Krohmer, Anton   II-614
Kulikov, Alexander S.   I-481
Kulkarni, Raghav   II-159
Künnemann, Marvin   I-859, II-552
Kupec, Martin   I-432
Kuperberg, Denis   II-197, II-299
Kurpisz, Adam   I-872
Kutten, Shay   II-675

Lahav, Ori   II-311
Lapinskas, John   I-654
Laplante, Sophie   I-506
Larsen, Kim G.   II-69
Laura, Luigi   I-605
Laurière, Mathieu   I-506
Lazić, Ranko   II-260
Lee, Troy   I-761
Lengler, Johannes   II-688
Leppänen, Samuli   I-872
Leroux, Jérôme   II-324
Li, Jerry   I-886
Li, Jian   I-898, I-910
Liew, Vincent   I-103
Lin, Chengyu   II-552
Lingas, Andrzej   I-785
Lohrey, Markus   II-337
Loitzenbauer, Veronika   I-713
Lokshtanov, Daniel   I-494, I-629,
    I-922, I-935
Lotker, Zvi   II-492
Lübbecke, Elisabeth   I-406

Mahajan, Meena   I-180
Mamouras, Konstantinos   II-286
Maneth, Sebastian   II-209, II-337
Manthey, Bodo   I-859
Maria, Clément   I-281
Markakis, Evangelos   I-39
Martin, Barnaby   I-256
Mastrolilli, Monaldo   I-872
Mathieu, Claire   I-773
Mazza, Damiano   II-350
Mehta, Ruta   I-554

Meunier, Pierre-Étienne   I-1022
Miao, Peihan   II-552
Michalewski, Henryk   II-362
Mihajlin, Ivan   I-481
Miles, Eric   I-749
Mio, Matteo   II-362
Misra, Pranabendu   I-922
Mitchell, Joseph S.B.   I-947
Molinaro, Marco   I-960
Mömke, Tobias   I-973
Moseley, Benjamin   I-78, I-737
Mottet, Antoine   I-256
Mouawad, Amer E.   I-985
Movahedi, Mahnush   II-575
Mozes, Shay   I-580
Mukherjee, Anish   II-159
Murlak, Filip   II-427
Muscholl, Anca   II-11

Nahum, Yinon   II-492
Nanongkai, Danupon   I-725
Nayyeri, Amir   I-997
Nicholson, Patrick K.   I-593
Nielsen, Jesper Buus   I-456
Nikolov, Aleksandar   I-1010
Nikzad, Afshin   I-39
Nishimura, Naomi   I-985
Nowak, Thomas   II-528

Ochremiak, Joanna   I-846
Otop, Jan   II-121

Panolan, Fahad   I-494, I-922
Park, Sunoo   II-663
Parotsidis, Nikos   I-605
Paskin-Cherniavsky, Anat   I-1
Pathak, Vinayak   I-985
Patitz, Matthew J.   I-1022
Pătrașcu, Mihai   I-103
Pecatte, Timothée   I-810
Peebles, John   I-886
Peleg, David   II-492
Peternek, Fabian   II-337
Petrișan, Daniela   II-286
Pietrzak, Krzysztof   I-1046
Polishchuk, Valentin   I-947
Post, Ian   I-469

Quanrud, Kent   I-318

Rabani, Yuval   I-78
Raman, Venkatesh   I-985
Ramanujan, M.S.   I-935
Raykov, Pavel   II-701
Reynier, Pierre-Alain   II-209
Richerby, David   I-642
Riely, James   II-247
Roland, Jérémie   I-506
Rotbart, Noy   II-564
Rothblum, Ron D.   I-666
Rothenberger, Ralf   II-516
Rubin, Sasha   II-375
Rudra, Atri   II-540

Saberi, Amin   I-39
Sachdeva, Sushant   I-193
Saha, Chandan   I-810
Saia, Jared   II-575
Saket, Rishi   I-822
Salavatipour, Mohammad R.   I-116
Sanyal, Swagato   I-1035
Sarma, Jayalal   I-834
Sauerwald, Thomas   II-516
Saurabh, Saket   I-494, I-629, I-922, I-935
Schewe, Sven   II-388
Schmitz, Sylvain   II-260
Schwentick, Thomas   II-159
Sekar, Shreyas   I-52
Seki, Shinnosuke   I-1022
Senellart, Pierre   II-56
Shen, Alexander   I-219
Shinkar, Igor   I-343
Shukla, Anil   I-180
Sidiropoulos, Anastasios   I-997
Sikora, Jamie   I-617
Silva, Alexandra   II-286
Siminelakis, Paris   II-467
Sivakumar, Rohit   I-116
Skórski, Maciej   I-1046
Skrzypczak, Michał   II-197, II-299
Sledneu, Dzmitry   I-785
Smid, Michiel   I-678
Sorkin, Gregory B.   I-567
Spegni, Francesco   II-375
Spirakis, Paul G.   I-393
Spreer, Jonathan   I-281
Srba, Jiří   II-69
Sreejith, A.V.   II-146
Staton, Sam   II-401
Steger, Angelika   II-688

Stehn, Fabian   I-678
Stein, Cliff   I-167
Stewart, Alistair   II-184
Sunil, K.S.   I-834
Sutre, Grégoire   II-324
Swernofsky, Joseph   II-414
Sysikaski, Mikko   I-947
Szabados, Michal   II-273

Talbot, Jean-Marc   II-209
Tarjan, Robert E.   I-689
Ta-Shma, Amnon   I-419
Terui, Kazushige   II-350
Thaler, Justin   I-268
Thapper, Johan   I-1058
Totzke, Patrick   II-324
Trivedi, Ashutosh   II-388

Uijlen, Sander   II-401
Uznański, Przemysław   I-393

Vafeiadis, Viktor   II-311
Vákár, Matthijs   II-31
Vardi, Moshe Y.   II-108
Varghese, Thomas   II-388
Vazirani, Vijay V.   I-554
Venturi, Daniele   I-456
Vidick, Thomas   I-355
Viola, Emanuele   I-749

Wagner, Lisa   II-504
Wang, Haitao   I-947
Wang, Yajun   I-1070
Wehar, Michael   II-414
Weimann, Oren   I-580
Weinstein, Omri   I-1082
Wiese, Andreas   I-973
Witkowski, Adam   II-427
de Wolf, Ronald   I-761
Wong, Sam Chiu-wai   I-1070
Woodruff, David P.   I-960, I-1082
Wootters, Mary   I-701

Yamaguchi, Yutaro   I-797
Yannakakis, Mihalis   II-184
Yaroslavtsev, Grigory   I-960
Yazdanbod, Sadra   I-554
Young, Maxwell   II-575
Yu, Huacheng   I-1094

Zehavi, Meirav   I-243
Zetzsche, Georg   II-440
Zeume, Thomas   II-159
Zhang, Shengyu   I-129
Zhou, Hang   I-773

Zimmermann, Martin   II-452
Zuleger, Florian   II-375
Zwick, Uri   I-689
Živný, Stanislav   I-517, I-1058

Printed in the United States
By Bookmasters